PRINCIPLES OF ELECTRICAL MACHINES

[FOR DEGREE, A.M.I.E., DIPLOMA AND OTHER ENGINEERING EXAMINATIONS]

Principles of
ELECTRICAL MACHINES

[FOR DEGREE, A.M.I.E., DIPLOMA AND OTHER ENGINEERING EXAMINATIONS]

V.K. MEHTA
ROHIT MEHTA

S Chand And Company Limited
(ISO 9001 Certified Company)

S Chand And Company Limited

(ISO 9001 Certified Company)

Head Office: D-92, Sector–2, Noida – 201301, U.P. (India), Ph. 91-120-4682700

Registered Office: A-27, 2nd Floor, Mohan Co-operative Industrial Estate, New Delhi – 110 044, Phone: 011-49731800

www.**schandpublishing.com;** e-mail: **info@schandpublishing.com**

Marketing Offices:

Chennai	:	Ph: 23632120; chennai@schandpublishing.com
Guwahati	:	Ph: 2738811, 2735640; guwahati@schandpublishing.com
Hyderabad	:	Ph: 40186018; hyderabad@schandpublishing.com
Jalandhar	:	Ph: 4645630; jalandhar@schandpublishing.com
Kolkata	:	Ph: 23357458, 23353914; kolkata@schandpublishing.com
Lucknow	:	Ph: 4003633; lucknow@schandpublishing.com
Mumbai	:	Ph: 25000297; mumbai@schandpublishing.com
Patna	:	Ph: 4011400; patna@schandpublishing.com

First Edition 2002
Reprints 2005, 2006 (Twice), 2007 (Twice), 2008; Second Edition 2009
Reprints 2009 (Twice); 2010, 2011 (Twice), 2012, 2013 (Twice), 2014 (Twice), 2015, 2016, 2017, 2018 (Twice), 2019, 2020, 2021, 2022 (Twice), 2023 (Twice), 2024, 2025 (Twice)

Reprint 2026

ISBN: 978-81-219-2191-6 **Product Code:** H3PEM52BEEL10ENAB09O

PRINTED IN INDIA

By Shri Shyamlal Printing Press Private Limited, Plot 20/4, Site-IV, Industrial Area Sahibabad, Ghaziabad – 201 010 and Published by S Chand And Company Limited, A-27, 2nd Floor, Mohan Co-operative Industrial Estate, New Delhi – 110 044.

PREFACE TO THE SECOND EDITION

The general response to the first edition of the book was very encouraging. Authors feel that their work has been amply rewarded and wish to express their deep sense of gratitude, in common to the large number of readers who have used it, and in particular to those of them who have sent helpful suggestions from time to time for the improvement of the book.

In the present revised edition, we have made sincere efforts to make the book up-to-date. This edition has three notable features. First, three new chapters viz. ***Electromechanical Energy Conversion, Circle Diagrams*** and ***Special-purpose Electric Machines*** have been added. Secondly, a large number of solved examples have been added to enhance the utility of the book. Thirdly, the book has been made design oriented keeping in view the growing demand of the industry. It is hoped that these features will make the book more useful.

Authors wish to thank their colleagues and friends who have contributed many valuable suggestions regarding the scope and content-sequence of the book. Authors are also indebted to M/s S. Chand & Company Ltd., New Delhi for bringing out this revised edition in a short time and pricing the book moderately inspite of heavy cost of paper and printing.

Errors might have crept in despite utmost care to avoid them. Authors shall be grateful if these are pointed out along with other suggestions for the improvement of the book.

V.K. MEHTA

ROHIT MEHTA

CONTENTS

1

Electromechanical Energy Conversion

INTRODUCTION

We daily use many devices that convert one form of energy into another form. For example, a heater converts electrical energy into heat energy while an electric bulb converts electrical energy into light energy. However, electromechanical conversion devices (*i.e.,* devices that convert electrical energy into mechanical energy or vice – versa) find wide practical applications. For example, an electric motor converts electrical energy into mechanical energy. On the other hand, an electric generator converts mechanical energy into electrical energy. A major reason for the widespread use of electromechanical energy conversion devices is that they are relatively efficient and permit an easy control. In this chapter, we shall discuss the basic principles of electromechanical energy conversion.

1.1 ELECTROMECHANICAL ENERGY CONVERSION

The conversion of electrical energy into mechanical energy or vice versa is known as **electromechanical energy conversion.**

Electromechanical energy conversion involves the interchange of energy between an electrical system and a mechanical system through the medium of a coupling electric field or magnetic field. Therefore, an electromechanical conversion system has three essential parts *viz., an electrical system, a mechanical system* and *a coupling field* (electric or magnetic). Fig. 1.1 shows the block diagram of an electromechanical energy conversion system. Note that from left to right, the system represents conversion from electrical to mechanical . However, from right to left, it will represent conversion from mechanical to electrical.

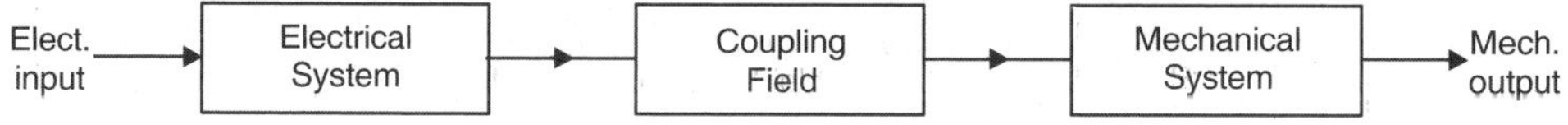

Fig. 1.1 Electromechanical Energy Conversion System

(*i*) **Electric field as coupling medium.** Electromechanical energy conversion can take place when electric field is used as the medium. Consider two oppositely charged plates of a capacitor which are separated by a dielectric medium. A force of attraction exists between the two plates that tends to move them together. If we allow one plate to move in the direction of the force, we are converting electrical energy into mechanical energy. On the other hand, if we apply an external force on one plate and try to increase the separation between them, we are then converting mechanical energy into electrical energy. Electrostatic microphones and electrostatic voltmeters use electrostatic fields for energy conversion.

(*ii*) **Magnetic field as coupling medium.** Electromechanical energy conversion can also take place more effectively when magnetic field is used as the medium. Consider the case of a current-carrying conductor placed in a magnetic field. The conductor experiences a force that tends to move it. If the conductor is free to move in the direction of the magnetic force, the magnetic field helps the conversion of electrical energy into mechanical energy. This is essentially the principle of operation of all electric motors. On the other hand, if an externally applied force moves the conductor in a direction opposite to the magnetic force, mechanical energy is converted into electrical energy. The generator action is based on this principle. Note that in both cases, the magnetic field acts as a medium for energy conversion.

It is important to note that the quantity of energy that can be converted by a device using electric field as a medium is relatively small. It is because the amount of force developed by an electric system is usually very small even when the applied voltage is high and the physical dimensions of the system are quite large. However, when magnetic field is used as a medium, a system with the same physical dimensions develops a much larger force than a system using an electric field as a medium. For this reason, the use of electric field as a medium for energy conversion has limited applications.

1.2 ELECTROMECHANICAL ENERGY CONVERSION DEVICES

Electromechanical energy conversion takes place through electric field or magnetic field as the medium. Although the various conversion devices operate on common set of physical principles, the structures of the devices depend on their function. Electromechanical energy conversion devices can be divided into the following three categories :

(*i*) **Transducers.** These conversion devices are used for measurement and control. They generally operate under linear input-output conditions and with relatively small signals. Examples include microphones, pick ups and loudspeakers.

(*ii*) **Force-producing devices.** These conversion devices are meant for producing force or torque with limited mechanical motion. Examples include relays, solenoid actuators and electromagnets.

(*iii*) **Continuous energy conversion devices.** These devices continuously convert electrical energy into mechanical energy or vice versa. They are used for bulk energy conversion and utilisation. Motors and generators are the examples of such conversion devices.

It may be noted that magnetic field is most suited as a medium for electromechanical energy conversion. *Therefore, in this book, we shall deal with magnetic field as the medium of energy conversion.*

1.3 FEATURES OF ELECTROMECHANICAL ENERGY CONVERSION

Electromechanical energy conversion takes place through the medium of magnetic field. The following salient features are worth noting in this energy conversion :

(*i*) As with any energy conversion system, the *principle of conservation of energy* holds good in case of electromechanical energy conversion. That is energy can neither be created nor destroyed; it can only be changed from one form to another.

(*ii*) During electromechanical energy conversion, various losses occur in the system. This is illustrated in Fig. 1.2 which shows the conversion of electrical energy into mechanical energy.

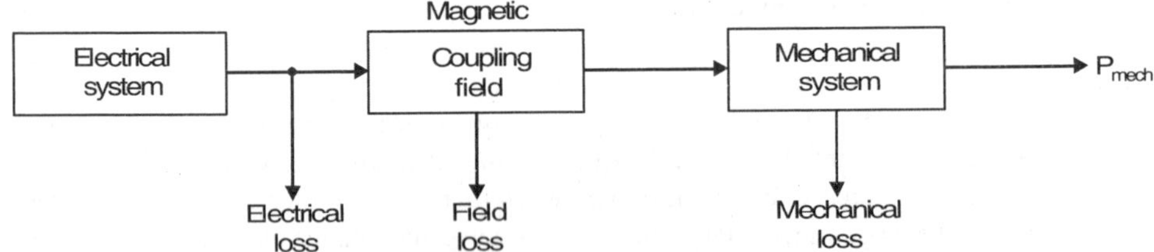

Fig. 1.2 Electromechanical Energy Conversion System

The electrical energy loss ($i^2 R$) is due to current (i) flowing in the winding (having resistance R) of the energy converter. The field loss is the core loss due to changing magnetic field in the magnetic core. The mechanical loss is the friction and windage loss due to the motion of the moving components. All these losses are converted into heat and raise the temperature of the energy conversion system.

(*iii*) Electromechanical energy conversion is a reversible process except for the losses in the system. The term reversible means that the energy can be transferred back and forth between the electrical and the mechanical systems. However, each time we go through an energy conversion process, some of the energy is used up to meet the losses in the conversion process. These losses are converted into heat and are lost from the system forever.

(*iv*) Electromechanical conversion devices are built with air gaps in the magnetic circuit to separate the fixed and moving parts. Most of the m.m.f. of the windings is required to overcome the air-gap reluctance so that most of the energy is stored in the air gap and is returned to the electric source when the field is reduced.

(*v*) The electromechanical energy conversion system can be analysed by using principle of conservation of energy, laws of electric and magnetic field, electric circuits and Newtonian mechanics.

(*vi*) The rotating electrical machines (motors and generators) continuously convert electrical energy into mechanical energy or vice versa. Fig. 1.3 shows the block diagram of electromechanical energy conversion in an electrical machine. The primary quantities involved in the mechanical system are torque (T) and speed (ω_m) while the analogous quantities in the electrical system are voltage (e) and current (i) respectively.

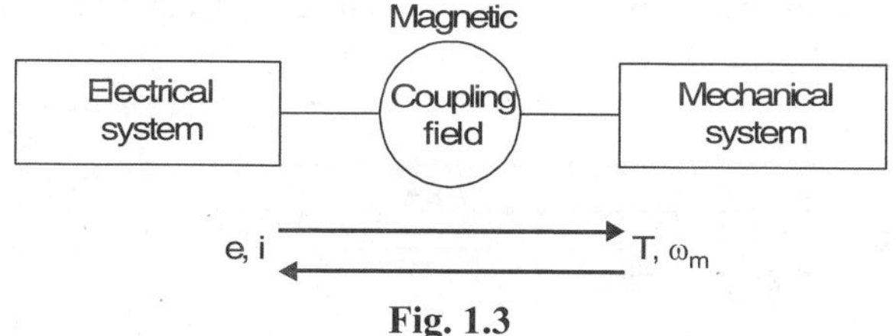

Fig. 1.3

1.4 ENERGY BALANCE EQUATION

An electromechanical energy conversion system has three essential parts *viz.,* an electrical system, a mechanical system and a coupling magnetic field as shown in Fig. 1.4. Since conversion of energy from one form into another form satisfies the principle of conservation of energy, the energy transfer equation is as under:

$$\begin{pmatrix}\text{Electrical energy}\\ \text{input from source}\end{pmatrix} = \begin{pmatrix}\text{Mechanical}\\ \text{energy output}\end{pmatrix} + \begin{pmatrix}\text{Increase in energy}\\ \text{stored in coupling field}\end{pmatrix} + \begin{pmatrix}\text{Energy}\\ \text{losses}\end{pmatrix} \quad ...(i)$$

Eq. (*i*) is applicable to all conversion devices. For motor action, the electrical and mechanical energy terms have positive values. For generator action, the electrical and mechanical energy terms have negative values.

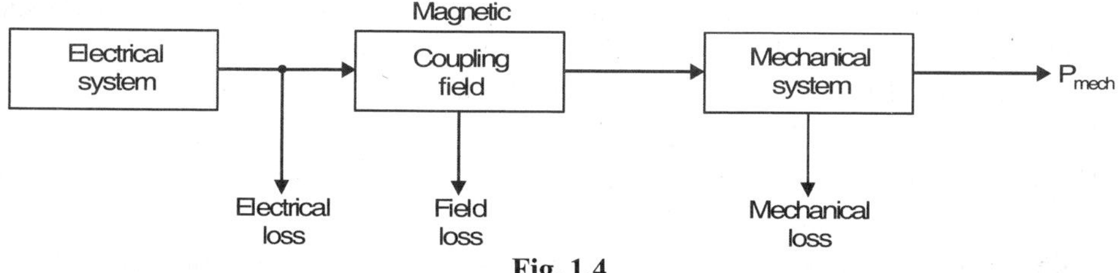

Fig. 1.4

During this energy conversion, energy loss occurs due to three causes *viz.,* (*i*) i^2R loss in the winding of the energy converter (*ii*) core or field loss due to changing magnetic field and (*iii*) mechanical loss is the friction and windage loss due to the motion of moving parts. All these losses are converted to heat. If the energy losses in the electrical system, the coupling magnetic field and the mechanical system are grouped with the corresponding terms in eq. (*i*) above, the energy balance equation can be written as under :

$$\begin{pmatrix}\text{Electrical energy}\\ \text{input } \textit{minus}\\ \text{resistance losses}\end{pmatrix} = \begin{pmatrix}\text{Mechanical energy}\\ \text{output } \textit{plus} \text{ friction}\\ \text{and windage losses}\end{pmatrix} + \begin{pmatrix}\text{Increase in stored}\\ \text{field energy } \textit{plus}\\ \text{core losses}\end{pmatrix} \quad ...(ii)$$

Now consider a differential time dt during which an increment of electrical energy dW_{elect} (excluding i^2R loss) flows to the system. During this time dt, let dW_{fld} be the energy supplied to the field (either stored or lost, or part stored and part lost) and dW_{mech} the energy converted to mechanical form (in useful form or as loss, or part useful and part as loss). In differential form, eq. (*ii*) can be expressed as

$$dW_{elect} = dW_{mech} + dW_{fld} \quad ...(iii)$$

If no mechanical work is done [*i.e.* $dW_{mech} = 0$], then eq. (*iii*) becomes :

$$dW_{elect} = dW_{fld}$$

In this case, electrical energy input is stored in the magnetic field (neglecting core losses).

1.5 ENERGY IN MAGNETIC SYSTEM

Consider singly–excited magnetic system shown in Fig. 1.5. It is the magnetic system of an attracted armature relay. Here a coil of N turns wound on the magnetic core is connected to an electric source. Let us assume that the armature is held stationary at some air gap and the current is increased from zero to some value i. As a result, flux ϕ will be established in the magnetic system.

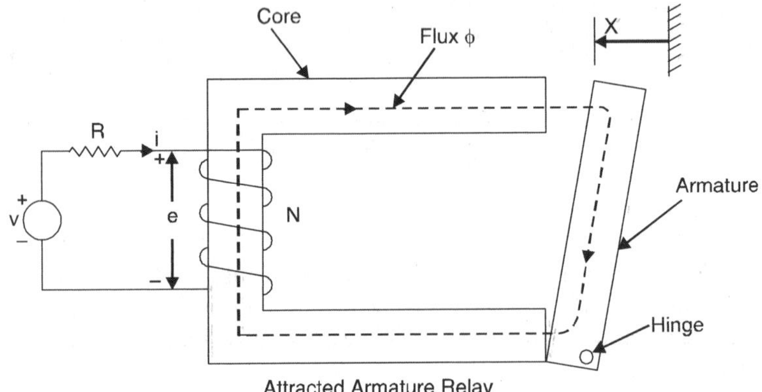

Attracted Armature Relay

Fig. 1.5

Total flux linkages, $\lambda = N\phi$

Induced e.m.f., $e = \dfrac{Nd\phi}{dt} = \dfrac{d}{dt}(N\phi) = \dfrac{d\lambda}{dt}$

For the coupling device to absorb energy from the electric circuit, the coupling field must produce a reaction in the circuit. This reaction is the e.m.f. e produced by the magnetic field.

The incremental electrical energy (dW_{elect}) due to the flow of current i in time dt is

$$dW_{elect} = e\,i\,dt$$

The energy balance equation in differential form is

$$dW_{elect} = dW_{mech} + dW_{fld}$$

Since we have assumed that the armature is held stationary, there is no mechanical output *i.e.*, $dW_{mech} = 0$. Therefore, *all the incremental electrical input energy is stored as incremental field energy *i.e.*

$$dW_{elect} = dW_{fld}$$

or $$e\,i\,dt = dW_{fld}$$

or $$\frac{d\lambda}{dt}\,i\,dt = dW_{fld}$$

$\therefore$ $$dW_{fld} = i\,d\lambda = N\,i\,d\phi \qquad (\because\ d\lambda = Nd\phi)$$

* Neglecting core loss.

The relationship between coil flux linkages λ and current i for a particuler air-gap length is shown in Fig. 1.6. The incremental field energy dW_{fld} is shown as crosshatched area in this figure. When the flux linkage is increased from zero to λ (or flux from zero to ϕ), the energy stored in the field is

$$W_{fld} = \int_0^\lambda i d\lambda = N \int_0^\phi i d\phi$$

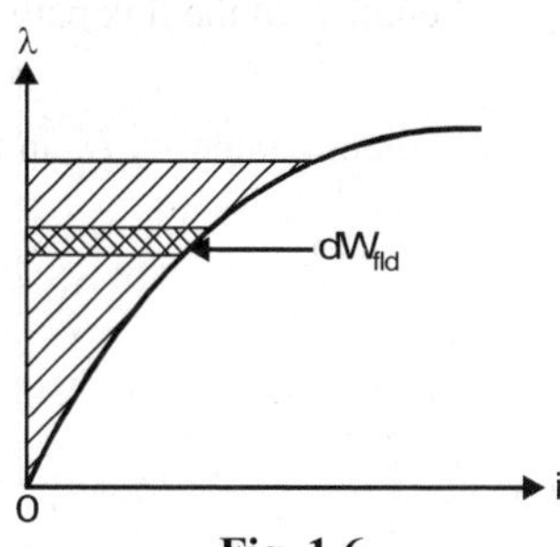

Fig. 1.6

The integral repesents the area between the λ axis and the λ-i characteristic and is equal to the entire shaded area shown in Fig. 1.6.

We can also derive another useful expression for the energy stored in the magnetic field. If l and A are the length and area of cross – section of the magnetic circuit respectively and B is the magnetic flux density, then,

$$Ni = Hl \text{ and } d\phi = A\, dB$$

Here H is the magnetic field intensity.

$$\therefore \qquad W_{fld} = N \int_0^\phi i d\phi = \int_0^B HlAdB = Al \int_0^B HdB$$

or

$$W_{fld} = Al \int_0^B HdB$$

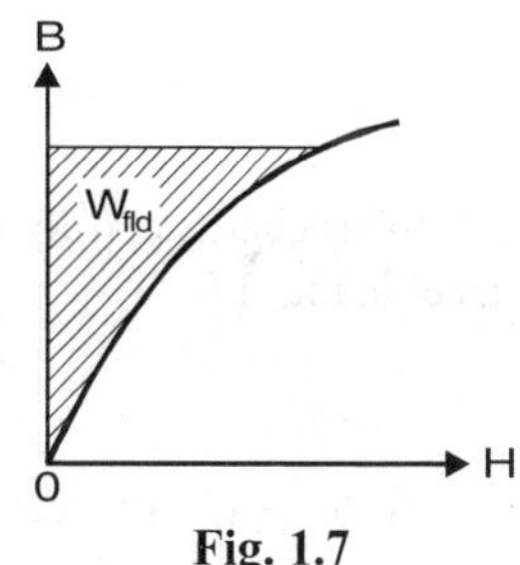

Fig. 1.7

Figure 1.7 shows B–H curve for the magnetic circuit.

Example 1.1. *Fig. 1.8 shows electromechanical system along with dimensions. The magnetic core is made of cast steel whose B–H characteristic is shown in Fig. 1.9. The coil has 250 turns and the coil resistance is 5Ω. For a fixed air gap length g = 5 mm, a d.c. source is connected to the coil to produce a flux density of 1 tesla in the air gap.*

(*i*) *Find the voltage of d.c. source.* (*ii*) *Find the stored field energy.*

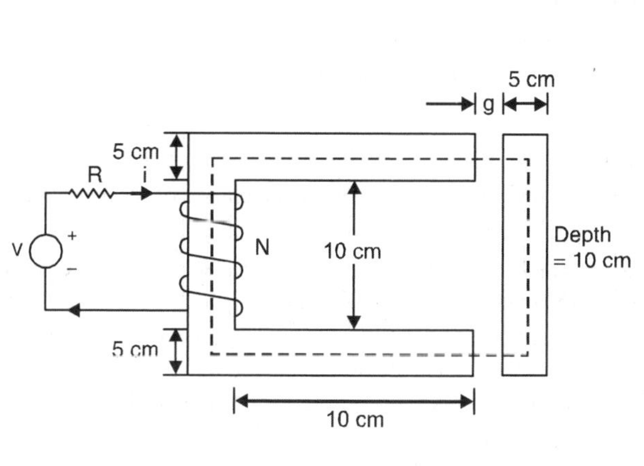

Fig. 1.8

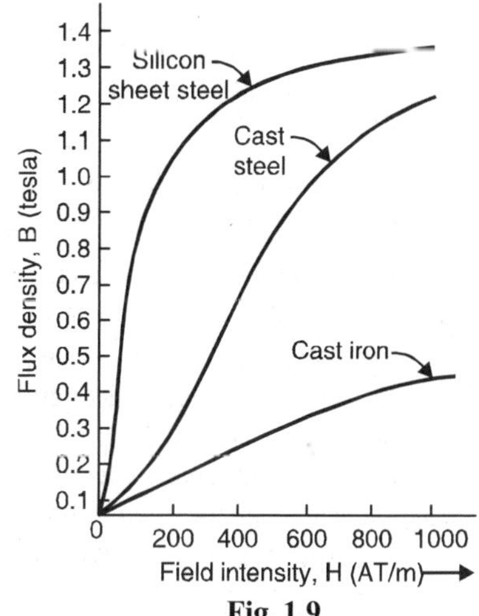

Fig. 1.9

Solution.

(*i*) From Fig. 1.9, the magnetic intensity H_c in the core material (cast steel) for a flux density of 1T is

$$H_c = 670 \text{ AT/m}$$

Length l_c of the flux path in the core is given by ;

$$l_c \simeq 2(10+5)+2(10+5) = 60 \text{ cm} = 0.6 \text{ m}$$

Magnetic intensity H_g in the air gap is given by ;

$$H_g = \frac{B_g}{\mu_0} = \frac{1}{4\pi \times 10^{-7}} = 795.8 \times 10^3 \text{ AT/m}$$

$$\begin{aligned} \text{Total m.m.f. required} &= H_c l_c + H_g \times 2g \\ &= (670 \times 0.6) + (795.8 \times 10^3 \times 2 \times 5 \times 10^{-3}) \\ &= 402 + 7958 = 8360 \text{ AT} \end{aligned}$$

$$\therefore \quad Ni = 8360 \quad \text{or} \quad i = 8360/N = 8360/250 = 33.44 \text{ A}$$

$$\therefore \quad \text{Voltage of dc source, } V = iR = 33.44 \times 5 = \mathbf{167.2 \ V}$$

(*ii*) Energy density in the core is given by ;

$$w_{fc} = \int_0^1 HdB$$

This energy density is given by the area enclosed between the B-axis and the B–H curve for cast steel in Fig. 1.9.

$$\therefore \quad w_{fc} = \frac{1}{2} \times B \times H = \frac{1}{2} \times 1 \times 670 = 335 \text{ J/m}^3$$

Volume of cast steel core is given by ;

$$V_c = 2(0.05 \times 0.1 \times 0.2) + 2(0.05 \times 0.1 \times 0.1) = 0.003 \text{ m}^3$$

∴ Field energy stored in the core is given by;

$$W_{fc} = w_{fc} \times V_c = 335 \times 0.003 = 1 \text{ J}$$

Energy density in the air gap is given by ;

$$w_{fg} = \frac{B^2}{2\mu_0} = \frac{(1)^2}{2 \times 4\pi \times 10^{-7}} = 398 \times 10^3 \text{ J/m}^3$$

$$\text{Volume of air gaps, } V_g = 2(0.05 \times 0.1 \times 0.005) = 0.05 \times 10^{-3} \text{ m}^3$$

∴ Field energy stored in the air gaps is given by ;

$$W_{fg} = w_{fg} \times V_g = 398 \times 10^3 \times 0.05 \times 10^{-3} = 19.9 \text{ J}$$

∴ Total stored field energy is given by ;

$$W_{fld} = W_{fc} + W_{fg} = 1 + 19.9 = \mathbf{20.9 \ J}$$

Note that most of the field energy is stored in the air gap because the reluctance of air gap is very large as compared to the iron part.

Note. We know that B–H curve or λ–i curve for air gap is a straight line ($B = \mu_0\mu_r H = \mu_0 H$ or $B \propto H$). A practical magnetic circuit of electromagnetic devices (*e.g.*, relays, motors, generators etc.) consists of an iron part and an air gap. The reluctance of air gap is very large as compared to the reluctance of the iron part. Therefore, we can neglect the reluctance of iron part so that B–H or λ–i curve of a practical magnetic circuit may be considered linear. This assumption leads to reasonable accuracy.

Example 1.2. *The relation between the total flux linkages and the current in the coil for the magnetic circuit shown in Fig. 1.10 is given by ;*

$$\lambda = \frac{6i}{2i+1} \text{ weber-turns}$$

Determine the energy stored in the magnetic field for $0 \le \lambda \le 2$ *Wb-turns.*

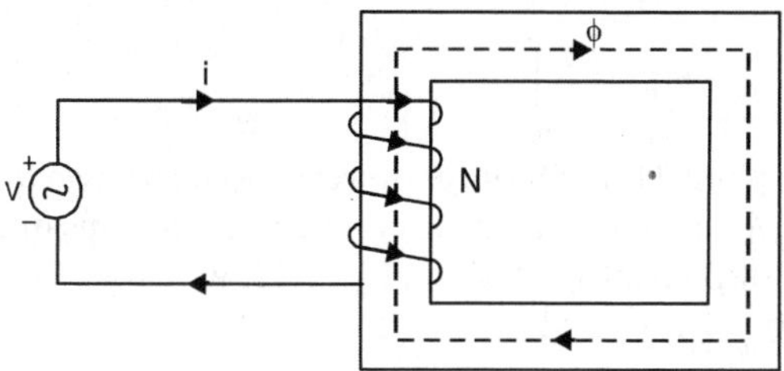

Fig. 1.10

Solution.

$$\lambda = \frac{6i}{2i+1}$$

or

$$\lambda(2i+1) = 6i$$

or

$$i = \frac{\lambda}{6-2\lambda}$$

∴ Energy stored in the magnetic field is given by ;

$$W_{fld} = \int_0^{\lambda} i\,d\lambda = \int_0^{2} \frac{\lambda}{6-2\lambda}\,d\lambda = \textbf{*0.648 J}$$

1.6 ENERGY AND COENERGY

The λ–i characteristic of an electromagnetic system depends on the air gap length and B–H characteristic of the magnetic material. These λ–i characteristics are shown in Fig. 1.11 for three values of air gap length. If the air gap length is large, the characteristic is essentially linear. The characteristic becomes nonlinear as the air gap length decreases.

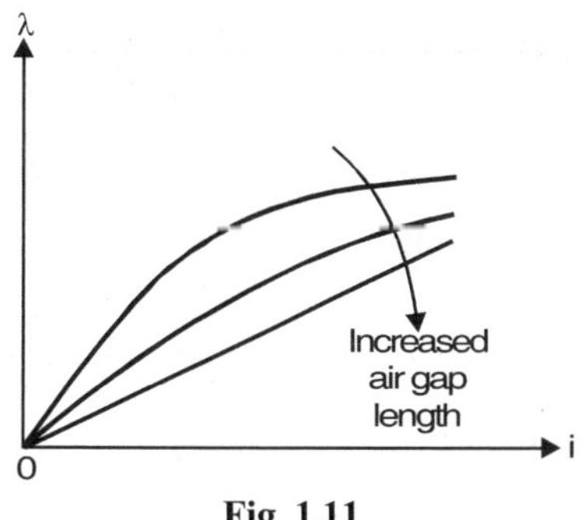

Fig. 1.11

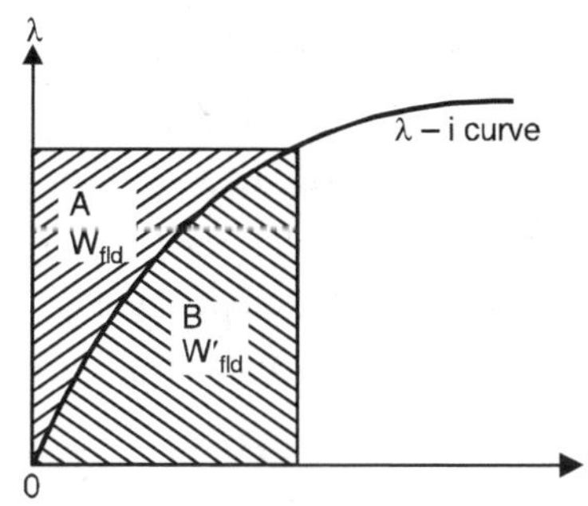

Fig. 1.12

For a particular value of air gap length, the energy stored in the field is represented by the area A between the λ-axis and the λ–i curve as shown in Fig. 1.12. The area B between the i-axis and the λ–i curve is known as the *coenergy* and is defined as :

---

* $$\int_0^2 \frac{\lambda}{6-2\lambda}\,d\lambda = -\frac{1}{2}\int_0^2 \frac{-2\lambda}{6-2\lambda}\,d\lambda = -\frac{1}{2}\int_0^2 \frac{6-2\lambda-6}{6-2\lambda}\,d\lambda$$

$$= -\frac{1}{2}\int_0^2 \left(\frac{6-2\lambda}{6-2\lambda} - \frac{6}{6-2\lambda}\right)d\lambda = -\frac{1}{2}\int_0^2 \left(1-\frac{6}{6-2\lambda}\right)d\lambda$$

$$= -\frac{1}{2}\left[\lambda + 6\log_e\left(\frac{6-2\lambda}{2}\right)\right]_0^2 = -\frac{1}{2}\left[\lambda + 3\log_e(6-2\lambda)\right]_0^2 = 0.648\text{ J}$$

$$W'_{fld} = \int_0^i \lambda \, di$$

This quantity has no physical significance. However, it can be used to find force developed in an electromagnetic system. We use coenergy to determine force in terms of the current i and energy to determine force in terms of flux linkage λ. From Fig. 1.12, we have,

$$W_{fld} + W'_{fld} = \lambda i$$

If the λ–i characteristic is linear, $W_{fld} = W'_{fld}$. However, if λ–i characteristic is non-linear, $W'_{fld} > W_{fld}$.

Example 1.3. *The λ–i characteristics of a magnetic circuit are described by the straight-line segments shown in Fig. 1.13. Find the energy W_{fld} and coenergy W'_{fld} for the magnetic circuit at (i) point 'a' and (ii) at point 'b'.*

Solution. (*i*) **At point 'a'.** Referring to Fig. 1.13,

Energy stored in field, W_{fld} = Area between the λ-axis and the λ–i curve (oa)

$$= \text{Area } oa1 = \frac{1}{2} \times 1 \times 1 = \mathbf{0.5\ J}$$

Coenergy, W'_{fld} = Area between the i-axis and the λ–i curve

$$= \text{Area } oa1 = \frac{1}{2} \times 1 \times 1 = \mathbf{0.5\ J}$$

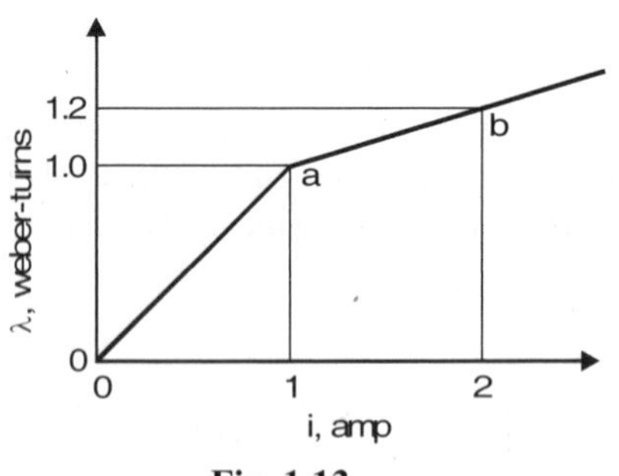

Fig. 1.13

Fig. 1.14

Note that energy (W_{fld}) and coenergy (W'_{fld}) at point 'a' are equal.

(*ii*) **At point 'b'.** Referring to Fig. 1.14,

Energy stored in field, W_{fld} = W_{fld} at point 'a' + Area between ab and λ-axis (See Fig. 1.14)

$$= 0.5 + \left(0.2 \times 1 + \frac{1}{2} \times 0.2 \times 1\right)$$

$$= 0.5 + (0.2 + 0.1) = \mathbf{0.8\ J}$$

Coenergy, W'_{fld} = Coenergy at point 'a' + Area between the curve ab and the i-axis (See Fig. 1.14)

$$= 0.5 + \left(1 \times 1 + \frac{1}{2} \times 1 \times 0.2\right)$$

$$= 0.5 + (1 + 0.1) = \mathbf{1.6\ J}$$

Thus the coenergy at point 'b' is 1.6 J and is larger than the energy of 0.8 J at the same point (*i.e.* point b).

1.7 FIELD ENERGY AND MECHANICAL FORCE

Consider the electromagnetic system shown in Fig. 1.15. Let the current through the coil be i when a voltage source v is applied across its terminals. The current i sets up magnetic flux ϕ in the magnetic circuit. The flux linkages induce an e.m.f. e in the coil.

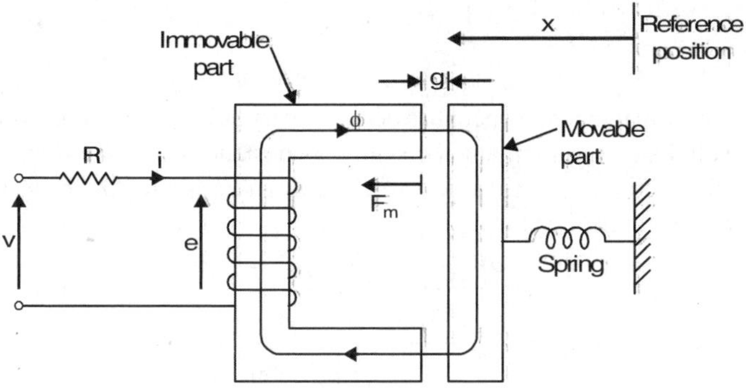

Fig. 1.15

Suppose the movable part moves from one position (say $x = x_1$) to another position ($x = x_2$) so that at the end of the movement, the air gap decreases. The λ–i characteristics of the system for these two positions are shown in Fig. 1.16(*i*). Note that operating points of the system are *a* when $x = x_1$ and *b* when $x = x_2$. The current $i(= v/R)$ will remain the same at both the positions in the steady state.

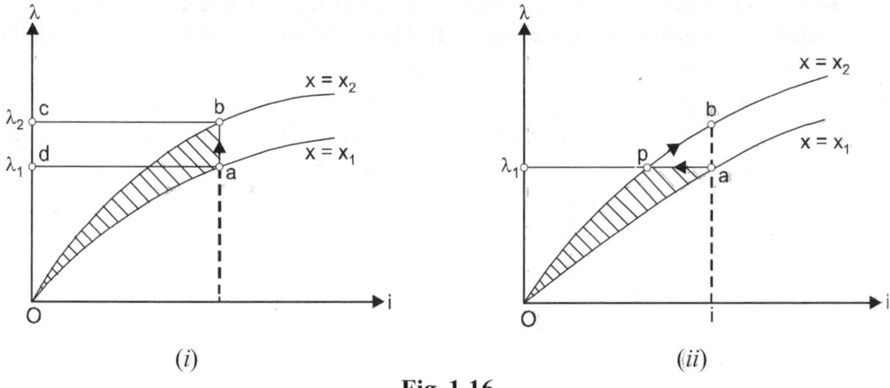

Fig. 1.16

If the movable part has moved slowly, the current has remained essentially constant during the motion. Therefore, the operating point moves upward from point *a* to point *b* as shown in Fig. 1.16(*i*). During the motion,

$$dW_{elect} = \int e\, i\, dt = \int_{\lambda_1}^{\lambda_2} i\, d\lambda = \text{Area } abcd$$

$$dW_{fld} = \text{Area } obc - \text{Area } oad$$

Now,

$$\begin{aligned} dW_{mech} &= dW_{elect} - dW_{fld} \\ &= (\text{Area } abcd) - (\text{Area } obc - \text{Area } oad) \\ &= \text{Area } abcd + \text{Area } oad - \text{Area } obc \\ &= \text{Area } oab \end{aligned}$$

Thus if the motion occurs under constant-current conditions, the mechanical work done (dW_{mech}) is represented by the shaded area in Fig. 1.16(*i*). In fact, this area is the increase in the coenergy.

$\therefore$ $$dW_{mech} = dW'_{fld}$$

If f_m is the mechanical force causing the differential displacement dx, then,

$$f_m dx = dW_{mech} = dW'_{fld}$$

or
$$f_m = \left.\frac{\partial W'_{fld}(i,x)}{\partial x}\right|_{i\,=\,\text{constant}} \qquad ...(i)$$

Let us now consider that the movement has occurred very quickly. It may be assumed that during the motion, the flux linkages have remained essentially constant as shown in Fig. 1.16(*ii*). It can be shown that during the motion, the mechanical work done is represented by the shaded area *oap*. In fact, this area is decrease in field energy.

$$\therefore \qquad f_m dx = dW_{mech} = -dW_{fld}$$

or
$$f_m = -\left.\frac{\partial W_{fld}(\lambda,x)}{\partial x}\right|_{\lambda\,=\,\text{constant}} \qquad ...(ii)$$

The following points may be noted from this analysis :

(*i*) Eqs. (*i*) and (*ii*) express the mechanical force of electrical origin in terms of partial derivatives of coenergy and energy functions *viz.* $W'_{fld}(i, x)$ and $W_{fld}(\lambda, x)$.

(*ii*) In the limit when the differential displacement *dx* is small, the area *oab* in Fig. 1.16(*i*) and area *oap* in Fig. 1.16(*ii*) will be essentially the same. Therefore, the force computed from eq. (*i*) and eq. (*ii*) will be the same. The selection of the energy or coenergy function as a basis for analysis is a matter of convenience. The choice depends upon the initial description of the system and the desired variables in the result.

(*iii*) The algebraic signs in eq. (*i*) and (*ii*) show that the mechanical force acts in a direction to increase the coenergy at constant current or to decrease the magnetic field stored energy at constant flux.

Example 1.4. *The λ–i relationship for an electromagnetic system is given by ;*

$$i = \left(\frac{\lambda g}{0.09}\right)^2$$

which is valid for the limits 0 < i < 4A and 3 < g < 10 cm.

For current i = 3A and air gap length g = 5 cm, find the mechanical force on the moving part using energy and coenergy of the field.

Solution.
$$i = \left(\frac{\lambda g}{0.09}\right)^2$$

or
$$\lambda = \frac{0.09 i^{1/2}}{g}$$

Using coenergy of the field

The coenergy of the system is given by ;

$$W'_{fld} = \int_0^i \lambda di = \int_0^i \frac{0.09 i^{1/2}}{g} di$$

$$= \frac{0.09}{g}\frac{2}{3}(i)^{3/2} \text{ joules}$$

The mechanical force in terms of coenergy function is given by ;

$$f_m = \left.\frac{\partial W'_{fld}(i,g)}{\partial g}\right|_{i\,=\,\text{constant}}$$

$$= -0.09 \times \frac{2}{3} i^{3/2} \times \frac{1}{g^2}$$

For $g = 0.05$ m and $i = 3$A, we have,

$$f_m = -0.09 \times \frac{2}{3} \times 3^{3/2} \times \frac{1}{(0.05)^2} = \mathbf{-124.7\ N}$$

Using energy of the field

The energy of the system is given by ;

$$W_{fld} = \int_0^{\lambda} i d\lambda = \int_0^{\lambda} \left(\frac{\lambda g}{0.09}\right)^2 d\lambda$$

$$= \frac{g^2}{(0.09)^2} \times \frac{\lambda^3}{3}$$

The mechanical force in terms of energy function is given by;

$$f_m = -\left.\frac{\partial W_{fld}(\lambda, g)}{\partial g}\right|_{\lambda = \text{constant}}$$

$$= -\frac{\lambda^3}{3} \times \frac{2g}{(0.09)^2}$$

For $g = 0.05$ m and $i = 3$A, the value of λ is given by;

$$\lambda = \frac{0.09 \times 3^{1/2}}{0.05} = 3.12 \text{ Wb-turns}$$

$$\therefore \quad f_m = -\frac{(3.12)^3}{3} \times \frac{2 \times 0.05}{(0.09)^2} = \mathbf{-124.7\ N}$$

Note that the value of mechanical force is the same whether the determination is made on the basis of coenergy or energy. Also the negative sign for the force indicates that the force acts in such a direction so as to decrease the air gap length.

1.8 LINEAR SYSTEM

Refer back to the electromagnetic system shown in Fig. 1.15. If the reluctance of the magnetic core path is negligible as compared to that of the air gap path, then λ–i relation becomes linear.

For this idealised system, we have,

$$\lambda = L(x)i$$

where $L(x)$ is the inductance of the coil whose value depends on the air gap length (x).

The field energy of the system is given by;

$$W_{fld} = \int_0^{\lambda} i\, d\lambda$$

$$= \int_0^{\lambda} \frac{\lambda}{L(x)} d\lambda \qquad \left(\because i = \frac{\lambda}{L(x)}\right)$$

$$= \frac{\lambda^2}{2L(x)} = \frac{[L(x)i]^2}{2L(x)} = \frac{1}{2}L(x)i^2$$

The mechanical force in terms of energy function is

$$f_m = -\left.\frac{\partial}{\partial x}\left(\frac{\lambda^2}{2L(x)}\right)\right|_{\lambda = \text{constant}}$$

$$= \frac{\lambda^2}{2L^2(x)} \frac{dL(x)}{dx}$$

$$\therefore \quad f_m = \frac{1}{2} i^2 \frac{dL(x)}{dx} \quad ...(i) \qquad \left[\because \frac{\lambda^2}{2L(x)} = \frac{1}{2} L(x) i^2 \right]$$

For a linear system, energy and coenergy are equal *i.e.*,

$$W_{fld} = W'_{fld}$$

$\therefore$ Mechanical force in terms of coenergy function is

$$f_m = \frac{\partial}{\partial x} \left(\frac{1}{2} L(x) i^2 \right) \Bigg|_{i = \text{constant}}$$

or

$$f_m = \frac{1}{2} i^2 \frac{dL(x)}{dx} \quad ...(ii)$$

Eqs. (*i*) and (*ii*) show that the same expressions are obtained for force whether analysis is based on energy or coenergy functions.

Another useful expression for f_m. If the reluctance of the magnetic core path in Fig. 1.15 (Refer back) is neglected, then,

$$Ni = H_g \times 2g = \frac{B_g}{\mu_0} \times 2g$$

Energy stored in the field is given by ;

$$W_{fld} = A_g l_g \int_0^{B_g} H_g dB_g = A_g l_g \int_0^{B_g} \frac{B_g}{\mu_0} dB_g$$

[Here l_g and A_g are the length and area of air gap]

$$= \frac{B_g^2}{2\mu_0} \times \text{volume of air gap}$$

$$= \frac{B_g^2}{2\mu_0} \times A_g \times 2g \qquad (\because l_g = 2g)$$

$\therefore$ Mechanical force in terms of energy function is

$$f_m = \frac{-\partial}{\partial g} \left(\frac{B_g^2}{2\mu_0} \times A_g \times 2g \right)$$

or

$$f_m = -\frac{B_g^2}{2\mu_0} (2A_g)$$

Note that total area of cross-section of the air gap is $2A_g$.

Example 1.5. *In the electromagnetic relay of Fig. 1.17 excited from a voltage source, the current and flux linkages are related as :*

$$i = \lambda^2 + 2\lambda (1 - x)^2 \; ; \; x < 1$$

Find the force on the armature as a function of λ.

Solution.

$$W_{fld} = \int_0^{\lambda} i d\lambda = \int_0^{\lambda} [\lambda^2 + 2\lambda(1 - x)^2] d\lambda$$

$$= \frac{\lambda^3}{3} + \lambda^2 (1 - x)^2$$

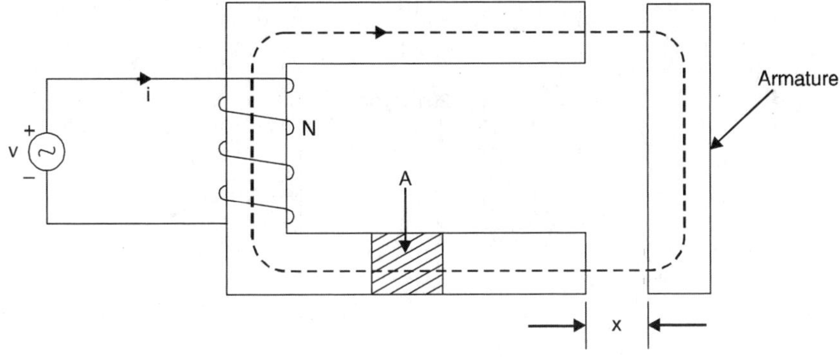

Fig. 1.17

Now $$f_m = -\frac{\partial W_{fld}}{\partial x} = \mathbf{2\lambda^2(1-x)}$$

Example 1.6. *An electromagnet is made using a horse shoe core as shown in Fig. 1.18. The core has an effective length of 600 mm and a cross-sectional area of 500 mm². A rectangular block of iron is held by electromagnetic force and a force of 20 N is required to free it. The magnetic circuit through the block is 200 mm long and the effective cross-sectional area is again 500 mm². The relative permeability of both core and block is 700. If the magnet is energised by a coil of 100 turns, find the coil current.*

Solution. Fig. 1.18 shows the conditions of the problem. There are two air gaps in the magnetic circuit. Therefore, the force to free the rectangular block is double that at any one air gap.

$$f_m = 2 \times \frac{B^2 A}{2\mu_0}$$

or $$20 = \frac{B^2 A}{\mu_0}$$

$\therefore$ $$B^* = \sqrt{\frac{20 \times \mu_0}{A}} = \sqrt{\frac{20 \times 4\pi \times 10^{-7}}{500 \times 10^{-6}}} = 0.224\text{T}$$

$$H = \frac{B}{\mu_0 \mu_r} = \frac{0.224}{4\pi \times 10^{-7} \times 700} = 250 \text{ AT/m}$$

$$\text{m.m.f.} = H \times l = 250 \times (600 + 200) \times 10^{-3}$$

$$= 200 \text{ AT}$$

or $$Ni = 200$$

$\therefore$ $$i = 200/N = 200/100 = \mathbf{2A}$$

N=100

Core

20 N

Rectangular block

Fig. 1.18

Example 1.7. *A solenoid relay is operated from a 110 V d.c. supply and the 5000-turn coil resistance is 5.5 kΩ. The core diameter of the relay is 20 mm and the gap length is 1.5 mm, the armature being stationary. The gap faces may be taken as parallel and the permeability of the ferro-magnetic parts as very high. Find (i) the gap flux density (ii) the coil inductance and (iii) the pull on the armature.*

Solution. Fig. 1.19 shows the conditions of the problem.

---

* Note that B is the same in the magnetic circuit.

(*i*) $$i = \frac{V}{R} = \frac{110}{5.5 \times 10^3} = 20 \times 10^{-3}\text{ A}$$

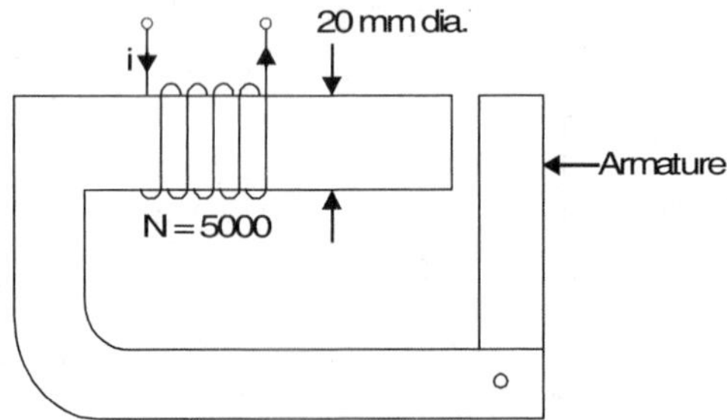

Fig. 1.19

$$\text{m.m.f.} = Ni = 5000 \times 20 \times 10^{-3} = 100\text{ AT}$$

$$H = \frac{\text{m.m.f.}}{l_g} = \frac{100}{1.5 \times 10^{-3}} = 0.67 \times 10^5\text{ AT/m}$$

$$B = \mu_0 H = 4\pi \times 10^{-7} \times 0.67 \times 10^5 = 84 \times 10^{-3}\text{ T} = \mathbf{84\text{ mT}}$$

(*ii*) $$\phi = B \times A = 84 \times 10^{-3} \times (\pi \times 10^2 \times 10^{-6}) = 26.3 \times 10^{-6}\text{ Wb}$$

$\therefore$ $$L = \frac{N\phi}{i} = \frac{5000 \times 26.3 \times 10^{-6}}{20 \times 10^{-3}} = \mathbf{6.56\text{ H}}$$

(*iii*) The inductance is inversely proportional to the air gap length (x).

$\therefore$ $$L = 6.56 \times \frac{1.5}{x} = \frac{9.84}{x}$$

$\therefore$ $$\frac{dL}{dx} = \frac{d}{dx}\left(\frac{9.84}{x}\right) = -\frac{9.84}{x^2} = -\frac{9.84}{(1.5)^2} = -4.37\text{ H/mm} = -4370\text{ H/m}$$

The negative sign shows that inductance increases with the decrease in air gap length (x).

$\therefore$ Magnitude of pull on the armature is

$$f_m = \frac{1}{2}i^2\frac{dL}{dx} = \frac{1}{2} \times (20 \times 10^{-3})^2 \times 4370 = \mathbf{0.88\text{ N}}$$

Example 1.8. *A plunger relay is shown in Fig. 1.20. When it is energised, the current remains constant at 2A and the plunger moves such that the air gap is reduced in length from 20 mm to 10 mm. If losses are neglected, calculate the average force experienced by the plunger.*

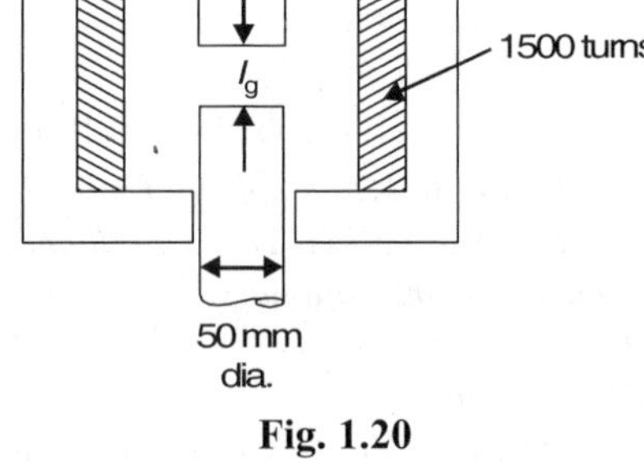

Fig. 1.20

Solution. For air gap length l_g = 10 mm

$$\text{m.m.f.} = Ni = 1500 \times 2 = 3000\text{ AT}$$

$$H = \frac{\text{m.m.f.}}{l_g} = \frac{3000}{10 \times 10^{-3}} = 3 \times 10^5\text{ AT/m}$$

$$B = \mu_0 H = (4\pi \times 10^{-7}) \times (3 \times 10^5) = 0.377\text{ T}$$

$$\phi = B \times A = (0.377) \times [\pi \times (25)^2 \times 10^{-6}] = 0.74 \times 10^{-3}\text{ Wb}$$

$\therefore$ $$L = \frac{N\phi}{i} = \frac{1500 \times 0.74 \times 10^{-3}}{2} = 0.56\text{ H}$$

For air gap length $l_g = 20$ mm

$$L = \frac{1}{2} \times 0.56 = 0.28\,\text{H}$$

$$\Delta L = 0.56 - 0.28 = 0.28\,\text{H}; \quad \Delta x = 20\,\text{mm} - 10\,\text{mm} = 10\,\text{mm} = 0.01\,\text{m}$$

∴ Average force on the plunger is given by ;

$$f_m = \frac{1}{2} i^2 \frac{\Delta L}{\Delta x} = \frac{1}{2} \times (2)^2 \times \frac{0.28}{0.01} = \mathbf{56\ N}$$

Example 1.9. *A smooth core armature working in a 4-pole field magnet has a gap (iron to iron) of 0.5 cm. The area of the surface of each pole is 0.1 m². The ampere-turns absorbed by each pole are 3000. Calculate (i) the mechanical force exerted by each pole on the armature (ii) energy stored in the four air gaps.*

Solution. (*i*) *AT* per gap = Flux × Reluctance of air gap $= (B \times A) \times \left(\frac{l_g}{A\mu_0}\right) = \frac{Bl_g}{\mu_0}$

$$\frac{B}{\mu_0} = \frac{\text{AT per gap}}{l_g} = \frac{3000}{0.5 \times 10^{-2}} = 6 \times 10^5$$

or
$$B = \mu_0 \times 6 \times 10^5 = (4\pi \times 10^{-7}) \times 6 \times 10^5 = 0.75\ \text{Wb/m}^2$$

Mechanical force exerted by each pole is given by ;

$$f_m = \frac{B^2 A}{2\mu_0} = \frac{(0.75)^2 \times 0.1}{2 \times 4\pi \times 10^{-7}} = \mathbf{22381\ N}$$

(*ii*) Volume of 4 air gaps $= 4\,A\,l_g = 4 \times 0.1 \times 0.5 \times 10^{-2} = 0.002\ \text{m}^3$

$$\text{Energy stored in air gaps} = \frac{B^2}{2\mu_0} \times \text{volume of 4 air-gaps}$$

$$= \frac{(0.75)^2}{2 \times 4\pi \times 10^{-7}} \times 0.002 = \mathbf{448\ J}$$

Example 1.10. *An iron ring having a mean circumference of 30 cm and a cross-sectional area of 1 cm² has two radial saw cuts at diameterically opposite points. A brass plate is inserted in each gap (thickness of each gap being 0.1 mm). If the ring is wound with 200 turns, calculate the magnetising current to exert a pull of 5 kg between the two halves. Assume the magnetic data for the iron to be :*

B (Wb/m²)	*0.79*	*1.0*	*1.3*
H (AT/m)	*250*	*350*	*520*

Solution. The total force of attraction at the two separations is = 5 × 9.80 = 49 N. Therefore, force of attraction at each separation, $f_m = 49/2 = 24.5$ N.

Now,
$$f_m = \frac{B^2 A}{2\mu_0} \quad \therefore \quad B = \sqrt{\frac{2 \times 4\pi \times 10^{-7} \times 24.5}{1 \times 10^{-4}}} = 0.79\ \text{Wb/m}^2$$

Corresponding to B = 0.79 Wb/m², we have, H = 250 AT/m

Length of iron path = 30 cm = 0.3 m

AT for iron path = 250 × 0.3 = 75

H for brass $= B/\mu_0 = 0.79/4\pi \times 10^{-7} = 628662$ AT/m

Thickness of brass plates $= 0.1 + 0.1 = 0.2\ \text{mm} = 0.2 \times 10^{-3}$ m

AT for brass paths $= 628662 \times 0.2 \times 10^{-3} = 125.73$

Total *AT* required = 75 + 125.73 = 200.73

∴ Magnetising current required = 200.73/200 = **1 A**

Example 1.11. *The arm of a starter is held in the "ON" position by means of an electromagnet. The torque exerted by the spring is 5 Nm and the effective radius at which the force is exerted is 10 cm. Area of each pole face is 2.5 cm² and each air gap is 0.4 mm. Find the minimum number of AT required to keep the arm in the "ON" position.*

Solution. Fig. 1.21 shows the whole arrangement. Let f_m newtons be the force exerted by the electromagnet.

$$\text{Torque} = \text{Force} \times \text{radius}$$

or $\quad 5 = f_m \times 0.1 \qquad \therefore \qquad f_m = 5/0.1 = 50 \text{ N}$

The force exerted at each pole of the magnet, f'_m 50/2 = 25 N.

Fig. 1.21

Now $$f'_m = \frac{B^2 A}{2\mu_0}$$

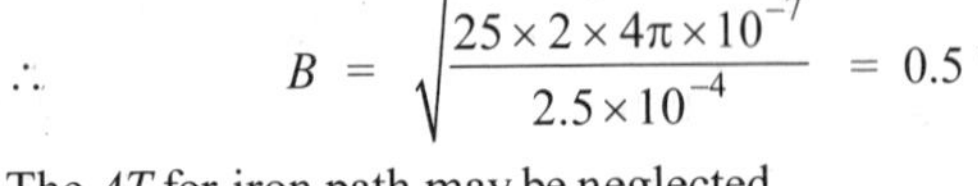

$$\therefore \qquad B = \sqrt{\frac{25 \times 2 \times 4\pi \times 10^{-7}}{2.5 \times 10^{-4}}} = 0.5 \text{ Wb/m}^2$$

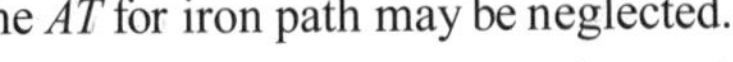

The *AT* for iron path may be neglected.

$$H \text{ in air gap} = B/\mu_0 = 0.5/4\pi \times 10^{-7} = 397887 \text{ AT/m}$$

$$\text{Total air gap length} = 2 \times 0.4 \times 10^{-3} = 0.8 \times 10^{-3} \text{ m}$$

$$AT \text{ required} = 397887 \times 0.8 \times 10^{-3} = \mathbf{318.3}$$

TUTORIAL PROBLEMS

1. Find the pull exerted on the plunger of an electromagnet when the total flux uniformly distributed is 500 μWb. Diameter of the plunger is 2.54 cm. **[196.3 N]**
2. A horse shoe magnet has two poles, each of area 5 cm². Find the pull between the poles and the keeper when the flux density at the contact surface is 1 Wb/m². **[398 N]**
3. The core material for use in an electromagnet should not have a flux density more than 1.5 Wb/m². How much area each of the two poles should have if the magnet is to lift 200 kg ? **[10.95 cm²]**
4. A circular crane magnet has an iron cross-section of 200 cm² and a mean magnetic path of 80 cm. Assuming the total length of each air gap to be 1.5 mm, calculate (*i*) the *AT* to produce a gap flux of 0.025 Wb (*ii*) the force to separate the contact surface, assuming no leakage or fringing.

B(Wb/m²)	1.0	1.2	1.4
H (AT/m)	900	1230	2100

[(*i*) 4080 AT (*ii*) 2500 kg (force)]

5. Each of the two pole faces of a lifting magnet has an area of 150 cm² and this may also be taken as the cross-sectional area of the 40 cm long flux path in the magnet. Determine the *AT* needed on the magnet if it is to lift a 900 kg iron block separated by 0.5 mm from the pole faces. Assume the magnetic leakage factor to be 1.2. Neglect fringing of the gap flux and reluctance of the flux path in the iron block.

H(AT/m)	400	600	800	1200	1600
B(Wb/m²)	0.81	0.98	1.1	1.24	1.35

[955]

1.9 ROTATING ELECTRICAL MACHINES

Rotating electrical machines continuously convert electrical energy into mechanical energy or vice-versa. This interchange of energy between an electrical system and a mechanical system takes place through the medium of coupling magnetic field. When the conversion takes place from electrical to mechanical form, the machine is called a *motor.* When mechanical energy is converted to electrical energy, the machine is called a *generator.* When the electrical system is characterised by direct current, the devices are called d.c. motors and d.c. generators. Similarly, when the electrical system is characterised by alternating current, the conversion devices are called a.c. motors and a.c. generators.

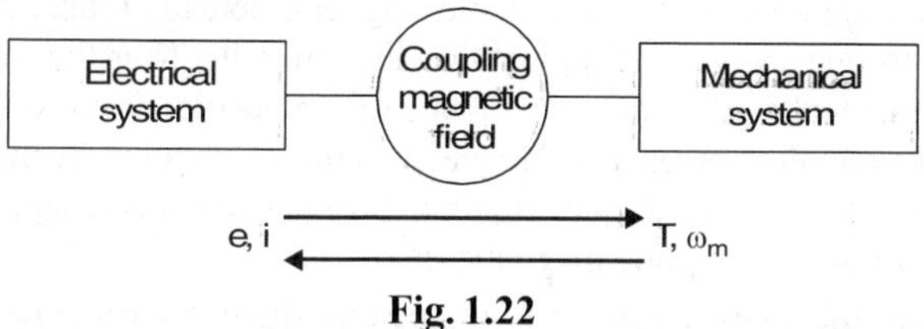

Fig. 1.22

Fig. 1.22 shows the block diagram of electromechanical energy conversion in a rotating electrical machine. The primary quantities involved in the mechanical system are torque (*T*) and angular speed (ω_m) while the analogous quantities in the electrical system are voltage (*e*) and current (*i*) respectively.

Principles of Operation. The operation of all rotating electrical machines are based on the following two fundamental electromagnetic laws :

(*i*) *When the magnetic flux linking a conductor or coil changes, an e.m.f. is induced in it whose magnitude is given by ;*

$$e = \frac{Nd\phi}{dt} \qquad \text{...Faraday's Law}$$

This is the basic equation that governs the induced voltage in the armature winding whether the electrical machine is d.c. or a.c. The ultimate forms of the induced voltage equations appear different for the two types of machines only because the details of mechanical construction differ.

(*ii*) *When a current-carrying conductor of length l is placed at right angles to a uniform magnetic field (of flux density B), the conductor experiences a mechanical force (F) whose magnitude is given by ;*

$$F = Bil \qquad \text{...Maxwell's Law}$$

This basic equation of force is used to find the developed torque (*T*) whether the electrical machine is d.c. or a.c. The ultimate forms of the torque equations appear different for the two types of machines only because the details of mechanical construction differ.

1.10 MOTORS AND GENERATORS

In motoring action, the driving torque is developed because current-carrying conductors are placed in the magnetic field. In generating action, e.m.f. is induced because revolving conductors cut through the magnetic field. In a practical machine (motor or generator), both these effects occur simultaneously.

(*i*) **Motor action.** Motor action takes place when the electrical system causes a current *i* to flow through the conductors that are placed in a magnetic field. As a result, a force acts on each conductor and if the conductors are located on a structure that is free to rotate (rotor), an electromagnetic torque *T* is produced which rotates the rotor with an angular speed ω_m. With the development of this motor action, the revolving conductors cut through the magnetic field and e.m.f. *e* (called reaction e.m.f. or back e.m.f.) is induced in the conductors. This back e.m.f. *e* opposes the applied voltage originating from the electrical system. The electrical work done in the electrical system to overcome this opposition is converted into mechanical energy. Note that it is the coupling magnetic field which is responsible for the production of driving torque *T* as well as the back e.m.f. *e*. Therefore, conversion of electrical energy to mechanical energy takes place through the coupling magnetic field.

(*ii*) **Generator action.** In the case of generator action, the reverse process takes place. Here the rotor (armature containing conductors) is driven by a prime mover (steam turbine, diesel engine etc.) in a magnetic field. Consequently, e.m.f. *e* (called generated e.m.f.) appears across the armature winding terminals. When electrical load (*e.g.* electric bulb, heater etc.) is connected to these terminals, a current *i* is made to flow, delivering electrical power to the load. The flow of this current *i* through the armature conductors interacts with the magnetic field to produce a reaction or opposing torque. This reaction torque opposes the applied torque originating from the prime mover (mechanical system). The mechanical work done to overcome this opposition is converted into electrical energy. Note that it is the coupling magnetic field which is responsible for the production of generated e.m.f. e and the reaction or opposing torque. Therefore, conversion of mechanical energy to electrical energy takes place through the coupling magnetic field.

Conclusion. From the above discussion, we see that the two basic effects (*i.e.* motoring and generating) are inseparable during electromechanical energy conversion in an electrical machine. In a *motor*, the machine draws power from electrical supply and converts it to mechanical power. This conversion is due to the setting up of generating action that opposes the applied voltage. In a *generator*, the power is supplied mechanically and is converted to the electrical form. This conversion is due to setting up of motor action that opposes the driving torque applied at the generator shaft.

1.11 BASIC CONSTRUCTION OF ELECTRICAL MACHINES

The principles of operation of an electrical machine (generator or motor) suggest that its structure should be such so that rotary motion can take place between a magnetic field and current-carrying conductors. Therefore, an electrical machine has two major members, stator and rotor, separated by an air gap as shown in Fig. 1.23.

Stator : This part of the electrical machine is stationary and normally is the outer frame of the machine.

Rotor : This part of the electrical machine is free to rotate and normally is the inner part of the machine.

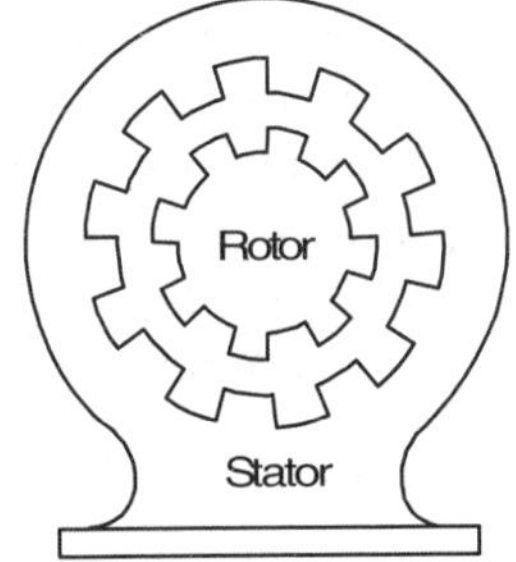

Fig. 1.23

(*i*) The function of one member (stator or rotor) is to provide a magnetic field and is called the field structure. The function of the other member (called armature) is to provide an assembly of conductors (armature winding) so that e.m.f. is induced in them due to the magnetic field produced by the first member.

(*ii*) Both stator and rotor are made of ferromagnetic materials so that reluctance of each is negligible compared with the reluctance of the air gap. Therefore, almost all the m.m.f. in the magnetic circuit of a rotating electrical machine is used up to set up the required flux in the air gap.

(*iii*) In most electrical machines, slots are cut on the inner periphery of the stator and outer periphery of the rotor as shown in Fig. 1.23. Conductors are placed in these slots. The conductors placed in the slots of stator or rotor are interconnected to form windings. The winding in which voltage is induced is called the *armature winding.* The winding through which current is passed to produce the primary source of magnetic flux is called the *field winding.*

(*iv*) So far as the basic functioning of the electrical machine is concerned, it makes no difference which member (*i.e.* member containing armature winding or field winding) is the revolving member. In d.c. machines, armature is made the rotating member due to practical considerations. In a.c. machines, both types of construction are employed.

(*v*) The practical method of producing magnetic field is by means of current-carrying windings called field windings. There are two types of field construction *viz. salient-pole field* and *non-salient-pole field.* In salient-pole construction, the field windings are wound on the projecting iron parts as shown in Fig. 1.24. In non-salient pole construction, the field windings are embedded in slots as shown in Fig. 1.25.

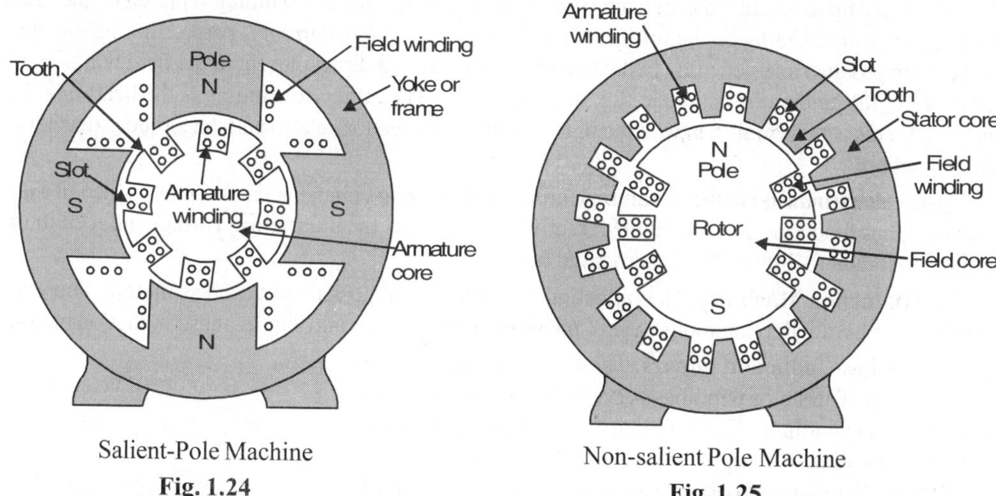

Salient-Pole Machine
Fig. 1.24

Non-salient Pole Machine
Fig. 1.25

D.C. machines always employ salient-pole field winding construction. However, both salient-pole and non-salient pole field winding construction is employed for a.c. machines.

(*vi*) The armature winding (whether on stator or rotor) always consists of a winding embedded in slots in cylindrical core. Although the armature winding is not provided for the purpose of producing a magnetic field, nevertheless the current in the armature winding will also produce magnetic flux (called armature flux). The armature flux will greatly affect the flux conditions in the machine. Considered from its field producing potentiality, the armature winding is non-salient-pole type winding. The number of magnetic poles that it tends to produce in the structure depends upon the distribution of the direction of current through the conductors as one progresses around the armature.

1.12 COMMONLY USED ELECTRICAL MACHINES

The most commonly used rotating electrical machines are (*i*) D.C. machines (*ii*) Induction machines and (*iii*) Synchronous machines. In these machines, conversion of energy from electrical to mechanical form or vice-versa takes place through magnetic field as the medium.

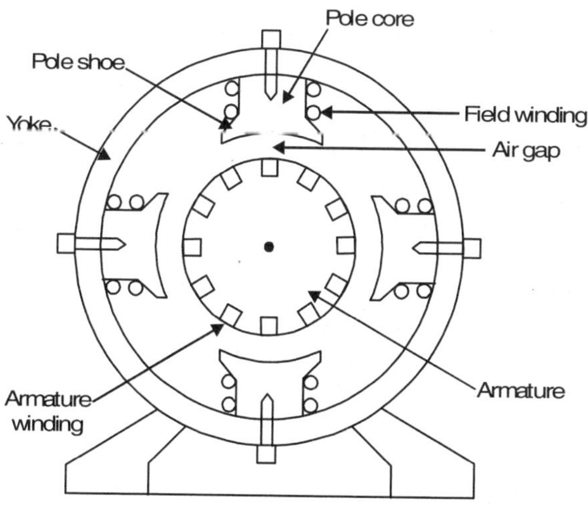

Fig. 1.26

(*i*) **D.C. Machines.** In a d.c. machine (generator or motor), the field winding is placed on the stator and the armature winding on the rotor as shown in Fig. 1.26. A d.c. current is passed through the field winding to produce magnetic flux in the machine. By changing direct current in the field winding, we can control the amount of magnetic flux in the machine. The armature winding is the heart of a d.c. machine. This is the winding in which e.m.f. is induced (generator action) and the torque is developed (motor action).

The voltage induced in the armature winding is alternating voltage. A commutator is used in a d.c. machine that functions as a mechanical rectifier. It converts the alternating voltage induced in the armature coils into direct voltage across the brushes.

(*ii*) **Induction Machines.** The induction machines are almost always used as motors. There are two basic types of induction motors *viz.* 3-phase induction motors and single-phase induction motors.

(*a*) **3-phase induction motors.** These motors operate on 3-phase supply. The stator windings serve as both armature windings and field windings. The rotor can have a winding similar to the stator winding or a cage-type winding. The latter is formed by placing aluminium or copper bars in the rotor slots and shorting them at the ends by means of rings. Fig. 1.27 shows the structure of a 3-phase induction motor.

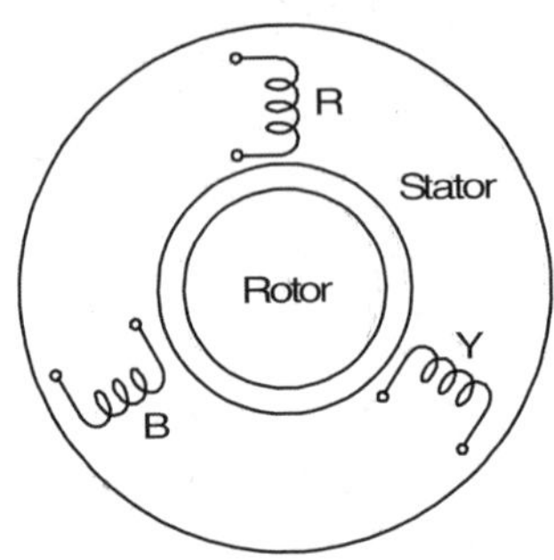

Fig. 1.27

When the stator windings are connected to 3-phase supply, flux is produced in the air gap and revolves at a fixed speed known as *synchronous speed.* This revolving flux induces voltage in the stator windings as well as rotor windings. Since the rotor circuit is closed, current flows in the rotor windings and reacts with the revolving flux to produce torque. Under steady-state conditions, the speed of the rotor is very close to synchronous speed.

(*b*) **Single-phase induction motors.** These motors operate on 1-phase supply. The stator carries a single-phase winding and the rotor is always of squirrel-cage type. A single-phase supply does not produce revolving magnetic field so that a single-phase motor is not a self-starting machine. To make a single-phase induction motor self-starting, we should somehow produce a revolving stator magnetic field. This is achieved by converting a single-phase supply into two-phase supply through the use of an additional winding.

(*iii*) **Synchronous machines.** The most common synchronous machines are 3-phase type. In these machines, the rotor carries the field winding and the stator carries the armature winding as shown in Fig. 1.28.

The field winding is excited by direct current to produce flux in the air gap. A synchronous machine (generator or motor) operates at *synchronous speed i.e.* at the speed at which the magnetic field created by the field coils rotates. When the machine converts mechanical power into 3-phase power, it is called *synchronous generator* or *alternator.* On the other hand, when the machine converts 3-phase power into mechanical power, it is called *synchronous motor.* A synchronous motor develops torque only when running at synchronous speed. However, synchronous motors are not used so much because they run at constant speed (*i.e.* synchronous speed) but because they possess other unique electrical properties.

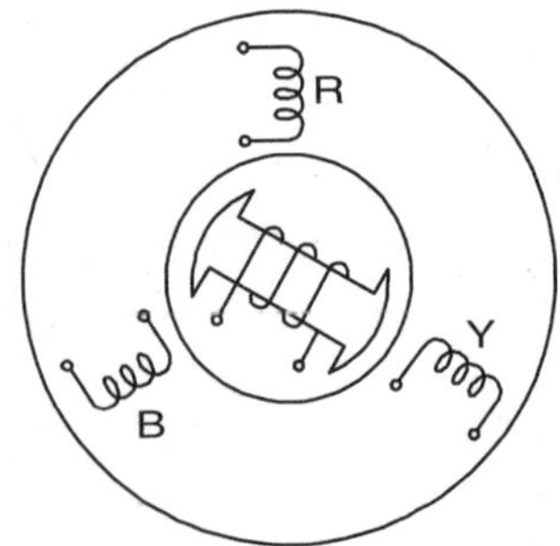

Fig. 1.28

1.13 MULTIPLY–EXCITED MAGNETIC FIELD SYSTEM

So far we have considered singly-excited magnetic systems *i.e.* systems in which there is only one source of excitation. Examples of singly-excited magnetic systems are electromagnets, relays, moving iron instruments etc. Most of the electromagnetic energy conversion devices belong to doubly-excited or multiply-excited magnetic systems. A doubly-excited magnetic system is one which has more than one independent sources of excitation *e.g.* synchronous machines, d.c. shunt machines, tachometers etc.

Fig. 1.29 shows a simple doubly-excited *rotating electromagnetic system. The fixed part of the magnetic system is called the *stator* and the moving part is called the *rotor.* The latter is mounted on a shaft and is free to rotate between the poles of the stator. Both stator and rotor have windings that carry currents *i.e.* it is a doubly-excited magnetic system. The current can be fed into the rotor circuit through fixed brushes and rotor-mounted slip rings.

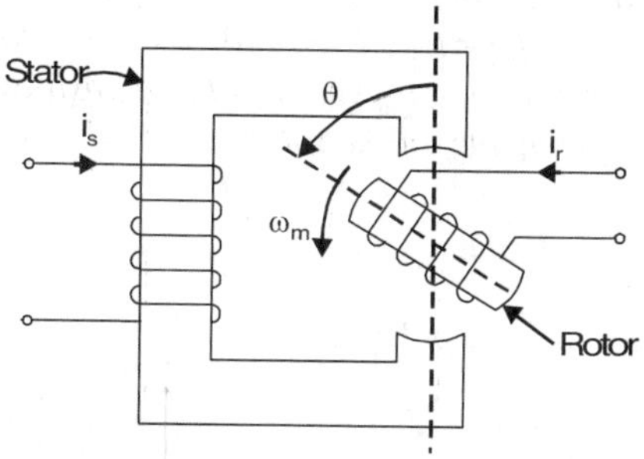

Fig. 1.29

The total flux linkages λ_s with stator winding can be expressed as :

$$\lambda_s = L_{ss} i_s + L_{sr} i_r$$

where L_{ss} = Self-inductance of the stator winding

L_{sr} = Mutual inductance between stator and rotor windings

Similarly, the total flux linkages with rotor winding can be expressed as :

$$\lambda_r = L_{rr} i_r + L_{rs} i_s$$

where L_{rr} = Self-inductance of the rotor winding

L_{rs} = Mutual inductance between rotor and stator windings

The total energy W_{fld} stored in the coupling magnetic field is

$$W_{fld} = \frac{1}{2} L_{ss} i_s^2 + \frac{1}{2} L_{rr} i_r^2 + L_{sr} i_s i_r$$

Following the procedure used to determine an expression for force developed in a translational actuator, it may be shown that the torque developed in a rotational electromagnetic system is

$$T = \left. \frac{\partial W'_{fld}(i, \theta)}{\partial \theta} \right|_{i = \text{constant}}$$

In a linear magnetic system, energy and coenergy are the same *i.e.* $W_{fld} = W'_{fld}$.

$$\therefore \quad T = \frac{1}{2} i_s^2 \frac{dL_{ss}}{d\theta} + \frac{1}{2} i_r^2 \frac{dL_{rr}}{d\theta} + i_s i_r \frac{dL_{sr}}{d\theta} \quad ...(i)$$

The first two terms on the right-hand side of eq. (*i*) represent torques produced in the machine because of variation of self-inductance with rotor position. This component of torque is called the *reluctance torque*. The third term represents torque produced by the variation of the mutual inductance between the stator and rotor windings.

Cylindrical machines. In cylindrical machines, the rotor as well as stator is round with uniform air gap. If the effects of slots on the stator and rotor are neglected, the reluctance of the magnetic path is independent of the position of the rotor. Therefore, self-inductances L_{ss} and L_{rr} are constant and no

* Most of energy converters, particularly the higher-power ones produce rotational motion.

reluctance torques are produced. The mutual inductance L_{sr} varies with rotor position and the torque produced in a cylindrical machine is given by ;

$$T = i_s i_r \frac{dL_{sr}}{d\theta}$$

Example 1.12. *For the electromagnetic system shown in Fig. 1.30, the rotor has no winding and the inductance of the stator as a function of the rotor position θ is $L_{ss} = L_0 + L_2 \cos 2\theta$ (See Fig. 1.30). The stator current is $i_s = I_{sm} \sin \omega t$. Find an expression for the torque acting on the rotor.*

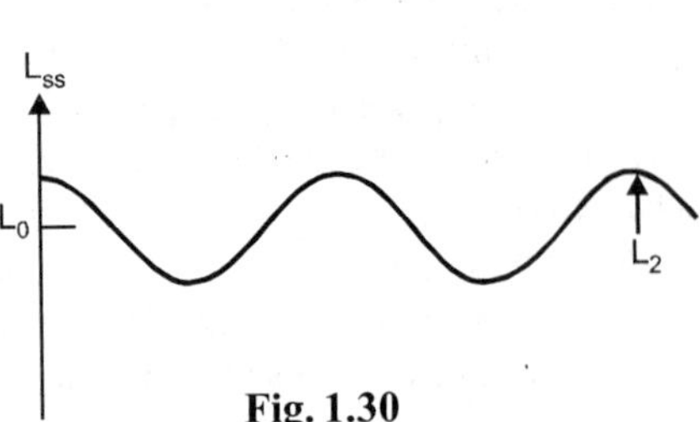

Fig. 1.30

Solution.
$$T = \frac{1}{2} i_s^2 \frac{dL_{ss}}{d\theta} + \frac{1}{2} i_r^2 \frac{dL_{rr}}{d\theta} + i_s i_r \frac{dL_{sr}}{d\theta}$$

Since rotor has no winding, $i_r = 0$.

$$\therefore \quad T = \frac{1}{2} i_s^2 \frac{dL_{ss}}{d\theta}$$

Here $\quad i_s = I_{sm} \sin \omega t \,; \qquad L_{ss} = L_0 + L_2 \cos 2\theta$

$$\therefore \quad T = \frac{1}{2} \times (I_{sm} \sin \omega t)^2 \times \frac{d}{d\theta}(L_0 + L_2 \cos 2\theta)$$

$$= -\frac{1}{2} I_{sm}^2 \sin^2 \omega t \times L_2 \sin 2\theta \times 2$$

or
$$\mathbf{T = -I_{sm}^2 L_2 \sin 2\theta \sin^2 \omega t \ Nm}$$

Example 1.13. *For the system shown in Fig. 1.31, the various inductances are :*

$$L_{11} = (4 + \cos 2\theta) \times 10^{-3} H$$
$$L_{12} = 0.15 \cos \theta H$$
$$L_{22} = (20 + 5 \cos 2\theta) H$$

Find the torque developed if $i_1 = 1$ A and $i_2 = 0.02$ A.

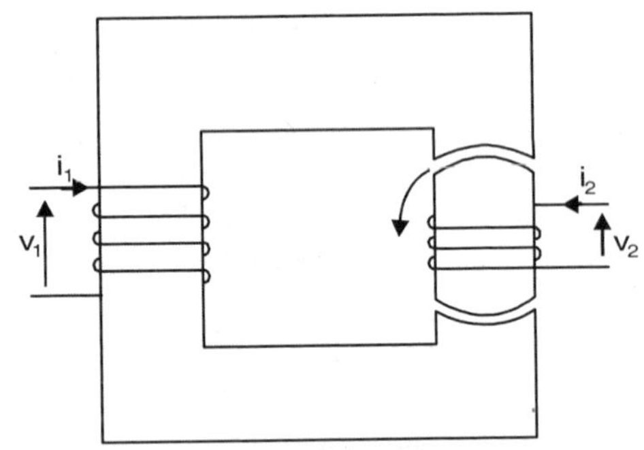

Fig. 1.31

Solution. Since the expression for the torque should be a function of the currents i_1 and i_2 rather than the flux linkages, we will use the coenergy of the system as the state function. The coenergy of the system is given by ;

$$W'_{fld} = \frac{1}{2} i_1^2 L_{11} + \frac{1}{2} i_2^2 L_{22} + i_1 i_2 L_{12}$$

$$= \frac{1}{2} i_1^2 (4 + \cos 2\theta) \times 10^{-3} + \frac{1}{2} i_2^2 (20 + 5 \cos 2\theta) + i_1 i_2 (0.15 \cos \theta)$$

$$\therefore \quad T = \frac{\partial W'_{fld}}{\partial \theta}$$

$$= -\sin 2\theta \times 10^{-3} i_1^2 - 5 \sin 2\theta \, i_2^2 - 0.15 \sin \theta \, i_1 i_2$$

$$= -\sin 2\theta \times 10^{-3} \times (1)^2 - 5 \sin 2\theta \times (0.02)^2 - 0.15 \sin \theta \times (1 \times 0.02)$$

or
$$\mathbf{T = -3 \times 10^{-3} \sin 2\theta - 3 \times 10^{-3} \sin \theta \ Nm}$$

Example 1.14. *The dynamometer type of instrument illustrated in Fig. 1.32 consists of a fixed coil 1 and a moving coil 2 suspended on a pivot acting against a spring. Find an expression for the average torque when the instrument is used as an a.c. ammeter.*

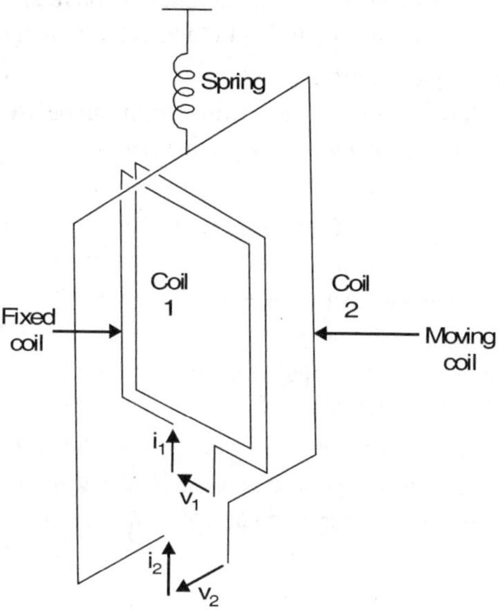

Fig. 1.32

Solution. The self-inductances of the two coils are independent of the angular displacement θ between them. The mutual inductance L_{12} between the coils will be maximum when $\theta = 0°$ and zero when $\theta = 90°$ so that $L_{12} = M \cos \theta$.

Now $$T = \frac{1}{2} i_1^2 \frac{dL_{11}}{d\theta} + \frac{1}{2} i_2^2 \frac{dL_{22}}{d\theta} + i_1 i_2 \frac{dL_{12}}{d\theta}$$

Now L_{11} and L_{22} are constant so that torque is given by ;

$$T = i_1 i_2 \frac{dL_{12}}{d\theta} = i_1 i_2 \frac{d}{d\theta}(M \cos\theta) = -i_1 i_2 M \sin\theta$$

When the instrument is used as an a.c. ammeter,

$$i_1 = i_2 = I_m \cos \omega t$$

$$\therefore \quad T = -(I_m \cos \omega t)(I_m \cos \omega t) M \sin\theta$$

$$= -I_m^2 M \sin\theta \cos^2 \omega t$$

$$\therefore \quad T = -\frac{I_m^2 M \sin\theta}{2}(1 + \cos 2\omega t)$$

The average torque (T_{av}) of this torque over the complete cycle is given by;

$$T_{av} = \frac{1}{2\pi} \int_0^{2\pi} -\frac{I_m^2 M \sin\theta}{2}(1 + \cos 2\omega t)\, d(\omega t)$$

$$\therefore \quad \boldsymbol{T_{av} = -\frac{I_m^2 M \sin\theta}{2}} \textbf{ Nm}$$

Thus the deflection of the moving coil is directly proportional to $(\text{current})^2$ for small angular displacements.

SHORT ANSWER QUESTIONS

Q. 1. *What are the essential parts of an electromechanical energy conversion system ?*

Ans. The three essential parts of electromechanical energy conversion system are (*i*) electrical system (*ii*) mechanical system (*iii*) coupling field (electric or magnetic field).

Q. 2. *What is the significance of coenergy ?*

Ans. The coenergy has no physical significance. However, it can be used to derive expressions for force (or torque) developed in an electromagnetic system.

Q. 3. *What are actuators ?*

Ans. The energy conversion devices that are used to produce translational forces are known as actuators. Examples include solenoids, relays and electromagnets.

Q. 4. *How does electromechanical energy conversion take place ?*

Ans. The process of electromechanical energy conversion takes place through the medium of the electric field or magnetic field. Although the various conversion devices operate on similar physical principles, the structures of the devices depend on their function.

Q. 5. *What is the fundamental principle involved in electromechanical energy conversion ?*

Ans. The electromechanical energy conversion is based on the principle of conservation of energy which states that energy can neither be created nor destroyed; it can only be changed from one form to another.

Q. 6. *What are the dimensions of coenergy ?*

Ans. The dimensions of coenergy are the same as that of energy.

Q. 7. *What is the drawback of electric field as a medium for electromechanical energy conversion ?*

Ans. When electric field is used as a medium for electromechanical energy conversion, the amount of force developed is usually very small.

Q. 8. *What is the advantage of magnetic field as a medium for electromechanical energy conversion ?*

Ans. When magnetic field is used as a medium for electromechanical energy conversion, the amount of force developed is very large as compared to the case when electric field is used as the medium.

Q. 9. *Why do we neglect magnetic non-linearity and core losses in electromagnetic energy conversion devices ?*

Ans. Electromagnetic energy conversion devices are built with air gaps in the magnetic circuit to separate the fixed and moving parts. Most of the m.m.f. of the windings is required to overcome the air-gap reluctance. Therefore, we can neglect the reluctance of the iron part of the magnetic circuit. Moreover, the core losses are very small and can be neglected without affecting too much accuracy. However, these effects may be taken into account if required in the design of the device.

Q. 10. *What are the losses that occur during electromechanical energy conversion ?*

Ans. The losses that occur during electromechanical energy conversion are :

(*i*) i^2R losses in the windings.

(*ii*) friction and windage losses.

(*iii*) core losses.

MULTIPLE-CHOICE QUESTIONS

1. The losses that occur during electromechanical energy conversion are converted into...............
   (*i*) heat (*ii*) sound
   (*iii*) light (*iv*) none of above
2. In a linear magnetic circuit, the relation between energy (W_{fld}) and coenergy (W'_{fld}) is...............
   (*i*) $W_{fld} > W'_{fld}$ (*ii*) $W_{fld} = W'_{fld}$
   (*iii*) $W_{fld} < W'_{fld}$ (*iv*) none of above
3. In a non-linear magnetic circuit, the relation between energy (W_{fld}) and coenergy (W'_{fld}) is...............
   (*i*) $W_{fld} = W'_{fld}$ (*ii*) $W_{fld} > W'_{fld}$
   (*iii*) $W_{fld} < W'_{fld}$ (*iv*) none of above
4. ....is the example of singly excited magnetic field system.
   (*i*) Relay
   (*ii*) D.C. shunt motor
   (*iii*) Synchronous motor
   (*iv*) None of above
5. ....is the example of multiply excited magnetic field system.
   (*i*) Relay
   (*ii*) Solenoid
   (*iii*) D.C. shunt motor
   (*iv*) None of above
6. The examples of continuous energy conversion devices are...............
   (*i*) motors and generators
   (*ii*) relays
   (*iii*) solenoids
   (*iv*) none of above
7. For air, the curve between flux linkages (λ) and current (i) is...............
   (*i*) a parabola (*ii*) a hyperbola
   (*iii*) a straight line (*iv*) none of above
8. Electromagnetic energy conversion devices are built with air gaps in the magnetic circuit to...............
   (*i*) increase reluctance of magnetic path
   (*ii*) decrease reluctance of magnetic path
   (*iii*) separate the fixed and moving parts
   (*iv*) none of above.
9. For iron, the curve between flux linkages (λ) and current (i) is...............
   (*i*) linear
   (*ii*) nonlinear but not circle
   (*iii*) circle
   (*iv*) none of above
10. When electric field is used as the medium for electromechanical energy conversion, the force developed is...............
    (*i*) zero (*ii*) very small
    (*iii*) very large (*iv*) none of above

ANSWERS TO MULTIPLE-CHOICE QUESTIONS

1. (*i*)	**2.** (*ii*)	**3.** (*iii*)	**4.** (*i*)	**5.** (*iii*)
6. (*i*)	**7.** (*iii*)	**8.** (*iii*)	**9.** (*ii*)	**10.** (*ii*)

2

D.C. Generators

INTRODUCTION

Although a far greater percentage of the electrical machines in service are a.c. machines, the d.c. machines are of considerable industrial importance. The principal advantage of the d.c. machine, particularly the d.c. motor, is that it provides a fine control of speed. Such an advantage is not claimed by any a.c. motor. However, d.c. generators are not as common as they used to be, because direct current, when required, is mainly obtained from an a.c. supply by the use of rectifiers. Nevertheless, an understanding of d.c. generator is important because it represents a logical introduction to the behaviour of d.c. motors. Indeed many d.c. motors in industry actually operate as d.c. generators for a brief period. In this chapter, we shall deal with various aspects of d.c. generators.

2.1 GENERATOR PRINCIPLE

An electric generator is a machine that converts mechanical energy into electrical energy.

An electric generator is based on the principle that whenever flux is cut by a conductor, an e.m.f. is induced which will cause a current to flow if the conductor circuit is closed. The direction of induced e.m.f. (and hence current) is given by *Fleming's right hand rule. Therefore, the essential components of a generator are :

(*a*) a magnetic field

(*b*) conductor or a group of conductors

(*c*) motion of conductor *w.r.t.* magnetic field

2.2 SIMPLE LOOP GENERATOR

Consider a single turn loop *ABCD* rotating clockwise in a uniform magnetic field with a constant speed as shown in Fig. 2.1 As the loop rotates, the flux linking the coil sides *AB* and *CD* changes continuously. Hence the e.m.f. induced in these coil sides also changes but the e.m.f. induced in one coil side **adds to that induced in the other.

(*i*) When the loop is in position no. 1 [See Fig. 2.1], the generated e.m.f. is zero because the coil sides (*AB* and *CD*) are cutting no flux but are moving parallel to it.

(*ii*) When the loop is in position no. 2, the coil sides are moving at an angle to the flux and, therefore, a low e.m.f. is generated as indicated by point 2 in Fig. 2.2.

(*iii*) When the loop is in position no. 3, the coil sides (*AB* and *CD*) are at right angle to the flux and are, therefore, cutting the flux at a maximum rate. Hence at this instant, the generated e.m.f. is maximum as indicated by point 3 in Fig. 2.2.

(*iv*) At position 4, the generated e.m.f. is less because the coil sides are cutting the flux at an angle.

---

* **Fleming's Right hand rule.** Stretch the thumb, fore-finger and middle finger of your right hand so that they are at right angles to each other. If the fore-finger points in the direction of field, thumb in the direction of motion of the conductor, then middle finger will point in the direction of induced e.m.f.

** It is because the coil sides always remain under the influence of opposite poles *i.e.,* if one coil side is under the influence of the *N*-pole, then the other coil side will be under the influence of *S*-pole and *vice-versa*.

(v) At position 5, no magnetic lines are cut and hence induced e.m.f. is zero as indicated by point 5 in Fig. 2.2.

(vi) At position 6, the coil sides move under a pole of opposite polarity and hence the direction of generated e.m.f. is reversed. The maximum e.m.f. in this direction (*i.e.,* reverse direction, See Fig. 2.2) will be when the loop is at position 7 and zero when at position 1. This cycle repeats with each revolution of the coil.

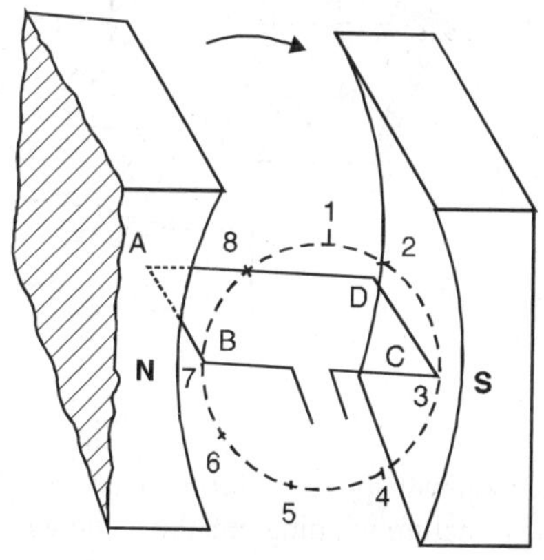

Fig. 2.1

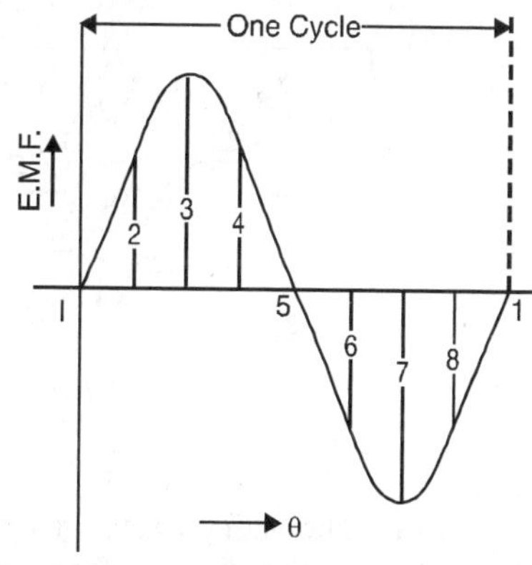

Fig. 2.2

Note that e.m.f. generated in the loop is alternating one. It is because any coil side, say *AB* has e.m.f. in one direction when under the influence of *N*-pole and in the other direction when under the influence of *S*-pole. If a load is connected across the ends of the loop, then alternating current will flow through the load. The alternating voltage generated in the loop can be converted into direct voltage by a device called *commutator*. We then have the d.c. generator. In fact, a commutator is a mechanical rectifier.

2.3 ACTION OF COMMUTATOR

If, somehow, connection of the coil side to the external load is reversed at the same instant the current in the coil side reverses, the current through the load will be direct current. This is what a commutator does. Fig. 2.3 shows a commutator having two segments C_1 and C_2. It consists of a cylindrical metal ring cut into two halves or segments C_1 and C_2 respectively separated by a thin sheet of mica. The commutator is mounted on but insulated from the rotor shaft. The ends of coil sides *AB* and *CD* are connected to the segments C_1 and C_2 respectively as shown in Fig. 2.4. Two stationary carbon brushes rest on the commutator and lead current to the external load. With this arrangement, the commutator at all times connects the coil side under *S*-pole to the +*ve* brush and that under *N*-pole to the –*ve* brush.

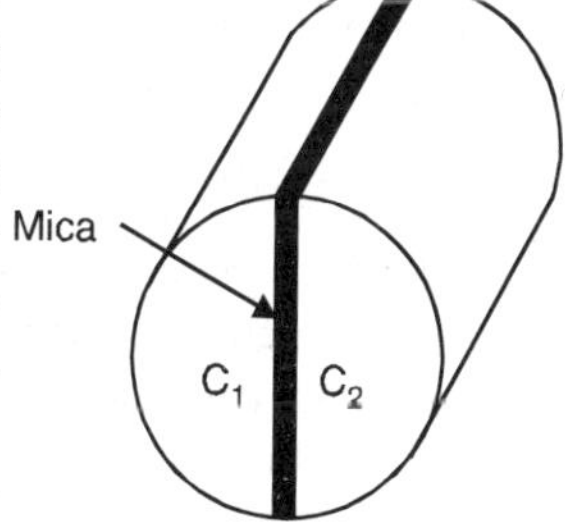

Fig. 2.3

(i) In Fig. 2.4, the coil sides *AB* and *CD* are under *N*-pole and *S*-pole respectively. Note that segment C_1 connects the coil side *AB* to point *P* of the load resistance *R* and the segment C_2 connects the coil side *CD* to point *Q* of the load. Also note the direction of current through load. It is from *Q* to *P*.

(ii) After half a revolution of the loop (*i.e.,* 180° rotation), the coil side *AB* is under *S*-pole and the coil side *CD* under *N*-pole as shown in Fig. 2.5. The currents in the coil sides now flow in the reverse direction but the segments C_1 and C_2 have also moved through 180° *i.e.,* segment C_1 is now in contact with +*ve* brush and segment C_2 in contact with –*ve* brush.

Note that commutator has reversed the coil connections to the load *i.e.*, coil side *AB* is now connected to point *Q* of the load and coil side *CD* to the point *P* of the load. Also note the direction of current through the load. It is *again from *Q* to *P*.

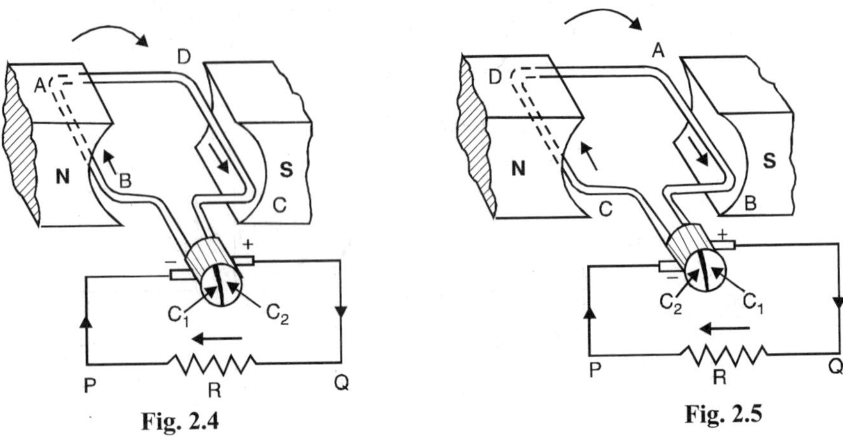

Fig. 2.4 Fig. 2.5

Thus the alternating voltage generated in the loop will appear as direct voltage across the brushes. The reader may note that e.m.f. generated in the armature winding of a d.c. generator is alternating one. It is by the use of commutator that we convert the generated alternating e.m.f. into direct voltage. The purpose of brushes is simply to lead current from the rotating loop or winding to the external stationary load.

The variation of voltage across the brushes with the angular displacement of the loop will be as shown in Fig. 2.6. This is not a steady direct voltage but has a pulsating character. It is because the voltage appearing across the brushes varies from zero to maximum value and back to zero twice for each revolution of the loop. A pulsating direct voltage such as is produced by a single loop is not suitable for many commercial uses. What we require is the steady direct voltage. This can be achieved by using a large number of coils connected in series. The resulting arrangement is known as *armature winding*.

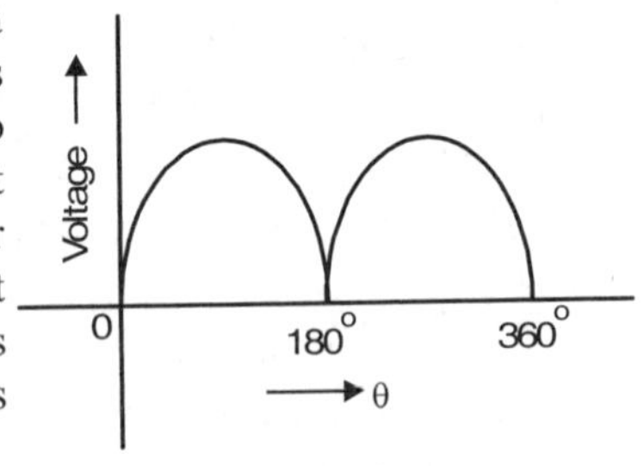

Fig. 2.6

2.4 CONSTRUCTION OF D.C. GENERATOR

The d.c. generators and d.c. motors have the same general construction. In fact, when the machine is being assembled, the workmen usually do not know whether it is a d.c. generator or motor. Any d.c. generator can be run as a d.c. motor and *vice-versa*. All d.c. machines have five principal components *viz.*, (*i*) field system (*ii*) armature core (*iii*) armature winding (*iv*) commutator (*v*) brushes [See Fig. 2.7].

(*i*) **Field system.** The function of the field system is to produce uniform magnetic field within which the armature rotates. It consists of a number of salient poles (of course, even number) bolted to the inside of circular frame (generally called *yoke*). The yoke is usually made of solid cast steel whereas the pole pieces are composed of stacked laminations. Field coils are mounted on the poles and carry the d.c. exciting current. The field coils are connected in such a way that adjacent poles have opposite polarity.

* The situation is that currents in the coil sides are reversed and at the same time the connections of the coil sides to the external load are reversed. This means that current will flow in the same direction through the load.

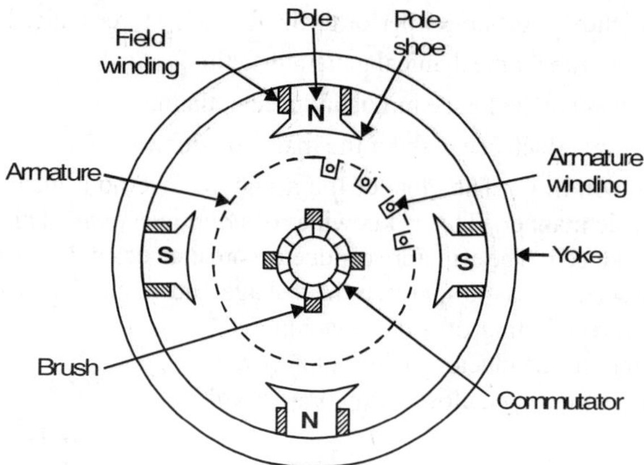

Fig. 2.7

The m.m.f. developed by the field coils produces a magnetic flux that passes through the pole pieces, the air gap, the armature and the frame (See Fig. 2.8). Practical d.c. machines have air gaps ranging from 0·5 mm to 1·5 mm. Since armature and field systems are composed of materials that have high permeability, most of the m.m.f. of field coils is required to set up flux in the air gap. By reducing the length of air gap, we can reduce the size of field coils (*i.e.* number of turns).

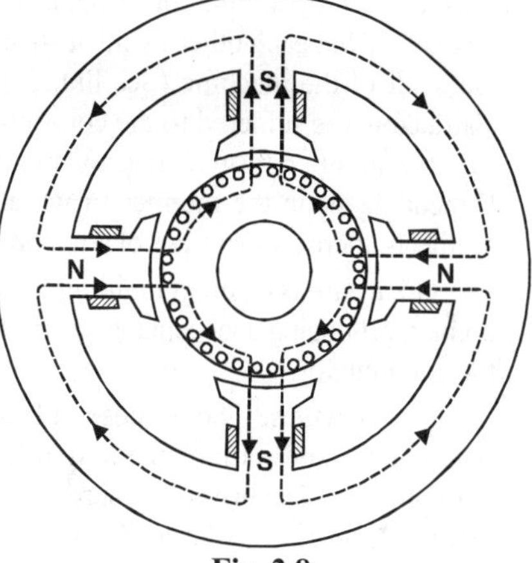

Fig. 2.8

The yoke not only serves as a support for field poles, interpoles (not shown), compensating windings (not shown) etc. but it also provides return path for the magnetic flux created by the field windings.

(*ii*) **Armature core.** The armature core is keyed to the machine shaft and rotates between the field poles. It consists of slotted soft-iron laminations (about 0·4 to 0·6 mm thick) that are stacked to form a cylindrical core as shown in Fig 2.9. The laminations (See Fig. 2.10) are individually coated with a thin insulating film so that they do not come in electrical contact with each other. The purpose of laminating the core is to reduce the eddy current loss. The laminations are slotted to accommodate and provide mechanical security to the armature winding and to give shorter air gap for the flux to cross between the pole face and the armature "teeth".

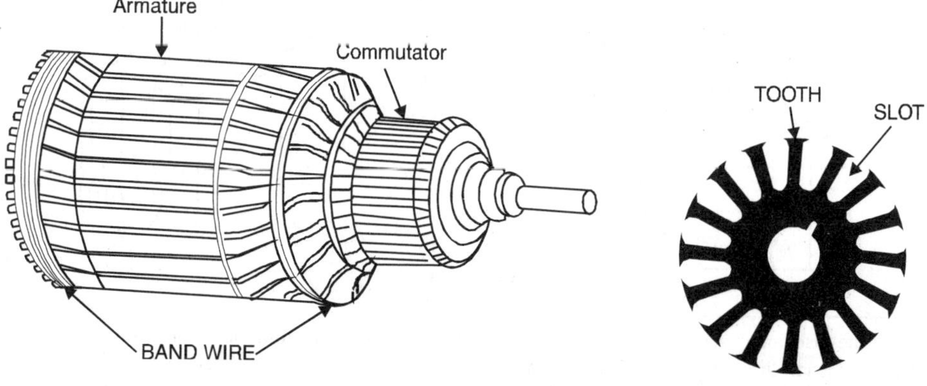

Fig. 2.9

Fig. 2.10

The armature of the d.c. generator performs the following three major functions :

(*a*) It permits rotation for mechanical generator action.

(*b*) Since it houses conductors, e.m.f. is induced in them.

(*c*) It provides low reluctance path for the magnetic flux.

(*iii*) **Armature winding.** The slots of the armature core hold insulated conductors that are connected in a suitable manner. This is known as *armature winding.* This is the winding in which "working" e.m.f. is induced. The armature conductors are connected in series-parallel; the conductors being connected in series so as to increase the voltage and in parallel paths so as to increase the current. The armature winding of a d.c. machine is a closed-circuit winding; the conductors being connected in a symmetrical manner forming a closed loop or series of closed loops.

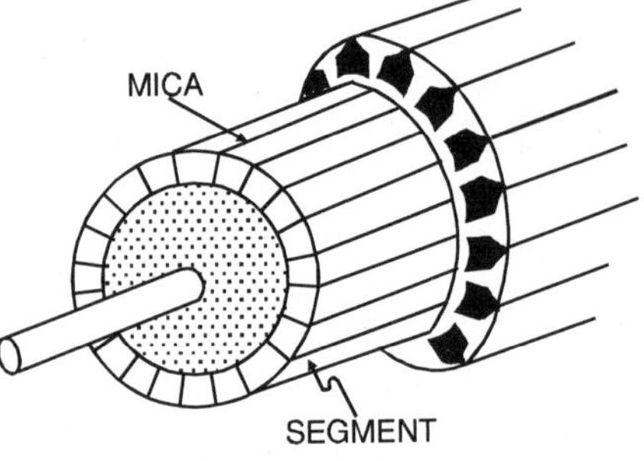

Fig. 2.11

(*iv*) **Commutator.** A commutator is a mechanical rectifier which converts the alternating voltage generated in the armature winding into direct voltage across the brushes. The commutator is made of copper segments insulated from each other by mica sheets and mounted on the shaft of the machine (See Fig. 2.11). The armature conductors are soldered to the commutator segments in a suitable manner to give rise to the armature winding. Depending upon the manner in which the armature conductors are connected to the commutator segments, there are two types of armature winding in a d.c. machine *viz.,* (*a*) lap winding (*b*) wave winding.

Great care is taken in building the commutator because any eccentricity will cause the brushes to bounce, producing unacceptable sparking. The sparks may burn the brushes and overheat and carbonise the commutator.

(*v*) **Brushes.** The purpose of brushes is to ensure *electrical connections between the rotating commutator and stationary external load circuit. The brushes are made of **carbon and rest on the commutator. The brush pressure is adjusted by means of adjustable springs (See Fig. 2.12). If the brush pressure is very large, the friction produces heating of the commutator and the brushes. On the other hand, if it is too weak, the imperfect contact with the commutator may produce sparking.

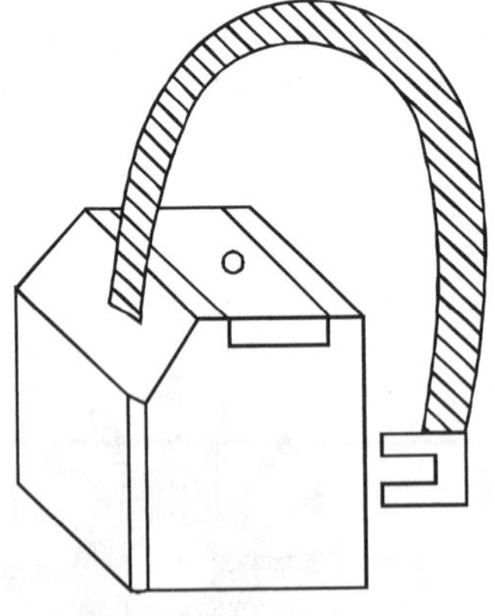

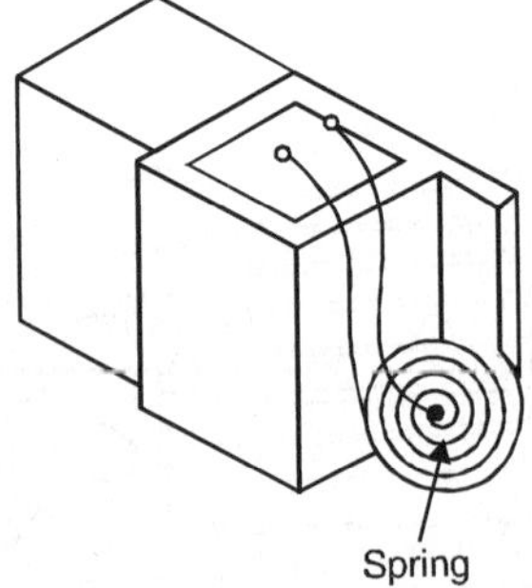

Fig. 2.12

* When the machine is acting as a generator, the brushes carry current from the commutator to the external stationary load. In case the machine is acting as a motor, they feed the supply current to the commutator.

** Because carbon has good electrical conductivity and is soft enough not to scratch the commutator. To improve the conductivity, sometimes we add a small amount of copper.

Multipole machines have as many brushes as they have poles. For example, a 4-pole machine has 4 brushes. As we go round the commutator, the successive brushes have positive and negative polarities. Brushes having the same polarity are connected together so that we have two terminals *viz.*, the +*ve* terminal and the −*ve* terminal.

2.5 GENERAL FEATURES OF D.C. ARMATURE WINDINGS

(*i*) A d.c. machine (generator or motor) generally employs windings distributed in slots over the circumference of the armature core. Each conductor lies at right angles to the magnetic flux and to the direction of its movement. Therefore, the induced e.m.f. in the conductor is given by ;

$$e = B\,l\,v \text{ volts}$$

where B = magnetic flux density in Wb/m^2

l = length of the conductor in metres

v = velocity (in m/s) of the conductor

(*ii*) The armature conductors are connected to form *coils*. The basic component of all types of armature windings is the armature coil. Fig. 2.13 (*i*) shows a single-turn coil. It has two *conductors* or *coil sides* connected at the *back of the armature. Fig. 2.13 (*ii*) shows a 4-turn coil which has 8 conductors or coil sides.

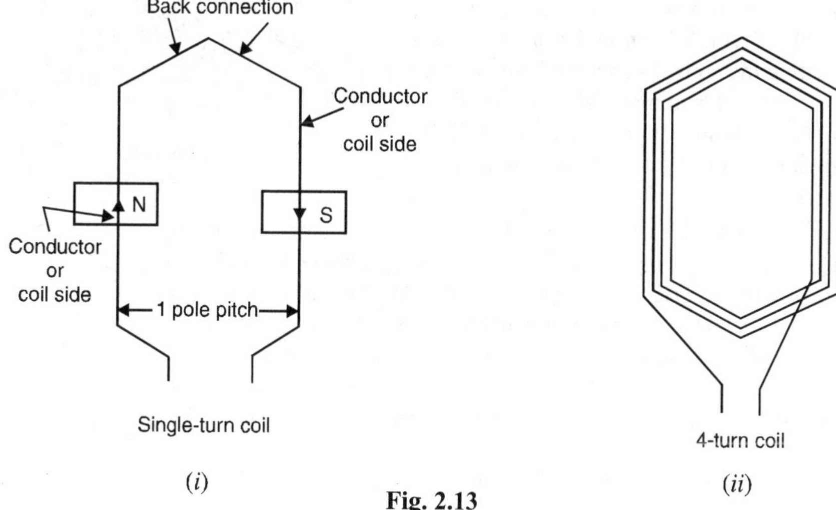

Fig. 2.13

The coil sides of a coil are placed a pole span apart i.e., one coil side of the coil is under *N*-pole and the other coil side is under the next *S*-pole at the corresponding position as shown in Fig. 2.13 (*i*). Consequently the e.m.f.s of the coil sides **add together. If the e.m.f. induced in one conductor is 2·5 volts, then the e.m.f. of a single-turn coil will be = 2 × 2·5 = 5 volts. For the same flux and speed, the e.m.f. of a 4-turn coil will be = 8 × 2·5 = 20 V.

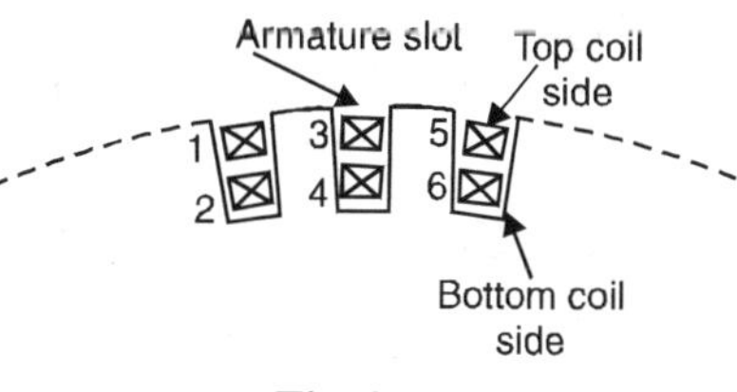

Fig. 2.14

(*iii*) Most of d.c. armature windings are †**double layer windings** *i.e.,* there are two coil sides per slot as shown in Fig. 2.14. One coil side of a coil lies at the top of a slot and the other coil side lies at the bottom of some other slot. The coil ends will then lie side by side. In two-layer

* That part of the coil that crosses at the back of the armature (back connections) does not have a voltage induced in it and is inactive in generating e.m.f.

** As the armature rotates, the induced e.m.f.s in the conductors separated by a pole pitch apart are always in opposite directions so that they aid each other.

† Coils are arranged in two layers.

winding, it is desirable to number the coil sides rather than the slots. The coil sides are numbered as indicated in Fig. 2.14. The coil sides at the top of slots are given odd numbers and those at the bottom are given even numbers. The coil sides are numbered in order round the armature.

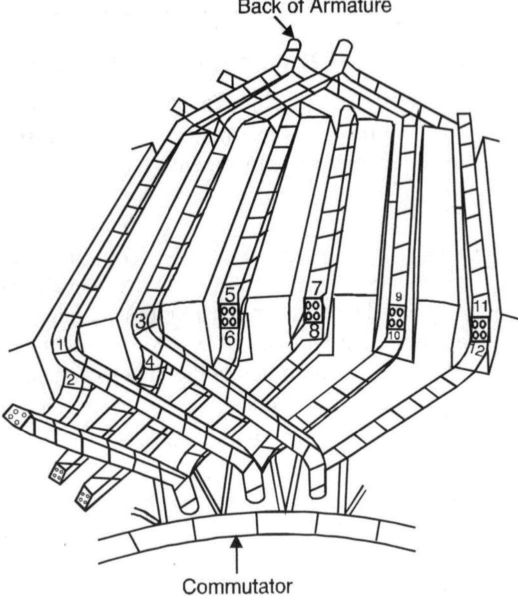

Fig. 2.15

As discussed above, each coil has one side at the top of a slot and the other side at the bottom of another slot ; the coil sides are nearly a pole pitch apart. In connecting the coils, it is ensured that top coil side is joined to the bottom coil side and *vice-versa*. This is illustrated in Fig. 2.15. The coil side 1 at the top of a slot is joined to coil side 10 at the bottom of another slot about a pole pitch apart. The coil side 12 at the bottom of a slot is joined to coil side 3 at the top of another slot. How coils are connected at the back of the armature and at the front (commutator end) will be discussed in later sections. It may be noted that as far as connecting the coils is concerned, the number of turns per coil is immaterial. For simplicity, then, the coils in winding diagrams will be represented as having only *one turn (*i.e.,* two conductors).

(*iv*) The coil sides are connected through commutator segments in such a manner as to form a series-parallel system ; a number of conductors are connected in series so as to increase the voltage and two or more such series-connected paths in parallel to share the current. Fig. 2.16 shows how the two coils connected through commutator segments (A, B, C etc) have their e.m.f.s added together. If voltage induced in each conductor is 2·5 V, then voltage between segments A and $C = 4 \times 2{\cdot}5 = 10$ V. It may be noted here that in the conventional way of representing a developed armature winding, full lines represent top coil sides (*i.e.,* coil sides lying at the top of a slot) and dotted lines represent the bottom coil sides (*i.e.,* coil sides lying at the bottom of a slot).

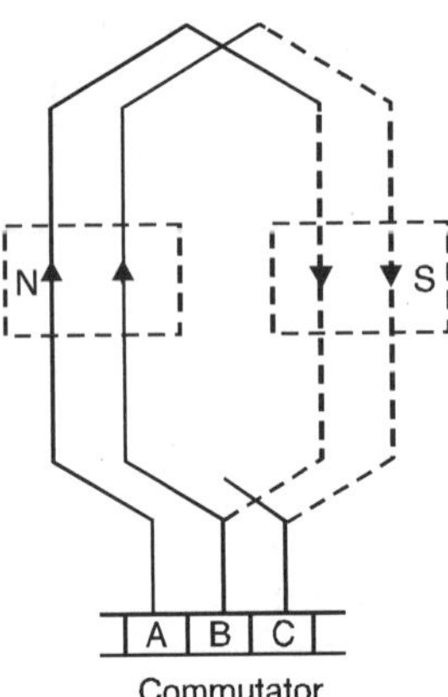

Fig. 2.16

(*v*) The d.c. armature winding is a closed circuit winding. In such a winding, if one starts at some point in the winding and traces through the winding, one will come back to the starting point without passing through any external connection. D.C. armature windings must be of the closed type in order to provide for the commutation of the coils.

2.6 COMMUTATOR PITCH (Y_C)

The commutator pitch is the number of commutator segments spanned by each coil of the winding. It is denoted by Y_C.

In Fig. 2.17, one side of the coil is connected to commutator segment 1 and the other side connected to commutator segment 2. Therefore, the number of commutator segments spanned by the coil is 1 *i.e.,* $Y_C = 1$. In Fig. 2.18, one side of the coil is connected to commutator segment 1 and the other side to commutator segment 8. Therefore, the number of commutator segments spanned by the coil = 8 –

---

* For single turn coils, there will be as many conductors as the coil sides.

1 = 7 segments *i.e.,* $Y_C = 7$. The commutator pitch of a winding is always a whole number. Since each coil has two ends and as two coil connections are joined at each commutator segment,

∴ Number of coils = Number of commutator segments

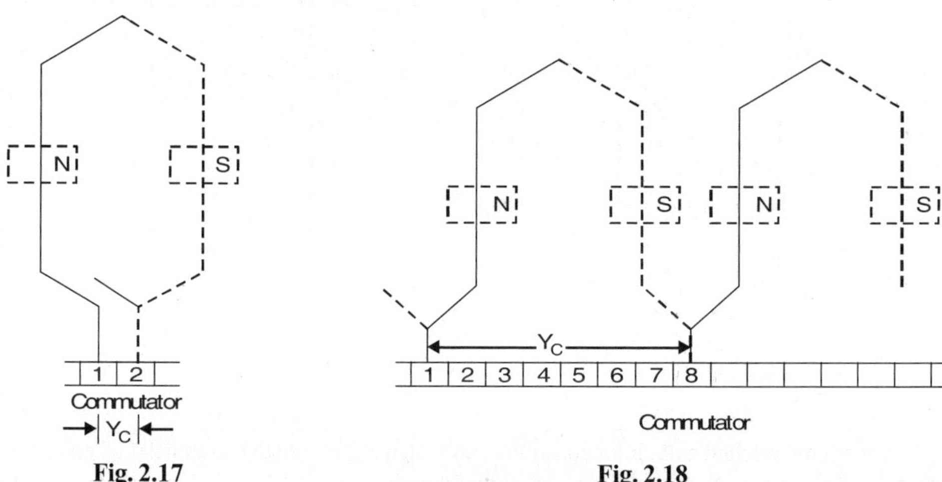

Fig. 2.17 **Fig. 2.18**

For example, if an armature has 30 conductors, the number of coils will be 30/2 = 15. Therefore, number of commutator segments is also 15. *Note that commutator pitch is the most important factor in determining the type of d.c. armature winding.*

2.7 POLE-PITCH

It is the distance measured in terms of number of armature slots (or armature conductors) per pole. Thus if a 4-pole generator has 16 coils, then number of slots = 16.

$$\therefore \quad \text{Pole pitch} = \frac{16}{4} = 4 \text{ slots}$$

Also
$$\text{Pole pitch} = \frac{\text{No. of conductors}}{\text{No. of poles}} = \frac{16 \times 2}{4} = 8 \text{ conductors}$$

The pole pitch is an important factor in the design of armature winding.

2.8 COIL SPAN OR COIL PITCH (Y_S)

It is the distance measured in terms of the number of armature slots (or armature conductors) spanned by a coil. Thus if the coil span is 9 slots, it means one side of the coil is in slot 1 and the other side in slot 10.

2.9 FULL-PITCHED COIL

If the coil span or coil pitch is equal to pole pitch (*i.e.* distance between the two sides of a coil is 180° electrical), it is called *full-pitched coil.* In other words, a full-pitched coil means that when one coil side is under the centre of *N*-pole, the other coil side must be under the centre of next *S*-pole as shown in Fig. 2.19. In this case, the e.m.f.s in the coil sides are additive and have a phase difference of 0°.If the e.m.f. induced in one coil side is 2.5V, then e.m.f. across the coil terminals = 2 × 2.5 = 5V. The e.m.f. induced in a full-pitched coil is maximum. Therefore, coil span or coil pitch should always be one pole pitch unless there is a good reason for making it shorter.

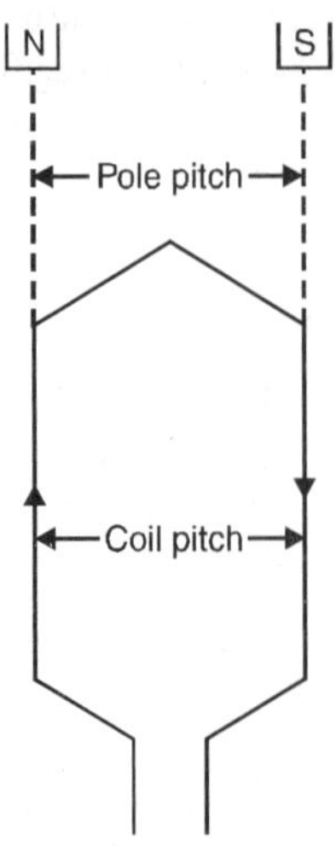

Fig. 2.19

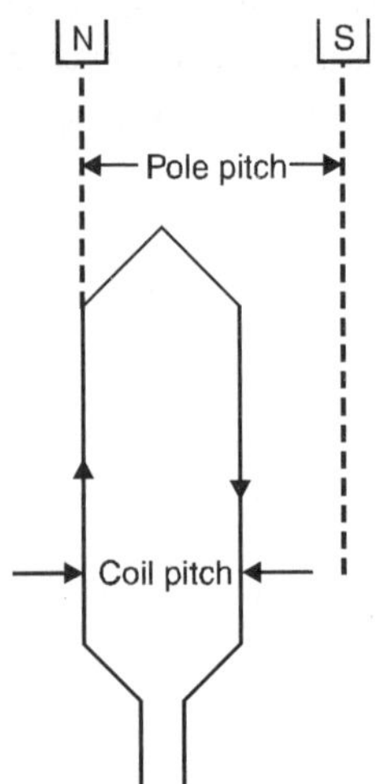

Fig. 2.20

Fractional pitched coil. If the coil span or coil pitch is less than the pole pitch (*i.e.* coil pitch is less than 180° electrical), then it is called *fractional pitched coil* (See Fig. 2.20). In this case, the phase difference between the e.m.f.s in the two coil sides will not be zero so that the e.m.f. in the coil will be less compared to full-pitched coil. Fractional pitch winding requires less copper but if the coil pitch is too small, an appreciable reduction in the generated e.m.f. results.

There are many situations where we cannot use full-pitched coils. For example, for a 2-pole machine, it is very difficult to place full-pitched coils. Similarly, if the number of slots in not *exactly divisible by the number of poles, we cannot use full-pitched coils. Under such situations, we have to resort to fractional pitch winding. In this case, the maximum possible pitch may be used as the pitch of the coil.

Example 2.1. *The armature of a d.c. generator has 10 slots. Calculate the coil pitch for a (i) 2-pole winding and (ii) 4-pole winding. Show the placement of coils in each case.*

Solution. A 10-slot armature using double-layer winding requires 10 coils.

(*i*) Number of slots, $S = 10$

Number of poles, $P = 2$

$$\therefore \quad \text{Number of slots/pole} = \frac{S}{P} = \frac{10}{2} = 5$$

Since the number of slots is exactly divisible by the number of poles, it is possible to use full-pitch coils. Therefore, coil pitch, Y_S = **5 slots.** The placement of coils in the armature is shown in Fig. 2.21.

Here one side of coil 1 is placed in slot 1 so that the other side of this coil is placed in slot 6. The coil 2 goes in slots 2 and 7, the coil 3 in slots 3 and 8 and so on. Since there are 5 slots per pole and a pole spans 180° electrical, the slot span = 180°/5 = 36°electrical. It means that the angle from the centre of one slot to the centre of the next slot is 36° electrical. Since the number of slots is equal to the number of teeth, each coil spans 5 teeth.

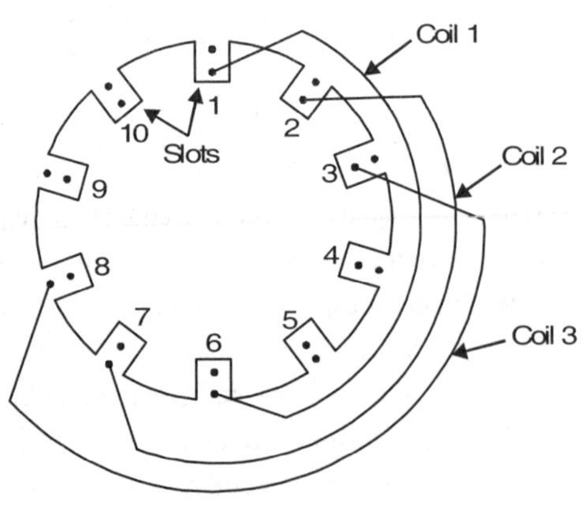

Fig. 2.21

* Remember that we always use double-layer winding.

(*ii*) Number of slots, $S = 10$

Number of poles, $P = 4$

$$\therefore \quad \text{Number of slots/pole} = \frac{S}{P} = \frac{10}{4} = 2.5$$

Since the number of slots is not exactly divisible by the number of poles, full-pitched coils cannot by used. The maximum number of slots that the coil can span is 2. Therefore, the coil pitch, Y_S = **2 slots.** The placement of coils in the armature is shown in Fig. 2.22.

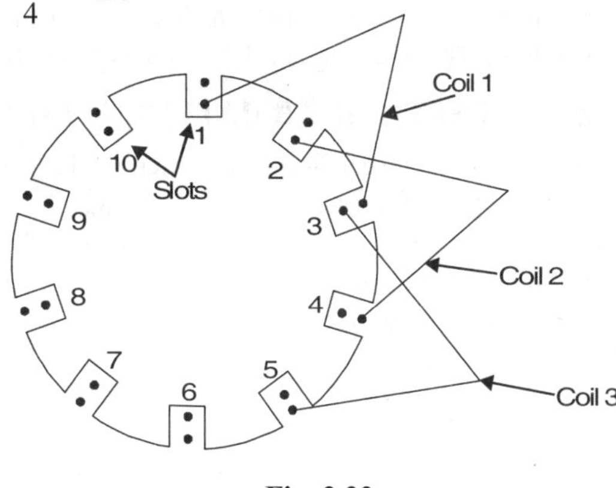

Fig. 2.22

Here one side of coil 1 is placed in slot 1 so that the other side of this coil is placed in slot 3. The coil 2 goes in slots 2 and 4, the coil 3 in slots 3 and 5 and so on. Note that in this case, slot span = 180°/2.5 = 72° electrical. Therefore, the coil pitch = 2 × 72° = 144° electrical. Note that for full-pitched coil, the coil pitch is 180° electrical.

2.10 TYPES OF D.C. ARMATURE WINDINGS

The different armature coils in a d.c. armature winding must be connected in series with each other by means of end connections (back connection and front connection) in a manner so that the generated voltages of the respective coils will aid each other in the production of the terminal e.m.f. of the winding. Two basic methods of making these end connections are :

1. Simplex lap winding
2. Simplex wave winding

1. Simplex lap winding. For a simplex lap winding, the commutator pitch $Y_C = 1$ and coil span $Y_S \simeq$ pole pitch. Thus the ends of any coil are brought out to adjacent commutator segments and the result of this method of connection is that all the coils of the armature are in sequence with the last coil connected to the first coil. Consequently, closed circuit winding results. This is illustrated in Fig. 2.23 where a part of the lap winding is shown. Only two coils are shown for simplicity. The name lap comes from the way in which successive coils overlap the preceding one.

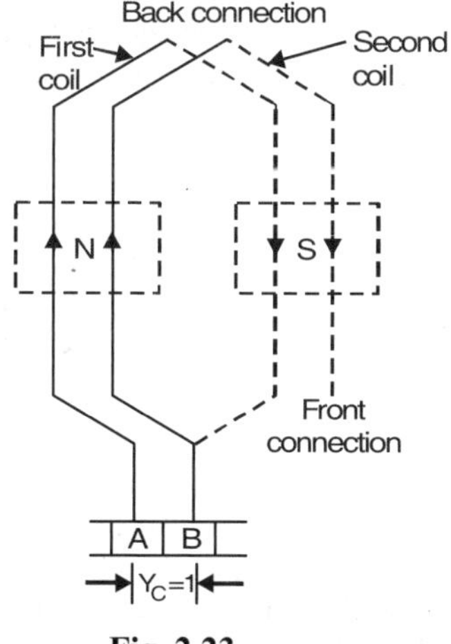

Fig. 2.23

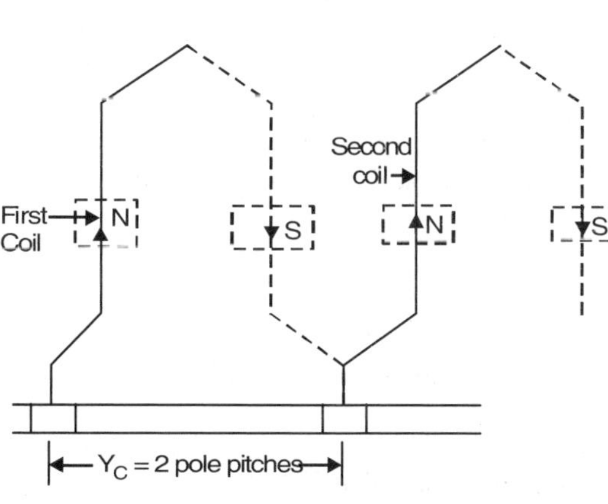

Fig. 2.24

2. Simplex wave winding. For a simplex wave winding, the commutator pitch $Y_C \simeq 2$ pole pitches and coil span $\simeq$ pole pitch. The result is that the coils under consecutive pole pairs will be joined together in series thereby adding together their e.m.f.s [See Fig. 2.24]. After passing once around the armature, the winding falls in a slot to the left or right of the starting point and thus connecting up another circuit. Continuing in this way, all the conductors will be connected in a single closed winding. This winding is called wave winding from the appearance (wavy) of the end connections.

2.11 FURTHER ARMATURE WINDING TERMINOLOGY

Apart from the terms discussed earlier, the following terminology requires discussion :

(*i*) **Back Pitch (Y_B).** It is the distance measured in terms of armature conductors between the two sides of a coil at the back of the armature (See Fig. 2.25). It is denoted by Y_B. For example, if a coil is formed by connecting conductor 1 (upper conductor in a slot) to conductor 12 (bottom conductor in another slot) at the back of the armature, then back pitch is $Y_B = 12 - 1 = 11$ conductors.

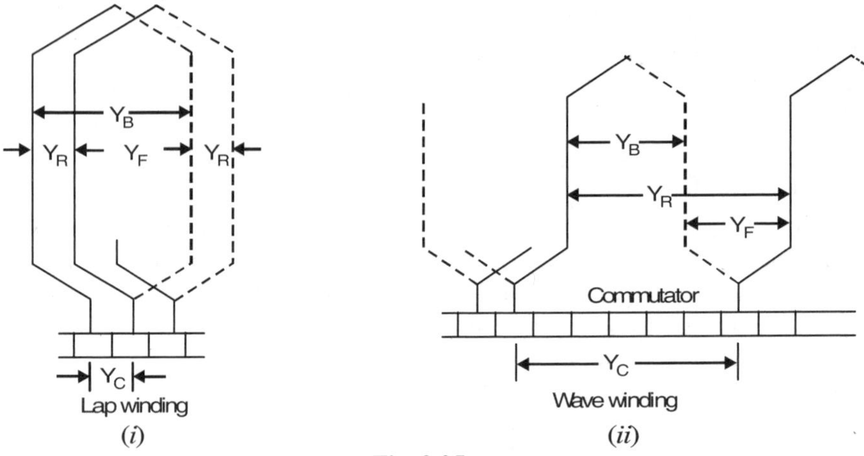

Fig. 2.25

(*ii*) **Front Pitch (Y_F).** It is the distance measured in terms of armature conductors between the coil sides attached to any one commutator segment [See Fig. 2.25]. It is denoted by Y_F. For example, if coil side 12 and coil side 3 are connected to the same commutator segment, then front pitch is $Y_F = 12 - 3 = 9$ conductors.

(*iii*) **Resultant Pitch (Y_R).** It is the distance (measured in terms of armature conductors) between the beginning of one coil and the beginning of the next coil to which it is connected (See Fig. 2.25). It is denoted by Y_R. Therefore, the resultant pitch is the algebraic sum of the back and front pitches.

(*iv*) **Commutator Pitch (Y_C).** It is the number of commutator segments spanned by each coil of the armature winding.

For simplex lap winding, $Y_C = 1$

For simplex wave winding, $Y_C \simeq 2$ pole pitches (segments)

(*v*) **Progressive Winding.** A progressive winding is one in which, as one traces through the winding, the connections to the commutator will progress around the machine in the same direction as is being traced along the path of each individual coil.

Fig. 2.26 (*i*) shows progressive lap winding. Note that $Y_B > Y_F$ and $Y_C = +1$.

(*vi*) **Retrogressive Winding.** A retrogressive winding is one in which, as one traces through the winding, the connections to the commutator will progress around the machine in the opposite direction to that which is being traced along the path of each individual coil. Fig. 2.26 (*ii*) shows retrogressive lap winding. Note that $Y_F > Y_B$ and $Y_C = -1$. A retrogressive winding is seldom used because it requires more copper.

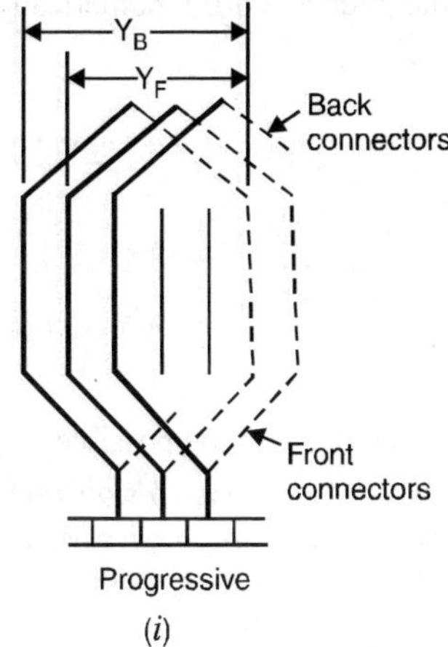

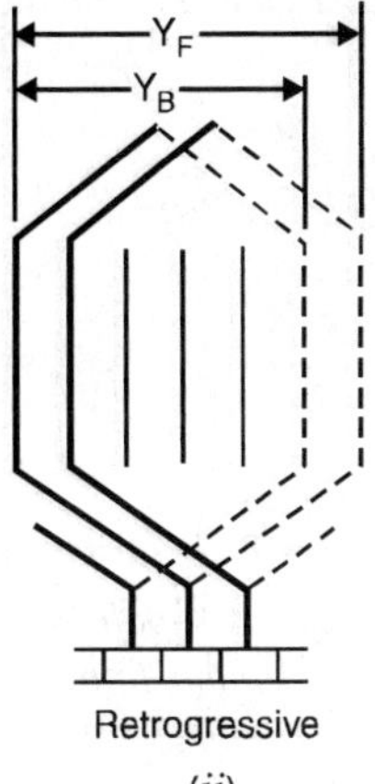

Fig. 2.26

2.12 GENERAL RULES FOR D.C. ARMATURE WINDINGS

In the design of d.c. armature winding (lap or wave), the following rules may be followed :

(*i*) The back pitch (Y_B) as well as front pitch (Y_F) should be nearly equal to pole pitch. This will result in increased e.m.f. in the coils.

(*ii*) Both pitches (Y_B and Y_F) should be odd. This will permit all end connections (back as well as front connection) between a conductor at the top of a slot and one at the bottom of a slot.

(*iii*) The number of commutator segments is equal to the number of slots or coils (or half the number of conductors).

No. of commutator segments = No. of slots = No. of coils

It is because each coil has two ends and two coil connections are joined at each commutator segment.

(*iv*) The winding must close upon itself *i.e.*, it should be a closed circuit winding.

2.13 RELATIONS BETWEEN PITCHES FOR SIMPLEX LAP WINDING

In a simplex lap winding, the various pitches should have the following relation :

(*i*) The back and front pitches are odd and are of opposite signs. They differ numerically by 2.

$$\therefore \quad Y_B = Y_F \pm 2$$

$$Y_B = Y_F + 2 \quad \text{...for progressive winding}$$

$$Y_B = Y_F - 2 \quad \text{...for retrogressive winding}$$

(*ii*) Both Y_B and Y_F should be nearly equal to pole pitch.

(*iii*) Average pitch $= \dfrac{Y_B + Y_F}{2}$. It equals pole pitch $(= Z/P)$.

(*iv*) Commutator pitch, $Y_C = \pm 1$

$Y_C = +1$ for progressive winding

$Y_C = -1$ for retrogressive winding

(*v*) The resultant pitch (Y_R) is even, being the arithmetical difference of two odd numbers *viz.*, Y_B and Y_F.

(*vi*) If Z = number of armature conductors and P = number of poles, then,

$$\text{Pole-pitch} = \frac{Z}{P}$$

Since Y_B and Y_F both must be about one pole pitch and differ numerically by 2,

$$\left.\begin{aligned} Y_B &= \frac{Z}{P}+1 \\ Y_F &= \frac{Z}{P}-1 \end{aligned}\right\} \text{For progressive winding}$$

$$\left.\begin{aligned} Y_B &= \frac{Z}{P}-1 \\ Y_F &= \frac{Z}{P}+1 \end{aligned}\right\} \text{For retrogressive winding}$$

It is clear that Z/P must be an even number to make the winding possible.

Example 2.2. *A 4-pole, simplex lap-wound armature contains 16 slots and has two coil sides per slot. Find back pitch, front pitch and commutator pitch for (i) progressive winding (ii) retrogressive winding.*

Solution. Pole pitch $= \frac{Z}{P} = \frac{16 \times 2}{4} =$ **8 conductors**

(*i*) *For progressive winding*

Back pitch, $Y_B = \frac{Z}{P}+1 = 8+1 =$ **9 conductors**

Front pitch, $Y_F = \frac{Z}{P}-1 = 8-1 =$ **7 conductors**

Commutator pitch, $Y_C =$ **+ 1 segment**

(*ii*) *For retrogressive winding*

Back pitch, $Y_B = \frac{Z}{P}-1 = 8-1 =$ **7 conductors**

Front pitch, $Y_F = \frac{Z}{P}+1 = 8+1 =$ **9 conductors**

Commutator pitch, $Y_C =$ **–1 segment**

Example 2.3. *Design a 4-pole, simplex lap winding suitable for an armature containing 20 slots. Assume single turn coils with 2 conductors per slot.*

Solution.

Number of commutator segments = Number of slots = 20

Number of armature coils = Number of slots = 20

Number of armature conductors = 2 × 20 = 40

Pole pitch $= \frac{Z}{P} = \frac{40}{4} = 10$ conductors

Let us design the progressive lap winding. For progressive winding :

Back pitch, $Y_B = \frac{Z}{P}+1 = 10+1 = 11$ conductors

Front pitch, $Y_F = \frac{Z}{P}-1 = 10-1 = 9$ conductors

Winding table. Fig. 2.27 shows the winding table.

Back connections		*Front connections*
1 to (1 + 11) = 12	⟶	12 to (12 – 9) = 3
3 to (3 + 11) = 14	⟶	14 to (14 – 9) = 5
5 to (5 + 11) = 16	⟶	16 to (16 – 9) = 7
7 to (7 + 11) = 18	⟶	18 to (18 – 9) = 9

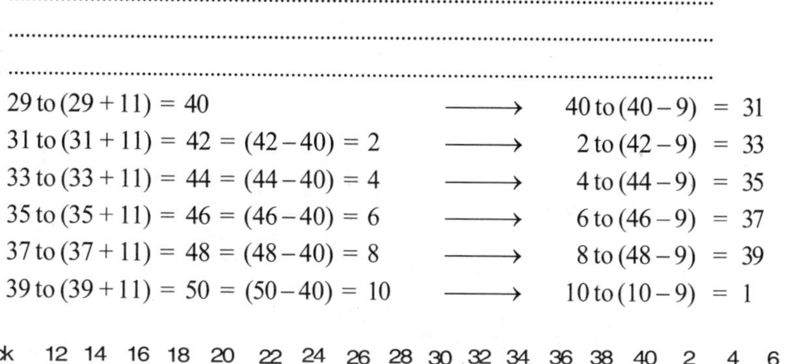

....

....

....

$$29 \text{ to } (29+11) = 40 \longrightarrow 40 \text{ to } (40-9) = 31$$
$$31 \text{ to } (31+11) = 42 = (42-40) = 2 \longrightarrow 2 \text{ to } (42-9) = 33$$
$$33 \text{ to } (33+11) = 44 = (44-40) = 4 \longrightarrow 4 \text{ to } (44-9) = 35$$
$$35 \text{ to } (35+11) = 46 = (46-40) = 6 \longrightarrow 6 \text{ to } (46-9) = 37$$
$$37 \text{ to } (37+11) = 48 = (48-40) = 8 \longrightarrow 8 \text{ to } (48-9) = 39$$
$$39 \text{ to } (39+11) = 50 = (50-40) = 10 \longrightarrow 10 \text{ to } (10-9) = 1$$

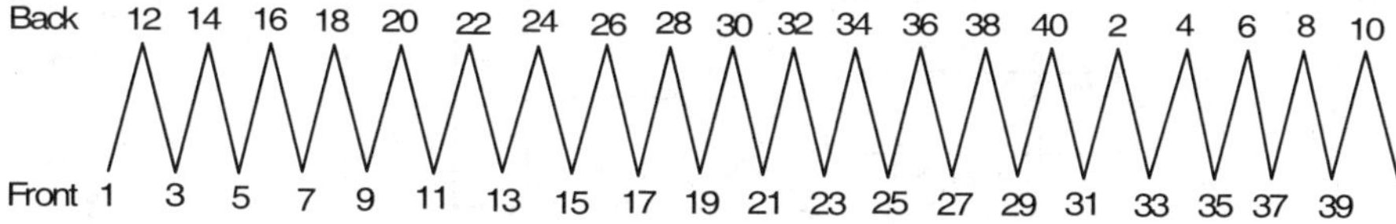

Fig. 2.27

Note that last conductor 10 joins to the free end of conductor 1 at which the winding started. Therefore, the winding closes upon itself.

Developed diagram. Developed diagram is obtained by imagining the cylindrical surface of the armature to be cut by an axial plane and then flattened out. Fig. 2.28 (*i*) shows the developed diagram of the winding. Note that *full lines represent the top coil sides (or conductors) and dotted lines represent the bottom coil sides (or conductors).

The winding goes from commutator segment 1 by conductor 1 across the back to conductor 12 and at the front to commutator segment 2, thus forming a coil. Then from commutator segment 2, through conductors 3 and 14 back to commutator segment 3 and so on till the winding returns to commutator segment 1 after using all the 40 conductors.

Position and number of brushes. We now turn to find the position and the number of brushes required. The brushes, like field poles, remain fixed in space as the commutator and winding revolve. It is very important that brushes are in correct position relative to the field poles. The arrowhead marked "rotation" in Fig. 2.28 (*i*) shows the direction of motion of the conductors. By right-hand rule, the direction of e.m.f. in each conductor will be as shown.

In order to find the position of brushes, the *ring diagram* shown in Fig. 2.28 (*ii*) is quite helpful. A positive brush will be placed on that commutator segment where the currents in the coils are meeting to flow *out* of the segment. A negative brush will be placed on that commutator segment where the currents in the coils are meeting to flow *in*. Referring to Fig. 2.28 (*i*), there are four brushes — two positive and two negative. *Therefore, we arrive at a very important conclusion that in a simplex lap winding, the number of brushes is equal to the number of poles.* If the brushes of the same polarity are connected together, then all the armature conductors are connected in four parallel paths ; each path containing an equal number of conductors in series. This is illustrated in Fig. 2.29.

Since segments 6 and 16 are connected together through positive brushes and segments 11 and 1 are connected together through negative brushes, there are four parallel paths, each containing 10 conductors in series. *Therefore, in a simplex lap winding, the number of parallel paths is equal to the number of poles.*

---

* The two conductors which lie in the same slot are drawn nearer to each other than to those in the other slots.

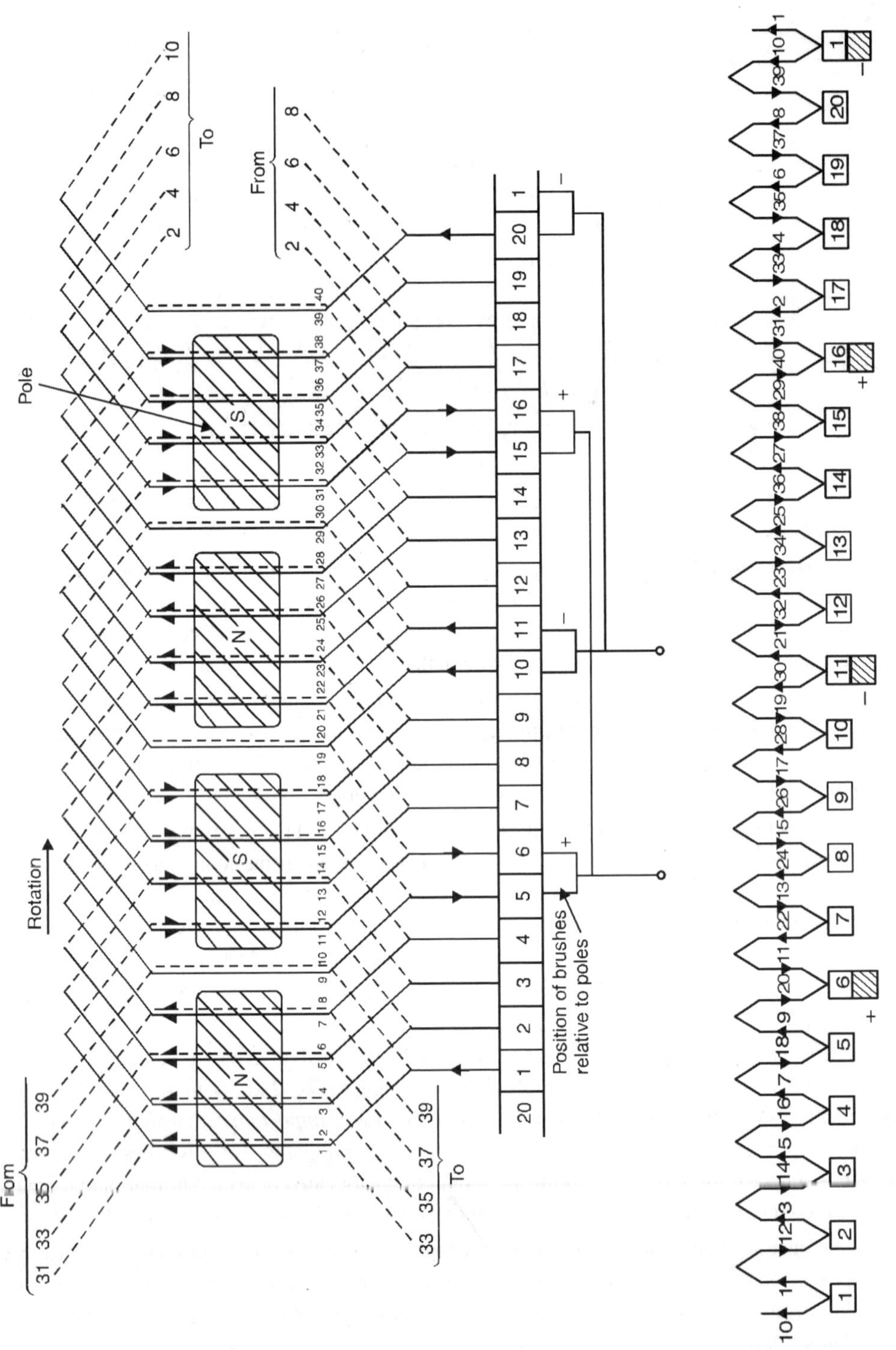

Fig. 2.28 (*i*)

Fig. 2.28 (*ii*)

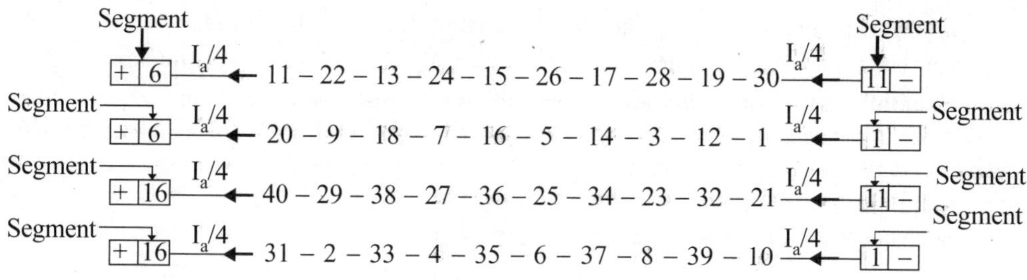

Fig. 2.29

Conclusions. From the above discussion, the following conclusions can be drawn :

(*i*) The total number of brushes is equal to the number of poles.

(*ii*) *The armature winding is divided into as many parallel paths as the number of poles.* If the total number of armature conductors is Z and P is the number of poles, then,

Number of conductors/path $= Z/P$

In the present case, there are 40 armature conductors and 4 poles. Therefore, the armature winding has 4 parallel paths, each consisting of 10 conductors in series.

(*iii*) E.M.F. generated = E.M.F. per parallel path

$$= \text{average e.m.f. per conductor} \times \frac{Z}{P}$$

(*iv*) Total armature current, $I_a = P \times$ current per parallel path

(*v*) The armature resistance can be found as under :

Let l = length of each conductor ; a = cross-sectional area

A = number of parallel paths = P ....for simplex lap winding

$$\text{Resistance of whole winding, } R = \frac{\rho l}{a} \times Z$$

$$\text{Resistance per parallel path} = \frac{R}{A} = \frac{\rho l Z}{a \times A}$$

Since there are A (= P) parallel paths, armature resistance R_a is given by :

$$R_a = \frac{\text{Resistance per parallel path}}{A} = \frac{1}{A}\left(\frac{\rho l Z}{a \times A}\right)$$

$$\therefore \quad R_a = \frac{\rho l Z}{a A^2}$$

Example 2.4. *A 4-pole, simplex progressive lap-wound armature contains 21 slots and has two coil sides per slot. Find (i) slot span (ii) the back pitch (iii) commutator pitch (iv) the front pitch.*

Solution. The slot pitch or slot span is the number of slots per pole. If there are S slots on the armature and the number of poles is P, then,

$$\text{Slot span, } Y_S = S/P$$

(*i*) $$\text{Slot span, } Y_S = \frac{S}{P} = \frac{21}{4} = 5\frac{1}{4} \text{ slots}$$

The most convenient way of expressing the span of a coil (or coil pitch) is in terms of the slots. Thus in this case, the coil pitch should be $5\frac{1}{4}$ slots. Since a coil can span only a whole number of slots, in solving this problem, the fraction 1/4 is dropped.

$$\therefore \quad Y_S = \mathbf{5 \text{ slots}}$$

(*ii*) The back pitch is equal to coil sides per slot (C_S) times the slot span (Y_S) plus 1.

$$\therefore \quad Y_B = C_S\, Y_S + 1 = 2 \times 5 + 1 = \mathbf{11 \text{ conductors}}$$

(*iii*) For a progressive simplex lap winding, commutator pitch, $Y_C = +1$ **segment**

(*iv*) For a progressive simplex lap winding, $Y_F = Y_B - 2 = 11 - 2 =$ **9 conductors**

Example 2.5. *A 4-pole, lap-wound armature has 28 slots and 28 segments. Each slot has two coil sides. If the winding is progressive, determine (i) the back pitch, (ii) the commutator pitch, (iii) the front pitch.*

Solution.

Number of armature conductors, $Z = 2 \times 28 = 56$

$$\text{Pole pitch} = \frac{Z}{P} = \frac{56}{4} = 14 \text{ conductors}$$

(*i*) For a progressive winding,

$$\text{Back pitch, } Y_B = \frac{Z}{P} + 1 = 14 + 1 = \mathbf{15\ conductors}$$

(*ii*) Commutator pitch, $Y_C = +1$ **segment**

(*iii*) $$\text{Front pitch, } Y_F = \frac{Z}{P} - 1 = 14 - 1 = \mathbf{13\ conductors}$$

Example 2.6. *An 8-pole, 240 V, simplex lap-wound armature delivers 125 A. Determine (i) the voltage per path (ii) the current per path and (iii) the current per brush.*

Solution.

(*i*) Voltage per path = Generator terminal voltage = **240 V**

(*ii*) $$\text{Current per path} = \frac{I_a}{A} = \frac{I_a}{P} = \frac{125}{8} = \mathbf{15{\cdot}62\ A}$$

(*iii*) $$\text{Current per brush} = \frac{125}{2} = \mathbf{62{\cdot}5\ A}$$

TUTORIAL PROBLEMS

1. A 4-pole simplex lap-wound armature contains 21 slots and has two coil sides per slot. Design the armature winding.
2. Design the armature winding for a 4-pole simplex lap winding with 15 slots and 15 segments. Each slot has two coil sides.
3. Draw a developed diagram of a simplex 2-layer lap winding of a 4-pole generator with 24 single-turn coils.

2.14 SIMPLEX WAVE WINDING

The essential difference between a lap winding and a wave winding is in the commutator connections. In a simplex lap winding, the coils approximately pole pitch apart are connected in series and the commutator pitch $Y_C = \pm 1$ segment. As a result, the coil voltages add. This is illustrated in Fig. 2.30. In a simplex wave winding, the coils approximately pole pitch apart are connected in series and the commutator pitch $Y_C \simeq$ *2 pole pitches (segments). Thus in a wave winding, successive coils "wave" forward under successive poles instead of "lapping" back on themselves as in the lap winding. This is illustrated in Fig. 2.31.

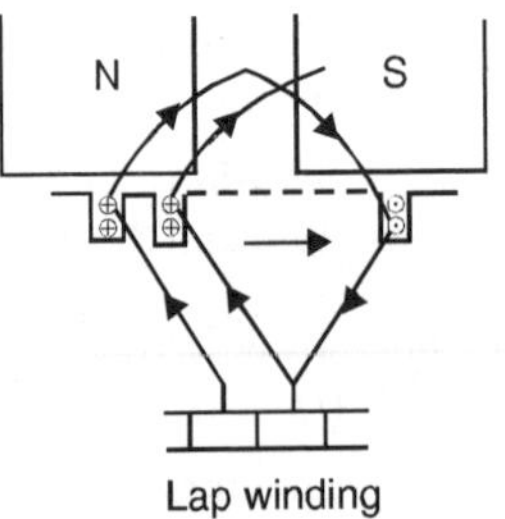

Lap winding

Fig. 2.30

The simplex wave winding must not close after it passes once around the armature but it must connect to a commutator segment adjacent to the first and the next coil must be adjacent to the first as indicated in Fig. 2.31. This is repeated each time

* measured in terms of coils or slots.

around until connections are made to all the commutator segments and all the slots are occupied after which the winding automatically returns to the starting point. If, after passing once around the armature, the winding connects to a segment to the left of the starting point, the winding is *retrogressive* [See Fig. 2.31 (*i*)]. If it connects to a segment to the right of the starting point, it is *progressive* [See Fig. 2.31 (*ii*)]. This type of winding is called wave winding because it passes around the armature in a wave-like form.

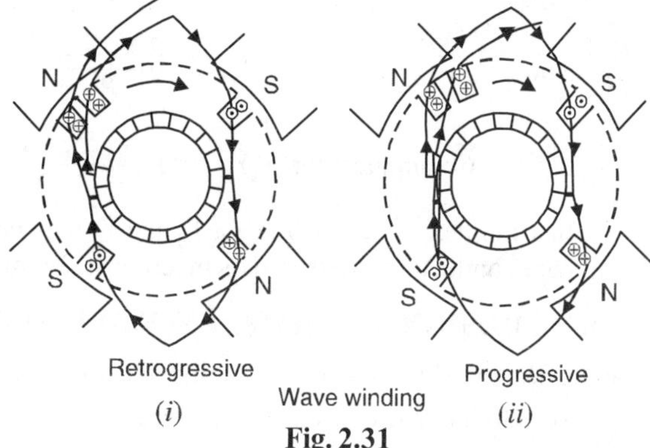

Retrogressive (*i*) Wave winding Progressive (*ii*)

Fig. 2.31

Various pitches. The various pitches in a wave winding are defined in a manner similar to lap winding.

(*i*) The distance measured in terms of armature conductors between the two sides of a coil at the back of the armature is called **back pitch Y_B** (See Fig. 2.32). *The Y_B must be an odd integer so that a top conductor and a bottom conductor will be joined.*

(*ii*) The distance measured in terms of armature conductors between the coil sides attached to any one commutator segment is called **front pitch Y_F** (See Fig. 2.32). *The Y_F must be an odd integer so that a top conductor and a bottom conductor will be joined.*

(*iii*) Resultant pitch, $Y_R = Y_B + Y_F$ (See Fig. 2.32)

The resultant pitch must be an even integer since Y_B and Y_F are odd. Further Y_R is approximately two pole pitches because Y_B as well as Y_F is approximately one pole pitch.

(*iv*) Average pitch, $Y_A = \dfrac{Y_B + Y_F}{2}$

When one tour of armature has been completed, the winding should connect to the next top conductor (progressive) or to the preceding top conductor (retrogressive). In either case, the difference will be of 2 conductors or one slot. If P is the number of poles and Z is the total number of armature conductors, then,

$$P \times Y_A = Z \pm 2$$

or

$$Y_A = \frac{Z \pm 2}{P} \qquad ...(i)$$

Fig. 2.32

Since P is always even and $Z = PY_A \pm 2$, Z must be even. It means that $Z \pm 2/P$ must be an integer. In eq. (*i*), plus sign will give progressive winding and the negative sign retrogressive winding.

(*v*) The number of commutator segments spanned by a coil is called **commutator pitch (Y_C)** (See Fig. 2.32). Suppose in a simplex wave winding,

P = Number of poles ; N_C = Number of commutator segments ;

Y_C = Commutator pitch

$\therefore$ Number of pair of poles = $P/2$

If $Y_C \times P/2 = N_C$, then the winding will close on itself in passing once around the armature. In order to connect to the adjacent conductor and permit the winding to proceed,

$$Y_C \times \frac{P}{2} = N_C \pm 1$$

or $$Y_C = \frac{2N_C \pm 2}{P} = \frac{N_C \pm 1}{P/2} = \frac{\text{No. of commutator seg. } \pm 1}{\text{Number of pair of poles}}$$

Now $$Y_C = \frac{2N_C \pm 2}{P} = \frac{Z \pm 2}{P} = Y_A \qquad (\because 2N_C = Z)$$

$$\therefore \quad \text{Commutator pitch, } Y_C = Y_A = \frac{Y_B + Y_F}{2}$$

In a simplex wave winding Y_B, Y_F and Y_C may be equal. Note that Y_B, Y_F and Y_A are in terms of armature conductors whereas Y_C is in terms of commutator segments.

2.15 DESIGN OF SIMPLEX WAVE WINDING

In the design of simplex wave winding, the following points may be kept in mind :

(*i*) Both pitches Y_B and Y_F are odd and are of the same sign.

(*ii*) $$\text{Average pitch, } Y_A = \frac{Z \pm 2}{P} \qquad ...(i)$$

(*iii*) Both Y_B and Y_F are nearly equal to pole pitch and *may be equal or differ by 2*. If they differ by 2, they are one more and one less than Y_A.

(*iv*) Commutator pitch is given by ;

$$Y_C = Y_A = \frac{\text{Number of commutator segments } \pm 1}{\text{Number of pair of poles}}$$

The plus sign for progressive winding and negative for retrogressive winding.

(*v*) $$Y_A = \frac{Z \pm 2}{P}$$

Since Y_A must be a whole number, there is a restriction on the value of Z. With $Z = 180$, this winding is impossible for a 4-pole machine because Y_A is not a whole number.

(*vi*) $$Z = PY_A \pm 2$$

$$\therefore \quad \text{Number of coils} = \frac{Z}{2} = \frac{PY_A \pm 2}{2}$$

Example 2.7. *A 4-pole armature has 30 armature conductors. The armature is to be simplex wave-wound with single-turn coils. Determine for a retrogressive winding (i) back pitch (ii) front pitch (iii) commutator pitch.*

Solution.

(*i*) $$\text{Average pitch, } Y_A = \frac{Z \pm 2}{P}$$

For a retrogressive winding, $$Y_A = \frac{Z - 2}{P} = \frac{30 - 2}{4} = 7 \text{ conductors}$$

$$\therefore \quad Y_B = \mathbf{7\ conductors}$$

(*ii*) $$Y_A = \frac{Y_B + Y_F}{2}$$

or $$7 = \frac{7 + Y_F}{2} \qquad \therefore \quad Y_F = \mathbf{7\ conductors}$$

(*iii*) $$Y_C = Y_A = \mathbf{7\ commutator\ segments}$$

Alternatively, for a retrogressive winding,

$$Y_C = \frac{\text{No. of commutator segments} - 1}{\text{No. of pair of poles}} = \frac{15 - 1}{2} = \mathbf{7\ segments}$$

Example 2.8. *A 4-pole, simplex wave-wound armature has 21 slots and 21 segments. Determine the commutator pitch.*

Solution.

$$\text{Commutator pitch, } Y_C = \frac{2N_C \pm 2}{P} = \frac{2 \times 21 \pm 2}{4} = \mathbf{10} \textit{ or } \mathbf{11 \text{ commutator segments}}$$

Example 2.9. *Design a 4-pole wave winding for an armature with 21 slots. Assume single turn coils and two conductors per slot.*

Solution.

$$\text{Average pitch, } Y_A = \frac{Z \pm 2}{P} = \frac{42+2}{4} = 11 \text{ conductors}$$

We have taken + 2 on the assumption that progressive winding is desired. Since Y_A is a whole number, the wave winding is possible. Taking $Y_B = 11$ conductors, we have $Y_F = 11$ conductors.

$$\text{Commutator pitch, } Y_C = Y_A = 11 \text{ segments}$$

Winding Table. The winding table for the winding is given below :

Back connections		*Front connections*
1 to (1 + 11) = 12	⟶	12 to (12 + 11) = 23
23 to (23 + 11) = 34	⟶	34 to (34 + 11) = 45 = 45 – 42 = 3
3 to (3 + 11) = 14	⟶	14 to (14 + 11) = 25
25 to (25 + 11) = 36	⟶	36 to (36 + 11) = 47 = 47 – 42 = 5
......		......
......		......
......		......
19 to (19 + 11) = 30	⟶	30 to (30 + 11) = 41
41 to (41 + 11) = 52 = 52 – 42 = 10	⟶	10 to (10 + 11) = 21
21 to (21 + 11) = 32	⟶	32 to (32 + 11) = 43 = 43 – 42 = 1

Since we have come back to conductor No. 1 from where we started, the winding gets closed at this stage.

Developed diagram. Fig. 2.33 (*i*) shows the developed diagram for the winding. Note that full lines represent the top coil sides (or conductors) and dotted lines represent the bottom coil sides (or conductors). The two conductors which lie in the same slot are drawn nearer to each other than to those in the other slots.

Referring to Fig. 2.33 (*i*), conductor 1 connects at the back to conductor 12 (1 + 11) which in turn connects at the front to conductor 23 (12 + 11) and so on round the armature until the winding is complete. Note that the commutator pitch $Y_C = 11$ segments. This means that the number of commutator segments spanned between the start end and finish end of any coil is 11 segments.

Position and number of brushes. We now turn to find the position and the number of brushes. The arrowhead marked "rotation" in Fig. 2.33 (*i*) shows the direction of motion of the conductors. By right hand rule, the direction of e.m.f. in each conductor will be as shown.

In order to find the position of brushes, the ring diagram shown in Fig. 2.33 (*ii*) is quite helpful. It is clear that only two brushes — one positive and one negative — are required (though two positive and two negative brushes can also be used). We find that there are two parallel paths between the positive brush and the negative brush. This is illustrated in Fig. 2.34.

Therefore, we arrive at a very important conclusion that *in a simplex wave winding, the number of parallel paths is two irrespective of the number of poles*. Note that the first parallel path has 11 coils (or 22 conductors) while the second parallel path has 10 coils (or 20 conductors). This fact is

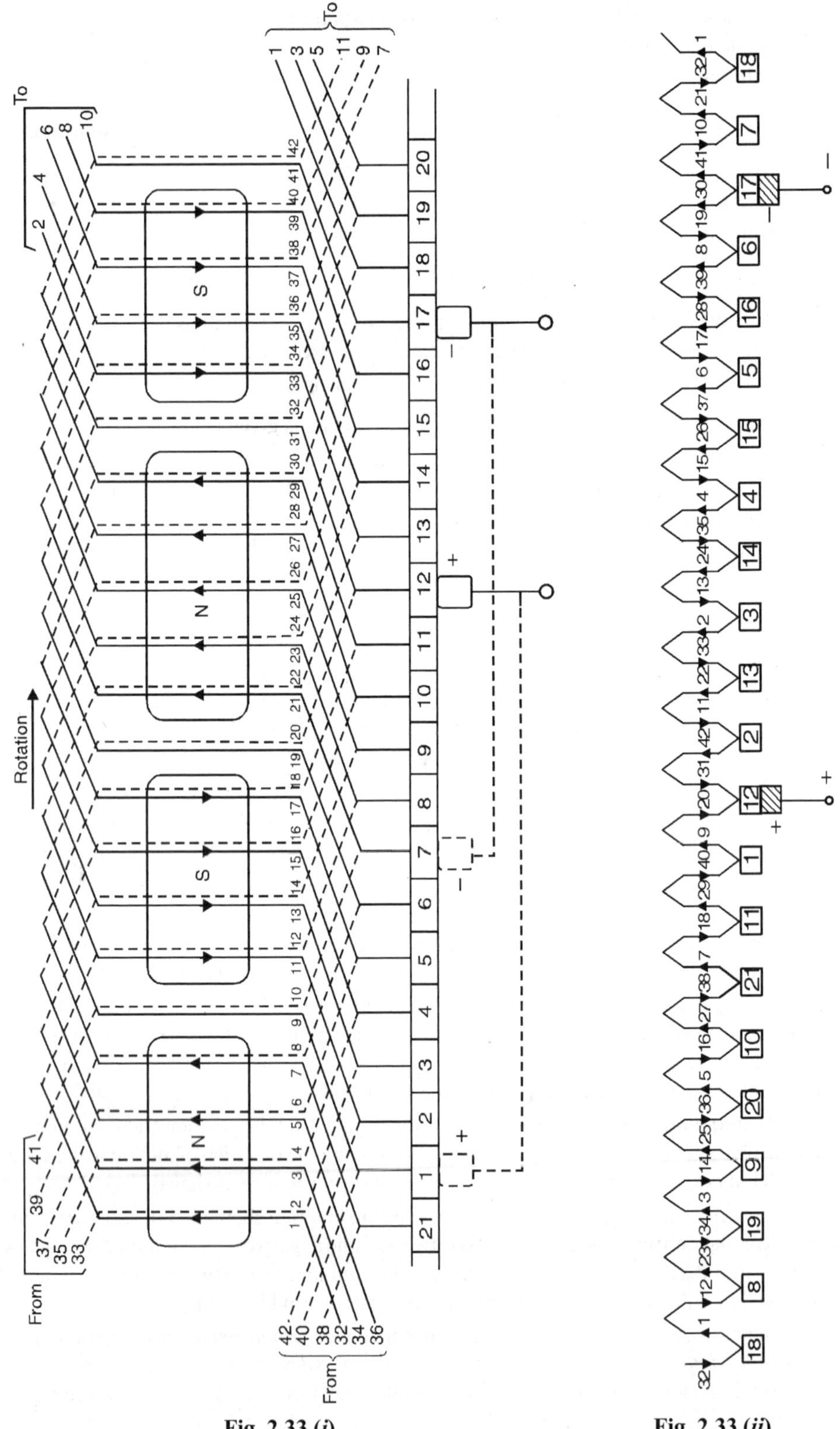

Fig. 2.33 (*i*)

Fig. 2.33 (*ii*)

not important as it may appear at first glance. The coils in the smaller group should supply less current to the external circuit. But the identity of the coils in either parallel path is rapidly changing from moment to moment. Therefore, the average value of current through any particular coil is the same.

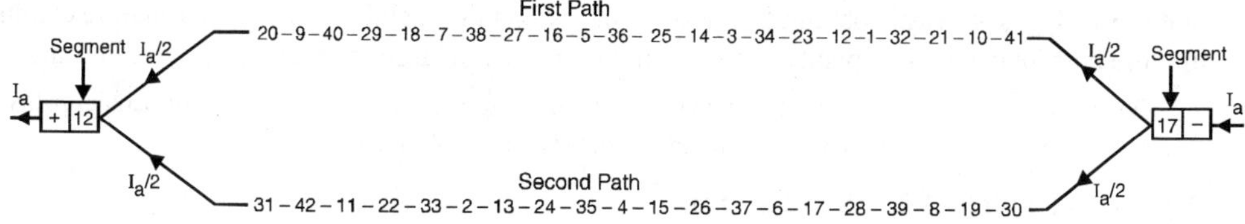

Fig. 2.34

Conclusions. From the above discussion, the following conclusions can be drawn :

(*i*) Only two brushes are necessary but as many brushes as there are poles may be used.

(*ii*) *The armature winding is divided into two parallel paths irrespective of the number of poles.* If the total number of armature conductors is Z and P is the number of poles, then,

$$\text{Number of conductors/path} = \frac{Z}{2}$$

(*iii*)
$$\text{E.M.F. generated} = \text{E.M.F. per parallel path}$$
$$= \text{Average e.m.f. per conductor} \times \frac{Z}{2}$$

(*iv*) Total armature current, $I_a = 2 \times$ current per parallel path

(*v*) The armature can be wave-wound if Y_A or Y_C is a whole number.

2.16 DUMMY COILS

In a simplex wave winding, the average pitch Y_A (or commutator pitch Y_C) should be a whole number. Sometimes the standard armature punchings available in the market have slots that do not satisfy the above requirement so that more coils (usually only one more) are provided than can be utilised. These extra coils are called *dummy or dead coils*. The dummy coil is inserted into the slots in the same way as the others to make the armature *dynamically balanced but it is not a part of the armature winding.

Let us illustrate the use of dummy coils with a numerical example. Suppose the number of slots is 22 and each slot contains 2 conductors. The number of poles is 4. For simplex wave wound armature,

$$Y_A = \frac{Z \pm 2}{P} = \frac{2 \times 22 \pm 2}{4} = \frac{44 \pm 2}{4} = 11\frac{1}{2} \text{ or } 10\frac{1}{2}$$

Since the results are not whole numbers, the number of coils (and hence segments) must be reduced. If we make one coil dummy, we have 42 conductors and

$$Y_A = \frac{42 \pm 2}{4} = 11 \text{ or } 10$$

This means that armature can be wound only if we use 21 coils and 21 segments. The extra coil or dummy coil is put in the slot. One end of this coil is taped and the other end connected to the unused commutator segment (segment 22) for the sake of appearance. Since only 21 segments are required, the two (21 and 22 segments) are connected together and considered as one.

2.17 APPLICATIONS OF LAP AND WAVE WINDINGS

In multipolar machines, for a given number of poles (P) and armature conductors (Z), a wave winding has a higher terminal voltage than a lap winding because it has more conductors in series. On the

* An armature having some slots without windings would be out of balance mechanically.

other hand, the lap winding carries more current than a wave winding because it has more parallel paths.

In small machines, the current-carrying capacity of the armature conductors is not critical and in order to achieve suitable voltages, wave windings are used. On the other hand, in large machines suitable voltages are easily obtained because of the availability of large number of armature conductors and the current carrying capacity is more critical. Hence in large machines, lap windings are used.

Note : In general, a high-current armature is lap-wound to provide a large number of parallel paths and a low-current armature is wave-wound to provide a small number of parallel paths.

2.18 MULTIPLEX WINDINGS

A simplex lap-wound armature has as many parallel paths as the number of poles. A simplex wave-wound armature has two parallel paths irrespective of the number of poles. In case of a 10-pole machine, using simplex windings, the designer is restricted to either two parallel circuits (wave) or ten parallel circuits (lap). Sometimes it is desirable to increase the number of parallel paths. For this purpose, *multiplex windings* are used. The sole purpose of multiplex windings is to increase the number of parallel paths enabling the armature to carry a large total current. The degree of multiplicity or plex determines the number of parallel paths in the following manner :

(*i*) A lap winding has pole times the degree of plex parallel paths.

Number of parallel paths, $A = P \times$ plex

Thus a *duplex lap winding* has $2P$ parallel paths, *triplex lap winding* has $3P$ parallel paths and so on. If an armature is changed from simplex lap to duplex lap without making any other change, the number of parallel paths is doubled and each path has half as many coils. The armature will then supply twice as much current at half the voltage.

(*ii*) A wave winding has two times the degree of plex parallel paths.

Number of parallel paths, $A = 2 \times$ plex

Note that the number of parallel paths in a multiplex wave winding depends upon the degree of plex and not on the number of poles. Thus a *duplex wave winding* has 4 parallel paths, *triplex wave winding* has 6 parallel paths and so on.

2.19 FUNCTION OF COMMUTATOR AND BRUSHES

The e.m.f. generated in the armature winding of a d.c. generator is alternating one. The commutator and brushes cause the alternating e.m.f. of the armature conductors to produce a p.d. always in the same direction between the terminals of the generator. In lap as well as wave winding, it will be observed that currents in the coils to a brush are either all directed towards the brush (positive brush) or all directed away from the brush (negative brush). Further, the direction of current in coil reverses as it passes the brush. Thus when the coil approaches the contact with the brush, the current through the coil is in one direction ; when the coil leaves the contact with the brush, the current has been reversed. *This reversal of current in the coil as the coil passes a brush is called* **commutation** *and takes place while the coil is short-circuited by the brush.* These changes occur in every coil in turn. If, at the instant when the brush breaks contact with the commutator segment connected to the coil undergoing commutation, the current in the coil has not been reversed, the result will be sparking between the commutator segments and the brush.

The criterion of good commutation is that it should be sparkless. In order to have sparkless commutation, the brushes on the commutator should be placed at points known as *neutral points* where no voltage exists between adjacent segments. The conductors connected to these segments lie between the poles in position of zero magnetic flux which is termed as *magnetic neutral axis* (M.N.A.).

2.20 E.M.F. EQUATION OF A D.C. GENERATOR

We shall now derive an expression for the e.m.f. generated in a d.c. generator.

Let
$$\phi = \text{flux/pole in Wb}$$
$$Z = \text{total number of armature conductors}$$
$$P = \text{number of poles}$$
$$A = \text{number of parallel paths} = 2 \text{ .... for wave winding}$$
$$= P \text{ .... for lap winding}$$
$$N = \text{speed of armature in r.p.m.}$$
$$E_g = \text{e.m.f. of the generator} = \text{e.m.f./parallel path}$$

Flux cut by one conductor in one revolution of the armature,
$$d\phi = P\phi \text{ webers}$$

Time taken to complete one revolution,
$$dt = 60/N \text{ second}$$

$$\text{e.m.f. generated/conductor} = \frac{d\phi}{dt} = \frac{P\phi}{60/N} = \frac{P\phi N}{60} \text{ volts}$$

$$\text{e.m.f. of generator, } E_g = \text{e.m.f. per parallel path}$$
$$= (\text{e.m.f./conductor}) \times \text{No. of conductors in series per parallel path}$$
$$= \frac{P\phi N}{60} \times \frac{Z}{A}$$

$$\therefore \quad E_g = \frac{P\phi Z N}{60\,A} \qquad \dots(i)$$

where $A = 2$ ... *for wave winding*

$= P$ ... *for lap winding*

The following points may be noted :

(*i*) Eq. (*i*) is also valid for a d.c. motor. In case of a d.c. generator, induced e.m.f. in the armature is called **generated e.m.f.** or **generated voltage** (E_g). When the d.c. machine operates as a motor, the induced e.m.f. in the armature is called **counter e.m.f.** or **back e.m.f.** E_b ($= P\phi ZN/60A$).

(*ii*) For a given d.c. machine (generator or motor), Z, P and A are constant so that E_g (or E_b) $\propto \phi N$. *Therefore, for a given d.c. machine, the induced voltage in the armature is directly proportional to flux per pole* (ϕ) *and speed of rotation (N).*

Note. In deriving eq. (*i*) above, we have assumed that the poles cover the entire armature periphery. This is virtually impossible. In practical machines, the poles cover 60% to 80% of the armature periphery. This fact must be taken into consideration while computing flux per pole (ϕ).

2.21 ARMATURE RESISTANCE (R_a)

The resistance offered by the armature circuit is known as armature resistance (R_a) and includes :

(*i*) resistance of armature winding

(*ii*) resistance of brushes

Thc armature resistance depends upon the construction of machine. Except for small machines, its value is generally less than 1Ω.

2.22 TYPES OF D.C. GENERATORS

The magnetic field in a d.c. generator is normally produced by electromagnets rather than permanent magnets. Generators are generally classified according to their methods of field excitation. On this basis, d.c. generators are divided into the following two classes :

(*i*) Separately excited d.c. generators (*ii*) Self-excited d.c. generators

The behaviour of a d.c. generator on load depends upon the method of field excitation adopted.

2.23 SEPARATELY EXCITED D.C. GENERATORS

A d.c. generator whose field magnet winding is supplied from an independent external d.c. source (*e.g.*, a battery etc.) is called a *separately excited generator*. Fig. 2.35 shows the connections of a separately excited generator. The voltage output depends upon the speed of rotation of armature and the field current ($E_g = P\phi ZN/60A$). The greater the speed and field current, greater is the generated e.m.f. It may be noted that separately excited d.c. generators are rarely used in practice. The d.c. generators are normally of self-excited type.

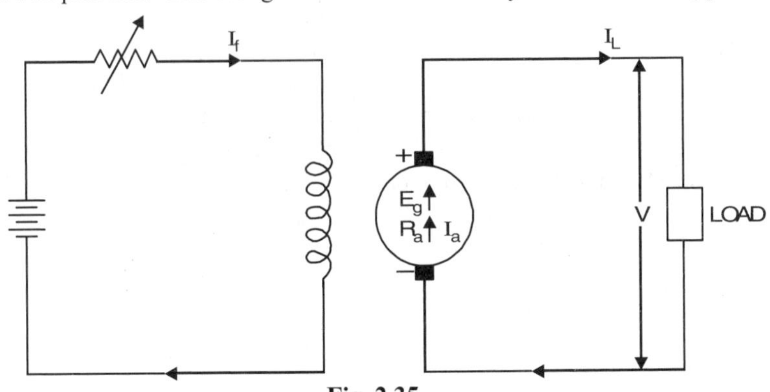

Fig. 2.35

$$\text{Armature current, } I_a = I_L$$

$$\text{Terminal voltage, } V = E_g - I_a R_a$$

$$\text{Electric power developed} = E_g I_a$$

$$\text{Power delivered to load} = E_g I_a - I_a^2 R_a = I_a (E_g - I_a R_a) = V I_a$$

2.24 SELF-EXCITED D.C. GENERATORS

A d.c. generator whose field magnet winding is supplied current from the output of the generator itself is called a **self-excited generator*. There are three types of self-excited generators depending upon the manner in which the field winding is connected to the armature, namely;

(*i*) Series generator ; (*ii*) Shunt generator ; (*iii*) Compound generator

(*i*) **Series generator.** In a series wound generator, the field winding is connected in series with armature winding so that whole armature current flows through the field winding as well as the load. Fig. 2.36 shows the connections of a series wound generator. Since the field winding carries the whole of load current, it has a few turns of thick wire having low resistance. Series generators are rarely used except for special purposes *e.g.*, as boosters.

Fig. 2.36

$$\text{Armature current, } I_a = I_{se} = I_L = I \text{ (say)}$$

$$\text{Terminal voltage, } V = E_g - I(R_a + R_{se})$$

$$\text{Power developed in armature} = E_g I_a$$

$$\text{Power delivered to load} = E_g I_a - I_a^2 (R_a + R_{se}) = I_a [E_g - I_a (R_a + R_{se})] = V I_a \text{ or } V I_L$$

* How is self-excitation achieved ? When the armature is rotated, a small voltage is induced in the armature winding due to residual flux in the poles. This voltage produces a small field current in the field winding and causes the flux per pole to increase. The increased flux increases the induced voltage which further increases the field current. These events take place rapidly and the generator builds up to the rated generated voltage.

(*ii*) **Shunt generator.** In a shunt generator, the field winding is connected in parallel with the armature winding so that terminal voltage of the generator is applied across it. The shunt field winding has many turns of fine wire having high resistance. Therefore, only a part of armature current flows through shunt field winding and the rest flows through the load. Fig. 2.37 shows the connections of a shunt-wound generator.

$$\text{Shunt field current, } I_{sh} = V/R_{sh}$$
$$\text{Armature current, } I_a = I_L + I_{sh}$$
$$\text{Terminal voltage, } V = E_g - I_a R_a$$
$$\text{Power developed in armature} = E_g I_a$$
$$\text{Power delivered to load} = V I_L$$

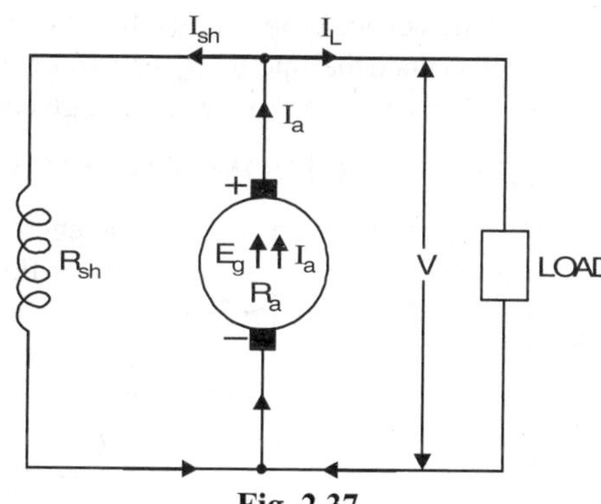

Fig. 2.37

(*iii*) **Compound generator.** In a compound-wound generator, there are *two sets of field windings on each pole—one is in series and the other in parallel with the armature. A compound wound generator may be :

(*a*) **Short Shunt** in which only shunt field winding is in parallel with the armature winding [See Fig. 2.38 (*i*)].

(*b*) **Long Shunt** in which shunt field winding is in parallel with both series field and armature winding [See Fig. 2.38 (*ii*)].

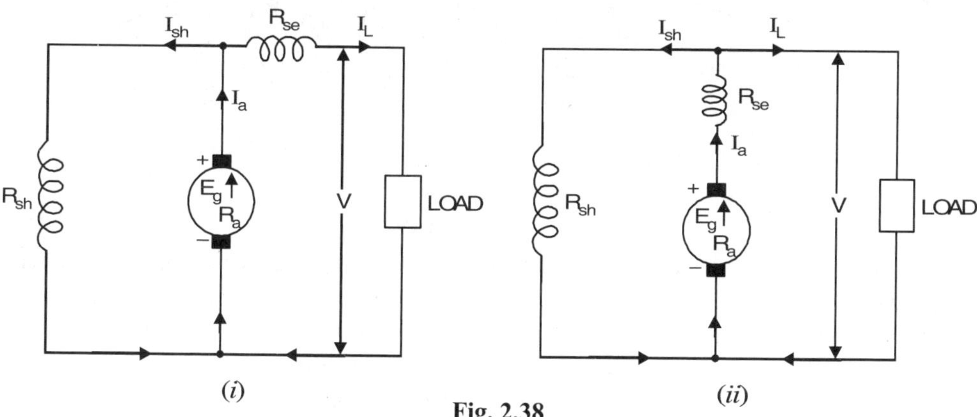

Fig. 2.38

Short shunt

$$\text{Series field current, } I_{se} = I_L$$
$$\text{Shunt field current, } I_{sh} = \frac{V + I_{se} R_{se}}{R_{sh}}$$
$$\text{Terminal voltage, } V = E_g - I_a R_a - I_{se} R_{se}$$
$$\text{Power developed in armature} = E_g I_a$$
$$\text{Power delivered to load} = V I_L$$

Long shunt

$$\text{Series field current, } I_{se} = I_a = I_L + I_{sh}$$
$$\text{Shunt field current, } I_{sh} = V/R_{sh}$$

* Normally, the majority of m.m.f. is provided by the shunt field. The two windings may be connected to aid each other (*cumulative compounding*) or they may oppose each other (*differential compounding*).

$$\text{Terminal voltage, } V = E_g - I_a(R_a + R_{se})$$
$$\text{Power developed in armature} = E_g I_a$$
$$\text{Power delivered to load} = V I_L$$

2.25 BRUSH CONTACT DROP

It is the voltage drop over the brush contact resistance when current flows. Obviously, its value will depend upon the amount of current flowing and the value of contact resistance. This drop is generally small and may be neglected if not given.

2.26 ELECTROMECHANICAL ENERGY CONVERSION IN A GENERATOR

A generator converts mechanical energy into electrical energy. Let us see how it happens. The load current delivered by a d.c. generator flows through all the armature conductors. The current always flows in the same direction in those conductors that are momentarily under a *N*-pole. The same is true for the conductors that are momentarily under a *S*-pole. However, currents under the *N*-pole flow opposite to those under the *S*-pole as shown in Fig. 2.39.

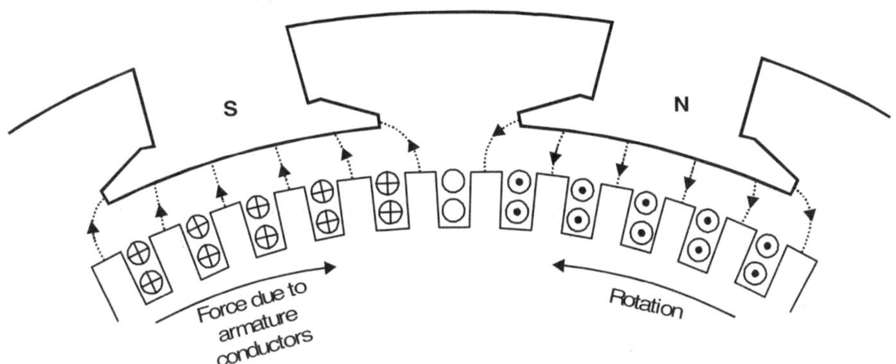

Fig. 2.39

Since all the conductors lie in a magnetic field, they all experience a force. By Fleming's left hand rule, we find that the forces on the conductors all act in the same direction. As can be seen, the conductors produce a torque which acts opposite to the direction in which the generator is driven. To keep the machine going, we must exert a torque on the shaft to overcome this opposing electromagnetic torque, the resulting mechanical energy is converted into electrical energy which is delivered to the generator load. In Fig. 2.39, the generator is driven in anticlockwise direction by the prime mover. The armature conductors under the *S*-pole carry currents that flow *into* the page, away from the reader. On the other hand, the armature currents under the *N*-pole flow *out* of the page, towards the reader.Therefore, by Fleming's left hand rule, the force on every conductor acts towards the right, producing a net clockwise torque. This braking torque acts opposite to the direction of rotation.

Example 2.10. *Calculate the e.m.f. generated by a 4-pole wave-wound generator having 65 slots with 12 conductors per slot when driven at 1200 r.p.m. The flux per pole is 0·02 Wb.*

Solution.
$$E_g = \frac{P \phi Z N}{60 A}$$

Here $P = 4$; $\phi = 0{\cdot}02$ Wb ; $N = 1200$ r.p.m. ; $Z = 12 \times 65 = 780$; $A = 2$

$$\therefore \quad E_g = \frac{4 \times 0{\cdot}02 \times 780 \times 1200}{60 \times 2} = \textbf{624 volts}$$

Example 2.11. *A 6-pole lap-wound d.c. generator has 600 conductors on its armature. The flux per pole is 0.02 Wb. Calculate (i) the speed at which the generator must be run to generate 300 V. (ii) What would be the speed if the generator were wave-wound ?*

Solution.

(*i*) **Lap wound** $$E_g = \frac{P\phi ZN}{60\,A}$$

$$\therefore \quad N = \frac{E_g \times 60\,A}{P\phi Z} = \frac{300 \times 60 \times 6}{6 \times 0{\cdot}02 \times 600} = \mathbf{1500\ r.p.m.}$$

(*ii*) **Wave wound** $$N = \frac{E_g \times 60\,A}{P\phi Z} = \frac{300 \times 60 \times 2}{6 \times 0{\cdot}02 \times 600} = \mathbf{500\ r.p.m.}$$

Example 2.12. *An 8-pole, lap-wound armature rotated at 350 r.p.m. is required to generate 260 V. The useful flux per pole is 0·05 Wb. If the armature has 120 slots, calculate the number of conductors per slot.*

Solution. $$E_g = \frac{P\phi ZN}{60\,A}$$

$$\therefore \quad Z = \frac{E_g \times 60\,A}{P\phi N} = \frac{260 \times 60 \times 8}{8 \times 0{\cdot}05 \times 350} = 890$$

$\therefore$ No. of conductors/slot $= 890/120 = 7.41$

This value must be an even number.

Hence, conductors/slot = **8**

Example 2.13. *The armature of a 6-pole, 600 r.p.m. lap-wound generator has 90 slots. If each coil has 4 turns, calculate the flux per pole required to generate an e.m.f. of 288 volts.*

Solution. Each turn has two active conductors and 90 coils are required to fill 90 slots.

$$\therefore \quad Z = 90 \times 4 \times 2 = 720$$

$$E_g = \frac{P\phi ZN}{60\,A}$$

or $$\phi = \frac{E_g \times 60\,A}{PZN} = \frac{288 \times 60 \times 6}{6 \times 720 \times 600} = \mathbf{0{\cdot}04\ Wb}$$

Example 2.14. *The armature of a d.c. generator has 81 slots and the commutator has 243 segments. It is wound to give lap winding having 1 turn per coil. If the flux per pole is 30 mWb, calculate the generated e.m.f. at a speed of 1200 r.p.m. Number of poles is 6.*

Solution. The number of coils is equal to the number of commutator segments. Each turn has 2 active conductors.

$$\therefore \quad Z = 243 \times 2 = 486\ ;\ \phi = 30 \times 10^{-3}\ \text{Wb} = 0{\cdot}03\ \text{Wb}$$

$$E_g = \frac{P\phi ZN}{60\,A} = \frac{6 \times 0{\cdot}03 \times 486 \times 1200}{60 \times 6} = \mathbf{291{\cdot}6\ volts}$$

Example 2.15. *The armature of a 2-pole, 200 V generator has 400 conductors and runs at 300 r.p.m. If the number of turns in each field coil is 1200, what is the average value of e.m.f. induced in each coil on breaking the field if the flux dies away completely in 0·1 s ?*

Solution. $$E_g = \frac{P\phi ZN}{60\,A} = \left(\frac{P}{A}\right) \times \frac{\phi ZN}{60}$$

Here $P/A = 1$; $Z = 400$; $N = 300$ r.p.m. ; $E_g = 200$ V

$$\therefore \qquad 200 = 1 \times \frac{\phi \times 400 \times 300}{60} \quad \text{or} \quad \phi = 0.1 \text{ Wb}$$

If n (= 1200) is the number of turns of each field coil, then magnitude of e.m.f. induced on breaking the field

$$= n\frac{d\phi}{dt} = 1200 \times \frac{0\cdot1}{0\cdot1} = \mathbf{1200\ V}$$

Example 2.16. *A 4-pole, lap-connected d.c. machine has an armature resistance of 0.15 Ω. Find the armature resistance of the machine if rewound for wave connection.*

Solution. Let r ohm be the resistance of each parallel path. Then $r/4 = 0.15$. Therefore, $r = 0.6\ \Omega$. When the machine is rewound for wave connection, then there will be two parallel paths each of resistance = $2r = 2 \times 0.6 = 1.2\ \Omega$. Therefore, armature resistance for wave wound machine = $1.2 \parallel 1.2 = 1.2/2 =$ **0.6 Ω.**

Example 2.17. *A generator has a rated armature current of 250 A. What is the current per path of the armature if the armature is simplex wave wound or simplex lap wound ? The machine has 12 poles.*

Solution. For simplex wave wound, number of parallel paths is $A = 2$. Therefore, current per parallel path $= I_a/A = 250/2 =$ **125*A***. For simplex lap wound, $A = P = 12$. Therefore, current per parallel path $= I_a/A = 250/12 =$ **20.83*A***.

Example 2.18. *A 4-pole generator has an induced e.m.f. of 262V when driven at a speed of 400 r.p.m. The armature is lap wound and has 652 conductors, its resistance being 0.15Ω. The bore of the pole shoe is 42 cm diameter, the pole subtends an angle of 60° and is 20 cm long. Calculate the flux density in the air-gap.*

Solution. Fig. 2.40. shows the conditions of the problem.

$$E_g = \frac{P\phi Z N}{60A}$$

Putting the values of E_g, P, Z, N and A, we have, flux/pole, $\phi = 0.0603$ Wb.

$$\text{Pole shoe arc} = \pi \times \text{Armature dia.} \times \theta/360°$$
$$= \pi \times 0.42 \times 60°/360° = 0.07\pi \text{ m}$$
$$\text{Pole shoe area} = \text{Pole shoe arc} \times \text{Pole shoe length}$$
$$= 0.07\pi \times 0.2 = 0.044 \text{ m}^2$$
$$\text{Air-gap flux density} = 0.0603/0.044 = \mathbf{1.37\ Wb/m^2}$$

This example shows that poles do not cover the entire armature periphery. This fact must be taken into account while computing flux per pole.

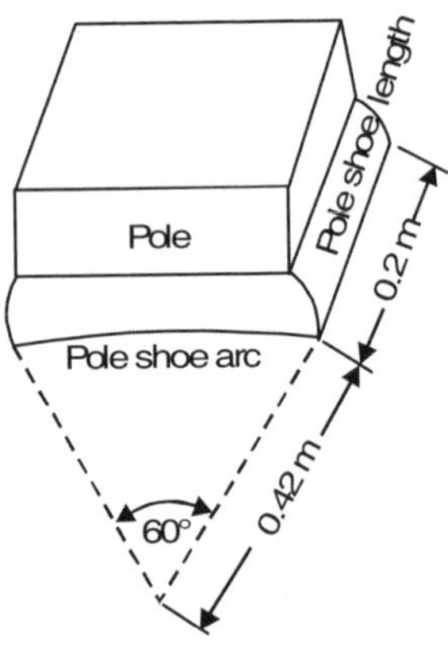

Fig. 2.40

Example 2.19. *A 4-pole lap wound d.c. armature has a bore diameter of 0.7 m. It has 560 conductors and the ratio of pole arc/pole pitch is 0.63. The armature is running at 600 r.p.m. and the flux density in the air gap in 1.2 Wb/m². Determine the induced e.m.f. in the armature if the effective length of armature conductor is 20 cm.*

Solution.

$$\text{Pole pitch} = \frac{\pi \times \text{bore diameter}}{\text{Number of poles}} = \frac{\pi \times 0.7}{4} = 0.55\text{m}$$

$$\text{Pole shoe arc} = \text{Pole pitch} \times \frac{\text{Pole arc}}{\text{Pole pitch}} = 0.55 \times 0.63 = 0.3465\text{m}$$

$$\text{Area of pole face} = \text{Pole shoe arc} \times \text{axial length} = 0.3465 \times 0.2 = 0.0693\text{m}^2$$

$$\text{Flux per pole, } \phi = 1.2 \times 0.0693 = 0.08316 \text{ Wb}$$

Now $$E_g = \frac{P\phi ZN}{60A}$$

Putting the values of P, ϕ, Z, N and A, we have, E_g = **465.7 V**

Example 2.20. *A 6-pole d.c. generator runs at 1200 r.p.m. on no-load and has a generated e.m.f. of 250 V. Its armature diameter is 350 mm and the radial air-gap between the field poles and the armature is 3 mm. The axial length of the field poles is 260 mm and the field pole effective coverage is 80% including fringing. If the armature has 96 coils having 3 turns per coil and is wound duplex lap, calculate (i) flux per pole (ii) effective pole arc length and (iii) average air-gap flux density.*

Solution. Fig. 2.41. depicts the conditions of the problem showing two poles only.

(*i*) $$E_g = \frac{P\phi ZN}{60A}$$

Here $Z = (96 \times 3) \times 2 = 576 ; N = 1200 \text{ r.p.m.}$

$A = P \times \text{plex} = 6 \times 2 = 12 ; E_g = 250\text{V}$

Putting the various values, ϕ = **0.0434 Wb**

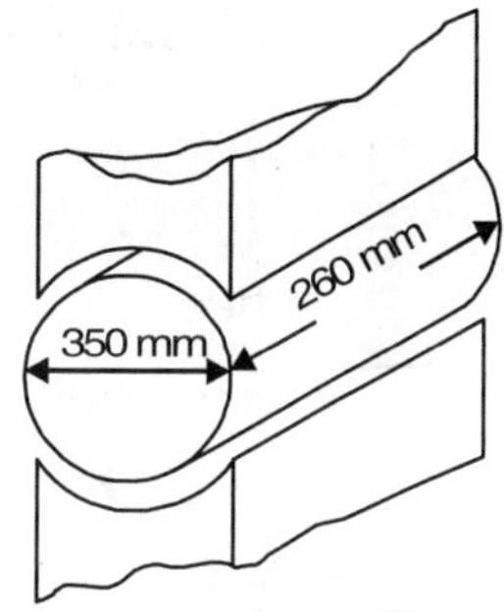

Fig. 2.41

(*ii*) The inner diameter of pole shoe arc, $D = 350 + 3 + 3 = 356$ mm. Therefore, pole pitch $= \pi D/6 = \pi \times 0.356/6$ m.

∴ Effective pole arc length = 80% of $(\pi \times 0.356/6)$ = **0.149 m**

(*iii*) Area of pole face = Pole shoe arc × axial length of the pole $= 0.149 \times 0.260 = 0.03874\,\text{m}^2$

$$\therefore \text{Air-gap flux density} = \frac{\text{Flux/pole}}{\text{Pole face area}} = \frac{0.0434}{0.03874} = \mathbf{1.12\ Wb/m^2}$$

Example 2.21. *The armature of a d.c. generator consists of 40 coils and each coil has 20 turns. When the armature is rotated at 200 rad/s in a 4-pole field structure having a flux of 5 mWb/pole and there are four paths in the armature, calculate (i) number of armature conductors (ii) the voltage between brushes generated by the armature.*

Solution. (*i*) Number of armature conductors is

$$Z = \text{No. of coils} \times \text{No. of turns/coil} \times 2^*$$
$$= (40) \times (20) \times 2 = \mathbf{1600}$$

(*ii*) $$\omega = \frac{2\pi N}{60} \quad \text{or} \quad N = \frac{\omega \times 60}{2\pi}$$

$$\text{Voltage between brushes, } E_g = \frac{P\phi ZN}{60A} = \frac{P\phi Z}{60A} \times \frac{60 \times \omega}{2\pi}$$

$$\therefore \quad E_g = \frac{P\phi Z\omega}{2\pi A}$$

Here $P = 4 ; \phi = 5 \times 10^{-3}$ Wb ; $Z = 1600 ; \omega = 200$ rad/s ; $A = 4$

$$\therefore \quad E_g = \frac{4 \times (5 \times 10^{-3}) \times 1600 \times 200}{2\pi \times 4} = \mathbf{254.6\ V}$$

* Each turn has 2 conductors.

Example 2.22. *A 24-slot, 2-pole d.c. generator has 18 turns per coil. The average flux density per pole is 1T. The effective length of the machine is 20 cm and the radius of the armature is 10 cm. The magnetic poles are designed to cover 80% of armature periphery. If the armature angular velocity is 183.2 rad/s, determine the induced e.m.f. in the armature winding. What is the induced e.m.f. per turn ?*

Solution. *No of coils = 24 ; Effective length of machine, l = 20 cm

Radius of armature, r = 10 cm ; Angular velocity, ω = 183.2 rad/s

Number of total armature conductors is

$$Z = 2 \times \text{No. of turns/coil} \times \text{No. of coils}$$
$$= 2 \times 18 \times 24 = 864$$

$$\text{Actual pole area, } A_P = \frac{2\pi r l}{P} = \frac{2\pi \times 0.1 \times 0.2}{2} = 0.063 \text{ m}^2$$

$$\text{Effective pole area} = 80\% \text{ of } A_P = 0.8 \times 0.063 = 0.05 \text{ m}^2$$

$$\therefore \quad \text{Effective flux/pole, } \phi = B \times 0.05 = 1 \times 0.05 = 0.05 \text{ Wb}$$

Induced e.m.f. in the armature winding is given by;

$$E_g = \frac{P\phi ZN}{60\,A}$$

Here
$$N = \frac{60 \times \omega}{2\pi} = \frac{60 \times 183.2}{2\pi} = 1750 \text{ r.p.m.} ; A = 2^{**}$$

$$\therefore \quad E_g = \frac{2 \times 0.05 \times 864 \times 1750}{60 \times 2} = \mathbf{1260\ V}$$

The induced e.m.f. per parallel path is 1260 V. Since there are two parallel paths, the number of coils in each path = 24/2 = 12. Therefore, induced e.m.f./coil = 1260/12 = 105V. Since there are 18 turns per coil, the induced e.m.f./turn = 105/18 = **5.83 V**.

Example 2.23. *A 100 kW, 240 V shunt generator has a field resistance of 55 Ω and armature resistance of 0·067 Ω. Find the full-load generated voltage.*

Solution. Fig. 2.42 shows the shunt generator circuit.

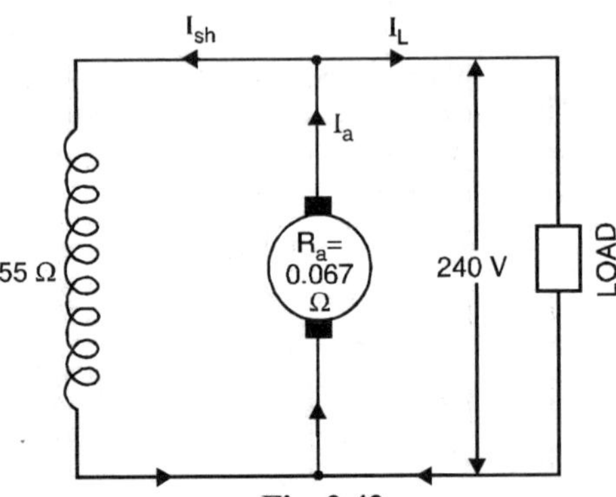

Fig. 2.42

$$I_L = \frac{100 \times 10^3}{240} = 416{\cdot}7 \text{ A}$$
$$I_{sh} = 240/55 = 4.36 \text{ A}$$
$$I_a = I_L + I_{sh}$$
$$= 416{\cdot}7 + 4{\cdot}36 = 421{\cdot}1 \text{ A}$$
$$E_g = V + I_a R_a$$
$$= 240 + 421{\cdot}1 \times 0{\cdot}067 = \mathbf{268{\cdot}2\ V}$$

Example 2.24. *A 4-pole d.c. shunt generator with a wave-wound armature has to supply a load of 500 lamps each of 100 W at 250 V. Allowing 10 V for the voltage drop in the connecting leads between the generator and the load and drop of 1 V per brush, calculate the speed at which the generator should be driven. The flux per pole is 30 mWb and the armature and shunt field resistances are respectively 0·05 Ω and 65 Ω. The number of armature conductors is 390.*

---

* For a double layer winding, the number of coils is the same as the number of armature slots.

** Since it is a 2-pole machine, the number of parallel paths is 2 whether the armature has simplex lap or simplex wave winding.

Solution. Load current, $I_L = \dfrac{500 \times 100}{250} = 200\text{ A}$

$$\text{Voltage across shunt} = 250 + 10 = 260\text{ V}$$

$$I_{sh} = 260/65 = 4\text{ A}$$

$$I_a = I_L + I_{sh} = 200 + 4 = 204\text{ A}$$

$$\text{Brush drop} = 2 \times 1 = 2\text{ V}$$

$$E_g = \text{Voltage across shunt} + 2 + I_a R_a$$

$$= 260 + 2 + 204 \times 0{\cdot}05 = 272{\cdot}2\text{ V}$$

Now, $$E_g = \frac{P \phi Z N}{60\, A}$$

$\therefore$ $$N = \frac{E_g \times 60\, A}{P \phi Z} = \frac{272{\cdot}2 \times 60 \times 2}{4 \times 30 \times 10^{-3} \times 390} = \mathbf{698\ r.p.m.}$$

[For wave winding, $A = 2$]

Example 2.25. *A 30 kW, 300 V d.c. shunt generator has armature and field resistances of 0·05 Ω and 100 Ω respectively. Calculate the total power developed by the armature when it delivers full load output.*

Solution. Fig. 2.43 shows the shunt generator circuit.

$$I_L = \frac{30 \times 10^3}{300} = 100\text{ A}$$

$$I_{sh} = 300/100 = 3\text{A}$$

$$I_a = I_L + I_{sh} = 100 + 3 = 103\text{ A}$$

$$E_g = V + I_a R_a$$

$$= 300 + 103 \times 0{\cdot}05 = 305{\cdot}15\text{ V}$$

Power developed by armature $= E_g I_a$

$$= 305{\cdot}15 \times 103$$

$$= 31{\cdot}43 \times 10^3\text{ W} = \mathbf{31.43\ kW}$$

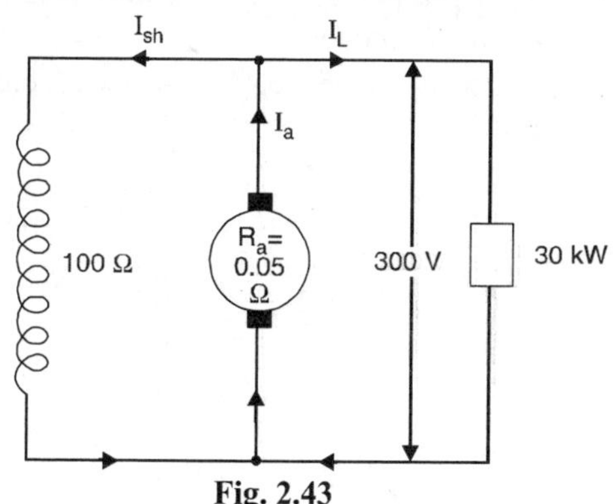

Fig. 2.43

Example 2.26. *A 4-pole lap-wound d.c. shunt generator has a useful flux per pole of 0·07 Wb. The armature winding consists of 220 turns, each of 0·004 Ω resistance. Calculate the terminal voltage when running at 900 r.p.m. if the armature current is 50 A.*

Solution. $$E_g = \frac{P \phi Z N}{60\, A}$$

Here $Z = 220 \times 2 = 440$; $\phi = 0{\cdot}07$ Wb ; $N = 900$ r.p.m. ; $A = P = 4$

$\therefore$ $$E_g = \frac{4 \times 0{\cdot}07 \times 440 \times 900}{60 \times 4} = 462\text{ V}$$

No. of turns per parallel path $= 220/4 = 55$

Resistance per parallel path $= 0{\cdot}004 \times 55 = 0{\cdot}22\ \Omega$

$\therefore$ $$R_a = 0{\cdot}22/A = 0{\cdot}22/4 = 0{\cdot}055\ \Omega$$

Now $$E_g = V + I_a R_a$$

or $$462 = V + 50 \times 0{\cdot}055$$

$\therefore$ $$V = \mathbf{459{\cdot}25\ volts}$$

Example 2.27. *A d.c. shunt generator runs at 400 r.p.m. and delivers 500 kW to bus-bars having a constant voltage of 400 V. Assuming the field excitation to be constant at 5 A, calculate the speed at which the generator must run if the load on it is to be reduced to 300 kW. The armature resistance is 0·015 Ω.*

Solution. $*E_g \propto \phi N$

In the present problem, ϕ is given to be constant. Therefore, $E_g \propto N$.

Initial conditions

$$\text{Load current, } I_1 = 500 \times 10^3/400 = 1250 \text{ A}$$
$$\text{Armature current} = 1250 + 5 = 1255 \text{ A}$$
$$\therefore \quad \text{Generated e.m.f., } E_1 = 400 + 1255 \times 0{\cdot}015 = 418{\cdot}8 \text{ V}$$

Final conditions

$$\text{Load current, } I_2 = 300 \times 10^3/400 = 750 \text{ A}$$
$$\text{Armature current} = 750 + 5 = 755 \text{ A}$$
$$\therefore \quad \text{Generated e.m.f., } E_2 = 400 + 755 \times 0{\cdot}015 = 411{\cdot}3 \text{ V}$$
$$E_2/E_1 = N_2/N_1$$
$$\therefore \quad N_2 = (E_2/E_1) \times N_1 = (411{\cdot}3/418{\cdot}8) \times 400 = \mathbf{393 \text{ r.p.m.}}$$

Example 2.28. *A d.c. shunt generator has an induced voltage on open circuit of 127 V. When the machine is on load, the terminal voltage is 120 V. Find the load current if the field resistance be 15 Ω and the armature resistance 0·02 Ω. Ignore armature reaction.*

Solution. Fig. 2.44 shows the conditions of the problem.

$$E_g = V + I_a R_a$$
$$\text{or} \quad I_a R_a = E_g - V = 127 - 120 = 7 \text{ V}$$
$$\therefore \quad I_a = \frac{7}{R_a} = \frac{7}{0{\cdot}02} = 350 \text{ A}$$
$$I_{sh} = \frac{V}{R_{sh}} = \frac{120}{15} = 8\text{A}$$
$$\therefore \quad I_L = I_a - I_{sh} = 350 - 8 = \mathbf{342 \text{ A}}$$

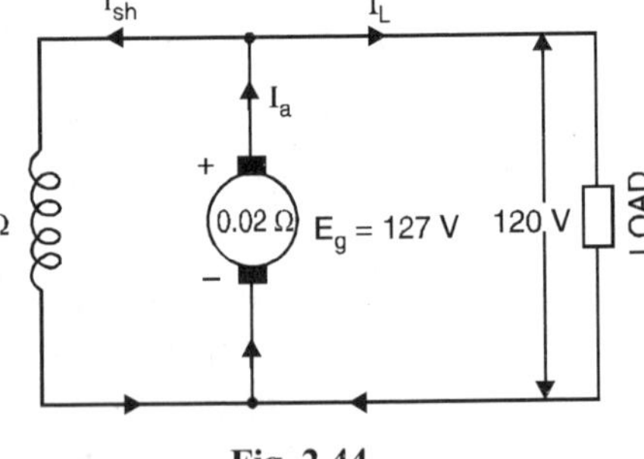

Fig. 2.44

Example 2.29. *A separately excited generator when running at 1200 r.p.m. supplies 200 A at 125 V to a circuit of constant resistance. What will be the load current when the speed is dropped to 1000 r.p.m. if the field current is unaltered? Armature resistance = 0·04 Ω and total drop at brushes = 2V.*

Solution. Fig. 2.45 shows the conditions of the problem.

Fig. 2.45

$$\text{Load resistance, } R_L = \frac{V}{I_L} = \frac{125}{200} = 0{\cdot}625 \ \Omega$$
$$E_{g_1} = V + I_L R_a + 2 = 125 + 200 \times 0{\cdot}04 + 2 \text{ (Brush drop)}$$
$$= 135 \text{ V} \qquad \text{(at } N_1 = 1200 \text{ r.p.m.)}$$

Now $$\frac{E_{g_2}}{E_{g_1}} = \frac{N_2}{N_1} \qquad [\because E_g \propto \phi N \text{ or } E_g \propto N (\because \phi \text{ is constant})]$$

---

* $E_g = \dfrac{P \phi Z N}{60 A}$. As P, Z and A are constant, $E_g \propto \phi N$.

$$\therefore \quad E_{g_2} = E_{g_1} \times \frac{N_2}{N_1} = 135 \times \frac{1000}{1200} = 112{\cdot}5 \text{ V}$$

Voltage across series combination of R_L and R_a

$$V' = E_{g_2} - \text{brush drop} = 112{\cdot}5 - 2 = 110{\cdot}5 \text{ V}$$

$$\text{New load current, } I_L' = \frac{V'}{R_a + R_L} = \frac{110{\cdot}5}{0{\cdot}04 + 0{\cdot}625} = \mathbf{166\ A}$$

Example 2.30. *Estimate the reduction in speed of a generator with constant excitation on bus-bars to decrease its load from 500 kW to 250 kW. The resistance between terminals is 0·015 Ω. The bus bar voltage is 500 V.*

Solution.

$$\text{For the first case, load current, } I_1 = \frac{500 \times 10^3}{500} = 1000 \text{ A}$$

$$E_{g_1} = V + I_1 R_a = 500 + 1000 \times 0{\cdot}015 = 515 \text{ V}$$

$$\text{For the second case, load current, } I_2 = \frac{250 \times 10^3}{500} = 500 \text{ A}$$

$$E_{g_2} = V + I_2 R_a = 500 + 500 \times 0{\cdot}015 = 507{\cdot}5 \text{ V}$$

Now $$\frac{E_{g_2}}{E_{g_1}} = \frac{N_2}{N_1} \qquad (\because \phi \text{ is constant})$$

$$\therefore \quad \frac{N_2}{N_1} = \frac{E_{g_2}}{E_{g_1}} = \frac{507{\cdot}5}{515} = 0{\cdot}9855$$

$$\text{\% age reduction in speed} = 1 - 0{\cdot}9855 = 0{\cdot}0145 = \mathbf{1{\cdot}45\%}$$

Example 2.31. *A 4-pole, 900 r.p.m. d.c. machine has a terminal voltage of 220 V and an induced voltage of 240 V at rated speed. The armature resistance is 0·2 Ω. Is the machine operating as a generator or a motor ? Find the armature current and the number of armature coils if the air-gap flux per pole is 10 mWb and the armature turns per coil are 8. The armature is wave-wound.*

Solution. Since the induced voltage E_g (= 240 V) is more than the terminal voltage V (= 220 V), the machine is operating as a generator.

$$V = E_g - I_a R_a$$

or $$220 = 240 - I_a \times 0{\cdot}2 \quad \therefore \quad I_a = \frac{240 - 220}{0{\cdot}2} = \mathbf{100\ A}$$

$$E_g = \frac{P \phi Z N}{60 A} = \frac{\phi Z N}{60} \times \frac{P}{A}$$

$$\therefore \quad 240 = \frac{(10 \times 10^{-3}) \times Z \times 900}{60} \times \frac{4}{2} \quad \therefore \quad Z = 800$$

Since there are 8 turns per coil, the number of conductors per coil = 2 × 8 = 16.

$$\therefore \quad \text{Number of coils} = \frac{Z}{16} = \frac{800}{16} = \mathbf{50}$$

Example 2.32. *A 4-pole d.c. shunt generator with a shunt field resistance of 100 Ω and armature resistance of 1 Ω has 378 wave-connected conductors in its armature. The flux per pole is 0·02 Wb. If a load resistance of 10 Ω is connected across the armature terminals and the generator is driven at 1000 r.p.m., calculate the power absorbed by the load.*

Solution. Fig. 2.46 shows the generator connections.

$$E_g = \frac{P \phi Z N}{60 A} = \frac{4 \times 0{\cdot}02 \times 378 \times 1000}{60 \times 2} = 252 \text{ V}$$

If V is the terminal voltage of the generator, then,

$$I_L = \frac{V}{10} \; ; \; I_{sh} = \frac{V}{100}$$

$$\therefore \quad I_a = I_L + I_{sh} = \frac{V}{10} + \frac{V}{100} = \frac{11V}{100}$$

Now $\quad V = E_g - I_a R_a$

or $\quad V = 252 - \frac{11V}{100} \quad \therefore \quad V = 227$ volt

$$\therefore \quad I_L = \frac{V}{10} = \frac{227}{10} = 22{\cdot}7 \text{ A}$$

Fig. 2.46

Power absorbed in load $= VI_L = 227 \times 22{\cdot}7 =$ **5153 W**

Example 2.33. *A compound generator is to supply a load of 250 lamps, each rated at 100 W, 250 V. The armature, series and shunt windings have resistances of 0·06 Ω, 0·04 Ω and 50 Ω respectively. Determine the generated e.m.f. when the machine is connected in* (i) *long shunt* (ii) *short shunt. Take drop per brush as 1 V.*

Solution. (*i*) Fig. 2.47 shows the connections of the machine when connected in long shunt.

$$I_L = \frac{250 \times 100}{250} = 100 \text{ A}$$

$$I_{sh} = 250/50 = 5 \text{ A}$$

$$I_a = I_L + I_{sh} = 100 + 5 = 105 \text{ A}$$

$$E_g = V + I_a (R_a + R_{se}) + \text{Total brush drop}$$

$$= 250 + 105\,(0{\cdot}06 + 0{\cdot}04) + 2 \times 1 = \mathbf{262{\cdot}5\ V}$$

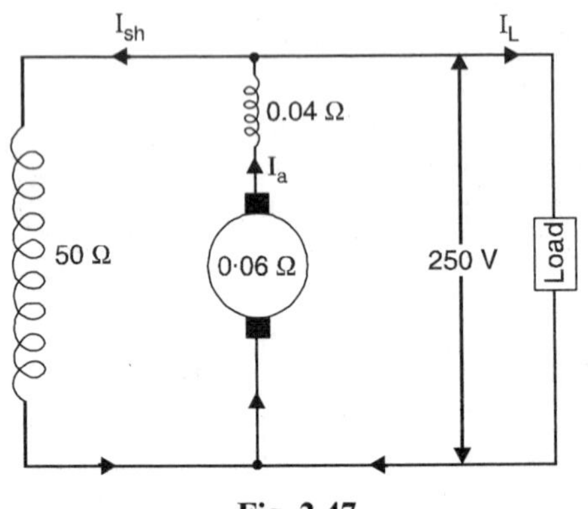

Fig. 2.47

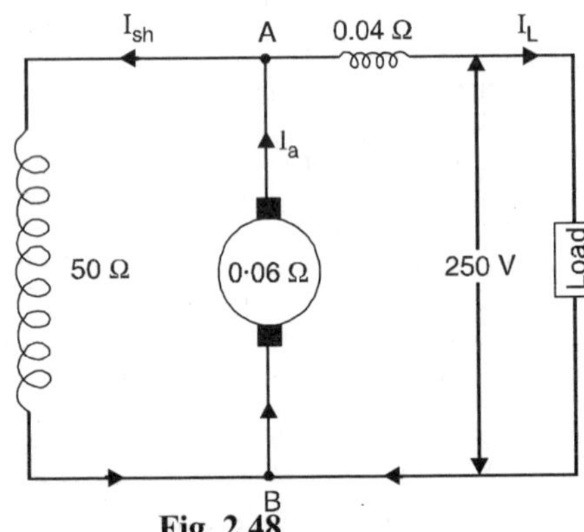

Fig. 2.48

(*ii*) Fig. 2.48 shows the connections when the machine is connected in short shunt.

$$I_L = 100 \text{ A} \qquad \textit{... same as before}$$

Voltage across shunt, $V_{AB} = 250 + 100 \times 0{\cdot}04 = 254$ V

$$I_{sh} = 254/50 = 5{\cdot}08 \text{ A}$$

$$I_a = I_L + I_{sh} = 100 + 5{\cdot}08 = 105{\cdot}08 \text{ A}$$

$$E_g = V_{AB} + I_a R_a + \text{Total brush drop}$$

$$= 254 + 105{\cdot}08 \times 0{\cdot}06 + 2 \times 1 = \mathbf{262{\cdot}3\ V}$$

Example 2.34. *A long-shunt compound generator delivers a load current of 50 A at 500 V and has armature, series field and shunt field resistances of 0·05 Ω, 0·03 Ω and 250 Ω respectively. Calculate the armature current and the generated e.m.f. Allow 1 V per brush for contact drop.*

Solution. Fig. 2.49 shows the conditions of the problem.

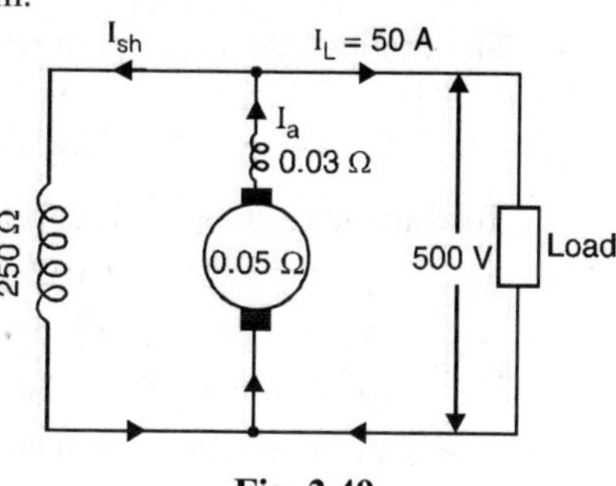

Fig. 2.49

$$I_{sh} = \frac{500}{250} = 2\text{A}$$

$$I_a = I_L + I_{sh} = 50 + 2 = \mathbf{52\ A}$$

Current through series field winding

$$= I_a = 52\text{ A}$$

Voltage drop in series field winding

$$= I_a R_{se} = 52 \times 0{\cdot}03 = 1{\cdot}56\text{ V}$$

$$\text{Armature voltage drop} = I_a R_a = 52 \times 0{\cdot}05 = 2{\cdot}6\text{ V}$$

$$\text{Brush drop} = 2 \times 1 = 2\text{ V}$$

$$\text{Generated voltage}, E_g = V + I_a R_a + I_a R_{se} + \text{Brush drop}$$

$$= 500 + 2{\cdot}6 + 1{\cdot}56 + 2 = \mathbf{506{\cdot}16\ V}$$

Example 2.35. *A short-shunt d.c. compound generator supplies 200A at 100V. The resistances of armature, series field and shunt field windings are 0.04, 0.03 and 60 ohms respectively. Find the e.m.f. generated. Also find the e.m.f. generated if the same machine is connected as long-shunt machine.*

Solution. Fig. 2.50 (*i*) shows short-shunt connection. Armature terminal voltage, $V_a = 100 + I_L R_{se} = 100 + 200 \times 0.03 = 106$ V

$I_{sh} = V_a/R_{sh} = 106/60 = 1.767\text{A}$; $I_a = I_L + I_{sh} = 200 + 1.767 = 201.767$ A

$\therefore$ Generated e.m.f., $E_g = V_a + I_a R_a = 106 + 201.767 \times 0.04 =$ **114.07 V**

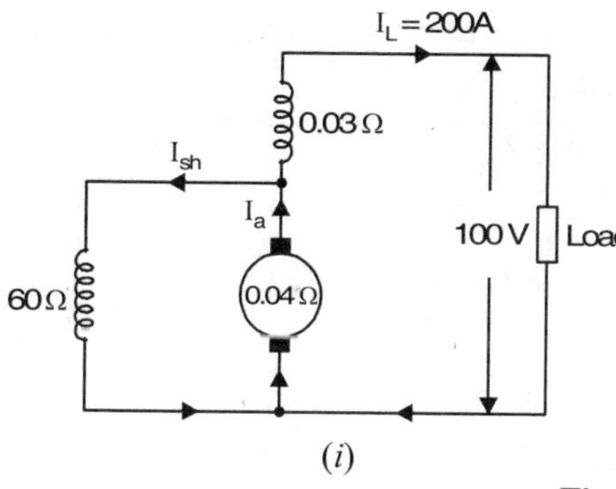

(*i*)

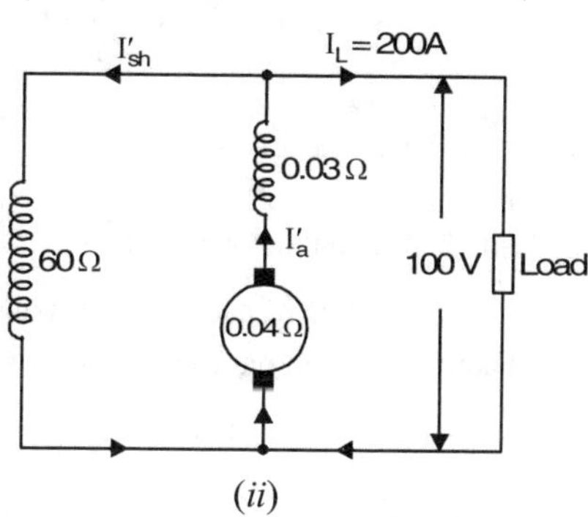

(*ii*)

Fig. 2.50

Fig. 2.50 (*ii*) shows the same machine connected as a long-shunt machine.

$I'_{sh} = 100/60 = 1.667$ A ; $I'_a = I_L + I'_{sh} = 200 + 1.667 = 201.667$ A

Drop in armature and series winding $= I'_a(R_a + R_{se}) = 201.667\,(0.04 + 0.03) - 14.12$V

$\therefore$ Generated e.m.f., $E'_g = 100 + 14.12 =$ **114.12 V**

Example 2.36. *In a long-shunt compound generator, the terminal voltage is 230 V when generator delivers 150 A. Determine (i) induced e.m.f. (ii) total power generated (iii) distribution of this power. Given that shunt field, series field, diverter and armature resistances are 92 Ω, 0·015 Ω, 0·03 Ω and 0·032 Ω respectively.*

Solution. Fig. 2.51 shows the conditions of the problem.

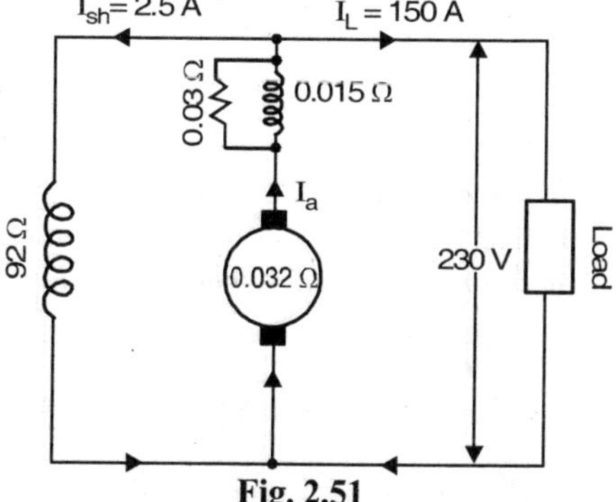

Fig. 2.51

$$I_{sh} = \frac{230}{92} = 2{\cdot}5\text{ A}$$

$$I_a = I_L + I_{sh} = 150 + 2{\cdot}5 = 152{\cdot}5\text{ A}$$

Since series field and diverter resistances are connected in parallel, their combined resistance

$$= \frac{0{\cdot}03 \times 0{\cdot}015}{0{\cdot}03 + 0{\cdot}015} = 0{\cdot}01\,\Omega$$

Total armature circuit resistance

$$= 0.032 + 0{\cdot}01 = 0{\cdot}042\,\Omega$$

$$\text{Voltage drop} = 152{\cdot}5 \times 0{\cdot}042 = 6{\cdot}4\text{ V}$$

(*i*) Induced voltage, $E_g = 230 + 6{\cdot}4 = \mathbf{236{\cdot}4\ V}$

(*ii*) Total power generated in the armature $= E_g I_a = 236{\cdot}4 \times 152{\cdot}5 = \mathbf{36051\ W}$

(*iii*) Power dissipated in armature $= I_a^2 R_a = (152{\cdot}5)^2 \times 0{\cdot}032 = \mathbf{744\ W}$

Power dissipated in series field and diverter $= (152{\cdot}5)^2 \times 0{\cdot}01 = \mathbf{232\ W}$

Power dissipated in shunt winding $= V I_{sh} = 230 \times 2{\cdot}5 = \mathbf{575\ W}$

Power delivered to load $= 230 \times 150 = \mathbf{34500\ W}$

Note that sum of all these powers dissipated is equal to the total power generated in the armature.

Example 2.37. *Calculate the armature resistance of a 6-pole lap-wound armature winding from the following data : Number of slots = 150 ; Conductors per slot = 8 ; Mean length of one turn = 250 cm ; Cross-section of each conductor = 10 mm × 2·5 mm. The resistance of 1 metre of copper wire 1 mm^2 in cross-section is 0·0213 Ω.*

Solution. Mean length of each conductor is

$$l = \frac{250}{2} = 125\text{ cm} = 1{\cdot}25\text{ m}$$

$$\text{Resistance of each conductor} = \frac{\rho\, l}{a} = \frac{0{\cdot}0213 \times 1{\cdot}25}{10 \times 2{\cdot}5} = 1{\cdot}065 \times 10^{-3}\,\Omega$$

$$\text{Number of conductors in series} = \frac{Z}{A} = \frac{150 \times 8}{6} = 200$$

∴ Resistance of each parallel path $= 200 \times 1{\cdot}065 \times 10^{-3} = 0{\cdot}213\,\Omega$

There are 6 such parallel paths.

$$\therefore \quad \text{Armature resistance, } R_a = \frac{0{\cdot}213}{6} = \mathbf{0{\cdot}0355\,\Omega}$$

Example 2.38. *A 4-pole d.c. generator runs at 750 r.p.m. and generates an e.m.f. of 240 V. The armature is wave-wound and has 792 conductors. If the total flux per pole is 0·0145 Wb, what is the leakage coefficient ?*

Solution.

$$E_g = \frac{P\,\phi\,Z\,N}{60\,A}$$

$$\therefore \quad \text{Useful flux/pole, } \phi = \frac{60\,A \times E_g}{P\,Z\,N} = \frac{60 \times 2 \times 240}{4 \times 792 \times 750} = 0{\cdot}0121\text{ Wb}$$

$$\therefore \quad \text{Leakage coefficient, } \lambda = \frac{\text{Total flux/pole}}{\text{Useful flux/pole}} = \frac{0{\cdot}0145}{0{\cdot}0121} = \mathbf{1.2}$$

Example 2.39. *A 1500 kW, 550 V, 10-pole generator runs at 150 r.p.m. There are 2500 lap-connected conductors and the full-load copper losses are 25 kW. The air-gap flux density has a uniform value of 0.9 Wb/m^2. Calculate* (*i*) *the no-load terminal voltage and* (*ii*) *the area of pole shoe.*

Solution.

$$(i) \quad \text{Full-load current, } I = \frac{1500 \times 10^3}{550} = 2727\text{ A}$$

$$\text{Full-load Cu loss} = I^2R = 25 \times 10^3 \text{ W (given)}$$

$$\therefore \quad \text{Resistance drop} = IR = \frac{I^2R}{I} = \frac{25 \times 10^3}{2727} = 9{\cdot}2 \text{ V}$$

$$\therefore \quad \text{Generated voltage, } E_g = \text{No-load terminal voltage}$$

$$= 550 + 9{\cdot}2 = \mathbf{559{\cdot}2 \text{ V}}$$

(*ii*)

$$E_g = \frac{P \phi Z N}{60 A}$$

$$\therefore \quad \text{Flux/pole, } \phi = \frac{60 A \times E_g}{P Z N} = \frac{60 \times 10 \times 559{\cdot}2}{10 \times 2500 \times 150} = 0{\cdot}0895 \text{ Wb}$$

$$\therefore \quad \text{Pole area} = \frac{\text{Flux}}{\text{Flux density}} = \frac{0{\cdot}0895}{0{\cdot}9} = 0{\cdot}0994 \text{ m}^2 = \mathbf{994 \text{ cm}^2}$$

Example 2.40. *In a 120 V compound generator, the resistance of the armature, shunt and series windings are 0·06 Ω, 25 Ω and 0·04 Ω respectively. The load current is 100 A at 120 V. Find the armature current and induced e.m.f. when the machine is connected as (i) long shunt (ii) short shunt. How will the ampere-turns of the series field be changed in (i) if a diverter of 0.1 Ω be connected in parallel with the series winding ? Neglect brush contact drop and ignore armature reaction.*

Solution.

(*i*) **Long Shunt** [See Fig. 2.52 (*i*)].

$$I_{sh} = \frac{120}{25} = 4.8 \text{ A} \; ; \; I_L = 100 \text{ A} \; ; \; I_a = 100 + 4{\cdot}8 = \mathbf{104{\cdot}8 \text{ A}}$$

$$\text{Voltage drop in series winding} = I_a R_{se} = 104{\cdot}8 \times 0{\cdot}04 = 4.19 \text{ V}$$

$$\text{Armature voltage drop} = I_a R_a = 104{\cdot}8 \times 0{\cdot}06 = 6{\cdot}29 \text{ V}$$

$$\text{Generated voltage, } E_g = 120 + I_a R_a + I_a R_{se} = 120 + 6{\cdot}29 + 4{\cdot}19 = \mathbf{130{\cdot}5 \text{ V}}$$

(*ii*) **Short Shunt** [See Fig. 2.52 (*ii*)].

$$\text{Voltage across series winding} = I_L R_{se} = 100 \times 0{\cdot}04 = 4 \text{ V}$$

$$\text{Voltage across shunt winding} = 120 + 4 = 124 \text{ V}$$

$$\therefore \quad I_{sh} = 124/25 = 5\text{A} \; ; \; I_a = 100 + 5 = \mathbf{105 \text{ A}}$$

$$\text{Armature drop} = I_a R_a = 105 \times 0{\cdot}06 = 6{\cdot}3 \text{ V}$$

$$\text{Generated voltage, } E_g = 120 + 6.3 + 4 = \mathbf{130{\cdot}3 \text{ V}}$$

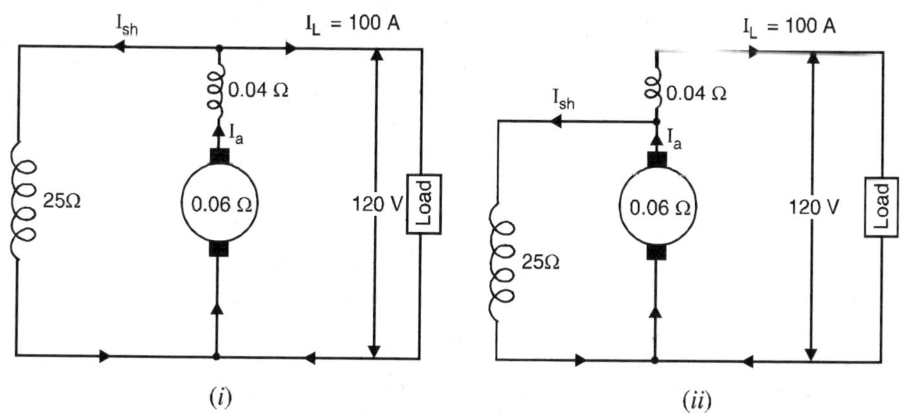

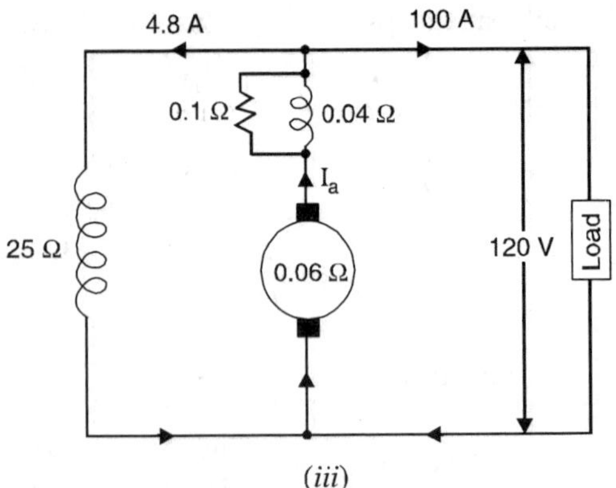

(*iii*)

Fig. 2.52

When a 0.1 Ω diverter is connected across the series field of long-shunt generator [See Fig. 2.52 (*iii*)], then by current divider rule,

$$\text{Current through series winding} = I_a \times \frac{R_{div}}{R_{div} + R_{se}} = 104{\cdot}8 \times \frac{0{\cdot}1}{0{\cdot}1 + 0{\cdot}04} = 74{\cdot}86 \text{ A}$$

Since the number of series turns is the same, change in series field ampere-turns is the same as the change of current in the series winding.

∴ Decrease in series field ampere-turns

$$= \frac{104{\cdot}8 - 74{\cdot}86}{104{\cdot}8} \times 100 = \mathbf{28{\cdot}6\%}$$

Example 2.41. *A separately excited d.c. generator has armature circuit resistance of 0.1Ω and the total brush drop is 2V. When running at 1000 r.p.m., it delivers a current of 100A at 250 V to a load of constant resistance. If the generator speed drops to 700 r.p.m. with field current unaltered, find the current delivered to load.*

Solution. Load resistance, $R_L = 250/100 = 2.5\ \Omega$

For the first case, $E_g = V + I_a R_a + \text{Brush drop}$

$= 250 + 100 \times 0.1 + 2 = 262$ V

For the second case, $N' = 700$ r.p.m. and generated e.m.f. E'_g is given by ;

$$\frac{E'_g}{E_g} = \frac{N'}{N} \quad \text{or} \quad E'_g = \frac{N'}{N} \times E_g = \frac{700}{1000} \times 262 = 183.4 \text{ volts}$$

Let the new load current be I'_a. Then terminal voltage of the generator is $2.5\ I'_a$.

Now $2.5\ I'_a = E'_g - I'_a R_a - \text{Brush drop}$

or $2.5\ I'_a = 183.4 - I'_a (0.1) - 2$

∴ $I'_a = \mathbf{69.77\ A}$

TUTORIAL PROBLEMS

1. A 4-pole shunt generator with wave-wound armature has 41 slots, each having 12 conductors. The armature resistance is 0·05 Ω and shunt field resistance is 200 Ω. The flux per pole is 25 mWb. If a load resistance of 10 Ω is connected across the armature terminals, calculate the voltage across the load when generator is driven at 1000 r.p.m. **[389·5 V]**
2. A long shunt compound generator has full load output of 100 kW at 250 volts. The armature, series and shunt windings have resistances of 0·05 Ω, 0·03 Ω and 55 Ω respectively. Find the armature current and generated e.m.f. **[404·5 A, 282·3 V]**

3. A shunt generator supplies 75 A at 200 V through feeders of resistance 0·08 Ω. The armature and shunt field windings have resistances of 0·04 Ω and 80 Ω respectively. Find the terminal voltage and generated e.m.f. **[206 V, 209·1 V]**

4. A 4-pole short shunt compound generator has armature, shunt field and series field resistances of 0·4 Ω, 160 Ω and 0·2 Ω respectively. The armature is lap connected with 440 conductors and is driven at 600 r.p.m. Calculate the flux per pole when the machine is delivering 120 A at 400 V. **[0.108 Wb]**

5. The resistance of field circuit of a shunt-excited d.c. generator is 200 Ω. When the output of the generator is 100 kW, the terminal voltage is 500 V and the generated e.m.f. 525 V. Calculate (*i*) the armature resistance (*ii*) the value of generated e.m.f. when the output is 60 kW, if the terminal voltage is then 520 V. **[(*i*) 0·123 Ω (*ii*) 534·5 V]**

6. A 4-pole long-shunt lap-wound generator supplies 25 kW at a terminal voltage of 500 V. The armature resistance is 0·03 Ω, series field resistance is 0·04 Ω and shunt field resistance is 200 Ω. The brush drop may be taken as 1V. Determine the e.m.f. generated. Calculate also the number of conductors if the speed is 1200 r.p.m. and flux per pole is 0·02 Wb. Neglect armature reaction. **[505·67 V ; 1264 conductors]**

7. A separately excited d.c. generator when running at 1200 r.p.m. supplies 200 A at 125 V to a circuit of constant resistance. What will be the generated voltage when the speed is dropped to 1000 r.p.m. and the field current is reduced to 80% ? Armature resistance, 0·04 Ω and total drop at brushes, 2V. Ignore saturation and armature reaction. **[90 V]**

8. A short-shunt compound generator supplies a current of 100 A at 220 V. The resistance of the shunt field is 50 Ω, of the series winding 0·025 Ω and that of armature 0·05 Ω. Find the e.m.f. generated. **[227·72 V]**

9. A short-shunt compound generator delivers a load current of 30 A at 220 V and has armature, series field and shunt field resistances of 0·05 Ω, 0·3 Ω and 200 Ω respectively. Calculate the armature current and induced e.m.f. Allow 1 V per brush for contact drop. **[31·145 A ; 232·56 V]**

10. In a 110 V compound generator, the resistance of the armature, shunt and series windings are 0·06 Ω, 25 Ω and 0·04 Ω respectively. The load consists of 200 lamps each rated at 55 W, 110 V. Find the armature current and e.m.f. generated when the machine is connected (*i*) long shunt, (*ii*) short shunt. (*iii*) How will the ampere-turns of the series winding be changed if in (*i*) a diverter of 0·1 Ω is connected in parallel with the series field winding ? Ignore armature reaction and brush contact drop. **[(*i*) 104·4 A ; 120·4 V (*ii*) 104·6 A ; 120·3 V (*iii*) 28·5% *decrease*]**

2.27 LOSSES IN A D.C. MACHINE

The losses in a d.c. machine (generator or motor) may be divided into three classes *viz* (*i*) copper losses (*ii*) iron or core losses and (*iii*) mechanical losses. All these losses appear as heat and thus raise the temperature of the machine. They also lower the efficiency of the machine.

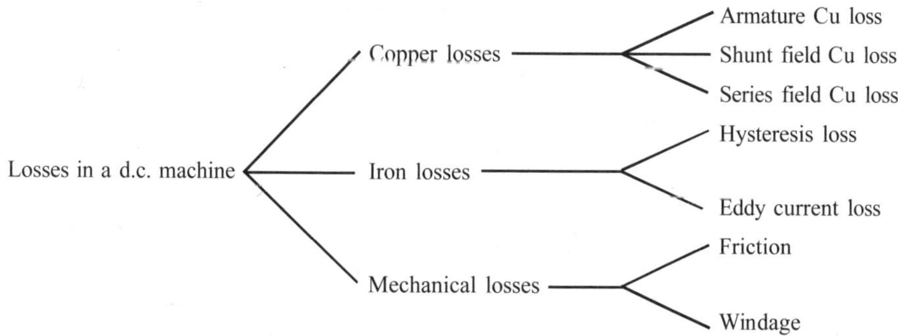

1. **Copper losses.** These losses occur due to currents in the various windings of the machine.

(*i*) Armature copper loss $= I_a^2 R_a$

(*ii*) Shunt field copper loss $= I_{sh}^2 R_{sh}$

(*iii*) Series field copper loss $= I_{se}^2 R_{se}$

Copper losses also occur in interpole field winding and compensating field winding and must be accounted for if required. As we shall see, these windings are connected in series with armature of the d.c. machine and, therefore, carry armature current (I_a).

Note. There is also brush contact loss due to brush contact resistance (*i.e.*, resistance between the surface of brush and surface of commutator). This loss is generally included in armature copper loss.

2. Iron or Core losses. These losses occur in the armature of a d.c. machine and are due to the rotation of armature in the magnetic field of the poles. They are of two types *viz.*, (*i*) hysteresis loss (*ii*) eddy current loss.

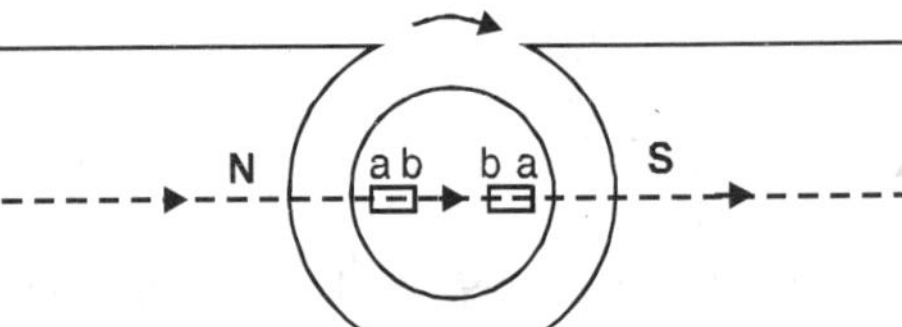

Fig. 2.53

(*i*) **Hysteresis loss.** Hysteresis loss occurs in the armature of the d.c. machine since any given part of the armature is subjected to magnetic field reversals as it passes under successive poles. Fig. 2.53 shows an armature rotating in two-pole machine. Consider a small piece *ab* of the armature. When the piece *ab* is under *N*-pole, the magnetic lines pass from *a* to *b*. Half a revolution later, the same piece of iron is under *S*-pole and magnetic lines pass from *b* to *a* so that magnetism in the iron is reversed. In order to reverse continuously the molecular magnets in the armature core, some amount of power has to be spent which is called hysteresis loss. It is given by **Steinmetz** formula. This formula is

$$\text{Hysteresis loss, } P_h = \eta B_{max}^{1.6} f V \text{ watts}$$

where
B_{max} = Maximum flux density in armature
f = Frequency of magnetic reversals
= $NP/120$ where N is in r.p.m.
V = Volume of armature in m^3
η = Steinmetz hysteresis co-efficient

In order to reduce this loss in a d.c. machine, armature core is made of such materials which have a low value of steinmetz hysteresis co-efficient *e.g.*, silicon steel.

(*ii*) **Eddy current loss.** In addition to the voltages induced in the armature conductors, there are also voltages induced in the armature core. These voltages produce circulating currents in the armature core as shown in Fig. 2.54. These are called *eddy currents* and power loss due to their flow is called eddy current loss. The eddy current loss appears as heat which raises the temperature of the machine and lowers its efficiency.

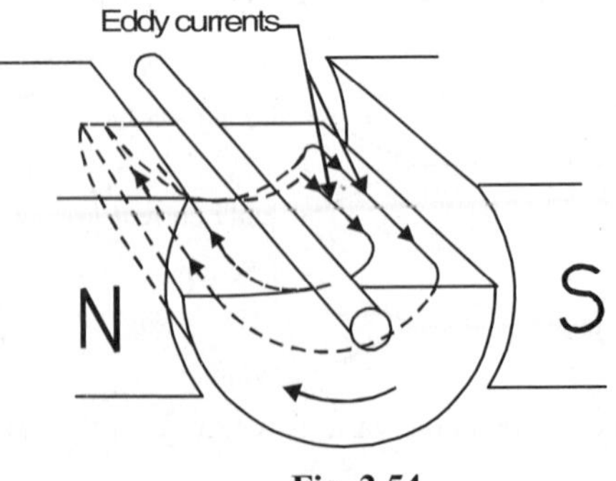

Fig. 2.54

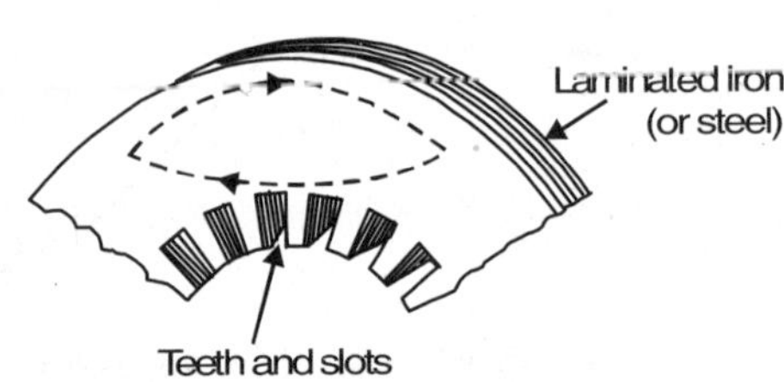

Fig. 2.55

If a continuous solid iron core is used, the resistance to eddy current path will be small due to large cross-sectional area of the core. Consequently, the magnitude of eddy current and hence eddy current loss* will be large. The magnitude of eddy current can be reduced by making core resistance as high as practical. The core resistance can be greatly increased by constructing the core of thin, round iron sheets called *laminations* [See Fig. 2.55]. The laminations are insulated from each other with a coating of varnish. The insulating coating has a high resistance, so very little current flows from one lamination to the other. Also, because each lamination is very thin, the resistance to current flowing through the width of a lamination is also quite large. Thus laminating a core increases the core resistance which decreases the eddy current and hence the eddy current loss.

$$\text{Eddy current loss, } P_e = K_e B_{max}^2 f^2 t^2 V \text{ watts}$$

where K_e = Constant depending upon the electrical resistance of core and system of units used

B_{max} = Maximum flux density in Wb/m^2

f = Frequency of magnetic reversals in Hz

t = Thickness of lamination in m

V = Volume of core in m^3

It may be noted that eddy current loss depends upon the square of lamination thickness. For this reason, lamination thickness should be kept as small as possible.

3. Mechanical losses. These losses are due to friction and windage.

(*i*) friction loss *e.g.*, bearing friction, brush friction etc.

(*ii*) windage loss *i.e.*, air friction of rotating armature.

These losses depend upon the speed of the machine. But for a given speed, they are practically constant.

Note. Iron losses and mechanical losses together are called **rotational losses**.

2.28 CONSTANT AND VARIABLE LOSSES

The losses in a d.c. generator (or d.c. motor) may be sub-divided into (*i*) constant losses (*ii*) variable losses.

(*i*) **Constant losses.** Those losses in a d.c. generator which remain constant at all loads are known as *constant losses*. The constant losses in a d.c. generator are :

(*a*) iron losses, (*b*) mechanical losses, (*c*) shunt field losses

(*ii*) **Variable losses.** Those losses in a d.c. generator which vary with load are called *variable losses*. The variable losses in a d.c. generator are :

(*a*) Copper loss in armature winding ($I_a^2 R_a$)

(*b*) Copper loss in series field winding ($I_{se}^2 R_{se}$)

$$\text{Total losses} = \text{Constant losses} + \text{Variable losses}$$

Note. Field Cu loss is constant for shunt and compound generators.

2.29 POWER STAGES

The various power stages in a d.c. generator are represented diagrammatically in Fig. 2.56.

$A - B$ = Iron and friction losses

$B - C$ = Copper losses

---

* Eddy current power loss = Induced voltage × eddy current
The induced voltage cannot be changed by modifying the core material. Therefore, eddy current loss can be reduced by decreasing the magnitude of eddy current.

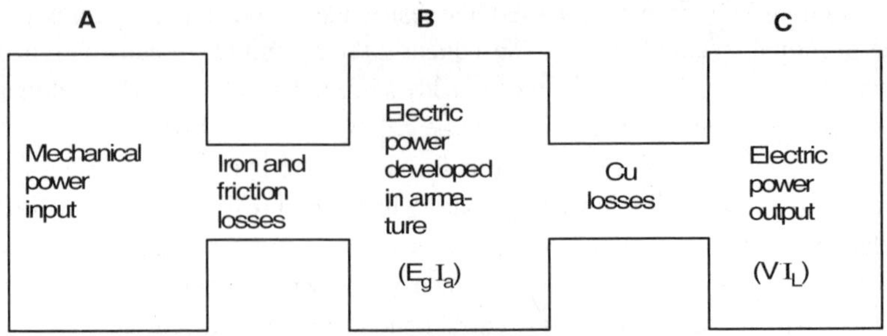

Fig. 2.56

(*i*) **Mechanical efficiency**

$$\eta_m = \frac{B}{A} = \frac{E_g I_a}{\text{Mechanical power input}}$$

(*ii*) **Electrical efficiency**

$$\eta_e = \frac{C}{B} = \frac{V I_L}{E_g I_a}$$

(*iii*) **Commercial** or **overall efficiency**

$$\eta_c = \frac{C}{A} = \frac{V I_L}{\text{Mechanical power input}}$$

Clearly, $\eta_c = \eta_m \times \eta_e$

Unless otherwise stated, commercial efficiency is always understood.

Now, commercial efficiency, $\eta_c = \dfrac{C}{A} = \dfrac{\text{output}}{\text{input}} = \dfrac{\text{input} - \text{losses}}{\text{input}}$

2.30 CONDITION FOR MAXIMUM EFFICIENCY

The efficiency of a d.c. generator is not constant but varies with load. Consider a shunt generator delivering a load current I_L at a terminal voltage V.

$$\begin{aligned}
\text{Generator output} &= V I_L \\
\text{Generator input} &= \text{Output} + \text{Losses} \\
&= V I_L + \text{Variable losses} + \text{Constant losses} \\
&= V I_L + I_a^2 R_a + W_C \\
&= V I_L + (I_L + I_{sh})^2 R_a + W_C \qquad [\because I_a = I_L + I_{sh}]
\end{aligned}$$

The shunt field current I_{sh} is generally small as compared to I_L and, therefore, can be neglected.

$\therefore$ Generator input $= V I_L + I_L^2 R_a + W_C$

Now
$$\eta = \frac{\text{output}}{\text{input}} = \frac{V I_L}{V I_L + I_L^2 R_a + W_C}$$

$$= \frac{1}{1 + \left(\dfrac{I_L R_a}{V} + \dfrac{W_C}{V I_L}\right)} \qquad ...(i)$$

The efficiency will be maximum when the denominator of eq. (*i*) is minimum *i.e.*,

$$\frac{d}{dI_L}\left(\frac{I_L R_a}{V} + \frac{W_C}{V I_L}\right) = 0$$

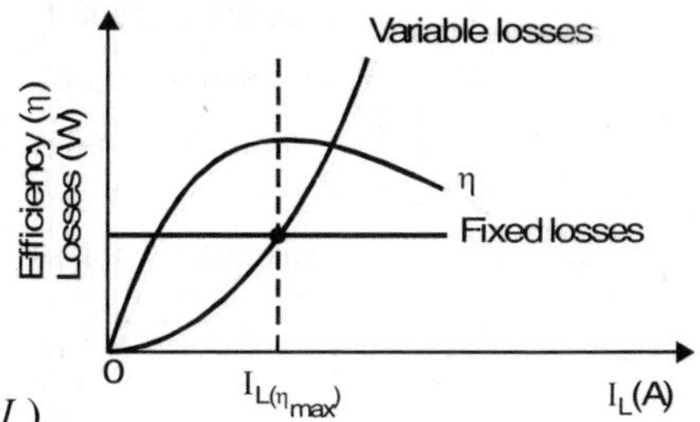

$$\text{or} \quad \frac{R_a}{V} - \frac{W_C}{V I_L^2} = 0$$

$$\text{or} \quad \frac{R_a}{V} = \frac{W_C}{V I_L^2}$$

$$\text{or} \quad I_L^2 R_a = W_C$$

i.e. **Variable loss = Constant loss** ($\because I_L \simeq I_a$)

The load current corresponding to maximum efficiency is given by ;

$$I_L = \sqrt{\frac{W_C}{R_a}}$$

Variable losses
η
Fixed losses
Efficiency (η)
Losses (W)
0
$I_{L(\eta_{max})}$
I_L(A)
Load current

Fig. 2.57

Hence, the efficiency of a d.c. generator will be maximum when the load current is such that variable loss is equal to the constant loss. Fig. 2.57 shows the graph between efficiency (η) and losses (W) versus load current (I_L). The rotational losses (Iron losses + Mechanical losses) and shunt field losses do not change with load current and are called constant or fixed losses. These are represented by a horizontal line parallel to load current axis. The variable losses (copper losses in various windings) vary with load current. The point at which the variable losses are equal to the fixed losses determines the maximum efficiency. The value of load current corresponding to maximum efficiency can be easily determined from the graph.

Example 2.42. *A shunt generator supplies 96 A at a terminal voltage of 200 volts. The armature and shunt field resistances are 0·1 Ω and 50 Ω respectively. The iron and frictional losses are 2500 W. Find (i) e.m.f. generated (ii) copper losses (iii) commercial efficiency.*

Solution. Fig. 2.58 shows the connections of shunt generator.

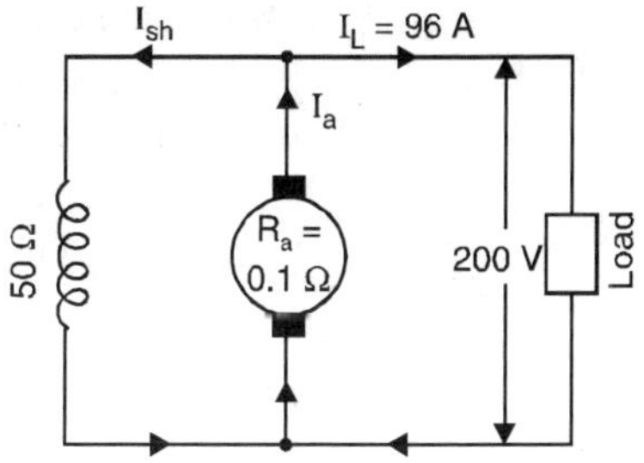

(*i*) $I_{sh} = 200/50 = 4 \text{ A}$

$I_a = I_L + I_{sh} = 96 + 4 = 100 \text{ A}$

$E_g = V + I_a R_a$

$= 200 + 100 \times 0{\cdot}1 = \mathbf{210 \text{ V}}$

(*ii*) Armature Cu loss $= I_a^2 R_a = (100)^2 \times 0.1$

$= 1000 \text{ W}$

Shunt Cu loss $= I_{sh}^2 R_{sh} = (4)^2 \times 50 = 800 \text{ W}$

Total Cu loss $= 1000 + 800 = \mathbf{1800 \text{ W}}$

(*iii*) Total losses = Rotational losses + Cu losses

$= 2500 + 1800 = 4300 \text{ W}$

Output power $= 96 \times 200 = 19200 \text{ W}$

Input power $= 19200 + 4300 = 23500 \text{ W}$

$$\therefore \quad \eta_c = \frac{19200}{23500} \times 100 = \mathbf{81.7\%}$$

I_{sh}
I_L = 96 A
I_a
50 Ω
R_a = 0.1 Ω
200 V
Load

Fig. 2.58

Example 2.43. *The shunt generator delivers full load current of 200 A at 240 V. The shunt field resistance is 60 Ω and full load efficiency is 90%. The rotational losses are 800 W. Find (i) armature resistance (ii) current at which maximum efficiency occurs.*

Solution.

(*i*) Generator output $= 240 \times 200 = 48000 \text{ W}$

Generator input $= 48000/0{\cdot}9 = 53333 \text{ W}$

Total losses $= 53333 - 48000 = 5333 \text{ W}$

Shunt field current, $I_{sh} = 240/60 = 4 \text{ A}$

$$\text{Armature current, } I_a = 200 + 4 = 204 \text{ A}$$

$$\text{Shunt Cu loss} = I_{sh}^2 R_{sh} = (4)^2 \times 60 = 960 \text{ W}$$

$$\text{Constant losses} = \text{Rotational losses + Shunt Cu loss}$$

$$= 800 + 960 = 1760 \text{ W}$$

$$\therefore \quad \text{Armature Cu loss} = \text{Total losses} - \text{Constant losses}$$

$$= 5333 - 1760 = 3573 \text{ W}$$

i.e., $I_a^2 R_a = 3573$

or $R_a = 3573/I_a^2 = 3573/(204)^2 = \mathbf{0{\cdot}0858\ \Omega}$

(*ii*) For maximum efficiency, the condition is :

$$\text{Variable losses} = \text{Constant losses}$$

or $^*I_L^2 R_a = 1760$

$$\therefore \quad I_L = \sqrt{\frac{1760}{0{\cdot}0858}} = \mathbf{143.22\ A}$$

Example 2.44. *A 10 kW d.c. shunt generator has the following losses at full load :*

Mechanical losses = 290 W ; Iron losses = 420 W ; Shunt Cu loss = 120 W ;

Armature Cu loss = 595 W

Calculate the efficiency at (*i*) *no load* (*ii*) *25% of full-load.*

Solution.

(*i*) **At no load.** The rotational losses remain constant at no load or at any other load.

$$\text{Rotational losses} = \text{Mechanical losses + Iron losses} = 290 + 420 = 710 \text{ W}$$

$$\text{Total losses at no load} = 710 + \text{Shunt Cu loss} = 710 + 120 = 830 \text{ W}$$

The efficiency at no load is **zero** because there is no output.

(*ii*) **At 25% of full load.**

$$\text{Constant losses} = 830 \text{ W} \qquad \textit{...same as before}$$

$$\text{Armature Cu loss} = {}^{**}(1/4)^2 \text{ of F.L. value} = (1/4)^2 \times 595 = 37 \text{ W}$$

$$\text{Total losses} = 830 + 37 = 867 \text{ W}$$

$$\text{Output} = (1/4) \times 10 \text{ kW} = 2{\cdot}5 \text{ kW} = 2500 \text{ W}$$

$$\text{Input} = 2500 + 867 = 3367 \text{ W}$$

$$\therefore \quad \eta = \frac{2500}{3367} \times 100 = \mathbf{74\%}$$

Example 2.45. *A 75 kW shunt generator is operated at 230 V. The rotational losses are 1810 W and shunt field circuit draws 5·35 A. The armature circuit has a resistance of 0·035 Ω and brush drop is 2·2 V. Calculate* (*i*) *total losses* (*ii*) *output of prime mover* (*iii*) *efficiency at rated load.*

Solution.

(*i*) $\text{Load current, } I_L = 75 \times 10^3/230 = 326{\cdot}1 \text{ A}$

$$\text{Armature current, } I_a = I_L + I_{sh} = 326{\cdot}1 + 5{\cdot}35 = 331{\cdot}5 \text{ A}$$

$$\text{Armature Cu loss} = I_a^2 R_a = (331{\cdot}5)^2 \times 0{\cdot}035 = 3846 \text{ W}$$

$$\text{Shunt Cu loss} = V I_{sh} = 230 \times 5{\cdot}35 = 1230 \text{ W}$$

$$\text{Brush power loss} = \text{Brush drop} \times I_a = 2{\cdot}2 \times 331{\cdot}5 = 729 \text{ W}$$

$$\therefore \quad \text{Total losses} = 1810 + 3846 + 1230 + 729 = \mathbf{7615\ W}$$

(*ii*) $\text{Output power} = 75 \text{ kW} = 75000 \text{ W}$

---

* Variable losses in armature are $I_a^2 R_a$. In the derivation of expression for maximum efficiency, I_{sh} was neglected so that $I_a = I_L$.

** When the generator is loaded to 25% of its normal rating, armature current is 25% (or 1/4) of its full load value. Since Cu loss varies as square of current, we have $(1/4)^2$ term here.

$$\text{Output of prime mover} = 75000 + 7615 = \mathbf{82615\ W}$$

(*iii*)
$$\eta = \frac{75000}{82615} \times 100 = \mathbf{90.8\ \%}$$

Example 2.46. *A long-shunt compound generator running at 1000 r.p.m. supplies 22 kW at a terminal voltage of 220 V. The resistances of the armature, shunt field and the series field are 0·05 Ω, 110 Ω and 0·06 Ω respectively. The overall efficiency at the above load is 88%. Find (i) Cu losses (ii) iron and friction losses (iii) torque exerted by the prime mover.*

Solution. Fig. 2.59 shows the conditions of the problem.

Fig. 2.59

$$I_L = \frac{22 \times 10^3}{220} = 100\text{ A} ; I_{sh} = \frac{220}{110} = 2\text{A}$$

$$I_a = I_L + I_{sh} = 100 + 2 = 102\text{ A}$$

Voltage drop in series field winding

$$= I_a R_{se} = 102 \times 0{\cdot}06 = 6{\cdot}12\text{ V}$$

(*i*)
$$\text{Armature Cu loss} = I_a^2 R_a = (102)^2 \times 0{\cdot}05 = 520{\cdot}2\text{ W}$$

$$\text{Series field Cu loss} = I_a^2 R_{se} = (102)^2 \times 0{\cdot}06 = 624{\cdot}3\text{ W}$$

$$\text{Shunt field Cu loss} = V I_{sh} = 220 \times 2 = 440\text{ W}$$

$$\text{Total Cu losses} = 520{\cdot}2 + 624{\cdot}3 + 440 = \mathbf{1584{\cdot}5\ W}$$

(*ii*)
$$\text{Output} = 22\text{ kW} = 22000\text{ W} ; \text{Input} = 22000/0{\cdot}88 = 25000\text{ W}$$

$$\text{Total losses} = \text{Input} - \text{Output} = 25000 - 22000 = 3000\text{ W}$$

$$\therefore \quad \text{Iron and friction losses} = \text{Total losses} - \text{Total Cu losses}$$

$$= 3000 - 1584{\cdot}5 = \mathbf{1415{\cdot}5\ W}$$

(*iii*) Let T N-m be the torque exerted by the prime mover.

$$\text{Input power} = T \times \text{Angular velocity}$$

or
$$25000 = T \times \frac{2\pi N}{60}$$

$$\therefore \quad T = \frac{25000 \times 60}{2\pi \times 1000} = \mathbf{238{\cdot}74\ N\text{-}m}$$

Example 2.47. *A short-shunt compound generator supplies a current of 100 A at 220 V. The resistance of the shunt field is 50 Ω, of the series winding 0·025 Ω and of the armature 0·05 Ω. Iron and friction losses amount to 1 kW. Find (i) the e.m.f. generated (ii) the copper losses (iii) the b.h.p. of the prime mover (iv) commercial efficiency.*

Solution. Fig. 2.60 shows the generator circuit.

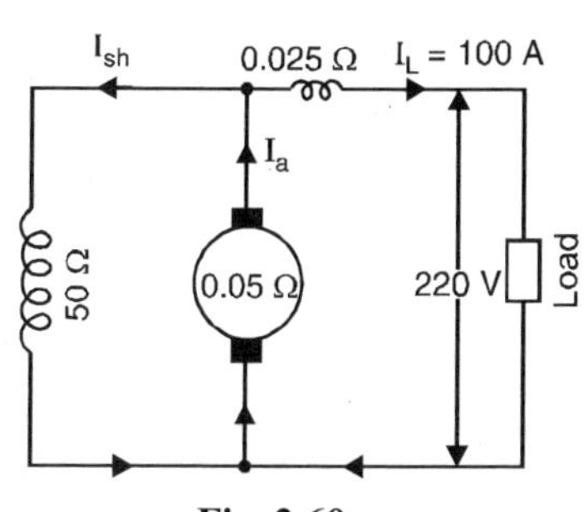

Fig. 2.60

(*i*) Voltage drop in series winding

$$= I_L R_{se} = 100 \times 0{\cdot}025 - 2{\cdot}5\text{ V}$$

Voltage across shunt winding = 220 + 2·5 = 222·5 V

$$I_{sh} = \frac{222{\cdot}5}{50} = 4{\cdot}45\text{ A}$$

$$I_a = I_L + I_{sh} = 100 + 4{\cdot}45 = 104{\cdot}45\text{ A}$$

$$\text{Voltage drop in armature} = I_a R_a = 104{\cdot}45 \times 0{\cdot}05$$

$$= 5{\cdot}22\text{ V}$$

$$\therefore \quad \text{Generated e.m.f., } E_g = V + I_a R_a + I_L R_{se} = 220 + 5{\cdot}22 + 2{\cdot}5 = \mathbf{227{\cdot}72\ V}$$

(*ii*) Electric power output $= VI_L = 220 \times 100 = 22000$ W

Electric power developed in the armature

$$= E_g I_a = 227{\cdot}72 \times 104{\cdot}45 = 23790 \text{ W}$$

$\therefore$ Cu losses $= 23790 - 22000 =$ **1790 W**

(*iii*) Total losses = Cu losses + Iron and friction losses

$$= 1790 + 1000 = 2790 \text{ W}$$

Mechanical input power = Output + Total losses

$$= 22000 + 2790 = 24790 \text{ W}$$

$$\therefore \quad \text{b.h.p. of prime mover} = \frac{24790}{746} = \mathbf{33{\cdot}24\ h.p.}$$

$$(iv)\ \text{Commercial efficiency, } \eta_c = \frac{\text{Output}}{\text{Input}} \times 100 = \frac{22000}{24790} \times 100 = \mathbf{88{\cdot}74\%}$$

Example 2.48. *A long-shunt compound wound generator gives 240 V at full-load output of 100 A. The resistances of the various windings of the machine are : armature (including brush contact) 0·1 Ω, series field 0·02 Ω, interpole 0·025 Ω, shunt field (including regulating resistance) 100 Ω. The iron losses at F.L. are 1000 W, windage and frictional losses amount to 500 W. Calculate the F.L. efficiency of the machine.*

Solution. Fig.2.61 shows the generator circuit.

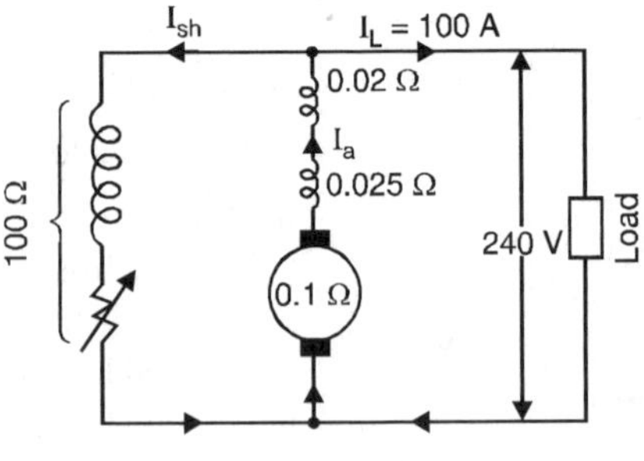

Output power $= 240 \times 100 = 24000$ W

Total armature circuit resistance

$$= R_a + R_{se} + R_{inter}$$
$$= 0{\cdot}1 + 0{\cdot}02 + 0{\cdot}025 = 0{\cdot}145\ \Omega$$

$$I_{sh} = \frac{240}{100} = 2{\cdot}4 \text{ A}$$

$$I_a = I_L + I_{sh} = 100 + 2{\cdot}4 = 102{\cdot}4 \text{ A}$$

Fig. 2.61

Armature circuit Cu loss $= I_a^2 \times 0{\cdot}145$

$$= (102{\cdot}4)^2 \times 0{\cdot}145 = 1521 \text{ W}$$

Shunt field Cu loss $= VI_{sh} = 240 \times 2{\cdot}4 = 576$ W

Iron loss = 1000 W ; Windage and friction loss = 500 W

Total losses $= 1521 + 576 + 1000 + 500 = 3597$ W

$$\text{Efficiency} = \frac{\text{Output}}{\text{Output + Losses}} \times 100 = \frac{24000}{24000 + 3597} \times 100 = \mathbf{87\%}$$

Example 2.49. *The hysteresis and eddy current losses in a d.c. machine running at 1000 r.p.m. are 250 W and 100 W respectively. If the flux remains constant, at which speed the total iron losses be halved ?*

Solution.

With constant flux, $P_h \propto$ speed or $250 = K_1 \times 1000$ $\quad \therefore K_1 = 0{\cdot}25$

With constant flux, $P_e \propto (\text{speed})^2$ or $100 = K_2 \times (1000)^2$ $\quad \therefore K_2 = 10^{-4}$

Suppose *N* r.p.m. is the speed at which total iron losses become half of the losses at 1000 r.p.m.

$$\therefore \quad \text{New losses} = \frac{250 + 100}{2} = 175 \text{ W}$$

Now $\quad K_1 N + K_2 N^2 = 175$

or $\quad 0{\cdot}25\,N + 10^{-4} N^2 = 175$

or $\quad N^2 + 2500\,N - 175 \times 10^4 = 0$

$$\therefore \quad N = \frac{-2500 \pm \sqrt{(2500)^2 + 175 \times 4 \times 10^4}}{2} = \mathbf{{}^{*}570\ r.p.m.}$$

* The negative value is mathematically absurd and hence rejected.

Example 2.50. *A shunt generator supplies 100 A at a terminal voltage of 200 V. The prime mover is developing 32 b.h.p. Shunt field resistance = 50 Ω ; armature resistance = 0·1 Ω. Find (i) the iron and friction losses (ii) the copper losses (iii) the commercial efficiency.*

Solution. Fig. 2.62 shows the generator circuit.

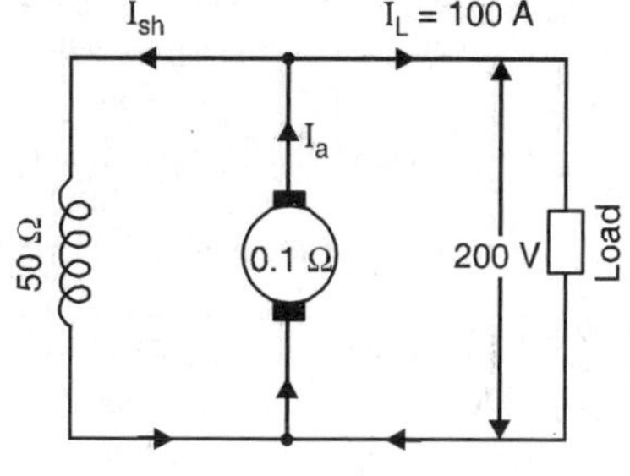

Fig. 2.62

$$I_{sh} = \frac{200}{50} = 4 \text{ A}$$

$$I_a = I_L + I_{sh} = 100 + 4 = 104 \text{ A}$$

Generated e.m.f., $E_g = V + I_a R_a = 200 + 104 \times 0{\cdot}1 = 210{\cdot}4$ V

(*i*) Mechanical input power = 32 × 746 = 23872 W

Electrical power developed in armature

$= E_g I_a = 210{\cdot}4 \times 104 = 21882$ W

∴ Iron and friction losses = 23872 – 21882 = **1990 W**

(*ii*) Electrical power output = 200 × 100 = 20000 W

∴ Copper losses = 21882 – 20000 = **1882 W**

(*iii*) Commercial efficiency, $\eta_c = \frac{\text{Output}}{\text{Input}} \times 100 = \frac{20000}{23872} \times 100 =$ **83.8%**

Example 2.51. *A shunt generator delivers 195 A at a terminal p.d. of 250 V. The armature resistance and shunt field resistance are 0·02 Ω and 50 Ω respectively. The iron and friction losses equal 950 W. Find (i) e.m.f. generated (ii) Cu losses (iii) output of the prime mover (iv) mechanical, electrical and commerical efficiencies.*

Solution. Fig. 2.63 shows the generator circuit.

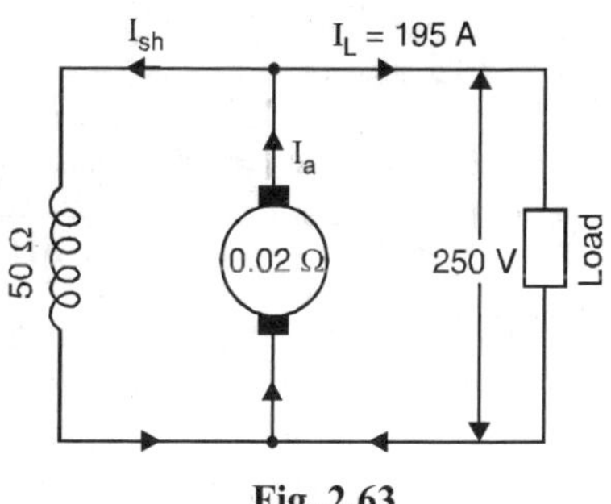

Fig. 2.63

(*i*) $I_{sh} = \frac{250}{50} = 5\text{A}$; $I_a = 195 + 5 = 200$ A

∴ Generated e.m.f., $E_g = V + I_a R_a = 250 + 200 \times 0{\cdot}02 =$ **254 V**

(*ii*) Armature Cu loss = $I_a^2 R_a = (200)^2 \times 0{\cdot}02$ = 800 W

Shunt Cu loss = $VI_{sh} = 250 \times 5 = 1250$ W

∴ Total Cu loss = 800 + 1250 = **2050 W**

(*iii*) Total losses = 2050 + 950 = 3000 W

Output = 250 × 195 = 48750 W

Input = 48750 + 3000 = 51750 W

∴ Output of prime mover = **51750 W**

(*iv*) Electrical power produced in armature

= 51750 – 950 = 50800 W

Mechanical efficiency, $\eta_m = \frac{50800}{51750} \times 100 =$ **98·2%**

Electrical efficiency, $\eta_e = \frac{48750}{50800} \times 100 =$ **95·9%**

Overall efficiency, $\eta_c = \frac{48750}{51750} \times 100 =$ **94·2 %**

TUTORIAL PROBLEMS

1. A shunt generator supplies 195 A at 220 V. Armature resistance = 0.02 Ω ; shunt field resistance = 44 Ω. If the iron and frictional losses amount to 1600 W, find (*i*) e.m.f. generated (*ii*) Cu losses (*iii*) b.h.p. of the engine (*iv*) electrical and commercial efficiencies.
[(*i*) 224 V (*ii*) 1900 W (*iii*) 62·25 h.p. (*iv*) 95·8% ; 92·5%]

2. The output of a shunt generator is 24 kW at a terminal voltage of 200 V. Armature resistance = 0·05 Ω and shunt field resistance = 40 Ω. If the iron and frictional losses equal the copper losses at this load, find (*i*) b.h.p. of the engine (*ii*) overall efficiency. **[(*i*) 36·9 h.p. (*ii*) 87·1 %]**

3. A shunt generator has a full-load current of 196 A at 220 V. The rotational losses are 720 W and the shunt field coil resistance is 55 Ω. If it has a F.L. efficiency of 88%, find the armature resistance. Also find the load current corresponding to maximum efficiency. **[0·107 Ω ; 122·3 A]**

4. The output of a shunt generator is 66 A at 100 V. Armature resistance = 0·04 Ω, shunt field resistance = 25 Ω. If the overall efficiency at this load is 88%, find (*i*) Cu losses (*ii*) iron and frictional losses (*iii*) b.h.p. of the driving engine. **[(*i*) 596 W (*ii*) 304 W (*iii*) 10 h.p.]**

5. A 4-pole d.c. shunt generator is delivering 20 A to a load of 10 Ω. If the armature resistance = 0·5 Ω and the shunt field resistance = 50 Ω, find the induced e.m.f. and efficiency of the machine. Allow a drop of 1V per brush. **[214 V ; 77·9%]**

6. A short-shunt compound generator supplies a load of 10 kW at 200 V through a pair of feeders of total resistance 0·1 Ω. Armature resistance = 0·01 Ω ; shunt field resistance = 52·5 Ω ; series field resistance = 0·1 Ω. Iron and frictional losses amount to 631 W. Find (*i*) terminal p.d. (*ii*) e.m.f. generated (*iii*) Cu losses (*iv*) b.h.p. of the driving engine (*v*) commercial efficiency.
[(*i*) 205 V (*ii*) 210·5 V (*iii*) 1119 W (*iv*) 16·1 h.p. (*v*) 85·4%]

7. In a d.c. machine, the total iron loss is 8 kW at its rated speed and excitation. If excitation remains the same but speed is reduced by 25%, the total iron loss is found to be 5 kW. Calculate the hysteresis and eddy current losses at (*i*) full-speed (*ii*) half the rated speed.
[(*i*) P_h = 2·67 kW ; P_e = 5·33 kW (*ii*) P_h = 1·335 kW ; P_e = 1·332 kW]

8. A short-shunt compound generator supplies 100 A at 110 V. Armature resistance = 0·05 Ω; shunt field resistance = 57 Ω ; series field resistance = 0·04 Ω. Iron and frictional losses amount to 852 W. Find (*i*) generated e.m.f. (*ii*) total Cu losses (*iii*) overall efficiency. **[(*i*) 119·1 V (*ii*) 1148·1 W (*iii*) 84·7%]**

SHORT ANSWER QUESTIONS

Q. 1. *What is the importance of d.c. machines ?*

Ans. D. C. generators are not as common as they used to be, because direct current when required is mainly produced by electronic rectifiers. However d.c. motors are widely used in industry due to the following reasons :

(*i*) Speed of a d.c. motor can be changed easily over a wide range.

(*ii*) Speed regulation of d.c. motors is very good.

(*iii*) The torque-speed characteristics of d.c. motors can be varied over a wide range while retaining high efficiency.

Due to above reasons, d.c. motors are used to drive such devices as hoists, fans, pumps, punch-presses etc.

Q. 2. *What purpose is served by the pole shoe in a d.c. machine ?*

Ans. The pole shoes serve the following purposes :

(*i*) They spread out the flux in the air gaps.

(*ii*) Since they are of larger cross-section, the reluctance of the magnetic path is reduced.

(*iii*) They support the field coils.

Q. 3. *What are the advantages and disadvantages of carbon brushes ?*

Ans. *Advantages* :

(*i*) They lubricate and polish the commutator.

(*ii*) If sparking occurs, they damage the commutator less than with the copper brushes.

(*iii*) They provide good commutation.

Disadvantages :

(*i*) The contact resistance is high and causes a loss of about 2V. Hence they are unsuited for low voltage machines in which this forms a larger percentage loss.

(*ii*) Due to high voltage drop, the commutator must be made larger than for copper brushes.

(*iii*) Their low current density necessitates large brush holders.

Q. 4. *Why is the armature of a d.c. machine made of silicon steel ?*

Ans. The most important function of an armature is to provide a low reluctance path for the flux through the armature from the *N*-Pole-shoes to the *S* ones. It is, therefore, made of a material having high permeability *e.g.*, silicon steel. An additional advantage of silicon steel is that it has low hysteresis loss.

Q. 5. *Why do we use slotted armature in a d.c. machine ?*

Ans. The slotted armature has the following advantages :

(*i*) The air-gap can be reduced to a value required for mechanical clearance. This reduces the reluctance of the magnetic circuit.

(*ii*) The conductors can be properly placed so that there is little fear of them being displaced.

(*iii*) The drag on the conductors is greatly reduced so that their insulation is in no danger of damage by pressure.

Q. 6. *Why is armature winding placed on the rotor of a d.c. machine ?*

Ans. The armature winding of a d.c. machine is placed on the rotor to facilitate commutation *i.e.*, to convert the alternating voltage produced in the winding into direct voltage at the brushes.

Q. 7. *What factors determine the number of poles in a d.c. machine ?*

Ans. The number of poles in a d.c. machine is governed by :

(*i*) frequency (*ii*) current per brush arm

The upper limit to the number of poles is imposed by frequency which is of the order 20-30 Hz for large machines and not exceeding 50 Hz for small machines. The lower limit is due to current per brush arm which should not exceed 400 A otherwise an excessively long commutator will be necessary.

Q. 8. *Why is armature winding of a d.c. machine always double layer winding ?*

Ans. When the coil sides are placed in one layer, it is difficult to arrange their ends so that they will pass each other. In order to avoid this difficulty, the coils are generally arranged in two layers. Each coil has one of its sides at the top of the slot and the other at the bottom of some other slot. The coil ends will then lie side by side.

Q. 9. *What is the essential difference between lap and wave windings ?*

Ans. The essential difference between a lap winding and a wave winding is in the commutator connections. In a lap winding, the commutator pitch is 1 whereas for a wave winding it is about twice the pole pitch.

Q. 10. *The armature winding of a d.c. machine is a single closed loop. How do we get parallel paths ?*

Ans. Whether the winding be lap or wave, when the brushes are lifted, the armature circuit is a single closed loop. However, when brushes are lowered on to the commutator, two or more parallel paths are formed depending upon the type of winding. Each parallel path consists of same number of coils in series.

(*i*) With a lap winding, the number of parallel paths formed is equal to the number of poles.

(*ii*) With a wave winding, the number of parallel paths formed is 2 irrespective of the number of poles.

Q. 11. *Why is lap winding used in large multipolar d.c. generators ?*

Ans. For a given number of armature conductors, a lap winding can carry a heavier current than a wave winding because it has more parallel paths. Therefore, lap winding is used in large multipolar d.c. generators to avoid having conductors of large cross-section.

Q. 12. *Why are "equalisers" not used in wave winding ?*

Ans. There is no need for "equalisers" in a wave-wound machine because conductors in each of the two parallel paths pass under all the N and S poles successively.

Q. 13. *What is the need of multiplex winding ?*

Ans. With a simplex wave winding, there are only 2 parallel paths whereas for a lap winding, the number of parallel paths is equal to the number of poles. Thus in case of a 10-pole machine, using simplex winding, the designer is restricted to either two parallel circuits (wave) or ten parallel circuits (lap). However, sometimes, it is desired to increase the number of parallel paths, enabling the armature to carry large current and at the same time reducing the conductor current. For this purpose, multiplex windings are used. Thus with duplex wave winding, the above 10-pole machine will have 4 parallel paths. Likewise, with duplex lap winding, the machine will have 20 parallel paths.

Q. 14. *Discuss the nomenclature of multiplex winding.*

Ans. The nomenclature of multiplex winding is according to the number of parallel paths formed.

(*i*) The number of parallel paths in a multiplex lap winding is given by :

$$\text{No. of parallel paths} = \text{No. of poles } (P) \times \text{Degree of plex}$$

Thus, a duplex lap winding is one in which number of parallel paths is $2P$. The triplex lap winding is one in which number of parallel paths is $3P$.

(*ii*) The number of parallel paths in a multiplex wave winding is given by :

$$\text{No. of parallel paths} = 2 \times \text{Degree of plex}$$

Thus, a duplex wave winding is one in which the number of parallel paths $= 2 \times 2 = 4$. A triplex wave winding will have $2 \times 3 = 6$ parallel paths.

Q. 15. *What are the advantages of using a large number of poles in a d.c. machine ?*

Ans. The advantages of using a large number of poles in a d.c. machine are :

(*i*) The flux per pole is reduced. This reduces the thickness of armature core and yoke and hence the weight of the machine.

(*ii*) The length of the magnetic circuit is reduced, resulting in the reduction of field ampere turns.

(*iii*) The length of the armature end connections is reduced.

(*iv*) The current per brush arm is reduced. This reduces the length of the commutator.

(*v*) Too small a number of poles means a large number of armature ampere-turns per pole and a high value of armature reaction, which requires a longer air-gap.

Q. 16. *What are the disadvantages of using a large number of poles in a d.c. machine ?*

Ans. The disadvantages of a large number of poles in a d.c. machine are :

(*i*) More commutator segments are needed or alternatively there must be more volts per segment for a given commutator diameter and thickness of segment.

(*ii*) The brush arms are closer together, increasing the danger of flash-over.

(*iii*) Frequency is proportional to number of poles and core loss depends upon frequency.

(*iv*) The armature diameter is increased.

Q. 17. *What decides the upper and lower limit to the number of poles in a d.c. machine ?*

Ans. The upper limit to the number of poles is imposed by frequency which is of the order 20-30 Hz for large machines and not exceeding 50 Hz for small ones. The lower limit is decided by the current per brush arm which should not exceed 400 A otherwise an excessively long commutator will be necessary. These limits usually restrict the choice to about two alternatives.

Q. 18. *What are the advantages and disadvantages of a separately excited generator over a self-excited generator ?*

Ans. The advantages of a separately excited generator over a self-excited generator are :

(*i*) The field ampere-turns can be independently controlled.

(*ii*) It will operate in a stable condition with any field excitation.

(*iii*) We can have a wide range of output voltages.

The disadvantages of a separately excited generator as compared to a self-excited generator are :

(*i*) A separate d.c. source is required.

(*ii*) It is an inconvenient arrangement.

Note : D. C. generators are rarely separately excited. But practically all a.c. generators are separately excited.

Q. 19. *Upon what factors the armature resistance (R_a) of a d.c. machine depends?*

Ans. The armature resistance of a d.c. machine depends upon :

(*i*) number, size and connection of armature coils.

(*ii*) contact resistance between the carbon brush and the copper commutator.

Q. 20. *What does the name plate of d.c. generator generally indicate?*

Ans. The name plate of a d.c. generator generally indicates :

(*i*) output in kW (*ii*) output voltage

(*iii*) speed (*iv*) permissible temperature rise.

MULTIPLE-CHOICE QUESTIONS

1. The armature of a d.c. machine is made of...............

(*i*) silicon steel (*ii*) wrought iron

(*iii*) cast steel (*iv*) soft iron

2. A 4-pole d.c. machine has .... magnetic circuits.

(*i*) 2 (*ii*) 8

(*iii*) 4 (*iv*) none of the above

3. The greatest eddy current loss occurs in the .... of a d.c. machine.

(*i*) field poles (*ii*) yoke

(*iii*) commutating poles

(*iv*) armature

4. The commutator pitch for a simplex lap winding is equal to ....

(*i*) number of poles of the machine

(*ii*) pole pairs

(*iii*) 1 (*iv*) none of the above

5. In a d.c. machine, the number of commutator segments is equal to ....

(*i*) number of conductors

(*ii*) twice the number of poles

(*iii*) number of coils
(*iv*) none of the above

6. The nature of armature winding of a d.c. machine is decided by ....
(*i*) front pitch (*ii*) commutator pitch
(*iii*) back pitch (*iv*) none of the above

7. An 8-pole duplex lap winding will have ...... parallel paths
(*i*) 8 (*ii*) 4
(*iii*) 32 (*iv*) 16

8. A triplex wave winding will have .... parallel paths.
(*i*) 6 (*ii*) 2
(*iii*) 4 (*iv*) none of the above

9. A 6-pole lap-wound generator has 300 conductors; the e.m.f. induced per conductor being 5 V. The generated voltage of the generator is ....
(*i*) 60 V (*ii*) 1500 V
(*iii*) 360 V (*iv*) 250 V

10. The armature of a d.c. machine having 10 poles is rotating at a speed of 10 revolutions per second. The number of hysteresis loops formed per second is ....
(*i*) 100 (*ii*) 10
(*iii*) 150 (*iv*) 50

11. The greatest percentage of heat loss in a d.c. machine is due to ....
(*i*) eddy current loss (*ii*) hysteresis loss
(*iii*) copper loss (*iv*) frictional loss

12. A shunt generator delivers 195 A at a terminal p.d. of 250 V. The armature resistance and shunt field resistance are 0·02 Ω and 50 Ω respectively. What is the value of generated e.m.f. ?
(*i*) 246 V (*ii*) 270 V
(*iii*) 254 V (*iv*) 282 V

13. In the above question, what is the value of copper losses ?
(*i*) 825 W (*ii*) 2050 W
(*iii*) 1025 W (*iv*) 960 W

14. If W_c is the constant loss and R_a is the armature resistance of a d.c. generator, then load current I_L corresponding to maximum efficiency is
(*i*) $I_L = \sqrt{\frac{R_a}{W_c}}$ (*ii*) $I_L = \frac{W_c}{\sqrt{R_a}}$
(*iii*) $I_L = \frac{R_a}{\sqrt{W_c}}$ (*iv*) $I_L = \sqrt{\frac{W_c}{R_a}}$

15. Rotational losses consist of
(*i*) magnetic and mechanical losses
(*ii*) magnetic and copper losses
(*iii*) mechanical and copper losses
(*iv*) none of the above

16. A 30 kW, 300 V d.c. shunt generator has armature and field resistances of 0·05 Ω and 100 Ω respectively. What is the generated e.m.f. ?
(*i*) 295·25 V (*ii*) 311·25 V
(*iii*) 305·15 V (*iv*) none of the above

17. In the above question, what is the total power developed when the generator delivers full output power ?
(*i*) 16·34 kW (*ii*) 20·62 kW
(*iii*) 12·46 kW (*iv*) 31·43 kW

18. In Fig. 2.64, the electrical efficiency is

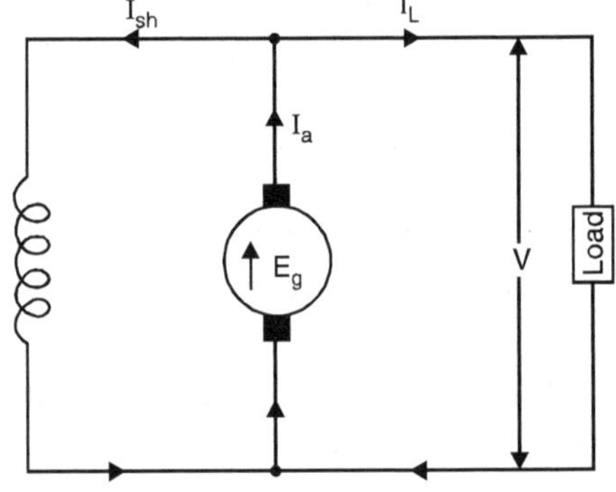

Fig. 2.64

(*i*) $\frac{V I_a}{E_g I_a}$ (*ii*) $\frac{V I_L}{E_g I_a}$
(*iii*) $\frac{E_g I_a}{V I_a}$ (*iv*) none of the above

19. The efficiency of a d.c. generator means its
(*i*) electrical efficiency
(*ii*) overall efficiency
(*iii*) mechanical efficiency
(*iv*) none of the above

20. A 4-pole, lap-wound d.c. shunt generator has an armature winding consisting of 220 turns each of 0·004 Ω. The armature resistance is
(*i*) 0·5 Ω (*ii*) 1 Ω
(*iii*) 0·025 Ω (*iv*) 0·055 Ω

ANSWERS TO MULTIPLE-CHOICE QUESTIONS

1. (*i*)	**2.** (*iii*)	**3.** (*iv*)	**4.** (*iii*)	**5.** (*iii*)
6. (*ii*)	**7.** (*iv*)	**8.** (*i*)	**9.** (*iv*)	**10.** (*iv*)
11. (*iii*)	**12.** (*iii*)	**13.** (*ii*)	**14.** (*iv*)	**15.** (*i*)
16. (*iii*)	**17.** (*iv*)	**18.** (*ii*)	**19.** (*ii*)	**20.** (*iv*)

HINTS TO MULTIPLE-CHOICE QUESTIONS

1. It is desired that armature should provide a path of low reluctance for the flux. For this reason, armature is made of material having high permeability *e.g.*, silicon steel.

2. In a d.c. machine, there are as many magnetic circuits as there are poles. Thus, a 4-pole d.c. machine will have 4 magnetic circuits. However, m.m.f. acting in each circuit is due to the ampere-turns of a pair of poles.

3. Since every part of the armature sees cyclic reversals in magnetic flux direction, it is potentially subjected to more severe eddy current-losses than the field pole shoes. Hence armature is invariably laminated to reduce the eddy current loss.

4. The commutator pitch for a simplex lap winding is given by ;

$$Y_c = \pm 1$$

The plus sign applies if, as is usual, the winding is *progressive* ; the minus sign, if it is *retrogressive*.

5. In a d.c. armature winding, the two ends of a coil are connected to two different commutator segments and two coils are connected to each segment so that number of commutator segments is equal to the number of coils.

6. It is the commutator pitch that determines the grouping of the coils in series and parallel. In other words, it is the commutator pitch that determines as to how many parallel paths, each consisting of equal number of coils in series, will be formed. Hence the nature of winding is determined by the commutator pitch.

7. With a n-plex lap winding, the number of parallel paths is $P\,n$ where

P = number of poles

n = 2, 3 etc and stands for duplex, triplex etc.

Hence 8-pole duplex lap winding will have $8 \times 2 = 16$ parallel paths.

8. With a n-plex wave winding, the number of parallel paths is $2n$ where $n = 2, 3$ etc. and stands for duplex, triplex etc.

Hence a triplex wave winding will have $2 \times 3 = 6$ parallel paths.

9. Number of conductors per parallel path = 300/6 = 50. Hence the generator voltage = 50×5 = 250 V.

10. One hysteresis loop is formed as the armature traverses a pair of poles. Hence number of loops formed = $5 \times 10 = 50$ loops per second.

11. The armature current of a d.c. machine is very large. Since copper loss is proportional to the square of current, hence a greater part of power loss occurs in the armature.

12. $I_{sh} = 250/50 = 5$ A $\quad \therefore \quad I_a = I_{sh} + I_L = 5 + 195 = 200$ A

$$\text{Armature drop} = I_a R_a = 200 \times 0{\cdot}02 = 4 \text{ V}$$

$$\therefore \quad \text{Generated e.m.f.} = V + I_a R_a = 250 + 4 = 254 \text{ V}$$

13.

$$\text{Armature Cu loss} = I_a^2 R_a = (200)^2 \times 0{\cdot}02 = 800 \text{ W}$$

$$\text{Shunt Cu loss} = V I_{sh} = 250 \times 5 = 1250 \text{ W}$$

$$\text{Total Cu loss} = 1250 + 800 = 2050 \text{ W}$$

14. The efficiency of a d.c. generator is maximum when

$$\text{Variable loss} = \text{Constant loss}$$

or $$I_L^2 R_a = W_c \quad \text{or} \quad I_L = \sqrt{\frac{W_c}{R_a}}$$

15. Usually magnetic and mechanical losses are collectively called rotational losses.

16. $$I_L = \frac{30 \times 10^3}{300} = 100\,\text{A}\,;\; I_{sh} = \frac{V}{R_{sh}} = \frac{300}{100} = 3\text{A}$$

$\therefore$ $$I_a = I_L + I_{sh} = 100 + 3 = 103\,\text{A}$$

$$\text{Generated e.m.f., } E_g = V + I_a R_a = 300 + 103 \times 0{\cdot}05 = 300 + 5{\cdot}15 = 305{\cdot}15\,\text{V}$$

17. Power developed in armature $= E_g I_a = \dfrac{305{\cdot}15 \times 103}{1000} = 31{\cdot}43$ kW

18. $$\text{Electrical efficiency, } \eta_e = \frac{\text{Power in load}}{\text{Total power generated}} = \frac{V I_L}{E_g I_a}$$

19. Unless stated otherwise, efficiency of a d.c. generator means its overall or commercial efficiency.

20. Total resistance of 220 turns $= 220 \times 0{\cdot}004 = 0{\cdot}88\,\Omega$

Since there are 4 parallel paths in armature,

Resistance of each path $= 0{\cdot}88/4 = 0{\cdot}22\,\Omega$

Now, there are four such resistances in parallel each of value $0{\cdot}22\,\Omega$.

$\therefore$ Armature resistance, $R_a = 0{\cdot}22/4 = 0{\cdot}055\,\Omega$

3

Armature Reaction and Commutation

INTRODUCTION

In a d.c. generator, the purpose of field winding is to produce magnetic field (called *main flux*) whereas the purpose of armature winding is to carry armature current. Although the armature winding is not provided for the purpose of producing a magnetic field, nevertheless the current in the armature winding will also produce magnetic flux (called *armature flux*). The armature flux distorts and weakens the main flux posing problems for the proper operation of the d.c. generator. The action of armature flux on the main flux is called *armature reaction.*

In the previous chapter (Art. 2.19), it was hinted that current in the coil is reversed as the coil passes a brush. This phenomenon is termed as *commutation*. The criterion for good commutation is that it should be sparkless. In order to have sparkless commutation, the brushes should lie along magnetic neutral axis. In this chapter, we shall discuss the various aspects of armature reaction and commutation in a d.c. generator.

3.1 ARMATURE REACTION

So far we have assumed that the only flux acting in a d.c. machine is that due to the main poles called *main flux*. However, current flowing through armature conductors also creates a magnetic flux (called *armature flux*) that distorts and weakens the flux coming from the poles. This distortion and field weakening takes place in both generators and motors. The action of armature flux on the main flux is known as *armature reaction*.

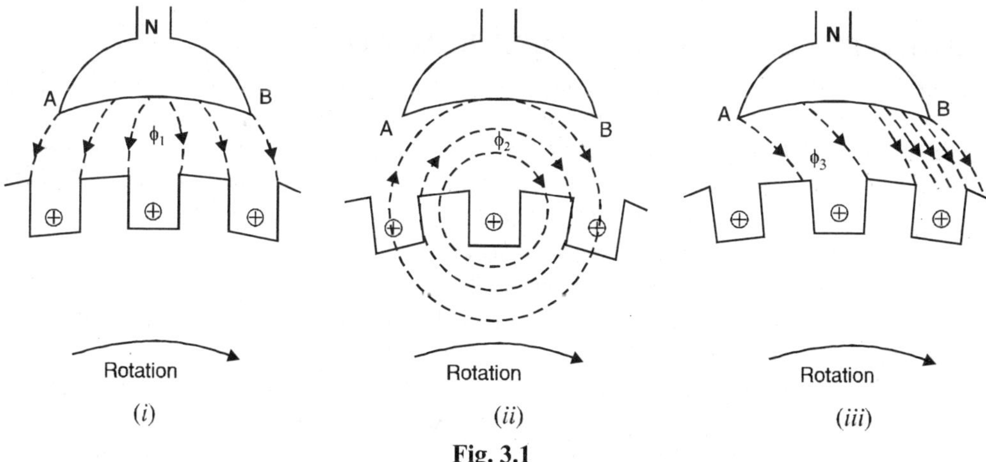

Fig. 3.1

The phenomenon of armature reaction in a d.c. generator is shown in Fig. 3.1. Only one pole is shown for clarity. When the generator is on no-load, a small current flowing in the armature does not appreciably affect the main flux ϕ_1 coming from the pole [See Fig. 3.1 (*i*)]. When the generator is loaded, the current flowing through armature conductors sets up flux ϕ_2. Fig. 3.1 (*ii*) shows flux due to armature current alone. By superimposing ϕ_1 and ϕ_2, we obtain the resulting flux ϕ_3 as shown in Fig. 3.1 (*iii*). Referring to Fig. 3.1 (*iii*), it is clear that flux density at the trailing pole tip (point *B*) is increased while at the *leading pole tip (point A), it is decreased. This unequal field distribution produces the following two effects :

(*i*) The main flux is distorted.

(*ii*) Due to higher flux density at pole tip B, saturation sets in. Consequently, the increase in flux at pole tip B is less than the decrease in flux under pole tip A. Flux ϕ_3 at full load is, therefore, less than flux ϕ_1 at no load. As we shall see, the weakening of flux due to armature reaction depends upon the position of brushes.

3.2 GEOMETRICAL AND MAGNETIC NEUTRAL AXES

(*i*) The *geometrical neutral axis* (G.N.A.) is the axis that bisects the angle between the centre line of adjacent poles [See Fig. 3.2 (*i*)]. Clearly, it is the axis of symmetry between two adjacent poles.

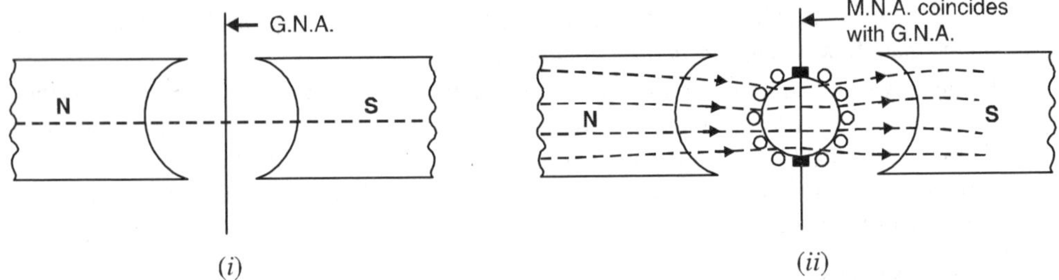

Fig. 3.2

(*ii*) The *magnetic neutral axis* (M. N. A.) is the axis drawn perpendicular to the mean direction of the flux passing through the centre of the armature. Clearly, no e.m.f. is produced in the armature conductors along this axis because then they cut no flux. With no current in the armature conductors, the M.N.A. coincides with G. N. A. as shown in Fig. 3.2. (*ii*). *In order to achieve sparkless commutation, the brushes must lie along M. N. A.*

3.3 EXPLANATION OF ARMATURE REACTION

With no current in armature conductors, the M.N.A. coincides with G.N.A. However, when current flows in armature conductors, the combined action of *main flux* and *armature flux* shifts the M.N.A. from G.N.A. In case of a generator, the M.N.A. is shifted in the direction of rotation of the machine. In order to achieve sparkless commutation, the brushes have to be moved along the new M.N.A. Under such a condition, the armature reaction produces the following **two effects :

1. *It demagnetises or weakens the main flux.*
2. *It cross-magnetises or distorts the main flux.*

Let us discuss these effects of armature reaction by considering a 2-pole generator (though the following remarks also hold good for a multipolar generator).

(*i*) Fig. 3.3 (*i*) shows the flux due to main poles (*main flux*) when the armature conductors carry no current. The flux across the air gap is uniform. The m.m.f. producing the main flux is represented in magnitude and direction by the vector OF_m in Fig. 3.3 (*i*). Note that OF_m is perpendicular to G.N.A.

---

* As the conductors rotate, the pole tip they first meet is the leading pole tip and the other trailing pole tip.

** If the brushes remain along G.N.A. (*i.e.*, they are not shifted along new M.N.A.), *only* cross-magnetising effect (*i.e.*, distortion of main flux) of armature reaction occurs. In case of a generator, the flux is weakened at the leading pole tip and strengthened at the trailing pole tip. Clearly, it is the distortion of the main flux. Since the decrease of flux at one pole tip is equal to the increase in flux at the other tip, the flux per pole remains the same.

(*ii*) Fig. 3.3 (*ii*) shows the flux due to current flowing in armature conductors *alone* (main poles unexcited). The armature conductors to the left of G.N.A. carry current "in" (×) and those to the right carry current "out" (•). The direction of magnetic lines of force can be found by cork screw rule. It is clear that armature flux is directed downward parallel to the brush axis. The m.m.f. producing the armature flux is represented in magnitude and direction by the vector OF_A in Fig. 3.3 (*ii*).

(*iii*) Fig. 3.3 (*iii*) shows the flux due to the main poles and that due to current in armature conductors acting together. The resultant m.m.f. OF is the vector sum of OF_m and OF_A as shown in Fig. 3.3 (*iii*). Since M.N.A. is always perpendicular to the resultant m.m.f., the M.N.A. is shifted through an angle θ. Note that M.N.A. is shifted in the direction of rotation of the generator.

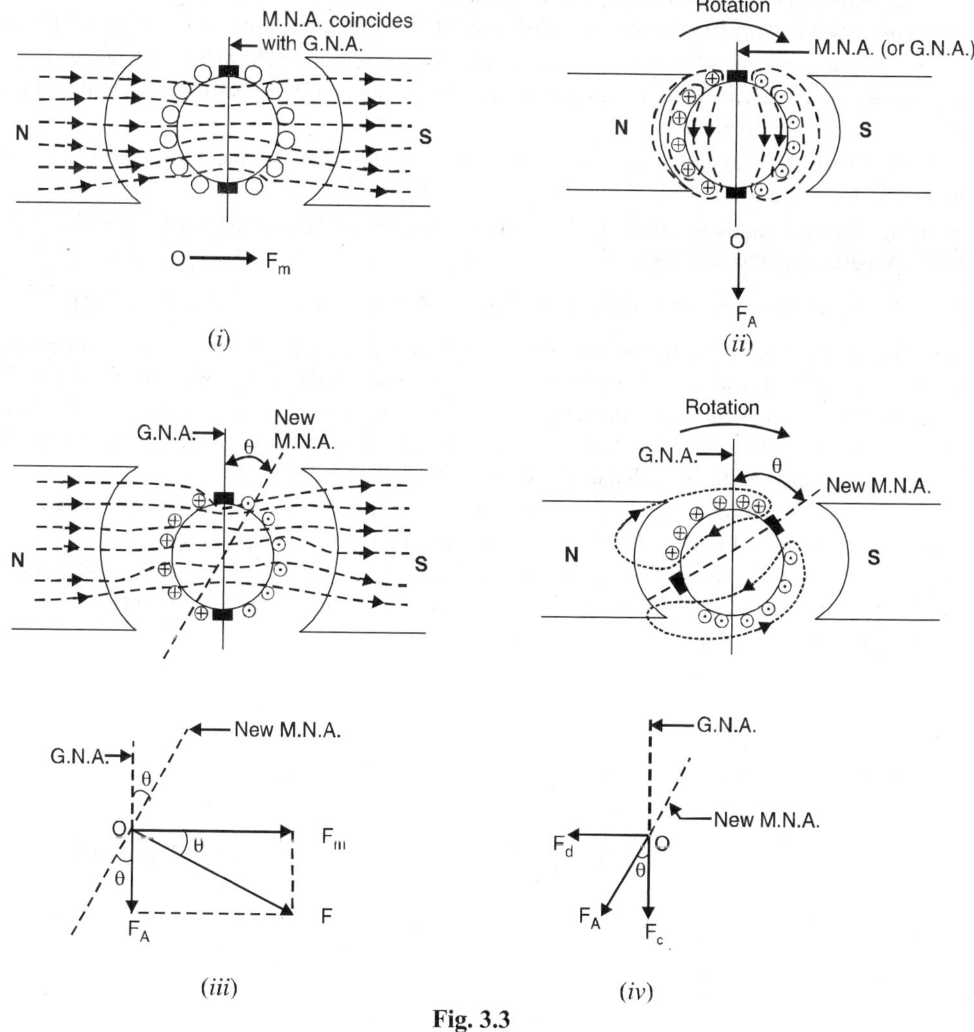

Fig. 3.3

(*iv*) In order to achieve sparkless commutation, the brushes must lie along the M.N.A. Consequently, the brushes are shifted through an angle θ so as to lie along the new M.N.A. as shown in Fig. 3.3 (*iv*). Due to brush shift, the m.m.f. F_A of the armature is also rotated through the same angle θ. It is because some of the conductors which were earlier under *N*-pole now come under *S*-pole and *vice-versa*. The result is that armature m.m.f. F_A will no longer be vertically downward but will be rotated in the direction of rotation through an angle θ as shown in Fig. 3.3 (*iv*). Now F_A can be resolved into rectangular components F_c and F_d.

(*a*) The component F_d is in direct opposition to the m.m.f. OF_m due to main poles. It has a demagnetising effect on the flux due to main poles. For this reason, it is called the demagnetising or weakening component of armature reaction.

(*b*) The component F_c is at right angles to the m.m.f. OF_m due to main poles. It distorts the main field. For this reason, it is called the cross-magnetising or distorting component of armature reaction.

It may be noted that with the increase of armature current, both demagnetising and distorting effects will increase.

Conclusions

(*i*) With brushes located along G.N.A. (*i.e.*, $\theta = 0°$), there is no demagnetising component of armature reaction ($F_d = 0$). There is only distorting or cross-magnetising effect of armature reaction.

(*ii*) With the brushes shifted from G.N.A., armature reaction will have both demagnetising and distorting effects. Their relative magnitudes depend on the amount of shift. This shift is directly proportional to the armature current.

(*iii*) The demagnetising component of armature reaction weakens the main flux. On the other hand, the distorting component of armature reaction distorts the main flux.

(*iv*) The demagnetising effect leads to reduced generated voltage while cross-magnetising effect leads to sparking at the brushes.

3.4 DEMAGNETISING AND CROSS-MAGNETISING CONDUCTORS

With the brushes in the G.N.A. position, there is *only* cross-magnetising effect of armature reaction. However, when the brushes are shifted from the G.N.A. position, the armature reaction will have both demagnetising and cross-magnetising effects. Consider a 2-pole generator with brushes shifted (lead) θ_m mechanical degrees from G.N.A. We shall identify the armature conductors that produce demagnetising effect and those that produce cross-magnetising effect.

(*i*) The armature conductors θ_m^o on either side of G.N.A. produce flux in direct *opposition to main flux as shown in Fig. 3.4 (*i*). Thus the conductors lying within angles $AOC = BOD = 2\theta_m$ at the top and bottom of the armature produce demagnetising effect. These are called demagnetising armature conductors and constitute the demagnetising ampere-turns of armature reaction (Remember two conductors constitute a turn).

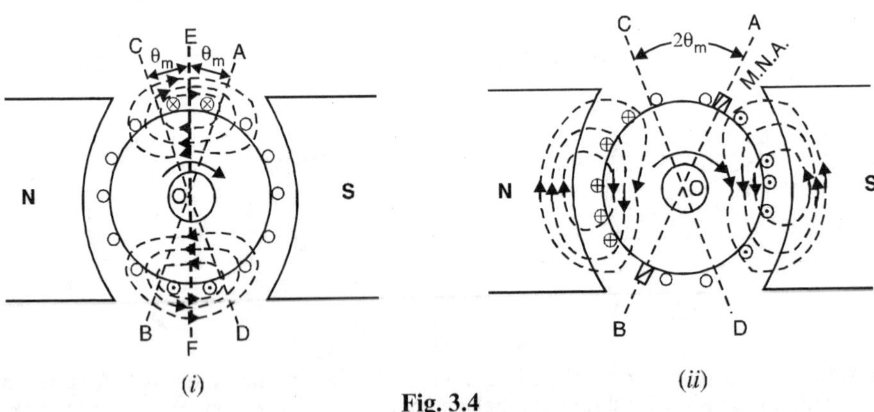

Fig. 3.4

(*ii*) The axis of magnetisation of the remaining armature conductors lying between angles *AOD* and *COB* is at right angles to the main flux as shown in Fig. 3.4 (*ii*). These conductors produce the

---

* By cork screw rule.

cross-magnetising (or distorting) effect *i.e.*, they produce uneven flux distribution on each pole. Therefore, they are called cross-magnetising conductors and constitute the cross-magnetising ampere-turns of armature reaction.

3.5 CALCULATION OF DEMAGNETISING AMPERE-TURNS PER POLE (AT_d/POLE)

It is sometimes desirable to neutralise the demagnetising ampere-turns of armature reaction. This is achieved by adding extra ampere-turns to the main field winding. We shall now calculate the demagnetising ampere-turns per pole (AT_d/pole).

Let Z = total number of armature conductors

I = current in each armature conductor

$= I_a/2$ .... for simplex wave winding

$= I_a/P$ .... for simplex lap winding

θ_m = forward lead in mechanical degrees

Referring to Fig. 3.4 (*i*) above, we have,

Total demagnetising armature conductors

= Conductors in angles *AOC* and *BOD*

$$= \frac{4\theta_m}{360} \times Z$$

Since two conductors constitute one turn,

∴ Total demagnetising ampere-turns

$$= \frac{1}{2}\left[\frac{4\theta_m}{360} \times Z\right] \times I = \frac{2\theta_m}{360} \times ZI$$

These demagnetising ampere-turns are due to a pair of poles.

$$\therefore \quad \text{Demagnetising ampere-turns/pole} = \frac{\theta_m}{360} \times ZI$$

i.e.,

$$AT_d/\text{pole} = \frac{\theta_m}{360} \times ZI$$

As mentioned above, the demagnetising ampere-turns of armature reaction can be neutralised by putting extra turns on each pole of the generator.

$$\therefore \quad \text{No. of extra turns/pole} = \frac{AT_d}{I_{sh}} \quad \text{.... for a shunt generator}$$

$$= \frac{AT_d}{I_a} \quad \text{.... for a series generator}$$

Note. When a conductor passes a pair of poles, one cycle of voltage is generated. We say one cycle contains 360 electrical degrees. Suppose there are *P* poles in a generator. In one revolution, there are 360 mechanical degrees and 360 × *P*/2 electrical degrees.

$$\therefore \quad 360^\circ \text{ mechanical} = 360 \times \frac{P}{2} \text{ electrical degrees}$$

or

$$1^\circ \text{ mechanical} = \frac{P}{2} \text{ electrical degrees}$$

$$\therefore \quad \theta \text{ (mechanical)} = \frac{\theta \text{ (electrical)}}{\text{Pair of poles}}$$

or

$$\theta_m = \frac{\theta_e}{P/2} \quad \therefore \quad \theta_m = \frac{2\theta_e}{P}$$

3.6 CROSS-MAGNETISING AMPERE-TURNS PER POLE (AT_c/POLE)

We now calculate the cross-magnetising ampere-turns per pole (AT_c/pole).

Total armature reaction ampere-turns per pole

$$= \frac{Z/2}{P} \times I = \frac{Z}{2P} \times I \qquad (\because \text{ two conductors make one turn})$$

Demagnetising ampere-turns per pole is given by ;

$$AT_d/\text{pole} = \frac{\theta_m}{360} \times Z I \qquad \text{(found above)}$$

$\therefore$ Cross-magnetising ampere-turns/pole are

$$AT_c/\text{pole} = \frac{Z}{2P} \times I - \frac{\theta_m}{360} \times Z I$$

$$= Z I \left(\frac{1}{2P} - \frac{\theta_m}{360} \right)$$

$$\therefore \qquad AT_c/\text{pole} = Z I \left(\frac{1}{2P} - \frac{\theta_m}{360} \right)$$

3.7 COMPENSATING WINDINGS

The cross-magnetising effect of armature reaction may cause trouble in d.c. machines subjected to large fluctuations in load. In order to neutralise the cross-magnetising effect of armature reaction, a compensating winding is used.

A compensating winding is an auxiliary winding embedded in slots in the pole faces as shown in Fig. 3.5. It is connected in series with armature in a manner so that the direction of current through the compensating conductors in any one pole face will be opposite to the direction of the current through the adjacent armature conductors [See Fig. 3.5]. Let us now calculate the number of compensating conductors/pole face.

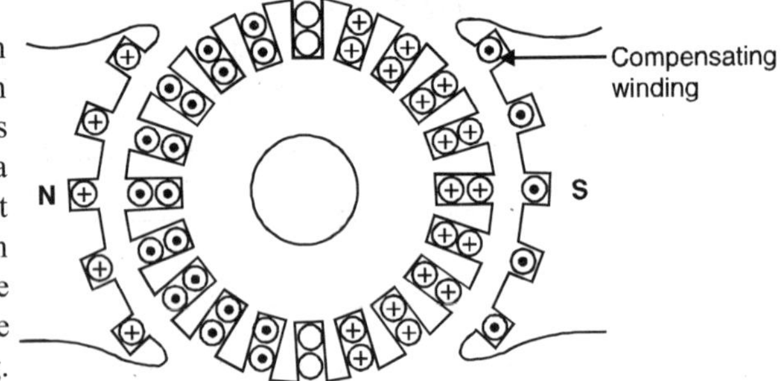

Fig. 3.5

In calculating the conductors per pole face required for the compensating winding, it should be remembered that the current in the compensating conductors is the armature current I_a whereas the current in armature conductors is I_a/A where A is the number of parallel paths.

Let Z_c = No. of compensating conductors/pole face

Z_a = No. of active armature conductors

I_a = Total armature current

I_a/A = Current in each armature conductor

$$\therefore \qquad Z_c I_a = Z_a \times \frac{I_a}{A}$$

$$\text{or} \qquad Z_c = \frac{Z_a}{A}$$

The use of a compensating winding considerably increases the cost of a machine and is justified only for machines intended for severe service *e.g.*, for high speed and high voltage machines.

3.8 AT/POLE FOR COMPENSATING WINDING

Only the cross-magnetising ampere-turns produced by conductors under the pole face are effective in producing the distortion in the pole cores. If Z is the total number of armature conductors and P is the number of poles, then,

$$\text{No. of armature conductors/pole} = \frac{Z}{P}$$

$$\text{No. of armature turns/pole} = \frac{Z}{2P}$$

$$\text{No. of armature turns under pole face} = \frac{Z}{2P} \times \frac{\text{Pole arc}}{\text{Pole pitch}}$$

If I is the current through each armature conductor, then,

$AT/$pole required for compensating winding

$$= \frac{ZI}{2P} \times \frac{\text{Pole arc}}{\text{Pole pitch}}$$

$$= \text{Armature } AT/\text{pole} \times \frac{\text{Pole arc}}{\text{Pole pitch}}$$

Example 3.1. *A 4-pole generator has a wave-wound armature with 722 conductors and it delivers 100 A on full-load. If the brush lead is 8°, calculate the armature demagnetising and cross-magnetising ampere-turns per pole.*

Solution. $I = I_a/2 = 100/2 = 50 \text{ A} \quad ; \quad Z = 722 \quad ; \quad \theta_m = 8°$

$$AT_d/\text{pole} = ZI \times \frac{\theta_m}{360} = 722 \times 50 \times \frac{8}{360} = \mathbf{802}$$

$$AT_c/\text{pole} = ZI\left(\frac{1}{2P} - \frac{\theta_m}{360}\right) = 722 \times 50 \times \left(\frac{1}{2 \times 4} - \frac{8}{360}\right) = \mathbf{3710}$$

Example 3.2. *A 4-pole generator supplies a current of 143 A. It has 492 conductors (i) wave connected (ii) lap connected. When delivering full load, the brushes are given an actual lead of 10°. Calculate the demagnetising ampere-turns/pole. The field winding is shunt connected and takes 10 A. Calculate the number of extra shunt field turns necessary to neutralise this demagnetisation.*

Solution. $Z = 492 \quad ; \quad \theta_m = 10° \quad ; \quad I_a = 143 + 10 = 153 \text{ A}$

(*i*) **Wave-connected**

$$I = \frac{I_a}{2} = \frac{153}{2} \text{ A}$$

$$AT_d/\text{pole} = ZI \times \frac{\theta_m}{360} = 492 \times \frac{153}{2} \times \frac{10}{360} = \mathbf{1046 \ AT}$$

$$\text{Extra shunt field turns/pole} = \frac{AT_d/\text{pole}}{I_{sh}} = \frac{1046}{10} \simeq \mathbf{105}$$

(*ii*) **Lap-connected**

$$I = \frac{I_a}{P} = \frac{153}{4} \text{ A}$$

$$AT_d/\text{pole} = ZI \times \frac{\theta_m}{360} = 492 \times \frac{153}{4} \times \frac{10}{360} = \mathbf{523 \ AT}$$

$$\text{Extra shunt field turns/pole} = \frac{523}{10} \simeq \mathbf{52}$$

Example 3.3. *An 8-pole lap-connected d.c. shunt generator delivers an output of 240 A at 500 V. The armature has 1408 conductors and 160 commutator segments. If the brushes are given a lead of 4 segments from the no-load neutral axis, find the demagnetising and cross-magnetising AT/pole.*

Solution. $I = I_a/P = 240/8 = 30\text{ A}$; $Z = 1408$; $\theta_m = \dfrac{4 \times 360}{160} = 9^\circ$

$$\therefore \quad AT_d/\text{pole} = ZI \times \frac{\theta_m}{360} = 1408 \times 30 \times \frac{9}{360} = \mathbf{1056\ AT}$$

$$AT_c/\text{pole} = ZI\left(\frac{1}{2P} - \frac{\theta_m}{360}\right) = 1408 \times 30\left(\frac{1}{2 \times 8} - \frac{9}{360}\right) = \mathbf{1584\ AT}$$

Example 3.4. *An 8-pole wave-wound d.c. generator has 480 armature conductors. The armature current is 200 A. Find the armature reaction demagnetising and cross-magnetising ampere-turns per pole if (i) brushes are on G.N.A. (ii) brushes are shifted 6° electrical from G.N.A.*

Solution. $Z = 480$; $I = I_a/2 = 200/2 = 100\text{ A}$

(*i*) **When brushes are on G.N.A.** When the brushes are on G.N.A., the whole armature reaction has cross-magnetising effect.

$$AT_c/\text{pole} = \text{Armature } AT/\text{pole} = \frac{ZI}{2P}$$

$$= \frac{480 \times 100}{2 \times 8} = \mathbf{3000\ AT}$$

$$AT_d/\text{pole} = \textbf{zero}$$

(*ii*) **When brushes are shifted 6° electrical from G.N.A.**

$$\theta_m = \frac{2\theta_e}{P} = \frac{2 \times 6}{8} = 1{\cdot}5^\circ$$

$$AT_d/\text{pole} = ZI \times \frac{\theta_m}{360} = 480 \times 100 \times \frac{1{\cdot}5}{360} = \mathbf{200\ AT}$$

$$\therefore \quad AT_c/\text{pole} = 3000 - 200 = \mathbf{2800\ AT}$$

Example 3.5. *A 500 V, wave-wound, 750 r.p.m. shunt generator supplies a load current of 195 A. The armature has 720 conductors and shunt field resistance is 100 Ω. Find the demagnetising ampere-turns/pole if the brushes are advanced through 3 segments at this load. Also calculate the extra shunt field turns required to neutralise this demagnetisation.*

Solution. $I_L = 195\text{ A}$; $I_{sh} = 500/100 = 5\text{ A}$ $\therefore$ $I_a = I_L + I_{sh} = 195 + 5 = 200\text{ A}$

Current in each conductor, $I = I_a/A = 200/2 = 100\text{ A}$

No. of commutator segments $= 720/2 = 360$

$$\theta_m = \frac{3}{360} \times 360 = 3^\circ \quad ; \quad Z = 720$$

$$\therefore \quad AT_d/\text{pole} = ZI \times \frac{\theta_m}{360} = 720 \times 100 \times \frac{3}{360} = \mathbf{600\ AT}$$

$$\text{Extra shunt turns/pole} = \frac{AT_d/\text{pole}}{I_{sh}} = \frac{600}{5} = \mathbf{120}$$

Example 3.6. *Determine per pole the number (i) of cross-magnetising ampere-turns (ii) of back ampere turns and (iii) series turns to balance the back ampere-turns in case of a generator having the following data :*

500 conductors, total current 200 A, 6 poles, 2-circuit wave winding, angle of lead = 10°, leakage co-efficient = 1·3

Solution. $I = I_a/A = 200/2 = 100\text{ A}$; $Z = 500$; $\theta_m = 10^\circ$

(*i*)
$$AT_c/\text{pole} = ZI\left(\frac{1}{2P} - \frac{\theta_m}{360}\right) = 500 \times 100 \times \left(\frac{1}{2 \times 6} - \frac{10}{360}\right) = \mathbf{2778\ AT}$$

(*ii*) $$AT_d/\text{pole} = ZI \times \frac{\theta_m}{360} = 500 \times 100 \times \frac{10}{360} = \mathbf{1390\ AT}$$

(*iii*) Series turns required to balance the demagnetising ampere-turns

$$= \lambda \times \frac{AT_d/\text{pole}}{I_a} = 1{\cdot}3 \times \frac{1390}{200} = \mathbf{9}$$

Example 3.7. *A 500 kW, 500 V, 10 pole d.c. generator has a lap wound armature with 800 conductors. Calculate the number of pole-face conductors in each pole of a compensating winding if the pole face covers 75 per cent of pole pitch.*

Solution.

$$\text{Load current supplied} = \frac{500 \times 10^3}{500} = 1000\ \text{A}$$

Current in each armature conductor, $I = 1000/10 = 100\ \text{A}$

$AT/$pole for compensating winding

$$= \text{Armature } AT/\text{pole} \times \frac{\text{Pole arc}}{\text{Pole pitch}}$$

$$= \frac{ZI}{2P} \times 0{\cdot}75 = \frac{800 \times 100}{2 \times 10} \times 0{\cdot}75 = 3000\ \text{AT}$$

$$\text{Turns per pole} = \frac{3000}{I_a} = \frac{3000}{1000} = 3$$

$\therefore$ No. of compensating conductors/pole $= 3 \times 2 = \mathbf{6}$

Example 3.8. *A 250 V, 14.9 kW, 8–pole d.c. motor has single turn coils. The armature is wave-wound with 94 commutator segments. If the brushes are given a lead of 2 commutator segments at full load, calculate (i) total armature reaction ampere – turns per pole (ii) cross – magnetising ampere – turns per pole. Assume an efficiency of 80%.*

Solution. $$\text{Motor input} = \frac{\text{Output}}{\text{Efficiency}} = \frac{14.9 \times 10^3}{0.8} = 18625\ \text{W}$$

Motor input current, $I_L = 18625/250 = 74.5\ \text{A}$

Since motor type is not given, we shall assume that armature current is equal to the input motor current, *i.e.*,

$$I_a = I_L = 74.5\ \text{A}$$

Current in each armature conductor is

$$I = I_a/2 = 74.5/2 = 37.25\ \text{A}$$

Number of armature conductors is

$$Z = 94 \times 2 = 188$$

$$\text{Angle of shift, } \theta_m = \frac{2}{94} \times 360° = (360/47)°$$

(*i*) Total armature reaction ampere – turns/pole

$$= \frac{ZI}{2P} = \frac{188 \times 37.25}{2 \times 8} = \mathbf{438}$$

(*ii*) Demagnetising ampere–turns per pole is

$$AT_d/\text{pole} = ZI \times \frac{\theta_m}{360°} = 188 \times 37.25 \times \frac{(360/47)°}{360°} = 149$$

$\therefore$ $$AT_c/\text{pole} = 438 - 149 = \mathbf{289}$$

Example 3.9. *A 400V, 1000A, lap-wound d.c. machine has 10 poles and 860 armature conductors. Calculate the number of conductors in the pole face to give full compensation if the pole face covers 70% of pole span.*

Solution. The current in each armature conductor is

$$I = \frac{1000}{10} = 100\,A$$

AT/pole for compensating winding

$$= \text{Armature } AT\,/\,\text{pole} \times \frac{\text{Pole arc}}{\text{Pole pitch}}$$

$$= \frac{ZI}{2P} \times 0.7 = \frac{860 \times 100}{2 \times 10} \times 0.7 = \mathbf{3010}$$

Example 3.10. *Determine the number of turns on each commutating pole of a 6 – pole machine if the flux density in the air – gap of the commutating pole = 0.5 Wb/m² at full – load and the effective length of the air-gap is 4 mm. The full-load current is 500 A and the armature is lap – wound with 540 conductors. Assume the ampere – turns required for the remainder of the magnetic circuit to be one – tenth of that of the air – gap.*

Solution. The *commutating poles* (or *interpoles*) are narrow poles that are located in the interpolar region centred along the mechanical neutral axis of the generator as shown in Fig. 3.6. The commutating pole windings are connected in series with the armature so that they carry armature current (I_a). The *compole winding must provide enough m.m.f. to oppose armature m.m.f. and set up magnetic flux for compole air gap and its magnetic circuit. The current in each armature conductor is

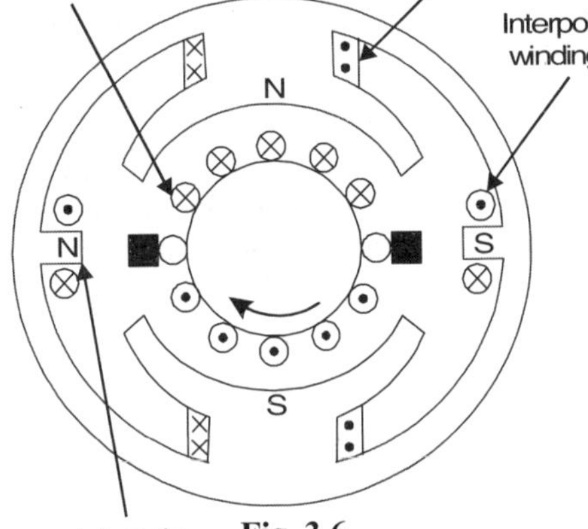

Fig. 3.6

$$I = I_a/A = 500\,/\,6\,\text{A}$$

$$\text{Armature m.m.f./pole} = \frac{ZI}{2P}$$

$$= \frac{540 \times (500/6)}{2 \times 6} = 3750 \text{ AT}$$

$$\text{Compole air–gap m.m.f.} = \frac{B}{\mu_0} \times l_g = \frac{0.5}{4\pi \times 10^{-7}} \times 0.004 = 1592 \text{ AT}$$

M.M.F. for rest of magnetic circuit = 10 % of 1592 = 159 AT

∴ Required compole m.m.f. = 3750 + 1592 + 159 = 5501 AT

Let N_{cp} be the number of turns of the compole winding.

$$\therefore \quad N_{cp} I_a = 5501$$

$$\text{or} \quad N_{cp} = 5501/500 = \mathbf{11}$$

Example 3.11. *Find the number of turns required on each commutating pole of a 5 kW, 230V, 4–pole separately excited d.c. generator having a wave connected armature with 564 conductors. The flux density in the commutating pole air–gap is 0.15 Wb / m² at full-load and the effective length of the gap is 0.25cm. Neglect the ampere–turns required for the iron parts of the commutating pole magnetic circuit.*

---
* Shorthand name for commutating pole.

Solution. Full–load current, $I_L = 5 \times 10^3/230 = 21.74\,\text{A}$

Since the generator is separtely excited, $I_a = I_L = 21.74\,\text{A}$

Current in each armature conductor, $I = I_a/A = 21.74/2\,\text{A}$

$$\text{Armature m.m.f./pole} = \frac{ZI}{2P} = \frac{564 \times (21.74/2)}{2 \times 4} = 766\,\text{AT}$$

$$\text{Compole air-gap m.m.f.} = \frac{B}{\mu_0} \times l_g = \frac{0.15}{4\pi \times 10^{-7}} \times 0.25 \times 10^{-2} = 298\,\text{AT}$$

$$\therefore \quad \text{Required compole m.m.f.} = 766 + 298 = 1064\,\text{AT}$$

Now $N_{cp} \times I_a = 1064$

$$\therefore \quad N_{cp} = 1064/I_a = 1064/21.74 = \mathbf{49}$$

TUTORIAL PROBLEMS

1. An 8-pole d.c. generator has 480 armature conductors and a wave winding. The armature current is 200 A. Find the armature reaction demagnetising and cross-magnetising ampere-turns per pole if brushes are shifted by 6° mechanical from G.N.A. **[800 AT ; 2200 AT]**
2. An 8-pole generator has an output of 200 A at 500 V, the lap-connected armature has 1280 conductors, 160 commutator segments. If the brushes are advanced 4-segments from the no-load neutral axis, estimate the armature demagnetising and cross-magnetising ampere-turns per pole. **[800 AT ; 1200 AT]**
3. A lap wound d.c. generator has 100 armature conductors and 10 poles. The rated armature current is 800 A. Find (*i*) the cross-magnetising ampere-turns/pole if the brushes are set along G.N.A. (*ii*) the demagnetising AT/pole if the brushes are shifted 5° electrical from G.N.A. **[(*i*) 4000 AT ; (*ii*) 222 AT]**
4. A 4-pole, 50 kW, 250 V, wave-wound shunt generator has 400 armature conductors. Brushes are given a lead of 4 commutator segments. Calculate the demagnetising ampere-turns per pole if the shunt field resistance is 50 Ω. Also calculate extra shunt field turns/pole to neutralise the demagnetisation. **[820 AT ; 164]**
5. A 4-pole, wave-wound armature has 880 conductors and delivers 120 A. The brushes have been displaced through 3 angular degrees from the G.N.A. Calculate the additional field current for neutralising the demagnetisation if the field winding has 1100 turns/pole. **[0·4 A]**
6. A 250 V, 25 kW, 4-pole d.c. generator has 328 wave-connected armature conductors. When the machine is delivering full-load, the brushes are given a lead of 7·2 electrical degrees. Calculate the cross-magnetising ampere-turns per pole. **[1886 AT]**

3.9 COMMUTATION

Fig. 3.7 shows the schematic diagram of 2-pole lap-wound generator. There are two parallel paths between the brushes. Therefore, each coil of the winding carries one half ($I_a/2$ in this case) of the total current (I_a) entering or leaving the armature.

Note that the currents in the coils connected to a brush are either all towards the brush (positive brush) or all directed away from the brush (negative brush). Therefore, current in a coil will reverse as the coil passes a brush. This reversal of current as the coil passes a brush is called *commutation*.

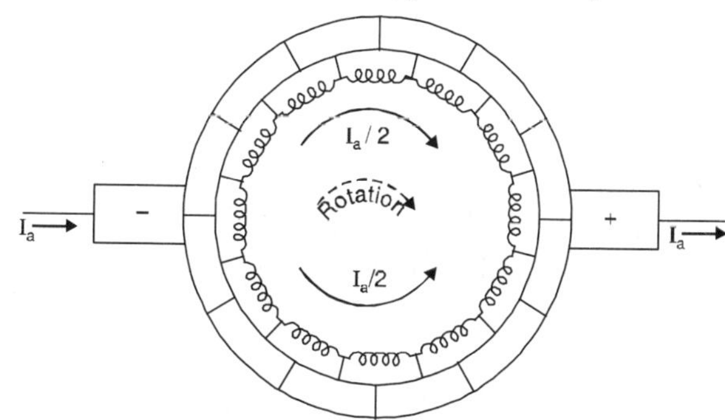

Fig. 3.7

The reversal of current in a coil as the coil passes the brush axis is called **commutation.**

When commutation takes place, the coil undergoing commutation is short-circuited by the brush. The brief period during which the coil remains short-circuited is known as *commutation period* T_c. If the current reversal is completed by the end of commutation period, it is called *ideal commutation.* If the current reversal is not completed by that time, then sparking occurs between the brush and the commutator which results in progressive damage to both.

Ideal commutation. Let us discuss the phenomenon of ideal commutation (*i.e.,* coil has no inductance) in one coil in the armature winding shown in Fig. 3.7 above. For this purpose, we consider the coil *A*. The brush width is equal to the width of one commutator segment and one mica insulation. Suppose the total armature current is 40 A. Since there are *two parallel paths, each coil carries a current of 20 A.

(*i*) In Fig. 3.8 (*i*), the brush is in contact with segment 1 of the commutator. The commutator segment 1 conducts a current of 40 A to the brush ; 20 A from coil *A* and 20 A from the adjacent coil as shown. The coil *A* has yet to undergo commutation.

(*ii*) As the armature rotates, the brush will make contact with segment 2 and thus short-circuits the coil *A* as shown in Fig. 3.8 (*ii*). There are now two parallel paths into the brush as long as the short-circuit of coil *A* exists. Fig. 3.8 (*ii*) shows the instant when the brush is one-fourth on segment 2 and three-fourth on segment 1. For this condition, the resistance of the path through segment 2 is three times the resistance of the path through segment 1 (∵ contact resistance varies inversely as the area of contact of brush with the segment). The brush again conducts a current of 40 A ; 30 A through segment 1 and 10 A through segment 2. Note that current in coil *A* (the coil undergoing commutation) is reduced from 20 A to 10 A.

(*iii*) Fig. 3.8 (*iii*) shows the instant when the brush is one-half on segment 2 and one-half on segment 1. The brush again conducts 40 A ; 20 A through segment 1 and 20 A through segment 2 (∵ now the resistances of the two parallel paths are equal). Note that now current in coil *A* is zero.

(*iv*) Fig. 3.8 (*iv*) shows the instant when the brush is three-fourth on segment 2 and one-fourth on segment 1. The brush conducts a current of 40 A ; 30 A through segment 2 and 10 A through segment 1. Note that current in coil A is 10 A but in the reverse direction to that before the start of commutation. The reader may see the action of the commutator in reversing the current in a coil as the coil passes the brush axis.

(*v*) Fig. 3.8 (*v*) shows the instant when the brush is in contact only with segment 2. The brush again conducts 40 A ; 20 A from coil *A* and 20 A from the adjacent coil to coil *A*. Note that now

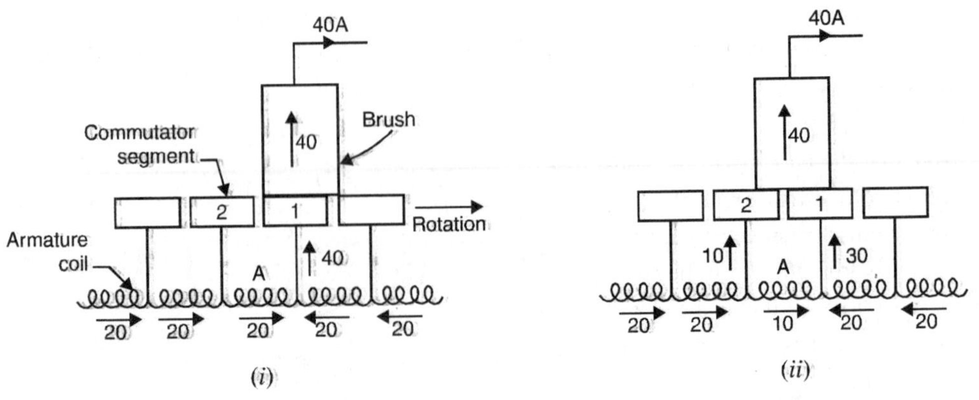

(*i*) (*ii*)

* We are considering 2-pole lap winding.

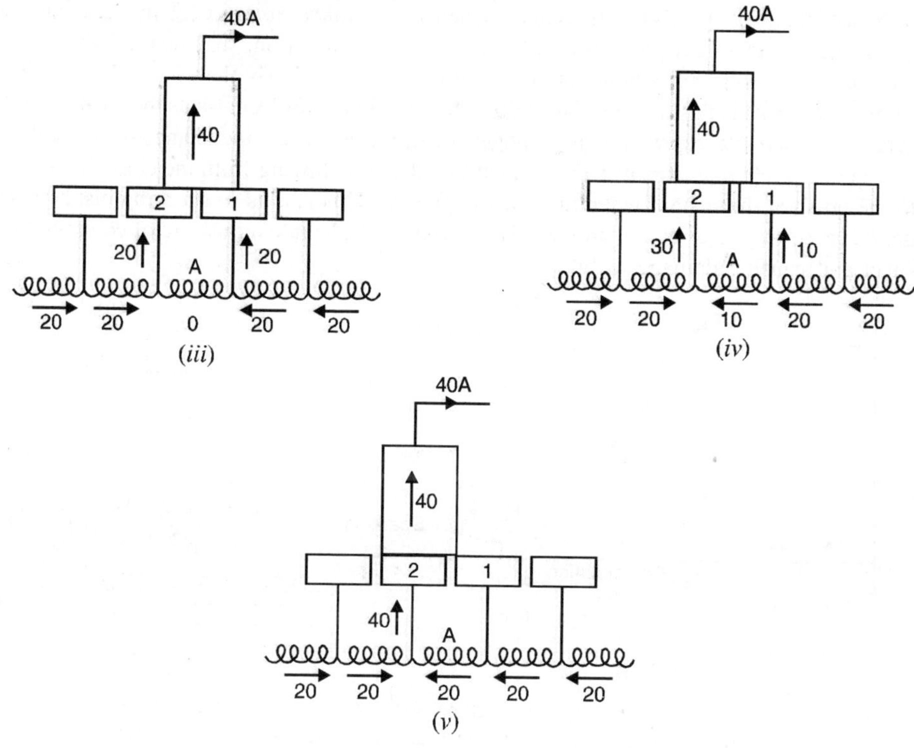

Fig. 3.8

current in coil *A* is 20 A but in the reverse direction. Thus the coil *A* has undergone commutation. Each coil undergoes commutation in this way as it passes the brush axis. Note that during commutation, the coil under consideration remains short-circuited by the brush.

Fig. 3.9 shows the current-time graph for the coil A undergoing commutation. The horizontal line *AB* represents a constant current of 20 A upto the beginning of commutation. From the finish of commutation, it is represented by another horizontal line *CD* on the opposite side of the zero line and at the same distance from it as *AB i.e.,* the current has exactly reversed (–20 A). The way in which current changes from *B* to *C* depends upon the conditions under which the coil undergoes commutation. If the current changes at a uniform rate (*i.e., BC* is a straight line), then it is called ideal commutation as shown in Fig. 3.9. Under such conditions, no sparking will take place between the brush and the commutator.

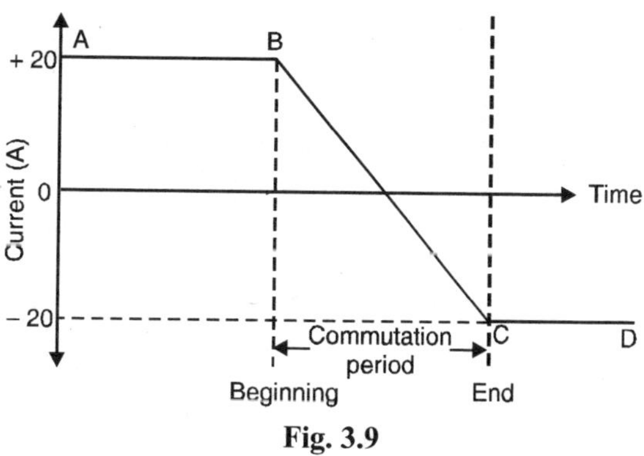

Fig. 3.9

Practical difficulties. The ideal commutation (*i.e.,* straight line change of current) cannot be attained in practice. This is mainly due to the fact that the armature coils have appreciable inductance. When the current in the coil undergoing commutation changes, *self-induced e.m.f.

* When current is flowing in any coil, it produces magnetic lines linked with it. Any change in current alters the number of lines and an e.m.f. is produced in the coil. This e.m.f. is called self-induced e.m.f. because it is caused by the change of current in the coil itself.

is produced in the coil. This is generally called *reactance voltage*. This reactance voltage opposes the change of current in the coil undergoing commutation. The result is that the change of current in the coil undergoing commutation occurs more slowly than it would be under ideal commutation. This is illustrated in Fig. 3.10. The straight line BC represents the ideal commutation whereas the curve BE represents the change in current when self-inductance of the coil is taken into account. Note that current CE (= 8 A in Fig. 3.10) is flowing from the commutator segment 1 to the brush at the instant when they part company. This results in sparking just as when any other current-carrying circuit is broken. The sparking results in overheating of commutator-brush contact and causing damage to both.

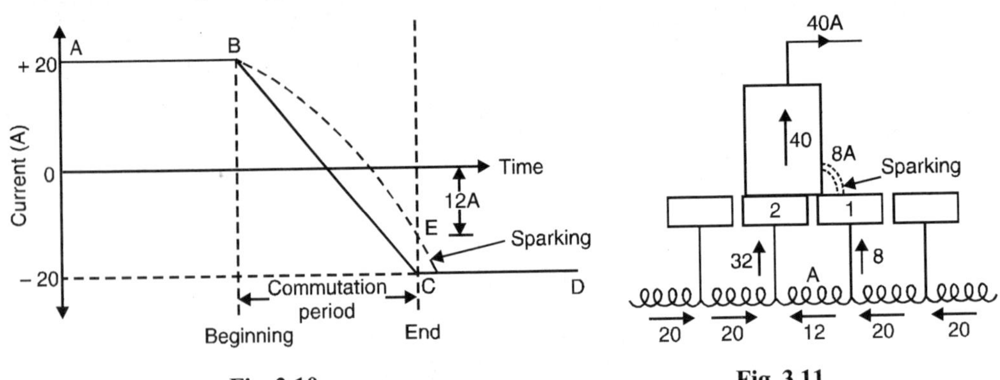

Fig. 3.10 **Fig. 3.11**

Fig. 3.11 illustrates how sparking takes place between the commutator segment and the brush. At the end of commutation or short-circuit period, the current in coil A is reversed to a value of 12 A (instead of 20 A) due to inductance of the coil. When the brush breaks contact with segment 1, the remaining 8 A current jumps from segment 1 to the brush through air causing *sparking between segment 1 and the brush.

3.10 CALCULATION OF REACTANCE VOLTAGE

Reactance voltage = Coefficient of self-inductance × Rate of change of current

When a coil undergoes commutation, two commutator segments remain short-circuited by the brush. Therefore, the time of short circuit (or commutation period T_c) is equal to the time required by the commutator to move a distance equal to the circumferential thickness of the brush minus the thickness of one insulating strip of mica.

Let W_b = brush width in cm ; W_m = mica thickness in cm

v = peripheral speed of commutator in cm/s

$$\therefore \text{ Commutation period, } T_c = \frac{W_b - W_m}{v} \text{ seconds}$$

The commutation period is very small, say of the order of 1/500 second.

Let the current in the coil undergoing commutation change from $+I$ to $-I$ (amperes) during the commutation. If L is the inductance of the coil, then reactance voltage is given by ;

$$\text{Reactance voltage, } E_R = L \times \frac{2I}{T_c} \quad \text{.... for linear commutation}$$

---

* At the end of commutation period, 8 A still flows to the commutator segment 1. However, there is no direct connection to the brush. Therefore, this current (*i.e.,* 8 A) takes a path through the air between the segment 1 and the brush and sparking results.

Example 3.12. *A 4-pole, wave-wound d.c. machine running at 1500 r.p.m. has a commutator of 30 cm diameter. If the armature current is 150 A, thickness of brush 1·25 cm and the self-inductance of each armature coil is 0·07 mH, calculate the average e.m.f. induced in each coil during commutation. Assume linear commutation and neglect mica thickness.*

Solution.

$$\text{Peripheral speed of the commutator, } v = \frac{\pi D N}{60} = \frac{\pi \times 30 \times 1500}{60} = 2356 \text{ cm/s}$$

$$\text{Commutation period, } T_c = \frac{W_b - W_m}{v} = \frac{1{\cdot}25 - 0}{2356} = 5{\cdot}3 \times 10^{-4} \text{ s}$$

$$\text{Current in each coil, } I = 150/2 = 75 \text{ A} \quad \text{.... wave-wound}$$

$$\text{Reactance voltage, } E_R = L \times \frac{2I}{T_c} = 0{\cdot}07 \times 10^{-3} \times \frac{2 \times 75}{5{\cdot}3 \times 10^{-4}} = \mathbf{19{\cdot}8 \text{ V}}$$

Example 3.13. *Calculate the reactance voltage for a machine having the following particulars :*
Revolution per minute = 900
Number of commutator segments = 55
Brush width in commutator segments = 1·74
Coefficient of self induction = 153 × 10^{-6} H
Assume linear commutation and neglect mica thickness.

Solution.

$$\text{Peripheral speed of the commutator, } v = 55 \times \frac{900}{60} = 825 \text{ segments/s}$$

$$\text{Commutation period, } T_c = \frac{W_b - W_m}{v} = \frac{1{\cdot}74 - 0}{825} = 2{\cdot}11 \times 10^{-3} \text{ s}$$

$$\text{Reactance voltage, } E_R = L \times \frac{2I}{T_c} = 153 \times 10^{-6} \times \frac{2 \times 27}{2{\cdot}11 \times 10^{-3}} = \mathbf{3{\cdot}91 \text{ V}}$$

Example 3.14. *A commutator having a diameter of 76 cm rotates at a speed of 600 r.p.m. Find the time of commutation if the brush width is 1·5 cm. Neglect mica thickness.*

Solution.

$$\text{Speed of commutator, } v = \frac{\pi D N}{60} = \frac{\pi \times 76 \times 600}{60} = 760\,\pi \text{ cm/s}$$

$$\text{Period of commutation, } T_c = \frac{W_b - W_m}{v} = \frac{1{\cdot}5 - 0}{760\,\pi} = \mathbf{0{\cdot}628 \times 10^{-3} \text{ s}}$$

3.11 METHODS OF IMPROVING COMMUTATION

Improving commutation means to make current reversal in the short-circuited coil as sparkless as possible. The following are the two principal methods of improving commutation :

(*i*) Resistance commutation
(*ii*) E.M.F. commutation

We shall discuss each method in turn.

3.12 RESISTANCE COMMUTATION

The reversal of current in a coil (*i.e.*, commutation) takes place while the coil is short-circuited by the brush. Therefore, there are two parallel paths for the current as long as the short circuit exists. If the contact resistance between the brush and the commutator is made large, then current would divide in the inverse ratio of contact resistances (as for any two resistances in parallel). This is the key point in improving commutation. This is achieved by using carbon brushes (instead of Cu brushes) which have high contact resistance. This method of improving commutation is called resistance commutation.

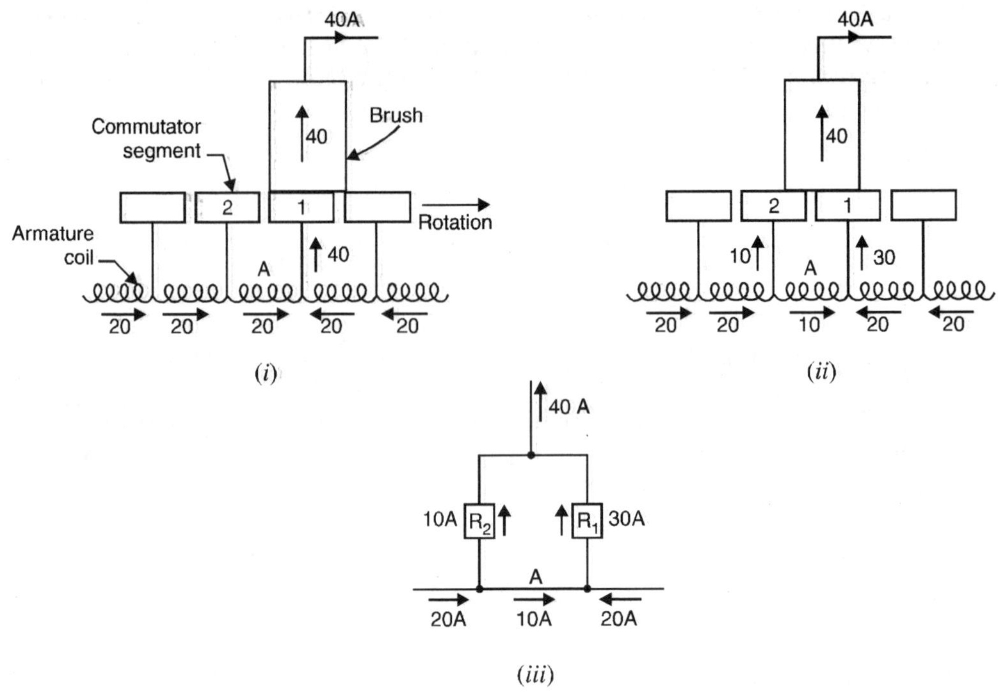

Fig. 3.12

Figs. 3.12 and 3.13 illustrates how high contact resistance of carbon brush improves commutation (*i.e.*, reversal of current) in coil *A*. In Fig. 3.12 (*i*), the brush is entirely on segment 1 and, therefore, the current in coil *A* is 20 A. The coil *A* is yet to undergo commutation. As the armature rotates, the brush short-circuits the coil *A* and there are two parallel paths for the current into the brush. Fig. 3.12 (*ii*) shows the instant when the brush is one-fourth on segment 2 and three-fourth on segment 1. The equivalent electric circuit is shown in Fig. 3.12 (*iii*) where R_1 and R_2 represent the brush contact resistances on segments 1 and 2. A resistor is not shown for coil *A* since it is assumed that the coil resistance is negligible as compared to the brush contact resistance. The values of current in the parallel paths of the equivalent circuit are determined by the respective resistances of the paths. For the condition shown in Fig. 3.12 (*ii*), resistor R_2 has three times the resistance of resistor R_1. Therefore, the current distribution in the paths will be as shown. Note that current in coil *A* is reduced from 20 A to 10 A due to division of current in the inverse ratio of contact resistances. If the Cu brush is used (which has low contact resistance), $R_1 \simeq R_2$ and the current in coil *A* would not have reduced to 10 A.

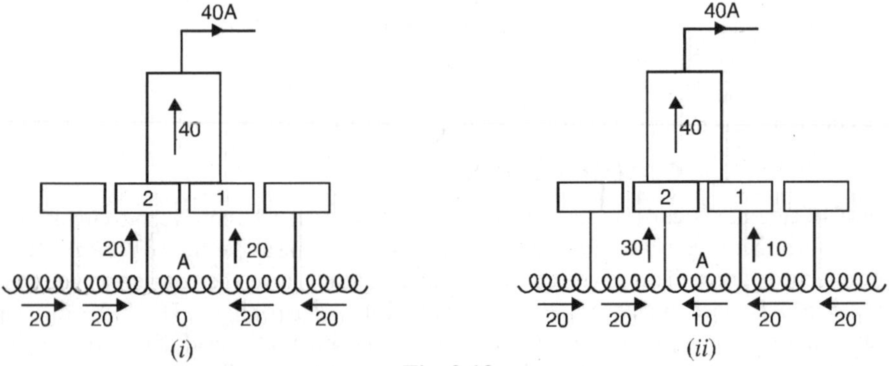

Fig. 3.13

As the carbon brush passes over the commutator, the contact area with segment 2 increases and that with segment 1 decreases *i.e.,* R_2 decreases and R_1 increases. Therefore, more and more current passes to the brush through segment 2. This is illustrated in Figs. 3.13 (*i*) and 3.13 (*ii*). When the break between the brush and the segment 1 finally occurs [See Fig. 3.13 (*iii*)], the current in the coil is reversed and commutation is achieved.

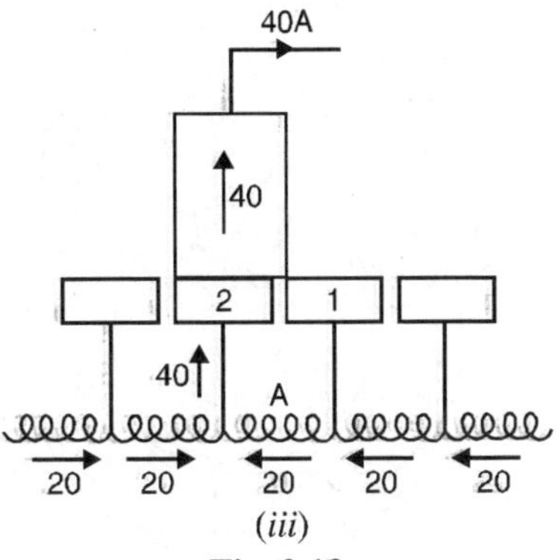

Fig. 3.13

It may be noted that the main cause of sparking during commutation is the production of reactance voltage and carbon brushes cannot prevent it. Nevertheless, the carbon brushes do help in improving commutation. The other minor advantages of carbon brushes are :

(*i*) The carbon lubricates and polishes the commutator.

(*ii*) If sparking occurs, it damages the commutator less than with copper brushes and the damage to the brush itself is of little importance.

3.13 E.M.F. COMMUTATION

In this method, an arrangement is made to neutralise the reactance voltage by producing a reversing voltage in the coil undergoing commutation. The reversing voltage acts in opposition to the reactance voltage and neutralises it to some extent. If the reversing voltage is equal to the reactance voltage, the effect of the latter is completely wiped out and we get sparkless commutation. The reversing voltage may be produced in the following two ways :

(*i*) By brush shifting

(*ii*) By using interpoles or *compoles

(*i*) *By brush shifting.* In this method, the brushes are given sufficient forward lead (for a generator) to bring the short-circuited coil (*i.e.,* coil undergoing commutation) under the influence of the next pole of opposite polarity. Since the short-circuited coil is now in the reversing field, the reversing voltage produced cancels the reactance voltage. This method suffers from the following drawbacks :

(*a*) The reactance voltage depends upon armature current. Therefore, the brush shift will depend on the magnitude of armature current which keeps on changing. This necessitates frequent shifting of brushes.

(*b*) The greater the armature current, the greater must be the forward lead for a generator. This increases the demagnetising effect of armature reaction and further weakens the main field.

(*ii*) *By using interpoles or compoles.* The best method of neutralising reactance voltage is by using interpoles or compoles. This method is discussed in Art. 3.14.

3.14 INTERPOLES OR COMPOLES

The best way to produce reversing voltage to neutralise the reactance voltage is by using interpoles or compoles. These are small poles fixed to the yoke and spaced mid-way between the main poles (See Fig. 3.14). They are wound with comparatively few turns and connected in series with the armature so that they carry armature current. Their polarity is the same as the next main pole ahead in the direction of rotation for a generator (See Fig. 3.14). Connections for a d.c. generator with interpoles is shown in Fig. 3.15.

* Interpoles are also called commutating poles. The latter are usually abbreviated to compoles.

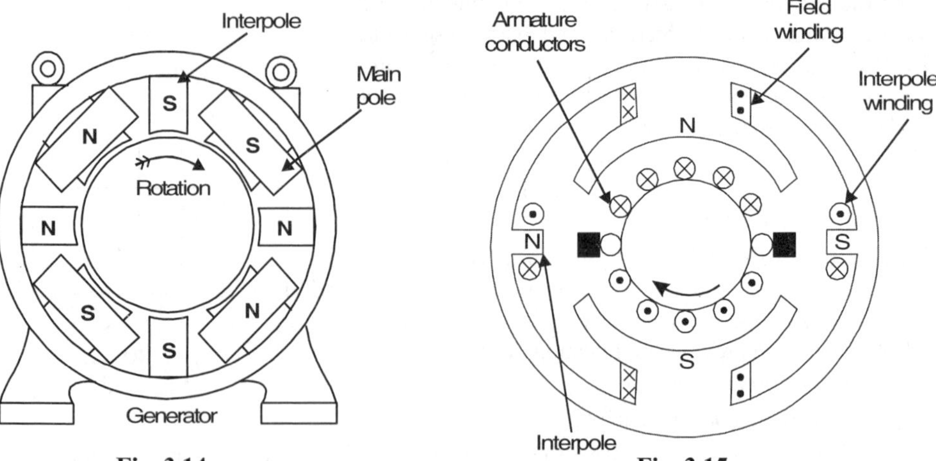

Fig. 3.14 Fig. 3.15

Functions of Interpoles. The machines fitted with interpoles have their brushes set on geometrical neutral axis (no lead). The interpoles perform the following two functions :

(*i*) As their polarity is the same as the main pole ahead (for a generator), they induce an e.m.f. in the coil (undergoing commutation) which opposes reactance voltage. This leads to sparkless commutation. The e.m.f. induced by compoles is known as *commutating* or *reversing e.m.f.* Since the interpoles carry the armature current and the reactance voltage is also proportional to armature current, the neutralisation of reactance voltage is automatic.

(*ii*) The m.m.f. of the compoles neutralises the cross-magnetising effect of armature reaction in small region in the space between the main poles. It is because the two m.m.f.s oppose each other in this region.

Fig. 3.16 shows the circuit diagram of a shunt generator with commutating winding and compensating winding. Both these windings are connected in series with the armature and so they carry the armature current. However, the functions they perform must be understood clearly. The main function of commutating winding is to produce reversing (or commutating) e.m.f. in order to cancel the reactance voltage. In addition to this, the m.m.f. of the commutating winding neutralises the cross-magnetising ampere-turns in the space between the main poles. The compen-sating winding neutralises the cross-magnetising effect of armature reaction under the pole faces.

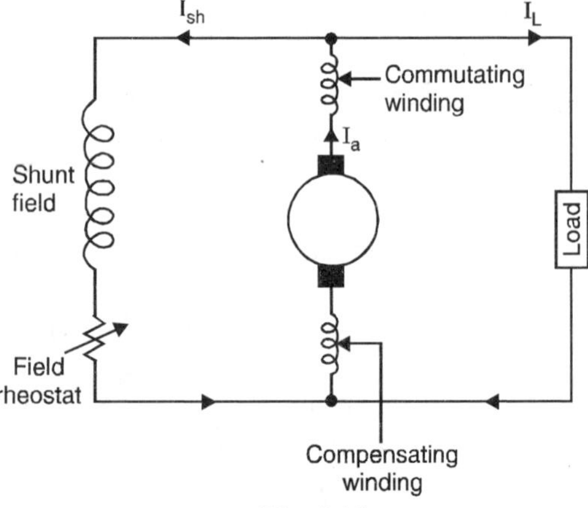

Fig. 3.16

Example 3.15. *Calculate the ampere-turns for each commutating pole of an 8 pole generator with 107 slots, each containing 1000 ampere conductors. The inter-pole air-gap is 1·2 cm. The flux density in the gap is to be 0·32 Wb/m². Neglect iron parts of the circuit and leakage.*

Solution. Ampere-turns for each commutating pole is equal to the sum of armature ampere-turns per pole and ampere-turns required for the gap (neglecting the effect of iron parts and of leakage).

$$\text{Armature ampere-conductors/pole} = \frac{107 \times 1000}{8}$$

$$\text{Armature ampere-turns/pole} = \frac{107 \times 1000}{8 \times 2} = 6688 \text{ AT}$$

$$AT \text{ required for air gap} = \frac{B}{\mu_0} \times l_g = \frac{0{\cdot}32}{4\pi \times 10^{-7}} \times 1{\cdot}2 \times 10^{-2} = 3056 \text{ AT}$$

$\therefore$ AT required on each commutating pole $= 6688 + 3056 =$ **9744 AT**

Example 3.16. *Estimate the number of turns needed on each commutating pole of a 6-pole generator delivering 200 kW at 200 V. Given that number of lap-connected conductors = 540, interpole gap = 1 cm, flux density in the interpolar air-gap = 0·3 Wb/m². Ignore the effect of iron parts of the circuit and of leakage.*

Solution.

$$\text{Armature current, } I_a = \frac{200 \times 10^3}{200} = 1000 \text{ A}$$

Current in each conductor, $I = (1000/6)\text{A}$ .... lap-wound

$$\text{Armature } AT/\text{pole} = \frac{ZI}{2P} = \frac{540 \times 1000}{2 \times 6 \times 6} = 7500 \text{ AT}$$

$$AT \text{ required for air-gap} = \frac{B}{\mu_0} \times l_g = \frac{0{\cdot}3}{4\pi \times 10^{-7}} \times 1 \times 10^{-2} = 2387 \text{ AT}$$

AT for each commutating pole $= 7500 + 2387 = 9887$ AT

$\therefore$ Number of turns on each commutating pole

$$= 9887/I_a = 9887/1000 \simeq \mathbf{10}$$

3.15 EQUALISING CONNECTIONS

We know that the armature circuit in lap winding of a multipolar machine has as many parallel paths as the number of poles. Because of wear in the bearings, and for other reasons, the air gaps in a generator become unequal and, therefore, the flux in some poles becomes greater than in others. This causes the voltages of the different paths to be unequal. With unequal voltages in these parallel paths, circulating current will flow even if no current is supplied to an external load. If these currents are large, some of the brushes will be required to carry a greater current at full load than they were designed to carry and this will cause sparking. To relieve the brushes of these circulating currents, points on the armature that are at the same potential are connected together by means of copper bars called *equaliser rings* This is achieved by connecting to the same equaliser ring the coils that occupy the same positions relative to the poles (See Fig. 3.17). Thus referring to Fig. 3.17, the coil consisting of conductor 1 and conductor 8 occupies the same position relative to the poles as the coil consisting of conductors 13 and 20. Therefore, the two coils are connected to the same equaliser ring. The equalisers provide a low resistance path for the circulating current. As a result, the circulating current due to the slight differences in the voltages of the various parallel paths passes through the equaliser rings instead of passing through the brushes. This reduces sparking.

Equaliser rings should be used only on windings in which the number of coils is a multiple of the number of poles. For best results, each coil should be connected to an equaliser ring but this is seldom done. Satisfactory results are obtained by connecting about every third coil to an equaliser ring. In order to distribute the connections to the equaliser rings equally, the number of coils per pole must be divisible by the connection pitch.

Note. Equaliser rings are not used in wave winding because there is no imbalance in the voltages of the two parallel paths. This is due to the fact that conductors in each of the two paths pass under all N and S poles successively (unlike a lap winding where all conductors in any parallel path lie under one pair of poles). Therefore, even if there are inequalities in pole flux, they will affect each path equally.

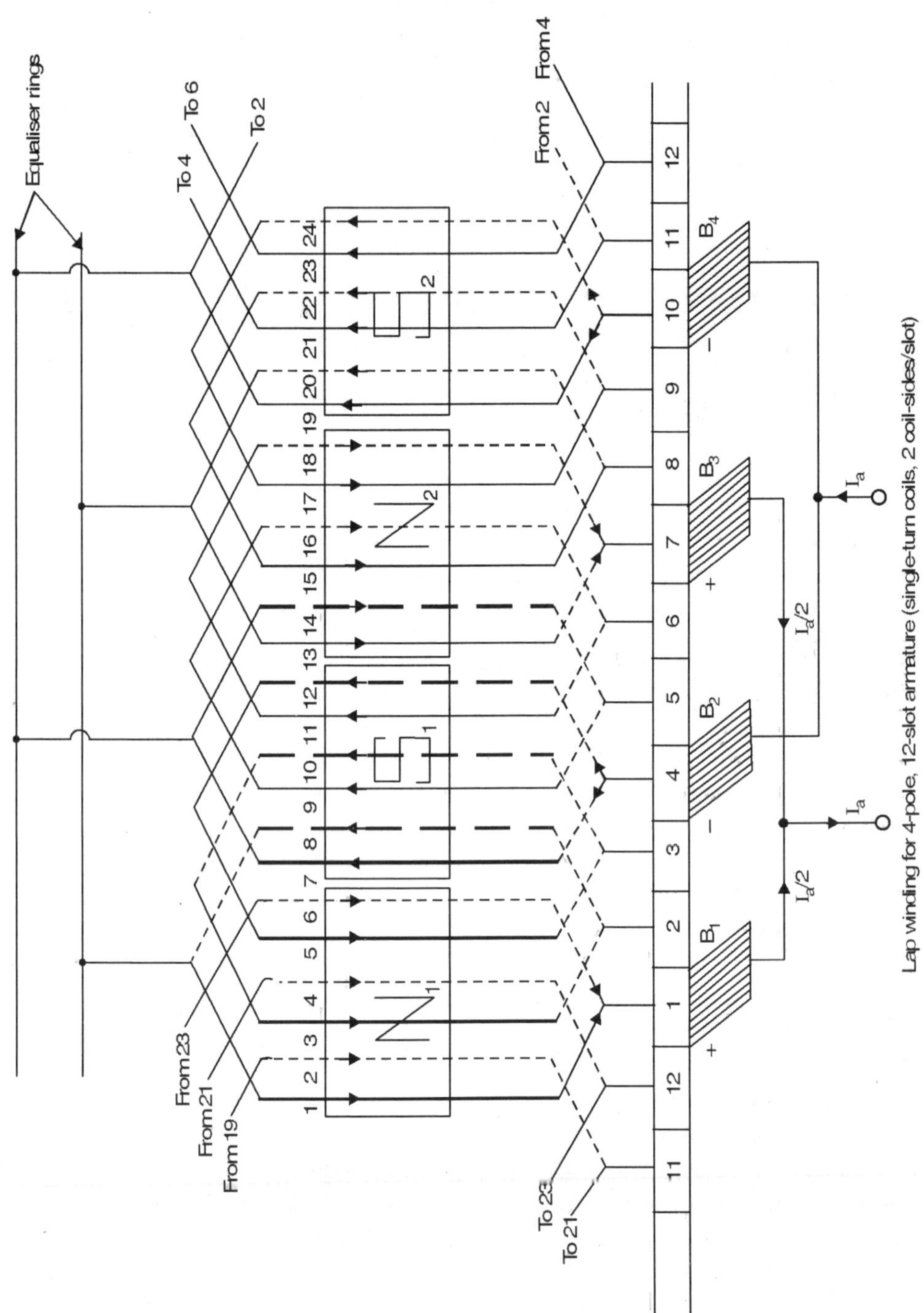
Equaliser rings
To 4
To 6
To 2
From 2
From 4
From 23
From 21
From 19
To 23
To 21
B_1
B_2
B_3
B_4
I_a
$I_a/2$
Lap winding for 4-pole, 12-slot armature (single-turn coils, 2 coil-sides/slot)

Fig. 3.17

3.16 IMPORTANT FACTS ABOUT D.C. GENERATORS

While studying d.c. generators, the following facts should be kept in mind :

(*i*) The d.c. generator is driven at a constant speed by the prime mover.

(*ii*) The most commonly used winding for a d.c. generator is the double layer winding. For a double layer winding, the number of coils is equal to the number of armature slots.

(*iii*) How we connect the armature conductors to the commutator determines the type of winding. There are two general types of windings; the **lap winding** and the **wave winding.**

(*iv*) The lap winding is used for low-voltage and high-current machines. On the other hand, wave winding is employed for high-voltage and low-current machines.

(*v*) The armature winding has finite resistance, however small it may be. Therefore, terminal voltage (V) of a d.c. generator is always less than the induced e.m.f. (E_g).

(*vi*) If the generator is not compensated for the armature reaction, there is less flux/pole under load than at no load.

(*vii*) The *torque developed by the armature conductors is equal and opposite to the torque applied by the prime mover *i.e.* torque developed *opposes* the armature rotation (Lenz's law). The mechanical energy is supplied to the prime mover to overcome this opposition. As a result, mechanical energy is converted into electrical energy during generator action and the law of conservation of energy holds good.

(*viii*) If the specific information regarding the adverse effect of armature reaction on the generator performance is not known, we assume that either the armature reaction is negligible or the generator is appropriately compensated for it.

(*ix*) We generally use the term **load** in a d.c. generator to signify the load current. Thus "no load" means an open circuit *i.e.* the generator is delivering no current. On the other hand, "full-load" means that the generator is delivering rated load current at the rated terminal voltage.

(*x*) The efficiency of a d.c. generator is maximum when **variable losses** are equal to **fixed** or **constant losses**.

SHORT ANSWER QUESTIONS

Q. 1. *What is geometrical neutral axis (G.N.A.) ?*

Ans. The geometrical neutral axis (G.N.A.) is the axis that bisects the angle between the centre line of two adjacent poles. Clearly, it is the axis of symmetry between two adjacent poles.

Q. 2. *What is magnetic neutral axis (M.N.A.) ?*

Ans. The magnetic neutral axis (M.N.A.) is the axis drawn perpendicular to the mean direction of the flux passing through the centre of armature. Clearly, no e.m.f. is produced in the armature conductors along this axis because then they cut no flux.

Q. 3. *What is armature reaction ?*

Ans. The armature reaction in a d.c. generator means the effect of magnetic field set up by armature current on the distribution of flux under the main poles of the machine.

Q. 4. *What are the effects of armature reaction in a d.c. generator ?*

Ans. The armature reaction in a d.c. generator has the following two effects :

(*i*) It demagnetises or weakens the main flux.

(*ii*) It cross-magnetises or distorts the main flux.

---

* The armature conductors carry current and are placed in the magnetic field. Therefore, torque is developed by them.

Q. 5. *With brushes along G.N.A., what is the effect of armature reaction?*

Ans. With brushes along G.N.A., there is no *demagnetising effect of armature reaction. There is only distorting or cross-magnetising effect of armature reaction.

Q. 6. *With brushes along M.N.A., what is the effect of armature reaction?*

Ans. With brushes along M.N.A., the armature reaction will have both demagnetising and distorting effects. Their relative magnitudes depend upon the amount of shift of M.N.A. from G.N.A.

Q. 7. *For sparkless commutation, brushes should be shifted along M.N.A. Is it a satisfactory method ?*

Ans. The shifting of M.N.A. from G.N.A. is directly proportional to the armature current. Then, with a changing armature current which generally takes place in d.c. generators, the position of the M.N.A. likewise changes. Then it is not possible to set the brushes in any one position. Consequently, shifting of brushes is not an entirely satisfactory method of eliminating sparking. This method is generally used in small machines.

Q. 8. *Should the brushes of a loaded d.c. generator be placed along G.N.A. ?*

Ans. No. The brushes must be advanced by a certain angle in the direction of rotation of the machine. This is called brush lead and depends upon armature current.

Q. 9. *What is the function of interpoles ?*

Ans. The interpoles are much narrower than main poles and are placed between the main poles. The interpole flux neutralises not only the field in the G.N.A. due to armature reaction but also produces proper flux to generate an e.m.f. in the short-circuited coil, equal and opposite to the e.m.f. of self-inductance. This produces an ideal condition for commutation.

Q. 10. *What is the chief application of compensating windings ?*

Ans. The compensating windings are not used on many d.c. machines because of the cost of manufacturing them. Their chief use is on high speed and high voltage generators of large capacity.

MULTIPLE-CHOICE QUESTIONS

1. In d.c. generators armature reaction is produced by
   (*i*) the field current (*ii*) armature current
   (*iii*) both field and armature currents
   (*iv*) none of the above
2. Flux distribution due to armature reaction causes the M.N.A. in a generator to
   (*i*) remain stationary
   (*ii*) move in the direction of rotation
   (*iii*) move opposite to the direction of rotation
   (*iv*) none of the above
3. In a clockwise rotating loaded d.c. generator, brushes have to be shifted
   (*i*) clockwise
   (*ii*) anticlockwise
   (*iii*) either clockwise or anticlockwise
   (*iv*) none of the above
4. Armature reaction is increased when
   (*i*) the field current increases
   (*ii*) the armature current decreases
   (*iii*) the armature current increases
   (*iv*) none of the above
5. In a d.c. generator, the brushes are shifted from G.N.A. in the direction of rotation to
   (*i*) decrease speed
   (*ii*) increase speed
   (*iii*) increase voltage
   (*iv*) reduce sparking

* Actually there will be a slight decrease in the main flux. In case of a generator, the flux will be weakened at leading pole tip and strengthened at the trailing pole tip.

6. Interpoles are used in a d.c. generator to
   (*i*) limit the current to a safe value
   (*ii*) stabilise the voltage
   (*iii*) raise the voltage with load
   (*iv*) reduce sparking at the brushes
7. When a d.c. generator carries no armature current,
   (*i*) M.N.A. coincides with G.N.A.
   (*ii*) M.N.A. is behind G.N.A.
   (*iii*) M.N.A. is ahead of G.N.A.
   (*iv*) none of the above
8. Due to armature reaction in a d.c. generator, the flux under leading pole tip
   (*i*) is decreased
   (*ii*) is increased
   (*iii*) may decrease or increase
   (*iv*) remains unchanged
9. In order to have sparkless commutation, the brushes should be placed
   (*i*) along G.N.A.
   (*ii*) along M.N.A.
   (*iii*) along G.N.A. or M.N.A.
   (*iv*) none of above
10. With respect to the direction of rotation, the interpoles on a d.c. generator have the same polarity as the main poles
    (*i*) behind them
    (*ii*) ahead of them
    (*iii*) either (*i*) or (*ii*)
    (*iv*) neither (*i*) nor (*ii*)
11. The interpoles are connected
    (*i*) in parallel with armature
    (*ii*) in series with armature
    (*iii*) either in series or in parallel
    (*iv*) both in series and parallel with armature
12. The conductors of compensating winding are housed
    (*i*) in the slots in the pole faces
    (*ii*) in the slots in the armature
    (*iii*) partly in armature slots and partly in pole faces
13. In a d.c. generator, the effect of armature reaction on the main pole flux is to
    (*i*) reduce it
    (*ii*) distort it
    (*iii*) both reduce and distort it
    (*iv*) reverse it
14. The compensating winding is connected
    (*i*) in parallel with armature winding
    (*ii*) in series with armature winding
    (*iii*) partly in series and partly in parallel with armature winding
    (*iv*) none of the above
15. During commutation, the coil undergoing commutation .... by the brush
    (*i*) remains short-circuited
    (*ii*) remains open-circuited
    (*iii*) either (*i*) or (*ii*)
    (*iv*) none of the above
16. The current-time graph for ideal commutation is
    (*i*) a straight line (*ii*) parabola
    (*iii*) hyperbola (*iv*) none of the above
17. The commutation period is of the order of
    (*i*) 1/10 second (*ii*) 1/50 second
    (*iii*) 1/20 second (*iv*) 1/500 second
18. The interpole winding is connected
    (*i*) in series with armature winding
    (*ii*) in parallel with armature winding
    (*iii*) partly in series and partly in parallel with armature winding
    (*iv*) none of the above
19. When the armature current increases, the forward lead for a generator
    (*i*) decreases
    (*ii*) increases
    (*iii*) remains unchanged
    (*iv*) insufficient data
20. The reactance voltage in a d.c. shunt generator depends upon
    (*i*) armature current
    (*ii*) field current
    (*iii*) both (*i*) and (*ii*)
    (*iv*) neither (*i*) nor (*ii*)

ANSWERS TO MULTIPLE-CHOICE QUESTIONS

1. (*ii*)	**2.** (*ii*)	**3.** (*i*)	**4.** (*iii*)	**5.** (*iv*)
6. (*iv*)	**7.** (*i*)	**8.** (*i*)	**9.** (*ii*)	**10.** (*ii*)
11. (*ii*)	**12.** (*i*)	**13.** (*iii*)	**14.** (*ii*)	**15.** (*i*)
16. (*i*)	**17.** (*iv*)	**18.** (*i*)	**19.** (*ii*)	**20.** (*i*)

HINTS TO SELECTED MULTIPLE-CHOICE QUESTIONS

2. The flux distribution due to armature reaction causes the M.N.A. in a generator to move in the direction of rotation.

4. Armature reaction depends upon armature current. The greater the armature current, the greater is the armature reaction and *vice-versa.*

5. In order to have sparkless commutation, the brushes should lie along M.N.A.

4

D.C. Generator Characteristics

INTRODUCTION

The speed of a d.c. machine operated as a generator is fixed by the prime mover. For general-purpose operation, the prime mover is equipped with a speed governor so that the speed of the generator is practically constant. Under such condition, the generator performance deals primarily with the relation between excitation, terminal voltage and load. These relations can be best exhibited graphically by means of curves known as generator characteristics. These characteristics show at a glance the behaviour of the generator under different load conditions.

4.1 D.C. GENERATOR CHARACTERISTICS

The following are the three most important characteristics of a d.c. generator :

1. Open Circuit Characteristic (O.C.C.). This curve shows the relation between the generated e.m.f. at no-load (E_0) and the field current (I_f) at constant speed. It is also known as magnetic characteristic or no-load saturation curve. Its shape is practically the same for all generators whether separately or self-excited. The data for O.C.C. curve are obtained experimentally by operating the generator at no load and constant speed and recording the change in terminal voltage as the field current is varied.

2. Internal or Total characteristic (E/I_a). This curve shows the relation between the generated e.m.f. on load (E) and the armature current (I_a). The e.m.f. E is less than E_0 due to the demagnetising effect of armature reaction. Therefore, this curve will lie below the open circuit characteristic (O.C.C.). The internal characteristic is of interest chiefly to the designer. It cannot be obtained directly by experiment. It is because a voltmeter cannot read the e.m.f. generated on load due to the voltage drop in armature resistance. The internal characteristic can be obtained from external characteristic if winding resistances are known because armature reaction effect is included in both characteristics.

3. External characteristic (V/I_L). This curve shows the relation between the terminal voltage (V) and load current (I_L). The terminal voltage V will be less than E due to voltage drop in the armature circuit. Therefore, this curve will lie below the internal characteristic. This characteristic is very important in determining the suitability of a generator for a given purpose. It can be obtained by making simultaneous measurements of terminal voltage and load current (with voltmeter and ammeter) of a loaded generator.

4.2 OPEN CIRCUIT CHARACTERISTIC OF A D.C. GENERATOR

The O.C.C. for a d.c. generator is determined as follows. The field winding of the d.c. generator (series or shunt) is disconnected from the machine and is separately excited from an external d.c. source as shown in Fig. 4.1 (*ii*). The generator is run at fixed speed (*i.e.*, normal speed). The *field current (I_f) is increased from zero in steps and the corresponding values of generated e.m.f. (E_0) read

* Generated e.m.f. $= \dfrac{P\phi ZN}{60A} = \left(\dfrac{PZ}{60\,A}\right) \times \phi N$

For a given generator, the quantities in the bracket are constant.

$\therefore$ Generated e.m.f. $\propto \phi N$

Thus the generated e.m.f. in a generator is directly proportional to the flux/pole and to the speed.

off on a voltmeter connected across the armature terminals. On plotting the relation between E_0 and I_f, we get the open circuit characteristic as shown in Fig. 4.1 (*i*).

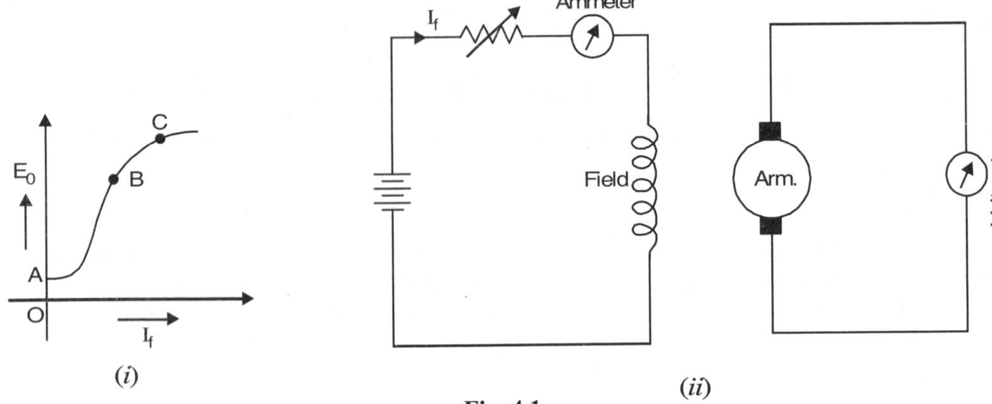

Fig. 4.1

The following points may be noted from O.C.C. :

(*i*) When the field current is zero, there is some generated e.m.f. *OA*. This is due to the residual magnetism in the field poles.

(*ii*) Over a fairly wide range of field current (upto point *B* in the curve), the curve is linear. It is because in this range, *reluctance of iron is negligible as compared with that of air gap. The air gap reluctance is constant and hence linear relationship.

(*iii*) After point *B* on the curve, the reluctance of iron also comes into picture. It is because at higher flux densities, μ_r for iron decreases and reluctance of iron is no longer negligible. Consequently, the curve deviates from linear relationship.

(*iv*) After point *C* on the curve, the magnetic saturation of poles begins and E_0 tends to level off.

The reader may note that the O.C.C. of even self-excited generator is obtained by running it as a separately excited generator.

4.3 CHARACTERISTICS OF A SEPARATELY EXCITED D.C. GENERATOR

The obvious disadvantage of a separately excited d.c. generator is that we require an external d.c. source for **excitation. But since the output voltage may be controlled more easily and over a wide range (from zero to a maximum), this type of excitation finds many applications.

(*i*) **Open circuit characteristic.** The O.C.C. of a separately excited generator is determined in a manner described in Art. 4.2. Fig. 4.2 shows the variation of generated e.m.f. on no load with field current for various fixed speeds. Note that if the value of constant speed is increased, the steepness of the curve also increases. When the field current is zero, the residual magnetism in the poles will give rise to the small initial e.m.f. as shown.

Generated e.m.f.
N increasing
N_3
N_2
N_1
A
O
Field current

Fig. 4.2

(*ii*) **Internal and External Characteristics.** The external characteristic of a separately excited generator is the curve between the terminal voltage (V) and the load current I_L (which is the same as armature current in this case). In order to determine the external characteristic, the circuit set up is as shown in

* At low flux density, μ_r for iron is high and hence iron path offers negligible reluctance. Remember μ_r for iron depends upon flux density. As flux density increases, μ_r decreases.

** Excitation is the term applied to the production of a magnetic field in the machine.

Fig. 4.3 (*i*). As the load current increases, the terminal voltage falls due to two reasons :

(*a*) The armature reaction weakens the main flux so that actual e.m.f. generated E on load is less than that generated (E_0) on no load.

(*b*) There is voltage drop across armature resistance ($= I_L R_a = I_a R_a$).

Due to these reasons, the external characteristic is a drooping curve [curve 3 in Fig. 4.3 (*ii*)]. Note that in the absence of armature reaction and armature drop, the generated e.m.f. would have been E_0 (curve 1).

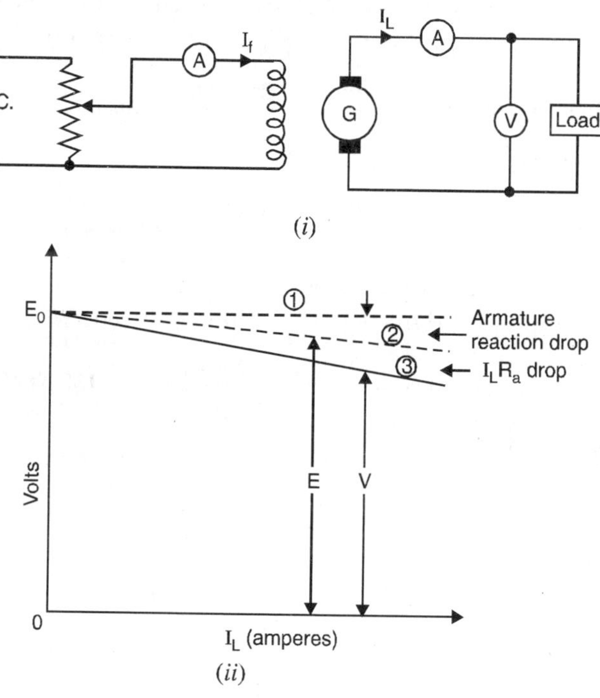

Fig. 4.3

The internal characteristic can be determined from external characteristic by adding $I_L R_a$ drop to the external characteristic. It is because armature reaction drop is included in the external characteristic. Curve 2 is the internal characteristic of the generator and should obviously lie above the external characteristic.

Example 4.1. *A 240 kW, 240V, 6–pole, 600 r.p.m. separately excited d.c. generator is delivering the rated load current at the rated voltage. The generator has the following specifications:*

Armature resistance, R_a = 0.01 Ω; Field winding resistance, R_f = 30 Ω; Voltage of external source, V_f = 120 V; Number of turns/pole for field winding, N_f = 500 ; Rotational losses, P_r = 10 kW. The demagnetising m.m.f. due to armature reaction is 25 % of the armature current. Its magnetising curve

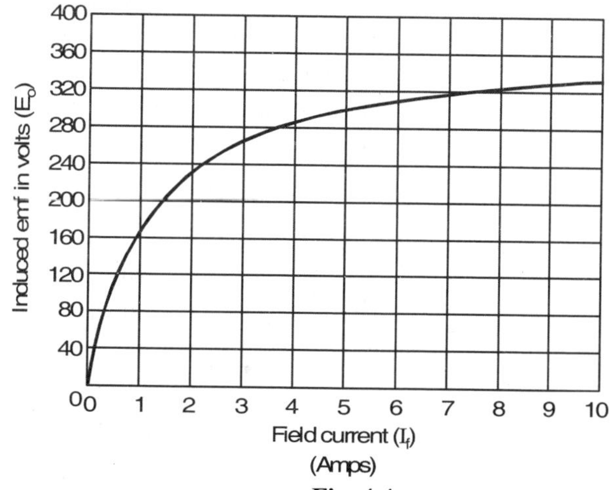

Fig. 4.4

is given in Fig. 4.4. Determine (i) induced e.m.f. at full load (ii) power developed (iii) torque developed (iv) applied torque (v) efficiency (vi) the external resistance in the field winding.

Solution. Fig. 4.5 shows the connections of the separately excited d.c. generator.

$$F.L.\ \text{current},\ I_L = I_a = \frac{240\times10^3}{240} = 1000\ \text{A}$$

(*i*) The induced e.m.f. at full load is

$$E_g = V + I_a R_a = 240 + 1000\times0.01 = \mathbf{250\ V}$$

(*ii*) Power developed, $P_d = E_g I_a = 250\times1000\ \text{W} = \mathbf{250\ kW}$

(*iii*) Angular velocity, $\omega = \dfrac{2\pi N}{60} = \dfrac{2\pi\times600}{60} = 20\pi\ \text{rad}/s$

∴ Torque developed at full load is given by ;

$$T_d = \frac{P_d}{\omega} = \frac{250\times10^3}{20\pi} = \mathbf{3979\ Nm}$$

(*iv*) Mechanical power input of full load is

$$P_m(in) = P_d + P_r = 250 + 10 = 260\ \text{kW}$$

∴ Applied torque $= \dfrac{P_m(in)}{\omega} = \dfrac{260\times10^3}{20\pi} = \mathbf{4138\ Nm}$

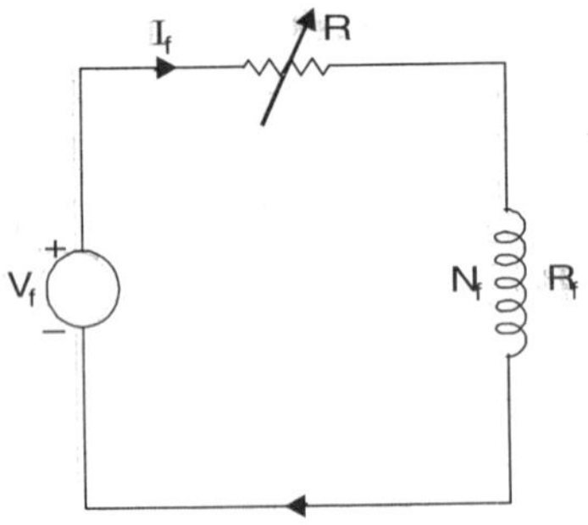

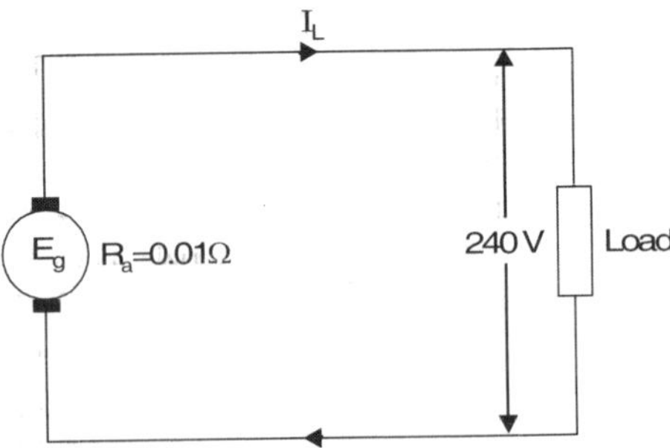

Fig. 4.5

(*v*) At full load, the induced e.m.f. is 250 *V* and the *effective field current* required is 2.5 *A* (from magnetisation curve in Fig. 4.4). This is the field current that must flow in the field winding and the corresponding m.m.f. = 500 × 2.5 = 1250 AT/pole. The demagnetising m.m.f. due to armature reaction = 0.25 × 1000 = 250 *AT*/ pole. Therefore, total m.m.f. to be provided by the field winding = 1250 + 250 = 1500 AT/pole.

∴ Actual field current at full load is

$$I_f = 1500/500 = 3\ \text{A}$$

Power loss in the field winding, $P_f = V_f I_f = 120\times3 = 360\ \text{W} = 0.36\ \text{kW}$

∴ Total input power, $P_i = 260 + 0.36 = 260.36\ \text{kW}$

∴ Generator efficiency, $\eta = \dfrac{240}{260.36}\times100 = \mathbf{92.2\%}$

(*vi*) Total resistance in the field winding is

$$R_{Tf} = 120/3 = 40\Omega$$

$$\therefore\quad R = 40 - R_f = 40 - 30 = \mathbf{10\Omega}$$

4.4 VOLTAGE BUILD-UP IN A SELF-EXCITED GENERATOR

Let us see how voltage builds up in a self-excited generator.

(*i*) **Shunt generator.** Consider a shunt generator. If the generator is run at a constant speed, some e.m.f. will be generated due to residual magnetism in the main poles. This small e.m.f. circulates a field current which in turn produces additional flux to reinforce the original residual flux (provided field winding connections are correct). This process continues and the generator builds up the normal generated voltage following the O.C.C. shown in Fig. 4.6 (*i*).

The field resistance R_f can be represented by a straight line passing through the origin as shown in Fig. 4.6 (*ii*). The two curves can be shown on the same diagram as they have the same ordinate [See Fig. 4.6 (*iii*)].

Since the field circuit is inductive, there is a delay in the increase in current upon closing the field circuit switch. The rate at which the current increases depends upon the voltage available for increasing it. Suppose at any instant, the field current is i (= OA) and is increasing at the rate di/dt. Then,

$$E_0 = i R_f + L \frac{di}{dt}$$

where R_f = total field circuit resistance

L = inductance of field circuit

At the considered instant, the total e.m.f. available is AC [See Fig. 4.6 (*iii*)]. An amount AB of the e.m.f. AC is absorbed by the voltage drop $i R_f$ and the remainder part BC is available to overcome $L\, di/dt$. Since this surplus voltage is available, it is possible for the field current to increase above the value OA. However, at point D, the available voltage is OM and is all absorbed by $i R_f$ drop. Consequently, the field current cannot increase further and the generator build up stops.

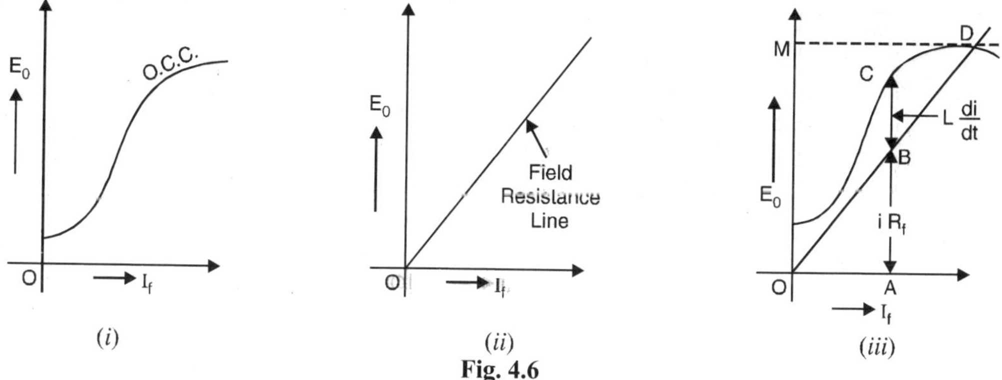

Fig. 4.6

We arrive at a very important conclusion that the voltage build up of the generator is given by the point of intersection of O.C.C. and field resistance line. Thus in Fig. 4.6 (*iii*), D is point of intersection of the two curves. Hence the generator will build up a voltage OM.

(*ii*) **Series generator.** During initial operation, with no current yet flowing, a residual voltage will be generated exactly as in the case of a shunt generator. The residual voltage will cause a current to flow through the whole series circuit when the circuit is closed. There will then be voltage build up to an equilibrium point exactly analogous to the build up of a shunt generator. The voltage build up graph will be similar to that of shunt generator except that now *load current* (instead of field current for shunt generator) will be taken along x-axis.

(*iii*) **Compound generator.** When a compound generator has its series field flux aiding its shunt field flux, the machine is said to be *cumulative compound.* When the series field is connected in reverse so that its field flux opposes the shunt field flux, the generator is then *differential compound.*

The easiest way to build up voltage in a compound generator is to start under no load conditions. At no load, only the shunt field is effective. When no-load voltage build up is achieved, the generator is loaded. If under load, the voltage rises, the series field connection is cumulative. If the voltage drops significantly, the connection is differential compound.

4.5 CRITICAL FIELD RESISTANCE FOR A SHUNT GENERATOR

We have seen above that voltage build up in a shunt generator depends upon field circuit resistance. If the field circuit resistance is R_1 (line OA), then generator will build up a voltage OM as shown in *Fig. 4.7. If the field circuit resistance is **increased to R_2 (line OB), the generator will build up a voltage OL, slightly less than OM. As the field circuit resistance is increased, the slope of resistance line also increases. When the field resistance line becomes tangent (line OC) to O.C.C., the generator would *just* excite. If the field circuit resistance is increased beyond this point (say line OD), the generator will ***fail to excite. The field circuit resistance represented by line OC (tangent to O.C.C.) is called *critical field resistance* R_C for the shunt generator. It may be defined as under :

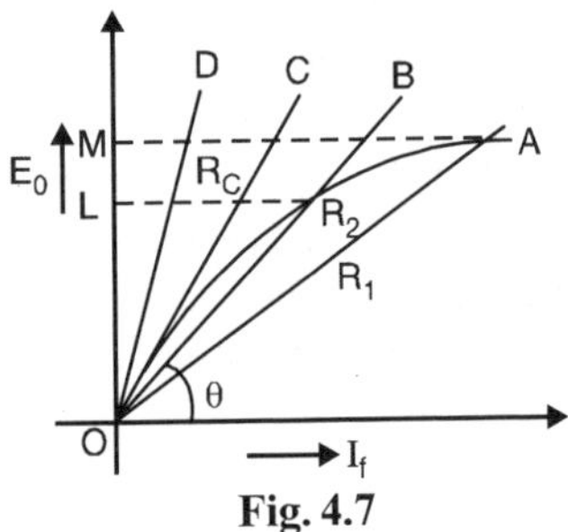

Fig. 4.7

The maximum field circuit resistance (for a given speed) with which the shunt generator would just excite is known as its **critical field resistance.**

It should be noted that shunt generator will build up voltage only if field circuit resistance is less than critical field resistance.

4.6 CRITICAL RESISTANCE FOR A SERIES GENERATOR

Fig. 4.8 shows the voltage build up in a series generator. Here R_1, R_2 etc. represent the total circuit resistance (load resistance and field winding resistance). If the total circuit resistance is R_1, then series generator will build up a voltage OL. The line OC is tangent to O.C.C. and represents the critical resistance R_C for a series generator. If the total resistance of the circuit is more than R_C (say line OD), the generator will fail to build up voltage. Note that Fig. 4.7 is similar to Fig. 4.7 with the following differences :

(*i*) In Fig. 4.7, R_1, R_2 etc. represent the total field circuit resistance. However, R_1, R_2 etc. in Fig. 4.8 represent the total circuit resistance (load resistance and series field winding resistance etc.).

(*ii*) In Fig. 4.7, field current alone is represented along X-axis. However, in Fig. 4.8 load current I_L is represented along X-axis. Note that in a series generator, field current = load current I_L.

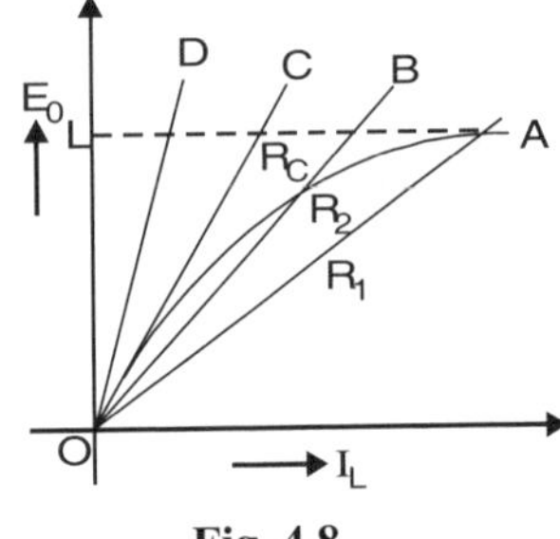

Fig. 4.8

4.7 CHARACTERISTICS OF SERIES GENERATOR

Fig. 4.9 (*i*) shows the connections of a series wound generator. Since there is only one current (that which flows through the whole machine), the load current is the same as the exciting current.

(*i*) **O.C.C.** Curve 1 shows the open circuit characteristic (O.C.C.) of a series generator. It can be obtained experimentally by disconnecting the field winding from the machine and exciting it from a separate d.c. source as discussed in Art. 4.2.

---

* The effect of residual magnetism has been neglected in drawing O.C.C. so that it starts from the origin.

** Resistance = slope of line = tan θ

Here θ is the angle which the line makes with x-axis. As θ increases, tan θ and hence resistance increases.

*** Since now all the available e.m.f. is immediately absorbed by $i\,R_f$ drop.

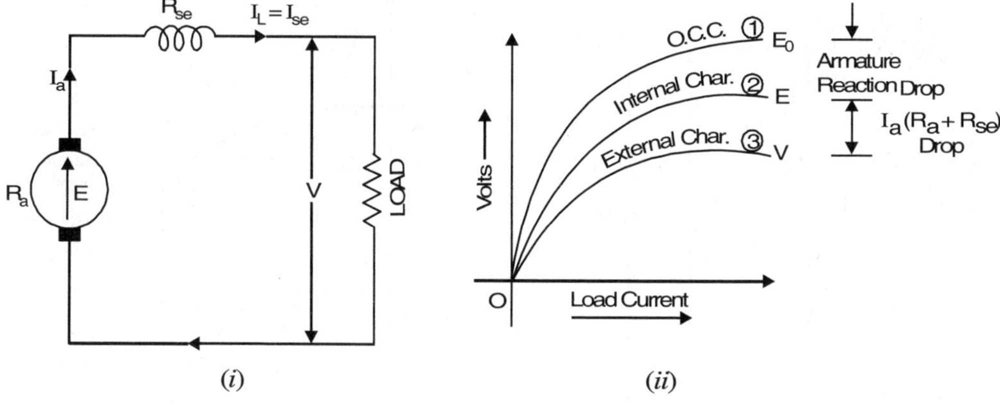

Fig. 4.9

(*ii*) **Internal characteristic.** Curve 2 shows the total or internal characteristic of a series generator. It gives the relation between the generated e.m.f. E on load and *armature current. Due to armature reaction, the flux in the machine will be less than the flux at no load. Hence, e.m.f. E generated under load conditions will be less than the e.m.f. E_0 generated under no load conditions. Consequently, internal characteristic curve lies below the O.C.C. curve ; the difference between them representing the effect of armature reaction [See Fig. 4.9 (*ii*)].

(*iii*) **External characteristic.** Curve 3 shows the external characteristic of a series generator. It gives the relation between terminal voltage V and load current I_L.

$$V = E - I_a (R_a + R_{se})$$

Therefore, external characteristic curve will lie below internal characteristic curve by an amount equal to ohmic drop [*i.e.*, $I_a (R_a + R_{se})$] in the machine as shown in Fig. 4.9 (*ii*).

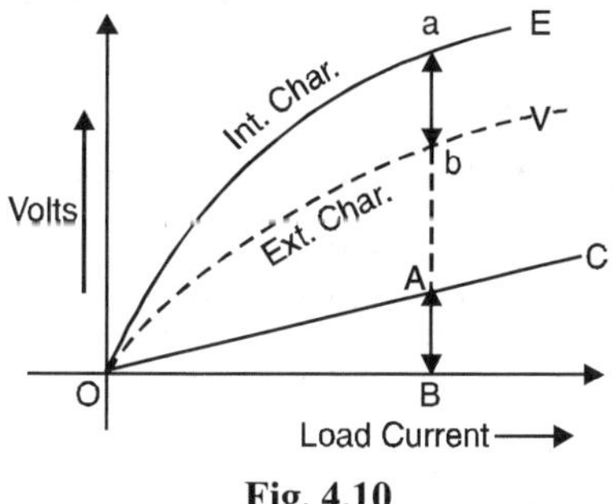

Fig. 4.10

The internal and external characteristics of a d.c. series generator can be plotted from one another as shown in Fig. 4.10. Suppose we are given the internal characteristic of the generator. Let the line OC represent the resistance of the whole machine *i.e.* $R_a + R_{se}$. If the load current is OB, drop in the machine is AB *i.e.*

$$AB = \text{Ohmic drop in the machine}$$
$$= OB\,(R_a + R_{se})$$

Now raise a perpendicular from point B and mark a point b on this line such that $ab = AB$. Then point b will lie on the external characteristic of the generator. Following similar procedure, other points of external characteristic can be located. It is easy to see that we can also plot internal characteristic from the external characteristic.

4.8 CHARACTERISTICS OF A SHUNT GENERATOR

Fig. 4.11 (*i*) shows the connections of a shunt wound generator. The armature current I_a splits up into two parts; a small fraction I_{sh} flowing through shunt field winding while the major part I_L goes to the external load.

(*i*) **O.C.C.** The O.C.C. of a shunt generator is similar in shape to that of a series generator as shown in Fig. 4.11 (*ii*). The line OA represents the shunt field circuit resistance. When the generator is run at normal speed, it will build up a voltage OM. At no-load, the terminal voltage of the generator will be constant (= OM) represented by the horizontal dotted line MC.

* In case of series generator, $I_a = I_{se} = I_L$.

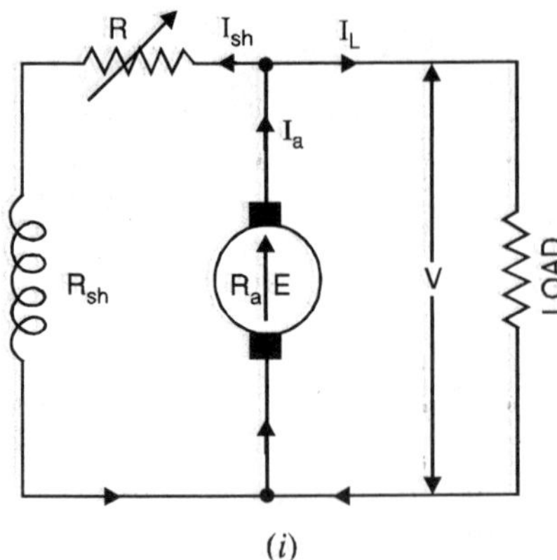

(i)

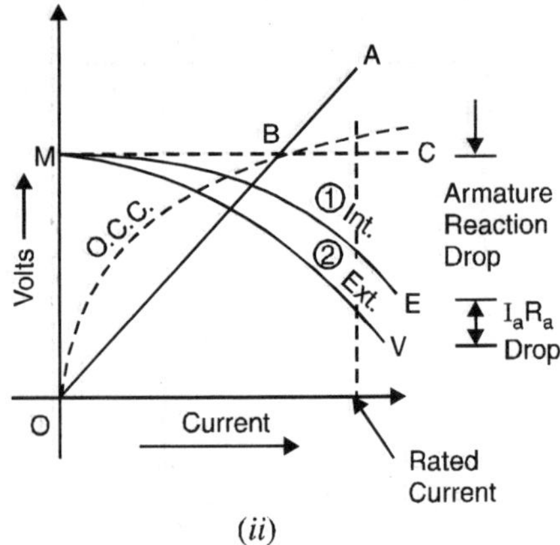

(ii)

Fig. 4.11

(*ii*) **Internal characteristic.** When the generator is loaded, flux per pole is reduced due to armature reaction. Therefore, e.m.f. E generated on load is less than the e.m.f. generated at no load. As a result, the internal characteristic *(E/I_a) drops down slightly as shown in Fig. 4.11 (*ii*).

(*iii*) **External characteristic.** Curve 2 shows the external characteristic of a shunt generator. It gives the relation between terminal voltage V and load current I_L.

$$V = E - I_a R_a$$
$$= E - (I_L + I_{sh}) R_a$$

Therefore, external characteristic curve will lie below the internal characteristic curve by an amount equal to drop in the armature circuit [*i.e.*, $(I_L + I_{sh}) R_a$] as shown in Fig. 4.11 (*ii*).

Note. It may be seen from the external characteristic that change in terminal voltage from no-load to full load is small. The terminal voltage can always be maintained constant by adjusting the field rheostat R automatically.

4.9 CRITICAL EXTERNAL RESISTANCE FOR SHUNT GENERATOR

If the load resistance across the terminals of a shunt generator is decreased, then load current increases. However, there is a limit to the increase in load current with the decrease of load resistance. Any decrease of load resistance beyond this point, instead of increasing the current, ultimately results in **reduced current. Consequently, the external characteristic turns back (dotted curve) as shown in Fig. 4.12. The tangent OA to the curve represents the minimum external resistance required to excite the shunt generator on load and is called *critical external resistance.* If the resistance of the external circuit is less than the critical external resistance (represented by tangent OA in Fig. 4.12), the machine will refuse to excite or will de-excite if already running. This means that external resistance is so low as virtually to short circuit the machine and so doing away with its excitation.

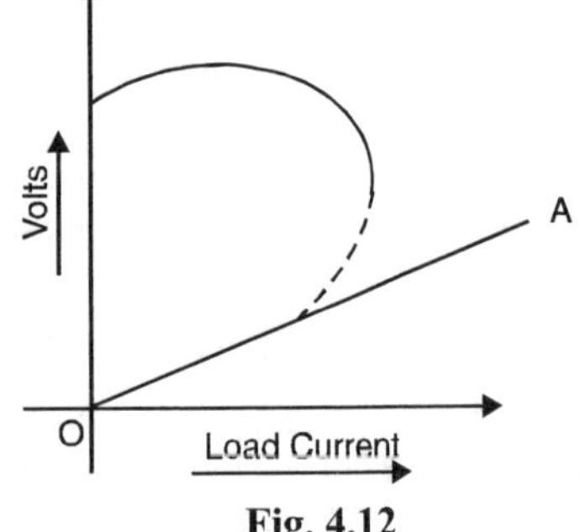

Fig. 4.12

* $I_a = I_L + I_{sh}$. As I_{sh} is small, therefore, $I_a \simeq I_L$.

** Actually on reducing load resistance, the current momentarily increases but this results in the reduction of terminal voltage due to (*i*) armature reaction effect and (*ii*) $I_a R_a$ drop. This decreases the field current and hence the machine flux. By the time steady conditions are obtained, the load current is actually less than before.

Note. There are two critical resistances for a shunt generator *viz*., (*i*) critical field resistance (*ii*) critical external resistance. For the shunt generator to build up voltage, the former should not be exceeded and the latter must not be gone below.

4.10 HOW TO DRAW O.C.C. AT DIFFERENT SPEEDS ?

If we are given O.C.C. of a generator at a constant speed N_1, then we can easily draw the O.C.C. at any other constant speed N_2. Fig. 4.13 illustrates the procedure. Here we are given O.C.C. at a constant speed N_1. It is desired to find the O.C.C. at constant speed N_2 (it is †assumed that $N_2 < N_1$). For constant excitation, $E \propto N$.

$$\therefore \quad \frac{E_2}{E_1} = \frac{N_2}{N_1}$$

or
$$E_2 = E_1 \times \frac{N_2}{N_1}$$

As shown in Fig. 4.13, for $I_f = OH$, $E_1 = HC$. Therefore, the new value of e.m.f. (E_2) for the same I_f but at N_2 is

$$E_2 = HC \times \frac{N_2}{N_1} = HD$$

Fig. 4.13

This locates the point D on the new O.C.C. at N_2. Similarly, other points can be located taking different values of I_f. The locus of these points will be the O.C.C. at N_2.

4.11 CRITICAL SPEED (N_C)

The critical speed of a shunt generator is the minimum speed below which it fails to excite. Clearly, it is the speed for which the given shunt field resistance represents the critical resistance. In Fig. 4.14, curve 2 corresponds to critical speed because the shunt field resistance (R_{sh}) line is tangential to it. If the generator runs at full speed N, the new O.C.C. moves upward and the R'_{sh} line represents critical resistance for this speed.

$$\therefore \quad \text{Speed} \propto \text{Critical resistance}$$

In order to find critical speed, take any convenient point C on excitation axis and erect a perpendicular so as to cut R_{sh} and R'_{sh} lines at points B and A respectively. Then,

$$\frac{BC}{AC} = \frac{N_C}{N}$$

or
$$N_C = N \times \frac{BC}{AC}$$

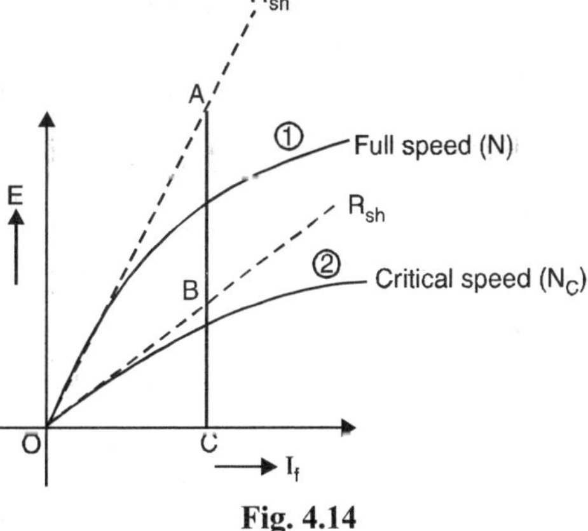

Fig. 4.14

4.12 CONDITIONS FOR VOLTAGE BUILD-UP OF A SHUNT GENERATOR

The necessary conditions for voltage build-up in a shunt generator are :

(*i*) There must be some residual magnetism in generator poles.

(*ii*) The connections of the field winding should be such that the field current strengthens the residual magnetism.

(*iii*) The resistance of the field circuit should be less than the critical resistance. In other words, the speed of the generator should be higher than the critical speed.

† The treatment is equally valid for $N_2 > N_1$.

Example 4.2. *The O.C.C. of a separately excited d.c. generator driven at 400 r.p.m. is as follows* :

Field current (A)	:	*2*	*3*	*4*	*5*	*6*	*7*	*8*	*9*
E.M.F. (V)	:	*110*	*155*	*186*	*212*	*230*	*246*	*260*	*271*

The machine is connected as a shunt generator and driven at 400 r.p.m. Find :

(i) The e.m.f. to which the machine will excite when the field circuit resistance is 34 Ω

(ii) The additional resistance required in the field circuit to reduce e.m.f. to 220 V

(iii) The critical value of shunt field resistance

(iv) The critical speed when the field circuit resistance is 34 Ω

(v) The lowest speed at which an e.m.f. of 225 V can be obtained with field circuit resistance of 34 Ω

Solution. If the e.m.f. for zero field current is not given, then curve is drawn from the origin. Plot the O.C.C. from the given data as shown in Fig. 4.15. Draw the 34 Ω-*line *OA* as shown.

(*i*) The field resistance line cuts O.C.C. at point *J*. Therefore, the voltage to which the machine will excite = *OM* = **255 V**.

(*ii*) From point *N* on the *y*-axis (*ON* = 220 V), draw a horizontal line to cut O.C.C. at point *B*. The resistance represented by line *OB* = 220/5·5 = 40 Ω.

∴ Additional resistance required = 40·0 – 34 = **6 Ω**

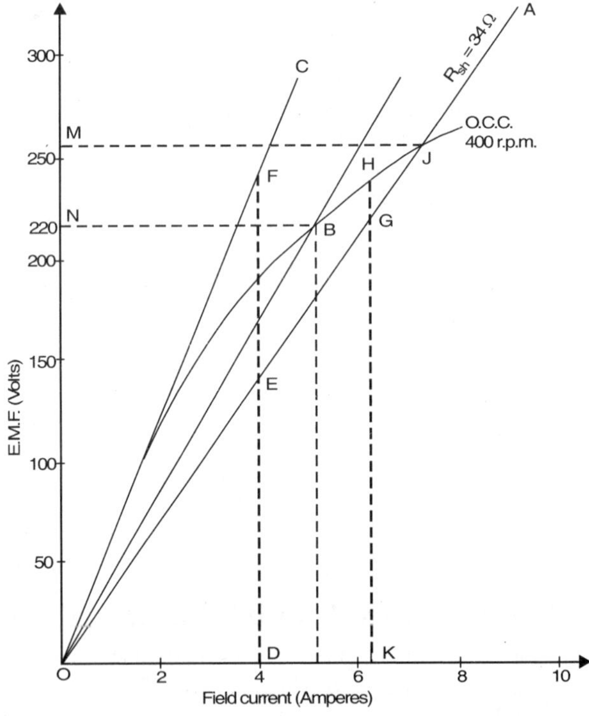

Fig. 4.15

* To draw a straight line, we need two points, one point can be the origin (0, 0) and the other any point such that volts/amperes = R_{sh} (= 34 Ω).

(*iii*) Draw the line *OC* tangential at the origin to O.C.C. This line represents the critical resistance.

$$\text{Critical resistance} = \text{Slope of line } OC = \frac{240}{4} = \mathbf{60\ \Omega}$$

(*iv*) At any point convenient (point *D* on *x*-axis), erect a perpendicular which cuts both *OA* and *OC*. Then,

$$\frac{DE}{DF} = \frac{N_C}{N}$$

or
$$N_C = \frac{DE}{DF} \times N = \frac{135}{240} \times 400 = \mathbf{225\ r.p.m.}$$

(*v*) From *G*(= 225 V), draw a perpendicular line *GK* cutting the O.C.C. at point *H*. If *N* is the lowest speed possible for getting 225 V with 34 Ω shunt circuit resistance, then,

$$\frac{GK}{HK} = \frac{N'}{400}$$

or
$$N' = 400 \times \frac{GK}{HK} = 400 \times \frac{225}{241} = \mathbf{373\ r.p.m.}$$

Example 4.3. *A shunt generator is to be converted into a level compounded generator by the addition of a series field winding. From the test on the machine with shunt excitation only, it is found that shunt current is 3·1 A to give 400 V on no load and 4·8 A to give the same voltage when the machine is supplying its full load of 200 A. The shunt winding has 1200 turns per pole. Find the number of series turns required per pole.*

Solution. Shunt field AT/pole required to produce 400 V on no-load = 3·1 × 1200 = 3720 AT

Shunt field AT/pole required to produce 400 V on full-load = 4·8 × 1200 = 5760 AT

Additional AT/pole required due to demagnetising effect of load current

$$= 5760 - 3720 = 2040 \text{ AT}$$

If *N* is the number of series turns required per pole when load current is 200 A, then,

$$N \times 200 = 2040 \quad \text{or} \quad N = \frac{2040}{200} = \mathbf{10{\cdot}2}$$

Example 4.4. *The following figures were obtained from an open circuit test on a shunt generator driven at 1000 r.p.m. :*

Field current (A) :	*1*	*1·5*	*2*	*2·5*	*3*
O.C. Voltage (V) :	*104*	*119*	*130*	*138*	*145*

If the field resistance is 50 Ω, find graphically :

(*i*) *The terminal voltage on open circuit when the speed is 1000 r.p.m.*

(*ii*) *The terminal voltage on open-circuit when the speed is 750 r.p.m.*

(*iii*) *The additional field resistance required so that the machine builds up to 89 V when the speed is 750 r.p.m.*

Solution. Plot the O.C.C. from the given data as shown in Fig. 4.16. The 50 Ω line *OA* is drawn as usual.

(*i*) The point where the resistance line *OA* intersects the O.C.C. gives the open circuit voltage of the generator at 1000 r.p.m.

$$\therefore \quad \text{Open circuit voltage} = OM = \mathbf{143\ V}$$

(*ii*)
$$E \propto \phi N$$

For constant excitation, $E \propto N$. Consider $I_f = 1\ A$. From the graph, the e.m.f. generated at 1000 r.p.m. is 104 V.

$$\therefore \quad \frac{E_{750}}{E_{1000}} = \frac{750}{1000}$$

$$\text{or} \qquad E_{750} = E_{1000} \times \frac{750}{1000} = 104 \times \frac{750}{1000} = 78 \text{ V}$$

Repeating this procedure for each field current, the following table is obtained :

Field current (A)	:	1	1·5	2	2·5	3
O.C. Voltage (V)	:	78	89	97·5	103·5	108·7

Draw the new O.C.C. from this data. This gives the O.C.C. at 750 r.p.m. The point where the 50 Ω line intersects this O.C.C. gives the open circuit voltage at 750 r.p.m.

Open-circuit voltage $= OD =$ **96 V**

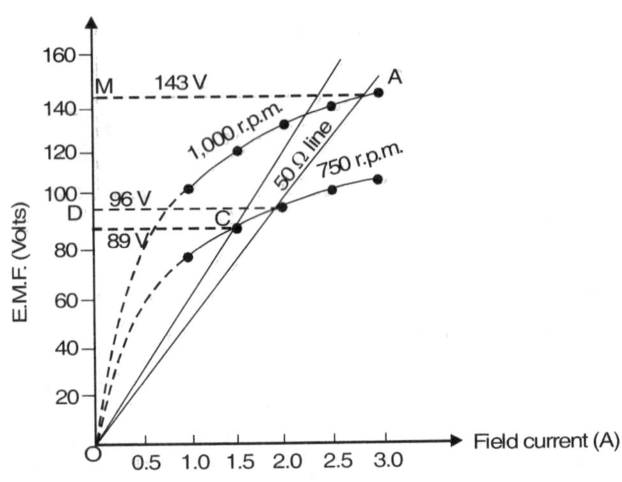

Fig. 4.16

(*iii*) Locate the point C on the O.C.C. for 750 r.p.m. corresponding to 89 V. The corresponding field current is 1·5 A.

$$\therefore \text{ Field circuit resistance } = \frac{89}{1 \cdot 5} = 59 \cdot 3\ \Omega$$

Additional field resistance required

$$= 59 \cdot 3 - 50 = \mathbf{9 \cdot 3\ \Omega}$$

Example 4.5. *A shunt generator gave the following open-circuit characteristic :*

Field current (A)	:	*0.5*	*1*	*1·5*	*2*	*2·5*	*3*	*3·5*
O.C. e.m.f. (V)	:	*54*	*107*	*152*	*185*	*210*	*230*	*245*

The armature and field resistances are 0·1 Ω and 80 Ω respectively. Calculate :

(*i*) *The voltage to which the machine will excite when run as a shunt generator at the same speed*

(*ii*) *The volts lost due to armature reaction when 100 A are passing in the armature at a terminal voltage of 175 V*

(*iii*) *The percentage reduction in speed for the machine to fail to excite on open circuit*

Solution. Plot the O.C.C. from the given data as shown in Fig. 4.17. The 80 Ω line OA is drawn as usual.

(*i*) The point where the resistance line OA intersects the O.C.C. represents the voltage to which the generator will build up.

∴ O.C. voltage to which the generator builds up $= OM =$ **222 V**

(*ii*) When the terminal voltage of the generator is 175 V, then,

$$\text{Shunt field current, } I_{sh} = \frac{175}{R_{sh}} = \frac{175}{80} = 2 \cdot 2 \text{ A}$$

Corresponding to this field current (*i.e.*, 2·2 A), the open circuit voltage is $OC = 195$ V.

Voltage lost due to armature reaction and armature drop

$$= 195 - 175 = 20\text{ V}$$

$$\text{Armature drop} = I_a R_a = 100 \times 0{\cdot}1 = 10\text{ V}$$

$\therefore$ Volts lost due to armature reaction = 20 – 10 = **10 V**

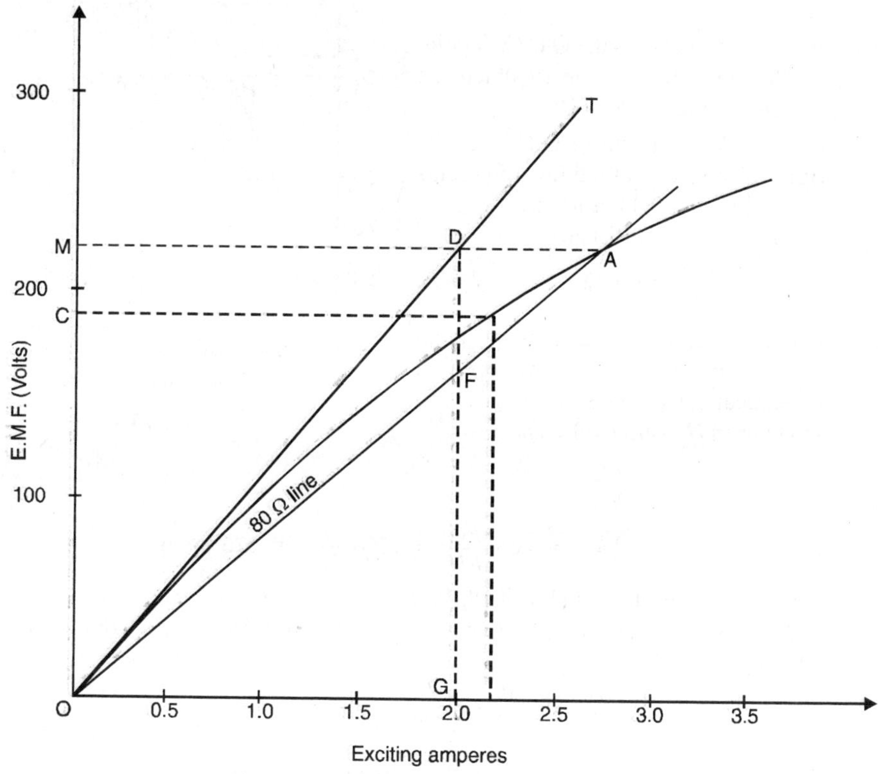

Fig. 4.17

(*iii*) Draw line *OT* tangential to O.C.C. at the origin. At any convenient point *G* on *x*-axis, erect a perpendicular *DFG*. Then,

$$\frac{N_C}{N} = \frac{FG}{DG}$$

or $$\frac{N - N_C}{N} = \frac{DG - FG}{DG}$$

or $$\frac{N - N_C}{N} = \frac{220 - 160}{220} = 0{\cdot}273$$

$\therefore$ Percentage reduction in speed $= 0{\cdot}273 \times 100 =$ **27·3%**

Example 4.6. *A d.c. shunt generator gave the following open-circuit characteristic when driven at 750 r.p.m.*

Field Current (A)	:	*0.5*	*1*	*1.5*	*2*	*2.5*
E.M.F. (V)	:	*50*	*84*	*105*	*120*	*131*

(*i*) *If the machine is run as a shunt generator at 750 r.p.m., to what voltage will it excite with shunt field resistance equal to 70 Ω ?*

(*ii*) *What is the critical value of shunt field resistance ?*

(*iii*) *What is the critical speed when the shunt field resistance is 70 Ω ?*

(iv) *With shunt field resistance equal to 55 Ω, what reduction in speed be made to make the open-circuit voltage equal to 100 V ?*

Solution. Plot the O.C.C. from the given data as shown in Fig. 4.18. The line *OA* represents 70 Ω line.

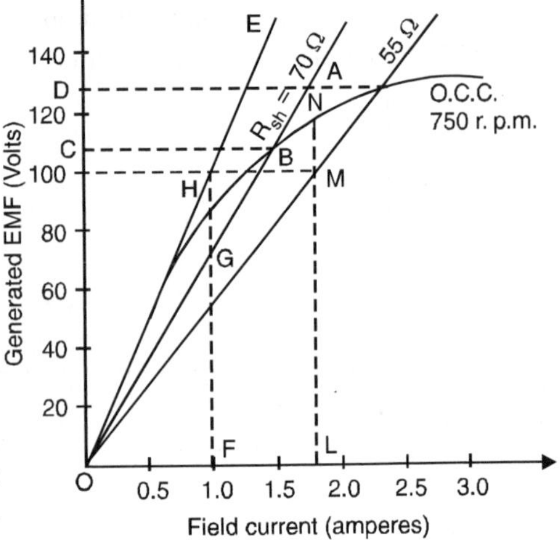

Fig. 4.18

(*i*) The line *OA* intersects the O.C.C. at point *B*. Therefore, the voltage to which the generator excites = *OC* = **105 V.**

(*ii*) Draw line *OE* tangent to O.C.C. at the origin. Then slope of the line represents the critical shunt field resistance.

∴ Critical shunt field resistance

$$= \frac{100\ V}{1\ A} = \mathbf{100\ \Omega}$$

(*iii*) To find critical speed, take any convenient point *F* on the *x*-axis and erect a perpendicular which cuts the lines *OA* and *OE* at *G* and *H* respectively. Then,

$$\frac{FG}{FH} = \frac{N_C}{N}$$

or $$N_C = N \times \frac{FG}{FH} = 750 \times \frac{70}{100} = \mathbf{525\ r.p.m.}$$

(*iv*) Draw 55 Ω line as shown in Fig. 4.18. With 55 Ω shunt field resistance, open-circuit voltage = *OD* = 128 V. To generate 100 V with field resistance of 55 Ω, the operating point should be at *M*. If *N′* is the speed, then,

$$\frac{LM}{LN} = \frac{N'}{N}$$

or $$N' = N \times \frac{LM}{LN} = 750 \times \frac{100}{115} = 652 \text{ r.p.m.}$$

∴ Reduction in speed = 750 – 652 = **98 r.p.m.**

Example 4.7. *The magnetisation characteristic for a 4-pole, 110 V, 1000 r.p.m. shunt generator is as follows :*

Field current (A)	:	*0*	*0.5*	*1*	*1.5*	*2*	*2.5*	*3*
O.C. voltage (V)	:	*5*	*50*	*85*	*102*	*112*	*126*	*120*

The armature is lap connected with 144 conductors. Field resistance is 45 Ω. Determine the residual flux per pole.

Solution. There is no need to draw the O.C.C. The answer can be obtained from the table. When the field current is zero, the induced e.m.f. of 5 V is due to the residual magnetism of generator poles.

$$E_g = \frac{P\,\phi\,Z\,N}{60\ A}$$

Here E_g = 5V ; *P* = 4 ; *Z* = 144 ; *N* = 1000 r.p.m. ; *A* = *P* = 4

$$\therefore \quad 5 = \frac{4 \times \phi \times 144 \times 1000}{60 \times 4}$$

or $$\phi = 2{\cdot}08 \times 10^{-3}\ \text{Wb} = \mathbf{2{\cdot}08\ mWb}$$

TUTORIAL PROBLEMS

1. The saturation curve of a 50 kW, 220 V shunt generator driven at 800 r.p.m. is as follows :

Field current (*A*)	:	0	0·5	1	2	3	4	5
E.M.F. (*V*)	:	10	50	100	175	220	245	262

(*i*) What value of field resistance will give on open circuit 220 V at 800 r.p.m. ?
(*ii*) What is the critical resistance of the shunt field at 800 r.p.m. ?
(*iii*) To what voltage the machine will excite when the field resistance is 55 Ω and the speed 800 r.p.m. ?
[(*i*) 73 Ω (*ii*) 100 Ω (*iii*) 257 V]

2. The relationship between the field current and open circuit voltage of a d.c. shunt generator driven at constant speed is as under :

Field current (A)	:	0·5	1	1·5	2	2·5	3	3·5
E.M.F. (*V*)	:	60	120	138	145	149	151	152

The resistance of the armature circuit is 0·1 Ω. If the field resistance is adjusted to 53 Ω and the generator is run at the same speed as before, calculate :
(*i*) The open circuit voltage
(*ii*) The load current when the terminal voltage is 100 V
(*iii*) The terminal voltage when the load has a resistance of 0·4 Ω
Neglect the effect of armature reaction. **[(*i*) 150 V (*ii*) 428·9 A (*iii*) 117 V]**

3. The magnetisation curve of a d.c. shunt generator driven at 800 r.p.m. is as under :

Field current (A)	:	0	0·5	1	2	3	4	5
E.M.F. (*V*)	:	10	60	120	199	232	248	258

If the resistance of the shunt field is 50 Ω, determine the additional resistance required in the shunt field circuit to give 240 V at a speed of 1000 r.p.m. **[64·3 Ω]**

4. The O.C.C. of a separately excited generator driven at 500 r.p.m. is as follows :

Field current (A)	:	0.2	0·4	0.6	0.8	1	1.2	1.4	1.6
E.M.F. (*V*)	:	40	66	86	101	112	121	128	133

The machine is connected as a shunt generator and driven at 500 r.p.m. Find (*i*) the open circuit voltage when the field circuit resistance is 94 Ω (*ii*) the additional resistance required in the field circuit to reduce the e.m.f. to 110 V (*iii*) critical value of shunt field resistance.
[(*i*) 126 V (*ii*) 22·7 Ω (*iii*) 200 Ω]

5. The magnetisation characteristic of a shunt generator driven at 1000 r.p.m. is as follows :

Field current (*A*)	:	0	0·5	1	1·5	2	2·5	3
O.C. voltage (*V*)	:	5	50	85	102	112	116	120

The field resistance is 45 Ω. Determine (*i*) voltage the machine will build up at no load (*ii*) the critical field resistance (*iii*) the critical speed. **[(*i*) 118 V (*ii*) 100 Ω (*iii*) 445 r.p.m.]**

4.13 COMPOUND GENERATOR CHARACTERISTICS

In a compound generator, both series and shunt excitation are combined as shown in Fig. 4.19. The shunt winding can be connected either across the armature only (*short-shunt connection S*) or across armature plus series field (*long-shunt connection G*). The compound generator can be *cumulatively compounded or differentially compounded generator. The latter is rarely used in practice. Therefore, we shall discuss the characteristics of cumulatively-compounded generator. It may be noted that external characteristics of long and short shunt compound generators are almost identical.

External characteristic. Fig. 4.20 shows the external characteristics of a cumulatively compounded generator. The series excitation aids the shunt excitation. The degree of compounding depends upon the increase in series excitation with the increase in load current.

* (*i*) If the series and shunt excitation help each other, it is called cumulatively-compounded generator.
(*ii*) If shunt and series excitation oppose each other, it is called differentially-compounded generator.

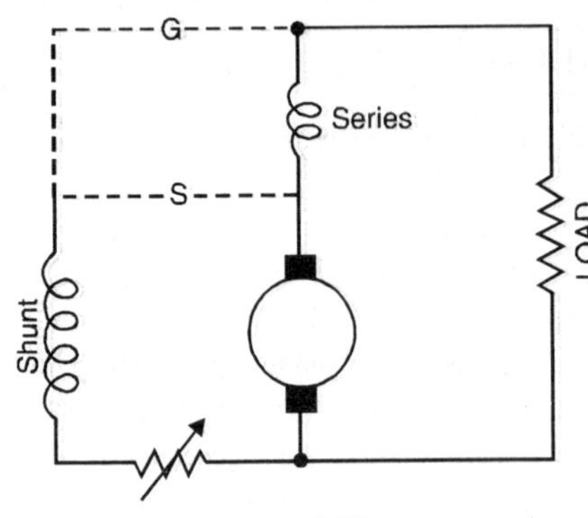

Fig. 4.19

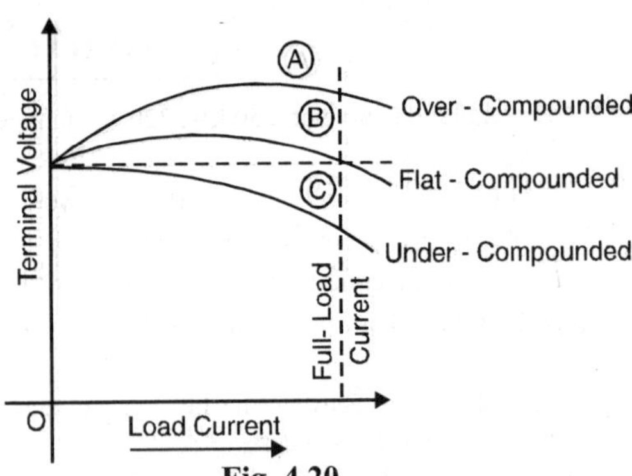

Fig. 4.20

(*i*) If series winding turns are so adjusted that with the increase in load current the terminal voltage increases, it is called *over-compounded generator*. In such a case, as the load current increases, the series field m.m.f. increases and tends to increase the flux and hence the generated voltage. The increase in generated voltage is greater than the $I_a R_a$ drop so that instead of decreasing, the terminal voltage increases as shown by curve A in Fig. 4.20.

(*ii*) If series winding turns are so adjusted that with the increase in load current, the terminal voltage substantially remains constant, it is called *flat-compounded generator*. The series winding of such a machine has lesser number of turns than the one in over-compounded machine and, therefore, does not increase the flux as much for a given load current. Consequently, the full-load voltage is nearly equal to the no-load voltage as indicated by curve B in Fig. 4.20.

(*iii*) If series field winding has lesser number of turns than for a flat-compounded machine, the terminal voltage falls with increase in load current as indicated by curve C in Fig. 4.20. Such a machine is called *under-compounded generator*.

4.14 VOLTAGE REGULATION

The change in terminal voltage of a generator between full and no load (at constant speed) is called the *voltage regulation,* usually expressed as a percentage of the voltage at full-load.

$$\%\text{ Voltage regulation} = \frac{V_{NL} - V_{FL}}{V_{FL}} \times 100$$

where V_{NL} = Terminal voltage of generator at no load

V_{FL} = Terminal voltage of generator at full load

Note that voltage regulation of a generator is determined with field circuit and speed held constant. If the voltage regulation of a generator is 10%, it means that terminal voltage increases 10% as the load is changed from full load to no load.

Note. For an ideal generator (*i.e.* constant-voltage generator), the voltage regulation is zero. A positive voltage regulation means that terminal voltage at no load is higher than at full load. On the other hand, a negative voltage regulation means that the terminal voltage at full load is higher than at no load.

Example 4.8. *A shunt generator is rated at 240 V full load. At no load, the voltage is 252 V. What is the percentage voltage regulation ?*

Solution.

$$\%\text{ Voltage regulation} = \frac{V_{NL} - V_{FL}}{V_{FL}} \times 100 = \frac{252 - 240}{240} \times 100 = \mathbf{5\%}$$

Note. If $V_{NL} - V_{FL}$ is small, the generator is said to have good regulation. On the other hand, if this difference is large, the regulation is poor.

Example 4.9. *A shunt generator is converted into a short-shunt compound generator by the addition of a series field winding. From the test on the machine with shunt excitation only, it is found that a field current of 5A gives 440 V on no-load and that 6A gives 440 V at full load current of 200 A. The shunt winding has 1600 turns per pole. Find the number of series turns required per pole.*

Solution. Shunt field AT/pole required to produce 440 V at no load $= 5 \times 1600 = 8000$ AT

Shunt field AT/pole required to produce 440 V on load $= 6 \times 1600 = 9600$ AT

Increase in AT/pole $= 9600 - 8000 = 1600$ AT

The same increase in AT/pole can be obtained by adding N series turns/pole.

$\therefore \quad N \times 200 = 1600$ or $N =$ **8 turns/pole**

Example 4.10. *A 60 kW d.c. shunt generator has 1600 turns/pole in its shunt field winding. A shunt field current of 1.25 A is required to generate 125 V at no-load and 1.75 A to generate 150 V at full load. Calculate :*

(i) The minimum number of series turns/pole needed to produce the required no-load and full-load voltages as a short-shunt compound generator

(ii) If the generator is equipped with 3 series turns per pole having a resistance of 0·02 Ω, calculate the diverter resistance required to produce the desired compounding

(iii) Voltage regulation of the compound generator

Solution.

(*i*) Extra shunt field *AT* required $= 1600\,(1{\cdot}75 - 1{\cdot}25) = 800$

$$I_{se} = I_L = \frac{60 \times 10^3}{150} = 400 \text{ A}$$

No. of series turns/pole required $= \dfrac{800}{I_{se}} = \dfrac{800}{400} = \mathbf{2}$

(*ii*) Fig. 4.21 shows the generator connections. The actual number of series turns per pole is 3. Therefore, series current required is

$$I'_{se} \times 3 = 800$$

or $\quad I'_{se} = 800/3$ A

Current in diverter, $I_d = 400 - 800/3 = 400/3$ A

The voltages across series winding and the diverter are equal.

$$\therefore \quad 0{\cdot}02 \times I'_{se} = I_d R_d$$

or
$$R_d = \frac{0{\cdot}02 \times I'_{se}}{I_d}$$

$$= \frac{0{\cdot}02 \times 800/3}{400/3} = \mathbf{0{\cdot}04\ \Omega}$$

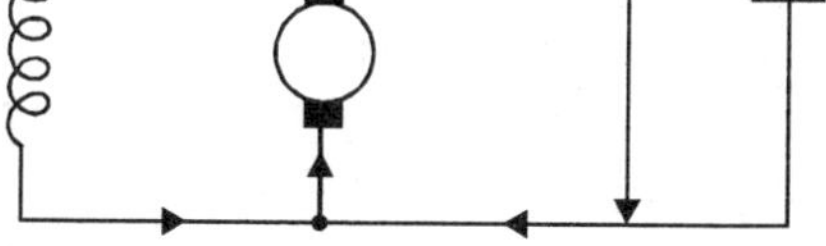

Fig. 4.21

(*iii*) Voltage regulation $= \dfrac{V_{NL} - V_{FL}}{V_{FL}} \times 100 = \dfrac{125 - 150}{150} \times 100 = \mathbf{-16{\cdot}7\%}$

Example 4.11. *In a 110 V compound generator, the resistances of the armature, shunt and series windings are 0·06 Ω, 25 Ω and 0·04 Ω respectively. The load consists of 200 lamps each rated at 55 W, 110 V. Find the e.m.f. generated and armature current when the machine is connected (i) long shunt (ii) short shunt (iii) How will the ampere-turns of series winding be changed if in (i) a diverter of resistance 0·1 Ω be connected in parallel with the series winding ? Ignore armature reaction and brush contact drop.*

Solution.

(*i*) **Long-shunt connection** [See Fig. 4.22 (*i*)]

$$\text{Load current, } I_L = \frac{200 \times 55}{110} = 100 \text{ A}$$

$$\text{Shunt field current, } I_{sh} = \frac{110}{25} = 4{\cdot}4 \text{ A}$$

$$\text{Armature current, } I_a = I_L + I_{sh} = 100 + 4{\cdot}4 = \mathbf{104{\cdot}4\ A}$$

$$\text{Generated e.m.f., } E_g = 110 + I_a (R_a + R_{se})$$

$$= 110 + 104{\cdot}4\,(0{\cdot}06 + 0{\cdot}04) = \mathbf{120{\cdot}4\ V}$$

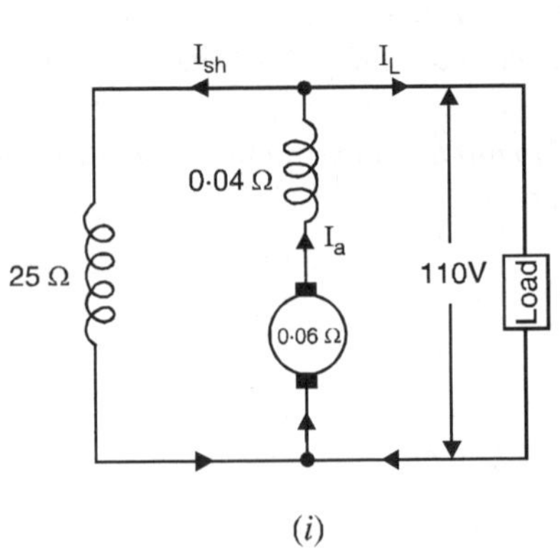

(*i*)

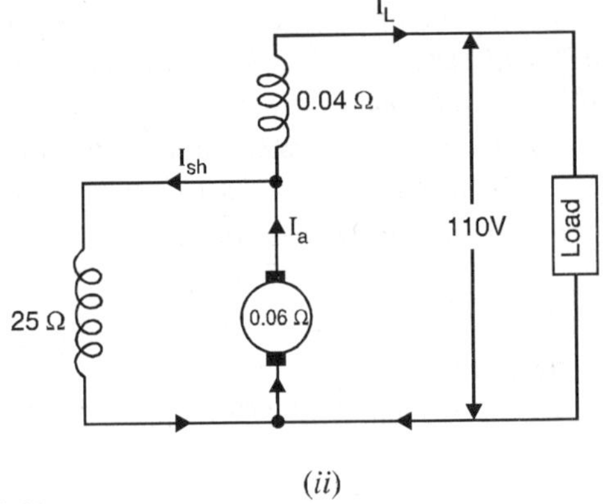

(*ii*)

Fig. 4.22

(*ii*) **Short-shunt connection** [See Fig. 4.22 (*ii*)]

$$I_L = I_{se} = 100 \text{ A} \qquad \text{(as calculated above)}$$

$$\text{Voltage across shunt winding, } V' = 110 + I_{se} R_{se} = 110 + 100 \times 0{\cdot}04 = 114 \text{ V}$$

$$\therefore \quad I_{sh} = \frac{114}{25} = 4{\cdot}56 \text{ A}$$

$$\text{Armature current, } I_a = I_L + I_{sh} = 100 + 4{\cdot}56 = \mathbf{104{\cdot}56\ A}$$

$$\text{Generated e.m.f., } E_g = V' + I_a R_a = 114 + 104{\cdot}56 \times 0{\cdot}06 = \mathbf{120{\cdot}3\ V}$$

(*iii*) Fig. 4.23 shows a diverter of resistance 0·1 Ω connected in parallel with the series winding of long-shunt connection. As calculated in part (*i*), I_a = 104·4 A. The current through series winding (current divider rule) is

$$I'_{se} = 104{\cdot}4 \times \frac{0{\cdot}1}{0{\cdot}04 + 0{\cdot}1} = 74{\cdot}57 \text{ A}$$

Reduction of current in series winding

$$= \frac{104{\cdot}4 - 74{\cdot}57}{104{\cdot}4} \times 100 = 28{\cdot}57\%$$

Since series ampere-turns are proportional to series current (turns are same),

Reduction in series ampere-turns = **28·57 %**

Fig. 4.23

Example 4.12. *A long shunt compound generator has a shunt field winding of 1000 turns per pole and series field winding of 4 turns per pole and resistance of 0·05 Ω. In order to obtain the rated voltage both at no-load and full load as shunt generator, it is necessary to increase field current by 0·2 A. The full-load armature current of the compound generator is 80 A. Calculate the diverter resistance connected in parallel with series field to obtain flat compound operation.*

Solution. Fig. 4.24 shows the generator connections. Additional shunt field AT required to maintain rated voltage at no-load and full-load

$$= 1000 \times 0{\cdot}2 = 200 \text{ AT}$$

No. of series turns/pole = 4

Current required in series winding to produce 200 AT

$$= \frac{200}{4} = 50 \text{ A}$$

It is given that armature current $I_a = 80$ A. Therefore, 80 – 50 = 30 A must pass through diverter resistance.

Now $\quad 0{\cdot}05 \times 50 = 30 \times R_d$

$$\therefore \quad R_d = \frac{0{\cdot}05 \times 50}{30} = \mathbf{0{\cdot}0833\ \Omega}$$

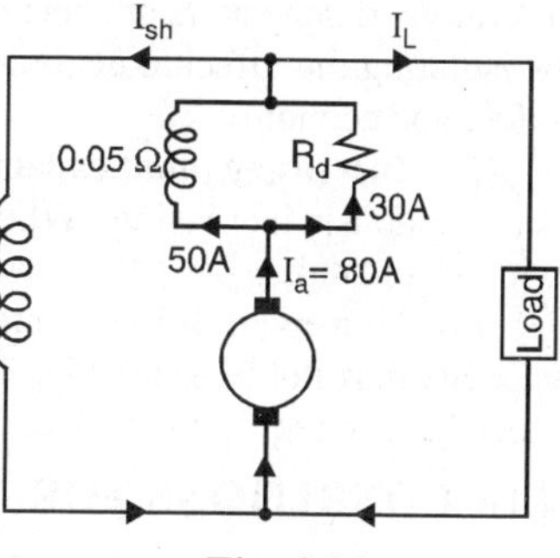

Fig. 4.24

Example 4.13. *A shunt generator is to be compounded so that the no-load terminal voltage is equal to that at full-load when the output is 400 A. Number of shunt turns/pole = 640. With series field cut out and shunt field excited from an external source, it is found that this can be accomplished by increasing the field current from 4·2 A at no-load to 6·2 A at full-load. Long shunt connections are used.*

(*i*) *How many series turns are required ?*

(*ii*) *If the generator is equipped with 5.5 turns, find the ratio of diverter to series field resistance. Neglect series AT at no-load and the drop in the series winding at full-load.*

Solution.

(*i*) Additional shunt field AT required to maintain rated voltage both at no-load and full-load

$$= (6{\cdot}2 - 4{\cdot}2) \times 640 = 1280 \text{ AT}$$

∴ Number of series turns required to produce 1280 AT

$$= \frac{1280}{400} = \mathbf{3{\cdot}2}$$

(*ii*) Now the actual number of series turns is 5.5. Current required to produce 1280 AT by series winding is [See Fig. 4.25]

$$5{\cdot}5 \times I_{se} = 1280$$

$$\therefore \quad I_{se} = 1280/5.5 = 232{\cdot}7 \text{ A}$$

Current thro' diverter, $I_d = 400 - 232{\cdot}7 = 167{\cdot}3$ A

Now $\quad I_d \times R_d = I_{se} \times R_{se}$

$$\therefore \quad \frac{R_d}{R_{se}} = \frac{I_{se}}{I_d} = \frac{232{\cdot}7}{167{\cdot}3} = \mathbf{1{\cdot}39 : 1}$$

Fig. 4.25

4.15 NEED FOR PARALLEL OPERATION OF D.C. GENERATORS

In a d.c. power plant, power is usually supplied from several generators of small ratings connected in parallel instead of from one large generator. This is due to the following reasons :

(*i*) **Continuity of service.** If a single large generator is used in the power plant, then in case of its breakdown, the whole plant will be shut down. However, if power is supplied from a number of small units operating in parallel, then in case of failure of one unit, the continuity of supply can be maintained by other healthy units.

(*ii*) **Efficiency.** Generators run most efficiently when loaded to their rated capacity. Electric power costs less per kWh when the generator producing it is efficiently loaded. Therefore, when load demand on power plant decreases, one or more generators can be shut down and the remaining units can be efficiently loaded.

(*iii*) **Maintenance and repair.** Generators generally require routine maintenance and repair. Therefore, if generators are operated in parallel, the routine or emergency operations can be performed by isolating the affected generator while load is being supplied by other units. This leads to both safety and economy.

(*iv*) **Increasing plant capacity.** In the modern world of increasing population, the use of electricity is continuously increasing. When *added capacity* is required, the new unit can be simply paralleled with the old units.

(*v*) **Non-availability of single large unit.** In many situations, a single unit of desired large capacity may not be available. In that case a number of smaller units can be operated in parallel to meet the load requirement. Generally a single large unit is more expensive.

4.16 CONDITIONS FOR PARALLEL OPERATION OF SHUNT GENERATORS

There are two main conditions to be met when d.c. generators (shunt, compound or series generators) are to be paralleled *viz.*

(*i*) *All polarities must be correct.* The positive terminals of all generators must be connected to the positive bus bar and the negative terminals must be connected to the negative bus bar.

(*ii*) *The terminal voltages of the generators being paralleled must be the same (ideally).*

Once shunt generators are put in parallel and adjusted so that they share the load satisfactorily, they will continue to operate properly even if the load changes. The natural tendency is for the terminal voltage of a shunt generator to fall when the load increases. This property tends to ensure stable operation.

4.17 CONNECTING SHUNT GENERATORS IN PARALLEL

The generators in a power plant are connected in parallel through bus-bars. The bus-bars are heavy thick copper bars and they act as +ve and –ve terminals. The positive terminals of the generators are connected to the +ve side of bus-bars and negative terminals to the negative side of bus-bars.

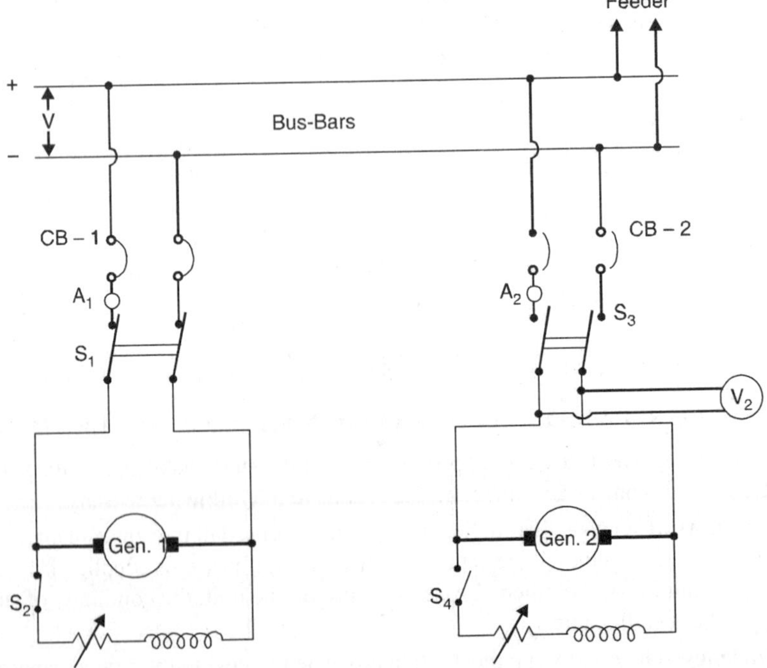

Fig. 4.26

Fig. 4.26 shows shunt generator G_1 connected to the bus-bars and supplying load. When the load on the power plant increases beyond the capacity of this generator, the second shunt generator G_2 is connected in parallel with the first to meet the increased load demand. The procedure for paralleling generator G_2 with generator G_1 is as under :

(*i*) The prime mover of generator G_2 is brought up to the rated speed. Now switch S_4 in the field circuit of the generator G_2 is closed.

(*ii*) Next circuit breaker *CB*-2 is closed and the excitation of generator G_2 is adjusted till it generates voltage equal to the bus-bars voltage. This is indicated by voltmeter V_2.

(*iii*) Now the generator G_2 is ready to be paralleled with generator G_1. The main switch S_3 is closed, thus putting generator G_2 in parallel with generator G_1. Note that generator G_2 is not supplying any load because its *generated e.m.f.* is equal to bus-bars voltage. The generator is said to be "*floating*" (*i.e.*, not supplying any load) on the bus-bars.

(*iv*) If generator G_2 is to deliver any current, then its generated voltage E should be greater than the bus-bars voltage V. In that case, current supplied by it is $I = (E - V)/R_a$ where R_a is the resistance of the armature circuit. By increasing the field current (and hence induced e.m.f. E), the generator G_2 can be made to supply proper amount of load.

(*v*) The load may be shifted from one shunt generator to another merely by adjusting the field excitation. Thus if generator G_1 is to be shut down, the whole load can be shifted onto generator G_2 provided it has the capacity to supply that load. In that case, reduce the current supplied by generator G_1 to zero (This will be indicated by ammeter A_1), open *CB*-1 and then open the main switch S_1.

4.18 LOAD SHARING BY SHUNT GENERATORS IN PARALLEL

The load sharing between shunt generators in parallel can be easily regulated because of their *drooping characteristics. The load may be shifted from one generator to another merely by adjusting the field excitation. Let us discuss the load sharing of two generators which have unequal no-load voltages.

Let E_1, E_2 = no-load voltages of the two generators

R_1, R_2 = their armature resistances

V = common terminal voltage (Bus-bars voltage)

Then,

$$I_1 = \frac{E_1 - V}{R_1} \quad \text{and} \quad I_2 = \frac{E_2 - V}{R_2}$$

Thus current output of the generators depends upon the values of E_1 and E_2. These values may be changed by field rheostats. The common terminal voltage (or bus-bars voltage) will depend upon (*i*) the e.m.f.s of individual generators and (*ii*) the total load current supplied. It is generally desired to keep the bus-bars voltage constant. This can be achieved by adjusting the field excitations of the generators operating in parallel.

4.19 PARALLEL OPERATION OF COMPOUND GENERATORS

The compound generators will not have stable operation if we just meet the requirements of correct polarity and equal terminal voltages. This is illustrated in Fig. 4.27. We shall now show that connections of Fig. 4.27 lead to unstable parallel operation of compound generators.

Suppose for some reason, the speed of one of the machines say generator G_1 increases momentarily.

* With drooping characteristics, any disturbance on the electrical system that might cause the load carried by any one machine to increase will immediately decrease the terminal voltage of the machine. This will destroy the steady-state equilibrium conditions and immediately result in this machine drooping some of its load. If a generator has a rising external characteristic, any disturbance which might cause its load to increase would tend to increase its terminal voltage. This would cause the machine to take even more load. This increased load would result in further voltage increase. The reactions are cumulative and, therefore, result in instability.

This will momentarily increase the generated e.m.f. of G_1 and this generator will supply more share of load current. This increase in armature current of G_1, in turn, increases the current in the series field winding of G_1. As a result, the generated e.m.f. of G_1 further increases, causing G_1 to supply more share of load current. At the same time, the armature current supplied by generator G_2 is decreased, thereby decreasing current in its series field winding and causing it to supply smaller and smaller portion of load current. This process continues until generator G_1 is supplying the entire load current and is attempting to operate machine G_2 as a motor. This causes the circuit breakers to open, disconnecting the generators from the load.

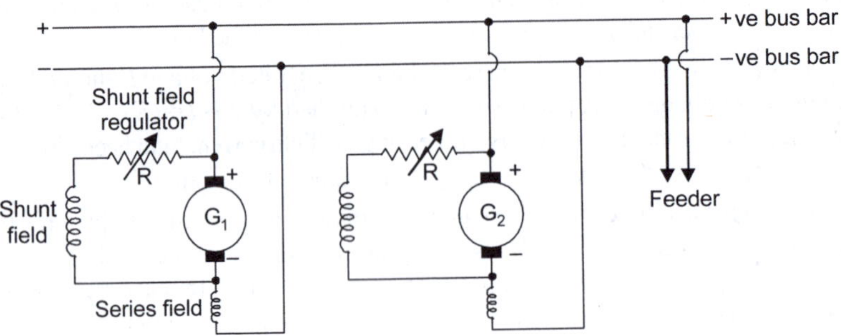

Fig. 4.27

This instability of the compound generators in parallel is overcome by using an equaliser connection as shown in Fig. 4.28. The equaliser is a bus bar that connects the two series field coils in parallel. Under this condition, the voltage across the series field coils of the two machines and their currents remain the same. As a result, the series field currents of the two machines remain constant regardless of the distribution of current in the two armatures. Thus the flux produced by the two series fields remains the same for a given load current on the bus.

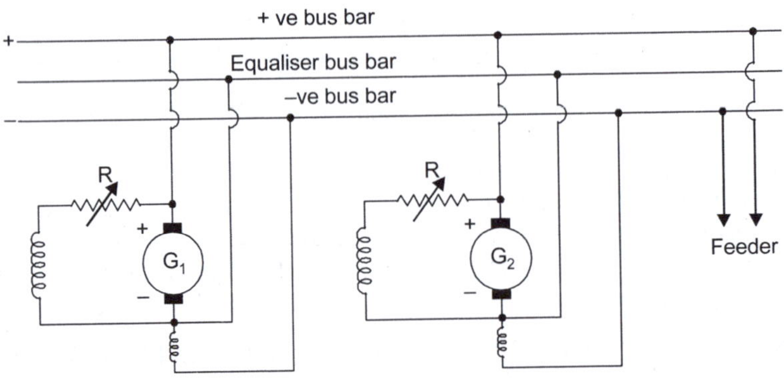

Fig. 4.28

Thus for stable parallel operation of compound generators, in addition to the requirments of correct polarity and equal terminal voltages, they require an equaliser connection before being paralleled. Compound generators are put in service and taken out of service in the same manner as shunt generators. Load is shifted from one compound generator to the other by the adjustment of shunt–field rheostats.

Note. In connecting a compound generator on to the bus bars, it is necessary that first series field windings have equaliser connection. Then shunt field of the incoming machine is adjusted so that the voltage generated is exactly equal to the bus bar voltage. The incoming machine is now connected to the bus bars.

4.20 PARALLEL OPERATION OF SERIES GENERATORS

Although d.c. series generators are not usually used for supplying power, d.c. series motors are frequently arranged to operate as d.c. series generators during electric braking. For electric braking of electric trains, the d.c. series motors are disconnected from the supply mains and are connected in parallel across resistors. Due to kinetic energy of the train, the motors continue to run and act as self–excited d.c. series generators.

Fig.4.29 shows two identical d.c. series generators G_1 and G_2 connected in parallel. Since the generators are identical, $E_1 = E_2 = E$ (say) and they supply equal currents *i.e.* $I_1 = I_2 = I$ (say). Suppose due to any reason, E_1 increases slightly so that $E_1 > E_2$. Under such conditions, I_1 becomes greater than I_2. As a result, the field of generator G_1 is strengthened thus further increasing E_1 while the field of generator G_2 is weakened thus decreasing E_2 further. The process continues until a stage is reached when generator G_1 is supplying the entire load current and is attempting to operate machine G_2 as a motor. At this stage, the two generators will form a short–circuited loop and the current will be dangerously high. This causes the circuit breakers to open, disconnecting the generators from the load. Therefore, this method of parallel connection will lead to unstable operation.

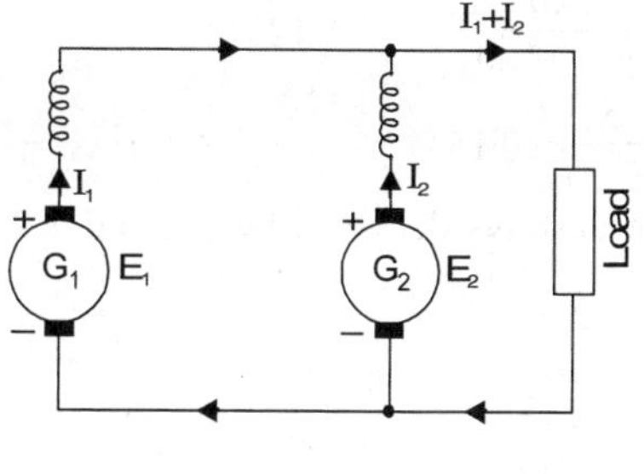

Fig. 4.29

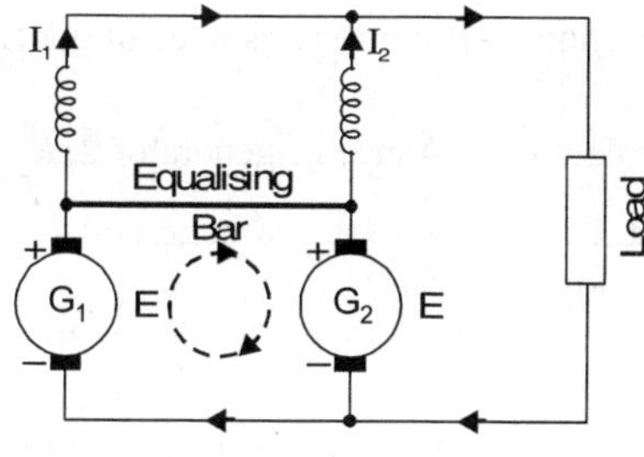

Fig. 4.30

The above condition can be prevented by using **equalising bar** (a thick copper bar of negligible resistance) across the two machines as shown in Fig. 4.30. Note that the equalising bar brings the series field windings in parallel. With the field coils connected in parallel, the value of flux is determined solely by the total load current. The two similar generators will supply nearly equal currents. Any slight difference between the two currents is confined to the loop formed by the armatures and the equalising bar (dotted circle).

Example 4.14. *Two shunt generators operating in parallel are excited so that their no-load voltages are equal. The generators have straight line drooping external characteristics. How can combined characteristic be obtained ?*

Solution. The way in which two or more generators in parallel share a given load depends on the external characteristics of the generators. Fig. 4.31 shows the external characteristics of two shunt generators that have the same no-load voltages. Here *AB* is the external characteristic of machine 1 and *AD* that of machine 2. For a common terminal voltage *V*, the generator 1 delivers I_1 amperes (= *VS*) while generator 2 delivers I_2 amperes (= *VQ*). Note that generator 1 having more drooping curve delivers less current. The total current is $I = I_1 + I_2$ (*VS* + *VQ* = *VP*). Then *P* is a point on the total characteristic. Other points may be obtained in a similar manner and so the total or combined characteristic *AT* may be drawn. If the characteristics can be represented with sufficient accuracy by straight lines, then results can be obtained by calculation instead of graphically.

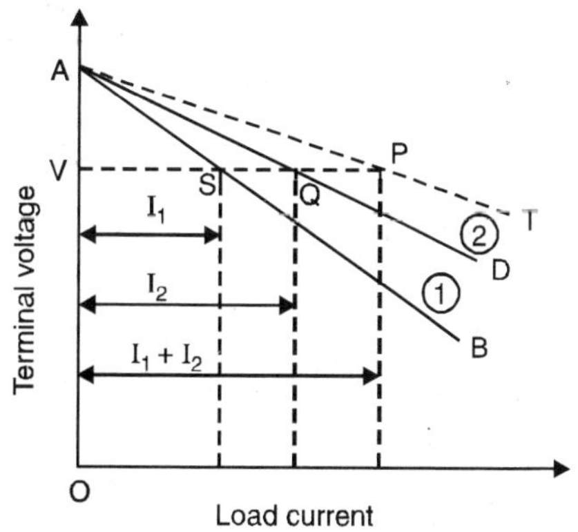

Fig. 4.31

Example 4.15. *A number of shunt generators are operating in parallel. Explain their behaviour.*

Solution.

(*i*) If the generated voltage (E) of the generator is *above* the bus-bars voltage (V), *generator action* is taking place and the unit is delivering current to the load.

(*ii*) If the generated voltage of the generator is equal to the bus-bars voltage, no effective generating action is taking place and no current is flowing to the load. The generator is said to be *floating* on the line. It is neither contributing nor drawing current and is still being rotated by its own prime mover.

(*iii*) If the generator is set so as to develop generated voltage less than the bus-bars voltage, it will draw current *from* the bus-bars and will be operating as a *motor*. In this case, the unit is drawing power (rather than contributing) from the bus-bars.

Example 4.16. *Two d.c. shunt generators operating in parallel supply a total load current of 200 A. The terminal voltage of one generator falls uniformly from 240 V to 225 V when delivering 120 A. The terminal voltage of the second generator falls uniformly from 230 V to 215 V when delivering 100 A. Find the load current shared by each generator and the bus-bars voltage.*

Solution. Armature resistance of generator 1, $R_1 = \dfrac{240-225}{120} = 0{\cdot}125\ \Omega$

Armature resistance of generator 2, $R_2 = \dfrac{230-215}{100} = 0{\cdot}15\ \Omega$

Let V be the bus-bars voltage and I_1 and I_2 be the currents delivered by generators 1 and 2 respectively.

$$\therefore \quad I_1 + I_2 = 200 \quad \text{...}(i)$$

$$V = E_1 - I_1 R_1 = E_2 - I_2 R_2$$

$$\therefore \quad 240 - I_1 \times 0{\cdot}125 = 230 - I_2 \times 0{\cdot}15$$

$$\text{or} \quad 240 - 0{\cdot}125\, I_1 = 230 - (200 - I_1) \times 0{\cdot}15 \quad [\because I_2 = 200 - I_1]$$

On solving, $I_1 = \mathbf{145{\cdot}5\ A}$ and $I_2 = 200 - 145{\cdot}5 = \mathbf{54{\cdot}5\ A}$

$$V = E_1 - I_1 R_1 = 240 - 145{\cdot}5 \times 0{\cdot}125 = \mathbf{221{\cdot}8\ volts}$$

Example 4.17. *Two shunt wound generators running in parallel supply a total load current of 4000 A. Each generator has an armature resistance of 0·02 Ω and a field resistance of 40 Ω. The fields are excited so that the e.m.f. induced in one generator is 480 V and in another 490 V. Find the bus-bars voltage and kilowatt output of each machine.*

Solution. Fig. 4.32 shows the connections of the generators.

Let V = bus-bars voltage

I_1 = current output of generator 1

I_2 = current output of generator 2

$I_1 + I_2 = 4000$

$I_{sh} = \dfrac{V}{R_{sh}}$

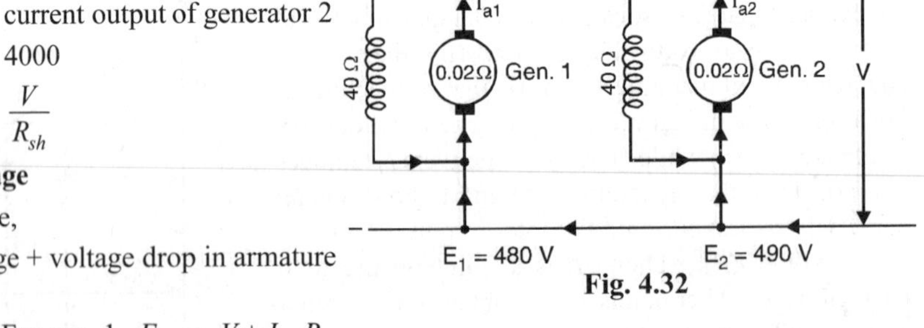

Fig. 4.32

Bus-bars voltage

In each machine,

Bus-bars voltage + voltage drop in armature = E.M.F. generated

For gen. 1, $E_1 = V + I_{a1} R_a$

$$\text{or} \quad 480 = V + \left(I_1 + \frac{V}{40}\right) \times 0{\cdot}02 \quad \text{...}(i)\ (\because I_{a1} = I_1 + I_{sh})$$

$$\text{For gen. 2,} \quad 490 = V + \left(I_2 + \frac{V}{40}\right) \times 0{\cdot}02 \quad \text{...}(ii)$$

Substracting (*i*) from (*ii*), we have,

$$0{\cdot}02\,I_2 - 0{\cdot}02\,I_1 = 10$$

or $$I_2 - I_1 = 500$$

or $$(4000 - I_1) - I_1 = 500 \qquad (\because I_1 + I_2 = 4000)$$

$\therefore$ $$I_1 = 1750\text{ A} \quad \text{and} \quad I_2 = 4000 - 1750 = 2250\text{ A}$$

Now $$480 = V + \left(I_1 + \frac{V}{40}\right) \times 0\cdot 02$$

or $$480 = V + \left(1750 + \frac{V}{40}\right) \times 0\cdot 02$$

On solving, V = **444·8 volts**

kilowatt ouput

$$\text{kW output of gen. 1} = \frac{V I_1}{1000} = \frac{444\cdot 8 \times 1750}{1000} = \mathbf{778{\cdot}4\ kW}$$

$$\text{kW output of gen. 2} = \frac{V I_2}{1000} = \frac{444\cdot 8 \times 2250}{1000} = \mathbf{1001\ kW}$$

Example 4.18. *Two shunt generators running in parallel supply a total load current of 3000 A. The generators have armature resistances 0·05 Ω and 0·03 Ω, field resistances 30 Ω and 25 Ω and induced e.m.f.s 400 V and 380 V respectively. Calculate (i) current supplied by each generator (ii) bus-bars voltage and (iii) kW ouput of each generator.*

Solution. Fig. 4.33 shows the circuit connections.

Let V = bus-bars voltage

I_1 = current output of generator 1

I_2 = current output of generator 2

$$I_{sh1} = V/30 \quad ; \quad I_{sh2} = V/25$$

$$I_{a1} = I_1 + \frac{V}{30} \quad ; \quad I_{a2} = I_2 + \frac{V}{25}$$

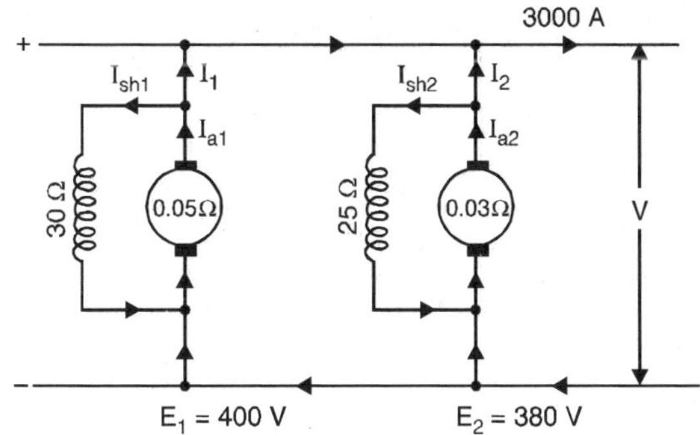

Fig. 4.33

$$I_1 + I_2 = 3000 \quad \therefore \quad I_2 = 3000 - I_1$$

In each machine, $E = V + \text{Armature drop}$

$\therefore$ $$400 = V + \left(I_1 + \frac{V}{30}\right) \times 0{\cdot}05$$

or $$400 = 1{\cdot}0016\,V + 0{\cdot}05\,I_1$$

$\therefore$ $$0{\cdot}05\,I_1 = 400 - 1{\cdot}0016\,V \qquad ...(i)$$

Also $$380 = V + \left(I_2 + \frac{V}{25}\right) \times 0{\cdot}03$$

or $$380 = 1{\cdot}0012\,V + 0{\cdot}03\,I_2 = 1{\cdot}0012\,V + (3000 - I_1) \times 0{\cdot}03$$

or $$0{\cdot}03\,I_1 = -290 + 1{\cdot}0012\,V \quad ...(ii)$$

Dividing eq. (*i*) by eq. (*ii*), we have,

$$\frac{0\cdot05\,I_1}{0\cdot03\,I_1} = \frac{400 - 1\cdot0016\,V}{-290 + 1\cdot0012\,V}$$

or $$\frac{5}{3} = \frac{400 - 1\cdot0016\,V}{-290 + 1\cdot0012\,V}$$

On solving, V = **331 volts**

Now $$0{\cdot}05\,I_1 = 400 - 1{\cdot}0016\,V$$

or $$0{\cdot}05\,I_1 = 400 - 1{\cdot}0016 \times 331$$

$\therefore$ I_1 = **1373·7 A**

$I_2 = 3000 - 1373{\cdot}7$ = **1626·3 A**

$$\text{kW output of generator 1} = \frac{V I_1}{1000} = \frac{331 \times 1373\cdot7}{1000} = \mathbf{454{\cdot}7\ kW}$$

$$\text{kW output of generator 2} = \frac{V I_2}{1000} = \frac{331 \times 1626\cdot3}{1000} = \mathbf{538{\cdot}3\ kW}$$

Example 4.19. *Two shunt generators operating in parallel deliver a total current of 250 A. One of the generator is rated 50 kW and the other 100 kW. The voltage rating of both machines is 500 V and have regulations 6% (smaller) and 4%. Assuming linear characteristics, determine (i) the current delivered by each machine (ii) terminal voltage.*

Solution.

50 kW generator

F.L. voltage drop $= 500 \times 0{\cdot}06 = 30$ V

F.L. current $= 50 \times 10^3/500 = 100$ A

Drop per ampere $= 30/100 = 3/10$ V/A

100 kW generator

F.L. voltage drop $= 500 \times 0{\cdot}04 = 20$ V

F.L. current $= 100 \times 10^3/500 = 200$ A

Drop per ampere $= 20/200 = 1/10$ V/A

Let I_1 and I_2 be the currents supplied by two generators and V be the terminal voltage. Then,

$$V = 500 - \frac{3}{10} \times I_1 \quad \text{...first generator}$$

$$V = 500 - \frac{1}{10} \times I_2 \quad \text{... second generator}$$

$$\therefore \quad 500 - \frac{3}{10} \times I_1 = 500 - \frac{1}{10} \times I_2$$

or $$3\,I_1 = I_2 \quad ...(i)$$

Also $$I_1 + I_2 = 250 \quad ...(ii)$$

(*i*) From eqs. (*i*) and (*ii*), we have, I_1 = **62·5 A** ; I_2 = **187·5 A**

(*ii*) Terminal voltage, $V = 500 - \frac{3}{10} \times I_1 = 500 - \frac{3}{10} \times 62{\cdot}5$ = **481·25 V**

Example 4.20. *Four compound generators operating in parallel supply a load of 1000 A. The resistances of the series windings are given in Fig. 4.34. If the machines are excited so that the e.m.f.s generated are as indicated, find (i) current output of each machine (ii) current in each section of the equaliser and (iii) bus-bars voltage. Neglect the shunt field currents.*

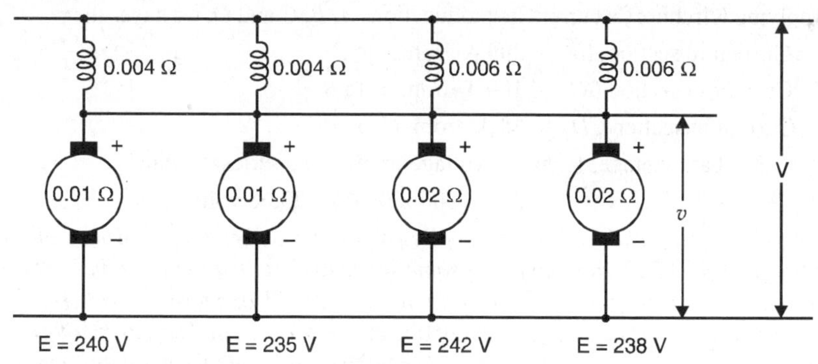

Fig. 4.34

Solution.

(*i*) Let V = bus-bars voltage

v = voltage between equaliser and negative terminal

I_1, I_2, I_3, I_4 = current outputs of generators [See Fig. 4.35]

Now $I_1 + I_2 + I_3 + I_4 = 1000$

or $$\frac{240-v}{0\cdot01} + \frac{235-v}{0\cdot01} + \frac{242-v}{0\cdot02} + \frac{238-v}{0\cdot02} = 1000$$

Multiplying throughout by 1/100, we get,

$$240 - v + 235 - v + 121 - \frac{v}{2} + 119 - \frac{v}{2} = 10$$

$\therefore$ v = 235 volts

The current outputs of generators are :

$$I_1 = \frac{240-235}{0\cdot01} = \textbf{500 A} \;;\quad I_2 = \frac{235-235}{0\cdot01} = \textbf{0 A}$$

$$I_3 = \frac{242-235}{0\cdot02} = \textbf{350 A} \;;\quad I_4 = \frac{238-235}{0\cdot02} = \textbf{150 A}$$

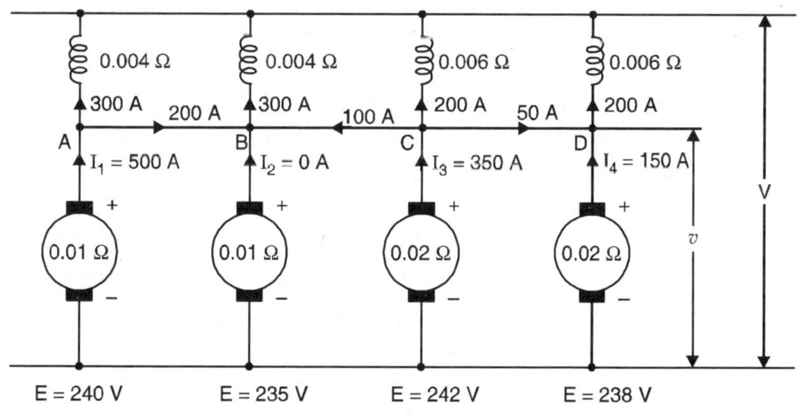

Fig. 4.35

(*ii*) The total current of 1000 A divides among series windings in the ratio :

$$\frac{1}{0\cdot004}:\frac{1}{0\cdot004}:\frac{1}{0\cdot006}:\frac{1}{0\cdot006} \quad i.e., 3:3:2:2$$

∴ Currents in series windings will be 300 A, 300 A, 200 A and 200 A as shown in Fig. 4.35. Applying Kirchhoff's current law to junctions *A*, *B*, *C* and *D*, we have,

Current in section AB = **200 A** from *A* to *B*

Current in section BC = **100 A** from *C* to *B*

Current in section CD = **50 A** from *C* to *D*

(*iii*) Bus-bars voltage, $V = v -$ voltage drop in one series winding

$= 235 - 300 \times 0\cdot004 =$ **233·8 volts**

Example 4.21. *In a certain sub-station, there are 5 shunt generators in parallel each having an armature resistance of 0·1 Ω, running at the same speed and excited to give equal induced e.m.f.s. Each generator supplies an equal share of a total load of 250 kW at a terminal voltage of 500 V into a load of fixed resistance. If the field current of one generator is raised by 4% the others remaining unchanged, calculate the power output of each machine and their terminal voltage under these conditions. Assume that speed remains constant and flux is proportional to field current.*

Solution. Fig. 4.36 shows the connections of the generators prior to the increased excitation of one.

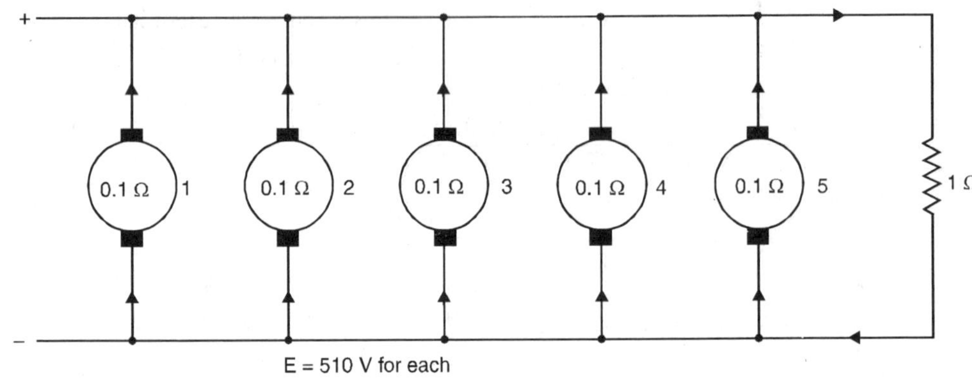

Fig. 4.36

Prior to increased excitation of one

Load supplied by each generator = 250/5 = 50 kW

Output current of each generator = $50 \times 10^3/500$ = 100 A

Armature drop in each generator = $100 \times 0\cdot1$ = 10 V

Induced e.m.f. of each generator = 500 + 10 = 510 V

Load resistance, R_L = 500/500 = 1 Ω

After increased excitation of one. When field excitation of one is increased by 4%, its induced e.m.f. is increased by 4%. Now 4% of 510 = 20·4 V.

Let I_1 = Current supplied by the generator with increased excitation

I_2 = Current supplied by each of other 4 generators

V = New voltage of bus-bars

∴ $I = I_1 + 4I_2$

Now $V = 530\cdot4 - 0\cdot1\, I_1$ ...(*i*)

And $V = 510 - 0\cdot1 I_2$ ...(*ii*)

Also $V = 1 \times (I_1 + 4I_2)$ ...(iii)

$\therefore$ $530 \cdot 4 - 0 \cdot 1 I_1 = 510 - 0 \cdot 1 I_2$

or $I_1 - I_2 = 204$ ...(iv)

Also $1 \times (I_1 + 4I_2) = 510 - 0 \cdot 1 I_2$

or $I_1 + 4 \cdot 1\, I_2 = 510$ ...(v)

Solving eqs. (*iv*) and (*v*), we get, $I_1 = 264\text{ A}\,;\, I_2 = 60\text{ A}$

New bus-bars voltage, $V = I_1 + 4I_2 = 264 + 4 \times 60 = \textbf{504 volts}$

Output of the generator with increased excitation

$$= VI_1 = \frac{504 \times 264}{1000} \text{ kW} = \textbf{133 kW}$$

Output of each of the other 4 generators

$$= VI_2 = \frac{504 \times 60}{1000} \text{ kW} = \textbf{30·24 kW}$$

Example 4.22. *Two separately excited d.c. generators running in parallel each supply a current of 200 A at 200 V. The armature resistance of each machine is 0·05 Ω. Calculate (i) the percentage change in the generated e.m.f. of one machine necessary to reduce its load to zero (ii) the resulting percentage drop in terminal voltage, assuming that the total load current and the e.m.f. of the other machine remain unchanged.*

Solution. Initially generated e.m.f. of each machine $= 200 + 0 \cdot 05 \times 200 = 210$ V

Total load current $= 200 + 200 = 400$ A

When one machine is supplying no current (*i.e.*, it is floating), the other machine will supply the total load of 400 A.

$\therefore$ New terminal voltage $= 210 - 400 \times 0 \cdot 05 = 190$ V

$\therefore$ Generated e.m.f. of floating machine $= 190$ V

(*i*) % age change (reduction) in the generated e.m.f. of floating machine

$$= \frac{210 - 190}{210} \times 100 = \textbf{9·5\%}$$

(*ii*) % age reduction in the terminal voltage

$$= \frac{200 - 190}{200} \times 100 = \textbf{5\%}$$

Example 4.23. *A 110 V battery is connected to the same bus-bars as a d.c. generator. Battery resistance is 0·025 Ω and generator resistance is 0·1 Ω. The battery floats when the load current is 100 A. Find the generator output (i) when the battery floats (ii) when the load current is 50 A.*

Solution.

(*i*) **When the battery floats.** When the battery floats (*i.e.*, it supplies no current), the bus-bars terminal voltage is equal to the battery e.m.f. (= 110 V).

$\therefore$ Generated e.m.f. of the generator

$$= 110 + 100 \times 0 \cdot 1 = 120 \text{ V}$$

$\therefore$ Generator output $=$ Bus-bars voltage $\times$ current supplied

$= 110 \times 100 = 11 \times 10^3\text{W} = \textbf{11 kW}$

(*ii*) **When the load current is 50 A**

Let I_1 = Current supplied by generator

I_2 = Current supplied by battery

V = New terminal or bus-bars voltage

$$\therefore \quad I_1 + I_2 = 50$$

$$I_1 = \frac{120 - V}{0 \cdot 1} = 1200 - 10\,V$$

$$I_2 = \frac{110 - V}{0 \cdot 025} = 4400 - 40\,V$$

$$\therefore \quad (1200 - 10\,V) + (4400 - 40\,V) = 50$$

or
$$V = 111 \text{ volts}$$

$$I_1 = \frac{120 - 111}{0 \cdot 1} = 90 \text{ A}$$

$$\therefore \quad \text{Generator output} = V I_1 = 111 \times 90 = 9{\cdot}9 \times 10^3 \text{W} = \mathbf{9{\cdot}9\ kW}$$

Example 4.24. *The terminal voltage of a generator falls from 250 V on open-circuit to 238 V when delivering 60 A. It is connected to a load in parallel with a battery having a constant e.m.f. of 245 V and a resistance of 0·1 Ω. Find the current supplied by the generator when the total load current is (i) 50 A (ii) 100 A (iii) zero.*

Solution. Internal resistance of generator $= \dfrac{250 - 238}{60} = 0{\cdot}2\ \Omega$

Let
I = total load current
I_1 = current supplied by generator
I_2 = current supplied by battery
V = terminal voltage

$$\therefore \quad I_1 = \frac{250 - V}{0 \cdot 2} \quad \text{and} \quad I_2 = \frac{245 - V}{0 \cdot 1}$$

Now
$$I_1 + I_2 = I$$

or
$$\frac{250 - V}{0 \cdot 2} + \frac{245 - V}{0 \cdot 1} = I$$

or
$$V = \frac{3700 - I}{15}$$

(*i*) **When total load current I = 50 A**

$$V = \frac{3700 - 50}{15} = \frac{3650}{15} = 243{\cdot}3 \text{ volts}$$

$$\therefore \quad I_1 = \frac{250 - V}{0 \cdot 2} = \frac{250 - 243 \cdot 3}{0 \cdot 2} = \mathbf{33{\cdot}5\ A}$$

(*ii*) **When total load current I = 100 A**

$$V = \frac{3700 - 100}{15} = \frac{3600}{15} = 240 \text{ volts}$$

$$\therefore \quad I_1 = \frac{250 - V}{0 \cdot 2} = \frac{250 - 240}{0 \cdot 2} = \mathbf{50\ A}$$

(*iii*) **When total load current I = 0**

$$V = \frac{3700 - 0}{15} = 246{\cdot}7 \text{ volts}$$

$$\therefore \quad I_1 = \frac{250 - V}{0 \cdot 2} = \frac{250 - 246 \cdot 7}{0 \cdot 2} = \mathbf{16{\cdot}5\ A}$$

Example 4.25. *Two 220 V d.c. generators operate in parallel. One machine has a terminal voltage of 270 V on no-load and 220 V at a load current of 35 A. The other has a voltage of 280 V on no-load and 220 V at 50 A. The external characteristics are linear. Calculate the bus-bars voltage and the output current of each machine when the total load is (i) 60 A (ii) 20 A.*

Solution.

$$\text{Armature resistance of generator 1, } R_1 = \frac{270-220}{35} = 1{\cdot}43\ \Omega$$

$$\text{Armature resistance of generator 2, } R_2 = \frac{280-220}{50} = 1{\cdot}2\ \Omega$$

Let

I_1 = current supplied by generator 1

I_2 = current supplied by generator 2

V = bus-bars voltage

I = total load current

Now $$I_1 + I_2 = I$$

or $$\frac{270-V}{1{\cdot}43} + \frac{280-V}{1{\cdot}2} = I$$

or $$1{\cdot}533\,V = 422 - I$$

or $$V = 275{\cdot}3 - 0{\cdot}652\,I$$

(*i*) **When total load current I = 60 A**

$$V = 275{\cdot}3 - 0{\cdot}652 \times 60 = \mathbf{236{\cdot}2\ volts}$$

$$I_1 = \frac{270-236{\cdot}2}{1{\cdot}43} = \mathbf{23{\cdot}6\ A}$$

$$I_2 = \frac{280-236{\cdot}2}{1{\cdot}2} = \mathbf{36{\cdot}4\ A}$$

Alternatively, $$I_2 = I - I_1 = 60 - 23{\cdot}6 = \mathbf{36{\cdot}4\ A}$$

(*ii*) **When total load current I = 20 A**

$$V = 275{\cdot}3 - 0{\cdot}652 \times 20 = \mathbf{262{\cdot}4\ volts}$$

$$I_1 = \frac{270-262{\cdot}4}{1{\cdot}43} = \mathbf{5{\cdot}3\ A}$$

$$I_2 = I - I_1 = 20 - 5{\cdot}3 = \mathbf{14{\cdot}7\ A}$$

Example 4.26. *Two shunt generators and a battery are working in parallel. The open-circuit voltage, armature and field resistances of generators are 250V, 0.24 Ω, 100 Ω and 248V, 0.12 Ω and 100 Ω respectively. If the generators supply the same current when the load on the bus-bars is 40A, calculate the e.m.f. of the battery if its internal resistance is 0.172 Ω.*

Solution. Fig. 4.37 shows the conditions of the problem. Let current supplied by each generator be I and that supplied by battery be I_b. Let the bus-bar voltage be V.

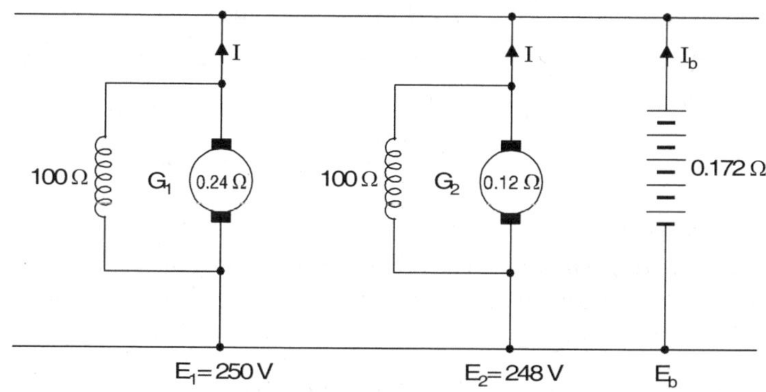

Fig. 4.37

For generator G_1 : $V + *\left(I + \frac{V}{100}\right) \times 0.24 = 250$ ...(*i*)

For generator G_2 : $V + \left(I + \frac{V}{100}\right) \times 0.12 = 248$ ...(*ii*)

Subtracting eq. (*ii*) from eq. (*i*), we have,

$$\left(I + \frac{V}{100}\right) \times 0.12 = 2 \qquad ...(iii)$$

Putting this value in eq. (*ii*), we get,

$$V + 2 = 248 \quad or \quad V = 248 - 2 = 246 \text{ volts}$$

Putting $V = 246$ volts in eq. (*iii*), we have,

$$\left(I + \frac{246}{100}\right) \times 0.12 = 2 \qquad \therefore \qquad I = 14.2 \text{ A}$$

It is given that load on the bus-bars is 40 A.

$$\therefore \qquad I + I + I_b = 40 \quad or \quad I_b = 40 - 2I = 40 - 2 \times 14.2 = 11.6 \text{ A}$$

$$\therefore \qquad \text{Battery e.m.f., } E_b = V + I_b R_i = 246 + (11.6 \times 0.172) = \mathbf{248\ V}$$

Example 4.27. *Two similar generators operating in parallel are excited to the same voltage and they carry total load of 400 A. If the armature resistance of each is 0.05Ω and the terminal voltage is 200V, find*

(i) load shared by each machine.

(ii) percentage change in no-load voltage of one so that it does not deliver any current

(iii) the terminal voltage corresponding to (ii)

Solution. (*i*) It is given that the generators (G_1 and G_2) are excited to the same voltage (say E volts) and have the same armature resistance ($R_a = 0.05\ \Omega$) and the same terminal voltage ($V = 200$ volts). Therefore, they will share the load equally**.

∴ Current supplied by each generator is

$$I = 400/2 = \mathbf{200\ A}$$

$$\text{Excitation voltage, } E = V + IR_a = 200 + 200 \times 0.05 = 210 \text{ V}$$

(*ii*) One of the two generators, say G_1, will stop delivering current when its excitation voltage is reduced to the terminal voltage of G_2 while supplying a total load of 400 A.

$$\text{Terminal voltage of } G_2, \quad V_2 = E - I_2 R_a = 210 - 400 \times 0.05 = 190 \text{ V}$$

Therefore, generator G_1 will stop delivering current when its excitation voltage is reduced to 190 V.

∴ Percentage change in no-load voltage of G_1

$$= \frac{210 - 190}{210} \times 100 = \mathbf{9.52\ \%}$$

(*iii*) Terminal voltage of G_1, $V_1 = \mathbf{190\ V}$

Example 4.28. *Two d.c. generators A and B are connected to a common load. The generator A has a constant e.m.f. of 400 V and internal resistance of 0.25Ω while generator B has a constant e.m.f. of 410 V and internal resistance of 0.4Ω.*

---

* $V + (I + I_{sh}) R_a = E_1$. Here $I_{sh} = V/100$; $R_a = 0.24\ \Omega$ and $E_1 = 250$ V.

** Current supplied by generator, $I = \frac{E - V}{R_a}$

(*i*) *Calculate the current and power output of each generator if the terminal voltage is 390 V.*
(*ii*) *What would be the current and power from each and the terminal voltage if the load is open-circuited ?*

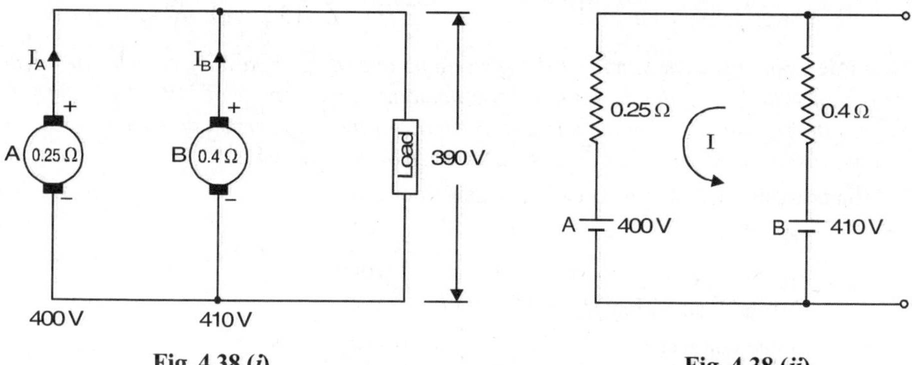

Fig. 4.38 (*i*) **Fig. 4.38 (*ii*)**

Solution. Fig. 4.38 (*i*) shows the conditions of the problem. The terminal voltage V = 390 volts.

(*i*) Current supplied by $A, I_A = \dfrac{E_A - V}{R_A} = \dfrac{400-390}{0.25} = \mathbf{40\ A}$

Current supplied by $B, I_B = \dfrac{E_B - V}{R_B} = \dfrac{410-390}{0.4} = \mathbf{50\ A}$

Power output from $A, P_A = VI_A = \dfrac{390 \times 40}{1000}$ kW = **15.6 kW**

Power output from $B, P_B = VI_B = \dfrac{390 \times 50}{1000}$ kW = **19.5 kW**

(*ii*) If the load is open–circuited as shown in Fig. 4.38 (*ii*), the two generators are put in series and the circuit current is determined by the net circuit voltage and total circuit resistance. Note that machine *B* (having higher e.m.f.) acts as a generator and drives machine *A* as a motor.

Net voltage in the circuit = 410 – 400 = 10V

Total circuit resistance = $R_A + R_B$ = 0.25 +0.4 = 0.65Ω

Circuit current, I = 10/0.65 = **15.4 A**

∴ Terminal voltage, *V = $E_A + IR_A$ = 400 + 15.4 × 0.25 = **403.8 V**

Power taken by A from B = VI = 403.8 × 15.4 = **6219 W**

Example 4.29. *Four d.c. generators of open circuit voltages 220V, 230V, 240V and 250V and internal resistances 0.1Ω, 0.1Ω, 0.25Ω and 0.2Ω respectively are joined in parallel across a 5Ω resistor. Calculate the terminal voltage and power delivered to the load.*

Solution.

$$\text{Terminal voltage, } V = \frac{\text{**Sum of short circuit currents}}{\text{Terminal conductances}}$$

$$= \frac{(220/0.1)+(230/0.1)+(240/0.25)+(250/0.2)}{(1/0.1)+(1/0.1)+(1/0.25)+(1/0.2)+(1/5)}$$

---

* Also $V = E_B - IR_B$.

** Millman's theorem.

$$= \frac{6710}{29.2} = \textbf{229.8 volts}$$

$$\text{Power delivered to load} = \frac{V^2}{R_L} = \frac{(229.8)^2}{5} = 10.5 \times 10^3 \text{ W} = \textbf{10.5 kW}$$

Example 4.30. *A station contains 6 d.c. generators operating in parallel, each having an armature current of 1000A, an induced e.m.f. of 250V and a terminal voltage of 240V. The external load has a constant resistance. Calculate the new terminal voltage and current of each generator if the e.m.f. of one generator is raised by 5% ; the others remaining unaltered.*

Solution. Resistance of armature of each generator is

$$R_a = \frac{E-V}{I_a} = \frac{250-240}{1000} = 0.01\Omega$$

$$\text{Total load current, } I_L = 6 \times 1000 = 6000 \text{ A}$$

$$\text{Load conductance, } G_L = I_L/V = 6000/240 = 25\text{S}$$

$$\text{E.M.F. of affected machine, } E' = 250 + 5\% \text{ of } 250$$

$$= 250 + 12.5 = 262.5 \text{ volts}$$

$$\therefore \quad \text{New terminal voltage, } V' = \frac{\text{Sum of short circuit currents}}{\text{Terminal conductances}}$$

$$\text{Terminal conductances} = {}^*(6/0.01) + 25 = 625\text{S}$$

$$\text{Sum of S.C. currents} = 5 \times (250/0.01) + (262.5/0.01) = 151250 \text{ A}$$

$$\therefore \quad V' = 151250/625 = \textbf{242 volts}$$

$$\text{Total load current, } I'_L = V' \times G_L = 242 \times 25 = 6050 \text{ A}$$

Current supplied by the generator with increased voltage

$$= \frac{E-V'}{R_a} = \frac{262.5-242}{0.01} = \textbf{2050 A}$$

Total current supplied by other 5 generators

$$= 6050 - 2050 = 4000 \text{ A}$$

$$\therefore \quad \text{Current supplied by each} = 4000/5 = \textbf{800 A}$$

TUTORIAL PROBLEMS

1. Three identical 5 kW shunt generators *A*, *B* and *C* are driven by a single prime mover. Each machine has a field resistance of 60 Ω and an armature circuit resistance of 0·18 Ω. The three machines are set to no-load voltages of (*A*) 124 V, (*B*) 120 V and (*C*) 115 V. They are connected to bus-bars that is maintained at 120 V by other machines. Determine armature current for each machine.

   [*A* = 24·2 A ; *B* = 2 A (floating) ; *C* = – 25·8 A (acting as a motor)]

2. Two d.c. generators operating in parallel supply a total load current of 150 A. The terminal voltage of one machine falls uniformly from 230 V on open circuit to 218 V when delivering 80 A; the terminal voltage of the second falls uniformly from 236 V to 220 V when delivering 80 A. Find the current output of each generator and the bus-bars voltage. **[*First* : 68·6 A ; Second : 81·4 A ; 219·7 V]**

3. Two shunt generators operating in parallel supply a total load of 3600 A. Each machine has a field resistance of 60 Ω and an armature resistance of 0·01 Ω. The e.m.f.s, generated are 455 V and 460 V respectively. Calculate the bus-bars voltage and the kW output of each machine.

   [439·5 V ; 681 kW ; 901 kW]

---

* Conductance of one generator = 1/0.01S. Therefore, conductance of 6 generators = 6 × 1/0.01 = 6/0.01S.

4. A shunt generator has external characteristic such that the terminal voltage falls uniformly from 450 V on no-load to 440 V when delivering 100 A. It is in parallel with a battery of 224 cells each of e.m.f. 2 V and internal resistance of 0·02 Ω. Find how a load of 400 A would be shared.

[*Generator* : 83·3 A ; *Battery* : 316·7 A]

5. A station consists of 6 generators operating in parallel each having an armature current of 100 A, an induced e.m.f. 250 V and a terminal voltage of 240 V. The external load has a constant resistance. Calculate the new terminal voltage and current of each generator if the e.m.f. of one generator is raised by 5%, the e.m.f.s of others remaining the same. **[*One generator* : 2050 A ; *Each of other 5* : 800 A]**

6. Two short-shunt compound generators *A* and *B* running in parallel supply a load current of 140 A at a terminal voltage 100 V. An equalising bar connects the two machines. The data regarding the machines are :

   Generator *A* : Armature resistance 0·02 Ω ; shunt field resistance 80 Ω ; series field resistance 0·02 Ω.

   Generator *B* : Armature resistance 0·05 Ω ; shunt field resistance 100 Ω ; series field resistance 0.05 Ω; e.m.f. of generator *B*, 105 V.

   Calculate (*i*) current in series windings (*ii*) armature currents (*iii*) current in equaliser (*iv*) e.m.f. generated by generator *A*. **[(*i*) 100 A ; 40 A (*ii*) 82·3 A ; 60 A (*iii*) 19 A (*iv*) 103·6 V]**

7. Five generators are running in parallel, the armature current of each being 400 A and the e.m.f. induced 230 V ; the bus-bars voltage is 220 V. If the e.m.f. of one generator is raised by 4%, find the new bus-bars voltage and current in each armature.

[221·7 V ; *In one armature* 700 A ; *each of four others* 332 A]

4.21 D.C. GENERATOR SPECIFICATIONS

The nominal characteristics of a d.c. generator are found on the nameplate that is secured to the yoke or stator of the machine. These specifications or nominal characteristics are the values guaranteed by the manufacturer. A typical 100 kW compound generator may have the following information engraved or stamped on its nameplate :

Voltage	250 V	**Type**	Compound
Load current	400 A	**Exciting current**	20 A
Output power	100 kW	**Temperature rise**	50°C
Speed	1200 r.p.m.	**Insulation Class**	A

From the nameplate data, we can infer a number of interesting conclusions :

(*i*) The d.c. generator can deliver continuously a power of 100 kW at a voltage of 250 V without exceeding a temperature rise of 50°C.

(*ii*) The shunt field current is 20 A.

(*iii*) The armature current is 420 A.

(*iv*) If the compound generator is connected short-shunt, the series field current is 400 A.

(*v*) The class *A* designation refers to the class of insulation used in the machine.

4.22 FRAME SIZE VERSUS SPEED OF ELECTRICAL MACHINES

The physical size of an electrical machine depends upon its power and speed of rotation. Consider a 100 kW, 250V, 2000 r.p.m. d.c. generator. Suppose we have to build another generator having the same power and voltage, but running at half the speed. To generate the same voltage at half the speed, we either have to double the number of conductors on the armature or double the flux from the poles. Consequently, we must either increase the size of the armature or increase the size of the poles. In practice, we increase both. Therefore, *for a given power output, a low-speed machine is always bigger than a high-speed machine.* This is true for both d.c. and a.c. machines. Therefore, low-speed motors are more costly than high-speed motors of equal power. Consequently, for low-speed applications, it is often cheaper to use a high-speed motor with a gear box than to use a low-speed motor directly.

In addition to speed, the size of an electrical machine is also affected by its type of enclosure, its class of insulation, its rated torque and the type of motor.

SHORT ANSWER QUESTIONS

Q. 1. *Why is the resistance of the field winding of a d.c. shunt generator kept below critical field resistance?*

Ans. For satisfactory working, the machine (shunt generator) should operate with field resistance appreciably below the critical value. If the field circuit resistance is more than the critical field resistance, the generator will fail to build up any voltage.

Q. 2. *Why does the field winding of a series d.c. machine has less number of turns than that of d.c. shunt machine?*

Ans. The resistance of the shunt field winding is large so that current in the winding is small compared to the rated armature current of the machine. To produce the necessary m.m.f., the winding consists of many turns. The resistance of the series field winding is made as low as possible so that voltage drop across the winding is minimum. To produce the same flux at rated conditions, the turns of a series winding are fewer than those of a shunt winding since the series winding carries the rated current of the machine.

Q. 3. *What is the necessity of a compound generator?*

Ans. In practice, power circuits require a constant voltage, especially lighting circuits. Shunt and series generators cannot be used in such situations. As the load current of a shunt generator increases, its terminal voltage falls. In the series generator, on the other hand, an increase in the load current is accompanied by rise in the terminal voltage.

In a compound generator, both series and shunt windings are placed on each field pole. The series winding is so connected that flux produced by it aids the flux produced by shunt winding. The generator so connected is said to be *cumulatively-compounded.* As the load current increases, the flux produced by the series winding compensates for the loss of flux due to armature reaction. Consequently, the terminal voltage of the generator remains constant.

Q. 4. *Why do we use over-compounded generators in power stations?*

Ans. In power stations, there is a considerable distance between the generator and the load. We have to compensate not only for armature reaction and the armature voltage drop, but also for IR drop in the line between the generator and the load. In such situations, we place a few additional turns on the series winding of the cumulatively compounded generator so that the terminal voltage increases as the load current rises. Such a generator is called *over-compounded generator* and compensates for the voltage drop in the line to maintain constant voltage at the load terminals.

Q. 5. *How does sudden changes in load in a d.c. machine cause "flash-over"?*

Ans. As the load on a generator changes suddenly, the armature reaction causes severe distortion of flux in the air-gap. This causes irregular voltage distribution around the commutator. The high value of voltage between segments over a portion of the commutator may cause flash-over, particularly in high voltage machines where the average voltage per segment is necessarily high.

Q. 6. *Can a d.c. machine be operated under conditions different from those on the nameplate?*

Ans. The nameplate values of a generator imply that the machine will perform satisfactorily during a long machine life if operated under the limits and conditions specified on the nameplate. It must not be inferred that the machine cannot be operated under conditions different from those on the nameplate. However, operation that exceeds any of the nameplate ratings may result in premature failure of the machine.

Q. 7. *What happens if a d.c. machine is operated at a speed below the rated speed?*

Ans. If a d.c. machine is operated at a speed below the rated value, the machine may be overheated due to :

(*i*) increased field current necessary to produce rated voltage

(*ii*) decrease in fanning action because of decrease in speed

Q. 8. *How can you improve the design of a d.c. machine ?*

Ans. The design of a d.c. machine can be improved by one of the following two ways :

(*i*) By using a low-cost winding material with less resistance than copper. Of the pure metals, only silver has a lower resistivity.

(*ii*) By using a low-cost magnetic material with a higher permeability than present magnetic materials.

Q. 9. *Show that for the same power output, a low-speed machine is always bigger than a high-speed machine.*

Ans. Suppose we have a 100 kW, 250 V, 2000 r.p.m. generator. Now, it is desired to build another generator having the same power and voltage, but running at half the speed.

To generate the same voltage at half the speed, we either have to double the number of conductors on the armature or double the flux from the poles. Consequently, we must either increase the size of the armature or increase the size of the poles. In practice, we increase both. It follows, therefore, that for a given power output, a low-speed machine is always bigger than a high-speed machine. This is true for both d.c. and a.c. machines.

Q. 10. *What are the two requirements to be met while connecting shunt generators in parallel?*

Ans. The following two main requirements are to be met while connecting shunt generators in parallel :

(*i*) The positive terminals of all the machines must be connected to the positive bus-bars and the negative terminals must be connected to the negative bus-bars.

(*ii*) The terminal voltages of the machines being paralleled must be the same.

Q. 11. *How will you disconnect one of the two shunt generators operating in parallel ?*

Ans. In order to disconnect one of the two shunt generators operating in parallel, the following procedure is adopted :

(*i*) The load of the generator to be disconnected is shifted over to the other by gradually reducing its field current and increasing the field current of the other generator.

(*ii*) When the current output of the machine to be disconnected is reduced almost to zero, the main switch is opened.

Q. 12. *In a power station, we have several small shunt generators in parallel instead of a single large shunt generator. Why?*

Ans. It is generally more desirable in a generating station to have several small generators that may be paralleled than to have one large generator to supply all the load. This is due to the following reasons :

(*i*) The load supplied by a generating station is usually variable. As machines operate at greatest efficiency at or near full-load, it is possible with several machines to keep each machine either fully loaded or shut down entirely.

(*ii*) A generator that breaksdown may be removed from the circuit for repairs without interrupting the power supply.

(*iii*) Generators may be added to the system as the power demand increases.

Q. 13. *What is critical resistance of a shunt generator? How is it determined?*

Ans. The maximum value of shunt field resistance (at a particular speed) which would just excite the shunt generator is called critical resistance of the generator for that speed. It is given by the slope of tangent to the O.C.C. of the generator at a particular speed.

Q. 14. *What do you mean by critical speed of a shunt generator?*

Ans. Critical speed of a shunt generator is that speed for which the given shunt field resistance represents critical resistance.

Q. 15. *What is the location of series field winding in a compound generator?*

Ans. A compound generator has series as well as shunt field windings. The generator may be connected in long shunt or short shunt. Regardless of these connections, both shunt and series field windings are located on the same poles pieces.

MULTIPLE-CHOICE QUESTIONS

1. For the same rating, a d.c. machine has .... an a.c. machine.
   (*i*) the same weight as
   (*ii*) more weight than
   (*iii*) less weight than
   (*iv*) none of the above

2. The field winding of a d.c. shunt machine usually carries .... of the rated current of the machine.
   (*i*) 2% to 5% (*ii*) 15% to 20%
   (*iii*) more than 20 % (*iv*) less than 0·5%

3. A separately excited d.c. generator is not used because ....
   (*i*) it is costly
   (*ii*) voltage drops considerably with load
   (*iii*) separate d.c. source is required for field circuit
   (*iv*) none of the above

4. A d.c. compound generator having full-load terminal voltage equal to the no-load voltage is called .... generator.
   (*i*) under-compounded
   (*ii*) over-compounded
   (*iii*) flat-compounded
   (*iv*) none of the above

5. The main drawback of a d.c. shunt generator is that ....
   (*i*) terminal voltage drops considerably with load
   (*ii*) shunt field circuit has high resistance
   (*iii*) generated voltage is small
   (*iv*) it is expensive

6. D.C. machines which are subjected to abrupt changes of load are provided with
   (*i*) interpole windings
   (*ii*) compensating windings
   (*iii*) equalisers
   (*iv*) copper brushes

7. Fig. 4.39 shows the external characteristic of .... generator.
   (*i*) over-compounded
   (*ii*) series
   (*iii*) flat-compounded
   (*iv*) shunt

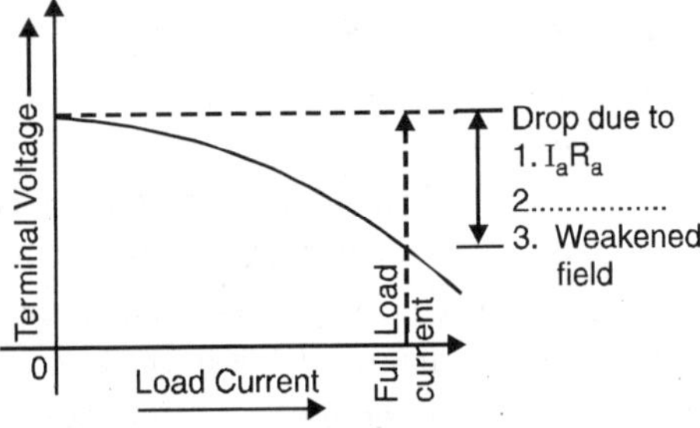

Fig. 4.39

8. Referring to Fig. 4.39, the voltage drop at no. 2 is due to ....
   (*i*) friction
   (*ii*) armature reaction
   (*iii*) field circuit
   (*iv*) none of the above

9. Fig. 4.40 shows the external characteristic of a .... generator.

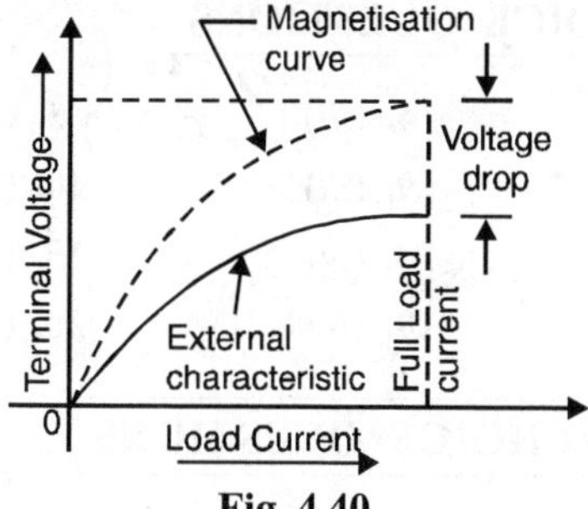

Fig. 4.40

(*i*) shunt (*ii*) flat-compounded
(*iii*) series (*iv*) over-compounded

10. Fig. 4.41 shows the external characteristics of three types of generators having the same rating. Curve 2 represents for .... generator.

(*i*) shunt (*ii*) separately-excited
(*iii*) compound (*iv*) none of the above

11. Curve 3 in Fig. 4.41 represents the external characteristic of a .... generator.

(*i*) compound (*ii*) separately-excited
(*iii*) shunt (*iv*) none of the above

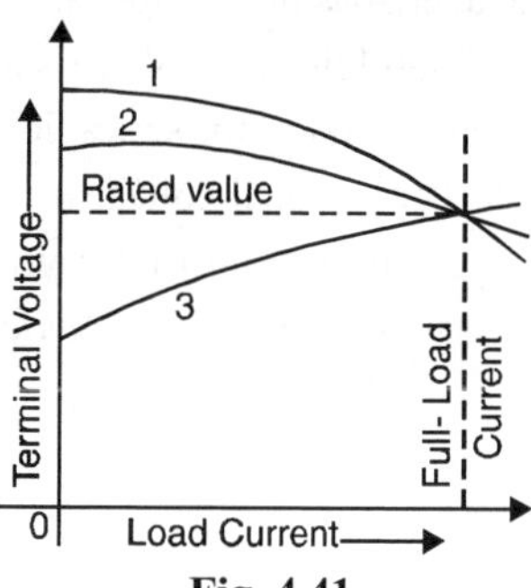

Fig. 4.41

12. Curve 1 in Fig. 4.41 represents the external characteristic of a .... generator.

(*i*) shunt
(*ii*) series
(*iii*) flat-compounded
(*iv*) none of the above

13. The curve *B* in Fig. 4.42 is the external characteristic of a .... generator.

(*i*) shunt
(*ii*) series
(*iii*) flat-compounded
(*iv*) under-compounded

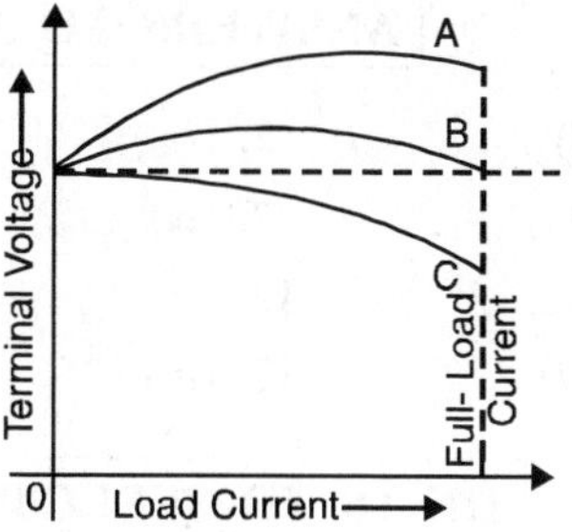

Fig. 4.42

14. The curve *A* in Fig. 4.42 represents the external characteristic of .... generator.

(*i*) shunt (*ii*) flat-compounded
(*iii*) under-compounded
(*iv*) over-compounded

15. The O.C.C. of a d.c. generator is also called its .... characteristic.

(*i*) magnetic (*ii*) internal
(*iii*) external (*iv*) performance

16. The line representing the critical resistance of a d.c. generator .... its O.C.C.

(*i*) intersects (*ii*) runs parallel to
(*iii*) just touches (*iv*) none of the above

17. Which of the following generator provides approximately constant voltage from no-load to full-load ?

(*i*) series (*ii*) shunt
(*iii*) flat-compounded
(*iv*) over-compounded

18. The level of compounding in a cumulatively compounded d.c. generator is generally adjusted by

(*i*) changing series field current
(*ii*) changing shunt field current
(*iii*) connecting it long-shunt
(*iv*) connecting it short-shunt

19. The d.c. generator that has poorest voltage regulation is

(*i*) shunt generator
(*ii*) over-compounded generator
(*iii*) flat-compounded generator
(*iv*) series generator

20. A shunt generator gives its greatest voltage at

(*i*) no-load (*ii*) full-load
(*iii*) half full-load (*iv*) none of the above

ANSWERS TO MULTIPLE-CHOICE QUESTIONS

1. (*ii*)	**2.** (*i*)	**3.** (*iii*)	**4.** (*iii*)	**5.** (*i*)
6. (*i*)	**7.** (*iv*)	**8.** (*ii*)	**9.** (*iii*)	**10.** (*ii*)
11. (*i*)	**12.** (*i*)	**13.** (*iii*)	**14.** (*iv*)	**15.** (*i*)
16. (*iii*)	**17.** (*iii*)	**18.** (*i*)	**19.** (*iv*)	**20.** (*i*)

HINTS TO SELECTED MULTIPLE-CHOICE QUESTIONS

3. The main disadvantage of a separately excited d.c. generator lies in the inconvenience and expense of providing a separate d.c. source for the field circuit. For this reason, the use of this type of generator is largely confined to experiments and in testing laboratories where such a source is available and a wide variation of output voltage is desirable.

5. The performance of a shunt generator is similar to that of a separately excited generator except that the voltage of the shunt generator falls off more rapidly with the increase in load. As in the separately excited generator, armature reaction and $I_a R_a$ drop in the armature circuit cause the terminal voltage of a shunt generator to decrease. Since the field excitation is obtained from the generator itself, the decrease in terminal voltage, because of above effects, also decreases the voltage applied to the field circuit. This causes a decrease in field current and field flux which in turn causes a further decrease in the terminal voltage.

17. The cumulative-compound generator combines the characteristics of the shunt and the series generators. At no-load, the shunt winding provides all the field flux since there is no current in the series field winding. As the load is increased on the generator, the series field adds an increasing amount of field flux to that of the shunt field. The terminal voltage of the loaded generator will depend on relative strengths of the two fields. If the full-load terminal voltage is equal to the no-load terminal voltage, the generator is said to be *flat-compounded.*

5

D.C. Motors

INTRODUCTION

D. C. motors are seldom used in ordinary applications because all electric supply companies furnish alternating current. However, for special applications such as in steel mills, mines and electric trains, it is advantageous to convert alternating current into direct current in order to use d.c. motors. The reason is that speed/torque characteristics of d.c. motors are much more superior to that of a.c. motors. Therefore, it is not surprising to note that for industrial drives, d.c. motors are as popular as 3-phase induction motors. Like d.c. generators, d.c. motors are also of three types *viz.*, series-wound, shunt-wound and compound-wound. The use of a particular motor depends upon the mechanical load it has to drive.

5.1 D.C. MOTOR PRINCIPLE

A machine that converts d.c. power into mechanical power is known as a d.c. motor. Its operation is based on the principle that when a current carrying conductor is palced in a magnetic field, the conductor experiences a mechanical force. The direction of this force is given by Fleming's left hand rule and magnitude is given by ;

$$F = BIl \text{ newtons}$$

Basically, there is no constructional difference between a d.c. motor and a d.c. generator. The same d.c. machine can be run as a generator or motor.

5.2 WORKING OF D.C. MOTOR

Consider a part of a multipolar d.c. motor as shown in Fig. 5.1. When the terminals of the motor are connected to an external source of d.c. supply :

(*i*) the field magnets are excited developing alternate *N* and *S* poles.

(*ii*) the armature conductors carry *currents. All conductors under *N*-pole carry currents in one direction while all the conductors under *S*-pole carry currents in the opposite direction.

Suppose the conductors under *N*-pole carry currents into the plane of the paper and those under *S*-pole carry currents out of the plane of the paper as shown in Fig. 5.1. Since each armature conductor is carrying current and is placed in the magnetic field, mechanical force acts on it. Referring to Fig. 5.1 and applying Fleming's left hand rule, it is clear that force on each conductor is tending to rotate the armature in anticlockwise direction. All these forces add together to produce a driving torque which sets the armature rotating. When the conductor moves from one side of a brush to the other, the current in that conductor is reversed and at the same time it comes under the influence of next pole which is of opposite polarity. Consequently, the direction of force on the conductor remains the same.

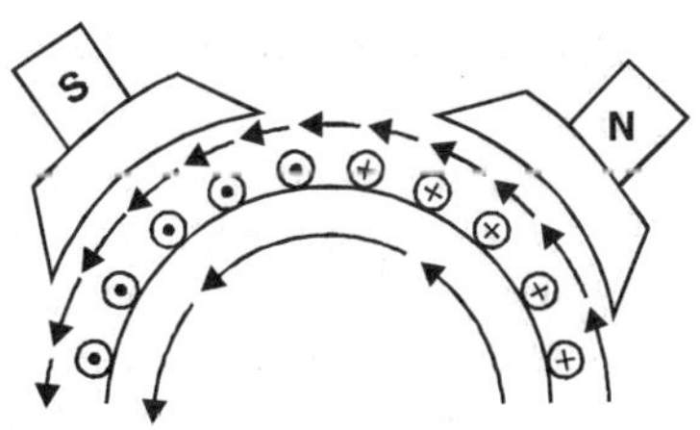

Fig. 5.1

* Since the armature of a motor is the same as that of generator, the current from the supply line must divide and pass through several paths of the armature—2 for wave wound armature and *P* for lap wound armature. As in a generator, currents through all conductors under a pole will be in the same direction.

5.3 BACK OR COUNTER E.M.F.

When the armature of a d.c. motor rotates under the influence of the driving torque, the armature conductors move through the magnetic field and hence e.m.f. is induced in them as in a generator. The induced e.m.f. acts in opposite direction to the applied voltage V (*Lenz's law) and is known as *back or counter e.m.f.* E_b. The back e.m.f. $E_b (= P\phi ZN/60\,A)$ is *always* less than the applied voltage V, although this difference is small when the motor is running under normal conditions.

Consider a shunt wound motor shown in Fig. 5.2. When d.c. voltage V is applied across the motor terminals, the field magnets are excited and armature conductors are supplied with current. Therefore, driving torque acts on the armature which begins to rotate. As the armature rotates, back e.m.f. E_b is induced which opposes the applied voltage V. The applied voltage V has to force current through the armature against the back e.m.f. E_b. The electric work done in overcoming and causing the current to flow against E_b is converted into mechanical energy developed in the armature. It follows, therefore, that energy conversion in a d.c. motor is only possible due to the production of back e.m.f. E_b.

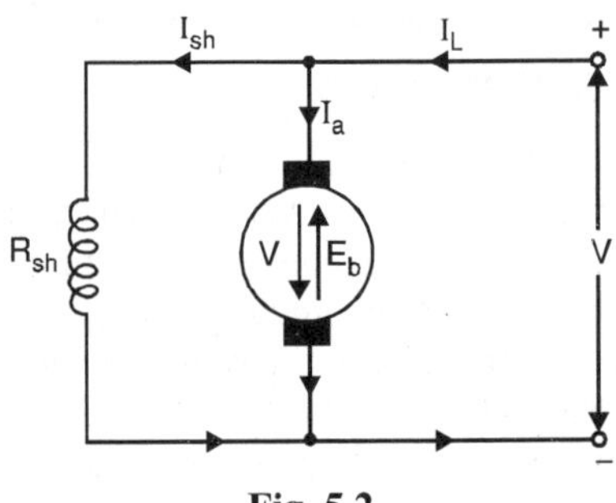

Fig. 5.2

Net voltage across armature circuit $= V - E_b$

If R_a is the armature circuit resistance, then, $I_a = \dfrac{V - E_b}{R_a}$

Since V and R_a are usually fixed, the value of E_b will determine the current drawn by the motor. If the speed of the motor is high, then back e.m.f. $E_b (= P\phi ZN/60\,A)$ is large and hence the motor will draw less armature current and *vice-versa*.

5.4 SIGNIFICANCE OF BACK E.M.F.

The presence of back e.m.f. makes the d.c. motor a *self-regulating* machine *i.e.*, it makes the motor to draw as much armature current as is just sufficient to develop the torque required by the load.

$$\text{Armature current, } I_a = \frac{V - E_b}{R_a}$$

(*i*) When the motor is running on no load, small torque is required to overcome the friction and windage losses. Therefore, the armature current I_a is small and the back e.m.f. is nearly equal to the applied voltage.

(*ii*) If the motor is suddenly loaded, the first effect is to cause the armature to slow down. Therefore, the speed at which the armature conductors move through the field is reduced and hence the back e.m.f. E_b falls. The decreased back e.m.f. allows a larger current to flow through the armature and larger current means increased driving torque. Thus, the driving torque increases as the motor slows down. The motor will stop slowing down when the armature current is just sufficient to produce the increased torque required by the load.

(*iii*) If the load on the motor is decreased, the driving torque is momentarily in excess of the requirement so that armature is accelerated. As the armature speed increases, the back e.m.f. E_b also increases and causes the armature current I_a to decrease. The motor will stop accelerating when the armature current is just sufficient to produce the reduced torque required by the load.

---

* According to Lenz's law, the direction of induced e.m.f. is such that it opposes the cause producing it. The cause producing the back e.m.f. E_b is the applied voltage V. Hence E_b opposes the applied voltage V.

It follows, therefore, that back e.m.f. in a d.c. motor regulates the flow of armature current i.e., it automatically changes the armature current to meet the load requirement.

5.5 VOLTAGE EQUATION OF D.C. MOTOR

Let in a d.c. motor (See Fig. 5.3),

V = applied voltage

E_b = back e.m.f.

R_a = armature resistance

I_a = armature current

Since back e.m.f. E_b acts in opposition to the applied voltage V, the net voltage across the armature circuit is $V - E_b$. The armature current I_a is given by ;

$$I_a = \frac{V - E_b}{R_a}$$

or

$$V = E_b + I_a R_a \quad ...(i)$$

This is known as voltage equation of the d.c. motor.

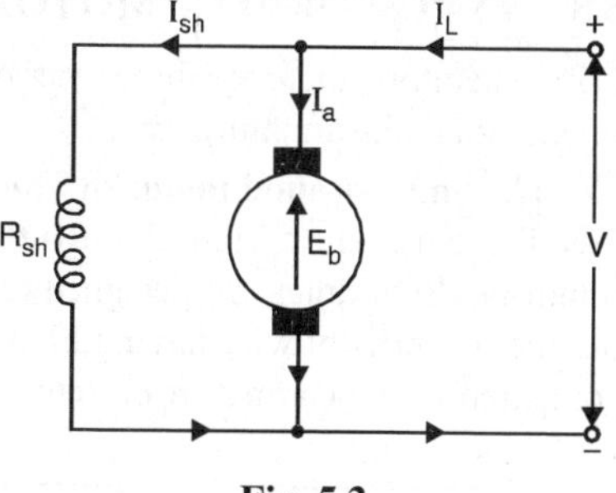

Fig. 5.3

5.6 POWER EQUATION

If eq. (*i*) above is multiplied by I_a throughout, we get,

$$VI_a = E_b I_a + I_a^2 R_a$$

This is known as power equation of the d.c. motor.

VI_a = electric power supplied to armature (*armature input*)

$E_b I_a$ = *power developed by armature (*armature output*)

$I_a^2 R_a$ = electric power wasted in armature (*armature Cu loss*)

Thus out of the armature input, a small portion (about 5%) is wasted as $I_a^2 R_a$ and the remaining portion $E_b I_a$ is converted into mechanical power within the armature.

5.7 CONDITION FOR MAXIMUM POWER

The mechanical power developed by the motor is $P_m = E_b I_a$

Now,

$$P_m = VI_a - I_a^2 R_a$$

Since, V and R_a are fixed, power developed by the motor depends upon armature current. For maximum power, dP_m/dI_a should be zero.

$$\therefore \quad \frac{dP_m}{dI_a} = V - 2I_a R_a = 0$$

or

$$I_a R_a = V/2$$

Now,

$$V = E_b + I_a R_a = E_b + V/2 \qquad [\because I_a R_a = V/2]$$

$$\therefore \quad E_b = V/2$$

Hence mechanical power developed by the motor is maximum when back e.m.f. is equal to half the applied voltage.

Limitations : In practice, we never aim at achieving maximum power due to the following reasons :

(*i*) The armature current under this condition is very large — much excess of rated current of the machine.

---

* This will be the mechanical power output in watts.

(*ii*) Half of the input power is wasted in the armature circuit. In fact, if we take into account other losses (iron and mechanical), the efficiency will be well below 50%.

5.8 TYPES OF D.C. MOTORS

Like generators, there are three types of d.c. motors characterised by the connections of field winding in relation to the armature *viz.* :

(*i*) **Shunt-wound motor** in which the field winding is connected in parallel with the armature [See Fig. 5.4]. The current through the shunt field winding is not the same as the armature current. Shunt field windings are designed to produce the necessary m.m.f. by means of a relatively large number of turns of wire having high resistance. Therefore, shunt field current is relatively small compared with the armature current.

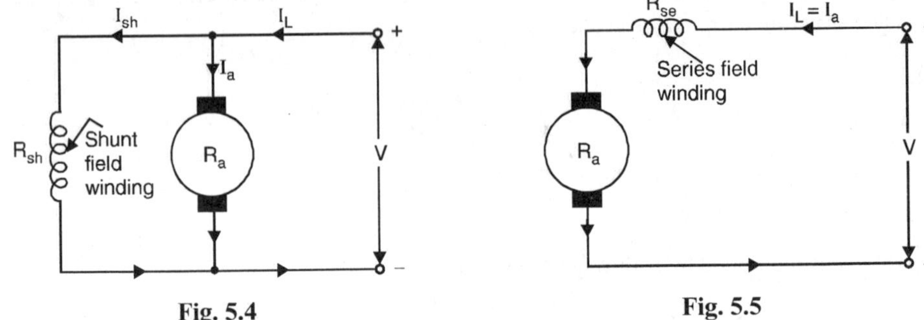

Fig. 5.4 **Fig. 5.5**

(*ii*) **Series-wound motor** in which the field winding is connected in series with the armature [See Fig. 5.5]. Therefore, series field winding carries the armature current. Since the current passing through a series field winding is the same as the armature current, series field windings must be designed with much fewer turns than shunt field windings for the same m.m.f. Therefore, a series field winding has a relatively small number of turns of thick wire and, therefore, will possess a low resistance.

(*iii*) **Compound-wound motor** which has two field windings ; one connected in parallel with the armature and the other in series with it. There are two types of compound motor connections (like generators). When the shunt field winding is directly connected across the armature terminals [See Fig. 5.6], it is called *short-shunt connection.* When the shunt winding is so connected that it shunts the series combination of armature and series field [See Fig. 5.7], it is called *long-shunt connection.*

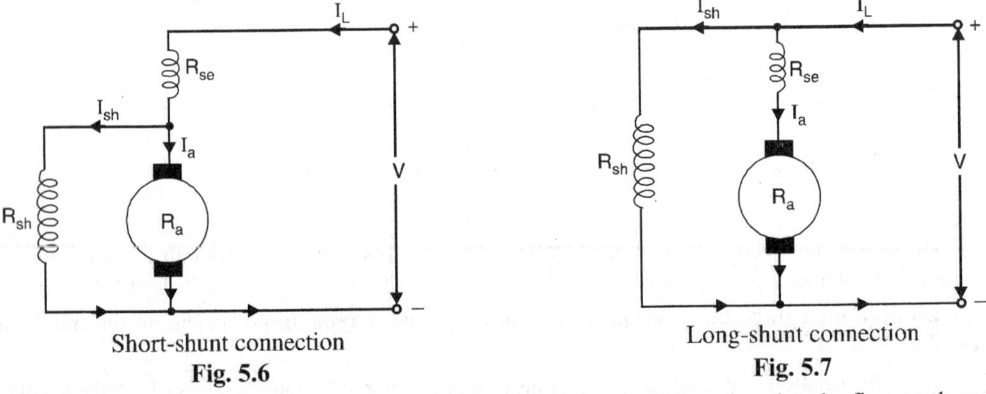

Short-shunt connection
Fig. 5.6

Long-shunt connection
Fig. 5.7

The compound machines (generators or motors) are always designed so that the flux produced by shunt field winding is considerably larger than the flux produced by the series field winding. Therefore, shunt field in compound machines is the basic dominant factor in the production of the magnetic field in the machine.

Example 5.1. *A 250 V shunt motor takes a total current of 20 A. The shunt field and armature resistances are 200 Ω and 0·3 Ω respectively. Determine (i) value of back e.m.f. (ii) gross mechanical power in the armature.*

Solution.

(*i*)

$$\text{Shunt current, } I_{sh} = 250/200 = 1{\cdot}25 \text{ A}$$

$$\text{Armature current, } I_a = 20 - 1{\cdot}25 = 18{\cdot}75 \text{ A}$$

$$\text{Back e.m.f., } E_b = V - I_a R_a = 250 - 18{\cdot}75 \times 0{\cdot}3 = \mathbf{244{\cdot}4\ V}$$

(*ii*)

$$\text{Mechanical power developed} = E_b I_a = 244{\cdot}4 \times 18{\cdot}75 = \mathbf{4582{\cdot}5\ W}$$

Example 5.2. *A 230 V motor has an armature circuit resistance of 0·6 Ω. If the full-load armature current is 30 A and no-load armature current is 4A, find the change in back e.m.f. from no load to full-load.*

Solution.

$$E_b = V - I_a R_a$$

$$\text{At no-load, } E_b = 230 - 4 \times 0{\cdot}6 = 227{\cdot}6 \text{ V}$$

$$\text{At full-load, } E_b = 230 - 30 \times 0{\cdot}6 = 212 \text{ V}$$

$$\text{Change in back e.m.f.} = 227{\cdot}6 - 212 = \mathbf{15{\cdot}6\ V}$$

Example 5.3. *A 4-pole, 500 V shunt motor has 720 wave-connected conductors in the armature. The full-load armature current is 60 A and the flux per pole is 0·03 Wb. The armature resistance is 0·2 Ω and the contact drop is 1 V per brush. Calculate the full-load speed of the motor.*

Solution.

$$V = E_b + I_a R_a + \text{Brush drop}$$

$$500 = E_b + 60 \times 0{\cdot}2 + 2 \times 1$$

$$\therefore \quad E_b = 500 - 60 \times 0{\cdot}2 - 2 \times 1 = 486 \text{ V}$$

Now

$$E_b = \frac{P \phi Z N}{60 A}$$

$$\therefore \quad N = \frac{E_b \times 60 A}{P \phi Z} = \frac{486 \times 60 \times 2}{4 \times 0{\cdot}03 \times 720} = \mathbf{675\ r.p.m.}$$

Example 5.4. *The counter e.m.f. of a shunt motor is 227 V, the field resistance is 160 Ω and field current is 1·5 A. If the line current is 39·5 A, find the armature resistance. Also find the armature current when the motor is stationary.*

Solution.

$$\text{Applied voltage, } V = I_{sh} R_{sh} = 1{\cdot}5 \times 160 = 240 \text{ volts}$$

$$\text{Armature current, } I_a = I_L - I_{sh} = 39{\cdot}5 - 1{\cdot}5 = 38 \text{ A}$$

$$V = E_b + I_a R_a$$

$$\therefore \quad R_a = \frac{V - E_b}{I_a} = \frac{240 - 227}{38} = \mathbf{0{\cdot}342\ \Omega}$$

At the moment of start-up, the armature is stationary so that $E_b = 0$.

$$\therefore \quad I_a = \frac{V}{R_a} = \frac{240}{0{\cdot}342} = \mathbf{701{\cdot}5\ A}$$

Example 5.5. *A 440 V shunt motor has armature resistance of 0·8 Ω and field resistance of 200 Ω. Determine the back e.m.f. when giving an output of 7·46 kW at 85% efficiency.*

Solution.

$$\text{Motor input power, } P_i = \frac{7{\cdot}46 \times 10^3}{\eta} = \frac{7{\cdot}46 \times 10^3}{0{\cdot}85} \text{ W}$$

$$\text{Motor line current, } I_L = \frac{P_i}{V} = \frac{7{\cdot}46 \times 10^3}{0{\cdot}85 \times 440} = 19.95 \text{ A}$$

$$\text{Shunt field current, } I_{sh} = \frac{V}{R_{sh}} = \frac{440}{200} = 2{\cdot}2 \text{ A}$$

$$\text{Armature current, } I_a = I_L - I_{sh} = 19{\cdot}95 - 2{\cdot}2 = 17{\cdot}75 \text{ A}$$

Now, $$E_b = V - I_a R_a = 440 - 17{\cdot}75 \times 0{\cdot}8 = \mathbf{425{\cdot}8\ V}$$

Example 5.6. *A 20 kW, 250V d.c. shunt generator has armature and field resistances of 0·1 Ω and 125 Ω respectively. Calculate the total armature power developed when running (i) as a generator delivering 20 kW output (ii) as a motor taking 20 kW input.*

Solution.

(*i*) **As a generator.** Fig. 5.8 (*i*) shows the connections of shunt generator.

$$I_L = 20 \times 10^3/250 = 80 \text{ A}$$
$$I_{sh} = 250/125 = 2 \text{ A}$$
$$I_a = I_L + I_{sh} = 80 + 2 = 82 \text{ A}$$
$$E_g = V + I_a R_a = 250 + 82 \times 0{\cdot}1 = 258{\cdot}2 \text{ V}$$

∴ Power developed in the armature

$$= E_g I_a = 258{\cdot}2 \times 82 = 21{\cdot}17 \times 10^3 \text{ W} = \mathbf{21{\cdot}17\ kW}$$

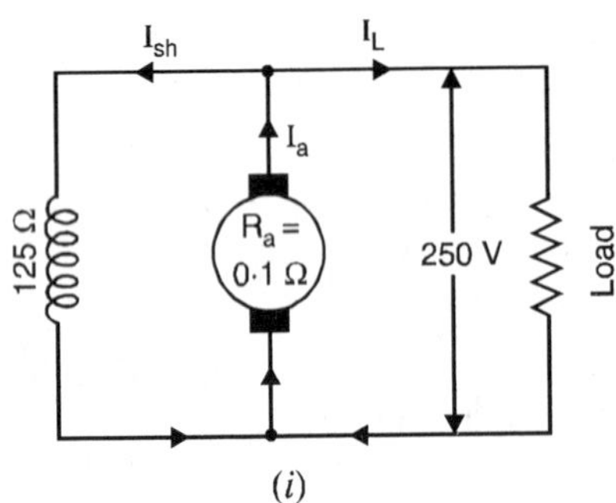

(*i*)

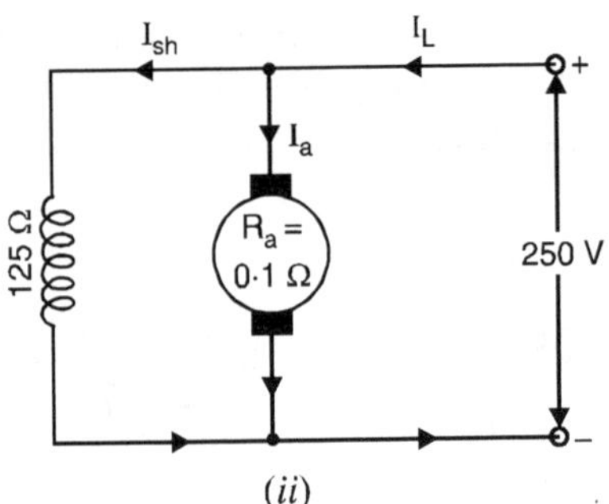

(*ii*)

Fig. 5.8

(*ii*) **As a motor.** Fig. 5.8 (*ii*) shows the connections of shunt motor.

$$I_L = 20 \times 10^3/250 = 80 \text{ A}$$
$$I_{sh} = 250/125 = 2 \text{ A}$$
$$I_a = I_L - I_{sh} = 80 - 2 = 78 \text{ A}$$
$$E_b = V - I_a R_a = 250 - 78 \times 0{\cdot}1 = 242{\cdot}2 \text{ V}$$

Power developed in the armature $= E_b I_a = 242{\cdot}2 \times 78 = 18{\cdot}9 \times 10^3 \text{ W} = \mathbf{18{\cdot}9\ kW}$

Example 5.7. *Find the useful flux per pole on no-load of 250 V, 6-pole shunt motor having a wave-connected armature winding with 110 turns. The armature resistance is 0.2 Ω. The armature current is 13·3 A at the no-load speed of 908 r.p.m.*

Solution.

$$E_b = V - I_a R_a = 250 - 13.3 \times 0.2 = 247{\cdot}34 \text{ V}$$

Now $$E_b = \frac{P \phi Z N}{60\ A}$$

∴ $$\text{Flux/pole, } \phi = \frac{E_b \times 60\ A}{P Z N}$$

Here $E_b = 247{\cdot}34$ V; $A = 2$; $P = 6$; $Z = 110 \times 2 = 220$; $N = 908$ r.p.m.

∴ $$\phi = \frac{247{\cdot}34 \times 60 \times 2}{6 \times 220 \times 908} = 24{\cdot}8 \times 10^{-3} \text{ Wb} = \mathbf{24.8\ mWb}$$

Example 5.8. *A 250 V shunt motor on no-load runs at 1000 r.p.m. and takes 5A. The total armature and shunt field resistances are 0.2 Ω and 250 Ω respectively. Calculate the speed when loaded and taking current of 50 A if armature reaction weakens the field by 3%.*

Solution.

At no-load. At no-load (See Fig. 5.9),

$$I_{sh} = \frac{250}{250} = 1\text{ A} ; I_{a1} = 5 - 1 = 4\text{ A} ; N_1 = 1000 \text{ r.p.m.}$$

$\therefore$ Back e.m.f. at no-load, $E_{b1} = V - I_{a1} R_a = 250 - 4 \times 0{\cdot}2 = 249{\cdot}2\text{V}$

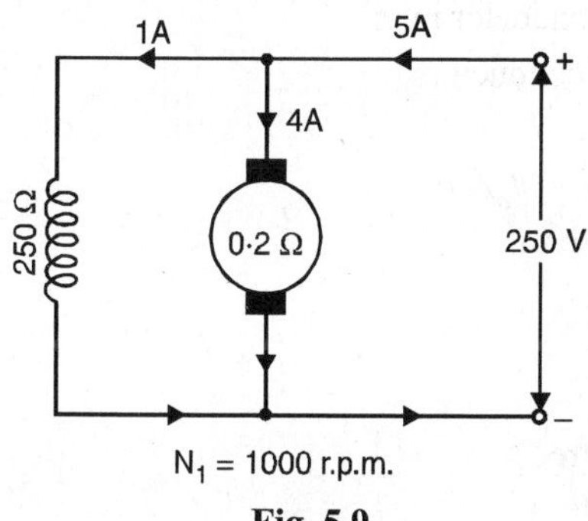

Fig. 5.9

1A 50A + 49 A 250 Ω 0·2 Ω 250 V –

N_2 = ?

Fig. 5.10

At load. At load (See Fig. 5.10),

$$I_{sh} = \frac{250}{250} = 1\text{ A} ; I_{a2} = 50 - 1 = 49\text{ A} ; N_2 = ?$$

$\therefore$ Back e.m.f. at load, $E_{b2} = V - I_{a2} R_a = 250 - 49 \times 0{\cdot}2 = 240{\cdot}2\text{V}$

Now $$\frac{E_{b2}}{E_{b1}} = \frac{N_2 \phi_2}{N_1 \phi_1} \qquad (\because E_b \propto \phi N)$$

$$\therefore \quad N_2 = \frac{E_{b2}}{E_{b1}} \times \frac{N_1 \phi_1}{\phi_2}$$

$$= \frac{240{\cdot}2}{249{\cdot}2} \times 1000 \times \frac{1}{0{\cdot}97} = \textbf{994 r.p.m.} \qquad (\because \phi_2 = 0{\cdot}97\phi_1)$$

TUTORIAL PROBLEMS

1. A 220 V shunt motor takes a total current of 20 A. The shunt field resistance is 250 Ω while armature resistance is 0·3 Ω. Calculate the back e.m.f. **[214.26V]**
2. The armature of a d.c. machine has a resistance of 0·1 Ω and is connected to a 230 V supply. Calculate the back e.m.f. when it is running (*i*) as a generator giving 80 A (*ii*) as a motor taking 80 A. **[(*i*) 238 V (*ii*) 222 V]**
3. A 4-pole d.c. motor is connected to a 500 V d.c. supply and takes an armature current of 80 A. The resistance of the armature circuit is 0·4 Ω. The armature is wave wound with 522 conductors and useful flux per pole is 0·025 Wb. Determine the speed of the motor. **[1075 r.p.m.]**
4. A 4-pole motor is fed at 440 V and takes an armature current of 50 A. The resistance of the armature circuit is 0·28 Ω. The armature winding is wave-connected with 888 conductors and useful flux per pole is 0·023 Wb. Calculate the speed of the motor. **[626 r.p.m.]**
5. A 220 V d.c. machine has an armature resistance of 0·5 Ω. If the full-load armature current is 20 A, find the induced e.m.f. when the machine acts as a generator and as a motor. **[230 V ; 210V]**

5.9 ARMATURE TORQUE OF D.C. MOTOR

Torque is the turning moment of a force about an axis and is measured by the product of force (F) and radius (r) at right angle to which the force acts *i.e.*

$$T = F \times r$$

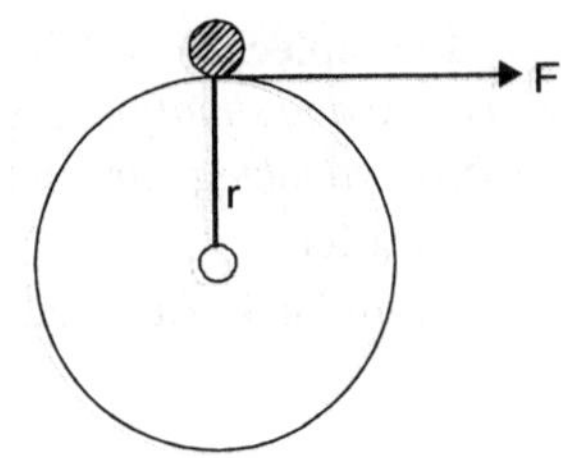

Fig. 5.11

In a d.c. motor, each conductor is acted upon by a circumferential force F at a distance r, the radius of the armature (Fig. 5·11). Therefore, each conductor exerts a torque, tending to rotate the armature. The sum of the torques due to all armature conductors is known as *gross or armature torque* (T_a).

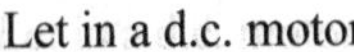

Let in a d.c. motor

r = average radius of armature in m

l = effective length of each conductor in m

Z = total number of armature conductors

A = number of parallel paths

i = current in each conductor = I_a/A

B = average flux density in Wb/m^2

ϕ = flux per pole in Wb

P = number of poles

Force on each conductor, $F = B\,i\,l$ newtons

Torque due to one conductor $= F \times r$ newton- metre

Total armature torque, $T_a = Z F r$ newton-metre

$= Z B i l r$

Now $i = I_a/A$, $B = \phi/a$ where a is the x-sectional area of flux path per pole at radius r. Clearly, $a = 2\pi r\, l/P$.

$$\therefore \quad T_a = Z \times (\phi/a) \times (I_a/A) \times l \times r$$

$$= Z \times \frac{\phi}{2\pi r\, l/P} \times \frac{I_a}{A} \times l \times r$$

$$= \frac{Z \phi I_a P}{2\pi A} \text{ N-m}$$

or $\quad T_a = 0{\cdot}159\, Z \phi I_a (P/A)$ N-m ...(*i*)

Since Z, P and A are fixed for a given machine,

$$\therefore \quad T_a \propto \phi I_a$$

Hence *torque in a d.c. motor is directly proportional to flux per pole and armature current.*

(*i*) For a *shunt motor,* flux ϕ is practically constant.

$$\therefore \quad T_a \propto I_a$$

(*ii*) For a *series motor,* flux ϕ is directly proportional to armature current I_a provided magnetic saturation does not take place.

$\therefore \quad T_a \propto I_a^2$ ...upto magnetic saturation

Alternative expression for T_a

$$E_b = \frac{P \phi Z N}{60\, A}$$

$$\therefore \quad \frac{P \phi Z}{A} = \frac{60 \times E_b}{N}$$

From eq. (*i*), we get the expression of T_a as :

$$T_a = 0{\cdot}159 \times \left(\frac{60 \times E_b}{N}\right) \times I_a$$

or $$T_a = 9.55 \times \frac{E_b I_a}{N} \text{ N-m}$$

Note that developed torque or gross torque means armature torque T_a.

5.10 SHAFT TORQUE (T_{sh})

The torque which is available at the motor shaft for doing useful work is known as *shaft torque*. It is represented by T_{sh}. Fig. 5.12 illustrates the concept of shaft torque.

The total or gross torque T_a developed in the armature of a motor is not available at the shaft because a part of it is lost in overcoming the iron and frictional losses in the motor. Therefore, shaft torque T_{sh} is somewhat less than the armature torque T_a. The difference $T_a - T_{sh}$ is called lost torque.

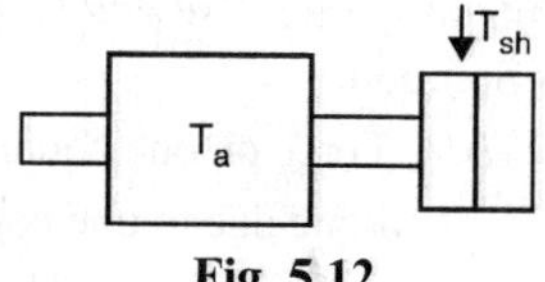

Fig. 5.12

Clearly, $$T_a - T_{sh} = 9{\cdot}55 \times \frac{\text{Iron and frictional losses}}{N}$$

For example, if the iron and frictional losses in a motor are 1600 W and the motor runs at 800 r.p.m., then,

$$T_a - T_{sh} = 9{\cdot}55 \times \frac{1600}{800} = 19{\cdot}1 \text{ N-m}$$

As stated above, it is the shaft torque T_{sh} that produces the useful output. If the speed of the motor is N r.p.m., then,

$$\text{Output in watts} = \frac{2\pi N T_{sh}}{60}$$

or $$T_{sh} = \frac{\text{Output in watts}}{2\pi N/60} \text{ N-m}$$

or $$T_{sh} = 9{\cdot}55 \times \frac{\text{Output in watts}}{N} \text{ N-m} \qquad (\because 60/2\pi = 9{\cdot}55)$$

5.11 BRAKE HORSE POWER (b.h.p.)

The horse power developed by the shaft torque is known as *brake horse power* (*b.h.p.*). If the motor is running at N r.p.m. and the shaft torque is T_{sh} newton-metres, then,

$$\text{W.D./revolution} = \text{force} \times \text{distance moved in 1 revolution}$$

$$= F \times 2\pi r = 2\pi \times T_{sh} \text{ J}$$

$$\text{W.D./minute} = 2\pi N T_{sh} \text{ J}$$

$$\text{W.D./sec.} = \frac{2\pi N T_{sh}}{60} \text{ Js}^{-1} \text{ or watts} = \frac{2\pi N T_{sh}}{60 \times 746} \text{ h.p.}$$

$\therefore$ $$\text{Useful output power} = \frac{2\pi N T_{sh}}{60 \times 746} \text{ h.p.}$$

or $$\text{b.h.p.} = \frac{2\pi N T_{sh}}{60 \times 746}$$

Example 5.9. *Calculate the value of torque established by the armature of a 4-pole motor having 774 conductors, two paths in parallel, 24 mWb flux per pole, when the total armature current is 50 A.*

Solution.

Armature torque, $$T_a = 0{\cdot}159\, Z \phi I_a (P/A)$$

Here, $$Z = 774\,;\ \phi = 24 \times 10^{-3} \text{ Wb}\,;\ I_a = 50 \text{ A}\,;\ P = 4\,;\ A = 2$$

$\therefore$ $$T_a = 0{\cdot}159 \times 774 \times 24 \times 10^{-3} \times 50 \times (4/2) = \mathbf{295{\cdot}35 \text{ N-m}}$$

Example 5.10. *An armature of a 6-pole machine 75cm in diameter has 664 conductors each having an effective length of 30 cm and carrying a current of 100 A. If 70% of total conductors lie simultaneously in the field of average flux density 0·85 Wb/m², calculate (i) armature torque (ii) horse power output at 250 r.p.m.*

Solution.

(*i*) Force on one conductor, $F = B\,i\,l = 0{\cdot}85 \times 100 \times 0{\cdot}3 = 25{\cdot}5$ N

Torque due to one conductor $= F \times r = 25{\cdot}5 \times 0{\cdot}375 = 9{\cdot}56$ N-m

No. of effective conductors, $Z = 664 \times 70/100 = 465$

∴ Total armature torque, $T_a = 9{\cdot}56 \times 465 =$ **4445·4 N-m**

(*ii*) Power output $= \dfrac{2\pi N T_a^*}{60 \times 746}$ h.p.

$$= \frac{2\,\pi \times 250 \times 4445{\cdot}4}{60 \times 746} = \mathbf{156\ h.p.}$$

Example 5.11. *A 230 V d.c. shunt motor takes a current of 40 A and runs at 1100 r.p.m. If armature and shunt field resistances are 0·25 Ω and 230 Ω respectively, find the torque developed by the armature.*

Solution.

Shunt field current, $I_{sh} = 230/230 = 1$A

Armature current, $I_a = 40 - 1 = 39$ A

Back e.m.f., $E_b = V - I_a R_a = 230 - 39 \times 0{\cdot}25 = 220{\cdot}25$ V

Now $T_a = 9{\cdot}55 \times E_b I_a / N$ ...See Art. 5·9

$= 9{\cdot}55 \times (220{\cdot}25 \times 39/1100)$ N-m $=$ **74·66 N-m**

Example 5.12. *A d.c. motor takes an armature current of 110 A at 480 V. The armature circuit resistance is 0·2 Ω. The machine has 6 poles and the armature is lap-connected with 864 conductors. The flux per pole is 0·05 Wb. Calculate (i) the speed and (ii) the gross torque developed by the motor.*

Solution.

(*i*) $E_b = V - I_a R_a = 480 - 110 \times 0{\cdot}2 = 458$ volts

Now $E_b = \dfrac{P\,\phi\,Z\,N}{60\,A}$

or $458 = \dfrac{6 \times 0{\cdot}05 \times 864 \times N}{60 \times 6}$

∴ $N =$ **636 r.p.m.**

(*ii*) Gross torque means armature torque T_a.

$$T_a = 0{\cdot}159\,\phi\,Z\,I_a \times \left(\frac{P}{A}\right) = 0{\cdot}159 \times 0{\cdot}05 \times 864 \times 110 \times \left(\frac{6}{6}\right) = \mathbf{756{\cdot}3\ N\text{-}m}$$

Example 5.13. *A 100 h.p., 500 V shunt motor has 4 poles and a 2 circuit wave winding with 492 armature conductors. The flux is 50 mWb per pole and the full-load efficiency 92%. The armature and commutating field windings have a total resistance of 0·1 Ω. The shunt field resistance is 250 Ω. Calculate for full load (i) the speed (ii) the useful torque.*

Solution.

Fig. 5.13 shows the shunt motor connections. The commutating field winding is connected in series with the armature winding so that it carries the armature current.

(*i*) Motor input power $= \dfrac{100 \times 746}{0{\cdot}92} = 8{\cdot}1 \times 10^4$ W

---

* Iron and frictional losses are neglected so that $T_a = T_{sh}$.

$$\text{Line current, } I_L = \frac{8{\cdot}1\times10^4}{500} = 162\text{ A}$$

$$I_{sh} = \frac{500}{250} = 2\text{ A}$$

$$I_a = I_L - I_{sh}$$
$$= 162 - 2 = 160\text{ A}$$

Now $E_b = V - I_a R_a$
$$= 500 - 160\times0{\cdot}1$$
$$= \mathbf{484\ volts}$$

Fig. 5.13

Also $E_b = \dfrac{P\,\phi\,Z\,N}{60\,A}$

or $$484 = \frac{4\times50\times10^{-3}\times492\times N}{60\times2}$$

$\therefore$ $N = \mathbf{590\ r.p.m.}$

(*ii*) $\text{Output power in watts} = \dfrac{2\pi\,N\,T_{sh}}{60}$

or $$100\times746 = \frac{2\pi\times590\times T_{sh}}{60}$$

$\therefore$ $$T_{sh} = \frac{100\times746\times60}{590\times2\,\pi} = \mathbf{1200\ N\text{-}m}$$

Example 5.14. *A 220 V d.c. shunt motor runs at 500 r.p.m. when the armature current is 50 A. Calculate the speed if the torque is doubled. The armature resistance of the motor is 0·2 Ω.*

Solution.

$$T \propto \phi I_a. \text{ Since } \phi \text{ is constant, } T \propto I_a.$$

Therefore, when torque is doubled, armature current drawn by the motor is also doubled.

$\therefore$ $I_{a1} = 50\text{ A and } I_{a2} = 2\times50 = 100\text{ A}$

Now $E_{b1} = V - I_{a1}R_a = 220 - 50\times0{\cdot}2 = 210\text{ volts}$

and $E_{b2} = V - I_{a2}R_a = 220 - 100\times0{\cdot}2 = 200\text{ volts}$

Now $\dfrac{E_{b2}}{E_{b1}} = \dfrac{N_2}{N_1}$ ($\because$ ϕ is constant)

$\therefore$ $$N_2 = \frac{E_{b2}}{E_{b1}}\times N_1 = \frac{200}{210}\times500 = \mathbf{476\ r.p.m.}$$

Example 5.15. *A 240 V, 4-pole shunt motor running at 1000 r.p.m. gives 15 h.p. with an armature current of 50 A and field current of 1 A. The armature winding is wave-connected and has 540 conductors. The armature resistance is 0·1 Ω and the drop at each brush is 1 V. Find (i) the useful torque (ii) the total torque (iii) useful flux per pole and (iv) iron and frictional losses.*

Solution. Fig. 5.14 shows the shunt motor connections.

(*i*) Useful power $= 15\times746 = 11190\text{ W}$

Now $\dfrac{2\pi\,N\,T_{sh}}{60} = 11190$

$\therefore$ $$T_{sh} = \frac{11190\times60}{2\pi N}$$
$$= \frac{11190\times60}{2\pi\times1000} = \mathbf{106\ N\text{-}m}$$

(*ii*) Back e.m.f., $E_b = V - I_a R_a - \text{Brush drop}$

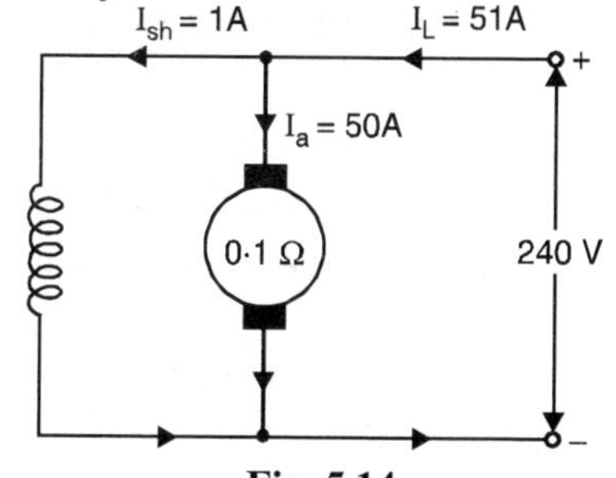

Fig. 5.14

$$= 240 - 50 \times 0.1 - 2 \times 1 = 233 \text{ volts}$$

$$\text{Total armature torque, } T_a = 9{\cdot}55 \times \frac{E_b I_a}{N} = 9{\cdot}55 \times \frac{233 \times 50}{1000} = \mathbf{110\ N\text{-}m}$$

(*iii*)
$$E_b = \frac{P \phi Z N}{60 A}$$

or
$$233 = \frac{4 \times \phi \times 540 \times 1000}{60 \times 2} \quad \text{or} \quad \phi = 13 \times 10^{-3} \text{ Wb} = \mathbf{13\ mWb}$$

(*iv*) Iron and frictional losses $= E_b I_a -$ Useful power

$$= 233 \times 50 - 11190 = \mathbf{460\ W}$$

Example 5.16. *A cutting tool exerts a tangential force of 400 N on a steel bar of diameter 10 cm which is being turned in a simple lathe. The lathe is driven by a chain at 840 r.p.m. from a 220 V d.c. motor which runs at 1800 r.p.m. Calculate the current taken by the motor if its efficiency is 80%. What size is the motor pulley if the lathe pulley has a diameter of 24 cm ?*

Solution. Useful torque, T_{sh} = Tangential force × radius = 400 × 0·05 = 20 N-m

$$\text{Output power} = \frac{2\pi N T_{sh}}{60} = \frac{2\pi \times 840 \times 20}{60} = 1760 \text{ W}$$

$$\text{Motor input power} = \frac{\text{Output}}{\eta} = \frac{1760}{0{\cdot}8} = 2200 \text{ W}$$

Current taken by motor = 2200/220 = **10 A**

Suppose N_1 and D_1 are the speed and diameter of the driver pulley respectively and N_2 and D_2 are the values for the lathe pulley. Then,

$$N_1 D_1 = N_2 D_2$$

$$\therefore \quad D_1 = \frac{N_2 D_2}{N_1} = \frac{840 \times 0{\cdot}24}{1800} = \mathbf{11{\cdot}2\ cm}$$

Example 5.17. *A shunt motor takes 5 A at 100 V when running light. Shunt field resistance is 50 Ω and armature resistance is 0·2 Ω. Find the iron and frictional losses. If the motor is loaded and takes 52 A, what is the b.h.p. ?*

Solution. When a shunt motor is running and doing no useful work, it is said to be "running light". Under these conditions, the power developed by the armature is entirely used up in overcoming the iron and frictional losses in the motor. For a motor, these losses are nearly constant at all loads.

$$I_{sh} = 100/50 = 2\text{A}\,;\; I_a = I_L - I_{sh} = 5 - 2 = 3 \text{ A}$$

$$\text{Back e.m.f., } E_b = V - I_a R_a = 100 - 3 \times 0{\cdot}2 = 99{\cdot}4 \text{ volts}$$

$$\text{Power developed in armature} = E_b I_a = 99{\cdot}4 \times 3 = \mathbf{298{\cdot}2\ W}$$

Since the motor is running light, output is zero.

∴ Iron and frictional losses = **298·2 W**

When the motor is loaded and taking a current of 52 A (= I_L), we have,

$$\text{Armature current, } I_a = I_L - I_{sh} = 52 - 2 = 50 \text{ A}$$

$$\text{Back e.m.f., } E_b = V - I_a R_a = 100 - 50 \times 0{\cdot}2 = 90 \text{ volts}$$

Mechanical power developed in the armature

$$= E_b I_a = 90 \times 50 = 4500 \text{ W}$$

$$\text{Useful output} = 4500 - \text{Iron and frictional losses}$$

$$= 4500 - 298{\cdot}2 = 4201{\cdot}8 \text{ W}$$

$$\therefore \quad \text{b.h.p.} = \frac{4201 \cdot 8}{746} = \mathbf{5{\cdot}65\ h.p.}$$

Example 5.18. *The armature of a 6-pole, 6 circuit shunt motor takes 400 A at a speed of 350 r.p.m. The flux per pole is 80 mWb, the number of armature turns is 600 and 3% of the torque is lost in windage, friction and iron losses. Calculate the b.h.p.*

Solution.

$$\text{Back e.m.f., } E_b = \frac{P \phi Z N}{60 A} = \frac{6 \times 80 \times 10^{-3} \times (600 \times 2) \times 350}{60 \times 6} = 560 \text{ V}$$

Since 3% torque is lost in windage, friction and iron losses,

$$\therefore \quad \text{Useful output power} = 0{\cdot}97 \times E_b I_a = 0{\cdot}97 \times 560 \times 400 = 2{\cdot}17 \times 10^5 \text{ W}$$

$$\therefore \quad \text{b.h.p.} = \frac{2{\cdot}17 \times 10^5}{746} = \mathbf{291\ h.p.}$$

Example 5.19. *A 4-pole series motor has 944 wave-connected armature conductors. At a certain load, the flux per pole is 34·6 mWb and the total mechanical torque developed is 209 N-m. Calculate the line current taken by the motor and the speed at which it will run. The applied voltage is 500 V and total motor resistance is 3 Ω.*

Solution. Fig. 5.15 shows the series motor connections. It is given that $^*R_a + R_{se} = 3\ \Omega$.

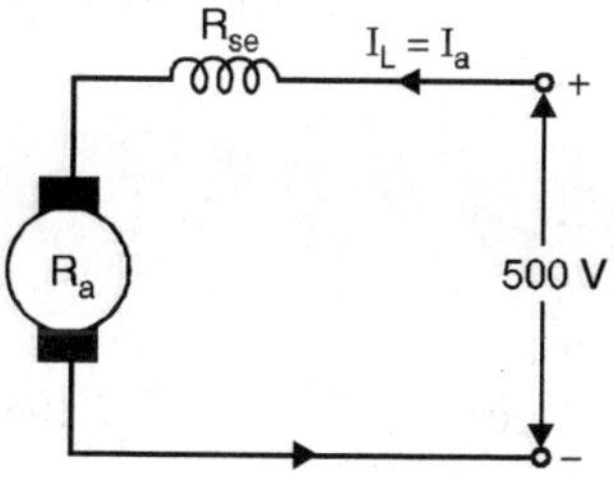

Fig. 5.15

$$T_a = 0{\cdot}159\, \phi Z I_a \left(\frac{P}{A}\right)$$

or $$209 = 0{\cdot}159 \times 34{\cdot}6 \times 10^{-3} \times 944 \times I_a \times \left(\frac{4}{2}\right)$$

$$\therefore \quad I_a = I_L = \mathbf{20\ A}$$

Now $$E_b = V - I_a (R_a + R_{se}) = 500 - 20 \times 3 = 440 \text{ volts}$$

But $$E_b = \frac{P \phi Z N}{60 A}$$

or $$N = \frac{E_b \times 60 A}{P \phi Z} = \frac{440 \times 60 \times 2}{4 \times 34{\cdot}6 \times 10^{-3} \times 944} = \mathbf{403\ r.p.m.}$$

Example 5.20. *A 460 V d.c. series motor runs at 500 r.p.m. taking a current of 40 A. Calculate the speed and percentage change in torque if the load is reduced so that the motor is taking 30 A. The total resistance of the armature and field circuits is 0·8 Ω. Assume that flux is proportional to the field current.*

Solution. $T \propto \phi I_a$. Since $\phi \propto I_a$, $T \propto I_a^2$.

For the first case, $T_1 \propto (40)^2$; For the second case, $T_2 \propto (30)^2$

$$\therefore \quad \frac{T_2}{T_1} = \left(\frac{(30)^2}{(40)^2}\right) = \frac{9}{16}$$

or $$\frac{T_1 - T_2}{T_1} = \frac{16 - 9}{16} = 0{\cdot}4375 = \mathbf{43{\cdot}75\%}$$

$$E_{b1} = V - I_{a1} (R_a + R_{se}) = 460 - 40 \times 0{\cdot}8 = 428 \text{ volts}$$

$$E_{b2} = V - I_{a2} (R_a + R_{se}) = 460 - 30 \times 0{\cdot}8 = 436 \text{ volts}$$

---

* It will also include compole winding resistance.

Now $$\frac{E_{b2}}{E_{b1}} = \frac{N_2 \phi_2}{N_1 \phi_1} = \frac{N_2 I_{a2}}{N_1 I_{a1}}$$

$\therefore$ $$N_2 = \left(\frac{E_{b2}}{E_{b1}} \times \frac{I_{a1}}{I_{a2}}\right) N_1 = \left(\frac{436}{428} \times \frac{40}{30}\right) \times 500 = \mathbf{679\ r.p.m.}$$

Example 5.21. *A 250 V, 4-pole series motor has a 2 circuit wave-wound winding with 105 slots, each containing 12 conductors. The gap flux per pole is 0·02 Wb when the motor is on full-load and taking a current of 45 A. The armature and field resistances are 0·2 Ω and 0·1 Ω respectively and iron and frictional losses total 700 W. Calculate (i) speed (ii) shaft-torque and (iii) b.h.p.*

Solution.

(*i*) Total circuit resistance, $R = R_a + R_{se} = 0{\cdot}2 + 0{\cdot}1 = 0{\cdot}3\ \Omega$

$\therefore$ Mechanical power developed in the armature (P_m)

$$= V I_a - I_a^2 R = 250 \times 45 - (45)^2 \times 0{\cdot}3 = 10642{\cdot}5 \text{ W}$$

Armature torque, $$T_a = 0{\cdot}159\, \phi Z I_a \left(\frac{P}{A}\right)$$

$$= 0{\cdot}159 \times 0{\cdot}02 \times (105 \times 12) \times 45 \times \left(\frac{4}{2}\right) = 361 \text{ N-m}$$

$$P_m = \frac{2\pi N T_a}{60}$$

or $$10642{\cdot}5 = \frac{2\pi \times N \times 361}{60}$$

$\therefore$ $N = \mathbf{282\ r.p.m.}$

(*ii*) Useful power $= P_m -$ Friction and iron losses

$= 10642{\cdot}5 - 700 = 9942{\cdot}5$ W

$\therefore$ $$9942{\cdot}5 = \frac{2\pi N T_{sh}}{60}$$

or $$9942{\cdot}5 = \frac{2\pi \times 282 \times T_{sh}}{60} \quad \text{or} \quad T_{sh} = \mathbf{336\ N\text{-}m}$$

(*iii*) $$\text{b.h.p.} = \frac{9942{\cdot}5}{746} = \mathbf{13{\cdot}3\ h.p.}$$

Example 5.22. *A 500 V, 37.3 kW, 1000 r.p.m. d.c. shunt motor has on full-load an efficiency of 90%. The armature circuit resistance is 0.24 Ω and there is total voltage drop of 2V at the brushes. The field current is 1.8 A. Determine (i) full-load line current (ii) full-load shaft torque (iii) total resistance in motor starter to limit the starting current to 1.5 times the full-load current.*

Solution.

(*i*) Motor input $= (37.3 \times 10^3)/0.9 = 41444$ W

$\therefore$ F.L. line current $= (41444)/500 = \mathbf{82.9\ A}$

(*ii*) Shaft torque, $$T_{sh} = 9.55 \times \frac{\text{Output power}}{N}$$

$$= 9.55 \times \frac{37.3 \times 10^3}{1000} = \mathbf{356\ Nm}$$

(*iii*) Starting line current $= 1.5 \times I_{F.L.} = 1.5 \times 82.9 = 124.3$ A

Armature current at starting $= 124.3 - 1.8 = 122.5$ A

If R is the starter resistance (which is in series with the armature), then,

$$122.5(R + 0.24) = 500 \quad \therefore \quad R = \mathbf{3.842\ \Omega}$$

TUTORIAL PROBLEMS

1. A 250 V, 4-pole, wave-wound series motor has 782 conductors on its armature. It has armature and series field resistance of 0·75 Ω. The motor takes a current of 40 A. Find its speed and gross torque developed if it has a flux per pole of 25 mWb. **[338 r.p.m. ; 249 N-m]**
2. A 220 V shunt motor takes 105 A. The armature resistance is 0·08 Ω and shunt field resistance is 44 Ω. The motor runs at 950 r.p.m. If iron and friction losses are equal to 2 kW, find (*i*) b.h.p. (*ii*) total torque (*iii*) shaft torque. **[(*i*) 25·74 h.p. (*ii*) 212·5 N-m (*iii*) 192·8 N-m]**
3. The armature winding of a 200 V, 4-pole series motor is lap-connected. There are 280 slots and each slot has 4 conductors. The current is 45 A and the flux per pole is 18 mWb. The field resistance is 0·3 Ω ; the armature resistance 0·5 Ω and the iron and friction losses total 800 W. The pulley diameter is 0·41 m. Find the pull in newton at the rim of the pulley. **[634 N]**
4. The output of a certain shunt motor is 5 b.h.p. when taking 50 A at 100 V and running at 750 r.p.m. Armature resistance is 0·06 Ω, shunt field resistance is 50 Ω. Find (*i*) iron and friction losses (*ii*) armature torque (*iii*) shaft torque. **[(*i*) 932 W (*ii*) 59·2 N-m (*iii*) 47·4 N-m]**
5. A 500 V, 37·3 kW, 1000 r.p.m. d.c. shunt motor has on full-load an efficiency of 90%. The armature circuit resistance is 0·24 Ω and there is a total voltage drop of 2 V at the brushes. The field current is 1·8 A. Determine (*i*) full-load line current (*ii*) full-load shaft torque (*iii*) the total resistance in motor starter to limit the starting current to 1·5 times the full-load current. **[(*i*) 82·9 A (*ii*) 356 N-m (*iii*) 3·825 Ω]**
6. A 200 V shunt motor - armature resistance 0·2 Ω, shunt field resistance 100 Ω – takes 7 A when running light. Find the iron and friction losses. Assuming these losses to remain constant, find the b.h.p. when taking a current of 52 A. **[995 W ; 11·4 h.p.]**
7. A 250 V shunt motor giving 20 h.p. at 1000 r.p.m. takes an armature current of 75 A. The armature resistance is 0·25 Ω and the load torque remains constant. If the flux is reduced by 20% of its normal value before the speed changes, find the instantaneous value of armature current and the torque. Determine the final value of armature current and speed. **[260 A ; 459 N-m ; 93·7A ; 1225 r.p.m.]**
8. A d.c. series motor takes a current of 100 A and runs at 1000 r.p.m. The total resistance of the motor is 0·1 Ω and the field is unsaturated. Calculate the speed and the percentage change in torque if the load is so altered that the motor current is 50 A. **[2050 r.p.m. ; 75%]**
9. A 4-pole, 250 V d.c. shunt motor has a lap-connected armature with 960 conductors. The flux per pole is 2×10^{-2} Wb. Calculate the torque developed by the armature and the useful torque when current taken by the motor is 30 A. The armature resistance is 0·12 Ω and the field resistance is 125 Ω. The rotational losses amount to 825 W. **[85·5 N-m ; 75·3 N-m]**
10. A 20 kW d.c. shunt generator delivering rated output at 1000 r.p.m. has a terminal voltage of 500 V. The armature resistance is 0·1 Ω, voltage drop per brush is 1 V and field resistance is 500 Ω. Calculate the speed at which the machine will run as a motor taking an input of 20 kW from a 500 V d.c. supply. **[976·1 r.p.m.]**

5.12 SPEED OF A D.C. MOTOR

$$E_b = V - I_a R_a$$

But $$E_b = P\phi ZN/60A$$

∴ $$\frac{P\phi ZN}{60A} = V - I_a R_a$$

or $$N = \frac{(V - I_a R_a)}{\phi}\frac{60A}{PZ}$$

or $$N = K\frac{(V - I_a R_a)}{\phi} \quad \text{where } K = 60A/PZ$$

But $$V - I_a R_a = E_b$$

∴ $$N = K\frac{E_b}{\phi}$$

or $$N \propto \frac{E_b}{\phi}$$

Therefore, in a d.c. motor, speed is directly proportional to back e.m.f. E_b and inversely proportional to flux per pole ϕ.

5.13 SPEED RELATIONS

If a d.c. motor has initial values of speed, flux per pole and back e.m.f. as N_1, ϕ_1 and E_{b1} respectively and the corresponding final values are N_2, ϕ_2 and E_{b2}, then,

$$N_1 \propto \frac{E_{b1}}{\phi_1} \quad \text{and} \quad N_2 \propto \frac{E_{b2}}{\phi_2}$$

$$\therefore \quad \frac{N_2}{N_1} = \frac{E_{b2}}{E_{b1}} \times \frac{\phi_1}{\phi_2}$$

(*i*) For a **shunt motor**, flux practically remains constant so that $\phi_1 = \phi_2$.

$$\therefore \quad \frac{N_2}{N_1} = \frac{E_{b2}}{E_{b1}}$$

(*ii*) For a **series motor**, $\phi \propto I_a$ prior to saturation.

$$\therefore \quad \frac{N_2}{N_1} = \frac{E_{b2}}{E_{b1}} \times \frac{I_{a1}}{I_{a2}}$$

where I_{a1} = initial armature current

I_{a2} = final armature current

5.14 SPEED REGULATION

The **speed regulation** *of a motor is the change in speed from full-load to no-load and is expressed as a percentage of the speed at full-load i.e.*

$$\% \text{ Speed regulation} = \frac{N.L.\text{ speed} - F.L.\text{ speed}}{F.L.\text{ speed}} \times 100$$

$$= \frac{N_0 - N}{N} \times 100$$

where N_0 = No - load speed

N = Full - load speed

5.15 TORQUE AND SPEED OF A D.C. MOTOR

For any motor, the torque and speed are very important factors. When the torque increases, the speed of a motor increases and *vice-versa*. We have seen that for a d.c. motor,

$$N = K\frac{(V - I_a R_a)}{\phi} = \frac{K E_b}{\phi} \qquad ...(i)$$

$$T_a \propto \phi I_a \qquad ...(ii)$$

If the flux decreases, from eq. (*i*), the motor speed increases but from eq. (*ii*) the motor torque decreases. This is not possible because the increase in motor speed must be the result of increased torque. Indeed, it is so in this case. When the flux decreases slightly, the armature current increases to a large value. As a result, inspite of the weakened field, the torque is momentarily increased to a high value and will exceed considerably the value corresponding to the load. The surplus torque available causes the motor to accelerate and back e.m.f. ($E_b = P \phi Z N/60 A$) to rise. Steady conditions of speed will ultimately be achieved when back e.m.f. has risen to such a value that armature current $[I_a = (V - E_b)/R_a]$ develops torque just sufficient to drive the load.

Illustration. Let us illustrate the above point with a numerical example. Suppose a 400 V shunt motor is running at 600 r.p.m., taking an armature current of 50 A. The armature resistance is 0.28 Ω. Let us see the effect of sudden reduction of flux by 5% on the motor.

Initially (prior to weakening of field), we have,

$$E_b = V - I_a R_a = 400 - 50 \times 0{\cdot}28 = 386 \text{ volts}$$

We know that $E_b \propto \phi N$. If the flux is reduced suddenly, $E_b \propto \phi$ because inertia of heavy armature prevents any rapid change in speed. It follows that when the flux is reduced by 5%, the generated e.m.f. must follow suit. Thus at the instant of reduction of flux, $E'_b = 0{\cdot}95 \times 386 = 366{\cdot}7$ volts.

Instantaneous armature current is

$$I'_a = \frac{V - E'_b}{R_a} = \frac{400 - 366{\cdot}7}{0{\cdot}28} = 118{\cdot}9 \text{ A}$$

Note that a sudden reduction of 5% in the flux has caused the armature current to increase about 2·5 times the initial value. This will result in the production of high value of torque. However, soon the steady conditions will prevail. This will depend on the system inertia ; the more rapidly the motor can alter the speed, the sooner the e.m.f. rises and the armature current falls.

Example 5.23. *A 230 V d.c. shunt motor takes 5 A at no-load and runs at 1000 r.p.m. Calculate the speed when loaded and taking a current of 30 A. The armature and field resistances are 0·2 Ω and 230 Ω respectively.*

Solution.

Let N_2 r.p.m. be the speed under loaded conditions.

No-load conditions

$$I_{sh} = 230/230 = 1\text{A} ; I_{a1} = I_L - I_{sh} = 5 - 1 = 4 \text{ A}$$

$$E_{b1} = V - I_{a1} R_a = 230 - 4 \times 0{\cdot}2 = 229{\cdot}2 \; V; \; N_1 = 1000 \text{ r.p.m.}$$

Loaded conditions

$$I_{a2} = I_L - I_{sh} = 30 - 1 = 29 \text{ A}$$

$$E_{b2} = V - I_{a2} R_a = 230 - 29 \times 0{\cdot}2 = 224{\cdot}2 \text{ V}$$

Since it is a shunt motor,

$$\therefore \quad \frac{N_2}{N_1} = \frac{E_{b_2}}{E_{b_1}} \qquad (\because \phi_1 = \phi_2 = \text{constant})$$

or

$$N_2 = \left(\frac{E_{b_2}}{E_{b_1}}\right) N_1 = \frac{224{\cdot}2}{229{\cdot}2} \times 1000 = \mathbf{978 \text{ r.p.m.}}$$

Example 5.24. *A 6-pole lap wound shunt motor has 500 conductors. The armature and shunt field resistances are 0·05 Ω and 25 Ω respectively. Find the speed of the motor if it takes 120 A from a d.c. supply of 100 V. Flux per pole is 20 mWb.*

Solution.

$$I_{sh} = 100/25 = 4 \text{ A}$$

$$I_a = I_L - I_{sh} = 120 - 4 = 116 \text{ A}$$

$$E_b = V - I_a R_a = 100 - 116 \times 0{\cdot}05 = 94{\cdot}2 \text{ V}$$

Now

$$E_b = \frac{P \phi Z N}{60 \, A}$$

$$\therefore \quad N = \frac{E_b \times 60 \, A}{P \phi Z} = \frac{94{\cdot}2 \times 60 \times 6}{6 \times 20 \times 10^{-3} \times 500} = \mathbf{565 \text{ r.p.m.}}$$

Example 5.25. *A d.c. series motor operates at 800 r.p.m. with a line current of 100 A from 230 V mains. Its armature circuit resistance is 0·15 Ω and its field resistance is 0·1 Ω. Find the speed at which the motor runs at a line current of 25 A, assuming the flux at this current is 45% of flux at 100 A.*

Solution.

$$E_{b1} = V - I_{a1}(R_a + R_{se}) = 230 - 100\,(0{\cdot}15 + 0{\cdot}1) = 205 \text{ volts}$$

$$E_{b2} = V - I_{a2}(R_a + R_{se}) = 230 - 25\,(0{\cdot}15 + 0{\cdot}1) = 223{\cdot}75 \text{ volts}$$

Now
$$\frac{N_2}{N_1} = \frac{E_{b2}}{E_{b1}} \times \frac{\phi_1}{\phi_2}$$

or
$$\frac{N_2}{800} = \frac{223{\cdot}75}{205} \times \frac{1}{0{\cdot}45} \qquad (\because \phi_2 = 0{\cdot}45\,\phi_1)$$

$\therefore$
$$N_2 = \mathbf{1940\ r.p.m.}$$

Example 5.26. *A d.c. shunt generator delivers an output of 100 kW at 500 V when running at 800 r.p.m. The armature and field resistances are 0·1 Ω and 100 Ω respectively. Calculate the speed of the same machine when running as a shunt motor and taking 100 kW input at 500 V. Allow 1 volt per brush for contact drop.*

Solution.

As a generator. Fig. 5·16 (*i*) shows the connections. The speed N_1 = 800 r.p.m.

$$I_L = 100 \times 10^3/500 = 200 \text{ A}$$

$$I_{sh} = 500/100 = 5 \text{ A}$$

$$I_{a1} = I_L + I_{sh} = 200 + 5 = 205 \text{ A}$$

Generated e.m.f.,
$$E_1 = V_1 + I_{a1} R_a + \text{Brush drop}$$
$$= 500 + 205 \times 0{\cdot}1 + 2 \times 1 = 522{\cdot}5 \text{ V}$$

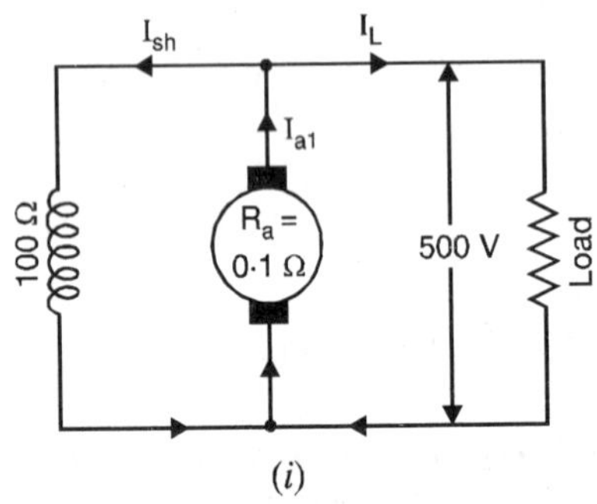

(*i*)

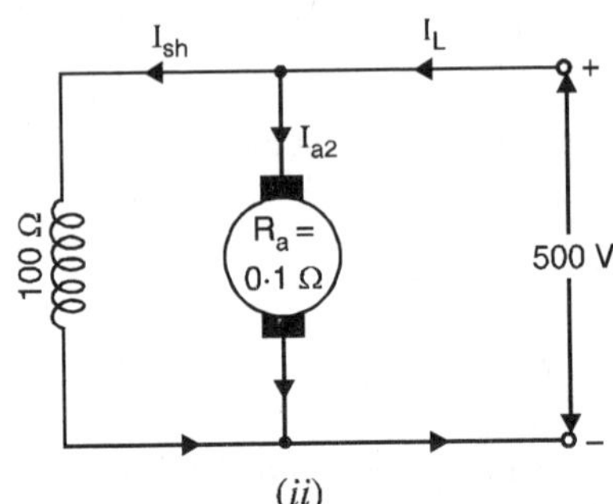

(*ii*)

Fig. 5.16

As a motor. Fig. 5·16 (*ii*) shows the connections.

$$I_{a2} = I_L - I_{sh} = 200 - 5 = 195 \text{ A}$$

Back e.m.f.,
$$E_2 = V - I_{a2} R_a - \text{Brush drop}$$
$$= 500 - 195 \times 0{\cdot}1 - 2 \times 1 = 478{\cdot}5 \text{ V}$$

Let N_2 be the speed of the machine in r.p.m. when working as a motor.

For a shunt machine,
$$\frac{N_2}{N_1} = \frac{E_2}{E_1} \qquad (\because \phi_2 = \phi_1 = \text{constant})$$

$\therefore$
$$N_2 = \left(\frac{E_2}{E_1}\right) N_1 = \frac{478{\cdot}5}{522{\cdot}5} \times 800 = \mathbf{732{\cdot}6\ r.p.m.}$$

Example 5.27. *A 100 kW, 500 V shunt generator was run as a motor on no load at its rated voltage and speed. The total current taken was 9.5 A including a shunt field current of 2.5 A. The resistance of the armature circuit is 0.1 Ω. Calculate the efficiency of the generator at (i) full-load (ii) half full-load.*

Solution.

At no-load as motor. Fig. 5.17 shows the conditions when the generator was run as a motor on no load. From the no-load data, we can find iron, friction and windage losses which remain constant from no-load to full-load.

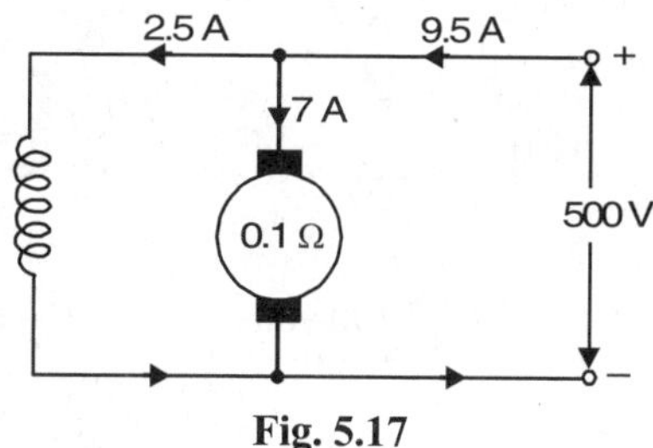

Fig. 5.17

$$\text{Input power at no-load} = 500 \times 9.5 = 4750 \text{ W}$$

$$\text{Armature Cu loss} = (7)^2 \times 0.1 = 4.9 \text{ W}$$

$$\text{Shunt field Cu loss} = 500 \times 2.5 = 1250 \text{ W}$$

$\therefore$ Iron, friction and windage losses = 4750 – 4.9 – 1250 = 3495 W

(*i*) **As generator at full-load**

$$\text{Output current at F.L.} = \frac{100 \times 10^3}{500} = 200 \text{ A}$$

$$\text{Armature current at F.L.} = 200 + 2.5 = 202.5 \text{ A}$$

$$\text{F.L. armature Cu loss} = (202.5)^2 \times 0.1 = 4100 \text{ W}$$

$$\text{F.L. shunt field Cu loss} = 500 \times 2.5 = 1250 \text{ W}$$

Total losses at F.L. = F.L. Arm. Cu Loss + F.L. Shunt field Cu Loss + Iron, friction and windage losses

= (4100) + (1250) + (3495) = 8845 W

$$\therefore \quad \text{F.L. efficiency} = \frac{100 \times 10^3}{100 \times 10^3 + 8845} \times 100 = \mathbf{91.9\%}$$

(*ii*) **As generator at half full-load**

$$\text{Output current} = 200/2 = 100 \text{ A}$$

$$\text{Armature current} = 100 + 2.5 = 102.5 \text{ A}$$

$$\text{Armature Cu loss} = (102.5)^2 \times 0.1 = 1050 \text{ W}$$

All other losses in the generator remain the same.

$$\therefore \quad \text{Total losses} = 1050 + 1250 + 3495 = 5795 \text{ W}$$

$$\therefore \quad \text{Efficiency at half full-load} = \frac{50 \times 10^3}{50 \times 10^3 + 5795} \times 100 = \mathbf{89.6\%}$$

Example 5.28. *A 200 V series motor takes a current of 100 A and runs at 1000 r.p.m. The total resistance of the motor is 0·1 Ω and the field is unsaturated. Calculate :*

(*i*) *the percentage change in torque and speed if the load is so changed that motor current is 500 A*

(*ii*) *the motor current and speed if the torque is halved*

Solution.

(*i*) Since the field is unsaturated, $T \propto \phi\, I_a$ or $T \propto I_a^2$ $\quad (\because \phi \propto I_a)$

For the first case, $T_1 \propto (100)^2$; For the second case, $T_2 \propto (50)^2$

$$\therefore \quad \frac{T_2}{T_1} = \frac{(50)^2}{(100)^2} = 0{\cdot}25$$

$$\therefore \quad \frac{T_1 - T_2}{T_1} = 1 - 0{\cdot}25 = \mathbf{0{\cdot}75} \text{ or } \mathbf{75\%}$$

Now

$$\frac{N_2}{N_1} = \frac{E_{b2}}{E_{b1}} \times \frac{I_{a1}}{I_{a2}} \quad \text{.... prior to saturation}$$

$$E_{b1} = V - I_{a1}(^*R_a + R_{se}) = 200 - 100 \times 0{\cdot}1 = 190 \text{ volts}$$
$$E_{b2} = V - I_{a2}(R_a + R_{se}) = 200 - 50 \times 0{\cdot}1 = 195 \text{ volts}$$
$$\therefore \quad \frac{N_2}{1000} = \frac{195}{190} \times \frac{100}{50} \quad \text{or} \quad N_2 = \mathbf{2052\ r.p.m.}$$

(*ii*)
$$\frac{I_{a2}^2}{I_{a1}^2} = \frac{T_2}{T_1} \quad \text{or} \quad \frac{I_{a2}^2}{(100)^2} = \frac{1}{2} \quad \text{or} \quad I_{a2} = 100 \times \sqrt{\frac{1}{2}} = \mathbf{70{\cdot}7\ A}$$
$$[\because I_{a1} = 100 \text{ A and } T_2/T_1 = 1/2]$$

$$E_{b1} = 190 \text{ volts}; \; E_{b2} = V - I_{a2}(R_a + R_{se}) = 200 - 70{\cdot}7 \times 0{\cdot}1 = 192{\cdot}93 \text{ volts}$$
$$\frac{N_2}{N_1} = \frac{E_{b2}}{E_{b1}} \times \frac{I_{a1}}{I_{a2}}$$
or
$$\frac{N_2}{1000} = \frac{192{\cdot}93}{190} \times \frac{100}{70{\cdot}7} \quad \text{or} \quad N_2 = \mathbf{1436\ r.p.m.}$$

Example 5.29. *Find (i) the no-load speed (ii) full-load speed and (iii) speed regulation for a 4-pole, 220 V, 24 h.p. shunt motor having field current = 5A ; armature resistance = 0·04 Ω ; flux per pole = 40 mWb; armature conductors = 160 ; 2 circuit wave-connection ; full-load current = 95 A; no-load current = 9 A. Neglect armature reaction.*

Solution. Fig. 5·18 shows the connections of shunt motor.

(*i*) **At no-load**

$$I_{a0} = I_{L0} - I_{sh} = 9 - 5 = 4 \text{ A}$$
$$E_{b0} = V - I_{a0} R_a$$
$$= 220 - 4 \times 0{\cdot}04 = 219{\cdot}84 \text{ volts}$$

Now
$$E_{b0} = \frac{P \phi Z N_0}{60 A}$$
or
$$219{\cdot}84 = \frac{4 \times 40 \times 10^{-3} \times 160 \times N_0}{60 \times 2}$$
$$\therefore \quad N_0 = \mathbf{1030\ r.p.m.}$$

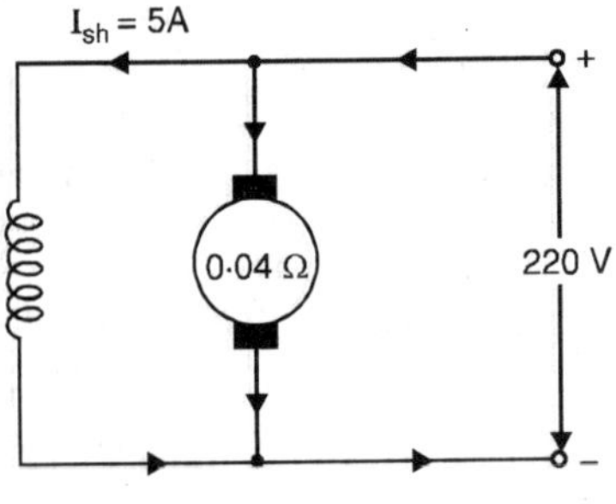

I_{sh} = 5A

0·04 Ω

220 V

Fig. 5.18

(*ii*) **At full-load**

$$I_a = I_L - I_{sh} = 95 - 5 = 90 \text{ A}$$
$$E_b = V - I_a R_a = 220 - 90 \times 0{\cdot}04 = 216{\cdot}4 \text{ volts}$$

Now
$$E_b = \frac{P \phi Z N}{60 A}$$
or
$$216{\cdot}4 = \frac{4 \times 40 \times 10^{-3} \times 160 \times N}{60 \times 2}$$
$$\therefore \quad N = \mathbf{1014\ r.p.m.}$$

(*iii*)
$$\text{Speed regulation} = \frac{N_0 - N}{N} \times 100 = \frac{1030 - 1014}{1014} \times 100 = \mathbf{1{\cdot}58\%}$$

Example 5.30. *A 250 V shunt motor giving 20 h.p. at 1000 r.p.m. takes an armature current of 75 A. The armature resistance is 0·25 Ω and the load torque remains the same. If the flux is reduced by 20% of its normal value before the speed changes, find the instantaneous value of the armature current and the torque. Determine the final value of the armature current and the speed.*

---

* Given that $R_a + R_{se} = 0{\cdot}1\ \Omega$.

Solution. Initially (*i.e.*, prior to the weakening of the field),

$$E_{b1} = V - I_{a1} R_a = 250 - 75 \times 0{\cdot}25 = 231{\cdot}25 \text{ volts}$$

We know that $E_b \propto \phi N$. If the flux is reduced suddenly, $E_b \propto \phi$ because the inertia of heavy armature prevents any rapid change in speed. It follows that when the flux is reduced by 20%, E_b is also reduced by 20%.

$\therefore$ Instantaneous back e.m.f., $(E_b)_{inst} = 0{\cdot}8 \times 231{\cdot}25 = 185$ volts

$$\text{Instantaneous armature current, } (I_a)_{inst} = \frac{V - (E_b)_{inst}}{R_a} = \frac{250 - 185}{0{\cdot}25} = \mathbf{260\ A}$$

$$\text{Instantaneous torque, } (T_a)_{inst} = 9{\cdot}55 \times \frac{(E_b)_{inst} \times (I_a)_{inst}}{N}$$

$$= 9{\cdot}55 \times \frac{185 \times 260}{1000} = \mathbf{459\ N\text{-}m}$$

Steady conditions. Since torque remains the same,

$$\phi_1 I_{a1} = \phi_2 I_{a2}$$

or $$I_{a2} = \frac{\phi_1}{\phi_2} \times I_{a1} = \frac{\phi_1}{0{\cdot}8\,\phi_1} \times 75 = \mathbf{93{\cdot}7\ A}$$

$\therefore$ $$E_{b2} = V - I_{a2} R_a = 250 - 93{\cdot}7 \times 0{\cdot}25 = 226{\cdot}6 \text{ volts}$$

Now $$\frac{N_2}{N_1} = \frac{E_{b2}}{E_{b1}} \times \frac{\phi_1}{\phi_2}$$

or $$\frac{N_2}{1000} = \frac{226{\cdot}6}{231{\cdot}25} \times \frac{1}{0{\cdot}8} \qquad (\because \phi_2 = 0{\cdot}8\,\phi_1)$$

$\therefore$ $$N_2 = \mathbf{1225\ r.p.m.}$$

Example 5.31. *A 60 kW, 400 V d.c. shunt motor has 4 poles and a 2-circuit wave-connected armature winding with 450 conductors. The flux per pole is 45 × 10^{-3} Wb. The full-load efficiency of the motor is 90·5%. The armature and interpole windings have total resistance of 0·1 Ω. The shunt field resistance is 200 Ω. Calculate* (*i*) *the line current* (*ii*) *full-load speed and* (*iii*) *armature torque.*

Solution. Fig. 5·19 shows the connections of shunt motor. Here R (= 0·1 Ω) is the combined resistance of the armature and interpole windings.

(*i*) $$\text{Input power to motor} = \frac{60 \times 10^3}{0{\cdot}905}$$

$$\text{Line current, } I_L = \frac{60 \times 10^3}{0{\cdot}905 \times 400} = \mathbf{165{\cdot}8\ A}$$

Fig. 5.19

(*ii*) $$I_{sh} = 400/200 = 2\text{ A}$$

$$I_a = I_L - I_{sh} = 165{\cdot}8 - 2 = 163{\cdot}8\text{ A}$$

$$\text{Back e.m.f., } E_b = V - I_a R$$

$$= 400 - 163{\cdot}8 \times 0{\cdot}1 = 383{\cdot}62 \text{ volts}$$

Now $$E_b = \frac{P \phi Z N}{60\ A}$$

or $$383{\cdot}62 = \frac{4 \times 45 \times 10^{-3} \times 450 \times N}{60 \times 2}$$

$\therefore$ $$N = \mathbf{568\ r.p.m.}$$

(*iii*) $$\text{Armature torque, } T_a = 0{\cdot}159\,\phi Z I_a (P/A)$$

$$= 0{\cdot}159 \times 45 \times 10^{-3} \times 450 \times 163{\cdot}8 \times (4/2) = \mathbf{1054.8\ N\text{-}m}$$

Example 5.32. *A d.c. shunt machine generates 250 V on open circuit at 1000 r.p.m. Effective armature resistance is 0·5 Ω, field resistance is 250 Ω, input to the machine running at no-load is 4 A at 250 V. Calculate the speed of the machine as a motor taking 40 A at 250 V. Armature reaction weakens field by 4%.*

Solution.

No-load speed as motor

$I_{sh} = 250/250 = 1$ A ; $I_{a0} = I_{L0} - I_{sh} = 4 - 1 = 3$ A

$$E_{b0} = V - I_{a0} R_a = 250 - 3 \times 0{\cdot}5 = 248{\cdot}5 \text{ volts}$$

The machine generated 250 volts while running at 1000 r.p.m. To generate 248·5 volts, speed is

$$N_0 = \frac{1000}{250} \times 248{\cdot}5 = 994 \text{ r.p.m.}$$

Speed at load

$$I_a = I_L - I_{sh} = 40 - 1 = 39 \text{ A}$$

$$E_b = V - I_a R_a = 250 - 39 \times 0{\cdot}5 = 230{\cdot}5 \text{ volts}$$

Now $$\frac{N}{N_0} = \frac{E_b}{E_{b0}} \times \frac{\phi_0}{\phi}$$

or $$\frac{N}{994} = \frac{230{\cdot}5}{248{\cdot}5} \times \frac{1}{0{\cdot}96} \qquad (\because \phi = 0{\cdot}96\, \phi_0)$$

$\therefore$ $$N = \mathbf{960 \text{ r.p.m.}}$$

Example 5.33. *A 400 V d.c. motor running at 1200 r.p.m. takes an armature current of 32·8 A. The armature resistance is 0·5 Ω. If the load torque increases by 25% and the flux increases by 10%, find (i) armature current (ii) speed and (iii) output of the machine. Neglect iron and friction losses.*

Solution.

(*i*) $$T_a \propto \phi I_a$$

For the first case, $T_a \propto \phi \times 32{\cdot}8$; For the second case, $1{\cdot}25\, T_a \propto 1{\cdot}1\, \phi \times I_{a2}$

$\therefore$ $$\frac{T_a}{1{\cdot}25\, T_a} = \frac{\phi \times 32{\cdot}8}{1{\cdot}1\, \phi \times I_{a2}} \quad \text{or} \quad I_{a2} = \mathbf{37{\cdot}27 \text{ A}}$$

(*ii*) $$E_{b1} = V - I_{a1} R_a = 400 - 32{\cdot}8 \times 0{\cdot}5 = 383{\cdot}6 \text{ volts}$$

$$E_{b2} = V - I_{a2} R_a = 400 - 37{\cdot}27 \times 0{\cdot}5 = 381{\cdot}4 \text{ volts}$$

Now $$\frac{N_2}{N_1} = \frac{E_{b2}}{E_{b1}} \times \frac{\phi}{1{\cdot}1\, \phi}$$

or $$\frac{N_2}{1200} = \frac{381{\cdot}4}{383{\cdot}6} \times \frac{1}{1{\cdot}1} \qquad \therefore N_2 = \mathbf{1084{\cdot}6 \text{ r.p.m.}}$$

(*iii*) Output of machine $$= \frac{E_{b2} I_{a2}}{1000} \text{ kW} = \frac{381{\cdot}4 \times 37{\cdot}27}{1000} = \mathbf{14{\cdot}2 \text{ kW}}$$

TUTORIAL PROBLEMS

1. A 220 V shunt motor takes 6 A on no-load and runs at 750 r.p.m. Resistance of shunt field is 110 Ω and that of armature is 0·25 Ω. Calculate the speed when loaded and taking a current of 52 A. Assume that armature reaction weakens the flux by 4%. **[740 r.p.m.]**
2. A 230 V shunt motor has an armature resistance of 0·5 Ω and field resistance of 115 Ω. At no-load, the speed is 1200 r.p.m. and the armature current is 2·5 A. On application of rated load, the speed drops to 1120 r.p.m. Determine the line current and the power input when the motor delivers rated load. **[33 A ; 8050 W]**

3. A 4-pole, 200 V d.c. shunt motor has 360 wave connected conductors on its armature. The full-load armature current is 30 A and flux per pole is 0·03 Wb. The armature resistance is 0·2 Ω and the contact drop is 1 V per brush. Determine the full-load speed of the motor. **[672 r.p.m.]**
4. A 4-pole, 460 V shunt motor has its armature wave-connected with 880 conductors. The useful flux per pole is 0·02 Wb and armature resistance is 0·7 Ω. If the armature current is 40 A, calculate (*i*) speed and (*ii*) torque. **[(*i*) 730 r.p.m. (*ii*) 226 N-m]**
5. A 230 V d.c. shunt motor takes a no-load current of 2 A and runs at 1200 r.p.m. If the full-load current is 40 A, find (*i*) speed on full-load (*ii*) percentage speed regulation. Resistance of armature is 0·25 Ω. **[(*i*) 1150 r.p.m. (*ii*) 4·35%]**
6. A 230 V shunt motor runs at 1000 r.p.m. and takes 5 A. The armature resistance of the motor is 0·25 Ω and shunt field resistance is 230 Ω. Calculate the drop in speed when the motor takes a line current of 41 A. **[40 r.p.m.]**
7. A shunt generator has an output of 10 kW at 500 V; the speed being 1000 r.p.m. The armature circuit resistance is 0·5 Ω and the field resistance is 250 Ω. Calculate the speed when running as a shunt motor taking 50 kW at 500 V. **[961 r.p.m.]**
8. The armature circuit resistance of a 18·65 kW, 250 V d.c. series motor is 0·1 Ω, the brush voltage drop is 3 V and the series field resistance is 0·05 Ω. When the motor takes 80 A, speed is 600 r.p.m. Calculate the speed when the current is 100 A. **[474 r.p.m.]**
9. A 4-pole d.c. shunt motor has a flux per pole of 0·04 Wb and the armature is lap-wound with 720 conductors. The shunt field resistance is 240 Ω and the armature resistance is 0·2 Ω. Brush contact drop is 1 V per brush. Determine the speed of the machine when running (*i*) as a motor taking 60 A (*ii*) as a generator supplying 120 A. The terminal voltage in each case is 480 V. **[(*i*) 972 r.p.m. (*ii*) 1055 r.p.m.]**
10. A 110 V d.c. series motor takes 50 A when running at 800 r.p.m. The armature resistance is 0·25 Ω and the series field resistance is 0·15 Ω. Iron and friction losses amount to 800 W. Find (*i*) the b.h.p. (*ii*) the total torque (*iii*) the shaft torque. **[(*i*) 5 h.p. (*ii*) 53·6 N-m (*iii*) 44 N-m]**

5.16 ARMATURE REACTION IN D.C. MOTORS

As in a d.c. generator, armature reaction also occurs in a d.c. motor. This is expected because when current flows through the armature conductors of a d.c. motor, it produces flux (armature flux) which acts on the flux produced by the main poles. For a motor with the same polarity and direction of rotation as is for generator, the direction of armature reaction field is reversed.

(*i*) In a generator, the armature current flows in the direction of the induced e.m.f. (*i.e.* generated e.m.f. E_g) whereas in a motor, the armature current flows against the induced e.m.f. (*i.e.* back e.m.f. E_b). Therefore, it should be expected that for the same direction of rotation and field polarity, the armature flux of the motor will be in the opposite direction to that of the generator. Hence instead of the main flux being distorted in the direction of rotation as in a generator, it is distorted opposite to the direction of rotation. We can conclude that :

Armature reaction in a d.c. generator weakens the flux at leading pole tips and strengthens the flux at trailing pole tips while the armature reaction in a d.c. motor produces the opposite effect.

(*ii*) In case of a d.c. generator, with brushes along G.N.A. and no commutating poles used, the brushes must be shifted in the direction of rotation (forward lead) for satisfactory commutation. However, in case of a d.c. motor, the brushes are given a negative lead *i.e.,* they are shifted against the direction of rotation.

With no commutating poles used, the brushes are given a forward lead in a d.c. generator and backward lead in a d.c. motor.

(*iii*) By using commutating poles (compoles), a d.c. machine can be operated with fixed brush positions for all conditions of load. Since commutating poles windings carry the armature current, then, when a machine changes from generator to motor (with consequent reversal of current), the polarities of commutating poles must be of opposite sign.

Therefore, in a d.c. motor, the commutating poles must have the same polarity as the main poles directly back of them. This is the opposite of the corresponding relation in a d.c. generator.

5.17 COMMUTATION IN D.C. MOTORS

Since the armature of a motor is the same as that of a generator, the current from the supply line must divide and pass through the paths of the armature windings. In order to produce unidirectional force (or torque) on the armature conductors of a motor, the conductors under any pole must carry the current in the same direction at all times. This is illustrated in Fig. 5·20. In this case, the current flows away from the observer in the conductors under the *N*-pole and towards the observer in the conductors under the *S*-pole. Therefore, when a conductor moves from the influence of *N*-pole to that of *S*-pole, the direction of current in the conductor must be reversed. This is termed as commutation. The function of the commutator and the brush gear in a d.c. motor is to cause the reversal of current in a conductor as it moves from one side of a brush to the other. For good commutation, the following points may be noted :

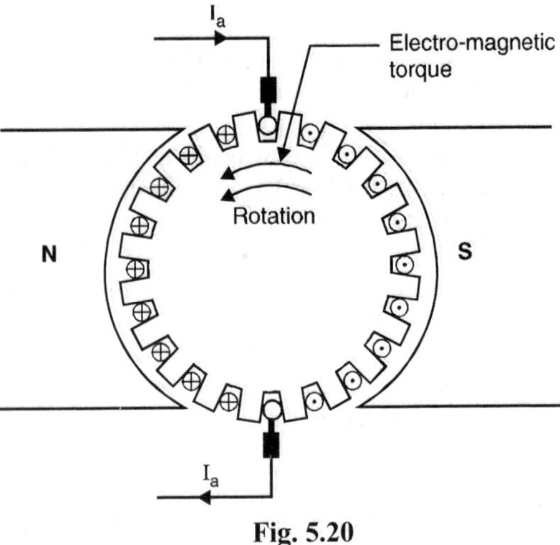

Fig. 5.20

(*i*) If a motor does not have commutating poles (compoles), the brushes must be given a negative lead *i.e.*, they must be shifted from G.N.A. against the direction of rotation of the motor.

(*ii*) By using interpoles, a d.c. motor can be operated with fixed brush positions for all conditions of load. For a d.c. motor, the commutating poles must have the same polarity as the main poles directly back of them. This is the opposite of the corresponding relation in a d.c. generator.

Note. A d.c. machine may be used as a motor or a generator without changing the commutating poles connections. When the operation of a d.c. machine changes from generator to motor, the direction of the armature current reverses. Since commutating poles winding carries armature current, the polarity of commutating pole reverses automatically to the correct polarity.

5.18 LOSSES IN A D.C. MOTOR

The losses occurring in a d.c. motor are the same as in a d.c. generator [See Art. 2.27]. These are [See Fig. 5·21]:

(*i*) copper losses (*ii*) iron losses or magnetic losses

(*iii*) mechanical losses

As in a generator, these losses cause (*a*) an increase of machine temperature and (*b*) reduction in the efficiency of the d.c. motor.

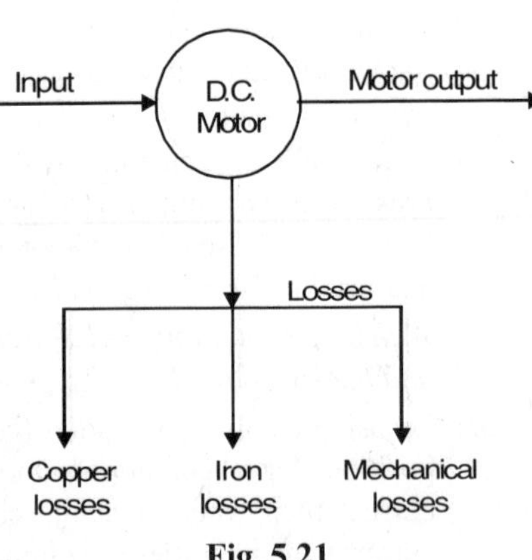

Fig. 5.21

The following points may be noted :

(*i*) Apart from armature Cu loss, field Cu loss and brush contact loss, Cu losses also occur in interpoles (commutating poles) and compensating windings. Since these windings carry armature current (I_a),

Loss in interpole winding $= I_a^2 \times$ Resistance of interpole winding

Loss in compensating winding $= I_a^2 \times$ Resistance of compensating winding

(*ii*) Since d.c. machines (generators or motors) are generally operated at constant flux density and constant speed, the iron losses are nearly constant.

(*iii*) The mechanical losses (*i.e.* friction and windage) vary as the cube of the speed of rotation of the d.c. machine (generator or motor). Since d.c. machines are generally operated at constant speed, mechanical losses are considered to be constant.

5.19 EFFICIENCY OF A D.C. MOTOR

Like a d.c. generator, the efficiency of a d.c. motor is the ratio of output power to the input power *i.e.*

$$\text{Efficiency, } \eta = \frac{\text{output}}{\text{input}} \times 100 = \frac{\text{output}}{\text{output + losses}} \times 100$$

As for a generator (See Art. 2.30), the efficiency of a d.c. motor will be maximum when :

$$\text{Variable losses} = \text{Constant losses}$$

Therefore, the efficiency curve of a d.c. motor is similar in shape to that of a d.c. generator (Refer back to Fig. 2.57).

5.20 POWER STAGES

The power stages in a d.c. motor are represented diagrammatically in Fig. 5.22.

$$A - B = \text{Copper losses}$$

$$B - C = \text{Iron and friction losses}$$

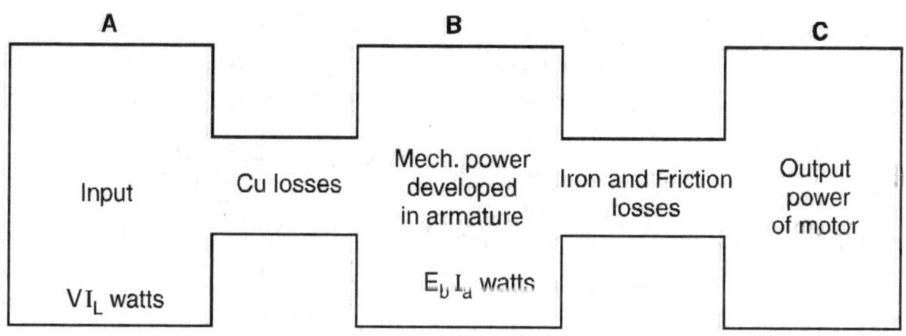

Fig. 5.22

$$\text{Overall efficiency, } \eta_c = C/A$$

$$\text{Electrical efficiency, } \eta_e = B/A$$

$$\text{Mechanical efficiency, } \eta_m = C/B$$

Example 5.34. *A 230 V shunt motor takes 5 A at no-load. The resistances of the armature and field circuit are 0·25 Ω and 115 Ω respectively. If the motor is loaded so as to carry 40 A, determine (i) iron and friction losses and (ii) efficiency.*

Solution.

(*i*) **No-load conditions**

$$\text{Input power} = VI_L = 230 \times 5 = 1150 \text{ W}$$

At no-load, the total input power supplied is used up in overcoming (*a*) small armature Cu loss (*b*) shunt field loss and (*c*) iron and friction losses.

$$I_{sh} = 230/115 = 2 \text{ A}; \; I_a = 5 - 2 = 3 \text{ A}$$

$$\text{Armature Cu loss} = I_a^2 R_a = (3)^2 \times 0{\cdot}25 = 2{\cdot}25 \text{ W}$$

$$\text{Field Cu loss} = I_{sh}^2 R_{sh} = (2)^2 \times 115 = 460 \text{ W}$$

$$\therefore \quad \text{Iron and friction losses} = 1150 - (2{\cdot}25 + 460) = \mathbf{687{\cdot}75 \text{ W}}$$

Alternatively. At no-load, the power developed in the armature ($E_b I_a$) is entirely used up in overcoming iron and friction losses.

$\therefore$ Iron and friction losses $= E_b I_a = (V - I_a R_a) I_a$
$= (230 - 3 \times 0{\cdot}25) \times 3 = \mathbf{687{\cdot}75\ W}$

(*ii*) **Load conditions**

Armature current, $I_a = 40 - 2 = 38$ A
Armature Cu loss $= I_a^2 R_a = (38)^2 \times 0{\cdot}25 = 361$ W
*Constant losses $= 687{\cdot}75 + 460 = 1147{\cdot}75$ W
Total losses $= 361 + 1147{\cdot}75 = 1508{\cdot}75$ W
Input power $= 230 \times 40 = 9200$ W
Output power $= 9200 - 1508{\cdot}75 = 7691{\cdot}25$ W
$\therefore$ $\eta = 7691{\cdot}25/9200 = \mathbf{0{\cdot}836}$ or $\mathbf{83{\cdot}6\ \%}$

Alternatively. Under loaded conditions, the power developed in the armature ($E_b I_a$) consists of (*i*) output power and (*ii*) iron and friction losses.

Power developed in armature $= E_b I_a = (V - I_a R_a) I_a$
$= (230 - 38 \times 0{\cdot}25)\ 38 = 8379$ W
Output power $= 8379 - 687{\cdot}75 = 7691{\cdot}25$ W
$\therefore$ $\eta = 7691{\cdot}25/9200 = \mathbf{0{\cdot}836}$ or $\mathbf{83{\cdot}6\%}$

Example 5.35. *A 230 V d.c. shunt motor has an armature resistance of 0.5 Ω and field resistance of 115 Ω. At no load, the speed is 1200 r.p.m. and the armature current 2.5 A. On application of rated load, the speed drops to 1120 r.p.m. Determine the line current and input power when the motor delivers rated load.*

Solution. $N_0 = 1200$ r.p.m. ; $E_{b0} = 230 - 2.5 \times 0.5 = 228.75$ V
$N = 1120$ r.p.m. ; $E_b = 230 - I_a \times 0.5$

Now $\dfrac{N}{N_0} = \dfrac{E_b}{E_{b0}}$ or $\dfrac{1120}{1200} = \dfrac{230 - 0.5 I_a}{228.75}$ $\therefore$ $I_a = 33$ A

Line current drawn by motor $= I_a + I_{sh} = 33 + (230/115) = \mathbf{35\ A}$
Power input at rated load $= 230 \times 35 = \mathbf{8050\ W}$

Example 5.36. *A 220 V shunt motor takes a total current of 80 A and runs at 800 r.p.m. Shunt field resistance and armature resistance are 50 Ω and 0·1 Ω respectively. If iron and friction losses amount to 1600 W, find* (i) *copper losses* (ii) *armature torque* (iii) *shaft torque* (iv) *efficiency.*

Solution.

Shunt field current, $I_{sh} = 220/50 = 4{\cdot}4$ A
Armature current, $I_a = 80 - 4{\cdot}4 = 75{\cdot}6$ A
Back e.m.f., $E_b = V - I_a R_a = 220 - 75{\cdot}6 \times 0{\cdot}1 = 212{\cdot}44$ V

(*i*) Input power $= V I_L = 220 \times 80 = 17600$ W
Power developed in armature $= E_b I_a = 212{\cdot}44 \times 75{\cdot}6 = 16060$ W
$\therefore$ Copper losses $= 17600 - 16060 = \mathbf{1540\ W}$

(*ii*) Armature torque, $T_a = 9{\cdot}55 \times E_b I_a/N = \dfrac{9{\cdot}55 \times 212{\cdot}44 \times 75{\cdot}6}{800} = \mathbf{192\ N\text{-}m}$

(*iii*) Output power $= 16060 - 1600 = 14460$ W
Shaft torque, $T_{sh} = 9{\cdot}55 \times \text{output}/N = 9{\cdot}55 \times 14460/800 = \mathbf{172{\cdot}6\ N\text{-}m}$

(*iv*) Efficiency $= \dfrac{14460}{17600} \times 100 = \mathbf{82{\cdot}1\%}$

---

* Constant losses = Iron and friction losses + Shunt field losses

Example 5.37. *A 250 V d.c. series motor runs at 500 r.p.m. The shaft torque is 130 N-m and the efficiency at this load is 88%. Find the current taken by the motor.*

Solution. Let I_L amperes be the current taken by the motor.

$$\text{Input power to motor} = 250 \times I_L \text{ watts}$$

$$\text{Output power of motor} = 0{\cdot}88 \times 250 \times I_L \text{ watts}$$

Also
$$\text{Output} = \frac{2\pi N T_{sh}}{60} = \frac{2\pi \times 500 \times 130}{60} = 6806 \text{ W}$$

$\therefore$
$$0{\cdot}88 \times 250 \times I_L = 6806$$

or
$$I_L = \frac{6806}{0{\cdot}88 \times 250} = \mathbf{30{\cdot}94\ A}$$

Example 5.38. *A 400 V d.c. shunt motor takes 5A at no-load. Its armature and field resistances are 0·5 Ω and 200 Ω respectively. Calculate the efficiency when the motor takes 50 A at full-load. Also calculate the percentage change in speed from no-load to full-load.*

Solution.

No-load conditions [See Fig. 5·23 (*i*)].

$$\text{Total motor input} = \text{Total no-load losses} = 400 \times 5 = 2000 \text{ W}$$

$$I_{sh} = 400/200 = 2 \text{ A} ; I_{a0} = I_{L0} - I_{sh} = 5 - 2 = 3 \text{ A}$$

Armature Cu loss = $(3)^2 \times 0{\cdot}5 = 4{\cdot}5$ W ; Field Cu loss = $400 \times 2 = 800$ W

$\therefore$ Iron and friction losses = $2000 - 4{\cdot}5 - 800 = 1195{\cdot}5$ W

The iron and friction losses will be assumed constant.

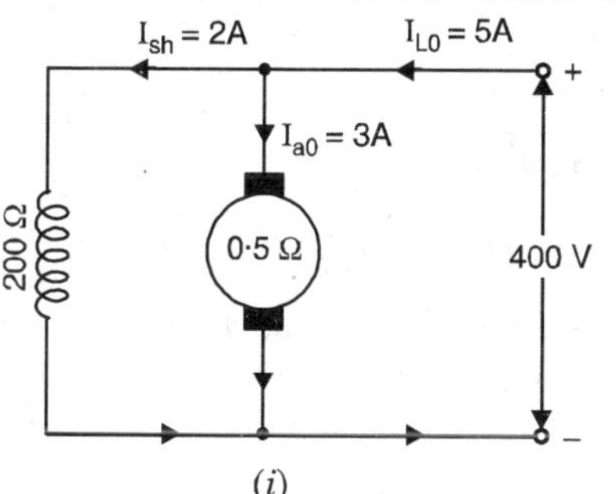

(*i*)

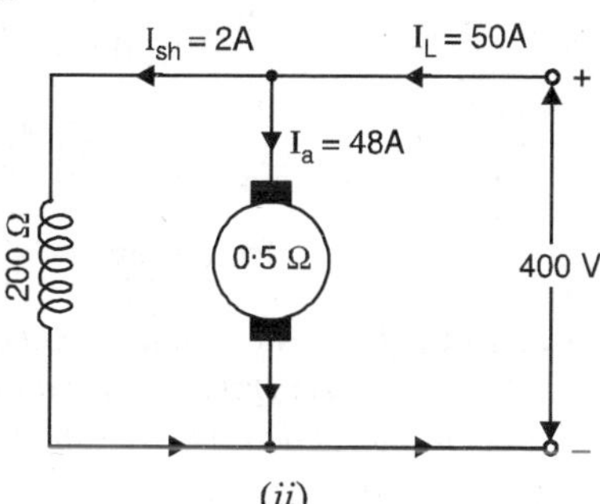

(*ii*)

Fig. 5.23

Under full-load [See Fig. 5·23 (*ii*)].

$$I_a = I_L - I_{sh} = 50 - 2 = 48 \text{ A}$$

$$\text{Armature Cu loss} = I_a^2 R_a = (48)^2 \times 0{\cdot}5 = 1152 \text{ W}$$

$$\text{Iron and friction losses} = 1195{\cdot}5 \text{ W} \qquad \text{...... assumed constant}$$

$$\text{Field Cu loss} = 400 \times 2 = 800 \text{ W}$$

$$\text{Total losses} = 1152 + 1195{\cdot}5 + 800 = 3147{\cdot}5 \text{ W}$$

$$\text{Input to motor} = 400 \times 50 = 20000 \text{ W}$$

$\therefore$
$$\text{F.L. efficiency} = \frac{\text{Input} - \text{Losses}}{\text{Input}} \times 100 = \frac{20000 - 3147{\cdot}5}{20000} \times 100 = \mathbf{84{\cdot}2\%}$$

$$E_{b0} = V - I_{a0} R_a = 400 - 3 \times 0{\cdot}5 = 398{\cdot}5 \text{ volts}$$

$$E_b = V - I_a R_a = 400 - 4{\cdot}8 \times 0{\cdot}5 = 376 \text{ volts}$$

Now
$$\frac{N}{N_0} = \frac{E_b}{E_{b0}} \qquad (\because \phi \text{ is constant})$$

or
$$\frac{N_0 - N}{N_0} = \frac{E_{b0} - E_b}{E_{b0}} = \frac{398{\cdot}5 - 376}{398{\cdot}5} = \mathbf{0{\cdot}0564} \text{ or } \mathbf{5{\cdot}64\%}$$

Example 5.39. *A 4-pole, 250 V series motor has a wave-connected armature with 1254 conductors. The flux per pole is 22 mWb when the motor is taking 50 A. Iron and friction losses amount to 1 kW. Armature resistance is 0·2 Ω and series field resistance is 0·2 Ω. Calculate (i) the speed (ii) the b.h.p. (iii) the shaft torque (iv) the efficiency at this load.*

Solution. Fig. 5·24 shows the connections of d.c. series motor.

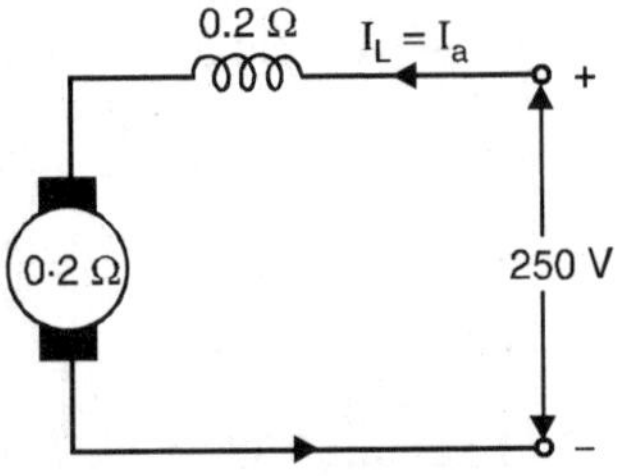

(*i*) $E_b = V - I_a(R_a + R_{se})$

$= 250 - 50 \times 0{\cdot}4 = 230$ volts

Also $E_b = \dfrac{P\phi ZN}{60A}$

or $230 = \dfrac{4 \times 22 \times 10^{-3} \times 1254 \times N}{60 \times 2}$

∴ $N =$ **250 r.p.m.**

Fig. 5.24

(*ii*) Mechanical power developed $= E_b I_a = 230 \times 50 = 11500$ W

Output of motor $= E_b I_a -$ Iron and friction losses

$= 11500 - 1000 = 10500$ W

b.h.p. $= \dfrac{10500}{746} =$ **14·08 h.p.**

(*iii*) Output power $= \dfrac{2\pi N T_{sh}}{60}$

or $10500 = \dfrac{2\pi \times 250 \times T_{sh}}{60}$

∴ $T_{sh} =$ **401·3 N-m**

(*iv*) Input power $= 250 \times 50 = 12500$ W

∴ Efficiency $= \dfrac{\text{Output power}}{\text{Input power}} \times 100 = \dfrac{10500}{12500} \times 100 =$ **84%**

Example 5.40. *A d.c. shunt machine when run as a motor on no-load takes 440 W and runs at 1000 r.p.m. The field current and armature resistance are 1 A and 0·5 Ω respectively. Calculate the efficiency of the machine when (i) running as a generator delivering 40 A at 220 V and (ii) as a motor taking 40 A from a 220 V supply.*

Solution.

Under no-load

Total motor input = Total no-load losses = 440 W

$I_{sh} = 1$ A ; $I_{L0} = 440/220 = 2$ A ∴ $I_{a0} = 2 - 1 = 1$ A

Armature Cu loss $= I_{a0}^2 \times R_a = (1)^2 \times 0{\cdot}5 = 0{\cdot}5$ W

Field Cu loss $= 220 \times 1 = 220$ W

∴ Iron and friction loss $= 440 - 220 - 0{\cdot}5 = 219{\cdot}5$ W

These losses will be assumed constant.

(*i*) **As a generator**

$I_L = 40$ A ; $I_a = I_L + I_{sh} = 40 + 1 = 41$ A

Armature Cu loss $= I_a^2 R_a = (41)^2 \times 0{\cdot}5 = 840{\cdot}5$ W

Field Cu loss $= 220 \times 1 = 220$ W

Iron and friction losses $= 219{\cdot}5$ W .... assumed constant

Total losses $= 840{\cdot}5 + 220 + 219{\cdot}5 = 1280$ W

Output of generator $= 220 \times 40 = 8800$ W

∴ Efficiency $= \dfrac{\text{Output}}{\text{Output + Losses}} \times 100 = \dfrac{8800}{8800 + 1280} \times 100 =$ **87·3%**

(*ii*) **As a motor**

$$I_L = 40\text{ A};\ I_a = I_L - I_{sh} = 40 - 1 = 39\text{ A}$$

$$\text{Armature Cu loss} = I_a^2 R_a = (39)^2 \times 0{\cdot}5 = 760{\cdot}5\text{ W}$$

$$\text{Field Cu loss} = 220 \times 1 = 220\text{ W}$$

$$\text{Iron and friction losses} = 219{\cdot}5\text{ W}$$

$$\text{Total losses} = 760{\cdot}5 + 220 + 219{\cdot}5 = 1200\text{ W}$$

$$\text{Motor input} = 220 \times 40 = 8800\text{ W}$$

$$\therefore \quad \text{Efficiency} = \frac{\text{Input} - \text{Losses}}{\text{Input}} \times 100 = \frac{8800 - 1200}{8800} \times 100 = \mathbf{86{\cdot}4\%}$$

Example 5.41.

(*i*) *A d.c. generator is connected to 220 V d.c. mains. The current delivered by the generator to the mains is 100 A. The armature resistance is 0.1 Ω. The generator is driven at a speed of 400 r.p.m. Calculate the induced e.m.f., electromagnetic torque and mechanical input power to the armature neglecting iron, windage and friction losses.*

(*ii*) *If the speed drops to 350 r.p.m., calculate the power input and output of the armature. State whether the machine is generating or motoring. Assume constant flux.*

Solution.

(*i*) Fig. 5.25 shows the conditions when the machine is operating as a generator and supplying current I_a (= 100 A) to the d.c. mains.

$$\text{Induced e.m.f., } E_1 = V + I_a R_a = 220 + 100 \times 0.1 = \mathbf{230\ volts}$$

$$\text{Torque, } {}^*T = \frac{E_1 I_a}{\omega} = \frac{230 \times 100}{2\pi \times (400/60)} = \mathbf{549\ Nm}$$

Since iron, windage and friction losses are neglected,

$$\therefore \quad \text{Mechanical power input} = E_1 I_a = 230 \times 100 = 23000\text{ W} = \mathbf{23\ kW}$$

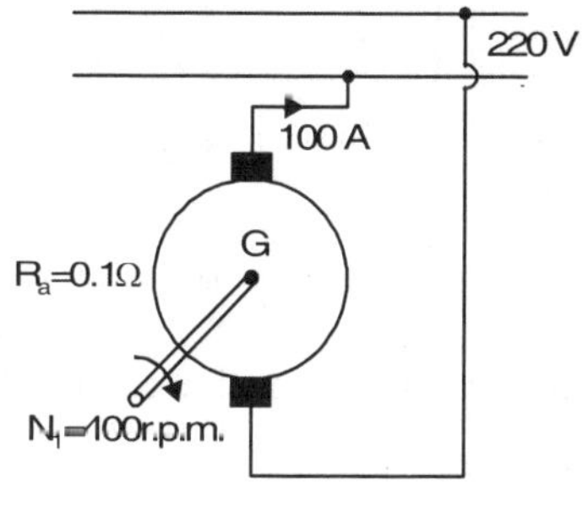

Fig. 5.25

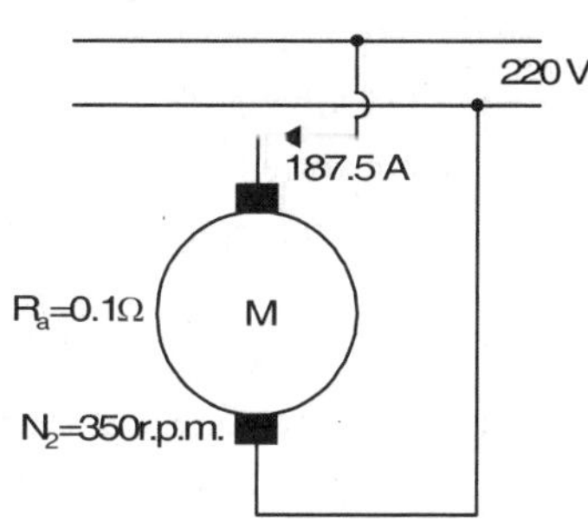

Fig. 5.26

(*ii*) When the speed of the machine drops to 350 r.p.m., the induced e.m.f. E_2 is given by ;

$$\frac{E_2}{E_1} = \frac{N_2}{N_1} \qquad (\because\ \phi \text{ is constant})$$

$$\therefore \quad E_2 = \frac{N_2}{N_1} \times E_1 = \frac{350}{400} \times 230 = 201.25\text{ volts}$$

---

* $\frac{2\pi NT}{60}$ = Mech. output power. Here N is in r.p.m. Now, $T\omega = T \times \frac{2\pi N}{60}$

Since the induced e.m.f. E_2 (= 201.25 volts) is less than the terminal voltage V(= 220 volts), the **machine will operate as a motor** (See Fig. 5.26) and draw current from the mains.

$$\text{Current drawn from mains, } I'_a = \frac{V - E_2}{R_a} = \frac{220 - 201.25}{0.1} = 187.5 \text{ A}$$

$$\text{Input power, } P_{in} = VI'_a = 220 \times 187.5 \text{ W} = \mathbf{41.25 \text{ kW}}$$

$$\text{Output power, } P_{out} = E_2 I'_a = 201.25 \times 187.5 \text{ W} = \mathbf{37.73 \text{ kW}}$$

Example 5.42. *A 7·46 kW, 250 V shunt motor takes a line current of 5 A when running light. Calculate the efficiency as a motor when delivering full-load output, if armature and field resistances are 0·5 Ω and 250 Ω respectively. At what output power will the efficiency be maximum ?*

Solution.

When running light

$$\text{Total motor input} = \text{Total no-load losses} = 250 \times 5 = 1250 \text{ W}$$

$$I_{sh} = 250/250 = 1 \text{ A}; \; I_{a0} = I_{L0} - I_{sh} = 5 - 1 = 4 \text{ A}$$

$$\text{Armature Cu loss} = I_{a0}^2 R_a = (4)^2 \times 0{\cdot}5 = 8 \text{ W}$$

$$\text{Field Cu loss} = 250 \times 1 = 250 \text{ W}$$

$$\text{Iron and friction losses} = 1250 - 8 - 250 = 992 \text{ W}$$

The iron and friction losses will be assumed constant.

When delivering full-load output. Let I_a be the full-load armature current (See Fig. 5·27).

$$\text{Armature input} = (250 \times I_a) \text{ W}$$

$$\text{F.L. output} = 7{\cdot}46 \times 1000 = 7460 \text{ W}$$

The losses in the armature are :

(*i*) Armature Cu loss $= (I_a^2 \times 0{\cdot}5)$ W

(*ii*) Iron and friction losses $= 992$ W

$$\therefore \quad 250\, I_a = 7460 + 992 + I_a^2 \times 0{\cdot}5$$

$$\text{or} \quad 0{\cdot}5\, I_a^2 - 250\, I_a + 8452 = 0$$

$$\text{On solving, } I_a = 36{\cdot}5 \text{ A}$$

$$\text{F.L. input current, } I_L = I_a + I_{sh} = 36{\cdot}5 + 1 = 37{\cdot}5 \text{ A}$$

$$\text{Motor input} = 250 \times 37{\cdot}5 = 9375 \text{ W}$$

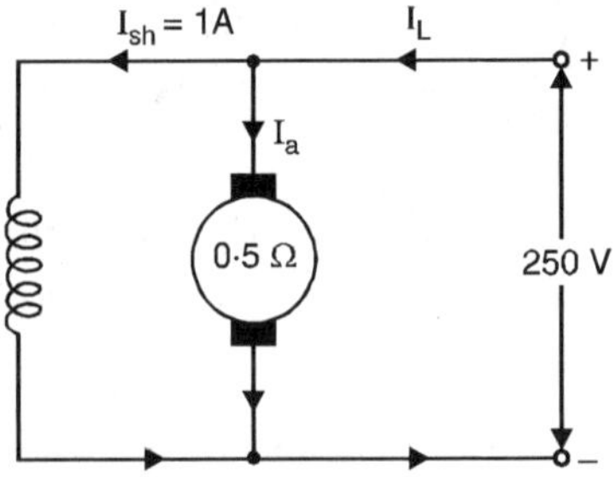

$$\text{F.L. efficiency} = \frac{\text{F.L. output}}{\text{F.L. input}} \times 100 = \frac{7460}{9375} \times 100 = \mathbf{79{\cdot}6\%}$$

Fig. 5.27

Now efficiency of the motor is maximum when armature Cu loss is equal to constant loss *i.e.*

$$I_a^2 R_a = {}^*1250 - 8$$

$$\text{or} \quad I_a = \sqrt{\frac{1242}{R_a}} = \sqrt{\frac{1242}{0{\cdot}5}} = 49{\cdot}84 \text{ A}$$

$$\text{Armature input} = 250\, I_a = 250 \times 49{\cdot}84 = 12460 \text{ W}$$

$$\text{Armature Cu loss} = I_a^2 R_a = (49{\cdot}84)^2 \times 0{\cdot}5 = 1242 \text{ W}$$

$$\text{Friction and iron losses} = 992 \text{ W} \quad \text{....... assumed constant}$$

$$\text{Armature output} = 12460 - 1242 - 992 = \mathbf{10226 \text{ W}}$$

* Constant losses = Iron and friction losses + Shunt field Cu loss = 1250 – 8 = 1242 W

Example 5.43. *The input to a 220 V d.c. shunt motor is 11 kW. Calculate (i) the torque developed (ii) the efficiency (iii) the speed at this load. The particulars of the motor are :*

No-load current = 5 A ; No-load speed = 1150 r.p.m.; Armature resistance = 0·5 Ω ; Shunt field resistance = 110 Ω.

Solution.

At no-load

$$\text{Total motor input} = \text{Total no-load losses} = 220 \times 5 = 1100 \text{ W}$$

$$I_{sh} = 220/110 = 2 \text{ A} ; I_{a0} = I_{L0} - I_{sh} = 5 - 2 = 3 \text{ A}$$

$$\text{Armature Cu loss} = (I_{a0})^2 \times R_a = (3)^2 \times 0{\cdot}5 = 4{\cdot}5 \text{ W}$$

$$\text{Constant losses} = 1100 - 4{\cdot}5 = 1095{\cdot}5 \text{ W}$$

When input is 11 kW

$$\text{Input current, } I_L = 11 \times 10^3/220 = 50 \text{ A}$$

$$\text{Armature current, } I_a = 50 - 2 = 48 \text{ A}$$

$$\text{Armature Cu loss} = I_a^2 R_a = (48)^2 \times 0{\cdot}5 = 1152 \text{ W}$$

$$\text{Total losses} = \text{Armature Cu loss} + \text{Constant losses} = 1152 + 1095{\cdot}5 = 2248 \text{ W}$$

$$\text{Output} = \text{Input} - \text{Losses} = 11000 - 2248 = 8752 \text{ W}$$

(*ii*) $$\text{Efficiency} = \frac{\text{Output}}{\text{Input}} \times 100 = \frac{8752}{11000} \times 100 = \mathbf{79{\cdot}6\%}$$

(*iii*) $$\text{Back e.m.f. at no load, } E_{b0} = V - I_{a0} R_a = 220 - 3 \times 0{\cdot}5 = 218{\cdot}5 \text{ volts}$$

$$\text{Back e.m.f. at given load, } E_b = V - I_a R_a = 220 - 48 \times 0{\cdot}5 = 196 \text{ volts}$$

∴ $$\frac{N}{N_0} = \frac{E_b}{E_{b0}} \qquad (\because \phi \text{ is constant})$$

or $$\frac{N}{1150} = \frac{196}{218{\cdot}5}$$

∴ $$N = \mathbf{1031 \text{ r.p.m.}}$$

(*i*) $$T_a = 9{\cdot}55 \times \frac{E_b I_a}{N} = 9{\cdot}55 \times \frac{196 \times 48}{1031} = \mathbf{87{\cdot}1 \text{ N-m}}$$

Example 5.44. *A 220 V series motor takes 50 A. The armature resistance is 0·1 Ω and the series field resistance is 0·08 Ω. If iron and friction losses are equal to copper losses at this load, find (i) the b.h.p. (ii) the efficiency.*

Solution. Fig. 5·28 shows the connections of the motor.

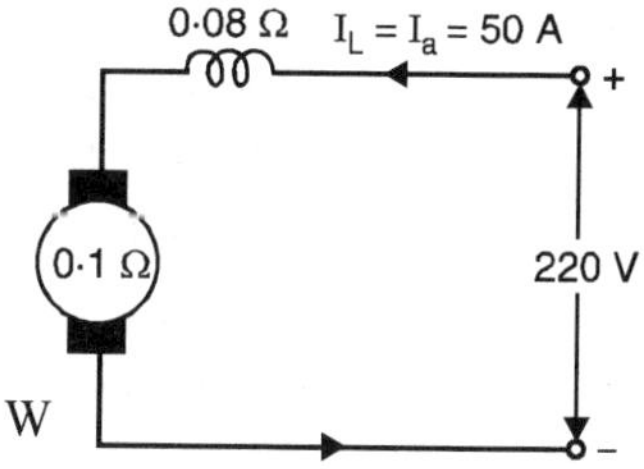

$$\text{Back e.m.f., } E_b = V - I_a (R_a + R_{se}) = 220 - 50\,(0{\cdot}1 + 0{\cdot}08) = 211 \text{ volts}$$

$$\text{Motor input} = 220 \times 50 = 11000 \text{ W}$$

$$\text{Mechanical power developed} = E_b I_a = 211 \times 50 = 10550 \text{ W}$$

$$\text{Copper losses} = I_a^2 (R_a + R_{se}) = (50)^2 \times (0{\cdot}1 + 0{\cdot}08) = 450 \text{ W}$$

0·08 Ω $I_L = I_a = 50$ A
+
0·1 Ω
220 V
–

Fig. 5.28

∴ $$\text{Iron and friction losses} = 450 \text{ W}$$

(*i*) $$\text{Output power} = E_b I_a - \text{Iron and friction losses} = 10550 - 450 = 10100 \text{ W}$$

∴ $$\text{b.h.p.} = \frac{10100}{746} = \mathbf{13{\cdot}54 \text{ h.p.}}$$

(*ii*) $$\text{Efficiency} = \frac{\text{Output}}{\text{Input}} \times 100 = \frac{10100}{11000} \times 100 = \mathbf{91{\cdot}8\%}$$

Example 5.45. *A 500 V d.c. shunt motor has an armature resistance of 0·1 Ω and a field resistance of 250 Ω. When running at no-load, the motor takes a current of 8 A. Find approximately the maximum efficiency and the load current at which it occurs.*

Solution.

$$\text{No-load input} = \text{Total no-load losses} = 500 \times 8 = 4000 \text{ W}$$

$$I_{sh} = 500/250 = 2 \text{ A} ; I_{a0} = 8 - 2 = 6 \text{ A}$$

$$\text{Armature Cu loss at no-load} = I_{a0}^2 R_a = (6)^2 \times 0{\cdot}1 = 3{\cdot}6 \text{ W}$$

As armature Cu loss is negligible, 4000 W represents the constant losses in the motor.

The maximum efficiency occurs when the armature Cu loss is equal to the constant loss. Let I_a be the armature current at which it occurs. Then,

$$I_a^2 R_a = 4000$$

or

$$I_a = \sqrt{\frac{4000}{R_a}} = \sqrt{\frac{4000}{0{\cdot}1}} = \mathbf{200 \text{ A}}$$

At this load, we have,

$$\text{Armature Cu loss} = 4000 \text{ W} ; \text{ Constant losses} = 4000 \text{ W}$$

$$\text{Total losses} = 4000 + 4000 = 8000 \text{ W}$$

$$\text{Input power} = 500 \times 200 = 100000 \text{ W}$$

$$\therefore \quad \text{Efficiency} = \frac{\text{Input} - \text{Losses}}{\text{Input}} \times 100 = \frac{100000 - 8000}{100000} \times 100 = \mathbf{92\%}$$

The expected efficiency will be slightly less than this value.

Example 5.46. *A 200 V shunt motor develops 23 h.p. when taking 20·2 kW. The field resistance is 50 Ω and the armature resistance is 0·06 Ω. What is the efficiency and power input when the output is 10 h.p. ?*

Solution.

When motor output is 23 h.p.

$$\text{Motor output} = 23 \text{ h.p.} = 23 \times 746 = 17158 \text{ W}$$

$$\text{Motor input} = 20{\cdot}2 \text{ kW} = 20200 \text{ W}$$

$$\text{Total losses} = 20200 - 17158 = 3042 \text{ W}$$

$$\text{Input current, } I_L = 20200/200 = 101 \text{ A}$$

$$I_{sh} = 200/50 = 4 \text{ A} ; I_a = I_L - I_{sh} = 101 - 4 = 97 \text{ A}$$

$$\text{Armature Cu loss} = I_a^2 R_a = (97)^2 \times 0{\cdot}06 = 565 \text{ W}$$

$$\text{Field Cu loss} = 200 \times 4 = 800 \text{ W}$$

$$\text{Total Cu losses} = 565 + 800 = 1365 \text{ W}$$

$$\text{Constant losses} = 3042 - 1365 = 1677 \text{ W}$$

When motor output is 10 h.p.

$$\text{Armature Cu loss} = {}^*565 \times \left(\frac{10}{23}\right)^2 = 107 \text{ W}$$

$$\text{Field Cu loss} = 200 \times 4 = 800 \text{ W}$$

$$\text{Constant losses} = 1677 \text{ W}$$

$$\text{Total losses} = 107 + 800 + 1677 = 2584 \text{ W}$$

$$\therefore \quad \text{Efficiency} = \frac{\text{Output}}{\text{Output} + \text{Losses}} \times 100 = \frac{7460}{7460 + 2584} \times 100 = \mathbf{74{\cdot}3\%}$$

---

* Armature Cu loss is directly proportional to the square of load.

$$\text{Input power} = 7460 + 2584 = 10044 \text{ W} \simeq \mathbf{10 \text{ kW}}$$

Example 5.47. *A 200 V d.c. shunt motor takes 4 A at no-load when running at 700 r.p.m. The field resistance is 100 Ω. The resistance of the armature at standstill gives a drop of 6 V across armature terminals when 10 A were passed through it. Calculate (i) the speed on load (ii) the armature torque (iii) efficiency. The normal input of motor is 8 kW.*

Solution.

At no-load

$$\text{Total motor input} = \text{Total no-load losses} = 200 \times 4 = 800 \text{ W}$$

$$I_{sh} = 200/100 = 2 \text{ A}; R_a = 6/10 = 0{\cdot}6\ \Omega;\ I_{a0} = 4 - 2 = 2 \text{ A}$$

$$\text{Armature Cu loss} = I_{a0}^2 \times R_a = (2)^2 \times 0{\cdot}6 = 2{\cdot}4 \text{ W}$$

$$\text{Constant losses} = 800 - 2{\cdot}4 = 797{\cdot}6 \text{ W}$$

$$\text{No-load } E_{b0} = V - I_{a0} R_a = 200 - 2 \times 0{\cdot}6 = 198{\cdot}8 \text{ volts}$$

At full-load

$$\text{F.L. line current, } I_L = 8 \times 10^3/200 = 40 \text{ A}$$

$$\text{F.L. armature current, } I_a = I_L - I_{sh} = 40 - 2 = 38 \text{ A}$$

$$\text{F.L. armature Cu loss} = I_a^2 R_a = (38)^2 \times 0{\cdot}6 = 866{\cdot}4 \text{ W}$$

(*i*) $$\text{Full-load } E_b = V - I_a R_a = 200 - 38 \times 0{\cdot}6 = 177{\cdot}2 \text{ volts}$$

Now $$\frac{E_b}{E_{b0}} = \frac{N}{N_0}$$

or $$\frac{177{\cdot}2}{198{\cdot}8} = \frac{N}{700} \qquad \therefore N = \mathbf{623{\cdot}9 \text{ r.p.m.}}$$

(*ii*) $$\text{Armature torque, } T_a = 9{\cdot}55 \times \frac{E_b I_a}{N} = 9{\cdot}55 \times \frac{177{\cdot}2 \times 38}{623{\cdot}9} = \mathbf{103 \text{ N-m}}$$

(*iii*) $$\text{Total F.L. losses} = \text{F.L. armature Cu loss} + \text{Constant losses}$$

$$= 866{\cdot}4 + 797{\cdot}6 = 1664 \text{ W}$$

$$\text{F.L. output} = 8000 - 1664 = 6336 \text{ W}$$

$$\therefore \quad \text{F.L. efficiency, } \eta = \frac{\text{F.L. output}}{\text{F.L. input}} \times 100 = \frac{6336}{8000} \times 100 = \mathbf{79{\cdot}2\%}$$

Example 5.48. *A 10 h.p., 230 V shunt motor takes a full-load current of 40 A. The armature and field resistances are 0·25 Ω and 230 Ω respectively. The total brush drop is 2 V and the core and friction losses are 380 W. Calculate the efficiency of the motor. Assume that *stray load loss is 1% of the output.*

Solution. Fig. 5·29 shows the connections of shunt motor.

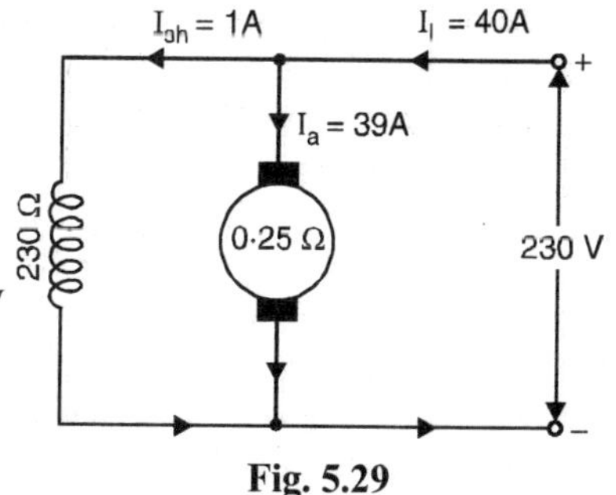

Fig. 5.29

$$\text{Motor Input} = 230 \times 40 = 9200 \text{ W}$$

$$I_{sh} = 230/230 = 1 \text{ A}$$

$$I_a = 40 - 1 = 39 \text{ A}$$

$$\text{Armature Cu loss} = (39)^2 \times 0{\cdot}25 = 380 \text{ W}$$

$$\text{Field Cu loss} = 230 \times 1 = 230 \text{ W}$$

$$\text{Core and friction loss} = 380 \text{ W}$$

$$\text{Brush contact loss} = 2 \times 39 = 78 \text{ W}$$

$$\text{Stray-load loss} = 1\% \text{ of Output}$$

---

* The stray load loss is due to leakage fluxes in teeth, slot wedges, pole faces etc. This loss is generally taken as 1% of output of the d.c. machine.

$$= \frac{1}{100} \times (10 \times 746) = 75 \text{ W}$$

$\therefore$ Total losses $= 380 + 230 + 380 + 78 + 75 = 1143$ W

Output power $= 9200 - 1143 = 8057$ W

$$\text{Efficiency} = \frac{\text{Output power}}{\text{Input power}} \times 100 = \frac{8057}{9200} \times 100 = \mathbf{87{\cdot}6\%}$$

Example 5.49. *The armature winding of a 200 V, 4-pole d.c. series motor is lap-connected. There are 280 slots and each slot has 4 conductors. The current is 45 A and the flux per pole is 18 mWb. The field resistance is 0.3 Ω; the armature resistance 0.5 Ω and the iron and friction losses total 800 W. The pulley diameter is 0.41 m. Find the pull in newton at the rim of the pulley.*

Solution. $E_b = V - I_a(R_a + R_{se}) = 200 - 45(0.5 + 0.3) = 164$ volts

Now $$E_b = \frac{P\phi Z N}{60 A}$$

Here $E_b = 164$ volts ; $P = 4$; $\phi = 18 \times 10^{-3}$ Wb ; $A = 4$; $Z = 280 \times 4$

$$\therefore \quad 164 = \frac{4 \times (18 \times 10^{-3}) \times (280 \times 4) \times N}{60 \times 4} \quad \text{or} \quad N = 488 \text{ r.p.m.}$$

Input power $= 200 \times 45 = 9000$ W

Cu losses $= I_a^2(R_a + R_{se}) = (45)^2 \times (0.5 + 0.3) = 1620$ W

Iron + Friction losses $= 800$ W (given)

$\therefore$ Total losses $= 1620 + 800 = 2420$ W

Output power $=$ Input power $-$ losses $= 9000 - 2420 = 6580$ W

$$\therefore \quad \text{Shaft torque, } T_{sh} = 9.55 \times \frac{\text{Output power}}{N} = 9.55 \times \frac{6580}{488} = 128 \text{ Nm}$$

The radius of the pulley, $r = 0.41/2 = 0.205$ m. Let F be the pull in newton at the rim of the pulley.

$\therefore$ $F \times r = T_{sh}$ or $F \times 0.205 = 128$ $\therefore$ $F =$ **624 N**

TUTORIAL PROBLEMS

1. A 220 V shunt motor is taking a current of 30 A. Armature and shunt field resistances are 0·2 Ω and 100 Ω respectively. Iron and friction losses amount to 500 W. Find the efficiency of the motor. **[82·4%]**
2. A shunt motor running at 600 r.p.m. takes 80 A at 250 V. The armature and shunt field resistances are 0·1 Ω and 50 Ω respectively. Iron and friction losses amount to 2188 W. Find (*i*) armature torque (*ii*) shaft torque (*iii*) copper losses (*iv*) efficiency. **[(*i*) 288 N-m (*ii*) 254 N-m (*iii*) 1812 W (*iv*) 80%]**
3. The output of certain shunt motor is 3·73 kW when taking 50 A at 100 V and running at 750 r.p.m. Armature resistance is 0·06 Ω and shunt field resistance is 50 Ω. Find (*i*) iron and friction losses (*ii*) armature torque (*iii*) shaft torque. **[(*i*) 932 W (*ii*) 59·2 N-m (*iii*) 47·4 N-m]**
4. The output of a shunt motor is 5 b.h.p. when taking 50 A at 100 V and running at 750 r.p.m. The armature resistance is 0·06 Ω and shunt field resistance is 50 Ω. Find (*i*) iron and friction losses (*ii*) armature torque (*iii*) shaft torque. **[(*i*) 923 W (*ii*) 59·2 N-m (*iii*) 47·4 N-m]**
5. A 50 h.p. (57·3 kW), 460 V d.c. shunt motor running light takes a current of 4 A and runs at a speed of 660 r.p.m. The resistance of the armature circuit is 0·3 Ω and that of the shunt field circuit is 270 Ω. When the motor is running at full-load, determine (*i*) the input current (*ii*) the speed (*iii*) the armature current at which efficiency is maximum. Ignore the effect of armature reaction. **[(*i*) 90·2 A (*ii*) 624 r.p.m. (*iii*) 78·33 A]**

6. A d.c. shunt motor running on no-load takes 5 A at 200 V. The field resistance is 100 Ω and the armature resistance is 0·1 Ω. Find the approximate maximum efficiency and the load current at which it occurs.

[90% ; 100 A]

7. A 220 V series motor takes a current of 35 A and runs at 500 r.p.m. The armature resistance is 0·25 Ω and the series field resistance is 0·3 Ω. If iron and friction losses amount to 600 W, find (*i*) armature torque (*ii*) shaft torque (*iii*) overall efficiency. **[(*i*) 134 N-m (*ii*) 122·5 N-m (*iii*) 83·4%]**

8. A shunt machine generates 220 V at 800 r.p.m. on open-circuit. Armature resistance is 0·4 Ω and field resistance is 160 Ω. The machine takes 5 A running as a motor on no-load at 220 V. Find the efficiency and speed of the machine as a motor taking 45 A at 220 V. Armature weakens the field by 3%.

[81·2% ; 759 r.p.m.]

9. One of the two similar 500 V shunt machines *A* and *B* running light takes 3 A. When *A* is mechanically coupled to *B*, the input to *A* is 3·5 A with *B* unexcited and 4·5 A when *B* is separately excited to generate 500 V. Calculate the friction and windage loss and core loss of each machine. **[250 W ; 500 W]**

10. A 100 V shunt motor is developing 6 b.h.p. and the overall efficiency at this load is 86%. Armature resistance is 0·06 Ω and shunt field resistance is 50 Ω. Find (*i*) current taken (*ii*) copper losses (*iii*) iron and friction losses. **[(*i*) 52·05 A (*ii*) 350 W (*iii*) 379 W]**

5.21 D.C. MOTOR CHARACTERISTICS

There are three principal types of d.c. motors *viz.*, shunt motors, series motors and compound motors. Both shunt and series types have only one field winding wound on the core of each pole of the motor. The compound type has two separate field windings wound on the core of each pole. The performance of a d.c. motor can be judged from its characteristic curves known as motor characteristics. Following are the three important characteristics of a d.c. motor :

(*i*) **Torque and Armature current characteristic (T_a/I_a)**

It is the curve between armature torque T_a and armature current I_a of a d.c. motor. It is also known as *electrical characteristic* of the motor.

(*ii*) **Speed and armature current characteristic (N/I_a)**

It is the curve between speed N and armature current I_a of a d.c. motor. It is very important characteristic as it is often the deciding factor in the selection of the motor for a particular application.

(*iii*) **Speed and torque characteristic (N/T_a)**

It is the curve between speed N and armature torque T_a of a d.c. motor. It is also known as *mechanical characteristic.*

5.22 CHARACTERISTICS OF SHUNT MOTORS

Fig. 5·30 shows the connections of a d.c. shunt motor. The field current I_{sh} is constant since the field winding is directly connected to the supply voltage V which is assumed to be constant. Hence, the flux in a shunt motor is *approximately constant.

(*i*) **T_a/I_a Characteristic.** We know that in a d.c. motor,

$$T_a \propto \phi I_a$$

Since the motor is operating from a constant supply voltage, flux ϕ is constant (neglecting armature reaction).

$$\therefore \qquad T_a \propto I_a$$

Hence T_a/I_a characteristic is a straight line passing through the origin as shown in Fig. 5·31. The shaft torque (T_{sh}) is less than T_a and is shown by a dotted line. It is clear from the curve that a very large current is required to start a heavy load. Therefore, a shunt motor should not be started on heavy load.

---

* Due to armature reaction, flux decreases a little but the decrease in flux is usually negligible under normal conditions.

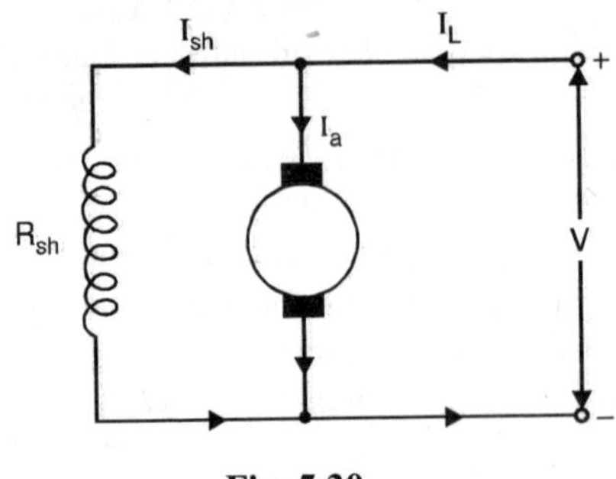

Fig. 5.30

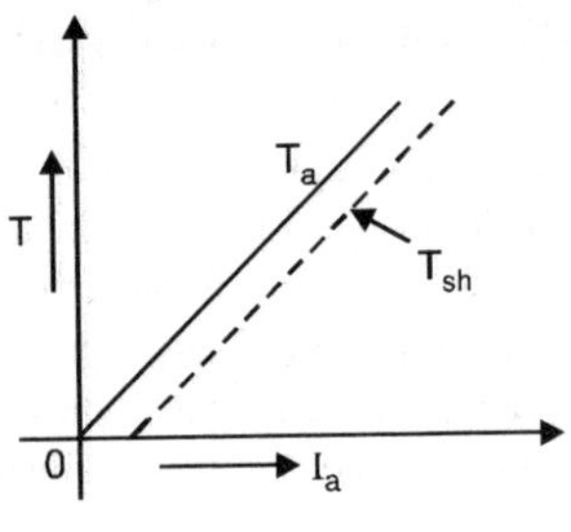

Fig. 5.31

(*ii*) **N/I_a Characteristic.** The speed N of a d.c. motor is given by ;

$$N \propto \frac{E_b}{\phi}$$

The flux ϕ and back e.m.f. E_b in a shunt motor are almost constant under normal conditions. Therefore, speed of a shunt motor will remain constant as the armature current varies (dotted line AB in Fig. 5·32). Strictly speaking, when load is increased, E_b $(= V - I_a R_a)$ and ϕ decrease due to the armature resistance drop and armature reaction respectively. However, E_b decreases slightly more than ϕ so that the *speed of the motor decreases slightly with load (line AC in Fig. 5.32).

(*iii*) **N/T_a Characteristic.** The curve is obtained by plotting the values of N and T_a for various armature currents (See Fig. 5·33). It may be seen that speed falls somewhat as the load torque increases.

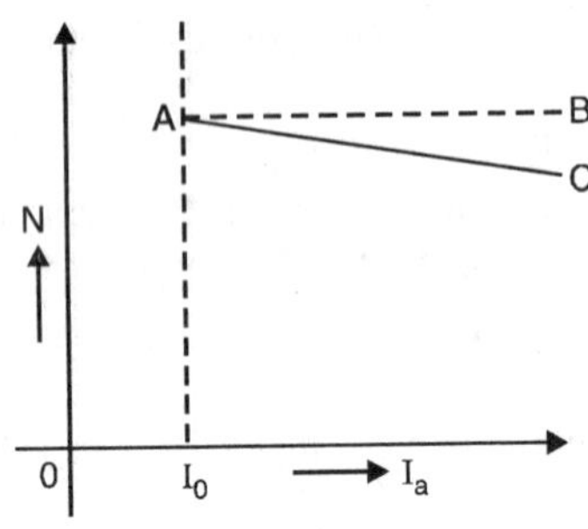

Fig. 5.32

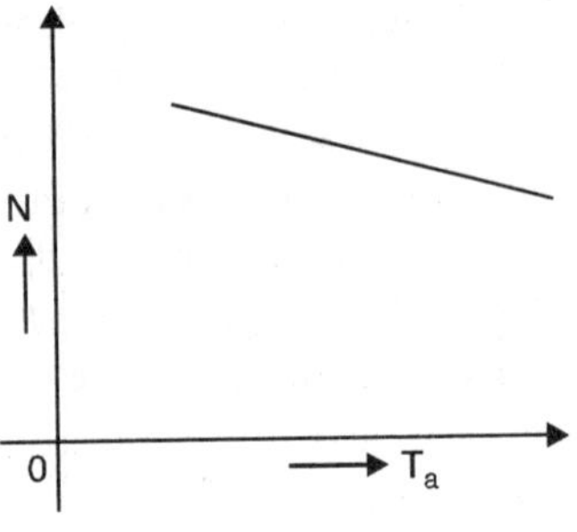

Fig. 5.33

Conclusions : Following two important conclusions are drawn from the above characteristics :

(*i*) There is slight change in the speed of a shunt motor from no-load to full-load. Hence, it is *essentially* a constant-speed motor.

(*ii*) The starting torque is not high because $T_a \propto I_a$.

5.23 CHARACTERISTICS OF SERIES MOTORS

Fig. 5·34 shows the connections of a series motor. Note that current passing through the field winding is the same as that in the armature. If the mechanical load on the motor increases, the armature current also increases. Hence, the flux in a series motor increases with the increase in armature current and *vice-versa.*

---

* It may be noted that characteristic does not have a point of zero armature current because a small current (no-load current I_0) is necessary to maintain rotation of the motor at no-load.

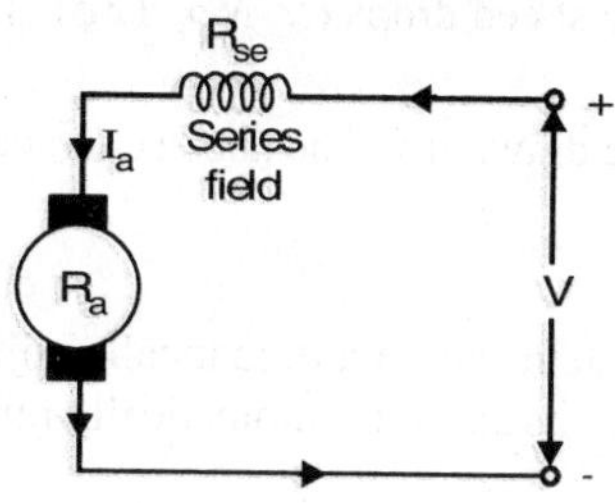

Fig. 5.34

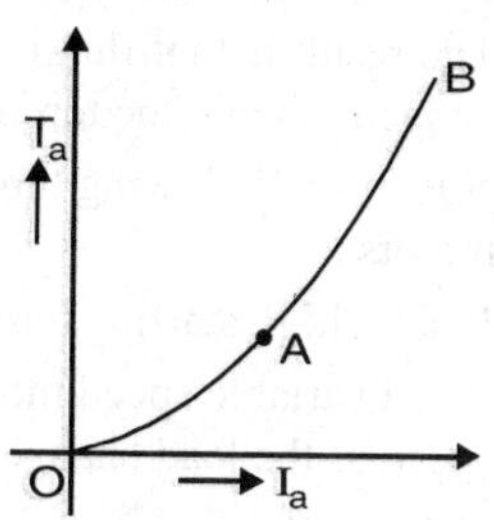

Fig. 5.35

(*i*) **T_a/I_a Characteristic.** We know that :

$$T_a \propto \phi I_a$$

Upto magnetic saturation, $\phi \propto I_a$ so that $T_a \propto I_a^2$

After magnetic saturation, ϕ is constant so that $T_a \propto I_a$

Thus upto magnetic saturation, the armature torque is directly proportional to the square of armature current. If I_a is doubled, T_a is almost quadrupled. Therefore, T_a/I_a curve upto magnetic saturation is a parabola (portion OA of the curve in Fig. 5·35). However, after magnetic saturation, torque is directly proportional to the armature current. Therefore, T_a/I_a curve after magnetic saturation is a straight line (portion AB of the curve in Fig. 5.35).

It may be seen that in the initial portion of the curve (*i.e.* upto magnetic saturation), $T_a \propto I_a^2$. This means that starting torque of a d.c. series motor will be very high as compared to a shunt motor (where $T_a \propto I_a$).

(*ii*) **N/I_a Characteristic.** The speed N of a series motor is given by ;

$$N \propto \frac{E_b}{\phi} \quad \text{where } E_b = V - I_a(R_a + R_{se})$$

When the armature current increases, the back e.m.f. E_b decreases due to $I_a(R_a + R_{se})$ drop while the flux ϕ increases. However, $I_a(R_a + R_{se})$ drop is quite small under normal conditions and may be neglected.

$$\therefore \qquad N \propto \frac{1}{\phi}$$

$$\propto \frac{1}{I_a} \text{ upto magnetic saturation}$$

Thus, upto magnetic saturation, the N/I_a curve follows the hyperbolic path as shown in Fig. 5·36. After saturation, the flux becomes constant and so does the speed.

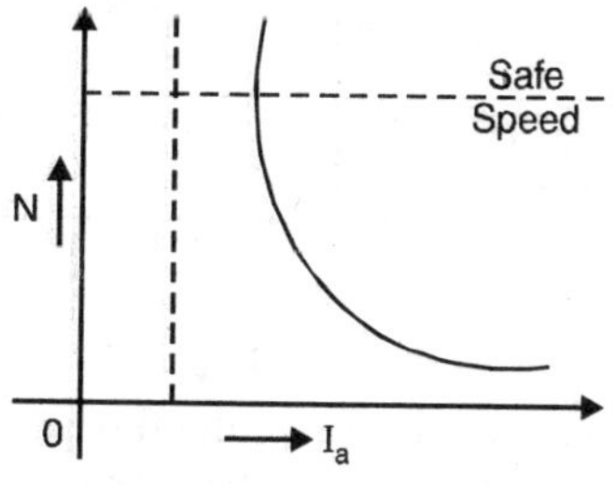

Fig. 5.36

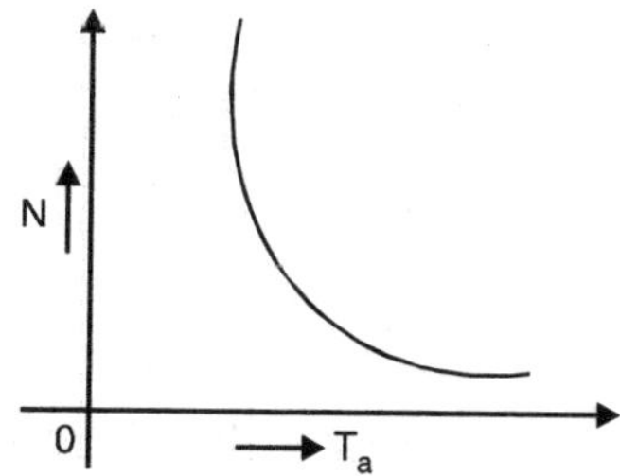

Fig. 5.37

(*iii*) **N/T_a Characteristic.** The N/T_a characteristic of a series motor is shown in Fig. 5.37. It is clear that series motor develops high torque at low speed and *vice-versa.* It is because an increase in torque requires an increase in armature current, which is also the field current.

The result is that flux is strengthened and hence the speed drops ($\because N \propto 1/\phi$). Reverse happens should the torque be low.

Conclusions : Following three important conclusions are drawn from the above characteristics of series motors :

(*i*) It has a high starting torque because initially $T_a \propto I_a^2$.

(*ii*) It is a variable speed motor (See N/I_a curve in Fig. 5.36) *i.e.,* it automatically adjusts the speed as the load changes. Thus if the load decreases, its speed is automatically raised and *vice-versa.*

(*iii*) At no-load, the armature current is very small and so is the flux. Hence, the speed rises to an excessive high value ($\because N \propto 1/\phi$). This is dangerous for the machine which may be destroyed due to centrifugal forces set up in the rotating parts. *Therefore, a series motor should never be started on no-load.* However, to start a series motor, mechanical load is first put and then the motor is started.

Note. The minimum load on a d.c. series motor should be great enough to keep the speed within limits. If the speed becomes dangerously high, then motor must be disconnected from the supply.

5.24 COMPOUND MOTORS

A compound motor has both series field and shunt field. The shunt field is always stronger than the series field. Compound motors are of two types :

(*i*) *Cumulative-compound motors* in which series field aids the shunt field.

(*ii*) *Differential-compound motors* in which series field opposes the shunt field.

Differential compound motors are rarely used due to their poor torque characteristics at heavy loads.

5.25 CHARACTERISTICS OF CUMULATIVE COMPOUND MOTORS

Fig. 5.38 shows the connections of a cumulative-compound motor. Each pole carries a series as well as shunt field winding ; the series field aiding the shunt field.

(*i*) **T_a/I_a Characteristic.** As the load increases, the series field increases but shunt field strength remains constant. Consequently, total flux is increased and hence the armature torque ($\because T_a \propto \phi I_a$). It may be noted that torque of a cumulative-compound motor is greater than that of shunt motor for a given armature current due to series field [See Fig. 5.39].

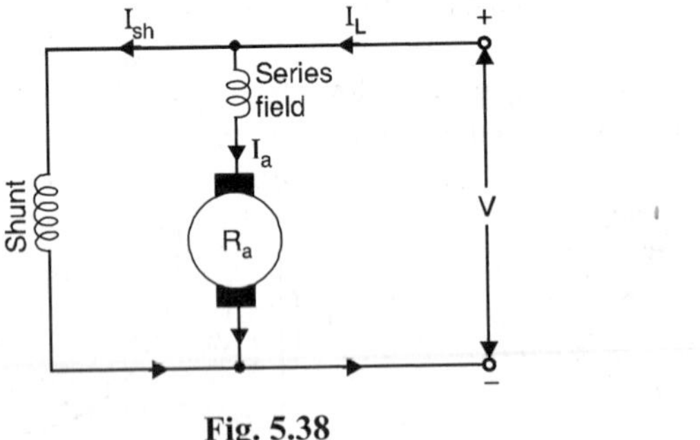

Fig. 5.38

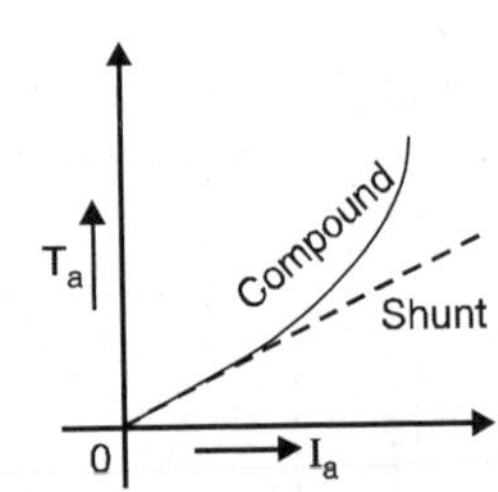

Fig. 5.39

(*ii*) **N/I_a Characteristic.** As explained above, as the load increases, the flux per pole also increases. Consequently, the speed ($N \propto 1/\phi$) of the motor falls as the load increases (See Fig. 5.40). It may be noted that as the load is added, the increased amount of flux causes the speed to decrease more than does the speed of a shunt motor. Thus the speed regulation of a cumulative compound motor is poorer than that of a shunt motor.

Note : Due to shunt field, the motor has a definite no load speed and can be operated safely at no-load.

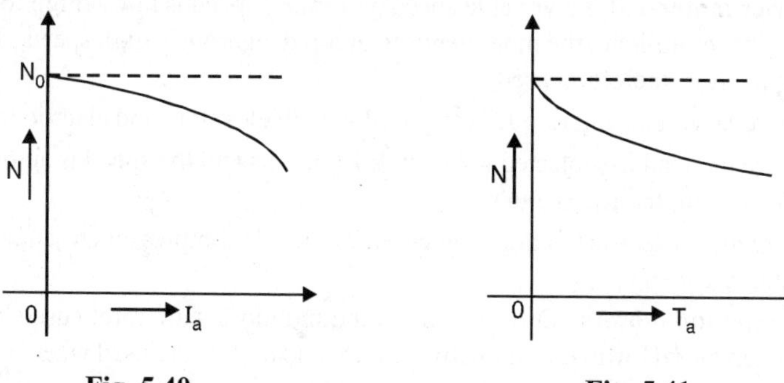

Fig. 5.40 Fig. 5.41

(*iii*) **N/T_a Characteristic.** Fig. 5.41 shows N/T_a characteristic of a cumulative compound motor. For a given armature current, the torque of a cumulative compound motor is more than that of a shunt motor but less than that of a series motor.

Conclusions : A cumulative compound motor has characteristics intermediate between series and shunt motors.

(*i*) Due to the presence of shunt field, the motor is prevented from running away at no-load.

(*ii*) Due to the presence of series field, the starting torque is increased.

5.26 COMPARISON OF THREE TYPES OF MOTORS

(*i*) The speed regulation of a shunt motor is better than that of a series motor.

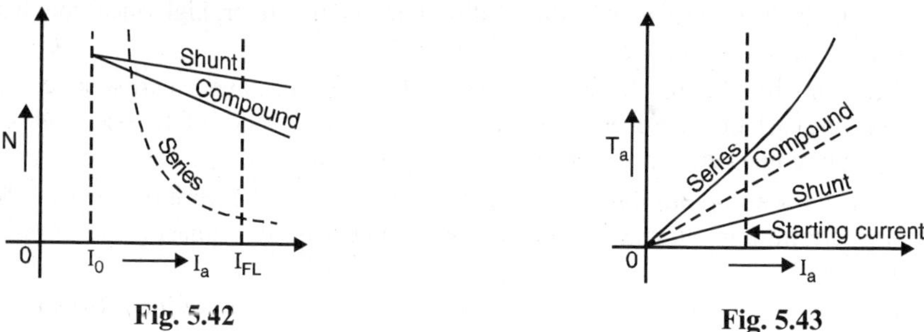

Fig. 5.42 Fig. 5.43

However, speed regulation of a cumulative compound motor lies between shunt and series motors (See Fig. 5.42).

(*ii*) For a given armature current, the starting torque of a series motor is more than that of a shunt motor. However, the starting torque of a cumulative compound motor lies between series and shunt motors (See Fig. 5.43).

(*iii*) Both shunt and cumulative compound motors have definite no-load speed. However, a series motor has dangerously high speed at no-load.

5.27 APPLICATIONS OF D.C. MOTORS

1. Shunt motors. The characteristics of a shunt motor reveal that it is an approximately constant speed motor. It is, therefore, used

(*i*) where the speed is required to remain almost constant from no-load to full-load

(*ii*) where the load has to be driven at a number of speeds and any one of which is required to remain nearly constant

Industrial use : Lathes, drills, boring mills, shapers, spinning and weaving machines etc.

2. Series motors. It is a variable speed motor *i.e.,* speed is low at high torque and *vice-versa.* However, at light or no-load, the motor tends to attain dangerously high speed. The motor has a high starting torque. It is, therefore, used

(*i*) where large starting torque is required *e.g.,* in elevators and electric traction

(*ii*) where the load is subjected to heavy fluctuations and the speed is automatically required to reduce at high torques and *vice-versa*

Industrial use : Electric traction, cranes, elevators, air compressors, vacuum cleaners, hair drier, sewing machines etc.

3. Compound motors. Differential-compound motors are rarely used because of their poor torque characteristics. However, cumulative-compound motors are used where a fairly constant speed is required with irregular loads or suddenly applied heavy loads.

Industrial use : Presses, shears, reciprocating machines etc.

5.28 TROUBLES IN D.C. MOTORS

Several troubles may arise in a d.c. motor and a few of them are discussed below :

1. Failure to start. This may be due to (*i*) ground fault (*ii*) open or short-circuit fault (*iii*) wrong connections (*iv*) too low supply voltage (*v*) frozen bearing or (*vi*) excessive load.

2. Sparking at brushes. This may be due to (*i*) troubles in brushes (*ii*) troubles in commutator (*iii*) troubles in armature or (*iv*) excessive load.

(*i*) Brush troubles may arise due to insufficient contact surface, too short a brush, too little spring tension or wrong brush setting.

(*ii*) Commutator troubles may be due to dirt on the commutator, high mica, rough surface or eccentricity.

(*iii*) Armature troubles may be due to an open armature coil. An open armature coil will cause sparking each time the open coil passes the brush. The location of this open coil is noticeable by a burnt line between segments connecting the coil.

3. Vibrations and pounding noises. These may be due to (*i*) worn bearings (*ii*) loose parts (*iii*) rotating parts hitting stationary parts (*iv*) armature unbalanced (*v*) misalignment of machine (*vi*) loose coupling etc.

4. Overheating. The overheating of motor may be due to (*i*) overloads (*ii*) sparking at the brushes (*iii*) short-circuited armature or field coils (*iv*) too frequent starts or reversals (*v*) poor ventilation (*vi*) incorrect voltage.

SHORT ANSWER QUESTIONS

Q. 1. *Why is the speed of a shunt motor practically constant ?*

Ans. The speed of a d.c. motor is given by :

$$N \propto \frac{V - I_a R_a}{\phi}$$

In a shunt motor, V, R_a and ϕ are practically constant. At no-load, the armature current is very small and $I_a R_a$ drop is negligible as compared to V. At full-load, $I_a R_a$ drop is ordinarily about 5% of V. Consequently, the full-load speed of the motor is about 95% of the no-load value. This fall in speed is reduced slightly due to armature reaction. In some cases, armature reaction is sufficient to cause the speed to remain nearly constant. For these reasons, a shunt motor is considered a constant speed motor, though the speed decreases slightly with the increase in load.

Q. 2. *What is the chief advantage of a d.c. series motor?*

Ans. The outstanding characteristic of a series motor is that it has high torque at low speeds and *vice-versa.* It is well-suited for traction purposes such as in electric trains. Acceleration is rapid because the torque is high at low speeds. Furthermore, the series motor automatically slows down as the train goes up an incline surface, yet turns at top speed on the flat ground. The power of a series motor tends to be constant because high torque is accompanied by low speed and *vice-versa.* Series motors are also used in electric cranes and hoists ; light loads are lifted quickly and heavy loads more slowly.

Q. 3. *Why is the actual operating efficiency of a series motor less than that of a shunt motor?*

Ans. Although the efficiency of a series motor varies about the same as that of a shunt motor, the actual operating efficiency is usually considerably lower. This is due to the fact that a series motor seldom runs at its rated load, but is continuously accelerating and decelerating.

Q. 4. *Can a d.c. series motor be operated on a.c. supply?*

Ans. If current direction through a series motor is reversed, the polarities of the field poles and the armature conductors are both reversed, resulting in a pushing effort between them in the same direction. Because of this, a series motor will operate on a.c. supply, its direction of rotation being the same for both half cycles. The lower the frequency of a.c., the greater the torque that can be produced. If the frequency is high, the inductive reactance of the coil limits the current that can flow and the torque of the motor.

Note. When d.c. is applied to a series motor, the current flowing has nothing to oppose it except the resistance in the circuit. This results in high starting current and torque.

Q. 5. *Will a d.c. shunt motor operate on an a.c. supply?*

Ans. The shunt winding has a large number of turns so that it has appreciable inductance. When a.c. is applied to a shunt motor, the large inductive reactance of shunt winding will reduce the field current too much. Consequently, shunt motor will not usually run on a.c. supply.

Q. 6. *Why is a d.c. series motor used to start heavy loads?*

Ans. In a d.c. motor, $T \propto \phi I_a$. For a d.c. series motor, at first, $\phi \propto I_a$ so that $T_a \propto I_a^2$. If current increases two-fold, the torque will increase four-fold. Consequently, the starting torque of a d.c. series motor is very high. For this reason, it is used to start heavy loads.

Q. 7. *Why is the torque of a cumulatively-compounded motor more than that of a shunt motor?*

Ans. The current-torque characteristic of a cumulatively-compounded motor is intermediate between those of a shunt motor and a series motor. When armature current is zero, the field strength is only due to shunt winding. As the load is applied, the series turns increase the flux, causing the torque $T_a \propto \phi I_a$ for any given current to be greater than it would be for a shunt motor.

Q. 8. *Why does not cumulatively-compounded motor run at dangerous speed at light loads?*

Ans. At no-load, the flux due to series field of a cumulatively-compounded motor is almost zero. However, the flux due to shunt winding prevents the motor from running at dangerously high speed.

Q. 9. *What will happen if a shunt motor running at no-load has its shunt winding opened accidentally?*

Ans. If a shunt motor, running at no-load, has its shunt field suddenly opened or burnt out, the back e.m.f. will fall to nearly zero. The armature current will continuously increase the speed of the motor until its parts fly apart.

Q. 10. *What will happen if the shunt field winding of a loaded shunt motor accidentally breaks?*

Ans. If the shunt winding of a loaded motor accidentally breaks, the loss of shunt field current would result in stopping of the motor or burn out the armature or fuse.

Q. 11. *How is the speed of a compound-wound motor controlled?*

Ans. The speed of a compound motor is controlled in the following ways :

(*i*) In order to reduce the speed of a compound motor, a resistance is inserted in the armature.

(*ii*) The speed can be increased either by a resistance in series with the shunt winding or a resistance in parallel with the series winding.

Q. 12. *What is the chief advantage of a cumulatively-compounded motor?*

Ans. The speed-torque characteristic of a cumulatively-compounded motor may approach those of the shunt or series motors, depending upon the strength of the two fields.

(*i*) If the motor contains a light series winding, it will have better torque characteristics than a shunt motor, but will largely retain the good regulation or efficient speed control of the shunt motor.

(*ii*) If the motor has a light shunt winding (one sufficient to keep the no-load speed to safe value) and a heavy series winding, it will have the characteristics of a series motor, but will not run at a dangerous speed with light loads.

Q. 13. *Why is a differentially-compounded motor rarely used?*

Ans. In a differentially-compounded motor, the flux due to series winding opposes the flux due to shunt winding. The series winding may be made to weaken the field sufficiently to make the speed at some particular load the same as at no-load. However, speed at other loads will be slightly different. This implies that a differentially-compounded motor has good speed regulation. The shunt motor also has good speed regulation with the added advantage that it has superior speed-torque characteristic as compared to differentially-compounded motor. For this reason, the latter is rarely used in practice.

MULTIPLE-CHOICE QUESTIONS

1. A d.c. motor is still used in industrial applications because it ....

(*i*) is cheap

(*ii*) is simple in construction

(*iii*) provides fine speed control

(*iv*) none of the above

2. Carbon brushes are preferable to copper brushes because ....

(*i*) they have longer life

(*ii*) they reduce armature reaction

(*iii*) they have lower resistance

(*iv*) they reduce sparking

3. The back e.m.f. in a d.c. motor ....

(*i*) opposes the applied voltage

(*ii*) aids the applied voltage

(*iii*) aids the armature current

(*iv*) none of the above

4. The value of back e.m.f. (E_b) in a d.c. motor is maximum at ....

(*i*) full-load (*ii*) no-load

(*iii*) half full-load (*iv*) none of above

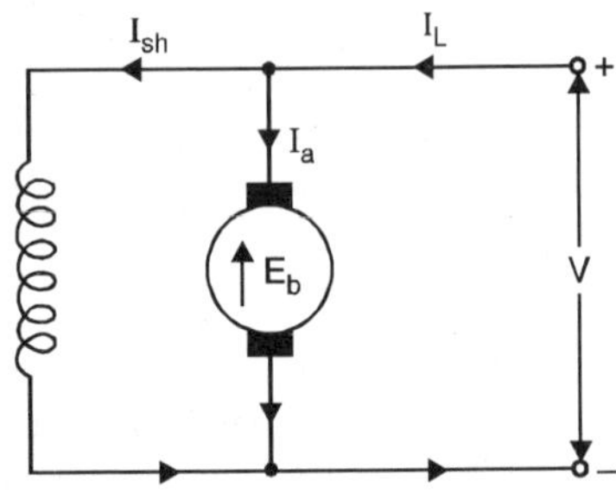

Fig. 5.44 Shunt motor on no-load

5. Fig. 5·44 shows the d.c. shunt motor on no-load. The quantity $E_b I_a$ represents ....

(*i*) input power to armature

(*ii*) copper losses in armature

(*iii*) core losses

(*iv*) total friction, windage and core losses

6. In Fig. 5·44, the quantity VI_a represents ....

(*i*) the mechanical output of the motor

(*ii*) core losses (*iii*) all losses

(*iv*) friction and windage losses

7. The mechanical power developed in a d.c. motor is maximum when back e.m.f. (E_b) is equal to .... the applied voltage (V).
   (*i*) twice (*ii*) half
   (*iii*) one-third (*iv*) none of above
8. The shaft torque (T_{sh}) in a d.c. motor is less than total armature torque (T_a) because of .... in the motor.
   (*i*) Cu losses (*ii*) field losses
   (*iii*) iron and friction losses
   (*iv*) none of the above
9. The amount of back e.m.f. of a shunt motor will increase when ....
   (*i*) the load is increased
   (*ii*) the field is strengthened
   (*iii*) the field is weakened
   (*iv*) none of the above
10. In a d.c. motor, the brushes are shifted from the mechanical neutral plane in a direction opposite to rotation to ....
    (*i*) reduce sparking
    (*ii*) increase speed
    (*iii*) decrease speed
    (*iv*) produce flat characteristics
11. The running speed of a d.c. series motor is basically determined by ....
    (*i*) field excitation (*ii*) load
    (*iii*) armature resistance
    (*iv*) none of the above
12. .... motor has the best speed regulation.
    (*i*) series
    (*ii*) cumulatively-compounded
    (*iii*) shunt
    (*iv*) differentially-compounded
13. The deciding factor in the selection of a d.c. motor for a particular application is its .... characteristic.
    (*i*) speed-torque
    (*ii*) speed-armature current
    (*iii*) torque-armature current
    (*iv*) none of the above
14. A d.c. series motor will overspeed when ....
    (*i*) the load is increased
    (*ii*) the field is opened
    (*iii*) the armature circuit is opened
    (*iv*) load is removed
15. D.C. shunt motors are used in those applications where .... is required.
    (*i*) high starting torque
    (*ii*) practically constant speed
    (*iii*) high no-load speed
    (*iv*) variable speed
16. D.C. series motors are used in those applications where .... is required.
    (*i*) high starting torque
    (*ii*) constant speed
    (*iii*) low no-load speed
    (*iv*) none of the above
17. Cumulatively-compounded motors are used where we require ....
    (*i*) variable speed
    (*ii*) poor speed regulation
    (*iii*) sudden heavy loads for short duration
    (*iv*) none of the above
18. For the same rating, ... motor has the highest starting torque.
    (*i*) shunt
    (*ii*) cumulatively-compounded
    (*iii*) differentially-compounded
    (*iv*) series
19. When load is removed, the motor that will run at the highest speed is the .... motor.
    (*i*) shunt (*ii*) series
    (*iii*) cumulatively-compounded
    (*iv*) differentially-compounded
20. For the same rating, ... motor has the least starting torque.
    (*i*) cumulatively-compounded
    (*ii*) series (*iii*) shunt
    (*iv*) data insufficient
21. A 440 V d.c. shunt motor has an armature resistance of 0·8 Ω and a field resistance of 200 Ω. Find the back e.m.f. when giving an output of 7·46 kW at 85% efficiency.
    (*i*) 222·4 V (*ii*) 425·8 V
    (*iii*) 312·6 V (*iv*) 392·7 V

22. A 250 V d.c. shunt motor takes a total current of 20 A. Resistance of shunt field winding is 200 Ω and that of armature is 0·3 Ω. What is the current in the armature ?

(*i*) 9·65 A (*ii*) 11·25 A
(*iii*) 18·75 A (*iv*) 16·62 A

23. In the above question, what is the approximate value of back e.m.f. ?

(*i*) 212 V (*ii*) 232 V
(*iii*) 227 V (*iv*) 244 V

24. A 220 V d.c. shunt motor takes a total current of 80 A and runs at 800 r.p.m. Resistance of shunt field is 50 Ω and that of armature is 0·1 Ω. The iron and friction losses amount to 1600 W. What is the driving power of the motor ?

(*i*) 16050 W (*ii*) 14500 W
(*iii*) 12600 W (*iv*) 18500 W

25. In question 24, what is b.h.p. ?

(*i*) 15·75 h.p. (*ii*) 11·78 h.p.
(*iii*) 14·28 h.p. (*iv*) 19·37 h.p.

26. In question 24, find the copper losses.

(*i*) 1120 W (*ii*) 1550 W
(*iii*) 1775 W (*iv*) 1140 W

27. A 220 V d.c. series motor is taking a current of 40 A. Resistance of armature = 0·5 Ω and resistance of series field = 0·25 Ω. What is the voltage at the brushes ?

(*i*) 190 V (*ii*) 210 V
(*iii*) 180 V (*iv*) none of the above

28. In the above question, what is the back e.m.f.?

(*i*) 170 V (*ii*) 210 V
(*iii*) 190 V (*iv*) none of the above

29. The armature resistance of a 6-pole lap wound generator is 0·05 Ω. If the armature is rewound using a wave winding, the armature resistance will be

(*i*) 0·15 Ω (*ii*) 0·1 Ω
(*iii*) 0·3 Ω (*iv*) 0·45 Ω

30. The armature of a 4-pole shunt motor has a lap winding accommodated in 60 slots, each containing 20 conductors. If the useful flux per pole is 23 mWb, the armature torque developed when the armature current is 50 A will be ....

(*i*) 176 N-m (*ii*) 192 N-m
(*iii*) 256 N-m (*iv*) 218 N-m

ANSWERS TO MULTIPLE-CHOICE QUESTIONS

1. (*iii*)	**2.** (*iv*)	**3.** (*i*)	**4.** (*ii*)	**5.** (*iv*)
6. (*iii*)	**7.** (*ii*)	**8.** (*iii*)	**9.** (*iii*)	**10.** (*i*)
11. (*ii*)	**12.** (*iii*)	**13.** (*i*)	**14.** (*iv*)	**15.** (*ii*)
16. (*i*)	**17.** (*iii*)	**18.** (*iv*)	**19.** (*ii*)	**20.** (*iii*)
21. (*ii*)	**22.** (*iii*)	**23.** (*iv*)	**24.** (*i*)	**25.** (*iv*)
26. (*ii*)	**27.** (*ii*)	**28.** (*iii*)	**29.** (*iv*)	**30.** (*iv*)

HINTS TO SELECTED MULTIPLE-CHOICE QUESTIONS

1. One potent reason for the widespread use of d.c. motors in industry is that their speed can be changed over a wide range by simple methods. Such a fine speed control is not easily possible with a.c. motors.

2. Since the contact resistance of carbon brushes is very high, sparking at the brushes can be prevented.

5. At no-load, there is no output. Consequently, the entire power developed in the armature is used up in overcoming friction, windage and core losses.

6. At no-load, there is no output. Consequently, the motor input (VI_a) is used up to overcome the losses.

8. The total torque developed in the armature is not available at the shaft as some percentage of it is used up in overcoming mechanical losses (friction) and iron losses in the motor.

16. In a d.c. motor, $T_a \propto \phi I_a$. At first (*i.e.*, at starting), $\phi \propto I_a$ in a series motor so that $T_a \propto I_a^2$. Consequently, starting torque is very high.

21.
$$\text{Motor input power} = \frac{7{\cdot}46 \times 1000}{0{\cdot}85}\ \text{W}$$
$$\text{Motor input current, } I_L = \frac{7{\cdot}46 \times 1000}{0{\cdot}85 \times 440} = 19{\cdot}95\ \text{A}$$
$$I_{sh} = 440/200 = 2{\cdot}2\ \text{A};\ I_a = I_L - I_{sh} = 19{\cdot}95 - 2{\cdot}2 = 17{\cdot}75\ \text{A}$$
$$\therefore \quad \text{Back e.m.f., } E_b = V - I_a R_a = 440 - 17{\cdot}75 \times 0{\cdot}8 = 425{\cdot}8\ \text{V}$$

22. $I_{sh} = 250/200 = 1{\cdot}25\ \text{A};\ I_a = 20 - 1{\cdot}25 = 18{\cdot}75\ \text{A}$

23.
$$E_b = V - I_a R_a = 250 - 18{\cdot}75 \times 0{\cdot}3 = 244\ \text{V}$$

24. $I_{sh} = 220/50 = 4{\cdot}4\ \text{A};\ I_a = 80 - 4{\cdot}4 = 75{\cdot}6\ \text{A}$
$$E_b = V - I_a R_a = 220 - 75{\cdot}6 \times 0{\cdot}1 = 212{\cdot}44\ \text{V}$$
$$\therefore \quad \text{Driving power} = E_b I_a = 212{\cdot}44 \times 75{\cdot}6 \simeq 16050\ \text{W}$$

25.
$$\text{Output power} = \text{Driving power} - \text{Iron and friction losses}$$
$$= 16050 - 1600 = 14450\ \text{W}$$
$$\therefore \quad \text{b.h.p.} = \frac{14450}{746} = 19{\cdot}37\ \text{h.p.}$$

26.
$$\text{Input power} = 220 \times 80 = 17600\ \text{W}$$
$$\text{Copper losses} = \text{Input power} - \text{driving power} = 17600 - 16050 = 1550\ \text{W}$$

27.
$$\text{Voltage at brushes} = 220 - I_a R_{se} = 220 - 40 \times 0{\cdot}25 = 210\ \text{V}$$

28.
$$\text{Motor resistance, } R_m = R_a + R_{se} = 0{\cdot}5 + 0{\cdot}25 = 0{\cdot}75\ \Omega$$
$$\text{Back e.m.f., } E_b = V - I_a R_m = 220 - 40 \times 0{\cdot}75 = 190\ \text{V}$$

29. In lap wound, resistance of each path $= 6 \times 0{\cdot}05 = 0{\cdot}3\ \Omega$. When wave-wound, there will be two paths, each of resistance $= 3 \times 0{\cdot}3 = 0{\cdot}9\ \Omega$. Therefore, armature resistance $= 0{\cdot}9/2 = 0{\cdot}45\Omega$.

30. If T is the armature torque in N-m and N is the number of revolutions per minute, then,
$$T \times \frac{2\pi N}{60} = \text{Total mechanical power developed}$$
$$= E_b I_a = \left[\frac{\phi Z N}{60} \times \frac{P}{A}\right] I_a$$
or
$$T = \frac{1}{2\pi} \times \phi Z \left(\frac{P}{A}\right) I_a = \frac{1}{2\pi} \times 23 \times 10^{-3} \times 60 \times 20 \times \frac{4}{4} \times 50$$
$$= 218\ \text{N-m}$$

6

Speed Control of D.C. Motors

INTRODUCTION

Although a far greater percentage of electric motors in service are a.c. motors, the d.c. motor is of considerable industrial importance. The principal advantage of a d.c. motor is that its speed can be changed over a wide range by a variety of simple methods. Such a fine speed control is generally not possible with a.c. motors. In fact, fine speed control is one of the reasons for the strong competitive position of d.c. motors in the modern industrial applications. In this chapter, we shall discuss the various methods of speed control of d.c. motors.

6.1 SPEED CONTROL OF D.C. MOTORS

The speed of a d.c. motor is given by :

$$N \propto \frac{E_b}{\phi}$$

or $$N = K\frac{(V - I_a R)}{\phi} \text{ r.p.m.} \qquad ...(i)$$

where $R = R_a$ ...*for shunt motor*

$= R_a + R_{se}$ ...*for series motor*

From exp. (*i*), it is clear that there are three main methods of controlling the speed of a d.c. motor, namely:

(*i*) By varying the flux per pole (ϕ). This is known as *flux control method.*

(*ii*) By varying the resistance in the armature circuit. This is known as *armature control method.*

(*iii*) By varying the applied voltage V. This is known as *voltage control method.*

6.2 SPEED CONTROL OF D.C. SHUNT MOTORS

The speed of a shunt motor can be changed by (*i*) flux control method (*ii*) armature control method (*iii*) voltage control method. The first method (*i.e.* flux control method) is frequently used because it is simple and inexpensive.

1. Flux control method. It is based on the fact that by varying the flux ϕ, the motor speed ($N \propto 1/\phi$) can be changed and hence the name flux control method. In this method, a variable resistance (known as *shunt field rheostat*) is placed in series with shunt field winding as shown in Fig. 6.1.

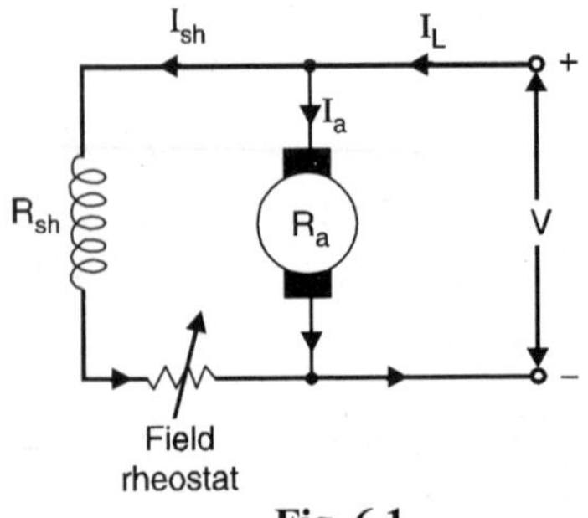

Fig. 6.1

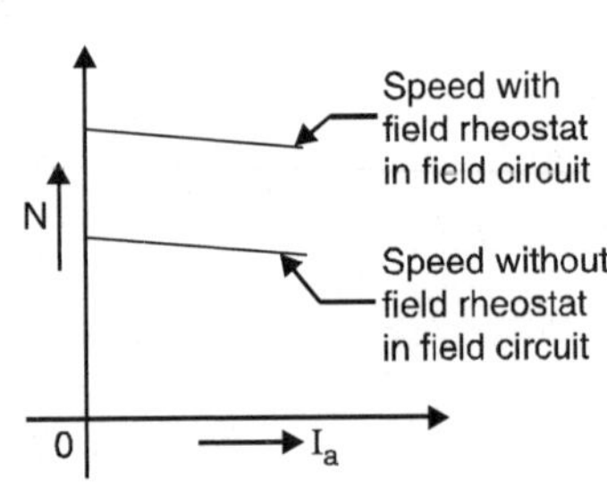

Fig. 6.2

The shunt field rheostat reduces the shunt field current I_{sh} and hence the flux ϕ. Therefore, we can only *raise the speed of the motor above the normal speed (See Fig. 6.2). Generally, this method permits to increase the speed in the ratio 3 : 1. Wider speed ranges tend to produce instability and poor commutation.

Advantages

(i) This is an easy and convenient method.

(ii) It is an inexpensive method since very little power is wasted in the shunt field rheostat due to relatively small value of I_{sh}.

(iii) The speed control exercised by this method is independent of load on the machine.

Disadvantages

(i) Only speeds higher than the normal speed can be obtained since the total field circuit resistance cannot be reduced below R_{sh} – the shunt field winding resistance.

(ii) There is a limit to the maximum speed obtainable by this method. It is because if the flux is too much weakened, commutation becomes poorer.

Note. *The field of a shunt motor in operation should never be opened because its speed will increase to an extremely high value.*

2. Armature control method. This method is based on the fact that by varying the voltage available across the armature, the back e.m.f. and hence the speed of the motor can be changed. This is done by inserting a variable resistance R_C (known as *controller resistance*) in series with the armature as shown in Fig. 6.3.

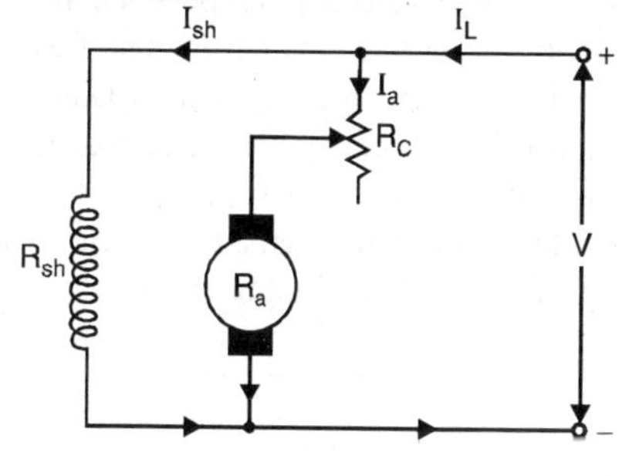

Fig. 6.3

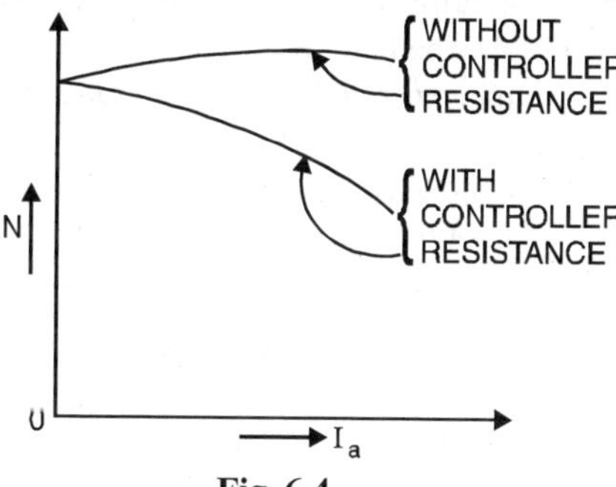

Fig. 6.4

$$N \propto V - I_a(R_a + R_C)$$

where R_C = controller resistance

Due to voltage drop in the controller resistance, the back e.m.f. (E_b) is †decreased. Since $N \propto E_b$, the speed of the motor is reduced. The highest speed obtainable is that corresponding to $R_C = 0$ *i.e.*, normal speed. Hence, this method can only provide speeds below the normal speed (See Fig. 6.4).

Disadvantages

(i) A large amount of power is wasted in the controller resistance since it carries full armature current I_a.

---

* When resistance in the field circuit is increased, the current (I_{sh}) in the field circuit decreases. This causes reduction in the field flux. Thus fewer magnetic lines are cut by the armature and, therefore, less back e.m.f. (E_b) is generated. As a result, more current (I_a) flows through the armature. The increase in armature current is greater than the decrease is flux. Therefore, the torque ($T \propto \phi I_a$) is greater and the armature speed increases. In doing so, the back e.m.f. increases and reduces the armature current. The speed increases until the developed torque becomes equal to load torque and then the motor again runs at a constant speed. Therefore, peculiar as it may seem, the speed of a shunt motor is increased by increasing the resistance of the field circuit.

† $E_b = V - I_a(R_a + R_C)$. Since V is constant, E_b will decrease.

(*ii*) The speed varies *widely with load since the speed depends upon the voltage drop in the controller resistance and hence on the armature current demanded by the load.

(*iii*) The output and efficiency of the motor are reduced.

(*iv*) This method results in poor speed regulation.

Due to above disadvantages, this method is seldom used to control the speed of shunt motors.

Note. The armature control method is a very common method for the speed control of d.c. series motors. The disadvantage of poor speed regulation is not important in a series motor which is used only where varying speed service is required.

3. Voltage control method. In this method, the voltage source supplying the field current is different from that which supplies the armature. This method avoids the disadvantages of poor speed regulation and low efficiency as in armature control method. However, it is quite expensive. Therefore, this method of speed control is employed for large size motors where efficiency is of great importance.

(*i*) **Multiple voltage control.** In this method, the shunt field of the motor is connected permanently across a fixed voltage source. The armature can be connected across several different voltages through a suitable switchgear. In this way, voltage applied across the armature can be changed. The speed will be approximately proportional to the voltage applied across the armature. Intermediate speeds can be obtained by means of a shunt field regulator.

(*ii*) **Ward-Leonard system.** In this method, the adjustable voltage for the armature is obtained from an adjustable-voltage generator while the field circuit is supplied from a separate source. This is illustrated in Fig. 6.5. The armature of the shunt motor *M* (whose speed is to be controlled) is connected directly to a d.c. generator *G* driven by a constant-speed a.c. motor *A*. The field of the shunt motor is supplied from a constant-voltage exciter *E*. The field of the generator *G* is also supplied from the exciter *E*. The voltage of the generator *G* can be varied by means of its field regulator. By reversing the field current of generator *G* by controller *FC*, the voltage applied to the motor may be reversed. Sometimes, a field regulator is included in the field circuit of shunt motor *M* for additional speed adjustment. With this method, the motor may be operated at **any speed upto its maximum speed.

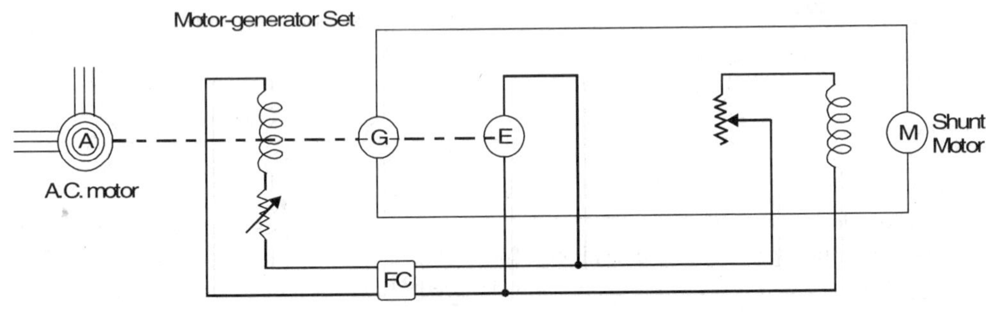

Ward-Leonard speed-control system

Fig. 6.5

Advantages

(*a*) The speed of the motor can be adjusted through a wide range without resistance losses which results in high efficiency.

---

* A more stable operation can be obtained by placing a diverter across the armature in addition to the controller resistance. A change in armature current (due to change in load torque) will not now be so effective in changing the voltage drop across the armature (See Example 6.28).

** Since we can vary the voltage of the generator *G* applied to motor *M*. Note that field circuit of the motor is supplied form a separate source. Therefore, the field circuit current is not affected by the variation of generator voltage.

(*b*) The motor can be brought to a standstill quickly, simply by rapidly reducing the voltage of generator *G*. When the generator voltage is reduced below the back e.m.f. of the motor, this back e.m.f. sends current through the generator armature, establishing dynamic braking. While this takes place, the generator *G* operates as a motor driving motor *A* which returns power to the line.

(*c*) This method is used for the speed control of large motors when a d.c. supply is not available.

The disadvantage of the method is that a special motor-generator set is required for each motor and the losses in this set are high if the motor is operating under light loads for long periods.

Applications. The Ward-Leonard system of speed control is expensive but is used where an unusually wide and very sensitive speed control is desired. This arrangement offers an excellent stepless speed control and is well suited for such applications as passenger elevators, electric excavators etc.

Example 6.1. *A 120 V shunt motor has an armature resistance of 0.3 Ω and is used as a separately excited motor in a Ward-Leonard system. The field current of the motor is adjusted at 1.2A. The output voltage of the exciter is 120 V. The d.c. generator has 200 turns per pole and is running at a speed of 1200 r.p.m. The magnetisation curve for the generator is shown in Fig. 6.6. The field winding resistance of the generator is 30Ω and its armature resistance is 0.2Ω. Determine the external resistance in the field winding circuit of the generator when the motor develops (i) a torque of 30 Nm at 2000 r.p.m. and takes 50A and (ii) the same torque but at a reduced speed of 715 r.p.m. Neglect the demagnetising effect due to armature reaction.*

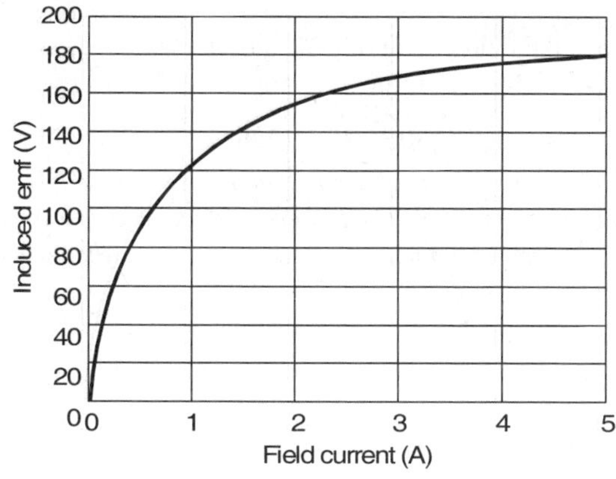

Fig. 6.6

Solution.

(*i*) The angular speed of the motor is

$$\omega_m = \frac{2\pi \times N}{60} = \frac{2\pi \times 2000}{60} = 209.44 \text{ rad/s}$$

The power developed by the motor is

$$P_d = \omega_m T = 209.44 \times 30 = 6283.2 \text{ W}$$

Back e.m.f. of motor, $$E_b = \frac{P_d}{I_a} = \frac{6283.2}{50} = 125.6 \text{ V}$$

The applied voltage across armature terminals of motor is

$$V_m = E_b + I_a{}^*R_{am} = 125.6 + 50 \times 0.3 = 140.6 \text{ V}$$

* R_{am} is the armature resistance of motor and R_{ag} is the armature resistance of generator.

The induced e.m.f. of the generator is

$$E_g = V_m + I_a R_{ag} = 140.6 + 50 \times 0.2 = 150.6 \text{ V}$$

From the magentisation curve of the generator in Fig. 6.6, the field current corresponding to E_g = 150.6 V is 1.75 A.

∴ Total resistance in the field circuit of generator is

$$R_{ft} = \frac{\text{Exciter voltage}}{1.75} = \frac{120}{1.75} = 68.57\Omega$$

∴ $$\text{External resistance} = R_{ft} - 30 = 68.57 - 30 = \mathbf{38.57\ \Omega}$$

(*ii*) Since the torque developed is the same, armature current of the motor remains the same.

$$\text{New back e.m.f. of motor, } E'_b = 125.6 \times \frac{715}{2000} = 44.9 \text{ V}$$

[$E_b \propto \phi N$. Since ϕ is fixed, $E_b \propto N$]

∴ New voltage across the armature terminals of motor is

$$V'_m = E'_b + I_a R_{am} = 44.9 + 50 \times 0.3 = 59.9 \text{ V}$$

∴ New induced e.m.f. of the generator is

$$E'_g = V'_m + I_a R_{ag} = 59.9 + 50 \times 0.2 = 69.9 \text{ V}$$

From the magnetisation curve of the generator in Fig. 6.6., the field current corresponding to E'_g = 69.9 V is 0.4 A.

∴ Total resistance in the field circuit of generator is

$$R'_{ft} = 120/0.4 = 300\Omega$$

∴ $$\text{External resistance} = 300 - 30 = \mathbf{270\ \Omega}$$

Example 6.2. *A 2000 kW, 500 V, variable speed motor is supplied by a 2500 kW generator using Ward-Leonard system shown in Fig. 6.7. The total resistance of the motor and generator armature circuit is 10 mΩ. The motor turns at a nominal speed of 300 r.p.m. when back e.m.f. is 500 V. Find the motor speed and torque when E_g = 400 V and E_b = 380 V.*

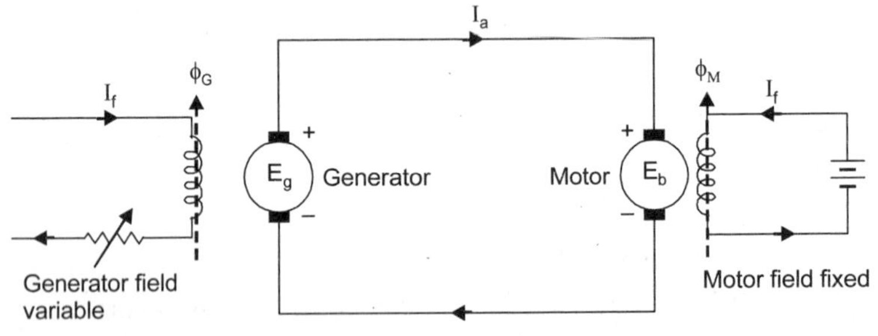

Fig. 6.7

Solution

$$\text{Armature current, } I_a = \frac{E_g - E_b}{\text{Total resistance}} = \frac{400V - 380V}{10\text{m}\Omega} = 2000 \text{ A}$$

$$\text{Power developed by motor, } P_d = E_b I_a = 380 \times 2000 \text{ W} = 760 \text{ kW}$$

∴ $$\text{Motor speed, } N = \frac{380}{500} \times 300 = \mathbf{228\ r.p.m.}$$

$$\text{Motor torque, } T = 9.55 \times \frac{P_d}{N} = 9.55 \times \frac{760 \times 10^3}{228} = \mathbf{31.8 \times 10^3\ Nm}$$

Example 6.3. *A 500 V shunt motor runs at its normal speed of 250 r.p.m. when the armature current is 200 A. The resistance of the armature is 0·12 Ω. Calculate the speed when a resistance is inserted in the field reducing the shunt field to 80% of normal value and armature current is 100 A.*

Solution

$$E_{b1} = V - I_{a1} R_a = 500 - 200 \times 0{\cdot}12 = 476 \text{ volts}$$

$$E_{b2} = V - I_{a2} R_a = 500 - 100 \times 0{\cdot}12 = 488 \text{ volts}$$

Now $N_1 = 250$ r.p.m. ; $\phi_2 = 0{\cdot}8\,\phi_1$; $N_2 = ?$

$$\therefore \quad \frac{N_2}{N_1} = \frac{E_{b2}}{E_{b1}} \times \frac{\phi_1}{\phi_2}$$

or

$$\frac{N_2}{250} = \frac{488}{476} \times \frac{\phi_1}{0{\cdot}8\,\phi_1}$$

$$\therefore \quad N_2 = \mathbf{320{\cdot}4 \text{ r.p.m.}}$$

Example 6.4. *A 220 V d.c. shunt motor having an armature resistance of 0·25 Ω carries an armature current of 50 A and runs at 600 r.p.m. If the flux is reduced by 10% by field regulator, find the speed assuming load torque remains the same.*

Solution.

Initial conditions

$$N_1 = 600 \text{ r.p.m.} \;;\; I_{a1} = 50 \text{ A}$$

$$E_{b1} = V - I_{a1} R_a = 220 - 50 \times 0{\cdot}25 = 207{\cdot}5 \text{ V}$$

Final conditions

$$\phi_2 = 0{\cdot}9\,\phi_1 \text{ so that } \phi_2/\phi_1 = 0{\cdot}9$$

As the load torque remains unchanged, $\phi_1 I_{a1} = \phi_2 I_{a2}$

$$\therefore \quad I_{a2} = \left(\frac{\phi_1}{\phi_2}\right) I_{a1} = \left(\frac{1}{0{\cdot}9}\right) 50 = 55{\cdot}6 \text{ A}$$

$$\therefore \quad E_{b2} = V - I_{a2} R_a = 220 - 55{\cdot}6 \times 0{\cdot}25 = 206{\cdot}1 \text{ V}$$

Using the formula,

$$\frac{N_2}{N_1} = \frac{E_{b2}}{E_{b1}} \times \frac{\phi_1}{\phi_2}$$

or

$$\frac{N_2}{600} = \frac{206{\cdot}1}{207{\cdot}5} \times \frac{1}{0{\cdot}9}$$

$$\therefore \quad N_2 = \mathbf{662 \text{ r.p.m.}}$$

Example 6.5. *A 4-pole, 500 V shunt motor has a total of 720 armature conductors which are wave connected. The full-load armature current is 60 A and the flux per pole is 0·03 Wb. The armature resistance is 0·2 Ω. The voltage drop across a brush is 1V. Calculate the full-load speed of the motor.*

Solution.

$$\text{Full-load } E_b = V - I_a R_a - \text{Brush drop}$$

$$= 500 \quad 60 \times 0{\cdot}2 - 2 = 486 \text{ volts}$$

Now

$$E_b = \frac{P \phi Z N}{60 A}$$

$$\therefore \quad N = \frac{E_b \times 60 A}{P \phi Z} = \frac{486 \times 60 \times 2}{4 \times 0{\cdot}03 \times 720} = \mathbf{675 \text{ r.p.m.}}$$

Example 6.6. *A 200 V shunt motor having armature resistance of 0·4 Ω and shunt field resistance of 100 Ω drives a load at 500 r.p.m. taking 27A. It is desired to run the motor at 700 r.p.m.*

Assuming the load torque to be constant, find the value of resistance to be used as field regulator. Neglect saturation effect.

Solution.

Initial conditions

$$N_1 = 500 \text{ r.p.m.} ; I_L = 27 \text{ A} ; I_{sh1} = 200/100 = 2 \text{ A}$$

$$I_{a1} = I_L - I_{sh1} = 27 - 2 = 25 \text{ A}$$

$$E_{b1} = 200 - 25 \times 0{\cdot}4 = 190 \text{ V}$$

Final conditions

$$E_{b2} = V - I_{a2} R_a = 200 - 0{\cdot}4\, I_{a2}$$

As the load torque is constant, $\phi_1 I_{a1} = \phi_2 I_{a2}$. But $\phi \propto I_{sh}$ so that we have,

$$I_{sh1} I_{a1} = I_{sh2} I_{a2}$$

$$\therefore \quad I_{sh2} = \frac{I_{sh1} I_{a1}}{I_{a2}} = \frac{2 \times 25}{I_{a2}} = \frac{50}{I_{a2}}$$

Now
$$\frac{N_1}{N_2} = \frac{E_{b1}}{E_{b2}} \times \frac{I_{sh2}}{I_{sh1}}$$

or
$$\frac{500}{700} = \frac{190}{200 - 0{\cdot}4\, I_{a2}} \times \frac{(50/I_{a2})}{2}$$

or
$$I_{a2}^2 - 500\, I_{a2} + 16625 = 0$$

Solving this equation, we get, $I_{a2} = 35$ A

$$\therefore \quad I_{sh2} = 50/I_{a2} = 50/35 = 1{\cdot}43 \text{ A}$$

$$\therefore \quad R_{sh2} = V/I_{sh2} = 200/1{\cdot}43 = 139{\cdot}86\ \Omega$$

$$\text{Field rheostat resistance} = 139{\cdot}86 - 100 = \mathbf{39{\cdot}86\ \Omega}$$

Example 6.7. *A 240 V d.c. shunt motor runs at 800 r.p.m. and takes armature current of 2 A. Find the resistance required in series with the shunt winding so that the motor may run at 950 r.p.m. when taking an armature current of 28 A. Assume flux is proportional to field current. Shunt field resistance is 160 Ω, armature resistance is 0·4 Ω.*

Solution.

$$E_{b1} = V - I_{a1} R_a = 240 - 2 \times 0{\cdot}4 = 239{\cdot}2 \text{ volts}$$

$$E_{b2} = V - I_{a2} R_a = 240 - 28 \times 0{\cdot}4 = 228{\cdot}8 \text{ volts}$$

Let R ohms be the resistance to be connected in series with shunt field winding.

$I_{sh1} = 240/160 = 1{\cdot}5$ A ; $I_{sh2} = 240/(160 + R)$ A ; $N_1 = 800$ r.p.m. ; $N_2 = 950$ r.p.m.

Now
$$\frac{N_2}{N_1} = \frac{E_{b2}}{E_{b1}} \times \frac{I_{sh1}}{I_{sh2}} \qquad (\because \text{ flux} \propto \text{field current})$$

or
$$\frac{950}{800} = \frac{228{\cdot}8}{239{\cdot}2} \times \frac{1{\cdot}5}{240/(160 + R)}$$

$$\therefore \quad 160 + R = 198{\cdot}63$$

or
$$R = 198{\cdot}63 - 160 = \mathbf{38{\cdot}63\ \Omega}$$

Example 6.8. *A 250 V shunt motor has a shunt field resistance of 250 Ω and an armature resistance of 0·25 Ω. For a given load torque and no additional resistance included in the shunt field circuit, the motor runs at 1500 r.p.m. drawing an armature current of 20 A. If a resistance of 250 Ω is inserted in series with the field, the load torque remaining the same, find the new armature current and the speed. Assume the magnetisation curve to be linear.*

Solution. New armature current. Since the magnetisation curve is linear, flux ∝ field current. As the load torque is constant,

$$\therefore \quad \phi_1 I_{a1} = \phi_2 I_{a2}$$

$$\text{or} \quad I_{sh1} I_{a1} = I_{sh2} I_{a2}$$

Now $I_{sh1} = 250/250 = 1\text{ A}$; $I_{sh2} = 250/(250 + 250) = 0{\cdot}5\text{ A}$

$$\therefore \quad 1 \times 20 = 0{\cdot}5 \times I_{a2} \quad \text{or} \quad I_{a2} = \mathbf{40\ A}$$

New Speed

$$E_{b1} = V - I_{a1} R_a = 250 - 20 \times 0{\cdot}25 = 245 \text{ volts}$$

$$E_{b2} = V - I_{a2} R_a = 250 - 40 \times 0{\cdot}25 = 240 \text{ volts}$$

$$\text{Now} \quad \frac{N_2}{N_1} = \frac{E_{b2}}{E_{b1}} \times \frac{I_{sh1}}{I_{sh2}}$$

$$\text{or} \quad \frac{N_2}{1500} = \frac{240}{245} \times \frac{1}{0{\cdot}5} \qquad \therefore N_2 = \mathbf{2938\ r.p.m.}$$

Example 6.9. *A 20 h.p., 250 V shunt motor has an armature circuit resistance (including brushes and interpoles) of 0·22 Ω and a field resistance of 170 Ω. At no-load and rated voltage, the speed is 1200 r.p.m. and the armature current is 3A. At full-load and rated voltage, the line current is 55 A and the flux is reduced by 6% (due to effects of armature reaction) from its value at no-load. What is the full-load speed ?*

Solution.

At no-load $\quad E_{b0} = V - I_{a0} R_a = 250 - 3 \times 0{\cdot}22 = 249{\cdot}3$ volts

No-load speed, $N_0 = 1200$ r.p.m. ; No-load flux $= \phi_0$

At full-load

$$I_{sh} = 250/170 = 1{\cdot}47 \text{ A} \; ; \; I_a = I_L - I_{sh} = 55 - 1{\cdot}47 = 53{\cdot}53 \text{ A}$$

$$E_b = V - I_a R_a = 250 - 53{\cdot}53 \times 0{\cdot}22 = 238{\cdot}2 \text{ volts}$$

At full-load, $\phi = 0{\cdot}94\, \phi_0$ or $\phi_0/\phi = 1/0{\cdot}94$

$$\text{Now} \quad \frac{N}{N_0} = \frac{E_b}{E_{b0}} \times \frac{\phi_0}{\phi}$$

$$\text{or} \quad \frac{N}{1200} = \frac{238{\cdot}2}{249{\cdot}3} \times \frac{1}{0{\cdot}94} \qquad \therefore N = \mathbf{1220\ r.p.m.}$$

Example 6.10. *A 220 V shunt motor has an armature resistance of 0·5 Ω and takes armature current of 40 A at full-load. By how much must the main flux be reduced to raise the speed by 50% if the developed torque is constant ?*

Solution. $$\frac{N_2}{N_1} = \frac{E_{b2}}{E_{b1}} \times \frac{\phi_1}{\phi_2}$$

$$\text{Now} \quad \phi_1 I_{a1} = \phi_2 I_{a2} \qquad \text{...torque remains constant}$$

$$\therefore \quad I_{a2} = \left(\frac{\phi_1}{\phi_2}\right) I_{a1} = 40\,x \quad \text{where } x = \frac{\phi_1}{\phi_2}$$

$$E_{b1} = V - I_{a1} R_a = 220 - 40 \times 0{\cdot}5 = 200 \text{ volts}$$

$$E_{b2} = V - I_{a2} R_a = 220 - 40\,x \times 0{\cdot}5 = (220 - 20\,x) \text{ volts}$$

$$N_2/N_1 = 3/2 \qquad \text{...given}$$

$$\therefore \quad \frac{3}{2} = \frac{220 - 20\,x}{200} \times x \quad \text{or} \quad x^2 - 11x + 15 = 0$$

On solving this equation, we have, $x = 1{\cdot}6$

$$\therefore \quad \frac{\phi_1}{\phi_2} = 1{\cdot}6 \quad \text{or} \quad \frac{\phi_1 - \phi_2}{\phi_1} = \frac{1{\cdot}6 - 1}{1{\cdot}6} = \mathbf{\frac{3}{8}} \text{ or } \mathbf{37{\cdot}5\%}$$

Example 6.11. *A 10 kW, 250 V shunt generator having an armature resistance of 0·1 Ω and field resistance of 250 Ω delivers full-load at rated voltage and 800 r.p.m. The machine is now run as a motor while taking 10 kW at 250 V. What is the speed of the motor ? Neglect brush contact drop.*

Solution.

As a generator

$$I_L = \frac{10 \times 10^3}{250} = 40 \text{ A} \; ; \; I_{sh} = \frac{250}{250} = 1 \text{ A}$$

$$\therefore \quad I_a = I_L + I_{sh} = 40 + 1 = 41 \text{ A}$$

$$E_g = V + I_a R_a = 250 + 41 \times 0{\cdot}1 = 254{\cdot}1 \text{ volts}$$

As a motor

$$I_L = \frac{10 \times 10^3}{250} = 40 \text{ A} \; ; \; I_{sh} = \frac{250}{250} = 1 \text{ A}$$

$$I_a = I_L - I_{sh} = 40 - 1 = 39 \text{ A}$$

$$\therefore \quad E_b = V - I_a R_a = 250 - 39 \times 0{\cdot}1 = 246{\cdot}1 \text{ volts}$$

Now $$\frac{N_m}{N_g} = \frac{E_b}{E_g}$$ ...since flux remains constant

or $$\frac{N_m}{800} = \frac{246{\cdot}1}{254{\cdot}1} \quad \therefore \; N_m = \mathbf{774{\cdot}8 \text{ r.p.m.}}$$

Example 6.12. *A 220 V, 15 kW, 850 r.p.m. motor draws 72·2 A when operating at rated conditions. The resistances of the armature and shunt field are 0·25 Ω and 100 Ω respectively. Determine the percentage reduction in the field flux in order to obtain a speed of 1650 r.p.m. when the armature current drawn is 40 A.*

Solution.

At rated conditions

$$I_{sh} = 220/100 = 2{\cdot}2 \text{ A} \; ; \; I_{a1} = 72{\cdot}2 - 2{\cdot}2 = 70 \text{ A}$$

$$E_{b1} = V - I_{a1} R_a = 220 - 70 \times 0{\cdot}25 = 202{\cdot}5 \text{ volts}$$

When armature current is 40 A

$$E_{b2} = V - I_{a2} R_a = 220 - 40 \times 0{\cdot}25 = 210 \text{ volts}$$

Now $$\frac{N_2}{N_1} = \frac{E_{b2}}{E_{b1}} \times \frac{\phi_1}{\phi_2}$$

or $$\frac{1650}{850} = \frac{210}{202{\cdot}5} \times \frac{\phi_1}{\phi_2} \quad \text{or} \quad \frac{\phi_1}{\phi_2} = \frac{1}{0{\cdot}534}$$

$$\therefore \quad \text{\% Reduction in flux} = \frac{\phi_1 - 0{\cdot}534\,\phi_1}{\phi_1} \times 100 = \mathbf{46{\cdot}6\%}$$

Example 6.13. *A 440 V shunt motor runs at its normal speed of 300 r.p.m. when armature current is 160 A. Find the speed when the armature current is 80 A and a resistance of 0·5 Ω is connected in series with the armature, the shunt field remaining constant. Resistance of armature is 0·2 Ω and of interpole winding is 0·05 Ω.*

Solution. Figure 6.8 shows the motor connections.

$$\frac{N_2}{N_1} = \frac{E_{b2}}{E_{b1}} \times \frac{\phi_1}{\phi_2}$$

$$E_{b1} = V - I_{a1} R$$

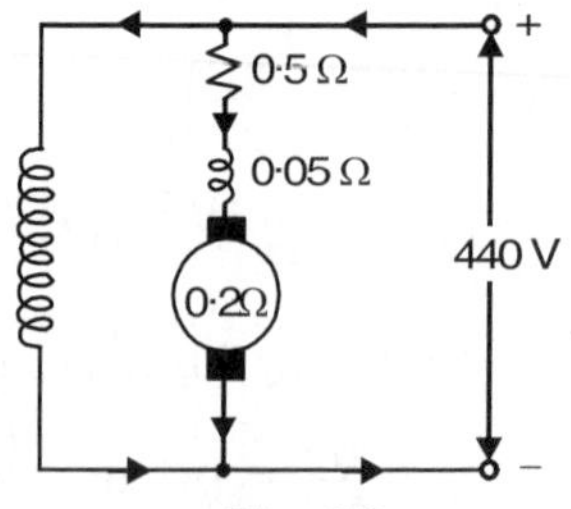

Fig. 6.8

$$= 440 - 160 \times (0{\cdot}2 + 0{\cdot}05) = 400 \text{ volts}$$

$$E_{b2} = V - I_{a2} R$$

$$= 440 - 80 \times (0{\cdot}2 + 0{\cdot}05 + 0{\cdot}5) = 380 \text{ volts}$$

Now $\phi_1 = \phi_2$; $N_1 = 300$ r.p.m. ; $N_2 = ?$

$$\therefore \quad \frac{N_2}{300} = \frac{380}{400} \times \frac{\phi_1}{\phi_1} \qquad \therefore N_2 = \mathbf{285 \text{ r.p.m.}}$$

Example 6.14. *A 230 V d.c. shunt motor takes an armature current of 20 A on a particular load. The armature circuit resistance is 0.5 Ω. Find the resistance required in series with the armature to reduce the speed to 50% if (i) the load torque is constant (ii) the load torque is proportional to the square of the speed.*

Solution. (*i*) The constant load torque means that here armature current remains the same.

For the first case, $E_1 = V - I_a R_a = 230 - 20 \times 0.5 = 220$ volts

For the second case, $N_2 = N_1/2$ so that the back e.m.f. E_2 is given by ;

$$\frac{E_2}{E_1} = \frac{N_2}{N_1} \text{ or } E_2 = \frac{N_2}{N_1} \times E_1 = \frac{1}{2} \times 220 = 110 \text{ volts}$$

Let R be the required resistance in series with the armature.

$$\therefore \qquad E_2 = V - I_a (R_a + R)$$

$$\text{or} \qquad 110 = 230 - 20\,(0.5 + R) \qquad \therefore R = \mathbf{5.5\ \Omega}$$

(*ii*) It is given that $T \propto N^2$. For a shunt motor, $T \propto I_a$ ($\because \phi$ is constant). Therefore, $I_a \propto N^2$. When speed is halved, the armature current becomes one-fourth. Therefore, new armature current $I'_a = I_a/4 = 20/4 = 5$A.

$$\therefore \qquad E_2 = V - I'_a (R_a + R')$$

$$\text{or} \qquad 110 = 230 - 5\,(0.5 + R') \qquad \therefore R' = \mathbf{23.5\ \Omega}$$

Example 6.15. *A 14.92 kW, 400 V, 400 r.p.m. d.c. shunt motor draws a current of 40 A when running at full-load. The moment of inertia of the rotating system in 7.5 kg m². If the starting current is 1.2 times the full-load current, calculate (i) the full-load torque (ii) the time required for the motor to attain the rated speed against full-load.*

Solution.

(*i*) F.L. output power = 14920 W ; Speed, N = 400 r.p.m.

$$\text{Now } \frac{2\pi NT}{60} = 14920 \text{ or } \frac{2\pi \times 400 \times T}{60} = 14920 \quad \therefore T = \mathbf{356 \text{ Nm}}$$

(*ii*) During starting perio d, thc torque available for accelerating the motor $= 1.2\,T - T = 0.2\,T = 0.2 \times 356 = 71.2$ Nm.

$$\text{Now, Torque} = I\frac{d\omega}{dt} \text{ or } 71.2 = 7.5 \times \frac{{}^*2\pi(20/3)}{dt} \quad \therefore dt = \mathbf{4.41 \text{ seconds}}$$

Example 6.16. *A 200 V d.c. shunt motor takes 4A at no-load when running at 700 r.p.m. The field resistance is 100Ω. The armature at standstill gives a drop of 6 volts across armature terminals when 10A were passed through it. Calculate (i) speed on load (ii) torque in Nm and (iii) efficiency. The normal input of the motor is 8 kW.*

$$* \quad \omega = \frac{2\pi N}{60} = 2\pi\left(\frac{N}{60}\right) = 2\pi\left(\frac{400}{60}\right) = 2\pi(20/3)$$

Solution.

(*i*) Armature resistance, $R_a = 6/10 = 0.6\Omega$; $I_{sh} = 200/100 = 2\text{A}$

F.L. line current, $I_L = 8 \times 10^3/200 = 40\text{A}$; $I_a = 40 - 2 = 38\text{A}$

$E_{b0} = 200 - 2 \times 0.6 = 198.8\text{ V}$; $E_b = 200 - 38 \times 0.6 = 177.2\text{ V}$

Now $$\frac{N}{N_0} = \frac{E_b}{E_{b0}} \text{ or } \frac{N}{700} = \frac{177.2}{198.8} \quad \therefore N = \mathbf{624\text{ r.p.m.}}$$

(*ii*) $$T_a = \frac{9.55 \times E_b I_a}{N} = \frac{9.55 \times 177.2 \times 38}{624} = \mathbf{103\text{ Nm}}$$

(*iii*) Constant losses = (No-load input power) – (No-load Arm. Cu loss)

$= (200 \times 4) - (2^2 \times 0.6) = 797.6\text{ W}$

F.L. Arm. Cu loss $= 38^2 \times 0.6 = 866.4\text{ W}$

Total F.L. losses $= 797.6 + 866.4 = 1664\text{ W}$

$$\therefore \quad \text{F.L. efficiency} = \frac{\text{Output}}{\text{Input}} = \frac{8000 - 1664}{8000} = \mathbf{0.792} \text{ or } \mathbf{79.2\%}$$

Example 6.17. *A belt-driven 100 kW, shunt generator running at 300 r.p.m. on 220 V bus-bars continues to run as a motor when the belt breaks, then taking 10 kW. What will be its speed ? Given armature resistance = 0.025 Ω, field resistance = 60 Ω and contact drop under each brush = 1V. Ignore armature reaction.*

Solution. As Generator

$$I_L = \frac{100 \times 10^3}{220} = 454.55\text{ A} ; \; I_{sh} = \frac{220}{60} = 3.67\text{ A} ; \; I_a = I_L + I_{sh} = 458.2\text{ A}$$

$$E_{b1} = 220 + 458.2 \times 0.025 + 2 \times 1 = 233.45\text{ V}$$

As motor

$$I_L = \frac{10 \times 10^3}{220} = 45.45\text{ A} ; \; I_{sh} = 3.67\text{ A} ; \; I_a = I_L - I_{sh} = 41.78\text{ A}$$

$$E_{b2} = 220 - 41.78 \times 0.025 - 2 \times 1 = 216.96\text{ V}$$

Now $$\frac{N_2}{N_1} = \frac{E_{b2}}{E_{b1}} \text{ or } \frac{N_2}{300} = \frac{216.96}{233.45} \quad \therefore N_2 = \mathbf{279\text{ r.p.m.}}$$

Example 6.18. *A d.c. shunt machine while running as generator develops a voltage of 250 V at 1000 r.p.m. at no-load. It has armature resistance of 0.5 Ω and field resistance of 250 Ω. When the machine runs as a motor, input to it at no-load is 4A, 250 V. Calculate the speed and efficiency of the machine when it runs as a motor taking 40 A at 250 V. Armature reaction weakens the field by 4%.*

Solution. Motor Speed

When the machine runs as a generator at 1000 r.p.m., it develops a voltage of 250 V. Clearly, if the machine is run as a motor at 1000 r.p.m., it will have a back e.m.f. of 250 V.

$\therefore \quad E_{b1} = 250\text{ V}$; $N_1 = 1000\text{ r.p.m.}$

When running as motor and drawing 40 A, we have,

$E_{b2} = 250 - (40 - 1)\, 0.5 = 230.5\text{ V}$ $\quad [\because I_a = I_L - I_{sh} = (40 - 1)]$

Now $$\frac{N_2}{N_1} = \frac{E_{b2}}{E_{b1}} \times \frac{\phi_1}{\phi_2}$$

or $$\frac{N_2}{1000} = \frac{230.5}{250} \times \frac{\phi_1}{0.96\phi_1} \qquad (\because \phi_2 = 0.96\,\phi_1)$$

$\therefore$ $$N_2 = 1000 \times \frac{230.5}{250} \times \frac{1}{0.96} = \textbf{960 r.p.m.}$$

Efficiency

Total losses at no-load $= 250 \times 4 = 1000$ W

Armature Cu loss at no-load $= I_{a0}^2 R_a = 3^2 \times 0.5 = 4.5$ W

$\therefore$ Constant losses $= 1000 - 4.5 = 995.5$ W

When the motor draws 40 A, its armature current $= 40 - 1 = 39$ A.

$\therefore$ Armature Cu loss $= 39^2 \times 0.5 = 760.5$ W

Total losses $= 995.5 + 760.5 = 1756$ W

Input power $= 250 \times 40 = 10{,}000$ W

Output power $= 10{,}000 - 1756 = 8244$ W

$\therefore$ $$\text{Efficiency} = \frac{8244}{10{,}000} \times 100 = \textbf{82.44\%}$$

TUTORIAL PROBLEMS

1. A 200 V shunt motor having an armature resistance of 0·2 Ω carries an armature current of 50 A and runs at 960 r.p.m. If the flux is reduced by 10%, find the speed, assuming load torque remains the same. **[1059 r.p.m.]**
2. A 240 V shunt motor has an armature resistance of 0·4 Ω and a field resistance of 120 Ω. It runs at 460 r.p.m. and the armature current is 25 A. What resistance must be placed in the shunt field circuit in order to raise the speed to 560 r.p.m., the torque remaining the same ? Assume the flux to be proportional to the field current. **[27·6 Ω]**
3. A 220 V shunt motor takes 64 A and runs at 500 r.p.m. Armature resistance is 0·25 Ω and field resistance is 55 Ω. Find the speed when a resistor of 20 Ω is connected in series with the field circuit, the load torque remaining the same. Assume the flux to be proportional to the field current. **[664 r.p.m.]**
4. A shunt motor connected across 440 V supply takes an armature current of 20 A and runs at 500 r.p.m. The armature resistance is 0·6 Ω. If the magnitude of flux is reduced by 30% and the torque developed by the armature increases by 40%, what is the speed of the motor ? **[695 r.p.m.]**
5. A 230 V shunt motor runs at 800 r.p.m. and takes armature current of 50 A. Find the resistance to be added to the field circuit to increase the speed to 1000 r.p.m. at an armature current of 80 A. Assume flux is proportional to field current. Armature resistance = 0·15 Ω and field winding resistance = 250 Ω. **[69 Ω]**
6. A shunt motor runs at 460 r.p.m. when taking 240 A at 100 V. Assuming flux proportional to the field current, what will be the speed when taking 90 A at 220 V ? Armature resistance = 0·032 Ω and shunt field resistance = 46 Ω. **[491 r.p.m.]**
7. A 250 V d.c. shunt motor has an armature resistance of 0·5 Ω and a field resistance of 250 Ω. When driving a load of constant torque at 600 r.p.m., the armature current is 20 A. If it is desired to raise the speed from 600 r.p.m. to 800 r.p.m., what resistance should be inserted in shunt field circuit ? Assume that the magnetic circuit is unsaturated. **[337·5 Ω]**
8. A 500 V d.c. shunt motor has an armature resistance of 0·1 Ω and a field resistance of 250 Ω. When

driving a load at 450 r.p.m., the armature current is 200 A. The speed of the machine is to be raised to 500 r.p.m. by inserting a resistance in the field circuit. The load torque is constant. Calculate the value of the resistance if the magnetisation curve is assumed to be a straight line. **[25 Ω]**

9. A 220 V shunt motor takes 64 A and runs at 500 r.p.m. Armature resistance is 0·25 Ω and field resistance is 55 Ω. Find the speed when a resistor of 20 Ω is included in the shunt field circuit and load torque is increased by 25%. Assume the flux to be proportional to the field current. **[647 r.p.m.]**

10. A 220 V shunt motor has armature resistance of 0·5 Ω and takes an armature current of 40 A on a certain load. By how much the main flux be reduced to raise the speed by 50% if the developed torque is constant ? Neglect saturation and armature reaction. **[37·3%]**

Example 6.19. *A shunt motor supplied at 230 V runs at 900 r.p.m. while taking armature current of 30 A ; the resistance of armature circuit being 0·4 Ω. Calculate the resistance required in series with the armature circuit to reduce the speed to 500 r.p.m., assuming that the armature current is 25 A.*

Solution.

Initial conditions

$$N_1 = 900 \text{ r.p.m.} ; \; I_{a1} = 30 \text{ A}$$

$$E_{b1} = V - I_{a1} R_a = 230 - 30 \times 0{\cdot}4 = 218 \text{ V}$$

Final conditions

$$N_2 = 500 \text{ r.p.m.} ; \; I_{a2} = 25 \text{ A} ; \; R_t = R_a + R_C$$

$$E_{b2} = V - I_{a2} R_t = 230 - 25 R_t$$

Since excitation is unchanged, $\phi_1 = \phi_2$.

$$\therefore \quad \frac{N_2}{N_1} = \frac{E_{b2}}{E_{b1}}$$

or $$\frac{500}{900} = \frac{230 - 25 R_t}{218} \qquad \therefore \quad R_t = 4{\cdot}356 \, \Omega$$

∴ Additional series resistance required in the armature is

$$R_C = R_t - R_a = 4{\cdot}356 - 0{\cdot}4 = \mathbf{3{\cdot}956 \, \Omega}$$

Example 6.20. *A 250 V shunt motor has an armature current of 20 A when running at 1000 r.p.m. against full-load torque. The armature resistance is 0·5 Ω. What resistance must be inserted in series with the armature to reduce the speed to 500 r.p.m. at the same torque ? What will be the speed if the load torque is halved with this resistance in the circuit ? Assume the flux to remain constant throughout and neglect brush contact drop.*

Solution. Figure 6.9 shows shunt motor connections.

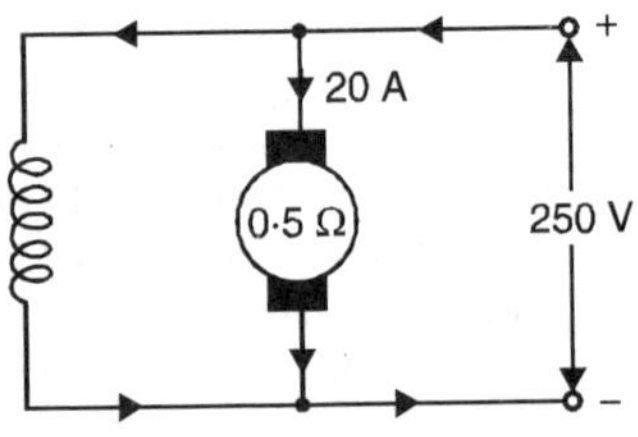

Fig. 6.9

At full-load torque. Let R_C be the resistance to be inserted in series with the armature to reduce the speed to 500 r.p.m. Since load torque remains the same and $\phi_1 = \phi_2$,

$$T_a = \phi_1 I_{a1} = \phi_2 I_{a2}$$

$$\therefore \quad I_{a1} = I_{a2} = 20 \text{ A}$$

$$E_{b1} = V - I_{a1} R_a = 250 - 20 \times 0{\cdot}5 = 240 \text{ volts}$$

$$E_{b2} = V - I_{a2} R_t = 250 - 20 \times R_t \quad \text{where } R_t = R_a + R_C$$

Now $$\frac{N_2}{N_1} = \frac{E_{b2}}{E_{b1}} \times \frac{\phi_1}{\phi_2}$$

or $$\frac{500}{1000} = \frac{250 - 20 R_t}{240} \qquad (\because \phi_1 = \phi_2)$$

$$\therefore \quad R_t = 6{\cdot}5 \, \Omega \quad \text{or} \quad R_C = R_t - R_a = 6{\cdot}5 - 0{\cdot}5 = \mathbf{6 \, \Omega}$$

At half full-load. Since load torque is halved and flux remains the same, the armature current is also halved *i.e.*, $I_{a3} = 20/2 = 10$ A.

$$E_{b3} = V - I_{a3} R_t = 250 - 10 \times 6{\cdot}5 = 185 \text{ volts}$$

Now $$\frac{N_3}{N_1} = \frac{E_{b3}}{E_{b1}} \qquad (\because \phi_1 = \phi_3)$$

or $$\frac{N_3}{1000} = \frac{185}{240} \qquad \therefore N_3 = \mathbf{771\ r.p.m.}$$

Example 6.21. *A 500 V, 10 h.p. shunt motor has a full-load efficiency of 85%. With the same shunt field and armature current, it is desired to reduce the speed by 30% by inserting a resistance in the armature circuit. Calculate the value of the inserted resistance. The resistances of the field and armature are 400 Ω and 0·25 Ω respectively.*

Solution. Let R_C be the required resistance.

$$\text{Input to motor} = \frac{\text{Output}}{\eta} = \frac{10 \times 746}{0{\cdot}85} = 8776 \text{ W}$$

$$\text{Input current} = \frac{8776}{500} = 17{\cdot}55 \text{ A}$$

$$\text{Shunt field current} = \frac{500}{400} = 1{\cdot}25 \text{ A}$$

$$\text{Armature current, } I_{a1} = 17{\cdot}55 - 1{\cdot}25 = 16{\cdot}3 \text{ A}$$

$$E_{b1} = V - I_{a1} R_a = 500 - 16{\cdot}3 \times 0{\cdot}25 = 495{\cdot}9 \text{ volts}$$

Now $$\frac{N_2}{N_1} = \frac{E_{b2}}{E_{b1}} \times \frac{\phi_1}{\phi_2}$$

Here $N_2 = 0{\cdot}7\, N_1$ and $\phi_1 = \phi_2$.

∴ $$\frac{0{\cdot}7\, N_1}{N_1} = \frac{E_{b2}}{495{\cdot}9} \quad \text{or} \quad E_{b2} = 347{\cdot}1 \text{ volts}$$

Now $$E_{b2} = V - I_{a2} R_t \quad \text{where } R_t = R_a + R_C$$

or $$347{\cdot}1 = 500 - 16{\cdot}3 \times R_t \qquad (\because I_{a2} = I_{a1} = 16{\cdot}3 \text{ A})$$

∴ $$R_t = 9{\cdot}37\ \Omega \quad \text{or} \quad R_C = R_t - R_a = 9{\cdot}37 - 0{\cdot}25 = \mathbf{9{\cdot}12\ \Omega}$$

Example 6.22. *A 220 V shunt motor with an armature resistance of 0·5 Ω is excited to give constant main field. At full-load, the motor runs at 500 r.p.m. and takes an armature current of 30 A. If a resistance of 1Ω is placed in the armature circuit, find the speed at (i) full-load torque (ii) double full-load torque. (iii) Also find the stalling torque.*

Solution.

(*i*) **At full-load torque.** With no additional resistance in the armature circuit,

$$N_1 = 500 \text{ r.p.m.} \quad ; \quad I_{a1} = 30 \text{ A}$$

∴ $$E_{b1} = V - I_{a1} R_a = 220 - 30 \times 0{\cdot}5 = 205 \text{ volts}$$

When 1 Ω resistance is added to the armature circuit, it is given that torque (= full-load torque) remains the same. Since flux is constant ($T \propto \phi I_a$), the armature current remains the same *i.e.*, $I_{a2} = I_{a1} = 30$ A.

∴ $$E_{b2} = V - I_{a2}(R_a + 1) = 220 - 30\,(0{\cdot}5 + 1) = 175 \text{ volts}$$

Now $$\frac{N_2}{N_1} = \frac{E_{b2}}{E_{b1}} \times \frac{\phi_1}{\phi_2}$$

or $$\frac{N_2}{500} = \frac{175}{205} \qquad (\because \phi_1 = \phi_2)$$

$\therefore \qquad N_2 = \frac{175}{205} \times 500 = \mathbf{427\ r.p.m.}$

(*ii*) **At double full-load torque.** With 1 Ω resistance in the armature circuit, the load torque is doubled. Since flux is constant, the armature current will be doubled *i.e.* $I_{a3} = 2I_{a1} = 2 \times 30 = 60$ A.

$\therefore \qquad E_{b3} = V - I_{a3}(R_a + 1) = 220 - 60\,(0{\cdot}5 + 1) = 130$ volts

Now $\qquad \frac{N_3}{N_1} = \frac{E_{b3}}{E_{b1}} \qquad (\because\ \phi_1 = \phi_3)$

or $\qquad \frac{N_3}{500} = \frac{130}{205} \qquad \therefore\ N_3 = \mathbf{317\ r.p.m.}$

(*iii*) **Stalling torque.** The stalling torque is the maximum torque *i.e.*, the value of load torque plus frictional torque which the motor will just fail to provide.

At stalling torque, $E_b = 0$.

Therefore, armature current I_{a4} is given by :

$$I_{a4} = \frac{V - E_b}{(R_a + 1)} = \frac{220 - 0}{(0{\cdot}5 + 1)} \simeq 150 \text{ A}$$

$$\simeq 5 \times 30 \text{ A (full-load current)}$$

Therfore, the stalling torque is *about five times the full-load torque*.

Example 6.23. *A 230 V d.c. shunt motor takes an armature current of 20 A on a particular load. The armature circuit resistance is 0·5 Ω. Find the resistance required in series with the armature to reduce the speed by 50% if (i) the load torque is constant (ii) the load torque is proportional to the square of speed.*

Solution.

(*i*) **Load torque constant**. With no additional resistance in the armature circuit,

$$E_{b1} = V - I_{a1} R_a = 230 - 20 \times 0{\cdot}5 = 220 \text{ volts}$$

Let R_C be the resistance added to the armature to reduce the speed by 50%. It is given that load torque remains the same. Since flux is constant ($T \propto \phi I_a$), the armature current remains the same *i.e.* $I_{a2} = 20$ A.

$\therefore \qquad E_{b2} = V - I_{a2}(R_a + R_C) = 230 - 20\,(0{\cdot}5 + R_C)$

Now $\qquad \frac{N_2}{N_1} = \frac{E_{b2}}{E_{b1}} \times \frac{\phi_1}{\phi_2}$

or $\qquad \frac{1}{2} = \frac{230 - 20\,(0{\cdot}5 + R_C)}{220} \qquad \left[\because \phi_1 = \phi_2 \text{ and } N_2 = \frac{N_1}{2}\right]$

$\therefore \qquad R_C = \mathbf{5{\cdot}5\ \Omega}$

(*ii*) **Load torque $\propto N^2$**

$T_a \propto \phi I_a \qquad$ or $\qquad T_a \propto I_a \qquad (\because\ \phi$ is constant)

Also $\qquad T_a \propto N^2$ (given) $\qquad \therefore \qquad I_a \propto N^2$

If the speed is halved, armature current will become one-fourth *i.e.* $I_{a3} = \frac{1}{4} I_{a2} = \frac{1}{4} \times 20 = 5$ A

$\therefore \qquad E_{b3} = V - I_{a3}(R_a + R_C) = 230 - 5\,(0{\cdot}5 + R_C)$

Since speed is halved and flux is constant, $E_{b3} = E_{b1}/2 = 220/2 = 110$ volts

$\therefore \qquad 110 = 230 - 5\,(0{\cdot}5 + R_C)$

or $\qquad R_C = \mathbf{23{\cdot}5\ \Omega}$

Example 6.24. *The speed of a 500 V shunt motor when running light is 1000 r.p.m. Its armature resistance including brushes is 2 Ω. To reduce the speed, the armature voltage is tapped off a 50 Ω resistor across the mains. With armature voltage tapped off 25 Ω, at what speed the motor will run when the supply current is 15 A ? Ignore field current and assume constant flux.*

Solution. When the motor is running light, $E_{b0} = 500$ V ; $N_0 = 1000$ r.p.m. To reduce the speed of the motor, the connections are made as shown in Fig. 6.10.

$$25 \times 15 + 25 \times I = 500$$

$$\therefore \qquad I = 5\text{A}$$

Armature current, $I_a = 15 - I = 15 - 5 = 10$ A

$$V_{BC} = I \times 25 = 5 \times 25 = 125 \text{ volts}$$

$$E_b = V - I_a R_a = 125 - 10 \times 2 = 105 \text{ volts}$$

Now $\dfrac{N}{N_0} = \dfrac{E_b}{E_{b0}}$ (∵ ϕ is constant)

$$\therefore \qquad N = \frac{E_b}{E_{b0}} \times N_0 = \frac{105}{500} \times 1000$$

$$= \textbf{210 r.p.m.}$$

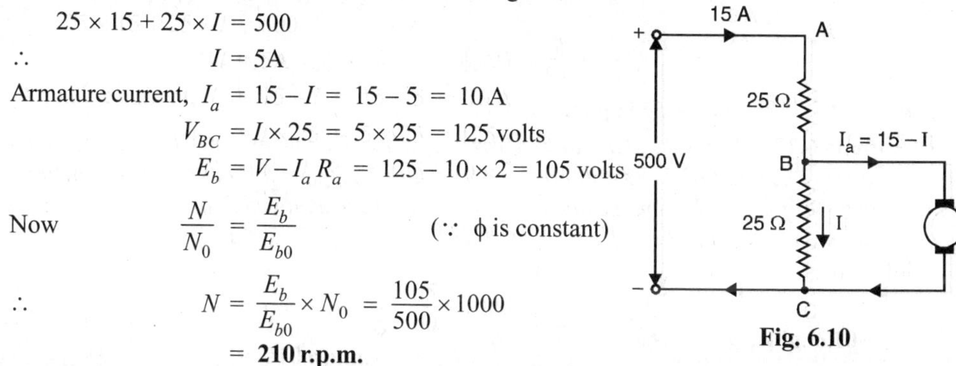

Fig. 6.10

Example 6.25. *A 250 V shunt motor with a constant main field drives a load, the torque of which varies as the cube of the speed. When running at 500 r.p.m., it takes 40 A. Find the speed at which it will run if a 25 Ω resistor is connected in series with the armature. Neglect motor losses.*

Solution. Since motor losses are neglected, the armature resistance is negligible.

When the motor takes 40 A at 500 r.p.m.,

$E_{b1} = 250$ V (armature resistance negligible) ; $N_1 = 500$ r.p.m.

$T_a \propto \phi I_a$ or $T_a \propto I_a$ (∵ ϕ is constant)

Also $T_a \propto N^3$ (given) ∴ $I_a \propto N^3$

$$\therefore \qquad \frac{I_{a2}}{I_{a1}} = \frac{N_2^3}{N_1^3}$$

or
$$I_{a2} = \frac{N_2^3}{(500)^3} \times 40 = \frac{40 N_2^3}{(500)^3}$$

Now $\dfrac{E_{b2}}{E_{b1}} = \dfrac{N_2}{N_1}$ (∵ ϕ is constant)

or
$$E_{b2} = \frac{N_2}{N_1} \times E_{b1} = \frac{N_2}{500} \times 250 = \frac{N_2}{2}$$

Also
$$E_{b2} = V - 25\, I_{a2} = 250 - 25\, I_{a2}$$

or $\dfrac{N_2}{2} = 250 - 25 \times \dfrac{40\, N_2^3}{(500)^3}$ (Putting value of I_{a2})

or
$$2000\, N_2^3 + (500)^3 N_2 - (500)^4 = 0$$

On solving, $N_2 =$ **250 r.p.m.**

Example 6.26. *A shunt motor fed from a 400 V direct current supply takes an armature current of 100 A when running at 800 r.p.m. If the total torque developed remains unchanged, find the speed at which the motor will run if the flux is increased to 120% of its original value and a resistance of 0·8 Ω is connected in series with the armature. The armature resistance is 0·2 Ω.*

Solution.
$$\frac{N_2}{N_1} = \frac{E_{b2}}{E_{b1}} \times \frac{\phi_1}{\phi_2}$$

Since torque remains constant, we have,

$$\phi_1 I_{a1} = \phi_2 I_{a2} \text{ or } I_{a2} = \left(\frac{\phi_1}{\phi_2}\right) I_{a1} = \frac{1}{1\cdot2} \times 100 = \frac{250}{3}\text{ A} \qquad (\because \phi_2 = 1\cdot2\,\phi_1)$$

$$E_{b1} = V - I_{a1} R_a = 400 - 100 \times 0\cdot2 = 380 \text{ volts}$$

$$E_{b2} = V - I_{a2}(R_a + 0\cdot8) = 400 - \frac{250}{3}(0\cdot2 + 0\cdot8) = \frac{950}{3} \text{ volts}$$

$$\therefore \quad \frac{N_2}{800} = \frac{950/3}{380} \times \frac{1}{1\cdot2} \qquad (\because \phi_2 = 1\cdot2\,\phi_1 \text{ or } \phi_1/\phi_2 = 1/1\cdot2)$$

or $N_2 = \textbf{555 r.p.m.}$

Example 6.27. *The full-load armature current of a 440 V shunt motor is 120 A. The armature resistance is 0·2 Ω and the speed 800 r.p.m. What will be the speed if the torque on the motor is reduced to 60% of its full-load value and a resistor of 1·5 Ω is included in the armature circuit, the field strength remaining the same ?*

Solution.

$$\frac{N_2}{N_1} = \frac{E_{b2}}{E_{b1}} \times \frac{\phi_1}{\phi_2} \text{ or } \frac{N_2}{N_1} = \frac{E_{b2}}{E_{b1}} \qquad (\because \phi_1 = \phi_2)$$

Now

$$\frac{T_2}{T_1} = \frac{\phi_2 I_{a2}}{\phi_1 I_{a1}} = \frac{I_{a2}}{I_{a1}} \qquad (\because \phi_1 = \phi_2)$$

$$\therefore \quad I_{a2} = \left(\frac{T_2}{T_1}\right) \times I_{a1} = 0\cdot6 \times 120 = 72 \text{ A} \qquad (\because T_2 = 0\cdot6\,T_1)$$

$$E_{b1} = V - I_{a1} R_a = 440 - 120 \times 0\cdot2 = 416 \text{ volts}$$

$$E_{b2} = V - I_{a2}(R_a + 1\cdot5) = 440 - 72(0\cdot2 + 1\cdot5) = 317\cdot6 \text{ volts}$$

$$\therefore \quad \frac{N_2}{800} = \frac{317\cdot6}{416} \text{ or } N_2 = \textbf{610 r.p.m.}$$

Example 6.28. *A 220 V shunt motor is taking a current of 50 A. Resistance of shunt field is 44 Ω and of the armature is 0·02 Ω. There is a resistance of 0·6 Ω in series with the armature and the speed is 800 r.p.m. What changes should be made in the armature circuit to raise the speed to 850 r.p.m., the torque remaining the same ? Also find :*

(*i*) *At what speed will the motor run if the load torque is reduced so that the armature current is 10 A ?*

(*ii*) *Suppose now that a diverter of 5 Ω resistance is placed across the armature and the series resistance adjusted so that the speed is again 800 r.p.m. with an armature current of 45 A. Find the value of this series resistance. Find also the speed when the armature current falls to 10 A.*

Solution. The motor circuit is shown in Fig. 6.11. To raise the speed, armature circuit resistance should be reduced.

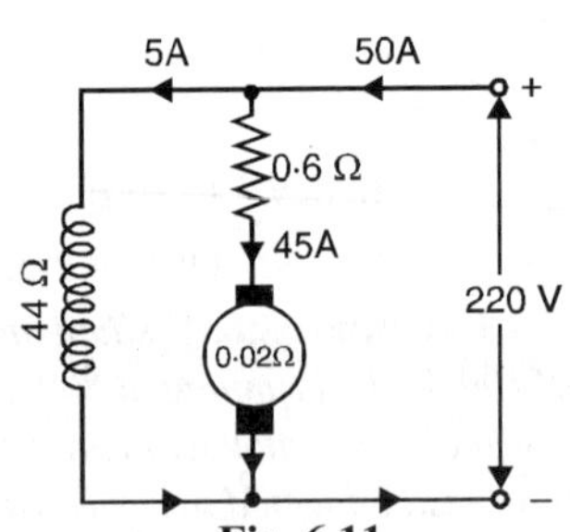

Fig. 6.11

Now

$$\frac{N_2}{N_1} = \frac{E_{b2}}{E_{b1}} \times \frac{\phi_1}{\phi_2}$$

Since flux remains constant,

$$\frac{N_2}{N_1} = \frac{E_{b2}}{E_{b1}}$$

$$I_{sh} = 220/44 = 5\text{A} \;;\; I_{a1} = 50 - 5 = 45 \text{ A}$$

Since torque remains the same and flux is constant in the two cases, the armature current remains the same *i.e.*, $I_{a2} = 45$ A.

$$E_{b1} = V - I_{a1}(R_a + 0\cdot6) = 220 - 45(0\cdot02 + 0\cdot6) = 192\cdot1 \text{ volts}$$

$$\therefore \quad \frac{850}{800} = \frac{E_{b2}}{192 \cdot 1} \qquad \therefore \quad E_{b2} = 204 \text{ volts}$$

Let R_t be the new resistance of the armature circuit to raise the speed from 800 r.p.m. to 850 r.p.m.

$$\therefore \quad E_{b2} = V - I_{a2} R_t$$

$$\text{or} \quad 204 = 220 - 45 \times R_t \qquad \therefore \quad R_t = 0 \cdot 356 \ \Omega$$

$\therefore$ Resistance of the armature circuit must be reduced by

$$= 0 \cdot 62 - 0 \cdot 356 = \mathbf{0 \cdot 264 \ \Omega}$$

(*i*) $I_{a1} = 45$ A ; $E_{b1} = 192 \cdot 1$ volts ; $N_1 = 800$ r.p.m.

If $I_{a3} = 10$ A ; $E_{b3} = V - I_{a3} (0 \cdot 02 + 0 \cdot 6) = 220 - 10 (0 \cdot 62) = 213 \cdot 8$ volts

$$\therefore \quad \frac{N_3}{N_1} = \frac{E_{b3}}{E_{b1}} \text{ or } \frac{N_3}{800} = \frac{213 \cdot 8}{192 \cdot 1} \qquad \therefore \ N_3 = \mathbf{890 \text{ r.p.m.}}$$

(*ii*) **Diverter in parallel with armature.** When a diverter of 5 Ω resistance is connected across the armature, the circuit connections become as shown in Fig. 6.12.

Let R = new series resistance

I = current through R

Since the speed is to be 800 r.p.m., the back e.m.f. should be $E_{b4} = E_{b1} = 192 \cdot 1$ volts.

$$\therefore \quad E_{b4} = V - IR - 45 \times R_a$$

$$\text{or} \quad 192 \cdot 1 = 220 - IR - 45 \times 0 \cdot 02$$

$$\therefore \quad IR = 27 \text{ volts}$$

Current through diverter,

$$I_d = \frac{V - IR}{5} = \frac{220 - 27}{5} = 38 \cdot 6 \text{ A}$$

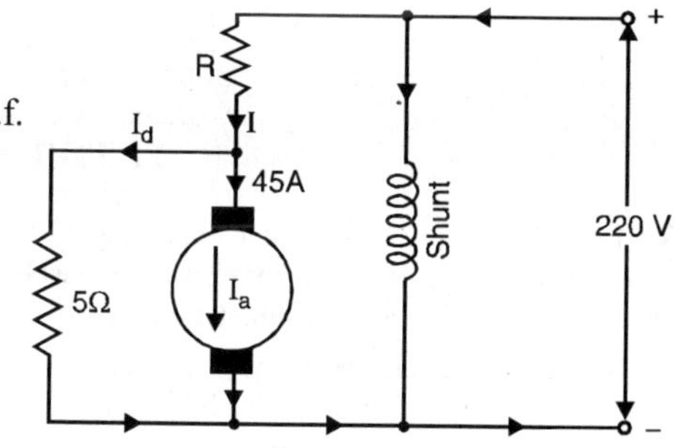

Fig. 6.12

$$\therefore \quad \text{Current through } R, I = I_a + I_d = 45 + 38 \cdot 6 = 83 \cdot 6 \text{ A}$$

$$\therefore \quad \text{Value of } R = \frac{IR}{I} = \frac{27}{83 \cdot 6} = \mathbf{0 \cdot 323 \ \Omega}$$

$$\text{When } I_a = 10 \text{ A} \ ; \ I_d = I - 10$$

$$\text{Voltage across diverter} = 5 (I - 10) = 220 - 0 \cdot 323 I$$

$$\text{or} \quad 5 I - 50 = 220 - 0 \cdot 323 I$$

$$\therefore \quad I = 50 \cdot 7 \text{ A}$$

$$E_{b5} = V - IR - 10 \times R_a$$

$$= 220 - 50 \cdot 7 \times 0 \cdot 323 - 10 \times 0 \cdot 02 = 203 \cdot 4 \text{ volts}$$

$$\therefore \quad \frac{N_5}{N_1} = \frac{E_{b5}}{E_{b1}} \text{ or } \frac{N_5}{800} = \frac{203 \cdot 4}{192 \cdot 1} \qquad \therefore \ N_5 = \mathbf{847 \text{ r.p.m.}}$$

Note the role of the diverter. The speed without diverter is 890 r.p.m. and with diverter, it is 847 r.p.m.

Example 6.29. *A 100 kW, belt-driven shunt generator running at 300 r.p.m. on 220 V bus-bars continues to run as a motor when the belt breaks, taking 10 kW. What will be its speed ? Armature resistance = 0·025 Ω, shunt field resistance = 60 Ω. Contact drop under each brush = 1 V. Ignore armature reaction.*

Solution.

As generator

$$\text{Load current, } I_L = \frac{100 \times 10^3}{220} = 455 \text{ A}$$

$$\text{Shunt field current, } I_{sh} = \frac{220}{60} = 3{\cdot}7\text{ A}$$

$$\text{Armature current, } I_a = I_L + I_{sh} = 455 + 3{\cdot}7 = 458{\cdot}7\text{ A}$$

$$\text{Generated e.m.f., } E_g = V + I_a R_a + \text{Brush drop}$$

$$= 220 + 458{\cdot}7 \times 0{\cdot}025 + 2 \times 1 = 233{\cdot}5 \text{ volts}$$

As Motor

$$\text{Line current, } I_L = \frac{10 \times 10^3}{220} = 45{\cdot}5\text{ A}$$

$$\text{Armature current, } I_a = I_L - I_{sh} = 45{\cdot}5 - 3{\cdot}7 = 41{\cdot}8\text{ A}$$

$$\text{Back e.m.f., } E_b = V - I_a R_a - \text{Brush drop}$$

$$= 220 - 41{\cdot}8 \times 0{\cdot}025 - 2 \times 1 = 217 \text{ volts}$$

Now $$\frac{N_m}{N_g} = \frac{E_b}{E_g}$$ ....flux is constant

or $$\frac{N_m}{300} = \frac{217}{233{\cdot}5} \qquad \therefore \; N_m = \mathbf{279\ r.p.m.}$$

TUTORIAL PROBLEMS

1. A 500 V d.c. shunt motor running at 720 r.p.m. takes an armature current of 50 A. Effective armature resistance is 0·4 Ω. What resistance must be placed in series with the armature to reduce the speed to 600 r.p.m., the torque remaining constant ? **[1·6 Ω]**
2. A 250 V shunt motor with armature resistance of 0·5 Ω runs at 600 r.p.m. on full-load and takes an armature current of 20 A. If a resistance of 0·1 Ω is placed in the armature circuit, find the speed at (*i*) full-load torque (*ii*) half full-load torque. **[(*i*) 550 r.p.m. (*ii*) 587·5 r.p.m.]**
3. A 200 V d.c. shunt motor having an armature circuit resistance of 0·2 Ω and a field-circuit resistance of 100 Ω draws a line current of 50 A at full-load at a speed of 1500 r.p.m. Calculate its speed at half full-load. **[1539 r.p.m.]**
4. A 240 V shunt motor has an armature current of 15 A when running at 800 r.p.m. against full-load torque. The armature resistance is 0·6 Ω. What resistance must be inserted in series with the armature to reduce the speed to 400 r.p.m. at the same torque ? What will be the speed if the load torque is halved with this resistance in the circuit ? Assume the flux to remain constant throughout. **[7·7 Ω ; 614·7 r.p.m.]**
5. A 230 V shunt motor having an armature resistance of 0·05 Ω and a field resistance of 75 Ω draws a line current of 7A while running light at 1120 r.p.m. For a load at which the line current is 46 A, determine the speed of the motor. **[1110·5 r.p.m.]**
6. A d.c. shunt motor drives a centifugal pump whose torque varies as the square of the speed. The motor is fed from 200 V supply and takes 50 A when running at 1000 r.p.m. What resistance must be inserted in the armature circuit in order to reduce the speed to 800 r.p.m. ? The armature and field resistances of the motor are 0·1 Ω and 100 Ω respectively. **[1·32 Ω]**

6.3 SPEED CONTROL OF D.C. SERIES MOTORS

The speed control of d.c. series motors can be obtained by (*i*) flux control method (*ii*) armature-resistance control method. The latter method is mostly used.

1. Flux control method. In this method, the flux produced by the series motor is varied and hence the speed. The variation of flux can be achieved in the following ways :

(*i*) **Field diverters.** In this method, a variable resistance (called *field diverter*) is connected in parallel with series field winding as shown in Fig. 6.13. Its effect is to shunt some portion of the line current from the series field winding, thus weakening the field and increasing the speed ($\because N \propto 1/\phi$). The lowest speed obtainable is that corresponding to zero current in the diverter (*i.e.*, diverter is open). Obviously, the lowest speed obtainable is the normal speed of the motor. Consequently, this method can only provide speeds above the normal speed. The series field diverter method is often employed in traction work.

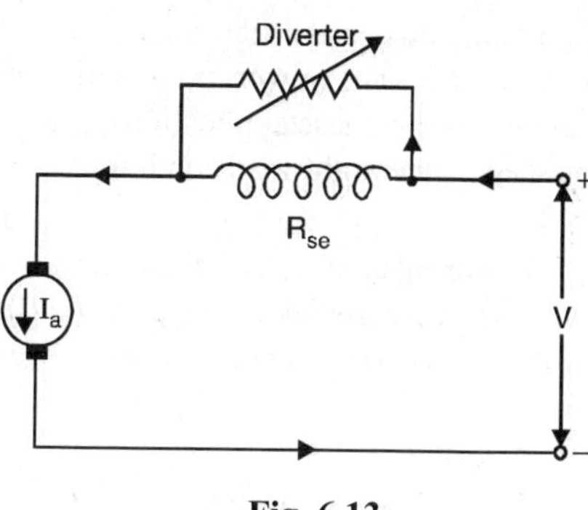

Fig. 6.13

(*ii*) **Armature diverter.** In order to obtain speeds below the normal speed, a variable resistance (called *armature diverter*) is connected in parallel with the armature as shown in Fig. 6.14. The diverter shunts some of the line current, thus reducing the armature current. Now for a given load, if I_a is decreased, the flux ϕ must increase ($T \propto \phi I_a$). Since $N \propto 1/\phi$, the motor speed is decreased. By adjusting the armature diverter, any speed lower than the normal speed can be obtained.

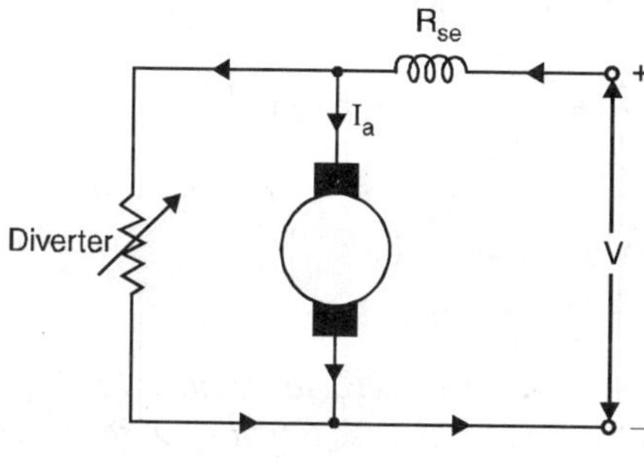
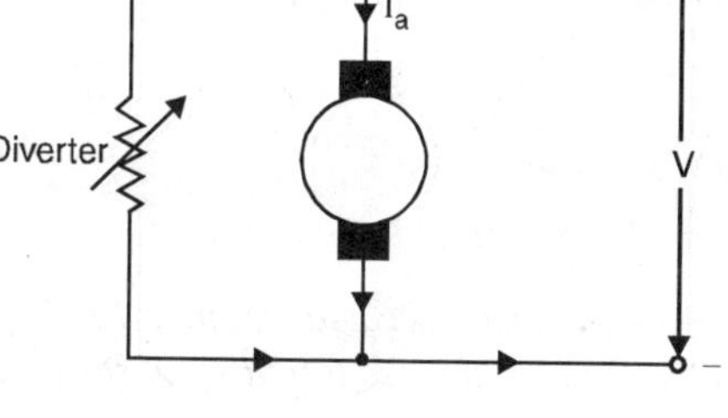

Fig. 6.14

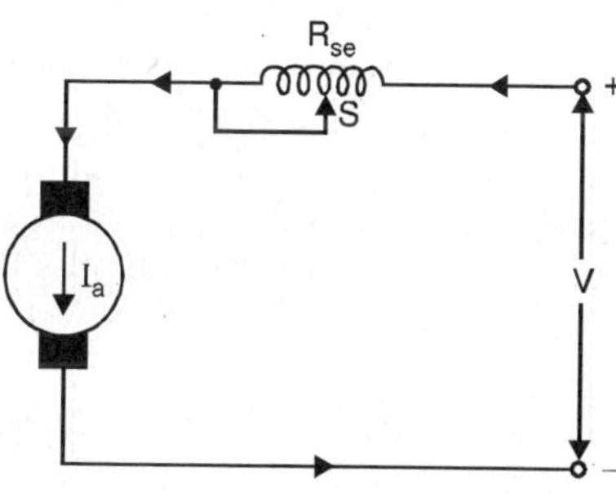

Fig. 6.15

(*iii*) **Tapped field control.** In this method, the flux is reduced (and hence speed is increased) by decreasing the number of turns of the series field winding as shown in Fig. 6.15. The switch *S* can short circuit any part of the field winding, thus decreasing the flux and raising the speed. With full turns of the field winding, the motor runs at normal speed and as the field turns are cut out, speeds higher than normal speed are achieved.

(*iv*) **Paralleling field coils.** This method is usually employed in the case of fan motors. By regrouping the field coils as shown in Fig. 6.16, several fixed speeds can be obtained.

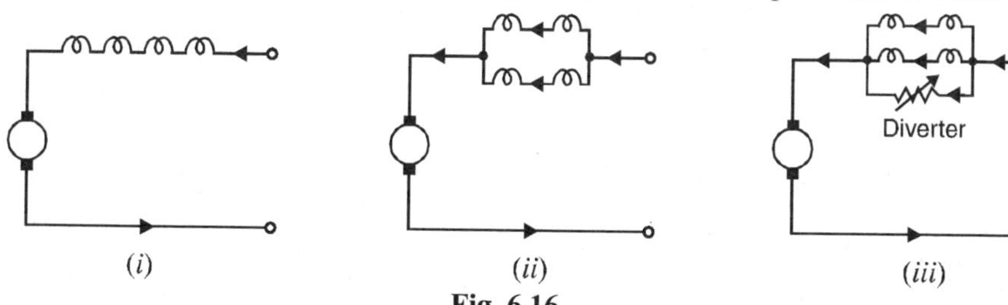

Fig. 6.16

2. Armature-resistance control. In this method, a variable resistance is directly connected in series with the supply to the complete motor as shown in Fig. 6.17. This reduces the voltage available across the armature and hence the speed falls. By changing the value of variable resistance, any speed below the normal speed can be obtained. This is the most common method employed to control the speed of d.c. series motors. Although this method has poor speed

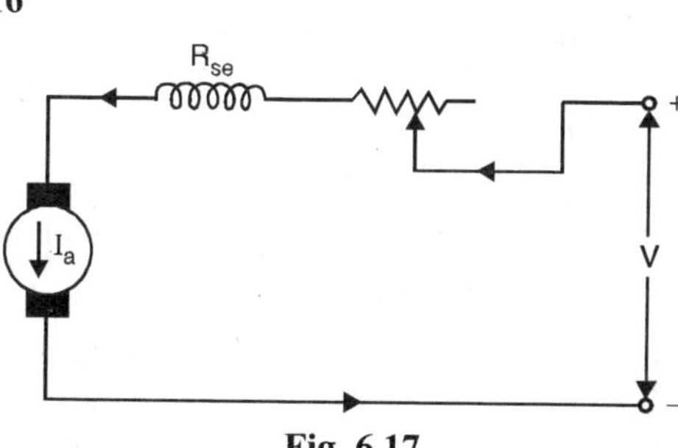

Fig. 6.17

regulation, this has no significance for series motors because they are used in varying speed applications. The loss of power in the series resistance for many applications of series motors is not too serious since in these applications, the control is utilised for a large portion of the time for reducing the speed under light-load conditions and is only used intermittently when the motor is carrying full-load.

Example 6.30. *A 200 V d.c. series motor runs at 800 r.p.m. when taking a line current of 15 A. The armature resistance and series field resistance are 0·6 Ω and 0·4 Ω respectively. Find the speed at which it will run when connected in series with a 5 Ω resistance and taking the same current at the same voltage.*

Solution.

$$R_a + R_{se} = 0{\cdot}6 + 0{\cdot}4 = 1\,\Omega \;;\; N_1 = 800 \text{ r.p.m.} \;;\; I_{a1} = I_{a2} = 15\text{ A} \;;\; N_2 = ?$$

Without 5 Ω resistance in series, we have,

$$E_{b1} = V - I_{a1}(R_a + R_{se}) = 200 - 15\,(0{\cdot}6 + 0{\cdot}4) = 185 \text{ volts}$$

With 5 Ω resistance in series, we have,

$$E_{b2} = V - I_{a2}(R_a + R_{se} + 5) = 200 - 15\,(0{\cdot}6 + 0{\cdot}4 + 5) = 110 \text{ volts}$$

Now $$\frac{N_2}{N_1} = \frac{E_{b2}}{E_{b1}} \qquad (\because \text{ flux is the same})$$

or $$\frac{N_2}{800} = \frac{110}{185} \qquad \therefore\; N_2 = \mathbf{476 \text{ r.p.m.}}$$

Example 6.31. *A 240 V d.c. series motor takes 40 A when giving its rated output at 1500 r.p.m. Its resistance is 0·3 Ω. Find what resistance must be added to obtain rated torque (i) at starting (ii) at 1000 r.p.m.*

Solution. Since torque remains the same ($T \propto I_a^2$), the armature current is same (*i.e.*, 40 A) in the two cases.

(*i*) Let R be the series resistance added at starting to obtain the rated torque. Then,

$$40 = \frac{V}{R + 0{\cdot}3} = \frac{240}{R + 0{\cdot}3}$$

$$\therefore \qquad R = \mathbf{5{\cdot}7\,\Omega}$$

(*ii*)
$$E_{b1} = V - I_{a1} R_a = 240 - 40 \times 0{\cdot}3 = 228 \text{ volts} \;;\; N_1 = 1500 \text{ r.p.m.}$$
$$E_{b2} = V - I_{a1}(R_a + R) = 240 - 40\,(0{\cdot}3 + R) \;;\; N_2 = 1000 \text{ r.p.m.}$$

Now $$\frac{N_2}{N_1} = \frac{E_{b2}}{E_{b1}} \qquad (\because \text{ flux is the same})$$

or $$\frac{1000}{1500} = \frac{240 - 40\,(0{\cdot}3 + R)}{228} \qquad \therefore\; R = \mathbf{1{\cdot}9\,\Omega}$$

Example 6.32. *A d.c. series motor connected to 440 V supply runs at 600 r.p.m. when taking a current of 50 A. Calculate the value of resistance which when inserted in series with the motor will reduce the speed to 400 r.p.m., the gross torque then being half the previous value. The resistance of the motor is 0·2 Ω. Assume flux to be proportional to field current.*

Solution.

Without the insertion of extra resistance, we have,

$$E_{b1} = V - I_{a1} R_m = 440 - 50 \times 0{\cdot}2 = 430 \text{ volts} \;;\; N_1 = 600 \text{ r.p.m.}$$

Let R be the resistance added to reduce the speed to 400 r.p.m. at half the load torque.

$$\frac{T_2}{T_1} = \frac{I_{a2}^2}{I_{a1}^2}$$

$$\therefore \quad I_{a2} = \sqrt{\frac{T_2}{T_1}} \times I_{a1} = \frac{1}{\sqrt{2}} \times 50 = 35{\cdot}35 \text{ A} \qquad \left(\because T_2 = \frac{T_1}{2}\right)$$

$$\therefore \quad E_{b2} = V - I_{a2}(R_m + R) = 440 - 35{\cdot}35\,(0{\cdot}2 + R)\ ;\ N_2 = 400 \text{ r.p.m.}$$

Now
$$\frac{N_2}{N_1} = \frac{E_{b2}}{E_{b1}} \times \frac{\phi_1}{\phi_2}$$

or
$$\frac{400}{600} = \frac{440 - 35{\cdot}35\,(0{\cdot}2 + R)}{430} \times \frac{50}{35{\cdot}35} \qquad (\because \phi \propto I_a)$$

$$\therefore \quad R = \mathbf{6{\cdot}52\ \Omega}$$

Example 6.33. *A 440 V d.c. series motor of negligible resistance and with unsaturated magnetic circuit takes 50 A when running at a certain speed on a given load. If the load torque varies as the cube of the speed, find the resistance necessary to reduce the speed by 50%. What is then the current ?*

Solution.

$$T_a \propto \phi I_a \quad \text{or} \quad T_a \propto I_a^2 \qquad (\because \phi \propto I_a)$$

Also $T_a \propto N^3$ (given) $\quad \therefore \quad I_a^2 \propto N^3$

For the first case, $I_{a1}^2 \propto N_1^3$; For the second case, $I_{a2}^2 \propto N_2^3$

$$\therefore \quad \left(\frac{I_{a2}}{I_{a1}}\right)^2 = \left(\frac{N_2}{N_1}\right)^3 = \left(\frac{1}{2}\right)^3 = \frac{1}{8} \qquad \left(\because \frac{N_2}{N_1} = \frac{1}{2}\right)$$

or
$$I_{a2} = I_{a1}\sqrt{\frac{1}{8}} = 50 \times \sqrt{\frac{1}{8}} = \mathbf{17{\cdot}68\ A}$$

Now
$$E_{b1} = V - I_{a1} \times 0 = 440 \text{ volts}$$

Let R be the series resistance inserted to reduce the speed by 50%.

$$\therefore \quad E_{b2} = V - I_{a2}R = (440 - 17{\cdot}68 \times R) \text{ volts}$$

Now
$$\frac{N_2}{N_1} = \frac{E_{b2}}{E_{b1}} \times \frac{I_{a1}}{I_{a2}} \qquad (\because \phi \propto I_a)$$

or
$$\frac{1}{2} = \frac{440 - 17{\cdot}68\,R}{440} \times \frac{50}{17{\cdot}68}$$

$$\therefore \quad R = \mathbf{20{\cdot}5\ \Omega}$$

Example 6.34. *A 4-pole, series-wound fan motor runs normally at 600 r.p.m. on a 250 V d.c. supply taking 20 A. The field coils are connected all in series. Estimate the speed and current taken by the motor if the coils are reconnected in two parallel groups of two in series. The load torque increases as the square of the speed. Assume that flux is directly proportional to the current and ignore losses.*

Solution. There are four field coils and the two types of field connections are shown in Fig. 6.18.

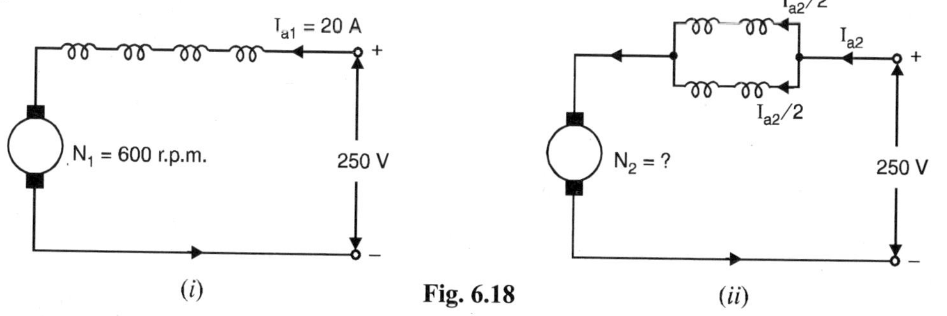

Fig. 6.18

$$T_a \propto \phi I_a \quad \text{and} \quad T_a \propto N^2 \text{ (given)} \qquad \therefore \; \phi I_a \propto N^2$$

For the first case, $\phi_1 I_{a1} \propto N_1^2$; For the second case, $\phi_2 I_{a2} \propto N_2^2$

$$\therefore \qquad \left(\frac{N_2}{N_1}\right)^2 = \frac{\phi_2 I_{a2}}{\phi_1 I_{a1}}$$

Since losses are negligible, resistance of all windings (field as well as armature) are negligible. This means that back e.m.f. in each case equals supply voltage *i.e.* $E_{b1} = E_{b2} = 250$ volts.

$$\text{Now} \qquad \frac{N_2}{N_1} = \frac{E_{b2}}{E_{b1}} \times \frac{\phi_1}{\phi_2}$$

$$\text{or} \qquad \frac{N_2}{N_1} = \frac{\phi_1}{\phi_2} \qquad (\because E_{b1} = E_{b2} = 250 \text{ volts})$$

$$\text{or} \qquad \left(\frac{N_2}{N_1}\right)^2 = \left(\frac{\phi_1}{\phi_2}\right)^2$$

$$\text{or} \qquad \frac{\phi_2 I_{a2}}{\phi_1 I_{a1}} = \left(\frac{\phi_1}{\phi_2}\right)^2 \qquad \left[\because \left(\frac{N_2}{N_1}\right)^2 = \frac{\phi_2 I_{a2}}{\phi_1 I_{a1}}\right]$$

$$\text{Now} \qquad \phi_1 \propto 20 \quad \text{and} \quad \phi_2 \propto I_{a2}/2$$

$$\therefore \qquad \frac{I_{a2}}{20} = \left(\frac{20}{I_{a2}/2}\right)^3$$

$$\therefore \qquad I_{a2} = 20 \times 2^{3/4} = \mathbf{33 \cdot 64 \ A}$$

$$\text{Also} \qquad \frac{N_2}{N_1} = \frac{\phi_1}{\phi_2} = \frac{I_{a1}}{I_{a2}/2} = \frac{2I_{a1}}{I_{a2}} = \frac{2 \times 20}{33 \cdot 64}$$

$$\therefore \qquad N_2 = N_1 \times \frac{2 \times 20}{33 \cdot 64} = 600 \times \frac{2 \times 20}{33 \cdot 64} = \mathbf{714 \ r.p.m.}$$

Example 6.35. *A 100 V series motor when taking 40 A runs at 200 r.p.m. The armature resistance is 0·1 Ω and there are four field coils, each of 0·01 Ω in series with it. Find the speed if the load torque be doubled and the coils arranged in series-parallel (two in series, two sets in parallel).*

Solution.

Coils in series [See Fig. 6.19 (*i*)].

$$\text{Resistance of motor} = 0 \cdot 1 + 4 \times 0 \cdot 01 = 0 \cdot 14 \ \Omega$$

$$\therefore \qquad E_{b1} = 100 - 40 \times 0 \cdot 14 = 94 \cdot 4 \text{ volts}$$

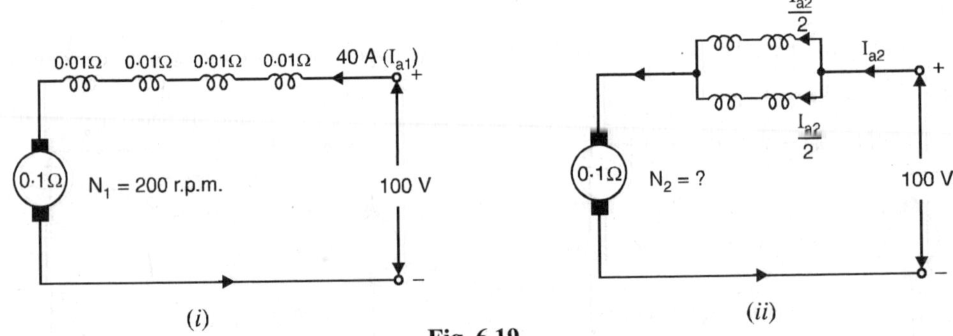

Fig. 6.19

Coils in series-parallel [See Fig. 6.19 (*ii*)]. In this case, the field current is half the new armature current (*i.e.* $I_{a2}/2$).

$$\text{Now} \quad \frac{T_1}{T_2} = \frac{\phi_1 I_{a1}}{\phi_2 I_{a2}}$$

$$\text{or} \quad \frac{1}{2} = \frac{40 \times 40}{I_{a2}/2 \times I_{a2}} \qquad (\because \phi \propto I_a)$$

$$\text{or} \quad I_{a2}^2 = 6400 \quad \therefore \; I_{a2} = 80 \text{ A}$$

$$\therefore \quad \text{Field current} = I_{a2}/2 = 80/2 = 40 \text{ A}$$

$$\text{Resistance of motor} = 0{\cdot}1 + \frac{0{\cdot}02}{2} = 0{\cdot}11\ \Omega$$

$$\therefore \quad E_{b2} = 100 - 80 \times 0{\cdot}11 = 91{\cdot}2 \text{ volts}$$

$$\text{Now} \quad \frac{N_2}{N_1} = \frac{E_{b2}}{E_{b1}} \times \frac{\phi_1}{\phi_2}$$

$$\text{or} \quad \frac{N_2}{200} = \frac{91{\cdot}2}{94{\cdot}4} \times \frac{40}{40} \quad \therefore \; N_2 = \mathbf{193 \text{ r.p.m.}}$$

Example 6.36. *A d.c. series motor having a resistance of 1 Ω drives a fan for which the torque varies as the square of the speed. At 220 V, the set runs at 350 r.p.m. and takes 25 A. The speed is to be raised to 500 r.p.m. by increasing the voltage. Determine the necessary voltage and the corresponding current assuming the field to be unsaturated.*

Solution.

$$T_a \propto \phi I_a \quad \text{or} \quad T_a \propto I_a^2 \qquad (\because \phi \propto I_a)$$

$$\text{Also} \quad T_a \propto N^2 \text{ (given)} \quad \therefore \; I_a \propto N$$

$$\therefore \quad \frac{I_{a2}}{I_{a1}} = \frac{N_2}{N_1} = \frac{500}{350}$$

$$\text{or} \quad I_{a2} = I_{a1} \times \frac{500}{350} = 25 \times \frac{500}{350} = \mathbf{\frac{250}{7} \text{ A}}$$

$$E_{b1} = 220 - I_{a1} R_m = 220 - 25 \times 1 = 195 \text{ volts}$$

Let V volts be the applied voltage to increase the speed of the motor from 350 r.p.m. to 500 r.p.m.

$$E_{b2} = V - I_{a2} R_m = V - \frac{250}{7} \times 1 = \frac{7V - 250}{7} \text{ volts}$$

$$\text{Now} \quad \frac{N_2}{N_1} = \frac{E_{b2}}{E_{b1}} \times \frac{\phi_1}{\phi_2}$$

$$\text{or} \quad \frac{500}{350} = \frac{\frac{7V - 250}{7}}{195} \times \frac{25}{250/7} \qquad (\because \phi \propto I_a)$$

$$\therefore \quad V = \mathbf{433{\cdot}7 \text{ volts}}$$

Example 6.37. *A d.c. series motor takes 20 A at 100 V and runs at 950 r.p.m. Armature resistance is 0.05 Ω and series field resistance is 0.2 Ω. At what speed will it run developing same torque but with a 10 Ω resistance placed in parallel with the armature ? Assume that the flux is proportional to the field current.*

Solution.

Prior to resistance connection. Figure 6.20 (*i*) shows the motor circuit *before* 10 Ω resistance is placed in parallel with the armature.

$$\text{Resistance of motor,} \quad R_m = 0{\cdot}2 + 0{\cdot}05 = 0{\cdot}25\ \Omega$$

$$\therefore \quad E_{b1} = V - I_{a1} R_m = 100 - 20 \times 0{\cdot}25 = 95 \text{ volts}$$

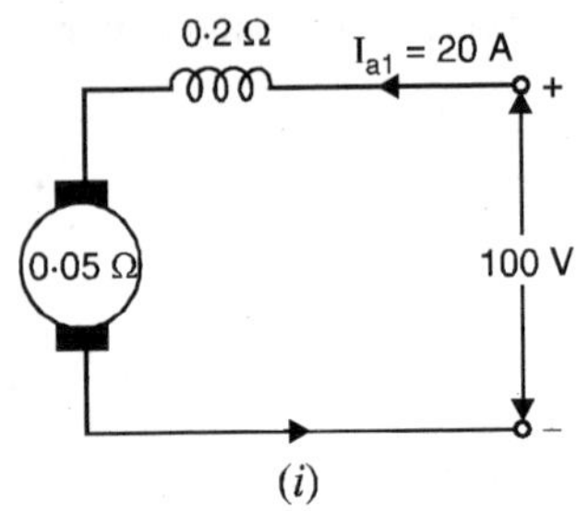

(i)

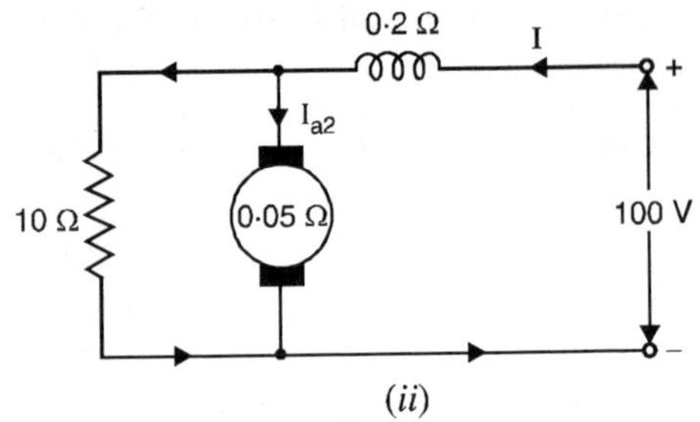

(ii)

Fig. 6.20

After resistance connection. Figure 6.20 (*ii*) shows the motor circuit *after* placing 10 Ω resistance in parallel with armature. Let I be the line current of the motor.

$$\text{Voltage drop in series field} = 0{\cdot}2\,I$$

$$\text{P.D. at the brushes} = 100 - 0{\cdot}2\,I$$

$$\text{Current in } 10\,\Omega \text{ resistor} = \frac{100 - 0{\cdot}2I}{10}$$

$$\therefore \quad \text{Armature current, } I_{a2} = I - \frac{100 - 0{\cdot}2I}{10} = \frac{10{\cdot}2I - 100}{10}$$

Since torque in the two cases is the same,

$$\therefore \quad \phi_1 I_{a1} = \phi_2 I_{a2}$$

$$\text{or} \quad 20 \times 20 = I \times \frac{10{\cdot}2I - 100}{10}$$

$$\text{or} \quad 10{\cdot}2\,I^2 - 100\,I - 4000 = 0$$

$$\text{On solving, we have, } I = 25{\cdot}3 \text{ A}$$

$$\therefore \quad I_{a2} = \frac{10{\cdot}2 \times 25{\cdot}3 - 100}{10} = 15{\cdot}8 \text{ A}$$

$$\text{P.D. at the brushes} = 100 - 0{\cdot}2 \times 25{\cdot}3 = 94{\cdot}94 \text{ volts}$$

$$\therefore \quad E_{b2} = \text{P.D. at brushes} - I_{a2} R_a = 94{\cdot}94 - 15{\cdot}8 \times 0{\cdot}05 = 94{\cdot}15 \text{ volts}$$

$$\text{Now} \quad \frac{N_2}{N_1} = \frac{E_{b2}}{E_{b1}} \times \frac{\phi_1}{\phi_2}$$

$$\text{or} \quad \frac{N_2}{950} = \frac{94{\cdot}15}{95} \times \frac{20}{25{\cdot}3} \qquad \therefore \quad N_2 = \mathbf{744 \text{ r.p.m.}}$$

Example 6.38. *A d.c. series motor runs at 1000 r.p.m. when taking 20 A at 200 V. Armature resistance is 0.5 Ω and series field resistance is 0.2 Ω. Find the speed for a total current of 20 A when a diverter of 0·2 Ω resistance is used across the series field. Flux for a field current of 10 A is 70 % of that for 20 A.*

Solution.

Prior to diverter connections. Figure 6.21 (*i*) shows the motor connections before diverter is connected.

$$\text{Total motor resistance} = 0{\cdot}5 + 0{\cdot}2 = 0{\cdot}7\,\Omega$$

$$E_{b1} = V - I_{a1} \times 0{\cdot}7 = 200 - 20 \times 0{\cdot}7 = 186 \text{ volts}$$

After diverter connections. Figure 6.21 (*ii*) shows a diverter of 0·2 Ω connected across the series field winding. Since the diverter and series field have the same resistance, the input current of 20 A will divide equally between them. It is given that flux for a field current of 10 A is 70% of that for 20 A.

Therefore, $\phi_2 = 0{\cdot}7\,\phi_1$ or $\phi_1/\phi_2 = 1/0{\cdot}7$.

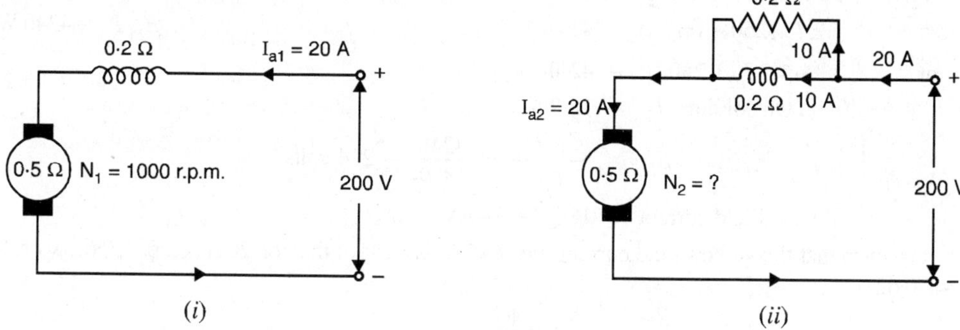

Fig. 6.21

$$\text{Total motor resistance} = 0{\cdot}5 + \frac{0{\cdot}2}{2} = 0{\cdot}6\,\Omega$$

$$E_{b2} = V - I_{a2} \times 0{\cdot}6 = 200 - 20 \times 0{\cdot}6 = 188 \text{ volts}$$

Now
$$\frac{N_2}{N_1} = \frac{E_{b2}}{E_{b1}} \times \frac{\phi_1}{\phi_2}$$

or
$$\frac{N_2}{1000} = \frac{188}{186} \times \frac{1}{0{\cdot}7} \qquad \therefore \; N_2 = \mathbf{1444\ r.p.m.}$$

Example 6.39. *A 250 V series motor runs at 1000 r.p.m., gives 5 h.p. and takes a current of 20 A when the two sections of the field winding are connected in series. What current the motor will take for the same output when the two sections are connected in parallel and what are the corresponding torque and speed ? The resistance of the armature is 1.1Ω and that of each section of the field winding is 0.4 Ω. Given that flux due to new field current is 62% of that for 20 A.*

Solution.

Field windings in series. Figure 6.22 (*i*) shows the two sections of the field winding is series.

$$\text{Total motor resistance} = 1{\cdot}1 + 0{\cdot}4 + 0{\cdot}4 = 1{\cdot}9\,\Omega\,; \; N_1 = 1000 \text{ r.p.m.}$$

$$E_{b1} = V - I_{a1} \times 1{\cdot}9 = 250 - 20 \times 1{\cdot}9 = 212 \text{ volts}$$

$$\text{Motor output} = E_{b1} \times I_{a1} = 212 \times 20 = 4240 \text{ W}$$

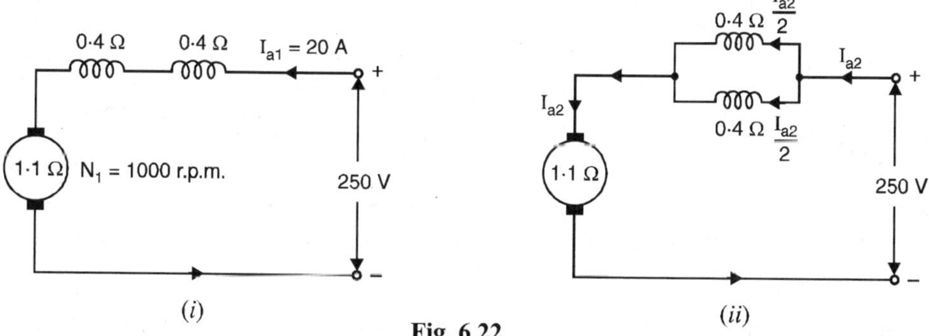

Fig. 6.22

Field windings in parallel. Figure 6.22 (*ii*) shows the motor connections when the two sections of field winding are in parallel. For the same output, let I_{a2} be the armature current.

$$\text{Total motor resistance} = 1{\cdot}1 + \frac{0{\cdot}4}{2} = 1{\cdot}3\,\Omega$$

$$E_{b2} = V - I_{a2} \times 1{\cdot}3 = 250 - 1{\cdot}3\, I_{a2}$$

$$\text{Motor output} = E_{b2} I_{a2} = (250 - 1\cdot3\, I_{a2})\, I_{a2}$$

Since motor output is the same in the two cases,

$$\therefore \quad I_{a2}(250 - 1\cdot3\, I_{a2}) = 4240 \qquad (\because E_{b1} I_{a1} = 4240\text{ W})$$

or
$$-1\cdot3\, I_{a2}^2 + 250\, I_{a2} = 4240$$

$$\text{On solving, } I_{a2} = \mathbf{18\cdot9\ A}$$

$$E_{b2} = \frac{E_{b2}\, I_{a2}}{I_{a2}} = \frac{4240}{18\cdot9} = 224 \text{ volts}$$

$$\text{Field current} = 18\cdot9/2 = 9\cdot45\text{ A}$$

It is given that flux at new field current (*i.e.* 9·45 A) is 62 % of that for 20 A *i.e.*, $\phi_2 = 0\cdot62\, \phi_1$ or $\phi_1/\phi_2 = 1/0\cdot62$.

Now
$$\frac{N_2}{N_1} = \frac{E_{b2}}{E_{b1}} \times \frac{\phi_1}{\phi_2}$$

or
$$\frac{N_2}{1000} = \frac{224}{212} \times \frac{1}{0\cdot62} \qquad \therefore \quad N_2 = \mathbf{1700\ r.p.m.}\ (\textit{nearly})$$

$$\text{Torque} = 9\cdot55 \times \frac{\text{Output power}}{N_2} = 9\cdot55 \times \frac{5 \times 746}{1700} = \mathbf{21\ N\text{-}m}$$

TUTORIAL PROBLEMS

1. A 220 V d.c. series motor runs at 900 r.p.m. when taking a line current of 40 A. The armature resistance and series field resistance are 0·06 Ω and 0·04 Ω respectively. If current taken remains the same, calculate the series resistance required to reduce the speed to 600 r.p.m. **[1·63 Ω]**
2. A d.c. series motor is taking 30 A from a 230 V d.c. supply. The total resistance of the armature and field circuits is 0·6 Ω. Calculate the resistance required in series with the armature to reduce the speed by 40% if the torque is proportional to the square of the speed and flux is proportional to the current. **[7·94 Ω]**
3. A d.c. series motor runs at 500 r.p.m. when taking a current of 60 A at 460 V. The resistance of the armature circuit is 0·2 Ω and that of the field winding is 0·1 Ω. Calculate the speed when a 0·15 Ω diverter is connected in parallel with the field winding. Assume the torque to remain unaltered and the flux to be proportional to the field current. **[643 r.p.m.]**
4. A 220 V d.c. series motor takes 40 A when running at 700 r.p.m. Calculate the speed at which the motor will run and the current taken from the supply if the field is shunted by a resistance equal to the field resistance and the load torque is increased by 50%. Armature resistance = 0·15 Ω and field resistance = 0·1 Ω. Assume that flux is proportional to field current. **[794 r.p.m. ; 69·3A]**
5. A 230 V d.c. series motor runs at 900 r.p.m. when the current is 30 A. The total resistance of the armature and field circuits is 0·8 Ω. Calculate the value of the additional resistance required in series with the machine to reduce the speed to 500 r.p.m. if the gross torque is (*i*) constant (*ii*) proportional to speed (*iii*) proportional to the square of the speed. Assume the magnetic circuit to be saturated.
[(*i*) 3·05 Ω (*ii*) 5·67 Ω (*iii*) 9·2 Ω]
6. A 4-pole, 230 V d.c. series motor runs at 1000 r.p.m. when the load current is 12 A. The series field resistance is 0·8 Ω and the armature resistance is 1 Ω. The series field coils are now regrouped from all in series to two in series with two parallel paths. The line current is now 20 A. If the corresponding weakening of the field is 15%, calculate the speed of the motor. **[1163 r.p.m.]**
7. A 210 V d.c. series motor takes a current of 70 A and runs at 800 r.p.m. when developing full-load torque. The armature and field resistances are 0·11 Ω and 0.09 Ω respectively. Estimate the speed when the motor is developing the same torque but with a 0·15 Ω diverter in parallel with the field winding. Assume that flux per pole is directly proportional to the current in the field winding. **[1010 r.p.m.]**
8. A 220 V d.c. series motor takes 10 A and runs at 600 r.p.m., its total resistance is 0·8 Ω. At what speed will it run when a 5-ohm resistance is connected in series, the motor is taking the same current at the same supply voltage ? **[458 r.p.m.]**

9. A d.c. series motor with unsaturated field and negligible resistance, when running at a certain speed on a given load takes 60 A at 460 V. If the load torque varies as the cube of the speed, calculate the series resistance required to reduce the speed by 25%. **[6 Ω]**

10. A d.c. series motor is connected to a 220 V d.c. supply and runs at 500 r.p.m. taking a current of 50 A. Calculate the value of series resistance which will reduce the speed to 300 r.p.m. ; the load torque then being one-half of its previous value. Resistance of motor is 0·1 Ω. Assume the flux to be proportional to field current. **[3·54 Ω]**

6.4 SERIES-PARALLEL CONTROL

Another method used for the speed control of d.c. series motors is the series-parallel method. In this system which is widely used in traction system, two (or more) similar d.c. series motors are mechanically coupled to the same load.

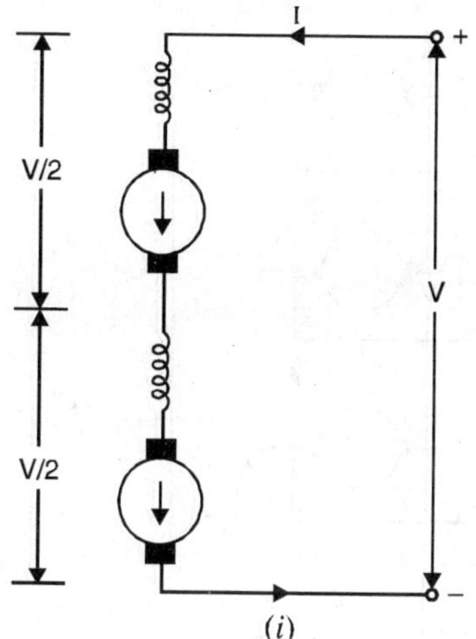

(*i*)

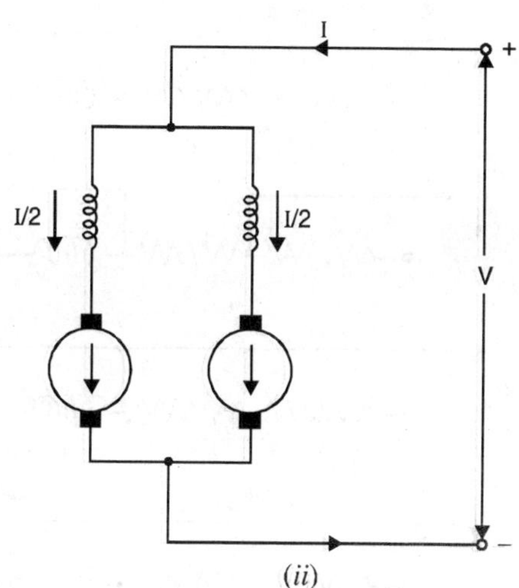

(*ii*)

Fig. 6.23

When the motors are connected in series [See Fig. 6.23 (*i*)], each motor armature will receive one -half the normal voltage. Therefore, the speed will be low. When the motors are connected in parallel, each motor armature receives the normal voltage and the speed is high [See Fig. 6.23 (*ii*)]. Thus we can obtain two speeds. Note that for the same load on the pair of motors, the system would run approximately *four times the speed when the machines are in parallel as when they are in series.

Series-parallel and resistance control. In electric traction, series-parallel method is usually combined with resistance method of control. In the simplest case, two d.c. series motors are coupled mechanically and drive the same vehicle.

(*i*) At standstill, the motors are connected is series via a starting rheostat. The motors are started up in series with each other and starting resistance is cut out step by step to increase the speed. When all the resistance is cut out (See Fig. 6.24), the voltage applied to each motor is about one-half of the line voltage. The speed is then about one-half of what it would be if the full line voltage were applied to each motor.

(*ii*) To increase the speed further, the two motors are connected in parallel and at the same time the starting resistance is connected in series with the combination (See Fig. 6.24). The starting resistance is again cut out step by step until full speed is attained. Then field control is introduced.

---

* When is parallel : speed $\propto E_b/\phi \propto \dfrac{E_b}{I/2} \propto \dfrac{2V}{I}$ $\qquad (\because E_b \simeq V)$

When is series : speed $\propto E_b/\phi \propto E_b/I \propto \dfrac{V/2}{I}$

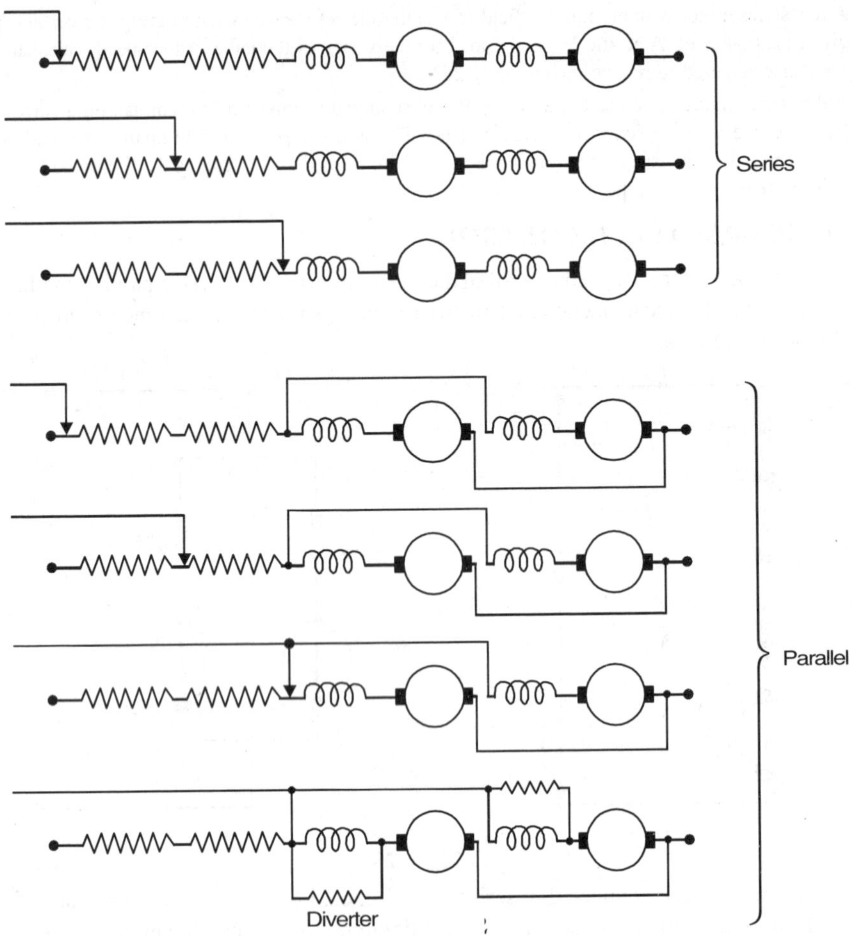

Fig. 6.24

Example 6.40. *Two series motors are exactly alike in construction except that the air gap of one is slightly greater than that of the other. When taking 70 A at 600 V, one motor runs at 800 r.p.m. and the other at 850 r.p.m. If the two motors are mechanically coupled and connected in series across 600 V, at what speed will they run when the current is 70 A? Total resistance of each motor is 0·28 Ω.*

Solution.

First motor $E_{b1} = 600 - 70 \times 0{\cdot}28 = 580{\cdot}4$ volts ; $N_1 = 800$ r.p.m.

$E_{b1} \propto N_1 \phi_1$ or $E_{b1} = k N_1 \phi_1$ $\quad \therefore \quad k\phi_1 = \dfrac{E_{b1}}{N_1} = \dfrac{580{\cdot}4}{800}$

Second motor $E_{b2} = 600 - 70 \times 0{\cdot}28 = 580{\cdot}4$ volts ; $N_2 = 850$ r.p.m.

$E_{b2} \propto N_2 \phi_2$ or $E_{b2} = k N_2 \phi_2$ $\quad \therefore \quad k\phi_2 = \dfrac{E_{b2}}{N_2} = \dfrac{580 \cdot 4}{850}$

Motors in series $E_b' = 600 - 70 \times (0{\cdot}28 + 0{\cdot}28) = 560{\cdot}8$ volts

Now $E_b' = E'_{b1} + E'_{b2}$

or $560{\cdot}8 = k\phi_1 N + k\phi_2 N$ where N = common speed

or $560{\cdot}8 = \dfrac{580 \cdot 4}{800} \times N + \dfrac{580 \cdot 4}{850} \times N$

$\therefore$ $N =$ **398 r.p.m.**

6.5 ELECTRIC BRAKING

Sometimes it is desirable to stop a d.c. motor quickly. This may be necessary in case of emergency or to save time if the motor is being used for frequently repeated operations. The motor and its load may be brought to rest by using either (*i*) mechanical (friction) braking or (*ii*) electric braking. In mechanical braking, the motor is stopped due to the friction between the moving parts of the motor and the brake shoe *i.e.* kinetic energy of the motor is dissipated as heat. Mechanical braking has several disadvantages including non-smooth stop and greater stopping time.

In electric braking, the kinetic energy of the moving parts (*i.e.*, motor) is converted into electrical energy which is dissipated in a resistance as heat or alternatively, it is returned to the supply source (Regenerative braking). For d.c. shunt as well as series motors, the following three methods of electric braking are used :

(*i*) Rheostatic or Dynamic braking

(*ii*) Plugging

(*iii*) Regenerative braking

It may be noted that electric braking cannot hold the motor stationary and mechanical braking is necessary. However, the main advantage of using electric braking is that it reduces the wear and tear of mechanical brakes and cuts down the stopping time considerably due to high braking retardation.

(*i*) **Rheostatic or Dynamic braking.** In this method, the armature of the running motor is disconnected from the supply and is connected across a variable resistance *R*. However, the field winding is left connected to the supply. The armature, while slowing down, rotates in a strong magnetic field and, therefore, operates as a generator, sending a large current through resistance *R*. This causes the energy possessed by the rotating armature to be dissipated quickly as heat in the resistance. As a result, the motor is brought to standstill quickly.

Fig. 6.25 (*ii*) shows dynamic braking of a shunt motor. The braking torque can be controlled by varying the resistance *R*. If the value of *R* is decreased as the motor speed decreases , the braking torque may be maintained at a high value. At a low value of speed, the braking torque becomes small and the final stopping of the motor is due to friction. This type of braking is used extensively in connection with the control of elevators and hoists and in other applications in which motors must be started, stopped and reversed frequently.

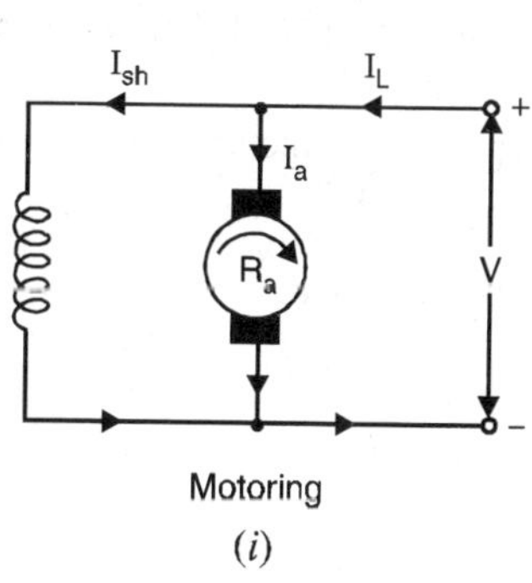

Motoring

(*i*)

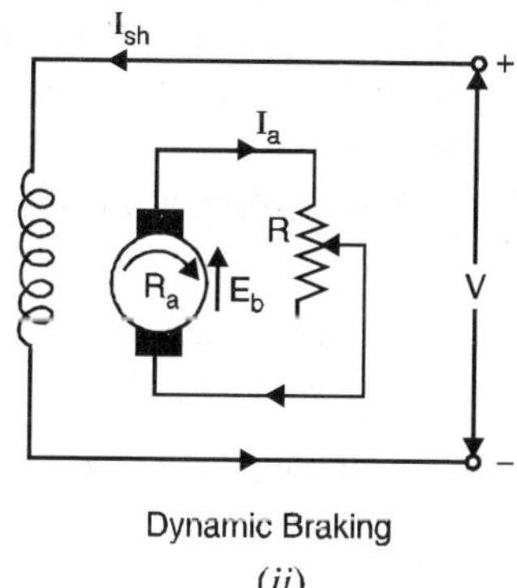

Dynamic Braking

(*ii*)

Fig. 6.25

We now investigate how braking torque depends upon the speed of the motor. Referring to Fig. 6.25 (*ii*),

$$\text{Armature current, } I_a = \frac{E_b}{R+R_a} = \frac{k_1 N \phi}{R+R_a} \qquad (\because E_b \propto \phi N)$$

$$\text{Braking torque, } T_B = k_2 I_a \phi = k_2 \phi \left(\frac{k_1 N \phi}{R+R_a}\right) = k_3 N \phi^2$$

where k_2 and k_3 are constants

For a shunt motor, ϕ is constant.

$\therefore$ Braking torque, $T_B \propto N$

Therefore, braking torque decreases as the motor speed decreases.

(*ii*) **Plugging.** In this method, connections to the armature are reversed so that motor tends to rotate in the opposite direction, thus providing the necessary braking effect. When the motor comes to rest, the supply must be cut off otherwise the motor will start rotating in the opposite direction.

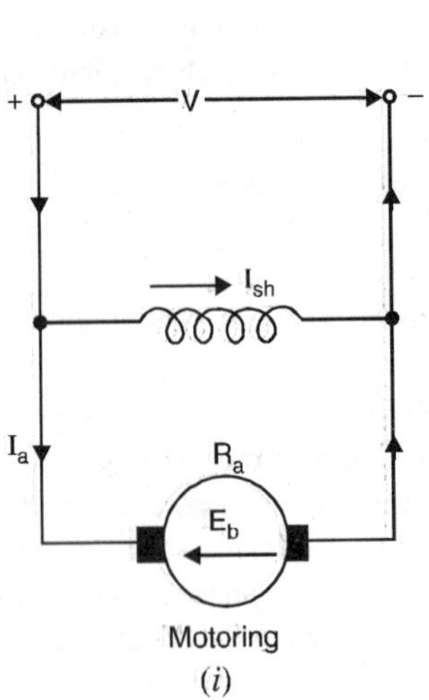

Motoring
(*i*)

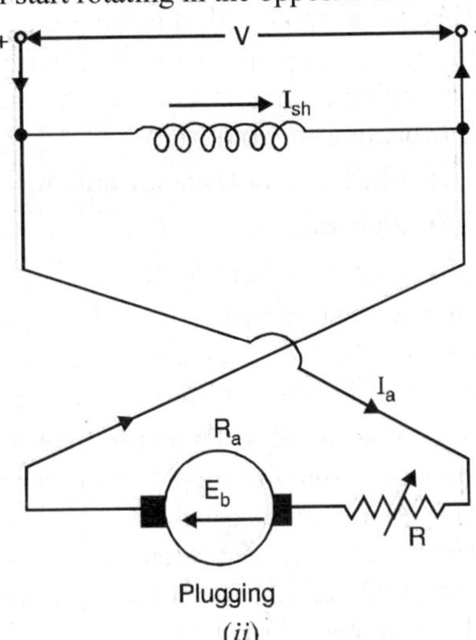

Plugging
(*ii*)

Fig. 6.26

Fig. 6.26 (*ii*) shows plugging of a d.c. shunt motor. Note that armature connections are reversed while the connections of the field winding are kept the same. As a result the current in the armature reverses. During the normal running of the motor [See Fig. 6.26 (*i*)], the back e.m.f. E_b opposes the applied voltage V. However, when armature connections are reversed, back e.m.f. E_b and V act *in the same direction* around the circuit. Therefore, a voltage equal to $V + E_b$ is impressed across the armature circuit. Since $E_b \simeq V$, the impressed voltage is approximately $2V$. In order to limit the current to safe value, a variable resistance R is inserted in the circuit at the time of changing armature connections.

We now investigate how braking torque depends upon the speed of the motor. Referring to Fig. 6.26 (*ii*),

$$\text{Armature current, } I_a = \frac{V + E_b}{R + R_a} = \frac{V}{R + R_a} + \frac{k_1 N \phi}{R + R_a} \qquad (\because\ E_b \propto \phi N)$$

$$\text{Braking torque, } T_B = k_2 I_a \phi = k_2 \phi \left[\frac{V}{R + R_a} + \frac{k_1 N \phi}{R + R_a}\right]$$

$$= k_3 \phi + k_4 N \phi^2$$

For a shunt motor, ϕ is constant.

$\therefore$ Braking torque, $T_B = k_5 + k_6 N$

Thus braking torque decreases as the motor slows down. Note that there is some braking torque ($T_B = k_5$) even when the motor speed is zero.

(*iii*) **Regenerative braking.** In the regenerative braking, the motor is run as a generator. As a result, the kinetic energy of the motor is converted into electrical energy and returned to the supply. Fig. 6.27 shows two methods of regenerative braking for a shunt motor.

(*a*) In one method, field winding is disconnected from the supply and field current is increased by exciting it from another source [See Fig. 6.27 (*i*)]. As a result, induced e.m.f. E exceeds the supply voltage V and the machine feeds energy into the supply. Thus braking torque is provided upto the speed at which induced e.m.f. and supply voltage are equal. As the machine slows down, it is not possible to maintain induced e.m.f. at a higher value than the supply voltage. Therefore, this method is possible only for a limited range of speed.

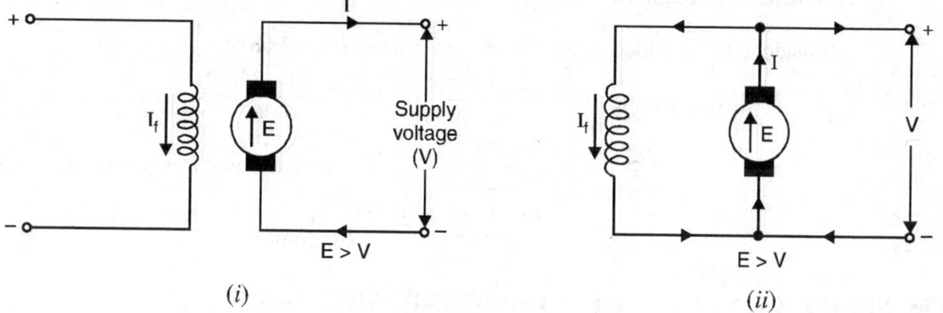

Fig. 6.27

(*b*) In a second method, the field excitation does not change but the load causes the motor to run above the normal speed (*e.g.*, descending load on a crane). As a result, the induced e.m.f. E becomes greater than the supply voltage V [See Fig. 6.27 (*ii*)]. The direction of armature current I, therefore, reverses but the direction of shunt field current I_f remains unaltered. Hence the torque is reversed and the speed falls until E becomes less than V.

Example 6.41. *A 20 h.p., 220 V shunt motor takes a full-load current of 82 A, speed 1000 r.p.m., armature resistance 0·1 Ω, shunt field resistance 110 Ω. It is to be braked by plugging. What resistance must be placed in series to limit the current to 120 A ? Find also the initial value of the braking torque.*

Solution.

$$\text{Field current, } I_{sh} = \frac{220}{110} = 2\text{A}$$

$$\therefore \quad \text{Full-load armature current, } I_a = I_L - I_{sh} = 82 - 2 = 80 \text{ A}$$

$$E_b = V - I_a R_a = 220 - 80 \times 0{\cdot}1 = 212 \text{ volts}$$

$$\text{Voltage across armature when braking} = V + E_b = 220 + 212 = 432 \text{ volts}$$

∴ Total resistance required to limit current to 120 A

$$= \frac{432}{120} = 3{\cdot}6\ \Omega$$

$$\text{Resistance to be added} = 3{\cdot}6 - 0{\cdot}1 = \mathbf{3{\cdot}5\ \Omega}$$

$$\text{Full-load torque} = 9{\cdot}55 \times \frac{\text{Output}}{N} = 9{\cdot}55 \times \frac{20 \times 746}{1000} = 142{\cdot}5 \text{ N-m}$$

$$\therefore \quad \text{Initial braking torque} = \frac{120}{80} \times 142{\cdot}5 = \mathbf{213{\cdot}8 \text{ N-m}}$$

Example 6.42. *A 20 h.p., 220 V d.c. series motor takes a full-load current of 70 A and runs at 600 r.p.m. Resistance of the motor is 0·25 Ω. It is to be braked by plugging. What resistance must be placed in series to limit the initial current to 120 A ? Find also the initial braking torque. The magnetisation curve at 600 r.p.m. is as follows :*

Field current (A)	:	*20*	*40*	*60*	*80*	*100*	*120*
E.M.F. (V)	:	*88*	*160*	*200*	*230*	*242*	*252*

Solution. From the magnetisation curve, when the field current is 120 A, the induced e.m.f. is 252 volts.

$\therefore$ Voltage across the armature circuit when braking

$$= V + E_b = 220 + 252 = 472 \text{ volts}$$

$$\text{Total resistance required} = \frac{472}{120} = 3{\cdot}93\ \Omega$$

$$\text{Resistance to be added} = 3{\cdot}93 - R_m = 3{\cdot}93 - 0{\cdot}25 = \mathbf{3{\cdot}68\ \Omega}$$

$$\text{Full-load torque} = 9{\cdot}55 \times \frac{\text{Output}}{N} = 9{\cdot}55 \times \frac{20 \times 746}{600} = 237 \text{ N-m}$$

$$\therefore \quad \frac{237}{T_B} = \frac{70 \times 220}{120 \times 252} \qquad (\because T_a \propto \phi I_a \text{ and } E \propto \phi \text{ at constant speed})$$

or

$$T_B = \frac{237 \times 120 \times 252}{70 \times 220} = \mathbf{465 \text{ N-m}}$$

6.6 SPEED CONTROL OF COMPOUND MOTORS

Speed control of compound motors may be obtained by any one of the methods described for shunt motors. Speed control cannot be obtained through adjustment of the series field since such adjustment would radically change the performance characteristics of the motor.

6.7 NECESSITY OF D.C. MOTOR STARTER

At starting, when the motor is stationary, there is no back e.m.f. in the armature. Consequently, if the motor is directly switched on to the mains, the armature will draw a heavy current ($I_a = V/R_a$) because of small armature resistance. As an example, 5 h.p., 220 V shunt motor has a full-load current of 20 A and an armature resistance of about 0·5 Ω. If this motor is directly switched on to supply, it would take an armature current of 220/0·5 = 440 A which is 22 times the full-load current. This high starting current may result in :

(*i*) burning of armature due to excessive heating effect,

(*ii*) damaging the commutator and brushes due to heavy sparking,

(*iii*) excessive voltage drop in the line to which the motor is connected. The result is that the operation of other appliances connected to the line may be impaired and in particular cases, they may refuse to work.

In order to avoid excessive current at starting, a variable resistance (known as *starting resistance*) is inserted in series with the armature circuit. This resistance is gradually reduced as the motor gains speed (and hence E_b increases) and eventually it is cut out completely when the motor has attained full speed. The value of starting resistance is generally such that starting current is limited to 1·25 to 2 times the full-load current.

6.8 TYPES OF D.C. MOTOR STARTERS

The starting operation of a d.c. motor consists in the insertion of external resistance into the armature circuit to limit the starting current taken by the motor and the removal of this resistance in steps as the motor accelerates. When the motor attains the normal speed, this resistance is totally cut out of the armature circuit. It is very important and desirable to provide the starter with protective devices to enable the starter arm to return to OFF position

(*i*) when the supply fails, thus preventing the armature being directly across the mains when this voltage is restored. For this purpose, we use *no-volt release coil.*

(*ii*) when the motor becomes overloaded or develops a fault causing the motor to take an excessive current. For this purpose, we use *overload release coil.*

There are two principal types of d.c. motor starters *viz.*, **three-point starter** and **four-point starter.** As we shall see, the two types of starters differ only in the manner in which the no-volt release coil is connected.

6.9 THREE-POINT STARTER

This type of starter is widely used for starting shunt and compound motors.

Schematic diagram. Fig. 6.28 shows the schematic diagram of a three-point starter for a shunt motor with protective devices. It is so called because it has three terminals *L*, *Z* and *A*. The starter consists of starting resistance divided into several sections and connected in series with the armature. The tapping points of the starting resistance are brought out to a number of studs. The three terminals *L*, *Z* and *A* of the starter are connected respectively to the positive line terminal, shunt field terminal and armature terminal. The other terminals of the armature and shunt field windings are connected to the negative terminal of the supply. The *no-volt release coil* is connected in the shunt field circuit. One end of the handle is connected to the terminal *L* through the *over-load release coil*. The other end of the handle moves against a spiral spring and makes contact with each stud during starting operation, cutting out more and more starting resistance as it passes over each stud in clockwise direction.

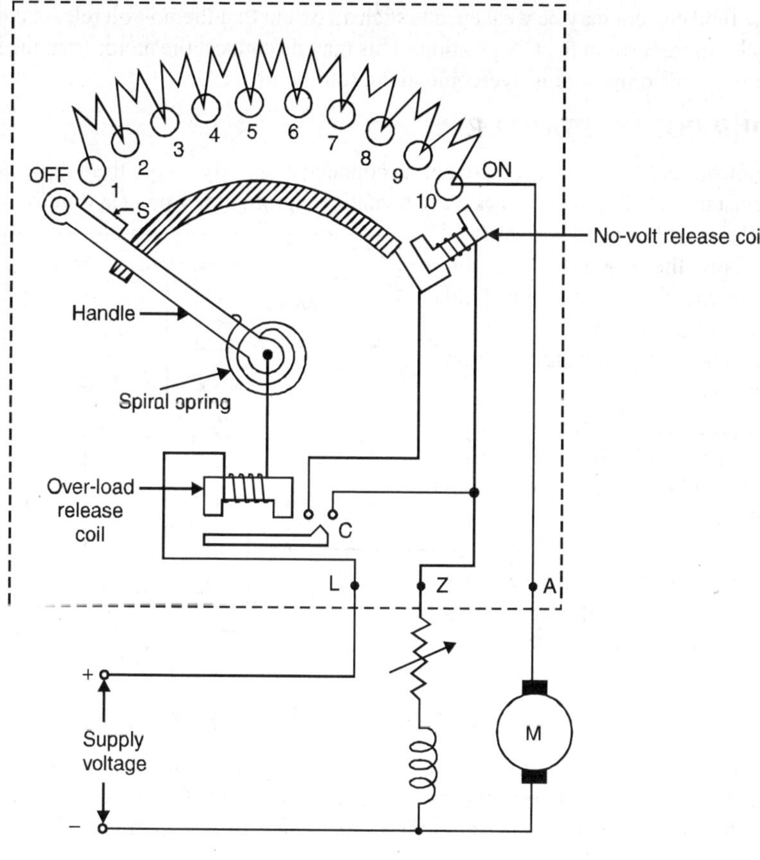

Fig. 6.28

Operation

(*i*) To start with, the d.c. supply is switched on with handle in the OFF position.

(*ii*) The handle is now moved clockwise to the first stud. As soon as it comes in contact with the first stud, the shunt field winding is directly connected across the supply, while the whole starting resistance is inserted in series with the armature circuit.

(*iii*) As the handle is gradually moved over to the final stud, the starting resistance is cut out of the armature circuit in steps. The handle is now held magnetically by the *no-volt release coil which is energised by shunt field current.

(*iv*) If the supply voltage is suddenly interrupted or if the field excitation is accidentally cut, the no-volt release coil is demagnetised and the handle goes back to the OFF position under the pull of the spring. If no-volt release coil were not used, then in case of failure of supply, the handle would remain on the final stud. If then supply is restored, the motor will be directly connected across the supply, resulting in an excessive armature current.

(*v*) If the motor is over-loaded (or a fault occurs), it will draw excessive current from the supply. This current will increase the ampere-turns of the over-load release coil and pull the armature *C*, thus short-circuiting the no-volt release coil. The no-volt coil is demagnetised and the handle is pulled to the OFF position by the spring. Thus, the motor is automatically disconnected from the supply.

Drawback. In a three-point starter, the no-volt release coil is connected in series with the shunt field circuit so that it carries the shunt field current. While excercising speed control through field regulator, the field current may be weakened to such an extent that the no-volt release coil may not be able to keep the starter arm in the ON position. This may disconnect the motor from the supply when it is not desired. This drawback is overcome in the four point starter.

6.10 FOUR-POINT STARTER

In a four-point starter, the no-volt release coil is connected directly across the supply line through a protective resistance *R*. Fig. 6.29 shows the schematic diagram of a 4-point starter for a shunt motor (over-load release coil omitted for clarity of the figure). Now the no-volt release coil circuit is **independent of the shunt field circuit. Therefore, proper speed control can be exercised without affecting the operation of no-volt release coil.

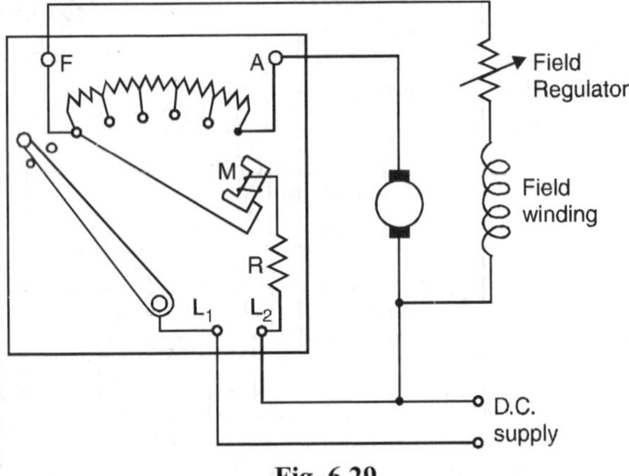

Fig. 6.29

Note that the only difference between a three-point starter and a four-point starter is the manner in which no-volt release coil is connected. However, the working of the two starters is the same. It may be noted that the three-point starter also provides protection against an open-field circuit. This protection is not provided by the four-point starter.

6.11 GRADING OF STARTING RESISTANCE — SHUNT MOTORS

For starting the motor satisfactorily, the starting resistance is divided into a number of sections in such a way that current fluctuates between maximum (I_m) and minimum (I) values. The upper limit is that value established as the maximum permissible for the motor ; it is generally 1·5 times the full-load

* In the ON or running position, the soft iron piece *S* attached to handle is held by the no-volt release coil.

** The reader may note that line current divides into three parts ; one part passes through starting resistance and motor armature ; second part passes through field regulator and field winding ; third part passes through no-volt release coil and protective resistance *R*.

current of the motor. The lower limit is the value set as a minimum for starting operation ; it may be equal to full-load current of the motor or some predetermined value. Fig. 6.30 shows shunt-wound motor with starting resistance divided into three sections between four studs. The resistances of these sections should be so selected that current during starting remains between I_m and I as shown in Fig. 6.31.

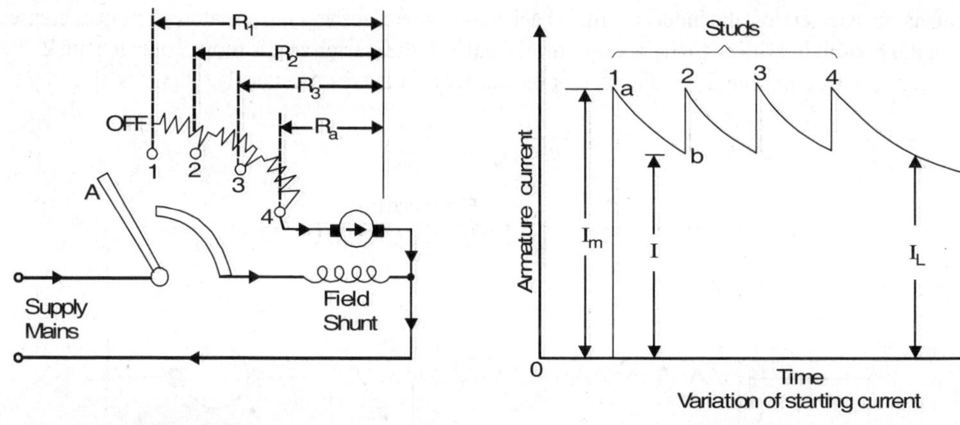

Fig. 6.30 **Fig. 6.31**

(*i*) When arm *A* is moved from OFF position to stud 1, field and armature circuits are energised and whole of the starting resistance is in series with the armature. The armature current jumps to maximum value given by ;

$$I_m = \frac{V}{R_1}$$

where R_1 = Resistance of starter and armature

(*ii*) As the armature accelerates, the generated e.m.f. increases and the armature current decreases as indicated by curve *ab*. When the current has fallen to *I*, arm *A* is moved over to stud 2, cutting out sufficient resistance to allow the current to rise to I_m again. This operation is repeated until the arm *A* is on stud 4 and the whole of the starting resistance is cut out of the armature circuit.

(*iii*) Now the motor continues to accelerate and the current decreases until it settles down at some value I_L such that torque due to this current is just sufficient to meet the load requirement.

6.12 STARTER STEP CALCULATIONS FOR D.C. SHUNT MOTOR

Fig. 6.32 shows a d.c. shunt motor starter with *n* resistance sections and (*n* + 1) studs.

Let R_1 = Total resistance in the armature circuit when the starter arm is on stud no. 1 (See Fig. 6.32)

R_2 = Total resistance in the armature circuit when the starter arm is on stud no. 2 and so on

I_m = Upper current limit

I = Lower current limit

n = Number of sections in the starter resistance

V = Applied voltage

R_a = Armature resistance

On stud 1. When the starter arm moves to stud 1, the total resistance in the armature circuit is R_1 and the circuit current jumps to maximum values I_m given by ;

$$I_m = \frac{V}{R_1} \quad ...(i)$$

Since torque $\propto \phi I_a$, it follows that the maximum torque acts on the armature to accelerate it. As the armature accelerates, the induced e.m.f. (back e.m.f.) increases and the armature current decreases. When the current has fallen to the predetermined value I, the starter arm is moved over to stud 2. Let the value of back e.m.f. be E_{b1} at the instant the starter arm leaves the stud 1. Then I is given by ;

$$I = \frac{V - E_{b1}}{R_1} \quad ...(ii)$$

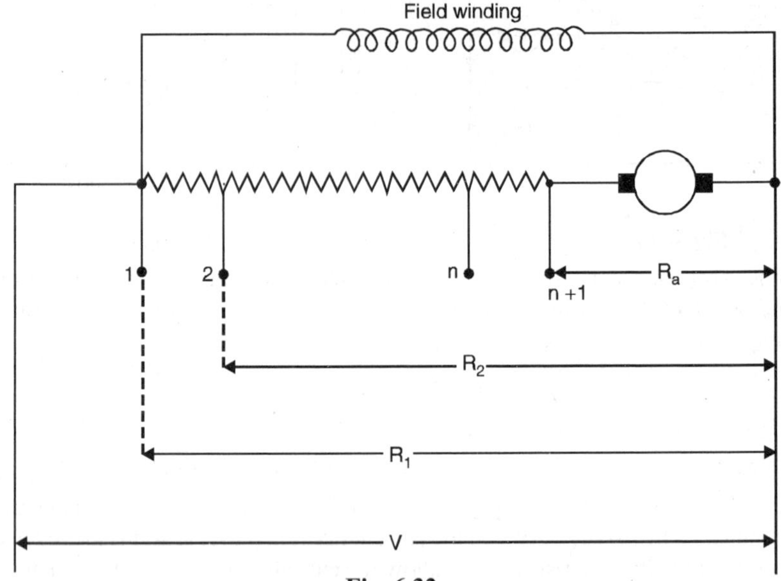

Fig. 6.32

On stud 2. As the starter arm moves over to stud 2, sufficient resistance is cut out (now total circuit resistance is R_2) and current rises to maximum value I_m once again given by ;

$$I_m = \frac{V - {}^{*}E_{b1}}{R_2} \quad ...(iii)$$

The acceleration continues and the back e.m.f. increases and the armature current decreases. When the current has fallen to the predetermined value I, the starter arm is moved over to stud 3. Let E_{b2} be the value of back e.m.f. at the instant the starter arm leaves the stud 2. Then,

$$I = \frac{V - E_{b2}}{R_2} \quad ...(iv)$$

On stud 3.

As the starter arm moves to stud 3, $I_m = \dfrac{V - E_{b2}}{R_3}$ ...(v)

As the starter arm leaves stud 3, $I = \dfrac{V - E_{b3}}{R_3}$ ...(vi)

---

* Due to inertia, the speed of the motor cannot change so that at the instant the starter arm comes in contact with stud 2, the back e.m.f. remains E_{b1}.

On *n*th stud.

As the starter arm leaves *n*th stud, $I = \dfrac{V - E_{bn}}{R_n}$

On (*n* + 1) th stud. When the starter arm moves over to (*n* + 1) th stud, all the external starting resistance is cut out, leaving only the armature resistance R_a.

$$\therefore \qquad I_m = \frac{V - E_{bn}}{R_a} \quad \text{and} \quad I = \frac{V - E_b}{R_a}$$

Dividing eq. (*ii*) by eq. (*iii*), we get,

$$\frac{I}{I_m} = \frac{R_2}{R_1}$$

Dividing eq. (*iv*) by eq. (*v*), we get,

$$\frac{I}{I_m} = \frac{R_3}{R_2}$$

Continuing these divisions, we have finally,

$$\frac{I}{I_m} = \frac{R_a}{R_n}$$

$$\text{Let } \frac{I}{I_m} = k. \quad \text{Then } \frac{R_2}{R_1} = \frac{R_3}{R_2} = \ldots\ldots = \frac{R_a}{R_n} = k$$

If we multiply these *n* equal ratios together, then,

$$\frac{R_2}{R_1} \times \frac{R_3}{R_2} \times \frac{R_4}{R_3} \ldots\ldots \times \frac{R_a}{R_n} = k^n$$

$$\therefore \qquad \frac{R_a}{R_1} = k^n$$

Thus we can calculate the values of R_2, R_3, R_4 etc. if the values of R_1, R_a and n are known.

Example 6.43. *A 220 V shunt motor has an armature resistance of 0·4 Ω. The armature current at starting must not exceed 40 A. If the number of sections is 6, calculate the values of resistor steps to be used in this starter.*

Solution.

$$R_1 = 220/40 = 5{\cdot}5\ \Omega \quad ; \quad n = 6 \quad ; \quad \frac{R_a}{R_1} = \frac{0{\cdot}4}{5{\cdot}5} = k^n = k^6$$

$$\therefore \qquad k = \left(\frac{0{\cdot}4}{5{\cdot}5}\right)^{1/6} = 0{\cdot}6462$$

$R_2 = k R_1 = 0{\cdot}6462 \times 5{\cdot}5 = 3{\cdot}554\ \Omega$; $R_3 = k R_2 = 0{\cdot}6462 \times 3{\cdot}554 = 2{\cdot}296\ \Omega$

$R_4 = k R_3 = 0{\cdot}6462 \times 2{\cdot}296 = 1{\cdot}484\ \Omega$; $R_5 = k R_4 = 0{\cdot}6462 \times 1{\cdot}484 = 0{\cdot}958\ \Omega$

$R_6 = k R_5 = 0{\cdot}6462 \times 0{\cdot}958 = 0{\cdot}619\ \Omega$; $R_7 = R_a = 0{\cdot}4\ \Omega$

Resistance of first section = $R_1 - R_2 = 5{\cdot}5 - 3{\cdot}554$ = **1·946 Ω**

Resistance of second section = $R_2 - R_3 = 3{\cdot}554 - 2{\cdot}296$ = **1·258 Ω**

Resistance of third section = $R_3 - R_4 = 2{\cdot}296 - 1{\cdot}484$ = **0·812 Ω**

Resistance of fourth section = $R_4 - R_5 = 1{\cdot}484 - 0{\cdot}958$ = **0·526 Ω**

Resistance of fifth section = $R_5 - R_6 = 0{\cdot}958 - 0{\cdot}619$ = **0·339 Ω**

Resistance of sixth section = $R_6 - R_a = 0{\cdot}619 - 0{\cdot}4$ = **0·219 Ω**

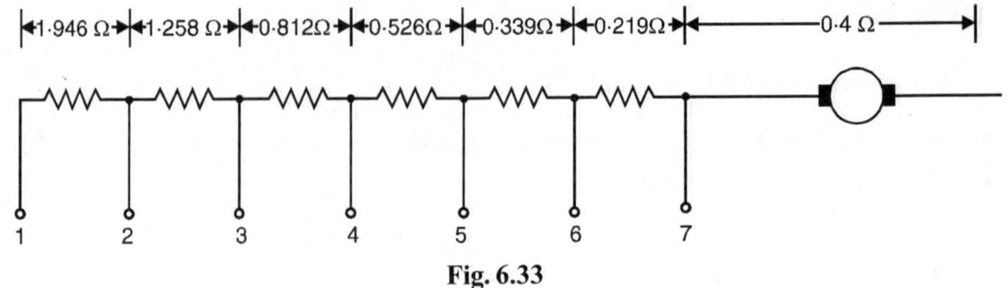

Fig. 6.33

The various section resistances are shown in Fig. 6.33.

Example 6.44. *A starter is required for a 220 V d.c. shunt motor. The maximum allowable current is 55 A and the minimum current is 35 A. Find the number of sections of starter resistance required and the resistance of each section. The armature resistance of the motor is 0·4 Ω.*

Solution.

$$R_a = 0{\cdot}4\,\Omega \quad ; \quad k = \frac{I}{I_m} = \frac{35}{55} = 0{\cdot}636 \quad ; \quad R_1 = \frac{220}{55} = 4\,\Omega$$

$$\frac{R_a}{R_1} = k^n \quad \text{or} \quad \frac{0{\cdot}4}{4} = (0{\cdot}636)^n \qquad \therefore \; n = 5{\cdot}1$$

Since n must be an integer, $n = \mathbf{5}$

$R_2 = k\,R_1 = 0{\cdot}636 \times 4 = 2{\cdot}544\,\Omega$; $R_3 = k\,R_2 = 0{\cdot}636 \times 2{\cdot}544 = 1{\cdot}618\,\Omega$

$R_4 = k\,R_3 = 0{\cdot}636 \times 1{\cdot}618 = 1{\cdot}028\,\Omega$; $R_5 = k\,R_4 = 0{\cdot}654\,\Omega$

Resistance of first section $= R_1 - R_2 = 4 - 2{\cdot}544 = \mathbf{1{\cdot}456\,\Omega}$

Resistance of second section $= R_2 - R_3 = 2{\cdot}544 - 1{\cdot}618 = \mathbf{0{\cdot}926\,\Omega}$

Resistance of third section $= R_3 - R_4 = 1{\cdot}618 - 1{\cdot}028 = \mathbf{0{\cdot}59\,\Omega}$

Resistance of fourth section $= R_4 - R_5 = 1{\cdot}028 - 0{\cdot}654 = \mathbf{0{\cdot}374\,\Omega}$

Resistance of fifth section $= R_5 - R_a = 0{\cdot}654 - 0{\cdot}4 = \mathbf{0{\cdot}254\,\Omega}$

Example 6.45. *There are 6 sections in the starter of a 5 h.p., 220 V shunt-wound motor. Starting current is not to exceed 1·5 times the full-load current. The motor efficiency is 85%. The armature Cu loss is half of the total loss. Find the values of step resistances. Neglect-shunt field current.*

Solution.

$$\text{Full-load current} = \frac{5 \times 746}{0{\cdot}85 \times 220} = 20\text{ A}$$

$$\text{Starting current} = 1{\cdot}5 \times 20 = 30\text{ A}$$

$$\text{Total losses} = \frac{15}{100} \times 220 \times 20 = 660\text{ W}$$

$$\text{Armature Cu loss} = \frac{660}{2} = 330\text{ W}$$

$$\therefore \quad \text{Armature resistance, } R_a = \frac{330}{(20)^2} = 0{\cdot}83\,\Omega$$

Total resistance of armature circuit at start is

$$R_1 = \frac{220}{30} = 7{\cdot}33\,\Omega$$

$$k^n = \frac{R_a}{R_1} \quad \text{or} \quad k^6 = \frac{0{\cdot}83}{7{\cdot}33} \qquad \therefore \; k = \left(\frac{0{\cdot}83}{7{\cdot}33}\right)^{1/6} = 0{\cdot}696$$

$R_2 = kR_1 = 0{\cdot}696 \times 7{\cdot}33 = 5{\cdot}10\,\Omega$; $R_3 = kR_2 = 0{\cdot}696 \times 5{\cdot}10 = 3{\cdot}55\,\Omega$

$R_4 = kR_3 = 0{\cdot}696 \times 3{\cdot}55 = 2{\cdot}47\,\Omega$; $R_5 = kR_4 = 0{\cdot}696 \times 2{\cdot}47 = 1{\cdot}72\,\Omega$

$R_6 = kR_5 = 0{\cdot}696 \times 1{\cdot}72 = 1{\cdot}20\,\Omega$; $R_7 = R_a = 0{\cdot}83\,\Omega$

∴ Resistance of first section $= R_1 - R_2 = 7{\cdot}33 - 5{\cdot}10 = \mathbf{2{\cdot}23\,\Omega}$

Resistance of second section $= R_2 - R_3 = 5{\cdot}10 - 3{\cdot}55 = \mathbf{1{\cdot}55\,\Omega}$

Resistance of third section $= R_3 - R_4 = 3{\cdot}55 - 2{\cdot}47 = \mathbf{1{\cdot}08\,\Omega}$

Resistance of fourth section $= R_4 - R_5 = 2{\cdot}47 - 1{\cdot}72 = \mathbf{0{\cdot}75\,\Omega}$

Resistance of fifth section $= R_5 - R_6 = 1{\cdot}72 - 1{\cdot}20 = \mathbf{0{\cdot}52\,\Omega}$

Resistance of sixth section $= R_6 - R_a = 1{\cdot}20 - 0{\cdot}83 = \mathbf{0{\cdot}37\,\Omega}$

Example 6.46. *Design the resistance sections of eight-stud starter for a 100 h.p., 500 V shunt motor. The lower current limit is to be the full-load current. The full-load efficiency is 94%. The total Cu losses are 3·7 % of input and shunt field resistance is 250 Ω.*

Solution. There are eight studs and seven sections of resistances.

$$\text{Input power} = \frac{\text{Output power}}{\eta} = \frac{100 \times 746}{0{\cdot}94} = 79362 \text{ W}$$

$$\text{Input current, } I_L = \frac{\text{Input power}}{500} = \frac{79362}{500} = 158{\cdot}7 \text{ A}$$

$$\text{Shunt field current, } I_{sh} = \frac{500}{250} = 2\text{A}$$

$$\text{Armature current, } I_a = 158{\cdot}7 - 2 = 156{\cdot}7 \text{ A}$$

$$\text{Total Cu losses} = 3{\cdot}7\% \text{ of input} = 0{\cdot}037 \times 79362 = 2936 \text{ W}$$

$$\text{Shunt field loss} = I_{sh}^2 R_{sh} = (2)^2 \times 250 = 1000 \text{ W}$$

$$\text{Armature Cu loss} = 2936 - 1000 = 1936 \text{ W}$$

$$\therefore \quad \text{Armature resistance, } R_a = \frac{1936}{I_a^2} = \frac{1936}{(156{\cdot}7)^2} = 0{\cdot}079\,\Omega$$

$$\text{Minimum current, } I = 156{\cdot}7 \text{ A}$$

$$\text{Maximum current, } I_m = \frac{500}{R_1} \text{ where } R_1 = \text{Total resistance of armature circuit on stud 1}$$

For a seven-section starter, we have,

$$\left(\frac{I}{I_m}\right)^7 = \frac{R_a}{R_1} \quad \text{or} \quad \left(\frac{156{\cdot}7 R_1}{500}\right)^7 = \frac{0{\cdot}079}{R_1}$$

$$\therefore \quad \left(\frac{156{\cdot}7}{500}\right)^7 \times R_1^8 = 0{\cdot}079 \quad \therefore \; R_1 = 2{\cdot}01\,\Omega$$

Now $$I_m = \frac{500}{2{\cdot}01} = 249 \text{ A} \quad \therefore \; k = \frac{I}{I_m} = \frac{156{\cdot}7}{249} = 0{\cdot}63$$

∴ $R_2 = kR_1 = 0{\cdot}63 \times 2{\cdot}01 = 1{\cdot}27\,\Omega$; $R_3 = kR_2 = 0{\cdot}63 \times 1{\cdot}27 = 0{\cdot}80\,\Omega$

$R_4 = kR_3 = 0{\cdot}63 \times 0{\cdot}80 = 0{\cdot}50\,\Omega$; $R_5 = kR_4 = 0{\cdot}63 \times 0{\cdot}50 = 0{\cdot}32\,\Omega$

$R_6 = kR_5 = 0{\cdot}63 \times 0{\cdot}32 = 0{\cdot}2\,\Omega$; $R_7 = kR_6 = 0{\cdot}63 \times 0{\cdot}2 = 0{\cdot}13\,\Omega$

$R_8 = R_a = 0{\cdot}079\,\Omega$

The various starter sections are shown in Fig. 6.34.

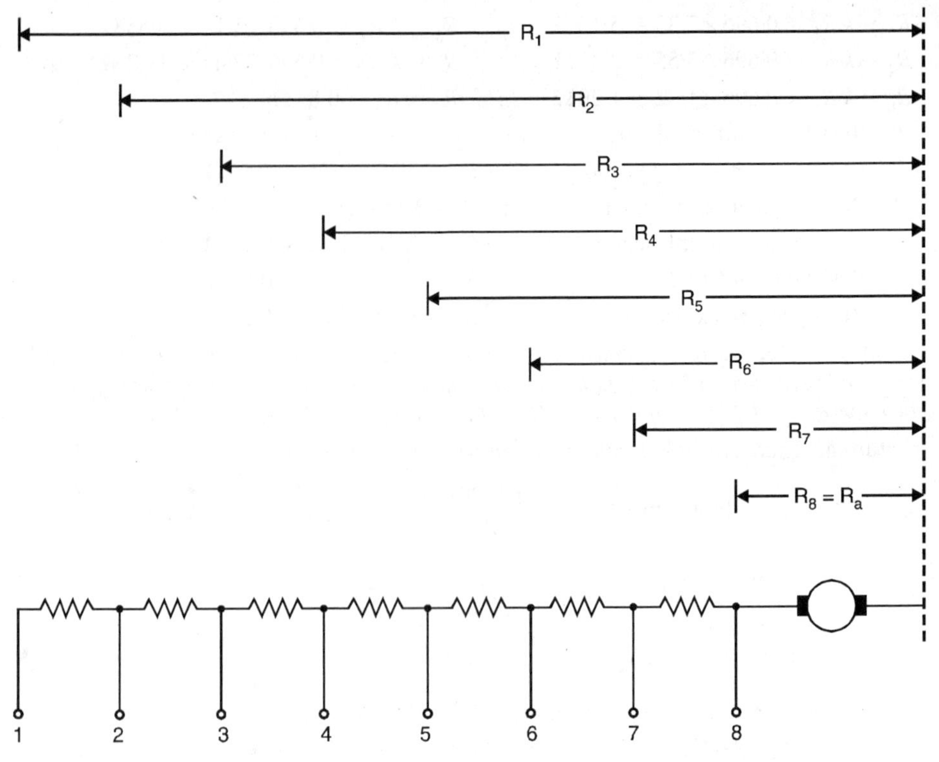

Fig. 6.34

Resistance of first section $= R_1 - R_2 = 2{\cdot}01 - 1{\cdot}27 = \mathbf{0{\cdot}74\ \Omega}$

Resistance of second section $= R_2 - R_3 = 1{\cdot}27 - 0{\cdot}80 = \mathbf{0{\cdot}47\ \Omega}$

Resistance of third section $= R_3 - R_4 = 0{\cdot}80 - 0{\cdot}50 = \mathbf{0{\cdot}30\ \Omega}$

Resistance of fourth section $= R_4 - R_5 = 0{\cdot}50 - 0{\cdot}32 = \mathbf{0{\cdot}18\ \Omega}$

Resistance of fifth section $= R_5 - R_6 = 0{\cdot}32 - 0{\cdot}20 = \mathbf{0{\cdot}12\ \Omega}$

Resistance of sixth section $= R_6 - R_7 = 0{\cdot}20 - 0{\cdot}13 = \mathbf{0{\cdot}07\ \Omega}$

Resistance of seventh section $= R_7 - R_a = 0{\cdot}13 - 0{\cdot}079 = \mathbf{0{\cdot}05\ \Omega}$

Example 6.47. *The 4-pole, lap-wound armature winding of a 500 V d.c. shunt motor is housed in a total number of 60 slots each slot containing 20 conductors. The armature resistance is 1·31 Ω. If during the period of starting, the minimum torque required is 218 N-m and the maximum torque is 1·5 times the minimum torque, find how many sections the starter should have and calculate the resistances of these sections. Take useful flux per pole to be 23 mWb.*

Solution. $T_a = 0{\cdot}159\ \phi Z I_a (P/A)$

Here $\phi = 23$ mWb $= 23 \times 10^{-3}$ Wb; $Z = 60 \times 20 = 1200$ conductors; $P = 4$; $A = 4$; $T_a = 218$ N-m

$\therefore \quad 218 = 0{\cdot}159 \times 23 \times 10^{-3} \times 1200 \times I_a \times (4/4) \quad \therefore\ I_a = 50$ A

Maximum current, $I_m = 1{\cdot}5 \times 50 = 75$ A ; Minimum current, $I = 50$ A

$$R_1 = 500/I_m = 500/75 = 6{\cdot}667\ \Omega$$

Let n be the number of sections of the starter. Then,

$$\left(\frac{I}{I_m}\right)^n = \frac{R_a}{R_1} \quad \text{or} \quad \left(\frac{50}{75}\right)^n = \frac{1{\cdot}31}{6{\cdot}667} \qquad \therefore\ n = \mathbf{4}$$

Hence there are 4 sections and 5 studs. Also $k = \frac{I}{I_m} = \frac{2}{3}$

$R_2 = kR_1 = \frac{2}{3} \times 6{\cdot}667 = 4{\cdot}44\ \Omega$; $R_3 = kR_2 = \frac{2}{3} \times 4{\cdot}44 = 2{\cdot}96\ \Omega$

$R_4 = kR_3 = \frac{2}{3} \times 2{\cdot}96 = 1{\cdot}98\ \Omega$; $R_5 = R_a = 1{\cdot}31\ \Omega$

Resistance of first section $= R_1 - R_2 = 6{\cdot}667 - 4{\cdot}44 = \mathbf{2{\cdot}23\ \Omega}$

Resistance of second section $= R_2 - R_3 = 4{\cdot}44 - 2{\cdot}96 = \mathbf{1{\cdot}48\ \Omega}$

Resistance of third section $= R_3 - R_4 = 2{\cdot}96 - 1{\cdot}98 = \mathbf{0{\cdot}98\ \Omega}$

Resistance of fourth section $= R_4 - R_a = 1{\cdot}98 - 1{\cdot}31 = \mathbf{0{\cdot}67\ \Omega}$

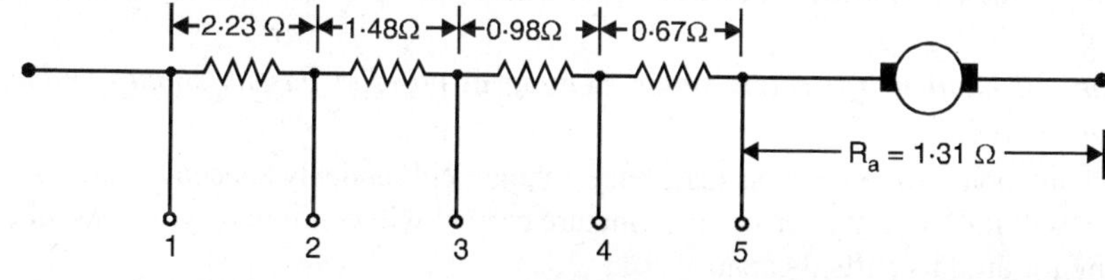

Fig. 6.35

The various sections of the starter are shown in Fig. 6.35.

TUTORIAL PROBLEMS

1. A starter is required for a 250 V shunt motor. The maximum current limit is to be 67 A and the minimum limit about 3/4 of this value. The armature resistance is 0·5 Ω. Find the number of sections of the starter and the resistance of each section.

[Sections = 7 ; 0·93 Ω ; 0·70 Ω ; 0·53 Ω ; 0·40 Ω ; 0·29 Ω ; 0·22 Ω ; 0·17 Ω]

2. Find the values of the step resistances in a 6-stud starter for a 5 h.p. (3·73 kW), 200 V shunt motor. The maximum current in the line is limited to twice the full-load value. The total Cu loss is 50% of the total loss. The normal field current is 0·6 A and the full-load efficiency is found to be 88%.

[1·628 Ω ; 1·075 Ω ; 0·708 Ω ; 0·468 Ω ; 0·309 Ω]

3. Design the resistance sections of a seven-stud starter for 36·775 kW, 400 V d.c. shunt motor. Full-load efficiency is 92% and total Cu losses are 5% of the input. Shunt field resistance is 200 Ω. The lower limit of the current through the armature is to be full-load value.

[0·973 Ω ; 0·593 Ω ; 0·36 Ω ; 0·218 Ω ; 0·133 Ω ; 0·081 Ω]

SHORT ANSWER QUESTIONS

Q. 1. *Why is the speed of a shunt motor practically constant ?*

Ans. The speed of a d.c. motor is given by :

$$N \propto \frac{V - I_a R_a}{\phi}$$

In a shunt motor, V, R_a and ϕ are practically constant. At no load, the armature current is very small and $I_a R_a$ drop is negligible as compared to V. At full-load, $I_a R_a$ drop is ordinarily about 5% of V. Consequently, the full-load speed of the motor is about 95% of the no-load value. This fall in speed is reduced slightly due to armature reaction. In some cases, armature reaction is sufficient to cause the speed to remain nearly constant. For these reasons, a shunt motor is considered a constant speed motor, though the speed decreases slightly with the increase in load.

Q. 2. *What is the chief advantage of a d.c. series motor ?*

Ans. The outstanding characteristic of a series motor is that it has high torque at low speeds and *vice-versa*. It is well-suited for traction purposes such as in electric trains. Acceleration is rapid because the torque is high at low speeds. Furthermore, the series motor automatically slows down as the train goes up an incline surface, yet turns at top speed on the flat ground. The power of a series motor tends to be constant because high torque is accompanied by low speed and *vice-versa*. Series motors are also used in electric cranes and hoists ; light loads are lifted quickly and heavy loads more slowly.

Q. 3. *Why does not cumulatively-compounded motor run at dangerous speed at light loads ?*

Ans. At no-load, the flux due to series field of a cumulatively-compounded motor is almost zero. However, the flux due to shunt winding prevents the motor from running at dangerously high speed.

Q. 4. *What will happen if a shunt motor running at no-load has its shunt winding opened accidentally?*

Ans. If a shunt motor, running at no-load, has its shunt field suddenly opened or burnt out, the back e.m.f. will fall to nearly zero. The armature current will continuously increase the speed of the motor until its parts fly apart.

Q. 5. *What will happen if the shunt field winding of a loaded shunt motor accidentally breaks ?*

Ans. If the shunt winding of a loaded motor accidentally breaks, the loss of shunt field current would result in stopping of the motor or burn out the armature or fuse.

Q. 6. *How is the speed of a compound-wound motor controlled ?*

Ans. The speed of a compound motor is controlled in the following ways :

(*i*) In order to reduce the speed of a compound motor, a resistance is inserted in the armature.

(*ii*) The speed can be increased either by a resistance in series with the shunt winding or a resistance in parallel with the series winding.

Q. 7. *What is the chief advantage of a cumulatively-compounded motor ?*

Ans. The speed-torque characteristic of a cumulatively-compounded motor may approach those of the shunt or series motors, depending upon the strength of the two fields.

(*i*) If the motor contains a light series winding, it will have better torque characteristics than a shunt motor but will largely retain the good regulation or efficient speed control of the shunt motor.

(*ii*) If the motor has a light shunt winding (one sufficient to keep the no-load speed to safe value) and a heavy series winding, it will have the characteristics of a series motor but will not run at a dangerous speed with light loads.

Q. 8. *Why is a differentially-compounded motor rarely used ?*

Ans. In a differentially-compounded motor, the flux due to series winding opposes the flux due to shunt winding. The series winding may be made to weaken the field sufficiently to make the speed at some particular load the same as at no-load. However, speed at other loads will be slightly different. This implies that a differentially-compounded motor has good speed regulation. The shunt motor also has good speed regulation with the added advantage that it has superior speed-torque characteristic as compared to differentially-compounded motor. For this reason, the latter is rarely used in practice.

Q. 9. *What is the necessity of a starter in a d.c. motor ?*

Ans. When we apply rated voltage (V) to a d.c. motor, the starting current in the armature is very high because of the absence of back e.m.f. The high starting current may result in :

(*i*) burning out the armature

(*ii*) damaging the commutator and brushes due to heavy sparking (*iii*) overloading the line to which the machine is connected.

In order to overcome above ill effects during starting of the d.c. motor, a starter is used with it.

Q. 10. *What factors may be controlled in the operation of a d.c. motor?*

Ans. The operation of a d.c. motor may require the control of any of the following factors :

(*i*) starting (*ii*) speed (*iii*) direction of rotation (*iv*) braking

Q. 11. *What steps are required in the starting of a d.c. motor ?*

Ans. The starting operation of a d.c. motor consists in

(*i*) the insertion of external resistance into the armature circuit to limit the starting current taken by the motor

(*ii*) the removal of this resistance in steps as the motor accelerates

The insertion and removal of resistance may be done manually or automatically.

Q. 12. *How are starters for d.c. motors rated ?*

Ans. Starters of d.c. motors are rated according to the rating of the motor with which they are to be used. Thus a 10 h.p., 230 V starter is designed to be used with a 10 h.p., 230 V motor. The resistance of the starter is such so as to permit a starting current in the range from 150% to 200% of the rated armature current ; the actual magnitude depends upon design of the motor.

Q. 13. *What does the nameplate of a d.c. motor indicate ?*

Ans. The nameplate data of a d.c. motor are :

(*i*) Horse power (*ii*) voltage (*iii*) current

(*iv*) speed (*v*) temperature rise

By nameplate the manufacturer guarantees that the motor will deliver the rated horsepower without exceeding the rated temperature rise, if it is operated at rated voltage and speed. Under these conditions, the line current taken by the motor is that specified on the nameplate.

Q. 14. *What is the difference between three-point and four-point starter?*

Ans. The two types of starters differ only in the manner in which the no-volt release coil is connected. In a three-point starter, the no-volt release coil is connected in series with the shunt field circuit so that it carries the shunt field current. In a four-point starter, the no-volt release coil is connected directly across the supply line through a protective resistance.

Q. 15. *If the load is removed from a d.c. series motor in operation, what will happen?*

Ans. If the load is removed from a d.c. series motor in operation, the speed increases rapidly, and at very light loads, the motor may run at far more than safe speed. In very small motors this is not true, since friction offers enough load to restrain the motor from running at dangerous speed. A large d.c. series motor should always be connected to the load by gears or other positive drives ; it should never be used with a belt drive or any other drive where it is possible for the load to be entirely removed.

MULTIPLE-CHOICE QUESTIONS

1. A d.c. motor runs at 1725 r.p.m. at full-load and 1775 r.p.m. at no-load. The speed regulation is

(*i*) 4·7 % (*ii*) 2·9 %

(*iii*) 7·6 % (*iv*) 1·5 %

2. A 4-pole, 500 V shunt motor has 720 wave-connected conductors on its armature. The full-load armature current is 60 A and flux per pole is 0·03 Wb. The armature resistance is 0·2 Ω and the contact drop per brush is 1 V. What is the full-load speed of the motor ?

(*i*) 675 r.p.m. (*ii*) 522 r.p.m.

(*iii*) 484 r.p.m. (*iv*) 576 r.p.m.

3. A shunt motor runs at 500 r.p.m. on a 200 V circuit. Its armature resistance is 0·5 Ω and the current taken is 30 A in addition to the field current. What resistance must be placed in series with the armature circuit to reduce the speed to 300 r.p.m., the current in armature remaining the same ?
   (*i*) 4·24 Ω (*ii*) 3·78 Ω
   (*iii*) 6·74 Ω (*iv*) 2·47 Ω
4. A 240 V series motor takes 40 A when giving its rated output at 1500 r.p.m. Its resistance is 0·3 Ω. Find what resistance must be added to obtain rated torque at starting ?
   (*i*) 2·4 Ω (*ii*) 1·5 Ω
   (*iii*) 3·2 Ω (*iv*) 5·7 Ω
5. In the above question, what resistance must be added to obtain rated torque at 1000 r.p.m. ?
   (*i*) 1·9 Ω (*ii*) 2·3 Ω
   (*iii*) 3·2 Ω (*iv*) 4·8 Ω
6. A series motor of resistance of 1 Ω between the terminals runs at 800 r.p.m. at 200 V with a current of 15 A. Find the speed at which it will run when connected in series with a 5 Ω resistance and taking the same current at the same supply voltage.
   (*i*) 476 r.p.m. (*ii*) 384 r.p.m.
   (*iii*) 572 r.p.m. (*iv*) 602 r.p.m.
7. What happens to speed when the flux is reduced by 10% in a 200 V d.c. shunt motor having an armature resistance of 0·2 Ω, carrying a current of 50 A and running at 960 r.p.m. prior to weakening of field ? The total torque may be assumed to be constant and iron and friction losses may be neglected.
   (*i*) 1250 r.p.m. (*ii*) 1058 r.p.m.
   (*iii*) 920 r.p.m. (*iv*) 576 r.p.m.
8. By flux control method of speed control of a d.c. shunt motor, we can obtain speeds
   (*i*) above the normal speed only
   (*ii*) below the normal speed only
   (*iii*) above as well as below the normal speed
   (*iv*) none of the above
9. By putting controller resistance in series with the armature of a d.c. motor, we can obtain speeds
   (*i*) above the normal speed only
   (*ii*) below the normal speed only
   (*iii*) above as well as below the normal speed
   (*iv*) none of the above
10. The speed of a d.c. motor can be controlled by changing
   (*i*) its flux
   (*ii*) armature circuit resistance
   (*iii*) applied voltage
   (*iv*) all of the above
11. Motor starters are essential for
   (*i*) accelerating the motor
   (*ii*) starting the motor
   (*iii*) avoiding excessive starting current
   (*iv*) preventing fuse blowing
12. The only disadvantage of field control method for controlling the speed of a d.c. shunt motor is that it
   (*i*) gives speed lower than the normal speed
   (*ii*) is wasteful
   (*iii*) needs a large rheostat
   (*iv*) adversely affects commutation
13. The rheostatic speed control method is very
   (*i*) economical (*ii*) efficient
   (*iii*) unsuitable for rapidly changing loads
   (*iv*) suitable for getting speeds above the normal
14. The value of the diverter resistance is of the order of
   (*i*) 25 Ω (*ii*) 100 Ω
   (*iii*) 0·1 Ω (*iv*) any of above
15. The difference between no-load and full-load speed of a d.c. shunt motor is of the order of
   (*i*) 25% (*ii*) 1%
   (*iii*) 10% (*iv*) 50%

ANSWERS TO MULTIPLE-CHOICE QUESTIONS

1. (*ii*)	**2.** (*i*)	**3.** (*iv*)	**4.** (*iv*)	**5.** (*i*)
6. (*i*)	**7.** (*ii*)	**8.** (*i*)	**9.** (*ii*)	**10.** (*iv*)
11. (*iii*)	**12.** (*iv*)	**13.** (*iii*)	**14.** (*iii*)	**15.** (*ii*)

HINTS TO SELECTED MULTIPLE-CHOICE QUESTIONS

1. Speed regulation $= \dfrac{N_0 - N_{FL}}{N_{FL}} \times 100 = \dfrac{1775 - 1725}{1725} \times 100 = 2{\cdot}9\%$

2.
$$E_b = V - I_a R_a - \text{brush drop} = 500 - 60 \times 0{\cdot}2 - 2 = 486 \text{ volts}$$
$$E_b = \frac{\phi ZN}{60} \times \frac{P}{A} \quad \text{or} \quad 486 = \frac{0{\cdot}03 \times 720 \times N}{60} \times \left(\frac{4}{2}\right) \qquad \therefore N = 675 \text{ r.p.m.}$$

3. $\dfrac{N_2}{N_1} = \dfrac{E_{b2}}{E_{b1}} \times \dfrac{\phi_1}{\phi_2}$. As field current remains same, $\phi_1 = \phi_2$.

$$\therefore \quad \frac{N_2}{N_1} = \frac{E_{b2}}{E_{b1}}$$
$$E_{b1} = 200 - 30 \times 0{\cdot}5 = 185 \text{ V}; \; E_{b2} = 200 - 30\,R_t$$
$$\therefore \quad \frac{300}{500} = \frac{200 - 30R_t}{185} \qquad \therefore R_t = 2{\cdot}97\,\Omega$$

Additional resistance required $= R_t - R_a = 2{\cdot}97 - 0{\cdot}5 = 2{\cdot}47\,\Omega$

4. Since torque is same, armature current will be the same. Motor resistance is $R_m = 0{\cdot}3\,\Omega$. Let R ohm be the desired resistance to be added in series.

At starting :
$$I_a = \frac{V - E_b}{R_m + R} = \frac{V}{R_m + R} \qquad (\because E_b = 0)$$
or
$$40 = \frac{240}{0{\cdot}3 + R} \qquad \therefore R = 5{\cdot}7\,\Omega$$

5. At 1000 r.p.m. ;
$$\frac{E_{b2}}{E_{b1}} = \frac{N_2}{N_1} = \frac{1000}{1500} = \frac{1}{1{\cdot}5} \qquad (\because \text{flux is same})$$
$$\therefore \quad E_{b2} = \frac{E_{b1}}{1{\cdot}5} = \frac{240 - 40 \times 0{\cdot}3}{1{\cdot}5} = 152 \text{ V}$$

If R is the resistance to be added, then,
$$I_a = \frac{V - E_{b2}}{R_m + R} \quad \text{or} \quad 40 = \frac{240 - 152}{0{\cdot}3 + R} \qquad \therefore R = 1{\cdot}9\,\Omega$$

6. $E_b \propto \phi N$ or $E_b \propto N$ (∵ Armature current and hence ϕ is the same)
$$E_{b1} = 200 - 15 \times 1 = 185 \text{ V}; \; E_{b2} = 200 - 15\,(5 + 1) = 110 \text{ V}$$
Now
$$\frac{E_{b2}}{E_{b1}} = \frac{N_2}{N_1} \quad \text{or} \quad \frac{110}{185} = \frac{N_2}{800} \qquad \therefore N_2 = 476 \text{ r.p.m.}$$

7. When the flux is decreased, speed ($N \propto E_b/\phi$) of the motor increases.

Now
$$\frac{N_2}{N_1} = \frac{E_{b2}}{E_{b1}} \times \frac{\phi_1}{\phi_2}$$

Also, $T_a \propto \phi I_a$. As torque is the same in the two cases,

$$\therefore \quad \phi_1 I_{a1} = \phi_2 I_{a2} \text{ or } \phi_1 \times 50 = 0{\cdot}9\, \phi_1 \times I_{a2} \qquad \therefore\ I_{a2} = 50/0{\cdot}9 = 55{\cdot}55 \text{ A}$$

$$E_{b1} = 200 - 50 \times 0{\cdot}2 = 190 \text{ V};$$

$$E_{b2} = 200 - 55{\cdot}55 \times 0{\cdot}2 = 188{\cdot}9 \text{ V}$$

$$\therefore \quad \frac{N_2}{960} = \frac{188{\cdot}9}{190} \times \frac{\phi_1}{0{\cdot}9\, \phi_1} \qquad \therefore\ N_2 = 1058 \text{ r.p.m.}$$

9. This method gives speeds below the normal speed, not above it because armature voltage can be decreased (not increased) by the controller resistance.

12. The only disadvantage of this method is that commutation becomes unsatisfactory because the effect of armature reaction is greater on a weakened field.

7

Testing of D.C. Machines

INTRODUCTION

There are several tests that are conducted upon a d.c. machine (generator or motor) to judge its performance. One important test is performed to measure the efficiency of a d.c. machine. The efficiency of a d.c. machine depends upon its losses. The smaller the losses, the greater is the efficiency of the machine and *vice-versa*. The consideration of losses in a d.c. machine is important for two principal reasons. First, losses determine the efficiency of the machine and appreciably influence its operating cost. Secondly, losses determine the *heating of the machine and hence the power output that may be obtained without undue deterioration of the insulation. In this chapter, we shall focus our attention on the various methods for the determination of the efficiency of a d.c. machine.

7.1 EFFICIENCY OF A D.C. MACHINE

The power that a d.c. machine receives is called the input and the power it gives out is called the output. Therefore, the efficiency of a d.c. machine, like that of any energy-transferring device, is given by;

$$\text{Efficiency} = \frac{\text{Output}}{\text{Input}} \qquad ...(i)$$

$$\text{Output} = \text{Input} - \text{Losses} \quad \text{and} \quad \text{Input} = \text{Output} + \text{Losses}$$

Therefore, the efficiency of a d.c. machine can also be expressed in the following forms :

$$\text{Efficiency} = \frac{\text{Input} - \text{Losses}}{\text{Input}} \qquad ...(ii)$$

$$\text{Efficiency} = \frac{\text{Output}}{\text{Output} + \text{Losses}} \qquad ...(iii)$$

The most obvious method of determining the efficiency of a d.c. machine is to directly load it and measure the input power and output power. Then we can use eq. (*i*) to determine the efficiency of the machine. This method suffers from three main drawbacks. First, this method requires the application of load on the machine . Secondly, for machines of large rating, the loads of the required sizes may not be available. Thirdly, even if it is possible to provide such loads, large power will be dissipated, making it an expensive method.

The most common method of measuring the efficiency of a d.c. machine is to determine its losses (instead of measuring the input and output on load). We can then use eq. (*ii*) or eq. (*iii*) to determine the efficiency of the machine. This method has the obvious advantage of convenience and economy.

7.2 EFFICIENCY BY DIRECT LOADING

In this method, the d.c. machine is loaded and output and input are measured to find the efficiency. For this purpose, two simple methods can be used.

(*i*) **Brake test.** In this method, a brake is applied to a water-cooled pulley mounted on the motor shaft as shown in Fig. 7.1. One end of the rope is fixed to the floor *via* a spring balance *S* and

* The losses that occur in a d.c. machine are (*i*) copper losses (*ii*) iron losses or magnetic losses and (*iii*) mechanical losses. All these losses appear as heat and will thus raise the temperature of the machine.

a known mass is suspended at the other end. If the spring balance reading is S kg-wt and the suspended mass has a weight of W kg-wt, then,

Net pull on the rope = $(W - S)$ kg-wt = $(W - S) \times 9{\cdot}81$ newtons

If r is the radius of the pulley in metres, then the shaft torque T_{sh} developed by the motor is

$$T_{sh} = (W - S) \times 9{\cdot}81 \times r \text{ N-m}$$

If the speed of the pulley is N r.p.m., then,

$$\text{Output power} = \frac{2\pi N T_{sh}}{60} = \frac{2\pi N \times (W - S) \times 9{\cdot}81 \times r}{60} \text{ watts}$$

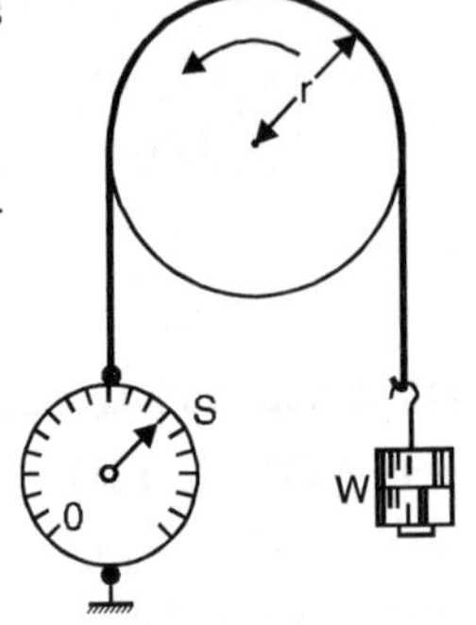

Fig. 7.1

Let V = Supply voltage in volts

I = Current taken by the motor in amperes

$\therefore$ Input to motor = VI watts

$$\therefore \quad \text{Efficiency} = \frac{2\pi N (W - S) \times r \times 9{\cdot}81}{60 \times VI}$$

(*ii*) In another method, the motor drives a calibrated generator *i.e.* one whose efficiency is known at all loads. The output of the generator is measured with the help of an ammeter and voltmeter.

$$\therefore \quad \text{Output of motor} = \frac{\text{Generator output}}{\text{Generator efficiency}}$$

Let V = Supply voltage is volts

I = Current taken by the motor is amperes

Input to motor = VI

Thus efficiency of the motor can be determined.

Because of several disadvantages (See Art. 7.1), direct loading method is used only for determining the efficiency of small machines.

Example 7.1. *In a brake-test on a small shunt motor, the speed was 1500 r.p.m., the load on one side of the brake band was 28·9 N and on the other side 1·67 N. The diameter of the brake pulley was 15·2 cm. If the input current was 2 A at 250 V, calculate (i) the torque (ii) brake horse power and (iii) the efficiency.*

Solution.

(*i*) $$\text{Torque, } T_{sh} = (W - S) \times r = (28{\cdot}9 - 1{\cdot}67) \times \frac{0{\cdot}152}{2} = \mathbf{2{\cdot}07 \text{ N-m}}$$

(*ii*) $$\text{Output power} = \frac{2\pi N T_{sh}}{60} = \frac{2\pi \times 1500 \times 2{\cdot}07}{60} = 325 \text{ W}$$

$$\text{Brake horse power} = 325/746 = \mathbf{0{\cdot}436 \text{ h.p.}}$$

(*iii*) $$\text{Input to motor} = VI = 250 \times 2 = 500 \text{ W}$$

$$\text{Efficiency} = \frac{\text{Output}}{\text{Input}} \times 100 = \frac{325}{500} \times 100 = \mathbf{65\%}$$

Example 7.2. *The following readings were obtained when doing a load test on a d.c. shunt motor using a brake drum :*

Spring balance readings = 10 kg and 35 kg ; Diameter of drum = 40 cm;

Speed of the motor = 950 r.p.m. ; Applied voltage = 200 V; Line current = 30 A

Calculate (i) the output power (ii) efficiency of the motor.

Solution.

(*i*) Force on the drum, $F = (35 - 10) = 25 \text{ kg-wt} = 25 \times 9{\cdot}8 \text{ N}$

Shaft torque, $T_{sh} = F \times r = 25 \times 9{\cdot}8 \times 0{\cdot}2 = 49 \text{ N-m}$

$$\text{Motor output} = \frac{2\pi N T_{sh}}{60} = \frac{2\pi \times 950 \times 49}{60} = \mathbf{4876\ W}$$

(*ii*) $$\text{Input to motor} = VI = 200 \times 30 = 6000\text{ W}$$

$$\therefore \quad \text{Efficiency} = \frac{\text{Output}}{\text{Input}} \times 100 = \frac{4876}{6000} \times 100 = \mathbf{81.3\%}$$

Example 7.3. *A 120 V, 0.75 h.p. motor is tested at 2400 r.p.m. using the prony brake test. The input current is 7A and the deflection force on the spring is 4.57 N. The effective length of the torque arm is 50 cm and its dead weight is 0.03 N. Determine the torque and the efficiency of the motor.*

Solution.

$$\text{Net force on the spring, } F = 4.57 - 0.03 = 4.54\text{ N}$$

$$\text{Shaft torque of motor, } T_{sh} = F \times 0.5 = 4.54 \times 0.5 = \mathbf{2.27\ Nm}$$

$$\text{Angular velocity of shaft, } \omega_m = \frac{2\pi N}{60} = \frac{2\pi \times 2400}{60} = 251.3\text{ rad/s}$$

$$\text{Output power, } P_{out} = \omega_m T_{sh} = 251.3 \times 2.27 = 570.52\text{ W}$$

$$\text{Input power, } P_{in} = VI = 120 \times 7 = 840\text{ W}$$

$$\therefore \quad \text{Motor efficiency, } \eta = \frac{P_{out}}{P_{in}} \times 100 = \frac{570.52}{840} \times 100 = \mathbf{67.9\%}$$

TUTORIAL PROBLEMS

1. In a brake test on a d.c. motor, the effective load on the brake drum was 23 kg-wt, the effective diameter of the drum 45 cm and the speed 960 r.p.m. The input to the motor was 28 A at 230 V. Calculate (*i*) the efficiency (*ii*) the brake horse power of the motor. **[(*i*) 79·3% (*ii*) 6·85 h.p.]**
2. The brake test on a d.c. shunt motor gave the following results : tensions 3·2 kg and 0·3 kg, radius of pulley 10 cm, speed 1500 r.p.m., V = 230 volts, line current = 2·8A. Find (*i*) output torque (*ii*) output power (*iii*) horse power and (*iv*) efficiency. **[(*i*) 2·845 N-m (*ii*) 446·9 W (*iii*) 0·608 h.p. (*iv*) 69·4%]**
3. In a brake test, the effective load on the pulley was 38·1 kg, the effective diameter of the pulley 63·5 cm and speed 720 r.p.m. The motor took 49 A at 220 V. Calculate the output power and the efficiency at this load. **[8945 W; 83%]**

7.3 SWINBURNE'S METHOD FOR DETERMINING EFFICIENCY

In this method, the d.c. machine (generator or motor) is run as a motor at no-load and losses of the machine are determined. Once the losses of the machine are known, its efficiency at any desired load can be determined in advance. *It may be noted that this method is applicable to those machines in which flux is practically *constant at all loads e.g., shunt and compound machines.* Let us see how the efficiency of a d.c. shunt machine (generator or motor) is determined by this method. The test consists of two steps :

(*i*) **Determination of hot resistances of windings.** The armature resistance and shunt field resistance are measured using a battery, voltmeter and ammeter. Since these resistances are measured when the machine is cold, they must be converted to values corresponding to the temperature at which the machine would work on full-load. Generally, these values are measured for a temperature rise of 40° C above the room temperature. Let the hot resistances of armature and shunt field be R_a and R_{sh} respectively.

---

* This method cannot be used for d.c. series machines for two reasons. First, this test requires the d.c. machine to be run as motor on no-load. A d.c. series motor should not be run at no-load because at no-load, its speed is very high and may result in damage to mechanical parts of the machine and its foundation. Secondly, in series d.c. machine, iron and friction losses are not constant but vary with load.

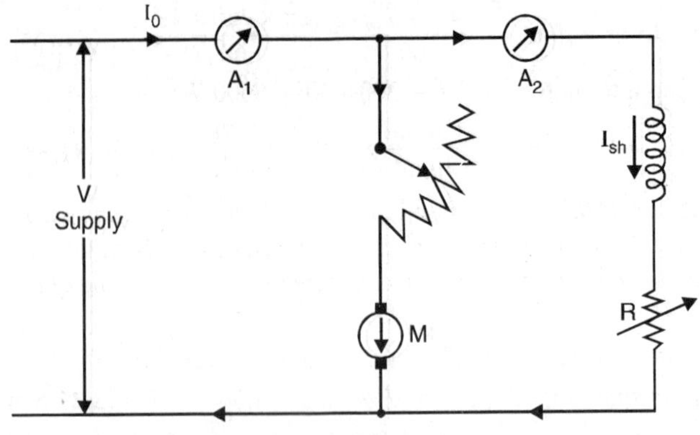

Fig. 7.2

(*ii*) **Determination of constant losses.** The machine is run as a motor on no-load with supply voltage adjusted to the rated voltage *i.e.* voltage stamped on the nameplate. The speed of the motor is adjusted to the rated speed with the help of field regulator R as shown is Fig. 7.2.

Let V = Supply voltage

I_0 = No-load current read by ammeter A_1

I_{sh} = Shunt-field current read by ammeter A_2

$\therefore$ No-load armature current, $I_{a0} = I_0 - I_{sh}$

No-load input power to motor $= V I_0$

No-load power input to armature $= V I_{a0} = V(I_0 - I_{sh})$

Since the output of the motor is zero, the no-load input power to the armature supplies (*a*) iron losses in the core (*b*) friction loss (*c*) windage loss (*d*) armature Cu loss $[I_{a0}^2 R_a$ or $(I_0 - I_{sh})^2 R_a]$.

*Constant losses, W_C = Input to motor – Armature Cu loss

or $W_C = V I_0 - (I_0 - I_{sh})^2 R_a$

Since constant losses are known, the efficiency of the machine at any other load can be determined. Suppose it is desired to determine the efficiency of the machine at load current I. Then,

Armature current, $I_a = I - I_{sh}$ ... if the machine is motoring

$= I + I_{sh}$ ... if the machine is generating

Efficiency when running as a motor

Input power to motor $= V I$

Armature Cu loss $= I_a^2 R_a = (I - I_{sh})^2 R_a$

Constant losses $= W_C$ ..... found above

Total losses $= (I - I_{sh})^2 R_a + W_C$

$$\therefore \quad \text{Motor efficiency, } \eta_m = \frac{\text{Input} - \text{Losses}}{\text{Input}} = \frac{V I - (I - I_{sh})^2 R_a - W_C}{V I}$$

Efficiency when running as a generator

Output of generator $= V I$

Armature Cu loss $= (I + I_{sh})^2 R_a$

Constant losses $= W_C$ ..... found above

Total losses $= (I + I_{sh})^2 R_a + W_C$

---

* Note that W_C = shunt field loss + friction and windage losses + iron losses. These losses remain the same at all loads and hence are called constant losses. Note that only loss which varies with load is the armature Cu loss.

$\therefore$ Generator efficiency, $\eta_g = \dfrac{\text{Output}}{\text{Output + Losses}} = \dfrac{V\,I}{V\,I + (I + I_{sh})^2 R_a + W_C}$

Advantages of Swinburne's test

(*i*) The power required to carry out the test is small because it is a no-load test. Therefore, this method is quite economical.

(*ii*) The efficiency can be determined at any load because constant losses are known.

(*iii*) This test is very convenient.

Disadvantages of Swinburne's test

(*i*) It does not take into account the *stray load losses that occur when the machine is loaded.

(*ii*) This test does not enable us to check the performance of the machine on full-load. For example, it does not indicate whether commutation on full-load is satisfactory and whether the temperature rise is within the specified limits.

(*iii*) This test does not give quite accurate efficiency of the machine. It is because iron losses under actual load are greater than those measured. This is mainly due to armature reaction distorting the field.

Example 7.4. *A 220 V d.c. shunt motor at no-load takes a current of 2·5 A. The resistances of the armature and shunt field are 0·8 Ω and 200 Ω respectively. Estimate the efficiency of the motor when input current is 32 A. State the assumptions made.*

Solution.

No-load input power $= V I_0 = 220 \times 2{\cdot}5 = 550$ W

$I_{sh} = 220/200 = 1{\cdot}1$ A ; No-load armature current, $I_{a0} = 2{\cdot}5 - 1{\cdot}1 = 1{\cdot}4$ A

No-load armature Cu loss $= I_{a0}^2 R_a = (1{\cdot}4)^2 \times 0{\cdot}8 = 1{\cdot}6$ W

Constant losses $= 550 - 1{\cdot}6 = 548{\cdot}4$ W

When input current is 32 A

Armature current, $I_a = 32 - 1{\cdot}1 = 30{\cdot}9$ A

Armature Cu loss $= I_a^2 R_a = (30{\cdot}9)^2 \times 0{\cdot}8 = 764$ W

Total losses $=$ Constant losses $+ I_a^2 R_a = 548{\cdot}4 + 764 = 1312{\cdot}4$ W

Input power $= 220 \times 32 = 7040$ W

Output power $= 7040 - 1312{\cdot}4 = 5727{\cdot}6$ W

Efficiency, $\eta = \dfrac{5727{\cdot}6}{7040} \times 100 = \mathbf{81.36\%}$

For assumptions, the reader is advised to refer to Art. 7.3.

Example 7.5. *A 100 kW, 460 V shunt generator was run as a motor on no-load at its rated voltage and speed. The total current taken was 9·8 A including a shunt current of 2·7 A. The resistance of the armature circuit (including compoles) at normal working temperature was 0·11 Ω. Calculate the efficiency at (i) full-load (ii) half full-load.*

Solution.

No-load input power $= V I_0 = 460 \times 9{\cdot}8 = 4508$ W

No-load armature current, $I_{a0} = 9{\cdot}8 - 2{\cdot}7 = 7{\cdot}1$ A

No-load armature Cu loss $= I_{a0}^2 R_a = (7{\cdot}1)^2 \times 0{\cdot}11 = 5{\cdot}5$ W

Constant losses $= 4508 - 5{\cdot}5 = 4502{\cdot}5$ W

---

* These are extra losses due to eddy currents in the conductors and due to changes in the magnetic field. These losses occur only on load.

(*i*) **Efficiency at full-load**

$$\text{Output current at full-load} = \frac{100 \times 10^3}{460} = 217{\cdot}5 \text{ A}$$

$$\text{Armature current at full-load, } I_a = 217{\cdot}5 + 2{\cdot}7 = 220{\cdot}2 \text{ A}$$

$$\text{Armature Cu loss at full-load} = I_a^2 R_a = (220{\cdot}2)^2 \times 0{\cdot}11 = 5325 \text{ W}$$

$$\text{Total losses at full-load} = \text{Constant losses} + \text{Arm. Cu loss at full-load}$$

$$= 4502{\cdot}5 + 5325 = 9827{\cdot}5 \text{ W}$$

$$= 9{\cdot}83 \text{ kW (app.)}$$

$$\text{Input power at full-load} = 100 + 9{\cdot}83 = 109{\cdot}83 \text{ kW}$$

$$\therefore \quad \text{Efficiency at full-load} = \frac{100}{109{\cdot}83} \times 100 = \mathbf{91{\cdot}1\%}$$

(*ii*) **Efficiency at half full-load**

$$\text{Output current at half full-load} = 217{\cdot}5/2 = 108{\cdot}75 \text{ A}$$

$$\text{Armature current at half full-load, } I_a = 108{\cdot}75 + 2{\cdot}7 = 111{\cdot}45 \text{ A}$$

$$\text{Armature Cu loss at half full-load} = I_a^2 R_a = (111{\cdot}45)^2 \times 0{\cdot}11 = 1365 \text{ W}$$

$$\text{Total losses at half full-load} = \text{Constant losses} + \text{Arm. Cu loss at half full-load}$$

$$= 4502{\cdot}5 + 1365 = 5867{\cdot}5 \text{ W} = 5{\cdot}87 \text{ kW (app.)}$$

$$\text{Input power at half full-load} = 50 + 5{\cdot}87 = 55{\cdot}87 \text{ kW}$$

$$\therefore \quad \text{Efficiency at half full-load} = \frac{50}{55{\cdot}87} \times 100 = \mathbf{89{\cdot}5\%}$$

Example 7.6. *A d.c. shunt motor gave the following results of Swinburne test :*

Voltage : 600 V ; Current : 8A; Armature resistance : 0·1 Ω; Field resistance : 300 Ω

Find the efficiency of the machine (i) as a generator delivering 100 A at 600 V (ii) as a motor having a line current of 100 A at 600 V. Neglect increase in resistance due to rise in temperature. Assume stray load losses as 1% of output.

Solution.

$$\text{No-load input power} = V I_0 = 600 \times 8 = 4800 \text{ W}$$

$$\text{Shunt field current, } I_{sh} = 600/300 = 2 \text{ A}$$

$$\text{No-load armature current, } I_{a0} = 8 - 2 = 6 \text{ A}$$

$$\text{No-load armature Cu loss} = I_{a0}^2 \times R_a = (6)^2 \times 0{\cdot}1 = 3{\cdot}6 \text{ W}$$

$$\text{Constant losses} = 4800 - 3{\cdot}6 = 4796{\cdot}4 \text{ W}$$

(*i*) **As a generator**

$$\text{Output current} = 100 \text{ A}$$

$$\text{Armature current, } I_a = 100 + 2 = 102 \text{ A}$$

$$\text{Armature Cu loss} = I_a^2 R_a = (102)^2 \times 0{\cdot}1 = 1040{\cdot}4 \text{ W}$$

$$\text{Stray load losses} = 1\% \text{ of output} = \frac{1}{100} \times 600 \times 100 = 600 \text{ W}$$

$$\text{Total losses} = \text{Constant losses} + \text{Stray load losses} + \text{Arm. Cu loss}$$

$$= 4796{\cdot}4 + 600 + 1040{\cdot}4 = 6436{\cdot}8 \text{ W}$$

$$\text{Output of generator} = 600 \times 100 = 60000 \text{ W}$$

$$\therefore \quad \text{Generator efficiency} = \frac{\text{Output}}{\text{Output + Losses}} \times 100 = \frac{60000}{60000 + 6436{\cdot}8} \times 100 = \mathbf{90{\cdot}3\%}$$

(*ii*) **As a motor**

$$\text{Input power} = 600 \times 100 = 60000 \text{ W}$$

$$\text{Armature current, } I_a = 100 - 2 = 98 \text{ A}$$

$$\text{Armature Cu loss} = I_a^2 R_a = (98)^2 \times 0{\cdot}1 = 960{\cdot}4 \text{ W}$$

$$\text{Output} = 60000 - 4796{\cdot}4 - 960{\cdot}4 = 54243{\cdot}4 \text{ W}$$

$$\text{Stray load losses} = \frac{1}{100} \times 54243{\cdot}4 = 542{\cdot}4 \text{ W}$$

$$\text{Net output} = 54243{\cdot}4 - 542{\cdot}4 = 53701 \text{ W}$$

$$\therefore \quad \text{Motor efficiency} = \frac{53701}{60000} \times 100 = \mathbf{89{\cdot}5\%}$$

Example 7.7. *The no-load test of a 44·76 kW, 220 V d.c. shunt motor gave the following figures: Input current = 13.25 A ; Field current = 2.55 A*
Resistance of armature at 75°C = 0·032 Ω ; brush drop = 2 V
Find (i) full-load current (ii) full-load efficiency.

Solution.

$$\text{No-load input power} = V I_0 = 220 \times 13{\cdot}25 = 2915 \text{ W}$$

$$\text{No-load armature current, } I_{a0} = 13{\cdot}25 - 2{\cdot}55 = 10{\cdot}7 \text{ A}$$

$$\text{No-load armature Cu loss} = I_{a0}^2 R_a = (10{\cdot}7)^2 \times 0{\cdot}032 = 3{\cdot}6 \text{ W}$$

$$\text{Loss due to brush drop} = \text{Brush drop} \times I_{a0} = 2 \times 10{\cdot}7 = 21{\cdot}4 \text{ W}$$

$$\therefore \quad \text{Constant losses} = 2915 - 3{\cdot}6 - 21{\cdot}4 = 2890 \text{ W}$$

Full-load conditions

Let I_a be the full-load armature current.

$$\therefore \quad \text{F.L. line current} = (I_a + 2{\cdot}55) \text{ A}$$

$$\text{F.L. input power to motor} = V \times (I_a + 2{\cdot}55) \text{ W}$$

$$\text{Total F.L. losses} = \text{Constant losses} + \text{Brush loss} + \text{Arm. Cu loss}$$

$$= 2890 + 2\,I_a + 0{\cdot}032\,I_a^2$$

$$\text{F.L. input power} = \text{F.L. output power} + \text{total F.L. losses}$$

$$\text{or} \quad V \times (I_a + 2{\cdot}55) = 44760 + 2890 + 2I_a + 0{\cdot}032\,I_a^2$$

$$\text{or} \quad 0{\cdot}032\,I_a^2 - 218\,I_a + 47090 = 0$$

$$\text{On solving, we get, } I_a = 223{\cdot}5 \text{ A}$$

$$(i) \quad \therefore \quad \text{F.L. line current} = I_a + I_{sh} = 223{\cdot}5 + 2{\cdot}55 = \mathbf{226 \text{ A}}$$

$$\text{F.L. input power} = 220 \times 226 = 49720 \text{ W}$$

$$(ii) \quad \text{F.L. efficiency} = \frac{44760}{49720} \times 100 = \mathbf{90\%}$$

Example 7.8. *A 50 kW, 440 V shunt generator having an armature circuit resistance including inter-pole winding of 0·15 Ω at normal working temperature was run as a shunt motor on no-load at rated voltage and speed. The total current drawn by the motor was 5 A including shunt field current of 1·5 A. Calculate the efficiency of the shunt generator at 3/4th full-load.*

Solution.

No-load conditions

$$\text{No-load input power} = V I_0 = 440 \times 5 = 2200 \text{ W}$$

$$\text{No-load armature current, } I_{a0} = 5 - 1{\cdot}5 = 3{\cdot}5 \text{ A}$$

$$\text{No-load armature Cu loss} = I_{a0}^2 R_a = (3{\cdot}5)^2 \times 0{\cdot}15 = 2 \text{ W}$$

$$\text{Constant losses} = 2200 - 2 = 2198 \text{ W}$$

3/4th Full-load conditions

$$\text{F.L. output current of generator} = \frac{50 \times 10^3}{440} = 113{\cdot}6 \text{ A}$$

Generator output current at 3/4th F.L. $= \frac{3}{4} \times 113 \cdot 6 = 85 \cdot 2$ A

Armature current, $I_a = 85{\cdot}2 + 1{\cdot}5 = 86{\cdot}7$ A

Armature Cu loss $= I_a^2 R_a = (86{\cdot}7)^2 \times 0{\cdot}15 = 1127{\cdot}5$ W

Total losses = Constant losses + Arm. Cu loss

$= 2198 + 1127{\cdot}5 = 3225{\cdot}5$ W

Generator output at 3/4th F.L. $= (3/4) \times 50 = 37{\cdot}5$ kW

Efficiency of the generator at 3/4th full-load

$$= \frac{\text{Output}}{\text{Output + Losses}} \times 100 = \frac{37 \cdot 5 \times 10^3}{37 \cdot 5 \times 10^3 + 3225 \cdot 5} \times 100$$

= **91·85%**

Example 7.9. *In a test on a d.c. shunt generator whose full-load output is 200 kW at 250 V, the following figures were obtained:*

(*i*) *When running light as a motor at full-speed, the line current was 36 A, the field current 12 A and the supply voltage 250 V.*

(*ii*) *With the machine at rest, a p.d. of 6 V produced a current of 400 A through the armature circuit.*

Explain how these results may be utilised to obtain the efficiency of the generator at full-load and half full-load. Neglect voltage drop at the brushes.

Solution.

At no-load

No-load input power $= VI_0 = 250 \times 36 = 9000$ W

No-load armature current, $I_{a0} = 36 - 12 = 24$ A

Armature resistance, $R_a = 6/400 = 0{\cdot}015\ \Omega$

No-load armature Cu loss $= I_{a0}^2 R_a = (24)^2 \times 0{\cdot}015 = 8{\cdot}64$ W

Constant losses $= 9000 - 8{\cdot}64 = 8991{\cdot}4$ W

At full-load

F.L. output current $= \frac{200 \times 10^3}{250} = 800$ A

F.L. armature current, $I_a = 800 + 12 = 812$ A

F.L. armature Cu loss $= I_a^2 R_a = (812)^2 \times 0{\cdot}015 = 9890$ W

F.L. total losses = Constant losses + F.L. armature Cu loss

$= 8991{\cdot}4 + 9890 = 18881$ W

∴ F.L. efficiency $= \frac{200 \times 10^3}{200 \times 10^3 + 18881} \times 100 =$ **91·4%**

At half full-load

Output current $= \frac{100 \times 10^3}{250} = 400$ A

Armature current $= 400 + 12 = 412$ A

Armature Cu loss $= (412)^2 \times R_a = (412)^2 \times 0{\cdot}015 = 2546$ W

Total losses $= 2546 + 8991{\cdot}4 = 11537$ W

∴ Efficiency $= \frac{100 \times 10^3}{100 \times 10^3 + 11537} \times 100 =$ **89·6%**

TUTORIAL PROBLEMS

1. The current taken by a 460 V shunt motor when running light is 7A. The resistance of the armature circuit is 0·15 Ω and that of the field circuit is 230 Ω. Calculate the output (in horsepower) and the efficiency when the current taken is 130 A. **[72·6 h.p.; 90·5%]**
2. A 200 V shunt motor develops an output of 17·158 kW when taking 20·2 kW. The field resistance is 50 Ω and armature resistance 0·06 Ω. What is the efficiency and power input when the output is 7·46 kW ? **[74·6%; 10 kW]**
3. A 230 V shunt motor running on no-load and at normal speed takes an armature current of 2·5 A from 230 V mains. The field circuit resistance is 230 Ω and the armature-circuit resistance is 0·3 Ω. Calculate the motor output and efficiency when total current taken from the mains is 35 A.

   If the motor is used as a 230 V shunt generator, find the efficiency and the input power for an output current of 35 A. [*Motor* : **9·25 h.p. ; 85·7%** *Generator*: **87·1%; 9·24 kW]**

7.4 REGENERATIVE OR HOPKINSON'S TEST

This method of determining the efficiency of a d.c. machine saves power and gives more accurate results. In order to carry out this test, we require two identical d.c. machines and a source of electrical power.

Principle. Two identical d.c. shunt machines are mechanically coupled and connected in parallel across the d.c. supply. By *adjusting the field excitations of the machines, one is run as a motor and the other as a generator. The electric power from the generator and electrical power from the d.c. supply are fed to the motor. The electric power given to the motor is mostly converted into mechanical power, the rest going to the various motor losses. This mechanical power is given to the generator. The electrical power of the generator is given to the motor except that which is wasted as generator losses. Thus the electrical power taken from the d.c. supply is the sum of motor and generator losses and this can be measured directly by a voltmeter and an ammeter. Since the power input from the d.c. supply is equal to the power required to supply the losses of the two machines, this test can be carried out with a small amount of power. By adjusting the field strengths of the machines, any load can be put on the machines. Therefore, we can measure the total loss of the machines at any load. Since the machines can be tested under full-load conditions (of course at the expense of power equal to the losses in the two machines), the temperature rise and commutation qualities of the machines can be observed.

Circuit. Fig. 7.3 shows the essential connections for Hopkinson's test. Two identical d.c. shunt machines are mechanically coupled and are connected in parallel across the d.c. supply. By **adjusting the field strenghts of the two machines, the machine M is made to run as a motor and machine G as a generator. The motor M draws current I_1 from the †generator G and current I_2 from the d.c. supply so that input current to motor M is $(I_1 + I_2)$. Power taken from the d.c. supply is VI_2 and is equal to the total motor and generator losses. The field current of motor M is I_4 and that of generator G is I_3.

* The machine whose field is strengthened acts as a generator while whose field is weakened acts as a motor.

** To start the test, machine M is started up as motor with paralleling switch S open. The speed of M is adjusted to normal value by means of its field regulator. The machine M drives the machine G as a generator. The excitation of G is adjusted (G is yet to be paralleled with M) till its generated voltage is equal to the supply voltage. Under such conditions, the voltmeter V_1 will read zero. In this condition, the machine G is floating *i.e.*, neither giving nor taking any current from the line. Now switch S is closed to parallel the machines. By increasing the field current of G or by decreasing the field current of M or by both adjustments, any desired load can be put on the generator G.

† Note that current I_1 supplied by generator G is much greater than the current I_2 supplied by the d.c. supply. It is because the output of the generator (= generator input – generator losses) is entirely supplied to the motor.

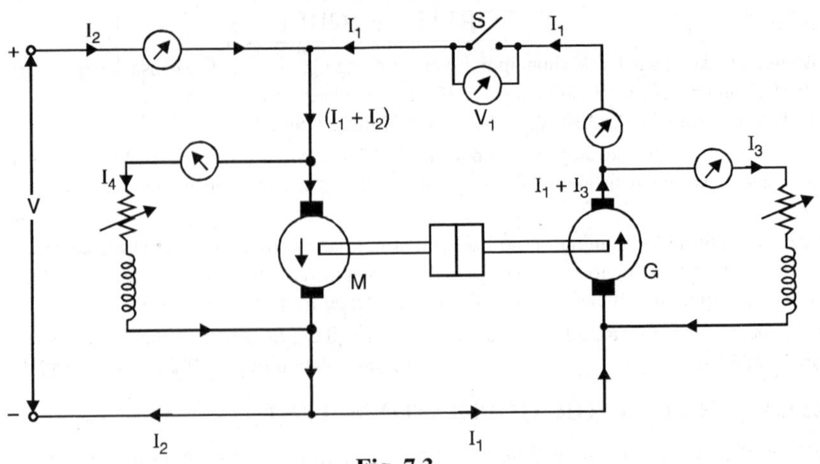

Fig. 7.3

Calculations. If V be the supply voltage, then,

$$\text{Motor input} = V(I_1 + I_2)$$

$$\text{Generator output} = VI_1$$

We shall find the efficiencies of the machines considering two cases *viz.* (*i*) assuming that both machines have the same efficiency η (*ii*) assuming iron, friction and windage losses are the same in both machines.

(*i*) **Assuming that both machines have the same efficiency η.**

Motor output $= \eta \times \text{motor input} = \eta V(I_1 + I_2) = \text{Generator input}$

Generator output $= \eta \times \text{generator input} = \eta \times \eta V(I_1 + I_2) = \eta^2 V(I_1 + I_2)$

But generator output is VI_1.

$$\therefore \quad \eta^2 V(I_1 + I_2) = VI_1$$

or

$$\eta = \sqrt{\frac{I_1}{I_1 + I_2}}$$

This expression gives the value of efficiency sufficiently accurate for a rough test. However, if greater accuracy is required, the efficiencies of the two machines should be calculated separately as indicated below.

(*ii*) **Assuming that iron, friction and windage losses are same in both machines.** It is not correct to assume that the two machines have the same efficiency. It is because armature and field losses in the two machines are not the same. However, iron, friction and windage losses in the two machines will be the same because the machines are identical. On this assumption, we can find the efficiency of each machine as under :

Let

R_a = armature resistance of each machine

I_3 = field current of generator G

I_4 = field current of motor M

Armature Cu loss in generator $= (I_1 + I_3)^2 R_a$

Armature Cu loss in motor $= (I_1 + I_2 - I_4)^2 R_a$

Shunt Cu loss in generator $= VI_3$

Shunt Cu loss in motor $= VI_4$

Power drawn from the d.c. supply is VI_2 and is equal to the *total losses* of the motor and generator

i.e.

$$VI_2 = \text{Total losses of motor and generator}$$

If we subtract armature and shunt Cu losses of the two machines from VI_2, we get iron, friction and windage losses of the two machines.

Iron, friction and windage losses of two machines (*M* and *G*)

$$= VI_2 - [(I_1 + I_3)^2 R_a + (I_1 + I_2 - I_4)^2 R_a + VI_3 + VI_4] = W \text{(say)}$$

∴ Iron, friction and windage losses of *each* machine = $W/2$

For generator

$$\text{Output of generator} = VI_1$$

$$\text{Total losses} = \frac{W}{2} + (I_1 + I_3)^2 R_a + VI_3 = W_g \text{ (say)}$$

$$\therefore \quad \text{Generator efficiency, } \eta_g = \frac{VI_1}{VI_1 + W_g}$$

For motor

$$\text{Input to motor} = V(I_1 + I_2)$$

$$\text{Total losses} = (I_1 + I_2 - I_4)^2 R_a + VI_4 + \frac{W}{2} = W_m \text{ (say)}$$

$$\therefore \quad \text{Motor efficiency, } \eta_m = \frac{\text{Input} - \text{Losses}}{\text{Input}} = \frac{V(I_1 + I_2) - W_m}{V(I_1 + I_2)}$$

7.5 ALTERNATE CONNECTIONS FOR HOPKINSON'S TEST

Fig. 7.4 shows the alternative *connections for Hopkinson's test. The main difference is that now the shunt field windings are directly connected across the lines. Therefore, the input line current is I_1, *excluding the field currents*. The power VI_1 drawn from the d.c. supply is equal to the total losses of the two machines *except* the shunt field losses of the two machines *i.e.*,

VI_1 = Total losses of the two machines except shunt field losses of the two machines

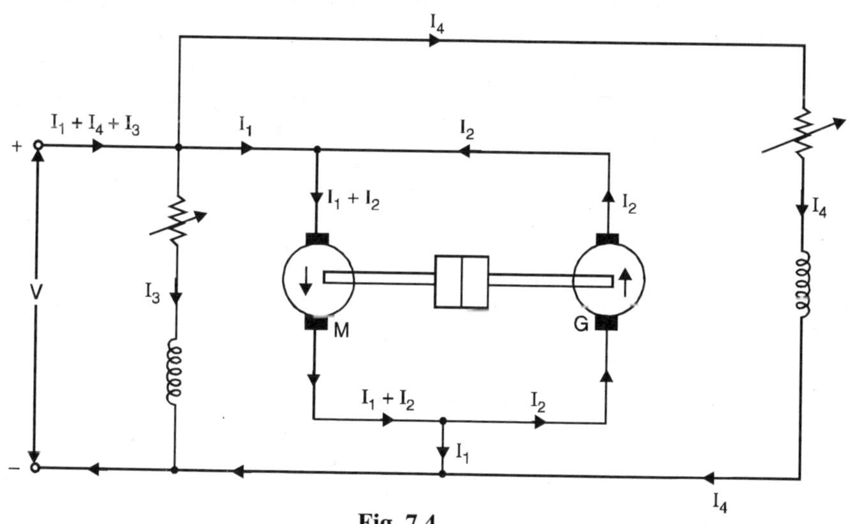

Fig. 7.4

$$\text{Motor armature Cu loss} = (I_1 + I_2)^2 R_a$$

$$\text{Generator armature Cu loss} = I_2^2 R_a$$

* Note that instruments and other controls have been omitted for the clarity of the figure.

Iron, friction and windage losses of the two machines are VI_1 minus armature Cu losses of the two machines *i.e.*

Iron, friction and windage losses of the two machines

$$= VI_1 - [(I_1 + I_2)^2 R_a + I_2^2 R_a] = W \text{ (say)}$$

Iron, friction and windage losses of each machine $= W/2$

Motor efficiency

$$\text{Motor input, } P_i = V(I_1 + I_2 + I_3)$$

$$\text{Motor losses} = (I_1 + I_2)^2 R_a + VI_3 + W/2 = W_m \text{ (say)}$$

$$\therefore \quad \text{Motor efficiency, } \eta_m = \frac{\text{Motor input} - \text{Losses}}{\text{Motor input}} = \frac{P_i - W_m}{P_i}$$

Generator efficiency

$$\text{Generator output} = VI_2$$

$$\text{Generator losses} = I_2^2 R_a + VI_4 + W/2 = W_g \text{ (say)}$$

$$\therefore \quad \text{Generator efficiency, } \eta_g = \frac{VI_2}{VI_2 + W_g}$$

7.6 ADVANTAGES OF HOPKINSON'S TEST

The advantages of Hopkinson's test are :

(*i*) The total power required to test the two machines is small compared with the full-load power of each machine.

(*ii*) The machines can be tested under full-load conditions so that commutation qualities and temperature rise can be checked.

(*iii*) It is more accurate to measure the loss directly than to measure it as the difference of the measured input and output.

(*iv*) All the measurements are electrical which are simpler and more accurate than mechanical measurements.

The main disadvantage is that two similar d.c. machines are required.

Example 7.10. *In a Hopkinson's test on two 440 V, 200 kW generators, the circulating current is equal to the full-load current, and in addition, 90 A are taken from the supply. Find the approximate efficiency of each machine.*

Solution.

$$\text{Full-load output of generator} = \frac{200 \times 10^3}{440} = 454{\cdot}5 \text{ A}$$

$$\therefore \quad \text{Circulating current, } I_1 = 454{\cdot}5 \text{ A}$$

$$\text{Current from supply, } I_2 = 90 \text{ A}$$

Approximate efficiency of each machine is

$$\eta = \sqrt{\frac{I_1}{I_1 + I_2}} = \sqrt{\frac{454 \cdot 5}{454 \cdot 5 + 90}} = \mathbf{0{\cdot}914} \textit{ or } \mathbf{91{\cdot}4\%}$$

Example 7.11. *The Hopkinson's test on two identical shunt machines gave the following results:*

Input voltage = 500 V ; Input current = 15 A

Output current of generator = 120 A ; Field current of generator = 4 A

Field current of motor = 3 A ; Armature resistance of each machine = 0·06 Ω

Find the efficiency of motor and generator.

Solution. Fig. 7.5 shows the motor-generator set. We shall find the armature Cu loss and shunt Cu loss for both machines.

Motor

$$\text{Armature current} = 120 + 15 - 3 = 132 \text{ A}$$

$$\text{Armature Cu loss} = (132)^2 \times 0{\cdot}06 = 1045{\cdot}44 \text{ W}$$

$$\text{Shunt Cu loss} = 500 \times 3 = 1500 \text{ W}$$

$$\text{Total Cu loss} = 1045{\cdot}44 + 1500 = 2545{\cdot}44 \text{ W}$$

Generator

$$\text{Armature current} = 120 + 4 = 124 \text{ A}$$

$$\text{Armature Cu loss} = (124)^2 \times 0{\cdot}06 = 922{\cdot}56 \text{ W}$$

$$\text{Shunt Cu loss} = 500 \times 4 = 2000 \text{ W}$$

$$\text{Total Cu loss} = 922{\cdot}56 + 2000 = 2922{\cdot}56 \text{ W}$$

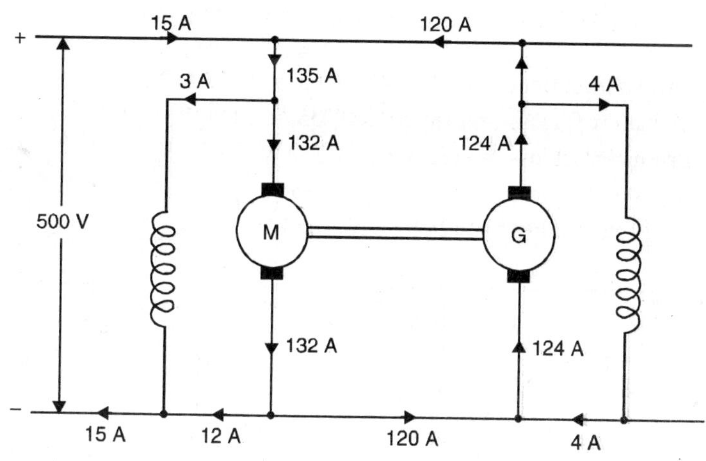

Fig. 7.5

For the set

$$\text{Total arm. and shunt Cu loss for the set} = 2545{\cdot}44 + 2922{\cdot}56 = 5468 \text{ W}$$

$$\text{Total input} = 500 \times 15 = 7500 \text{ W}$$

$$\therefore \quad \text{Iron, friction and windage losses} = 7500 - 5468 = 2032 \text{ W}$$

$$\text{Iron, friction and windage losses in each machine} = 2032/2 = 1016 \text{ W}$$

Motor efficiency

$$\text{Total motor losses} = \text{Total Cu losses + Iron, friction and windage losses}$$

$$= 2545{\cdot}44 + 1016 = 3561{\cdot}44 \text{ W}$$

$$\text{Motor input} = 500 \times 135 = 67500 \text{ W}$$

$$\therefore \quad \text{Motor efficiency} = \frac{67500 - 3561{\cdot}44}{67500} \times 100 = \mathbf{94{\cdot}7\%}$$

Generator efficiency

$$\text{Total generator losses} = \text{Total Cu losses + Iron, friction and windage losses}$$

$$= 2922{\cdot}56 + 1016 = 3938{\cdot}56 \text{ W}$$

$$\text{Generator output} = 500 \times 120 = 60000 \text{ W}$$

$$\therefore \quad \text{Generator efficiency} = \frac{60000}{60000 + 3938{\cdot}56} \times 100 = \mathbf{93{\cdot}8\%}$$

Example 7.12. *The Hopkinson's test on two similar d.c. shunt machines gave the following full-load data :*

Line voltage = 110 V ; *Line current = 48 A*

Motor armature current = 230 A ; *Field currents are 3 A and 3·5 A*

The armature resistance of each machine is 0·035 Ω. Calculate the efficiency of each machine assuming a brush contact drop of 1V per brush.

Solution. Fig. 7.6 shows the connections for Hopkinson's test on two similar d.c. shunt machines. The machine with greater excitation will act as a generator and the other as a motor. We shall find the armature Cu loss, brush contact loss and shunt Cu loss for both machines.

Motor

$$\text{Armature Cu loss} = (230)^2 \times 0{\cdot}035 = 1851{\cdot}5 \text{ W}$$

$$\text{Brush contact loss} = 230 \times 2 = 460 \text{ W}$$

$$\text{Shunt Cu loss} = 110 \times 3 = 330 \text{ W}$$

$$\text{Total Cu loss} = 1851{\cdot}5 + 460 + 330 = 2641{\cdot}5 \text{ W}$$

Generator

$$\text{Armature current} = (233 - 48) + 3{\cdot}5 = 188{\cdot}5 \text{ A}$$

$$\text{Armature Cu loss} = (188{\cdot}5)^2 \times 0{\cdot}035 = 1244 \text{ W}$$

$$\text{Brush contact loss} = 188{\cdot}5 \times 2 = 377 \text{ W}$$

$$\text{Shunt Cu loss} = 110 \times 3{\cdot}5 = 385 \text{ W}$$

$$\text{Total Cu loss} = 1244 + 377 + 385 = 2006 \text{ W}$$

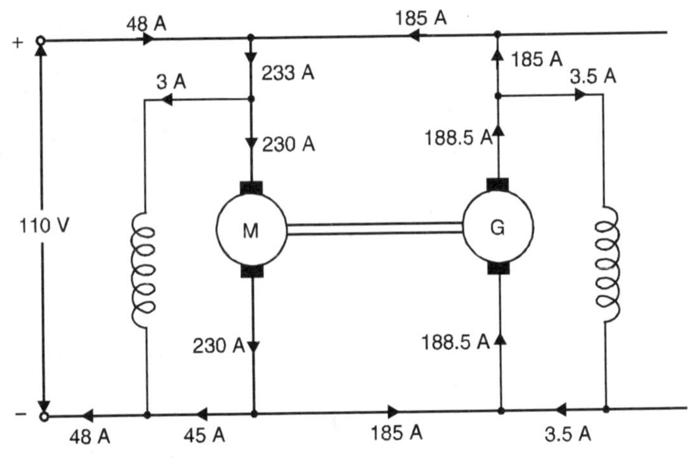

Fig. 7.6

For the Set

$$\text{Total arm. and shunt Cu loss for the set} = 2641{\cdot}5 + 2006 = 4647{\cdot}5 \text{ W}$$

$$\text{Total input} = 110 \times 48 = 5280 \text{ W}$$

$$\text{Iron, friction and windage losses} = 5280 - 4647{\cdot}5 = 632{\cdot}5 \text{ W}$$

$$\text{Iron, friction and windage losses in each machine} = 632{\cdot}5/2 = 316 \text{ W}$$

Motor efficiency

$$\text{Total motor losses} = \text{Total Cu losses + Iron, friction and windage losses}$$

$$= 2641{\cdot}5 + 316 = 2957{\cdot}5 \text{ W}$$

$$\text{Motor input} = 110 \times 233 = 25630 \text{ W}$$

$$\therefore \quad \text{Motor efficiency} = \frac{25630 - 2957{\cdot}5}{25630} \times 100 = \mathbf{88{\cdot}5\%}$$

Generator efficiency

$$\text{Total generator losses} = \text{Total Cu losses} + \text{Iron, friction and windage losses}$$
$$= 2006 + 316 = 2322 \text{ W}$$
$$\text{Generator output} = 110 \times 185 = 20350 \text{ W}$$
$$\therefore \quad \text{Generator efficiency} = \frac{20350}{20350 + 2322} \times 100 = \mathbf{89{\cdot}8\%}$$

Example 7.13. *Two 1000 kW, 500 V shunt machines were tested by Hopkinson method. When the output current of the generator was 2000 A, the input current from the supply mains was 380 A and shunt currents of the generator and motor were 22 A and 17 A respectively. The resistance of the armature circuit of each machine was 0·01 Ω. Calculate the efficiency of the generator at full-load by assuming (i) equal efficiencies (ii) equal iron and friction losses.*

Solution. Fig. 7.7 shows the connections of two shunt machines for Hopkinson's test.

(*i*) **Equal efficiencies**

$$\text{F.L. output current of generator} = \frac{1000 \times 10^3}{500} = 2000 \text{ A}$$
$$\therefore \quad \text{Current supplied by generator, } I_1 = 2000 \text{ A (\textit{i.e.} F.L. current)}$$
$$\text{Current from supply mains, } I_2 = 380 \text{ A}$$
$$\therefore \quad \text{F.L. efficiency of each machine} = \sqrt{\frac{I_1}{I_1 + I_2}} = \sqrt{\frac{2000}{2000 + 380}} = \mathbf{0{\cdot}9165} \textit{ or } \mathbf{91{\cdot}65\%}$$

(*ii*) **Equal iron and friction losses.** We shall first find armature Cu loss and shunt Cu loss for both machines.

Motor

$$\text{Armature current} = 2000 + 380 - 17 = 2363 \text{ A}$$
$$\text{Armature Cu loss} = (2363)^2 \times 0{\cdot}01 = 55850 \text{ W} = 55{\cdot}85 \text{ kW}$$
$$\text{Shunt Cu loss} = 500 \times 17 = 8500 \text{ W} = 8{\cdot}5 \text{ kW}$$
$$\text{Total Cu loss} = 55{\cdot}85 + 8{\cdot}5 = 64{\cdot}35 \text{ kW}$$

Generator

$$\text{Armature current} = 2000 + 22 = 2022 \text{ A}$$
$$\text{Armature Cu loss} = (2022)^2 \times 0{\cdot}01 = 40900 \text{ W} = 40{\cdot}9 \text{ kW}$$
$$\text{Shunt Cu loss} = 500 \times 22 = 11000 \text{ W} = 11 \text{ kW}$$
$$\text{Total Cu loss} = 40{\cdot}9 + 11 = 51{\cdot}9 \text{ kW}$$

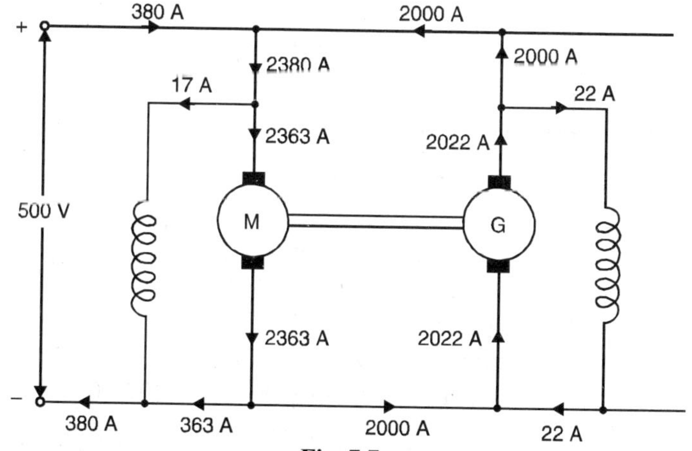

Fig. 7.7

For the set

$$\text{Total armature and shunt Cu loss for the set} = 64{\cdot}35 + 51{\cdot}9 = 116{\cdot}25 \text{ kW}$$

$$\text{Total input} = 500 \times 380 = 190000 \text{ W} = 190 \text{ kW}$$

$$\text{Iron and friction losses} = 190 - 116{\cdot}25 = 73{\cdot}75 \text{ kW}$$

$$\therefore \quad \text{Iron and friction losses in each machine} = 73{\cdot}75/2 = 36{\cdot}9 \text{ kW}$$

Generator efficiency

$$\text{F.L. total losses of generator} = \text{F.L. total Cu losses} + \text{Iron and friction losses}$$

$$= 51{\cdot}9 + 36{\cdot}9 = 88{\cdot}8 \text{ kW}$$

$$\text{F.L. generator output} = 1000 \text{ kW}$$

$$\therefore \quad \text{Generator efficiency} = \frac{1000}{1000 + 88{\cdot}8} \times 100 = \mathbf{91{\cdot}85\%}$$

Example 7.14. *In Hopkinson's test on 250 V machines, the line current was 50 A and the motor current 400 A not including the field currents of 6 A and 5 A. The armature resistance of each machine was 0·015 Ω. Calculate the efficiency of each machine.*

Solution. Fig. 7.8 shows the connections for Hopkinson's test on two d.c. shunt machines. Note that shunt field windings are directly connected across the lines.

$$\text{Motor armature Cu loss} = (400)^2 \times 0{\cdot}015 = 2400 \text{ W}$$

$$\text{Generator armature Cu loss} = (350)^2 \times 0{\cdot}015 = 1838 \text{ W}$$

Total iron, friction and windage losses for the two machines

$$= 250 \times 50 - (2400 + 1838) = 8262 \text{ W}$$

Iron, friction and windage losses per machine

$$= 8262/2 = 4131 \text{ W}$$

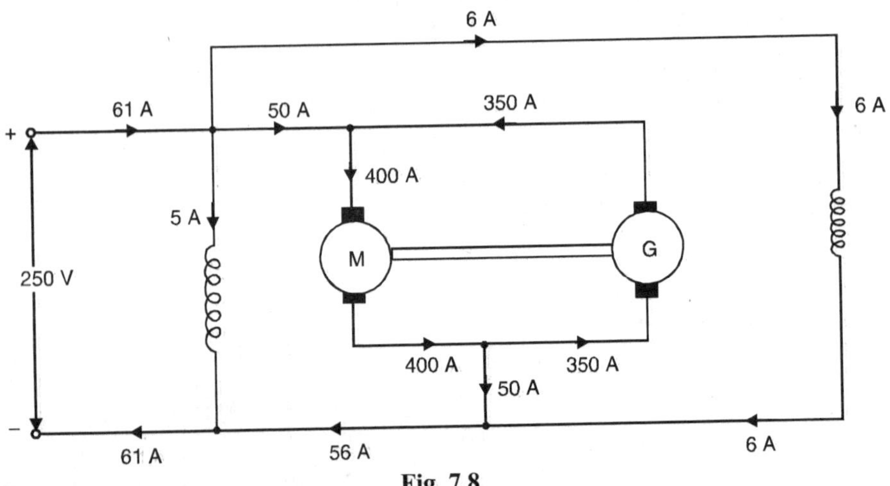

Fig. 7.8

Motor efficiency

$$\text{Motor armature Cu loss} = 2400 \text{ W} \qquad \text{... found above}$$

$$\text{Motor field Cu loss} = 250 \times 5 = 1250 \text{ W}$$

$$\text{Iron, friction and windage losses} = 4131 \text{ W} \qquad \text{... found above}$$

$$\text{Total motor losses} = 2400 + 1250 + 4131 = 7781 \text{ W}$$

$$\text{Motor input} = 250 \times (400 + 5) = 101250 \text{ W}$$

$$\therefore \quad \text{Motor efficiency} = \frac{101250 - 7781}{101250} \times 100 = \mathbf{92{\cdot}3\%}$$

Generator efficiency

Generator armature Cu loss = 1838 W ... found above

Generator field Cu loss = $250 \times 6 = 1500$ W

Iron, friction and windage losses = 4131 W ...found above

Total generator losses = $1838 + 1500 + 4131 = 7469$ W

Generator output = $250 \times 350 = 87500$ W

$$\therefore \quad \text{Generator efficiency} = \frac{87500}{87500 + 7469} \times 100 = \mathbf{92{\cdot}1\%}$$

TUTORIAL PROBLEMS

1. In Hopkinson test, the following readings were obtained :

   Line voltage = 100 V ; Motor current = 30 A

   Generator current = 25 A ; Armature resistance of each machine = 0·25 Ω

   Calculate the efficiency of each machine, ignoring the field currents and assuming that their iron and mechanical losses are the same. **[*Motor* : 90·5%; *Generator* : 92·05%]**
2. The following readings were obtained during Hopkinson test on 230 V similar d.c. shunt machines : Armature currents : 37 A, 30 A; Field currents : 0·85 A, 0·8 A. Calculate the efficiencies of the machines. The armature resistance of each machine is 0·33 Ω. **[*Motor* : 87·7% ; *Generator* : 87·9%]**
3. The Hopkinson's test on two shunt machines gave the following results for full-load : Line voltage = 250 V; Current taken from supply system excluding field currents = 50 A; Motor armature current = 380 A; Field currents = 5 A and 4·2 A. Calculate the efficiency of the machine working as a generator. Armature resistance of each machine is 0·02 Ω. **[92·02%]**

7.7 RETARDATION OR RUNNING DOWN TEST

This is the best and simplest method to find the efficiency of a constant-speed d.c. machine (*e.g.*, shunt generator and motor). In this method, we find the mechanical (friction and windage) and iron losses of the machine. Then knowing the armature and shunt Cu losses at any load, the efficiency of the machine can be calculated at that load.

Principle. Consider a d.c. shunt motor running at no-load.

(*i*) If the supply to the armature is cut off but field remains normally excited, the motor slows down gradually and finally stops. The kinetic energy of the armature is used up to overcome *friction, windage and iron losses.*

(*ii*) If the supply to the armature as well as field excitation is cut off, the motor again slows down and finally stops. Now the kinetic energy of the armature is used up to overcome *only the friction and windage losses*. This is expected because in the absence of flux, there will be no iron losses.

By carrying out the first test, we can find out the friction, windage and iron losses and hence the efficiency of the machine. However, if we perform the second test also, we can separate friction and windage losses from the iron losses.

Theory of retardation test. In the retardation test, the d.c. machine is run as a motor at a speed just above the normal. Then the supply to the armature is cut off while the field is normally excited. The speed is allowed to fall to some value just below normal. The time taken for this fall of speed is noted. From these observations, the *rotational losses (*i.e.*, friction, windage and iron losses) and hence the efficiency of the machine can be determined.

Let N = normal speed in r.p.m.

ω = normal angular velocity in rad/s = $2\pi N/60$

* Friction, windage and iron losses are collectively known as *rotational losses.*

$\therefore$ Rotational losses, W = Rate of loss of K.E. of armature

or $$W = \frac{d}{dt}\left(\frac{1}{2} I\omega^2\right) = I\omega \frac{d\omega}{dt}$$

Here I is the moment of inertia of the armature. As $\omega = 2\pi N/60$,

$$\therefore \quad W = I \times \frac{2\pi N}{60} \times \frac{d}{dt}\left(\frac{2\pi N}{60}\right) = \left(\frac{2\pi}{60}\right)^2 I N \frac{dN}{dt}$$

or $$W = 0{\cdot}011\, I N \frac{dN}{dt}$$

Let us illustrate the application of retardation test with a numerical example. Suppose the normal speed of a d.c. machine is 1000 r.p.m. When retardation test is performed, the time taken for the speed to fall from 1030 r.p.m. to 970 r.p.m. is 15 seconds with field normally excited. If the moment of inertia of the armature is 75 kg m^2, then,

Rotational losses, $$W = 0{\cdot}011\, I N \frac{dN}{dt}$$

Here $I = 75$ kg m^2; $N = 1000$ r.p.m.

$dN = 1030 - 970 = 60$ r.p.m. ; $dt = 15$ sec

$$\therefore \quad W = 0{\cdot}011 \times 75 \times 1000 \times \frac{60}{15} = 3300 \text{ watts}$$

The main difficulty with this method is the accurate determination of the speed which is continuously changing.

7.8 MOMENT OF INERTIA (I) OF THE ARMATURE

In retardation test, the rotational losses are given by;

$$W = 0{\cdot}011\, I N \frac{dN}{dt}$$

In order to find W, the value of I must be known. It is difficult to determine I directly or by calculation. Therefore, we perform another experiment by which either I is calculated or it is eliminated from the above expression.

(*i*) **First method.** It is a fly-wheel method in which the value of I is calculated. First, retardation test is performed with armature alone and dN/dt_1 is determined. Next, a fly-wheel of known moment of inertia I_1 is keyed on to the shaft of the machine. For the same change in speed, dN/dt_2 is noted. Since the addition of fly-wheel will not materially affect the rotational losses in the two cases,

$\therefore$ For the first case, $$W = 0{\cdot}011\, I N \frac{dN}{dt_1}$$

For the second case, $$W = 0{\cdot}011\, (I + I_1) N \frac{dN}{dt_2}$$

$$\therefore \quad 0{\cdot}011\, I N \frac{dN}{dt_1} = 0{\cdot}011\, (I + I_1) N \frac{dN}{dt_2}$$

or $$I \frac{dN}{dt_1} = (I + I_1) \frac{dN}{dt_2}$$

or $$\frac{I + I_1}{I} = \frac{dN/dt_1}{dN/dt_2} = \frac{dt_2}{dt_1}$$

or $$\frac{I_1}{I} = \frac{dt_2 - dt_1}{dt_1} = \frac{t_2 - t_1}{t_1}$$

or $$I = I_1 \times \frac{t_1}{t_2 - t_1}$$

Since the values of I_1, t_1 and t_2 are known, the moment of inertia I of the armature can be determined.

(*ii*) **Second method.** In this method, I is eliminated from the expression by an experiment. First, retardation test is performed with armature alone. The rotational losses are given by;

$$W = 0{\cdot}011\, I N \frac{dN}{dt_1}$$

Next the motor is loaded with a known amount of power W' with a *brake. For the same change in speed, dN/dt_2 is noted. Then,

$$W + W' = 0{\cdot}011\, I N \frac{dN}{dt_2}$$

$$\therefore \quad \frac{W + W'}{W} = \frac{dt_1}{dt_2} = \frac{t_1}{t_2}$$

or

$$\frac{W'}{W} = \frac{t_1 - t_2}{t_2}$$

$$\therefore \quad W = W' \times \frac{t_2}{t_1 - t_2}$$

Since the values of W', t_1 and t_2 are known, the value of W can be determined.

7.9 ELECTRIC LOADING IN RETARDATION TEST

In a retardation test, the rotational losses W are given by ;

$$W = 0{\cdot}011\, I N \frac{dN}{dt}$$

As discussed in Art. 7.8, we can eliminate I (moment of inertia of armature) from the above expression by applying either mechanical or electric loading to the armature. The electric loading is preferred because of convenience and reliability. Fig. 7.9 illustrates how electric loading is applied to slow down the armature. The double throw switch S is thrown to the supply and the machine is brought to full-load speed. Then the switch S is thrown to the other side connecting a non-inductive resistance R across the armature. The supply now is cut off and the power dissipated in R acts as a retarding torque to slow down the armature.

Let $\quad V'$ = *average* voltage across R

I'_a = *average* current through R

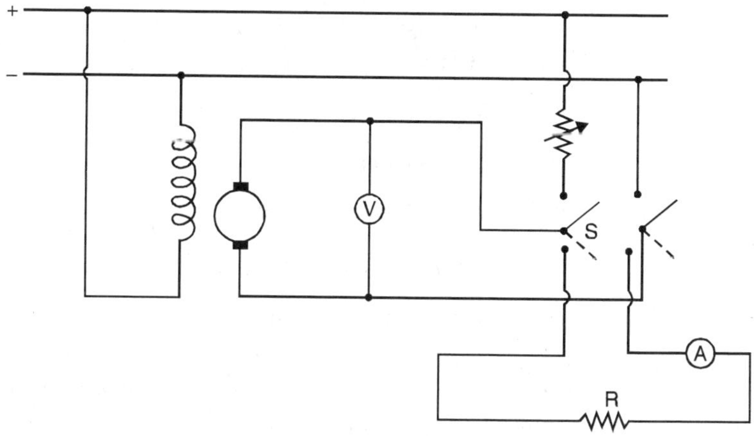

Fig. 7.9

* You can also put electrical load.

The electric loading W' (or extra power loss) is given by ;

$$W' = \text{average voltage} \times \text{average current} = V' I'_a$$

Example 7.15. *In a retardation test on a separately excited motor, the induced e.m.f. in the armature falls from 400 V to 380 V in 65 seconds on disconnecting the armature from the supply. The same fall takes place in 40 seconds if immediately after disconnection, armature is connected to a resistance which takes 10 A (average) during this fall. Calculate the rotational losses of the motor.*

Solution.

Electrical load, $W' = \text{Average voltage} \times \text{Average current}$

$$= \frac{380 + 400}{2} \times 10 = 3900 \text{ watts}$$

Rotational losses, $W = W' \times \dfrac{t_2}{t_1 - t_2}$

Here $W' = 3900$ watts; $t_1 = 65$ sec ; $t_2 = 40$ sec

$\therefore$ $W = 3900 \times \dfrac{40}{65 - 40} = \mathbf{6240\ watts}$

Example 7.16. *A retardation test is performed on a 1500 r.p.m. d.c. machine. The time taken for the speed to fall from 1524 r.p.m. to 1476 r.p.m. is*

(*a*) *40 seconds with no excitation*

(*b*) *25 seconds with full excitation at no-load*

(*c*) *20 seconds with full excitation and armature supplying an extra load of 1000 watts*

Calculate (i) the moment of inertia of the armature (ii) friction and windage losses i.e. mechanical losses (iii) core or iron losses at the mean speed of 1500 r.p.m.

Solution.

(*i*) Rotational losses, $W = W' \times \dfrac{t_2}{t_1 - t_2}$

Here $W' = 1000$ watts ; $t_1 = 25$ sec ; $t_2 = 20$ sec

$\therefore$ $W = 1000 \times \dfrac{20}{25 - 20} = 4000$ watts

Also $W = 0{\cdot}011\, I N \dfrac{dN}{dt}$

Here $W = 4000$ watts ; $N = 1500$ r.p.m.

$dN = 1524 - 1476 = 48$ r.p.m. ; $dt = 25$ sec

$\therefore$ $4000 = 0{\cdot}011 \times I \times 1500 \times \dfrac{48}{25}$

or $I = \mathbf{126{\cdot}26\ kg\ m^2}$

(*ii*) Mechanical losses, $W_m = 0{\cdot}011\, IN \dfrac{dN}{dt}$

$$= 0{\cdot}011 \times 126{\cdot}26 \times 1500 \times \frac{48}{40} = \mathbf{2500\ watts}$$

(*iii*) Iron losses $= 4000 - 2500 = \mathbf{1500\ watts}$

Example 7.17. *A retardation test is made on a separately-excited d.c. machine as a motor. The induced voltage falls from 240 V to 225 V in 25 seconds on opening the armature circuit and 6 seconds on suddenly changing the armature connection from supply to a load resistance taking 10 A (average). Find the efficiency of the machine when running as a motor and taking a current of 25 A on a supply of 250 V. The resistance of its armature is 0·4 Ω and that of its field winding is 250 Ω.*

Solution.

$$\text{Average voltage across load} = \frac{240 + 225}{2} = 232 \cdot 5 \text{ V}$$

$$\text{Power absorbed by load, } W' = \text{Average voltage} \times \text{Average current}$$
$$= 232 \cdot 5 \times 10 = 2325 \text{ watts}$$

$$\text{Rotational losses, } W = W' \times \frac{t_2}{t_1 - t_2}$$

Here $W' = 2325$ watts ; $t_1 = 25$ seconds ; $t_2 = 6$ seconds

$$\therefore \quad W = 2325 \times \frac{6}{25 - 6} = 734 \text{ watts}$$

$$\text{Line current, } I_L = 25 \text{ A}$$
$$\text{Shunt field current, } I_{sh} = 250/250 = 1 \text{ A}$$
$$\text{Armature current, } I_a = I_L - I_{sh} = 25 - 1 = 24 \text{ A}$$
$$\text{Armature Cu loss} = I_a^2 R_a = (24)^2 \times 0 \cdot 4 = 230 \cdot 4 \text{ watts}$$
$$\text{Shunt Cu loss} = 250 \times 1 = 250 \text{ watts}$$
$$\therefore \quad \text{Total losses} = 734 + 230 \cdot 4 + 250 = 1214 \cdot 4 \text{ watts}$$
$$\text{Input to motor} = 250 \times 25 = 6250 \text{ watts}$$
$$\therefore \quad \text{Motor efficiency, } \eta_m = \frac{\text{Input} - \text{Losses}}{\text{Input}} \times 100$$
$$= \frac{6250 - 1214 \cdot 4}{6250} \times 100 = \mathbf{80 \cdot 6\%}$$

TUTORIAL PROBLEMS

1. A retardation test is carried out on a 1500 r.p.m. d.c. machine. The time taken for the speed to fall from 1525 r.p.m. to 1475 r.p.m. is
   (*a*) 43 seconds with no excitation
   (*b*) 26 seconds with full excitation at no-load
   (*c*) 20 seconds with full excitation and the armature supplying a load of 1·1 kW
   Calculate (*i*) the moment of inertia of the armature and (*ii*) core loss for normal excitation at the mean speed of 1500 r.p.m. **[(*i*) 117 kg m^2 (*ii*) 1460 W]**
2. In a retardation test on a separately-excited d.c. motor, the induced e.m.f. in the armature falls from 220 V to 190 V in 30 seconds on disconnecting the armature from the supply. The same fall takes place in 20 seconds if, immediately after disconnection, armature is connected to a resistance which takes 10 A (average) during fall. Find rotational losses of the machine. **[4100 W]**
3. In a retardation test on a d.c. motor, with field normally excited, the speed fell from 1525 r.p.m. to 1475 r.p.m. is 25 seconds. With an average load of 1·0 kW supplied by the armature, the same speed drop occurred in 20 seconds. Find the moment of inertia of the rotating parts in kg m^2. **[121·6 kg m^2]**

SHORT ANSWER QUESTIONS

Q. 1. *How does higher permeability of magnetic material of a d.c. machine affect its performance?*

Ans. If the permeability of the magnetic material is large, a high value of flux density can be established. A higher flux density will allow less armature coil turns for the same performance. This would result in lower winding resistance and at the same time less coil inductance. Both results are highly beneficial in reducing losses and improving commutation.

Q. 2. *Show that for the same power output, a low-speed machine is always bigger than a high speed machine.*

Ans. Suppose we have a 100 kW, 250 V, 2000 r.p.m. generator. Now, it is desired to build another generator having the same power and voltage, but running at half the speed.

To generate the same voltage at half the speed, we either have to double the number of conductors on the armature or double the flux from the poles. Consequently, we must either increase the size of the armature or increase the size of the poles. In practice, we increase both. It follows, therefore, that for a given power output, a low-speed machine is always bigger than a high-speed machine. This is true for both d.c. and a.c. machines.

Q. 3. *Which losses occur in a d.c. machine?*

Ans. The losses in a d.c. machine consist of

(*i*) copper losses (*ii*) magnetic losses and (*iii*) mechanical losses

(*i*) **Copper losses**

(*a*) Armature copper loss $= I_a^2 R_a$

(*b*) Field Cu loss : Shunt field loss $= I_{sh}^2 R_{sh}$ (or $V I_{sh}$)

Series field loss $= I_{se}^2 R_{se}$

Interpole field loss $= I_a^2 R$

(*c*) Loss due to brush contact resistance. This is generally included in the armature copper loss.

(*ii*) **Magnetic losses.** These are also known as iron or core losses and consist of

(*a*) Hysteresis loss, $P_h \propto B_{max}^{1.6} f$

(*b*) Eddy current loss, $P_e \propto B_{max}^2 f^2$

These losses are practically constant for shunt and compound machines because field current in these machines is approximately constant.

(*iii*) **Mechanical losses.** These losses consist of

(*a*) frictional loss at the bearings and commutator

(*b*) brush frictional loss

(*c*) air friction loss due to rotating armature

The magnetic and mechanical losses are generally called **rotational losses**.

Q. 4. *Why is the efficiency of a d.c. generator not determined by direct loading?*

Ans. It is wasteful and difficult to measure the input and output of a d.c. generator by direct loading, especially in larger sizes. In practice, losses of the generator are measured and efficiency found from the following relation :

$$\text{Efficiency} = \frac{\text{output}}{\text{input}} = \frac{\text{output}}{\text{output} + \text{losses}}$$

Q. 5. *Why is the efficiency of a large d.c. motor not determined by brake test?*

Ans. The brake test is not suitable for large d.c. motors owing to the difficulty of dissipating the heat generated at the pulley.

Q. 6. *What is friction and windage loss in a d.c. machine equal to?*

Ans. The friction and windage loss is equal to the mechanical power required to drive the machine as a generator with the armature and field circuits open.

Q. 7. *What is the sum of mechanical loss and core loss in a d.c. machine equal to?*

Ans. The core loss in a d.c. machine is due to hysteresis and eddy currents in the armature core and pole faces. The sum of mechanical loss and core loss is equal to the electrical power developed in the armature when the machine is running as a motor with no mechanical load.

Q. 8. *Why is Swinburne's test preferred to determine the efficiency of a d.c. machine?*

Ans. This method involves purely electrical measurements which are capable of being carried out with a high degree of accuracy. It possesses the further advantage that the power required for testing is only a small fraction of the rated output of the machine.

Q. 9. *Show that eddy-current loss is directly proportional to the square of the thickness of laminations.*

Ans. Consider a solid iron armature core for a 2-pole d.c. machine as shown is Fig. 7.10 (*i*). If the armature core is rotated, e.m.f.s are generated in the core in exactly the same way as they are generated in conductors placed on the armature. These induced e.m.f.s circulate currents– known as *eddy currents*– in the core as shown dotted in Fig. 7.10 (*i*). Due to the very low resistance of the core, these eddy currents have large magnitude. This results in a large eddy-current loss and heating of the armature. If the core is made of laminations insulated from each other, the eddy currents are confined to the respective laminations as shown dotted in Fig.7.10 (*ii*). The result is that eddy-current loss is reduced. Let us illustrate this point. Suppose the core is split up into 5 laminations as shown in Fig. 7.10 (*ii*). The e.m.f. per lamination is only a fifth of that generated in the solid core. Moreover, the cross-section area per path is reduced a fifth so that resistance per path is five times that of the solid core. As a result, the current per path is 1/25 that of the solid core.

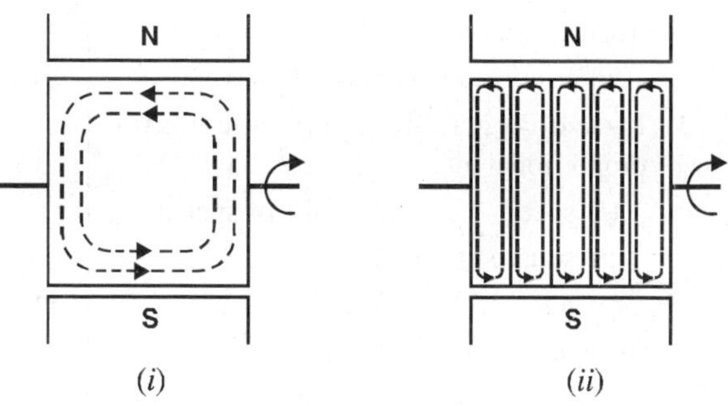

Fig. 7.10

$$\therefore \quad \frac{\text{Eddy – current loss } (I^2R) \text{ per lamination}}{\text{Eddy current loss } (I^2R) \text{ in solid core}} = \left(\frac{1}{25}\right)^2 \times 5 = \frac{1}{125}$$

Since there are 5 laminations, we have,

$$\frac{\text{Total eddy-current loss in laminated core}}{\text{Total eddy-current loss in solid core}} = \frac{5}{125} = \frac{1}{25} = \left(\frac{1}{5}\right)^2$$

Therefore, the eddy-current loss is directly proportional to the square of the thickness of the laminations.

Q. 10. *What are stray-load losses?*

Ans. Stray-load losses consist of the additional I^2R and core losses which are caused by the load current. The load current produces armature reaction which distorts the flux. This change in flux distribution with load will produce non-uniform distribution of current over the cross-section of the armature conductors. This results in greater I^2R loss. Change in flux distribution with load also results in some additional core losses over those existing under core-loss test conditions. The stray-load loss cannot be calculated and it is difficult to determine it by test. For d.c. machines, the stray-load loss is taken as 1% of the output.

MULTIPLE-CHOICE QUESTIONS

1. The efficiency of a large d.c. machine is not determined by direct loading because
   (*i*) it is difficult to provide means for operating a large machine under load
   (*ii*) it is an expensive method
   (*iii*) it requires the application of load
   (*iv*) all the above
2. In a d.c. machine, maximum losses occur due to....
   (*i*) copper losses
   (*ii*) iron losses
   (*iii*) mechanical losses
   (*iv*) friction and windage losses
3. In a d.c. machine, windage loss is proportional to ......
   (*i*) speed (*ii*) cube of speed
   (*iii*) square of speed
   (*iv*) none of the above
4. Swinburne test cannot be used for
   (*i*) shunt motor (*ii*) compound motor
   (*iii*) series motor (*iv*) insufficient data
5. Losses in compole winding are proportional to ....
   (*i*) armature current
   (*ii*) square of armature current
   (*iii*) cube of armature current
   (*iv*) none of the above
6. If the thickness of the armature laminations is made less than 0·4 mm, ...
   (*i*) the eddy-current loss is increased
   (*ii*) cost of construction is increased
   (*iii*) there are insulation problems
   (*iv*) none of the above
7. The most economical method of finding no-load losses of a large d.c. machine is..... test.
   (*i*) Swinburne's (*ii*) Hopkinson's
   (*iii*) Retardation (*iv*) none of above
8. The armature copper losses in a d.c. machine are about ..... of total full-load losses.
   (*i*) 5 – 10% (*ii*) 10 – 15%
   (*iii*) 30 – 40% (*iv*) less than 5%
9. Mechanical losses of a d.c. machine depend upon.......
   (*i*) armature resistance
   (*ii*) speed of machine
   (*iii*) load current
   (*iv*) none of above
10. Hopkinson's test can be used to determine ....
    (*i*) efficiency only
    (*ii*) temperature rise only
    (*iii*) efficiency as well as temperature rise
    (*iv*) none of above
11. The iron losses are not constant in a .... d.c. machine.
    (*i*) series (*ii*) shunt
    (*iii*) compound (*iv*) cannot say
12. The disadvantage of Hopkinson's test is that......
    (*i*) power required is large
    (*ii*) temperature rise cannot be found
    (*iii*) two identical machines are needed
    (*iv*) none of above
13. The mechanical losses ..... with speed.
    (*i*) decrease (*ii*) increase
    (*iii*) first decrease and then increase
    (*iv*) none of above
14. The efficiency of a d.c. machine is maximum when ....
    (*i*) Variable losses = Constant losses
    (*ii*) Variable losses = 1·5 × Constant losses
    (*iii*) Variable losses = (Constant losses)^2
    (*iv*) Variable losses = 0·5 × Constant losses
15. A d.c. machine has maximum efficiency near.....
    (*i*) half full-load (*ii*) full-load
    (*iii*) twice the full-load
    (*iv*) three-fourth full-load

ANSWERS TO MULTIPLE-CHOICE QUESTIONS

1. (*iv*)	**2.** (*i*)	**3.** (*ii*)	**4.** (*iii*)	**5.** (*ii*)
6. (*ii*)	**7.** (*i*)	**8.** (*iii*)	**9.** (*ii*)	**10.** (*iii*)
11. (*i*)	**12.** (*iii*)	**13.** (*ii*)	**14.** (*i*)	**15.** (*ii*)

HINTS TO SELECTED MULTIPLE-CHOICE QUESTIONS

5. The compole winding carries armature current.

6. It is desirable that the thickness of laminations should be as small as possible. However, if the thickness of lamination is made less than 0·4 mm, the reduction in eddy-current loss does not justify the extra cost of construction.

7. Swinburne's test is convenient and economical because power required to test a large machine is small *i.e.,* only no-load input power.

11. The iron losses are constant in a machine whose flux is constant *i.e.,* shunt and compound machines. However, iron losses vary in a d.c. series machine.

8

Transformer

INTRODUCTION

The transformer is probably one of the most useful electrical devices ever invented. It can change the magnitude of alternating voltage or current from one value to another. This useful property of transformer is mainly responsible for the widespread use of alternating currents rather than direct currents *i.e.,* electric power is generated, transmitted and distributed in the form of alternating current. Transformers have no moving parts, rugged and durable in construction, thus requiring very little attention. They also have a very high efficiency — as high as 99%. In this chapter, we shall study some of the basic properties of transformers.

8.1 TRANSFORMER

A transformer is a static piece of equipment used either for raising or lowering the voltage of an a.c. supply with a corresponding decrease or increase in current. It essentially consists of two windings, the primary and secondary, wound on a common laminated magnetic core as shown in Fig. 8.1. The winding connected to the a.c. source is called *primary winding* (or primary) and the one connected to load is called *secondary winding* (or secondary). The alternating voltage V_1 whose magnitude is to be changed is applied to the primary. Depending upon the number of turns of the primary (N_1) and secondary (N_2), an alternating e.m.f. E_2 is induced in the secondary. This induced e.m.f. E_2 in the secondary causes a secondary current I_2. Consequently, terminal voltage V_2 will appear across the load. If $V_2 > V_1$, it is called a **step up-transformer**. On the other hand, if $V_2 < V_1$, it is called a **step-down transformer**.

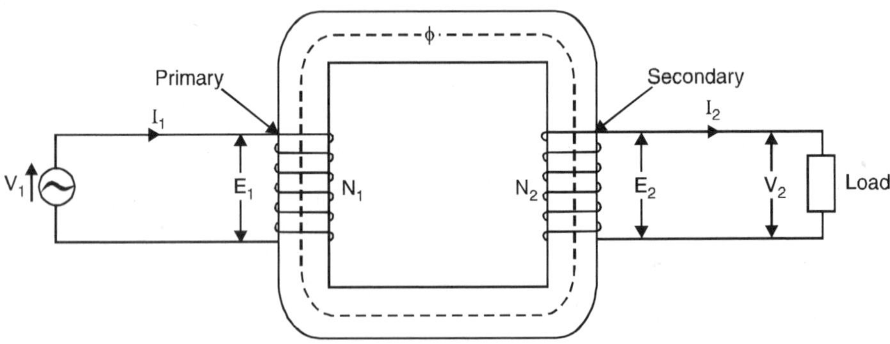

Fig. 8.1

Working. When an alternating voltage V_1 is applied to the primary, an alternating flux ϕ is set up in the core. This alternating flux links both the windings and induces e.m.f.s E_1 and E_2 in them according to Faraday's laws of electromagnetic induction. The e.m.f. E_1 is termed as primary e.m.f. and e.m.f. E_2 is termed as secondary e.m.f.

Clearly, $$E_1 = -N_1 \frac{d\phi}{dt}$$

and $$E_2 = -N_2 \frac{d\phi}{dt}$$

$$\therefore \quad \frac{E_2}{E_1} = \frac{N_2}{N_1}$$

Note that magnitudes of E_2 and E_1 *depend upon the number of turns on the secondary and primary respectively. If $N_2 > N_1$, then $E_2 > E_1$ (or $V_2 > V_1$) and we get a step-up transformer. On the other hand, if $N_2 < N_1$, then $E_2 < E_1$ (or $V_2 < V_1$) and we get a step-down transformer. If load is connected across the secondary winding, the secondary e.m.f. E_2 will cause a current I_2 to flow through the load. Thus, a transformer enables us to transfer a.c. power from one circuit to another with a change in voltage level.

The following points may be noted carefully :

(*i*) The transformer action is based on the laws of electromagnetic induction.

(*ii*) There is no electrical connection between the primary and secondary. The a.c. power is transferred from primary to secondary through magnetic flux.

(*iii*) There is no change in frequency *i.e.*, output power has the same frequency as the input power.

(*iv*) The losses that occur in a transformer are :

(*a*) core losses — eddy current and hysteresis losses

(*b*) copper losses — in the resistance of the windings

In practice, these losses are very small so that output power is nearly equal to the input primary power. In other words, a transformer has very high efficiency.

8.2 THEORY OF AN IDEAL TRANSFORMER

An ideal transformer is one that has

(*i*) no winding resistance

(*ii*) no leakage flux *i.e.*, the same flux links both the windings

(*iii*) no iron losses (*i.e.*, eddy current and hysteresis losses) in the core

Although ideal transformer cannot be physically realised, yet its study provides a very powerful tool in the analysis of a practical transformer. In fact, practical transformers have properties that approach very close to an ideal transformer.

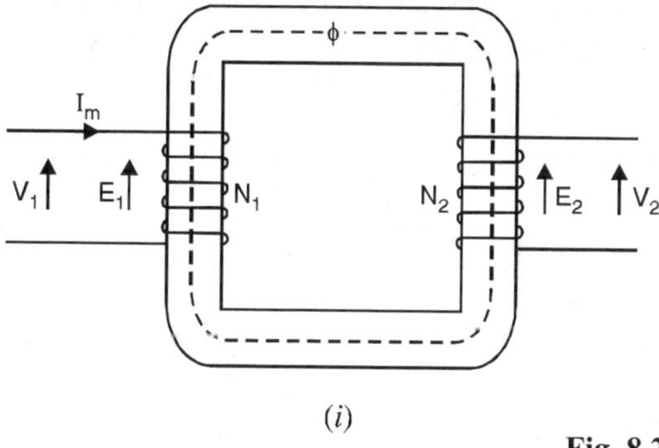

(*i*)

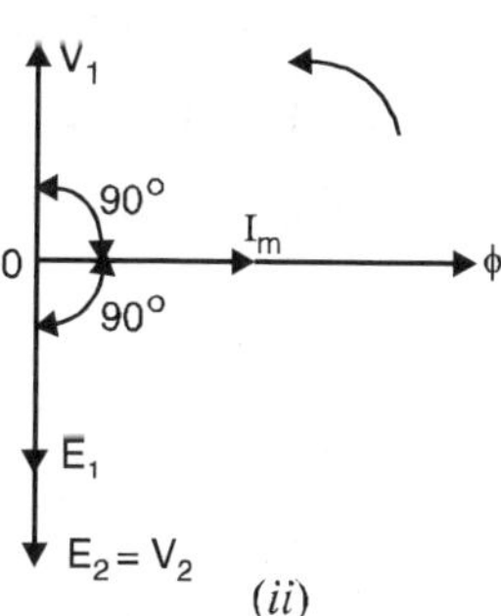

(*ii*)

Fig. 8.2

Consider an ideal transformer on no load *i.e.*, secondary is open-circuited as shown in Fig. 8.2 (*i*). Under such conditions, the primary is simply a coil of pure inductance. When an alternating voltage V_1 is applied to the primary, it draws a small magnetising current I_m which lags behind the applied voltage by **90°. This alternating current I_m produces an alternating flux ϕ which is proportional to and in †phase with it. The alternating flux ϕ links both the windings and induces e.m.f. E_1 in the

* Since $d\phi/dt$ is common for both the windings.

** In a pure inductive circuit, current lags behind the voltage by 90°.

† Since there are no iron losses, flux ϕ is in phase with I_m.

primary and e.m.f. E_2 in the secondary. The primary e.m.f. E_1 is, at every instant, equal to and in opposition to V_1 (Lenz's law). Both e.m.f.s E_1 and E_2 lag behind flux ϕ by 90° (See Art. 8.3). However, their magnitudes depend upon the number of primary and secondary turns.

Fig. 8.2 (*ii*) shows the phasor diagram of an ideal transformer on no load. Since flux ϕ is common to both the windings, it has been taken as the reference phasor. As shown in Art. 8.3, the primary e.m.f. E_1 and secondary e.m.f. E_2 lag behind the flux ϕ by 90°. Note that E_1 and E_2 are in phase. But E_1 is equal to V_1 and 180° out of phase with it.

8.3 E.M.F. EQUATION OF A TRANSFORMER

Consider that an alternating voltage V_1 of frequency f is applied to the primary as shown in Fig. 8.2 (*i*). The sinusoidal flux ϕ produced by the primary can be represented as :

$$\phi = \phi_m \sin \omega t$$

The instantaneous e.m.f. e_1 induced in the primary is

$$e_1 = -N_1 \frac{d\phi}{dt} = -N_1 \frac{d}{dt}(\phi_m \sin \omega t)$$

$$= -\omega N_1 \phi_m \cos \omega t = -2\pi f N_1 \phi_m \cos \omega t$$

$$= 2\pi f N_1 \phi_m \sin(\omega t - 90^\circ) \quad ...(i)$$

It is clear from the above equation that maximum value of induced e.m.f. in the primary is

$$E_{m1} = 2\pi f N_1 \phi_m$$

The r.m.s. value E_1 of the primary e.m.f. is

$$E_1 = \frac{E_{m1}}{\sqrt{2}} = \frac{2\pi f N_1 \phi_m}{\sqrt{2}}$$

or $$E_1 = 4{\cdot}44 f N_1 \phi_m$$

*Similarly, $$E_2 = 4{\cdot}44 f N_2 \phi_m$$

In an ideal transformer, $E_1 = V_1$ and $E_2 = V_2$.

Note. It is clear from exp. (*i*) above that e.m.f. E_1 induced in the primary lags behind the flux ϕ by 90°. Likewise, e.m.f. E_2 induced in the secondary lags behind flux ϕ by 90°.

8.4 VOLTAGE TRANSFORMATION RATIO (*K*)

From the above equations of induced e.m.f., we have (See Fig. 8.3),

$$\frac{E_2}{E_1} = \frac{N_2}{N_1} = K$$

The constant K is called *voltage transformation ratio*. Thus if $K = 5$ (*i.e.* $N_2/N_1 = 5$), then $E_2 = 5\,E_1$.

For an ideal transformer ;

(*i*) $E_1 = V_1$ and $E_2 = V_2$ as there is no voltage drop in the windings.

$$\therefore \quad \frac{E_2}{E_1} = \frac{V_2}{V_1} = \frac{N_2}{N_1} = K$$

(*ii*) There are no losses. Therefore, volt-amperes input to the primary are equal to the output volt-amperes *i.e.*

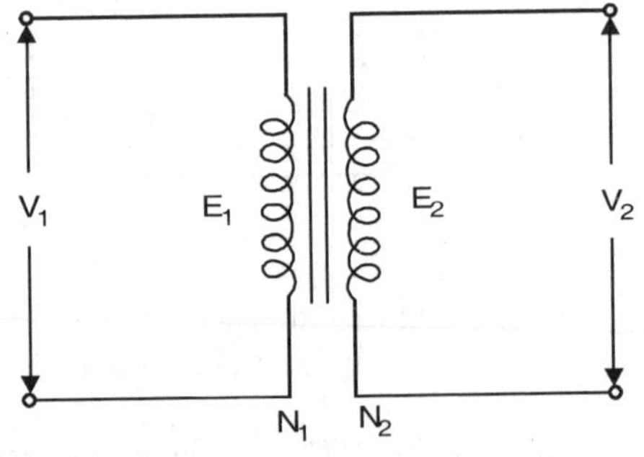

Fig. 8.3

* The e.m.f. E_2 induced in the secondary is produced by the same flux (*i.e.* $\phi = \phi_m \sin \omega t$) that causes e.m.f. E_1. Thus the only difference in the r.m.s. values of the two is the difference in the number of turns.

$$V_1 I_1 = V_2 I_2$$

or
$$\frac{I_2}{I_1} = \frac{V_1}{V_2} = \frac{1}{K}$$

Hence, currents are in the inverse ratio of voltage transformation ratio. This simply means that if we raise the voltage, there is a corresponding decrease of current.

Example 8.1. *A 2000/200 V, 20 kVA transformer has 66 turns in the secondary. Calculate (i) primary turns (ii) primary and secondary full-load currents. Neglect the losses.*

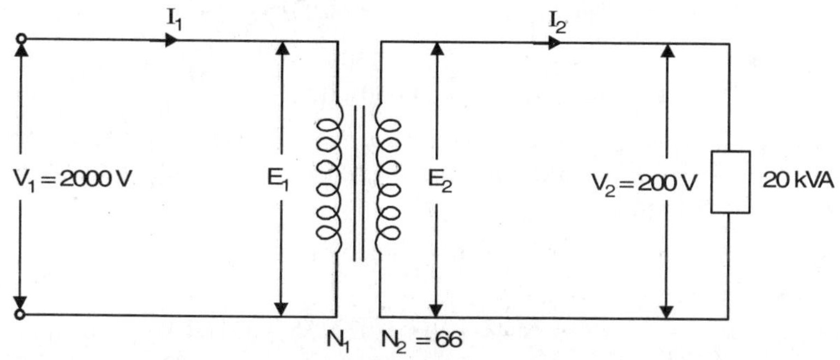

Fig. 8.4

Solution. Fig. 8.4 represents the conditions of the problem.

$$K = V_2/V_1 = 200/2000 = 1/10$$

(*i*)
$$\frac{N_2}{N_1} = K = \frac{1}{10}$$

$$\therefore \quad N_1 = N_2 \times 10 = 66 \times 10 = \textbf{660 turns}$$

(*ii*)
$$V_1 I_1 = V_2 I_2 = 20 \times 10^3$$

$$\therefore \quad I_2 = \frac{20 \times 10^3}{V_2} = \frac{20 \times 10^3}{200} = \textbf{100 A}$$

$$I_1 = \frac{20 \times 10^3}{V_1} = \frac{20 \times 10^3}{2000} = \textbf{10A}$$

Alternatively, $I_1 = K I_2 = (1/10)\ 100 = \textbf{10 A}$

Example 8.2. *An ideal 25 kVA transformer has 500 turns on the primary winding and 40 turns on the secondary winding. The primary is connected to 3000 V, 50 Hz supply. Calculate (i) primary and secondary currents on full-load (ii) secondary e.m.f. and (iii) the maximum core flux.*

Solution.

$$K = \frac{N_2}{N_1} = \frac{40}{500} = \frac{4}{50}$$

(*i*)
$$I_1 = \frac{\text{Volt-amperes}}{V_1} = \frac{25 \times 10^3}{3000} = \textbf{8·33 A}$$

$$I_2 = \frac{I_1}{K} = 8·33 \times \frac{50}{4} = \textbf{104·2 A}$$

(*ii*)
$$E_2/E_1 = K$$

$$\therefore \quad E_2 = K E_1 = (4/50) \times 3000 = \textbf{240 V}$$

(*iii*)
$$E_1 = 4·44 f N_1 \phi_m$$

or
$$3000 = 4·44 \times 50 \times 500 \times \phi_m$$

$$\therefore \quad \phi_m = \frac{3000}{4{\cdot}44 \times 50 \times 500} = 27 \times 10^{-3} \text{ Wb} = \mathbf{27\ mWb}$$

Example 8.3. *A single phase 2200/250 V, 50 Hz transformer has a net core area of 36 cm² and a maximum flux density of 6 Wb/m². Calculate the number of turns of primary and secondary.*

Solution. $E_1 = 4{\cdot}44\, f N_1\, \phi_m$

Here $E_1 = 2200$ V ; $f = 50$ Hz ; $\phi_m = B_m \times A = 6 \times 36 \times 10^{-4} = 0{\cdot}0216$ Wb

$$\therefore \quad N_1 = \frac{E_1}{4{\cdot}44\, f\, \phi_m} = \frac{2200}{4{\cdot}44 \times 50 \times 0{\cdot}0216} = \mathbf{459}$$

Also
$$N_2 = \frac{E_2}{4{\cdot}44\, f\, \phi_m} = \frac{250}{4{\cdot}44 \times 50 \times 0{\cdot}0216} = \mathbf{52}$$

Example 8.4. *A sinusoidal flux 0.02 Wb (maximum) links with 55 turns of a transformer secondary. Calculate the r.m.s. value of the induced e.m.f. in the secondary. The supply frequency is 50 Hz.*

Solution. $\phi_m = 0.02$ Wb ; $N_2 = 55$ turns ; $f = 50$ Hz

∴ R.M.S. value of induced e.m.f. in the secondary is

$$E_2 = 4.44\, f \phi_m N_2$$
$$= 4.44 \times 50 \times 0.02 \times 55 = \mathbf{244.2\ V}$$

Example 8.5. *A single phase 50 Hz transformer has square core of 20 cm side. The permissible maximum flux density in the core is 1 Wb/m². Calculate the number of turns per limb on the high and low voltage sides for a 3000/220 V ratio. To allow for insulation of stampings, assume the net iron length to be 0·9 × gross iron length.*

Solution. Net iron length $= 0{\cdot}9 \times 20 = 18$ cm

Net X-sectional area, $A = 18 \times 20 = 360 \text{ cm}^2 = 0{\cdot}036 \text{ m}^2$

Max. flux in core, $\phi_m = B_m \times A = 1 \times 0{\cdot}036 = 0{\cdot}036$ Wb

$$E_1 = 4{\cdot}44\, f N_1\, \phi_m$$

$$\therefore \quad N_1 = \frac{E_1}{4{\cdot}44\, f\, \phi_m} = \frac{3000}{4{\cdot}44 \times 50 \times 0{\cdot}036} = \mathbf{375}$$

Also
$$N_2 = \frac{E_2}{4{\cdot}44\, f\, \phi_m} = \frac{220}{4{\cdot}44 \times 50 \times 0{\cdot}036} = \mathbf{27}$$

Note that one length is affected due to stamping.

Example 8.6. *A single-phase core-type transformer has 400 primary turns. The net cross-sectional area of the core is 60 cm² and its magnetic length is 0·8 m. The primary voltage is 500 V, 50 Hz. What is the maximum flux density in the core ? Assuming a relative permeability of 2000 at this flux density, calculate the magnetising current.*

Solution. Fig. 8.5 shows the core of a core-type transformer. The core is built up of thin laminations such as are used in the armature of a d.c. generator. The magnetic length is l (= 0·8 m in the example).

$$E_1 = 4{\cdot}44\, B_m\, a f N_1$$

or
$$500 = 4{\cdot}44 \times B_m \times 60 \times 10^{-4} \times 50 \times 400$$

$$\therefore \quad B_m = \mathbf{0{\cdot}938\ Wb/m^2}$$

Now Primary m.m.f. = Flux × Reluctance

or
$$N_1 \times (I_m)_{max} = \phi_m \times \frac{l}{a\, \mu_0\, \mu_r} = \left(\frac{\phi_m}{a}\right) \times \frac{l}{\mu_0\, \mu_r}$$

or
$$N_1 \times (I_m)_{max} = B_m \times \frac{l}{\mu_0\, \mu_r}$$

or
$$400 \times (I_m)_{max} = 0{\cdot}938 \times \frac{0{\cdot}8}{4\pi \times 10^{-7} \times 2000}$$

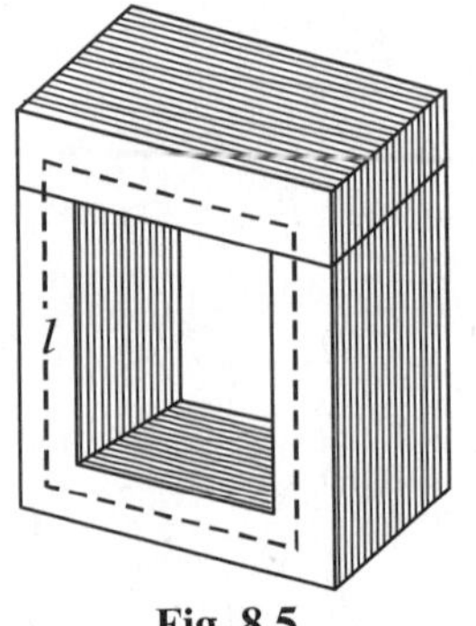

Fig. 8.5

$\therefore \quad (I_m)_{max} = 0{\cdot}747$ A

$\therefore$ R.M.S. value of magnetising current is

$$I_m\,(r.m.s.) = \frac{0{\cdot}747}{\sqrt{2}} = \mathbf{0{\cdot}528\ A}$$

Example 8.7. *A transformer having 90 turns on the primary and 2250 turns on the secondary is connected to 120 V, 60 Hz source. The coupling between the primary and secondary is perfect but the magnetising current is 4 A. Calculate*

(i) the peak voltage across the secondary terminals

(ii) the instantaneous voltage across the secondary when the instantaneous voltage across the primary is 37 V

Solution. $N_1 = 90$ turns ; $N_2 = 2250$ turns ; $K = N_2/N_1 = 25$; $E_1 = 120\ V$

(*i*) R.M.S. value of induced secondary voltage is

$$E_2 = (N_2/N_1)\,E_1 = 25 \times 120 = 3000 \text{ V}$$

The voltage varies sinusoidally so that peak secondary voltage is

$$E_{2(max)} = \sqrt{2} \times E_2 = \sqrt{2} \times 3000 = \mathbf{4242\ V}$$

(*ii*) Now $E_2 = KE_1 = 25\,E_1$. Therefore, the instantaneous secondary voltage is *always* 25 times greater than the instantaneous primary voltage.

$$\therefore \quad E_{2\,inst.} = 25 \times E_{1\,inst} = 25 \times 37 = \mathbf{925\ V}$$

TUTORIAL PROBLEMS

1. A 50 kVA, 6600/250 V transformer has 52 secondary turns. Find (*i*) the number of primary turns (*ii*) full-load primary and secondary currents. Neglect losses. **[(*i*) 1373 (*ii*) 7·58 A ; 200 A]**
2. The net cross-sectional area of the core of 400/3000 V, 50 Hz transformer is 600 cm^2. If the maximum flux density in the core is 1·3 Wb/m^2, find the number of turns on the primary and secondary. **[24 ; 180]**
3. In a certain 50 kVA transformer, the number of turns on the primary and secondary windings is 834 and 58 respectively. If primary is connected to a 3300 V supply, find (*i*) secondary voltage (*ii*) the primary and secondary currents when the transformer is fully loaded. Neglect the losses. **[(*i*) 230 V (*ii*) 15·2 A ; 218 A]**
4. A single phase transformer has a ratio of 1 : 10 and a secondary winding of 1000 turns. The primary winding is connected to a 25 V sinusoidal supply. If the maximum value of flux in the core is 2·15 mWb, find (*i*) the frequency of the supply (*ii*) the number of primary turns (*iii*) the secondary voltage on open-circuit. **[(*i*) 25 Hz (*ii*) 100 turns (*iii*) 250 V]**
5. A single phase 50 Hz transformer has 20 primary turns and 273 secondary turns. The net cross-sectional area of the core is 400 cm^2. If the primary winding is connected to 230 V supply, find (*i*) peak value of flux density in the core (*ii*) voltage induced in the secondary winding. **[(*i*) 1·24 Wb/m^2 (*ii*) 3003 V]**

8.5 PRACTICAL TRANSFORMER

A practical transformer differs from the ideal transformer in many respects. The practical transformer has (*i*) iron losses (*ii*) winding resistances and (*iii*) magnetic leakage, giving rise to leakage reactances.

(*i*) **Iron losses.** Since the iron core is subjected to alternating flux, there occurs eddy current and hysteresis loss in it. These two losses together are known as *iron losses* or *core losses*. The iron losses depend upon the supply frequency, maximum flux density in the core, volume of the core etc. It may be noted that magnitude of iron losses is quite small in a practical transformer.

(*ii*) **Winding resistances.** Since the windings consist of copper conductors, it immediately follows that both primary and secondary will have winding resistance. The primary resistance R_1 and secondary resistance R_2 act in series with the respective windings as shown in Fig. 8.6. When current flows through the windings, there will be power loss as well as a loss in voltage due to IR drop. This will affect the power factor and E_1 will be less than V_1 while V_2 will be less than E_2.

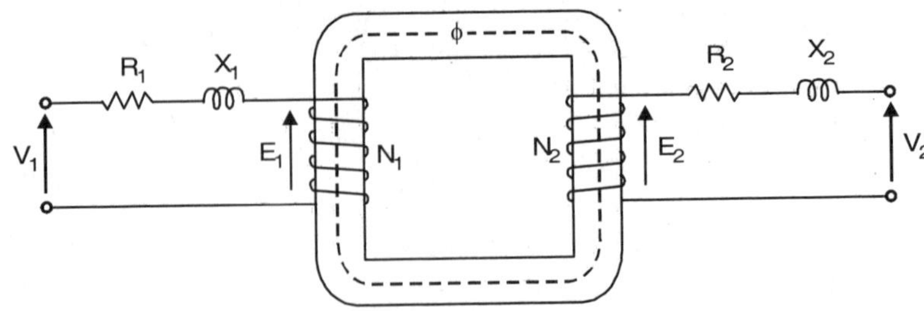

Fig. 8.6

(*iii*) **Leakage reactances.** Both primary and secondary currents produce flux. The flux ϕ which links both the windings is the useful flux and is called mutual flux. However, primary current would produce some flux ϕ_1 which would not link the secondary winding (See Fig. 8.7). Similarly, secondary current would produce some flux ϕ_2 that would not link the primary winding. The flux such as ϕ_1 or ϕ_2 which links only one winding is called *leakage flux*. The leakage flux paths are mainly through the air. The *effect of these leakage fluxes would be the same as though inductive reactances were connected in series with each winding of transformer that had no leakage flux as shown in Fig. 8.6. In other words, the effect of primary leakage flux ϕ_1 is to introduce an inductive reactance X_1 in series with the primary winding as shown in Fig. 8.6. Similarly, the secondary leakage flux ϕ_2 introduces an inductive reactance X_2 in series with the secondary winding. There will be no power loss due to leakage reactance. However, the presence of leakage reactance in the windings changes the power factor as well as there is voltage loss due to IX drop.

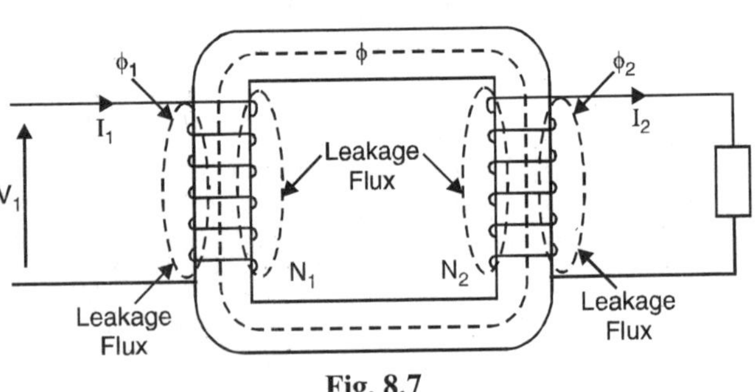

Fig. 8.7

Note. Although leakage flux in a transformer is quite small (about 5% of ϕ) compared to the mutual flux ϕ, yet it cannot be ignored. It is because leakage flux paths are through air of high reluctance and hence require considerable m.m.f. It may be noted that energy is conveyed from the primary winding to the secondary winding by mutual flux ϕ which links both the windings.

8.6 PRACTICAL TRANSFORMER ON NO LOAD

Consider a practical transformer on no load *i.e.*, secondary on open-circuit as shown in Fig. 8.8 (*i*). The primary will draw a small current I_0 to supply (*i*) the iron losses and (*ii*) a very **small amount of copper loss in the primary. Hence the primary no load current I_0 is not 90° behind the applied voltage V_1 but lags it by an angle $\phi_0 < 90°$ as shown in the phasor diagram in Fig. 8.8 (*ii*).

* The primary leakage flux ϕ_1 is alternating one and hence induces back e.m.f. e in the primary winding.

$$\therefore \quad \text{Primary leakage inductance, } L_1 = \frac{\text{Primary leakage flux linkages}}{\text{Primary current}} = \frac{N_1 \phi_1}{I_1}$$

$$\therefore \quad \text{Primary leakage reactance, } X_1 = 2\pi f L_1$$

Also secondary leakage reactance, $X_2 = 2\pi f L_2$ where $L_2 = N_2 \phi_2 / I_2$

** At no load, copper loss in the primary is $I_0^2 R_1$. Since I_0 is quite small, this loss may be neglected.

No load input power, $W_0 = V_1 I_0 \cos \phi_0$

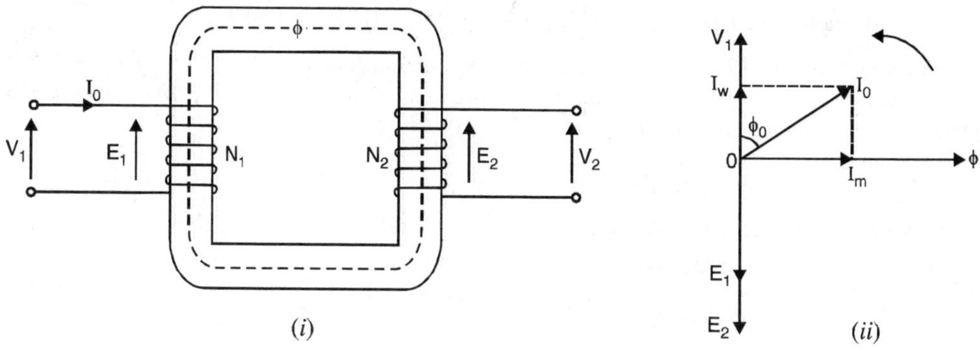

Fig. 8.8

As seen from the phasor diagram in Fig. 8.8 (*ii*), the no-load primary current I_0 can be resolved into two rectangular components *viz.*

(*i*) The component I_W in phase with the applied voltage V_1. This is known as *active* or *working* or *iron loss component* and *supplies the iron loss and a very small primary copper loss.

$$I_W = I_0 \cos \phi_0$$

(*ii*) The component I_m lagging behind V_1 by 90° and is known as *magnetising component*. It is this component which produces the mutual flux ϕ in the core.

$$I_m = I_0 \sin \phi_0$$

Clearly, I_0 is phasor sum of I_m and I_W.

$$\therefore \quad I_0 = \sqrt{I_m^2 + I_W^2}$$

$$\text{No load p.f., } \cos \phi_0 = I_W / I_0$$

It is emphasised here that no load primary copper loss (*i.e.* $I_0^2 R_1$) is very small and may be neglected. Therefore, the no load primary input power is practically equal to the iron loss in the transformer *i.e.*

No load input power, W_0 = Iron loss

Note. At no load, there is no current in the secondary so that $V_2 = E_2$. On the primary side, the drops in R_1 and X_1 due to I_0 are also very small because of the smallness of I_0. Hence, we can say that at no load, $V_1 = E_1$.

Example 8.8. *A transformer takes a current of 0·6 A and absorbs 64 W when primary is connected to its normal supply of 200 V, 50 Hz ; the secondary being on open circuit. Find the magnetising and iron loss currents.*

Solution.

No load primary power, $W_0 = V_1 I_0 \cos \phi_0$

Iron loss component, $I_W = I_0 \cos \phi_0 = W_0/V_1 = 64/200 = \mathbf{0 \cdot 32\ A}$

No load current, $I_0 = \sqrt{I_m^2 + I_W^2}$

$\therefore$ Magnetising current, $I_m = \sqrt{I_0^2 - I_W^2} = \sqrt{(0 \cdot 6)^2 - (0 \cdot 32)^2} = \mathbf{0 \cdot 507\ A}$

Example 8.9. *A 230/2300 V transformer takes no load current of 5 A at 0·25 power factor lagging. Find (i) the core loss and (ii) magnetising current.*

Solution. (*i*) Core loss, $W_0 = V_1 I_0 \cos \phi_0 = 230 \times 5 \times 0 \cdot 25 = \mathbf{287 \cdot 5\ W}$

---

* Remember that the component of current in phase with voltage produces the true or real power. However, component 90° out of phase with voltage produces reactive power *i.e.*, magnetic flux in this case.

(*ii*) Iron-loss current, $I_W = I_0 \cos \phi_0 = W_0/V_1 = 287{\cdot}5/230 = 1{\cdot}25$ A

$\therefore$ Magnetising current, $I_m = \sqrt{I_0^2 - I_W^2} = \sqrt{(5)^2 - (1{\cdot}25)^2} = \mathbf{4{\cdot}84\ A}$

Example 8.10. *A 230/2300 V transformer takes a no load current of 6·5 A and absorbs 187 W. If the resistance of primary is 0·06 Ω, find (i) the core loss (ii) no load p.f. (iii) active component of current and (iv) magnetising current.*

Solution.

No-load loss, $W_0 = 187$ W

Primary Cu loss due to no-load current $= I_0^2 R_1 = (6{\cdot}5)^2 \times 0{\cdot}06 = 2{\cdot}5$ W

(*i*) Iron loss $= 187 - 2{\cdot}5 = \mathbf{184{\cdot}5\ W}$

(*ii*) No load p.f., $\cos \phi_0 = \dfrac{W_0}{V_1 I_0} = \dfrac{187}{230 \times 6{\cdot}5} = \mathbf{0{\cdot}125}$ *lag*

(*iii*) Active component, $I_W = \dfrac{W_0}{V_1} = \dfrac{187}{230} = \mathbf{0{\cdot}81\ A}$

(*iv*) Magnetising current, $I_m = \sqrt{I_0^2 - I_W^2} = \sqrt{(6{\cdot}5)^2 - (0{\cdot}81)^2} = \mathbf{6{\cdot}4\ A}$

Example 8.11. *A single-phase transformer on no-load takes 4·5 A at a power factor of 0·25 lagging when connected to a 230 V, 50 Hz supply. The number of turns of the primary winding is 250. Calculate (i) the magnetising current (ii) the core loss and (iii) the maximum value of flux in the core.*

Solution. Fig. 8.9 shows the phasor diagram of the transformer on no load.

$V_1 = 230$ volts ; $N_1 = 250$ turns ; $\cos \phi_0 = 0{\cdot}25$ lag ; $I_0 = 4{\cdot}5$ A

(*i*) Magnetising current, $I_m = I_0 \sin \phi_0$

$= 4{\cdot}5 \times 0{\cdot}968 = \mathbf{4{\cdot}35\ A}$

(*ii*) Core loss, $W_0 = V_1 I_0 \cos \phi_0$

$= 230 \times 4{\cdot}5 \times 0{\cdot}25 = \mathbf{259\ W}$

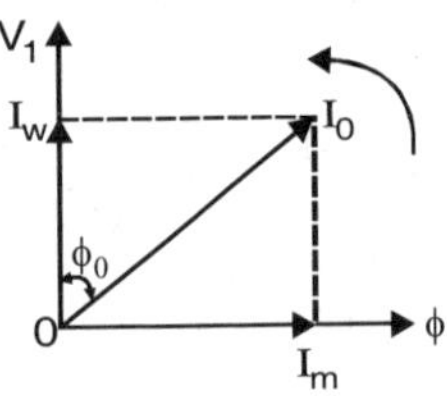

Fig. 8.9

(*iii*) Since no-load current I_0 is small, $E_1 \simeq V_1$.

$V_1 = 4{\cdot}44\ \phi_m f N_1$

or $230 = 4{\cdot}44 \times \phi_m \times 50 \times 250$

$\therefore$ $\phi_m = \mathbf{4{\cdot}14 \times 10^{-3}\ Wb}$

Example 8.12. *A single-phase transformer has 480 turns on the primary and 90 turns on the secondary. The mean length of the flux path in the iron core is 180 cm and the joints are equivalent to an air gap of 0·1 mm. If the maximum value of flux density is to be 1·1 Wb/m² when a p.d. of 2200 V at 50 Hz is applied to the primary, calculate (i) the cross-sectional area of the core (ii) the secondary voltage on no load (iii) the primary current and power factor on no load. Assume ampere-turns/metre for 1·1 Wb/m² in iron to be 400, the corresponding iron loss to be 2 W/kg at 50 Hz and density of iron to be 7·8 g/cm³.*

Solution.

(*i*) Since no-load current is small, $E_1 \simeq V_1$.

$\therefore$ $V_1 = 4{\cdot}44\ \phi_m f N_1$

or $2200 = 4{\cdot}44 \times \phi_m \times 50 \times 480$

$\therefore$ $\phi_m = 0{\cdot}0206$ Wb

$\therefore$ Cross-sectional area of transformer core

$$= \frac{\phi_m}{B_m} = \frac{0{\cdot}0206}{1{\cdot}1} = 0{\cdot}0187\ \text{m}^2 = \mathbf{187\ cm^2}$$

This is the net cross-sectional area of the core. The gross area will be about 10% greater than this value to allow for the insulation between the laminations.

(*ii*) No-load secondary voltage $= K E_1 = \dfrac{N_2}{N_1} \times E_1 = \dfrac{90}{480} \times 2200 = \mathbf{412 \cdot 5\ V}$

(*iii*)

$$AT/\text{m for iron} = 400$$

$$\therefore \quad AT \text{ for iron} = 400 \times 1{\cdot}8 = 720 \text{ AT}$$

$$AT \text{ for airgap} = H\,l = \frac{B_m}{\mu_0} \times l = \frac{1{\cdot}1}{4\pi \times 10^{-7}} \times 0{\cdot}1 \times 10^{-3} = 87{\cdot}5 \text{ AT}$$

$$\text{Total } AT \text{ at } B_m = 720 + 87{\cdot}5 = 807{\cdot}5 \text{ AT}$$

$$\therefore \quad \text{Max. value of magnetising current} = \frac{807{\cdot}5}{480} = 1{\cdot}682 \text{ A}$$

$$\therefore \quad I_m\,(r.m.s.) = 1.682/\sqrt{2} = 1{\cdot}19 \text{ A}$$

$$\text{Volume of iron} = 180 \times 187 \simeq 33700 \text{ cm}^3$$

$$\therefore \quad \text{Mass of iron} = 33700 \times 7{\cdot}8\ g = \frac{33700 \times 7 \cdot 8}{1000} \text{ kg} = 262{\cdot}86 \text{ kg}$$

$$\therefore \quad \text{Iron loss} = 2 \times 262{\cdot}86 = 525{\cdot}72 \text{ W}$$

Iron loss component of no-load current is

$$I_W = \frac{525 \cdot 72}{V_1} = \frac{525 \cdot 72}{2200} = 0{\cdot}24 \text{ A}$$

$$\text{No-load current, } I_0 = \sqrt{I_m^2 + I_W^2} = \sqrt{(1 \cdot 19)^2 + (0 \cdot 24)^2} = \mathbf{1{\cdot}21\ A}$$

$$\text{No-load power factor, } \cos\phi_0 = \frac{I_W}{I_0} = \frac{0 \cdot 24}{1 \cdot 21} = \mathbf{0{\cdot}198}\ \textit{lagging}$$

TUTORIAL PROBLEMS

1. A 2200/200 V transformer takes a no load current of 0·6 A and absorbs 400 watts. Find (*i*) the magnetising current and (*ii*) iron-loss current. [(*i*) **0·572 A** (*ii*) **0·18 A**]
2. A 2200/200 V transformer takes 0·5 A at a p.f. of 0·3 lagging at no load. Find (*i*) the magnetising current and (*ii*) iron-loss current. [(*i*) **0.477 A** (*ii*) **0·15 A**]
3. The no load current of a transformer is 5 A at 0·3 p.f. lagging when supplied at 230 V, 50 Hz. The number of turns on the primary winding is 200. Calculate (*i*) the maximum value of flux in the core (*ii*) the core loss and (*iii*) the magnetising current. [(*i*) **5·18 mWb** (*ii*) **345 W** (*iii*) **4·77 A**]

8.7 IDEAL TRANSFORMER ON LOAD

Let us connect a load Z_L across the secondary of an *ideal transformer as shown in Fig. 8.10 (*i*). The secondary e.m.f. E_2 will cause a current I_2 to flow through the load.

$$I_2 = \frac{{}^{**}E_2}{Z_L} = \frac{V_2}{Z_L}$$

The angle at which I_2 leads or lags V_2 (or E_2) depends upon the resistance and reactance of the load. In the present case, we have considered inductive load so that current I_2 lags behind V_2 (or E_2) by ϕ_2.

---

* Since practical transformers have full-load efficiencies well over 99%, they closely approach the ideal transformers.

** Since there is no voltage drop in an ideal transformer, $E_2 = V_2$.

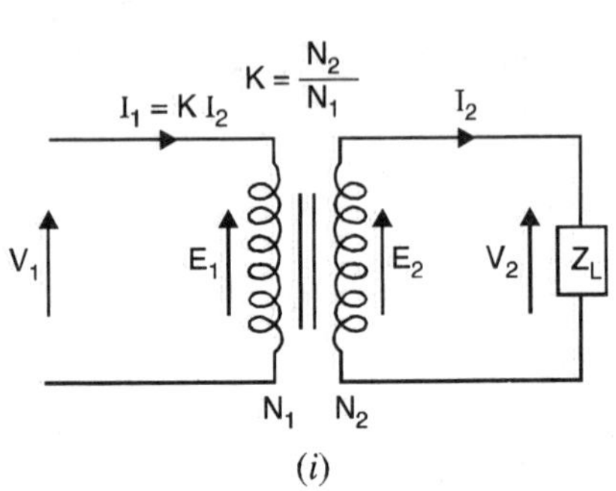

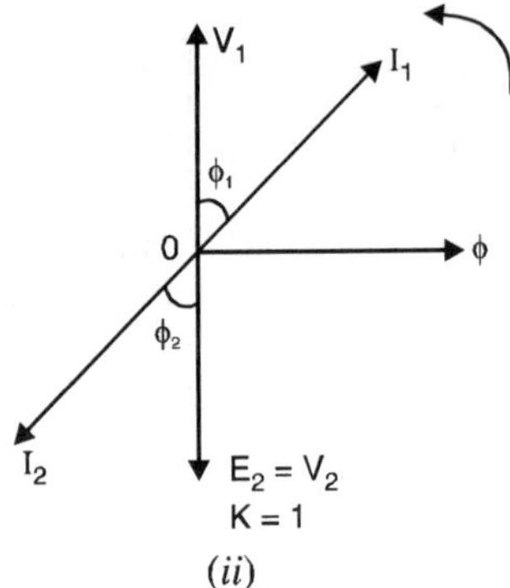

Fig. 8.10

The secondary current I_2 sets up an m.m.f. $N_2\ I_2$ which produces a flux in the *opposite direction to the flux ϕ originally set up in the primary by the magnetising current. This will change the flux in the core from the original value. However, the flux in the core should not **change from the original value. In order to fulfil this condition, the primary must develop an m.m.f. which exactly counterbalances the secondary m.m.f. $N_2\ I_2$. Hence a primary current I_1 must ***flow such that :

$$N_1 I_1 = N_2 I_2$$

or
$$I_1 = \frac{N_2}{N_1} I_2 = K I_2$$

Thus when a transformer is loaded and carries a secondary current I_2, then a current I_1 (= $K I_2$) must flow in the primary to maintain the m.m.f. balance. In other words, the primary must draw enough current to neutralise the demagnetising effect of secondary current so that mutual flux ϕ remains constant. Thus as the secondary current increases, the primary current I_1 (= $K I_2$) increases in unison and keeps the mutual flux ϕ constant. The power input, therefore, automatically increases with the output. For example if $K = 2$ and $I_2 = 2$ A, then primary will draw a current $I_1 = K I_2 = 2 \times 2 = 4$ A. If secondary current is increased to 4 A, then primary current will become $I_1 = K I_2 = 2 \times 4 = 8$ A.

Phasor diagram. Fig. 8.10 (*ii*) shows the phasor diagram of an ideal transformer on load. Note that in drawing the phasor diagram, the value of K has been assumed unity so that primary phasors are equal to secondary phasors. The secondary current I_2 lags behind V_2 (or E_2) by ϕ_2. It causes a primary current $I_1 = K I_2 = 1 \times I_2$ which is in *antiphase* with it.

(*i*)
$$\phi_1 = \phi_2$$

or
$$\cos \phi_1 = \cos \phi_2$$

Thus, power factor on the primary side is equal to the power factor on the secondary side.

(*ii*) Since there are no losses in an ideal transformer, input primary power is equal to the secondary output power *i.e.*,

---

* Lenz's law states that voltage induced in the secondary (*i.e.* E_2) acts in a direction so as to oppose what is causing it. This means that m.m.f. set up by the secondary current will oppose the mutual flux ϕ which is producing E_2.

** $E_1 = 4{\cdot}44\, f \phi_m N_1$

In an ideal transformer, $V_1 = E_1$. Since applied voltage V_1 is kept fixed, E_1 must remain unchanged. This is possible only if the flux remains fixed. Hence, mutual flux ϕ remains fixed whether a load is connected or not.

*** When the load is switched in, the secondary ampere-turns $N_2\ I_2$ tend to produce flux in opposition to that already established. This momentarily reduces the flux and the fall in primary e.m.f. E_1 allows a current I_1 to flow such that $I_1\ N_1$ (the load primary m.m.f.) restores the balance *i.e.*, the flux to its original level.

$$V_1 I_1 \cos \phi_1 = V_2 I_2 \cos \phi_2$$

Example 8.13. *An ideal transformer having 90 turns on the primary and 2250 turns on the secondary is connected to 200 V, 50 Hz supply. The load across the secondary draws a current of 2 A at a p.f. of 0·8 lagging. Calculate (i) the value of primary current and (ii) the peak value of flux linked with the secondary. Draw the phasor diagram.*

Solution. The conditions of the problem are represented in Fig. 8.11 (*i*).

$$K = N_2/N_1 = 2250/90 = 25$$

(*i*) $$I_1 = K I_2 = 25 \times 2 = \mathbf{50\ A}$$

(*ii*) $$E_1 = 4{\cdot}44 f N_1 \phi_m$$

or $$200 = 4{\cdot}44 \times 50 \times 90 \times \phi_m$$

$\therefore$ $$\phi_m = \frac{200}{4{\cdot}44 \times 50 \times 90} = \mathbf{0{\cdot}01\ Wb}$$

This flux links both the windings.

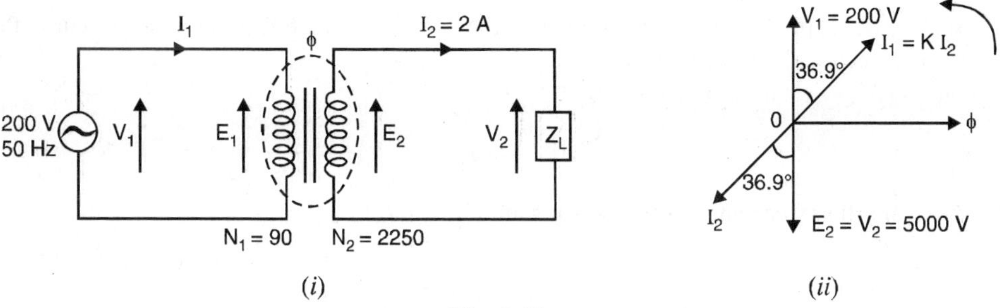

Fig. 8.11

Phasor diagram. Fig. 8.11 (*ii*) shows the phasor diagram (not to scale) of the transformer.

$$E_2 = K E_1 = 25 \times 200 = 5000 \text{ V}$$

$$\phi_2 = \text{phase angle between } V_2 \text{ and } I_2$$

$$= \cos^{-1} 0{\cdot}8 = 36{\cdot}9^\circ$$

The phase angle between V_1 and I_1 is also 36·9°.

Example 8.14. *An ideal transformer has 1000 turns on its primary and 500 turns on its secondary. The driving voltage on the primary side is 100 V and the load resistance is 5 Ω. Calculate V_2, I_1 and I_2.*

Solution.

$$K = N_2/N_1 = 500/1000 = 1/2$$

$$V_2 = K V_1 = (1/2) \times 100 = \mathbf{50\ V}$$

$$I_2 = V_2/R_2 = 50/5 = \mathbf{10\ A}$$

$$I_1 = K I_2 = (1/2) \times 10 = \mathbf{5\ A}$$

Example 8.15. *A single-phase transformer having a core of cross-sectional area of 150 cm² operates at a maximum flux density of 1.1 Wb/m² from a 50 Hz supply. If the secondary winding has 66 turns, determine the output in kVA when connected to a load of 4Ω impedance. Neglect any voltage drop in the secondary.*

Solution. Maximum value of flux in the core is

$$\phi_m = B_m \times a = (1.1) \times (150 \times 10^{-4}) = 0.0165 \text{ Wb}$$

Secondary induced e.m.f., $$E_2 = 4.44 f \phi_m N_2$$

$$= 4.44 \times 50 \times 0.0165 \times 66 = 242 \text{ V}$$

Since voltage drop in the secondary is neglected,

$\therefore$ $$V_2 = E_2 = 242 \text{ V}$$

$$\text{Secondary current, } I_2 = V_2/Z_L = 242/4 = 60.5 \text{ A}$$

$$\therefore \qquad \text{Output} = \frac{V_2 I_2}{1000} \text{ kVA} = \frac{242 \times 60.5}{1000} = \mathbf{14.6\ kVA}$$

8.8 PRACTICAL TRANSFORMER ON LOAD

We shall consider two cases (*i*) when such a transformer is assumed to have no winding resistance and leakage flux (*ii*) when the transformer has winding resistance and leakage flux.

(*i*) **No winding resistance and leakage flux**

Fig. 8.12 shows a practical transformer with the assumption that resistances and leakage reactances of the windings are negligible. With this assumption, $V_2 = E_2$ and $^*V_1 = E_1$. Let us take the usual case of inductive load which causes the secondary current I_2 to lag the secondary voltage V_2 by ϕ_2. The total primary current I_1 must meet two requirements *viz.*

(*a*) It must supply the no-load current I_0 to meet the iron losses in the transformer and to provide flux in the core.

(*b*) It must supply a current I_2' to counteract the demagnetising effect of secondary current I_2. The magnitude of I_2' will be such that :

$$N_1 I_2' = N_2 I_2$$

or

$$I_2' = \frac{N_2}{N_1} I_2 = K I_2$$

The total primary current I_1 is the phasor sum of I_2' and I_0 *i.e.*,

$$I_1 = I_2' + I_0 \qquad \textit{...phasor sum}$$

where

$$I_2' = -K I_2$$

Note that I_2' is 180° out of phase with I_2.

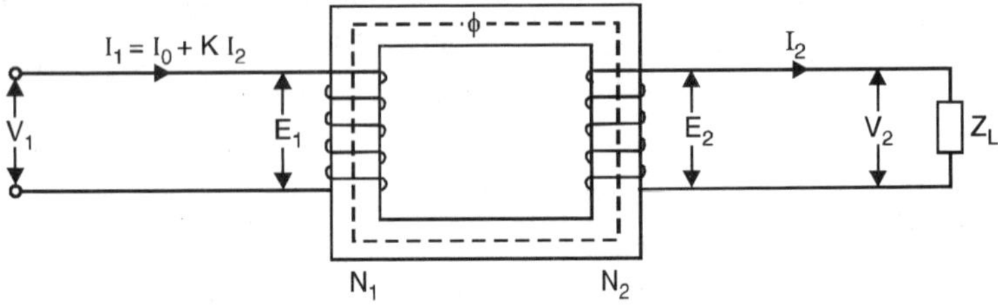

Fig. 8.12

Phasor diagram. Fig. 8.13 shows the phasor diagram for the usual case of inductive load. Both E_1 and E_2 lag behind the mutual flux ϕ by 90°. The current I_2' represents the primary current to neutralise the demagnetising effect of secondary current I_2. Now $I_2' = K I_2$ and is *antiphase* with I_2. I_0 is the no-load current of the transformer. The phasor sum of I_2' and I_0 gives the total primary current I_1. Note that in drawing the phasor diagram, the value of K is assumed to be unity so that primary phasors are equal to secondary phasors.

$$\text{Primary p.f.} = \cos\phi_1$$

$$\text{Secondary p.f.} = \cos\phi_2$$

$$\text{Primary input power} = V_1 I_1 \cos\phi_1$$

$$\text{Secondary output power} = V_2 I_2 \cos\phi_2$$

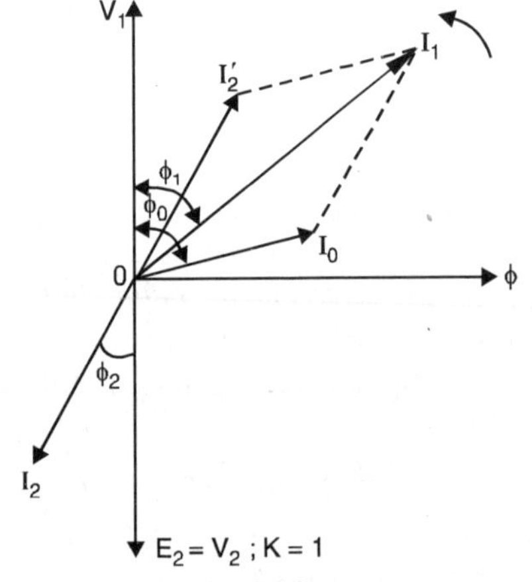

Fig. 8.13

* However, V_1 and E_1 are 180° out of phase.

Example 8.16. *A single-phase transformer with a ratio of 440/110 V takes no-load current of 5 A at 0·2 p.f. lagging. If the secondary supplies a current of 120 A at a p.f. 0·8 lagging, find the current taken by the primary.*

Solution. Fig. 8.14 shows the phasor diagram of the transformer.

$$\cos\phi_2 = 0{\cdot}8 \quad \therefore \quad \phi_2 = 36{\cdot}87^\circ$$

$$\cos\phi_0 = 0{\cdot}2 \quad \therefore \quad \phi_0 = 78{\cdot}46^\circ$$

Now

$$K = V_2/V_1 = 110/440 = 1/4$$

$$I_2' = K I_2 = (1/4) \times 120 = 30 \text{ A}$$

$$I_0 = 5 \text{ A}$$

Angle between I_2' and I_0 = $78{\cdot}46^\circ - 36{\cdot}87^\circ = 41{\cdot}59^\circ$

Using parallelogram method, we have,

$$I_1 = \sqrt{(I_0)^2 + (I_2')^2 + 2I_0\, I_2' \cos 41{\cdot}59^\circ}$$

$$= \sqrt{(5)^2 + (30)^2 + 2 \times 5 \times 30 \times \cos 41{\cdot}59^\circ} = \mathbf{3{\cdot}39\ A}$$

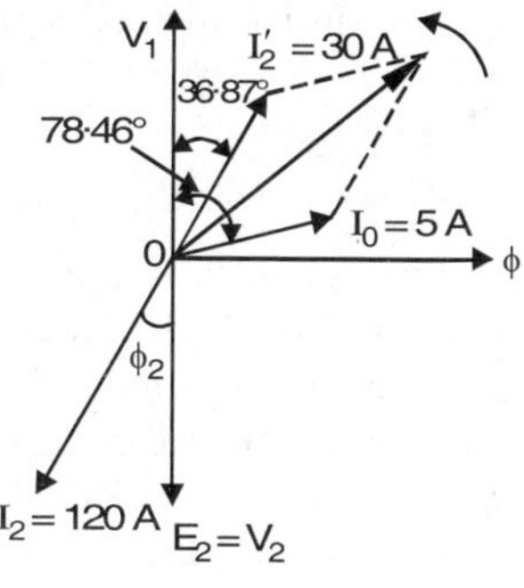

Fig. 8.14

Example 8.17. *A single-phase transformer has 1000 turns on the primary and 200 turns on the secondary. The no-load current is 3 A at a p.f. of 0·2 lagging. Calculate the primary current and power factor when the secondary current is 280 A at a p.f. of 0·8 lagging.*

Solution. Fig. 8.15 shows the phasor diagram of the transformer.

$$K = \frac{200}{1000} = \frac{1}{5}$$

$$\therefore \quad I_2' = K I_2 = \frac{1}{5} \times 280 = 56 \text{ A}$$

$$\cos\phi_2 = 0{\cdot}8 \quad \therefore \quad \sin\phi_2 = 0{\cdot}6$$

$$\cos\phi_0 = 0{\cdot}2 \quad \therefore \quad \sin\phi_0 = 0{\cdot}98$$

Referring to Fig. 8.15 and resolving the currents along rectangular axes, we have,

$$I_1 \cos\phi_1 = I_2' \cos\phi_2 + I_0 \cos\phi_0$$

$$= (56 \times 0{\cdot}8) + (3 \times 0{\cdot}2) = 45{\cdot}4 \text{ A}$$

$$I_1 \sin\phi_1 = I_2' \sin\phi_2 + I_0 \sin\phi_0$$

$$= (56 \times 0{\cdot}6) + (3 \times 0{\cdot}98)$$

$$= 36{\cdot}54 \text{ A}$$

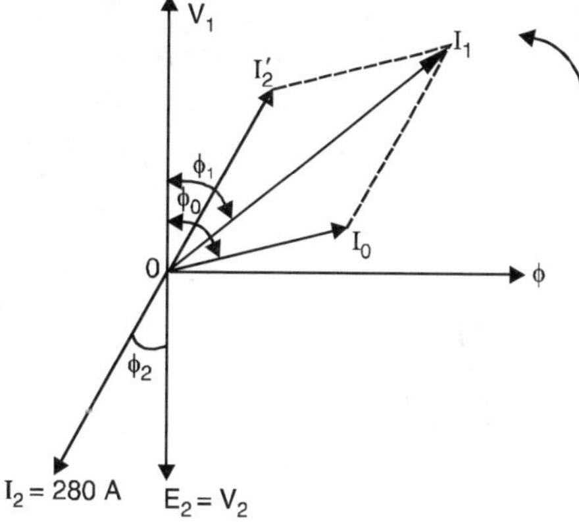

Fig. 8.15

Now,

$$I_1 = \sqrt{(I_1 \cos\phi_1)^2 + (I_1 \sin\phi_1)^2}$$

$$= \sqrt{(45{\cdot}4)^2 + (36{\cdot}54)^2} = \mathbf{58{\cdot}3\ A}$$

$$\tan\phi_1 = \frac{I_1 \sin\phi_1}{I_1 \cos\phi_1} = \frac{36{\cdot}54}{45{\cdot}4} = 0{\cdot}805$$

$$\therefore \quad \phi_1 = \tan^{-1} 0{\cdot}805 = 38{\cdot}83^\circ$$

$\therefore$ Primary p.f. $= \cos\phi_1 = \cos 38{\cdot}83^\circ =$ **0·78** *lag*

Example 8.18. *A single-phase transformer with a ratio of 6600/600 V has a load impedance of (4 + j3) ohms connected across the terminals of low voltage winding. Calculate the kW delivered to the load. Neglecting losses in the transformer, find the current taken by it from the supply lines.*

Solution.

Load impedance, $Z = \sqrt{4^2 + 3^2} = 5\ \Omega$

Load p.f., $\cos\phi_2 = R/Z = 4/5 = 0{\cdot}8$ *lag*

Load current, $I_2 = V_2/Z = 600/5 = 120$ A

$$\text{Power delivered to load} = V_2 I_2 \cos\phi_2 = 600 \times 120 \times 0{\cdot}8 = 57{\cdot}6 \times 10^3 \text{ W} = \mathbf{57{\cdot}6 \text{ kW}}$$

$$K = V_2/V_1 = 600/6600 = 1/11$$

Since losses are neglected, we have,

$$N_1 I_1 = N_2 I_2$$

or

$$I_1 = \left(\frac{N_2}{N_1}\right) I_2 = K I_2 = \left(\frac{1}{11}\right) \times 120 = \mathbf{10{\cdot}91 \text{ A}}$$

Example 8.19. *A single-phase transformer has a turns ratio of 144/432 and operates at a maximum flux of $7{\cdot}5 \times 10^{-3}$ Wb at 50 Hz. When on no load, the transformer takes 0·24 kVA at a power factor of 0·26 lagging from the supply. If the transformer supplies a load of 1·2 kVA at a power factor of 0·8 lagging, find (i) the magnetising current (ii) primary current (iii) the primary power factor.*

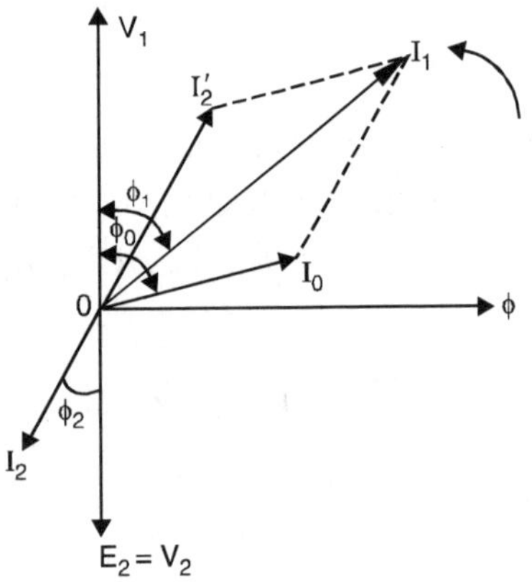

Solution. Fig. 8.16 shows the phasor diagram of the transformer. Now $K = 432/144$.

Fig. 8.16

(*i*)

$$V_1 \simeq E_1 = 4{\cdot}44\ \phi_m f N_1 = 4{\cdot}44 \times 7{\cdot}5 \times 10^{-3} \times 50 \times 144 = 240 \text{ V}$$

$$\text{No-load current, } I_0 = \frac{0{\cdot}24 \times 1000}{V_1} = \frac{0{\cdot}24 \times 1000}{240} = 1 \text{ A}$$

$$\phi_0 = \cos^{-1}(0{\cdot}26) = 75^\circ$$

$$\therefore \text{ Magnetising current, } I_m = I_0 \sin\phi_0 = 1 \times \sin 75^\circ = \mathbf{0{\cdot}97 \text{ A}}$$

(*ii*)

$$V_2 \simeq E_2 = K E_1 = \frac{432}{144} \times 240 = 720 \text{ V}$$

$$\therefore \quad I_2 = \frac{1{\cdot}2 \times 1000}{V_2} = \frac{1{\cdot}2 \times 1000}{720} = 1{\cdot}67 \text{ A}$$

$$I_2' = K I_2 = (432/144) \times 1{\cdot}67 = 5 \text{ A}$$

Referring to Fig. 8.16 and resolving the currents along rectangular axes, we have,

$$I_1 \cos\phi_1 = I_2' \cos\phi_2 + I_0 \cos\phi_0 = 5 \times 0{\cdot}8 + 1 \times 0{\cdot}26 = 4{\cdot}26 \text{ A}$$

$$I_1 \sin\phi_1 = I_2' \sin\phi_2 + I_0 \sin\phi_0 = 5 \times 0{\cdot}6 + 1 \times 0{\cdot}97 = 3{\cdot}97 \text{ A}$$

$$\therefore \quad I_1 = \sqrt{(I_1 \cos\phi_1)^2 + (I_1 \sin\phi_1)^2} = \sqrt{(4{\cdot}26)^2 + (3{\cdot}97)^2} = \mathbf{5{\cdot}82 \text{ A}}$$

(*iii*)

$$\text{Primary p.f., } \cos\phi_1 = \frac{I_2' \cos\phi_2 + I_0 \cos\phi_0}{I_1} = \frac{4{\cdot}26}{5{\cdot}82} = \mathbf{0{\cdot}73} \textit{ lagging}$$

Example 8.20. *A small substation has a single-phase 6600/240 V transformer supplying four feeders which take the following loads :*

(i) 10 kW at 0.8 p.f. lag (ii) 50 A at 0.7 p.f. lag (iii) 5 kW at unity p.f. (iv) 8 kVA at 0.6 p.f. lead. Determine the primary current and power factor. Neglect losses in the transformer.

Solution. We shall first determine the total current supplied by the secondary to the four loads. Let I_1, I_2, I_3 and I_4 be the currents supplied to the four loads at power factors $\cos\phi_1$, $\cos\phi_2$, $\cos\phi_3$ and $\cos\phi_4$ respectively.

$$I_1 = \frac{10 \times 1000}{240 \times 0.8} = 52.08 \text{ A at } \cos\phi_1 = 0.8 \text{ lag}$$

$$I_2 = 50 \text{ A at } \cos\phi_2 = 0.7 \text{ lag}$$

$$I_3 = \frac{5\times1000}{240\times1} = 20.83 \text{ A at } \cos\phi_3 = 1$$

$$I_4 = \frac{8\times1000}{240} = 33.33 \text{ A at } \cos\phi_4 = 0.6 \text{ lead}$$

Let I_S be the total current supplied by the secondary at a power factor of $\cos\phi_S$. Resolving all the currents along horizontal and vertical axes, we have,

$$\begin{aligned} I_S\cos\phi_S &= I_1\cos\phi_1 + I_2\cos\phi_2 + I_3\cos\phi_3 + I_4\cos\phi_4 \\ &= 52.08\times0.8 + 50\times0.7 + 20.83\times1 + 33.33\times0.6 \\ &= 41.66 + 35 + 20.83 + 20 = 117.49 \text{ A} \\ I_S\sin\phi_S &= I_1\sin\phi_1 + I_2\sin\phi_2 + I_3\sin\phi_3 - I_4\sin\phi_4 \\ &= 52.08\times0.6 + 50\times0.7 + 0 - 33.33\times0.8 \\ &= 31.25 + 35 + 0 - 26.66 = 39.59 \text{ A} \end{aligned}$$

$$\therefore \quad I_S = \sqrt{(I_S\cos\phi_S)^2 + (I_S\sin\phi_S)^2}$$

$$= \sqrt{(117.49)^2 + (39.59)^2} = 124 \text{ A}$$

$$\therefore \quad I_P = I_S\frac{V_S}{V_P} = 124\times\frac{240}{6600} = \mathbf{4.51\ A}$$

$$\tan\phi_S = \frac{39.59}{117.49} = 0.34 \quad \therefore \phi_S = 18.8°$$

$\therefore$ Secondary p.f. $= \cos\phi_S = \cos 18.8° =$ **0.94** *lag*

Since the transformer is ideal, primary p.f. is equal to the secondary p.f.

Example 8.21. *A single-phase transformer takes 10 A at no load at p.f. of 0.2 lagging. The turn ratio is 4:1 (step-down). If the load on the secondary is 200 A at a p.f. of 0.85 lagging, find the primary current and power factor. Neglect voltage drop in the windings.*

Solution. $K = N_2/N_1 = 1/4$; $\mathbf{I_0} = 10\angle-\phi_0^\circ$. where $\phi_0 = \cos^{-1} 0.2 = 78.5°$ lagging. The secondary current $\mathbf{I_2} = 200\angle-\phi_L$ where $\phi_L = \cos^{-1} 0.85 = 31.8°$ lagging. It is reflected in the primary as $\mathbf{I'_2}$ where

$$\mathbf{I'_2} = K\mathbf{I_2} = \frac{1}{4}\times 200 {}^{*}\angle-31.8° = 50\angle-31.8° \text{A}$$

The current $\mathbf{I'_2}$ lags behind the applied voltage by 31.8°.

$$\begin{aligned} \therefore \quad \text{Primary current, } \mathbf{I_1} &= \mathbf{I_0} + \mathbf{I'_2} = 10\angle-78.5° + 50\angle-31.8° \\ &= (2 - j\,9.8) + (42.5 - j\,26.35) \\ &= (44.5 - j\,36.15)\text{ A} = 57.33\angle-39.10° \text{ A} \end{aligned}$$

$\therefore$ Primary current $=$ **57.33 A** and Primary p.f. $= \cos(-39.10°) =$ **0.776** *lag*

TUTORIAL PROBLEMS

1. A 3300/240 V single phase transformer on no load takes 2 A at a p.f. of 0·25 lagging. Determine the primary current and power factor when the transformer is supplying a load of 60 A at p.f. 0·9 lagging. **[4·43 A, 1]**

---

* $\phi_L = \cos^{-1}(0.85) = 31.8°$

2. A 4 : 1 ratio step-down transformer takes 1A at 0·15 p.f. on no load. Determine the primary current and power factor when the transformer is supplying a load of 2 A at 0·8 p.f. lagging. Neglect voltage drops in the windings. **[7 A ; 0·735** *lag*]

3. A 400/200 V single phase transformer is supplying a load of 50 A at a p.f. of 0·866 lagging. The no-load current is 2 A at 0·208 p.f. lagging. Calculate the primary current and power factor. **[26·4 A ; 0·838** *lag*]

4. A transformer has a primary winding of 800 turns and a secondary winding of 200 turns. When the load current on the transformer is 80 A at 0·8 power factor lagging, the primary current is 25 A at 0·707 lagging. Determine the no-load current of the transformer and its phase w.r.t. the applied voltage. **[5·93 A ; ϕ_0 = 73·3°]**

5. The primary of a certain transformer takes 1A at a power factor of 0·4 when it is connected across 200 V, 50 Hz supply and the secondary is on open circuit. The number of turns on the primary is twice that on the secondary. A load taking 50 A at a lagging power factor of 0·8 is now connected across the secondary. What is now the primary current ? **[25·9 A]**

(*ii*) Transformer with resistance and leakage reactance

Fig. 8.17 shows a practical transformer having winding resistances and leakage reactances. These are the *actual* conditions that exist in a transformer. There is voltage drop in R_1 and X_1 so that primary e.m.f. E_1 is less than the applied voltage V_1. Similarly, there is voltage drop in R_2 and X_2 so that secondary terminal voltage V_2 is less than the secondary e.m.f. E_2. Let us take the usual case of inductive load which causes the secondary current I_2 to lag behind the secondary voltage V_2 by ϕ_2. The total primary current I_1 must meet two requirements *viz.*

(*a*) It must supply the no-load current I_0 to meet the iron losses in the transformer and to provide flux in the core.

(*b*) It must supply a current I_2' to counteract the demagnetising effect of secondary current I_2. The magnitude of I_2' will be such that :

$$N_1 I_2' = N_2 I_2$$

or

$$I_2' = \frac{N_2}{N_1} I_2 = K I_2$$

The total primary current I_1 will be the phasor sum of I_2' and I_0 *i.e.*,

$$\mathbf{I}_1 = \mathbf{I}_2' + \mathbf{I}_0 \text{ where } \mathbf{I}_2' = -K\,\mathbf{I}_2$$

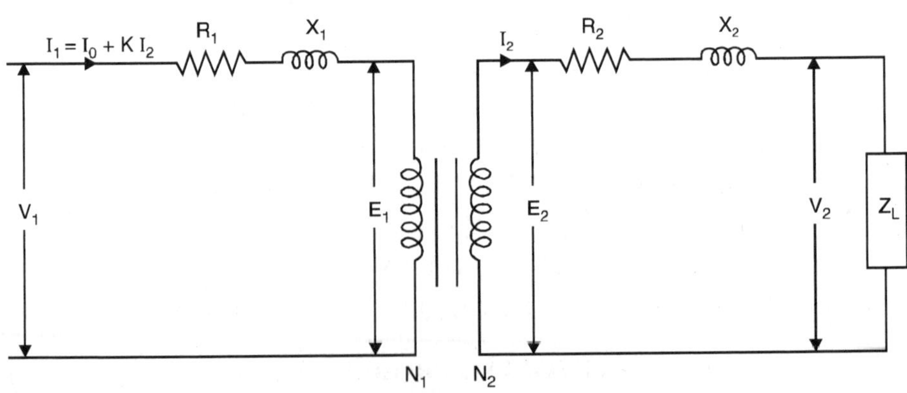

Fig. 8.17

$$\mathbf{V}_1 = -\mathbf{E}_1 + \mathbf{I}_1 (R_1 + jX_1) \text{ where } \mathbf{I}_1 = \mathbf{I}_0 + (-K\mathbf{I}_2) \quad \text{...phasor sum}$$

$$= -\mathbf{E}_1 + \mathbf{I}_1 \mathbf{Z}_1 \quad \text{...phasor sum}$$

$$\mathbf{V}_2 = \mathbf{E}_2 - \mathbf{I}_2 (R_2 + jX_2)$$

$$= \mathbf{E}_2 - \mathbf{I}_2 \mathbf{Z}_2$$

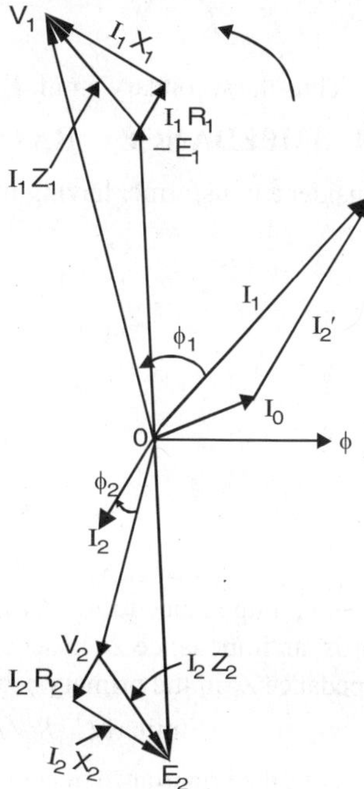

Fig. 8.18

Phasor diagram. Fig. 8.18 shows the phasor diagram of a practical transformer for the usual case of inductive load. Both E_1 and E_2 lag the mutual flux ϕ by 90°. The current I_2' represents the primary current to neutralise the demagnetising effect of secondary current I_2. Now $I_2' = K I_2$ and is opposite to I_2. Also I_0 is the no-load current of the transformer. The phasor sum of I_2' and I_0 gives the total primary current I_1.

Note that counter e.m.f. that opposes the applied voltage $\mathbf{V_1}$ is $-\mathbf{E_1}$. Therefore, if we add $\mathbf{I_1}R_1$ (in phase with $\mathbf{I_1}$) and $\mathbf{I_1}X_1$ (90° ahead of $\mathbf{I_1}$) to $-\mathbf{E_1}$, we get the applied primary voltage $\mathbf{V_1}$. The phasor $\mathbf{E_2}$ represents the induced e.m.f. in the secondary by the mutual flux ϕ. The secondary terminal voltage $\mathbf{V_2}$ will be what is left over after subtracting $\mathbf{I_2}R_2$ and $\mathbf{I_2}X_2$ from $\mathbf{E_2}$.

$$\text{Load power factor} = \cos\phi_2$$

$$\text{Primary power factor} = \cos\phi_1$$

$$\text{Input power to transformer, } P_1 = V_1 I_1 \cos\phi_1$$

$$\text{Output power of transformer, } P_2 = V_2 I_2 \cos\phi_2$$

Note : The reader may draw the phasor diagram of a loaded transformer for (*i*) unity p.f. and (*ii*) leading p.f. as an exercise.

Example 8.22. *The primary of a 1000/250 V transformer has a resistance of 0·15 Ω and leakage reactance of 0·8 Ω. Find the primary induced e.m.f. when the primary current is 60 A at 0·8 p.f. lagging.*

Solution.

$$\text{Primary impedance, } \mathbf{Z_1} = (0{\cdot}15 + j\,0{\cdot}8)\,\Omega = 0{\cdot}814\angle 79{\cdot}6^\circ\,\Omega$$

$$\text{Power factor angle, } \phi_1 = \cos^{-1} 0{\cdot}8 = 36{\cdot}9^\circ$$

Taking applied voltage as the reference phasor, we have, $\mathbf{V_1} = 1000\angle 0^\circ$ V.

Now,

$$\begin{aligned} -\mathbf{E_1} &= \mathbf{V_1} - \mathbf{I_1}\mathbf{Z_1} \\ &= 1000\angle 0^\circ - 60\angle -36{\cdot}9^\circ \times 0{\cdot}814\angle 79{\cdot}6^\circ \\ &= 1000\angle 0^\circ - 48{\cdot}8\angle 42{\cdot}7^\circ \\ &= 1000 - (36 + j33) \\ &= (964 - j\,33)\text{V} = 964{\cdot}5\angle -2^\circ\text{ V} \end{aligned}$$

$$\therefore \quad \text{Primary e.m.f., } \mathbf{E_1} = (-964 + j\,33)\text{V} = \mathbf{964{\cdot}5\angle 178^\circ\text{ V}}$$

Example 8.23. *The voltage on the secondary of a single phase transformer is 200 V when supplying a load of 8 kW at a p.f. of 0·8 lagging. The secondary resistance is 0·04 Ω and secondary leakage reactance is 0·8 Ω. Calculate the induced e.m.f. in the secondary.*

Solution. Taking secondary voltage as the reference phasor, we have, $\mathbf{V_2} = 200\angle 0^\circ$ V.

$$\text{Secondary current, } I_2 = \frac{8 \times 10^3}{200 \times 0{\cdot}8} = 50\text{ A}$$

$$\text{Power factor angle, } \phi_2 = \cos^{-1} 0{\cdot}8 = 36{\cdot}9^\circ$$

$$\therefore \quad \mathbf{I_2} = 50\angle -36{\cdot}9^\circ\text{ A}$$

$$\mathbf{Z_2} = (0{\cdot}04 + j\,0{\cdot}8)\,\Omega = 0{\cdot}8\angle 87{\cdot}14^\circ\,\Omega$$

$$\begin{aligned} \mathbf{E_2} &= \mathbf{V_2} + \mathbf{I_2}\mathbf{Z_2} \\ &= 200\angle 0^\circ + 50\angle -36{\cdot}9^\circ \times 0{\cdot}8\angle 87{\cdot}14^\circ \\ &= 200\angle 0^\circ + 40\angle 50{\cdot}24^\circ \end{aligned}$$

$$= (200 + j\,0) + (25{\cdot}58 + j\,30{\cdot}75)$$
$$= (225{\cdot}58 + j\,30{\cdot}75) \text{ volts} = \mathbf{227{\cdot}67\angle 7{\cdot}8^\circ \text{ volts}}$$

Thus the secondary e.m.f. $\mathbf{E}_2$ leads the secondary terminal voltage $\mathbf{V}_2$ by 7·8°.

8.9 IMPEDANCE RATIO

Consider a transformer having impedance Z_2 in the secondary as shown in Fig. 8.19.

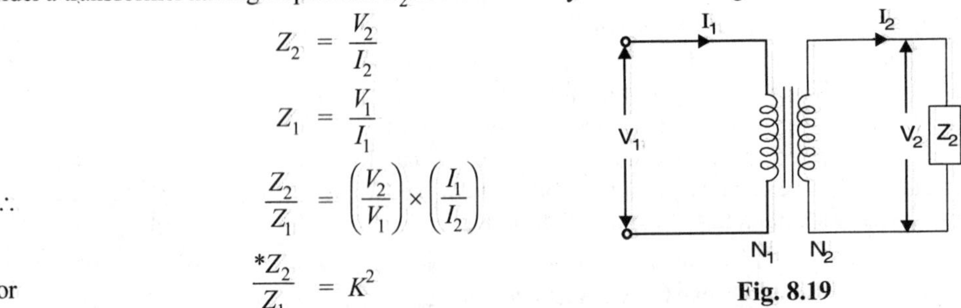

$$Z_2 = \frac{V_2}{I_2}$$

$$Z_1 = \frac{V_1}{I_1}$$

$$\therefore \quad \frac{Z_2}{Z_1} = \left(\frac{V_2}{V_1}\right) \times \left(\frac{I_1}{I_2}\right)$$

or
$$\frac{{}^*Z_2}{Z_1} = K^2$$

Fig. 8.19

i.e., impedance ratio (Z_2/Z_1) is equal to the square of voltage transformation ratio. In other words, an impedance Z_2 in secondary becomes Z_2/K^2 when transferred to primary. Likewise, an impedance Z_1 in the primary becomes $K^2\,Z_1$ when transferred to the secondary.

Similarly, $R_2/R_1 = K^2$ and $X_2/X_1 = K^2$.

Note the importance of above relations. We can transfer the parameters from one winding to the other. Thus :

(*i*) A resistance R_1 in the primary becomes $K^2\,R_1$ when transferred to the secondary.
(*ii*) A resistance R_2 in the secondary becomes R_2/K^2 when transferred to the primary.
(*iii*) A reactance X_1 in the primary becomes $K^2\,X_1$ when transferred to the secondary.
(*iv*) A reactance X_2 in the secondary becomes $X_2\,/K^2$ when transferred to the primary.

Note : It is important to remember that :
(*i*) When transferring resistance or reactance from primary to secondary, *multiply* it by K^2.
(*ii*) When transferring resistance or reactance from secondary to primary, *divide* it by K^2.
(*iii*) When transferring voltage or current from one winding to the other, only ** K is used.

8.10 SHIFTING IMPEDANCES IN A TRANSFORMER

Fig. 8.20 shows a transformer where resistances and reactances are shown external to the windings. The resistance and reactance of one winding can be transferred to the other by appropriately using the factor K^2. This makes the analysis of the transformer a simple affair because then we have to work in one winding only.

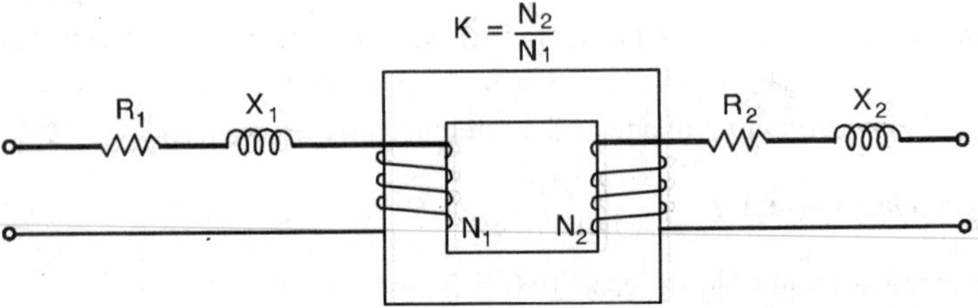

Fig. 8.20

* Note that it is an impedance ratio. Here Z_2 is the actual impedance of the secondary. However, when this impedance is transferred to primary, it becomes $Z_1 = Z_2/K^2$. It is obvious Z_1 in primary becomes $Z_2 = K^2\,Z_1$ when transferred to secondary.

** Thus any voltage V_1 in the primary becomes $K\,V_1$ in the secondary. On the other hand, any voltage V_2 in the secondary becomes V_2/K in the primary. Again a current I_1 in the primary becomes I_1/K in the secondary. Any current I_2 in the secondary becomes $K\,I_2$ in the primary.

(*i*) **Referred to primary.** When secondary resistance or reactance is transferred to the primary, it is **divided** by K^2. It is then called *equivalent secondary resistance or reactance referred to primary* and is denoted by R_2' or X_2'.

Equivalent resistance of transformer referred to primary is

$$R_{01} = R_1 + R_2' = R_1 + R_2/K^2$$

Equivalent reactance of transformer referred to primary is

$$X_{01} = X_1 + X_2' = X_1 + X_2/K^2$$

Equivalent impedance of transformer referred to primary is

$$Z_{01} = \sqrt{R_{01}^2 + X_{01}^2}$$

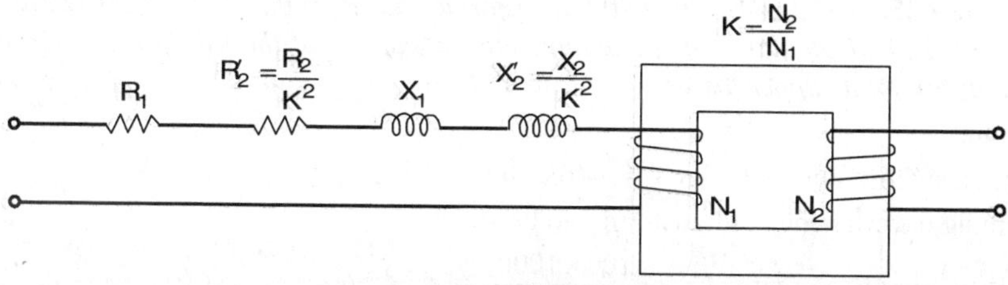

Fig. 8.21

Fig. 8.21 shows the resistance and reactance of the secondary referred to the primary. Note that secondary now has no resistance or reactance.

(*ii*) **Referred to secondary.** When primary resistance or reactance is transferred to the secondary, it is **multiplied** by K^2. It is then called *equivalent primary resistance or reactance referred to the secondary* and is denoted by R_1' or X_1'.

Equivalent resistance of transformer referred to secondary is

$$R_{02} = R_2 + R_1' = R_2 + K^2 R_1$$

Equivalent reactance of transformer referred to secondary is

$$X_{02} = X_2 + X_1' = X_2 + K^2 X_1$$

Equivalent impedance of transformer referred to secondary is

$$Z_{02} = \sqrt{R_{02}^2 + X_{02}^2}$$

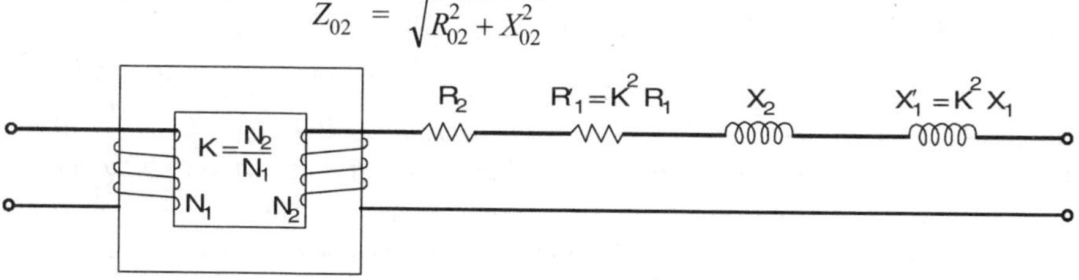

Fig. 8.22

Fig. 8.22 shows the resistance and reactance of the primary referred to the secondary. Note that primary now has no resistance or reactance.

Example 8.24. *A 10 kVA, 2000/400 V single-phase transformer has $R_1 = 5\ \Omega$; $X_1 = 12\ \Omega$; $R_2 = 0{\cdot}2\ \Omega$ and $X_2 = 0{\cdot}48\ \Omega$. Determine the equivalent impedance of the transformer referred to (i) primary side (ii) secondary side.*

Solution.

$$K = 400/2000 = 1/5$$

(*i*)
$$R_{01} = R_1 + \frac{R_2}{K^2} = 5 + \frac{0{\cdot}2}{(1/5)^2} = 10\ \Omega$$

$$X_{01} = X_1 + \frac{X_2}{K^2} = 12 + \frac{0 \cdot 48}{(1/5)^2} = 24\ \Omega$$

$$\therefore \qquad Z_{01} = \sqrt{R_{01}^2 + X_{01}^2} = \sqrt{10^2 + 24^2} = \mathbf{26\ \Omega}$$

(*ii*)
$$R_{02} = R_2 + K^2 R_1 = 0{\cdot}2 + (1/5)^2 \times 5 = 0{\cdot}4\ \Omega$$

$$X_{02} = X_2 + K^2 X_1 = 0{\cdot}48 + (1/5)^2 \times 12 = 0{\cdot}96\ \Omega$$

$$\therefore \qquad Z_{02} = \sqrt{R_{02}^2 + X_{02}^2} = \sqrt{(0{\cdot}4)^2 + (0{\cdot}96)^2} = \mathbf{1{\cdot}04\ \Omega}$$

Alternatively :
$$Z_{02} = K^2 Z_{01} = (1/5)^2 \times 26 = \mathbf{1{\cdot}04\ \Omega}$$

Example 8.25. *A 100 kVA, 2200/440 V transformer has $R_1 = 0{\cdot}3\ \Omega$; $X_1 = 1{\cdot}1\ \Omega$; $R_2 = 0{\cdot}01\ \Omega$ and $X_2 = 0{\cdot}035\ \Omega$. Calculate (i) the equivalent impedance of the transformer referred to the primary and (ii) total copper losses.*

Solution.

Voltage transformation ratio, $K = 440/2200 = 1/5$

Assuming the efficiency of transformer to be 100 %,

$$\text{Full-load primary current, } I_1 = \frac{100 \times 1000}{2200} = 45{\cdot}45 \text{ A}$$

$$\text{Full-load secondary current, } I_2 = \frac{100 \times 1000}{440} = 227{\cdot}25 \text{ A}$$

Alternatively ;
$$I_2 = I_1/K = 45{\cdot}45 \times 5 = 227{\cdot}25 \text{ A}$$

(*i*)
$$R_{01} = R_1 + \frac{R_2}{K^2} = 0 \cdot 3 + \frac{0 \cdot 01}{(1/5)^2} = 0 \cdot 55\ \Omega$$

$$X_{01} = X_1 + \frac{X_2}{K^2} = 1 \cdot 1 + \frac{0 \cdot 035}{(1/5)^2} = 1 \cdot 975\ \Omega$$

$$\therefore \qquad Z_{01} = \sqrt{R_{01}^2 + X_{01}^2} = \sqrt{(0 \cdot 55)^2 + (1 \cdot 975)^2} = \mathbf{2{\cdot}05\ \Omega}$$

(*ii*)
$$\text{Total Cu losses} = I_1^2 R_{01} = (45{\cdot}45)^2 \times 0{\cdot}55 = \mathbf{1136{\cdot}14\ W}$$

Alternatively :
$$\text{Total Cu losses} = I_1^2 R_1 + I_2^2 R_2$$
$$= (45{\cdot}45)^2 \times 0{\cdot}3 + (227{\cdot}25)^2 \times 0{\cdot}01 = \mathbf{1136{\cdot}14\ W}$$

Example 8.26. *A transformer with a 10 : 1 ratio and rated at 50 kVA, 2400/ 240 V, 50 Hz is used to step down the voltage of a distribution system. The low tension voltage is to be kept constant at 240 V.*

(i) What load impedance connected to low-tension side will be loading the transformer fully at 0·8 power factor (lag) ?

(ii) What is the value of this impedance referred to high tension side ?

(iii) What is the value of the current referred to high tension side ?

Solution.

(*i*)
$$\text{F.L. secondary current, } I_2 = \frac{kVA \times 10^3}{V_2} = \frac{50 \times 10^3}{240} = \frac{625}{3} \text{ A}$$

$$\text{Load impedance, } Z_2 = \frac{V_2}{I_2} = \frac{240}{625/3} = \mathbf{1{\cdot}152\ \Omega}$$

(ii) $$K = 240/2400 = 1/10$$

Secondary impedance referred to primary side is

$$Z_2' = \frac{Z_2}{K^2} = \frac{1 \cdot 152}{(1/10)^2} = 1 \cdot 152 \times 100 = \mathbf{115 \cdot 2\ \Omega}$$

(iii) Secondary current referred to primary is

$$I_2' = K I_2 = \frac{1}{10} \times \frac{625}{3} = \mathbf{20 \cdot 83\ A}$$

TUTORIAL PROBLEMS

1. A 100 kVA transformer has 400 turns on the primary and 80 turns on the secondary. The primary and secondary resistances are 0·3 Ω and 0·01 Ω respectively and the corresponding leakage reactances are 1·1 Ω and 0·035 Ω respectively. Calculate the equivalent impedance referred to primary circuit. **[2·05 Ω]**
2. A 10 kVA, 2000/400 V single-phase transformer has the following resistances and leakage reactances:
   Primary winding : Resistance 5 Ω ; leakage reactance 12 Ω
   Secondary winding : Resistance 0·2 Ω ; leakage reactance 0·48 Ω
   Calculate (i) the equivalent resistance referred to secondary (ii) the equivalent leakage reactance referred to secondary. **[(i) 0·4 Ω (ii) 0·06 Ω]**
3. A 40 kVA, 2000/250 V transformer has a primary resistance of 1·15 Ω and a secondary resistance of 0·0155 Ω. Calculate (i) the total resistance in terms of secondary winding (ii) the total resistance drop on full-load and (iii) the total copper loss on full-load. **[(i) 0·0334 Ω (ii) 5·35 V (iii) 855 W]**

8.11 IMPORTANCE OF SHIFTING IMPEDANCES

If we shift all the impedances from one winding to the other, the transformer is eliminated and we get an equivalent electrical circuit. Various voltages and currents can be readily obtained by solving this electrical circuit.

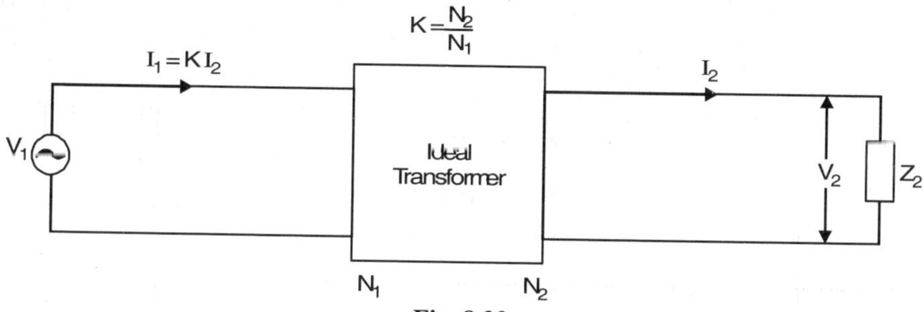

Fig. 8.23

Consider an *ideal transformer having an impedance Z_2 in the secondary as shown in Fig. 8.23.

(i) **Referred to primary.** When impedance Z_2 in the secondary is transferred to the primary, it becomes Z_2/K^2 as shown in Fig. 8.24 (i). Note that in Fig. 8.24 (i), the secondary of the ideal transformer is on open-circuit. Consequently, both primary and secondary currents are **zero. We can, therefore, remove the transformer, yielding the equivalent circuit shown in Fig. 8.24 (ii). The primary current can now be readily found out.

$$I_1 = \frac{V_1}{(Z_2 / K^2)}$$

* A practical transformer consists of an ideal transformer with various resistances and reactances external to it. The treatment will, therefore, be equally valid for a practical transformer.

** The ideal transformer is located between primary terminals *ab* and secondary terminals *cd* [See Fig. 8.24 (i)]. Terminals *c*, *d* are open and hence secondary current is zero. Since secondary current is zero, primary current will also be zero *i.e.*, no current will flow through terminals *a*, *b*. The ideal transformer can thus be removed.

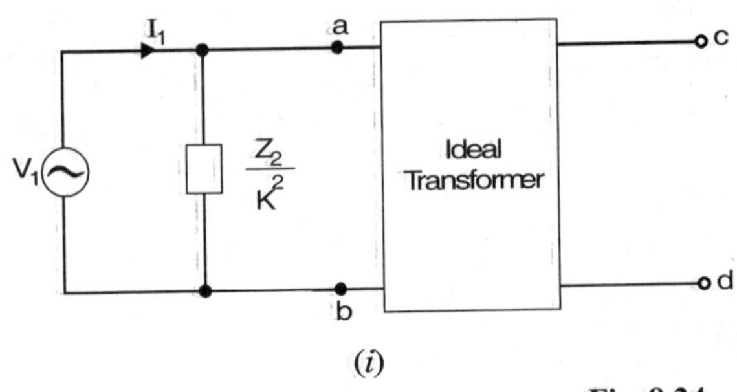

(i)

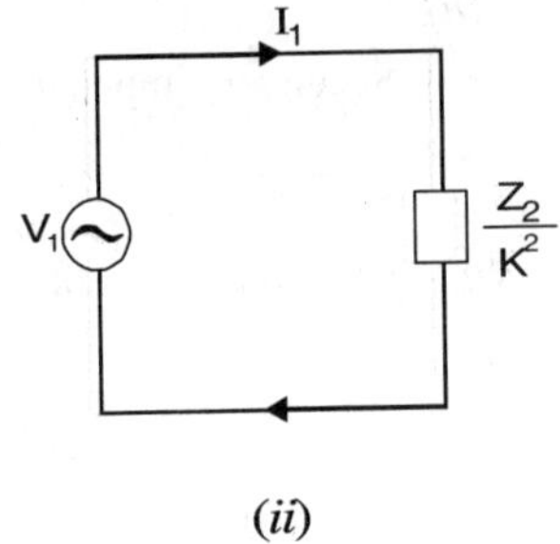

(ii)

Fig. 8.24

The circuits of Fig. 8.23 and Fig. 8.24 (*ii*) are electrically equivalent. Thus referring to Fig. 8.23,

$$I_1 = K I_2$$

Also if we refer to Fig. 8.24 (*ii*), we have,

$$I_1 = \frac{V_1}{(Z_2 / K^2)} = \frac{K^2 V_1}{Z_2} = \frac{K (K V_1)}{Z_2}$$

$$= K \frac{V_2}{Z_2} = K I_2 \qquad \left(\because I_2 = \frac{V_2}{Z_2}\right)$$

Thus the value of primary current I_1 is the same whether we use Fig. 8.23 or Fig. 8.24 (*ii*). Obviously, it is easier to use Fig. 8.24 (*ii*) as it contains no transformer.

(*ii*) **Referred to secondary.** Refer back to Fig. 8.23. There is no impedance on the primary side. However, voltage V_1 in the primary when transferred to the secondary becomes $K V_1$ as shown in Fig. 8.25 (*i*). Note that in Fig. 8.25 (*i*), the primary of the transformer is on open circuit. Consequently, both primary and secondary currents are zero. As before, we can remove the transformer, yielding the equivalent circuit shown in Fig. 8.25 (*ii*). The secondary current I_2 can be readily found out as :

$$I_2 = \frac{K V_1}{Z_2}$$

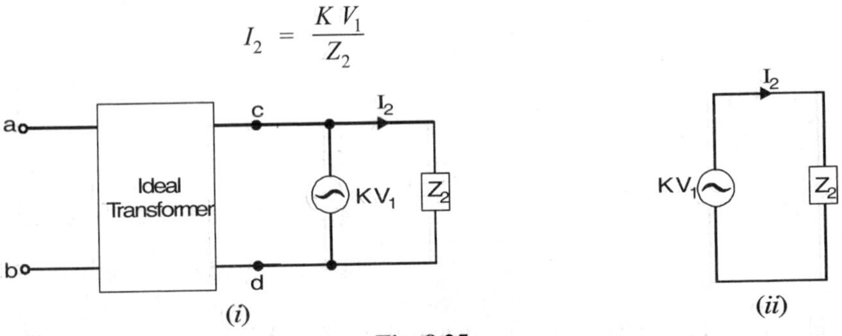

Fig. 8.25

The circuits of Fig. 8.23 and Fig. 8.25 (*ii*) are electrically equivalent. Thus referring back to Fig. 8.23, we have,

$$I_2 = V_2 / Z_2$$

Also if we refer to Fig. 8.25 (*ii*), we have,

$$I_2 = \frac{K V_1}{Z_2} = \frac{K (V_2 / K)}{Z_2} = \frac{V_2}{Z_2}$$

Thus the value of secondary current I_2 is the same whether we use Fig. 8.23 or Fig. 8.25 (*ii*). Obviously, it is easier to use Fig. 8.25 (*ii*) as it contains no transformer.

Example 8.27. *A single phase 400/2000 V transformer has a resistance of 1 Ω connected in series with primary winding and a 225 Ω resistor connected across its secondary winding. Calculate the primary current when the circuit is supplied at 400 V.*

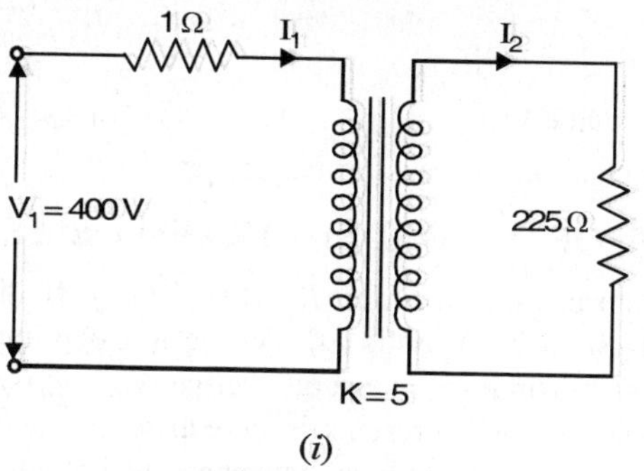

(i)

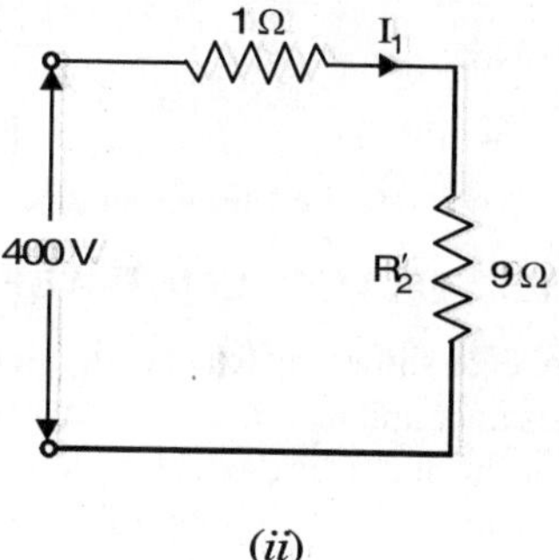

(ii)

Fig. 8.26

Solution. The conditions of the problem are represented in Fig. 8.26 (*i*).

$$K = 2000/400 = 5\,;R_1 = 1\,\Omega\,;R_2 = 225\,\Omega$$

Equivalent resistance of transformer referred to primary is

$$R_{01} = R_1 + R_2'$$

$$= R_1 + \frac{R_2}{K^2} = 1 + \frac{225}{(5)^2} = 10\,\Omega$$

The equivalent circuit of the transformer referred to primary is shown in Fig. 8.26 (*ii*). This is a simple series circuit.

$$\text{Primary current, } I_1 = \frac{V_1}{R_{01}} = \frac{V_1}{R_1 + R_2'} = \frac{400}{1+9} = \mathbf{40\ A}$$

Example 8.28. *Calculate the secondary voltage V and primary current I in the circuit shown in Fig. 8.27 (i). The transformer has primary to secondary turn ratio of 1 : 100.*

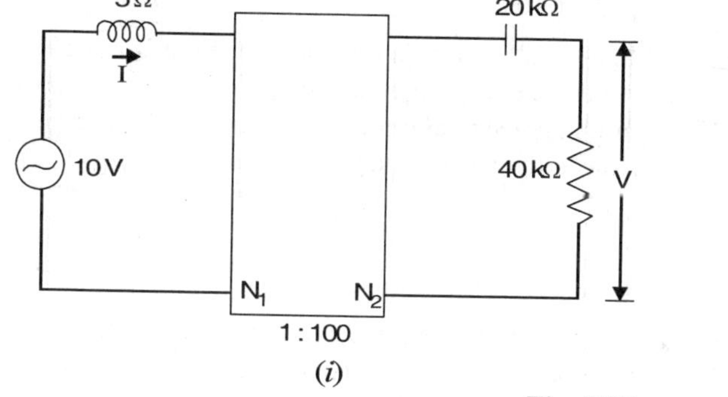

(i)

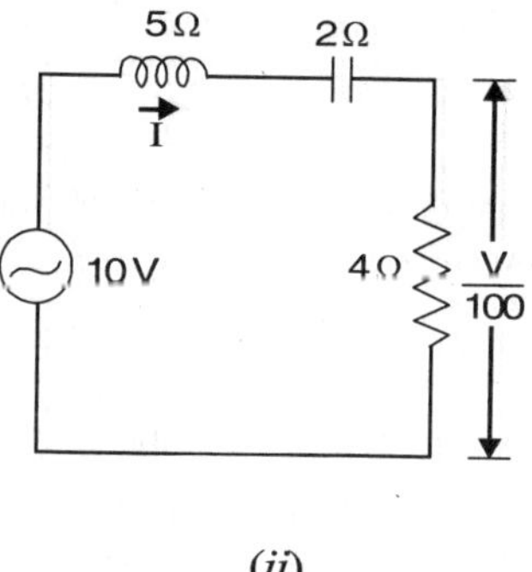

(ii)

Fig. 8.27

Solution.

$$K = 100/1 = 100\,;\mathbf{Z_2} = (40000 - j\,20000)\,\Omega$$

Equivalent impedance of secondary referred to primary is

$$\mathbf{Z_2'} = \frac{Z_2}{K^2} = \frac{1}{(100)^2}(40000 - j\,20000) = (4 - j\,2)\,\Omega$$

The equivalent circuit of the transformer referred to primary is shown in Fig. 8.27 (*ii*). Note that voltage across 40 kΩ resistance in Fig. 8.27 (*i*) is *V*. But when this resistance is transferred to the primary, the voltage across this resistance becomes V/K or $V/100$.

Referring to the equivalent circuit in Fig. 8.27 (*ii*), we have,

$$\text{Circuit impedance, } Z = \sqrt{R^2 + (X_L - X_C)^2} = \sqrt{4^2 + (5-2)^2} = 5\,\Omega$$

$$\therefore \quad \text{Primary current, } I = 10/Z = 10/5 = \mathbf{2\ A}$$

Now, $$\frac{V}{100} = I \times 4 = 2 \times 4$$

$$\therefore \quad V = (2 \times 4) \times 100 = \mathbf{800\ V}$$

Note how the equivalent circuit permits easy calculations of transformer problems.

8.12 EXACT EQUIVALENT CIRCUIT OF A LOADED TRANSFORMER

Fig. 8.28 shows the exact equivalent circuit of a transformer on load. Here R_1 is the primary winding resistance and R_2 is the secondary winding resistance. Similarly, X_1 is the leakage reactance of primary winding and X_2 is the leakage reactance of the secondary winding. The parallel circuit $R_0 - X_0$ is the no-load equivalent circuit of the transformer. The resistance R_0 represents the core losses (hysteresis and eddy current losses) so that current I_W which supplies the core losses is shown passing through R_0. The inductive reactance X_0 represents a loss-free coil which passes the magnetising current I_m. The phasor sum of I_W and I_m is the *no-load current I_0 of the transformer.

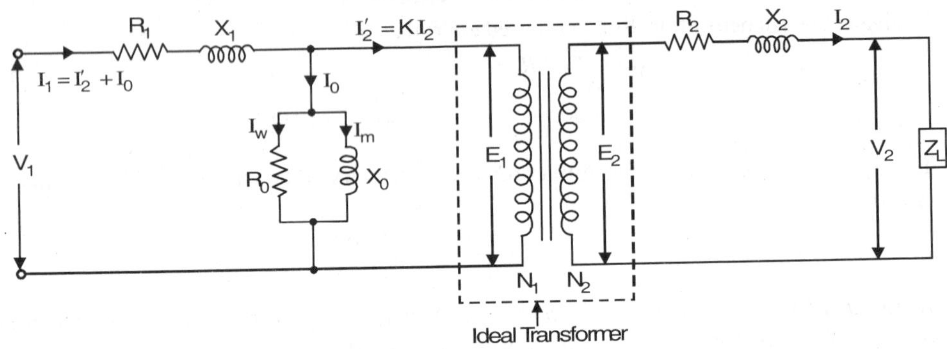

Fig. 8.28

Note that in the equivalent circuit shown in Fig. 8.28, the imperfections of the transformer have been taken into account by various circuit elements. Therefore, the transformer is now the ideal one. Note that equivalent circuit has created two normal electrical circuits separated only by an ideal transformer whose function is to change values according to the equation :

$$\frac{E_2}{E_1} = \frac{N_2}{N_1} = \frac{I_2'}{I_2}$$

The following points may be noted from the equivalent circuit :

(*i*) When the transformer is on *no-load* (*i.e.*, secondary terminals are open-circuited), there is no current in the secondary winding. However, the primary draws a small no-load current I_0. The no-load primary current I_0 is composed of (*a*) magnetising current (I_m) to create magnetic flux in the core and (*b*) the current I_W required to supply the core losses.

(*ii*) When the secondary circuit of a transformer is closed through some external load Z_L, the voltage E_2 induced in the secondary by mutual flux will produce a secondary current I_2. There will be $I_2 R_2$ and $I_2 X_2$ drops in the secondary winding so that load voltage V_2 will be less than E_2.

$$\mathbf{V_2} = \mathbf{E_2} - \mathbf{I_2}\,(R_2 + j\,X_2) = \mathbf{E_2} - \mathbf{I_2}\,\mathbf{Z_2}$$

(*iii*) When the transformer is loaded to carry the secondary current I_2, the primary current consists of two components :

(*a*) The no-load current I_0 to provide magnetising current and the current required to supply the core losses.

(*b*) The primary current I_2' ($= K\,I_2$) required to supply the load connected to the secondary.

* Note that no-load current of a transformer is independent of the load current I_2.

$\therefore$ Total primary current, $\mathbf{I}_1 = \mathbf{I}_0 + (-K\mathbf{I}_2)$ ....phasor sum

(*iv*) Since the transformer in Fig. 8.28 is now ideal, the primary induced voltage E_1 can be calculated from the relation :

$$\frac{E_1}{E_2} = \frac{N_1}{N_2}$$

If we add $I_1 R_1$ and $I_1 X_1$ drops to E_1, we get the primary input voltage V_1.

$$\mathbf{V}_1 = -\mathbf{E}_1 + \mathbf{I}_1 (R_1 + jX_1) = -\mathbf{E}_1 + \mathbf{I}_1 \mathbf{Z}_1 \quad \text{...phasor sum}$$

or

$$\mathbf{V}_1 = -\mathbf{E}_1 + \mathbf{I}_1 \mathbf{Z}_1 \quad \text{...phasor sum}$$

8.13 SIMPLIFIED EQUIVALENT CIRCUIT OF A LOADED TRANSFORMER

The no-load current I_0 of a transformer is small as compared to the rated primary current. Therefore, voltage drops in R_1 and X_1 due to I_0 are negligible. The equivalent circuit shown in Fig. 8.28 above can, therefore, be simplified by transferring the shunt circuit $R_0 - X_0$ to the input terminals as shown in Fig. 8.29. This modification leads to only slight loss of accuracy.

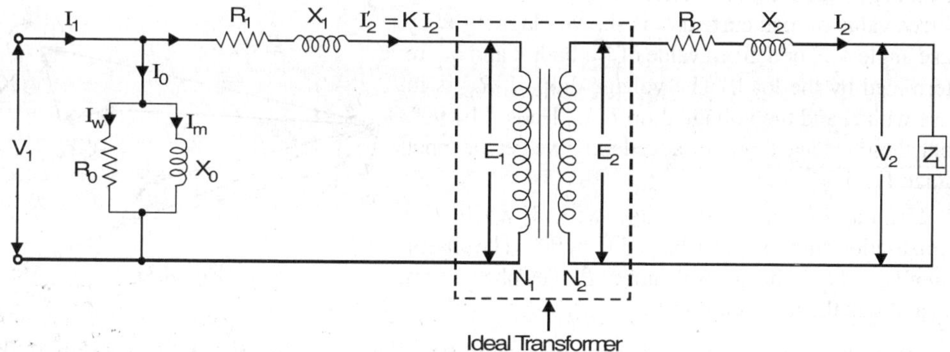

Fig. 8.29

(*i*) **Equivalent circuit referred to primary.** If all the secondary quantities are referred to the primary, we get the equivalent circuit of the transformer referred to the primary as shown in Fig. 8.30 (*i*). This further reduces to Fig. 8.30 (*ii*). Note that when secondary quantities are referred to primary, resistances/reactances/impedances are divided by K^2, voltages are divided by K and currents are multiplied by K.

$$\therefore \quad R_2' = \frac{R_2}{K^2} \; ; \; X_2' = \frac{X_2}{K^2} \; ; \; Z_L' = \frac{Z_L}{K^2} \; ; \; V_2' = \frac{V_2}{K} \; ; I_2' = K I_2$$

$$\mathbf{Z}_{01} = R_{01} + jX_{01}$$

where

$$R_{01} = R_1 + R_2' \; ; \; X_{01} = X_1 + X_2'$$

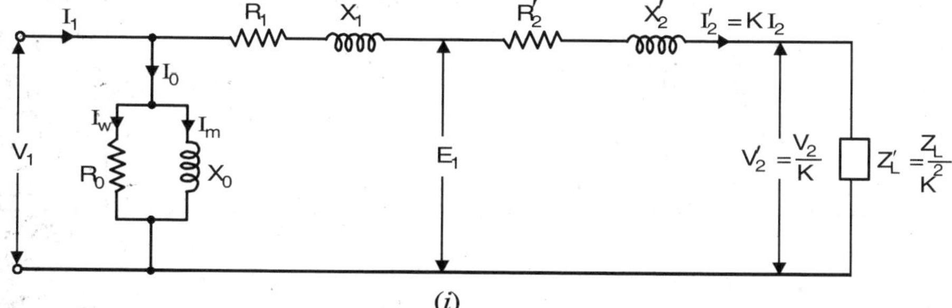

(*i*)

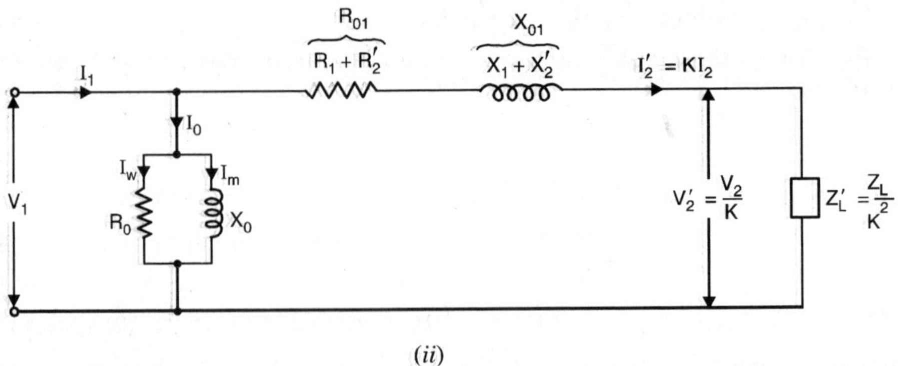

(ii)

Fig. 8.30

Phasor diagram. Fig. 8.31 shows the phasor diagram corresponding to the equivalent circuit shown in Fig. 8.30 (*ii*). The referred value of load voltage V_2' is chosen as the reference phasor. The referred value of load current I_2' is shown *lagging V_2' by phase angle ϕ_2. For a given value of V_2', both I_2' and ϕ_2 are determined by the load. The voltage drop $I_2'\, R_{01}$ is in phase with I_2' and the voltage drop $I_2'\, X_{01}$ leads I_2' by 90º. When these voltage drops are added to V_2', we get the input voltage V_1.

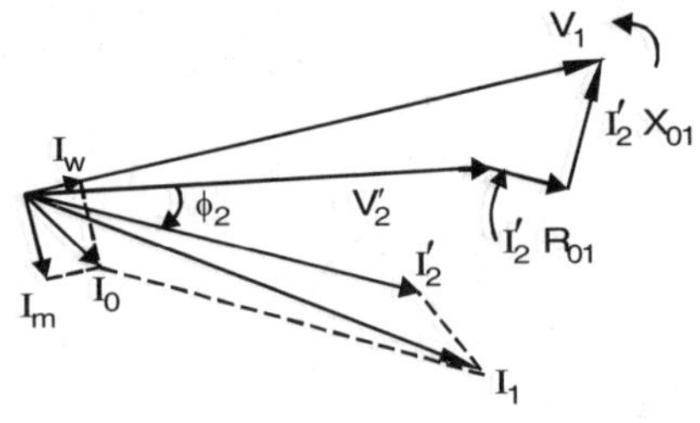

Fig. 8.31

The current I_W is in phase with V_1 while the magnetisation current I_m lags behind V_1 by 90º. The phasor sum of I_W and I_m is the no-load current I_0. The phasor sum of I_0 and I_2' is the input current I_1.

(*ii*) **Equivalent circuit referred to secondary.** If all the primary quantities are referred to secondary, we get the equivalent circuit of the transformer referred to secondary as shown in Fig. 8.32 (*i*). This further reduces to Fig. 8.32 (*ii*). Note that when primary quantities are referred to secondary resistances/ reactances/impedances are multiplied by K^2, voltages are multiplied by K and currents are divided by K.

$$\therefore \quad R_1' = K^2 R_1 \;;\; X_1' = K^2 X_1 \;;\; V_1' = K V_1 \;;\; I_1' = \frac{I_1}{K}$$

$$\mathbf{Z_{02}} = R_{02} + j\, X_{02}$$

where $$R_{02} = R_2 + R_1' \;;\; X_{02} = X_2 + X_1'$$

(*i*)

* The load is assumed to be inductive.

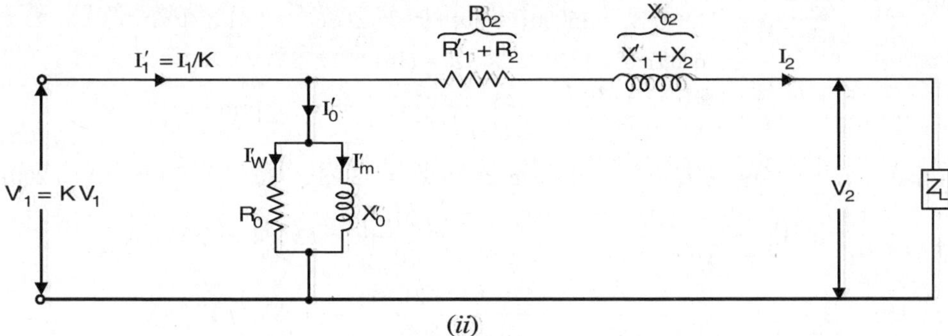

(ii)

Fig. 8.32

Phasor diagram. Fig. 8.33 shows the phasor diagram of the equivalent circuit shown in Fig. 8.32 (*ii*). The load voltage V_2 is chosen as the reference phasor. The load current I_2 is shown *lagging the load voltage V_2 by phase angle ϕ_2. The voltage drop $I_2 R_{02}$ is in phase with I_2 and the voltage drop $I_2 X_{02}$ leads I_2 by 90°. When these voltage drops are added to V_2, we get the referred primary voltage V'_1 $(= KV_1)$.

The current **I'_W is in phase with V'_1 while the magnetising current I'_m lags behind V'_1 by 90°. The phasor sum of I'_W and I'_m gives the referred value of no-load current I'_0. The phasor sum of I'_0 and load current I_2 gives the referred primary current I'_1 $(= I_1/K)$.

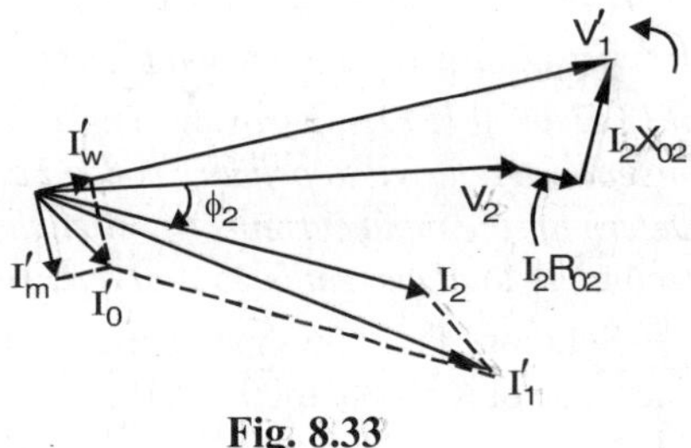

Fig. 8.33

Example 8.29. *A 4500/16000 V, 1500 kVA, 50 Hz transformer has the following parameters :*

$$R_1 = 0{\cdot}03\ \Omega,\quad R_2 = 0{\cdot}44\ \Omega,\quad R_0 = 1688\ \Omega$$
$$X_1 = 0{\cdot}092\ \Omega,\quad X_2 = 1{\cdot}34\ \Omega,\quad X_0 = 256\ \Omega$$

The transformer is supplying full-load at a power factor of 0·8 lagging. Using exact equivalent circuit, find the input current.

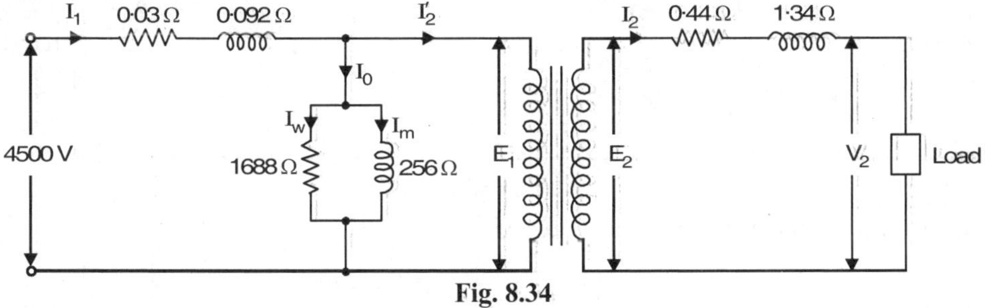

Fig. 8.34

Solution. Fig. 8.34 shows the exact equivalent circuit of the transformer. Here $K = 16000/4500 = 160/45$. Taking the load voltage as reference phasor, we have, $\mathbf{V_2} = 16000\ \angle 0°$ V.

$$\text{F.L. secondary current, } I_2 = \frac{\text{kVA} \times 10^3}{V_2} = \frac{1500 \times 10^3}{16000} = 93{\cdot}75 \text{ A}$$

$$\therefore \quad \mathbf{I_2} = 93{\cdot}75\ \angle -36{\cdot}9° \text{ A} \qquad (\because \text{ Load is at 0·8 p.f. lagging})$$

$$\mathbf{Z_2} = (0{\cdot}44 + j\,1{\cdot}34)\ \Omega = 1{\cdot}41\ \angle 71{\cdot}82°\ \Omega$$

* The load is assumed to be inductive.

** I'_W and I'_m are the values referred to secondary.

$$\mathbf{E}_2 = \mathbf{V}_2 + \mathbf{I}_2\mathbf{Z}_2 = 16000(1 + j0) + 93{\cdot}75\ \angle{-36{\cdot}9^\circ} \times 1{\cdot}41\ \angle 71{\cdot}82^\circ$$
$$= 16000\ \angle 0^\circ + 132{\cdot}19\ \angle 34{\cdot}92^\circ = (16108 + j\,75{\cdot}7)\text{ V}$$

Now $$\mathbf{E}_1 = \frac{\mathbf{E}_2}{K} = \frac{1}{K} \times \mathbf{E}_2 = \frac{45}{160} \times (16108 + j\,75{\cdot}7) = (4530{\cdot}4 + j\,21{\cdot}3)\text{ V}$$

and $$\mathbf{I}_2' = K\,\mathbf{I}_2 = \frac{160}{45} \times 93{\cdot}75\ \angle{-36{\cdot}9^\circ} = 333{\cdot}3\ \angle{-36{\cdot}9^\circ}\text{ A} = (266{\cdot}64 - j\,200)\text{ A}$$

$$\mathbf{I}_W = \frac{\mathbf{E}_1}{(1688 + j\,0)} = \frac{(4530{\cdot}4 + j\,21{\cdot}3)}{(1688 + j\,0)} = (2{\cdot}684 + j\,0{\cdot}013)\text{ A}$$

$$\mathbf{I}_m = \frac{\mathbf{E}_1}{(0 + j\,256)} = \frac{(4530{\cdot}4 + j\,21{\cdot}3)}{(0 + j\,256)} = (0{\cdot}083 - j\,17{\cdot}7)\text{ A}$$

$$\mathbf{I}_0 = \mathbf{I}_W + \mathbf{I}_m = (2{\cdot}684 + j\,0{\cdot}013) + (0{\cdot}083 - j\,17{\cdot}7) = (2{\cdot}77 - j\,17{\cdot}69)\text{ A}$$

$\therefore$ $$\mathbf{I}_1 = \mathbf{I}_2' + \mathbf{I}_0 = (266{\cdot}64 - j\,200) + (2{\cdot}77 - j\,17{\cdot}69)$$
$$= (269{\cdot}4 - j\,217{\cdot}7)\text{ A} = \mathbf{346{\cdot}4\ \angle{-38{\cdot}94^\circ}\ A}$$

Example 8.30. *A 4000/400 V, 10 kVA transformer has primary and secondary winding resistances of 13 Ω and 0·15 Ω respectively. The leakage reactance referred to primary is 45 Ω. The magnetising impedance referred to primary is 6 kΩ and the resistance corresponding to the core loss is 12 kΩ.. Determine the input current (i) when the secondary terminals are open-circuited and (ii) when the secondary load current is 25 A at a power factor of 0·8 lagging. Use simplified equivalent circuit.*

Solution. Fig. 8.35 shows the simplified equivalent circuit of the transformer referred to primary. The value of $K = 400/4000 = 1/10$.

$$R_{01} = R_1 + R'_2 = R_1 + \frac{R_2}{K^2} = 13 + \frac{0{\cdot}15}{(1/10)^2} = 28\ \Omega \quad ; \quad X_{01} = 45\ \Omega \quad \text{(given)}$$

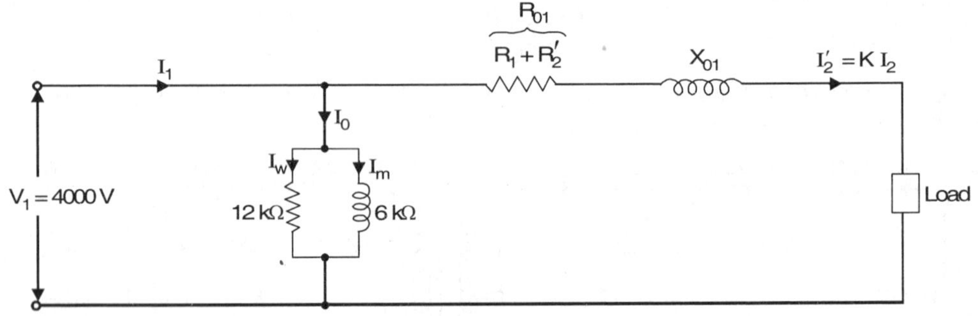

Fig. 8.35

(*i*) Core loss component of current is given by;

$$I_W = \frac{V_1}{R_0} = \frac{4000}{12 \times 10^3} = 0{\cdot}333\text{ A}$$

$$\text{Magnetising current, } I_m = \frac{V_1}{X_0} = \frac{4000}{6 \times 10^3} = 0{\cdot}667\text{ A}$$

$\therefore$ No-load current, $\mathbf{I}_0 = (0{\cdot}333 - j\,0{\cdot}667)\text{ A} = \mathbf{0{\cdot}745\ \angle{-63{\cdot}5^\circ}\ A}$

(*ii*) Secondary load current, $\mathbf{I}_2 = 25\ \angle{-36{\cdot}9^\circ}\text{ A}$

$$\mathbf{I}_2' = K\,\mathbf{I}_2 = \frac{1}{10} \times 25\ \angle{-36{\cdot}9^\circ} = 2{\cdot}5\ \angle{-36{\cdot}9^\circ}\text{ A} = (2 - j\,1{\cdot}5)\text{ A}$$

$$\text{Input current, } \mathbf{I}_1 = \mathbf{I}_2' + \mathbf{I}_0 = (2 - j\,1{\cdot}5) + (0{\cdot}333 - j\,0{\cdot}667)$$
$$= (2{\cdot}333 - j\,2{\cdot}167)\text{ A} = \mathbf{3{\cdot}18\ \angle{-43^\circ}\ A}$$

8.14 APPROXIMATE EQUIVALENT CIRCUIT OF LOADED TRANSFORMER

The no-load current I_0 in a transformer is only 1 – 3% of the rated primary current and may be neglected without any serious error. The transformer can then be shown as in Fig. 8.36. This is an

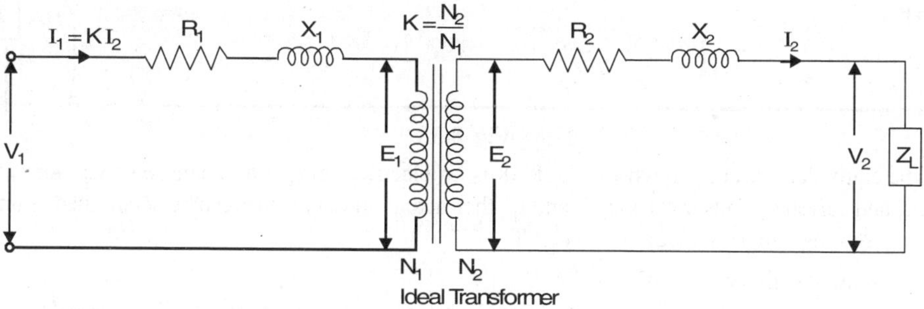

Fig. 8.36

approximate representation because no-load current has been neglected. Note that all the circuit elements have been shown external so that the transformer is an ideal one.

As shown in Art. 8.11, if we refer all the quantities to one side (primary or secondary), the ideal transformer stands removed and we get the equivalent circuit.

(*i*) **Equivalent circuit of transformer referred to primary.** If all the secondary quantities are referred to the primary, we get the equivalent circuit of the transformer referred to primary as shown in Fig. 8.37. Note that when secondary quantities are referred to primary, resistances/reactances are divided by K^2, voltages are divided by K and currents are multiplied by K.

The equivalent circuit shown in Fig. 8.37 is an electrical circuit and can be solved for various currents and voltages. Thus if we find V_2' and I_2', then actual secondary values can be determined as under :

Actual secondary voltage, $V_2 = K\,V_2'$

Actual secondary current, $I_2 = I_2'/K$

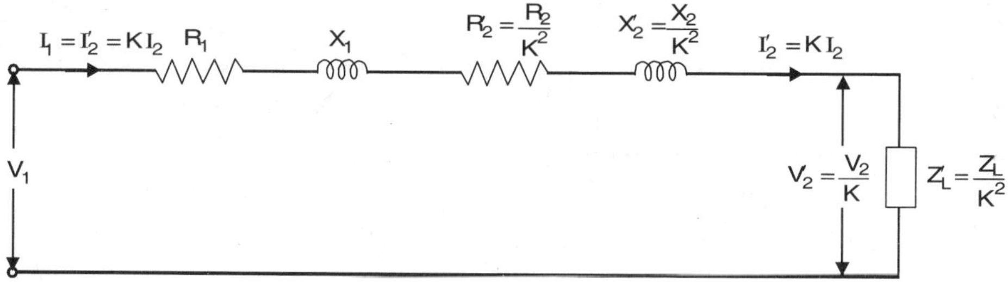

Fig. 8.37

(*ii*) **Equivalent circuit of transformer referred to secondary.** If all the primary quantities are referred to secondary, we get the equivalent circuit of the transformer referred to secondary as shown in Fig. 8.38. Note that when primary quantities are referred to secondary, resistances/reactances are multiplied by K^2, voltages are multiplied by K and currents are divided by K.

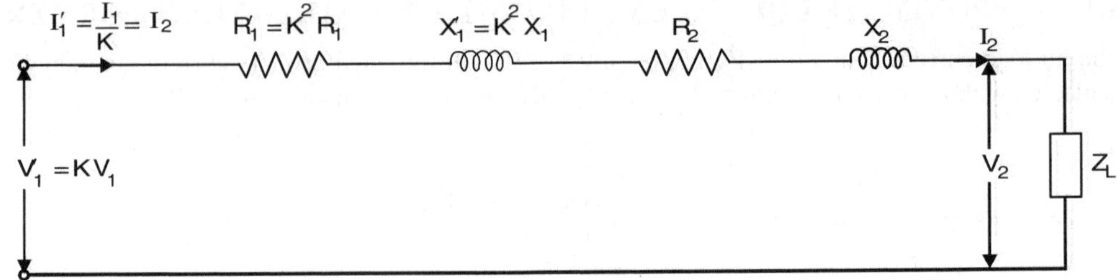

Fig. 8.38

The equivalent circuit shown in Fig. 8.38 is an electrical circuit and can be solved for various voltages and currents. Thus if we find V_1' and I_1', then actual primary values can be determined as under :

$$\text{Actual primary voltage, } V_1 = V_1'/K$$

$$\text{Actual primary current, } I_1 = K I_1'$$

Note : The same final answers will be obtained whether we use the equivalent circuit referred to primary or secondary. The use of a particular equivalent circuit would depend upon the conditions of the problem.

8.15 APPROXIMATE VOLTAGE DROP IN A TRANSFORMER

The approximate equivalent circuit of transformer referred to secondary is shown in Fig. 8.39. At no-load, the secondary voltage is $K V_1$. When a load having a lagging p.f. cos ϕ_2 is applied, the secondary carries a current I_2 and voltage drops occur in $(R_2 + K^2 R_1)$ and $(X_2 + K^2 X_1)$. Consequently, the secondary voltage falls from $K V_1$ to V_2. Referring to Fig. 8.39, we have,

$$\mathbf{V_2} = K\mathbf{V_1} - \mathbf{I_2}\,[(R_2 + K^2R_1) + j\,(X_2 + K^2 X_1)]$$

$$= K\mathbf{V_1} - \mathbf{I_2}\,(R_{02} + j\,X_{02})$$

$$= K\mathbf{V_1} - \mathbf{I_2}\,\mathbf{Z_{02}}$$

$$\text{Drop in secondary voltage} = K\mathbf{V_1} - \mathbf{V_2} = \mathbf{I_2}\,\mathbf{Z_{02}}$$

The phasor diagram is shown in Fig. 8.40. It is clear from the phasor diagram that drop in secondary voltage is $AC = I_2 Z_{02}$. It can be found as follows. With O as centre and OC as radius, draw an arc cutting OA produced at M. Then $AC = AM \simeq AN$. From B, draw BD perpendicular to OA produced. Draw CN perpendicular to OM and draw $BL \parallel OM$.

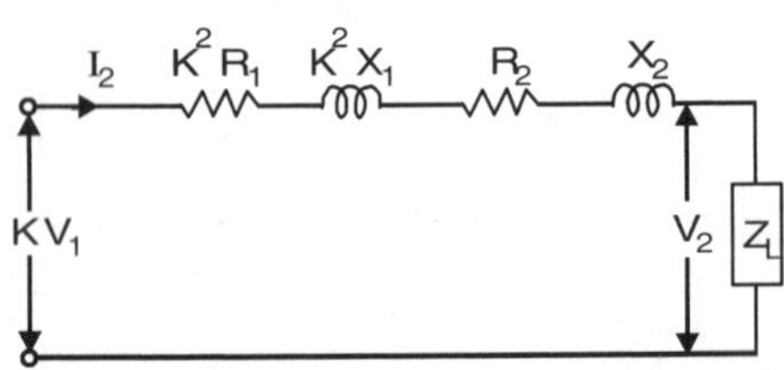

Fig. 8.39

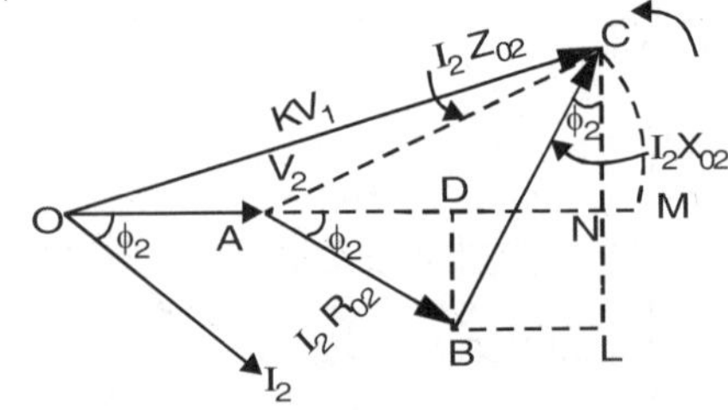

Fig. 8.40

Approximate drop in secondary voltage $= AN = AD + DN$

$$= AD + BL \qquad (\because BL = DN)$$

$$= I_2\,R_{02}\cos\phi_2 + I_2\,X_{02}\sin\phi_2$$

For a load having a leading p.f. cos ϕ_2, we have,

$$\text{Approximate voltage drop} = I_2\,R_{02}\cos\phi_2 - I_2\,X_{02}\sin\phi_2$$

Note : If the circuit is referred to primary, then it can be easily established that :

$$\text{Approximate voltage drop} = I_1\,R_{01}\,{}^*\cos\phi_2 \pm I_1\,X_{01}\sin\phi_2$$

* Assuming primary p.f., cos ϕ_1 = secondary p.f., cos ϕ_2.
This assumption is quite reasonable because the two values differ by a very small amount.

8.16 VOLTAGE REGULATION

The **voltage regulation** *of a transformer is the arithmetic difference (not phasor difference) between the no-load secondary voltage* (${}_0V_2$) *and the secondary voltage* V_2 *on load expressed as percentage of no-load voltage i.e.*

$$\text{\%age voltage regulation} = \frac{{}_0V_2 - V_2}{{}_0V_2} \times 100$$

where

$${}_0V_2 = \text{No-load secondary voltage} = K\,V_1$$

$$V_2 = \text{Secondary voltage on load}$$

As shown in Art. 8.15,

$${}_0V_2 - V_2 = I_2 R_{02} \cos\phi_2 \pm I_2 X_{02} \sin\phi_2$$

The +ve sign is for lagging p.f. and –ve sign for leading p.f.

It may be noted that %age voltage regulation of the transformer will be the same whether primary or secondary side is considered.

Example 8.31. *A 10 kVA, 2000/400 V single-phase transformer has the following data :*

$$R_1 = 5\,\Omega;\quad X_1 = 12\,\Omega;\quad R_2 = 0{\cdot}2\,\Omega;\quad X_2 = 0{\cdot}48\,\Omega$$

Determine the secondary terminal voltage at full-load, 0·8 p.f. lagging when the primary supply voltage is 2000 V.

Solution.

$$K = 400/2000 = 1/5$$

$$R_{02} = R_2 + K^2 R_1 = 0{\cdot}2 + (1/5)^2 \times 5 = 0{\cdot}4\,\Omega$$

$$X_{02} = X_2 + K^2 X_1 = 0{\cdot}48 + (1/5)^2 \times 12 = 0{\cdot}96\,\Omega$$

$$\text{F.L. secondary current, } I_2 = \frac{10 \times 1000}{400} = 25\text{ A}$$

$$\cos\phi_2 = 0{\cdot}8 \quad \therefore \quad \sin\phi_2 = 0{\cdot}6$$

$$\text{Voltage drop} = I_2 (R_{02} \cos\phi_2 + X_{02} \sin\phi_2) \qquad \text{...See Art. 8.15}$$

$$= 25\,(0{\cdot}4 \times 0{\cdot}8 + 0{\cdot}96 \times 0{\cdot}6) = 25\,(0{\cdot}32 + 0{\cdot}576) = 22{\cdot}4\text{ V}$$

∴ Secondary terminal voltage at full-load is

$$V_2 = 400 - 22{\cdot}4 = \mathbf{377{\cdot}6\ V}$$

Example 8.32. *The primary and secondary windings of a 40 kVA, 6600/250 V single-phase transformer have resistances of 10 Ω and 0·02 Ω respectively. The leakage reactance of the transformer referred to the primary side is 35 Ω. Calculate the percentage voltage regulation of the transformer when supplying full-load current at a p.f. of 0·8 lagging.*

Solution.

$$K = 250/6600 = 25/660$$

$$R_{01} = R_1 + \frac{R_2}{K^2} = 10 + 0{\cdot}02 \times (660/25)^2 = 23{\cdot}93\,\Omega,\quad X_{01} = 35\,\Omega\ (given)$$

$$\text{F.L. primary current, } I_1 = \frac{40 \times 1000}{6600} = 6{\cdot}06\text{ A}$$

$$\cos\phi_2 = 0{\cdot}8 \quad \therefore \quad \sin\phi_2 = 0{\cdot}6$$

$$\text{Voltage drop} = I_1 (R_{01} \cos\phi_2 + X_{01} \sin\phi_2)$$

$$= 6{\cdot}06\,(23{\cdot}93 \times 0{\cdot}8 + 35 \times 0{\cdot}6) = 243\text{ V}$$

$$\therefore \quad \text{\% age voltage regulation} = (243/6600) \times 100 = \mathbf{3{\cdot}7\ \%}$$

Example 8.33. *The primary and secondary windings of a 40 kVA, 6600/250 V single-phase transformer have resistances of 10 Ω and 0·02 Ω respectively. The leakage reactance of the transformer referred to the primary is 35 Ω. Calculate (i) the primary voltage required to circulate full-load current when the secondary is short-circuited (ii) the full-load regulation at 0·8 p.f. lagging. Neglect the no-load current.*

Solution.

$$K = 250/6600 = 25/660$$

(*i*) The resistance of secondary winding referred to primary is

$$R'_2 = \frac{R_2}{K^2} = \frac{0{\cdot}02}{(25/660)^2} = 0{\cdot}02 \times \left(\frac{660}{25}\right)^2 = 13{\cdot}9\,\Omega$$

$$\therefore \quad R_{01} = R_1 + R'_2 = 10 + 13{\cdot}9 = 23{\cdot}9\,\Omega\,; X_{01} = 35\,\Omega\ (given)$$

If the secondary is short circuited, the impedance of the load reflected in the primary is zero. Therefore, total impedance of transformer referred to primary is

$$\mathbf{Z_{01}} = R_{01} + j\,X_{01} = (23{\cdot}9 + j\,35)\,\Omega$$

$$\text{Magnitude, } Z_{01} = \sqrt{(23{\cdot}9)^2 + (35)^2} = 42{\cdot}24\,\Omega$$

$$\text{F.L. Primary current, } I_1 = 40 \times 10^3/6600 = 6{\cdot}06 \text{ A}$$

∴ Primary voltage required to circulate full-load current is

$$V_{SC} = I_1 \times Z_{01} = 6{\cdot}06 \times 42{\cdot}24 = \mathbf{256\ V}$$

(*ii*)

$$\text{Voltage drop} = I_1\,(R_{01}\cos\phi_2 + X_{01}\sin\phi_2)$$

$$= 6{\cdot}06\,(23{\cdot}93 \times 0{\cdot}8 + 35 \times 0{\cdot}6) = 243 \text{ V}$$

$$\therefore \quad \%\text{ voltage regulation} = \frac{243}{6600} \times 100 = \mathbf{3{\cdot}7\ \%}$$

Example 8.34. *Calculate the % voltage regulation of a transformer in which the percentage resistance drop is 1% and percentage reactance drop is 5% when the power factor is (i) 0·8 lagging (ii) unity and (iii) 0·8 leading.*

Solution.

$$\%\text{ voltage regulation} = \frac{I_2\,R_{02}\cos\phi_2 \pm I_2\,X_{02}\sin\phi_2}{{}_0V_2} \times 100$$

$$= \frac{100 \times I_2\,R_{02}}{{}_0V_2}\cos\phi_2 \pm \frac{100 \times I_2\,X_{02}}{{}_0V_2}\sin\phi_2$$

$$= v_r\cos\phi_2 \pm v_x\sin\phi_2$$

where

$$v_r = \frac{100 \times I_2\,R_{02}}{{}_0V_2} = \text{Percentage resistive drop}$$

$$v_x = \frac{100 \times I_2\,X_{02}}{{}_0V_2} = \text{Percentage reactive drop}$$

(*i*) When $\cos\phi_2 = 0{\cdot}8$ lag, regulation $= v_r\cos\phi_2 + v_x\sin\phi_2 = 1 \times 0{\cdot}8 + 5 \times 0{\cdot}6 =$ **3·8 %**

(*ii*) When $\cos\phi_2 = 1$, regulation $= 1 \times 1 + 5 \times 0 =$ **1%**

(*iii*) When $\cos\phi_2 = 0{\cdot}8$ lead, regulation $= 1 \times 0{\cdot}8 - 5 \times 0{\cdot}6 =$ **– 2·2%**

Example 8.35. *A 80 kVA, 2000/200 V, 50 Hz single-phase transformer has impedance drop of 8% and resistance drop of 4%. Calculate the % voltage regulation of the transformer at full-load 0·8 p.f. lagging.*

Solution.

$$\%\text{ impedance drop} = \frac{I_2\,Z_{02}}{{}_0V_2} \times 100 \quad \text{or} \quad 8 = \frac{I_2\,Z_{02}}{200} \times 100$$

$$\therefore \quad I_2\,Z_{02} = \frac{8 \times 200}{100} = 16 \text{ V}$$

$$\%\text{ resistance drop} = \frac{I_2 \times R_{02}}{{}_0V_2} \times 100$$

or $$4 = \frac{I_2 R_{02}}{200} \times 100 \quad \therefore \quad I_2 R_{02} = 8\text{ V}$$

$\therefore$ $$I_2 X_{02} = \sqrt{(I_2 Z_{02})^2 - (I_2 R_{02})^2} = \sqrt{(16)^2 - (8)^2} = 13{\cdot}86\text{ V}$$

$\therefore$ $$\%\text{ voltage regulation} = \frac{I_2 R_{02} \cos \phi_2 + I_2 X_{02} \sin \phi_2}{{}_0V_2} \times 100$$

$$= \frac{8 \times 0{\cdot}8 + 13{\cdot}86 \times 0{\cdot}6}{200} \times 100 = \mathbf{7{\cdot}35\ \%}$$

Example 8.36. *A 50 Hz, 1-ϕ transformer has a turn ratio of 6. The resistances are 0.9 Ω and 0.03 Ω and reactances are 5 Ω and 0.13 Ω for high voltage and low voltage windings respectively. Find (i) the voltage to be applied to the h.v. side to obtain full-load current of 200 A in the l.v. winding on short circuit (ii) the power factor on short circuit.*

Solution. The statement of the problem reveals that it is a step-down transformer *i.e.* primary is the h.v. side.

Now $$N_1/N_2 = 6 \quad \therefore K = N_2/N_1 = 1/6$$

Equivalent resistance of transformer referred to primary is

$$R_{01} = R_1 + R_2/K^2 = 0.9 + 0.03/(1/6)^2 = 1.98\ \Omega$$

Equivalent reactance of transformer referred to primary is

$$X_{01} = X_1 + X_2/K^2 = 5 + 0.13/(1/6)^2 = 9.68\ \Omega$$

Equivalent impedance of transformer referred to primary is

$$Z_{01} = \sqrt{(R_{01})^2 + (X_{01})^2} = \sqrt{(1.98)^2 + (9.68)^2} = 9.88\ \Omega$$

$$\text{Primary current, } I_1 = KI_2 = (1/6) \times 200 = 33.33\text{ A}$$

(*i*) Voltage to be applied to h.v. side to obtain full-load current on short-circuit of *l*.v. side is

$$V_{SC} = I_1 Z_{01} = 33.33 \times 9.88 = \mathbf{329.33\ V}$$

(*ii*) Power factor on short circuit is given by ;

$$\cos \phi_S = R_{01}/Z_{01} = 1.98/9.88 = \mathbf{0.2}\ \textit{lag}$$

Example 8.37. *A transformer has a reactive drop of 6% and a resistive drop of 2.5%. Determine the lagging power factor at which the voltage regulation is maximum and the value of this regulation.*

Solution.

$$\text{Voltage regulation} = \frac{I_2 R_{02} \cos \phi_2 + I_2 X_{02} \sin \phi_2}{V_2}$$

The voltage regulation will be maximum if

$$\frac{d}{d\phi_2}(\text{Regulation}) = 0$$

or $$\frac{d}{d\phi_2}\left(\frac{I_2 R_{02} \cos \phi_2 + I_2 X_{02} \sin \phi_2}{{}_0V_2}\right) = 0$$

or $$\frac{-I_2 R_{02}}{{}_0V_2} \sin \phi_2 + \frac{I_2 X_{02}}{{}_0V_2} \cos \phi_2 = 0$$

or $$\tan \phi_2 = \frac{X_{02}}{R_{02}}$$

Note that maximum regulation occurs at lagging power factor. Putting the given values, $\tan \phi_2 = 6/2.5 = 2.4$. Therefore, $\phi_2 = \tan^{-1} 2.4 = 67.38°$.

$\therefore$ $$\text{Power factor} = \cos \phi_2 = \cos 67.38° = \mathbf{0.385}\ \textit{lag}$$

$$\text{Maximum regulation} = \frac{I_2R_{02}}{{}_0V_2} \times 100 \times \cos\phi_2 + \frac{I_2X_{02}}{{}_0V_2} \times 100 \times \sin\phi_2$$

$$= (2.5) \times 0.385 + (6) \times 0.923 = \mathbf{6.5\%}$$

Example 8.38. *Calculate the percentage voltage drop for a transformer with a percentage resistance of 2.5% and a percentage reactance of 5% of rating 500 kVA when it is delivering 400 kVA at 0.8 p.f. lagging.*

Solution.

$$\% \text{ Voltage drop} = \frac{(\%R)I\cos\phi}{I_{F.L.}} + \frac{(\%X)I\sin\phi}{I_{F.L.}}$$

Here $I_{F.L.}$ is the full-load current and I is the actual current.

$$\therefore \quad \% \text{ Voltage drop} = \frac{^*(\%R)\,kW}{kVA \text{ rating}} + \frac{(\%X)\,kVAR}{kVA \text{ rating}}$$

In the present case, kW = 400 × 0.8 = 320 ; kVAR = 400 × 0.6 = 240

$$\therefore \quad \% \text{ Voltage drop} = \frac{2.5 \times 320}{500} + \frac{5 \times 240}{500} = \mathbf{4\%}$$

Example 8.39. *A 80 kVA, 2000/200 V, 50 Hz single-phase transformer has impedance drop of 8% and resistance drop of 4%. Calculate the regulation of the transformer at full-load 0.8 p.f. lagging. Also find the power factor at which voltage regulation will be zero.*

Solution.

$$\% \text{ Impedance drop} = \frac{I_2Z_{02}}{{}_0V_2} \times 100$$

or

$$8 = \frac{I_2Z_{02}}{200} \times 100 \qquad \therefore \qquad I_2Z_{02} = 16\text{V}$$

Similarly,

$$4 = \frac{I_2R_{02}}{200} \times 100 \qquad \therefore \qquad I_2R_{02} = 8\text{V}$$

Now

$$I_2X_{02} = \sqrt{(I_2Z_{02})^2 - (I_2R_{02})^2} = \sqrt{(16)^2 - (8)^2} = 13.86 \text{ V}$$

$$\therefore \quad \% \text{ Voltage regulation} = \frac{I_2R_{02}\cos\phi_2 + I_2X_{02}\sin\phi_2}{{}_0V_2} \times 100$$

$$= \frac{8 \times 0.8 + 13.86 \times 0.6}{200} \times 100 = \mathbf{7.35\%}$$

Condition for zero regulation

$$\text{Voltage regulation} = \frac{I_2R_{02}\cos\phi_2 + I_2X_{02}\sin\phi_2}{{}_0V_2}$$

The voltage regulation will be zero when the numerator is zero *i.e.*

$$I_2R_{02}\cos\phi_2 + I_2X_{02}\sin\phi_2 = 0$$

$$\therefore \quad \tan\phi_2 = -R_{02}/X_{02}$$

The negative sign shows that zero regulation occurs at a leading power factor. It's value is given by;

$$\text{Power factor} = \frac{R_{02}}{\sqrt{(R_{02})^2 + (X_{02})^2}} \qquad (\because \cos\phi_2 = R_{02}/Z_{02})$$

$$\text{In the present case, power factor} = \frac{8}{\sqrt{(8)^2 + (13.86)^2}} = \mathbf{0.5} \textit{ leading}$$

* Multiply the numerator and denominator by the rated voltage.

TUTORIAL PROBLEMS

1. A 500 kVA, 3800/400 V, 50 Hz single phase transformer has the following data :

$$R_1 = 0{\cdot}114\ \Omega\ ;\ X_1 = 0{\cdot}459\ \Omega\ ;\ R_2 = 0{\cdot}00102\ \Omega\ ;\ X_2 = 0{\cdot}00416\ \Omega$$

Find the percentage voltage regulation of the transformer when supplying full-load current at a p.f. of 0·8 lagging. **[2·2 %]**

2. A 100 kVA transformer has 400 turns on the primary and 80 turns on the secondary. The primary and secondary resistances are 0·3 Ω and 0·1 Ω respectively and the corresponding reactances are 1·1 Ω and 0·035 Ω respectively. The supply voltage is 2200 V. Calculate the voltage regulation and secondary terminal voltage for full-load having a p.f. of (*i*) 0·8 lagging (*ii*) 0·8 leading.

[(*i*) 3·36 % ; 425·2 V (*ii*) – 1·54 % ; 446·8 V]

3. The primary and secondary windings of a 30 kVA, 6000/230 V transformer have resistances of 10 Ω and 0·016 Ω respectively. The total reactance of the transformer referred to the primary is 23 Ω. Calculate the percentage voltage regulation of the transformer when supplying full-load current at a p.f. 0·8 lagging. **[2·55 %]**

4. The primary and secondary windings of a 30 kVA, 6000/230 V transformer have resistances of 10 Ω and 0·016 Ω respectively. The reactance of the transformer referred to primary is 34 Ω. Calculate (*i*) the primary voltage required to circulate full-load current when the secondary is short-circuited (*ii*) the percentage voltage regulation of the transformer for a load of 30 kVA having a p.f. of 0·8 lagging.

[(*i*) 200 V (*ii*) 3·1 %]

8.17 TRANSFORMER TESTS

The circuit *constants, efficiency and voltage regulation of a transformer can be determined by two simple tests (*i*) *open-circuit test* and (*ii*) *short-circuit test.* These tests are very convenient as they provide the required information without actually loading the transformer. Further, the power required to carry out these tests is very small as compared with full-load output of the transformer. These tests consist of measuring the input voltage, current and power to the primary first with secondary open-circuited (open-circuit test) and then with the secondary short-circuited (short circuit test).

8.18 OPEN-CIRCUIT OR NO-LOAD TEST

This test is conducted to determine the iron losses (or core losses) and parameters R_0 and X_0 of the transformer. In this test, the rated voltage is applied to the primary (usually **low-voltage winding) while the secondary is left open-circuited. The applied primary voltage V_1 is measured by the voltmeter, the no-load current I_0 by ammeter and no-load input power W_0 by wattmeter as shown in Fig. 8.41 (*i*). As the normal rated voltage is applied to the primary, therefore, normal iron losses will occur in the transformer core. Hence wattmeter will record the iron losses and small copper loss in the primary. Since no-load current I_0 is very small (usually 2 – 10 % of rated current), Cu losses in the primary under no-load condition are negligible as compared with iron losses. *Hence, wattmeter reading practically gives the iron losses in the transformer.* It is reminded that iron losses are the same at all loads. Fig. 8.41 (*ii*) shows the equivalent circuit of transformer on no-load.

---

* R_0, X_0, R_{01} (or R_{02}) and X_{01} (or X_{02}).

** The open-circuit test is always made on the low-voltage winding of the transformer. Of course, iron losses will be the same if measured on either winding. But if measurement is made on high-voltage winding, the current I_0 would be inconveniently small and the applied voltage inconveniently large.

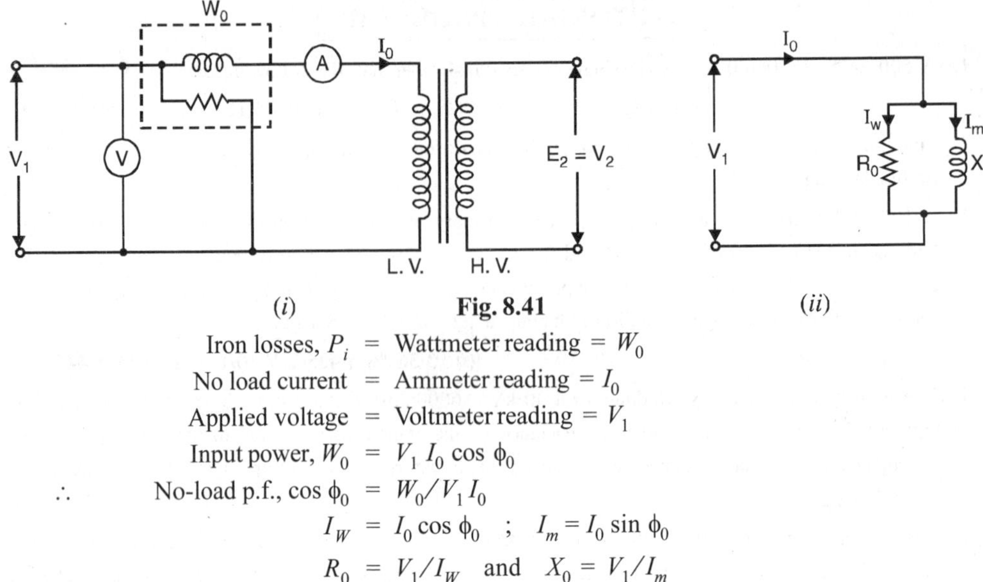

(i) **Fig. 8.41** (ii)

$$\text{Iron losses, } P_i = \text{Wattmeter reading} = W_0$$

$$\text{No load current} = \text{Ammeter reading} = I_0$$

$$\text{Applied voltage} = \text{Voltmeter reading} = V_1$$

$$\text{Input power, } W_0 = V_1 I_0 \cos \phi_0$$

$$\therefore \quad \text{No-load p.f., } \cos \phi_0 = W_0/V_1 I_0$$

$$I_W = I_0 \cos \phi_0 \quad ; \quad I_m = I_0 \sin \phi_0$$

$$R_0 = V_1/I_W \quad \text{and} \quad X_0 = V_1/I_m$$

Thus open-circuit test enables us to determine iron losses and parameters R_0 and X_0 of the transformer.

8.19 SHORT-CIRCUIT OR IMPEDANCE TEST

This test is conducted to determine R_{01} (or R_{02}), *X_{01} (or X_{02}) and full-load copper losses of the transformer. In this test, the secondary (usually **low-voltage winding) is short-circuited by a thick conductor and variable low voltage is applied to the primary as shown in Fig. 8.42 (*i*). The low input voltage is gradually raised till at voltage V_{SC}, full-load current I_1 flows in the primary. Then I_2 in the secondary also has full-load value since $I_1/I_2 = N_2/N_1$. Under such conditions, the copper loss in the windings is the same as that on full load.

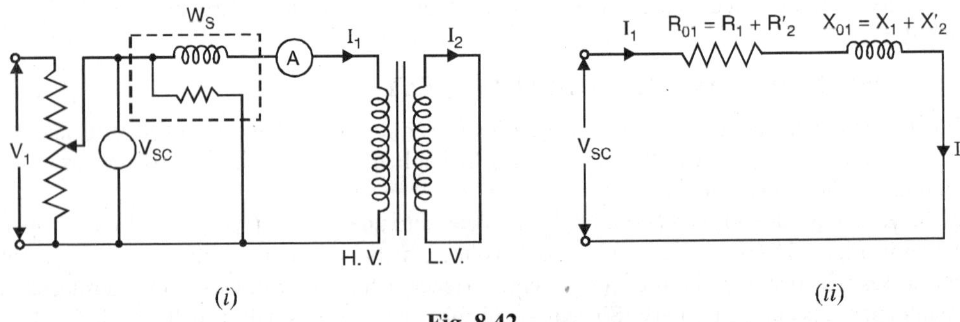

(i) (ii)

Fig. 8.42

There is no output from the transformer under short-circuit conditions. Therefore, input power is all loss and this loss is almost entirely copper loss. It is because iron loss in the core is negligibly

* R_1 and R_2 (and hence R_{01} or R_{02}) can be easily found by d.c. measurements. However, leakage reactances X_{01} or X_{02} cannot be found by any simple means. The short-circuit test is very convenient to find X_{01} or X_{02}.

** While making a short-circuit test, low-voltage winding is always short-circuited and measurements are made on the high-voltage winding. If measurements are made on the low-voltage winding (of course high-voltage winding being short-circuited), the voltage will be inconveniently low and the current would be inconveniently high.

†small since the voltage V_{SC} is very small. *Hence, the wattmeter will practically register the full-load copper losses in the transformer windings*. Fig. 8.42 (*ii*) shows the equivalent circuit of a transformer on short circuit as referred to primary ; the no-load current being neglected due to its smallness.

$$\text{Full load Cu loss, } P_C = \text{Wattmeter reading} = W_S$$

$$\text{Applied voltage} = \text{Voltmeter reading} = V_{SC}$$

$$\text{F.L. primary current} = \text{Ammeter reading} = I_1$$

$$P_C = I_1^2 R_1 + I_1^2 R'_2 = I_1^2 R_{01}$$

$$\therefore \quad R_{01} = P_C / I_1^2$$

where R_{01} is the total resistance of transformer referred to primary.

Total impedance referred to primary, $Z_{01} = V_{SC}/I_1$

Total leakage reactance referred to primary, $X_{01} = \sqrt{Z_{01}^2 - R_{01}^2}$

Short-circuit p.f., $\cos \phi_s = P_C / V_{SC} I_1$

Thus short-circuit test gives full-load Cu loss, R_{01} and X_{01}.

Note : The short-circuit test will give full-load Cu loss only if the applied voltage V_{SC} is such so as to circulate full-load currents in the windings. If in a short-circuit test, current value is other than full-load value, the Cu loss will be corresponding to that current value.

8.20 ADVANTAGES OF TRANSFORMER TESTS

The above two simple transformer tests offer the following advantages :

(*i*) The power required to carry out these tests is very small as compared to the full-load output of the transformer. In case of open-circuit test, power required is equal to the iron loss whereas for a short-circuit test, power required is equal to full-load copper loss.

(*ii*) These tests enable us to determine the efficiency of the transformer accurately at any load and p.f. without actually loading the transformer.

(*iii*) The short-circuit test enables us to determine R_{01} and X_{01} (or R_{02} and X_{02}). We can thus find the total voltage drop in the transformer as referred to primary or secondary. This permits us to calculate voltage regulation of the transformer.

Example 8.40. *A single-phase, 50 Hz transformer has a full-load secondary current of 500 A; the primary current being one-fifth of this value. The transformer has the following parameters :*

$$R_1 = 0{\cdot}6\ \Omega\ ;\ R_2 = 0{\cdot}03\ \Omega\ ;\ X_1 = 2\ \Omega \text{ and } X_2 = 0{\cdot}06\ \Omega$$

If the secondary is short-circuited, find the primary voltage required to circulate full-load current. Neglect the no-load current. What is the power factor on short circuit ?

Solution. Neglecting no-load current, the equivalent circuit of transformer on short-circuit as referred to primary will be as shown in Fig. 8.43. It is given that full-load $I_2 = 500$ A and full load $I_1 = 100$ A. Hence $K = I_1/I_2 = 1/5$.

$$R_{01} = R_1 + R_2/K^2 = 0{\cdot}6 + 0{\cdot}03 \times 5^2 = 1{\cdot}35\ \Omega$$

$$X_{01} = X_1 + X_2/K^2 = 2 + 0{\cdot}06 \times 5^2 = 3{\cdot}5\ \Omega$$

$$\therefore \quad Z_{01} = \sqrt{R_{01}^2 + X_{01}^2} = \sqrt{(1{\cdot}35)^2 + (3.5)^2} = 3{\cdot}75\ \Omega$$

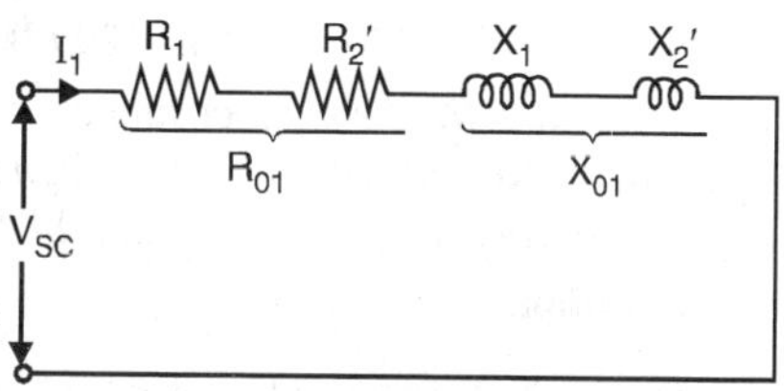

Fig. 8.43

Primary voltage required to circulate full-load current,

$$V_{SC} = I_1 Z_{01} = 100 \times 3{\cdot}75 = \mathbf{375\ V}$$

Short circuit p.f. is

$$\cos \phi_s = R_{01}/Z_{01} = 1{\cdot}35/3{\cdot}75 = \mathbf{0{\cdot}36}\ \textit{lag}$$

† Iron loss is proportional to the square of flux and hence to the square of applied voltage. Since applied voltage V_{SC} under short circuit conditions is about 1/10th of normal voltage, the iron loss will be 1/100th of the normal value. Consequently, the iron loss is negligible under short-circuit conditions.

Example 8.41. *The low-voltage winding of a 300 kVA, 11000/2200 V, 50 Hz transformer has 190 turns and a resistance of 0·06 Ω. The high voltage side has 910 turns and a resistance of 1.6 Ω. When the low-voltage winding is short-circuited, the full-load current is obtained with 550 V applied to h.v. side. Calculate the equivalent resistance and leakage reactance referred to high-voltage side. Assume full-load efficiency of 98·5 %.*

Solution.

$$V_{SC} = 550 \text{ V};\ K = 190/910 = 19/91\ ;\ \eta = 98{\cdot}5\ \%$$

$$\text{F.L. primary current, } I_1 = \frac{300 \times 10^3}{11000 \times 0{\cdot}985} = 27{\cdot}7 \text{ A}$$

$$Z_{01} = V_{SC}/I_1 = 550/27{\cdot}7 = 19{\cdot}8\ \Omega$$

$$R'_2 = R_2/K^2 = 0{\cdot}06/(19/91)^2 = 1{\cdot}38\ \Omega$$

$$R_{01} = R_1 + R'_2 = 1{\cdot}6 + 1{\cdot}38 = \mathbf{2{\cdot}98\ \Omega}$$

$$X_{01} = \sqrt{Z_{01}^2 - R_{01}^2} = \sqrt{(19{\cdot}8)^2 - (2{\cdot}98)^2} = \mathbf{19{\cdot}5\ \Omega}$$

Example 8.42. *In a no-load test of a single-phase transformer, the following test data were obtained :*

Primary voltage : 220 V ; Secondary voltage : 110 V

Primary current : 0·5 A ; Power input : 30 W

Find (i) the turns ratio (ii) the magnetising component of no-load current (iii) the iron-loss component of no-load current (iv) iron loss. The resistance of primary winding is 0·6 Ω.

Solution.

(*i*) $$\text{Turns ratio} = \frac{N_1}{N_2} = \frac{E_1}{E_2} \simeq \frac{V_1}{V_2} = \frac{220}{110} = \mathbf{2}$$

(*ii*) $$W_0 = V_1 I_0 \cos\phi_0$$

$\therefore$ $$\text{No-load p.f., } \cos\phi_0 = \frac{W_0}{V_1 I_0} = \frac{30}{220 \times 0{\cdot}5} = 0{\cdot}273\ ;\ \sin\phi_0 = 0{\cdot}962$$

$$\text{Magnetising current, } I_m = I_0 \sin\phi_0 = 0{\cdot}5 \times 0{\cdot}962 = \mathbf{0{\cdot}48\ A}$$

(*iii*) Component of no-load current corresponding to core loss is

$$I_W = I_0 \cos\phi_0 = 0{\cdot}5 \times 0{\cdot}273 = \mathbf{0{\cdot}1365\ A}$$

(*iv*) $$\text{Primary Cu loss} = I_0^2 R_1 = (0{\cdot}5)^2 \times 0{\cdot}6 = 0{\cdot}15 \text{ W}$$

$\therefore$ $$\text{Iron loss} = 30 - 0{\cdot}15 = \mathbf{29{\cdot}85\ W}$$

Example 8.43. *A 10 kVA, 200/400 V, 50 Hz single-phase transformer gave the following test results :*

O.C. test : 200 V, 1.3 A, 120 W .... on L.V. side

S.C. test : 22 V, 30 A, 200 W .... on H.V. side

Calculate (i) the magnetising current and the component corresponding to core loss at normal frequency and voltage (ii) the magnetising-branch impedances and (iii) percentage voltage regulation when supplying full-load at 0·8 p.f. leading.

Solution.

(*i*) $$\text{No-load p.f., } \cos\phi_0 = \frac{W_0}{V_1 I_0} = \frac{120}{200 \times 1{\cdot}3} = 0{\cdot}462$$

$$\text{Magnetising current, } I_m = I_0 \sin\phi_0 = 1{\cdot}3 \times 0{\cdot}886 = \mathbf{1{\cdot}15\ A}$$

Component of no-load current corresponding to core loss is

$$I_W = I_0 \cos\phi_0 = 1{\cdot}3 \times 0{\cdot}462 = \mathbf{0{\cdot}6\ A}$$

(*ii*)
$$R_0 = \frac{V_1}{I_0 \cos \phi_0} = \frac{200}{0 \cdot 6} = \mathbf{333\ \Omega}$$

$$X_0 = \frac{V_1}{I_0 \sin \phi_0} = \frac{200}{1 \cdot 15} = \mathbf{174\ \Omega}$$

(*iii*) Total impedance referred to H.V. side (secondary) is

$$Z_{02} = \frac{V_{SC}}{\text{sec. current}} = \frac{22}{30} = 0{\cdot}733\ \Omega$$

Total resistance referred to H.V. side (secondary) is

$$R_{02} = \frac{W_S}{(\text{sec. current})^2} = \frac{200}{(30)^2} = 0{\cdot}222\ \Omega$$

Total leakage reactance referred to H.V. side (secondary) is

$$X_{02} = \sqrt{Z_{02}^2 - R_{02}^2} = \sqrt{(0{\cdot}733)^2 - (0{\cdot}222)^2} = 0{\cdot}698\ \Omega$$

$$\text{F.L. current on H.V. side, } I_2 = \frac{10 \times 10^3}{400} = 25\text{ A}$$

$$\text{Approximate voltage drop} = I_2 (R_{02} \cos \phi_2 - X_{02} \sin \phi_2)$$
$$= 25(0{\cdot}222 \times 0.8 - 0{\cdot}698 \times 0{\cdot}6) = -6\text{ V}$$

Note that there is a voltage rise due to leading power factor.

$$\%\text{ age voltage regulation} = (-6/400) \times 100 = \mathbf{-1{\cdot}5\ \%}$$

Example 8.44. *A 200 kVA, 2000/440 V, 50 Hz single-phase transformer gave the following test results :*

O.C. test : 2000 V, 1·8 A, 1·75 kW .... on H.V. side

S.C. test : 13 V, 300 A, 1 kW .... on L.V. side

Obtain the equivalent circuit as referred to H.V. side.

Solution.

Component of no-load current corresponding to core loss is

$$I_W = \frac{W_0}{V_1} = \frac{1 \cdot 75 \times 10^3}{2000} = 0{\cdot}875\text{ A}$$

$$\text{Magnetising current, } I_m = \sqrt{I_0^2 - I_W^2} = \sqrt{(1 \cdot 8)^2 - (0 \cdot 875)^2} = 1{\cdot}57\text{ A}$$

$$\therefore \quad R_0 = \frac{V_1}{I_W} = \frac{2000}{0 \cdot 875} = 2286\ \Omega$$

$$X_0 = \frac{V_1}{I_m} = \frac{2000}{1 \cdot 57} = 1274\ \Omega$$

Total impedance referred to L.V. side (secondary) is

$$Z_{02} = V_{SC}/I_2 = 13/300 = 0{\cdot}0433\ \Omega$$

Total resistance referred to L.V. side (secondary) is

$$R_{02} = \frac{W_S}{I_2^2} = \frac{1 \times 10^3}{(300)^2} = 0{\cdot}011\ \Omega$$

$$\therefore \quad X_{02} = \sqrt{Z_{02}^2 - R_{02}^2} = \sqrt{(0 \cdot 0433)^2 - (0 \cdot 011)^2} = 0{\cdot}042\ \Omega$$

Total resistance and reactance referred to H.V. side (primary) are

$$R_{01} = \frac{R_{02}}{K^2} = \frac{0{\cdot}011}{(440/2000)^2} = 0{\cdot}011 \times \left(\frac{2000}{440}\right)^2 = 0{\cdot}227\ \Omega$$

$$X_{01} = \frac{X_{02}}{K^2} = \frac{0{\cdot}042}{(440/2000)^2} = 0{\cdot}042 \times \left(\frac{2000}{440}\right)^2 = 0{\cdot}868\ \Omega$$

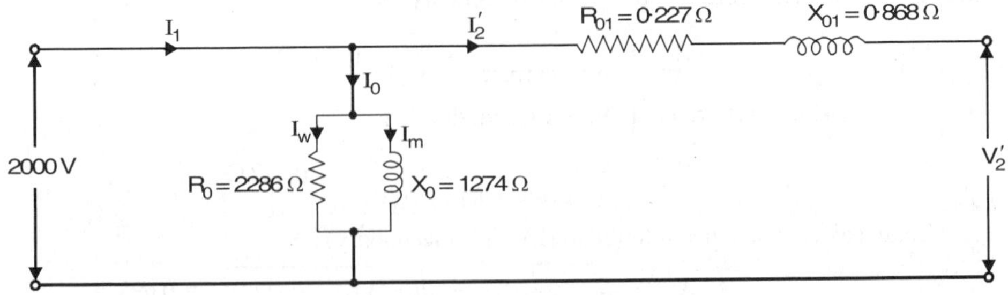

Fig. 8.44

Fig. 8.44 shows the required equivalent circuit as referred to H.V. side (primary).

Example 8.45. *A 4 kVA, 200/400 V, 50 Hz single-phase transformer gave the following test results :*

No-load : Low-voltage data 200 V, 0·7 A, 60 W

Short circuit : High-voltage data 9 V, 6 A, 21·6 W

Calculate (i) the magnetising current and the component corresponding to iron loss at normal frequency and voltage (ii) the efficiency on full-load at unity power factor (iii) the secondary terminal voltage on full-load at power factors of unity, 0·8 lagging and 0·8 leading.

Solution.

(*i*) From O.C. test, we have,

$$\text{No-load p.f., } \cos\phi_0 = \frac{W_0}{V_1 I_0} = \frac{60}{200 \times 0{\cdot}7} = 0{\cdot}43 \quad \therefore \sin\phi_0 = 0{\cdot}9$$

$$\text{Magnetising current, } I_m = I_0 \sin\phi_0 = 0{\cdot}7 \times 0{\cdot}9 = \mathbf{0{\cdot}63\ A}$$

$$\text{Iron loss component, } I_W = I_0 \cos\phi_0 = 0{\cdot}7 \times 0{\cdot}43 = \mathbf{0{\cdot}3\ A}$$

(*ii*) Copper loss at 6 A on 400 V side = 21·6 W

$$\text{F.L. secondary current (400 V side), } I_2 = \frac{4 \times 10^3}{400} = 10\ \text{A}$$

$$\therefore \quad \text{Total F.L. Cu losses} = 21{\cdot}6 \times (10/6)^2 = 60\ \text{W}$$

$$\text{Total F.L. losses} = 60 + 60 = 120\ \text{W}$$

$$\text{F.L. output at unity p.f.} = 4 \times 10^3 \times 1 = 4000\ \text{W}$$

$$\text{F.L. efficiency} = \frac{4000}{4000 + 120} \times 100 = \mathbf{97{\cdot}1\ \%}$$

(*iii*) From S.C. test (H.V. side), we have,

$$Z_{02} = \frac{V_{SC}}{\text{Secondary current}} = \frac{9}{6} = 1{\cdot}5\ \Omega$$

$$\cos\phi_s = \frac{\text{Watts}}{\text{Volt-amp}} = \frac{21{\cdot}6}{9 \times 6} = 0{\cdot}4 \ ; \ \sin\phi_s = 0{\cdot}92$$

$$\therefore \quad R_{02} = Z_{02} \cos\phi_s = 1{\cdot}5 \times 0{\cdot}4 = 0{\cdot}6\ \Omega$$

$$\text{and} \quad X_{02} = Z_{02} \sin\phi_s = 1{\cdot}5 \times 0{\cdot}92 = 1{\cdot}37\ \Omega$$

$$\text{Voltage drop in secondary} = I_2 (R_{02} \cos\phi_2 \pm X_{02} \sin\phi_2)$$

At unity p.f. ($\cos \phi_2 = 1$), voltage drop in secondary

$$= 10(0{\cdot}6 \times 1 + 0) = 6 \text{ V}$$

$$\therefore \quad \text{Load voltage} = 400 - 6 = \mathbf{394\ V}$$

At p.f. of 0·8 lagging ($\cos \phi_2 = 0{\cdot}8$), voltage drop in secondary

$$= 10(0{\cdot}6 \times 0{\cdot}8 + 1{\cdot}37 \times 0{\cdot}6) = 13 \text{ V}$$

$$\therefore \quad \text{Load voltage} = 400 - 13 = \mathbf{387\ V}$$

At p.f. of 0·8 leading, voltage drop in secondary

$$= 10(0{\cdot}6 \times 0{\cdot}8 - 1{\cdot}37 \times 0{\cdot}6) = -3{\cdot}4 \text{ V}$$

$$\therefore \quad \text{Load voltage} = 400 + 3{\cdot}4 = \mathbf{403{\cdot}4\ V}$$

Example 8.46. *A 100 kVA, 6600/330 V, 50 Hz, single-phase transformer took 10 A and 436 W at 100 V in a short-circuit test, the figures referring to the high voltage side. Calculate the voltage to be applied to the high voltage side on full-load at 0·8 p.f. lagging when the secondary load voltage is 330 V.*

Solution. From S.C. test, we have,

$$Z_{01} = \frac{V_{SC}}{\text{Primary current}} = \frac{100}{10} = 10\ \Omega$$

$$\cos \phi_p = \frac{\text{Watts}}{VA} = \frac{436}{100 \times 10} = 0{\cdot}436\ ;\ \sin \phi_p = 0{\cdot}9$$

$$\therefore \quad R_{01} = Z_{01} \cos \phi_p = 10 \times 0{\cdot}436 = 4{\cdot}36\ \Omega$$

$$X_{01} = Z_{01} \sin \phi_p = 10 \times 0{\cdot}9 = 9\ \Omega$$

$$\text{F.L. primary current, } I_1 = \frac{100 \times 10^3}{6600} = 15{\cdot}15 \text{ A}$$

F.L. voltage drop in primary when p.f. $\cos \phi_2 = 0{\cdot}8$

$$= I_1 (R_{01} \cos \phi_2 + X_{01} \sin \phi_2)$$

$$= 15{\cdot}15\,(4{\cdot}36 \times 0{\cdot}8 + 9 \times 0{\cdot}6) = 134 \text{ V}$$

$\therefore$ Voltage to be applied to the H.V. side (primary)

$$= 6600 + 134 = \mathbf{6734\ V}$$

Example 8.47. *The corrected instrument readings obtained from open and short-circuit tests on 10 kVA, 450/120 V, 50 Hz transformer are :*

O.C. test : $V_1 = 120$ V ; $I_1 = 4{\cdot}2$ A ; $W_1 = 80$ W ; W_1 and I_1 were read on the low-voltage side

S.C. test : $V_1 = 9{\cdot}65$ V ; $I_1 = 22{\cdot}2$ A ; $W_1 = 120$ W with low voltage winding short-circuited

Compute (i) the equivalent circuit (simplified) constants (ii) efficiency and voltage regulation for full-load at 0·8 p.f. lagging (iii) efficiency for half full-load and 0·8 p.f. lagging.

Solution. In O.C. test, the primary is open and secondary draws a no-load current of 4·2 A. Now $K = 120/450 = 4/15$.

(*i*) $\therefore$ No-load primary current, $I_0 = K \times 4{\cdot}2 = (4/15) \times 4{\cdot}2 = 1{\cdot}12$ A

$$\text{Primary core-loss component, } I_W = \frac{\text{Watts}}{450} = \frac{80}{450} = 0{\cdot}178 \text{ A}$$

$$\text{Primary magnetising current, } I_m = \sqrt{I_0^2 - I_W^2} = \sqrt{(1{\cdot}12)^2 - (0{\cdot}178)^2} = 1{\cdot}1 \text{ A}$$

$$\therefore \quad R_0 = \frac{450}{I_W} = \frac{450}{0{\cdot}178} = \mathbf{2530\ \Omega}\ ;\ X_0 = \frac{450}{I_m} = \frac{450}{1{\cdot}1} = \mathbf{409\ \Omega}$$

In S.C. test, secondary (120 V side) is short-circuited and instruments are placed in primary.

$$\therefore \quad Z_{01} = \frac{9{\cdot}65}{22{\cdot}2} = 0{\cdot}435\ \Omega$$

$$R_{01} = \frac{120}{(22 \cdot 2)^2} = \mathbf{0 \cdot 243\ \Omega}$$

$$X_{01} = \sqrt{Z_{01}^2 - R_{01}^2} = \sqrt{(0 \cdot 435)^2 - (0 \cdot 243)^2} = \mathbf{0 \cdot 361\ \Omega}$$

Thus the required four parameters of the equivalent circuit are determined (See Fig. 8.45).

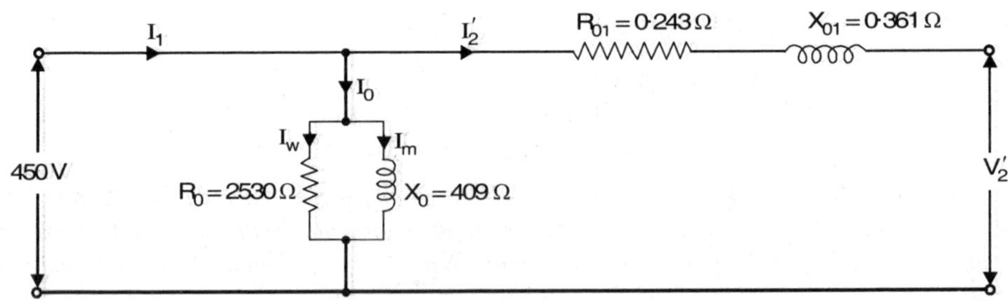

Fig. 8.45

(*ii*) F.L. primary current, $I_1 = \frac{10 \times 10^3}{450} = 22 \cdot 2$ A

F.L. voltage drop in primary $= I_1 (R_{01} \cos \phi_2 + X_{01} \sin \phi_2)$

$= 22 \cdot 2 (0 \cdot 243 \times 0 \cdot 8 + 0 \cdot 361 \times 0 \cdot 6) = 9 \cdot 2$ V

$\therefore$ F.L. voltage regulation $= \frac{9 \cdot 2}{450} \times 100 = \mathbf{2 \cdot 04\ \%}$

It is clear that S.C. test is conducted for full-load current. Therefore, S.C. test gives F.L. Cu losses (= 120 *W*).

Total F.L. losses $= 80 + 120 = 200$ W

F.L. output $= (10 \times 10^3) \times$ p.f. $= 10 \times 10^3 \times 0 \cdot 8 = 8000$ W

$\therefore$ F.L. efficiency $= \frac{8000}{8000 + 200} \times 100 = \mathbf{97 \cdot 57\ \%}$

(*iii*) Cu loss at half full-load $= \left(\frac{1}{2}\right)^2 \times 120 = 30$ W

Total losses at half full-load $= 80 + 30 = 110$ W

Output at half full-load $= 8000/2 = 4000$ W

$\therefore$ Half full-load efficiency $= \frac{4000}{4000 + 110} \times 100 = \mathbf{97 \cdot 34\ \%}$

Example 8.48. *A 20 kVA, single-phase, 50 Hz, 2200/200 V transformer gave the following test results :*

O.C. test : 2200 V applied to primary, power taken 220 W

S.C. test : Power required to circulate F.L. current in short-circuited secondary 240 W

Calculate the efficiency at full-load and half full-load at p.f. 0·8 lagging.

Solution.

Iron losses = 220 W ; F.L. Cu losses = 240 W

At full-load, 0·8 p.f. lagging

Output $= 20 \times 10^3 \times$ p.f. $= 20 \times 10^3 \times 0 \cdot 8 = 16000$ W

Total losses $= 220 + 240 = 460$ W

Input $= 16000 + 460 = 16460$ W

$\therefore$ F.L. efficiency $= \frac{16000}{16460} \times 100 = \mathbf{97 \cdot 21\ \%}$

At half full-load, 0·8 p.f. lagging

$$\text{Cu losses} = \left(\frac{1}{2}\right)^2 \times 240 = 60 \text{ W}$$

$$\text{Iron losses} = 220 \text{ W}$$

$$\text{Total losses} = 220 + 60 = 280 \text{ W}$$

$$\text{Output} = 10 \times 10^3 \times 0{\cdot}8 = 8000 \text{ W}$$

$$\text{Input} = 8000 + 280 = 8280 \text{ W}$$

$$\therefore \quad \text{Efficiency} = \frac{8000}{8280} \times 100 = \mathbf{96{\cdot}63\ \%}$$

Example 8.49. *A single-phase, 10 kVA, 500/250 V transformer has the following constants :*

Resistance : Primary 0·2 Ω ; Secondary 0·5 Ω

Reactance : Primary 0·4 Ω ; Secondary 0·1 Ω

Resistance of equivalent exciting circuit referred to primary, R_0 = 1500 Ω

Reactance of equivalent exciting circuit referred to primary, X_0 = 750 Ω

What would be the readings of the instruments when the transformer is connected for the open circuit and short-circuit tests ?

Solution. Fig. 8.46 (*i*) shows the connections of the transformer for open-circuit test.

$$I_W = \frac{V_1}{R_0} = 500/1500 = 1/3 \text{ A} \; ; I_m = \frac{V_1}{X_0} = \frac{500}{750} = \frac{2}{3} \text{ A}$$

$$\therefore \quad I_0 = \sqrt{I_W^2 + I_m^2} = \sqrt{(1/3)^2 + (2/3)^2} = 0{\cdot}745 \text{ A}$$

$$\text{No-load primary input} = V_1 I_W = 500 \times \frac{1}{3} = 167 \text{ W}$$

Therefore, readings of ammeter, voltmeter and wattmeter would be **0·745 A** ; **500 V** ; **167 W** respectively.

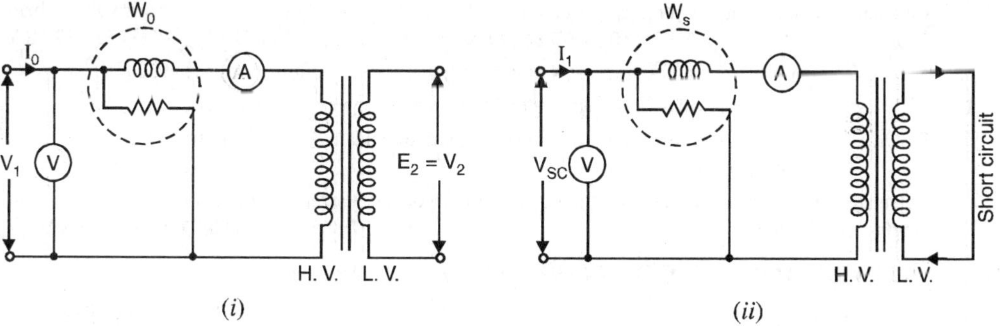

(*i*) (*ii*)

Fig. 8.46

Fig. 8.46 (*ii*) shows the connections of the transformer for short-circuit test. Here the secondary (*i.e.* L.V. winding) is short-circuited so that all instruments are placed in the primary. For the transformer, $K = 250/500 = 1/2$.

$$R_{01} = R_1 + R'_2 = R_1 + \frac{R_2}{K^2} = 0{\cdot}2 + \frac{0{\cdot}5}{(1/2)^2} = 2{\cdot}2\,\Omega$$

$$X_{01} = X_1 + X'_2 = X_1 + \frac{X_2}{K^2} = 0{\cdot}4 + \frac{0{\cdot}1}{(1/2)^2} = 0{\cdot}8\,\Omega$$

$$\therefore \quad Z_{01} = \sqrt{(R_{01})^2 + (X_{01})^2} = \sqrt{(2{\cdot}2)^2 + (0{\cdot}8)^2} = 2{\cdot}34\,\Omega$$

$$\text{F.L. primary current, } I_1 = \frac{10 \times 10^3}{500} = 20 \text{ A}$$

$$\therefore \quad V_{SC} = I_1 Z_{01} = 20 \times 2{\cdot}34 = 46{\cdot}8 \text{ V}$$

$$\text{Power absorbed} = I_1^2 R_{01} = (20)^2 \times 2{\cdot}2 = 880 \text{ W}$$

Therefore, ammeter, voltmeter and wattmeter will read **20 A**; **46·8 V** and **880 W** respectively.

TUTORIAL PROBLEMS

1. A 50 kVA, 6360/240 V transformer gave the following test results :

O.C. test : Primary voltage 6360 V; primary current 1 A ; power input 2 kW

S.C. test : Voltage across secondary winding 50 V ; current in secondary winding 175 A : power input 2 kW

Find the efficiency of the transformer when supplying full-load at a power factor of 0·8 lagging. **[89·2 %]**

2. A short-circuit test when performed on the H.V. side of a 10 kVA, 2000/400 V single-phase transformer gave the following data : 60 V, 4 A, 100 W. If the L.V. side is delivering full-load current at 0·8 p.f. lagging, find the voltage applied to H.V. side so that secondary voltage at on load is the same as at no-load. **[2067 V]**

3. A 5 kVA, 200/400 V single-phase transformer gave the following test results :

O.C. test : 200 V applied to primary ; power absorbed 72 W

S.C. test : 8V applied to primary circulated F.L. current in short-circuited secondary ; power absorbed 80 W

Calculate the efficiency at (*i*) full-load at 0·9 p.f. lagging (*ii*) half full-load at 0·9 p.f. lagging. **[(*i*) 96·72 % (*ii*) 94·86 %]**

4. Obtain the equivalent circuit of a 200/400 V, 50 Hz, single-phase transformer from the following test data:

O.C. test : 200 V, 0·7 A, 70 W .... on L.V. side

S.C. test : 15 V, 10 A, 85 W .... on H.V. side

Calculate the secondary voltage when delivering 5 kW at 0·8 p.f. lagging, the primary voltage being 200 V. **[R_0 = 57·14 Ω ; X_0 = 330 Ω ; R_{01} = 0·21 Ω ; X_{01} = 0·31 Ω ; 377·8 V]**

5. A 10 kVA, 500/2000 V, single-phase, 50 Hz transformer gave the following test results :

O.C. test : 500 V at normal frequency applied to primary ; power absorbed 120 W

S.C. test : 15 V at normal frequency applied to primary circulated full-load current in the short-circuited secondary ; power absorbed 100 W

Calculate (*i*) efficiency at full-load at unity p.f. (*ii*) secondary terminal voltage on F.L. at unity p.f., 0·8 lagging, 0·8 leading. **[(*i*) 97·86 % (*ii*) 1980 V ; 1950 V ; 2018 V]**

8.21 SEPARATION OF COMPONENTS OF CORE LOSSES

The core losses (or iron losses) consist of hysteresis loss and eddy current loss. Sometimes it is desirable to find the hysteresis loss component and eddy current loss component in the total core losses.

$$\text{Hysteresis loss, } P_h = k_h f B_m^{1{\cdot}6} \text{ watts/m}^3$$

$$\text{Eddy current loss, } P_e = k_e f^2 B_m^2 t^2 \text{ watts/m}^3$$

where B_m = maximum flux density ; f = frequency ; k_h, k_e = constants

For a given machine and maximum flux density (B_m),

$$P_h \propto f \text{ and } P_e \propto f^2$$

or

$$P_h = af \text{ and } P_e = bf^2$$

where a and b are constants.

$$\text{Total core loss, } P_i = af + bf^2$$

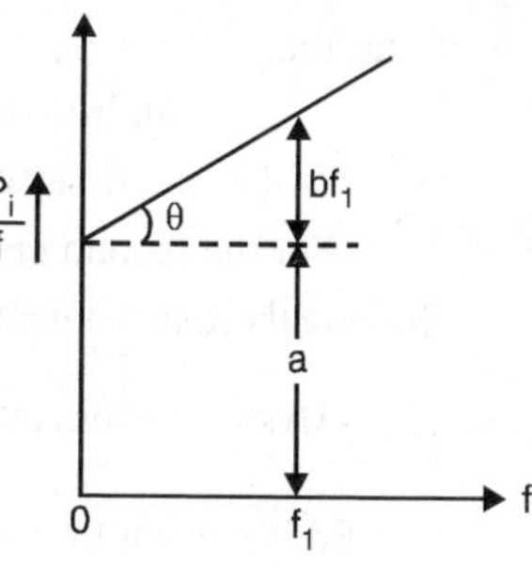

Fig. 8.47

Hence if the total core loss for given B_m is known at two frequencies, the constants a and b can be calculated. Knowing the values of a and b, the hysteresis loss component and eddy current loss component of the core loss can be determined.

P_i/f and f curve

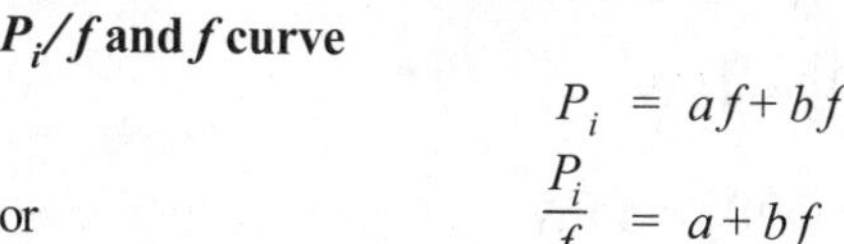

$$P_i = af + bf^2$$

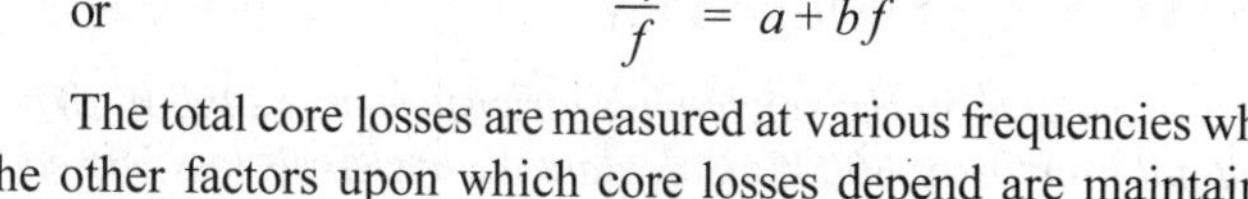

or $$\frac{P_i}{f} = a + bf$$

The total core losses are measured at various frequencies while the other factors upon which core losses depend are maintained constant. If a graph is plotted between P_i/f and f, it will be a straight line with slope $\tan\theta = b$ (See Fig. 8.47). Therefore, constants a and b can be evaluated. Hence the hysteresis and eddy current losses at a given frequency (say f_1) can be found out.

Example 8.50. *The iron loss in a certain transformer is 80 W at 25 Hz and 204 W at 60 Hz, the maximum flux density being the same. Calculate the total iron losses at 100 Hz at the same maximum flux density.*

Solution.

$$\text{Total iron losses, } P_i = af + bf^2 \qquad ...(i)$$

or $$\frac{P_i}{f} = a + bf$$

$$\text{At 25 Hz : } \frac{80}{25} = a + 25\,b \;;\; \text{At 60 Hz : } \frac{204}{60} = a + 60\,b$$

On solving these two equations, we have, $a = 107/35$; $b = 1/175$. Total iron loss at 100 Hz can be obtained by putting the values of a, b and f (= 100 Hz) in eq. (i).

$$\therefore \quad \text{Total iron loss, } P_i = \frac{107}{35} \times 100 + \frac{1}{175} \times (100)^2 = \mathbf{362{\cdot}85\ W}$$

Example 8.51. *In a test for the determination of losses of a 440 V, 50 Hz transformer, the total iron losses were found to be 2500 W at normal voltage and frequency. When the applied voltage and frequency were 220 V and 25 Hz, the iron losses were found to be 850 W. Calculate the eddy current loss at normal voltage and frequency.*

Solution. In a transformer, $\phi_m \propto V_1/f$. Since in the second case, voltage as well as frequency is halved, the ratio V_1/f remains the same. Therefore, maximum flux density in the two cases is the same.

$$\therefore \quad \text{Total iron losses, } P_i = af + bf^2$$

or $$\frac{P_i}{f} = a + bf$$

For the first case, $$\frac{2500}{50} = a + 50\,b$$

For the second case, $$\frac{850}{25} = a + 25\,b$$

Solving these two equations, we have, $b = 0{\cdot}64$.

$$\therefore \quad \text{Eddy current loss} = bf^2 = 0{\cdot}64 \times 50^2 = \mathbf{1600\ W}$$

We can also find the hysteresis loss by subtracting 1600 W from 2500 W.

Example 8.52. *The iron loss in a transformer core at normal flux density was measured at 30 Hz and 50 Hz, the results being 30 W and 54 W. Calculate the hysteresis and eddy current loss at 50 Hz.*

Solution.

$$\text{Total iron loss, } P_i = af + bf^2$$

$$\text{For the first case, } 30 = 30\,a + b \times (30)^2$$

$$\text{For the second case, } 54 = 50\,a + b \times (50)^2$$

Solving these two equations, we have, $a = 44/50$; $b = 1/250$.

$$\therefore \quad \text{Hysteresis loss at 50 Hz} = af = \frac{44}{50} \times 50 = \mathbf{44\ W}$$

$$\text{Eddy current loss at 50 Hz} = bf^2 = \frac{1}{250} \times (50)^2 = \mathbf{10\ W}$$

Example 8.53. *The total iron loss in a 460 V, 50 Hz, single-phase transformer is 2400 W. When a 230 V, 25 Hz supply is applied, the total iron loss is 800 W. Calculate the hysteresis loss and eddy current loss at normal voltage and frequency.*

Solution. In a transformer, $\phi_m \propto V_1/f$. Since in the second case, voltage as well as frequency is halved, the ratio V_1/f remains the same. Therefore, maximum flux density in the two cases is the same.

$$\therefore \quad \text{Total iron loss, } P_i = af + bf^2$$

$$\text{or} \quad \frac{P_i}{f} = a + bf$$

$$\text{For the first case, } \frac{2400}{50} = a + 50\,b$$

$$\text{For the second case, } \frac{800}{25} = a + 25\,b$$

Solving these two equations, we get, $a = 16$; $b = 16/25$.

$$\therefore \quad \text{Hysteresis loss at 50 Hz} = af = 16 \times 50 = \mathbf{800\ W}$$

$$\text{Eddy current loss at 50 Hz} = bf^2 = \frac{16}{25} \times (50)^2 = \mathbf{1600\ W}$$

8.22 WHY TRANSFORMER RATING IN kVA ?

An important factor in the design and operation of electrical machines is the relation between the life of the insulation and operating temperature of the machine. Therefore, temperature rise resulting from the losses is a determining factor in the rating of a machine. We know that copper loss in a transformer depends on current and iron loss depends on voltage. Therefore, the total loss in a transformer depends on the volt-ampere product only and not on the phase angle between voltage and current *i.e.*, it is independent of load *power factor. For this reason, the rating of a transformer is in kVA and not kW.

8.23 SUMPNER OR BACK-TO-BACK TEST

This test is conducted simultaneously on two identical transformers and provides data for finding the efficiency, regulation and **temperature rise. The main advantage of this test is that the transformers are tested under full-load conditions without much expenditure of power. The power required to conduct this test is equal to the losses of the two transformers. It may be noted that two identical transformers are needed to carry out this test.

* Rating in kW = kVA × load p.f.
The load p.f. depends upon the nature of load. Thus kW rating of transformer will be different for different load p.f.

** No doubt the open-circuit and short-circuit tests provide data for determining the efficiency and regulation of a transformer but they indicate no information regarding temperature rise. It is because in each of these tests, the power loss is either the core loss or copper loss but not both.

Circuit. Fig. 8.48 shows the connections for back-to-back test on two identical transformers T_1 and T_2. The primaries of the two transformers are connected in parallel across the rated voltage V_1 while the two secondaries are connected in phase opposition. Therefore, there will be no circulating current in the loop formed by the secondaries because their induced e.m.f.s are equal and in opposition. There is an auxiliary low-voltage transformer T which can be adjusted to give a variable voltage and hence current in the secondary loop circuit. A wattmeter W_1, an ammeter A_1 and voltmeter V_1 are connected to the input side. A wattmeter W_2 and ammeter A_2 are connected in the secondary circuit.

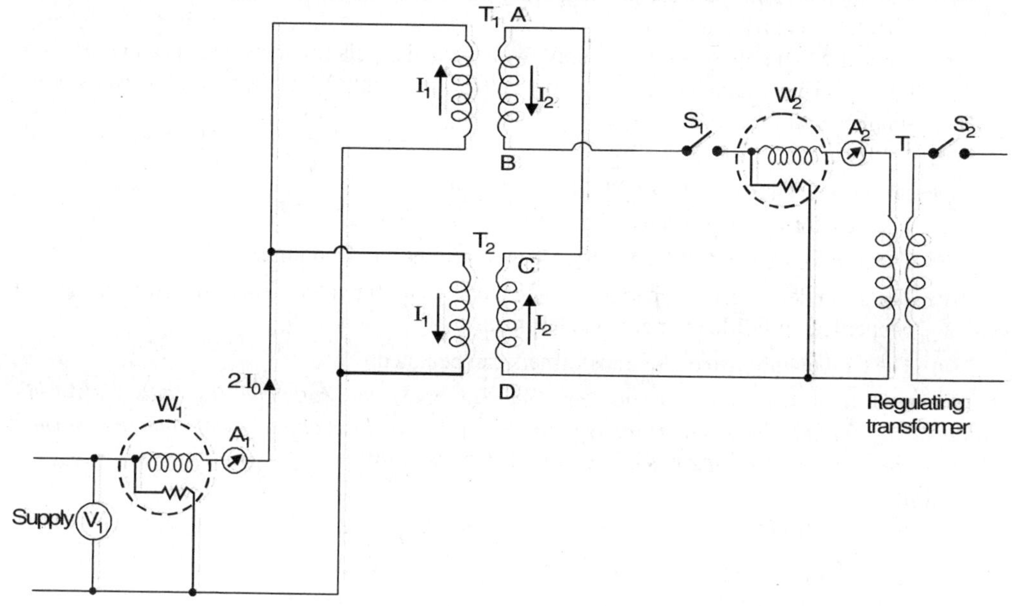

Fig. 8.48

Operation

(*i*) The secondaries of the transformers are in *phase opposition. With switch S_1 closed and switch S_2 open (*i.e.*, regulating transformer not in the circuit), there will be no circulating current ($I_2 = 0$) in the secondary loop circuit. It is because the induced e.m.f.s in the secondaries are equal and in opposition. This situation is just like an open-circuit test. Therefore, the current drawn from the supply is $2\ I_0$ where I_0 is the no-load current of each transformer. The reading of wattmeter W_1 will be equal to the core losses of the two transformers.

$$W_1 = \text{Core losses of the two transformers}$$

(*ii*) Now switch S_2 is also closed and output voltage of the regulating transformer is adjusted till full-load current I_2 flows in the secondary loop circuit. The full-load secondary current will cause full-load current I_1 ($= K\,I_2$) in the primary circuit. The primary current I_1 circulates in the primary winding **only and will not pass through W_1. Note that full-load currents are flowing through the primary and secondary windings. Therefore, reading of wattmeter W_2 will be equal to the full-load copper losses of the two transformers.

---

* This can be done by first connecting any two terminals (say A and C) together and a high-resistance voltmeter (of twice the rated secondary voltage) is connected to the remaining terminals B and D. For correct back-to-back connection, the voltmeter should give zero reading. If the voltmeter gives large reading, the terminal connections should be reversed.

** The primary windings are virtually short-circuited so far as I_1 is concerned. The injection of I_2 resembles the short-circuit test with the roles of primary and secondary windings reversed.

$$W_2 = \text{Full-load Cu losses of two transformers}$$

$$\therefore \quad W_1 + W_2 = \text{Total losses of two transformers at full load}$$

The following points may be noted :

(*a*) The wattmeter W_1 gives the core losses of the two transformers while wattmeter W_2 gives the full-load copper losses (or at any other load current I_2) of the two transformers. Therefore, power required to conduct this test is equal to the total losses of the two transformers.

(*b*) Although transformers are not supplying any load, yet full iron loss and full-load copper losses are occurring in them.

(*c*) There are two voltage sources (supply voltage and regulating transformer) and there is no interference between them. The supply voltage gives only $2\,I_0$ while regulating transformer supplies I_2 and hence I_1 $(= K\,I_2)$.

Advantages

(*i*) The power required to carry out the test is small.

(*ii*) The transformers are tested under full-load conditions.

(*iii*) The iron losses and full-load copper losses are measured simultaneously.

(*iv*) The secondary current I_2 can be adjusted to any current value. Therefore, we can find the copper loss at full-load or at any other load.

(*v*) The temperature rise of the transformers can be noted.

Example 8.54. *In a back-to-back test (See Fig. 8.48), the wattmeter W_1 reads 4 kW while wattmeter W_2 (at full-load current) reads 6 kW. Find the full-load efficiency of each transformer. The transformers are rated at 200 kVA. Assume p.f. to be unity.*

Solution.

$$\text{Iron loss of each transformer} = W_1/2 = 4/2 = 2\text{ kW}$$

$$\text{F.L. Cu loss of each transformer} = W_2/2 = 6/2 = 3\text{ kW}$$

$$\text{Total F.L. losses of each transformer} = 2 + 3 = 5\text{ kW}$$

$$\text{F.L. output of each transformer} = 200 \times 1 = 200\text{ kW}$$

$$\therefore \quad \text{F.L. efficiency of each transformer} = \frac{200}{200+5} \times 100 = \mathbf{97{\cdot}56\ \%}$$

Example 8.55. *Two similar 250 kVA, single-phase transformers gave the following results when tested by back-to-back method :*

Mains wattmeter, W_1 = *5 kW*

Primary series circuit wattmeter, W_2 = *7.5 kW (at full-load current)*

Find the efficiency of each transformer at 75 % full load and 0.8 p.f. lead.

Solution. Wattmeter W_1 gives the iron losses of the two transformers while wattmeter W_2 gives the full-load copper losses of the two transformers.

$\therefore$ Iron loss of each transformer, $P_i = W_1/2 = 5/2 = 2.5$ kW

F.L. Cu loss of each transformer, $P_C = W_2/2 = 7.5/2 = 3.75$ kW

Output of each transformer at 75% full-load and 0.8 p.f. lead

$$= 0.75 \times 250 \times 0.8 = 150\text{ kW}$$

Total losses of each transformer at 75% full-load

$$= P_i + (0.75)^2\,P_C = 2.5 + (0.75)^2 \times 3.75 = 4.61\text{ kW}$$

Input to each transformer at 75% full-load and 0.8 p.f. lead

$$= 150 + 4.61 = 154.61\text{ kW}$$

$\therefore$ Efficiency η of each transformer is given by ;

$$\eta = \frac{150}{154.61} \times 100 = \mathbf{97.02\%}$$

8.24 LOSSES IN A TRANSFORMER

The power losses in a transformer are of two types, namely;

1. Core or Iron losses 2. Copper losses

These losses appear in the form of heat and produce (*i*) an increase in temperature and (*ii*) a drop in efficiency.

1. Core or Iron losses (P_i). These consist of hysteresis and eddy current losses and occur in the transformer core due to the alternating flux. These can be determined by open-circuit test.

$$\text{Hysteresis loss} = k_h f B_m^{1.6} \text{ watts/m}^3$$

$$\text{Eddy current loss} = k_e f^2 B_m^2 t^2 \text{ watts/m}^3$$

Both hysteresis and eddy current losses depend upon (*i*) maximum flux density B_m in the core and (*ii*) supply frequency f. Since transformers are connected to constant-frequency, constant voltage supply, both f and B_m are constant. Hence, core or iron losses are practically the same at all loads.

$$\text{Iron or Core losses, } P_i = \text{Hysteresis loss + Eddy current loss}$$
$$= \text{Constant losses}$$

The hysteresis loss can be minimised by using steel of high silicon content whereas eddy current loss can be reduced by using core of thin laminations.

2. Copper losses (P_C). These losses occur in both the primary and secondary windings due to their ohmic resistance. These can be determined by short-circuit test.

$$\text{Total Cu losses, } P_C = I_1^2 R_1 + I_2^2 R_2$$
$$= I_1^2 R_{01} \text{ or } I_2^2 R_{02}$$

It is clear that copper losses vary as the square of load current. Thus if copper losses are 400 W at a load current of 10 A, then they will be $(1/2)^2 \times 400 = 100$ W at a load current of 5A.

$$\text{Total losses in a transformer} = P_i + P_C$$
$$= \text{Constant losses + Variable losses}$$

It may be noted that in a transformer, copper losses account for about 90% of the total losses.

8.25 EFFICIENCY OF A TRANSFORMER

Like any other electrical machine, the efficiency of a transformer is defined as the ratio of output power (in watts or kW) to input power (watts or kW) *i.e.*,

$$\text{Efficiency} = \frac{\text{Output power}}{\text{Input power}}$$

It may appear that efficiency can be determined by direclty loading the transformer and measuring the input power and output power. However, this method has the following drawbacks :

(*i*) Since the efficiency of a transformer is very high, even 1% error in each wattmeter (output and input) may give ridiculous results. This test, for instance, may give efficiency higher than 100%.

(*ii*) Since the test is performed with transformer on load, considerable amount of power is wasted. For large transformers, the cost of power alone would be considerable.

(*iii*) It is generally difficult to have a device that is capable of absorbing all of the output power.

(*iv*) The test gives no information about the proportion of various losses.

Due to these drawbacks, direct loading method is seldom used to determine the efficiency of a transformer. In practice, open-circuit and short-circuit tests are carried out to find the efficiency.

$$\text{Efficiency} = \frac{\text{Output}}{\text{Input}} = \frac{\text{Output}}{\text{Output + Losses}}$$

The losses can be determined by transformer tests.

8.26 EFFICIENCY FROM TRANSFORMER TESTS

$$\text{F.L. Iron loss} = P_i \quad \text{...from open-circuit test}$$
$$\text{F.L. Cu loss} = P_C \quad \text{...from short-circuit test}$$
$$\text{Total F.L. losses} = P_i + P_C$$

We can now find the full-load efficiency of the transformer at any p.f. without actually loading the transformer.

$$\text{F.L. efficiency, } \eta_{F.L.} = \frac{\text{Full-load } VA \times \text{p.f.}}{(\text{Full-load } VA \times \text{p.f.}) + P_i + P_C}$$

Also for *any load equal to $x \times$ full-load,

$$\text{Corresponding total losses} = P_i + x^2 P_C$$

$$\text{Corresponding } \eta_x = \frac{(x \times \text{Full-load } VA) \times \text{p.f.}}{(x \times \text{Full-load } VA \times \text{p.f.}) + P_i + x^2 P_C}$$

Note that iron loss remains the same at all loads.

8.27 CONDITION FOR MAXIMUM EFFICIENCY

$$\text{Output power} = V_2 I_2 \cos\phi_2$$

If R_{02} is the total resistance of the transformer referred to secondary, then,

$$\text{Total Cu loss, } P_C = I_2^2 R_{02}$$
$$\text{Total losses} = P_i + P_C$$

$$\therefore \quad \text{Transformer } \eta = \frac{V_2 I_2 \cos\phi_2}{V_2 I_2 \cos\phi_2 + P_i + I_2^2 R_{02}}$$

$$= \frac{V_2 \cos\phi_2}{V_2 \cos\phi_2 + P_i / I_2 + I_2 R_{02}} \quad \text{...(i)}$$

For a normal transformer, V_2 is approximately constant. Hence for a load of given p.f., efficiency depends upon load current I_2. It is clear from exp. (*i*) above that numerator is constant and for the efficiency to be maximum, the denominator should be minimum *i.e.*,

$$\frac{d}{dI_2}(\text{denominator}) = 0$$

or $$\frac{d}{dI_2}(V_2 \cos\phi_2 + P_i/I_2 + I_2 R_{02}) = 0$$

or $$0 - \frac{P_i}{I_2^2} + R_{02} = 0$$

or $$P_i = I_2^2 R_{02} \quad \text{...(ii)}$$

i.e., **Iron losses = Copper losses**

Hence efficiency of a transformer will be maximum when copper losses are equal to constant or iron losses.

From eq. (*ii*) above, the load current I_2 corresponding to maximum efficiency is given by ;

$$I_2 = \sqrt{\frac{P_i}{R_{02}}}$$

The relative value of these losses is in the control of the designer of the transformer according to the relative amount of copper and iron he uses. A transformer which is to operate continuously on full-load would, therefore, be designed to have maximum efficiency at full-load. However, distribution

* Thus if rated VA (*i.e.*, full-load VA) of a transformer is 1000, then half full-load VA = 500. Obviously, the value of $x = 1/2$.

transformers operate for long periods on light load. Therefore, their point of maximum efficiency is usually arranged to be between three-quarter and half full-load.

Note. In a transformer, iron losses are constant whereas copper losses are variable. In order to obtain maximum efficiency, the load current should be such that total Cu losses become equal to iron losses.

8.28 OUTPUT kVA CORRESPONDING TO MAXIMUM EFFICIENCY

Let

$$P_C = \text{Copper losses at full-load kVA}$$

$$P_i = \text{Iron losses}$$

$$x = \text{Fraction of full-load kVA at which efficiency is maximum}$$

$$\text{Total Cu losses} = x^2 P_C$$

$$\therefore \quad x^2 P_C = P_i \qquad \text{...for maximum efficiency}$$

or

$$x = \sqrt{\frac{P_i}{P_C}} = \sqrt{\frac{\text{Iron loss}}{\text{F.L. Cu loss}}}$$

∴ Output kVA corresponding to maximum efficiency

$$= x \times \text{Full-load kVA} = \text{Full-load kVA} \times \sqrt{\frac{\text{Iron loss}}{\text{F.L. Cu loss}}}$$

It may be noted that the value of kVA at which the efficiency is maximum is independent of p.f. of the load.

Example 8.56. *In a 50 kVA transformer, the iron loss is 500 W and full-load copper loss is 800 W. Find the efficiency at full-load and half full-load at 0·8 p.f. lagging.*

Solution.

Full-load, 0·8 p.f.

$$\text{Full-load output} = 50 \times 0{\cdot}8 = 40 \text{ kW}$$

$$\text{Total Full-load losses} = 500 + 800 = 1300 \text{ W} = 1{\cdot}3 \text{ kW}$$

$$\text{Full-load input} = 40 + 1{\cdot}3 = 41{\cdot}3 \text{ kW}$$

$$\text{Full-load } \eta = \frac{40}{41{\cdot}3} \times 100 = \mathbf{96{\cdot}85\ \%}$$

Half full-load, 0·8 p.f.

$$\text{Output at half full-load} = (50 \times 1/2) \times 0{\cdot}8 = 20 \text{ kW}$$

$$\text{Total loss at half full-load} = 500 + (1/2)^2 \times 800 = 700 \text{ W} = 0{\cdot}7 \text{ kW}$$

$$\text{Input at half full-load} = 20 + 0{\cdot}7 = 20{\cdot}7 \text{ kW}$$

$$\text{Half full-load } \eta = \frac{20}{20{\cdot}7} \times 100 = \mathbf{96{\cdot}6\ \%}$$

Example 8.57. *A 40 kVA transformer has iron loss of 450 W and full-load copper loss of 850 W. If the power factor of the load is 0·8 lagging, calculate (i) full-load efficiency (ii) the kVA load at which maximum efficiency occurs and (iii) the maximum efficiency.*

Solution.

(*i*)

$$\text{Total F.L. losses} = 450 + 850 = 1300 \text{ W} = 1{\cdot}3 \text{ kW}$$

$$\text{F.L. output} = 40 \times 0{\cdot}8 = 32 \text{ kW}$$

$$\text{F.L. Input} = 32 + 1{\cdot}3 = 33{\cdot}3 \text{ kW}$$

$$\text{F.L. efficiency} = \frac{32}{33{\cdot}3} \times 100 = \mathbf{96{\cdot}1\ \%}$$

(*ii*)

$$\text{Load for max. efficiency} = \text{F.L. kVA} \times \sqrt{\frac{\text{Iron loss}}{\text{F.L. Cu loss}}}$$

$$= 40 \times \sqrt{\frac{450}{850}} = \mathbf{29{\cdot}1\ kVA}$$

$$\text{Corresponding output kW} = 29{\cdot}1 \times 0{\cdot}8 = 23{\cdot}3\ \text{kW}$$

(*iii*) For maximum efficiency, Cu loss = Iron loss

$$\therefore \quad \text{Total losses} = 2 \times 450 = 900\ \text{W} = 0{\cdot}9\ \text{kW}$$

$$\text{Input} = 23{\cdot}3 + 0{\cdot}9 = 24{\cdot}2\ \text{kW}$$

$$\text{Maximum efficiency} = \frac{23{\cdot}3}{24{\cdot}2} \times 100 = \mathbf{96{\cdot}28\ \%}$$

Example 8.58. *The primary and secondary windings of a 50 kVA, 6600/220 V transformer have resistances of 7·8 Ω and 0·0085 Ω respectively. The transformer draws no-load current of 0·328 A at p.f. of 0·3 lagging. Calculate the efficiency at full-load if the p.f. of the load is 0·8 lagging.*

Solution.

$$K = 220/6600 = 1/30$$

$$\text{F.L. Iron losses} = 6600 \times 0{\cdot}328 \times 0{\cdot}3 = 650\ \text{W}$$

$$R_{01} = R_1 + R_2/K^2 = 7{\cdot}8 + (0{\cdot}0085) \times (30)^2 = 15{\cdot}45\ \Omega$$

$$\text{F.L. primary current, } I_1 = 50 \times 10^3/6600 = 7{\cdot}57\ \text{A}$$

$$\text{F.L. Cu losses} = I_1^2 R_{01} = (7{\cdot}57)^2 \times 15{\cdot}45 = 885{\cdot}4\ \text{W}$$

$$\text{Total F.L. losses} = 650 + 885{\cdot}4 = 1535{\cdot}4\ \text{W}$$

$$\text{F.L. output} = 50 \times 10^3 \times 0{\cdot}8 = 40{,}000\ \text{W}$$

$$\text{F.L. input} = 40{,}000 + 1535{\cdot}4 = 41535{\cdot}4\ \text{W}$$

$$\text{F.L. efficiency} = \frac{40{,}000}{41535{\cdot}4} \times 100 = \mathbf{96{\cdot}3\ \%}$$

Example 8.59. *A 440/110 V transformer has a primary resistance of 0·03 Ω and secondary resistance of 0·02 Ω. Its iron loss at normal input is 150 W. Determine the secondary current at which maximum efficiency will occur and the value of this maximum efficiency at a unity p.f. load.*

Solution.

$$K = 110/440 = 0{\cdot}25$$

$$R_{02} = R_2 + K^2 R_1 = 0{\cdot}02 + (0{\cdot}25)^2 \times 0{\cdot}03 = 0{\cdot}022\ \Omega$$

$$\text{Cu loss} = \text{Iron loss} \qquad \text{...for maximum } \eta$$

$$\text{or} \quad I_2^2 R_{02} = 150$$

$$\therefore \quad I_2 = \sqrt{\frac{150}{R_{02}}} = \sqrt{\frac{150}{0{\cdot}022}} = \mathbf{82{\cdot}58\ A}$$

$$\text{Output at unity p.f.} = 110 \times 82{\cdot}58 \times 1 = 9083{\cdot}8\ \text{W}$$

$$\text{Total losses} = 150 + 150 = 300\ \text{W} \qquad \text{...for maximum } \eta$$

$$\therefore \quad \text{Maximum efficiency} = \frac{9083{\cdot}8}{(9083{\cdot}8 + 300)} \times 100 = \mathbf{96{\cdot}8\ \%}$$

Example 8.60. *The efficiency of a 400 kVA, single-phase transformer is 98·77 % when delivering full-load at 0·8 p.f. lagging and 99·13% at half full-load at unity p.f. Calculate (i) iron loss and (ii) full-load copper loss.*

Solution.

Let P_i = Iron losses ...*constant at all loads*

P_C = F.L. copper losses

At full-load

$$\text{F.L. output at } 0{\cdot}8 \text{ p.f.} = 400 \times 0{\cdot}8 = 320\ \text{kW}$$

$$\text{F.L. input at } 0{\cdot}8 \text{ p.f.} = 320/0{\cdot}9877 = 324 \text{ kW}$$
$$\text{Total F.L. losses} = 324 - 320 = 4 \text{ kW}$$
i.e.,
$$P_i + P_C = 4 \text{ kW} \quad ...(i)$$

At half full-load

$$\text{Output at unity p.f.} = (1/2) \times 400 \times 1 = 200 \text{ kW}$$
$$\text{Input at unity p.f.} = 200/0{\cdot}9913 = 201{\cdot}8 \text{ kW}$$
$$\text{Total losses at half full-load} = 201{\cdot}8 - 200 = 1{\cdot}8 \text{ kW}$$
i.e.,
$$P_i + 0{\cdot}25\, P_C = 1{\cdot}8 \text{ kW} \quad ...(ii)$$

Solving eqs. (*i*) and (*ii*), P_i = **1·07 kW;** P_C = **2·93 kW**

Example 8.61. *A 20 kVA, 440/220 V, single-phase, 50 Hz transformer has iron loss of 324 W. The Cu loss is found to be 100 W when delivering half full-load current. Determine (i) the efficiency when delivering full-load current at 0·8 lagging p.f. and (ii) the percent of full-load kVA when the efficiency will be maximum.*

Solution.

$$\text{Iron loss} = 324 \text{ W} \quad ; \quad \text{F.L. Cu losses} = (2)^2 \times 100 = 400 \text{ W}$$

(*i*)
$$\text{Total F.L. losses} = 324 + 400 = 724 \text{ W} = 0{\cdot}724 \text{ kW}$$
$$\text{F.L. output at } 0{\cdot}8 \text{ p.f.} = 20 \times 0{\cdot}8 = 16 \text{ kW}$$
$$\therefore \quad \text{F.L. efficiency at } 0{\cdot}8 \text{ p.f.} = \frac{16}{16 + 0{\cdot}724} \times 100 = \mathbf{95{\cdot}67\ \%}$$

(*ii*)
$$\frac{\text{kVA for maximum } \eta}{\text{F.L. kVA}} = \sqrt{\frac{\text{Iron loss}}{\text{F.L. Cu loss}}} = \sqrt{\frac{324}{400}} = 0{\cdot}9$$
$$\therefore \quad \text{kVA for maximum } \eta = 0{\cdot}9 \times \text{F.L. kVA}$$

Therefore, efficiency will be maximum at 90% of full-load kVA.

Example 8.62. *A 600 kVA, single-phase transformer when working at unity p.f. has an efficiency of 92% at full-load and also at half full-load. Determine the efficiency when it operates at unity p.f. and 60% of full-load.*

Solution. It is given that the transformer has the same efficiency (*i.e.*, 92 %) at full-load and half full-load when the p.f. is unity. We shall use this data to find F.L. Cu loss and iron loss.

Let
$$P_C = \text{F.L. Cu loss}$$
$$P_i = \text{Iron loss} \quad \text{...constant at all loads}$$

At Full-load

$$\text{Output} = 600 \times 1 = 600 \text{ kW} \quad ; \quad \text{Input} = 600/0{\cdot}92 = 652{\cdot}2 \text{ kW}$$
$$\text{Total F.L. losses} = 652{\cdot}2 - 600 = 52{\cdot}2 \text{ kW}$$
or
$$P_C + P_i = 52{\cdot}2 \quad ...(i)$$

At Half full-load

$$\text{Output} = 300 \times 1 = 300 \text{ kW} \quad ; \quad \text{Input} = 300/0{\cdot}92 = 326{\cdot}1 \text{ kW}$$
$$\text{Total losses at half full-load} = 326{\cdot}1 - 300 = 26{\cdot}1 \text{ kW}$$
or
$$\frac{{}^*P_C}{4} + P_i = 26{\cdot}1 \quad ...(ii)$$

Solving eqs. (*i*) and (*ii*), we get, $P_C = 34{\cdot}8$ kW ; $P_i = 17{\cdot}4$ kW

Efficiency at 60 % Full-load, unity p.f.

$$\text{Output} = 600 \times 0{\cdot}6 \times 1 = 360 \text{ kW}$$

* At half full-load, Cu losses $= \left(\frac{1}{2}\right)^2 \times P_C = \frac{P_C}{4}$. However iron losses remain the same at all loads.

$$\text{Cu loss} = (0{\cdot}6)^2 \times \text{F.L. Cu loss} = (0{\cdot}6)^2 \times 34{\cdot}8 = 12{\cdot}53 \text{ kW}$$

$$\text{Total losses} = P_i + 12{\cdot}53 = 17{\cdot}4 + 12{\cdot}53 = 29{\cdot}93 \text{ kW}$$

$$\therefore \quad \text{Efficiency} = \frac{360}{360 + 29{\cdot}93} \times 100 = \mathbf{96{\cdot}3\ \%}$$

Example 8.63. *A 400 kVA transformer has an iron loss of 2 kW and the maximum efficiency at 0·8 p.f. occurs when the load is 240 kW. Calculate (i) the maximum efficiency at unity p.f. (ii) the efficiency on full-load at 0·71 p.f. lagging.*

Solution.

(*i*) **Maximum η at unity p.f.**

$$\text{kVA for maximum efficiency} = 240/0{\cdot}8 = 300 \text{ kVA}$$

$$\text{At maximum } \eta, \text{ copper loss} = \text{iron loss}$$

$$\therefore \quad \text{Copper loss at 300 kVA load} = 2 \text{ kW} \qquad (\because \text{ Iron loss} = 2 \text{ kW})$$

$$\text{Output} = 300 \times 1 = 300 \text{ kW} \; ; \; \text{Total losses} = 2 + 2 = 4 \text{ kW}$$

$$\therefore \quad \text{Maximum } \eta = \frac{300}{300 + 4} \times 100 = \mathbf{98{\cdot}6\%}$$

(*ii*) **F.L. efficiency at 0·71 p.f.**

$$\text{F.L. Cu loss} = \left(\frac{400}{300}\right)^2 \times 2 = 3{\cdot}55 \text{ kW} \; ; \; \text{Iron loss} = 2 \text{ kW}$$

$$\text{Total F.L. losses} = 3{\cdot}55 + 2 = 5{\cdot}55 \text{ kW}$$

$$\text{F.L. output} = 400 \times 0{\cdot}71 \text{ kW}$$

$$\therefore \quad \text{F.L. efficiency} = \frac{400 \times 0{\cdot}71}{400 \times 0{\cdot}71 + 5{\cdot}55} \times 100 = \mathbf{98\ \%}$$

Example 8.64. *A 10 kVA, 200/400 V, 50 Hz, single-phase transformer gave the following test results.*

O.C. test : 200 V ; 1·3 A ; 120 W ...on L.V. side

S.C. test : 22 V ; 30 A ; 200 W ...on H.V. side

(i) Calculate efficiency and secondary terminal voltage when supplying full load at 0·8 leading power factor .

(ii) Calculate the load at unity p.f. corresponding to maximum efficiency.

Solution.

$$(i) \quad \text{F.L. secondary current, } I_2 = \frac{10 \times 10^3}{400} = 25 \text{ A}$$

The S.C. test is carried out with 30 A flowing in the secondary.

$$\therefore \text{ F.L. Cu loss of transformer} = \left(\frac{25}{30}\right)^2 \times 200 = 140 \text{ W}$$

$$\text{Total F.L. losses} = 120 + 140 = 260 \text{ W} = 0{\cdot}26 \text{ kW}$$

$$\text{F.L. output} = 10 \times 0{\cdot}8 = 8 \text{ kW}$$

$$\therefore \quad \text{F.L. efficiency} = \frac{8}{8 + 0{\cdot}26} \times 100 = \mathbf{96{\cdot}85\ \%}$$

From S.C. test, we have,

$$Z_{02} = \frac{22}{30} = 0{\cdot}733\ \Omega \; ; \; R_{02} = \frac{200}{(30)^2} = 0{\cdot}222\ \Omega$$

$$\therefore \quad X_{02} = \sqrt{Z_{02}^2 - R_{02}^2} = \sqrt{(0{\cdot}733)^2 - (0{\cdot}222)^2} = 0{\cdot}698\ \Omega$$

Transformer voltage drop as referred to secondary

$$= I_2 (R_{02} \cos \phi_2 - X_{02} \sin \phi_2)$$

$$= 25 (0{\cdot}222 \times 0{\cdot}8 - 0{\cdot}698 \times 0{\cdot}6) = -6\text{ V}$$ [Voltage rise due to leading p.f.]

∴ Secondary voltage, $V_2 = 400 + 6 = \mathbf{406\ V}$

(*ii*) kVA for maximum efficiency $= \text{F.L. kVA} \times \sqrt{\dfrac{\text{Iron loss}}{\text{F.L. Cu loss}}}$

$$= 10 \times \sqrt{\frac{120}{140}} = 9{\cdot}25 \text{ kVA}$$

∴ Load of unity p.f. corresponding to η_{max}

$$= 9{\cdot}25 \times 1 = \mathbf{9{\cdot}25\ kW}$$

Example 8.65. *A 25 kVA, 2200/220 V, single-phase transformer has a primary resistance of 1 Ω and secondary resistance of 0·01 Ω. Find the full-load efficiency at 0·8 p.f. if the iron loss in the transformer is 200 W.*

Solution. $K = 220/2200 = 1/10 \; ; \; R_1 = 1\Omega \; ; \; R_2 = 0{\cdot}01\ \Omega$

$$R_{02} = R_2 + K^2 R_1 = 0{\cdot}01 + \frac{1}{100} \times 1 = 0{\cdot}02\ \Omega$$

F.L. secondary current, $I_2 = 25 \times 10^3/220 = 113{\cdot}6$ A

F.L. Cu loss $= I_2^2 R_{02} = (113{\cdot}6)^2 \times 0{\cdot}02 = 258$ W

Total F.L. losses $= 258 + 200 = 458$ W $= 0{\cdot}458$ kW

F.L. output $= 25 \times 0{\cdot}8 = 20$ kW

∴ F.L. efficiency $= \dfrac{20}{20 + 0.458} \times 100 = \mathbf{97{\cdot}8\%}$

Example 8.66. *A single-phase transformer is rated at 10 kVA, 230/100 V. When the secondary terminals are open circuited and the primary winding is supplied at normal voltage (230 V), the current input is 2·6 A at a p.f. of 0·3. When the secondary terminals are short-circuited, a voltage of 18 V applied to the primary causes the full-load current (100 A) to flow in the secondary ; the power input to the primary being 240 W. Calculate (i) the efficiency of the transformer at full-load unity p.f. (ii) the load at unity p.f. corresponding to maximum efficiency.*

Solution.

(*i*) Iron loss $=$ No-load primary input $= V_1 I_0 \cos \phi_0$

$$= 230 \times 2{\cdot}6 \times 0{\cdot}3 = 179{\cdot}4 \text{ W}$$

F.L. Cu loss $= 240$ W

Total F.L. losses $= 179{\cdot}4 + 240 = 419{\cdot}4$ W $= 0{\cdot}4194$ kW

F.L. output $= 10 \times 1 = 10$ kW

F.L. efficiency $= \dfrac{10}{10 + 0{\cdot}4194} \times 100 = \mathbf{95{\cdot}97\ \%}$

(*ii*) kVA for maximum η $= \text{F.L. kVA} \times \sqrt{\dfrac{\text{Iron loss}}{\text{F.L. Cu loss}}}$

$$= 10 \times \sqrt{\frac{179{\cdot}4}{240}} = 8{\cdot}65 \text{ kVA}$$

Load at unity p.f. corresponding to maximum η

$$= 8{\cdot}65 \times 1 = \mathbf{8{\cdot}65\ kW}$$

Example 8.67. *The maximum efficiency of a 500 kVA, 3300/500 V, 50 Hz, single-phase transformer is 97 % and occurs at 75 % of full-load at unity power factor. If the impedance is 10 %, calculate the voltage regulation at full-load, 0·8 power factor lagging.*

Solution.

$$\text{Output at maximum } \eta = 500 \times 0{\cdot}75 \times 1 = 375 \text{ kW}$$

$$\text{Total losses at max. } \eta = {}^*\text{Output}\left(\frac{1}{\eta} - 1\right) = 375\left(\frac{1}{0{\cdot}97} - 1\right) = 11{\cdot}6 \text{ kW}$$

At maximum efficiency, Cu loss = Iron loss

$$\therefore \quad \text{Cu loss} = \frac{\text{Total losses}}{2} = \frac{11{\cdot}6}{2} = 5{\cdot}8 \text{ kW}$$

$$\text{F.L. Cu loss} = 5{\cdot}8 \times \left(\frac{100}{75}\right)^2 = 10{\cdot}3 \text{ kW}$$

$$\% R^{**} = \frac{\text{F.L. Cu loss}}{V_2 I_2 \text{ rating}} \times 100 = \frac{10{\cdot}3 \times 10^3}{500 \times 10^3} \times 100 = 2{\cdot}06\,\% = v_r$$

$$\% Z = 10\,\% \quad (\textit{given})$$

$$\therefore \quad \% X = \sqrt{(10)^2 - (2{\cdot}06)^2} = 9{\cdot}78\,\% = v_x$$

$$\therefore \quad \text{Voltage regulation} = v_r \cos\phi_2 + v_x \sin\phi_2$$ ...See example 8.34

$$= 2{\cdot}06 \times 0{\cdot}8 + 9{\cdot}78 \times 0{\cdot}6 = \mathbf{7{\cdot}51\,\%}$$

TUTORIAL PROBLEMS

1. In a 100 kVA transformer, the iron loss is 1·2 kW and full-load copper loss is 2 kW. If the load p.f. is 0·8 lagging, find the efficiency at (*i*) full-load and (*ii*) half full-load. **[(*i*) 96·16 % (*ii*) 95·94 %]**

2. The primary and secondary windings of a 500 kVA, 6600/400 V transformer have resistances of 0·42 Ω and 0·0011 Ω respectively and the iron loss is 2·9 kW. Calculate the efficiency at (*i*) full-load and (*ii*) half full-load, assuming the power factor of the load to be 0·8 lagging. **[(*i*) 98·27 % (*ii*) 98·07 %]**

3. A 500/250 V, 1-phase transformer has constant losses of 150 W. When delivering an output of 8 kW to a load, the transformer takes 21 A at a p.f. of 0·8 lagging from 500 V supply. Calculate the variable losses of the transformer at this load.

4. A 100 kVA transformer has an iron loss of 1 kW and full-load Cu loss of 1·5 kW. Calculate the kVA loading at which the efficiency is maximum and its efficiency at this loading at 0·8 p.f. lagging. **[82·3 kVA ; 97·05 %]**

5. A 150 kVA transformer has iron-loss of 1·4 kW and full-load copper loss of 2·8 kW. Calculate (*i*) the efficiency of transformer at full-load (*ii*) the maximum efficiency of the transformer and (*iii*) the output power at the maximum level of efficiency. Assume unity power factor. **[(*i*) 97·2 % (*ii*) 98·13 % (*iii*) 106·05 kW]**

6. A 300 kVA, single-phase transformer is designed to have a resistance of 1·5 % and maximum efficiency occurs at a load of 173·2 kVA. Find its efficiency when supplying full-load at 0·8 p.f. lagging at normal voltage and frequency. **[97·6 %]**

---

* $\frac{1}{\eta} = \frac{\text{Output} + \text{Losses}}{\text{Output}} = 1 + \frac{\text{Losses}}{\text{Output}}$

** It is a usual practice to refer to the percentage resistance, reactance and impedance drops on full-load as simply % resistance, reactance and impedance of the transformer.

7. A 50 kVA, 6360/240 V transformer is tested on open and short circuit to obtain efficiency. The results of the tests are as follows :

   O.C. test : Primary voltage 6360 V, primary current 1 A, power input 2 kW

   S.C. test : Voltage across secondary winding 50 V, current in secondary winding 175 A, power input 2 kW.

   Find the efficiency of the transformer when supplying full-load at a power factor of 0·8 lagging. **[89·2 %]**

8. A 600 kVA, single-phase transformer has an efficiency of 92% both at full-load and half full-load at unity power factor. Determine its efficiency at 60% of full load at 0·8 power factor lag. **[85·9 %]**

9. The losses in a 80 kVA, 1100/250 V, single-phase transformer are as follows : Iron losses 800 W; copper loss with 160 A in low voltage winding is 200 W. Calculate the efficiency at full-load and half full-load at unity power-factor. **[98·04 % ; 97·57 %]**

10. A 4 kVA, 200/400 V, single-phase transformer has equivalent resistance and reactance referred to low-voltage side equal to 0·5 Ω and 1·5 Ω respectively. Find the terminal voltage on the high-voltage side when it supplies 3/4th full-load at power factor of 0·8 lagging, the supply voltage being 200 V. Hence find the output of the transformer and its efficiency if core losses are 100 W. **[361 V ; 2·4 kW ; 91·87 %]**

8.29 ALL-DAY (OR ENERGY) EFFICIENCY

The ordinary or commercial efficiency of a transformer is defined as the ratio of output power to the input power *i.e.*,

$$\text{Commercial efficiency} = \frac{\text{Output power}}{\text{Input power}}$$

There are certain types of transformers whose performance cannot be judged by this efficiency. For instance, distribution transformers used for supplying lighting loads have their primaries energised all the 24 hours in a day but the secondaries supply little or no load during the major portion of the day. It means that a constant loss (*i.e.*, iron loss) occurs during the whole day but copper loss occurs only when the transformer is loaded and would depend upon the magnitude of load. Consequently, the copper loss varies considerably during the day and the commercial efficiency of such transformers will vary from a low value (or even zero) to a high value when the load is high. The performance of such transformers is judged on the basis of energy consumption during the whole day (*i.e.*, 24 hours). This is known as all-day or energy efficiency.

The ratio of output in kWh to the input in kWh of a transformer over a 24-hour period is known as **all-day efficiency** *i.e.*,

$$\eta_{all\text{-}day} = \frac{\text{kWh output in 24 hours}}{\text{kWh input in 24 hours}}$$

All-day efficiency is of special importance for those transformers whose primaries are never open-circuited but the secondaries carry little or no load much of the time during the day. In the design of such transformers, efforts should be made to reduce the iron losses which continuously occur during the whole day.

Note. Efficiency of a transformer means commercial efficiency unless stated otherwise.

Example 8.68. *A 5 kVA distribution transformer has a full-load efficiency of 95 % at which Cu loss is equal to the iron loss. The transformer is loaded for 24 hours as under :*

No load for 10 hours, one-fourth full-load for 7 hours, half full-load for 5 hours and full-load for 2 hours.

Calculate the all-day efficiency of the transformer.

Solution. Since a distribution transformer supplies mostly lighting load, it has a load p.f. of unity.

$$\text{F.L. output} = 5 \times 1 = 5 \text{ kW}$$
$$\text{F.L. input} = 5/0{\cdot}95 = 5{\cdot}26 \text{ kW}$$
$$\text{Total F.L. losses} = 5{\cdot}26 - 5 = 0{\cdot}26 \text{ kW}$$
$$\text{Iron loss} = \text{F.L. Cu loss} = 0{\cdot}26/2 = 0{\cdot}13 \text{ kW}$$
$$\text{Cu loss at 1/4 F.L.} = (1/16) \times 0{\cdot}13 = 0{\cdot}008 \text{ kW}$$
$$\text{Cu loss at 1/2 F.L.} = (1/4) \times 0{\cdot}13 = 0{\cdot}03 \text{ kW}$$
$$\text{Iron loss in 24 hours} = 0{\cdot}13 \times 24 = 3{\cdot}12 \text{ kWh}$$
$$\text{Cu loss in 24 hours} = (0{\cdot}008 \times 7) + (0{\cdot}03 \times 5) + (0{\cdot}13 \times 2)$$
$$= 0{\cdot}056 + 0{\cdot}15 + 0{\cdot}26 = 0{\cdot}466 \text{ kWh}$$
$$\text{Total losses in 24 hours} = 3{\cdot}12 + 0{\cdot}466 = 3{\cdot}586 \text{ kWh}$$
$$\text{Total output in 24 hours} = (7 \times 5/4) + (5 \times 5/2) + 2 \times 5 = 31{\cdot}25 \text{ kWh}$$
$$\therefore \quad \text{All-day efficiency} = \frac{31{\cdot}25}{31{\cdot}25 + 3{\cdot}586} \times 100 = \mathbf{89{\cdot}7\ \%}$$

Note that commercial efficiency of this transformer is 95 % while its all-day efficiency is 89·7%.

Example 8.69. *A 40 kVA distribution transformer has iron loss of 500 W and full-load copper loss of 500 W. The transformer is supplying a lighting load (unity p.f.). The load cycle is as under :*

Full-load for 4 hours ; half full-load for 8 hours and no load for 12 hours.

Calculate the all-day efficiency of the transformer.

Solution.

$$\text{Iron loss} = 500 \text{ W} = 0{\cdot}5 \text{ kW}$$
$$\text{F.L. Cu loss} = 500 \text{ W} = 0{\cdot}5 \text{ kW}$$
$$\text{Cu loss at half full-load} = (1/4) \times 0{\cdot}5 = 0{\cdot}125 \text{ kW}$$
$$\text{Iron loss in 24 hours} = 0{\cdot}5 \times 24 = 12 \text{ kWh}$$
$$\text{Cu loss in 24 hours} = (0{\cdot}5 \times 4) + (0{\cdot}125 \times 8) = 3 \text{ kWh}$$
$$\text{Total loss in 24 hours} = 12 + 3 = 15 \text{ kWh}$$
$$\text{Output in 24 hours} = (40 \times 4) + (40 \times 1/2 \times 8) = 320 \text{ kWh}$$
$$\text{All-day efficiency} = \frac{320}{320 + 15} \times 100 = \mathbf{95{\cdot}52\ \%}$$

Example 8.70. *A transformer has its maximum efficiency of 98% at 15 kVA at unity power factor. During the day, it is loaded as under :*

12 hours : 2 kW at power factor 0·5 lagging
6 hours : 12 kW at power factor 0·8 lagging
6 hours : 18 kW at power factor 0·9 lagging

Find the all-day efficiency.

Solution.

$$\text{Output} = 15 \times 1 = 15 \text{ kW}; \text{Input} = 15/0{\cdot}98 = 15{\cdot}306 \text{ kW}$$
$$\text{Total losses at maximum } \eta = 15{\cdot}306 - 15 = 0{\cdot}306 \text{ kW} = 306 \text{ W}$$
$$\text{At maximum } \eta, \text{Cu loss} = \text{Iron loss}$$
$$\therefore \quad \text{Cu loss at 15 kVA} = 306/2 = 153 \text{ W}; \text{Iron loss} = 153 \text{ W}$$

kVA loads during the day : $2/0{\cdot}5 = 4$ kVA; $12/0{\cdot}8 = 15$ kVA ; $18/0{\cdot}9 = 20$ kVA

$$\text{Cu loss at 4 kVA} = (4/15)^2 \times 153 = 10{\cdot}9 \text{ W}$$
$$\text{Cu loss at 15 kVA} = (15/15)^2 \times 153 = 153 \text{ W}$$
$$\text{Cu loss at 20 kVA} = (20/15)^2 \times 153 = 272 \text{ W}$$

$$\text{Iron loss in 24 hrs} = \frac{153 \times 24}{1000} = 3{\cdot}672 \text{ kWh}$$

$$\text{Cu loss in 24 hrs} = \frac{1}{1000}[10{\cdot}9 \times 12 + 153 \times 6 + 272 \times 6] = 2{\cdot}681 \text{ kWh}$$

$$\text{Total losses in 24 hrs} = 3{\cdot}672 + 2{\cdot}681 = 6{\cdot}353 \text{ kWh}$$

$$\text{Total output in 24 hrs} = 2 \times 12 + 12 \times 6 + 18 \times 6 = 204 \text{ kWh}$$

$$\therefore \quad \text{All-day efficiency} = \frac{204}{204 + 6{\cdot}353} \times 100 = \mathbf{96{\cdot}98\ \%}$$

Example 8.71. *A 150 kVA transformer is loaded as under :*

Load increases from zero to 100 kVA in 3 hours from 7 a.m. to 10 a.m., stays at 100 kVA from 10 a.m. to 6 p.m. and then the transformer is disconnected till next day. Assuming the load to be resistive and core loss equal to full-load copper loss of 1 kW, determine the commercial and all-day efficiency of the transformer.

Solution. Since the load is resistive, its power factor is unity.

$$\text{Average load from 7 a.m. to 10 a.m.} = \frac{0 + 100}{2} = 50 \text{ kVA}$$

$$\text{Load from 10 a.m. to 6 p.m.} = 100 \text{ kVA}$$

$$\text{F.L. Cu loss} = 1 \text{ kW} ; \text{Iron loss} = 1 \text{ kW}$$

Commercial efficiency

For commercial efficiency of a transformer, the load cycle is not relevant.

$$\text{Output} = 150 \times 1 = 150 \text{ kW} ; \text{Total losses} = 1 + 1 = 2 \text{ kW}$$

$$\therefore \quad \text{Commercial efficiency} = \frac{150}{150 + 2} \times 100 = \mathbf{98{\cdot}68\ \%}$$

All-day efficiency

Iron loss remains the same at all loads.

$$\text{Cu loss at 50 kVA} = (50/150)^2 \times 1 = 1/9 \text{ kW}$$

$$\text{Cu loss at 100 kVA} = (100/150)^2 \times 1 = 4/9 \text{ kW}$$

$$\text{Cu loss in 24 hrs} = 1/9 \times 3 + 4/9 \times 8 = 3{\cdot}888 \text{ kWh}$$

$$\text{Iron loss in 24 hrs} = 1 \text{ kW} \times 24 \text{ hrs} = 24 \text{ kWh}$$

$$\text{Total losses in 24 hrs} = 3{\cdot}888 + 24 = 27{\cdot}888 \text{ kWh}$$

$$\text{Output in 24 hrs} = (50 \times 1) \times 3 + (100 \times 1)\, 8 = 950 \text{ kWh}$$

$$\therefore \quad \text{All-day efficiency} = \frac{950}{950 + 27{\cdot}888} \times 100 = \mathbf{97{\cdot}15\ \%}$$

Example 8.72. *A 10 kVA transformer is given an open-circuit test from which the iron loss is found to be 160 W. A short-circuit test shows the copper loss to be 240 W with full-load current flowing. Calculate the all-day efficiency if it operates at unity p.f. 6 hours per day on full-load, 2 hours on half full-load and the remainder of the time on no load.*

Solution.

$$\text{F.L. Cu loss} = 240 \text{ W} ; \text{Iron loss} = 160 \text{ W}$$

$$\text{Cu loss at half full-load} = \left(\frac{1}{2}\right)^2 \times 240 = 60 \text{ W}$$

$$\text{Iron loss in 24 hrs} = 0{\cdot}160 \times 24 = 3{\cdot}84 \text{ kWh}$$

$$\text{Cu loss in 24 hrs} = 0{\cdot}240 \times 6 + 0{\cdot}06 \times 2 = 1{\cdot}56 \text{ kWh}$$

$$\text{Total losses in 24 hrs} = 3{\cdot}84 + 1{\cdot}56 = 5{\cdot}4 \text{ kWh}$$

$$\text{Output in 24 hrs} = 10 \times 6 + 5 \times 2 = 70 \text{ kWh}$$

$$\therefore \quad \text{All-day efficiency} = \frac{70}{70 + 5{\cdot}4} \times 100 = \mathbf{92{\cdot}84\ \%}$$

TUTORIAL PROBLEMS

1. Find the all-day efficiency of a 500 kVA distribution transformer whose copper loss and iron loss at full-load are 4·5 kW and 3·5 kW respectively. During the day of 24 hours, it is loaded as under :

No. of hours	*Load in kW*	*Power factor*
6	400	0·8
10	300	0·75
4	100	0·8
4	0	–

[97·62 %]

2. A 100 kVA lighting transformer has a full-load loss of 3 kW, the losses being equally divided between iron and copper. During one day, the transformer operates on full-load for 3 hours, on half full-load for 4 hours, the output being negligible for the remainder of the day. Calculate the all-day efficiency. **[92·26 %]**

3. Find all-day efficiency of a transformer having a maximum efficiency of 98% at 15 kVA at unity power factor and loaded as follows :

12 hours : 2 kW at 0·5 p.f. lag
6 hours : 12 kW at 0·8 p.f. lag
6 hours : at no load **[95·3 %]**

8.30 CONSTRUCTION OF A TRANSFORMER

We usually design a power transformer so that it approaches the characteristics of an ideal transformer. To achieve this, following design features are incorporated :

(*i*) The core is made of silicon steel which has low hysteresis loss and high permeability. Further, core is laminated in order to reduce eddy current loss. These features considerably reduce the iron losses and the no-load current.

(*ii*) Instead of placing primary on one limb and secondary on the other, it is a usual practice to wind one-half of each winding on one limb. This ensures tight coupling between the two windings. Consequently, leakage flux is considerably reduced.

(*iii*) The winding resistances R_1 and R_2 are minimised to reduce I^2R loss and resulting rise in temperature and to ensure high efficiency.

8.31 TYPES OF TRANSFORMERS

Depending upon the manner in which the primary and secondary are wound on the core, transformers are of two types *viz.*, (*i*) core-type transformer and (*ii*) shell-type transformer.

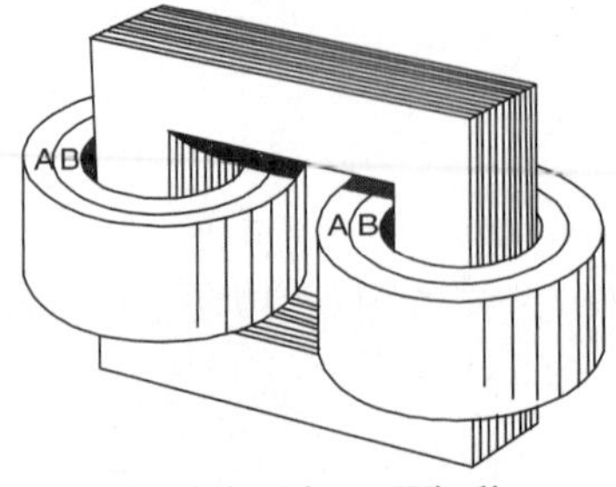

A High Voltage Winding
B Low Voltage Winding

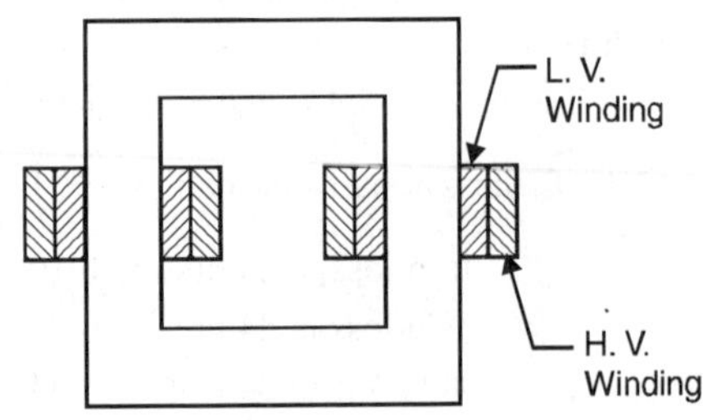

Core-type transformer

Fig. 8.49

(*i*) **Core-type transformer.** In a core-type transformer, half of the primary winding and half of the secondary winding are placed around each limb as shown in Fig. 8.49. This reduces the leakage flux. It is a usual practice to place the low-voltage winding below the high-voltage winding for mechanical considerations.

(*ii*) **Shell-type transformer.** This method of construction involves the use of a double magnetic circuit. Both the windings are placed around the central limb (See Fig. 8.50), the other two limbs acting simply as a low-reluctance flux path.

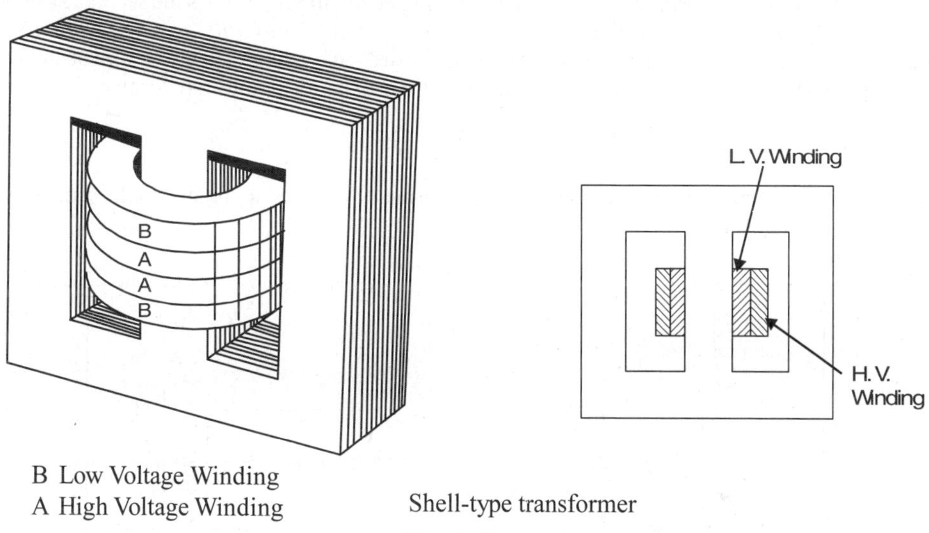

B Low Voltage Winding
A High Voltage Winding

Shell-type transformer

Fig. 8.50

The choice of type (whether core or shell) will not greatly affect the efficiency of the transformer. The core type is generally more suitable for high voltage and small output while the shell-type is generally more suitable for low voltage and high output.

8.32 COOLING OF TRANSFORMERS

In all electrical machines, the losses produce heat and means must be provided to keep the temperature low. In generators and motors, the rotating unit serves as a fan causing air to circulate and carry away the heat. However, a transformer has no rotating parts. Therefore, some other methods of cooling must be used. Heat is produced in a transformer by the iron losses in the core and I^2R loss in the windings. To prevent undue temperature rise, this heat is removed by cooling.

(*i*) In small transformers (below 50 kVA), natural air cooling is employed *i.e.*, the heat produced is carried away by the surrounding air.

(*ii*) Medium size power or distribution transformers are generally cooled by housing them in tanks filled with oil. The oil serves a double purpose, carrying the heat from the windings to the surface of the tank and insulating the primary from the secondary.

(*iii*) For large transformers, external radiators are added to increase the cooling surface of the oil filled tank. The oil circulates around the transformer and moves through the radiators where the heat is released to surrounding air. Sometimes cooling fans blow air over the radiators to accelerate the cooling process.

8.33 TRANSFORMER RATING

Transformers are rated in terms of the volt-ampere output under specified conditions of voltage and frequency. As in the rating of all electrical equipment, the operating temperature is the determining factor in the transformer rating. The nameplate of a transformer provides the following information :

(*i*) Voltage rating (for high-voltage and low-voltage winding)

(*ii*) kVA rating (*iii*) Frequency rating (*iv*) Temperature rise

The rated currents of the transformer windings are not usually given on the nameplate but can be calculated from the kVA rating and voltage ratings of the windings.

8.34 AUTOTRANSFORMER

An autotransformer has a single winding on an iron core and a part of winding is common to both the primary and secondary circuits. Fig. 8.51 (*i*) shows the connections of a step-down autotransformer whereas Fig. 8.51 (*ii*) shows the connections of a step-up autotransformer. In either case, the winding *ab* having N_1 turns is the primary winding and winding *bc* having N_2 turns is the secondary winding. *Note that the primary and secondary windings are connected electrically as well as magnetically.* Therefore, power from the primary is transferred to the secondary conductively as well as inductively (transformer action). The voltage transformation ratio *K* of an ideal autotransformer is

$$K = \frac{V_2}{V_1} = \frac{N_2}{N_1} = \frac{I_1}{I_2}$$

Note that in an autotransformer, secondary and primary voltages are related in the same way as in a 2-winding transformer.

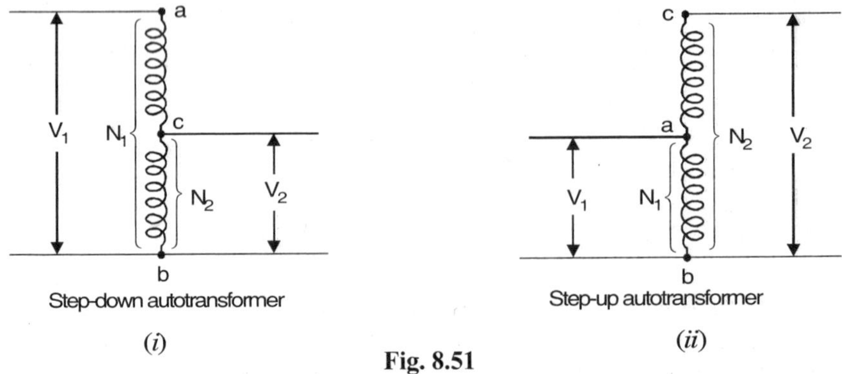

(*i*) (*ii*)

Fig. 8.51

Fig. 8.52 shows the connections of a loaded step-down as well as step-up autotransformer. In each case, I_1 is the input current and I_2 is the output or load current. Regardless of autotransformer connection (step-up or step-down), the current in the portion of the winding that is common to both the primary and the secondary is the difference between these currents (I_1 and I_2). The relative direction of the current through the common portion of the winding depends upon the connections of the autotransformer. It is because the type of connection determines whether input current I_1 or output current I_2 is larger. For step-down autotransformer $I_2 > I_1$ (as for 2-winding transformer) so that *$I_2 - I_1$ current flows through the common portion of the winding. For step-up autotransformer, $I_2 < I_1$. Therefore, $I_1 - I_2$ current flows in the common portion of the winding.

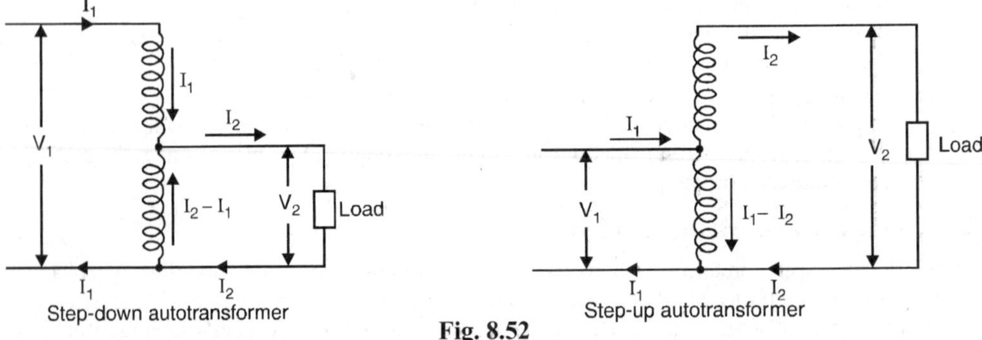

Fig. 8.52

* Note that secondary current [$(I_2 - I_1)$ or $(I_1 - I_2)$] in an autotransformer is less than the load current. This is unlike a two-winding transformer in which secondary current is equal to the load current.

In an ideal autotransformer, exciting current and losses are neglected. For such an autotransformer, as K approaches 1, the value of current in the common portion ($I_2 - I_1$ or $I_1 - I_2$) of the winding approaches *zero. Therefore, for value of K near unity, the common portion of the winding can be wound with wire of smaller cross-sectional area. For this reason, an autotransformer requires less copper.

8.35 THEORY OF AUTOTRANSFORMER

Fig. 8.53 (*i*) shows an ideal step-down autotransformer on load. Here winding 1-3 having N_1 turns is the primary winding while winding 2-3 having N_2 turns is the secondary winding. The input current is I_1 while the output or load current is I_2. Note that portion 1-2 of the winding has $N_1 - N_2$ turns and voltage across this portion of the winding is $V_1 - V_2$. The current through the common portion of the winding is $I_2 - I_1$.

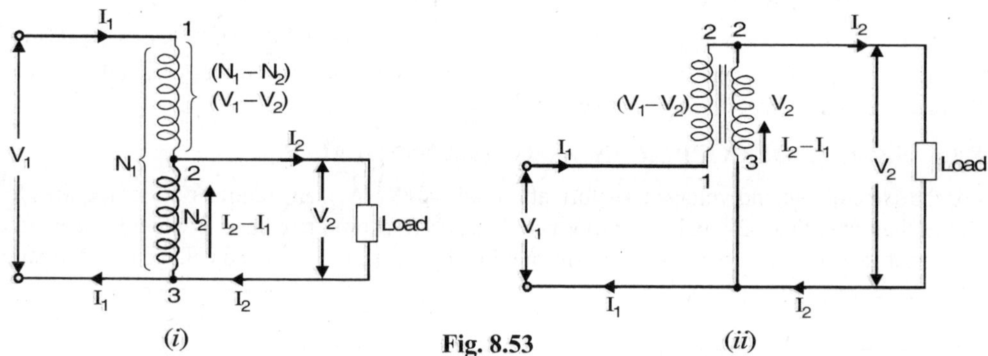

(*i*) **Fig. 8.53** (*ii*)

Fig. 8.53 (*ii*) shows the equivalent circuit of the autotransformer. From this equivalent circuit, we have,

$$\frac{V_2}{V_1 - V_2} = \frac{N_2}{N_1 - N_2}$$

$$V_2 (N_1 - N_2) = N_2 (V_1 - V_2)$$

or $$V_2 N_1 - V_2 N_2 = N_2 V_1 - N_2 V_2$$

or $$V_2 N_1 = N_2 V_1$$

$\therefore$ $$\frac{V_2}{V_1} = \frac{N_2}{N_1} = K \quad ...(i)$$

Also $$(V_1 - V_2) I_1 = (I_2 - I_1) V_2 \quad \text{...See Fig. 8.53 } (ii)$$

or $$V_1 I_1 - V_2 I_1 = V_2 I_2 - V_2 I_1$$

or $$V_1 I_1 - V_2 I_2$$

$\therefore$ $$\frac{V_2}{V_1} = \frac{I_1}{I_2} \quad ...(ii)$$

From eqs. (*i*) and (*ii*), $$\frac{V_2}{V_1} - \frac{N_2}{N_1} = \frac{I_1}{I_2} = K$$

Also $$V_1 I_1 = V_2 I_2 \quad \text{(Input apparent power = Output apparent power)}$$

Output. Since the primary and secondary windings of an autotransformer are connected magnetically as well as electrically, the power from primary is transferred to the secondary inductively (transformer action) as well as conductively (*i.e.*, conducted directly from source to the load).

$$\text{Output apparent power} = V_2 I_2$$

* $I_1 = K I_2$. As $K \to 1$, $I_1 \simeq I_2$.

$$\text{Apparent power transferred inductively} = V_2(I_2 - I_1) = V_2(I_2 - KI_2)$$
$$= V_2 I_2 (1-K) = V_1 I_1 (1-K)$$

$\therefore$ Power transferred inductively $=$ Input $\times (1-K)$

$\therefore$ Power transferred conductively $=$ Input $-$ Input $(1-K)$

$= \text{Input}\,[1-(1-K)]$

$= K \times \text{Input}$

Suppose the input power to an ideal autotransformer is 1000 W and its voltage transformation ratio $K = 1/4$. Then,

$$\text{Power transferred inductively} = \text{Input} \times (1-K) = 1000\left(1 - \frac{1}{4}\right) = 750 \text{ W}$$

$$\text{Power transferred conductively} = K \times \text{Input} = \frac{1}{4} \times 1000 = 250 \text{ W}$$

Note that input power to the autotransformer is 1000 W. Out of this, 750 W is transferred to the secondary by transformer action (inductively) while 250 W is *conducted* directly from the source to the load (*i.e.*, it is transferred conductively to the load).

8.36 SAVING OF COPPER IN AUTOTRANSFORMER

For the same output and voltage transformation ratio K (N_2/N_1), an autotransformer requires less copper than an ordinary 2-winding transformer. Fig. 8.54 (*i*) shows an ordinary 2-winding transformer whereas Fig. 8.54 (*ii*) shows an autotransformer having the same output and voltage transformation ratio K.

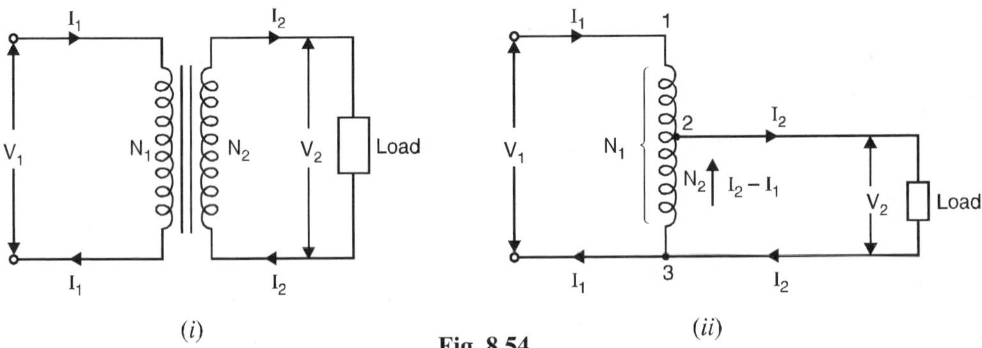

Fig. 8.54

The length of copper required in a winding is proportional to the number of turns and the area of cross-section of the winding wire is proportional to the current rating. Therefore, the volume and hence weight of copper required in a winding is proportional to current × turns *i.e.*,

Weight of Cu required in a winding $\propto$ current × turns

2-winding transformer

Weight of Cu required $\propto (I_1 N_1 + I_2 N_2)$

Autotransformer

Weight of Cu required in section 1-2 $\propto I_1 (N_1 - N_2)$

Weight of Cu required in section 2-3 $\propto (I_2 - I_1) N_2$

$\therefore$ Total weight of Cu required $\propto I_1 (N_1 - N_2) + (I_2 - I_1) N_2$

$$\frac{\text{Weight of Cu in autotransformer}}{\text{Weight of Cu in ordinary transformer}} = \frac{I_1 (N_1 - N_2) + (I_2 - I_1) N_2}{I_1 N_1 + I_2 N_2}$$

$$= \frac{N_1 I_1 - N_2 I_1 + N_2 I_2 - N_2 I_1}{N_1 I_1 + N_2 I_2}$$

$$= \frac{N_1 I_1 + N_2 I_2 - 2 N_2 I_1}{N_1 I_1 + N_2 I_2}$$

$$= 1 - \frac{2 N_2 I_1}{N_1 I_1 + N_2 I_2}$$

$$= 1 - \frac{2 N_2 I_1}{2 N_1 I_1} \qquad (\because N_2 I_2 = N_1 I_1)$$

$$= 1 - \frac{N_2}{N_1} = 1 - K$$

$\therefore$ Wt. of Cu in autotransformer (W_a) = $(1 - K) \times$ Wt. in ordinary transformer (W_o)

or $W_a = (1 - K) \times W_o$

$\therefore$ Saving in Cu $= W_o - W_a = W_o - (1 - K)\, W_o = K\, W_o$

or Saving in Cu $= K \times$ Wt. of Cu in ordinary transformer

Thus if $K = 0{\cdot}1$, the saving of Cu is only 10% but if $K = 0{\cdot}9$, saving of Cu is 90%. Therefore, the nearer the value of K of autotransformer is to 1, the greater is the saving of Cu.

8.37 ADVANTAGES AND DISADVANTAGES OF AUTOTRANSFORMERS

Advantages

(*i*) An autotransformer requires less Cu than a two-winding transformer of similar rating.

(*ii*) An autotransformer operates at a higher efficiency than a two-winding transformer of similar rating.

(*iii*) An autotransformer has better voltage regulation than a two-winding transformer of the same rating.

(*iv*) An autotransformer has smaller size than a two-winding transformer of the same rating.

(*v*) An autotransformer requires smaller exciting current than a two-winding transformer of the same rating.

It may be noted that these advantages of the autotransformer decrease as the ratio of transformation increases. Therefore, an autotransformer has marked advantages only for relatively low values of transformation ratio (*i.e.* values approaching 1).

Disadvantages

(*i*) There is a direct connection between the primary and secondary. *Therefore, the output is no longer d.c. isolated from the input.*

(*ii*) An autotransformer is not safe for stepping down a high voltage to a low voltage. As an illustration, Fig. 8.55 shows 11000/230 V step-down autotransformer. If an open circuit develops in the common portion 2-3 of the winding, then full-primary voltage (*i.e.*, 11000 V in this case) will appear across the load. In such a case, any one coming in contact with the secondary is subjected to high voltage. This could be dangerous to both the persons and equipment. For this reason, autotransformers are prohibited for general use.

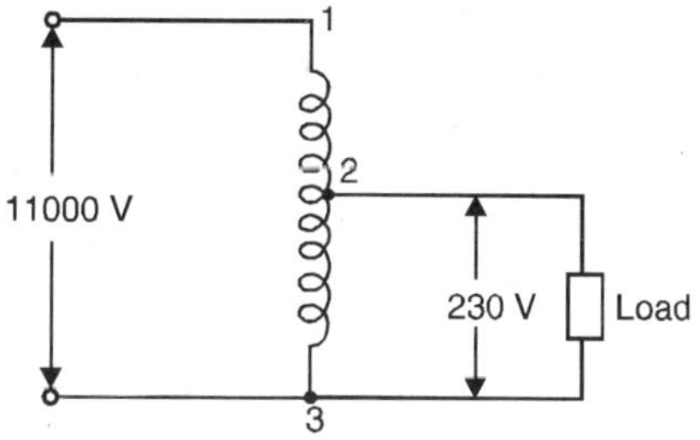

Fig. 8.55

(*iii*) The short-circuit current is much larger than for the two-winding transformer of the same rating. It can be seen from Fig. 8.55 that a short-circuited secondary causes part of the primary also to be short-circuited. This reduces the effective resistance and reactance.

8.38 APPLICATIONS OF AUTOTRANSFORMERS

(*i*) Autotransformers are used to compensate for voltage drops in transmission and distribution lines. When used for this purpose, they are known as *booster transformers*.

(*ii*) Autotransformers are used for reducing the voltage supplied to a.c. motors during the starting period.

(*iii*) Autotransformers are used for continuously variable supply.

Example 8.73. *Fig. 8.56 shows a step-down autotransformer. Find the currents in different parts of the winding.*

Solution. Load current, $I_2 = \frac{300}{30} = \mathbf{10\ A}$

$$V_1 I_1 = V_2 I_2$$

$$\therefore \quad I_1 = \frac{V_2 I_2}{V_1} = \frac{300 \times 10}{440} = \mathbf{6{\cdot}82\ A}$$

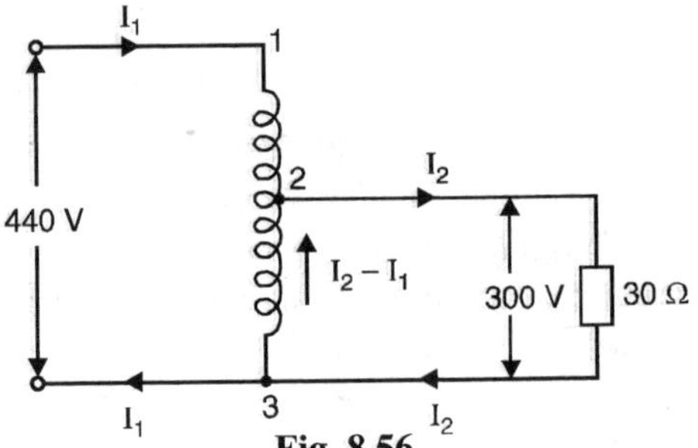

Fig. 8.56

Current in the common portion of the winding

$$= I_2 - I_1 = 10 - 6{\cdot}82 = \mathbf{3{\cdot}18\ A}$$

Note the direction of current in the common portion of the winding.

Example 8.74. *An autotransformer supplies a load of 2·5 kW at 110 V and at a power factor of unity. If the primary applied voltage is 220 V, calculate the power supplied to the load (i) inductively (ii) conductively.*

Solution. Neglecting losses, the input power is 2·5 kW. Now $K = 110/220 = 1/2$.

(*i*) Power transferred inductively $= \text{Input} \times (1 - K) = 2{\cdot}5 \left(1 - \frac{1}{2}\right) = \mathbf{1{\cdot}25\ kW}$

(*ii*) Power transferred conductively $= K \times \text{Input} = \frac{1}{2} \times 2{\cdot}5 = \mathbf{1{\cdot}25\ kW}$

Example 8.75. *An autotransformer is used to transform from 500 V to 440 V. The load is 20 kW at unity power factor. Neglecting losses and magnetising current, find the currents in the various parts of the winding.*

Solution. Fig. 8.57 shows the connections of autotransformer. Clearly, it is a step-down transformer.

Load current, $I_2 = \frac{20 \times 10^3}{440} = \mathbf{45{\cdot}45\ A}$

Now $V_1 I_1 = V_2 I_2$

or $500 \times I_1 = 440 \times 45{\cdot}45$

$\therefore \quad I_1 = \mathbf{40\ A}$

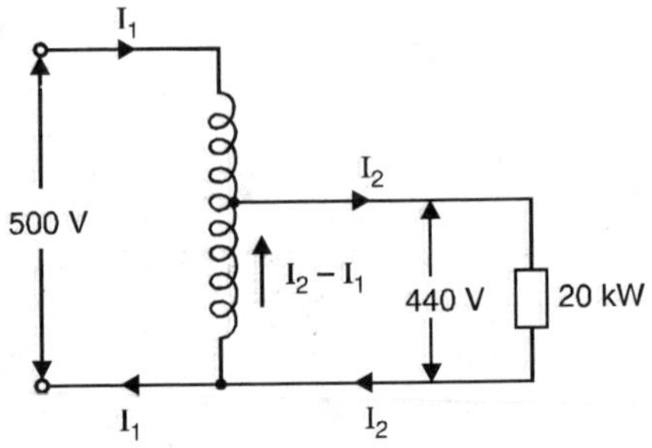

Fig. 8.57

Current in common portion of the winding

$$= I_2 - I_1 = 45{\cdot}45 - 40 = \mathbf{5{\cdot}45\ A}$$

Example 8.76. *The primary and secondary voltages of an autotransformer are 230 V and 75 V respectively. Calculate the currents in the different parts of the winding when load current is 200 A. Also calculate the saving of copper.*

Solution. Fig. 8.58 shows the autotransformer connections.

$$V_1 I_1 = V_2 I_2$$

or $$230 \times I_1 = 75 \times 200$$

$$\therefore \quad I_1 = \mathbf{65 \cdot 2\ A}$$

Current in the common portion of the winding

$$= I_2 - I_1 = 200 - 65 \cdot 2 = \mathbf{134 \cdot 8\ A}$$

$$\text{Saving of Cu} = K\,W_o = \frac{75}{230}\,W_o = 0 \cdot 326\,W_o$$

$$\therefore \%\ \text{saving of Cu} = 0 \cdot 326 \times 100 = \mathbf{32 \cdot 6\ \%}$$

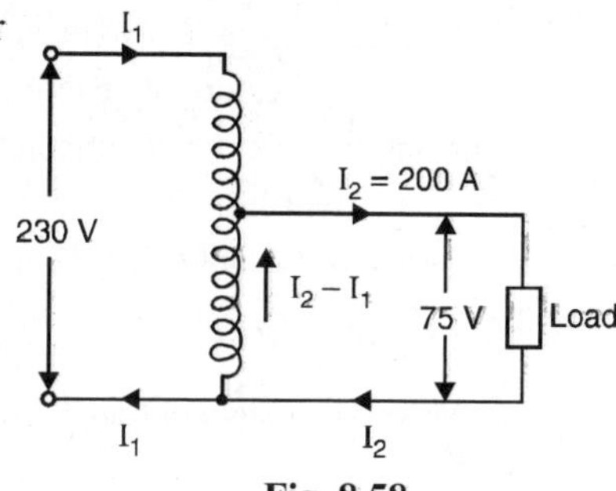

Fig. 8.58

Example 8.77. *An auto-transformer supplies a load of 5 kW at 125 V at unity power factor. If the primary voltage is 250 V, determine (i) transformation ratio (ii) secondary current (iii) primary current (iv) number of secondary turns if total number of turns is 250 (v) power transformed inductively (vi) power transformed conductively.*

Solution.

(*i*) Transformation ratio, $K = V_2/V_1 = 125/250 = 1/2 = \mathbf{0.5}$

(*ii*) Secondary current, $I_2 = \dfrac{5 \times 10^3}{V_2 \cos\phi_2} = \dfrac{5 \times 10^3}{125 \times 1} = \mathbf{40\ A}$

(*iii*) Primary current, $I_1 = K\,I_2 = (1/2) \times 40 = \mathbf{20\ A}$

(*iv*) Number of secondary turns, $N_2 = K\,N_1 = (1/2) \times 250 = \mathbf{125}$

(*v*) Power transferred inductively = *Input $(1-K)$ = $5(1-0.5)$ = **2.5 kW**

(*vi*) Power transferred conductively = $K \times$ Input = $(1/2) \times 5$ = **2.5 kW**

Example 8.78. *A 480/120 V, 5 kVA two winding transformer is to be used as an auto-transformer to supply power at 480 V from 600 V source. Draw the connection diagram and determine the kVA capacity as an auto-transformer.*

Solution. The rated output of two-winding transformer is 5 kVA.

$$\therefore \quad \text{Current rating of 480 V winding} = \frac{5 \times 10^3}{480} = 10.4\ \text{A}$$

$$\text{Current rating of 120 V winding} = \frac{5 \times 10^3}{120} = 41.67\ \text{A}$$

When this two-winding transformer is to be used as an auto-transformer, the circuit connections will be as shown in Fig. 8.59.

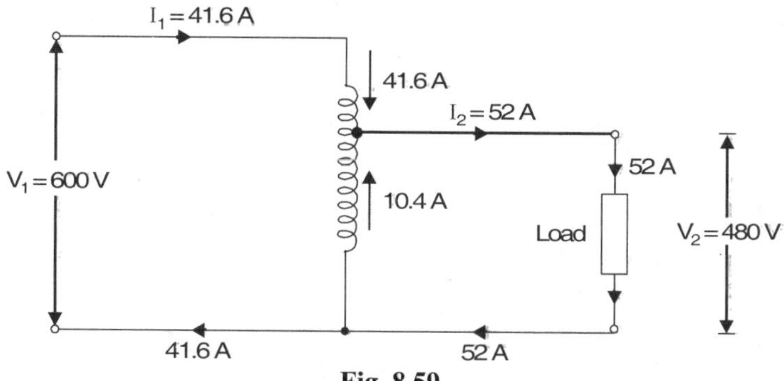

Fig. 8.59

* Transformer is assumed to be ideal so that output = input = 5 kW.

$$\text{Primary voltage, } V_1 = 600 \text{ V} \text{ ; Secondary voltage, } V_2 = 480 \text{ V}$$

$$\text{Transformation ratio, } K = V_2/V_1 = 480/600 = 4/5$$

Let the output current be I_2 amperes. Then primary current is $I_1 = K I_2 = (4/5) I_2 = 0.8 I_2$.

$$\text{Current thro' 480 V winding } = I_2 - I_1 = I_2 - 0.8 I_2 = 0.2 I_2$$

Since 480 V winding has the maximum current capacity of 10.4 A,

$$\therefore \quad 0.2 I_2 = 10.4 \quad \text{or} \quad I_2 = 10.4/0.2 = 52 \text{ A}$$

$$\text{Maximum output of transformer } = \frac{V_2 I_2}{1000} = \frac{480 \times 52}{1000} \text{ kVA} = \mathbf{25\ kVA}$$

TUTORIAL PROBLEMS

1. An autotransformer is used to step-down from 240 V to 200 V. The complete winding consists of 438 turns and the secondary delivers a load current of 15 A. Determine (*i*) secondary turns (*ii*) primary current (*iii*) current in the secondary winding. Neglect the effect of magnetising current. **[(*i*) 365 turns (*ii*) 12·5 A (*iii*) 2·5 A]**
2. An autotransformer supplies a load of 3 kW at 115 V at a unity power factor. If the applied primary voltage is 230 V, calculate the power transferred to the load (*i*) inductively (*ii*) conductively. **[(*i*) 1·5 kW (*ii*) 1·5 kW]**
3. An autotransformer is to be loaded with 5 kW at 100 V and at a power factor of unity. The voltage applied to the primary is 220 V. Calculate the power transferred to the load (*i*) inductively (*ii*) conductively. **[(*i*) 2750 W (*ii*) 2250 W]**

8.39 CONVERSION OF TWO-WINDING TRANSFORMER INTO AUTOTRANSFORMER

A two-winding transformer can be converted into an autotransformer, either step-up or step-down. Consider a 10 kVA, 2300/230 V two-winding transformer shown in Fig. 8.60 (*i*). If we want to convert it into autotransformer, the two windings of the transformer are connected in series. If we use the **additive polarity* as shown in Fig. 8.60 (*ii*), we get step-up autotransformer. The voltage rating of the autotransformer is now 2300/2530 V. If we use *subtractive polarity* as shown in Fig. 8.60 (*ii*), we get a step-down autotransformer. The voltage rating of the transformer is now 2300/2070 V.

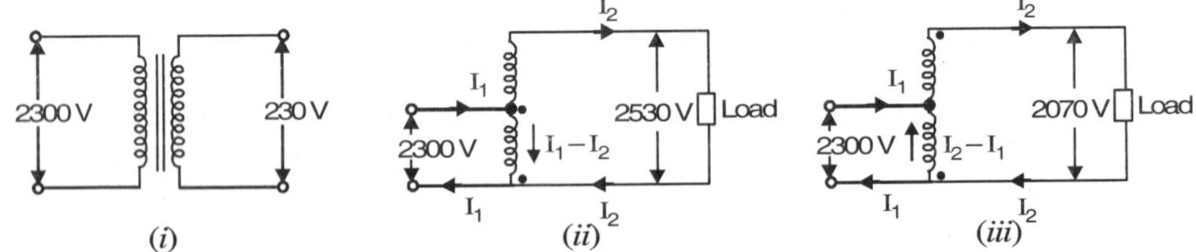

Fig. 8.60

When a two-winding transformer is converted into autotransformer, the kVA rating of the resulting autotransformer is greatly increased. This higher rating results from the conduction connection.

Example 8.79. *A single-phase, 100 kVA, 2000/200 V, two-winding transformer is converted into an autotransformer as shown in Fig. 8.61 such that more than 2000 V is obtained at the secondary. The portion ab is the 200 V winding and portion bc is the 2000 V winding. Find the kVA rating as autotransformer.*

* To ascertain additive or subtractive polarity, dot notation is used. Dots are placed on the primary and secondary terminals that go positive at the same time. For additive polarity, output voltage in this case is 2300 + 230 = 2530 V and for subtractive polarity, output voltage = 2300 – 230 = 2070 V.

Solution. The current ratings of the windings are :

$$I_{ab} = \frac{100 \times 10^3}{200} = 500 \text{ A}$$

$$I_{bc} = \frac{100 \times 10^3}{2000} = 50 \text{ A}$$

Therefore, for full-load operation of the transformer, the various currents are :

$$I_2 = 500 \text{ A} ; I_1 = 500 + 50 = 550 \text{ A}$$

Also $\quad V_2 = V_1 + 200 = 2000 + 200 = 2200 \text{ V}$

Fig. 8.61

$$\text{kVA rating of L.V. side} = \frac{2000 \times 550}{1000} = 1100 \text{ kVA}$$

$$\text{kVA rating of H.V. side} = \frac{2200 \times 500}{1000} = 1100 \text{ kVA}$$

Note that when a two-winding transformer is connected as an autotransformer, the kVA rating of the autotransformer is greatly increased. This higher rating as an autotransformer is the result of the fact that all of 1100 kVA does not have to be transferred by electromagnetic induction. In fact, all that the autotransformer has to do is to boost a current of 500 A through a potential rise of 200 V. In other words, 200 × 500/1000 = 100 kVA is actually transferred through transformer action (inductively). The other 1000 kVA is transferred through conduction.

8.40 PARALLEL OPERATION OF SINGLE-PHASE TRANSFORMERS

Two transformers are said to be connected in parallel if the primary windings are connected to supply busbars and secondary windings are connected to load busbars. Fig. 8.62 shows two transformers *A* and *B* in parallel. While connecting two or more than two transformers in parallel, it is essential that their terminals of similar polarities are joined to the same busbars as shown in Fig. 8.62. The wrong connections may result in a *dead short-circuit and transformers may be damaged unless protected by fuses or circuit breakers. There are three principal reasons for connecting transformers in parallel. Firstly, if one transformer fails, the continuity of supply can be maintained through other transformers. Secondly, when the load on the sub-station becomes more than the capacity of the existing transformers, another transformer can be added in parallel. Thirdly, any transformer can be taken out of the circuit for repair/routine maintenance without interrupting supply to the consumers.

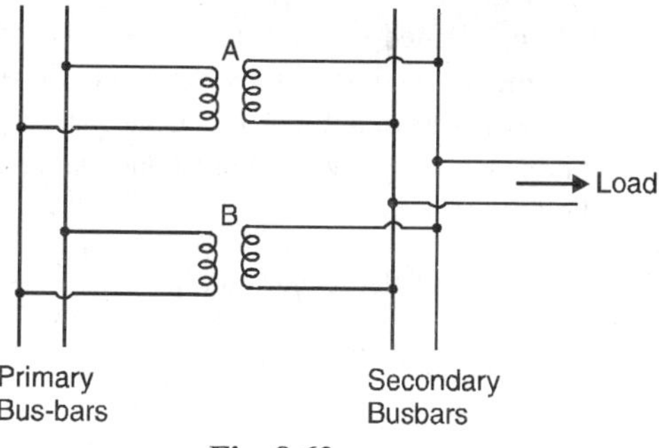

Fig. 8.62

Conditions for satisfactory parallel operation. In order that the transformers work satisfactorily in parallel, the following conditions should be satisfied :

(*i*) *Transformers should be properly connected with regard to their polarities.*

(*ii*) *The voltage ratings and voltage ratios of the transformers should be the same.*

(*iii*) *The per unit or percentage impedances of the transformers should be equal.*

(*iv*) *The reactance/resistance ratios of the transformers should be the same.*

Condition (*i*). Condition (*i*) is absolutely essential because wrong connections may result in

* This may happen when secondaries are so connected that their induced e.m.f.s are additive. Then, there will be a very large circulating current (short-circuit current) around the loop formed by the secondaries.

dead short-circuit. Fig. 8.63 (*i*) shows the correct method of connecting two single-phase transformers in parallel. It will be seen that around the loop formed by the secondaries, the two secondary e.m.f.s $\mathbf{E_A}$ and $\mathbf{E_B}$ oppose and there will be no circulating current.

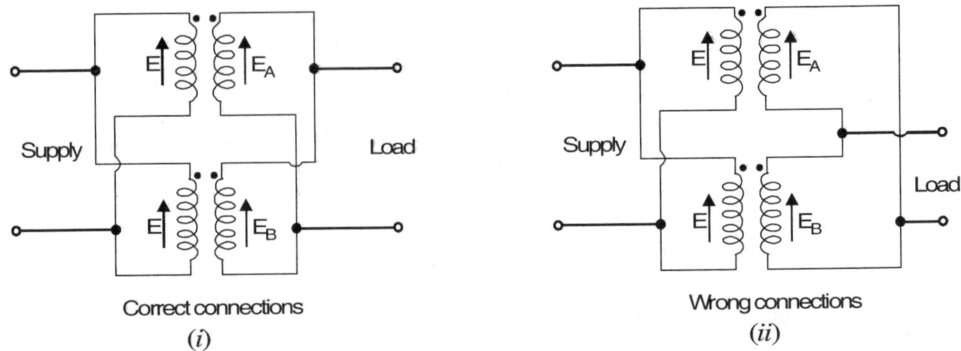

Fig. 8.63

Fig. 8.63 (*ii*) shows the wrong method of connecting two single-phase transformers in parallel. Here the two secondaries are so connected that their e.m.f.s $\mathbf{E_A}$ and $\mathbf{E_B}$ are additive. This may lead to short-circuit conditions and a very large circulating current will flow in the loop formed by the two secondaries. Such a condition may damage the transformers unless they are protected by fuses and circuit breakers.

Condition (*ii*). This condition is desirable for the satisfactory parallel operation of transformers. If this condition is not met, the secondary e.m.f.s will not be equal and there will be circulating current in the loop formed by the secondaries. This will result in the unsatisfactory parallel operation of transformers. Let us illustrate this point. Consider two single-phase transformers *A* and *B* operating in parallel as shown in Fig. 8.64. Let $\mathbf{E_A}$ and $\mathbf{E_B}$ be their no-load secondary voltages and $\mathbf{Z_A}$ and $\mathbf{Z_B}$ be their impedances referred to the secondary. Then at no-load, the circulating current in the loop formed by the secondaries is

$$\text{Circulating current, } \mathbf{I_C} = \frac{\mathbf{E_A} - \mathbf{E_B}}{\mathbf{Z_A} + \mathbf{Z_B}} \quad \text{...assuming } \mathbf{E_A} > \mathbf{E_B}$$

Even a small difference in the induced secondary voltages can cause a large circulating current in the secondary loop because impedances of the transformers are small. This secondary circulating current will cause current to be drawn from the supply by the primary of each transformer. These currents will cause copper losses in both primary and secondary. This creates heating with no useful output. When load is connected to the system, this circulating current will tend to produce unequal loading conditions *i.e.*, the transformers will not share the load according to their kVA ratings. It is because the circulating current will tend to make the terminal voltages of the same value for both transformers. Therefore, transformer with smaller voltage ratio will tend to carry more than its proper share of load. Thus, one transformer would tend to become overloaded than the other and the system could not be loaded to the summation of transformer ratings without overloading one transformer.

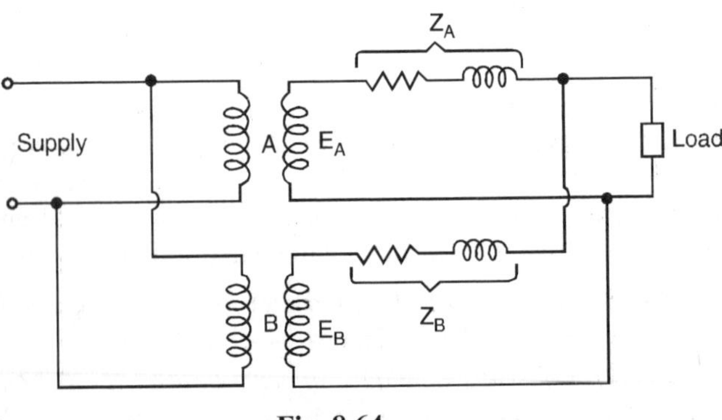

Fig. 8.64

Condition (*iii*). This condition is also desirable for proper parallel operation of transformers. If this condition is not met, the transformers will not share the load according to their kVA ratings. Sometimes this condition is not fulfilled by the design of the transformers. In that case, it can be corrected by inserting proper amount of resistance or reactance or both in series with either primary or secondary circuits of the transformers where the impedance is below the value required to fulfil condition (*iii*).

Condition (*iv*). If the reactance/resistance ratios of the two transformers are not equal, the power factor of the load supplied by the transformers will not be equal. In other words, one transformer will be operating with a higher and the other with a lower power factor than that of the load. Condition (*iii*) is much more important than condition (*iv*). Considerable deviation from condition (*iv*) will result in only a small reduction in the satisfactory degree of operation. When desired, condition (*iv*) may also be improved by inserting external impedance of proper value.

8.41 SINGLE-PHASE EQUAL VOLTAGE RATIO TRANSFORMERS IN PARALLEL

Fig. 8.65 shows two single-phase equal voltage ratio transformers *A* and *B* in parallel. The secondary e.m.f.s of the two transformers are equal (*i.e.*, $E_A = E_B = E$) because they have the same turns ratio and have their primaries connected to the same supply.

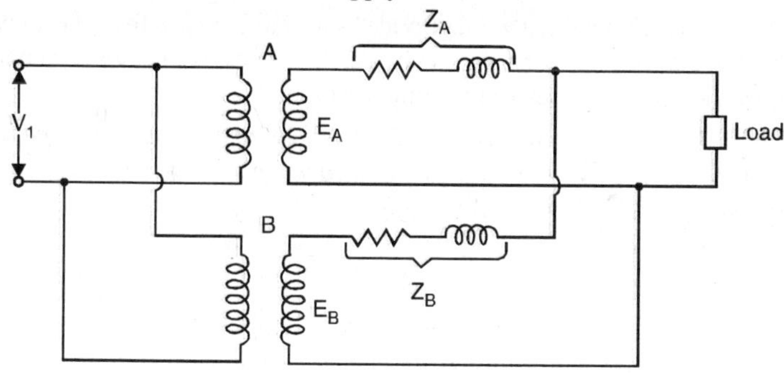

Fig. 8.65

If the magnetising current is ignored, the two transformers can be represented by their equivalent circuits referred to secondary as shown in Fig. 8.66. It is clear that the transformers will share total load in the same way as two impedances in parallel.

Let $\mathbf{Z_A}, \mathbf{Z_B}$ = Impedances of transformers referred to secondary

$\mathbf{I_A}, \mathbf{I_B}$ = Their respective currents

$\mathbf{V_2}$ = Common terminal voltage

$\mathbf{I}$ = Total load current

It is clear from Fig. 8.66 that :

$$\mathbf{I_A} + \mathbf{I_B} = \mathbf{I} \quad ...(i)$$

and $$\mathbf{I_A Z_A} = \mathbf{I_B Z_B}$$

$$\therefore \quad \mathbf{I_A} = \mathbf{I_B}\frac{\mathbf{Z_B}}{\mathbf{Z_A}}$$

$$\therefore \quad \mathbf{I_B}\frac{\mathbf{Z_B}}{\mathbf{Z_A}} + \mathbf{I_B} = \mathbf{I}$$

or $$\mathbf{I_B}\left(1 + \frac{\mathbf{Z_B}}{\mathbf{Z_A}}\right) = \mathbf{I}$$

Fig. 8.66

$$\therefore \quad \mathbf{I_B} = \mathbf{I}\frac{\mathbf{Z_A}}{\mathbf{Z_A} + \mathbf{Z_B}} \quad ...(ii)$$

Similarly, $\mathbf{I_A} = \mathbf{I}\dfrac{\mathbf{Z_B}}{\mathbf{Z_A} + \mathbf{Z_B}}$ ...(iii)

Thus the way in which the load current **I** is shared by the transformers is independent of load impedance and depends only on the transformer impedances.

kVA carried by each transformer

Let $\mathbf{S}$ = total load kVA = $\mathbf{V_2 I} \times 10^{-3}$

$\mathbf{S_A}$ = kVA carried by transformer *A*

$\mathbf{S_B}$ = kVA carried by transformer *B*

$$\therefore \quad \mathbf{S_A} = \mathbf{V_2 I_A} \times 10^{-3} = \mathbf{V_2 I} \times 10^{-3} \times \frac{\mathbf{Z_B}}{\mathbf{Z_A} + \mathbf{Z_B}} = \mathbf{S}\frac{\mathbf{Z_B}}{\mathbf{Z_A} + \mathbf{Z_B}}$$

or
$$\mathbf{S_A} = \mathbf{S}\frac{\mathbf{Z_B}}{\mathbf{Z_A} + \mathbf{Z_B}}$$

Also
$$\mathbf{S_B} = \mathbf{V_2 I_B} \times 10^{-3} = \mathbf{V_2 I} \times 10^{-3} \frac{\mathbf{Z_A}}{\mathbf{Z_A} + \mathbf{Z_B}} = \mathbf{S}\frac{\mathbf{Z_A}}{\mathbf{Z_A} + \mathbf{Z_B}}$$

or
$$\mathbf{S_B} = \mathbf{S}\frac{\mathbf{Z_A}}{\mathbf{Z_A} + \mathbf{Z_B}}$$

Therefore, $\mathbf{S_A}$ and $\mathbf{S_B}$ are obtained in magnitude as well as in phase from the above expressions. It may be noted that in these expressions, $\mathbf{Z_A}$ and $\mathbf{Z_B}$ can be expressed in ohms or in p.u. If p.u. values are to be used, they should be with respect to common base kVA.

Example 8.80. *Two 1-phase transformers with equal voltage ratios have impedances of (0·819 + j 2·503) Ω and (0·8 + j 2·31) Ω with respect to the secondary. If they operate in parallel, how they will share a total load of 2000 kW at p.f. 0·8 lagging?*

Solution.

$$\mathbf{Z_A} = (0{\cdot}819 + j\,2{\cdot}503)\ \Omega = 2{\cdot}633 \angle 71{\cdot}88^\circ\ \Omega$$

$$\mathbf{Z_B} = (0{\cdot}8 + j\,2{\cdot}31)\ \Omega = 2{\cdot}445 \angle 70{\cdot}9^\circ\ \Omega$$

$$\mathbf{Z_A} + \mathbf{Z_B} = (1{\cdot}619 + j\,4{\cdot}813)\ \Omega = 5{\cdot}078 \angle 71{\cdot}4^\circ\ \Omega$$

$$\text{Total load kVA} = 2000/0{\cdot}8 = 2500 \text{ kVA}$$

$$\therefore \quad \mathbf{S} = 2500 \angle -36{\cdot}87^\circ \text{ kVA}$$

$$\mathbf{S_A} = \mathbf{S}\frac{\mathbf{Z_B}}{\mathbf{Z_A} + \mathbf{Z_B}} = 2500 \angle -36{\cdot}87^\circ \times \frac{2\cdot 445 \angle 70\cdot 9^\circ}{5\cdot 078 \angle 71\cdot 4^\circ}$$

$= 1203{\cdot}7 \angle -37{\cdot}37^\circ$ kVA

= a load of 1203·7 cos 37·37° = **956·6 kW** *at 0·79 p.f. lagging*

$$\mathbf{S_B} = \mathbf{S}\frac{\mathbf{Z_A}}{\mathbf{Z_A} + \mathbf{Z_B}} = 2500 \angle -36{\cdot}87^\circ \times \frac{2\cdot 633 \angle 71\cdot 88^\circ}{5\cdot 078 \angle 71\cdot 4^\circ}$$

$= 1296{\cdot}3 \angle -36{\cdot}39^\circ$ kVA

= a load of 1296·3 cos 36·39° = **1043·5 kW** *at 0·805 p.f. lagging*

Example 8.81. *Two 1-phase transformers A and B rated at 250 kVA each are operated in parallel on both sides. Impedances for A and B are (1 + j 6) Ω and (1·2 + j 4·8) Ω respectively. Find the load shared by each when the total load is 500 kVA at 0·8 p.f. lagging.*

Solution.

$$\mathbf{S} = 500 \angle -36{\cdot}9^\circ \text{ kVA}$$

$$\frac{\mathbf{Z_A}}{\mathbf{Z_A} + \mathbf{Z_B}} = \frac{1 + j\,6}{2\cdot 2 + j\,10\cdot 8} = 0{\cdot}55 \angle 2{\cdot}1^\circ$$

$$\frac{\mathbf{Z_B}}{\mathbf{Z_A}+\mathbf{Z_B}} = \frac{1 \cdot 2 + j\,4 \cdot 8}{2 \cdot 2 + j\,10 \cdot 8} = 0{\cdot}45\,\angle{-2{\cdot}5^\circ}$$

$$\therefore \quad \mathbf{S_A} = \mathbf{S}\,\frac{\mathbf{Z_B}}{\mathbf{Z_A}+\mathbf{Z_B}} = 500\,\angle{-36{\cdot}9^\circ} \times 0{\cdot}45\,\angle{-2{\cdot}5^\circ} = \mathbf{225\,\angle{-39{\cdot}4^\circ}\ kVA}$$

$$\mathbf{S_B} = \mathbf{S}\,\frac{\mathbf{Z_A}}{\mathbf{Z_A}+\mathbf{Z_B}} = 500\,\angle{-36{\cdot}9^\circ} \times 0{\cdot}55\,\angle{2{\cdot}1^\circ} = \mathbf{275\,\angle{-34{\cdot}8^\circ}\ kVA}$$

Note that transformer *B* is overloaded by 10% [(275 – 250) × 100/250 = 10%].

Example 8.82. *Two single-phase transformers which have the same turns ratio are connected in parallel and supply a total load of 800 kW at 0·8 power factor lagging. Their ratings are as follows :*

Transformer	*Rating*	*p.u. resistance*	*p.u. reactance*
A	*400 kVA*	*0·02*	*0·04*
B	*600 kVA*	*0·01*	*0·05*

Determine the power output and the power factor of each transformer.

Solution.

$$\text{Total load kVA} = 800/0{\cdot}8 = 1000\ \text{kVA}$$

$$\therefore \quad \mathbf{S} = 1000\,(0{\cdot}8 - j\,0{\cdot}6) = (800 - j\,600)\ \text{kVA}$$

The p.u. resistances/reactances of the two transformers should be brought to the same kVA level, say 1000 kVA.

$$\therefore \quad R_A = 0{\cdot}02 \times \frac{1000}{400} = 0{\cdot}05\ ;\ X_A = 0{\cdot}04 \times \frac{1000}{400} = 0{\cdot}1$$

$$R_B = 0{\cdot}01 \times \frac{1000}{600} = 0{\cdot}0167\ ;\ X_B = 0{\cdot}05 \times \frac{1000}{600} = 0{\cdot}0833$$

$$\mathbf{S_B} = \mathbf{S}\,\frac{\mathbf{Z_A}}{\mathbf{Z_A}+\mathbf{Z_B}} = (800 - j\,600) \times \frac{0{\cdot}05 + j\,0{\cdot}1}{0{\cdot}05 + j\,0{\cdot}01 + 0{\cdot}0167 + j\,0{\cdot}0833}$$

$$= (800 - j\,600) \times \frac{0{\cdot}05 + j\,0{\cdot}1}{0{\cdot}0667 + j\,0{\cdot}1833} = (414 - j\,392)\ \text{kVA}$$

$$\mathbf{S_A} = \mathbf{S} - \mathbf{S_B} = (800 - j\,600) - (414 - j\,392) = (386 - j\,208)\ \text{kVA}$$

Transformer *A* : **386 kW, 440 kVA,** *0·878 p.f. lagging*

Transformer *B* : **414 kW, 570 kVA,** *0·726 p.f. lagging*

Example 8.83. *Two 2200/110 V single-phase transformers are connected in parallel to supply a common load of 125 kVA at 0·8 power factor lagging. Transformers are rated as follows :*

Transformer A : 100 kVA, 0·9 % resistance and 10 % reactance

Transformer B : 50 kVA, 1 % resistance and 5 % reactance

How will the two transformers share the common load ?

Solution. The percentage impedances of the two transformers should be brought to the same kVA level, say 100 kVA.

$$\mathbf{Z_A} = (0{\cdot}9 + j\,10)\,\Omega\ ;\ \mathbf{Z_B} = \frac{100}{50}\,(1 + j\,5) = (2 + j\,10)\,\Omega$$

$$\cos\phi = 0{\cdot}8 \ \therefore\ \phi = 37^\circ$$

$$\therefore \quad \text{Total load, } \mathbf{S} = 125\,\angle{-37^\circ}\ \text{kVA}$$

$$\frac{\mathbf{Z_A}}{\mathbf{Z_A}+\mathbf{Z_B}} = \frac{0{\cdot}9 + j\,10}{0{\cdot}9 + j\,10 + 2 + j\,10} = \frac{0{\cdot}9 + j\,10}{2{\cdot}9 + j\,20} = 0{\cdot}497\,\angle{3^\circ}$$

$$\frac{\mathbf{Z_B}}{\mathbf{Z_A}+\mathbf{Z_B}} = \frac{2 + j\,10}{2{\cdot}9 + j\,20} = 0{\cdot}5\angle{-3^\circ}$$

Load shared by transformer A is given by ;

$$\mathbf{S_A} = \mathbf{S}\frac{\mathbf{Z_B}}{\mathbf{Z_A}+\mathbf{Z_B}} = 125\angle -37^\circ \times 0{\cdot}5\angle -3^\circ = \mathbf{62{\cdot}5\angle -40^\circ\ kVA}$$

Load shared by transformer B is given by ;

$$\mathbf{S_B} = \mathbf{S}\frac{\mathbf{Z_A}}{\mathbf{Z_A}+\mathbf{Z_B}} = 125\angle -37^\circ \times 0{\cdot}497\angle 3^\circ = \mathbf{62{\cdot}12\angle -34^\circ\ kVA}$$

Example 8.84. *Two single-phase transformers A and B rated at 600 kVA and 500 kVA respectively are operated in parallel to supply a load of 1000 kVA at 0·8 lagging power factor. The resistance and reactance of transformer A are 3% and 6·5 % while that of transformer B are 1·5 % and 8%. Calculate the kVA loading and the power factor at which each transformer operates.*

Solution.

$$\mathbf{S} = 1000\,(0{\cdot}8 - j\,0{\cdot}6) = (800 - j\,600)\ \text{kVA}$$

Choosing 1000 kVA as the base kVA, we have,

$$\mathbf{Z_A}\,(\text{p.u.}) = \frac{1000}{600}\,(0{\cdot}03 + j\,0{\cdot}065) = 0{\cdot}05 + j\,0{\cdot}108$$

$$\mathbf{Z_B}\,(\text{p.u.}) = \frac{1000}{500}\,(0{\cdot}015 + j\,0{\cdot}08) = 0{\cdot}03 + j\,0{\cdot}16$$

$$\therefore \quad \mathbf{S_A} = \mathbf{S}\frac{\mathbf{Z_B}}{\mathbf{Z_A}+\mathbf{Z_B}} = (800 - j\,600) \times \frac{0{\cdot}03 + j\,0{\cdot}16}{0{\cdot}08 + j\,0{\cdot}268} = 584{\cdot}2\angle -30{\cdot}9^\circ\ \text{kVA}$$

Therefore, transformer A will carry a load of **584·2 kVA** *at a p.f. of 0·86 lagging.*

$$\mathbf{S_B} = \mathbf{S}\frac{\mathbf{Z_A}}{\mathbf{Z_A}+\mathbf{Z_B}} = (800 - j\,600) \times \frac{0{\cdot}05 + j\,0{\cdot}108}{0{\cdot}08 + j\,0{\cdot}268} = 426{\cdot}5\angle -45{\cdot}15^\circ\ \text{kVA}$$

Therefore, transformer B will carry a load of **426·5 kVA** *at a p.f. of 0·7 lagging.*

Example 8.85. *Two transformers A and B are running in parallel and supply a load of 2500 kVA at 0·8 lagging power factor. The ratings of the transformers are as under :*

	Transformer A	*Transformer B*
Voltage ratio	*4500/16000 V*	*4500/16000 V*
kVA rating	*1500*	*1500*
Total resistance (Ω)	*0·819*	*0·8*
Total reactance (Ω)	*2·503*	*2·31*

Find the current supplied by each transformer.

Solution.

$$\mathbf{Z_A} = (0{\cdot}819 + j\,2{\cdot}503)\,\Omega = 2{\cdot}633\angle 71{\cdot}88^\circ\,\Omega$$

$$\mathbf{Z_B} = (0{\cdot}8 + j\,2{\cdot}31)\,\Omega = 2{\cdot}445\angle 70{\cdot}9^\circ\,\Omega$$

$$\mathbf{Z_A} + \mathbf{Z_B} = (1{\cdot}619 + j\,4{\cdot}813)\,\Omega = 5{\cdot}078\angle 71{\cdot}4^\circ\,\Omega$$

$$\text{Total current, } I = \frac{2500 \times 10^3}{16000} = 156{\cdot}3\ \text{A}$$

Taking secondary terminal voltage as the reference phasor, we have,

$$\mathbf{I} = 156{\cdot}3\angle -36{\cdot}87^\circ\ \text{A} \qquad (\cos\phi = 0{\cdot}8\ \therefore\ \phi = 36{\cdot}87^\circ)$$

$$\mathbf{I_A} = \mathbf{I} \times \frac{\mathbf{Z_B}}{\mathbf{Z_A}+\mathbf{Z_B}} = 156{\cdot}3\angle -36{\cdot}87^\circ \times \frac{2{\cdot}445\angle 70{\cdot}9^\circ}{5{\cdot}078\angle 71{\cdot}4^\circ}$$

$$= \mathbf{75{\cdot}26\angle -37{\cdot}37^\circ\ A}$$

$$\mathbf{I_B} = \mathbf{I} \times \frac{\mathbf{Z_A}}{\mathbf{Z_A} + \mathbf{Z_B}} = 156{\cdot}3\ \angle{-36{\cdot}87^\circ} \times \frac{2{\cdot}633\ \angle 71{\cdot}88^\circ}{5{\cdot}078\ \angle 71{\cdot}4^\circ}$$

$$= \mathbf{81{\cdot}04\ \angle{-\ 36{\cdot}39^\circ}\ A}$$

Note that transformer *B* carries about 7% more load than transformer *A*. Therefore, working to the maximum capacity of 3000 kVA is impractical since one transformer must work above the full load.

TUTORIAL PROBLEMS

1. Two single-phase transformers with equal voltage ratios have impedances of (0·5 + *j* 3) ohm and (0·6 + *j* 10) ohm with respect to the secondary. If they operate in parallel, determine how they share a total load of 100 kW at p.f. 0·8 lagging. **[78·2 kW ; 21·8 kW]**
2. A 500 kVA transformer with 0·01 + *j* 0·05 p.u. impedance and a 250 kVA transformer with 0·015 + *j* 0·04 p.u. impedance are operating in parallel. The no-load secondary voltages of the two are equal. How will they share a 750 kVA, 0·8 lagging power factor load ? **[471 kVA** *at 0·762 lagging p.f.* ; **281 kVA** *at 0·856 lagging p.f.***]**
3. Two 1100/400 V, single-phase transformers have total equivalent impedances of (0·2 + *j* 0·5) Ω and (0·3 + *j* 0·6) Ω referred to the secondary side. They are connected in parallel and supply a load of 50 kW at a power factor of 0·85 lagging. Find (*i*) the magnitudes of the secondary currents and their angles relative to the secondary terminal voltages (*ii*) the power delivered by each transformer. **[(*i*) 81·6 ∠33·9° A ; 65·5 ∠29·1° A (*ii*) 27·1 kW ; 22·9 kW]**
4. Two 11000/400 V, single-phase transformers *A* and *B* are connected in parallel to supply a total load of 500 A at 0·8 power factor lagging at a terminal voltage of 400 V. The equivalent impedances of transformers referred to secondary side are (2 + *j* 3) ohm and (2·5 + *j* 5) ohm. Calculate the current supplied by each transformer. **[I_A = 304 A ; I_B = 197 A]**
5. A 650 kVA, single-phase transformer with 0·12 p.u. resistance and 0·06 p.u. reactance is connected in parallel with a 300 kVA transformer with 0·014 p.u. resistance and 0·045 p.u. reactance to share a load of 800 kVA at 0·8 p.f. lagging. If the transformers have the same voltage ratio, how will they share the load ? **[485 ∠–39° kVA ; 315 ∠–33° kVA]**

8.42 SINGLE- PHASE UNEQUAL VOLTAGE RATIO TRANSFORMERS IN PARALLEL

Fig. 8.67 shows two single-phase unequal voltage ratio transformers *A* and *B* in parallel. Since the voltage ratios of the transformers are unequal, their no-load secondary voltages will also be unequal. We shall calculate how load current is shared between the two transformers.

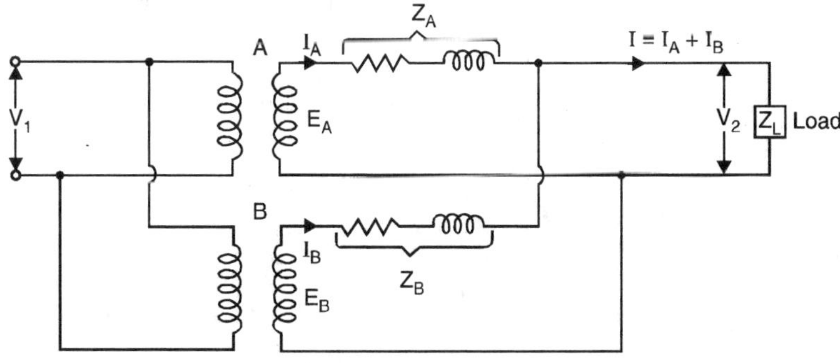

Fig. 8.67

Fig. 8.68 shows the equivalent circuit of the transformers referred to secondary in a simplified way.

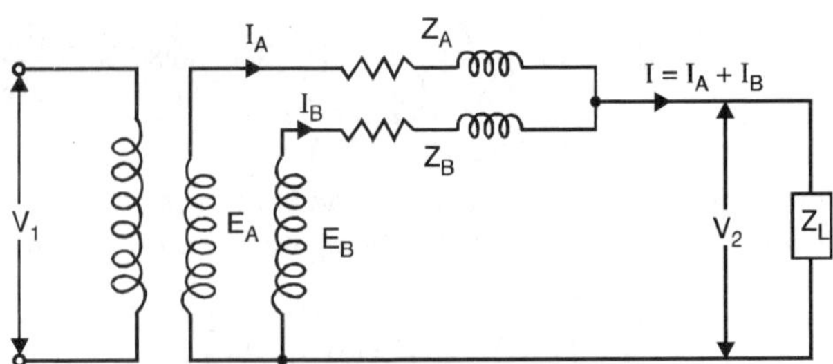

Fig. 8.68

Let $\mathbf{E_A, E_B}$ = no-load secondary voltages of the two transformers. It is assumed that $\mathbf{E_A > E_B}$.

$\mathbf{I_A, I_B}$ = their respective currents

$\mathbf{I}$ = load current

$\mathbf{Z_A, Z_B}$ = impedances of the transformers referred to secondary

$\mathbf{Z_L}$ = load impedance

$\mathbf{V_2}$ = load voltage

At no-load, the circulating current $\mathbf{I_C}$ is

$$\mathbf{I_C} = \frac{\mathbf{E_A - E_B}}{\mathbf{Z_A + Z_B}}$$

When the system is loaded, the load current **I** is shared by the two transformers. By Kirchhoff's voltage law,

$$\mathbf{E_A} = \mathbf{V_2 + I_A Z_A}$$

and $$\mathbf{E_B} = \mathbf{V_2 + I_B Z_B}$$

But $$\mathbf{V_2} = \mathbf{I Z_L} = \mathbf{(I_A + I_B) Z_L}$$

∴ $$\mathbf{E_A} = \mathbf{(I_A + I_B) Z_L + I_A Z_A} \qquad ...(i)$$

and $$\mathbf{E_B} = \mathbf{(I_A + I_B) Z_L + I_B Z_B} \qquad ...(ii)$$

Now $$\mathbf{E_A - E_B} = \mathbf{I_A Z_A - I_B Z_B}$$

or $$\mathbf{I_A} = \frac{\mathbf{(E_A - E_B) + I_B Z_B}}{\mathbf{Z_A}}$$

Putting this value of $\mathbf{I_A}$ in eq. (*ii*), we get,

$$\mathbf{E_B} = \left[\frac{\mathbf{(E_A - E_B) + I_B Z_B}}{\mathbf{Z_A}} + \mathbf{I_B}\right] \mathbf{Z_L + I_B Z_B}$$

On solving, $$\mathbf{I_B} = \frac{\mathbf{E_B Z_A - (E_A - E_B) Z_L}}{\mathbf{Z_A Z_B + Z_L (Z_A + Z_B)}} \qquad ...(iii)$$

From the symmetry of the expression, we get,

$$\mathbf{I_A} = \frac{\mathbf{E_A Z_B + (E_A - E_B) Z_L}}{\mathbf{Z_A Z_B + Z_L (Z_A + Z_B)}}$$

Also $$\mathbf{I} = \mathbf{I_A + I_B} = \frac{\mathbf{E_A Z_B + E_B Z_A}}{\mathbf{Z_A Z_B + Z_L (Z_A + Z_B)}}$$

$$\mathbf{V_2} = \mathbf{I Z_L} = \left[\frac{\mathbf{E_A Z_B + E_B Z_A}}{\mathbf{Z_A Z_B + Z_L (Z_A + Z_B)}}\right] \mathbf{Z_L}$$

or $$\mathbf{V}_2 = \frac{\mathbf{E}_A \mathbf{Z}_B + \mathbf{E}_B \mathbf{Z}_A}{\dfrac{\mathbf{Z}_A \mathbf{Z}_B}{\mathbf{Z}_L} + \mathbf{Z}_A + \mathbf{Z}_B}$$

Since the transformers have a common primary voltage, $\mathbf{E}_A$ and $\mathbf{E}_B$ will be in phase with each other.

Example 8.86. *Two transformers A and B are joined in parallel to the same load. Determine the current delivered by each transformer having given : open-circuit e.m.f. 6600 V for A and 6400 V for B. Equivalent leakage impedance in terms of secondary = (0·3 + j 3) ohm for A and (0·2 + j 1) ohm for B. The load impedance = (8 + j 6) ohm.*

Solution.

$$\mathbf{E}_A = 6600 \angle 0^\circ \text{ V} ; \mathbf{E}_B = 6400 \angle 0^\circ \text{ V} ; \mathbf{Z}_L = (8 + j\,6)\,\Omega ; \mathbf{Z}_A = (0{\cdot}3 + j\,3)\,\Omega$$
$$\mathbf{Z}_B = (0{\cdot}2 + j\,1)\,\Omega$$

Now $$\mathbf{I}_A = \frac{\mathbf{E}_A \mathbf{Z}_B + (\mathbf{E}_A - \mathbf{E}_B)\mathbf{Z}_L}{\mathbf{Z}_A \mathbf{Z}_B + \mathbf{Z}_L(\mathbf{Z}_A + \mathbf{Z}_B)}$$

$$= \frac{6600\,(0{\cdot}2 + j\,1) + (6600 - 6400)\,(8 + j\,6)}{(0{\cdot}3 + j\,3)\,(0{\cdot}2 + j\,1) + (8 + j\,6)\,(0{\cdot}3 + j\,3 + 0{\cdot}2 + j\,1)}$$

$$= \frac{2920 + j\,7800}{-22{\cdot}94 + j\,35{\cdot}9} = (117 - j\,156)\text{ A}$$

$\therefore$ Magnitude of $\mathbf{I}_A$, $I_A = \sqrt{(117)^2 + (156)^2} =$ **195 A**

Also $$\mathbf{I}_B = \frac{\mathbf{E}_B \mathbf{Z}_A - (\mathbf{E}_A - \mathbf{E}_B)\mathbf{Z}_L}{\mathbf{Z}_A \mathbf{Z}_B + \mathbf{Z}_L(\mathbf{Z}_A + \mathbf{Z}_B)}$$

$$= \frac{6400\,(0{\cdot}3 + j\,3) - (6600 - 6400)\,(8 + j\,6)}{(0{\cdot}3 + j\,3)\,(0{\cdot}2 + j\,1) + (8 + j\,6)\,(0{\cdot}3 + j\,3 + 0{\cdot}2 + j\,1)}$$

$$= \frac{320 + j\,18000}{-22{\cdot}94 + j\,35{\cdot}9} = (349 - j\,231)\text{ A}$$

$\therefore$ Magnitude of $\mathbf{I}_B$, $I_B = \sqrt{(349)^2 + (231)^2} =$ **420 A**

Example 8.87. *Two transformers A and B of ratings 500 kVA and 250 kVA are supplying a load of 750 kVA at 0·8 p.f. lagging. Their open-circuit voltages are 405 V and 415 V respectively. Transformer A has 0·01 p.u. resistance and 0·05 p.u. reactance and transformer B has 0·015 p.u. resistance and 0·04 p.u. reactance. Find (i) circulating current at no-load (ii) the load current shared by each transformer (iii) kVA and power factor of each transformer. Take the voltage across load as 400 V.*

Solution. We shall first convert p.u. values into ohmic values.

Rated current of transformer $A = \dfrac{500 \times 10^3}{400} = 1250$ A

Rated current of transformer $B = \dfrac{250 \times 10^3}{400} = 625$ A

$$\mathbf{Z}_A = (0{\cdot}01 + j\,0{\cdot}05)\text{ p.u.}$$

$$= (0{\cdot}01 + j\,0{\cdot}05) \times \frac{*400}{1250}\ \Omega = (0{\cdot}0032 + j\,0{\cdot}016)\,\Omega$$

$$= 0{\cdot}0163\ \angle 78{\cdot}7^\circ\,\Omega$$

$$\mathbf{Z_B} = (0{\cdot}015 + j\,0{\cdot}04)\text{ p.u.}$$

$$= (0{\cdot}015 + j\,0{\cdot}04) \times \frac{400}{625}\ \Omega = (0{\cdot}0096 + j\,0{\cdot}0256)\,\Omega$$

$$= 0{\cdot}0274\ \angle 69{\cdot}4^\circ\,\Omega$$

$$\mathbf{Z_A} + \mathbf{Z_B} = (0{\cdot}0128 + j\,0{\cdot}0416)\,\Omega = 0{\cdot}0436\ \angle 72{\cdot}9^\circ\,\Omega$$

The load is 750 kVA at 0·8 p.f. lagging. Therefore, load impedance $\mathbf{Z_L}$ is given by :

$$\frac{(\text{Load voltage})^2}{\mathbf{Z_L}} = 750 \times 10^3\ \angle -36{\cdot}9^\circ$$

or $$\frac{(400)^2}{\mathbf{Z_L}} = 750 \times 10^3\ \angle -36{\cdot}9^\circ$$

or $$\mathbf{Z_L} = 0{\cdot}214\ \angle 36{\cdot}9^\circ\,\Omega$$

(*i*) Circulating current at no-load is

$$\mathbf{I_C} = \frac{\mathbf{E_A} - \mathbf{E_B}}{\mathbf{Z_A} + \mathbf{Z_B}} = \frac{405 - 415}{0{\cdot}0436\ \angle 72{\cdot}9^\circ} = \mathbf{-\,230\ \angle -\,72{\cdot}9^\circ\ A}$$

(*ii*) $$\mathbf{I_A} = \frac{\mathbf{E_A\,Z_B} + (\mathbf{E_A} - \mathbf{E_B})\,\mathbf{Z_L}}{\mathbf{Z_A\,Z_B} + \mathbf{Z_L}\,(\mathbf{Z_A} + \mathbf{Z_B})}$$

$$= \frac{(405) \times (0{\cdot}0274\ \angle 69{\cdot}4^\circ) + (405 - 415) \times 0{\cdot}214\ \angle 36\cdot 9^\circ}{0{\cdot}0163\ \angle 78{\cdot}7^\circ \times 0{\cdot}0274\ \angle 69{\cdot}4^\circ + 0{\cdot}214\ \angle 36{\cdot}9^\circ \times 0{\cdot}0436\ \angle 72{\cdot}9^\circ}$$

$$= \frac{9{\cdot}36\ \angle 76{\cdot}4^\circ}{9{\cdot}72 \times 10^{-3}\ \angle 111{\cdot}6^\circ} = \mathbf{963\ \angle -\,35{\cdot}2^\circ\ A}$$

$$\mathbf{I_B} = \frac{\mathbf{E_B\,Z_A} - (\mathbf{E_A} - \mathbf{E_B})\,\mathbf{Z_L}}{\mathbf{Z_A\,Z_B} + \mathbf{Z_L}\,(\mathbf{Z_A} + \mathbf{Z_B})}$$

$$= \frac{415 \times 0{\cdot}0163\ \angle 78{\cdot}7^\circ - (405 - 415) \times 0{\cdot}214\ \angle 36{\cdot}9^\circ}{0{\cdot}0163\ \angle 78{\cdot}7^\circ \times 0{\cdot}0274\ \angle 69{\cdot}4^\circ + 0{\cdot}214\ \angle 36{\cdot}9^\circ \times 0{\cdot}0436\ \angle 72{\cdot}9^\circ}$$

$$= \frac{8{\cdot}54\ \angle 69{\cdot}4^\circ}{9{\cdot}72 \times 10^{-3}\ \angle 111{\cdot}6^\circ} = \mathbf{878\ \angle -\,42{\cdot}2^\circ\ A}$$

(*iii*) kVA of transformer *A* $= 400 \times 963 \times 10^{-3} =$ **385 kVA**

p.f. of transformer *A* $= \cos(-\,35{\cdot}2^\circ) =$ **0·82** *lagging*

kVA of transformer *B* $= 400 \times 878 \times 10^{-3} =$ **351 kVA**

p.f. of transformer *B* $= \cos(-\,42{\cdot}2^\circ) =$ **0·74** *lagging*

Example 8.88. *Two single-phase transformers A and B are connected in parallel to supply a load having a resistance of 5 Ω and inductive reactance of 2 Ω. The equivalent resistances referred to the secondary windings are 0·25 Ω and 0·3 Ω and equivalent reactances 1·5 Ω and 2 Ω respectively. The open circuit secondary voltages are in the ratio 100 : 98. Calculate the inductive reactance which when connected in series with the secondary of transformer B will cause the magnitudes of currents delivered by A and B to be in the ratio 1·8 : 1.*

---

* $$\text{p.u. } Z = \frac{I_{F.L.} \times Z}{V_{phase}}$$

Here $I_{F.L.}$ is the full-load or rated current and Z is the impedance in ohms.

Solution. Fig. 8.69 shows the equivalent circuit of the transformers *A* and *B* operating in parallel. Let *x* ohm be the total reactance of transformer *B* to meet the condition of the problem. Let $\mathbf{E_A} = 100 \angle 0°$ V. Then $\mathbf{E_B} = 98 \angle 0°$ V.

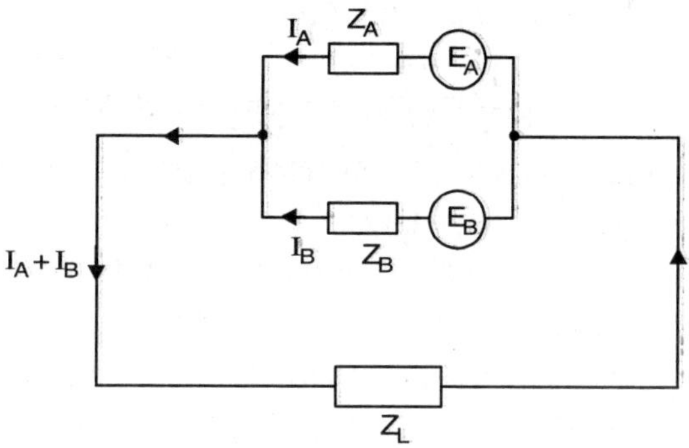

Fig. 8.69

$$\mathbf{Z_A} = (0{\cdot}25 + j\,1{\cdot}5)\,\Omega\,;\ \mathbf{Z_B} = (0{\cdot}3 + j\,x)\,\Omega\,;\ \mathbf{Z_L} = (5 + j\,2)\,\Omega$$

$$\mathbf{I_A} = \frac{\mathbf{E_A}\,\mathbf{Z_B} + (\mathbf{E_A} - \mathbf{E_B})\,\mathbf{Z_L}}{\mathbf{Z_A}\,\mathbf{Z_B} + \mathbf{Z_L}\,(\mathbf{Z_A} + \mathbf{Z_B})} = \frac{100\,\mathbf{Z_B} + (100 - 98)\,\mathbf{Z_L}}{\mathbf{Z_A}\,\mathbf{Z_B} + \mathbf{Z_L}\,(\mathbf{Z_A} + \mathbf{Z_B})}$$

$$\mathbf{I_B} = \frac{\mathbf{E_B}\,\mathbf{Z_A} - (\mathbf{E_A} - \mathbf{E_B})\,\mathbf{Z_L}}{\mathbf{Z_A}\,\mathbf{Z_B} + \mathbf{Z_L}\,(\mathbf{Z_A} + \mathbf{Z_B})} = \frac{98\,\mathbf{Z_A} - (100 - 98)\,\mathbf{Z_L}}{\mathbf{Z_A}\,\mathbf{Z_B} + \mathbf{Z_L}\,(\mathbf{Z_A} + \mathbf{Z_B})}$$

$$\therefore \quad \frac{\mathbf{I_A}}{\mathbf{I_B}} = \frac{100\,\mathbf{Z_B} + 2\,\mathbf{Z_L}}{98\,\mathbf{Z_A} - 2\,\mathbf{Z_L}} = \frac{\mathbf{Z_L} + 50\,\mathbf{Z_B}}{-\mathbf{Z_L} + 49\,\mathbf{Z_A}}$$

$$= \frac{(5 + j\,2) + 50\,(0{\cdot}3 + j\,x)}{-5 - j\,2 + 49\,(0{\cdot}25 + j\,1{\cdot}5)} = \frac{20 + j\,(2 + 50\,x)}{71{\cdot}9 \angle 84°}$$

It is given that $\left|\dfrac{\mathbf{I_A}}{\mathbf{I_B}}\right| = 1{\cdot}8$

$\therefore \quad (20)^2 + (2 + 50\,x)^2 = (1{\cdot}8)^2 \times (71{\cdot}9)^2$

or $\quad 400 + (2 + 50\,x)^2 = 16750$

or $\quad (2 + 50\,x)^2 = 16350$

or $\quad 2 + 50\,x = 127{\cdot}9 \qquad \therefore \quad x = 2{\cdot}52\,\Omega$

$\therefore$ Reactance to be added $= 2{\cdot}52 - 2 =$ **0·52 Ω**

Example 8.89. *Two transformers A and B are connected in parallel to a load of* $(2 + j\,1{\cdot}5)\,\Omega$. *Their impedances in secondary terms are* $\mathbf{Z}_A = (0{\cdot}15 + j\,0{\cdot}5)\,\Omega$ *and* $\mathbf{Z}_B = (0{\cdot}1 + j\,0{\cdot}6)\,\Omega$. *Their no-load terminal voltages are* $\mathbf{E}_A = 207 \angle 0°$ *V and* $\mathbf{E}_B = 205 \angle 0°$ *V. Find the power output and power factor of each transformer.*

Solution.

$$\mathbf{E_A} = 207 \angle 0°\,\text{V}\,;\ \mathbf{E_B} = 205 \angle 0°\,\text{V}\,;\ \mathbf{Z_A} = (0{\cdot}15 + j\,0{\cdot}5)\,\Omega$$

$$\mathbf{Z_B} = (0{\cdot}1 + j\,0{\cdot}6)\,\Omega\,;\ \mathbf{Z_L} = (2 + j\,1{\cdot}5)\,\Omega = 2{\cdot}5 \angle 36{\cdot}9°\,\Omega$$

Now

$$\mathbf{I_A} = \frac{\mathbf{E_A}\,\mathbf{Z_B} + (\mathbf{E_A} - \mathbf{E_B})\,\mathbf{Z_L}}{\mathbf{Z_A}\,\mathbf{Z_B} + \mathbf{Z_L}\,(\mathbf{Z_A} + \mathbf{Z_B})}$$

$$= \frac{207\,(0{\cdot}1 + j\,0{\cdot}6) + (207 - 205)\,(2 + j\,1{\cdot}5)}{(0{\cdot}15 + j\,0{\cdot}5)\,(0{\cdot}1 + j\,0{\cdot}6) + (2 + j\,1{\cdot}5)\,(0{\cdot}25 + j\,1{\cdot}1)}$$

$$= \frac{24{\cdot}7 + j\,127{\cdot}2}{-1{\cdot}435 + j\,2{\cdot}715} = \frac{129{\cdot}7 \angle 79°}{3{\cdot}07 \angle 117{\cdot}90°} = 42{\cdot}26 \angle -38{\cdot}9°\,\text{A}$$

$$= (32{\cdot}89 - j\,26{\cdot}55)\,\text{A}$$

Also

$$\mathbf{I_B} = \frac{\mathbf{E_B}\,\mathbf{Z_A} - (\mathbf{E_A} - \mathbf{E_B})\,\mathbf{Z_L}}{\mathbf{Z_A}\,\mathbf{Z_B} + \mathbf{Z_L}\,(\mathbf{Z_A} + \mathbf{Z_B})} = \frac{205\,(0{\cdot}15 + j\,0{\cdot}5) - 2\,(2 + j\,1{\cdot}5)}{3{\cdot}07\,\angle 117{\cdot}9^\circ}$$

$$= \frac{103\,\angle 75^\circ}{3{\cdot}07\,\angle 117{\cdot}9^\circ} = 33{\cdot}56\,\angle -42{\cdot}9^\circ\,\text{A} = (24{\cdot}58 - j\,22{\cdot}84)\,\text{A}$$

$$\therefore \quad \mathbf{I_A} + \mathbf{I_B} = (32{\cdot}89 - j\,26{\cdot}55) + (24{\cdot}58 - j\,22{\cdot}84) = (57{\cdot}47 - j\,49{\cdot}39)\,\text{A}$$

$$\text{Load voltage, } \mathbf{V_2} = (\mathbf{I_A} + \mathbf{I_B})\,\mathbf{Z_L} = (57{\cdot}47 - j\,49{\cdot}39)\,(2 + j\,1{\cdot}5)$$

$$= (189 - j\,12{\cdot}58)\text{ volt} = 189{\cdot}4\,\angle -3{\cdot}9^\circ\text{ volt}$$

$$\text{p.f. angle of transformer } A = -3{\cdot}9^\circ - (-38{\cdot}9^\circ) = 35^\circ$$

$$\therefore \quad \text{p.f. of transformer } A = \cos 35^\circ = \mathbf{0{\cdot}818}\ \textit{lag}$$

$$\text{p.f. of transformer } B = \cos\,[-3{\cdot}9^\circ - (-42{\cdot}9^\circ)] = \mathbf{0{\cdot}776}\ \textit{lag}$$

$$\text{Power output of transformer } A = V_2 I_A \cos\phi_A = 189{\cdot}4 \times 42{\cdot}26 \times 0{\cdot}818 = \mathbf{6548\ W}$$

$$\text{Power output of transformer } B = V_2 I_B \cos\phi_B = 189{\cdot}4 \times 33{\cdot}56 \times 0{\cdot}776 = \mathbf{4932\ W}$$

TUTORIAL PROBLEMS

1. A 500 kVA, 500 V, 1-phase transformer with a reactance of 4% and a resistance of 1% is connected in parallel on the high voltage side with a 250 kVA, 500 V transformer having drops of 6 % and 1·5 % respectively. The secondary terminal voltages on open circuit are 510 V in the first and 500 V in the second transformer. With the secondary windings connected in parallel, calculate (*i*) the cross-current in the secondaries on no-load (*ii*) the secondary current in each transformer when the total load is 700 kW at unity p.f. and (*iii*) terminal voltage at this load. **[(*i*) 121·5 A (*ii*) 1070 A ; 330 A (*iii*) 501 V]**
2. Two single-phase transformers, one of 100 kVA and the other of 50 kVA are connected in parallel to the same busbars on the primary side, their no-load secondary voltages being 1000 V and 950 V respectively. Their resistances are 2% and 2·5 % respectively and their reactances are 8% and 6% respectively. Calculate the no-load circulating current in the secondaries. **[25·1 A]**

8.43 THREE-PHASE TRANSFORMER

A three-phase system is used to generate and transmit electric power. Three-phase voltages are raised or lowered by means of three-phase transformers. A three-phase transformer can be built in two ways *viz.*, (*i*) by suitably connecting a bank of three single-phase transformers or (*ii*) by constructing a three-phase transformer on a common magnetic structure. In either case, the windings may be connected in $Y - Y$, $\Delta - \Delta$, $Y - \Delta$ or $\Delta - Y$.

(*i*) **Bank of three single-phase transformers.** Three similar single-phase transformers can be connected to form a three-phase transformer. The primary and secondary windings may be connected in star (Y) or delta (Δ) arrangement. Fig. 8.70 (*i*) shows a Y - Δ connection of a three-phase transformer. The primary windings are connected in star and the secondary windings are connected in delta. A more convenient way of showing this connection is illustrated in Fig. 8·70 (*ii*). The primary and secondary windings shown parallel to each other belong to the same single-phase transformer. The ratio of secondary phase voltage to primary phase voltage is the *phase transformation ratio K*.

$$\text{Phase transformation ratio, } K = \frac{\text{Secondary phase voltage}}{\text{Primary phase voltage}} = \frac{N_2}{N_1}$$

Referring to Fig. 8.70 (*ii*), primary line-to-line voltage is V and the primary line current is I. The phase transformation ratio is $K\ (= N_2/N_1)$. The secondary line voltage and line current are also shown.

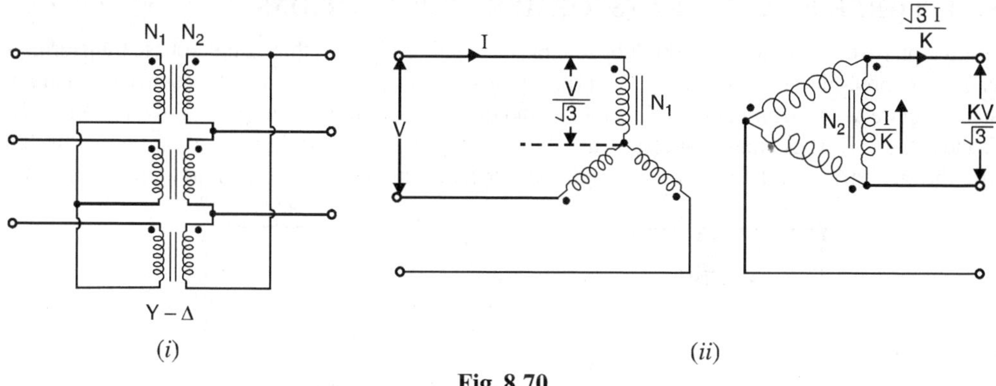

Fig. 8.70

(*ii*) **Three-phase transformer.** A three-phase transformer can be constructed by having three primary and three secondary windings on a common magnetic circuit. The basic principle of a 3-phase transformer is illustrated in Fig. 8.71 (*i*). The three single-phase core-type transformers, each with windings (primary and secondary) on only one leg have their unwound legs combined to provide a path for the returning flux. The primaries as well as secondaries may be connected in star or delta. If the primary is energised from a 3-phase supply, the central limb (*i.e.*, unwound limb) carries the fluxes produced by the 3-phase primary windings. Since the phasor sum of three primary currents at any instant is zero, the sum of three fluxes passing through the central limb must be zero. Hence no flux exists in the central limb and it may, therefore, be eliminated. This modification gives a three leg core-type 3-phase transformer. In this case, any two legs will act as a return path for the flux in the third leg. For example, if flux is ϕ in one leg at some instant, then flux is $\phi/2$ in the opposite direction through the other two legs at the same instant. All the connections of a 3-phase transformer are made inside the case and for delta-connected winding three leads are brought out while for star-connected winding four leads are brought out.

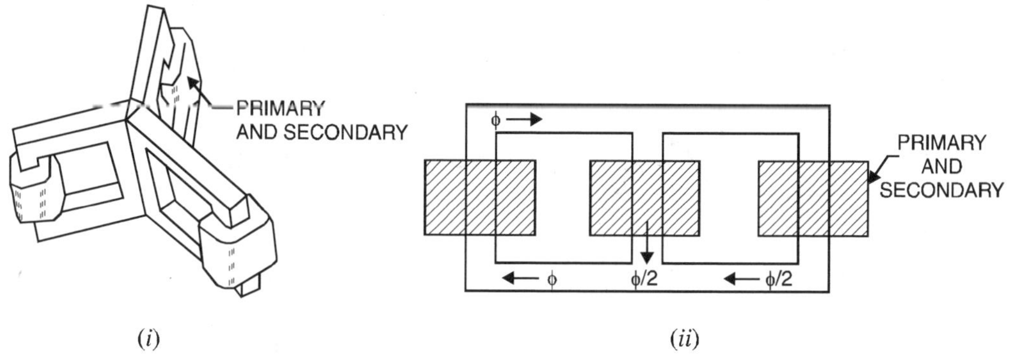

Fig. 8.71

For the same capacity, a 3-phase transformer weighs less, occupies less space and costs about 20% less than a bank of three single-phase transformers. Because of these advantages, 3-phase transformers are in common use, especially for large power transformations.

A disadvantage of the three-phase transformer lies in the fact that when one phase becomes defective, the entire three-phase unit must be removed from service. When one transformer in a bank of three single-phase transformers becomes defective, it may be removed from service and the other two transformers may be reconnected to supply service on an emergency basis until repairs can be made.

8.44 THREE-PHASE TRANSFORMER CONNECTIONS

A three-phase transformer can be built by suitably connecting a bank of three single-phase transformers or by one three-phase transformer. The primary or secondary windings may be connected in either star (Y) or delta (Δ) arrangement. The four most common connections are (*i*) Y - Y (*ii*) Δ - Δ (*iii*) Y - Δ and (*iv*) Δ - Y. These four connections are shown in Fig. 8.72. In this figure, the windings at the left are the primaries and those at the right are the secondaries. The primary and secondary

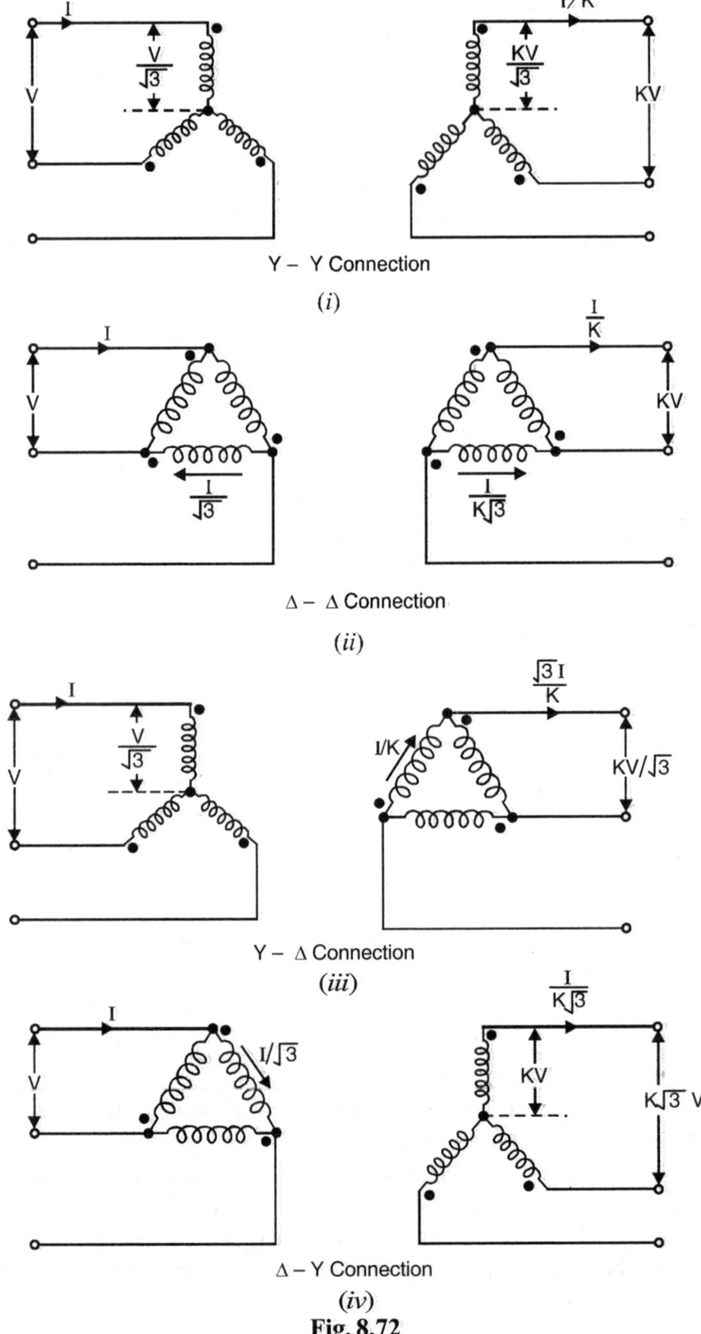

Fig. 8.72

voltages and currents are also shown. The primary line voltage is V and the primary line current is I. The phase transformation ratio K is given by ;

$$K = \frac{\text{Secondary phase voltage}}{\text{Primary phase voltage}} = \frac{N_2}{N_1}$$

(*i*) **Y - Y Connection.** In the Y - Y connection shown in Fig. 8.72 (*i*), 57·7% (or $1/\sqrt{3}$) of the line voltage is impressed upon each winding but full line current flows in each winding. Power circuits supplied from a Y - Y bank often create serious disturbances in communication circuits in their immediate vicinity. Because of this and other disadvantages, the Y - Y connection is seldom used.

(*ii*) **Δ - Δ Connection.** The Δ - Δ connection shown in Fig. 8.72 (*ii*) is often used for moderate voltages. An advantage of this connection is that if one transformer gets damaged or is removed from service, the remaining two can be operated in what is known as the open-delta or V-V connection. By being operated in this way, the bank still delivers three-phase currents and voltages in their correct phase relationships but the capacity of the bank is reduced to 57·7% of what it was with all three transformers in service.

(*iii*) **Y - Δ Connection.** The Y - Δ connection shown in Fig. 8.72 (*iii*) is suitable for stepping down a high voltage. In this case, the primaries are designed for 57·7% of the high-tension line voltages.

(*iv*) **Δ - Y Connection.** The Δ - Y connection shown in Fig. 8.72 (*iv*) is commonly used for stepping up to a high voltage.

Example 8.90. *A 3-phase, 50 Hz transformer has a delta-connected primary and star-connected secondary, the line voltages being 22000 V and 400 V respectively. The secondary has a star-connected balanced load at 0·8 power factor lagging. The line current on the primary side is 5A. Determine the current in each coil of the primary and in each secondary line. What is the output of the transformer in kW ?*

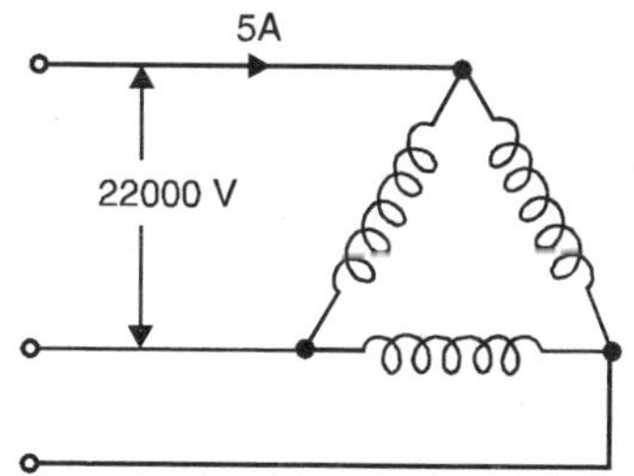

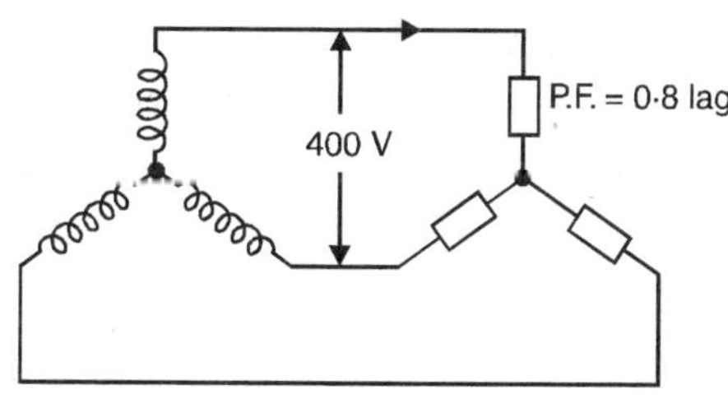

Fig. 8.73

Solution. Fig. 8.73 shows the conditions of the problem.

$$\text{Primary phase voltage} = 22000\text{ V}$$

$$\text{Secondary phase voltage} = 400 \div \sqrt{3} = 400/\sqrt{3}\text{ V}$$

$$\therefore \quad K = \frac{400/\sqrt{3}}{22000} = \frac{1}{55\sqrt{3}}$$

$$\text{Primary phase current} = 5 \div \sqrt{3} = \mathbf{5/\sqrt{3}\ A}$$

$$\text{Secondary phase current} = 5/\sqrt{3} \div K = \frac{5}{\sqrt{3}} \div \frac{1}{55\sqrt{3}} = 275\text{ A}$$

$$\text{Secondary line current} = \mathbf{275\ A}$$

$$\text{Output power} = \sqrt{3}\ V_L I_L \cos\phi$$

$$= \sqrt{3} \times 400 \times 275 \times 0{\cdot}8 = 152400\text{ W} = \mathbf{152{\cdot}4\ kW}$$

Example 8.91. *The input current to a 3-phase step down transformer connected to an 11 kV supply system is 14 A. Calculate the secondary line voltage and current for (i) star-star and (ii) delta-star if the phase turn ratio is 44.*

Solution.

$$K = \frac{\text{Secondary phase voltage}}{\text{Primary phase voltage}} = \frac{1}{44}$$

(*i*) Star-Star Connection

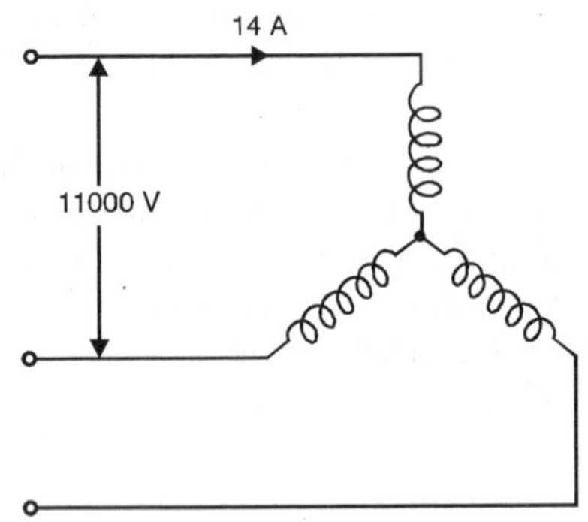

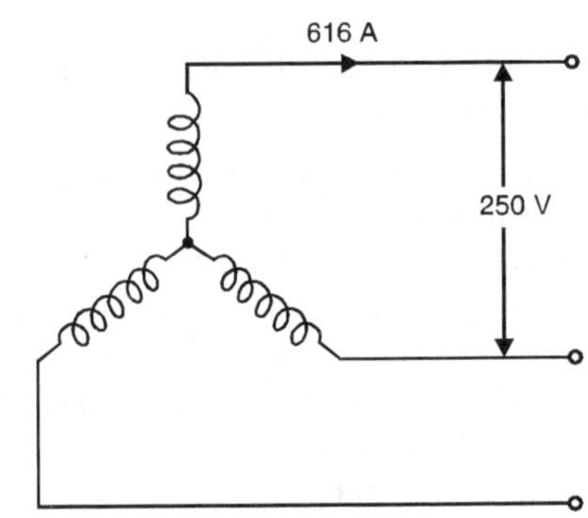

Fig. 8.74

Fig. 8.74 shows the star-star connection.

$$\text{Primary phase voltage} = 11000 \div \sqrt{3} = 6351 \text{ V}$$

$$\text{Secondary phase voltage} = 6351 \times K = 6351 \times 1/44 = 144{\cdot}34 \text{ V}$$

$$\text{Secondary line voltage} = 144{\cdot}34 \times \sqrt{3} = \mathbf{250 \text{ V}}$$

$$\text{Primary phase current} = 14 \text{ A}$$

$$\text{Secondary phase current} = 14 \div K = 14 \times 44 = \mathbf{616 \text{ A}}$$

(*ii*) Delta-Star Connection

$$\text{Primary phase voltage} = 11000 \text{ V}$$

$$\text{Secondary phase voltage} = 11000 \times K = 11000 \times 1/44 = 250 \text{ V}$$

$$\text{Secondary line voltage} = 250 \times \sqrt{3} = \mathbf{433 \text{ V}}$$

$$\text{Primary line current} = 14 \text{ A}$$

$$\text{Primary phase current} = 14 \div \sqrt{3} = 8{\cdot}08 \text{ A}$$

$$\text{Secondary phase current} = 8{\cdot}08 \div K = 8{\cdot}08 \times 44 = 355{\cdot}5 \text{ A}$$

$$\text{Secondary line current} = \mathbf{355{\cdot}5 \text{ A}}$$

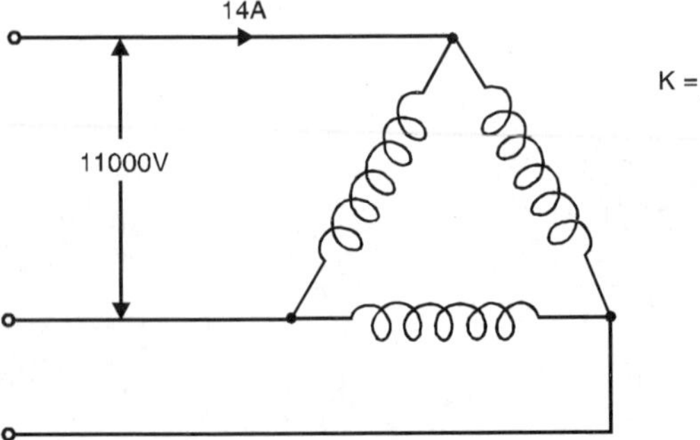

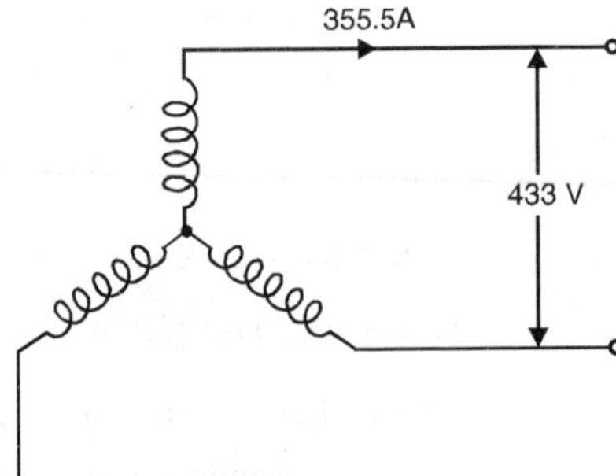

Fig. 8.75

Fig. 8.75 shows the delta-star connection.

Example 8.92. *Three single-phase, 50 kVA, 2300/230 V, 50 Hz transformers are connected to form a 3-phase, 4000/230 V transformer bank. The equivalent impedance of each transformer referred to low voltage side is (0·012 + j 0·016) Ω. The three-phase transformer supplies 3 - φ, 120 kVA, 230 V, 0·85 p.f. (lag) load.*

(i) Draw the schematic diagram showing the transformer connections.

(ii) Determine transformer winding currents.

(iii) Determine the primary voltage (line-to-line) required.

(iv) Determine the voltage regulation.

Solution. (*i*) The transformer should be connected in *Y* - Δ arrangement to meet the requirement. The connection diagram is shown in Fig.8.76 (*i*). Note that primary and secondary windings shown parallel to each other belong to the same single-phase transformer.

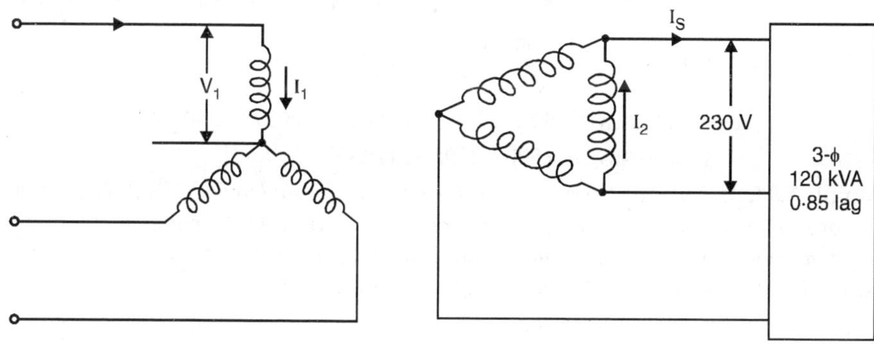

Fig. 8.76 (*i*)

(*ii*) Secondary line current, $I_S = \dfrac{120 \times 10^3}{\sqrt{3} \times 230} = 301{\cdot}24 \text{ A}$

Secondary phase current, $I_2 = I_S / \sqrt{3} = 301{\cdot}24/\sqrt{3} =$ **173·92 A**

Transformation ratio, $K = 230/2300 = 1/10$

Primary phase current, $I_1 = K I_2 = \dfrac{1}{10} \times 173{\cdot}92 =$ **17·39 A**

(*iii*) Calculations can be carried out on a per-phase basis.

Equivalent impedance of each transformer referred to primary is

$$\mathbf{Z_{01}} = \frac{\mathbf{Z_{02}}}{K^2} = (0{\cdot}012 + j\,0{\cdot}016)\,100$$

$$= (1{\cdot}2 + j\,1{\cdot}6)\,\Omega$$

$$\phi = \cos^{-1} 0{\cdot}85 = 31{\cdot}8^\circ$$

The primary equivalent circuit is shown in Fig. 8.76 (*ii*).

1.2 Ω 1.6 Ω 17.39 ∠–31.8° A

V_1 $V_2' = 2300$ volt

Fig. 8.76 (*ii*)

Primary phase voltage, $\mathbf{V_1} = 2300\angle 0^\circ + 17{\cdot}39\angle{-31{\cdot}8^\circ}(1{\cdot}2 + j\,1{\cdot}6)$

∴ Magnitude of $\mathbf{V_1}$, $V_1 = 2332{\cdot}4$ volt

∴ Primary line voltage $= \sqrt{3}\,V_1 = \sqrt{3} \times 2332{\cdot}4 =$ **4039·8 volt**

(*iv*) Voltage regulation $= \dfrac{2332{\cdot}4 - 2300}{2300} \times 100 =$ **1·41 %**

Example 8.93. *A 100 kVA, 3-phase, 50 Hz, 3300/400 V transformer is Δ - connected on the h.v. side and Y - connected on the l.v. side. The resistance of the h.v. winding is 3·5 Ω per phase and that of the l.v. winding 0·02 Ω per phase. Calculate the iron losses of the transformer at normal voltage and frequency if its full-load efficiency is 95·8 % at 0·8 p.f. (lag).*

Solution.

$$\text{Full-load output} = 100 \times 0{\cdot}8 = 80 \text{ kW}$$

$$\text{Full-load input} = 80/0{\cdot}958 = 83{\cdot}5 \text{ kW}$$

$$\text{Total F.L. losses} = 83{\cdot}5 - 80 = 3{\cdot}5 \text{ kW} = 3500 \text{ W}$$

$$\text{Transformation ratio, } K = \frac{\text{Secondary voltage/phase}}{\text{Primary voltage/phase}} = \frac{400/\sqrt{3}}{3300} = \frac{4}{33\sqrt{3}}$$

Equivalent resistance/phase referred to secondary

$$R_{02} = R_2 + K^2 R_1 = 0{\cdot}02 + (4/33 \times \sqrt{3})^2 \times 3{\cdot}5 = 0{\cdot}037\ \Omega$$

$$\text{F.L. secondary current, } I_2 = \frac{100 \times 10^3}{\sqrt{3} \times 400} = 144{\cdot}1 \text{ A}$$

$$\text{Total Cu losses} = 3 I_2^2 R_{02} = 3 \times (144{\cdot}1)^2 \times 0{\cdot}037 = 2305 \text{ W}$$

$$\therefore \quad \text{Iron losses} = 3500 - 2305 = \mathbf{1195\ W}$$

Example 8.94. *A 3-phase step-down transformer is connected to 6600 V on the primary side. The ratio of turns per phase is 12 and the line current drawn from the mains is 20 A. Find the secondary line voltage and secondary line current if the transformer is connected in (i) star-star (ii) delta-delta (iii) star-delta (iv) delta-star. Neglect losses.*

Solution. Since losses are neglected,

$$\text{Output in each case} = \frac{\sqrt{3} \times 6600 \times 20}{1000} = 228{\cdot}62 \text{ kVA}$$

$$\text{Transformation ratio, } K = N_2/N_1 = 1/12$$

(*i*) **Star-star connection**

$$\text{Primary phase voltage} = 6600/\sqrt{3} \text{ V}$$

$$\text{Secondary phase voltage} = K \times 6600/\sqrt{3} = \frac{1}{12} \times 6600/\sqrt{3}$$

$$\text{Secondary line voltage} = \sqrt{3} \times \frac{1}{12} \times 6600/\sqrt{3} = \mathbf{550\ V}$$

$$\text{Primary phase current} = 20 \text{ A}$$

$$\text{Secondary phase current} = 20/K = 20 \times 12 = \mathbf{240\ A}$$

$$\text{Secondary line current} = \mathbf{240\ A}$$

(*ii*) **Delta-delta connection**

$$\text{Primary phase voltage} = 6600 \text{ V}$$

$$\text{Secondary phase voltage} = K \times 6600 = \frac{1}{12} \times 6600 = 550 \text{ V}$$

$$\text{Secondary line voltage} = \mathbf{550\ V}$$

$$\text{Primary phase current} = 20/\sqrt{3} \text{ A}$$

$$\text{Secondary phase current} = 20/\sqrt{3} \times \frac{1}{K} = 20/\sqrt{3} \times 12$$

$$\text{Secondary line current} = \sqrt{3} \times \frac{20}{\sqrt{3}} \times 12 = \mathbf{240\ A}$$

(*iii*) **Star-delta connection**

$$\text{Primary phase voltage} = 6600/\sqrt{3} \text{ V}$$

$$\text{Secondary phase voltage} = K \times 6600/\sqrt{3} = \frac{1}{12} \times 6600/\sqrt{3} = 317{\cdot}55 \text{ V}$$

$$\text{Secondary line voltage} = \mathbf{317{\cdot}55 \ V}$$

$$\text{Primary phase current} = 20 \text{ A}$$

$$\text{Secondary phase current} = 20/K = 20 \times 12 = 240 \text{ A}$$

$$\text{Secondary line current} = \sqrt{3} \times 240 = \mathbf{415{\cdot}68 \ A}$$

(*iv*) **Delta-star connection**

$$\text{Primary phase voltage} = 6600 \text{ V}$$

$$\text{Secondary phase voltage} = K \times 6600 = \frac{1}{12} \times 6600 = 550 \text{ V}$$

$$\text{Secondary line voltage} = \sqrt{3} \times 550 = \mathbf{952{\cdot}6 \ V}$$

$$\text{Primary phase current} = 20/\sqrt{3} = 11{\cdot}55 \text{ A}$$

$$\text{Secondary phase current} = 11{\cdot}55/K = 11{\cdot}55 \times 12 = 138{\cdot}6 \text{ A}$$

$$\text{Secondary line current} = \mathbf{138{\cdot}6 \ A}$$

Example 8.95. *A small industrial plant draws an average load of 122 A at 0·73 p.f. lagging from the secondaries of its 2300/208 V, 60 kVA, Y - Δ transformer bank. Find :*

(i) the power used by the plant in kW

(ii) the total kVA used

(iii) the rated line currents available from the transformer bank

(iv) the rated transformer phase currents of the Δ secondaries

(v) percent of rated load on the transformers

(vi) primary line and phase currents

(vii) the kVA rating of each individual transformer

Solution. The circuit is shown in Fig. 8.77.

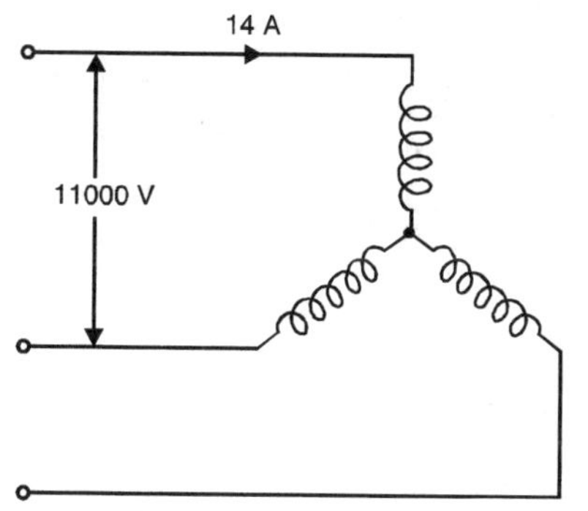

K = 1/44

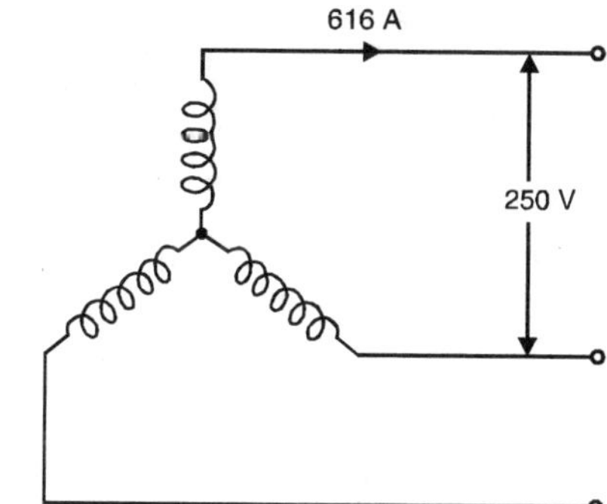

Fig. 8.77

$$\text{Transformation ratio, } K = \frac{\text{Secondary voltage/phase}}{\text{Primary voltage/phase}} = 208/2300/\sqrt{3}$$

(*i*) $$\text{Total power consumed} = \sqrt{3} \times 208 \times 122 \times 0{\cdot}73 = 32{\cdot}1 \times 10^3 \text{ W} = \mathbf{32{\cdot}1 \ kW}$$

(*ii*) $$\text{Total kVA used} = \frac{\text{kW}}{\text{p.f.}} = \frac{32{\cdot}1}{0{\cdot}73} = \mathbf{44 \ kVA}$$

(*iii*) $$\text{Rated line current} = \frac{60 \times 10^3}{\sqrt{3} \times 208} = \mathbf{167 \ A}$$

(*iv*) Rated phase current of Δ secondaries

$$= \frac{167}{\sqrt{3}} = \mathbf{96{\cdot}2\ A}$$

(*v*) $$\% \text{ of rated load} = \frac{44}{60} \times 100 = \mathbf{73{\cdot}3\ \%}$$

(*vi*) $$\text{Secondary phase current} = 122/\sqrt{3} = 70{\cdot}4\ \text{A}$$

$$\text{Primary phase current} = K \times 70{\cdot}4 = \frac{208}{2300/\sqrt{3}} \times 70{\cdot}4 = \mathbf{11\ A}$$

$$\text{Primary line current} = \mathbf{11\ A}$$

(*vii*) kVA rating of individual transformer

$$= 60/3 = \mathbf{20\ kVA}$$

Example 8.96. *A three-phase transformer, ratio 33/6·6 kV, Δ/Y, 2 MVA has a primary resistance of 8 Ω per phase and a secondary resistance of 0·08 Ω per phase. The percentage impedance is 7 %. Calculate the secondary load voltage with rated primary voltage and hence the regulation for full-load 0·75 p.f. lagging conditions.*

Solution.

$$\text{F.L. secondary current, } I_2 = \frac{2 \times 10^6}{\sqrt{3} \times 6{\cdot}6 \times 10^3} = 175\ \text{A}$$

$$\text{Transformation ratio, } K = \frac{\text{Secondary voltage/phase}}{\text{Primary voltage/phase}} = \frac{6{\cdot}6/\sqrt{3}}{33} = \frac{1}{8{\cdot}65}$$

$$R_{02} = R_2 + K^2 R_1 = 0{\cdot}08 + \left(\frac{1}{8{\cdot}65}\right)^2 \times 8 = 0{\cdot}187\ \Omega/\text{phase}$$

$$\text{Secondary impedance drop per phase} = \frac{7}{100} \times \frac{6600}{\sqrt{3}} = 266{\cdot}7\ \text{V}$$

$$\therefore \quad Z_{02} = 266{\cdot}7/175 = 1{\cdot}523\ \Omega \text{ per phase}$$

$$X_{02} = \sqrt{Z_{02}^2 - R_{02}^2} = \sqrt{(1{\cdot}523)^2 - (0{\cdot}187)^2} = 1{\cdot}51\ \Omega/\text{phase}$$

$$\text{Voltage drop per phase} = I_2 (R_{02} \cos \phi + X_{02} \sin \phi)$$

$$= 175\ (0{\cdot}187 \times 0{\cdot}75 + 1{\cdot}51 \times 0{\cdot}66) = 200\ \text{V}$$

$$\text{Secondary voltage/phase} = 6600/\sqrt{3} = 3810\ \text{V}$$

$$\text{Secondary load voltage/phase} = 3810 - 200 = 3610\ \text{V}$$

$$\therefore \quad \text{Secondary line voltage} = \sqrt{3} \times 3610 = \mathbf{6250\ V}$$

$$\text{Voltage regulation} = \frac{200}{3810} \times 100 = \mathbf{5{\cdot}23\ \%}$$

TUTORIAL PROBLEMS

1. A 3-phase transformer has 145 turns/phase on each secondary winding and 1382 on each primary winding. The primary side is connected to a 6600 V, 50 Hz, 3-phase supply. Calculate the no-load secondary line terminal P.D. when the windings are connected (*i*) star-mesh and (*ii*) mesh-star.
[(*i*) 400 V (*ii*) 1200 V]

2. A 50 kW, 440 V, three-phase a.c. motor has an efficiency of 90% and operates at 0·85 lagging power factor when on full-load. It is supplied by a 6600/440 V, mesh-star connected transformer. Ignoring the magnetising current, calculate the currents in the primary and secondary transformer phases on full-load.
[85·76 A; 3·3A]

3. A 5000 kVA, 3-phase transformer, 6·6/33 kV, Δ/*Y* has a no-load loss of 15 kW and a full-load loss of 50 kW. The impedance drop at full-load is 7%. Calculate the primary voltage when a load of 3200 kW at 0·8 p.f. lagging is delivered at 33 kV.**[6850 V]**

8.45 THREE-PHASE TRANSFORMATION WITH TWO SINGLE-PHASE TRANSFORMERS

It is possible to transform three-phase power by using two single-phase transformers. Two methods of doing this are :

(*i*) the connection of two identical single-phase transformers in *open delta* (or *V-V connection*).

(*ii*) the *T-T connection* (or *Scott connection*) of two nonidentical single-phase transformers.

Both of these methods of three-phase transformation result in slightly unbalanced output voltages under load because of unsymmetrical relations. The unbalance is negligible under usual conditions of operation.

8.46 OPEN-DELTA OR V-V CONNECTION

If one transformer breaks down in a star-star connected system of 3 single-phase transformers, three-phase power cannot be supplied until the defective transformer has been replaced or repaired. To eliminate this undesirable condition, single-phase transformers are generally connected in Δ-Δ. In this case, if one transformer breaks down, it is possible to continue supplying three-phase power with the other two transformers because this arrangement maintains correct voltage and phase relations on the secondary. However, with two transformers, the capacity of the bank is reduced to 57·7% of what it was with all three transformers in service (*i.e.,* complete Δ-Δ circuit).

Theory. If one transformer is removed in the Δ-Δ connection of three single-phase transformers, the resulting connection becomes *open delta or *V-V* connection. In complete Δ - Δ connection, the voltage of any one phase is equal and opposite to the sum of the voltages of the other two phases. Therefore, under no-load conditions if one transformer is removed, the other two will **maintain the same three line voltages on the secondary side. Under load conditions, the secondary line voltages will be slightly unbalanced because of the unsymmetrical relation of the impedance drops in the transformers.

Fig. 8.78 (*i*) shows open delta (or *V-V*) connection; one transformer shown dotted is removed. For simplicity, the load is considered to be star connected. Fig. 8.78 (*ii*) shows the phasor diagram for voltages and currents. Here V_{AB}, V_{BC} and V_{CA} represent the line-to-line voltages of the primary; V_{ab}, V_{bc} and V_{ca} represent line-to -line voltages of the secondary and V_{an}, V_{bn} and V_{cn} represent the phase voltages of the load. For inductive load, the load currents I_a, I_b and I_c will lag the corresponding voltages V_{an}, V_{bn} and V_{cn} by the load phase angle ϕ.

The transformer windings *ab* and *bc* will deliver power given by ;

$$P_{ab} = V_{ab} I_a \cos(30° + \phi)$$

$$P_{bc} = V_{cb} I_c \cos(30° - \phi)$$

Let $V_{ab} = V_{cb} = V$, the voltage rating of transformer secondary winding

$I_a = I_c = I$, current rating of the transformer secondary winding

p.f. $= 1$ *i.e.* $\phi = 0°$ ... For resistive load

∴ Power delivered to the resistive load by *V-V* connection is

$$P_V = P_{ab} + P_{bc} = VI\cos 30° + VI\cos 30° = 2\,VI\cos 30°$$

* It is called open delta because it is open or incomplete delta circuit. It is called *V-V* connection because of its appearance in the pictorial schematic.

** The basis of operation of the open-delta connection is the fact that the vector sum of any two of the line voltages in a balanced 3-phase system is equal to the third line voltage. Thus, even though a transformer has been removed, the voltage between the terminals to which it was connected remains unchanged.

With all the three transformers connected in delta, the power delivered to the resistive load is

$$P_{\Delta} = 3\,VI$$

$$\therefore \quad \frac{P_V}{P_{\Delta}} = \frac{2V\,I\cos 30^\circ}{3V\,I} = \frac{2\cos 30^\circ}{3} = 0{\cdot}577$$

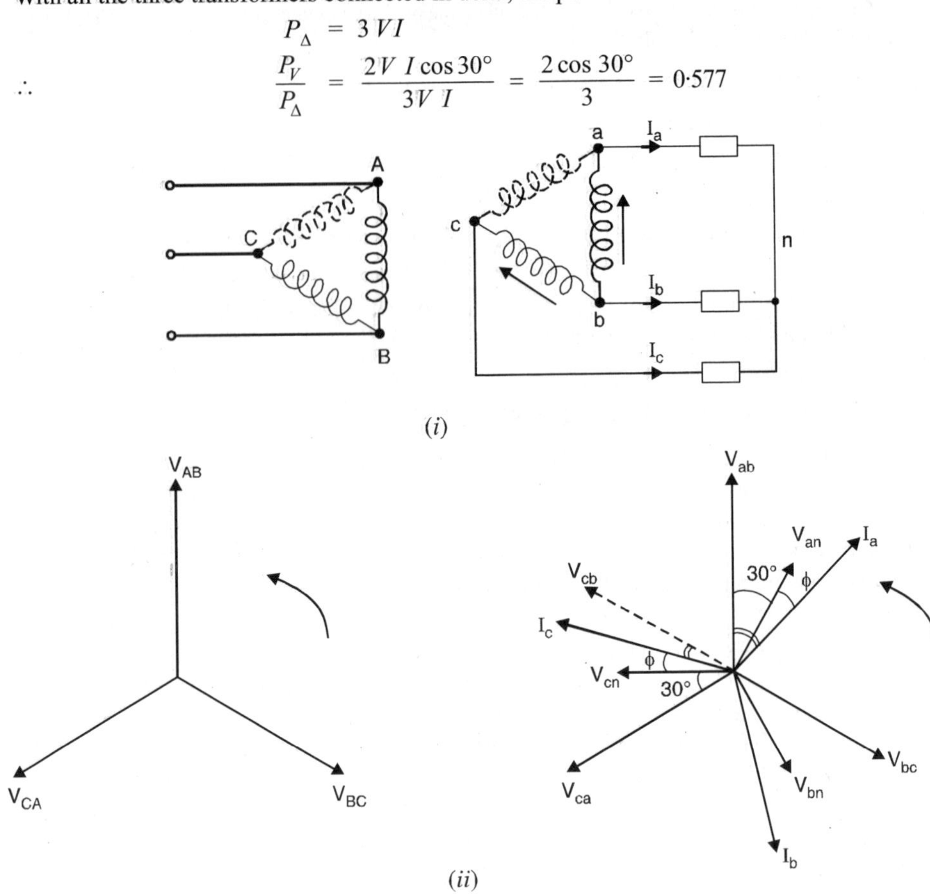

(ii)

Fig. 8.78

Hence the power-handling capacity of a V-V circuit (without overheating the transformers) is 57·7% of the capacity of a complete Δ-Δ circuit of the same transformers.

In a *V-V* circuit, only 86·6% of the rated capacity of the two transformers is available. This can be readily proved.

$$\frac{\text{Operating capacity}}{\text{Available capacity}} = \frac{2\,VI\cos 30^\circ}{2\,VI} = \cos 30^\circ = \frac{\sqrt{3}}{2} = 0{\cdot}866$$

Let us illustrate *V-V* connection with a numerical example. Suppose three identical single-phase transformers, each of capacity 10 kVA, are connected in Δ-Δ. The total rating of the three transformers is 30 kVA. When one transformer is removed, the system reverts to *V-V* circuit and can deliver 3-phase power to a 3-phase load. However, the kVA capacity of the *V-V* circuit is reduced to 30 × 0·577 = 17·3 kVA and not 20 kVA as might be expected. This reduced capacity can be determined in an alternate way. The available capacity of the two transformers is 20 kVA. When operating in *V-V* circuit, only 86·6% of the rated capacity is available *i.e.* 20 × 0·866 = 17·3 kVA.

8.47 POWER FACTOR OF TRANSFORMERS IN V-V CIRCUIT

When *V-V* circuit is delivering 3-phase power, the power factor of the two transformers is not the same (except at unity p.f.). Therefore, the voltage regulation of the two transformers will not be the same. If the load power factor angle is ϕ, then,

$$\text{p.f. of transformer 1} = \cos(30° - \phi)$$
$$\text{p.f. of transformer 2} = \cos(30° + \phi)$$

(*i*) *When load p.f.* $= 1$ *i.e.* $\phi = 0°$

In this case, *each* transformer will have a power factor of 0·866.

(*ii*) *When load p.f.* $= 0{\cdot}866$ *i.e.* $\phi = 30°$

In this case, one transformer will have a p.f. of cos (30° – 30°) = 1 and the other of cos (30° + 30°) = 0·5.

(*iii*) *When load p.f.* $= 0{\cdot}5$ *i.e.* $\phi = 60°$

In this case, one transformer will have a p.f. of cos (30° – 60°) = 0·866 and the other of cos (30° + 60°) = 0. Thus at a load p.f. of 0·5, one transformer delivers *all* the power at 0·866 p.f. and the other (*although still necessary to be in the circuit*) delivers *no power* at all.

8.48 APPLICATIONS OF OPEN-DELTA OR V-V CONNECTION

The *V-V* circuit has a number of features that are advantageous. A few applications are given below by way of illustration :

(*i*) The circuit can be employed in an emergency situation when one transformer in a complete Δ - Δ circuit must be removed for repair and continuity of service is required.

(*ii*) Upon failure of the primary or secondary of one transformer of a complete Δ - Δ circuit, the system can be operated as *V-V* circuit and can deliver 3-phase power (with reduced capacity) to a 3-phase load.

(*iii*) A circuit is sometimes installed as *V-V* circuit with the understanding that its capacity may be increased by adding one more transformer to form complete Δ-Δ circuit. As shown earlier,

$$\frac{P_V}{P_\Delta} = \frac{1}{\sqrt{3}} = 0{\cdot}577$$

$$\therefore \quad P_\Delta = \sqrt{3} \times P_V$$

Thus if a *V - V* circuit is changed to complete Δ - Δ circuit, the total capacity is increased by a factor of $\sqrt{3}$ (= 1·732). Thus by adding a third transformer to *V – V* connection, the capacity of the system is increased by 73.2%. However, the increase in cost is only 50% of two transformers. The supply companies take advantage of this fact by starting a 3-phase system using *V – V* connection in an area where the growth of power demand is slow. When the power demand increases, a third transformer is added to the *V – V* connection.

Example 8.97. *In an open delta transformer bank, the load power factor is 0·803. What are the individual transformer power factors ?*

Solution.

$$\text{Load p.f. angle, } \phi = \cos^{-1} 0{\cdot}803 = 36{\cdot}6°$$

In an open delta circuit, the two transformers have power factors of cos (30° – φ) and cos (30° + φ).

$$\therefore \quad \text{p.f. of transformer 1} = \cos(30° - 36{\cdot}6°) = \mathbf{0{\cdot}993}$$
$$\text{p.f. of transformer 2} = \cos(30° + 36{\cdot}6°) = \mathbf{0{\cdot}397}$$

Example 8.98.

(*i*) *Two identical 1-phase transformers are connected in open delta across a 3-phase mains and deliver a balanced load of 3000 kW at 11 kV and 0·8 p.f. lagging. Calculate the phase currents and the average power factor at which they operate.*

(*ii*) *If one more identical unit is added and the open delta is converted to close delta, calculate the additional load at the same power factor that can now be supplied for the same temperature rise. Also calculate the phase and line currents.*

Solution.

(*i*) If I is the line current, then,

$$I = \frac{3000 \times 10^3}{\sqrt{3} \times 11000 \times 0{\cdot}8} = 197 \text{ A}$$

Since line current has the same magnitude as the phase current,

$\therefore$ Secondary phase current $=$ **197 A**

The average power factor at which V - V bank operates is less than that of the load. This power factor is actually 86·6 % of the balanced load power factor.

$\therefore$ Transformer p.f. $=$ 86·6 % of 0·8 $=$ **0·693**

(*ii*) Additional load $=$ 72·5 % of 3000 kW $=$ **2175 kW**

Total load $=$ 3000 + 2175 $=$ 5175 kW

Now $\sqrt{3}\ V I_L \cos\phi = 5175 \times 10^3$

or
$$I_L = \frac{5175 \times 10^3}{\sqrt{3} \times 11000 \times 0{\cdot}8} = \textbf{340 A}$$

$\therefore$ Phase current $= I_L / \sqrt{3} = 340/\sqrt{3} =$ **196 A**

Example 8.99. *Two transformers, each rated at 250 kVA, 11/2 kV and 50 Hz are connected in open delta on both primary and secondary.*

(*i*) *Find the load kVA that can be supplied from this transformer connection.*

(*ii*) *A delta-connected three-phase load of 250 kVA, 0·8 p.f., 2 kV is connected to the L.V. terminals of this open-delta transformer. Determine the transformer currents on the 11 kV side of this transformer.*

Solution.

(*i*) Load kVA that can be supplied by the V - V bank

$=$ 0·866 × sum of kVA ratings of two transformers

$=$ 0·866 × 500 $=$ **433 kVA**

(*ii*) Let I be the secondary line current which is also the phase current in V - V connection.

Then, $\sqrt{3} \times 2 \times 10^3 \times I = 250 \times 10^3$

$$\therefore \quad I = \frac{250 \times 10^3}{\sqrt{3} \times 2 \times 10^3} = 72{\cdot}2 \text{ A}$$

Transformation ratio, $K = 2/11$

Line current on 11 kV side $= K \times 72{\cdot}2 = \frac{2}{11} \times 72{\cdot}2 =$ **13·12 A**

Phase current on 11 kV side $=$ **13·12 A**

Example 8.100. *A balanced 3-phase load of 30 kVA at a power factor of 0·866 lagging is connected to two transformers connected in open-delta to give 230 V three-phase system. Find the power delivered by each transformer.*

Solution. We know that the ratio of operating capacity to the available capacity in an open-delta is 0·866. Hence kVA capacity of each transformer is one-half of the total kVA load divided by 0·866.

$$\text{kVA of each transformer} = \frac{30/2}{0{\cdot}866} = 17{\cdot}32 \text{ kVA}$$

Load power factor angle, $\phi = \cos^{-1} 0{\cdot}866 = 30°$

The transformers in open-delta have power factors cos (30° – ϕ) and cos (30° + ϕ).

∴ Power supplied by one transformer is given by ;

$$P_1 = \text{kVA} \cos (30° - 30°) = 17{\cdot}32 \cos 0° = \mathbf{17{\cdot}32\ kW}$$

Power supplied by the other transformer is given by ;

$$P_2 = \text{kVA} \cos (30° + 30°) = 17{\cdot}32 \cos 60° = \mathbf{8{\cdot}66\ kW}$$

Example 8.101. *A balanced 3-phase load of 150 kW at 1000 V, 0.866 lagging power factor is supplied from 2000 V, 3-phase mains through single-phase transformers (ideal) connected in (i) delta-delta (ii) Vee-Vee. Find the current in the windings of each transformer and the power factor at which they operate in each case.*

Solution. (*i*) **Delta-Delta Connection**

$$I_L = \frac{150 \times 10^3}{\sqrt{3}\, V_L \cos\phi} = \frac{150 \times 10^3}{\sqrt{3} \times 1000 \times 0.866} = 100 \text{ A}$$

∴ Secondary line current = 100 A ; Secondary phase current = $100/\sqrt{3}$ = **57.7 A**

Primary phase current = K × secondary phase current

= (1000/2000) × 57.7 = **28.85 A**

Transformer power factor = **0.866** *lagging*

(*ii*) **Vee-Vee Connection**

Let I be the secondary line current which is also the phase current in *Vee-Vee* connection. Then,

$$I = \frac{150 \times 10^3}{\sqrt{3}\, V_L \cos\phi} = \frac{150 \times 10^3}{\sqrt{3} \times 1000 \times 0.866} = 100 \text{ A}$$

Secondary phase current = **100 A**

Primary phase current = KI_2 = (1000/2000) × 100 = **50 A**

Since load power factor is 0.866, $\phi = \cos^{-1}(0.866) = 30°$. Therefore, one transformer will have a p.f. cos (30° – 30°) = cos 0° = **1** and the other transformer will have a p.f. of cos (30° + 30°) = cos 60° = **0.5** *lagging*.

Example 8.102. *Three identical single-phase transformers are connected in Δ – Δ. Each transformer is rated at 20 kVA, 2300/230 V and the bank supplies a 40 kVA load at 0.7 p.f. lagging. If one defective transformer is removed for repair, calculate for the V – V connection,*

(*i*) *the kVA load carried by each transformer*

(*ii*) *%age rated load carried by each transformer*

(*iii*) *total kVA rating of the transformer bank in V – V*

(*iv*) *ratio of V – V to Δ – Δ bank transformer ratings*

(*v*) *%age increase in the load on each transformer when one transformer is removed. Comment on the results.*

Solution. (*i*) In *V – V* connection, each transformer delivers line (not phase) current. Therefore, kVA supplied by each transformer in *V – V* compared to the total 3-phase kVA is

$$\frac{\text{kVA/transformer}}{\text{Total 3-phase kVA}} = \frac{V_L I_L}{\sqrt{3}\, V_L I_L} = \frac{1}{\sqrt{3}}$$

∴ kVA load carried by each transformer

$$= 40/\sqrt{3} = \mathbf{23.1\ kVA}$$

(*ii*) % age rated load carried by each transformer

$$= \frac{\text{Load in kVA/transformer}}{\text{kVA rating/transformer}} \times 100$$

$$= \frac{23.1 \text{ kVA}}{20 \text{ kVA}} \times 100 = \mathbf{115.5\%}$$

(*iii*) Total kVA rating of $V-V$ connection

$$= 0.577 \times \text{Total kVA of } \Delta - \Delta \text{ bank}$$

$$= 0.577 \times (20 \times 3) = \mathbf{34.62 \text{ kVA}}$$

(*iv*) $$\text{Ratio of ratings} = \frac{V-V \text{ bank}}{\Delta - \Delta \text{ bank}} = \frac{34.62}{20 \times 3} = \mathbf{0.577}$$

(*v*) In $\Delta - \Delta$ connection, each transformer carries load $= 40/3 = 13.33$ kVA.

∴ %age increase in load on each transformer when one transformer is removed

$$= \frac{\text{kVA per transformer in } V-V}{\text{kVA per transformer in } \Delta - \Delta} \times 100$$

$$= \frac{23.1 \text{ kVA}}{13.33 \text{ kVA}} \times 100 = \mathbf{173.2\%}$$

Comments. When a $\Delta - \Delta$ connection is converted to $V-V$ connection, the following interesting information is obtained :

(*a*) The kVA rating of each transformer is 20 kVA. However, in $V-V$ connection, each transformer carries a load of 23.1 kVA. Thus each transformer in $V-V$ connection carries an overload = $(3.1/20) \times 100 = 15\%$. This overload may be sustained until such time the removed transformer is replaced.

(*b*) In $\Delta - \Delta$ connection, each transformer carries a load $= 40/3 = 13.33$ kVA. However, in $V-V$ connection, each transformer carries a load of 23.1 kVA. As a result of $V-V$ connection, load on each transformer increases by 173.2% [See part (*v*) of the example].

Example 8.103. *What should be the kVA rating of each transformer in a V – V bank when the 3-phase balanced load is 40 kVA ? If a third similar transformer is connected, what is the rated capacity ? What percentage increase in rating is affected in this way ?*

Solution. The kVA of each transformer has to be increased by 15%.

∴ kVA rating/ transformer $= (40/2) \times 1.15 = \mathbf{23 \text{ kVA}}$

$\Delta - \Delta$ bank rating $= 23 \times 3 = \mathbf{69 \text{ kVA}}$

$$\% \text{ age increase} = \frac{69 - 40}{40} \times 100 = \mathbf{72.5\%}$$

TUTORIAL PROBLEMS

1. If three transformers in a Δ - Δ transformer bank can handle 48 kW at a particular load, how much power can two of the transformers be expected to carry in open-delta or V -V if one transformer is removed ? **[27·7 kW]**
2. What should be the kVA rating of each transformer in a V - V bank when the three-phase load is 40 kVA? If a third similar transformer is connected for operation, what is the rated capacity ? What percentage increase in rating is affected in this way ? **[23 kVA ; 69 kVA ; 72.5 %]**
3. In an open-delta transformer bank, the load power factor is 0·866 lagging. What are the individual transformer power factors ? **[1 ; 0·5]**

8.49 SCOTT CONNECTION OR T-T CONNECTION

Although there are now no 2-phase transmission and distribution systems, a 2-phase supply is sometimes required. We can convert 3-phase supply into 2-phase supply through scott or T -T connection of two single-phase transformers. One is called the *main transformer* M which has a centre-tapped primary; the centre-tap being C [See Fig. 8.79 (*i*)]. The primary of this transformer has N_1 turns and is connected between the terminals B and Y of the 3-phase supply. The other transformer is called *teaser transformer* T and its primary has $0{\cdot}866\ N_1$ turns. One end of this primary is connected to centre-tap C and the other end to the terminal R of the 3-phase supply. The number of turns (N_2) of the secondary windings of the two transformers are equal. As we shall see, the voltages across the secondaries are equal in magnitude having a phase difference of 90º. Thus scott connection of two single-phase transformers enables us to convert 3-phase supply to 2-phase supply.

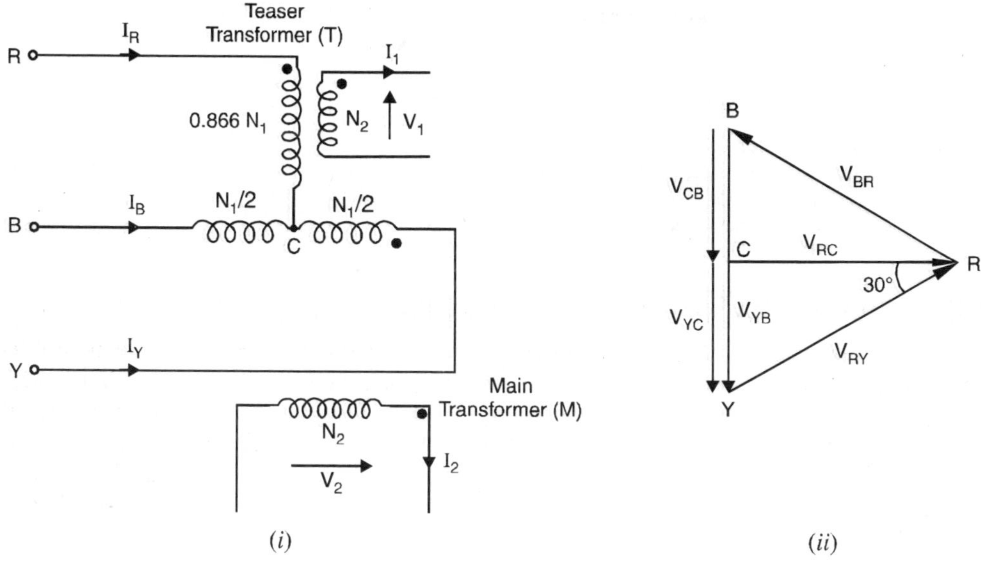

Fig. 8.79

Theory. Referring to Fig. 8.79 (*i*), the centre-tapped primary of the *main transformer* has line voltage $\mathbf{V_{YB}}$ applied to its terminals. The secondary terminal voltage $\mathbf{V_2}$ of the main transformer is

$$\mathbf{V_2} = \frac{N_2}{N_1}\mathbf{V_{YB}} = \frac{N_2}{N_1}V_L \qquad (V_L = \text{line voltage})$$

Fig. 8.79 (*ii*) shows the relevant phasor diagram. The line voltages of the 3-phase system $\mathbf{V_{RY}}$, $\mathbf{V_{YB}}$ and $\mathbf{V_{BR}}$ are balanced and are shown on the phasor diagram as a closed equilateral triangle. The voltages across the two halves of the centre tapped primary of the main transformer, $\mathbf{V_{CB}}$ and $\mathbf{V_{YC}}$ are equal and in phase with $\mathbf{V_{YB}}$. Clearly, $\mathbf{V_{RC}}$ leads $\mathbf{V_{YB}}$ by 90º. This voltage (*i.e.*, $\mathbf{V_{RC}}$) is applied to the primary of the *teaser transformer*. Therefore, the secondary voltage $\mathbf{V_1}$ of the teaser transformer will lead the secondary voltage $\mathbf{V_2}$ by 90º as shown in Fig. 8.80. We now show that magnitudes of $\mathbf{V_2}$ and $\mathbf{V_1}$ are equal.

Fig. 8.80

$$\mathbf{V_{RC}} = \mathbf{V_{RY}}\cos 30º = \frac{\sqrt{3}}{2}V_L = 0{\cdot}866\ V_L$$

$$V_1 = \frac{N_2}{0{\cdot}866\ N_1}V_{RC} = \frac{N_2}{0{\cdot}866\ N_1}\times 0{\cdot}866\ V_L = \frac{N_2}{N_1}V_L = V_2$$

Thus voltages $\mathbf{V}_1$ and $\mathbf{V}_2$ constitute balanced 2-phase system consisting of two voltages of equal magnitude having a phase difference of 90°.

Example 8.104. *Two transformers are required for a Scott connection operating from a 440 V, 3-phase supply for supplying two single-phase furnaces at 200 V on the two-phase side. If the total output is 150 kVA, calculate the secondary to primary turn ratio and the winding currents of each transformer.*

Solution. Main Transformer

$$\text{Primary voltage} = 440 \text{ V} \; ; \text{Secondary voltage} = 200 \text{ V}$$

$$\therefore \quad \frac{N_2}{N_1} = \frac{200}{440} = \frac{\mathbf{1}}{\mathbf{2.2}} (= K)$$

$$\text{Secondary current} = \frac{(150\times10^3)/2}{200} = \mathbf{375\ A}$$

$$\text{Primary current} = K I_2 = \frac{1}{2.2}\times 375 = \mathbf{170.5\ A}$$

Teaser Transformer

$$\text{Primary volts} = 0.866 \times 440 = 381 \text{ V}$$

$$\therefore \quad \frac{\text{Secondary turns}}{\text{Primary turns}} = \frac{200}{381} = \frac{\mathbf{1}}{\mathbf{1.095}}$$

8.50 3-PHASE TRANSFORMER CONNECTIONS AND PHASE SHIFT

A three-phase transformer can be built in two ways *viz* (*i*) by suitably connecting a bank of three single-phase transformers or (*ii*) by constructing a three-phase transformer on a common magnetic core. A noteable feature of a 3-phase transformer is that there may be a phase shift between corresponding primary and secondary *line* voltages. This phase shift depends upon the type of connection (star or delta) of primary and secondary windings. Depending upon the phase displacement between corresponding primary and secondary line voltages, three-phase transformers are divided into the following four groups :

Group 1 No phase displacement

Group 2 180° phase displacement

Group 3 –30° phase displacement

Group 4 +30° phase displacement

For example, the star-star or delta-delta connections have no phase displacement and belong to group 1. Only transformers in the same groups may be connected in parallel.

There is a phase shift of 30° between the corresponding primary and secondary *line voltages* in the Y–Δ bank and in the Δ–Y bank. There is no phase shift between corresponding primary and secondary line voltages in a $Y - Y$ bank, $\Delta - \Delta$ bank or a $V - V$ bank. *Because of the phase shift in $Y - \Delta$ and $\Delta - Y$ banks, they must *not be paralleled with $Y - Y$, $\Delta - \Delta$ or $V - V$ banks. Only the banks with the same phase shift should be operated in parallel.* Note that bank ratio (ratio of line voltages) for $Y - Y$, $\Delta - \Delta$ or $V - V$ banks is equal to the respective turns ratio.

Note : It is desirable that each transformer should have the same per unit resistance and per unit reactance. If this is not ensured, the transformers will not share the total load in proportion to their ratings.

---

* Any attempt to do so would cause large circulating currents and severe overheating of the windings even in the no-load conditions.

8.51 PARALLEL OPERATION OF 3-PHASE TRANSFORMERS

All the conditions for the successful parallel operation of single-phase transformers also apply to the parallel running of 3-phase transformers but with following additions :

(*i*) *The secondaries of all transformers must have the same phase sequence.*

(*ii*) *The phase displacement between primary and secondary line voltages must be the same for all transformers which are to be operated in parallel.*

(*iii*) *The secondaries of all transformers must have the same magnitude of line voltage.*

The above three conditions must be strictly observed. If these conditions are not compiled with, the secondaries will simply short-circuit one another and no output will be possible. The main difficulty arising from the parallel connection of 3-phase transformers is to ensure that condition (*ii*) is satisfied. For this, transformers in the same group should be connected in parallel.

Note : In dealing with 3-phase transformers, we consider balanced loading conditions. Therefore, calculations are made for one phase only.

Example 8.105. *Two three-phase transformers which have the same turns ratio are connected in parallel and supply a total load of 800 kW at 0.8 power factor lagging. Their ratings are as follows :*

Transformer	*Rating*	*p.u. resistance*	*p.u. reactance*
A	*400 kVA*	*0.02*	*0.04*
B	*600 kVA*	*0.01*	*0.05*

Determine the power output and the power factor of each transformer.

Solution.

$$\text{Total load kVA} = 800/0.8 = 1000 \text{ kVA}$$

$$\therefore \quad \mathbf{S} = 1000\,(0.8 - j\,0.6) = (800 - j\,600)\text{ kVA}$$

The p.u. resistances/reactances of the two transformers should be brought to the same kVA level, say 1000 kVA.

$$\therefore \quad R_A = 0.02 \times \frac{1000}{400} = 0.05 \;;\; X_A = 0.04 \times \frac{1000}{400} = 0.1$$

$$R_B = 0.01 \times \frac{1000}{600} = 0.0167 \;;\; X_B = 0.05 \times \frac{1000}{600} = 0.0833$$

$$\mathbf{S_B} = \mathbf{S}\frac{\mathbf{Z_A}}{\mathbf{Z_A} + \mathbf{Z_B}} = (800 - j\,600) \times \frac{0.05 + j\,0.1}{0.05 + j0.01 + 0.0167 + j0.0833}$$

$$= (800 - j\,600) \times \frac{0.05 + j\,0.1}{0.0667 + j\,0.1833} = (414 - j\,392)\text{ kVA}$$

$$\mathbf{S_A} = \mathbf{S} - \mathbf{S_B} = (800 - j\,600) - (414 - j\,392) = (386 - j\,208)\text{ kVA}$$

Transformer *A* : **386 kW, 440 kVA,** *0.878 p.f. lagging*

Transformer *B* : **414 kW, 570 kVA,** 0.726 *p.f. lagging*

Example 8.106. *A load of 1400 kVA at 0.866 p.f. lagging is supplied by two 3-phase transformers of 1000 kVA and 500 kVA capacity operating in parallel. The ratio of transformation is the same in both ; 6600/400 V delta-star. If the equivalent secondary impedances are (0.001 + j 0.003) ohm and (0.0028 + j 0.005) ohm per phase respectively, calculate the load and power factor of each transformer.*

Solution. On the base of 1000 kVA, we have,

$$\mathbf{Z_A} = (0.001 + j0.003)\Omega \;;\; \mathbf{Z_B} = (1000/500)(0.0028 + j0.005) = (0.0056 + j0.01)\Omega$$

Now $$\frac{\mathbf{Z_A}}{\mathbf{Z_A}+\mathbf{Z_B}} = \frac{(0.001+j\,0.003)}{(0.0066+j\,0.013)} = 0.2032\,\angle 8.5°$$

$$\mathbf{S} = 1400\,\angle\cos^{-1}0.866 = 1400\,\angle{-30°}\text{ kVA} = (1212-j\,700)\text{ kVA}$$

$$\mathbf{S_B} = \mathbf{S}\,\frac{\mathbf{Z_A}}{\mathbf{Z_A}+\mathbf{Z_B}} = 1400\,\angle{-30°}\times 0.2032\,\angle 8.5°$$

$$= 284.5\,\angle{-21.5°}\text{ kVA} = (265-j\,104)\text{ kVA}$$

$\therefore$ $$\mathbf{S_A} = \mathbf{S}-\mathbf{S_B} = (1212-j\,700)-(265-j\,104)$$

$$= (947-j\,596)\text{ kVA} = 1119\,\angle{-32.2°}\text{ kVA}$$

$$\cos\phi_A = \cos 32.2° = 0.846\text{ lag} \;;\; \cos\phi_B = \cos 21.5° = 0.93\text{ lag}$$

Transformer A : **1119 kVA** ; **0.846** *lag* ; Transformer B : **284.5 kVA** ; **0.93** *lag*

8.52 APPLICATIONS OF TRANSFORMERS

There are four principal applications of transformers *viz.*

(*i*) power transformers (*ii*) distribution transformers

(*iii*) autotransformers (*iv*) instrument transformers

(*i*) **Power Transformers.** They are designed to operate with an almost constant load which is equal to their rating. The maximum efficiency is designed to be at full-load. This means that full-load winding copper losses must be equal to the core losses.

(*ii*) **Distribution Transformers.** These transformers have variable load which is usually considerably less than the full-load rating. Therefore, these are designed to have their maximum efficiency at between 1/2 and 3/4 of full-load.

(*iii*) **Autotransformers.** An autotransformer has only one winding and is used in cases where the ratio of transformation (K), either step-up or step down, differs little from 1. For the same output and voltage ratio, an autotransformer requires less copper than an ordinary 2-winding transformer. Autotransformers are used for starting induction motors (reducing applied voltage during starting) and in boosters for raising the voltage of feeders.

(*iv*) **Instrument transformers.** Current and voltage transformers are used to extend the range of a.c. instruments.

8.53 INSTRUMENT TRANSFORMERS

For measuring a large current in a d.c. circuit, we use low-range ammeter with a suitable shunt. The measurement of high d.c. voltage is made using a low-range voltmeter with a multiplier. However, this method is not used for the measurement of high alternating currents and voltages for many good reasons. In order to measure high alternating currents and voltages, we employ specially designed transformers, called *instrument transformers.* These transformers facilitate the a.c. measurements with low-range a.c. instruments. There are two types of instrument transformers *viz.*

(*i*) Current transformers (*ii*) Potential transformers.

8.54 CURRENT TRANSFORMER (C.T.)

A current transformer (C.T.) is used to measure high alternating current in a power system. The primary of this transformer has a few turns of thick wire whereas the secondary has many turns of very fine wire as shown in Fig. 8.81. It is clear from the figure that a current transformer is simply a well-designed step-up transformer. Since voltage is stepped up, the current is stepped down which can be measured with a low-range a.c. ammeter.

The primary of the current transformer is connected in *series* with the line whose current is to be measured as shown in Fig. 8.81. The secondary of the transformer is connected across a low-range (0 – 5A) a.c. ammeter. The line current (I_P) and a.c. ammeter current (I_S) are *related as :

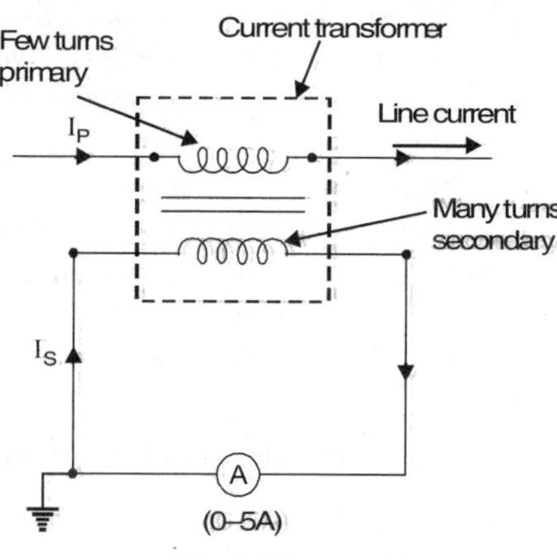

Fig. 8.81

$$N_P I_P = N_S I_S$$

or $$\frac{I_P}{I_S} = \frac{N_S}{N_P}$$

The primary to secondary current ratio (*i.e.* I_P/I_S) is called C.T. ratio (current transformation ratio).

$\therefore$ $$\frac{I_P}{I_S} = \text{C.T. ratio}$$

or $$I_P = I_S \times \text{C.T. ratio}$$

i.e., Line current (I_P) = A.C. ammeter reading × C.T. ratio

Thus if the reading of a.c. ammeter is 1A and the C.T. ratio is 100:1 (or 100/1), then line current = 1 × 100 = 100 A.

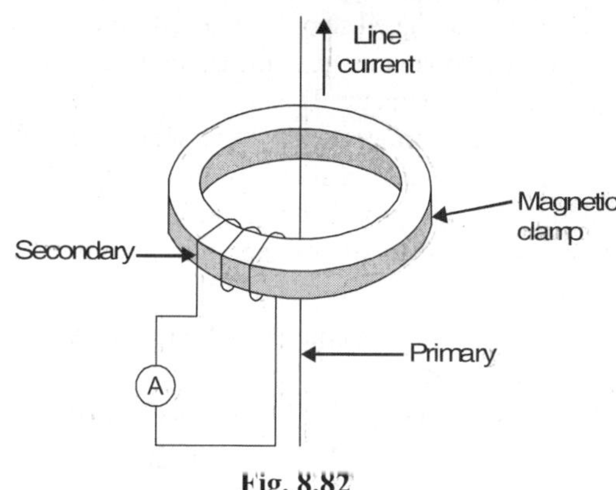

Fig. 8.82

Similarly, if the a.c. ammeter reads 2 A, the line current = 2 × 100 = 200 A. One most commonly used current transformer is the **clamp-on** or **clip-on type** shown in Fig. 8.82. It consists of a ring-shaped laminated core which carries the secondary winding. The current-carrying conductor itself acts as a one-turn primary that simply passes through the centre of the ring. The position of the primary is unimportant as long as it is more or less centred. This current transformer has the arrangement to open and close the ring-shaped core so that current can be measured without opening the line. The clamp-on current transformers are simple and inexpensive and are widely used in low-voltage (LV) and medium-voltage (MV) lines in the power system.

C.T. secondary should never be left open. Every precaution must be taken to *never* open the secondary circuit of a current transformer (C.T.) while current is flowing in the primary circuit. Opening of the secondary under these conditions removes the **opposing m.m.f. produced in the secondary and allows the flux in the core to reach a value determined only by the primary m.m.f. The result is that the flux in the core is greatly increased. This situation leads to the following harmful effects :

(*i*) The increased flux in the core may saturate it. The effect of saturation is to leave the transformer with a large value of residual flux. This will seriously ***impair the accuracy of the current transformer.

---

* The magnetising component of primary current is very small and can be neglected with reasonable accuracy.

** With secondary circuit closed, the secondary current sets up m.m.f. which opposes the primary m.m.f. As a result, the value of flux in the core is small.

*** Due to change in the magnetic characteristics of the core, the ratio of transformation of the transformer may change.

(*ii*) The flux in the core changes at a very high rate inducing a large voltage in the open secondary winding. The effect of excessive voltage may be to puncture the insulation of the transformer or to provide dangerous shock to the operator opening the circuit.

In view of the above, if we have to remove a meter in the secondary circuit of a C.T., we must first short-circuit the secondary *and then* remove the meter. The short circuit across the winding may be removed after the secondary circuit is again closed.

Note : Unlike an ordinary transformer, the primary current of C.T. is not controlled by secondary but remains constant at a value determined by the current in the main circuit. Therefore, short-circuiting a current transformer does not harm because the primary current remains unchanged and the secondary current depends only upon the turns ratio.

8.55 POTENTIAL TRANSFORMER (P.T.)

A potential transformer (P.T.) is used to measure high alternating potential difference (voltage) in a power system. The primary of this transformer has many turns while the secondary has few turns as shown in Fig. 8.83. It is clear from the figure that a potential transformer is simply a well-designed step-down transformer. The stepped down voltage is measured with a low-range a.c. voltmeter. The magnetic core of a potential transformer usually has a shell-type construction for better accuracy. It order to provide adequate protection to the operator, one end of the secondary winding is usually grounded.

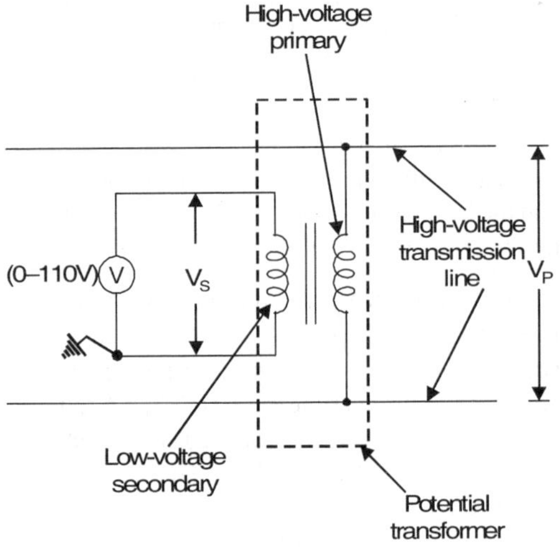

Fig. 8.83

The primary of the potential transformer is connected across the high-voltage line whose voltage is to be measured. A low-range (0 – 110 V) a.c. voltmeter is connected across the secondary. The line voltage (V_P) and a.c. voltmeter voltage (V_S) are related as :

$$\frac{V_P}{V_S} = \frac{N_P}{N_S}$$

The primary to secondary voltage ratio (*i.e.* V_P/V_S) is called P.T. ratio (potential transformation ratio).

$$\therefore \quad \frac{V_P}{V_S} = \text{P.T. ratio}$$

or $$V_P = V_S \times \text{P.T. ratio}$$

i.e. Line voltage (V_P) = A.C. voltmeter reading × P.T. ratio.

Thus if the reading of a.c. voltmeter is 50 V and the P.T. ratio is 100:1 (or 100/1), then line voltage = 50 × 100 = 5,000 V. Similarly, if the a.c. voltmeter reads 100 V, then line voltage = 100 × 100 = 10,000 V.

8.56 ADVANTAGES OF INSTRUMENT TRANSFORMERS

In order to measure high alternating currents and voltages in a power system, we prefer instrument transformers to shunts and multipliers for the following reasons :

(*i*) The errors due to stray inductance and capacitance in shunts, multipliers and their leads are eliminated.

(*ii*) The measuring circuit is isolated from the mains by the transformer.

(*iii*) We can use low-range and accurate a.c. instruments.

(*iv*) The length of the connecting leads from the transformer to the instrument is of lesser importance and leads may be of small cross-sectional area.

(*v*) By using a clip-on type of transformer core, the current in a heavy-current conductor can be measured without breaking the circuit.

Example 8.107. *Fig. 8.84 shows the connections of instrument transformers in a high-voltage circuit. The C.T. ratio is 80:5 while P.T. ratio is 100:1. If the ammeter, voltmeter and wattmeter read 4A, 110 V and 352 W respectively, determine (i) current (ii) voltage and (iii) the power supplied to the load.*

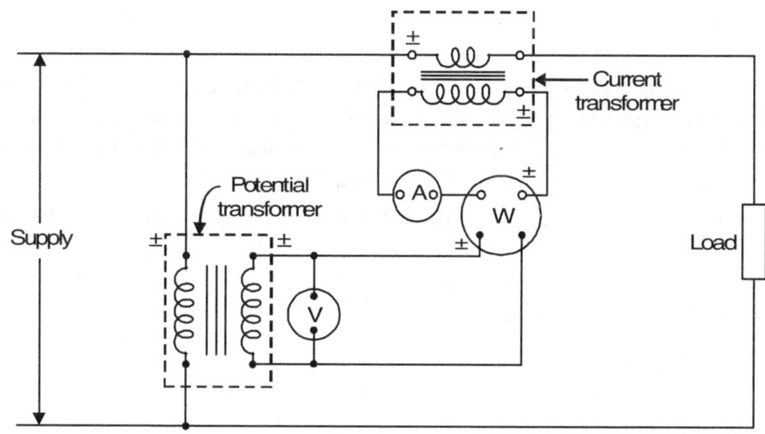

Fig. 8.84

Solution.

(*i*) Current, I = Ammeter reading × C.T. ratio

$= 4 \times (80/5) = \mathbf{64\ A}$

(*ii*) Voltage, V = Voltmeter reading × P.T. ratio

$= 110 \times (100/1) = \mathbf{11{,}000\ V}$

(*iii*) Power supplied, P = Wattmeter reading × C.T. ratio × P.T. ratio

$= 352 \times (80/5) \times (100/1) = \mathbf{563.2 \times 10^3\ W}$

Example 8.108. *A potential transformer of 14400/115 V and a current transformer of 75/5 A are used to measure voltage and current in a transmission line. If the voltmeter indicates 111 V and ammeter reads 3 A, what is the voltage and current in the line ?*

Solution.

Line voltage, V = Voltmeter reading × P.T. ratio

$= 111 \times (14400/115) = \mathbf{13900\ V}$

Line current, I = Ammeter reading × C.T. ratio

$= 3 \times (75/5) = \mathbf{45\ A}$

SHORT ANSWER QUESTIONS

Q. 1. *What is the principle of operation of a transformer ?*

Ans. A simple transformer consists of two windings placed on an iron core. When alternating voltage is applied to one winding (called the primary), it sets up alternating flux in the core. The alternating flux links the other winding (called the secondary) and induces voltage in it. When load is connected to the secondary, current flows through the load, thus delivering power to the load. In this way power is transformed from primary to secondary through the agency of magnetic flux.

Q. 2. *Why do we use iron-core in a transformer ?*

Ans. Transformer action demands only the existence of alternating flux linking the two windings. No doubt such action will be obtained if an air core is used but it will be obtained much more effectively if an iron core is used. It is because the flux is then substantially confined to a definite path (*i.e.*, iron path) having a much higher permeability than air.

Q.3. *What will happen if the primary of a transformer is connected to d.c. supply ?*

Ans. If the primary of a transformer is connected to a d.c. supply, the primary will draw a steady current and hence produce constant flux. Consequently, no back e.m.f. will be produced. The primary winding will draw excessive current due to low resistance of the primary. The result is that the primary will overheat and burn out or the fuses will blow. Care must be taken not to connect the primary of a transformer across the d.c. supply.

Q. 4. *Under what condition is d.c. supply applied safely to the primary of a transformer ?*

Ans. When the primary winding of a transformer is to be connected to a d.c. supply, a high resistance is connected in series with the primary. This series resistance limits the primary current to a safe d.c. value and thus prevents the primary from burning out.

Q. 5. *List four applications of transformers.*

Ans. Four important applications of transformer are :

(*i*) It can raise or lower the voltage or current in an a.c. circuit.

(*ii*) It can increase or decrease the value of a capacitor, an inductor or resistance in an a.c. circuit. It can thus act as an impedance transferring device.

(*iii*) It can isolate two circuits electrically.

(*iv*) It can be used to prevent d.c. from passing from one circuit to the other.

Q. 6. *Why is iron-cored transformer not used for high frequency applications ?*

Ans. At high frequencies, an iron-core transformer has the following disadvantages :

(*i*) high hysteresis and eddy current loss

(*ii*) iron becomes less effective as a magnetic circuit because at high frequencies, the flux penetrates to lesser extent in the core

Since air does not exhibit eddy current or hysteresis effects, the use of air-core transformers is common at high frequencies found in electronics and communications.

Q. 7. *What conditions are required for a transformer to operate satisfactorily at very low frequencies?*

Ans. A transformer intended to operate satisfactorily at low frequencies should meet the following requirements :

(*i*) A magnetic core of large-section.

(*ii*) A large number of turns. Clearly, this makes the transformer large in size.

Q. 8. *Where is a core-type construction suitable for a transformer ?*

Ans. In the core-type transformer, a core shaped as shown has primary and secondary windings wound on it. It is a usual practice to wind one half of each winding on one limb, low-voltage winding being innermost for mechanical strength. The core-type transformer affords better cooling surface for the winding than the shell-type core and is, therefore, more suitable for transformers which remain fully loaded.

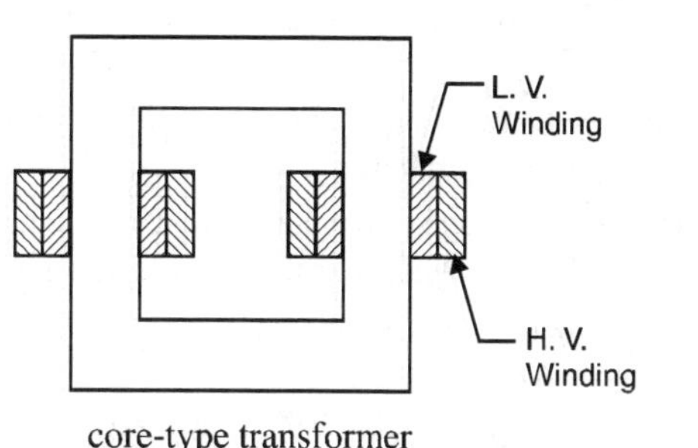

core-type transformer

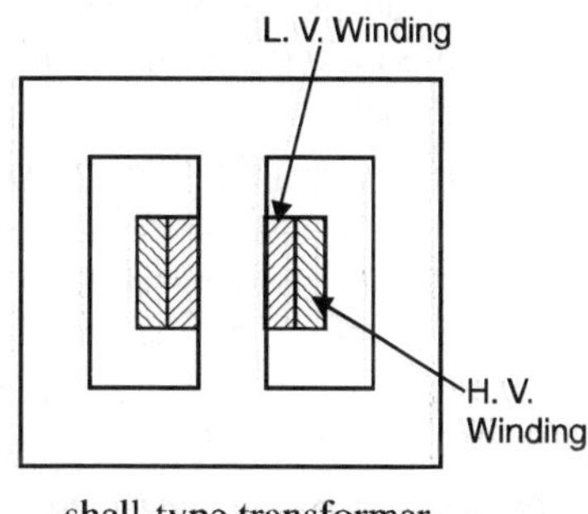

shell-type transformer

Fig. 8.85

Q. 9. *Where is shell-type construction suitable for a transformer ?*

Ans. In shell-type core, both the windings are wound on the central limb. This type of core is used for those transformers which work on poor power factor.

Q. 10. *Why are the cores of large transformers built up of circular cross-section ?*

Ans. The cross-section of the cores of small transformers are rectangular but for large transformers, it is a common practice to use approximately circular cross-section. It is because circular section has the smallest perimeter for a given area and, therefore, requires less copper than rectangular section.

Q. 11. *Why do we represent leakage flux in a transformer by inductive reactance ?*

Ans. In an actual transformer, there is primary leakage flux as well as secondary leakage flux. The primary leakage flux is the flux which links with the primary but not with the secondary. Similarly, the secondary leakage flux is the flux which links with the secondary but not with the primary winding. Since leakage flux links with only one winding, it produces an induced e.m.f. in that winding. It is, therefore, equivalent to a small inductance in series with that winding.

Q. 12. *What are the advantages of back to back test in determining the efficiency of a transformer ?*

Ans. In this method, the efficiency of a transformer is determined by putting it under full-load conditions and yet the power demand is small. This test requires two identical transformers (or even number of identical transformers). The primary windings are connected in parallel and supplied at normal voltage and frequency and the secondary windings are connected in series opposition and supplied through a variable voltage regulator. The method offers the following advantages :

(*i*) The transformers are tested under full-load conditions and yet the power demand is small.

(*ii*) The losses can be determined very accurately.

(*iii*) Two large transformers can be put under full-load conditions for several hours, so that the temperature rise can be measured, with an expenditure of energy equal to that required for losses only.

Q. 13. *Why are iron losses constant at all loads in a transformer ?*

Ans. Since the induced primary ampere-turns and secondary ampere-turns always neutralise one another, the flux in the core on load is the same as the flux on no load. Hence, the iron losses are constant and are independent of load.

Q. 14. *What is the difference between power transformers and distribution transformers ?*

Ans. Those transformers installed at the sending or receiving end of long high-voltage transmission lines are the power transformers. The distribution transformers (generally pole mounted) are those installed in the localities of the city to provide utilisation voltage at the consumer terminals.

(*i*) Power transformers generally operate at nearly full-load. However, distribution transformers operate at light loads during major part of the day.

(*ii*) The performance of a power transformer is generally judged from commercial efficiency whereas the performance of a distribution transformer is judged from all-day efficiency.

(*iii*) The rating of a power transformer is many times greater than that of a distribution transformer.

Q. 15. *What would happen if a power transformer designed for operation on 50 Hz were connected to a 500 Hz source of the same voltage ?*

Ans. Power transformers are made to operate on one particular frequency, usually 50 Hz. If the frequency is too high, the inductive reactance of the primary will prevent the primary from drawing sufficient power. The hysteresis and eddy current losses will be excessive.

Q. 16. *What would happen if a power transformer designed for operation on 50 Hz were connected to a 5 Hz source of the same voltage ?*

Ans. Power transformers are made to operate on one particular frequency, usually 50 Hz. If the frequency is too low, the primary will have insufficient reactance and too much primary current will flow, producing considerable copper losses. The transformer may start to smoke.

Q. 17. *If part of a primary winding of a transformer were accidentally short-circuited, what would be the immediate effect ?*

Ans. If a few turns of the primary of a transformer short out for some reason, a high current will be induced in the turns, producing excessive heat in the transformer. This is not only because of the 'shorted turns' heating but also because of the cancellation of the inductance of the primary by the magnetic field set up by the shorted turns. Cancellation of the inductance decreases the inductive reactance of the primary and excessive primary current flows. This causes excessive heat in the transformer.

Q. 18. *Why are autotransformers not safe for supplying a low-voltage from a high-voltage source?*

Ans. Autotransformers are not safe for supplying a low voltage from a high voltage source. It is because if the winding that is common to both primary and secondary accidentally becomes open-circuited, the full primary voltage will appear across the secondary. This may cause severe shock to the operating personnel.

Q. 19. *What functions are performed by instrument transformers ?*

Ans. Instrument transformers perform two functions *viz.*

(*i*) they act as *ratio devices*, making possible the use of standard low-voltage and low-current meters and instruments

(*ii*) they act as insulating devices to protect the apparatus and operating personnel from high voltages

There are two types of instrument transformers *viz.*, potential transformers and current transformers.

Q. 20. *What are the advantages and disadvantages of 3-phase transformers over 3 single-phase bank of transformers ?*

Ans. Advantages of 3-phase transformers over 3 single-phase bank of transformers are :

(*i*) less cost (*ii*) less weight

(*iii*) requires less space (*iv*) somewhat higher efficiency

Disadvantages of 3-phase transformers over 3 single-phase bank of transformers are :

(*i*) greater cost of standby units

(*ii*) increased cost and inconvenience of repairs

MULTIPLE-CHOICE QUESTIONS

1. A transformer will work on ....

(*i*) a.c. only (*ii*) d.c. only

(*iii*) a.c. as well as d.c.

(*iv*) none of the above

2. The primary and secondary of a transformer are .... coupled.

(*i*) electrically (*ii*) magnetically

(*iii*) electrically and magnetically

(*iv*) none of the above

3. A transformer transfers electrical energy from primary to secondary usually with a change in ....

(*i*) frequency (*ii*) power

(*iii*) voltage (*iv*) time period

4. The voltage per turn of the primary of a transformer is .... the voltage per turn of the secondary.

(*i*) the same as (*ii*) more than

(*iii*) less than

(*iv*) none of the above

5. The winding of the transformer with greater number of turns will be ....

(*i*) high-voltage winding

(*ii*) low-voltage winding

(*iii*) either high or low voltage winding

(*iv*) none of the above

6. A transformer does not possess .... changing property.

(*i*) impedance (*ii*) voltage

(*iii*) current (*iv*) power

7. The iron-core is used to .... of the transformer.

(*i*) increase the weight

(*ii*) provide tight magnetic coupling

(*iii*) reduce core losses

(*iv*) none of the above

8. The magnetic flux in the core of a single-phase transformer is ....

(*i*) purely alternating one

(*ii*) purely rotating one

(*iii*) partly alternating and partly rotating

(*iv*) none of the above

9. A transformer is so designed that primary and secondary have ....

(*i*) high leakage reactance

(*ii*) large resistance

(*iii*) tight magnetic coupling

(*iv*) good electric coupling

10. An ideal transformer is one which ....

(*i*) has no losses and leakage reactance

(*ii*) does not work

(*iii*) has the same number of primary and secondary turns

(*iv*) none of the above

11. Calculate the core-area required for a 1600 kVA, 6600/440 V, 50 Hz, single-phase core-type power transformer. Assume a maximum flux density of 1·2 Wb/m^2 and induced voltage per turn of 30 V.

(*i*) 975 cm^2 (*ii*) 1100 cm^2

(*iii*) 1125 cm^2 (*iv*) 1224 cm^2

12. An iron-cored transformer working at a maximum flux density of 0·8 Wb/m^2 is replaced by silicon steel working at a maximum flux density of 1·2 Wb/m^2. If the total flux is to remain unchanged, what is the reduction in volume expressed as a percentage of the original volume ? The frequency and voltage per turn are the same.

(*i*) 33 % (*ii*) 9 %

(*iii*) 22 % (*iv*) 11 %

13. A single-phase transformer has 400 primary and 1000 secondary turns. The net cross-sectional area of the core is 60 cm^2. If the primary is connected to 500 V, 50 Hz source, the peak value of flux density in the core is
(i) 1·2 Wb/m^2 (ii) 0·94 Wb/m^2
(iii) 2·1 Wb/m^2 (iv) 0·25 Wb/m^2

14. In the above question, what is the voltage induced in the secondary ?
(i) 950 V (ii) 1125 V
(iii) 840 V (iv) 1250 V

15. A single-phase, 50 Hz, core-type transformer has square core of 20 cm side. The permissible flux density is 1 Wb/m^2. If the stacking factor is 0·9, the voltage induced per turn is
(i) 6V (ii) 11V
(iii) 8V (iv) 14V

16. The no-load primary current I_0 in Fig. 8.86 is

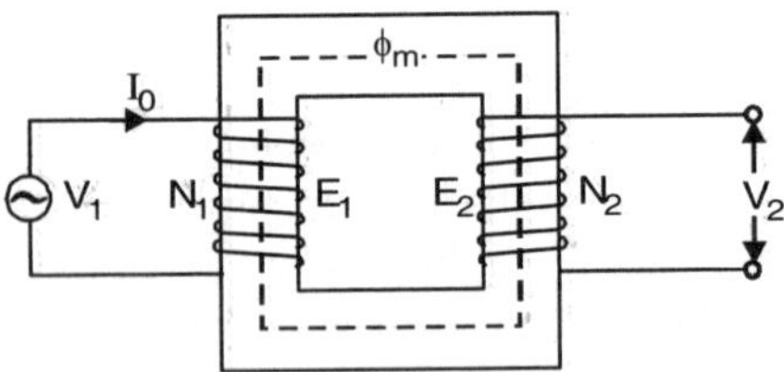

Fig. 8.86

(i) wholly reactive
(ii) wholly active
(iii) partly active and partly reactive
(iv) none of the above

17. The no-load input power to a transformer is practically equal to .... loss in the transformer.
(i) iron (ii) copper
(iii) eddy current
(iv) none of the above

18. In Fig. 8.86, the voltage transformation ratio is given by ....
(i) N_1/N_2 (ii) E_1/E_2
(iii) N_2/N_1
(iv) none of the above

19. A transformer takes a current of 0·6 A and absorbs 64 W when the primary is connected to its normal supply of 200 V, 50 Hz ; the secondary being on open circuit. The iron-loss current is ....
(i) 0·2 A (ii) 0·43 A
(iii) 1 A (iv) 0·32 A

20. In the above question, what is the magnetising current ?
(i) 0·15 A (ii) 0·507 A
(iii) 0·215 A (iv) 0·312 A

21. A 230/2300 V transformer takes no-load current of 5 A at 0·25 power factor lagging. The core loss is
(i) 300·2 W (ii) 192·5 W
(iii) 287·5 W (iv) 212·6 W

22. A 230/2300 V transformer takes no-load current of 6·5 A and absorbs 187 W. If the resistance of the primary is 0·06 Ω, what is the core loss ?
(i) 122·5 W (ii) 184·5 W
(iii) 206·4 W (iv) 191·3 W

23. In the above question, what is the no-load power factor ?
(i) 0·25 lag (ii) 0·35 lag
(iii) 0·125 lag (iv) 0·252 lag

24. Fig. 8.87 shows the equivalent circuit of a transformer on no-load. The element R_C accounts for ....
(i) core loss in the transformer
(ii) copper loss in the transformer
(iii) magnetic leakage in the transformer
(iv) none of the above

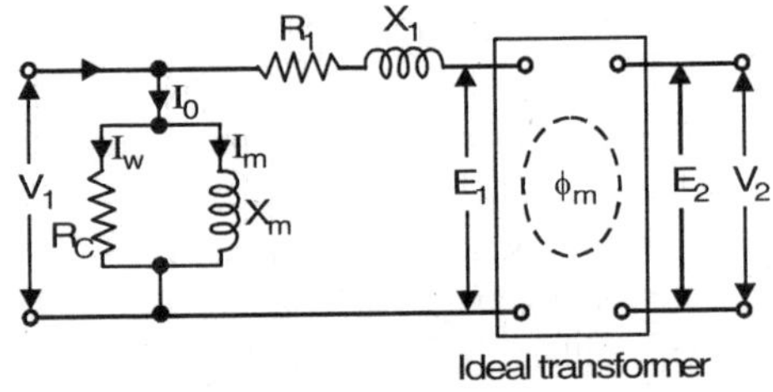

Fig. 8.87

25. In Fig. 8.87, the element X_m accounts for .... in the transformer.
(i) core loss (ii) copper loss
(iii) permeability of core
(iv) leakage reactance

26. In Fig. 8.87, V_1 = 200 V and wattmeter connected in the primary reads 400 W. The value of R_C is ....
(i) 200 Ω (ii) 50 Ω

(*iii*) 100Ω
(*iv*) none of the above

27. A transformer has full-load copper loss of 400 W. The copper loss at half full-load will be ....

(*i*) 100 W (*ii*) 200 W
(*iii*) 400 W
(*iv*) none of the above

28. A transformer has 200 W as iron loss at full-load. The iron loss at half full-load will be ....

(*i*) 50 W (*ii*) 100 W
(*iii*) 400 W (*iv*) 200 W

29. A transformer will have zero efficiency at ....

(*i*) full-load (*ii*) no-load
(*iii*) half full-load (*iv*) none of above

30. The approximate efficiency of a large transformer is ....

(*i*) 65 % (*ii*) 50 %
(*iii*) 80 % (*iv*) 95 %

ANSWERS TO MULTIPLE-CHOICE QUESTIONS

1. (*i*)	**2.** (*ii*)	**3.** (*iii*)	**4.** (*i*)	**5.** (*i*)
6. (*iv*)	**7.** (*ii*)	**8.** (*i*)	**9.** (*iii*)	**10.** (*i*)
11. (*iii*)	**12.** (*i*)	**13.** (*ii*)	**14.** (*iv*)	**15.** (*iii*)
16. (*iii*)	**17.** (*i*)	**18.** (*iii*)	**19.** (*iv*)	**20.** (*ii*)
21. (*iii*)	**22.** (*ii*)	**23.** (*iii*)	**24.** (*i*)	**25.** (*iii*)
26. (*iii*)	**27.** (*i*)	**28.** (*iv*)	**29.** (*ii*)	**30.** (*iv*)

HINTS TO SELECTED MULTIPLE-CHOICE QUESTIONS

1. Transformer action demands only the existence of alternating flux linking the two windings. Since alternating flux can be produced by a.c., a transformer will work on a.c. only.

4. Since the same mutual flux links both the windings, the voltage per turn of the primary is the same as the voltage per turn of the secondary.

7. Iron core ensures tight magnetic coupling between the two windings. The co-efficient of coupling between the windings of a practical transformer may be as high as 0·99.

10. An ideal transformer is one that has (*i*) no winding resistance (*ii*) no leakage flux and (*iii*) no core losses. An actual transformer has some losses, winding resistance and leakage reactance. It may be noted that losses in an actual transformer are quite small. Moreover, winding resistances and leakage reactances are also small. Consequently, the assumption of ideal transformer is satisfactory for most considerations.

11. $E = 4{\cdot}44 fN\phi_m$ or $E/N = 4{\cdot}44 f\phi_m$ or $30 = 4{\cdot}44 \times 50 \times \phi_m$

$\therefore \phi_m = 0{\cdot}135$ Wb ; Area of x-section $= \dfrac{\phi_m}{B_m} = \dfrac{0{\cdot}135}{1{\cdot}2} = 0{\cdot}1125 \text{ m}^2 = 1125 \text{ cm}^2$

12. Voltage per turn $= 4{\cdot}44 f\phi_m = 4{\cdot}44 fB_m a$

Since voltage per turn and frequency are the same, $B_m a$ = constant.

$\therefore a \propto \dfrac{1}{B_m}$; $a_1 \propto \dfrac{1}{0{\cdot}8} = 1{\cdot}25$ and $a_2 \propto \dfrac{1}{1{\cdot}2} = 0{\cdot}83$

$\therefore$ % reduction in core volume = % reduction in core area

$$= \frac{1{\cdot}25 - 0{\cdot}83}{1{\cdot}25} \times 100 = 33\ \%$$

13. $E_1 = 4{\cdot}44 fN_1\phi_m$ or $500 = 4{\cdot}44 \times 50 \times 400 \times \phi_m$ $\therefore \phi_m = 5{\cdot}63 \times 10^{-3}$ Wb

Maximum flux density, $B_m = \dfrac{\phi_m}{\text{core area}} = \dfrac{5{\cdot}63 \times 10^{-3}}{60 \times 10^{-4}} = 0{\cdot}94 \text{ Wb/m}^2$

14. Voltage induced in secondary, $E_2 = KE_1 = \dfrac{N_2}{N_1} \times E_1 = \dfrac{1000}{400} \times 500 = 1250$ V

17. The no-load input power ($V_1 I_0 \cos \phi_0$) supplies
   (*i*) iron losses of the transformer
   (*ii*) small quantity of primary Cu loss

   Since the no-load current I_0 is very small, primary Cu loss is negligible. Consequently, the no-load primary input power is practically equal to the iron loss in the transformer.

18. In a transformer, induced voltage per turn is the same for both primary and secondary.

$$\therefore \quad \frac{E_1}{N_1} = \frac{E_2}{N_2} \quad \text{or} \quad \frac{E_2}{E_1} = \frac{N_2}{N_1} = K$$

   The constant K is known as voltage transformation ratio.

   Note : In a transformer, $V_1 \simeq E_1$ and $V_2 \simeq E_2$ so that $\frac{E_2}{E_1} = \frac{V_2}{V_1} = \frac{N_2}{N_1} = K.$

19. No-load primary power, $W_0 = V_1 I_0 \cos \phi_0$

$$\text{Iron loss component, } I_w = I_0 \cos \phi_0 = \frac{W_0}{V_1} = \frac{64}{200} = 0{\cdot}32 \text{ A}$$

20. $$\text{Magnetising current, } I_m = \sqrt{I_0^2 - I_w^2} = \sqrt{(0\cdot6)^2 - (0\cdot32)^2} = 0{\cdot}507 \text{ A}$$

21. $$\text{Core loss, } W_0 = V_1 I_0 \cos \phi_0 = 230 \times 5 \times 0{\cdot}25 = 287{\cdot}5 \text{ W}$$

22. $$\text{No-load loss, } W_0 = 187 \text{ W}\ ;\ \text{Primary Cu loss} = I_0^2 R_1 = (6{\cdot}5)^2 \times 0{\cdot}06 = 2{\cdot}5 \text{ W}$$

$$\text{Iron loss} = W_0 - \text{Primary Cu loss} = 187 - 2{\cdot}5 = 184{\cdot}5 \text{ W}$$

23. $$\text{No-load p.f., } \cos \phi_0 = \frac{W_0}{V_1 I_0} = \frac{187}{230 \times 6\cdot5} = 0{\cdot}125 \text{ lag}$$

24. An ideal transformer has an infinite permeable core *i.e.*, the core has no eddy current and hysteresis loss. However, in an actual transformer, these losses do occur. We can represent these imperfections by two circuit elements R_c and X_m in parallel with the primary terminals of the ideal transformer.

   (*i*) The resistance R_c represents the iron losses and the heat they produce. The value of R_c is given by :

$$P_m = \frac{V_1^2}{R_c} \quad \text{or} \quad I_w^2 R_c$$

   where P_m = Iron losses

   (*ii*) The reactance X_m represents the permeability of the core. The value of X_m is given by ;

$$Q_m = \frac{V_1^2}{X_m}$$

   where Q_m = reactive power needed to set up mutual flux (ϕ_m)

26. $$R_c = \frac{V_1^2}{P_m} = \frac{(200)^2}{400} = 100\ \Omega$$

27. Since copper loss varies as the square of current, copper loss at half full-load will be

$$= (1/2)^2 \times \text{Full-load Cu loss} = (1/4) \times 400 = 100 \text{ W}$$

29. Since there is no output at no-load, the efficiency of transformer will be zero at no-load.

9

Three Phase Induction Motors

INTRODUCTION

The three-phase induction motors are the most widely used electric motors in industry. They run at essentially constant speed from no-load to full-load. However, the speed is frequency dependent and consequently these motors are not easily adapted to speed control. We usually prefer d.c. motors when large speed variations are required. Nevertheless, the 3-phase induction motors are simple, rugged, low-priced, easy to maintain and can be manufactured with characteristics to suit most industrial requirements. In this chapter, we shall focus our attention on the general principles of 3-phase induction motors.

9.1 THREE-PHASE INDUCTION MOTOR

Like any electric motor, a 3-phase induction motor has a stator and a rotor. The stator carries a 3-phase winding (called stator winding) while the rotor carries a short-circuited winding (called rotor winding). *Only the stator winding is fed from 3-phase supply. The rotor winding derives its voltage and power from the externally energised stator winding through electromagnetic induction and hence the name. The induction motor may be considered to be a transformer with a rotating secondary and it can, therefore, be described as a "transformer-type" a.c. machine in which electrical energy is converted into mechanical energy.

Advantages

(*i*) It has simple and rugged construction.

(*ii*) It is relatively cheap.

(*iii*) It requires little maintenance.

(*iv*) It has high efficiency and reasonably good power factor.

(*v*) It has self-starting torque.

Disadvantages

(*i*) It is essentially a constant speed motor and its speed cannot be changed easily.

(*ii*) Its starting torque is inferior to d.c. shunt motor.

9.2 CONSTRUCTION

A 3-phase induction motor has two main parts (*i*) stator and (*ii*) rotor. The rotor is separated from the stator by a small air-gap which ranges from 0·4 mm to 4 mm, depending on the power of the motor.

1. Stator. It consists of a steel frame which encloses a hollow, cylindrical core made up of thin laminations of silicon steel to reduce hysteresis and eddy current losses. A number of evenly spaced slots are provided on the inner periphery of the laminations [See Fig. 9.1]. The insulated conductors are placed in the stator slots and are suitably connected to form a

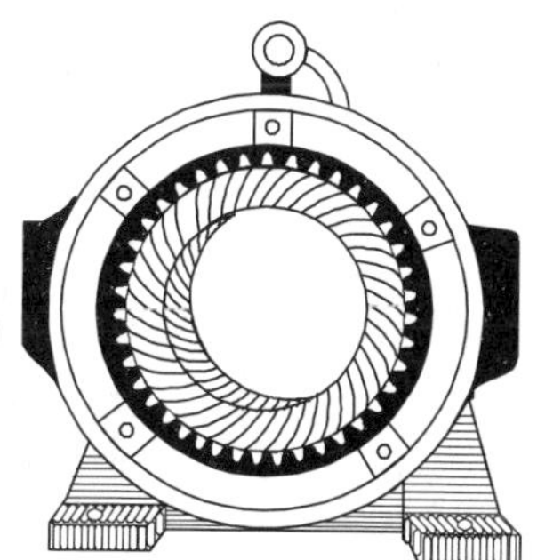

Fig. 9.1

* Unlike a d.c. motor where both stator winding (*i.e.*, field winding) and the rotor winding (*i.e.*, armature winding) are connected to a voltage source.

balanced 3-phase star or delta connected circuit. The 3-phase stator winding is wound for a definite number of poles as per requirement of speed. Greater the number of poles, lesser is the speed of the motor and *vice-versa.* When 3-phase supply is given to the stator winding, a rotating magnetic field (See Art. 9.3) of constant magnitude is produced. This rotating field induces currents in the rotor by electromagnetic induction.

2. Rotor. The rotor, mounted on a shaft, is a hollow laminated core having slots on its outer periphery. The winding placed in these slots (called rotor winding) may be one of the following two types :

(*i*) Squirrel cage type (*ii*) Wound type

(*i*) **Squirrel cage rotor.** It consists of a laminated cylindrical core having parallel slots on its outer periphery. One copper or aluminium bar is placed in each slot. All these bars are joined at each end by metal rings called end rings (See Fig. 9.2). This forms a permanently short-circuited winding which is indestructible. The entire construction (bars and end rings) resembles a squirrel cage and hence the name. The rotor is not connected electrically to the supply but has current induced in it by transformer action from the stator.

Those induction motors which employ squirrel cage rotor are called *squirrel cage induction motors*. Most of 3-phase induction motors use squirrel cage rotor as it has a remarkably simple and robust construction enabling it to operate in the most adverse circumstances. However, it suffers from the disadvantage of a low starting torque. It is because the rotor bars are permanently short-circuited and it is not possible to add any external resistance to the rotor circuit to have a large starting torque.

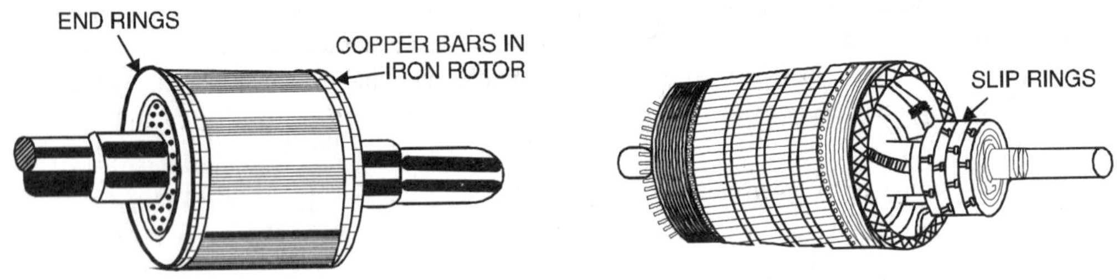

Fig. 9.2 **Fig. 9.3**

(*ii*) **Wound rotor.** It consists of a laminated cylindrical core and carries a 3-phase winding, similar to the one on the stator (See Fig. 9.3). The rotor winding is uniformly distributed in the slots and is usually star-connected. The open ends of the rotor winding are brought out and joined to three insulated slip rings mounted on the rotor shaft with one brush resting on each slip ring. The three brushes are connected to a 3-phase star-connected rheostat as shown in Fig. 9.4. At starting, the external resistances are included in the rotor circuit to give a large starting torque. These resistances are gradually reduced to zero as the motor runs up to speed.

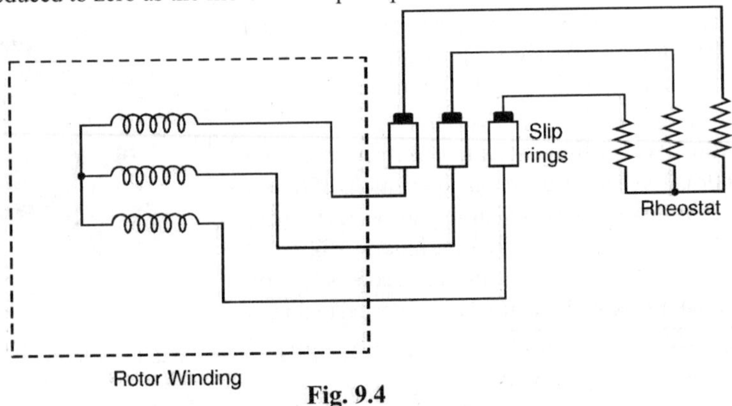

Fig. 9.4

The external resistances are used during starting period only. When the motor attains normal speed, the three brushes are short-circuited so that the wound rotor runs like a squirrel cage rotor.

9.3 ROTATING MAGNETIC FIELD DUE TO 3-PHASE CURRENTS

When a 3-phase winding is energised from a 3-phase supply, a rotating magnetic field is produced. This field is such that its poles do no remain in a fixed position on the stator but go on shifting their positions around the stator. For this reason, it is called a rotating field. It can be shown that magnitude of this rotating field is constant and is equal to $1{\cdot}5\ \phi_m$ where ϕ_m is the maximum flux due to any phase.

To see how rotating field is produced, consider a 2-pole, 3-phase *winding as shown in Fig. 9.6 (*i*). The three phases X, Y and Z are energised from a 3-phase source and currents in these phases are indicated as I_x, I_y and I_z [See Fig. 9.6 (*ii*)]. Referring to Fig. 9.6 (*ii*), the **fluxes produced by these currents are given by ;

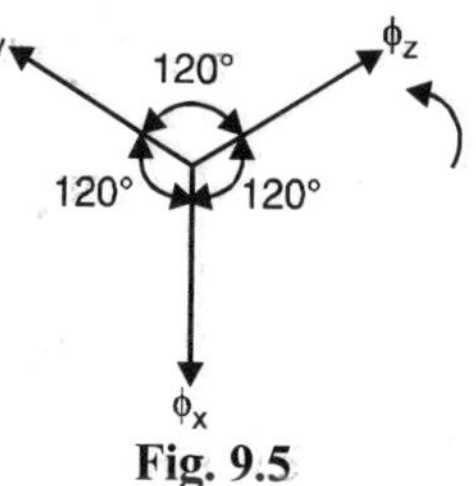

Fig. 9.5

$$\phi_x = \phi_m \sin \omega t$$

$$\phi_y = \phi_m \sin (\omega t - 120^\circ)$$

$$\phi_z = \phi_m \sin (\omega t - 240^\circ)$$

Here ϕ_m is the maximum flux due to any phase. Fig. 9.5 shows the phasor diagram of the three fluxes. We shall now prove that this 3-phase supply produces a rotating field of constant magnitude equal to $1{\cdot}5\ \phi_m$.

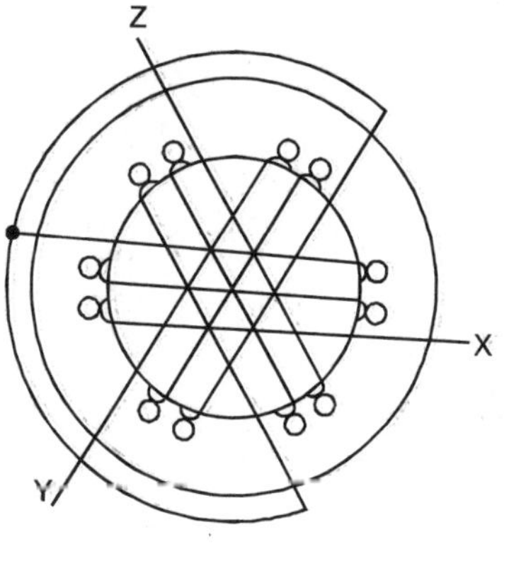

(*i*) 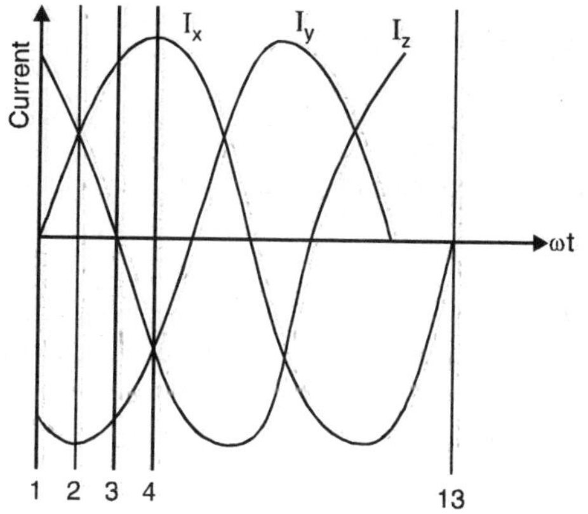 (*ii*)

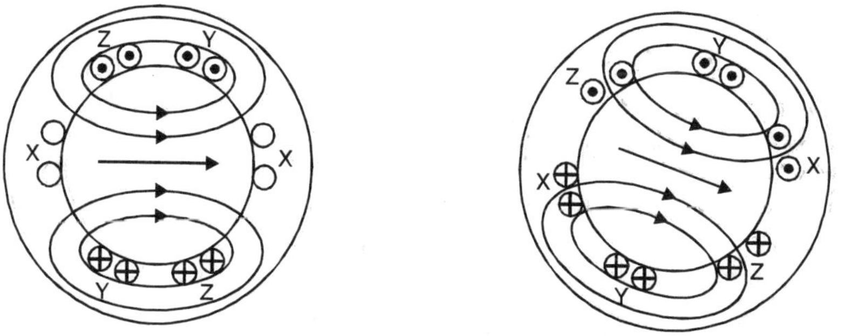

(1) (*iii*) (2)

---

* The three windings are 120° electrical apart.

** The flux produced by a current is in phase with the current that produces it.

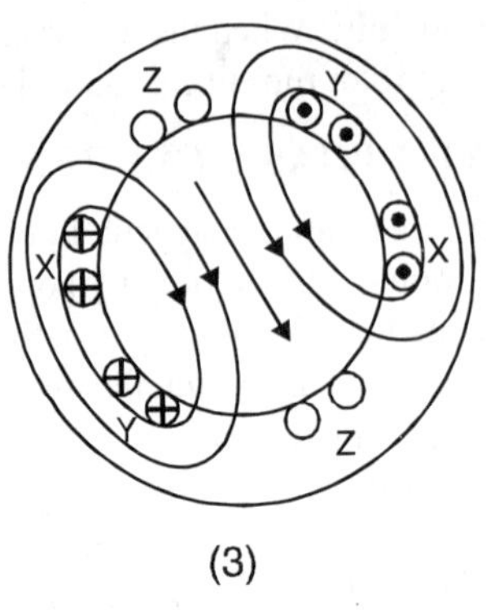

(3)

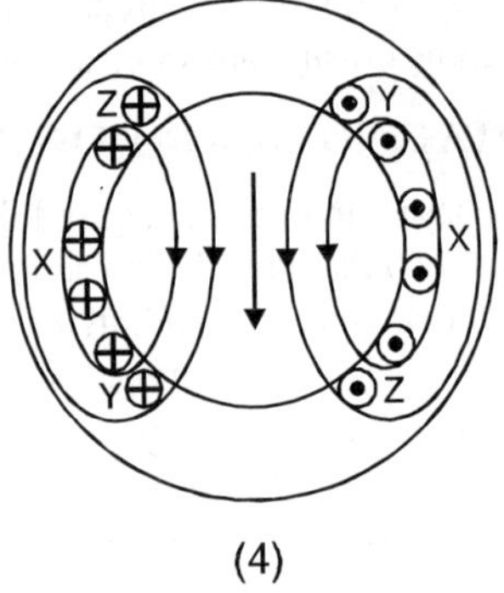

(4)

(iii)

Fig. 9.6

(i) At instant 1 [See Fig. 9.6 *(ii)* and Fig. 9.6 *(iii)*], the current in phase *X* is zero and currents in phases *Y* and *Z* are equal and opposite. The currents are flowing outward in the top conductors and inward in the bottom conductors. This establishes a resultant flux towards right. The magnitude of the resultant flux is constant and is equal to $1{\cdot}5\ \phi_m$ as proved under :

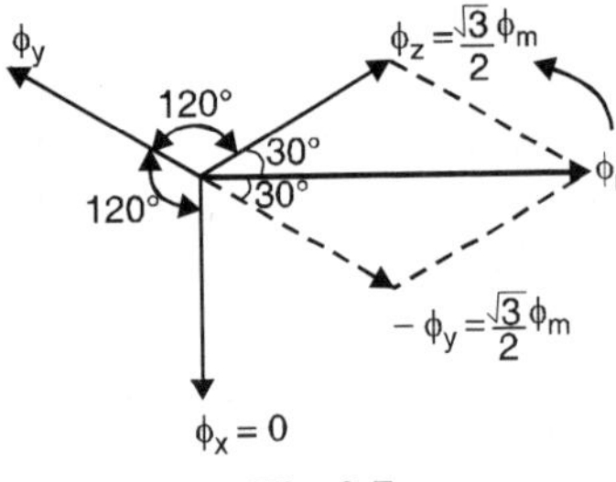

Fig. 9.7

At instant 1, $\omega t = 0°$. Therefore, the three fluxes are given by;

$$\phi_x = 0\ ;\ \phi_y = \phi_m \sin(-120°) = -\frac{\sqrt{3}}{2}\phi_m\ ;$$

$$\phi_z = \phi_m \sin(-240°) = \frac{\sqrt{3}}{2}\phi_m$$

The phasor sum of $-\phi_y$ and ϕ_z is the resultant flux ϕ_r (See Fig. 9.7). It is clear that :

$$\text{Resultant flux, } \phi_r = 2 \times \frac{\sqrt{3}}{2}\phi_m \cos\frac{60°}{2} = 2 \times \frac{\sqrt{3}}{2}\phi_m \times \frac{\sqrt{3}}{2} = 1{\cdot}5\ \phi_m$$

(ii) At instant 2, the current is maximum (negative) in phase *Y* and 0·5 maximum (positive) in phases *X* and *Z*. The magnitude of resultant flux is $1{\cdot}5\ \phi_m$ as proved under :

At instant 2, $\omega t = 30°$. Therefore, the three fluxes are given by ;

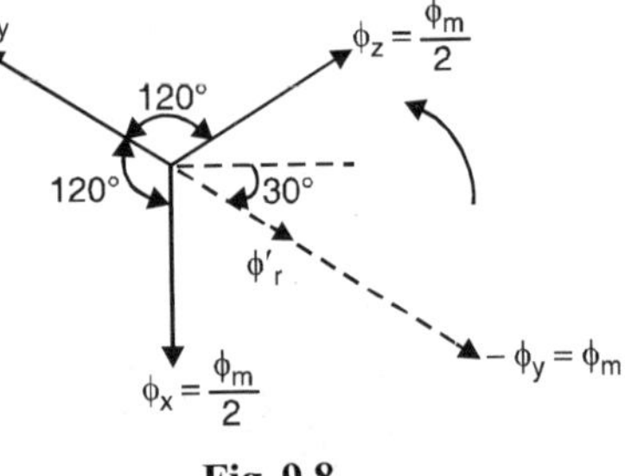

Fig. 9.8

$$\phi_x = \phi_m \sin 30° = \frac{\phi_m}{2}$$

$$\phi_y = \phi_m \sin(-90°) = -\phi_m$$

$$\phi_z = \phi_m \sin(-210°) = \frac{\phi_m}{2}$$

The phasor sum of ϕ_x, $-\phi_y$ and ϕ_z is the resultant flux ϕ_r.

Phasor sum of ϕ_x and ϕ_z, $\phi'_r = 2 \times \frac{\phi_m}{2}\cos\frac{120°}{2} = \frac{\phi_m}{2}$ ...along $-\phi_y$

Phasor sum of ϕ'_r and $-\phi_y$, $\phi_r = \frac{\phi_m}{2} + \phi_m = 1{\cdot}5\ \phi_m$ ...along $-\phi_y$

Note that resultant flux is displaced 30° clockwise from position 1.

(iii) At instant 3, current in phase Z is zero and the currents in phases *X* and *Y* are equal and opposite (currents in phases *X* and *Y* are 0·866 × max. value). The magnitude of resultant flux is $1{\cdot}5\ \phi_m$ as proved under :

At instant 3, $\omega t = 60°$. Therefore, the three fluxes are given by ;

$$\phi_x = \phi_m \sin 60^\circ = \frac{\sqrt{3}}{2}\,\phi_m$$

$$\phi_y = \phi_m \sin(-60^\circ) = -\frac{\sqrt{3}}{2}\,\phi_m$$

$$\phi_z = \phi_m \sin(-180^\circ) = 0$$

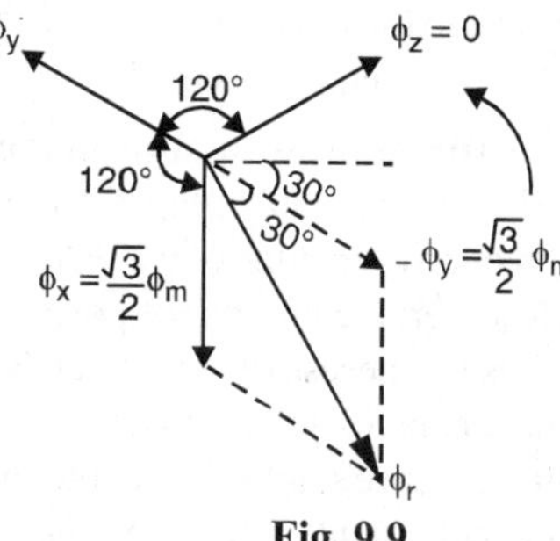

Fig. 9.9

The resultant flux ϕ_r is the phasor sum of ϕ_x and $-\phi_y$ ($\because \phi_z = 0$).

$$\therefore \qquad \phi_r = 2 \times \frac{\sqrt{3}}{2}\,\phi_m \cos\frac{60^\circ}{2} = 1{\cdot}5\,\phi_m$$

Note that resultant flux is displaced 60° clockwise from position 1.

(*iv*) At instant 4, the current in phase *X* is maximum (positive) and the currents in phases *Y* and *Z* are equal and negative (currents in phases *Y* and *Z* are 0·5 × max. value). This establishes a resultant flux downward as shown under :

At instant 4, ωt = 90°. Therefore, the three fluxes are given by;

$$\phi_x = \phi_m \sin 90^\circ = \phi_m$$

$$\phi_y = \phi_m \sin(-30^\circ) = -\frac{\phi_m}{2}$$

$$\phi_z = \phi_m \sin(-150^\circ) = -\frac{\phi_m}{2}$$

Fig. 9.10

The phasor sum of ϕ_x, $-\phi_y$ and $-\phi_z$ is the resultant flux ϕ_r.

Phasor sum of $-\phi_z$ and $-\phi_y$, $\quad \phi'_r = 2 \times \frac{\phi_m}{2} \cos\frac{120^\circ}{2} = \frac{\phi_m}{2}$ ...along $+\phi_x$

Phasor sum of ϕ'_r and ϕ_x, $\quad \phi_r = \frac{\phi_m}{2} + \phi_m = 1{\cdot}5\,\phi_m$ ...along $+\phi_x$

Note that the resultant flux is downward *i.e.*, it is displaced 90° clockwise from position 1.

It follows from the above discussion that a 3-phase supply produces a rotating field of constant value (= 1·5 ϕ_m where ϕ_m is the maximum flux due to any phase).

Speed of rotating magnetic field. The speed at which the rotating magnetic field revolves is called the *synchronous speed* (N_s). Referring to Fig. 9·6 (*ii*), the time instant 4 represents the completion of one-quarter cycle of alternating current I_x from the time instant 1. During this one quarter cycle, the field has rotated through 90°. At a time instant represented by 13* or one complete cycle of current I_x from the origin, the field has completed one revolution. Therefore, for a 2-pole stator winding, the field makes one revolution in one cycle of current. In a 4-pole stator winding, it can be shown that the rotating field makes one revolution in two cycles of current. In general, for *P* poles, the rotating field makes one revolution in *P*/2 cycles of current.

$$\therefore \qquad \text{Cycles of current} = \frac{P}{2} \times \text{revolutions of field}$$

$$\text{or} \qquad \text{Cycles of current per second} = \frac{P}{2} \times \text{revolutions of field per second}$$

Since revolutions per second is equal to the revolutions per minute (N_s) divided by 60 and the number of cycles per second is the frequency *f*,

$$\therefore \qquad f = \frac{P}{2} \times \frac{N_s}{60} = \frac{N_s P}{120}$$

$$\text{or} \qquad N_s = \frac{120\,f}{P}$$

The speed of the rotating magnetic field is the same as the speed of the alternator that is supplying

* At this instant, I_y and I_z also complete one cycle from the origin.

power to the motor if the two have the same number of poles. Hence the magnetic flux is said to rotate at synchronous speed.

Direction of rotating magnetic field. The phase sequence of the three-phase voltage applied to the stator winding in Fig. 9.6 (*ii*) is *X* - *Y* - *Z*. If this sequence is changed to *X* - *Z* - *Y*, it is observed that direction of rotation of the field is reversed *i.e.*, the field rotates counterclockwise rather than clockwise. However, the number of poles and the speed at which the magnetic field rotates remain unchanged. Thus it is necessary only to change the phase sequence in order to change the direction of rotation of the magnetic field. For a three-phase supply, this can be done by interchanging any two of the three lines. As we shall see, the rotor in a 3-phase induction motor runs in the same direction as the rotating magnetic field. Therefore, *the direction of rotation of a 3-phase induction motor can be reversed by interchanging any two of the three motor supply lines.*

9.4 ALTERNATE ANALYSIS FOR ROTATING MAGNETIC FIELD

We shall now use another useful method to find the *magnitude* and *speed* of the resultant flux due to three-phase currents. The three-phase sinusoidal currents produce fluxes ϕ_1, ϕ_2 and ϕ_3 which vary sinusoidally. The resultant flux at any instant will be the vector sum of all the three at that instant. The fluxes are represented by three variable magnitude vectors (See Fig. 9.11). In Fig. 9.11, the individual flux directions are fixed but their magnitudes vary sinusoidally as does the current that produces them. To find the magnitude of the resultant flux, resolve each flux into horizontal and vertical components and then find their vector sum.

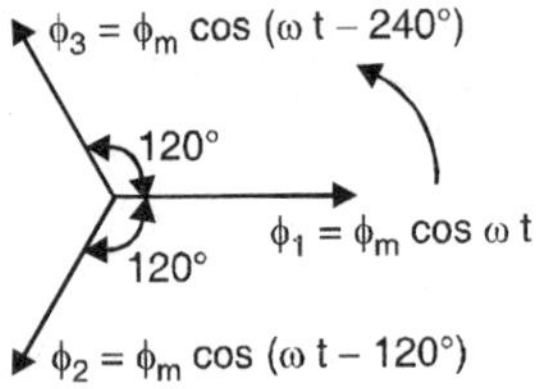

Fig. 9.11

$$\phi_h = {}^*\phi_m \text{ cost } \omega t - \phi_m \cos(\omega t - 120°) \cos 60° - \phi_m \cos(\omega t - 240°) \cos 60° = \frac{3}{2}\phi_m \text{ cost } \omega t$$

$$\phi_v = 0 - \phi_m \cos(\omega t - 120°) \sin 60° + \phi_m \cos(\omega t - 240°) \sin 60° = -\frac{3}{2}\phi_m \sin \omega t$$

The resultant flux is given by (See Fig. 9.12) ;

$$\phi_r = \sqrt{\phi_h^2 + \phi_v^2} = \frac{3}{2}\phi_m [\cos^2 \omega t + (-\sin \omega t)^2]^{1/2} = \frac{3}{2}\phi_m = 1{\cdot}5\,\phi_m = \text{Constant}$$

Thus the resultant flux has constant magnitude $(= 1{\cdot}5\,\phi_m)$ and does not change with time.

The angular displacement of ϕ_r relative to the OX axis is

$$\tan\theta = \frac{\phi_v}{\phi_h} = \frac{\frac{3}{2}\phi_m \sin \omega t}{\frac{3}{2}\phi_m \cos \omega t} = \tan \omega t$$

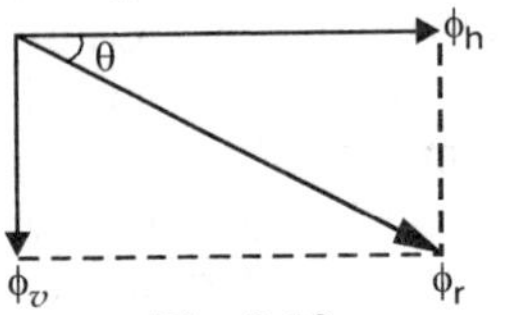

$$\therefore \qquad \theta = \omega t$$

Fig. 9.12

Thus the resultant magnetic field rotates at constant angular velocity ω $(= 2\pi f)$ rad/sec. For a *P*-pole machine, the **rotation speed (ω_m) is

$$\omega_m = \frac{2}{P}\,\omega \text{ rad/sec}$$

$$\text{or} \qquad \frac{2\pi N_s}{60} = \frac{2}{P} \times 2\pi f \qquad \ldots N_s \text{ is in r.p.m.}$$

$$\therefore \qquad N_s = \frac{120 f}{P}$$

Thus the resultant flux due to three-phase currents is of constant value (= 1·5 ϕ_m where ϕ_m is the

* ϕ_m is the maximum flux due to any phase. Note that treatment is equally valid for $\phi = \phi_m \sin \omega t$.

** We want to find the speed of rotating flux in r.p.m.

maximum flux in any phase) and this flux rotates around the stator winding at a synchronous speed of 120 f/P r.p.m.

For example, for a 6-pole, 50 Hz, 3-phase induction motor, $N_s = 120 \times 50/6 = 1000$ r.p.m. It means that flux rotates around the stator at a speed of 1000 r.p.m.

9.5 PRINCIPLE OF OPERATION

Consider a portion of 3-phase induction motor as shown in Fig. 9.13. The operation of the motor can be explained as under :

(*i*) When 3-phase stator winding is energised from a 3-phase supply, a rotating magnetic field is set up which rotates around the stator at synchronous speed N_s (= $120 f/P$).

(*ii*) The rotating field passes through the air gap and cuts the rotor conductors, which as yet, are stationary. Due to the relative speed between the rotating flux and the stationary rotor, e.m.f.s are induced in the rotor conductors. Since the rotor circuit is short-circuited, currents start flowing in the rotor conductors.

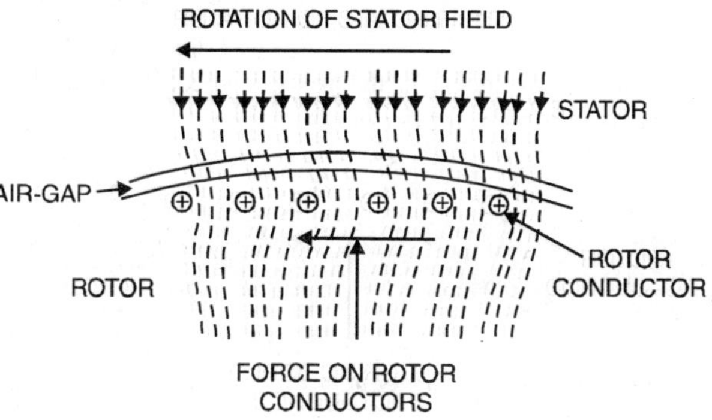

Fig. 9.13

(*iii*) The current-carrying rotor conductors are placed in the magnetic field produced by the stator. Consequently, mechanical force acts on the rotor conductors. The sum of the mechanical forces on all the rotor conductors produces a torque which tends to move the rotor in the *same direction* as the rotating field.

(*iv*) The fact that rotor is urged to follow the stator field (*i.e.*, rotor moves in the direction of stator field) can be explained by Lenz's law. According to this law, the direction of rotor currents will be such that they tend to oppose the cause producing them. Now, the cause producing the rotor currents is the relative speed between the rotating field and the stationary rotor conductors. Hence to reduce this relative speed, the rotor starts running in the same direction as that of stator field and tries to catch it.

9.6 SLIP

We have seen above that rotor rapidly accelerates in the direction of rotating field. In practice, the rotor can never reach the speed of stator flux. If it did, there would be no relative speed between the stator field and rotor conductors, no induced rotor currents and, therefore, no torque to drive the rotor. The friction and windage would immediately cause the rotor to slow down. Hence, the rotor speed (N) is always less than the stator field speed (N_s). This difference in speed depends upon load on the motor.

The difference between the synchronous speed N_s of the rotating stator field and the actual rotor speed N is called ***slip**. It is usually expressed as a percentage of synchronous speed *i.e.*,

$$\% \text{ age slip, } s = \frac{N_s - N}{N_s} \times 100$$

* The term slip is used because it describes what an observer riding on the stator field sees looking at the rotor – it appears to be slipping backward.

(*i*) The quantity $N_s - N$ is sometimes called slip speed.

(*ii*) When the rotor is stationary (*i.e.*, $N = 0$), slip, $s = 1$ or 100 %.

(*iii*) In an induction motor, the change in slip from no-load to full-load is hardly 0·1 % to 3 % so that it is essentially a constant-speed motor.

9.7 ROTOR CURRENT FREQUENCY

The frequency of a voltage or current induced due to the relative speed between a winding and a magnetic field is given by the general formula ;

$$\text{Frequency} = \frac{N\,P}{120}$$

where N = Relative speed between magnetic field and the winding

P = Number of poles

For a rotor speed N, the relative speed between the rotating flux and the rotor is $N_s - N$. Consequently, the rotor current frequency f' is given by ;

$$f' = \frac{(N_s - N)\,P}{120}$$

$$= \frac{s\,N_s\,P}{120} \qquad \left(\because s = \frac{N_s - N}{N_s}\right)$$

$$= s f \qquad (\because f = N_s P/120)$$

i.e., *Rotor current frequency* = *Fractional slip* × *Supply frequency*

(*i*) When the rotor is at standstill or stationary (*i.e.*, $s = 1$), the frequency of rotor current is the same as that of supply frequency ($f' = s f = 1 \times f = f$).

(*ii*) As the rotor picks up speed, the relative speed between the rotating flux and the rotor decreases. Consequently, the slip s and hence rotor current frequency decreases.

Note. The relative speed between the rotating field and stator winding is $N_s - 0 = N_s$. Therefore, the frequency of induced current or voltage in the stator winding is $f = N_s\, P/120$ — the supply frequency.

Example 9.1. *A 6-pole, 3-phase induction motor is connected to 50 Hz supply. If it is running at 970 r.p.m., find the slip.*

Solution.

$$\text{Synchronous speed, } N_s = 120 f/P = 120 \times 50/6 = 1000 \text{ r.p.m.}$$

$$\text{Slip, } s = \frac{N_s - N}{N_s} \times 100 = \frac{1000 - 970}{1000} \times 100 = \mathbf{3\ \%}\ \textit{or}\ \mathbf{0{\cdot}03}$$

Example 9.2. *A 3-phase induction motor is wound for 4 poles and is supplied from 50 Hz system. Calculate (i) the synchronous speed (ii) the speed of the motor when slip is 4 % and (iii) the rotor current frequency when the motor runs at 600 r.p.m.*

Solution.

(*i*) $$\text{Synchronous speed, } N_s = 120 f/P = 120 \times 50/4 = \mathbf{1500\ r.p.m.}$$

(*ii*) $$\text{Slip, } s = \frac{N_s - N}{N_s} \times 100$$

or $$4 = \frac{1500 - N}{1500} \times 100$$

$\therefore$ $$N = \mathbf{1440\ r.p.m.}$$

(*iii*) $$\text{When } N = 600 \text{ r.p.m., } s = \frac{1500 - 600}{1500} = 0{\cdot}6$$

$\therefore$ Rotor current frequency, $f' = s f = 0{\cdot}6 \times 50 =$ **30 Hz**

Example 9.3. *A 6-pole alternator running at 1000 r.p.m. supplies an 8-pole induction motor. Find the actual speed of the motor if the slip is 2·5 %.*

Solution. The frequency of 3-phase supply fed to the induction motor is determined from the speed of the alternator and its number of poles.

$$\text{Supply frequency, } f = NP/120 = 1000 \times 6/120 = 50 \text{ Hz}$$

$$\text{Synchronous speed, } N_s = 120f/P = 120 \times 50/8 = 750 \text{ r.p.m.}$$

$$\text{Slip, } s = \frac{N_s - N}{N_s} \times 100$$

or
$$2{\cdot}5 = \frac{750 - N}{750} \times 100$$

∴
$$N = \mathbf{731{\cdot}25\ r.p.m.}$$

Example 9.4. *A 50 Hz, 4-pole, 3-phase induction motor has a rotor current of frequency 2 Hz. Determine (i) the slip and (ii) speed of the motor.*

Solution.

(*i*)
$$f' = sf$$

or
$$s = f'/f = 2/50 = \mathbf{0{\cdot}04} \text{ } or \text{ } \mathbf{4\%}$$

(*ii*)
$$N_s = 120f/P = 120 \times 50/4 = 1500 \text{ r.p.m.}$$

Now
$$s = \frac{N_s - N}{N_s}$$

or
$$0{\cdot}04 = \frac{1500 - N}{1500}$$

∴
$$N = \mathbf{1440\ r.p.m.}$$

Example 9.5. *A 500 h.p., 3-phase, 440 V, 50 Hz induction motor has a speed of 950 r.p.m. on full-load. The machine has 6 poles. Calculate the full-load slip. How many cycles will the rotor voltage make per minute ?*

Solution.

$$\text{Synchronous speed, } N_s = \frac{120f}{P} = \frac{120 \times 50}{6} = 1000 \text{ r.p.m.}$$

$$\text{Full-load speed, } N = 950 \text{ r.p.m.}$$

∴
$$\text{Slip, } s = \frac{N_s - N}{N_s} = \frac{1000 - 950}{1000} = \mathbf{0.05} \text{ or } \mathbf{5\%}$$

$$\text{Frequency of rotor e.m.f., } f' = sf = 0.05 \times 50 = 2.5 \text{ Hz}$$

∴ Complete cycles made by rotor voltage per minute

$$= 2.5 \times 60 = \mathbf{150}$$

Example 9.6. *A slip-ring induction motor runs at 290 r.p.m. at full-load when connected to 50 Hz supply. Determine the number of poles and slip.*

Solution. The full-load speed of an induction motor is very close to synchronous speed. The full-load slip is less than 3%. Therefore, it is reasonable to assume that $N_s = 300$ r.p.m.

Now
$$N_s = \frac{120f}{P} \text{ or } 300 = \frac{120 \times 50}{P} \quad \therefore P = \mathbf{20}$$

$$\text{Slip, } s = \frac{N_s - N}{N_s} = \frac{300 - 290}{300} = \mathbf{0.0333} \text{ or } \mathbf{3.33\%}$$

Example 9.7. *A 10-pole, 3-phase alternator is coupled to an engine running at 600 r.p.m. It supplies 3-phase induction motor which has a full-load speed of 1440 r.p.m. Calculate the number of poles and slip of the motor.*

Solution. Frequency of the voltage generated by the alternator is

$$f = \frac{NP}{120} = \frac{600 \times 10}{120} = 50 \text{ Hz}$$

∴ Synchronous speed of the induction motor is

$$N_s = \frac{120 f}{P} = \frac{120 \times 50}{P}$$

For $P = 2$; $N_s = 3000$ r.p.m. and for $P = 4$; $N_s = 1500$ r.p.m. Since the full-load speed of the motor is close to the synchronous speed, probably $N_s = 1500$ r.p.m. Therefore, $P = \mathbf{4}$.

$$\text{Slip, } s = \frac{N_s - N}{N_s} = \frac{1500 - 1440}{1500} = \mathbf{0.04} \textit{ or } \mathbf{4\%}$$

TUTORIAL PROBLEMS

1. A 2-pole, 3-phase, 50 Hz induction motor is running on no load with a slip of 4 %. Calculate (*i*) the synchronous speed and (*ii*) speed of the motor. **[(*i*) 3000 r.p.m. (*ii*) 2880 r.p.m.]**
2. The frequency of e.m.f. in the stator of a 4-pole, 3-phase induction motor is 50 Hz and that in the rotor is 1·5 Hz. Determine (*i*) the slip and (*ii*) speed of the motor. **[(*i*) 3 % (*ii*) 1455 r.p.m.]**
3. A 3-phase, 50 Hz induction motor has 8 poles. If the full-load slip is 2·5 %, determine (*i*) synchronous speed (*ii*) rotor speed and (*iii*) rotor frequency. **[(*i*) 750 r.p.m. (*ii*) 731 r.p.m. (*iii*) 1·25 Hz]**

9.8 EFFECT OF SLIP ON THE ROTOR CIRCUIT

When the rotor is stationary, $s = 1$. Under these conditions, the per phase rotor e.m.f. E_2 has a frequency equal to that of supply frequency f. At any slip s, the relative speed between stator field and the rotor is decreased. Consequently, the rotor e.m.f. and frequency are reduced proportionally to $s\,E_2$ and $s\,f$ respectively. At the same time, per phase rotor reactance X_2, being frequency dependent, is reduced to $s\,X_2$.

Consider a 6-pole, 3-phase, 50 Hz induction motor. It has synchronous speed $N_s = 120\,f/P = 120 \times 50/6 = 1000$ r.p.m. At standstill, the relative speed between stator flux and rotor is 1000 r.p.m. and rotor e.m.f./phase = E_2 (say). If the full-load speed of the motor is 960 r.p.m., then,

$$s = \frac{1000 - 960}{1000} = 0{\cdot}04$$

(*i*) The relative speed between stator flux and the rotor is now only 40 r.p.m. Consequently, rotor e.m.f./phase is reduced to :

$$*E_2 \times \frac{40}{1000} = 0{\cdot}04\,E_2 \text{ or } s\,E_2$$

(*ii*) The frequency is also reduced in the same ratio to :

$$50 \times \frac{40}{1000} = 50 \times 0{\cdot}04 \quad \text{or} \quad s\,f$$

(*iii*) The per phase rotor reactance X_2 is likewise reduced to :

---

* If the relative speed between stator flux and rotor is 1000 r.p.m., the rotor e.m.f./phase = E_2. If the relative speed is now 40 r.p.m., rotor e.m.f./phase is

$$= \frac{40}{1000} \times E_2 \qquad \text{...unitary method}$$

$$X_2 \times \frac{40}{1000} = 0{\cdot}04 X_2 \text{ or } s X_2$$

Thus at any slip s,

$$\text{Rotor e.m.f./phase} = s E_2$$

$$\text{Rotor reactance/phase} = s X_2$$

$$\text{Rotor frequency} = sf$$

where E_2, X_2 and f are the corresponding values at standstill.

Example 9.8. *A 3-phase, 6-pole induction motor is connected to a 60 Hz supply. The voltage induced in the rotor bars is 4 V when the rotor is at standstill. Calculate the voltage and frequency induced in the rotor bars at 300 r.p.m.*

Solution.

$$\text{Synchronous speed, } N_s = 120 f/P = 120 \times 60/6 = 1200 \text{ r.p.m.}$$

$$\text{Slip, } s = \frac{N_s - N}{N_s} = \frac{1200 - 300}{1200} = 3/4$$

Corresponding to this slip, we have,

$$\text{Induced voltage} = 4 \times s = 4 \times 3/4 = \mathbf{3\ V}$$

$$\text{Frequency} = f \times s = 60 \times 3/4 = \mathbf{45\ Hz}$$

Note. *The subscript 1 (e.g. R_1, X_1, Z_1 etc.) is used for stator values while subscript 2 (e.g. R_2, X_2, Z_2 etc.) is used for rotor values at* **standstill**. *However superscript / (dash) alongwith subscript 2 (e.g. X_2', Z_2' etc.) is used for rotor values under* **running conditions**. Note that f indicates stator frequency and f' (= $s f$) indicates rotor frequency.

9.9 ROTOR CURRENT

Fig. 9.14 shows the circuit of a 3-phase induction motor at any slip s. The rotor is assumed to be of wound type and star connected. Note that rotor e.m.f./phase and rotor reactance/phase are $s E_2$ and

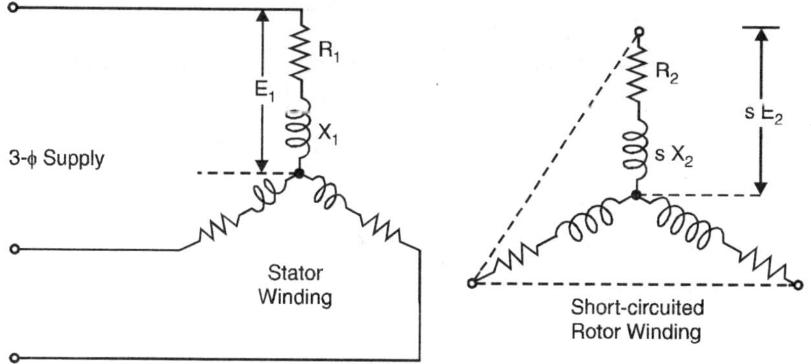

Fig. 9.14

$s X_2$ respectively. The rotor resistance/phase is R_2 and is independent of frequency and, therefore, does not depend upon slip. Likewise, stator winding values R_1 and X_1 do not depend upon slip.

Since the motor represents a balanced 3-phase load, we need consider one phase only ; the conditions in the other two phases being similar.

At standstill. Fig. 9.15 (*i*) shows one phase of the rotor circuit at standstill.

$$\text{Rotor current/phase, } I_2 = \frac{E_2}{Z_2} = \frac{E_2}{\sqrt{R_2^2 + X_2^2}}$$

$$\text{Rotor p.f., } \cos \phi_2 = \frac{R_2}{Z_2} = \frac{R_2}{\sqrt{R_2^2 + X_2^2}}$$

(*i*) (*ii*)

Fig. 9.15

When running at slip *s*. Fig. 9.15 (*ii*) shows one phase of the rotor circuit when the motor is running at slip *s*.

$$\text{Rotor current, } I'_2 = \frac{s\,E_2}{Z'_2} = \frac{s\,E_2}{\sqrt{R_2^2 + (s\,X_2)^2}}$$

$$\text{Rotor p.f., } \cos \phi'_2 = \frac{R_2}{Z'_2} = \frac{R_2}{\sqrt{R_2^2 + (s\,X_2)^2}}$$

Example 9.9. *A 3-phase, 400 V wound rotor motor has delta-connected stator winding and star-connected rotor-winding. The stator has 48 turns/phase while the rotor has 24 turns per phase. Find the standstill or open-circuited voltage across the slip rings.*

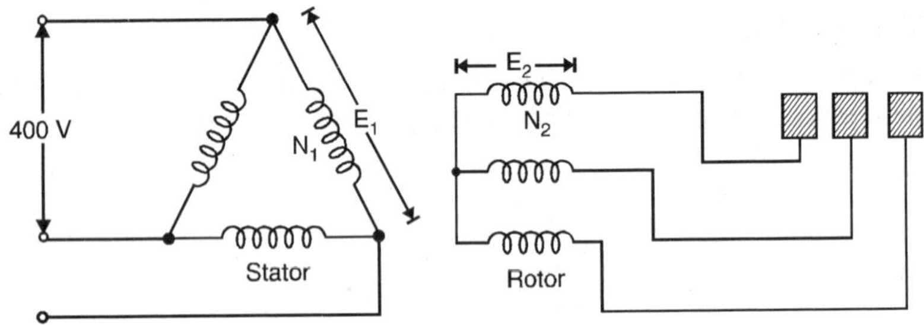

Fig. 9.16

Solution. Fig. 9.16 shows the conditions of the problem.

$$\text{Stator e.m.f./phase, } E_1 = 400 \text{ V}$$

$$\text{Stator turns/phase, } N_1 = 48$$

$$\text{Rotor turns/phase, } N_2 = 24$$

$$\therefore \quad K = N_2/N_1 = 24/48 = 1/2$$

$$\therefore \quad \text{Rotor e.m.f./phase} = K\,E_1 = (1/2) \times 400 = 200 \text{ V}$$

$$\text{Voltage between slip rings} = \text{Rotor line voltage} = \sqrt{3} \times 200 = \mathbf{346\ V}$$

Example 9.10. *A 6-pole, 3-phase, 50 Hz induction motor is running at full-load with a slip of 4%. The rotor is star-connected and its resistance and standstill reactance are 0·25 Ω and 1·5 Ω per phase. The e.m.f. between slip rings is 100 V. Find the rotor current per phase and p.f. assuming the slip rings are short-circuited.*

Solution.

$$\text{Rotor e.m.f./phase at standstill, } E_2 = 100/\sqrt{3} = 57{\cdot}7 \text{ V}$$

$$\text{Rotor e.m.f./phase at full-load} = s\,E_2 = 0{\cdot}04 \times 57{\cdot}7 = 2{\cdot}31 \text{ V}$$
$$\text{Rotor reactance/phase at full-load} = s\,X_2 = 0{\cdot}04 \times 1{\cdot}5 = 0{\cdot}06\ \Omega$$
$$\text{Rotor impedance/phase at full-load} = \sqrt{(0{\cdot}25)^2 + (0{\cdot}06)^2} = 0{\cdot}257\ \Omega$$
$$\text{Full-load rotor current/phase} = 2{\cdot}31/0{\cdot}257 = \mathbf{9\ A}$$
$$\text{Rotor power factor} = 0{\cdot}25/0{\cdot}257 = \mathbf{0{\cdot}97}\ \textit{lag}$$

Example 9.11. *A 4-pole, 3-phase, 50 Hz induction motor has a star-connected rotor. The rotor has a resistance of 0·1 Ω per phase and standstill reactance of 2Ω/phase. The induced e.m.f. between the slip rings is 100 V. If the full-load speed is 1460 r.p.m., calculate (i) the slip (ii) the e.m.f. induced in the rotor in each phase (iii) the rotor reactance per phase (iv) the rotor current and (v) rotor power factor. Assume slip rings are short-circuited.*

Soution.

(*i*)
$$N_s = \frac{120\,f}{P} = \frac{120 \times 50}{4} = 1500 \text{ r.p.m.}$$

$\therefore$
$$s = \frac{N_s - N}{N_s} \times 100 = \frac{1500 - 1460}{1500} \times 100 = \mathbf{2{\cdot}66\ \%}$$

(*ii*)
$$\text{Line e.m.f. of rotor at standstill} = 100 \text{ V}$$
$$\text{Phase e.m.f. of rotor at standstill, } E_2 = 100/\sqrt{3} = 57{\cdot}75 \text{ V}$$
$$\therefore \quad \text{Phase e.m.f. of rotor at full load} = s\,E_2 = 0{\cdot}0266 \times 57{\cdot}75 = \mathbf{1{\cdot}54\ V}$$

(*iii*) Rotor reactance/phase at standstill, $X_2 = 2\ \Omega$

$$\text{Rotor reactance/phase at full load} = s\,X_2 = 0{\cdot}0266 \times 2 = \mathbf{0{\cdot}0532\ \Omega}$$

(*iv*) Rotor impedance per phase at full-load is

$$Z_2' = \sqrt{R_2^2 + (s\,X_2)^2} = \sqrt{(0{\cdot}1)^2 + (0{\cdot}0532)^2} = 0{\cdot}1132\ \Omega$$

$\therefore$
$$\text{Rotor phase current} = \frac{1{\cdot}54}{0{\cdot}1132} = \mathbf{13{\cdot}57\ A}$$

(*v*)
$$\text{Rotor power factor} = \frac{R_2}{Z_2'} = \frac{0{\cdot}1}{0{\cdot}1132} = \mathbf{0{\cdot}883}\ \textit{lagging}$$

Example 9.12. *A 150 kW, 3000 V, 50 Hz, 6-pole star-connected induction motor has a star-connected slip ring rotor with a transformation ratio of 3·6 (stator/rotor). The rotor resistance is 0·1 Ω per phase and its per phase leakage inductance is 3·61 mH. The stator impedance may be neglected. Find the starting current on rated voltage with short-circuited slip rings.*

Solution.

$$\text{Transformation ratio, } K = \frac{\text{Rotor turns/phase}}{\text{Stator turns/phase}} = \frac{1}{3{\cdot}6}$$
$$X_2 = 2\pi f L_2 = 2\,\pi \times 50 \times 3{\cdot}61 \times 10^{-3} = 1{\cdot}13\ \Omega$$
$$\text{Rotor reactance/phase referred to stator} = X_2/K^2 = (3{\cdot}6)^2 \times 1{\cdot}13 = 14{\cdot}7\ \Omega$$
$$\text{Rotor resistance/phase referred to stator} = R_2/K^2 = (3{\cdot}6)^2 \times 0{\cdot}1 = 1{\cdot}3\ \Omega$$
$$\text{Stator voltage/phase} = 3000/\sqrt{3} \text{ V}$$

$\therefore$
$$\text{Starting current} = \frac{3000/\sqrt{3}}{\sqrt{(1{\cdot}3)^2 + (14{\cdot}7)^2}} = \mathbf{117{\cdot}4\ A}$$

Example 9.13. *A 3-phase slip ring motor gives a reading of 55 V across slip rings on open circuit when at standstill with normal stator voltage applied. The rotor is star-connected and has impedance of (0·7 + j 5) Ω per phase. Find the rotor current and p.f. when the machine is (i) at standstill with slip rings joined to a star-connected starter with a phase impedance of (4 + j 3) ohms and (ii) when running normally with a 5 % slip.*

Solution.

(*i*) **At standstill**

$$\text{Rotor e.m.f./phase, } E_2 = 55/\sqrt{3} = 31{\cdot}8 \text{ V}$$
$$\text{Rotor resistance/phase, } R_2 = 0{\cdot}7 + 4 = 4{\cdot}7\ \Omega$$
$$\text{Rotor reactance/phase, } X_2 = 5 + 3 = 8\ \Omega$$
$$\text{Rotor impedance/phase, } Z_2 = \sqrt{R_2^2 + X_2^2} = \sqrt{(4{\cdot}7)^2 + (8)^2} = 9{\cdot}28\ \Omega$$
$$\text{Rotor current/phase, } I_2 = \frac{E_2}{Z_2} = \frac{31{\cdot}8}{9{\cdot}28} = \mathbf{3{\cdot}43\ A}$$
$$\text{Rotor power factor, } \cos\phi_2 = R_2/Z_2 = 4{\cdot}7/9{\cdot}28 = \mathbf{0{\cdot}506}\ \textit{lagging}$$

(*ii*) **Running normally at 5 % slip**

$$\text{Rotor e.m.f./phase, } E'_2 = s\,E_2 = 0{\cdot}05 \times 31{\cdot}8 = 1{\cdot}59 \text{ V}$$
$$\text{Rotor impedance/phase, } Z'_2 = \sqrt{(0{\cdot}7)^2 + (s \times 5)^2} = \sqrt{(0{\cdot}7)^2 + (0{\cdot}05 \times 5)^2} = 0{\cdot}743\ \Omega$$
$$\text{Rotor current/phase, } I'_2 = \frac{E'_2}{Z'_2} = \frac{1{\cdot}59}{0{\cdot}743} = \mathbf{2{\cdot}14\ A}$$
$$\text{Rotor power factor, } \cos\phi'_2 = \frac{0{\cdot}7}{Z'_2} = \frac{0{\cdot}7}{0{\cdot}743} = \mathbf{0{\cdot}942}\ \textit{lagging}$$

TUTORIAL PROBLEMS

1. The star-connected rotor of a 3-phase induction motor has a resistance and standstill reactance of 0·4 Ω/phase and 2·5 Ω/phase respectively. The e.m.f. induced between slip rings at standstill is 80 V, the stator being connected to normal supply voltage. Find the rotor current and power factor at starting when the rings are (*i*) short-circuited (*ii*) joined to a star-connected resistance of 5Ω/phase.

   [(*i*) 18·25 A ; 0·16 (*ii*) 7·76 A ; 0·91]

2. A 1100 V, 50 Hz, delta-connected induction motor has a star-connected slip-ring rotor with a phase transformation ratio of 3·8. The rotor resistance and standstill leakage reactance are 0·012 Ω and 0·25 Ω per phase respectively. Neglecting stator impedance and magnetising current, determine (*i*) the rotor current at start with slip-rings shorted (*ii*) the rotor power factor at start with slip-rings shorted (*iii*) the rotor current at 4 % slip with slip-rings shorted (*iv*) the rotor p.f. at 4% slip with slip rings shorted (*v*) the external resistance per phase required to obtain a starting current of 100 A in the stator supply lines.

   [(*i*) 1157 A (*ii*) 0·048 *lag* **(*iii*) 742·3 A (*iv*) 0·77 (*v*) 0·707 Ω]**

3. A 4-pole, 3-phase, 50 Hz induction motor is running at full-load at 1400 r.p.m. The star-connected rotor has a resistance and standstill reactance of 0·3 Ω and 2 Ω per phase respectively. The e.m.f. between slip-rings at standstill is 240 V. Find for full load conditions (*i*) e.m.f. induced in each rotor phase (*ii*) rotor impedance/phase (*iii*) rotor current and p.f. assuming the slip-rings are short-circuited.

   [(*i*) 5·54 V (*ii*) 0·31 Ω (*iii*) 17·84 A ; 0·97 lag]

9.10 ROTOR TORQUE

The torque T developed by the rotor is directly proportional to :

(*i*) rotor current

(*ii*) rotor e.m.f.

(*iii*) power factor of the rotor circuit

∴ $$T \propto E_2 I_2 \cos\phi_2$$

or $$T = K E_2 I_2 \cos\phi_2$$

where I_2 = rotor current at standstill

E_2 = rotor e.m.f. at standstill

$\cos\phi_2$ = rotor p.f. at standstill

Note. The values of rotor e.m.f., rotor current and rotor power factor are taken for the given conditions.

9.11 STARTING TORQUE (T_S)

Let
$$E_2 = \text{rotor e.m.f. per phase at standstill}$$
$$X_2 = \text{rotor reactance per phase at standstill}$$
$$R_2 = \text{rotor resistance per phase}$$

Rotor impedance/phase, $Z_2 = \sqrt{R_2^2 + X_2^2}$ ...*at standstill*

Rotor current/phase, $I_2 = \dfrac{E_2}{Z_2} = \dfrac{E_2}{\sqrt{R_2^2 + X_2^2}}$ ... *at standstill*

Rotor p.f., $\cos\phi_2 = \dfrac{R_2}{Z_2} = \dfrac{R_2}{\sqrt{R_2^2 + X_2^2}}$ ...*at standstill*

∴ Starting torque, $T_s = K\,E_2 I_2 \cos\phi_2$

$$= K\,E_2 \times \frac{E_2}{\sqrt{R_2^2 + X_2^2}} \times \frac{R_2}{\sqrt{R_2^2 + X_2^2}}$$

$$= \frac{K\,E_2^2\,R_2}{R_2^2 + X_2^2}$$

Generally, the stator supply voltage V is constant so that flux per pole ϕ set up by the stator is also fixed. This in turn means that e.m.f. E_2 induced in the rotor will be constant.

$$\therefore \quad T_s = \frac{K_1\,R_2}{R_2^2 + X_2^2} = \frac{K_1\,R_2}{Z_2^2}$$

where K_1 is another constant.

It is clear that the magnitude of starting torque would depend upon the relative values of R_2 and X_2 *i.e.*, rotor resistance/phase and standstill rotor reactance/phase.

It can be shown that $K = 3/2\,\pi\,N_s$.

$$\therefore \quad T_s = \frac{3}{2\pi\,N_s} \cdot \frac{E_2^2\,R_2}{R_2^2 + X_2^2}$$

Note that here N_s is in r.p.s.

9.12 CONDITION FOR MAXIMUM STARTING TORQUE

It can be proved that starting torque will be maximum when rotor resistance/phase is equal to standstill rotor reactance/phase.

Now $$T_s = \frac{K_1\,R_2}{R_2^2 + X_2^2} \quad ...(i)$$

Differentiating eq. (*i*) *w.r.t.* R_2 and equating the result to zero, we get,

$$\frac{dT_s}{dR_2} = K_1\left[\frac{1}{R_2^2 + X_2^2} - \frac{R_2\,(2\,R_2)}{(R_2^2 + X_2^2)^2}\right] = 0$$

or $$R_2^2 + X_2^2 = 2\,R_2^2$$

or $$R_2 = X_2$$

Hence starting torque will be maximum when :

Rotor resistance/phase = Standstill rotor reactance/phase

Under the condition of maximum starting torque, $\phi_2 = 45°$ and rotor power factor is 0·707 lagging [See Fig. 9.17 (*ii*)].

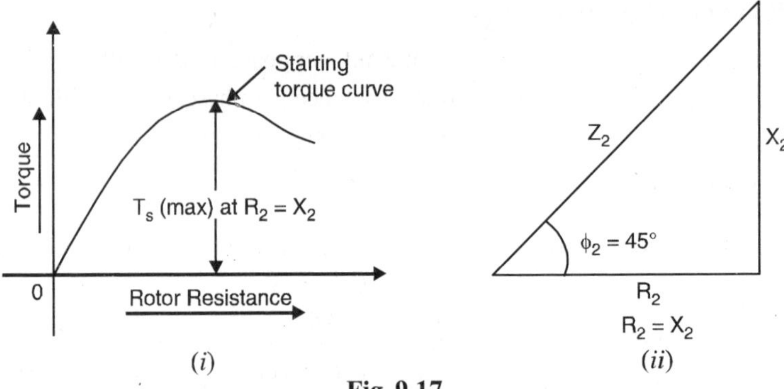

Fig. 9.17

Fig. 9.17 (*i*) shows the variation of starting torque with rotor resistance. As the rotor resistance is increased from a relatively low value, the starting torque increases until it becomes maximum when $R_2 = X_2$. If the rotor resistance is increased beyond this optimum value, the starting torque will decrease.

9.13 EFFECT OF CHANGE OF SUPPLY VOLTAGE

$$T_s = \frac{K\,E_2^2\,R_2}{R_2^2 + X_2^2}$$

Since $E_2 \propto$ Supply voltage V

$$\therefore \quad T_s = \frac{K_2\,V^2\,R_2}{R_2^2 + X_2^2}$$

where K_2 is another constant.

$$\therefore \quad T_s \propto V^2$$

Therefore, the starting torque is very sensitive to changes in the value of supply voltage. For example, a drop of 10 % in supply voltage will decrease the starting torque by about 20 %. This could mean the motor failing to start if it cannot produce a torque greater than the load torque plus friction torque.

9.14 STARTING TORQUE OF 3-PHASE INDUCTION MOTORS

The rotor circuit of an induction motor has low resistance and *high inductance. At starting, the rotor frequency is equal to the stator frequency (*i.e.*, 50 Hz) so that rotor reactance is large compared with rotor resistance. Therefore, rotor current lags the rotor e.m.f. by a large angle, the power factor is low and consequently the starting torque is small. When resistance is added to the rotor circuit, the rotor power factor is improved which results in improved starting torque. This, of course, increases the rotor impedance and, therefore, decreases the value of rotor current but the effect of improved power factor predominates and the starting torque is increased.

(*i*) **Squirrel-cage motors.** Since the rotor bars are permanently short-circuited, it is not possible to add any external resistance in the rotor circuit at starting. Consequently, the starting torque of such motors is low. Squirrel cage motors have starting torque of 1·5 to 2 times the full-load value with starting current of 5 to 9 times the full-load current.

(*ii*) **Wound rotor motors.** The resistance of the rotor circuit of such motors can be increased through the addition of external resistance. By inserting the proper value of external resistance (so that $R_2 = X_2$), maximum starting torque can be obtained. As the motor accelerates, the external resistance is gradually cut out until the rotor circuit is short-circuited on itself for running conditions.

* Because the rotor conductors are embedded in iron.

Example 9.14. *The 3-phase induction motor having star-connected rotor has an induced e.m.f. of 50 V between the slip rings at standstill on open-circuit. The rotor has a resistance and reactance per phase of 0·5 Ω and 4·5 Ω respectively. Find the current per phase and the power factor at starting when (i) the slip rings are short-circuited (ii) the slip rings are connected to a star-connected rheostat of 4 Ω per phase.*

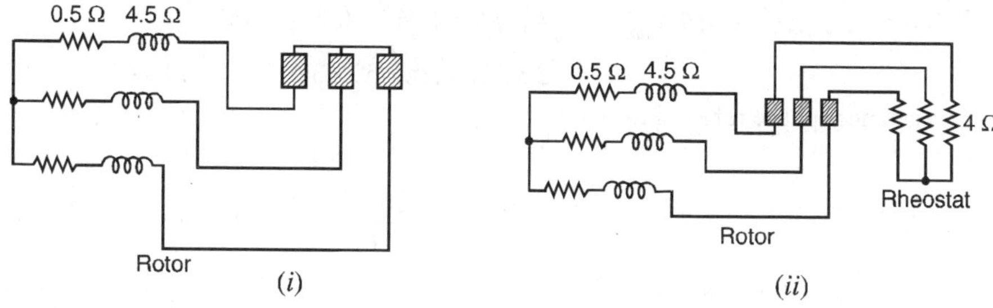

Fig. 9.18

Solution.

$$\text{Standstill rotor e.m.f./phase, } E_2 = 50/\sqrt{3} = 28{\cdot}87 \text{ V}$$

(*i*) **When slip rings are short-circuited** [See Fig. 9.18 (*i*)].

$$\text{Rotor impedance/phase} = \sqrt{(0{\cdot}5)^2 + (4{\cdot}5)^2} = 4{\cdot}53\ \Omega$$

$$\text{Rotor current/phase} = 28{\cdot}87/4{\cdot}53 = \mathbf{6{\cdot}38\ A}$$

$$\text{Rotor power factor} = 0{\cdot}5/4{\cdot}53 = \mathbf{0{\cdot}11}\ \textit{lag}$$

(*ii*) **When slip rings connected to Y-connected rheostat** [See Fig. 9.18 (*ii*)].

$$\text{Rotor resistance/phase} = 4 + 0{\cdot}5 = 4{\cdot}5\ \Omega$$

$$\text{Rotor impedance/phase} = \sqrt{(4{\cdot}5)^2 + (4{\cdot}5)^2} = 6{\cdot}36\ \Omega$$

$$\text{Rotor current/phase} = 28{\cdot}87/6{\cdot}36 = \mathbf{4{\cdot}54\ A}$$

$$\text{Rotor power factor} = 4{\cdot}5/6{\cdot}36 = \mathbf{0{\cdot}707}\ \textit{lag}$$

It is clear that by inserting additional resistance in the rotor circuit, the p.f. of the rotor circuit is improved, though the rotor current is decreased. However, the effect of improved p.f. predominates the decrease in rotor current. Hence, starting torque is increased.

Example 9.15. *A star-connected rotor winding of a 3-phase induction motor has a resistance of 0·2 Ω per phase and a standstill reactance of 1 Ω per phase. Find the value of external resistance per phase to be added to give maximum starting torque. What is the power factor of the rotor circuit then ?*

Solution.

Let R ohm be the external resistance per phase to be added to the rotor circuit to give maximum starting torque. For maximum starting torque,

$$\text{Rotor resistance/phase} = \text{Standstill rotor reactance/phase}$$

or
$$0{\cdot}2 + R = 1$$

∴
$$R = 1 - 0{\cdot}2 = \mathbf{0{\cdot}8\ \Omega}$$

$$\text{Rotor impedance/phase} = \sqrt{1^2 + 1^2} = \sqrt{2}$$

$$\text{Rotor power factor} = 1/\sqrt{2} = \mathbf{0{\cdot}707}\ \textit{lag}$$

Example 9.16. *A 150 kW, 3000 V, 50 Hz, 6-pole, star-connected induction motor has a star-connected slip-ring rotor with a transformation ratio of 3·6 (stator/rotor). The rotor resistance is 0·1 Ω/phase and its per phase leakage inductance is 3·61 mH. The stator impedance may be neglected. Find the starting torque on rated voltage with short-circuited slip rings.*

Solution.

$$\text{Transformation ratio, } K = \frac{\text{Rotor turns/phase}}{\text{Stator turns/phase}} = \frac{1}{3{\cdot}6}$$

Rotor resistance/phase referred to stator is

$$(R_2)_{stator} = R_2/K^2 = (3{\cdot}6)^2 \times 0{\cdot}1 = 1{\cdot}3\,\Omega$$

$$X_2 = 2\,\pi f L_2 = 2\pi \times 50 \times 3{\cdot}61 \times 10^{-3} = 1{\cdot}13\,\Omega$$

Rotor reactance/phase referred to stator is

$$(X_2)_{stator} = X_2/K^2 = 1{\cdot}13 \times (3{\cdot}6)^2 = 14{\cdot}7\,\Omega$$

$$N_s = \frac{120\,f}{P} = \frac{120 \times 50}{6} = 1000 \text{ r.p.m.} = 50/3 \text{ r.p.s.}$$

$$\text{Supply voltage/phase, } E_1 = 3000/\sqrt{3} \text{ volts}$$

$$\text{Starting torque, } T_s = \frac{3}{2\,\pi\,N_s} \times \frac{E_1^2\,(R_2)_{stator}}{(R_2)_{stator}^2 + (X_2)_{stator}^2}$$

$$= \frac{3}{2\,\pi\,(50/3)} \times \frac{(3000/\sqrt{3})^2 \times 1{\cdot}3}{(1{\cdot}3)^2 + (14{\cdot}7)^2} = \mathbf{513\ N\text{-}m}$$

Example 9.17. *The rotor resistance and standstill reactance per phase of a 3-phase induction motor are 0·02 Ω and 0·1 Ω respectively. What should be the value of the external resistance per phase to be inserted in the rotor circuit to give maximum torque at starting ?*

Solution. Let $R_x\,\Omega$/phase be the external resistance added to the rotor circuit to obtain maximum starting torque.

$$\text{Rotor resistance/phase, } R_2 = (0{\cdot}02 + R_x)\,\Omega$$

$$\text{Rotor reactance/phase at standstill, } X_2 = 0{\cdot}1\,\Omega$$

The starting torque will be maximum when :

Rotor resistance/phase = Standstill rotor reactance/phase

or $$0{\cdot}02 + R_x = 0{\cdot}1\,\Omega$$

$\therefore$ $$R_x = 0{\cdot}1 - 0{\cdot}02 = \mathbf{0{\cdot}08\,\Omega}$$

Example 9.18. *The rotor resistance and standstill reactance of a 3-phase induction motor are 0 ·2 Ω and 1·0 Ω per phase. The voltage between the slip rings with rotor locked and a full voltage on the stator is 110 V.*

(*i*) *Find the starting rotor current/phase when the rotor slip rings are short-circuited to form the normal running condition.*

(*ii*) *What should be the value of the external resistance per phase to be inserted in the rotor circuit to give maximum torque at starting ? Also find the rotor current/phase under this condition.*

Solution.

(*i*) Fig. 9.19 (*i*) shows the conditions of the problem.

$$\text{Rotor e.m.f./phase at standstill, } E_2 = 110/\sqrt{3} = 63{\cdot}5 \text{ V}$$

$$\text{Rotor impedance/phase at standstill, } Z_2 = \sqrt{R_2^2 + X_2^2} = \sqrt{(0{\cdot}2)^2 + (1)^2} = 1{\cdot}02\,\Omega$$

$$\text{Rotor phase current at standstill, } I_2 = E_2/Z_2 = 63{\cdot}5/1{\cdot}02 = \mathbf{62{\cdot}3\,A}$$

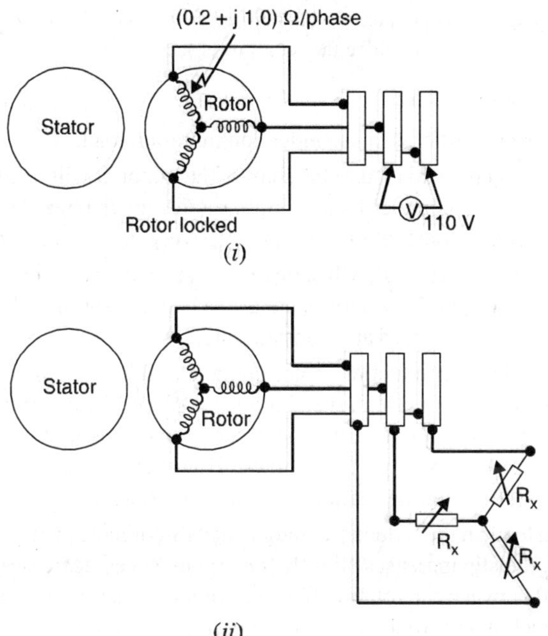

Fig. 9.19

(*ii*) Let R_x Ω/phase be the external resistance added to the rotor circuit [See Fig. 9.19 (*ii*)] to obtain the maximum starting torque.

The starting torque will be maximum when :

$$\text{Rotor resistance/phase} = \text{Standstill rotor reactance/phase}$$

or $$0{\cdot}2 + R_x = 1$$

or $$R_x = 1 - 0{\cdot}2 = \mathbf{0{\cdot}8\ \Omega/phase}$$

$$\text{Rotor impedance/phase} = \sqrt{(0{\cdot}8 + 0{\cdot}2)^2 + (1)^2} = 1{\cdot}414\ \Omega$$

$$\text{Rotor current/phase} = 63{\cdot}5/1{\cdot}414 = \mathbf{44{\cdot}7\ A}$$

Note that rotor current/phase is reduced by about 30 %. This will produce more than double the torque by improving the power factor of the rotor circuit. The efficiency has reduced considerably since the losses in the rotor have changed from $3I^2R = 3 \times (62{\cdot}3)^2 \times 0{\cdot}2 = 2328{\cdot}5$ W to $3 \times (44{\cdot}7)^2 \times 1 = 5994{\cdot}3$ W. Therefore, the external resistances must be removed as soon as the motor has reached the working speed.

TUTORIAL PROBLEMS

1. A 3-phase induction motor at standstill has a rotor voltage of 120 V between the slip rings on open-circuit. The rotor winding is star-connected and has reactance of 1 Ω per phase at standstill and a resistance of 0·2 Ω per phase. Find the rotor current per phase and power factor at start.

   [13·6 A ; 0·98 *lag***]**

2. A 3-phase induction motor with a star-connected rotor has an induced e.m.f. of 80 V between the slip rings at standstill on open-circuit. The resistance and standstill reactance of the rotor are 0·35 Ω and 2·3 Ω per phase respectively. Calculate the rotor current per phase and the power factor at start with an external resistance of 2 Ω per phase. **[14 A ; 0·71** *lag***]**

3. If the star-connected rotor winding of a 3-phase induction motor has a resistance of 0·01 Ω per phase and a standstill reactance of 0·08 Ω per phase, what must be the value of external resistance per phase to give maximum starting torque ? What will be the rotor power factor then ? **[0·07 Ω ; 0·707 *lag*]**

9.15 MOTOR UNDER LOAD

Let us now discuss the behaviour of 3-phase induction motor on load.

(*i*) When we apply mechanical load to the shaft of the motor, it will begin to slow down and the rotating flux will cut the rotor conductors at a higher and higher rate. The induced voltage and resulting current in rotor conductors will increase progressively, producing greater and greater torque.

(*ii*) The motor and mechanical load will soon reach a state of equilibrium when the motor torque is exactly equal to the load torque. When this state is reached, the speed will cease to drop any more and the motor will run at the new speed at a constant rate.

(*iii*) The drop in speed of the induction motor on increased load is small. It is because the rotor impedance is *low and a small decrease in speed produces a large rotor current. The increased rotor current produces a higher torque to meet the increased load on the motor. This is why induction motors are considered to be constant-speed machines. However, because they never actually turn at synchronous speed, they are sometimes called *asynchronous machines*.

Note that change in load on the induction motor is met through the adjustment of slip. When load on the motor increases, the slip increases slightly (*i.e.*, motor speed decreases slightly). This results in greater relative speed between the rotating flux and rotor conductors. Consequently, rotor current is increased, producing a higher torque to meet the increased load. Reverse happens should the load on the motor decrease.

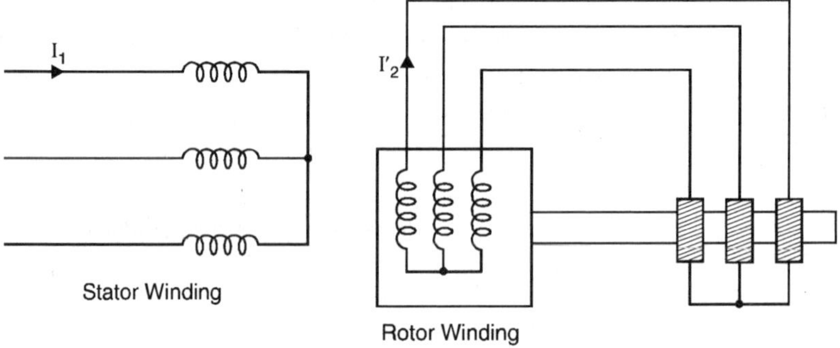

Fig. 9.20

(*iv*) With increasing load, the increased load currents I_2' are in such a direction so as to decrease the stator flux (Lenz's law), thereby decreasing the counter e.m.f. in the stator windings. The decreased counter e.m.f. allows motor stator currents (I_1) to increase, thereby increasing the power input to the motor. It may be noted that action of the induction motor in adjusting its stator or primary current with changes of current in the rotor or secondary is very much similar to the changes occurring in transformer with changes in load.

9.16 TORQUE UNDER RUNNING CONDITIONS

Let the rotor at standstill have per phase induced e.m.f. E_2, reactance X_2 and resistance R_2. Then under running conditions at slip s,

$$\text{Rotor e.m.f./phase, } E_2' = s\,E_2$$

* Rotor resistance is a fixed small value. The rotor frequency under running conditions is very small ($f' = sf$) and hence rotor reactance is also low. This results in low rotor impedance during running conditions.

$$\text{Rotor reactance/phase, } X_2' = s\,X_2$$

$$\text{Rotor impedance/phase, } Z_2' = \sqrt{R_2^2 + (s\,X_2)^2}$$

$$\text{Rotor current/phase, } I_2' = \frac{E_2'}{Z_2'} = \frac{s\,E_2}{\sqrt{R_2^2 + (s\,X_2)^2}}$$

$$\text{Rotor p.f., } \cos\phi_2' = \frac{R_2}{\sqrt{R_2^2 + (s\,X_2)^2}}$$

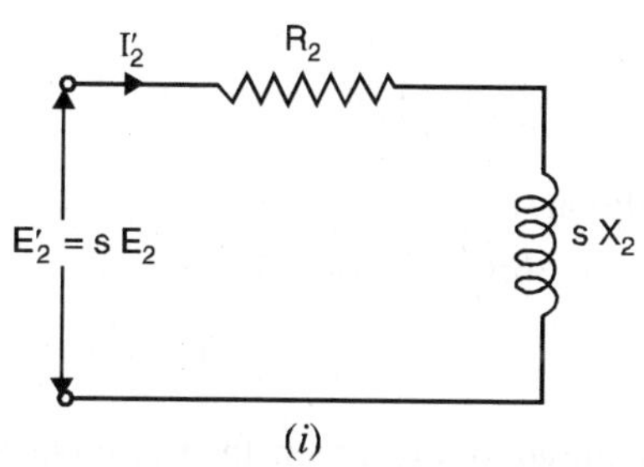

(*i*)

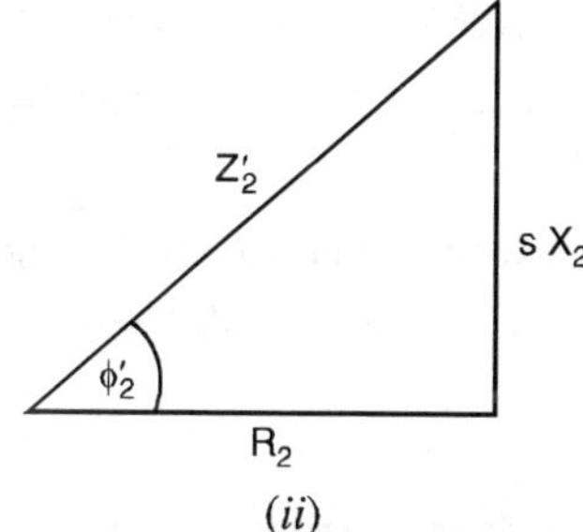

(*ii*)

Fig. 9.21

$$\text{Running Torque, } T_r \propto E_2'\,I_2'\cos\phi_2'$$

$$\propto \phi\,I_2'\cos\phi_2' \qquad (\because E_2' \propto \phi)$$

$$\propto \phi \times \frac{s\,E_2}{\sqrt{R_2^2 + (s\,X_2)^2}} \times \frac{R_2}{\sqrt{R_2^2 + (s\,X_2)^2}}$$

$$\propto \frac{\phi\,s\,E_2\,R_2}{R_2^2 + (s\,X_2)^2}$$

$$= \frac{K\,\phi\,s\,E_2\,R_2}{R_2^2 + (s\,X_2)^2}$$

$$= \frac{K_1\,s\,E_2^2\,R_2}{R_2^2 + (s\,X_2)^2} \qquad (\because E_2 \propto \phi)$$

If the stator supply voltage V is constant, then stator flux and hence E_2 will be constant.

$$\therefore \qquad T_r = \frac{K_2\,s\,R_2}{R_2^2 + (s\,X_2)^2}$$

where K_2 is another constant.

It may be seen that running torque is :

(*i*) directly proportional to slip *i.e.*, if slip increases (*i.e.*, motor speed decreases), the torque will increase and *vice-versa.*

(*ii*) directly proportional to square of supply voltage ($\because E_2 \propto V$).

It can be shown that value of $K_1 = 3/2\,\pi\,N_s$ where N_s is in r.p.s.

$$\therefore \qquad T_r = \frac{3}{2\,\pi\,N_s} \cdot \frac{s\,E_2^2\,R_2}{R_2^2 + (s\,X_2)^2} = \frac{3}{2\,\pi\,N_s} \cdot \frac{s\,E_2^2\,R_2}{(Z_2')^2}$$

At starting, $s = 1$ so that starting torque is

$$T_s = \frac{3}{2\,\pi\,N_s} \cdot \frac{E_2^2\,R_2}{R_2^2 + X_2^2} \qquad \text{... same derived in Art. 9.11}$$

9.17 MAXIMUM TORQUE UNDER RUNNING CONDITIONS

$$T_r = \frac{K_2\, s\, R_2}{R_2^2 + s^2 X_2^2} \qquad \text{...}(i)$$

In order to find the value of rotor resistance that gives maximum torque under running conditions, differentiate exp. (*i*) *w.r.t. s* and equate the result to zero *i.e.*,

$$\frac{dT_r}{ds} = \frac{K_2\,[R_2\,(R_2^2 + s^2X_2^2) - 2\,s\,X_2^2\,(s\,R_2)]}{(R_2^2 + s^2X_2^2)^2} = 0$$

or $\quad (R_2^2 + s^2X_2^2) - 2\,s^2\,X_2^2 = 0$

or $\quad R_2^2 = s^2\,X_2^2$

or $\quad R_2 = s\,X_2$

Thus for maximum torque (T_m) under running conditions :

Rotor resistance/phase = Fractional slip × Standstill rotor reactance/phase

Now $\quad T_r \propto \dfrac{s\,R_2}{R_2^2 + s^2X_2^2} \qquad$ *...from exp. (i) above*

For maximum torque, $R_2 = s\,X_2$. Putting $R_2 = s\,X_2$ in the above expression, the maximum torque T_m is given by ;

$$T_m \propto \frac{1}{2\,X_2}$$

Slip corresponding to maximum torque, $s = R_2/X_2$.

It can be shown that :

$$T_m = \frac{3}{2\,\pi\,N_s} \times \frac{E_2^2}{2\,X_2}\ \text{N-m} \qquad \text{... } N_s \text{ in r.p.s}$$

It is evident from the above equations that :

(*i*) The value of rotor resistance does not alter the value of the maximum torque but it only changes the value of the slip at which it occurs.

(*ii*) The maximum torque varies inversely as the standstill reactance. Therefore, it should be kept as small as possible.

(*iii*) The maximum torque varies directly with the square of the applied voltage.

(*iv*) To obtain maximum torque at starting ($s = 1$), the rotor resistance must be made equal to rotor reactance at standstill.

Example 9.19. *A 4-pole, 50 Hz, 3-phase induction motor has a rotor resistance of 0.024 Ω per phase and standstill reactance of 0.6 Ω per phase. Determine the speed at which maximum torque is developed.*

Solution.

Rotor resistance/phase, $R_2 = 0.024\ \Omega$

Standstill rotor reactance/phase, $X_2 = 0.6\ \Omega$

∴ Slip corresponding to max. torque, $s = \dfrac{R_2}{X_2} = \dfrac{0.024}{0.6\Omega} = 0.04$

∴ Speed corresponding to max. torque, $N = N_s\,(1 - s)$

$$= \frac{120f}{P}(1-s) = \frac{120\times 50}{4}(1-0.04)$$

$= \mathbf{1440\ r.p.m.}$

Example 9.20. *A 440 V, 3-phase, 50 Hz, 4-pole, Y-connected induction motor has a full-load speed of 1425 r.p.m. The rotor has an impedance of (0·4 + j 4) ohm per phase and rotor/stator turn ratio is 0·8. Calculate (i) full-load torque (ii) rotor current and (iii) full-load rotor Cu loss.*

Solution.

(*i*)
$$N_s = 120f/P = 120 \times 50/4 = 1500 \text{ r.p.m.} = 25 \text{ r.p.s.}$$

$$s = \frac{N_s - N}{N_s} = \frac{1500 - 1425}{1500} = \frac{75}{1500} = 0{\cdot}05$$

$$E_1 = 440/\sqrt{3} = 254 \text{ V/phase} \; ; \; E_2 = KE_1 = 0{\cdot}8 \times 254 = 203{\cdot}2 \text{ V}$$

$$\text{Full-load torque, } T_f = \frac{3}{2\pi N_s} \times \frac{s E_2^2 R_2}{R_2^2 + (sX_2)^2}$$

$$= \frac{3}{2\pi \times 25} \times \frac{0{\cdot}05 \times (203{\cdot}2)^2 \times 0{\cdot}4}{(0{\cdot}4)^2 + (0{\cdot}05 \times 4)^2} = \mathbf{78{\cdot}87 \text{ N-m}}$$

(*ii*)
$$\text{Rotor current, } I_2' = \frac{s E_2}{\sqrt{(R_2)^2 + (sX_2)^2}} = \frac{0{\cdot}05 \times 203{\cdot}2}{\sqrt{(0{\cdot}4)^2 + (0{\cdot}05 \times 4)^2}} = \mathbf{22{\cdot}73 \text{ A}}$$

(*iii*)
$$\text{Total rotor Cu loss} = 3 I_2'^2 R_2 = 3 \times (22{\cdot}73)^2 \times 0{\cdot}4 = \mathbf{620 \text{ W}}$$

Example 9.21. *A 3-phase induction motor has a 4-pole star-connected stator winding. The motor runs on a 50 Hz supply with 200 V between lines. The rotor resistance and standstill reactance per phase are 0·1 Ω and 0·9 Ω respectively. The ratio of rotor to stator turns is 0·67. Calculate (i) torque at 4 % slip (ii) maximum torque (iii) speed at maximum torque.*

Solution.

$$E_1 = 200/\sqrt{3} = 115{\cdot}5 \text{ V} \; ; E_2 = K E_1 = 115{\cdot}5 \times 0{\cdot}67 = 77{\cdot}4 \text{ V}$$

$$N_s = 120 f/P = 120 \times 50/4 = 1500 \text{ r.p.m.} = 25 \text{ r.p.s.}$$

(*i*) At 4 % slip, the torque is given by ;

$$T_r = \frac{3}{2\pi N_s} \times \frac{s E_2^2 R_2}{R_2^2 + (s X_2)^2} = \frac{3}{2\pi \times 25} \times \frac{0{\cdot}04 \times (77{\cdot}4)^2 \times 0{\cdot}1}{(0{\cdot}1)^2 + (0{\cdot}04 \times 0{\cdot}9)^2} = \mathbf{40{\cdot}5 \text{ N-m}}$$

(*ii*) The maximum torque occurs when $R_2 = s X_2$ or $s = R_2/X_2 = 1/9$.

$$\therefore \quad T_m = \frac{3}{2\pi N_s} \times \frac{s E_2^2 R_2}{R_2^2 + (s X_2)^2} = \frac{3}{2\pi \times 25} \times \frac{(1/9) \times (77{\cdot}4)^2 \times 0{\cdot}1}{(0{\cdot}1)^2 + (0{\cdot}1)^2} = \mathbf{63{\cdot}6 \text{ N-m}}$$

(*iii*) Slip at maximum torque, $s = R_2/X_2 = 1/9$

$$\therefore \text{ Speed at maximum torque} = (1 - s) N_s = \left(1 - \frac{1}{9}\right) \times 1500 = \mathbf{1333 \text{ r.p.m.}}$$

Example 9.22. *The resistance and standstill reactance of each phase of a 3-phase induction motor with star-connected rotor are 0·06 Ω and 0·4 Ω respectively. The full-load slip is 4 %. Calculate the resistance per phase of a star-connected rheostat which when connected to the rotor circuit will give a pull-out torque at one half of the full-load speed. What is then the power factor ?*

Solution. Let N_s be the synchronous speed.

$$\text{Full-load speed, } N_f = (1 - s_f) N_s = (1 - 0{\cdot}04) N_s = 0{\cdot}96 N_s$$

$$\text{One-half of full-load speed, } N = \frac{0{\cdot}96 N_s}{2} = 0{\cdot}48 N_s$$

$$\text{Slip at half of full load speed, } s = \frac{N_s - 0{\cdot}48 N_s}{N_s} = 0{\cdot}52$$

$$\text{Rotor resistance/phase, } R_2 = 0{\cdot}06 \, \Omega$$

$$\text{Rotor reactance/phase at standstill, } X_2 = 0{\cdot}4 \, \Omega$$

$$\text{Rotor reactance at slip } s = s X_2 = 0{\cdot}52 \times 0{\cdot}4 = 0{\cdot}208 \, \Omega$$

Let R_x ohm/phase be the resistance of the star-connected rheostat added to rotor circuit that will give pull-out (or maximum) torque at one-half of the full-load speed. Then,

$$0{\cdot}06 + R_x = 0{\cdot}208$$

or

$$R_x = 0{\cdot}208 - 0{\cdot}06 = \mathbf{0{\cdot}148\ \Omega}$$

$$\text{Power factor} = \frac{R_2 + R_x}{\sqrt{(R_2 + R_x)^2 + (s\,X_2)^2}} = \frac{0{\cdot}06 + 0{\cdot}148}{\sqrt{(0{\cdot}06 + 0{\cdot}148)^2 + (0{\cdot}208)^2}} = \mathbf{0{\cdot}707}\ \mathit{lag}$$

TUTORIAL PROBLEMS

1. A 3-phase induction motor has a 4-pole, star-connected stator winding and runs on a 220 V, 50 Hz supply. The rotor resistance is 0·1 Ω/phase and rotor reactance is 0·9 Ω per/phase. The ratio of stator to rotor turns is 1·75. The full-load slip is 5 %. Calculate (*i*) the full-load torque (*ii*) the maximum torque (*iii*) the speed at maximum torque. **[(*i*) 42 N-m (*ii*) 56 N-m (*iii*) 1330 r.p.m.]**
2. If the star-connected rotor winding of a three-phase induction motor has a resistance of 0·01 Ω/phase and a standstill leakage reactance of 0·08 Ω/phase, what must be the value of the resistance per phase of a starter to give maximum starting torque ? What is the percentage slip when the starting resistance has been reduced to 0·02 Ω/phase if the motor is still exerting the maximum torque ? **[0·07 Ω ; 37·5 %]**
3. A 6-pole, 3-phase, 50 Hz induction motor has a rotor resistance of 0·02 Ω/phase and standstill leakage reactance is 0.1 Ω/phase. Determine the speed to give maximum torque. What external resistance per phase is necessary in the rotor circuit to give half maximum torque at starting ? **[800 r.p.m. ; 0·07 Ω]**

9.18 TORQUE-SLIP CHARACTERISTICS

As shown in Art. 9.16, the motor torque under running conditions is given by ;

$$T = \frac{K_2\, s\, R_2}{R_2^2 + s^2 X_2^2}$$

If a curve is drawn between the torque and slip for a particular value of rotor resistance R_2, the graph thus obtained is called torque-slip characteristic. Fig. 9.22 shows a family of torque-slip characteristics for a slip-range from $s = 0$ to $s = 1$ for various values of rotor resistance.

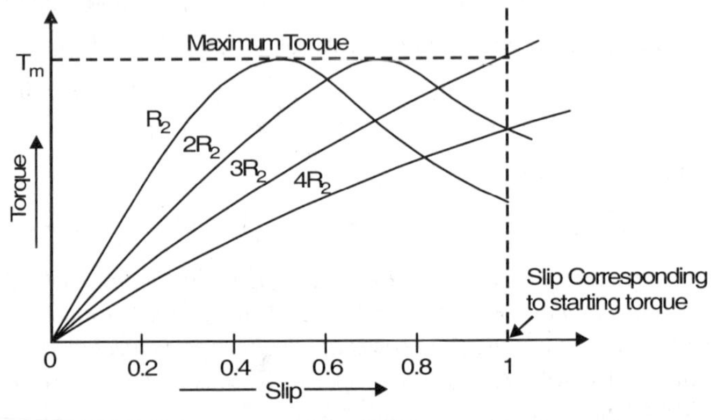

Fig. 9.22

The following points may be noted carefully :

(*i*) At $s = 0$, $T = 0$ so that torque-slip curve starts from the origin.

(*ii*) At normal speed, slip is small so that $s\,X_2$ is negligible as compared to R_2.

$$\therefore \quad T \propto s/R_2$$

$$\propto s \qquad \text{...as } R_2 \text{ is constant}$$

Hence torque-slip curve is a straight line from zero slip to a slip that corresponds to full-load.

(*iii*) As slip increases beyond full-load slip, the torque increases and becomes maximum at $s = R_2/X_2$. This maximum torque in an induction motor is called *pull-out torque* or *breakdown torque*. Its value is atleast twice the full-load value when the motor is operated at rated voltage and frequency.

(*iv*) When slip increases beyond that corresponding to maximum torque, the term $s^2X_2^2$ increases very rapidly so that R_2^2 may be neglected as compared to $s^2 X_2^2$.

$$\therefore \quad T \propto s/s^2 X_2^2$$

$$\propto 1/s \qquad \text{...as } X_2 \text{ is constant}$$

Thus the torque is now inversely proportional to slip. Hence torque-slip curve is a rectangular hyperbola.

(*v*) The maximum torque remains the same and is independent of the value of rotor resistance. *Therefore, the addition of resistance to the rotor circuit does not change the value of maximum torque but it only changes the value of slip at which maximum torque occurs.*

9.19 FULL-LOAD, STARTING AND MAXIMUM TORQUES

$$T_f \propto \frac{s R_2}{R_2^2 + (s X_2)^2} \qquad \text{...See Art. 9.16}$$

$$T_s \propto \frac{R_2}{R_2^2 + X_2^2} \qquad \text{...See Art. 9.11}$$

$$T_m \propto \frac{1}{2 X_2} \qquad \text{...See Art. 9.17}$$

Note that s corresponds to full-load slip.

(*i*) $\therefore$

$$\frac{T_m}{T_f} = \frac{R_2^2 + (s X_2)^2}{2 s R_2 X_2}$$

Dividing the numerator and denominator on R.H.S. by X_2^2, we get,

$$\frac{T_m}{T_f} = \frac{(R_2 / X_2)^2 + s^2}{2 s (R_2 / X_2)} = \frac{a^2 + s^2}{2 a s}$$

where

$$a = \frac{R_2}{X_2} = \frac{\text{Rotor resistance/phase}}{\text{Standstill rotor reactance/phase}}$$

(*ii*)

$$\frac{T_m}{T_s} = \frac{R_2^2 + X_2^2}{2 R_2 X_2}$$

Dividing the numerator and denominator on R.H.S. by X_2^2, we get,

$$\frac{T_m}{T_s} = \frac{(R_2 / X_2)^2 + 1}{2 (R_2 / X_2)} = \frac{a^2 + 1}{2 a}$$

where

$$a = \frac{R_2}{X_2} = \frac{\text{Rotor resistance/phase}}{\text{Standstill rotor reactance/phase}}$$

Example 9.23. *A 50 Hz, 8-pole induction motor has full-load slip of 4%. The rotor resistance and standstill reactance are 0·01 Ω and 0·1 Ω per phase respectively. Find (i) the speed at which maximum torque occurs and (ii) the ratio of maximum torque to full-load torque.*

Solution.

(*i*) Synchronous speed, $N_s = 120 f/P = 120 \times 50/8 = 750$ r.p.m.

Slip at which max. torque occurs $= R_2/X_2 = 0·01/0·1 = 0·1$

$\therefore$ Rotor speed at max. torque $=$ *$(1 - 0·1)\, N_s = (1 - 0·1)750 =$ **675 r.p.m.**

---

* $s = \dfrac{N_s - N}{N_s} = 1 - \dfrac{N}{N_s} \quad \therefore \; N = (1 - s)N_s$

(*ii*)
$$\frac{T_m}{T_f} = \frac{a^2 + s^2}{2as}$$ ...See Art. 9.19

where s = full-load slip = 0·04

$a = R_2/X_2 = 0·01/0·1 = 0·1$

$$\therefore \quad \frac{T_m}{T_f} = \frac{(0·1)^2 + (0·04)^2}{2 \times 0·1 \times 0·04} = \mathbf{1·45}$$

Example 9.24. *An 8-pole, 3-phase, 50 Hz induction motor has a rotor resistance of 0·025 Ω/phase and a rotor standstill reactance of 0·1 Ω/phase. At what speed is the torque maximum ? What proportion of maximum torque is the starting torque ?*

Solution.

$$N_s = 120f/P = 120 \times 50/8 = 750 \text{ r.p.m.}$$

$$R_2 = s X_2$$ ...*for maximum torque*

$$\therefore \quad s = R_2/X_2 = 0·025/0·1 = 0·25$$

Corresponding speed, $N = (1 - s) N_s = (1 - 0·25)\, 750 = \mathbf{562·5\ r.p.m.}$

$$\frac{T_s}{T_m} = \frac{2a}{a^2 + 1}$$ ...See Art. 9.19

$$a = R_2/X_2 = 0·025/0·1 = 0·25$$

$$\therefore \quad \frac{T_s}{T_m} = \frac{2 \times 0·25}{(0·25)^2 + 1} = \mathbf{0·47}$$

Example 9.25. *A 3-phase, 400/200 V, Y-Y connected wound-rotor induction motor has 0·06 Ω rotor resistance and 0·3 Ω standstill reactance per phase. Find the additional resistance required in the rotor circuit to make the starting torque equal to the maximum torque.*

Solution.

$$\frac{T_s}{T_m} = \frac{2a}{a^2 + 1}$$

Since $T_s = T_m$, $T_s/T_m = 1$

$$\therefore \quad 1 = \frac{2a}{a^2 + 1} \quad \text{or} \quad a = 1$$

Now
$$a = \frac{R_2 + R_x}{X_2}$$

where R_x = external resistance/phase added to the rotor circuit

$$\therefore \quad 1 = \frac{0·06 + R_x}{0·3}$$

or R_x = **0·24 Ω/phase**

Example 9.26. *An 8-pole, 50 Hz, 3-phase, slip-ring motor has an effective rotor resistance of 0·07 Ω/phase. Its stalling speed is 630 r.p.m. How much resistance must be inserted per phase to obtain the maximum torque at starting ? Ignore the magnetising current and the stator leakage impedance.*

Solution. The maximum torque is also the stalling torque. Therefore, stalling speed corresponds to maximum torque.

Synchronous speed, $N_s = \dfrac{120\,f}{P} = \dfrac{120 \times 50}{8} = 750$ r.p.m.

Stalling speed = 630 r.p.m.

$$\therefore \quad \text{Slip, } s = \frac{750 - 630}{750} = 0·16$$

Now $s = R_2/X_2 \quad \therefore \quad X_2 = R_2/s = 0{\cdot}07/0{\cdot}16 = 0{\cdot}44\,\Omega$

We know that $\dfrac{T_s}{T_m} = \dfrac{2a}{a^2+1}$

Since $T_s = T_m$, $1 = \dfrac{2a}{a^2+1}$ or $a = 1$

Let R_x ohm/phase be the external resistance added to the rotor circuit to obtain maximum torque at starting.

$$\therefore \quad a = \frac{R_2 + R_x}{X_2} \quad \text{or} \quad 1 = \frac{0{\cdot}07 + R_x}{0{\cdot}44}$$

$$\therefore \quad R_x = \mathbf{0{\cdot}37\,\Omega/phase}$$

Example 9.27. *A 4-pole, 3-phase, 50 Hz, slip-ring induction motor rotates at 1440 r.p.m. with slip-ring terminals short-circuited. The rotor resistance is 0·1 Ω/phase and standstill reactance/phase is 0·6 Ω. If an external resistance of 0·1 Ω/phase is added to the rotor circuit, what will be the new full-load speed ?*

Solution. Since the load on the motor is not changed, the full-load torque in the two cases is the same. However, due to the addition of extra resistance in the rotor circuit in the second case, the slip at which full-load torque develops in this case will change.

$$N_s = 120 f/P = 120 \times 50/4 = 1500 \text{ r.p.m.}$$

For the first case, $s_1 = \dfrac{1500 - 1440}{1500} = 0{\cdot}04$; For the second case, $s_2 = ?$

For the first case, $R_2 = 0{\cdot}1\,\Omega$; For the second case, $R'_2 = 0{\cdot}1 + 0{\cdot}1 = 0{\cdot}2\,\Omega$

Since the torque in the two cases is the same, we have,

$$\frac{3}{2\pi N_s} \times \frac{s_1 E_2^2 R_2}{R_2^2 + (s_1 X_2)^2} = \frac{3}{2\pi N_s} \times \frac{s_2 E_2^2 R'_2}{(R'_2)^2 + (s_2 X_2)^2}$$

or
$$\frac{s_1 R_2}{R_2^2 + (s_1 X_2)^2} = \frac{s_2 R'_2}{(R'_2)^2 + (s_2 X_2)^2}$$

or
$$\frac{0{\cdot}04 \times 0{\cdot}1}{(0{\cdot}1)^2 + (0{\cdot}04 \times 0{\cdot}6)^2} = \frac{s_2 \times (0{\cdot}2)}{(0{\cdot}2)^2 + (s_2 \times 0{\cdot}6)^2}$$

or
$$0{\cdot}378 = \frac{0{\cdot}2 \times s_2}{0{\cdot}04 + 0{\cdot}36\, s_2^2}$$

On solving, we have, $s_2 = 0{\cdot}08$

$\therefore$ New full-load speed $= N_s(1 - s_2) = 1500\,(1 - 0{\cdot}08) =$ **1380 r.p.m.**

Example 9.28. *A 6-pole, 50 Hz, 3-phase induction motor has rotor resistance and reactance per phase of 0·02 Ω and 0·1 Ω respectively. At what speed is the torque maximum ? What must be the value of external rotor resistance per phase to produce two-third of the maximum torque at starting?*

Solution.

At maximum torque, $s = R_2/X_2 = 0{\cdot}02/0{\cdot}1 = 0{\cdot}2$

Synchronous speed, $N_s = 120 f/P = 120 \times 50/6 = 1000$ r.p.m.

$\therefore$ Speed corresponding to maximum torque is

$$N = N_s(1 - s) = 1000\,(1 - 0{\cdot}2) = \mathbf{800 \text{ r.p.m.}}$$

It is given that $\dfrac{T_s}{T_m} = \dfrac{2}{3}$. Now $\dfrac{T_s}{T_m} = \dfrac{2a}{a^2+1}$

$$\therefore \quad \frac{2}{3} = \frac{2a}{a^2+1} \quad \text{or} \quad a^2 - 3a + 1 = 0$$

$$\therefore \quad a = 0{\cdot}38^*$$

Let $\quad R_x$ = additional rotor resistance per phase required

$$\therefore \quad a = \frac{R_2 + R_x}{X_2} \quad \text{or} \quad 0{\cdot}38 = \frac{0{\cdot}02 + R_x}{0{\cdot}1}$$

$$\therefore \quad R_x = \mathbf{0{\cdot}018\ \Omega/phase}$$

Example 9.29. *A 3-phase induction motor having a 6-pole, star-connected stator winding runs on 240 V, 50 Hz supply. The rotor resistance and standstill reactance are 0·12 Ω and 0·85 Ω per phase. The ratio of stator to rotor turns is 1·8 and full-load slip is 4 %. Calculate the developed torque at full-load, maximum torque and the speed at maximum torque.*

Solution.

$$\text{Transformation ratio, } K = \frac{\text{Rotor turns/phase}}{\text{Stator turns/phase}} = \frac{1}{1{\cdot}8}$$

$$\text{Rotor e.m.f./phase, } E_2 = K E_1 = (1/1{\cdot}8) \times 240/\sqrt{3} = 77 \text{ V}$$

$$\text{Full-load slip, } s = 0{\cdot}04$$

$$\text{Synchronous speed, } N_s = 120 f/P = 120 \times 50/6 = 1000 \text{ r.p.m.} = 50/3 \text{ r.p.s.}$$

$$\text{Full-load torque, } T_f = \frac{3}{2\pi N_s} \times \frac{s E_2^2 R_2}{R_2^2 + (s X_2)^2}$$

$$= \frac{3}{2\pi \times (50/3)} \times \frac{0{\cdot}04 \times (77)^2 \times 0{\cdot}12}{(0{\cdot}12)^2 + (0{\cdot}04 \times 0{\cdot}85)^2} = \mathbf{52{\cdot}4\ N\text{-}m}$$

$$\text{Slip at maximum torque, } s' = R_2/X_2 = 0{\cdot}12/0{\cdot}85 = 0{\cdot}14$$

$$\therefore \quad \text{Maximum torque, } T_m = \frac{3}{2\pi N_s} \times \frac{s' E_2^2 R_2}{R_2^2 + (s' X_2)^2}$$

$$= \frac{3}{2\pi \times (50/3)} \times \frac{0{\cdot}14 \times (77)^2 \times 0{\cdot}12}{(0{\cdot}12)^2 + (0{\cdot}14 \times 0{\cdot}85)^2} = \mathbf{99{\cdot}9\ N\text{-}m}$$

$$\text{Speed at maximum torque, } N = N_s(1 - s') = 1000\ (1 - 0{\cdot}14) = \mathbf{860\ r.p.m.}$$

Example 9.30. *The rotor of a 4-pole, 50 Hz, 3-phase slip-ring induction motor has a resistance of 0·25 Ω/phase and runs at 1440 r.p.m. on full load. Calculate the external resistance per phase which must be added to lower the speed to 1200 r.p.m., the torque remaining constant in both cases.*

Solution. The full-load torque of the motor remains the same but the slip at which full load torque will be developed with extra resistance in the rotor will change.

$$\text{Synchronous speed, } N_s = 120 f/P = 120 \times 50/4 = 1500 \text{ r.p.m.}$$

$$\text{For the first case, slip } s_1 = \frac{1500 - 1440}{1500} = 0{\cdot}04$$

$$\text{For the second case, slip } s_2 = \frac{1500 - 1200}{1500} = 0{\cdot}2$$

Let R_x ohm/phase be the resistance added to the rotor circuit to reduce the motor speed to 1200 r.p.m. Since torque remains the same in the two cases,

$$\therefore \quad \frac{3}{2\pi N_s} \times \frac{s_1 E_2^2 R_2}{R_2^2 + (s_1 X_2)^2} = \frac{3}{2\pi N_s} \times \frac{s_2 E_2^2 (R_2 + R_x)}{(R_2 + R_x)^2 + (s_2 X_2)^2}$$

$$\text{or} \quad \frac{s_1 R_2}{R_2^2 + (s_1 X_2)^2} = \frac{s_2 (R_2 + R_x)}{(R_2 + R_x)^2 + (s_2 X_2)^2}$$

* Larger value of a has been rejected.

For low value of s, s^2 is small so that we have,

$$\frac{s_1 R_2}{R_2^2} = \frac{s_2 (R_2 + R_x)}{(R_2 + R_x)^2}$$

or $$\frac{s_1}{R_2} = \frac{s_2}{(R_2 + R_x)}$$

or $$\frac{0{\cdot}04}{0{\cdot}25} = \frac{0{\cdot}2}{(0{\cdot}25 + R_x)}$$

$\therefore$ $$R_x = \mathbf{1\,\Omega/phase}$$

Example 9.31. *The resistance and standstill reactance of each phase of a 3-phase induction motor with star-connected rotor are 0.06 Ω and 0.4 Ω respectively. The full-load slip is 4%. Calculate the resistance per phase of a star-connected rheostat which when connected to the rotor will give a pull out torque at one-half of full-load speed. What is then power factor ?*

Solution. Let the synchronous speed of the motor be N_s.

Full-load slip, $s_f = 0.04$

Full-load speed, $N_f = N_s(1 - s_f) = N_s(1 - 0.04) = 0.96\,N_s$

One-half of full-load speed, $N = 0.5\,N_f = 0.5 \times 0.96\,N_s = 0.48\,N_s$

Slip at one-half of full-load speed, $s = \dfrac{N_s - 0.48N_s}{N_s} = 0.52$

Rotor resistance/phase, $R_2 = 0.06\,\Omega$

Standstill rotor reactance/phase, $X_2 = 0.4\,\Omega$

Rotor reactance/phase at slip s, $X'_2 = s\,X_2 = 0.52 \times 0.4 = 0.208\,\Omega$

Let r be the resistance/phase of the star-connected rheostat connected in the rotor. Pull out or maximum torque will occur at one-half of full-load speed if resistance per phase in the rotor circuit $(R_2 + r)$ is equal to the reactance per phase of the rotor at one-half of full-load spead $(= X'_2)$ *i.e.*

$$R_2 + r = X'_2 \quad \text{or} \quad 0.06 + r = 0.208 \qquad \therefore r = \mathbf{0.148\,\Omega}$$

$$\text{Power factor, } \cos\phi'_2 = \frac{R_2 + r}{\sqrt{(R_2 + r)^2 + (sX_2)^2}}$$

$$= \frac{0.06 + 0.148}{\sqrt{(0.06 + 0.148)^2 + (0.208)^2}} = \mathbf{0.707}\ \textit{lag}$$

TUTORIAL PROBLEMS

1. A 3000 V, 24-pole, 50 Hz, 3-phase, star-connected induction motor has a slip-ring rotor of resistance 0·016 Ω and standstill reactance 0·265 Ω per phase. Full-load torque is obtained at a speed of 247 r.p.m. Calculate (*i*) the ratio of maximum to full-load torque (*ii*) the speed at maximum torque. Neglect stator impedance. **[(*i*) 2·61 (*ii*) 235 r.p.m.]**
2. An 8-pole, 50 Hz, 3-phase induction motor has a full-load slip of 2·5 % and maximum torque of twice the full-load torque. At what value of slip does the maximum torque occur ? **[0·093]**
3. A 6-pole, wound-rotor induction motor operates from a 3-phase, 50 Hz supply. The rotor parameters are $R_2 = 0{\cdot}1\ \Omega$/phase and $X_2 = 0{\cdot}5\ \Omega$/phase. Calculate the speed at which maximum torque occurs and the value of external resistors necessary in the rotor circuit for this torque to occur at starting. **[800 r.p.m. ; 0·4 Ω/phase]**
4. For a 3-phase slip-ring induction motor, the maximum torque is 2·5 times the full-load torque and the starting torque is 1·5 times the full-load torque. Determine the percentage reduction in rotor circuit resistance to get full-load slip at 3 %. Neglect stator impedance. **[56·8 %]**

5. A 3-phase induction motor has a synchronous speed of 250 r.p.m. and 4 % slip at full-load. The rotor has a resistance of 0·02 Ω/phase and a standstill leakage reactance of 0·15 Ω/phase. Calculate (*i*) the ratio of maximum and full-load torque (*ii*) the speed at which the maximum torque develops. Neglect resistance and leakage reactance of the stator. **[(*i*) 1·82 (*ii*) 217 r.p.m.]**

9.20 TORQUE-SPEED CURVE OF 3-PHASE INDUCTION MOTOR

The torque of 3-phase induction motor depends upon its speed but the relationship between them cannot be expressed by a simple equation. Consequently, we prefer to show the relationship in the form of a curve. Fig. 9.23 shows the typical torque-speed curve of a conventional 3-phase induction motor. The curve reveals the following facts :

(*i*) The full-load torque is *T*. The starting torque (or *locked-rotor torque) is 1.5 *T* and the maximum torque (called breakdown torque) is 2.5 *T*.

(*ii*) At full load, the motor runs at a speed *N*. If the mechanical load increases, the speed will drop until the motor torque is again equal to the load torque. As soon as the two torques are equal, the motor will run at a constant but lower speed. However, if the torque exceeds 2.5 *T* (*i.e.* breakdown torque), the motor will suddenly stop.

(*iii*) The torque-speed curve is essentially a straight line between the no-load and full-load operating points. The slope of the line mainly depends upon the rotor resistance ; the higher the resistance, the sharper the slope.

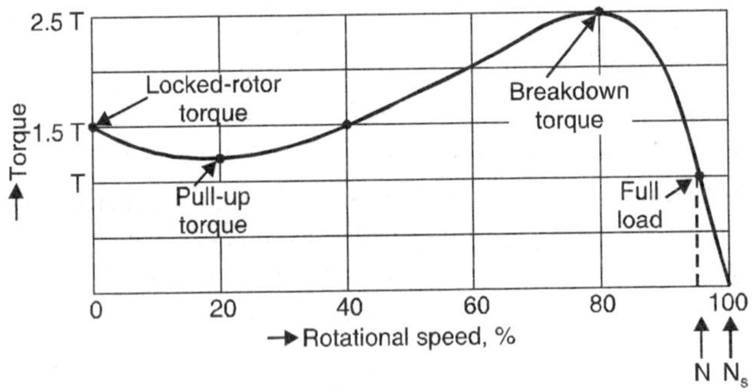

Fig. 9.23

Small motors (< 10 kW) develop their maximum torque at a speed of about 80% of synchronous speed. Big motors (> 1000 kW) attain their breakdown torque at about 98% of synchronous speed.

9.21 CHARACTERISTICS OF SQUIRREL-CAGE MOTORS

The 3-phase squirrel-cage induction motor is the most widely used motor in the industry. Therefore, it is desirable to study its characteristics under different load conditions.

(*i*) **Motor at no load.** When the motor runs at no load, the stator current is 30% to 50% of the full-load current. This no-load current consists of a *large* magnetising component which produces the revolving flux and a *small* active component that supplies the windage and friction losses in the rotor plus iron losses in the stator.

Since magnetising component in an induction motor is quite large (∵ there is an air-gap between stator and rotor), considerable reactive power is needed to create the revolving field. Therefore, the power factor at no-load is low ; it ranges from 0.2 lagging for small motors to 0.05 lagging for large motors. The efficiency is zero because the output power is zero.

* At starting, the rotor is stationary.

(*ii*) **Motor under load.** When the motor is loaded, both the magnetising current and reactive power (kVAR) remain about the same as at no load. However, the active power (kW) absorbed by the motor increases in direct proportion to the mechanical load. It follows that the power factor of the motor improves as the mechanical load increases. At full-load, it ranges from 0.7 lagging for small motors to 0.9 lagging for large motors. The efficiency at full load is particularly high ; it can attain 98% for very large motors.

(*iii*) **Locked rotor characteristics.** For testing purposes, we sometimes lock the rotor (*i.e.* hold the rotor stationary) of the motor while the normal voltage remains applied to the stator. This is called locked-rotor conditions. The locked-rotor current is 5 to 6 times the full-load current, making the $I^2 R$ losses 25 to 36 times higher than normal. Therefore, the rotor should never remain locked for more than a few seconds, otherwise the motor may be damaged.

Although the mechanical output power is zero under locked rotor conditions, the motor develops a strong torque. The power factor is low because considerable reactive power is needed to produce leakage flux in the rotor and stator windings. These leakage fluxes are much larger than in a transformer. It is because the stator and rotor windings are distributed in slots while primary and secondary windings of a transformer are tightly coupled.

9.22 SPEED REGULATION OF INDUCTION MOTORS

Like any other electrical motor, the speed regulation of an induction motor is given by ;

$$\% \text{ age speed regulation} = \frac{N_0 - N_{F.L.}}{N_{F.L.}} \times 100$$

where N_0 = no-load speed of the motor

$N_{F.L.}$ = full-load speed of the motor

If the no-load speed of the motor is 800 r.p.m. and its full-load speed in 780 r.p.m., then change in speed is 800 – 780 = 20 r.p.m. and percentage speed regulation = 20 × 100/780 = 2·56 %.

At no load, only a small torque is required to overcome the small mechanical losses and hence motor slip is small *i.e.*, about 1%. When the motor is fully loaded, the slip increases slightly *i.e.*, motor speed decreases slightly. It is because rotor impedance is low and a small decrease in speed produces a large rotor current. The increased rotor current produces a high torque to meet the full load on the motor. For this reason, the change in speed of the motor from no-load to full-load is small *i.e.*, the speed regulation of an induction motor is low. The speed regulation of an induction motor is 3 % to 5 %. Although the motor speed does decrease slightly with increased load, the speed regulation is low enough that the induction motor is classed as a constant-speed motor.

9.23 POWER FACTOR OF INDUCTION MOTOR

Like any other a.c. machine, the power factor of an induction motor is given by ;

$$\text{Power factor, } \cos\phi = \frac{\text{Active component of current } (I\cos\phi)}{\text{Total current } (I)}$$

The presence of air-gap between the stator and rotor of an induction motor greatly increases the reluctance of the magnetic circuit. Consequently, an induction motor draws a large magnetising current (I_m) to produce the required flux in the air-gap.

(*i*) At no load, an induction motor draws a large magnetising current and a small active component to meet the no-load losses. Therefore, the induction motor takes a *high no-load current lagging the applied voltage by a large angle. Hence the power factor of an induction motor on no load is low *i.e.*, about 0·1 lagging.

(*ii*) When an induction motor is loaded, the active component of current increases while the

* as compared to transformer no-load current

magnetising component remains about the same. Consequently, the power factor of the motor is increased. However, because of the large value of magnetising current, which is present regardless of load, the power factor of an induction motor even at full-load seldom exceeds 0·9 lagging.

9.24 MEASUREMENT OF SLIP

The slip (s) of an induction motor is an important factor in determining the *performance of the motor. It is because the slip (s) appears in all the voltage/power equations of the induction motor. Therefore, it is essential that its value is accurately known. There are many methods of finding the slip of an induction motor. However, we shall discuss the following methods to measure the slip of an induction motor (squirrel cage or slip-ring type) :

1. **By actual measurement of rotor speed.** The slip of an induction motor is given by ;

$$\text{Slip, } s = \frac{N_s - N}{N_s}$$

We measure N (speed of rotor) by a speedometer and synchronous speed N_s (= $120\,f/P$) is calculated from the knowledge of supply frequency (f) and the number of stator poles (P). Thus the slip of the motor can be calculated. This method is not accurate because slip speed ($N_s - N$) is very small as compared to N_s or N. Therefore, a slight error (2% error is quite common for a speedometer) in the measurement of N can cause a large error in the measurement of slip (s). For this reason, we rarely use this method to find s. Instead, we use those methods which *measure the slip directly.*

2. **By comparing stator and rotor frequencies.** We know that in an induction motor :

$$f' = sf$$

where f' = Rotor current frequency

f = Stator current frequency

The stator current frequency f (= 50 Hz) is generally known. Therefore, if rotor current frequency f' is measured by some device, the value of slip s can be determined. Generally, the rotor current frequency f' is quite small and can be easily measured by low-value centre zero d.c. moving-coil millivoltmeter. For example, consider the case of a 4-pole motor fed from 50 Hz and running at 1425 r.p.m. The synchronous speed of the motor, $N_s = 120\,f/P = 120 \times 50/4 = 1500$ r.p.m. so that the slip $s = 0.05$. Therefore, the frequency of rotor current is $f' = sf = 0.05 \times 50 = 2.5$ Hz. This is low enough frequency and can be easily measured.

(*i*) Fig. 9.24 shows the arrangement for measuring the rotor frequency of a slip-ring motor. The leads of the millivoltmeter are lightly pressed across the adjacent slip rings as they revolve.

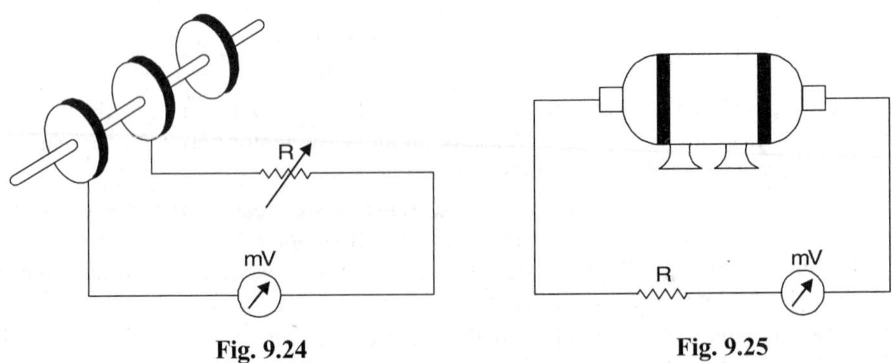

Fig. 9.24 **Fig. 9.25**

* For example, the approximate efficiency of an induction motor is $1 - s$ (See Page 406 of this book).

The current in the millivoltmeter follows the variations of the rotor current and hence the pointer oscillates about the zero position. The number of *oscillations (cycles) completed by the pointer over a given period can be counted. Hence the rotor **frequency f' can be determined.

$$\therefore \qquad \text{Slip, } s = \frac{\text{Rotor frequency}}{\text{Stator frequency}} = \frac{f'}{f}$$

(*ii*) Since a squirrel-cage motor does not have slip-rings, it is not possible to employ the millivoltmeter in the rotor circuit. In this case, it is sometimes possible to obtain an indication by connecting the millivoltmeter across the ends of the motor shaft as shown in Fig. 9.25.

We can use another method to find the slip of the motor. A large flat search coil of many turns is placed centrally against the end plate on the non-driving end of the motor. It is possible to pick up sufficient voltage by induction from the leakage fluxes to obtain the reading on the millivoltmeter. Although a large 50 Hz voltage will also be induced in the search coil, but this is too rapid to affect the millivoltmeter.

3. **Stroboscopic method.** This method employs the well-known principle that if a rotating object is illuminated by the light of an intermittent source, it will appear to be stationary provided the lamp flashes once and once only per revolution.

Fig. 9.26 shows the stroboscopic method for the measurement of slip (s) of an induction motor. In this method, a circular disc (known as ****stroboscopic disc*) painted with alternately black and white segments is rigidly attached to the shaft of the motor. The number of segments (both black and white) is equal to the number of poles of the motor. For a 6-pole motor, there will be six segments, three black and three white as shown in Fig. 9.26 (*i*). A neon glow lamp supplied from the motor line is arranged to illuminate the stroboscopic disc as shown in Fig. 9.26(*ii*).

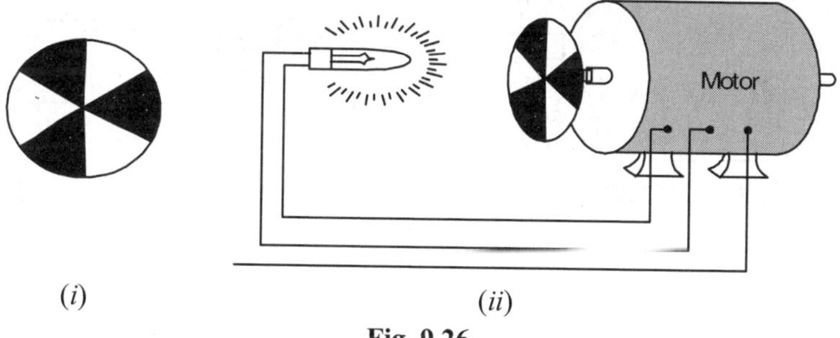

(*i*) (*ii*)

Fig. 9.26

The neon lamp used has the characteristic that it ceases to glow when voltage across it falls below a certain limit *i.e.* such a lamp glows twice in a cycle. If the disc were to rotate at synchronous speed, it would appear to be stationary. Since the speed of the rotor (and hence that of the stroboscopic disc) is less than the synchronous speed, the disc appears to rotate slowly backward in a direction opposite to the rotation of the motor. Counting the number of apparent revolutions of the disc in one minute gives the slip speed ($N_s - N$) in r.p.m. The slip s of the motor can be found from the following relation :

$$s = \frac{N_s - N}{N_s}$$

where $N_s - N$ = Apparent revolutions of the disc in one minute (r.p.m.)

$N_s = 120 f/P$ = Synchronous speed of motor in r.p.m.

* The pointer completes 1 oscillation (or 1 cycle) when it moves from zero to a maximum to the right, back to zero and on to a maximum to the left and then back to zero.

** For example, if the pointer completes 20 cycles in 10 seconds, then, $f' = 20/10 = 2\text{Hz}$.

*** Note that stroboscopic disc has a symmetrical pattern; a requirement for the method.

Note : In factories employing flourescent lights, the stroboscopic effect may sometimes produce the hazardous illusion that rapidly rotating machinery is rotating slowly or even standing still. This danger is eliminated by connecting various adjacent luminaries to different phases of a 3-phase supply. In this way, the illumination bursts are randomly distributed and no periodic bursts of illumination are produced.

9.25 POWER STAGES IN AN INDUCTION MOTOR

The input electric power fed to the stator of the motor is converted into mechanical power at the shaft of the motor. The various losses during the energy conversion are :

1. **Fixed losses**

(*i*) Stator iron loss (*ii*) Friction and windage loss

The rotor iron loss is negligible because the frequency of rotor currents under normal running condition is small.

2. **Variable losses**

(*i*) Stator copper loss (*ii*) Rotor copper loss

Fig. 9.27 shows how electric power fed to the stator of an induction motor suffers losses and finally converted into mechanical power.

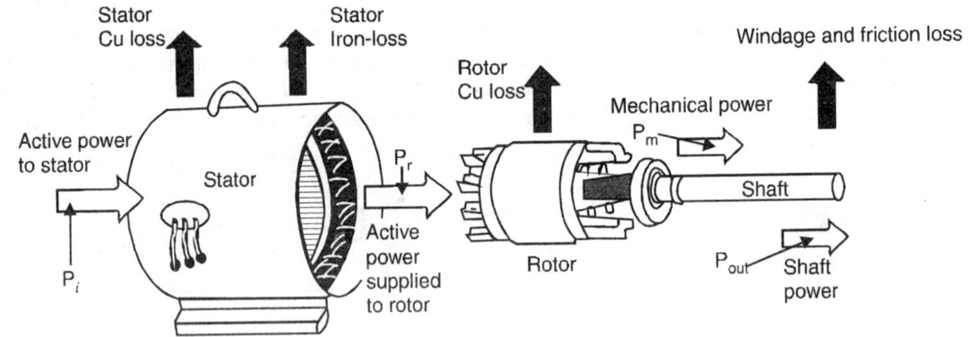

Fig. 9.27

The following points may be noted from the above diagram :

(*i*) Stator input, P_i = Stator output + Stator losses

= Stator output + Stator Iron loss + Stator Cu loss

(*ii*) Rotor input, P_r = Stator output

It is because stator output is entirely transferred to the rotor through air-gap by electromagnetic induction.

(*iii*) Mechanical power available, $P_m = P_r$ – Rotor Cu loss

This mechanical power available is the gross rotor output and will produce a gross torque T_g.

(*iv*) Mechanical power at shaft, $P_{out} = P_m$ – Friction and windage loss

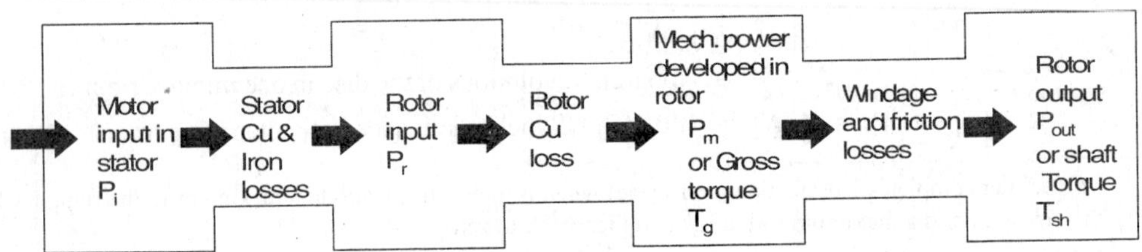

Fig. 9.28

Mechanical power available at the shaft produces a shaft torque T_{sh}.

$$\text{Clearly, } P_m - P_{out} = \text{Friction and windage loss}$$

Fig. 9.28 shows the block diagram of the power flow in a 3-phase induction motor. It may be noted that iron loss of the rotor is negligible and may be considerred zero. It is because under normal running conditions, the frequency of rotor currents is very small.

9.26 INDUCTION MOTOR TORQUE

The mechanical power P available from any electric motor can be expressed as :

$$P = \frac{2\,\pi\,N\,T}{60} \text{ watts}$$

where N = speed of the motor in r.p.m.

T = torque developed in N-m

$$\therefore \quad T = \frac{60}{2\pi}\frac{P}{N} = 9{\cdot}55\,\frac{P}{N} \text{ N-m}$$

If the gross output of the rotor of an induction motor is P_m and its speed is N r.p.m., then *gross torque T_g developed is given by :

$$T_g = 9{\cdot}55\,\frac{P_m}{N} \text{ N-m}$$

Similarly,

$$T_{sh} = 9{\cdot}55\,\frac{P_{out}}{N} \text{ N-m}$$

Note. Since windage and friction loss is small, $T_g = T_{sh}$. This assumption hardly leads to any significant error.

9.27 ROTOR OUTPUT

If T_g newton-metre is the gross torque developed and N r.p.m. is the speed of the rotor, then,

$$\text{Gross rotor output} = \frac{2\,\pi\,N\,T_g}{60} \text{ watts}$$

If there were no copper losses in the rotor, the output would equal rotor input and the rotor would run at synchronous speed N_s.

$$\therefore \quad \text{Rotor input} = \frac{2\,\pi\,N_s\,T_g}{60} \text{ watts}$$

$$\therefore \quad \text{Rotor Cu loss} = \text{Rotor input} - \text{Rotor output}$$

$$= \frac{2\,\pi\,T_g}{60}(N_s - N)$$

(i)
$$\frac{\text{Rotor Cu loss}}{\text{Rotor input}} = \frac{N_s - N}{N_s} = s$$

$$\therefore \quad \text{Rotor Cu loss} = s \times \text{Rotor input}$$

(ii)
$$\text{Gross rotor output, } P_m = \text{Rotor input} - \text{Rotor Cu loss} = \text{Rotor input} - s \times \text{Rotor input}$$

$$\therefore \quad P_m = \text{Rotor input}\,(1 - s)$$

(iii)
$$\frac{\text{Gross rotor output}}{\text{Rotor input}} = 1 - s = \frac{N}{N_s}$$

---

* Gross means total.

$$\text{Rotor gross output, } P_m = \text{Rotor input} - \text{Rotor Cu losses}$$

The rotor gross output is converted into mechanical energy and gives rise to gross torque. Out of this gross torque, some is lost due to windage and friction losses in the rotor and the rest appears as the useful or shaft torque T_{sh}.

(iv) $$\frac{\text{Rotor Cu loss}}{\text{Gross rotor output}} = \frac{s}{1-s}$$

It is clear that if the input power to rotor is P_r then $s\,P_r$ is lost as rotor Cu loss and the remaining $(1-s)\,P_r$ is converted into mechanical power. Consequently, induction motor operating at high slip has poor efficiency.

Note.

$$\frac{\text{Gross rotor output}}{\text{Rotor input}} = 1-s$$

If the stator losses as well as friction and windage losses are neglected, then,

$$\text{Gross rotor output} = \text{Useful output}$$

$$\text{Rotor input} = \text{Stator input}$$

$$\therefore \quad \frac{\text{Useful output}}{\text{Stator input}} = 1-s = \text{Efficiency}$$

Hence the approximate efficiency of an induction motor is $1-s$. Thus if the slip of an induction motor is 0·125, then its approximate efficiency is = 1 – 0·125 = 0·875 or 87·5 %.

9.28 INDUCTION MOTOR TORQUE EQUATION

The gross torque T_g developed by an induction motor is given by ;

$$T_g = \frac{\text{Rotor input}}{2\pi N_s} \qquad \text{...}N_s \text{ in r.p.s.}$$

$$= \frac{60 \times \text{Rotor input}}{2\pi N_s} \quad \text{(See Art. 9.27)} \qquad \text{...}N_s \text{ in r.p.m.}$$

Now, $$\text{Rotor input} = \frac{\text{Rotor Cu loss}}{s} = \frac{3\,(I'_2)^2\,R_2}{s} \qquad \text{...}(i)$$

As shown in Art. 9.16, under running conditions,

$$I'_2 = \frac{s\,E_2}{\sqrt{R_2^2 + (s\,X_2)^2}} = \frac{s\,K\,E_1}{\sqrt{R_2^2 + (s\,X_2)^2}}$$

where $$K = \text{Transformation ratio} = \frac{\text{Rotor turns/phase}}{\text{Stator turns/phase}}$$

$$\therefore \quad \text{Rotor input} = 3 \times \frac{s^2\,E_2^2\,R_2}{R_2^2 + (s\,X_2)^2} \times \frac{1}{s} = \frac{3\,s\,E_2^2\,R_2}{R_2^2 + (s\,X_2)^2} \quad \text{(Putting the value of } I'_2 \text{ in eq. } (i))$$

Also $$\text{Rotor input} = 3 \times \frac{s^2\,K^2 E_1^2\,R_2}{R_2^2 + (s\,X_2)^2} \times \frac{1}{s} = \frac{3\,s\,K^2\,E_1^2\,R_2}{R_2^2 + (s\,X_2)^2} \quad \text{(Putting the value of } I'_2 \text{ in eq. } (i))$$

$$\therefore \quad T_g = \frac{\text{Rotor input}}{2\pi N_s} = \frac{3}{2\pi N_s} \times \frac{s\,E_2^2\,R_2}{R_2^2 + (s\,X_2)^2} \qquad \text{...in terms of } E_2$$

$$= \frac{3}{2\pi N_s} \times \frac{s\,K^2\,E_1^2\,R_2}{R_2^2 + (s\,X_2)^2} \qquad \text{...in terms of } E_1$$

Note that in the above expressions of T_g, the values E_1, E_2, R_2 and X_2 represent the phase values.

Example 9.32. *An 8-pole, 3-phase, 50 Hz induction motor is running at a speed of 710 r.p.m. with an input power of 35 kW. The stator losses at this operating condition are known to be 1200 W while the rotational losses are 600 W. Find (i) the rotor Cu loss (ii) the gross mechanical power developed (iii) the gross torque developed (iv) mechanical power output (v) the net torque.*

Solution. Refer back to the block diagram (See Fig. 9.28) of power flow in a 3-phase induction motor.

$P_i = 35\,\text{kW}$; $P_r = 35 - 1.2 = 33.8\,\text{kW}$; $N = 710$ r.p.m.

$$\text{Synchronous speed, } N_s = \frac{120f}{P} = \frac{120 \times 50}{8} = 750 \text{ r.p.m.}$$

$$\text{Slip, } s = \frac{N_s - N}{N_s} = \frac{750 - 710}{750} = 0.0533$$

(*i*) Rotor Cu loss $= s\,P_r = 0.0533 \times 33.8 =$ **1.8 kW**

(*ii*) $P_m = P_r -$ Rotor Cu loss $= 33.8 - 1.8 =$ **32 kW**

(*iii*) $$T_g = 9.55 \times \frac{P_m}{N} = 9.55 \times \frac{32 \times 10^3}{710} = \mathbf{430.42\ Nm}$$

(*iv*) $P_{out} = P_m -$ Rotational losses $= 32000 - 600 =$ **31.4 kW**

(*v*) $$T_{sh} = 9.55 \times \frac{P_{out}}{N} = 9.55 \times \frac{31.4 \times 10^3}{710} = \mathbf{422.35\ Nm}$$

Example 9.33. *The input power to a 6-pole, 3-phase, 50 Hz induction motor is 42 kW ; the speed is 970 r.p.m. The stator losses are 1·2 kW and the friction and windage losses 1·8 kW. Find (i) the rotor Cu loss and (ii) the efficiency of the motor.*

Solution.

Synchronous speed, $N_s = 120f/P = 120 \times 50/6 = 1000$ r.p.m.

$$\text{Slip, } s = \frac{N_s - N}{N_s} = \frac{1000 - 970}{1000} = 0{\cdot}03$$

(*i*) Stator output = Stator input – Stator losses

= 42 – 1·2 = 40·8 kW

Rotor input = Stator output = 40·8 kW

Rotor Cu loss = $s \times$ Rotor input $= 0{\cdot}03 \times 40{\cdot}8 =$ **1·224 kW**

(*ii*) Useful output, P_{out} = Rotor input – Rotor Cu loss – Windage and friction loss

= 40·8 – 1·224 – 1·8 = 37·77 kW

$$\therefore \quad \text{Motor efficiency} = \frac{37{\cdot}77}{42} \times 100 = \mathbf{89{\cdot}9\ \%}$$

Example 9.34. *A 7·46 kW, 230 V, 3-phase, 50 Hz, 6-pole squirrel cage induction motor operates at full-load slip of 4 % when rated voltage and rated frequency are applied. Determine* (*i*) *speed of rotation of stator m.m.f.* (*ii*) *full-load speed* (*iii*) *frequency of rotor current under this condition and* (*iv*) *full-load torque in newton-metre.*

Solution.

(*i*) The stator m.m.f. revolves at synchronous speed.

$N_s = 120f/P = 120 \times 50/6 =$ **1000 r.p.m.**

(*ii*) $N = (1 - s)\,N_s = (1 - 0{\cdot}04)\,1000 =$ **960 r.p.m.**

(*iii*) $f' = sf = 0{\cdot}04 \times 50 =$ **2 Hz**

(*iv*) Neglecting windage and friction loss, $T_g = T_{sh}$.

$$\therefore \quad T_g = 9{\cdot}55\,\frac{P_{out}}{N} = 9{\cdot}55 \times \frac{7{\cdot}46 \times 10^3}{960} = \mathbf{74{\cdot}2\ N\text{-}m}$$

Example 9.35. *The full-load power input to a 3-phase induction motor is 50 kW and the slip is 3 %. Neglecting stator losses, calculate* (*i*) *the full-load rotor Cu losses per phase and* (*ii*) *total mechanical power developed.*

Solution.

(i) Rotor input $= 50\text{ kW}$ ...*Stator losses neglected*

Rotor Cu loss $= s \times \text{Rotor input} = 0{\cdot}03 \times 50 = 1{\cdot}5\text{ kW}$

Rotor Cu loss/phase $= 1{\cdot}5/3 = \mathbf{0{\cdot}5\ kW}$

(ii) Total mechanical power developed is

$$P_m = 50 - 1{\cdot}5 = \mathbf{48{\cdot}5\ kW}$$

Example 9.36. *A 6-pole, 50 Hz, 3-phase induction motor runs at 960 r.p.m. when the torque on the shaft is 200 N-m. If the stator losses are 1500 W and friction and windage losses are 500 W, find (i) rotor Cu loss and (ii) efficiency of the motor.*

Solution. $N_s = 120f/P = 120 \times 50/6 = 1000$ r.p.m.

$$s = \frac{N_s - N}{N_s} = \frac{1000 - 960}{1000} = 0{\cdot}04$$

(i) Rotor output, $P_{out} = \dfrac{2\pi\, N\, T_{sh}}{60} = \dfrac{2\pi \times 960 \times 200}{60} = 20096$ W

Gross rotor output, $P_m = 20096 + 500 = 20596$ W

$$\frac{\text{Rotor Cu loss}}{\text{Gross rotor output}} = \frac{s}{1-s}$$ ... See Art. 9.27

$\therefore$ Rotor Cu loss $= \dfrac{0{\cdot}04}{1 - 0{\cdot}04} \times 20596 = \mathbf{858\ W}$

(ii) Rotor input $=$ Stator output $= 20596 + 858 = 21454$ W

Stator input, $P_i = 21454 + 1500 = 22954$ W

$\therefore$ Efficiency $= \dfrac{P_{out}}{P_i} \times 100 = \dfrac{20096}{22954} \times 100 = \mathbf{87{\cdot}5\ \%}$

Example 9.37. *The power input to the rotor of a 440 V, 50 Hz, 6-pole, 3-phase induction motor is 100 kW. The rotor electromotive force is observed to make 120 cycles per minute. Calculate (i) the slip (ii) the rotor speed (iii) the mechanical power developed (iv) the rotor copper loss per phase (v) the rotor resistance per phase if the rotor current is 60 A.*

Solution.

Frequency of rotor e.m.f., $f' = 120/60 = 2$ Hz

(i) $f' = sf$

$\therefore$ Slip, $s = f'/f = 2/50 = \mathbf{0{\cdot}04}$

(ii) Rotor speed, $N = N_s(1-s) = 1000\,(1 - 0{\cdot}04) = \mathbf{960\ r.p.m.}$

$[\because N_s = 120f/P = 120 \times 50/6 = 1000 \text{ r.p.m.}]$

(iii) Mechanical power developed, $P_m = (1-s) \times$ Rotor input

$= (1 - 0{\cdot}04) \times 100 = \mathbf{96\ kW}$

(iv) Totol rotor Cu loss $= s \times$ Rotor input $= 0{\cdot}04 \times 100 = 4$ kW

Rotor Cu loss/phase $= 4/3 = \mathbf{1{\cdot}33\ kW}$

(v) If R_2 is the rotor resistance per phase, then,

$$R_2 = \frac{\text{Rotor Cu loss/phase}}{(I_2')^2} = \frac{1{\cdot}33 \times 10^3}{(60)^2} = \mathbf{0{\cdot}37\ \Omega/phase}$$

Example 9.38. *The input to a 3-phase induction motor is 50 kW. The stator losses amount to 800 W. Find the rotor copper loss per phase and the mechanical power developed if the slip is 3 %.*

Solution.

Rotor input $=$ Stator input $-$ Stator losses $= 50 - 0{\cdot}8 = 49{\cdot}2$ kW

$$\text{Total rotor Cu loss} = s \times \text{Rotor input} = 0{\cdot}03 \times 49{\cdot}2 = 1{\cdot}476 \text{ kW}$$

$$\therefore \quad \text{Rotor Cu loss/phase} = 1{\cdot}476/3 = 0.492 \text{ kW} = \mathbf{492 \ W}$$

$$\text{Mechanical power developed, } P_m = \text{Rotor input} - \text{Rotor Cu loss}$$

$$= 49{\cdot}2 - 1{\cdot}476 = \mathbf{47{\cdot}72 \ kW}$$

Example 9.39. *The power input to a 6-pole, 3-phase, 50 Hz induction motor is 42 kW; the speed is 970 r.p.m. The stator losses are 1·2 kW and the friction and windage losses are 1·8 kW. Find (i) the slip (ii) the rotor copper loss (iii) the b.h.p.. and (iv) the efficiency.*

Solution.

(*i*) Sychronous speed, $N_s = 120 f/P = 120 \times 50/6 = 1000$ r.p.m.

$$\text{Slip, } s = \frac{N_s - N}{N_s} = \frac{1000 - 970}{1000} = \mathbf{0{\cdot}03}$$

(*ii*) Rotor input = Stator input – Stator losses = 42 – 1·2 = 40·8 kW

Rotor Cu losses = $s \times$ Rotor input = 0·03 × 40·8 = **1·224 kW**

(*iii*) Useful output, P_{out} = Rotor input – Rotor Cu loss – Friction and windage loss

= 40·8 – 1·224 – 1·8 = 37·776 kW

$$\text{b.h.p.} = \frac{37{\cdot}776 \times 10^3}{746} = \mathbf{50{\cdot}6 \ h.p.}$$

(*iv*)
$$\text{Efficiency, } \eta = \frac{P_{out}}{\text{Stator input}} \times 100 = \frac{37{\cdot}776 \times 10^3}{42 \times 10^3} \times 100 = \mathbf{89{\cdot}9\%}$$

Example 9.40. *A 3-phase, 6-pole, 50 Hz induction motor develops 3·73 kW at 960 r.p.m. What will be the stator input if the stator loss is 280 W ?*

Solution.

$$\frac{\text{Mechanical power developed in Rotor}}{\text{Rotor input}} = \frac{N}{N_s}$$

Here N = 960 r.p.m. ; $N_s = 120 f/P = 120 \times 50/6$ = 1000 r.p.m.

Mechanical power developed in rotor = 3·73 kW = 3730 W

$$\therefore \quad \frac{3730}{\text{Rotor input}} = \frac{960}{1000}$$

$$\therefore \quad \text{Rotor input} = 3730 \times \frac{1000}{960} = 3885 \text{ W}$$

Stator input = Rotor input + Stator losses

= 3885 + 280 = **4165 W**

Example 9.41. *An 8-pole, 3-phase, 50 Hz induction motor running with a slip of 4% is taking 20 kW. Stator losses amount to 0·5 kW. If the mechanical torque lost in friction is 16·25 N-m, find (i) b.h.p. (ii) efficiency.*

Solution.

(*i*) Stator output = Rotor input = Stator input – Stator losses

= 20 – 0·5 = 19·5 kW

Rotor Cu loss = $s \times$ Rotor input = 0·04 × 19·5 = 0·78 kW = 780 W

$\therefore$ Gross rotor output, P_m = 19·5 – 0·78 = 18·72 kW

Synchronous speed, $N_s = 120 f/P = 120 \times 50/8$ = 750 r.p.m.

Rotor speed, $N = N_s(1 - s)$ = 750 (1 – 0·04) = 720 r.p.m.

$$\text{Gross rotor torque, } T_g = 9{\cdot}55 \times \frac{P_m}{N} = 9{\cdot}55 \times \frac{18{\cdot}72 \times 10^3}{720} = 248 \text{ N-m}$$

$$\text{Shaft torque, } T_{sh} = 248 - 16{\cdot}25 = 231{\cdot}75 \text{ N-m}$$

$$\therefore \quad \text{b.h.p.} = \frac{2\pi N T_{sh}}{60} = \frac{2\pi \times 720 \times 231{\cdot}75}{60} \text{ W}$$

$$= \frac{2\pi \times 720 \times 231{\cdot}75}{60 \times 746} \text{ h.p.} = \mathbf{23{\cdot}44 \text{ h.p.}}$$

(*ii*) $$\text{Efficiency} = \frac{23{\cdot}44 \times 746}{20 \times 10^3} \times 100 = \mathbf{87{\cdot}4\ \%}$$

Example 9.42. *A 6-pole, 3-phase, 50 Hz induction motor gives full-load output of 20 h.p. at 945 r.p.m. Stator losses amount to 1 kW. Allowing 13·65 N-m for mechanical torque lost in friction, find (i) rotor copper loss (ii) input to motor (iii) efficiency.*

Solution.

(*i*) $$\text{Synchronous speed, } N_s = 120 f/P = 120 \times 50/6 = 1000 \text{ r.p.m.}$$

$$\text{Rotor speed, } N = 945 \text{ r.p.m.}$$

Now, $$\text{h.p.} = \frac{2\pi N T_{sh}}{60 \times 746}$$

or $$20 = \frac{2\pi \times 945 \times T_{sh}}{60 \times 746}$$

$$\therefore \quad T_{sh} = \frac{20 \times 60 \times 746}{2\pi \times 945} = 151 \text{ N-m}$$

$$\text{Total torque developed, } T_g = T_{sh} + \text{Lost torque in friction}$$

$$= 151 + 13{\cdot}56 = 164{\cdot}56 \text{ N-m}$$

Now $$T_g = 9{\cdot}55 \times \frac{\text{Rotor input}}{N_s}$$

$$\therefore \quad \text{Rotor input} = \frac{T_g \times N_s}{9{\cdot}55} = \frac{164{\cdot}56 \times 1000}{9{\cdot}55} \simeq 17220 \text{ W}$$

$$\text{Slip, } s = \frac{N_s - N}{N_s} = \frac{1000 - 945}{1000} = 0{\cdot}055$$

$$\therefore \quad \text{Rotor Cu loss} = s \times \text{Rotor input} = 0{\cdot}055 \times 17220 = \mathbf{947 \text{ W}}$$

(*ii*) $$\text{Rotor input} = \text{Stator output} = 17220 \text{ W}$$

$$\text{Stator input} = \text{Stator output} + \text{Stator losses}$$

$$= 17220 + 1000 = \mathbf{18220 \text{ W}}$$

(*iii*) $$\text{Efficiency} = \frac{20 \times 746}{18220} \times 100 = \mathbf{81{\cdot}9\ \%}$$

Example 9.43. *A 25 h.p., 6-pole, 50 Hz slip-ring induction motor runs at 960 r.p.m. on full-load with a rotor current of 35 A. Allowing 250 W for the copper loss in the short-circuiting gear and 1000 W for mechanical losses, find the resistance per phase of the 3-phase rotor winding.*

Solution.

$$\text{Synchronous speed, } N_s = 120 f/P = 120 \times 50/6 = 1000 \text{ r.p.m.}$$

$$\text{Slip, } s = \frac{N_s - N}{N_s} = \frac{1000 - 960}{1000} = 0{\cdot}04$$

$$\text{Useful output, } P_{out} = 25 \times 746 = 18650 \text{ W}$$

$$\text{Total mechanical output, } P_m = P_{out} + \text{Losses} = 18650 + 250 + 1000 = 19900 \text{ W}$$

$$\therefore \quad \text{Rotor Cu losses} = P_m \times \frac{s}{1-s} = 19900 \times \frac{0 \cdot 04}{1 - 0 \cdot 04} = 830 \text{ W}$$

If R_2 is the rotor winding resistance per phase, then,

$$3 I'^2_2 R_2 = 830 - \text{Losses in short-circuiting gear}$$

or $$3 \times (35)^2 \times R_2 = 830 - 250 = 580$$

$$\therefore \quad R_2 = \frac{580}{3 \times (35)^2} = \mathbf{0{\cdot}158\ \Omega/phase}$$

Example 9.44 *A 3-phase, 500 V, 50 Hz induction motor with 6 poles develops 20 b.h.p. at 950 r.p.m. with a power factor of 0·86 lagging. The mechanical losses total 1 h.p.. Calculate for this load (i) the slip (ii) the rotor copper loss (iii) the input if the stator losses total 1500 W (iv) the line current.*

Solution.

(*i*) Synchronous speed, $N_s = 120 f/P = 120 \times 50/6 = 1000$ r.p.m.

$$\text{Slip, } s = \frac{N_s - N}{N_s} = \frac{1000 - 950}{1000} = \mathbf{0{\cdot}05}$$

(*ii*) Total mechanical power developed in the rotor is

$$P_m = 20 + 1 = 21 \text{ h.p.} = 21 \times 746 \text{ W} = 15666 \text{ W}$$

$$\text{Rotor input} = \frac{P_m}{1 - s} = \frac{15666}{1 - 0{\cdot}05} = 16491 \text{ W}$$

$$\text{Rotor Cu loss} = s \times \text{Rotor input} = 0{\cdot}05 \times 16491 = \mathbf{825\ W}$$

(*iii*) $$\text{Stator output} = \text{Rotor input} = 16491 \text{ W}$$

$$\text{Stator input} = \text{Stator output} + \text{Stator losses}$$

$$= 16491 + 1500 = \mathbf{17991\ W}$$

(*iv*) $$\sqrt{3}\ V_L I_L \cos\phi = 17991$$

$$\therefore \quad \text{Line current, } I_L = \frac{17991}{\sqrt{3} \times 500 \times 0{\cdot}86} = \mathbf{24{\cdot}16\ A}$$

Example 9.45. *An induction motor has an efficiency of 0.9 when delivering an output of 37 kW. At this load, the stator Cu loss and rotor Cu loss each equals the stator iron loss. The mechanical losses are one-third of the no-load loss. Calculate the slip.*

Solution.

$$\text{Motor input} = 37 \times 10^3/0.9 = 41{,}111 \text{ W}$$

$$\text{Total losses} = 41{,}111 - 37000 = 4111 \text{ W}$$

These include (*i*) stator Cu and iron losses (*ii*) rotor Cu loss (its iron loss is extremely small and neglected) and (*iii*) rotor mechanical losses (W_m). Note that stator iron loss and mechanical losses remain the same from no-load to full-load.

Let $$\text{Stator iron loss} = x$$

$$\therefore \quad \text{Stator Cu loss} = x$$

$$\text{Rotor Cu loss} = x$$

$$\text{Mechanical losses} = {}^*x/2$$

$$\therefore \quad x + x + x + x/2 = 4111 \quad \text{or } x = 1175 \text{ W}$$

$$\text{Rotor input} = \text{Gross output} + \text{Mechanical losses} + \text{Rotor Cu loss}$$

---

* It is given that no-load loss = 3 W_m.

Now, No-load loss = Stator iron loss + W_m [neglecting small stator Cu loss at no-load]

or $$3 W_m = x + W_m$$

$$\therefore \quad W_m = x/2$$

$$= 37{,}000 + x/2 + x = 37{,}000 + (1175/2) + 1175 = 38{,}762 \text{ W}$$

$$\therefore \quad \text{Slip, } s = \frac{\text{Rotor Cu loss}}{\text{Rotor input}} = \frac{1175}{38{,}762} = \mathbf{0.03} \text{ or } \mathbf{3\%}$$

Example 9.46. *An 18.65 kW, 4-pole, 50 Hz, 3-phase induction motor has friction and windage losses of 2.5% of the output. The full-load slip is 4%. Find for full-load (i) the rotor Cu loss (ii) the rotor input (iii) the shaft torque (iv) the gross electromagnetic torque.*

Solution.

$$\text{Motor output, } P_{out} = 18.65 \times 10^3 = 18650 \text{ W}$$

$$\text{Friction and Windage loss} = 2.5\% \text{ of } 18650 = 466 \text{ W}$$

$$\text{Rotor gross output, } P_m = 18650 + 466 = 19{,}116 \text{ W}$$

(i)

$$\frac{\text{Rotor Cu loss}}{\text{Rotor gross output}} = \frac{s}{1-s}$$

$$\therefore \quad \text{Rotor Cu loss} = \frac{s}{1-s} \times \text{Rotor gross output}$$

$$= \frac{0.04}{1-0.04} \times 19{,}116 = \mathbf{796.5 \text{ W}}$$

(ii)

$$\text{Rotor input, } P_r = \frac{\text{Rotor Cu loss}}{s} = \frac{796.5}{0.04} = \mathbf{19{,}912.5 \text{ W}}$$

(iii)

$$\text{Synchronous speed, } N_s = \frac{120f}{P} = \frac{120 \times 50}{4} = 1500 \text{ r.p.m.}$$

$$\text{Rotor speed, } N = N_s(1-s) = 1500(1-0.04) = 1440 \text{ r.p.m.}$$

$$\text{Shaft torque, } T_{sh} = 9.55 \frac{P_{out}}{N} = 9.55 \times \frac{18650}{1440} = \mathbf{123.7 \text{ Nm}}$$

(iv)

$$\text{Gross torque, } T_g = 9.55 \times \frac{P_m}{N} = 9.55 \times \frac{19{,}116}{1440} = \mathbf{126.8 \text{ Nm}}$$

Example 9.47. *A 3-phase, 60 Hz, 4-pole, star-connected induction motor having a nominal rating of 75 kW is excited by a 600 V source. The two-wattmeter method shows a total power consumption of 70 kW and an ammeter indicates a line current of 78 A. Precise measurements give a rotor speed of 1763 r.p.m. The following characterisitcs of the motor are known :*

Stator iron loss = 2 kW

Windage and friction loss = 1.2 kW

Resistance between two stator terminals = 0.34 Ω

Calculate (i) power supplied to the rotor (ii) rotor Cu losses (iii) motor output (iv) efficiency (v) gross torque.

Solution.

$$\text{Power input to stator} = 70 \text{ kW}$$

(i)

$$\text{Stator resistance/phase, } R_1 = 0.34/2 = 0.17\ \Omega$$

$$\text{Stator Cu loss} = 3I_1^2 R_1 = 3 \times (78)^2 \times 0.17 = 3.1 \text{ kW}$$

$$\text{Stator iron loss} = 2 \text{ kW (given)}$$

$$\text{Total stator losses} = 3.1 + 2 = 5.1 \text{ kW}$$

$$\text{Power supplied to rotor, } P_r = 70 - 5.1 = \mathbf{64.9 \text{ kW}}$$

(ii)

$$\text{Motor slip, } s = \frac{N_s - N}{N_s} = \frac{1800-1763}{1800} = 0.0205$$

$\therefore$ Rotor Cu loss $= s\,P_r = 0.0205 \times 64.9 = \mathbf{1.33\ kW}$

(*iii*) Mechanical power developed is given by ;

$$P_m = P_r - \text{Rotor Cu loss} = 64.9 - 1.33 = \mathbf{63.5\ kW}$$

$$\text{Motor output, } P_{out} = P_m - \text{windage and friction loss}$$

$$= 63.5 - 1.2 = 62.3 \text{ kW}$$

(*iv*) $$\text{Motor efficiency, } \eta = \frac{62.3}{70} \times 100 = \mathbf{89\%}$$

(*v*) $$\text{Gross torque, } T_g = 9.55 \times \frac{P_m}{N} = 9.55 \times \frac{63.5 \times 10^3}{1763} = \mathbf{344\ Nm}$$

Example 9.48. *A 3-phase, 60 Hz, 15 h.p., 460 V, 4-pole, 1728 r.p.m. induction motor delivers full output power to a load connected to its shaft. The windage and friction loss of the motor is 750 W. Determine (i) mechanical power developed (ii) air-gap power and (iii) rotor copper loss.*

Solution.

(*i*) $$\text{Full-load shaft power, } P_{out} = 15 \times 746 = 11{,}190 \text{ W}$$

$$\text{Mechanical power developed, } P_m = \text{Shaft power} + \text{Friction and windage loss}$$

$$= 11{,}190 + 750 = \mathbf{11{,}940\ W}$$

(*ii*) $$\text{Synchronous speed, } N_s = 120f/P = 120 \times 60/4 = 1800 \text{ r.p.m.}$$

$$\text{Slip, } s = \frac{N_s - N}{N_s} = \frac{1800 - 1728}{1800} = 0.04$$

$$\text{Air-gap power, } P_{ag} = \text{Input to rotor}$$

$$= \frac{P_m}{1-s} = \frac{11{,}940}{1-0.04} = \mathbf{12{,}437.5\ W}$$

(*iii*) $$\text{Rotor Cu loss} = s\,P_{ag} = 0.04 \times 12{,}437.5 = \mathbf{497.5\ W}$$

Example 9.49. *A 37.3 kW, 4-pole, 50 Hz induction motor has a friction and windage torque of 22 Nm. The stator losses equal the rotor circuit loss. Calculate the input power to the stator when the motor is delivering full-load output at a speed of 1440 r.p.m.*

Solution.

$$\text{Synchronous speed, } N_s = 120\,f/P = 120 \times 50/4 = 1500 \text{ r.p.m.}$$

$$\text{Slip, } s = \frac{N_s - N}{N_s} = \frac{1500 - 1440}{1500} = 0.04$$

$$\text{Windage and friction loss} = \frac{2\pi\,N\,T_w}{60} = \frac{2\pi \times 1440 \times 22}{60} = 3320 \text{ W}$$

$$\text{Mechanical power developed, } P_m = 37{,}300 + 3320 = 40{,}620 \text{ W}$$

$$\text{Rotor input, } P_r = \frac{P_m}{1-s} = \frac{40{,}620}{1-0.04} = 42{,}300 \text{ W}$$

$$\text{Stator losses} = \text{Rotor Cu loss} = s\,P_r = 0.04 \times 42{,}300 = 1690 \text{ W}$$

$\therefore$ $$\text{Stator input power} = 42{,}300 + 1690 = 44 \times 10^3 \text{ W} = \mathbf{44\ kW}$$

Example 9.50. *A 4-pole, 3-phase, 50 Hz induction motor at standstill has 120 V induced across its star-connected rotor terminals. The rotor resistance and standstill reactance per phase are 0.2 Ω and 1 Ω respectively.*

(i) Calculate the rotor speed when the rotor is drawing a current of 16 A at a particular load.

(ii) Find the speed at which the torque is maximum and the corresponding value of rotor input.

Solution. Standstill rotor e.m.f./phase, $E_2 = 120/\sqrt{3} = 69.3$ V

(*i*) Let s be the motor slip when the rotor is drawing a current of 16 A.

$$\text{Rotor current/phase, } I'_2 = \frac{sE_2}{\sqrt{R_2^2 + (sX_2)^2}}$$

or
$$16 = \frac{s \times 69.3}{\sqrt{(0.2)^2 + (s \times 1)^2}}$$

On solving, we have, $s = 0.048$

$$\text{Synchronous speed, } N_s = 120f/P = (120 \times 50)/4 = 1500 \text{ r.p.m.}$$

Now $s = \dfrac{N_s - N}{N_s}$ or $0.048 = \dfrac{1500 - N}{1500}$ $\therefore$ $N =$ **1428 r.p.m.**

(*ii*) Slip corresponding to maximum torque is given by ;

$$s' = R_2/X_2 = 0.2/1 = 0.2$$

Now $s' = \dfrac{N_s - N}{N_s}$ or $0.2 = \dfrac{1500 - N}{1500}$ $\therefore$ $N =$ **1200 r.p.m.**

Rotor current per phase at $s' = 0.2$ is given by ;

$$I'_2 = \frac{s' E_2}{\sqrt{R_2^2 + (s' X_2)^2}}$$

$$= \frac{0.2 \times 69.3}{\sqrt{(0.2)^2 + (0.2 \times 1)^2}} = 49 \text{ A}$$

$\therefore$ Rotor input for the three phases is given by ;

$$P_r = \frac{\text{Total Rotor Cu loss}}{s}$$

$$= \frac{3 I_2'^2 R_2}{s} = \frac{3 \times (49)^2 \times 0.2}{0.2} = 7203 \text{ W} = \mathbf{7.203 \text{ kW}}$$

TUTORIAL PROBLEMS

1. A 3-phase, 6-pole, 50 Hz induction motor develops 3·68 kW at 950 r.p.m. If the stator losses are 300 W, find the stator input. **[4·17 kW]**
2. The power input to a 3-phase induction motor is 40 kW. The stator losses total 1 kW and the mechanical losses are 1·6 kW. Determine (*i*) shaft power (*ii*) rotor Cu loss and (*iii*) efficiency of the motor with a slip of 3·5 %. **[(*i*) 36 kW (*ii*) 445 W (*iii*) 90 %]**
3. An 8-pole, 3-phase, 50 Hz induction motor is taking 50 kW and running at 725 r.p.m. Stator losses are 1·2 kW and frictional losses 1·8 kW. Find (*i*) rotor Cu loss and (*ii*) efficiency of the motor. **[(*i*) 1·627 kW (*ii*) 90·75 %]**
4. A 37·3 kW, 4-pole, 50 Hz induction motor has a friction and windage torque of 22 N-m. The stator losses equal rotor Cu loss. Calculate the power input to the stator when delivering full-load output at a speed of 1440 r.p.m. **[44 kW]**
5. A 50 Hz, 8-pole, 3-phase induction motor delivers an output of 3·1 kW at a slip of 0·02. The losses due to friction and windage are 100 W. Find the value of gross torque in N-m. **[41·58 N-m]**

9.29 PERFORMANCE CURVES OF SQUIRREL-CAGE MOTOR

The performance curves of a 3-phase induction motor indicate the variations of speed, power factor, efficiency, stator current and torque for different values of load. However, before giving the performance curves in one graph, it is desirable to discuss the variation of torque and stator current with slip.

(*i*) **Variation of torque and stator current with slip.** Fig. 9.29 shows the variation of torque and stator current with slip for a standard squirrel - cage motor. Generally, the rotor resistance is low so that full-load current occurs at low slip. Then even at full-load $f' (= sf)$ and, therefore, $X'_2 (= 2\pi f' L_2)$ are low. Between zero and full-load, rotor power factor (= $\cos \phi'_2$) and rotor impedance (= Z'_2) remain practically *constant. Therefore, rotor current $I'_2 (E'_2/Z'_2)$ and, therefore, torque (T_r) increase directly with the slip. Now stator current I_1 increases in proportion to I'_2. This is shown in Fig. 9.29 where T_r and I_1 are indicated as straight lines from no-load to full-load. As load and slip are increased beyond full-load, the increase in rotor reactance becomes appreciable. The increasing value of rotor impedance not only decreases the rotor power factor $\cos \phi'_2$ (= R_2/Z'_2) but also lowers the rate of increase of rotor current. As a result, the torque T_r and stator current I_1 do not increase directly with slip as indicated in Fig. 9.29. With the decreasing power factor and the lowered rate of increase in rotor current, the stator current I_1 and torque T_r increase at a lower rate. Finally, torque T_r reaches the maximum value at about 25 % slip in the standard squirrel cage motor. This maximum value of torque is called the *pull-out torque* or *breakdown torque*. If the load is increased beyond the breakdown point, the decrease in rotor power factor is greater than the increase in rotor current, resulting in a decreasing torque. The result is that motor slows down quickly and comes to a stop.

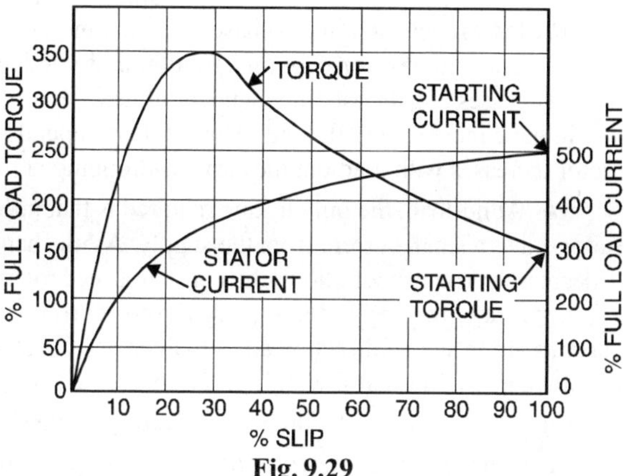

Fig. 9.29

In Fig. 9.29, the value of torque at starting (*i.e.*, $s = 100$ %) is 1·5 times the full-load torque. The starting current is about five times the full-load current. The motor is essentially a constant-speed machine having speed characteristics about the same as a d.c. shunt motor.

(*ii*) **Performance curves.** Fig. 9.30 shows the performance curves of 3-phase squirrel cage induction motor.

The following points may be noted:

(*a*) At no-load, the rotor lags behind the stator flux by only a small amount, since the only torque required is that needed to overcome the no-load losses. As mechanical load is added, the rotor speed decreases. A decrease in rotor speed allows the constant-speed rotating field to sweep across the rotor

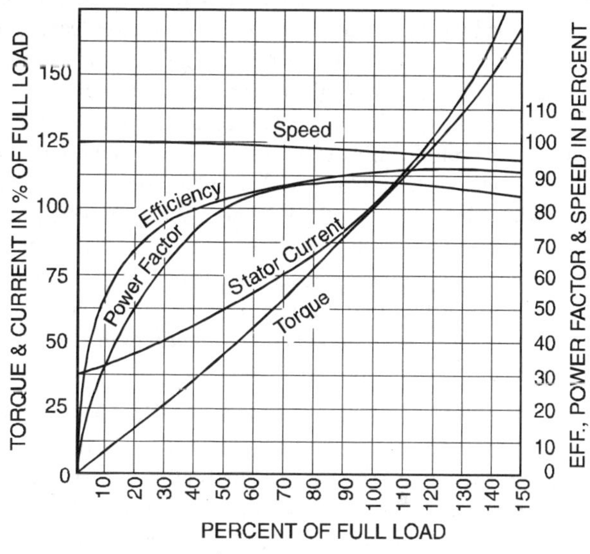

Fig. 9.30

* In a standard squirrel-cage motor, the change in slip is very small as the load is increased from zero to full-load. Therefore, the change of rotor impedance from no-load to full-load is negligible.

conductors at a faster rate, thereby inducing large rotor currents. This results in a larger torque output at a slightly reduced speed. This explains for speed-load curve in Fig. 9.30.

(*b*) At no-load, the current drawn by an induction motor is largely a magnetising current; the no-load current lagging the applied voltage by a large angle. Thus the power factor of a lightly loaded induction motor is very low. Because of the air gap, the reluctance of the magnetic circuit is high, resulting in a large value of no-load current as compared with a transformer. As load is added, the active or power component of current increases, resulting in a higher power factor. However, because of the large value of magnetising current, which is present regardless of load, the power factor of an induction motor even at full-load seldom exceeds 90%. Fig. 9.30 shows the variation of power factor with load of a typical squirrel-cage induction motor.

(*c*) $$\text{Efficiency} = \frac{\text{Output}}{\text{Output} + \text{Losses}}$$

The losses occurring in a 3-phase induction motor are Cu losses in stator and rotor windings, iron losses in stator and rotor core and friction and windage losses. The iron losses and friction and windage losses are almost *independent of load. Had I^2R been constant, the efficiency of the motor would have increased with load. But I^2R loss depends upon load. Therefore, the efficiency of the motor increases with load but the curve is dropping at high loads.

(*d*) At no-load, the only torque required is that needed to overcome no-load losses. Therefore, stator draws a small current from the supply. As mechanical load is added, the rotor speed decreases. A decrease in rotor speed allows the constant-speed rotating field to sweep across the rotor conductors at a faster rate, thereby inducing larger rotor currents. With increasing loads, the increased rotor currents are in such a direction so as to decrease the stator flux, thereby temporarily decreasing the counter e.m.f. in the stator winding. The decreased counter e.m.f. allows more stator current to flow.

(*e*) $$\text{Output} = \text{Torque} \times \text{Speed}$$

Since the speed of the motor does not change appreciably with load, the torque increases with increase in load.

9.30 PLUGGING OF AN INDUCTION MOTOR

In some industrial applications, it is desired to bring the running induction motor to a rapid stop. This can be done by simply interchanging the two stator leads. This process is called *plugging.* When we interchange two stator leads, the revolving field suddenly turns in the opposite direction to the rotor. During the *plugging period*, the motor acts as a *brake*. It absorbs kinetic energy from the still-revolving field, causing its speed to fall. The mechanical power associated with the rotor is entirely dissipated as heat in the rotor. Unfortunately, the rotor also continues to receive power from the stator which is also dissipated as heat. Consequently, plugging produces I^2R losses in the rotor which even exceed those when the rotor is locked. Motors should not be plugged too frequently because high rotor temperatures may melt the rotor bars or overheat the stator winding.

When very high intertia loads have to be brought to a stop, wound-rotor motors are recommended because most of the thermal energy absorbed by the rotor is dissipated by the external resistors. Furthermore, we can maintain a consistently high torque by gradually varying the rotor resistance during deceleration period.

9.31 BRAKING INDUCTION MOTOR WITH DIRECT CURRENT

An induction motor and its high-inertia load can be brought to a quick stop by circulating d.c. current in the stator winding. Any two terminals of the motor can be connected to the d.c. source. The direct current produces stationary *N*, *S* poles in the stator. The number of stationary poles created is equal to the number of poles which the motor develops normally. Thus, a 3-phase, 4-pole motor produces 4

* Iron losses in stator depend upon stator flux and supply frequency. Since both these factors are constant, the stator iron-loss is constant at all loads. Since rotor frequency is small, iron loss in rotor is small and may be neglected. As the speed of the motor does not vary appreciably with load, friction and windage losses may be assumed constant.

stationary poles, no matter how the motor terminals are connected to the d.c. source. When the rotor sweeps past the stationary field, an a.c. voltage is induced in the rotor bars. The voltage produces an a.c. current and the resulting rotor I^2R losses are dissipated at the expense of the kinetic energy stored in the revolving parts. The motor finally comes to rest when the entire kinetic energy has been dissipated as heat in the rotor.

The advantage of d.c. braking is that it produces far less heating than does plugging. The energy dissipated in the rotor is independent of the magnitude of the d.c. current. However, a smaller d.c. current increases the braking time, with the result that the temperature rise of the rotor is somewhat reduced. The d.c. current can be two to three times the rated current of the motor. Even larger values can be used provided the stator does not become too hot. The braking torque is directly proportional to the *square* of the d.c. braking current.

9.32 INDUCTION GENERATOR

If an induction motor whose stator windings are connected to a 3-phase line is driven by a prime-mover at a speed higher than synchronous speed, it acts as a generator. It converts the mechanical energy it receives from the prime-mover into electrical energy and this electrical energy is supplied to the mains. Such a machine is called an *induction generator* or *asynchronous generator.* When speed of the induction motor exceeds the synchronous speed, the slip (s) becomes *negative. Therefore, the relative motion between the rotor conductors and the rotating flux is reversed and as a result, the directions of rotor e.m.f. and rotor currents will also be reversed. However, as soon as this takes place, the motor develops a **counter torque which opposes the increase in speed. Thus generator operation occurs and mechanical energy of the prime-mover is converted into electrical energy which is supplied to the mains.

Fig. 9.31 shows an induction generator connected to a 3-phase line. The petrol engine is the prime-mover. As soon as the engine speed exceeds the synchronous speed, the motor becomes a generator, delivering active power P (kW) to the electrical systems (*i.e.* 3-phase line in this case) to which it is connected. However, to create its magnetic field, the motor has to absorb reactive power Q (kVAR). This power can only come from the supply lines. Consequently, the reactive power Q flows in the opposite direction to the active power P as shown in Fig. 9.31.

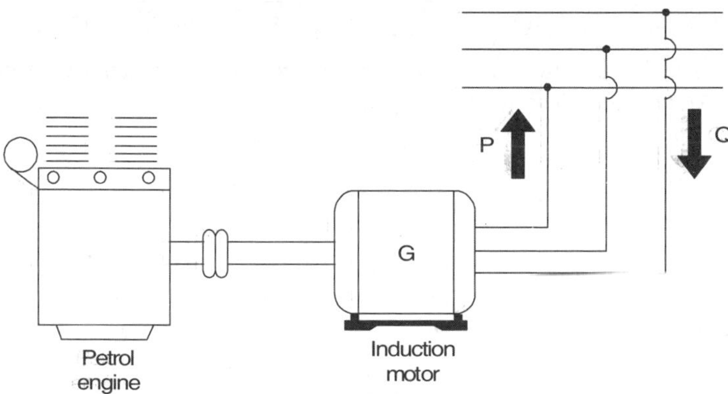

Fig. 9.31

The active power P (kW) delivered is directly proportional to the slip above synchronous speed. Thus, a higher engine speed produces a greater output. However, the rated output is reached at very small slip, generally less than 3%.

---

* Slip, $s = (N_s - N)/N_s$. If $N > N_s$, then s is negative.

** In an induction motor, $T \propto s$. If s is negative, T will be negative. The negative torque means that motor receives rather than delivers mechanical power and generator operation occurs.

It may be noted that an induction generator will deliver power only if it is supplied with proper reactive power to create its magnetic field. For this reason, an induction generator is generally connected to a 3-phase line. However, reactive power may be supplied by a group of capacitors connected to the terminals of the motor as shown in Fig. 9.32. In that case, the induction generator does not require external source for supply of reactive power. The frequency generated is slightly less than that corresponding to the rotation of the machine.

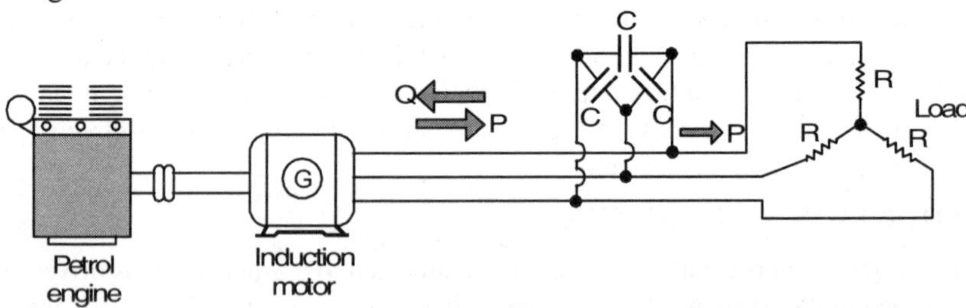

Fig. 9.32

The terminal voltage increases with the capacitance, but its magnitude is limited by saturation in the iron. If the capacitance is insufficient, the generator voltage will not build up. The capacitor bank must be large enough to supply the reactive power the machine normally absorbs when operating as a motor.

Example 9.51. *A 3-phase, 440 V, 4-pole, 1760 r.p.m., 30 kW squirrel-cage induction motor is to be used as an asynchronous generator. The rated current of the motor is 41 A and full-load power factor is 84%. (i) Calculate the capacitance required per phase if the capacitors are connected in delta. (ii) At what speed should the driving engine run to generate a frequency of 60 Hz ?*

Solutoin. (*i*) The apparent power *S* drawn by the machine when it operates as a motor is

$$S = \sqrt{3}\, V_L I_L = \sqrt{3} \times 440 \times 41 = 31.2 \times 10^3 \text{ VA} = 31.2 \text{ kVA}$$

The corresponding active power *P* absorbed is

$$P = S \cos\phi = 31.2 \times 0.84 = 26.2 \text{ kW}$$

The corresponding reactive power *Q* absorbed is

$$Q = \sqrt{S^2 - P^2} = \sqrt{(31.2)^2 - (26.2)^2} = 17 \text{ kVAR}$$

When the machine operates as an asynchronous generator, the capacitor bank must supply 17/3 = 5.7 kVAR per phase. The voltage per phase is 440 V because the capacitors are connected in delta.

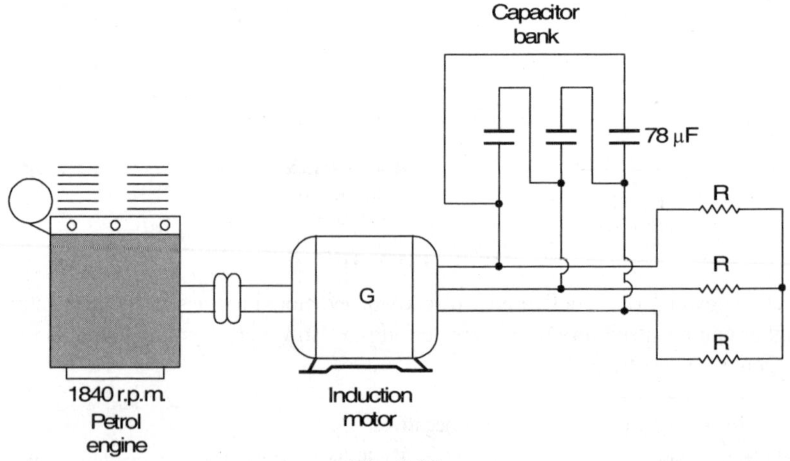

Fig. 9.33

∴ Capacitive current/phase, $I_C = Q/V_{ph} = (5.7 \times 10^3)/440 = 13\text{ A}$

Capacitive reactance/phase, $X_C = V_{ph}/I_C = 440/13 = 34\ \Omega$

$$\therefore \quad C = \frac{1}{2\pi f X_C} = \frac{1}{2\pi \times 60 \times 34}$$

$$= 78 \times 10^{-6}\text{ F} = \mathbf{78\ \mu F}$$

(*ii*) When the machine runs as a motor, its synchronous speed is $N_s = 120f/P = 120 \times 60/4 =$ 1800 r.p.m. and slip speed = 1800 – 1760 = 40 r.p.m. When the machine runs as a genterator, its slip speed should be equal to the slip speed when operating as a motor.

∴ Engine speed = 1800 + 40 = **1840 r.p.m.**

9.33 APPLICATIONS OF INDUCTION GENERATORS

The induction generator is *not* a self-excited generator. It is necessary to excite the stator with an external polyphase source at its rated voltage and frequency. It will generate only when it is connected to a source of fixed voltage and frequency and if it is then driven at a speed above the synchronous speed.

Thus the induction or asynchronous generator has limited applications. The most important use of the principle of the induction generator lies in automatic dynamic braking. For example, in a lift or crane driven by an induction motor, when the laden cage or hook is descending, the load torque on the motor acts in the direction of rotation. As a result, the motor speed exceeds the synchronous speed and the machine automatically becomes an induction generator and produces a *braking torque,* returning the energy of the descending load to the supply.

9.34 COMPLETE TORQUE/SLIP CURVE OF INDUCTION MACHINE

We have seen that a 3-phase induction machine may function (*i*) as a motor (*ii*) as a generator or (*iii*) as a brake. Fig. 9.34 shows the complete torque-slip characteristic of a 3-phase induction machine.

(*i*) **Motor mode.** In the motor mode, the stator terminals are connected to a 3-phase supply and the rotor rotates in the direction of the stator rotating magnetic field. The steady-state speed (N) of the machine is less than the synchronous speed (N_s). Most induction machines operate as motors supplying a torque that varies from zero to full-load torque. Between these limits, the torque-slip characterisitc is a straight line [See Fig. 9.34] *i.e.,* torque is directly proportional to the slip ($T \propto s$). As the slip decreases, the torque decreases and vice-versa. Note that in the motor mode, the slip varies from close to 0% at no-load to 100% at standstill. The linear relationship between torque and slip enables us to establish a very simple equation between the various parameters of the motor.

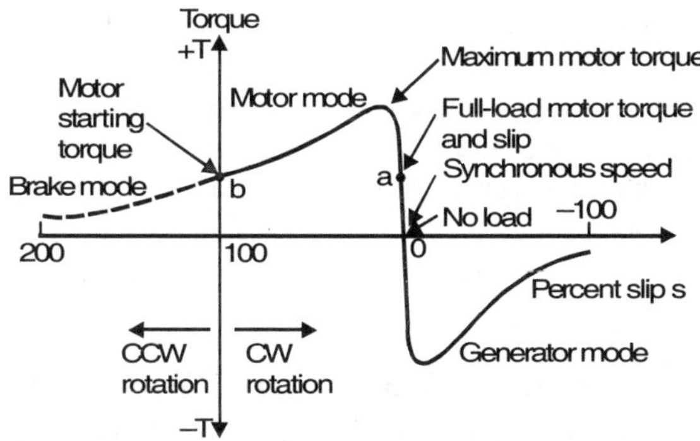

Fig. 9.34

(*ii*) **Generator mode.** If an induction motor whose stator windings are connected to the 3-phase supply is driven by a prime-mover at a speed higher than synchronous speed (*i.e.* at negative slip), the generator operation occurs. Since both slip and torque are negative, the motor receives rather than delivers mechanical energy. The mechanical energy that the machine receives is converted into electrical energy and is supplied to the mains. Induction machines are seldom used solely as generators.

(*iii*) **Brake mode.** Consider that the induction motor is operating at rated speed and rated load (point '*a*' in Fig.9.34) in a clockwise direction (CW). If we interchange any two stator leads, the rotating field suddenly turns in the opposite direction to the rotor and the motor acts as a brake. Consequently, the motor will slow down rapidly, pass through standstill (point '*b*' in Fig. 9.34) and accelerate the load in the counterclockwise (CCW) direction. This method of braking is called **plugging** (*i.e.* bringing induction motor to a rapid stop) and is used in many industrial applications. During plugging period, the kinetic energy of the revolving load is entirely dissipated as heat in the rotor. Moreover, the rotor also continues to receive power from the stator which is also dissipated as heat. Consequently, plugging produces severe heating of the rotor. In most applications, the stator is disconnected from the line terminals *before* the motor enters the braking mode at 100% slip.

Let us assume again that the motor is operating at rated speed and rated load in the clockwise (CW) direction. If for some *reason the load being driven accelerates the motor in the same direction (*i.e.* CW direction), the motor speed may exceed the synchronous speed. Under these circumstances, the motor acts as an induction generator.The mechanical energy of the moving parts is converted into electrical energy and is supplied to the mains. This tends to slow the motor down to its normal synchronous no-load speed. This type of braking is called **dynamic** *or* **regenerative braking.**

9.35 INDUCTION MOTOR AND TRANSFORMER COMPARED

An induction motor may be considered to be a transformer with a rotating short-circuited secondary. The stator winding corresponds to transformer primary and rotor winding to transformer secondary. However, the following differences between the two are worth noting :

(*i*) Unlike a transformer, the magnetic circuit of a 3-phase induction motor has an air gap. Therefore, the magnetising current in a 3-phase induction motor is much larger than that of the transformer. For example, in an induction motor, it may be as high as 30 – 50% of rated current whereas it is only 1 – 5% of rated current in a transformer.

(*ii*) In an induction motor, there is an air gap and the stator and rotor windings are distributed along the periphery of the air gap rather than concentrated on a core as in a transformer. Therefore, the leakage reactances of stator and rotor windings are quite large compared to that of a transformer.

(*iii*) In an induction motor, the inputs to the stator and rotor are electrical but the output from the rotor is mechanical. However, in a transformer, input as well as output is electrical.

(*iv*) *The main difference between the induction motor and transformer lies in the fact that the rotor voltage and its frequency are both proportional to slip s*. If f is the stator frequency, E_2 is the per phase rotor e.m.f. at standstill and X_2 is the standstill rotor reactance/phase, then at any slip s, these values are :

Rotor e.m.f./phase, $E'_2 = s\,E_2$, Rotor reactance/phase, $X'_2 = s\,X_2$

Rotor frequency, $f' = sf$

9.36 EQUIVALENT CIRCUIT OF 3-ϕ INDUCTION MOTOR AT ANY SLIP

In a 3-phase induction motor, the stator winding is connected to 3-phase supply and the rotor winding is short-circuited. The energy is transferred magnetically from the stator winding to the short-circuited

* The kinetic energy of the moving load (*e.g.*, a train descending a steep gradient or a descending lift) drives the motor above synchronous speed.

rotor winding. Therefore, an induction motor may be considered to be a transformer with a rotating secondary (short-circuited). The stator winding corresponds to transformer primary and the rotor winding corresponds to transformer secondary. In view of the similarity of the flux and voltage conditions to those in a transformer, one can expect that the equivalent circuit of an induction motor will be similar to that of a transformer. Fig. 9.35 shows the equivalent circuit (though not the only one) per phase for an induction motor. Let us discuss the stator and rotor circuits separately.

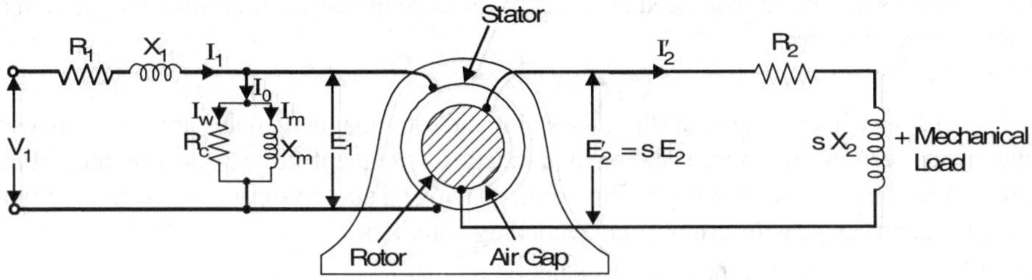

Fig. 9.35

Stator circuit. In the stator, the events are very similar to those in the transformer primary. The applied voltage per phase to the stator is V_1 and R_1 and X_1 are the stator resistance and leakage reactance per phase respectively. The applied voltage V_1 produces a magnetic flux which links the stator winding (*i.e.*, primary) as well as the rotor winding (*i.e.*, secondary). As a result, self-induced e.m.f. E_1 is induced in the stator winding and mutually induced e.m.f. E_2' $(= s\,E_2 = s\,K\,E_1$ where K is transformation ratio) is induced in the rotor winding. The flow of stator current I_1 causes voltage drops in R_1 and X_1.

$$\therefore \quad \mathbf{V_1} = -\mathbf{E_1} + \mathbf{I_1}(R_1 + jX_1) \quad \text{...phasor sum}$$

When the motor is at no-load, the stator winding draws a current I_0. It has two components viz., (*i*) I_w which supplies the no-load motor losses and (*ii*) magnetising component I_m which sets up magnetic flux in the core and the air-gap. The parallel combination of R_c and X_m, therefore, represents the no-load motor losses and the production of magnetic flux respectively.

$$\mathbf{I_0} = \mathbf{I_w} + \mathbf{I_m}$$

Rotor circuit. Here R_2 and X_2 represent the rotor resistance and standstill rotor reactance per phase respectively. At any slip s, the rotor reactance will be $s\,X_2$. The induced voltage/phase in the rotor is $E_2' = s\,E_2 = s\,K\,E_1$. Since the rotor winding is short-circuited, the whole of e.m.f. E_2' is used up in circulating the rotor current I_2'.

$$\therefore \quad \mathbf{E_2'} = \mathbf{I_2'}(R_2 + js\,X_2)$$

The rotor current $\mathbf{I_2'}$ is reflected as $\mathbf{I_2''}(= K\,\mathbf{I_2'})$ in the stator. The phasor sum of $\mathbf{I_2''}$ and $\mathbf{I_0}$ gives the stator current $\mathbf{I_1}$.

It is important to note that input to the primary and output from the secondary of a transformer are electrical. However, in an induction motor, the inputs to the stator and rotor are electrical but the output from the rotor is mechanical. To facilitate calculations, it is desirable and necessary to replace the mechanical load by an equivalent electrical load. We then have the transformer equivalent circuit of the induction motor.

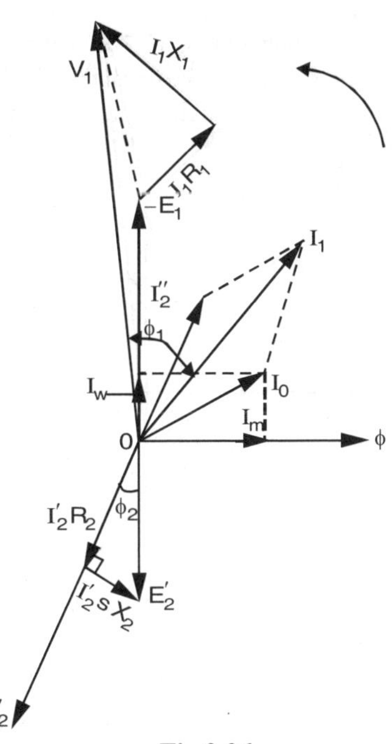

Fig.9.36

It may be noted that even though the frequencies of stator and rotor currents are different, yet the magnetic fields due to them rotate at synchronous speed N_s. The stator currents produce a magnetic flux which rotates at a speed N_s. At slip s, the speed of rotation of the rotor field relative to the rotor surface *in the direction of rotation of the rotor is*

$$= \frac{120\ f'}{P} = \frac{120\ s\ f}{P} = s\ N_s$$

But the rotor is revolving at a speed of N relative to the stator core. Therefore, the speed of rotor field relative to stator core

$$= s\ N_s + N = (N_s - N) + N = N_s$$

Thus no matter what the value of slip s, the stator and rotor magnetic fields are synchronous with each other when seen by an observer stationed in space. Consequently, the 3-phase induction motor can be regarded as being equivalent to a transformer having an air-gap separating the iron portions of the magnetic circuit carrying the primary and secondary windings.

Fig. 9.36 shows the phasor diagram of induction motor.

9.37 EQUIVALENT CIRCUIT OF THE ROTOR

We shall now see how mechanical load of the motor is replaced by the equivalent electrical load. Fig. 9.37 (*i*) shows the equivalent circuit per phase of the rotor at slip s. The rotor phase current is given by ;

$$I'_2 = \frac{s\ E_2}{\sqrt{R_2^2 + (s\ X_2)^2}}$$

Mathematically, this value is unaltered by writing it as :

$$I'_2 = \frac{E_2}{\sqrt{(R_2/s)^2 + (X_2)^2}}$$

As shown in Fig. 9.37 (*ii*), we now have a rotor circuit that has a fixed reactance X_2 connected in series with a variable resistance R_2/s and supplied with constant voltage E_2. Note that Fig. 9.37 (*ii*) transfers the variable to the resistance without altering power or power factor conditions.

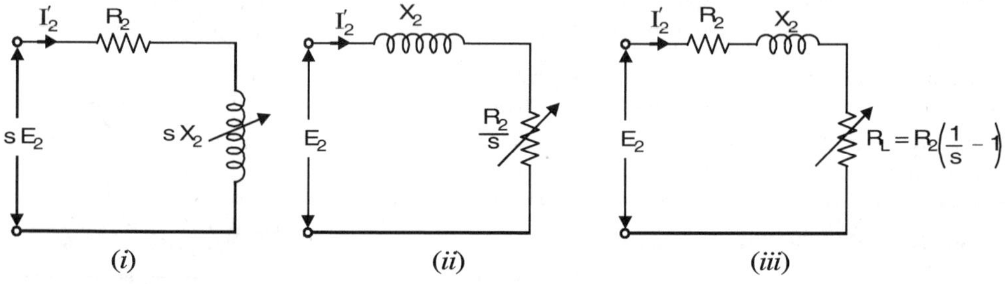

Fig. 9.37

The quantity R_2/s is greater than R_2 since s is a fraction. Therefore, R_2/s can be divided into a fixed part R_2 and a variable part $(R_2/s - R_2)$ *i.e.*,

$$\frac{R_2}{s} = R_2 + R_2\left(\frac{1}{s} - 1\right)$$

(*i*) The first part R_2 is the rotor resistance/phase and represents the rotor Cu loss.

(*ii*) The second part $R_2\left(\frac{1}{s} - 1\right)$ is a variable-resistance load. The power delivered to this load represents the total mechanical power developed in the rotor. Thus mechanical load on the induction motor can be replaced by a variable-resistance load of value $R_2\left(\frac{1}{s}-1\right)$. This is known as load resistance R_L.

$$\therefore \quad R_L = R_2\left(\frac{1}{s}-1\right)$$

Fig. 9.37 (*iii*) shows the equivalent rotor circuit along with load resistance R_L.

9.38 TRANSFORMER EQUIVALENT CIRCUIT OF INDUCTION MOTOR

Fig. 9.38 shows the equivalent circuit per phase of a 3-phase induction motor. Note that mechanical load on the motor has been replaced by an equivalent electrical resistance R_L given by ;

$$R_L = R_2\left(\frac{1}{s}-1\right) \quad ...(i)$$

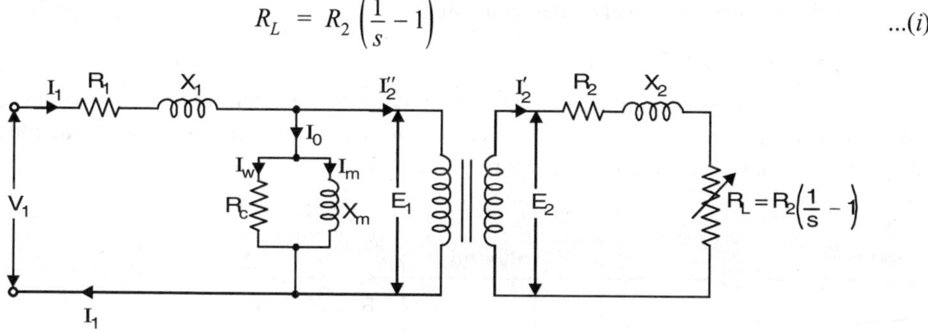

Fig. 9.38

Note that circuit shown in Fig. 9.38 is similar to the equivalent circuit of a transformer with secondary load equal to R_L given by eq. (*i*). The rotor e.m.f. in the equivalent circuit now depends *only* on the transformation ratio $K (= E_2/E_1)$.

Therefore, induction motor can be represented as an equivalent transformer connected to a variable-resistance load R_L given by eq. (*i*). The power delivered to R_L represents the total mechanical power developed in the rotor. Since the equivalent circuit of Fig. 9.38 is that of a transformer, the secondary (*i.e.*, rotor) values can be transferred to primary (*i.e.*, stator) through the appropriate use of transformation ratio K. Recall that when shifting resistance/reactance from secondary to primary, it should be *divided* by K^2 whereas current should be *multiplied* by K. The equivalent circuit of an induction motor referred to primary (stator) is shown in Fig. 9.39.

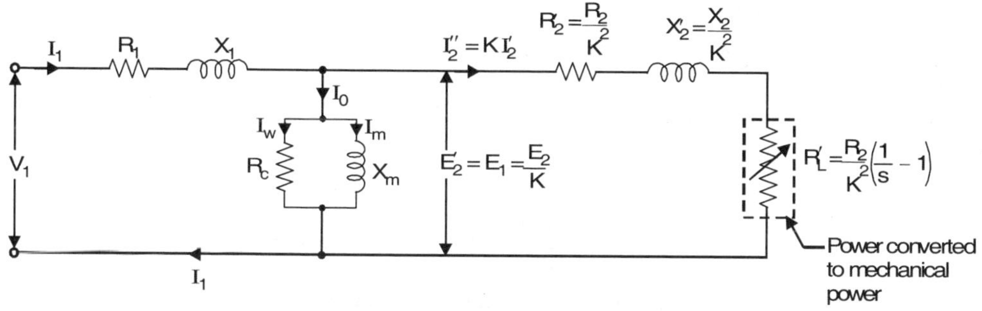

Fig. 9.39

Note that the element (*i.e.*, R_L') enclosed in the dotted box is the equivalent electrical resistance related to the mechanical load on the motor. The following points may be noted from the equivalent circuit of the induction motor :

(i) At no-load, the slip is practically zero and the load R'_L is infinite. This condition resembles that in a transformer whose secondary winding is open-circuited.

(ii) At standstill, the slip is unity and the load R'_L is zero. This condition resembles that in a transformer whose secondary winding is short-circuited.

(iii) When the motor is running under load, the value of R'_L will depend upon the value of the slip s. This condition resembles that in a transformer whose secondary is supplying variable and purely resistive load.

(iv) The equivalent electrical resistance R'_L related to mechanical load is slip or speed dependent. If the slip s increases, the load R'_L decreases and the rotor current increases and motor will develop more mechanical power. This is expected because the slip of the motor increases with the increase of load on the motor shaft.

9.39 POWER RELATIONS

The transformer equivalent circuit of an induction motor is quite helpful in analysing the various power relations in the motor. Fig. 9.40 shows the equivalent circuit per phase of an induction motor where all values have been referred to primary (*i.e.*, stator).

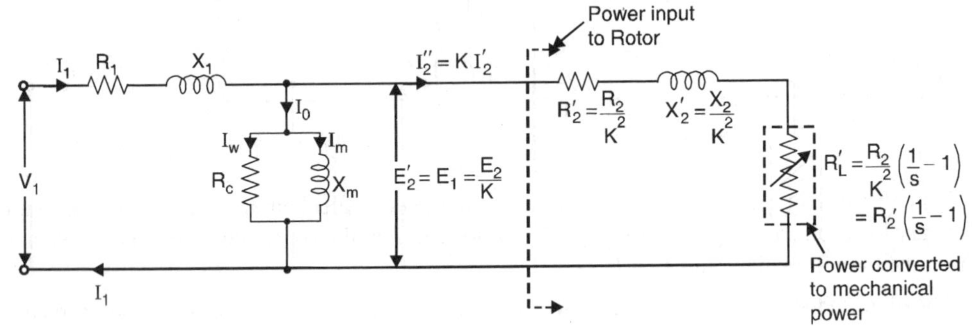

Fig. 9.40

(i) Total electrical load $= R'_2\left(\frac{1}{s}-1\right) + R'_2 = \frac{R'_2}{s}$

Power input to stator $= 3\, V_1\, I_1 \cos\phi_1$

There will be stator core loss and stator Cu loss. The remaining power will be the power transferred across the air-gap *i.e.*, input to the rotor.

(ii) Rotor input $= \dfrac{3\,(I''_2)^2\, R'_2}{s}$

Rotor Cu loss $= 3(I''_2)^2\, R'_2$

Total mechanical power developed by the rotor is

$$P_m = \text{Rotor input} - \text{Rotor Cu loss}$$

$$= \frac{3\,(I''_2)^2\, R'_2}{s} - 3\,(I''_2)^2\, R'_2 = 3\,(I''_2)^2\, R'_2\left(\frac{1}{s}-1\right)$$

This is quite apparent from the equivalent circuit shown in Fig. 9.40.

(iii) If T_g is the gross torque developed by the rotor, then,

$$P_m = \frac{2\pi\, N\, T_g}{60}$$

or

$$3\,(I''_2)^2\, R'_2\left(\frac{1}{s}-1\right) = \frac{2\pi\, N\, T_g}{60}$$

$$\text{or} \quad 3\,(I_2'')^2\,R_2'\left(\frac{1-s}{s}\right) = \frac{2\pi N T_g}{60}$$

$$\text{or} \quad 3\,(I_2'')^2\,R_2'\left(\frac{1-s}{s}\right) = \frac{2\pi N_s (1-s) T_g}{60} \qquad [\because N = N_s(1-s)]$$

$$\therefore \quad T_g = \frac{3\,(I_2'')^2\,R_2'/s}{2\pi N_s/60} \text{ N-m}$$

$$\text{or} \quad T_g = 9{\cdot}55\,\frac{3\,(I_2'')^2\,R_2'/s}{N_s} \text{ N-m}$$

Note that shaft torque T_{sh} will be less than T_g by the torque required to meet windage and frictional losses.

9.40 APPROXIMATE EQUIVALENT CIRCUIT OF INDUCTION MOTOR

As in case of a transformer, the approximate equivalent circuit of an induction motor is obtained by shifting the shunt branch (R_c - X_m) to the input terminals as shown in Fig. 9.41. This step has been taken on the assumption that voltage drop in R_1 and X_1 is small and the terminal voltage V_1 does not appreciably differ from the induced voltage E_1. Fig. 9.41 shows the approximate equivalent circuit per phase of an induction motor where all values have been referred to primary (*i.e.*, stator).

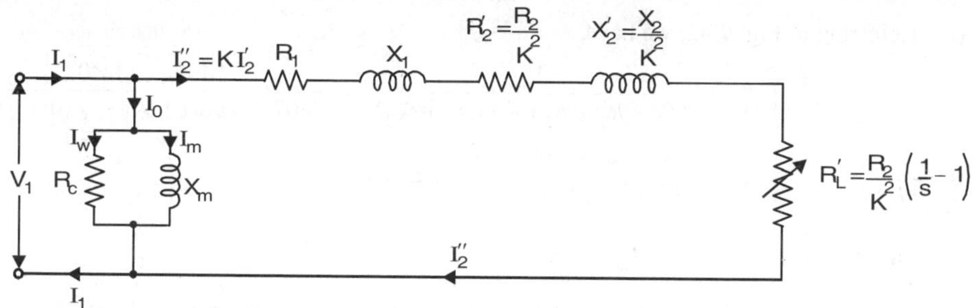

Fig. 9.41

The above approximate circuit of induction motor is not so readily justified as with the transformer. This is due to the following reasons :

(*i*) Unlike that of a power transformer, the magnetic circuit of the induction motor has an air-gap. Therefore, the exciting current of induction motor (30 to 40 % of full-load current) is much higher than that of the power transformer. Consequently, the exact equivalent circuit must be used for accurate results.

(*ii*) The relative values of X_1 and X_2 in an induction motor are larger than the corresponding ones to be found in the transformer. This fact does not justify the use of approximate equivalent circuit.

(*iii*) In a transformer, the windings are concentrated whereas in an induction motor, the windings are distributed. This affects the transformation ratio.

Inspite of the above drawbacks of approximate equivalent circuit, it yields results that are satisfactory for large motors. However, approximate equivalent circuit is not justified for small motors.

Example 9.52. *A 400 V, 50 Hz, 4-pole, 3-phase, star-connected induction motor has per phase stator impedance $Z_1 = (0{\cdot}07 + j\,0{\cdot}3)\ \Omega$ and per phase rotor impedance referred to stator side $Z'_2 = (0{\cdot}08 + j\,0{\cdot}3)\ \Omega$. The per phase magnetising reactance is 10 Ω and resistance representing core-loss is 50 Ω. The slip is 4 %. Using approximate equivalent circuit, calculate* (*i*) *stator current and stator power factor* (*ii*) *developed torque* (*iii*) *gross efficiency.*

Solution. As per the given data, we have,

$R_1 = 0{\cdot}07\,\Omega\,; X_1 = 0{\cdot}3\,\Omega\,; R'_2 = 0{\cdot}08\,\Omega\,; X'_2 = 0{\cdot}3\,\Omega$

Equivalent electrical resistance referred to stator is

$$R'_L = R'_2\left(\frac{1}{s}-1\right) = 0{\cdot}08\left(\frac{1}{0{\cdot}04}-1\right) = 1{\cdot}92\,\Omega$$

Fig. 9.42 shows the approximate equivalent circuit per phase with the circuit values. Here $V_1 = 400/\sqrt{3}$ volts.

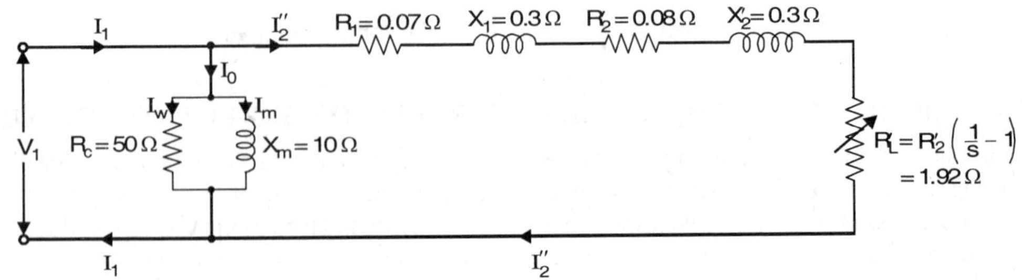

Fig. 9.42

(*i*) Referring to Fig. 9.42, we have,

$$I''_2 = \frac{\mathbf{V_1}}{(R_1+R'_2+R'_L)+j\,(X_1+X'_2)} = \frac{(400/\sqrt{3})\,\angle 0^0}{(0{\cdot}07+0{\cdot}08+1{\cdot}92)+j\,(0{\cdot}3+0{\cdot}3)}$$

$$= \frac{(400/\sqrt{3})\,\angle 0^0}{(2{\cdot}07+j\,0{\cdot}6)} = 107\,\angle{-16{\cdot}2^\circ}\text{ A}$$

No-load current, $\mathbf{I_0} = \mathbf{I_w}+\mathbf{I_m} = \dfrac{\mathbf{V_1}}{R_c}+\dfrac{\mathbf{V_1}}{j\,X_m}$

$$= \frac{(400/\sqrt{3})\,\angle 0^\circ}{50}+\frac{(400/\sqrt{3})\,\angle 0^\circ}{j10} = \frac{(400/\sqrt{3})}{50}-j\,\frac{(400/\sqrt{3})}{10}$$

$$= (4{\cdot}6-j\,23)\text{ A} = 23{\cdot}4\,\angle{-79^\circ}\text{ A}$$

Stator current, $\mathbf{I_1} = \mathbf{I''_2}+\mathbf{I_0} = 107\,\angle{-16{\cdot}2^\circ}+23{\cdot}4\,\angle{-79^\circ} = 119{\cdot}7\,\angle{-26{\cdot}3^\circ}\text{ A}$

∴ Magnitude of stator current = **119·7 A**

Stator power factor = $\cos(-26{\cdot}3^\circ)$ = **0·89** *lagging*

(*ii*) Synchronous speed, $N_s = 120f/P = 120\times 50/4 = 1500$ r.p.m.

Rotor speed, $N = N_s(1-s) = 1500\,(1-0{\cdot}04) = 1440$ r.p.m.

Gross rotor output, $P_m = 3\,(I''_2)^2\,R'_L = 3\times(107)^2\times 1{\cdot}92 = 65946$ W

If T_g is the gross torque in N-m, then,

$$T_g = 9{\cdot}55\times\frac{P_m}{N} = 9{\cdot}55\times\frac{65946}{1440} = \mathbf{437{\cdot}35\text{ N-m}}$$

(*iii*) Input to stator, $P_i = 3\,V_1\,I_1\cos\phi_1 = 3\times\dfrac{400}{\sqrt{3}}\times 119{\cdot}7\times 0{\cdot}89 = 73808$ W

∴ Gross efficiency $= \dfrac{P_m}{P_i}\times 100 = \dfrac{65946}{73808}\times 100$ = **89·4 %**

Example 9.53. *A 400 V, 4-pole, 50 Hz, 3-phase, star-connected induction motor has the following per phase parameters referred to stator :*

$R_1 = 0{\cdot}6\,\Omega\,; X_1 = 1{\cdot}1\,\Omega\,; R'_2 = 0{\cdot}3\,\Omega\,; X'_2 = 0{\cdot}5\,\Omega\,; X_m = 25\,\Omega$

The mechanical losses are 1000 W and stator core losses are 500 W. The slip is 3 %. Using approximate equivalent circuit, find (i) speed (ii) stator current (iii) stator power factor (iv) power input to rotor (v) gross torque (vi) shaft torque (vii) efficiency (viii) rotor copper loss/phase. Neglect R_c.

Solution.

Equivalent electrical resistance referred to stator is

$$R'_L = R'_2\left(\frac{1}{s}-1\right) = 0{\cdot}3\left(\frac{1}{0{\cdot}03}-1\right) = 9{\cdot}7\ \Omega$$

Fig. 9.43 shows the approximate equivalent circuit per phase with circuit values. Here $V_1 = 400/\sqrt{3}$ volts.

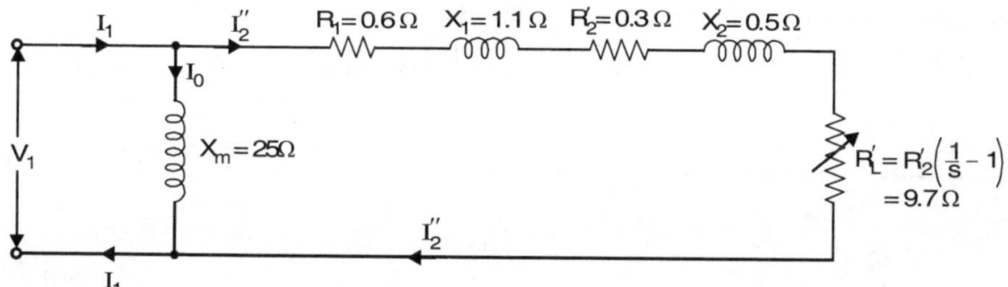

Fig. 9.43

(*i*) Synchronous speed, $N_s = 120 f/P = 120 \times 50/4 = 1500$ r.p.m.

Rotor speed, $N = N_s(1-s) = 1500\,(1-0{\cdot}03) =$ **1455 r.p.m.**

(*ii*) Referring to Fig. 9.43, we have,

$$\mathbf{I}''_2 = \frac{\mathbf{V}_1}{(R_1+R'_2+R'_L)+j\,(X_1+X'_2)} = \frac{(400/\sqrt{3})\angle 0^\circ}{(0{\cdot}6+0{\cdot}3+9{\cdot}7)+j\,(1{\cdot}1+0{\cdot}5)}$$

$$= \frac{231}{10{\cdot}6+j\,1{\cdot}6} = 21{\cdot}54\angle{-8{\cdot}6^\circ}\text{ A}$$

$$\mathbf{I}_0 = \frac{\mathbf{V}_1}{j\,X_m} = \frac{(400/\sqrt{3})\angle 0^\circ}{j\,25} = \frac{231}{j\,25} = 9{\cdot}24\angle{-90^\circ}\text{ A}$$

Stator current, $\mathbf{I}_1 = \mathbf{I}''_2 + \mathbf{I}_0 = 21{\cdot}54\angle{-8{\cdot}6^\circ} + 9{\cdot}24\angle{-90^\circ} = (21{\cdot}16 - j\,12{\cdot}46)$ A

$= 24{\cdot}56\angle{-30{\cdot}5^\circ}$ A

Magnitude of stator current = **24·56 A**

(*iii*) Stator p.f., $\cos\phi_1 = \cos(-30{\cdot}5^\circ) =$ **0·86** ***lagging***

(*iv*) Stator input, $P_i = 3\,V_1 I_1 \cos\phi_1 = 3\times(400/\sqrt{3})\times 24{\cdot}56\times 0{\cdot}86 = 14637$ W

Stator Cu losses $= 3\,I_1^2 R_1 = 3\times(24{\cdot}56)^2\times 0{\cdot}6 = 1086$ W

Stator core losses = 500 W

∴ Rotor input = 14637 – 1086 – 500 = **13051 W**

(*v*) Total mechanical power developed in the rotor is

$$P_m = 13051\,(1-s) = 13051\,(1-0{\cdot}03) = 12659\text{ W}$$

Gross torque, $T_g = 9{\cdot}55\times\dfrac{P_m}{N} = 9{\cdot}55\times\dfrac{12659}{1455} =$ **83·1 N-m**

(*vi*) Useful output, $P_{out} = P_m$ – Mechanical losses = 12659 – 1000 = 11659 W

Shaft torque, $T_{sh} = 9{\cdot}55\times\dfrac{P_{out}}{N} = 9{\cdot}55\times\dfrac{11659}{1455} =$ **76.52 N-m** (*vii*) Efficiency, $\eta = \frac{P_{out}}{P_i} \times 100 = \frac{11659}{14637} \times 100 = \mathbf{79{\cdot}65\ \%}$

(*viii*) Total rotor Cu loss $= s \times$ Rotor input $= 0{\cdot}03 \times 13051 = 391{\cdot}5$ W

Rotor Cu loss/phase $= 391{\cdot}5/3 = \mathbf{130{\cdot}5\ W}$

Example 9.54. *A 3-phase, star connected, 400 V, 50 Hz, 4-pole induction motor has the following per phase parameters referred to the stator :*

$R_1 = 0{\cdot}15\ \Omega\ ;\ X_1 = 0{\cdot}45\ \Omega\ ;\ R'_2 = 0{\cdot}12\ \Omega\ ;\ X'_2 = 0{\cdot}45\ \Omega\ ;\ X_m = 28{\cdot}5\ \Omega$

Calculate the stator current and power factor when the motor is operated at rated voltage and frequency with slip s = 0·04.

Solution. Fig. 9.44 shows the approximate equivalent circuit per phase with circuit values.

Equivalent electrical resistance referred to the stator is

$$R'_L = R'_2\left(\frac{1}{s}-1\right) = 0{\cdot}12\left(\frac{1}{0{\cdot}04}-1\right) = 2{\cdot}88\ \Omega$$

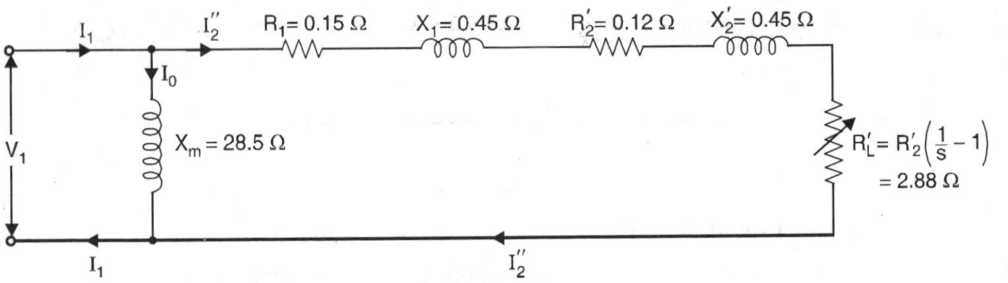

Fig. 9.44

$$\mathbf{I''_2} = \frac{\mathbf{V}_1}{(R_1 + R'_2 + R'_L) + j\,(X_1 + X'_2)}$$

$$= \frac{(400/\sqrt{3})\ \angle 0^\circ}{(0{\cdot}15 + 0{\cdot}12 + 2{\cdot}88) + j\,(0{\cdot}45 + 0{\cdot}45)}$$

$$= \frac{(400/\sqrt{3})}{3{\cdot}15 + j\,0{\cdot}9} = (67{\cdot}78 - j\,19{\cdot}36)\text{ A}$$

$$\mathbf{I}_0 = \frac{\mathbf{V}_1}{j\,X_m} = \frac{(400/\sqrt{3})\ \angle 0^\circ}{j\,28{\cdot}5} = -j\,8{\cdot}1\text{ A}$$

$\therefore$ Stator current, $\mathbf{I}_1 = \mathbf{I''_2} + \mathbf{I}_0 = (67{\cdot}78 - j\,19{\cdot}36) + (-j\,8{\cdot}1)$

$= (67{\cdot}78 - j\,27{\cdot}46)\text{ A} = 73{\cdot}13\ \angle{-22^\circ}\text{ A}$

Magnitude of stator current $=$ **73·13 A**

Stator power factor $= \cos(-22^\circ) =$ **0·927** *lag*

TUTORIAL PROBLEMS

1. A 440 V, 3-phase, 50 Hz, 8-pole, star-connected induction motor has the following per phase parameters referred to the stator :
   $R_1 = R'_2 = 0{\cdot}1\ \Omega\ ;\ X_1 = X'_2 = 0{\cdot}7\ \Omega\ ;\ R_c = 100\ \Omega\ ;\ X_m = 25\ \Omega$
   Mechanical losses are 1000 W and slip is 4%. Using approximate equivalent circuit, find (*i*) stator current (*ii*) stator power factor (*iii*) torque and (*iv*) efficiency.
   [(*i*) 93·3 A (*ii*) 0·838 lagging (*iii*) 706 N-m (*iv*) 87·5 %]
2. A 440 V, 3-phase, 50 Hz, 37·3 kW, star-connected induction motor has the following per phase parameters:
   $R_1 = 0{\cdot}1\ \Omega\ ;\ X_1 = 0{\cdot}4\ \Omega\ ;\ R'_2 = 0{\cdot}15\ \Omega\ ;\ X'_2 = 0{\cdot}44\ \Omega$

The motor has stator core loss of 1250 W and rotational loss of 1000 W. It draws a no-load line current of 20 A at a p.f. 0·09 lagging. When motor operates at a slip of 3 %, calculate (*i*) input line current and p.f. (*ii*) torque developed (*iii*) output and efficiency.

[(*i*) 57·4 A ; 0·875 lag (*ii*) 230 N-m (*iii*) 34075 W ; 88·7 %]

9.41 STARTING OF 3-PHASE INDUCTION MOTORS

The induction motor is fundamentally a transformer in which the stator is the primary and the rotor is short-circuited secondary. At starting, the voltage induced in the induction motor rotor is maximum ($\because s = 1$). Since the rotor impedance is low, the rotor current is excessively large. This large rotor current is reflected in the stator because of transformer action. This results in high starting current (4 to 10 times the full-load current) in the stator at *low power factor and consequently the value of starting torque is low. Because of the short duration, this value of large current does not harm the motor if the motor accelerates normally. However, this large starting current will produce large line-voltage drop. This will adversely affect the operation of other electrical equipment connected to the same lines. Therefore, it is desirable and necessary to reduce the magnitude of stator current at starting and several methods are available for this purpose.

9.42 METHODS OF STARTING 3-PHASE INDUCTION MOTORS

The method to be employed in starting a given induction motor depends upon the size of the motor and the type of the motor. The common methods used to start induction motors are :

(*i*) Direct-on-line starting
(*ii*) Stator resistance starting
(*iii*) Autotransformer starting
(*iv*) Star-delta starting
(*v*) Rotor resistance starting

Methods (*i*) to (*iv*) are applicable to both squirrel-cage and slip ring motors. However, method (*v*) is applicable only to slip ring motors. In practice, any one of the first four methods is used for starting squirrel cage motors, depending upon the size of the motor. But slip ring motors are invariably started by rotor resistance starting.

9.43 METHODS OF STARTING SQUIRREL-CAGE MOTORS

Except direct-on-line starting, all other methods of starting squirrel-cage motors employ reduced voltage across motor terminals at starting.

(*i*) **Direct-on-line starting.** This method of starting is just what the name implies – the motor is started by connecting it directly to 3-phase supply. The impedance of the motor at standstill is relatively low and when it is directly connected to the supply system, the starting current will be high (4 to 10 times the full-load current) and at a low power factor. Consequently, this method of starting is suitable for relatively small (upto 7·5 kW) machines.

Relation between starting and F.L. torques. We know that :

$$\text{Rotor input} = 2\pi N_s T = kT \qquad \ldots N_s \text{ in r.p.s. (See Art. 9.28)}$$

But $\quad \text{Rotor Cu loss} = s \times \text{Rotor input}$

$$\therefore \quad 3(I'_2)^2 R_2 = s \times kT$$

or $\quad T \propto (I'_2)^2/s \qquad \ldots$ for same R_2

or $\quad T \propto I_1^2/s \qquad (\because I'_2 \propto I_1)$

If I_{st} is the starting current, then starting torque (T_{st}) is

$$T_{st} \propto I_{st}^2 \qquad (\because \text{at starting } s = 1)$$

If I_f is the full-load current and s_f is the full-load slip, then,

---

* At starting ($s = 1$), the rotor reactance is high compared with the rotor resistance. This results in low power factor at starting.

$$T_f \propto I_f^2/s_f$$

$$\therefore \quad \frac{T_{st}}{T_f} = \left(\frac{I_{st}}{I_f}\right)^2 \times s_f$$

When the motor is started direct-on-line, the starting current is the short-circuit (blocked-rotor) current I_{sc}.

$$\therefore \quad \frac{T_{st}}{T_f} = \left(\frac{I_{sc}}{I_f}\right)^2 \times s_f$$

Let us illustrate the above relation with a numerical example. Suppose $I_{sc} = 5\,I_f$ and full-load slip $s_f = 0{\cdot}04$. Then,

$$\frac{T_{st}}{T_f} = \left(\frac{I_{sc}}{I_f}\right)^2 \times s_f = \left(\frac{5\,I_f}{I_f}\right)^2 \times 0{\cdot}04 = (5)^2 \times 0{\cdot}04 = 1$$

$$\therefore \quad T_{st} = T_f$$

Note that starting current is as large as five times the full-load current but starting torque is just equal to the full-load torque. Therefore, starting current is very high and the starting torque is comparatively low. If this large starting current flows for a long time, it may overheat the motor and damage the insulation.

(*ii*) **Stator resistance starting.** In this method, external resistances are connected in series with each phase of stator winding during starting [See Fig. 9.45]. This causes voltage drop across the resistances so that voltage available across motor terminals is reduced and hence the starting current. The starting resistances are gradually cut out in steps (two or more steps) from the stator circuit as the motor picks up speed. When the motor attains rated speed, the resistances are completely cut out and full line voltage is applied to the motor.

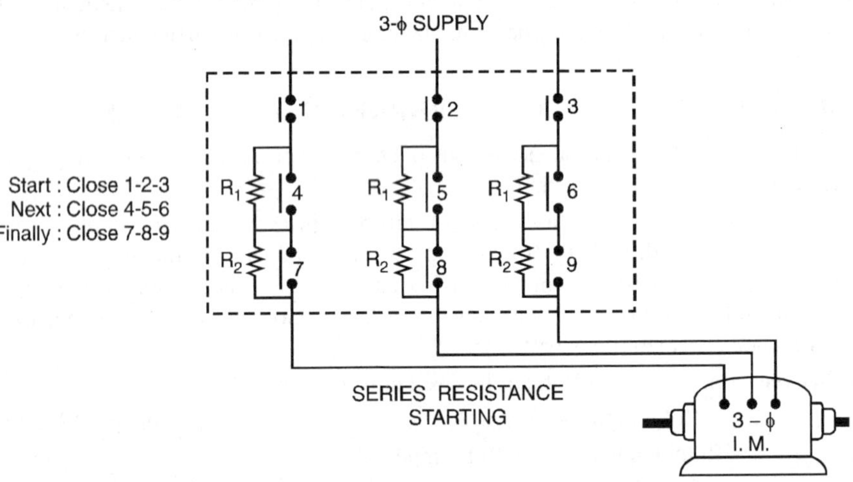

Fig. 9.45

This method suffers from two drawbacks. First, the reduced voltage applied to the motor during the starting period *lowers the starting torque and hence increases the accelerating time. Secondly, a lot of power is wasted in the starting resistances.

Relation between starting and F.L. torques. Let V be the rated voltage/phase. If the voltage is reduced by a fraction x by the insertion of resistors in the line, then voltage applied to the motor per phase will be xV.

* As shown in Art. 9.13, $T_s \propto V^2$ where V is the supply voltage.

$$\therefore \qquad I_{st} = x\,I_{sc}$$

Now
$$\frac{T_{st}}{T_f} = \left(\frac{I_{st}}{I_f}\right)^2 \times s_f$$

or
$$\frac{T_{st}}{T_f} = x^2\left(\frac{I_{sc}}{I_f}\right)^2 \times s_f$$

Thus while the starting current reduces by a fraction x of the rated-voltage starting current (I_{sc}), the starting torque is reduced by a fraction x^2 of that obtained by direct switching. The reduced voltage applied to the motor during the starting period lowers the starting current but at the same time increases the accelerating time because of the reduced value of the starting torque. Therefore, this method is used for starting small motors only.

(*iii*) **Autotransformer starting.** This method also aims at connecting the induction motor to a reduced supply at starting and then connecting it to the full voltage as the motor picks up sufficient speed. Fig. 9.46 shows the circuit arrangement for autotransformer starting. The tapping on the autotransformer is so set that when it is in the circuit, 65% to 80% of line voltage is applied to the motor.

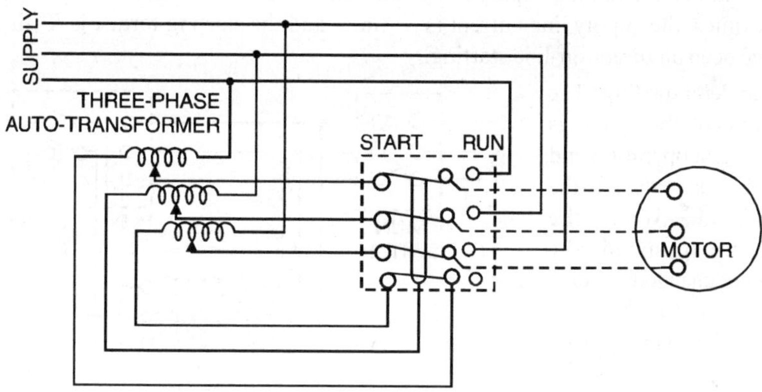

Fig. 9.46

At the instant of starting, the change-over switch is thrown to "start" position. This puts the autotransformer in the circuit and thus reduced voltage is applied to the circuit. Consequently, starting current is limited to safe value. When the motor attains about 80% of normal speed, the change-over switch is thrown to "run" position. This takes out the autotransformer from the circuit and puts the motor to full line voltage. Autotransformer starting has several advantages *viz* low power loss, low starting current and less radiated heat. For large machines (over 25 h.p.), this method of starting is often used. This method can be used for both star and delta connected motors.

Relation between starting and F.L. torques. Consider a star-connected squirrel-cage induction motor. If V is the line voltage, then voltage across motor phase on direct switching is $V/\sqrt{3}$ and starting current is $I_{st} = I_{sc}$. In case of autotransformer, if a tapping of transformation ratio K (a fraction) is used , then phase voltage across motor is $KV/\sqrt{3}$ and $I_{st} = K\,I_{sc}$.

Now
$$\frac{T_{st}}{T_f} = \left(\frac{I_{st}}{I_f}\right)^2 \times s_f = \left(\frac{K\,I_{sc}}{I_f}\right)^2 \times s_f = K^2\left(\frac{I_{sc}}{I_f}\right)^2 \times s_f$$

$$\therefore \qquad \frac{T_{st}}{T_f} = K^2\left(\frac{I_{sc}}{I_f}\right)^2 \times s_f$$

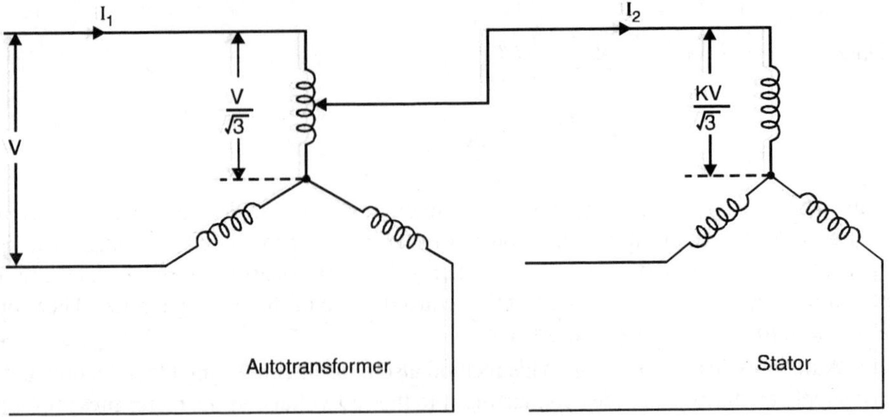

Fig. 9.47

The current taken from the supply or by autotransformer is $I_1 = KI_2 = K^2 I_{sc}$. Note that motor current is K times, the supply line current is K^2 times and the starting torque is K^2 times the value it would have been on direct-on-line starting.

(*iv*) **Star-delta starting.** The stator winding of the motor is designed for delta operation and is connected in star during the starting period. When the machine is up to speed, the connections are changed to delta. The circuit arrangement for star-delta starting is shown in Fig. 9.48.

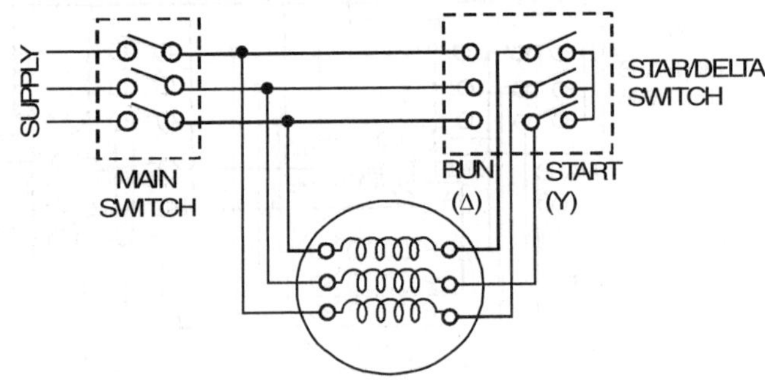

Fig. 9.48

The six leads of the stator windings are connected to the change-over switch as shown. At the instant of starting, the change-over switch is thrown to "Start" position which connects the stator windings in star. Therefore, each stator phase gets $V/\sqrt{3}$ volts where V is the line voltage. This reduces the starting current. When the motor picks up speed, the change-over switch is thrown to "Run" position which connects the stator windings in delta. Now each stator phase gets full line voltage V. The disadvantages of this method are :

(*a*) With star-connection during starting, stator phase voltage is $1/\sqrt{3}$ times the line voltage. Consequently, starting torque is $(1/\sqrt{3})^2$ or $1/3$ times the value it would have with Δ-connection. This is rather a large reduction in starting torque.

(*b*) The reduction in voltage is fixed.

This method of starting is used for medium-size machines (upto about 25 h.p.).

Relation between starting and F.L. torques. In direct delta starting,

$$\text{Starting current/phase, } I_{sc} = V/Z_{sc} \quad \text{where } V = \text{line voltage}$$

$$\text{Starting line current} = \sqrt{3}\, I_{sc}$$

In star starting, we have,

$$\text{Starting current/phase, } I_{st} = \frac{V/\sqrt{3}}{Z_{sc}} = \frac{1}{\sqrt{3}} I_{sc}$$

Now $$\frac{T_{st}}{T_f} = \left(\frac{I_{st}}{I_f}\right)^2 \times s_f = \left(\frac{I_{sc}}{\sqrt{3} \times I_f}\right)^2 \times s_f$$

or $$\frac{T_{st}}{T_f} = \frac{1}{3}\left(\frac{I_{sc}}{I_f}\right)^2 \times s_f$$

where I_{sc} = starting phase current (delta)

I_f = F.L. phase current (delta)

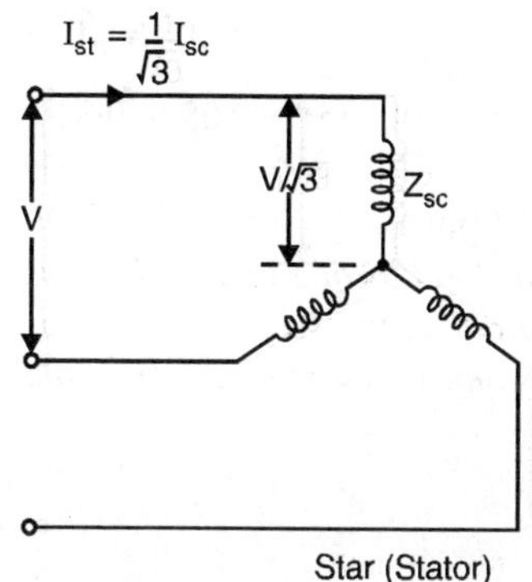

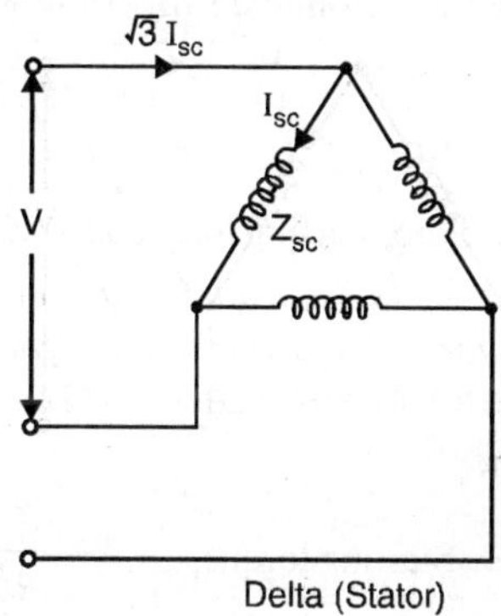

Fig. 9.49

Note that in star-delta starting, the starting line current is reduced to *one-third as compared to starting with the winding delta connected. Further, starting torque is reduced to one-third of that obtainable by direct delta starting. This method is cheap but limited to applications where high starting torque is not necessary *e.g.*, machine tools, pumps etc.

Example 9.55. *A 3-phase squirrel cage induction motor takes a starting current 6 times the full-load current. Find the starting torque as a percentage of full-load torque if the motor is started (i) direct-on-line (ii) through a star-delta starter. The full-load slip of the motor is 0·04.*

Solution.

$I_{sc} = 6\, I_f$ or $I_{sc}/I_f = 6$; $s_f = 0{\cdot}04$

(*i*) ***Direct-on-line starting***

$$\frac{T_{st}}{T_f} = \left(\frac{I_{sc}}{I_f}\right)^2 \times s_f = (6)^2 \times 0{\cdot}04 = 1{\cdot}44 = \mathbf{144\ \%}$$

Therefore, the starting torque is 144 % of the full-load torque.

(*ii*) ***Star-delta starting***

$$\frac{T_{st}}{T_f} = \frac{1}{3}\left(\frac{I_{sc}}{I_f}\right)^2 \times s_f = \frac{1}{3}(6)^2 \times 0{\cdot}04 = 0{\cdot}48 = \mathbf{48\%}$$

Therefore, starting torque is 48% of the full-load torque.

Example 9.56. *Determine approximately the starting torque of an induction motor in terms of full-load torque when started by means of (i) star-delta switch (ii) an autotransformer with 50% tapping. The short-circuit current of the motor at normal voltage is five times the full-load current and full-load slip is 5 %. Neglect magnetising current.*

---

* $$\frac{\text{Line current in star}}{\text{Line current in delta}} = \frac{I_{sc}}{\sqrt{3}} \div \sqrt{3}\, I_{sc} = \frac{I_{sc}}{\sqrt{3}} \times \frac{1}{\sqrt{3}\, I_{sc}} = \frac{1}{3}$$

Solution.

$I_{sc} = 5\,I_f$ or $I_{sc}/I_f = 5$; $s_f = 5\,\% = 0{\cdot}05$

(*i*) ***Star-delta starting***

$$\frac{T_{st}}{T_f} = \frac{1}{3}\left(\frac{I_{sc}}{I_f}\right)^2 \times s_f = \frac{1}{3} \times (5)^2 \times 0{\cdot}05 = 0{\cdot}42 = \mathbf{42\%}$$

Therefore, starting torque is 42 % of the full-load torque.

(*ii*) ***Autotransformer starting*** ($K = 50\,\% = 0{\cdot}5$)

$$\frac{T_{st}}{T_f} = K^2\left(\frac{I_{sc}}{I_f}\right)^2 \times s_f = (0{\cdot}5)^2 \times (5)^2 \times 0{\cdot}05 = 0{\cdot}31 = \mathbf{31\ \%}$$

Therefore, starting torque is 31% of the full-load torque.

Example 9.57. *The ratio of maximum torque to full-load torque in a 3-phase squirrel cage induction motor is 2·5. Calculate the ratio of starting torque to full-load torque for direct-on-line starting. The rotor resistance is 0·5 Ω/phase and rotor standstill reactance is 5 Ω/phase.*

Solution.

$$\text{Starting torque, } T_{st} = \frac{K\,E_2^2\,R_2}{R_2^2 + X_2^2}$$

$$\text{Maximum torque, } T_m = \frac{K\,E_2^2}{2\,X_2}$$

$$\text{Full-load torque, } T_f = \frac{T_m}{2{\cdot}5} \quad \text{...given}$$

$$= \frac{K\,E_2^2}{2{\cdot}5 \times 2\,X_2} = \frac{K\,E_2^2}{2{\cdot}5 \times 2 \times 5} = \frac{K\,E_2^2}{25}$$

$$\therefore \quad \frac{T_{st}}{T_f} = \frac{K\,E_2^2\,R_2}{R_2^2 + X_2^2} \times \frac{25}{K\,E_2^2} = \frac{25\,R_2}{R_2^2 + X_2^2}$$

$$\text{or} \quad \frac{T_{st}}{T_f} = \frac{25 \times 0{\cdot}5}{(0{\cdot}5)^2 + (5)^2} = \frac{12{\cdot}5}{25{\cdot}25} = \mathbf{0{\cdot}49}$$

Example 9.58. *Determine a suitable autotransformer ratio for starting a 3-phase induction motor with line current not exceeding twice the full-load current. The short circuit current is 4 times the full-load current and full-load slip is 2·5 %. Estimate the starting torque in terms of the full-load torque.*

Solution.

$I_{sc} = 4\,I_f$; $s_f = 2{\cdot}5\,\% = 0{\cdot}025$; $K = ?$

$$\text{Supply line current} = K^2\,I_{sc}$$

It is given that supply line current at start is equal to $2\,I_f$ and $I_{sc} = 4\,I_f$.

$$\therefore \quad 2\,I_f = K^2 \times 4\,I_f \quad \text{or} \quad K^2 = \frac{1}{2} \quad \therefore \quad K = \frac{1}{\sqrt{2}} = \mathbf{0{\cdot}707}$$

In case of autotransformer starting, we have,

$$\frac{T_{st}}{T_f} = K^2\left(\frac{I_{sc}}{I_f}\right)^2 \times s_f = \frac{1}{2} \times \left(\frac{4\,I_f}{I_f}\right)^2 \times 0{\cdot}025 = 0{\cdot}2 = \mathbf{20\%}$$

Therefore, starting torque is 20 % of full-load torque.

Example 9.59. *A 3 h.p., 3-phase induction motor with full-load efficiency and power factor of 0·83 and 0·8 respectively has a short-circuit current of 3·5 times the full-load current. Estimate the line current at the instant of starting the motor from a 500 V supply by means of a star-delta switch. Ignore the magnetising current.*

Solution.

$$\sqrt{3}\ V_L I_L \cos\phi = \frac{\text{Output}}{\eta}$$

or $$\sqrt{3} \times 500 \times I_L \times 0{\cdot}8 = \frac{3 \times 746}{0{\cdot}83}$$

∴ Full-load line current, $$I_L = \frac{3 \times 746}{\sqrt{3} \times 500 \times 0{\cdot}8 \times 0{\cdot}83} = 3{\cdot}89\ \text{A}$$

F.L. phase current, $$I_{ph} = 3{\cdot}89/\sqrt{3} = 2{\cdot}25\ \text{A}$$

∴ $$I_{sc} = 3{\cdot}5\, I_{ph} = 3{\cdot}5 \times 2{\cdot}25 = 7{\cdot}86\ \text{A}$$

In star-delta starter, $$I_{st} = \frac{I_{sc}}{\sqrt{3}} = \frac{7{\cdot}86}{\sqrt{3}} = \mathbf{4{\cdot}54\ A}$$

Example 9.60. *Find the ratio of starting to full-load current in a 15 h.p., 400 V, 3-phase induction motor with star/delta starter, given that full-load p.f. is 0·85, the full-load efficiency is 0·88, short-circuit current at 200 V is 40 A. Ignore magnetising current.*

Solution. The full-load line current I_L drawn by Δ-connected motor is

$$\sqrt{3}\ V_L I_L \cos\phi = \text{Output}/\eta$$

or $$\sqrt{3} \times 400 \times I_L \times 0{\cdot}85 = \frac{15 \times 746}{0{\cdot}88}$$

∴ $$I_L = 21{\cdot}6\ \text{A}$$

The line value of I_{sc} of Δ-connected motor is 40 A at 200 V.

∴ $$I_{sc}\ \text{(line value) at 400 V} = 40 \times 400/200 = 80\ \text{A}$$

$$I_{sc}\ \text{(phase value)} = 80/\sqrt{3}\ \text{A} = 46{\cdot}2\ \text{A}$$

With star-delta starting, $$I_{st}/\text{phase} = \frac{1}{\sqrt{3}} \times I_{sc}\ \text{per phase} = \frac{1}{\sqrt{3}} \times \frac{80}{\sqrt{3}} = \frac{80}{3}\ \text{A}$$

This is also the line current (star-connected) drawn by the motor during starting.

∴ $$\frac{\text{Line value of starting current}}{\text{Line value of F.L. current}} = \frac{80/3}{21{\cdot}6} = \mathbf{1{\cdot}24}$$

TUTORIAL PROBLEMS

1. A 3-phase induction motor is designed to run at 5 % slip at full-load. If motor draws 6 times the full-load current at starting at the rated voltage, find the ratio of starting to full-load torque. **[1·8]**
2. A 3-phase, 4-pole, 50 Hz, 208 V induction motor has a starting line current of 700 A and a starting torque of 225 N-m. If a reduced voltage of 120 V is applied to the stator at the time of starting, find the starting torque and starting line current. **[75 N-m ; 403 A]**
3. Find the ratio of starting to full-load current for a 7·46 kW, 400 V, 3-phase induction motor with star/delta starter, given that full-load efficiency is 0·87, the full-load p.f. is 0·85 and the short-circuit current is 15 A at 100 V. **[1·37]**
4. A 3-phase, 6-pole, 50 Hz induction motor takes 60 A at full-load speed of 940 r.p.m. and develops a torque of 150 N-m. The starting current at the rated voltage is 300 A. What is the starting torque ? If star/delta starter is used, determine the starting torque and starting current. **[225 N-m ; 100 A ; 75 N-m]**

5. The full-load slip of a 400 V, 3-phase, squirrel cage induction motor is 3·5 % and with locked rotor, full-load current is circulated when 92 V is applied between lines. Find the necessary tapping on an autotransformer to limit the starting current to twice the full-load current. **[67·8 %]**

9.44 STARTING OF SLIP-RING MOTORS

Slip-ring motors are invariably started by *rotor resistance starting*. In this method, a variable star-connected rheostat is connected in the rotor circuit through slip rings and full voltage is applied to the stator winding as shown in Fig. 9.50.

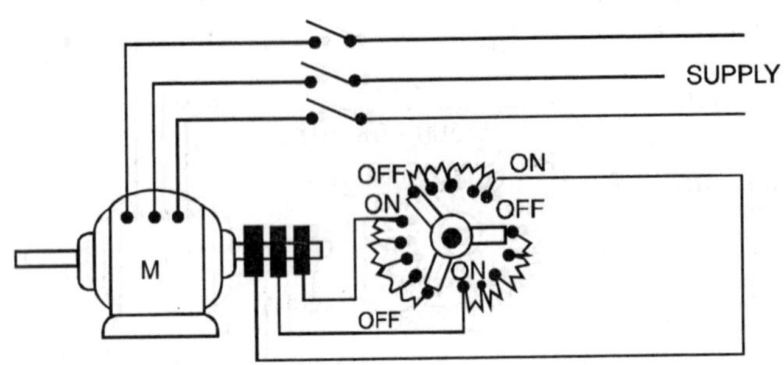

Fig. 9.50

(*i*) At starting, the handle of rheostat is set in the *OFF* position so that maximum resistance is placed in each phase of the rotor circuit. This reduces the starting current and at the same time starting torque is increased.

(*ii*) As the motor picks up speed, the handle of rheostat is gradually moved in clockwise direction and cuts out the external resistance in each phase of the rotor circuit. When the motor attains normal speed, the change-over switch is in the *ON* position and the whole external resistance is cut out from the rotor cricuit.

9.45 SLIP-RING MOTORS VERSUS SQUIRREL CAGE MOTORS

The slip-ring induction motors have the following advantages over the squirrel cage motors :

(*i*) High starting torque with low starting current.
(*ii*) Smooth acceleration under heavy loads.
(*iii*) No abnormal heating during starting.
(*iv*) Good running characterisitics after external rotor resistances are cut out.
(*v*) Adjustable speed.

The disadvantages of slip-ring motors are :

(*i*) The initial and maintenance costs are greater than those of squirrel cage motors.
(*ii*) The speed regulation is poor when run with resistance in the rotor circuit.

9.46 INDUCTION MOTOR RATING

The nameplate of a 3-phase induction motor provides the following information :

(*i*) Horsepower (*ii*) Line voltage (*iii*) Line current
(*iv*) Speed (*v*) Frequency (*vi*) Temperature rise

The horsepower rating is the mechanical output of the motor when it is operated at rated line voltage, rated frequency and rated speed. Under these conditions, the line current is that specified on the nameplate and the temperature rise does not exceed that specified.

The speed given on the nameplate is the actual speed of the motor at rated full-load; it is not the synchronous speed. Thus, the nameplate speed of the induction motor might be 1710 r.p.m. It is the rated full-load speed.

Example 9.61. *Calculate the full-load efficiency of a 415 V, 3-phase, 50 Hz delta-connected induction motor from the following test data :*

On no-load : Power intake = 1500 W at rated input voltage

On full-load : Line current = 50 A ; power factor = 0.85 lag, slip = 0.04, stator resistance per phase = 0.5 Ω

Assume the ratio of stator core loss to friction and windage loss as 3:2.

Solution. The no-load input power of 1500 W consists of (*i*) stator core loss (*ii*) friction and windage loss (neglecting small amount of stator I^2R loss). Since these losses are in the ratio 3:2,

∴ Stator core loss = (3/5) × 1500 = 900 W

Friction and windage loss = (2/5) × 1500 = 600 W

$$\text{Stator Cu loss at full-load} = 3I_1^2 R_1 = 3 \times (50/\sqrt{3})^2 \times 0.5 = 1250 \text{ W}$$

Total stator losses = 1250 + 900 = 2150 W

$$\text{Stator input} = \sqrt{3}\, V_L I_L \cos\phi = \sqrt{3} \times 415 \times 50 \times 0.85 = 30548 \text{ W}$$

Rotor input = Stator input – Stator losses

= 30548 – 2150 = 28398 W

Rotor Cu loss = s × Rotor input = 0.04 × 28398 = 1136 W

Rotor losses = Friction and windage loss + Rotor Cu loss

= 600 + 1136 = 1736 W

∴ Rotor output = 28398 – 1736 = 26,662 W

$$\therefore \quad \text{Efficiency} = \frac{26,662}{30548} \times 100 = \mathbf{87.28\%}$$

9.47 ABNORMAL OPERATION OF 3-PHASE INDUCTION MOTORS

The abnormal operation of a 3-phase induction motor may be due to *internal causes* (*e.g.* short-circuit in the stator, overheating of bearings etc) or due to *external causes.* External problems may be caused by :

(*i*) Mechanical overload (*ii*) Supply voltage changes (*iii*) Single-phasing (*iv*) Frequency changes

According to the national standards, a motor shall operate successfully on any voltage within ± 10% of the nominal voltage and for any frequency within ± 5% of the nominal frequency.

(*i*) **Mechanical overload.** Although standard induction motors can develop twice their rated power for short periods, they should not be allowed to run continuously beyond their rated capacity. Overloads cause overheating which deteriorates the insulation and reduces its useful life. In practice, thermal overload relays are provided in the motor circuit. The higher motor current causes these relays to trip, bringing the motor to a stop before its temperature gets too high.

(*iii*) **Line voltage changes.** We know that torque of an induction motor at any speed is directly proportional to the *square* of the applied voltage. Therefore, if the stator voltage *decreases by 10%, the torque will drop by 20%. On the other hand, if the line voltage increases, the flux per pole will also increase. This increases both the iron losses and the magnetising current. The result is that the temperature increases slightly and the power factor is somewhat reduced.

A slight *unbalance* of 3-phase voltages produces a serious unbalance of the three line currents. This condition increases the stator and rotor losses, resulting in higher temperature. A voltage unbalance of as little as 3.5% can cause the temperature to increase by 15°C.

* A line voltage drop is often produced during starting due to the heavy starting current drawn from the line.

(*iii*) **Single-phasing.** If one line of a 3-phase line is accidentally opened or if a fuse blows while the motor is running, the machine will continue to run as a single-phase motor. This condition is called *single-phasing.* The current drawn from the remaining two lines will almost double and the motor will begin to overheat. The thermal relays protecting the motor will eventually trip the circuit breaker, thereby disconnecting the motor from the line.

The torque-speed curve is seriously affected when a 3-phase motor operates on single-phase. The breakdown torque decreases to about 40% of its original value and the motor develops no starting torque at all. Consequently, a fully-loaded 3-phase motor may simply stop if one of its line is suddenly opened.

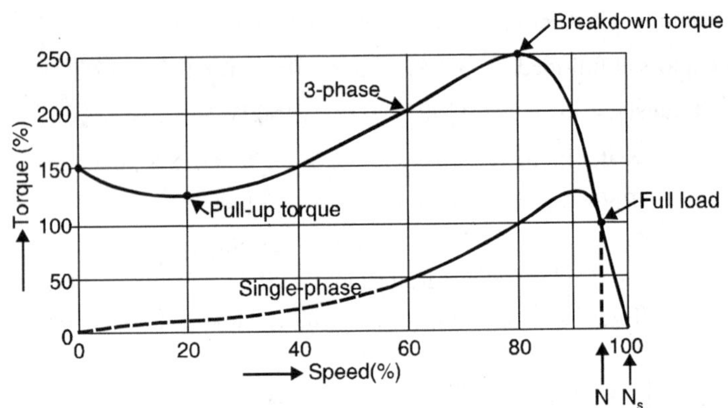

Fig. 9.51

Fig. 9.51 shows the typical torque-speed curves of a 3-phase motor when it runs normally and when it is single-phasing. Note carefully the effect on the torque-speed curve when a 3-phase motor is single-phasing.

(*iv*) **Frequency changes.** Significant frequency changes never take place on a large distribution system, except during a major disturbance. However, the frequency may vary considerably on isolated, low-power systems where electrical energy is generated by diesel engines or gas turbines. The most important effect of frequency change is the resulting change in motor speed. For example, if the frequency drops by 20%, the motor speed will drop by 20% ($N_s = 120\,f/P$). This change in motor speed may be unacceptable in many applications.

Example 9.62. *(i) Suggest a method to operate a 60 Hz motor on a 50 Hz line.*

(ii) Suggest a method to operate a 50 Hz motor on a 60 Hz line.

Solution. (*i*) Machine tools and other motor-driven equipment imported from countries where the frequency is 60 Hz may cause problems when they are connected to a 50 Hz system. Every motor will run 17% slower than the normal and this may not be acceptable in some applications.

A 60 Hz motor operates well on a 50 Hz line if its terminal voltage is reduced to 5/6 (or 83%) of its nameplate value. The breakdown torque and starting torque are then the same as before. The power factor, efficiency and temperature rise remain satisfactory.

(*ii*) A 50 Hz motor may cause problems when connected to a 60 Hz system. Every motor will run 20% faster than the normal and this may be objectionable in many cases.

A 50 Hz motor operates well on a 60 Hz line if its terminal voltage is increased to 6/5 (or 120%) of the nameplate rating. The new breakdown torque is then equal to the original breakdown torque and the starting torque is only slightly reduced. The power factor, efficiency and temperature rise remain satisfactory.

9.48 DOUBLE SQUIRREL-CAGE MOTORS

One of the advantages of the slip-ring motor is that resistance may be inserted in the rotor circuit to obtain high starting torque (at low starting current) and then cut out to obtain optimum running conditions. However, such a procedure cannot be adopted for a squirrel cage motor because its cage is permanently short-circuited. In order to provide high starting torque at low starting current, double-cage construction is used.

Construction. As the name suggests, the rotor of this motor has two squirrel-cage windings located one above the other as shown in Fig. 9.52 (*i*).

(*i*) The **outer winding** consists of bars of smaller cross-section short-circuited by end rings. Therefore, the resistance of this winding is high. Since the outer winding has relatively open slots and a poorer flux path around its bars [See Fig.9.52 (*ii*)], it has a low inductance. *Thus the resistance of the outer squirrel-cage winding is high and its inductance is low.*

(*ii*) The **inner winding** consists of bars of greater cross-section short-circuited by end rings. Therefore, the resistance of this winding is low. Since the bars of the inner winding are thoroughly buried in iron, it has a high inductance [See Fig. 9.52 (*ii*)]. *Thus the resistance of the inner squirrel-cage winding is low and its inductance is high.*

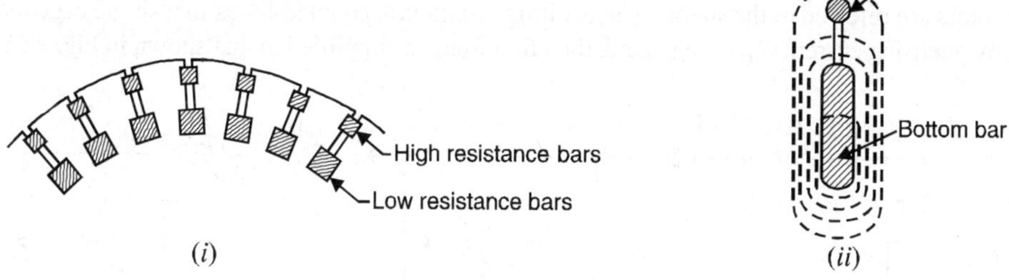

Fig. 9.52

Working. When a rotating magnetic field sweeps across the two windings, equal e.m.f.s are induced in each.

(*i*) At starting, the rotor frequency is the same as that of the line (*i.e.*, 50 Hz), making the reactance of the lower winding much higher than that of the upper winding. Because of the high reactance of the lower winding, nearly all the rotor current flows in the high-resistance outer cage winding. This provides the good starting characteristics of a high-resistance cage winding. Thus the outer winding gives high starting torque at low starting current.

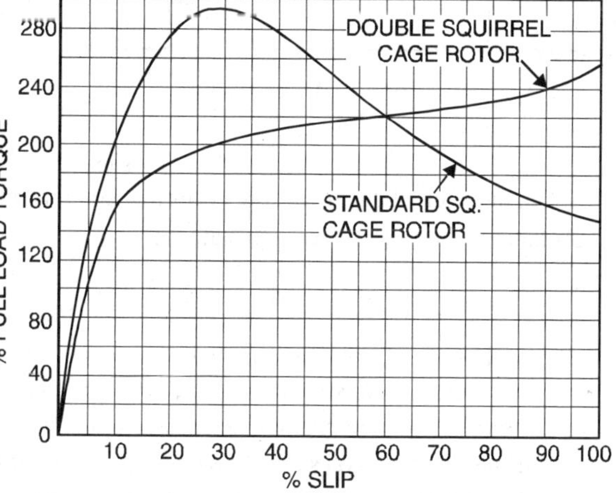

Fig. 9.53

(*ii*) As the motor accelerates, the rotor frequency decreases, thereby lowering the reactance of the inner winding, allowing it to carry a larger proportion of the total rotor current. At the normal operating speed of the motor, the rotor frequency is so low (2 to 3 Hz) that nearly all the rotor current flows in the low-resistance inner cage winding. This results in good operating efficiency and speed regulation.

Fig. 9.53 shows the operating characteristics of double squirrel-cage motor. The starting torque of this motor ranges from 200 to 250 percent of full-load torque with a starting current of 4 to 6 times the full-load value. It is classed as a high-torque, low starting current motor.

9.49 EQUIVALENT CIRCUIT OF DOUBLE SQUIRREL-CAGE MOTOR

Fig. 9.54 shows a section of the double squirrel cage motor. Here R_o and R_i are the per phase resistances of the outer cage winding and inner cage winding whereas X_o and X_i are the corresponding per phase standstill reactances. For the outer cage, the resistance is made intentionally high, giving a high starting torque. For the inner cage winding, the resistance is low and the leakage reactance is high, giving a low starting torque but high efficiency on load. Note that in a double squirrel cage motor, the outer winding produces the high starting and accelerating torque while the inner winding provides the running torque at good efficiency.

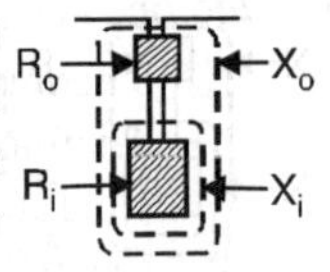

Fig. 9.54

Fig. 9.55 (*i*) shows the equivalent circuit for one phase of double cage motor referred to stator. The two cage impedances are effectively in parallel. The resistances and reactances of the outer and inner rotors are referred to the stator. The exciting circuit is accounted for as in a single cage motor. If the magnetising current (I_0) is neglected, then the circuit is simplified to that shown in Fig. 9.55 (*ii*).

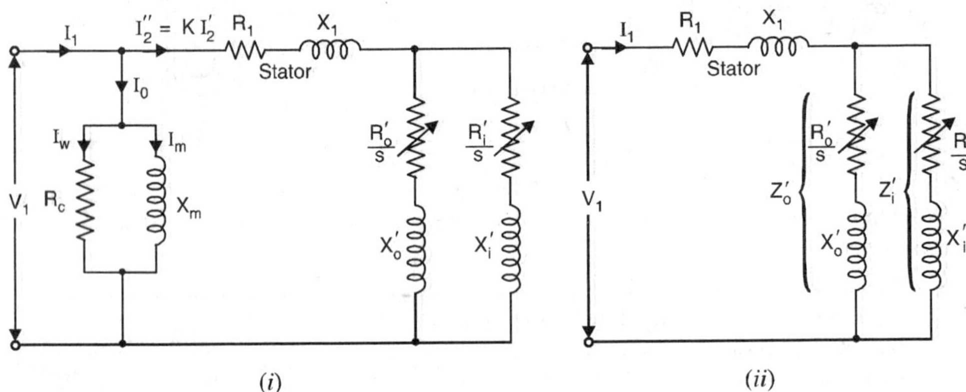

Fig. 9.55

From the equivalent circuit, the performance of the motor can be predicted.

Total impedance as referred to stator is

$$\mathbf{Z_{01}} = R_1 + jX_1 + \frac{1}{1/\mathbf{Z}'_i + 1/\mathbf{Z}'_o} = R_1 + j\,X_1 + \frac{\mathbf{Z}'_i\,\mathbf{Z}'_o}{\mathbf{Z}'_i + \mathbf{Z}'_o}$$

Example 9.63. *At standstill, the equivalent impedance/phase of the inner and outer cages of a double-cage rotor as referred to stator are (0·4 + j 2) Ω and (2 + j 0·4) Ω respectively. Calculate the ratio of torques produced by the two cages (i) at standstill (ii) at 5 % slip.*

Solution. Fig. 9.56 (*i*) shows the equivalent circuit for one phase referred to the stator. Let I_o and I_i be the currents in the outer and inner cages respectively.

(*i*) **At standstill** ($s = 1$)

$$\text{Impedance of inner cage, } Z'_i = \sqrt{(0{\cdot}4)^2 + (2)^2} = 2{\cdot}04\ \Omega$$

$$\text{Impedance of outer cage, } Z'_o = \sqrt{(2)^2 + (0{\cdot}4)^2} = 2{\cdot}04\,\Omega$$

$$\text{Power input to outer cage, } P_o = I_o^2 R'_o = 2\,I_o^2$$

$$\text{Power input to inner cage, } P_i = I_i^2 R'_i = 0{\cdot}4\,I_i^2$$

$$\therefore \quad \frac{\text{Torque of outer cage } (T_o)}{\text{Torque of inner cage } (T_i)} = \frac{P_o}{P_i} = \frac{2\,I_o^2}{0{\cdot}4\,I_i^2} = 5\left(\frac{I_o}{I_i}\right)^2$$

$$= 5\left(\frac{Z'_i}{Z'_o}\right)^2 = 5\left(\frac{2{\cdot}04}{2{\cdot}04}\right)^2 = 5$$

Therefore, $T_o/T_i = \mathbf{5}$

Note that at starting, the torque of the outer cage is 5 times that of the inner cage.

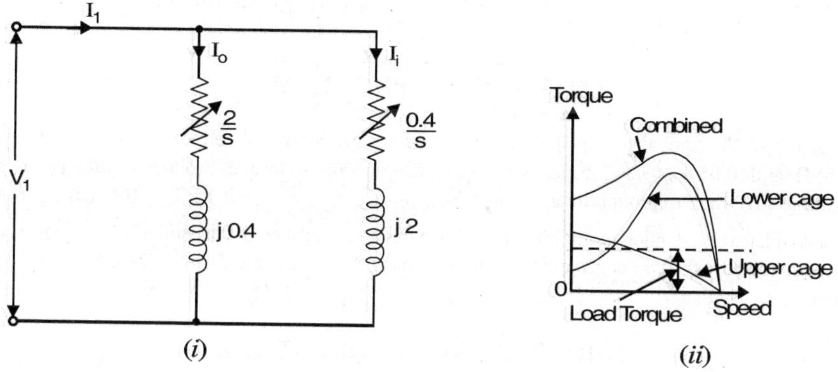

Fig. 9.56

(*ii*) **When slip $s = 0{\cdot}05$**

$$Z'_o = \sqrt{(2/0{\cdot}05)^2 + (0{\cdot}4)^2} = 40\,\Omega$$

$$Z'_i = \sqrt{(0{\cdot}4/0{\cdot}05)^2 + (2)^2} = 8{\cdot}25\,\Omega$$

$$\therefore \quad \frac{I_o}{I_i} = \frac{8{\cdot}25}{40} = 0{\cdot}206$$

$$P_o = \frac{I_o^2 R'_o}{s} = \frac{I_o^2 \times 2}{0{\cdot}05} = 40\,I_o^2 \;;\; P_i = \frac{I_i^2 R'_i}{s} = \frac{I_i^2 \times 0{\cdot}4}{0{\cdot}05} = 8\,I_i^2$$

$$\therefore \quad \frac{T_o}{T_i} = \frac{P_o}{P_i} = \frac{40\,I_o^2}{8\,I_i^2} = 5\left(\frac{I_o}{I_i}\right)^2 = 5 \times (0{\cdot}206)^2 = \mathbf{0{\cdot}21}$$

Note that during running conditions, the inner cage provides most of the torque. Fig. 9.56 (*ii*) shows torque-speed characteristics of the individual cages and the combined torque-speed characteristics.

Example 9.64. *In a double -cage induction motor, if the outer cage has an equivalent impedance at standstill of* $(2 + j\,2)$ *ohm and the inner cage an equivalent impedance of* $(0{\cdot}5 + j\,5)$ *ohm, determine the slip at which the two cages develop equal torques.*

Solution. Let s be the slip at which the two cages develop equal torques.

$$Z'_o = \sqrt{(2/s)^2 + (2)^2} \;;\; Z'_i = \sqrt{(0{\cdot}5/s)^2 + (5)^2}$$

$$\therefore \quad \left(\frac{I_o}{I_i}\right)^2 = \left(\frac{Z'_i}{Z'_o}\right)^2 = \frac{(0{\cdot}5/s)^2 + 25}{4/s^2 + 4}$$

Now $$P_o = I_o^2 \frac{R'_o}{s} = I_o^2 \times \frac{2}{s} = \frac{2\,I_o^2}{s} \; ; \; P_i = I_i^2 \frac{R'_i}{s} = \frac{0{\cdot}5\,I_i^2}{s}$$

$$\therefore \quad \frac{T_o}{T_i} = \frac{P_o}{P_i} = \left(\frac{I_o}{I_i}\right)^2 \times 4 = \frac{0{\cdot}25/s^2 + 25}{4/s^2 + 4} \times 4$$

But it is given that $T_o = T_i$.

$$\therefore \quad 1 = \frac{0{\cdot}25/s^2 + 25}{4/s^2 + 4} \times 4$$

or $$\frac{4}{s^2} + 4 = \frac{1}{s^2} + 100 \quad \therefore \quad s = 0{\cdot}177 = \mathbf{17{\cdot}7\ \%}$$

TUTORIAL PROBLEMS

1. At standstill, the equivalent impedance of inner and outer cages of a double-cage rotor are (0.01 + j 0·5) Ω and (0·05 + j 0·1) Ω respectively. Calculate the ratio of torques produced by the two cages (*i*) at starting (*ii*) when running with a slip of 5 %. **[(*i*) T_o/T_i = 100 (*ii*) T_o/T_i = 1/1·44]**
2. In a double-cage induction motor if the outer cage has an impedance at standstill of (2 + j 1) ohm, determine the slip at which the two cages develop equal torques if the inner cage has an impedance of (0·5 + j 3·5) ohm at standstill. **[25·1 %]**

SHORT ANSWER QUESTIONS

Q. 1. *What do you mean by synchronous speed of a 3-phase induction motor* ?

Ans. The speed at which the flux produced by 3-phase stator windings of induction motor rotates is called the synchronous speed of the motor. It is given by :

$$N_s = \frac{120\,f}{P}$$

where N_s = synchronous speed in r.p.m.

f = supply frequency in Hz

P = number of poles

Q. 2. *Why is the field producing winding of a 3-phase induction motor made stationary* ?

Ans. A 3-phase induction motor has two windings *viz. stator winding* supported by the stationary part of the machine and the *rotor winding* placed on the rotor. So far as the basic operation of the motor is concerned, it makes no difference which winding is located on the stator. The machine will function equally well with field producing winding as either the stationary or rotating element. Making the field winding stationary element eliminates the use of slip rings and brushes and, therefore, will result in a very trouble-free construction.

Q. 3. *Why does the rotor of a 3-phase induction motor rotate in the same direction as the rotating field* ?

Ans. When three-phase stator winding is fed from 3-phase supply, a rotating magnetic field is set up which cuts the rotor conductors. Since the rotor circuit is closed, currents start flowing in the rotor conductors. Now, rotor conductors are carrying currents and are in the magnetic field. Therefore, mechanical force acts on the rotor, tending to move it in the *same direction* as the stator field. The fact that the rotor is urged to follow the stator field (*i.e.,* rotor moves in the direction of stator field) can be explained by Lenz's law. According to Lenz's law, the direction of rotor currents will be such so as to oppose the cause producing them. Now, the cause producing the rotor currents is the relative speed between the rotating field and stationary rotor. Hence to reduce this relative speed, the rotor starts running in the same direction as the stator field and tries to catch it.

Q. 4. *Why cannot 3-phase induction motor run at synchronous speed* ?

Ans. As explained in Q.3 above, the rotor follows the stator field. In actual practice, rotor can never reach the speed of stator field (*i.e.* synchronous speed). If it did, there would be no relative movement between the stator field and rotor conductors and, therefore, no torque to drive the motor. Hence 3-phase induction motor can never run at synchronous speed.

Q. 5. *Why is the air gap between the rotor and stator of a 3-phase induction motor kept as short as possible* ?

Ans. The air gap between the rotor and stator of a 3-phase induction motor is made as small as possible in order that :

(*i*) mutual flux may be produced with a minimum exciting current.

(*ii*) leakage reactances are as small as possible.

Q. 6. *How is magnetising current kept small in a 3-phase induction motor* ?

Ans. The magnetising current drawn by a 3-phase induction motor is very large (30-50% of full-load stator winding current) owing to the presence of air-gap between the stator and rotor of the motor. It can be kept small by ;

(*i*) making the air-gap between stator and rotor very small.

(*ii*) using partially closed stator and rotor slots.

Q. 7. *What is the importance of slip in a 3-phase induction motor* ?

Ans. The rate at which the flux cuts the rotor conductors is directly proportional to the difference between the speed of rotating field (N_s) and that of the rotor (N). If the rotor speed were to become equal to the speed of the rotating field, there would be no generated e.m.f. (and hence current) in the rotor conductors. Consequently, there would be no motor action. Thus it is the slip of the rotor ($N_s - N$) r.p.m. which causes e.m.f.s to be generated and currents to flow in the rotor conductors. This is precisely the principle of operation of 3-phase induction motor.

Q. 8. *What is the difference between squirrel cage rotor and wound rotor* ?

Ans. Essentially, there is little difference between the squirrel cage and wound rotors. The object of the latter is merely to bring out the ends of polyphase rotor winding to slip rings in order that additional external resistance may be connected in series to improve the starting torque.

Q. 9. *What is the origin of the name squirrel cage rotor* ?

Ans. When this type of rotor first appeared, the common squirrel was a frequent house pet. The usual cage that housed him contained a rotating wheel that the animal could enter. This wheel afforded exercise and amusement to the pet. Since the rotor resembled squirrel's exercise cage, it was named squirrel cage rotor.

Q. 10. *Why is the power factor of a 3-phase induction motor low at no-load* ?

Ans. Because of air-gap, the reluctance of the magnetic circuit of 3-phase induction motor is very high. Consequently, the current drawn by the motor on no load is largely magnetising current; the no load current lagging the applied voltage by a large angle. For this reason, the p.f. of a lightly loaded 3-phase induction motor is very low.

Q. 11. *Why is the power factor of a fully loaded 3-phase induction motor not very high* ?

Ans. A 3-phase induction motor draws a large magnetising current because of high reluctance of the magnetic circuit; air-gap being the main reason. As load is added, the active component of current increases, resulting in higher power factor. However,because of the large value of magnetising current, which is present regardless of load, the p.f. of 3-phase induction motor even at full-load seldom exceeds 0·85.

Q. 12. *What are the advantages of skewed slots in the rotor of a squirrel cage motor ?*

Ans. It is a usual practice to use the rotor of a squirrel cage motor with skewed slots *i.e*,. slots that are not parallel to the shaft axis. This arrangement offers the following advantages :

(*i*) It reduces motor noise and vibration.

(*ii*) It increases the starting torque and reduces the starting current.

(*iii*) It increases the rotor resistance due to increased length of rotor bars.

Q. 13. *How will you design the rotor of a squirrel cage motor to have high starting torque ?*

Ans. When a high starting torque is required, a squirrel cage machine with specially designed rotor can be used without much sacrifice of efficiency but with some reduction of rated output. This can be achieved by either of the following two ways :

(*i*) The rotor bars can be made very deep so that owing to skin effect, their resistance will be high at starting when the rotor frequency is equal to the supply frequency.

(*ii*) By using double squirrel cage rotor.

Q. 14. *Why is the maximum torque of a squirrel cage induction motor called pull-out torque ?*

Ans. The maximum torque of a squirrel cage induction motor is also called pull-out torque because of the way the motor reacts to an overload. After the maximum torque point (which is three to four times the full-load torque), the decrease in rotor p.f. is greater than the increase in rotor current resulting in a decreasing torque and the motor quickly comes to stop.

It is generally experienced that a lathe will stall on a heavy cut. The machine will slow down as its cutting load is increased until suddenly it will stall and hum or growl loudly. This condition will persist until the load is removed or a fuse blows.

Q. 15. *When will rotor resistance dominate rotor reactance and vice-versa ?*

Ans. (*i*) When a 3-phase induction motor is in normal operation, the rotor frequency ($f' = sf$ where f is the supply frequency) is low and so is the rotor reactance. The current that flows is then largely limited by rotor resistance rather than reactance.

(*ii*) When the rotor is stationary (*i.e.*, at standstill), $s = 1$. This means that rotor sees the full-line frequency (*i.e.*, $f' = f$) and its reactance is dominant compared to its resistance.

MULTIPLE-CHOICE QUESTIONS

1. The stator of a 3-phase induction motor produces .... magnetic field.

(*i*) steady (*ii*) rotating
(*iii*) alternating (*iv*) none of the above

2. An induction motor is preferred to a d.c. motor because it ....

(*i*) provides high starting torque
(*ii*) provides fine speed control
(*iii*) has simple and rugged construction
(*iv*) none of the above

3. The air-gap between stator and rotor of a 3-phase induction motor ranges from ....

(*i*) 2 cm to 4 cm
(*ii*) 0·4 mm to 4 mm
(*iii*) 1 cm to 2 cm
(*iv*) 4 cm to 6 cm

4. The relation among synchronous speed (N_s), rotor speed (N) and slip (s) is ....

(*i*) $N = (s - 1) N_s$ (*ii*) $N = (1 - s) N_s$
(*iii*) $N = (1 + s) N_s$ (*iv*) $N = s N_s$

5. When a 3-phase induction motor is at no load, the slip is ....

(*i*) 1 (*ii*) 0·5
(*iii*) practically zero (*iv*) 0·2

6. When the rotor of a 3-phase induction motor is blocked, the slip is ....

(*i*) zero (*ii*) 0·5
(*iii*) 0·1 (*iv*) 1

7. In Fig. 9.57, the rotor frequency is .... when the motor is at standstill.

(*i*) zero (*ii*) 25 Hz
(*iii*) 50 Hz (*iv*) none of the above

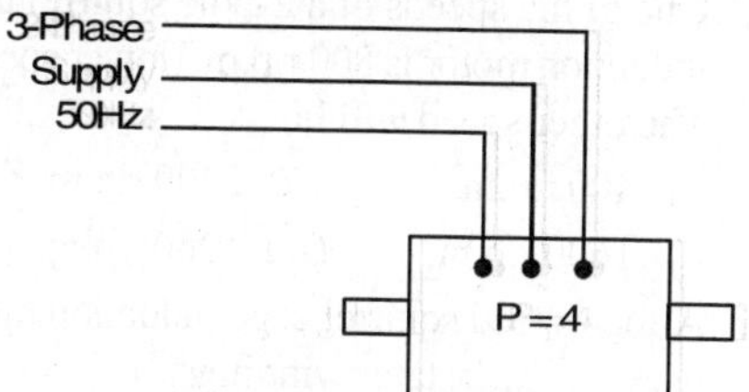

Fig. 9.57

8. If one of the lines in Fig. 9.57 is interchanged, then ....
   (i) flux speed decreases
   (ii) flux speed increases
   (iii) flux speed remains same but direction is reversed
   (iv) none of the above

9. The full-load slip of a 3-phase induction motor ranges from ....
   (i) 10% to 20% (ii) 20% to 30%
   (iii) 2% to 5% (iv) none of the above

10. The rotor winding of a 3-phase wound rotor is generally .... connected.
   (i) star (ii) delta
   (iii) partly star and partly delta
   (iv) none of the above

11. The torque characteristic of a 3-phase induction motor is similar to that of............
   (i) d.c. series motor
   (ii) d.c. shunt motor
   (iii) d.c. differentially compounded motor
   (iv) d.c. cumulatively compounded motor

12. The maximum voltage is induced in the rotor of a 3-phase induction motor when it ....
   (i) runs at no-load (ii) runs at full-load
   (iii) is blocked (iv) none of the above

13. The reactance of the rotor circuit of a 3-phase induction motor is maximum ....
   (i) at no-load (ii) at full-load
   (iii) at half full-load (iv) none of the above

14. When an induction motor is running at full-load, rotor reactance is .... rotor resistance.
   (i) comparable to
   (ii) very large compared to
   (iii) large compared to
   (iv) none of the above

15. If the slip of a 3-phase induction motor increases, the p.f. of the rotor circuit ....
   (i) is increased (ii) is decreased
   (iii) remains unchanged
   (iv) none of the above

16. The starting torque of a 3-phase induction motor is .... supply voltage.
   (i) independent of
   (ii) directly proportional to
   (iii) directly proportional to square of
   (iv) none of the above

17. The maximum torque of a 3-phase induction motor under running conditions is ....
   (i) inversely proportional to supply voltage
   (ii) inversely proportional to rotor reactance at standstill
   (iii) directly proportional to rotor resistance
   (iv) none of the above

18. If N_s and N are the speeds of rotating field and rotor respectively, the ratio rotor input/rotor output is equal to ....
   (i) N/N_s (ii) N_s/N
   (iii) $N_s - N$ (iv) $N - N_s$

19. For higher efficiency of 3-phase induction motor, the slip should be ....
   (i) large (ii) very large
   (iii) as small as possible
   (iv) 1

20. If a 3-phase induction motor is running at a slip s (in decimal), then approximate efficiency of the motor is ....
   (i) s^2 (ii) $1 + s$
   (iii) $s - 1$ (iv) $1 - s$

21. A 3-phase induction motor is running at 2% slip. If the input to rotor is 1000 W, then mechanical power developed by the motor is ....
   (i) 20 W (ii) 980 W
   (iii) 500 W (iv) 200 W

22. The approximate efficiency of a 3-phase, 50 Hz, 4-pole induction motor running at 1350 r.p.m. is ....

(*i*) 90% (*ii*) 40%
(*iii*) 65% (*iv*) none of the above

23. The conditions of an induction motor on no-load resemble those of a transformer whose secondary is ....
(*i*) short-circuited
(*ii*) open-circuited
(*iii*) supplying a variable resistive load
(*iv*) none of the above

24. The conditions of induction motor at standstill resemble those of a transformer whose secondary is ....
(*i*) short-circuited
(*ii*) open-circuited
(*iii*) supplying a variable resistive load
(*iv*) none of the above

25. The conditions of induction motor on load resemble those of a transformer whose secondary is ....
(*i*) open-circuited
(*ii*) short-circuited
(*iii*) supplying a variable resistive load
(*iv*) none of the above

26. One of the speeds of a 2-pole squirrel cage induction motor is 800 r.p.m. (lower speed). The other speed will be ....
(*i*) 400 r.p.m. (*ii*) 2400 r.p.m.
(*iii*) 1600 r.p.m. (*iv*) 1200 r.p.m.

27. A four-speed squirrel cage induction motor uses .... stator windings.
(*i*) four (*ii*) three
(*iii*) one (*iv*) two

28. An 8-pole alternator runs at 750 r.p.m. and supplies power to a 6-pole induction motor which has a full-load slip of 3%. The full-load speed of the motor is ....
(*i*) 1050 r.p.m. (*ii*) 970 r.p.m.
(*iii*) 960 r.p.m. (*iv*) 1250 r.p.m.

29. In the above question, what is the frequency of rotor e.m.f. ?
(*i*) 5 Hz (*ii*) 50 Hz
(*iii*) 1·5 Hz (*iv*) 6·5 Hz

30. What is the approximate efficiency of a 60 Hz, 6-pole, 3-phase induction motor running at 1050 r.p.m. ?
(*i*) 72·5% (*ii*) 81·2%
(*iii*) 76·8% (*iv*) 87·5%

ANSWERS TO MULTIPLE-CHOICE QUESTIONS

1. (*ii*)	**2.** (*iii*)	**3.** (*ii*)	**4.** (*ii*)	**5.** (*iii*)
6. (*iv*)	**7.** (*iii*)	**8.** (*iii*)	**9.** (*iii*)	**10.** (*i*)
11. (*ii*)	**12.** (*iii*)	**13.** (*iv*)	**14.** (*i*)	**15.** (*ii*)
16. (*iii*)	**17.** (*ii*)	**18.** (*ii*)	**19.** (*iii*)	**20.** (*iv*)
21. (*ii*)	**22.** (*i*)	**23.** (*ii*)	**24.** (*i*)	**25.** (*iii*)
26. (*iii*)	**27.** (*iv*)	**28.** (*ii*)	**29.** (*iii*)	**30.** (*iv*)

HINTS TO SELECTED MULTIPLE-CHOICE QUESTIONS

4.
$$s = \frac{N_s - N}{N_s} = \frac{N_s}{N_s} - \frac{N}{N_s} = 1 - \frac{N}{N_s}$$
$$\therefore \quad N = (1 - s)N_s$$

7. Rotor frequency, $f' = sf$ where f is supply frequency. At standstill, $s = 1$ so that $f' = f = 50$ Hz.

10. So that additional external resistance may be inserted in the rotor circuit at starting to increase the starting torque and decrease the starting current. As the motor gains speed, these external resistances are cut out of the rotor circuit.

13.
$$X'_2 = s\,X_2$$
where X'_2 = rotor reactance at slip s

$$X_2 = \text{rotor reactance at standstill}$$

Since $s = 1$ at starting, rotor reactance will be maximum under starting conditions.

15. $$\text{Rotor p.f., } \cos\phi_2' = \frac{R_2}{Z_2'}$$

An increase in slip causes an increase in rotor frequency and hence the total reactance. This in turn increases the total impedance Z_2'. Since rotor resistance (R_2) is constant, rotor p.f. is decreased.

16. $$\text{Starting torque, } T_s = \frac{K E_2^2 R_2}{R_2^2 + X_2^2} \qquad (\because s = 1)$$

Now $E_2 \propto V$, the supply voltage.

$$\therefore \qquad T_s = \frac{K_2 V^2 R_2}{R_2^2 + X_2^2}$$

or $$T_s \propto V^2$$

17. The maximum torque (T_{max}) of a 3-phase induction motor is given by,

$$T_{max} = \frac{3}{2\pi N_s} \times \frac{E_2^2}{2X_2}$$

where
N_s = synchronous speed
E_2 = rotor e.m.f./phase at standstill
X_2 = rotor reactance/phase at standstill

It is clear that $T_{max} \propto \dfrac{1}{X_2}$.

18. $$\frac{\text{Rotor output}}{\text{Rotor input}} = 1 - s = 1 - \left(\frac{N_s - N}{N_s}\right) = \frac{N}{N_s}$$

$$\therefore \qquad \frac{\text{Rotor input}}{\text{Rotor output}} = \frac{N_s}{N}$$

19. Approximate efficiency $= 1 - s$

where s = decimal slip

It is clear that for higher efficiency of 3-phase induction motor, the slip should be as small as possible.

20. $$\text{Turn ratio, } k = \frac{\text{Turns per phase on stator}}{\text{Turns per phase on rotor}}$$

Neglecting stator losses, power transferred to the rotor is given by:

$$P_r = \sqrt{3}\, E_s I_s \cos\phi_s$$

where E_s, I_s and $\cos\phi_s$ represent respectively the voltage, current and power factor of stator.

Rotor induced voltage at slip $s = sE_s/k$

Rotor current $= kI_s$

The stator and rotor power factors are approximately the same (say $\cos\phi$).

$$\text{Rotor losses} = \sqrt{3}\,\frac{s\,E_s}{k}\,k I_s \cos\phi = \sqrt{3}\, s E_s I_s \cos\phi$$

$$\text{Efficiency} = \frac{\text{Input} - \text{Loss}}{\text{Input}}$$

$$= \frac{\sqrt{3}\, E_s I_s \cos\phi - \sqrt{3}\, E_s I_s\, s\cos\phi}{\sqrt{3}\, E_s I_s \cos\phi}$$

$$= 1 - s \qquad \text{...}(i)$$

Since the stator losses are neglected, the actual efficiency is less than that given by eq. (*i*).

21. Mechanical power developed in 3-phase motor

$$= (1 - s) \times \text{Power input to rotor}$$
$$= (1 - 0{\cdot}02) \times 1000 = 980 \text{ W}$$

22.
$$N_s = \frac{120 f}{P} = 120 \times \frac{50}{4} = 1500 \text{ r.p.m.}$$
$$s = \frac{N_s - N}{N_s} = \frac{1500 - 1350}{1500} = 0{\cdot}1$$

∴ Approximate efficiency $= 1 - s = 1 - 0{\cdot}1 = 0{\cdot}9$ or 90%

23. The inputs to the stator and rotor are electrical but the output from the rotor is mechanical. We can represent the mechanical load by R_L – the electrical equivalent of the mechanical load.

$$\text{Rotor input} = \frac{(I_2')^2 R_2}{s} \quad \text{where } I_2' = \text{rotor current}$$
$$\text{Rotor loss} = (I_2')^2 R_2; \quad R_2 = \text{rotor resistance}$$

The difference between the two is the electrical power converted into mechanical power.

$$\text{Mechanical power} = \frac{(I_2')^2 R_2}{s} - (I_2')^2 R_2 = (I_2')^2 R_2\left(\frac{1}{s} - 1\right)$$
$$= (I_2')^2 R_2\left(\frac{1-s}{s}\right) = (I_2')^2 R_L$$

where R_L = Electrical equivalent of mechanical load on the motor

$$= R_2\left(\frac{1-s}{s}\right)$$

When 3-phase induction motor is on no load, slip s is practically zero.

∴ $$R_L = R_2\left(\frac{1-s}{s}\right) = \text{infinite (open circuited)}$$

Hence when induction motor is on no load, the condition is similar to that of a transformer whose secondary is open-circuited.

24. When the 3-phase induction motor is at standstill (*i.e.,* rotor is blocked), then, $s = 1$.

∴ $$R_L = R_2\left(\frac{1-s}{s}\right) = R_2\left(\frac{1-1}{1}\right) = 0$$

i.e. R_L is zero (*i.e.* short circuited).

Hence when induction motor is at standstill, the condition is similar to that of a transformer whose secondary is short-circuited.

25. When a 3-phase induction motor is running at slip s, the electrical equivalent of mechanical load is given by :

$$R_L = R_2\left(\frac{1-s}{s}\right)$$

The value of R_L is variable and depends upon the slip of the motor.

Thus when induction motor is running on load, the condition is similar to that of a transformer whose secondary is supplying a purely variable resistance load.

26. Multispeed squirrel cage motors are provided with stator windings that may be reconnected to form different number of poles. Two-speed motors usually have one stator winding that may be switched through suitable control equipment to provide two speeds; one of which is half of the other.

27. Four-speed squirrel-cage motors are equipped with two separate stator windings, each of which provides two speeds.

28. The supply frequency is found from the speed of the alternator and its number of poles.

$$\text{Supply frequency, } f = \frac{NP}{120} = \frac{750 \times 8}{120} = 50 \text{ Hz}$$

Synchronous speed of motor is

$$N_s = \frac{120\,f}{P} = \frac{120 \times 50}{6} = 1000 \text{ r.p.m.}$$

$$\%\text{ slip } s = \frac{N_s - N}{N_s} \times 100 \text{ or } 3 = \frac{1000 - N}{1000} \times 100 \qquad \therefore\ N = 970 \text{ r.p.m.}$$

29. Frequency of rotor e.m.f., $f' = sf = 0{\cdot}03 \times 50 = 1{\cdot}5$ Hz

30. $$\text{Synchronous speed, } N_s = \frac{120\,f}{P} = \frac{120 \times 60}{6} = 1200 \text{ r.p.m.}$$

$$\text{Fractional slip, } s = \frac{N_s - N}{N_s} = \frac{1200 - 1050}{1200} = 0{\cdot}125$$

$$\therefore\ \text{Approximate efficiency} = 1 - s = 1 - 0{\cdot}125 = 0{\cdot}875 \text{ or } 87{\cdot}5\%$$

10

Circle Diagrams

INTRODUCTION

In electrical engineering, we often use graphical methods to study the performance of a circuit because the relation between circuit variables are always most easily seen in a graph. For example, the resonance curve tells us more at a glance than could be learned from its description in words or in mathematics by long study. The shape of the graph depends upon the circuit conditions. For example, consider an $R - L$ a.c. series circuit with constant inductive reactance $X (= 2\pi f L)$ and variable resistance R connected to a constant voltage source V. If R is varied, the power factor angle ϕ as well as the magnitude of circuit current I changes. It can be shown that locus of the current phasor is a circle with diameter equal to V/X. Such a diagram is called a *circle diagram* (or more accurately current circle diagram). Although graphical methods do not give very accurate results, yet they provide very useful information at a glance. In this chapter, we shall study the circle diagram of induction motors and other topics including speed control of induction motors and rotor rheostat starter for slip-ring motors.

10.1 CIRCLE DIAGRAM FOR R – L SERIES CIRCUIT

Fig. 10.1(*i*) shows $R - L$ series a.c. circuit with a variable resistance R and constant inductive reactance $X (= 2\pi f L)$ connected to a constant voltage source V. The circuit current I lags behind the applied voltage V by an angle $\phi (= \tan^{-1} X/R)$ as shown in the impedance triangle in Fig. 10.1(*ii*). The magnitude of circuit current (I) is given by ;

$$I = \frac{V}{Z} = \frac{V}{\sqrt{R^2 + X^2}}$$

$$= \frac{V}{X} \times \frac{X}{\sqrt{R^2 + X^2}}$$

$$\therefore \quad I = \left(\frac{V}{X}\right) \sin\phi \qquad \left(\because \sin\phi = \frac{X}{Z} = \frac{X}{\sqrt{R^2 + X^2}}\right)$$

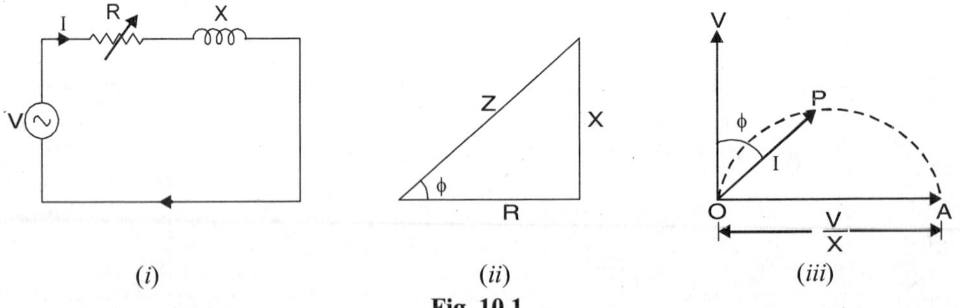

(*i*) (*ii*) (*iii*)

Fig. 10.1

The expression $I = (V/X) \sin\phi$ is the equation of a circle in polar form with diameter of the circle equal to V/X. As R is changed, ϕ as well as I changes and locus of the tip of the current phasor is a semi-circle OPA as shown in Fig. 10.1(*iii*). Note that voltage is represented by the vertical axis. This is the current circle diagram for the circuit. Note that polar coordinates of any point P on the circle diagram are I and ϕ. Let us discuss this circle diagram.

(*i*) When $R = 0$, *$\phi = 90°$ and the tip of the current phasor is at point A as shown in Fig. 10.1(*iii*). Under this condition, the magnitude of circuit current is maximum and is equal to **V/X (= OA = diameter of the circle). As R is increased from zero, the value of ϕ decreases and the tip of the current phasor moves towards the voltage phasor V. When $R \to \infty$, the tip of the current phasor is at point O and $I = 0$; $\phi = 0°$.

(*ii*) As R is increased from zero to infinity, the phase angle ϕ decrease from 90° to 0° and the magnitude of circuit current (I) decreases from the maximum value of V/X to zero. Thus the limits of the phase angle ϕ are 90° and 0°.

Note. For a lagging current, it is a usual practice to orient the circle such that its diameter is horizontal and the voltage phasor is vertical.

10.2 IMPORTANCE OF CIRCLE DIAGRAMS

A circle diagram is very useful because it conveys information that is not provided by an ordinary phasor diagram. A phasor diagram is constructed to show relationships between currents and voltages for one circuit condition. A circle diagram is effectively an infinite number of phasor diagrams drawn on one set of axes. There are a large number of examples of circle diagrams. One common example is the circle diagram of a 3-phase induction motor. This circle diagram enables us to determine output power, torque, power factor, efficiency, slip, speed, etc. of the motor graphically.

10.3 CIRCLE DIAGRAM OF 3-ϕ INDUCTION MOTOR

Fig. 10.2 shows the per phase approximate equivalent circuit of a 3-phase induction motor. It is clear that circuit to the right of points *ab* is a $R - L$ series circuit having constant reactance X_{01}, variable resistance ***$(R'_L + R_{01})$ and supplied from a constant voltage V_1. Therefore, the tip of the phasor current †I''_2 will lie on a circle with a diameter of V_1/X_{01}. This circle is shown dotted in Fig. 10.3. Note that maximum value of I''_2 is V_1/X_{01} and will occur when $(R_{01} + R'_L)$ is zero.

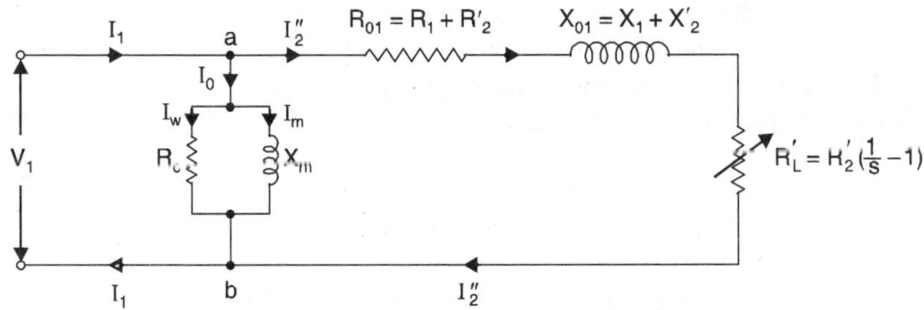

Fig. 10.2

The current I''_2 is the rotor current referred to the stator and its phase angle is ϕ_2 (lagging). The no-load current is I_0 and its phase angle is ϕ_0 (lagging). The total stator current I_1 is the phasor sum of I_0 and I''_2 and its phase angle is ϕ_1 (lagging).

---

* $\tan \phi = X/R = X/0 = \infty \quad \therefore \phi = 90°$.

** $I = \dfrac{V}{\sqrt{R^2 + X^2}} = \dfrac{V}{\sqrt{(0)^2 + X^2}} = \dfrac{V}{X}$

*** Now R_{01} is constant but R'_L is variable and its value depends upon the slip s of the motor. Therefore, $(R_{01} + R'_L)$ is variable.

† Note the notation we are following for rotor current in this book. The current I_2 is rotor current at standstill, I'_2 is rotor current at slip s and I''_2 is the rotor current (at slip s) referred to the stator side.

Now I_0 and $\cos \phi_0$ are constant. Therefore, the locus of stator phasor current I_1 is circle displaced from the origin by an amount I_0 as shown in Fig. 10.3. Its diameter is still V_1/X_{01} and is parallel to the horizontal axis *OC*. Hence we arrive at an important conclusion that as the load on induction motor changes, the locus of the stator phasor current I_1 is a circle. As we shall see, this circle diagram provides important information about the performance of the motor at a glance.

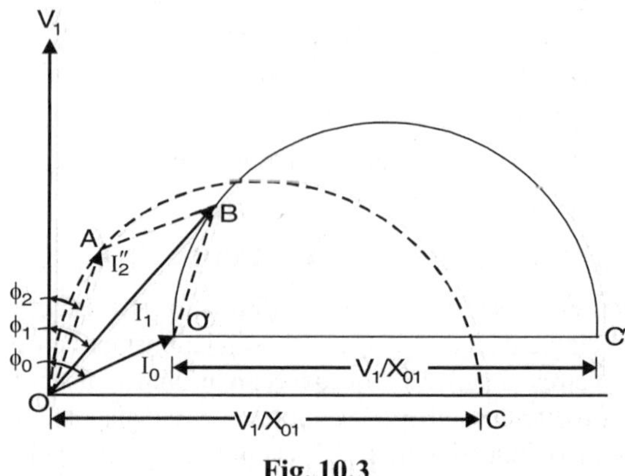

Fig. 10.3

10.4 DATA REQUIRED FOR CIRCLE DIAGRAM

We have seen that current locus of an induction motor is circular. Therefore, we can construct the circle diagram of the motor if the position of only *two points on the circle are known. These points should represent the measured values obtained from testing when the motor is connected to normal voltage. It is advisable that these two points should be as far as possible on the circle diagram. The obvious choice is for those points corresponding to synchronous speed and standstill condition. These points can be obtained from (*i*) **no-load test and (*ii*) blocked rotor test respectively.

The no-load test of the motor corresponds to open-circuit test of a transformer and provides the values of I_0 and $\cos \phi_0$. Therefore, the phasor current I_0 can be laid off lagging ϕ_0 behind the applied voltage V_1 as shown in Fig. 10.4. This locates the first point *O′* on the circle diagram. The †blocked rotor test corresponds to the short-circuit test of a transformer and provides I_{SN} (short circuit current corresponding to normal voltage) and $\cos \phi_s$ (power factor on short circuit). This locates the second point *A* on the circle diagram. The right bisector of chord *O′A* locates the centre *C* of the circle as shown in Fig. 10.4. With centre *C* and radius = *CO′*, the circle can be drawn.

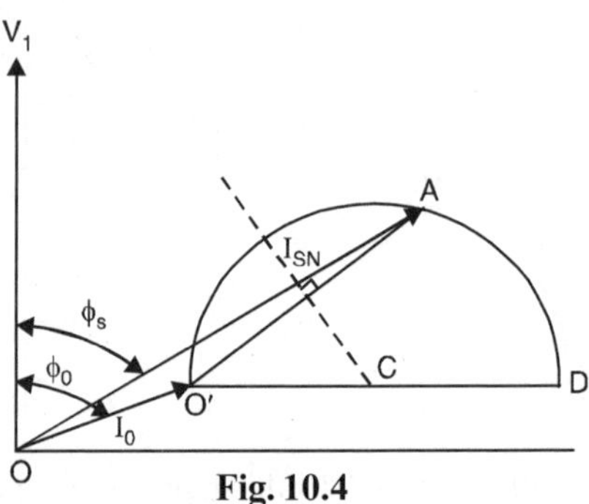

Fig. 10.4

10.5 TESTS TO OBTAIN DATA FOR CIRCLE DIAGRAM

In order to draw the circle diagram of an induction motor, we carry out two tests *viz* no-load test and blocked rotor test.

---

* The straight line joining these two points is the chord of the circle. The right bisector of this chord will pass through the centre of the circle.

** At no-load, an induction motor runs close to synchronous speed.

† In the blocked rotor test, the rotor is locked (*i.e.* it is held stationary) when the supply is switched on. Under this condition, the mechanical load on the motor is so heavy that it corresponds to the short-circuited secondary of the transformer. Therefore, the motor will draw very heavy current (6 to 10 times the full-load current).

1. No-load test. Fig. 10.5 shows the circuit connections for carrying out the no-load test on the induction motor. The stator is fed from 3-phase rated voltage and the motor is allowed to run without any external mechanical load on it. The motor input is measured by two-wattmeter method. The ammeter and voltmeter connected in the circuit measure current and voltage respectively.

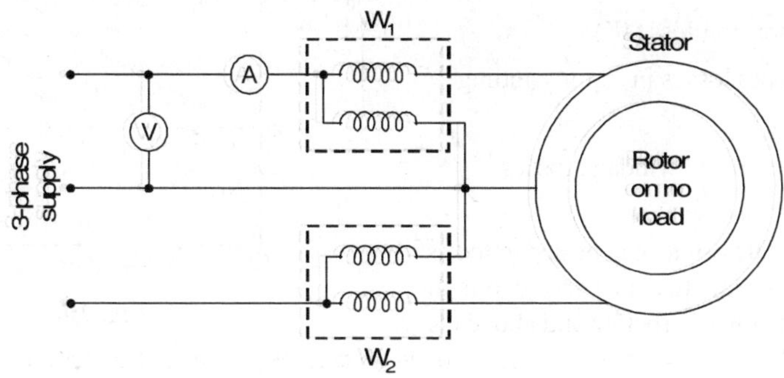

Fig. 10.5

Let

$$V_1 = \text{Applied voltage per phase}$$
$$I_0 = \text{Input no-load current per phase}$$
$$W_0 = \text{Total input power at no-load}$$
$$\cos\phi_0 = \text{No-load power factor}$$

$$\therefore \quad W_0 = 3\,V_1\,I_0\cos\phi_0$$

or

$$\cos\phi_0 = \frac{W_0}{3\,V_1 I_0}$$

Thus the values of I_0 and $\cos\phi_0$ are known from the no-load test. This enables us to locate one point on the circle diagram.

Additional information. This test can provide more information about the motor. At no-load, the speed of induction motor is very close to the synchronous speed so that for all practical purposes, we can consider that the slip s of the motor is zero at no-load. This means that quantity R'_L of the equivalent circuit (Refer back to Fig. 10.2) is *infinite. Therefore, the rotor circuit effectively behaves as an open circuit. As a result, the equivalent circuit of the motor at no-load will be as shown in Fig. 10.6. We can now obtain useful information from the no-load test.

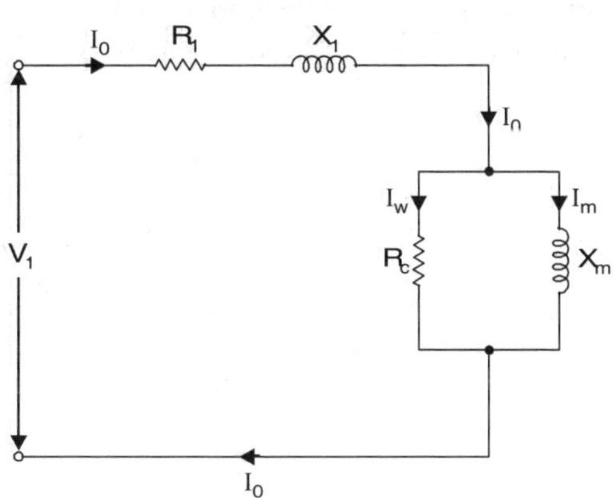

Fig. 10.6

(*i*) $I_w = I_0\cos\phi_0$; $I_m = I_0\sin\phi_0$

As the no-load current I_0 is very small, therefore, voltage across the shunt branch ($R_c - X_m$) is virtually V_1 (drop in R_1 and X_1 due to I_0 is negligible).

$$\therefore \quad R_c = \frac{V_1}{I_w} = \frac{V_1}{I_0\cos\phi_0}$$

---

* $R'_L = R'_2\left(\frac{1}{s} - 1\right)$. At $s = 0$, $R'_L \to \infty$.

and $$X_m = \frac{V_1}{I_m} = \frac{V_1}{I_0 \sin \phi_0}$$

(*ii*) The no-load input power W_0 is spent as :

(*a*) Stator iron loss (W_i).

(*b*) Copper losses in stator winding (W_c).

(*c*) Friction and windage losses (W_{F+W}).

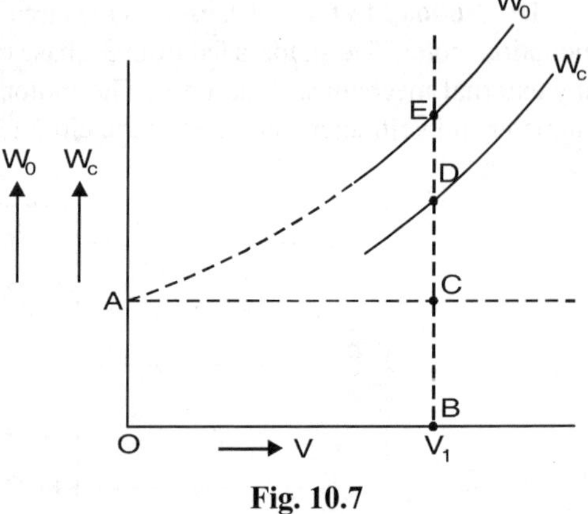

Fig. 10.7

These losses at no-load can be separated as follows. The no-load test is carried out at different applied voltages (below and above the normal voltage V_1). Corresponding to each applied voltage, W_0 is found. Then graph is drawn between W_0 and V as shown in Fig. 10.7. If we extend the curve to the left, it meets the vertical axis at point A. Similarly, curve between stator copper loss and applied voltage can be drawn. This is marked W_c in Fig. 10.7. At the rated voltage V_1, we have,

*$OA = BC$ = Friction and windage loss (W_{F+W})

CD = Stator copper loss (W_c)

DE = Iron loss (W_i)

Thus the losses at no-load stand separated.

2. Blocked rotor test. Fig. 10.8 shows the circuit connections for carrying out the blocked rotor test on the induction motor. In this test, the **rotor is blocked (*i.e.* rotor is held stationary) and † reduced voltage (15% to 20% of normal voltage) is applied to the stator terminals. The reduced voltage is so adjusted that 1.5 times full-load current flows in the stator. The motor input is measured by two wattmeters. The ammeter and voltmeter connected in the circuit measure current and voltage respectively. Note that blocked rotor test is just like the short-circuit test on a transformer.

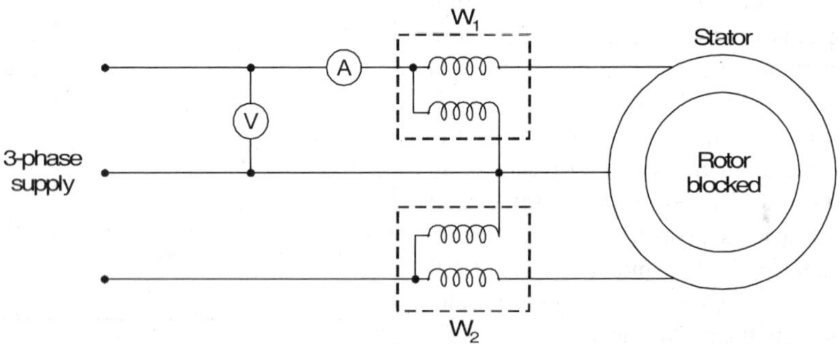

Fig. 10.8

Let V_s = Applied voltage per phase

I_s = Short-circuit current per phase with voltage V_s

* At point A, voltage is zero. Therefore, there is neither stator iron loss nor stator copper loss. This means OA is friction and windage loss.

** The rotor windings are short-circuited at the slip rings if the motor has wound rotor.

† If normal voltage is applied across the stator terminals with rotor blocked, about 6 to 10 times its full-load current will flow through the circuit. This may damage the motor.

$$W_s = \text{Total input power on short-circuit}$$

$$\cos\phi_s = \text{Power factor on short circuit}$$

$$\therefore \quad W_s = 3\,V_s I_s \cos\phi_s$$

or

$$\cos\phi_s = \frac{W_s}{3\,V_s I_s}$$

It is found that relation between short circuit current and applied voltage is approximately a straight line. For voltage V_s applied to the stator, short-circuit current is I_s. If normal voltage V_1 were applied to the stator, the short-circuit current would be :

$$I_{SN} = \frac{I_s}{V_s} \times V_1$$

where I_{SN} = short-circuit current **with normal voltage V_1**

Thus the values of I_{SN} and $\cos\phi_s$ are known from the blocked rotor test. This enables us to locate the second point on the circle diagram (the first point being located by the no-load test). The right bisector of the straight line joining these two points will pass through the centre of the circle. Thus the circle can be drawn.

Additional information. This test can provide more information about the motor. When the rotor is blocked, the rotor is stationary. Therefore, slip s of the motor is 1 and the quantity $*R'_L$ in the equivalent circuit (See Fig. 10.2) is zero. As a result, the equivalent circuit of the motor under blocked rotor condition becomes as shown in Fig. 10.9.

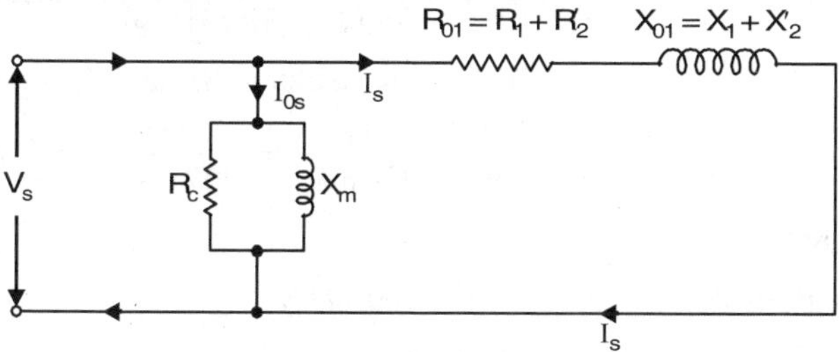

Fig. 10.9

(*i*) Under blocked rotor test, reduced voltage (V_s) is applied at the stator (input) terminals. The no-load current under short-circuit condition (I_{0s}) is very small and may be neglected.

$$\therefore \quad Z_{01} = V_s/I_s$$

Also $\quad R_{01} = Z_{01}\cos\phi_s \quad$ or $\quad R_1 + R'_2 = Z_{01}\cos\phi_s$

Now $\quad X_{01} = Z_{01}\sin\phi_s \quad$ or $\quad X_1 + X'_2 = Z_{01}\sin\phi_s$

Thus the blocked rotor test provides the values of $(R_1 + R'_2)$ and $(X_1 + X'_2)$.

(*ii*) The d.c. value of R_1 is measured by voltmeter-ammeter method by applying d.c. voltage to the stator winding. To get a.c. resistance, this d.c. value is multiplied by 1.5.

$$\therefore \quad R'_2 = R_{01} - R_1$$

Thus the values of R_1 and R'_2 (and hence R_2) can be found.

(*iii*) $X_{01} = X_1 + X'_2$. There is no simple method to find these reactances separately. It is a common practice to assume that $X_1 = X'_2$.

$$\therefore \quad X_1 = X'_2 = X_{01}/2$$

---

* $R'_L = R'_2\left(\frac{1}{s} - 1\right)$. When $s = 1$, $R'_L = 0$.

Example 10.1. *A 400 V, 40 h.p., 50 Hz, three-phase induction motor gave the following test data :*

No-load test : 400 V, 20 A, 1200 W

Blocked rotor test : 100 V, 45 A, 2750 W

The stator d.c. resistance per phase is 0.01 Ω. The ratio of a.c. to d.c. resistance is 1.5. The friction and windage loss is 300 W. Calculate the circuit elements of the approximate equivalent circuit of the motor.

Solution.

$$R_1(\text{a.c.}) = 1.5 \times R_1\,(\text{d.c.}) = 1.5 \times 0.01 = \mathbf{0.015\ \Omega/phase}$$

No-Load Test

$$\text{Total stator Cu loss, } W_c = 3\,I_0^2\,R_1 = 3 \times (20)^2 \times 0.015 = 18\text{ W}$$

$$\therefore \quad \text{Total iron loss (or core loss), } W_i = W_0 - W_c - \text{Friction and Windage loss}$$

$$= 1200 - 18 - 300 = 882\text{ W}$$

$$\therefore \quad \text{Iron loss/phase} = 882/3 = 294\text{ W}$$

$$\text{Applied voltage/phase, } V_1 = 400/\sqrt{3} = 231\text{ V}$$

Now $$\text{Core loss/phase} = V_1^2/R_c$$

$$\therefore \quad R_c = \frac{V_1^2}{\text{core loss/phase}} = \frac{(231)^2}{294} = \mathbf{181\Omega/phase}$$

Now $$\cos\phi_0 = \frac{1200}{3V_1I_0} = \frac{1200}{3\times 231\times 20} = 0.086$$

$$\therefore \quad \phi_0 = \cos^{-1}0.086 = 85° \text{ and } \sin\phi = 0.996$$

$$\therefore \quad X_m = \frac{V_1}{I_0\sin\phi_0} = \frac{231}{20\times 0.996} = \mathbf{11.6\Omega/phase}$$

Blocked rotor test

$$\text{Applied voltage/phase, } V_s = 100/\sqrt{3}\text{ V}$$

$$\text{Short circuit current/phase, } I_s = 45\text{ A}$$

$$\therefore \quad {}^*Z_{01} = V_s/I_s = (100/\sqrt{3})/45 = 1.28\Omega$$

Neglecting iron loss (or core loss) under blocked rotor condition, we have,

$$3\,I_s^2\,{}^{**}R_{01} = 2750$$

$$\therefore \quad R_{01} = \frac{2750}{3\times(45)^2} = 0.45\Omega$$

$$\therefore \quad {}^{\dagger}R'_2 = R_{01} - R_1 = 0.45 - 0.015 = \mathbf{0.435\ \Omega/phase}$$

Now $$X_{01} = \sqrt{Z_{01}^2 - R_{01}^2} = \sqrt{(1.28)^2 - (0.45)^2} = 1.2\ \Omega$$

Assuming that $X_1 = X'_2$, we have,

$$X_1 = X'_2 = X_{01}/2 = 1.2/2 = \mathbf{0.6\ \Omega/phase}$$

Example 10.2. *The tests on a 3-phase, star-connected, 400 V induction motor gave the following results :*

---

* Recall Z_{01} is the total impedance of motor referred to stator side.

** Recall that R_{01} is the total resistance of the motor referred to stator side.

† R'_2 is the rotor resistance (R_2) referred to the stator side.

No-load test : *Applied voltage 400 V, line current 20 A, wattmeter readigs 5000 W and –3200 W.*

Blocked rotor test : *Applied voltage 50 V, line current 60 A, wattmeter readings 2300 W and 750 W.*

D.C. test : *Applied voltage (across two windings in series) 18 V, current 60 A.*

The ratio of a.c. to d.c. resistance is 1.5. Calculate the elements of the approximate equivalent circuit of the motor.

Solution. D.C. test $R_1 \text{ (d.c.)} = \dfrac{V_{dc}}{2I_{dc}} = \dfrac{18}{2\times 60} = 0.15\ \Omega$

$\therefore$ $R_1\text{(a.c.)} = 0.15 \times 1.5 = \mathbf{0.225\ \Omega/phase}$

No-load test

Total no-load losses, $W_0 = (5000) + (-3200) = 1800$ W

Applied voltage/phase, $V_1 = 400/\sqrt{3} = 231$ V

No-load current/phase, I_0 = Line current = 20 A

Now $W_0 = 3\,V_1\,I_0 \cos\phi_0$

$\therefore$ $\cos\phi_0 = \dfrac{W_0}{3V_1I_0} = \dfrac{1800}{3\times 231\times 20} = 0.13$

Now $R_c = \dfrac{V_1}{I_0\cos\phi_0} = \dfrac{231}{20\times 0.13} = \mathbf{88.85\ \Omega/phase}$

and $X_m = \dfrac{V_1}{I_0\sin\phi_0} = \dfrac{231}{20\times 0.99}$ **11.67 Ω/phase**

Blocked rotor test

Total input power on short circuit, $W_s = (2300) + (750) = 3050$ W

Applied voltage/phase, $V_s = 50/\sqrt{3} = 28.87$ V

Short circuit current/phase, $I_s = 60$ A

Now $W_s = 3\,V_sI_s\cos\phi_s$

$\therefore$ P.F. on short-circuit, $\cos\phi_s = \dfrac{W_s}{3V_sI_s} = \dfrac{3050}{3\times 28.87\times 60} = 0.587$

Now $Z_{01} = \dfrac{V_s}{I_s} = \dfrac{28.87}{60}$

$\therefore$ $^*R_{01} = Z_{01}\cos\phi_s = \dfrac{28.87}{60}\times 0.587 = 0.282\ \Omega$

$\therefore$ $R_1 + R_2' = 0.282$ or $R_2' = 0.282 - R_1 = 0.282 - 0.225 = \mathbf{0.057\ \Omega/phase}$

Now $X_{01} = Z_{01}\sin\phi_s = \dfrac{28.87}{60}\times 0.81 = 0.39\ \Omega$ $\quad\therefore X_1 + X_2' = 0.39\ \Omega$

Assuming $X_1 = X_2'$, we have, $X_1 = X_2' = X_{01}/2 = 0.39/2 = \mathbf{0.195\ \Omega/phase}$

---

* We can also find R_{01} as follows. Neglecting core loss (*i.e.* iron loss) under blocked rotor condition, we have,

$3\,I_s^2\,R_{01} = 3050$ or $R_{01} = \dfrac{3050}{3\,I_s^2} = \dfrac{3050}{3\times(60)^2} = 0.282\ \Omega$

Example 10.3. *A 400 V, 50 Hz, 3-phase, star-connected, 4-pole induction motor has per phase stator impedance $Z_1 = (0.07 + j\,0.3)\ \Omega$ and rotor impedance referred to stator side, $Z_2' = (0.08 + j\,0.3)\ \Omega$. The magnetising reactance is 10 Ω and resistance representing core-loss is 50 Ω. Using approximate equivalent circuit, calculate (i) stator current (ii) motor power factor (iii) rotor current (iv) torque developed (v) gross efficiency. Assume a slip of 4%.*

Solution. Fig. 10.10 shows the approximate per phase equivalent circuit of the induction motor.

$$R_{01} = R_1 + R'_2 = 0.07 + 0.08 = 0.15\,\Omega$$
$$X_{01} = X_1 + X'_2 = 0.3 + 0.3 = 0.6\,\Omega$$
$$R'_L = R'_2\left(\frac{1-s}{s}\right) = 0.08\left(\frac{1-0.04}{0.04}\right) = 1.92\,\Omega$$

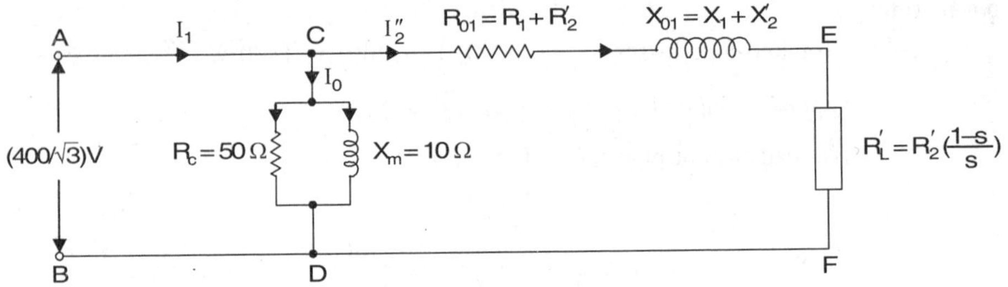

Fig. 10.10

(*i*) Total impedance Z of the circuit to the right of terminals CD is

$$Z = R_{01} + X_{01} + R'_L$$
$$= 0.15 + j\,0.6 + 1.92 = (2.07 + j\,0.6)\,\Omega$$

The voltage across terminals CD is $400/\sqrt{3}$ V so that :

$$\text{Rotor current, } I_2'' = \frac{400/\sqrt{3}}{Z} = \frac{400/\sqrt{3}}{(2.07 + j\,0.6)} = 107\,\angle{-16.5°}\text{ A}$$

$$\text{No-load current, } I_0 = \frac{V_1}{R_c} + \frac{V_1}{jX_m} = \frac{400/\sqrt{3}}{50} + \frac{400/\sqrt{3}}{j10}$$
$$= \frac{400}{\sqrt{3}\times 50} - \frac{j\,400}{\sqrt{3}\times 10} = (4.6 - j\,23)\text{ A}$$
$$= 23.4\,\angle{-79°}\text{ A}$$

$$\therefore \quad \text{Stator current, } I_1 = I_2'' + I_0 = 107\,\angle{-16.5°} + 23.4\,\angle{-79°}$$
$$= 119.7\,\angle{-26.5°}\text{ A}$$

∴ Magnitude of stator current is **119.7 A.**

(*ii*) Motor p.f. $= \cos\phi_1 = \cos 26.5° =$ **0.89** *lagging*

(*iii*) Rotor current referred to stator side is $I''_2 =$ **107 A.**

(*iv*) Synchronous speed, $N_s = 120f/P = 120 \times 50/4 = 1500$ r.p.m.

∴ Rotor speed, $N = N_s(1-s) = 1500\,(1-0.04) = 1440$ r.p.m.

$$\text{Rotor output} = 3\,I''^2_2 R'_L = 3\times(107)^2\times 1.92 = 65946\text{ W}$$

If T is the torque developed in Nm, then,

$$\text{Rotor output} = \frac{2\pi NT}{60}$$

or $$65946 = \frac{2\pi \times 1440 \times T}{60} \qquad \therefore\ T = \mathbf{437.3\ Nm}$$

(v) Input power to stator $= \sqrt{3}\, V_1 I_1 \cos \phi_1$

$$= \sqrt{3} \times 400 \times 119.7 \times 0.89 = 73808\ \text{W}$$

$\therefore$ $$\text{Gross efficiency} = \frac{\text{Gross output}}{\text{Input}} \times 100 = \frac{65946}{73808} \times 100 = \mathbf{89.4\%}$$

10.6 CONSTRUCTION OF CIRCLE DIAGRAM

We have seen that the circle diagram of an induction motor can be drawn using the data obtained from the no-load and blocked rotor tests on the motor. The construction of the circle diagram involves the following steps :

(i) From the no-load test, I_0 and $\cos \phi_0$ can be calculated. Selecting suitable current scale, phasor I_0 can be drawn lagging ϕ_0^0 behind the applied voltage V_1 as shown in Fig. 10.11. Thus one point O' of the circle diagram is located.

(ii) From the blocked rotor test, short-circuit current I_{SN} [$= (V_1/V_s)\, I_s$] corresponding to normal voltage and $\cos \phi_s$ can be found. With the selected current scale, the phasor I_{SN} can be drawn lagging ϕ_s^0 behind the applied voltage V_1 as shown in Fig. 10.11. This locates the second point A of the circle diagram. The phasor I''_2 represents the rotor current phasor referred to the stator side.

(iii) The right bisector of the line joining points O' and A intersects the horizontal line through O' at C. Therefore, C is the centre of the circle. With centre C and radius = CO', the circle can be drawn as shown in Fig. 10.11. The circular path $O'LAD$ is the locus of the stator input current I_1 (= phasor sum of I_0 and I''_2). For example for point L on the circle diagram, the stator current for the motor is I_1 and its power factor is $\cos \phi_1$. For point A, the stator current is I_{SN} and power factor $\cos \phi_s$.

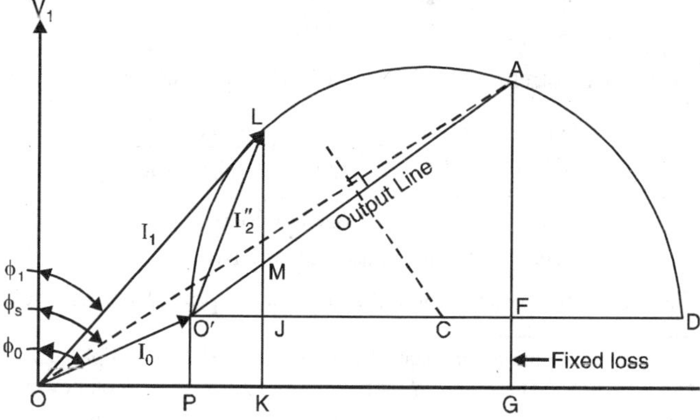

Fig. 10.11

Output line. The line $O'A$ is called the **output line.** It is because the vertical distance above this line up to the periphery of the circle (expressed in power scale) represents the output power of the motor. Thus for point L on the circle diagram, the output power of the motor is the vertical distance ML.

We shall now prove that line $O'A$ is rightly called the output line. Consider the point L on the circle diagram. Corresponding to point L, we have,

LK = *Input power to motor on power scale

JK = No-load input power = ** Fixed losses

MJ = Copper losses in rotor and stator (to be proved)

Now Motor output = Motor input – Fixed losses – Copper losses

$= LK - JK - MJ = LK - MK = LM$

∴ Motor output $= LM$

Thus for point L on the circle diagram, the motor output is LM *i.e.* vertical distance from point M upto the periphery of the circle. The line $O'A$ is the locus of such points as M and is called the *output line.*

To prove MJ = Copper losses. We now prove that MJ is approximately equal to copper loss (stator and rotor). For this purpose, we proceed as follows. Fig. 10.12 shows the circle portion of Fig. 10.11.

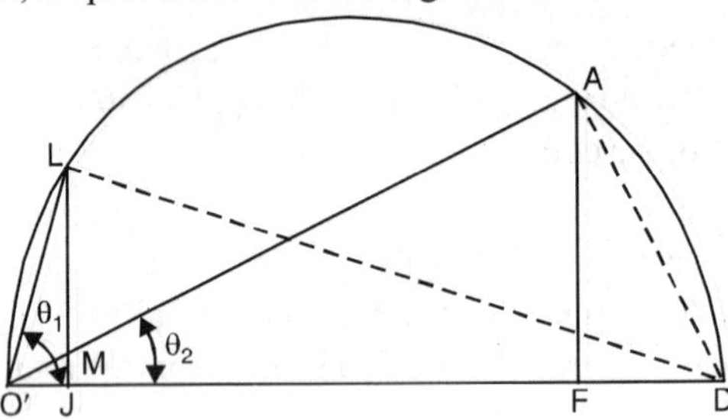

Fig. 10.12

Referring to Fig. 10.12, the triangles $O'MJ$ and $O'AF$ are similar.

$$\therefore \quad \frac{MJ}{AF} = \frac{O'J}{O'F} = \frac{\dagger O'L\cos\theta_1}{\dagger\dagger O'A\cos\theta_2}$$

In triangle $O'LD$, $\angle O'LD = 90°$ (property of circle). Therefore, $\cos\theta_1 = O'L/O'D$.

In triangle $O'AD$, $\angle O'AD = 90°$. Therefore, $\cos\theta_2 = O'A/O'D$.

$$\therefore \quad \frac{MJ}{AF} = \frac{(O'L)(O'L/O'D)}{(O'A)(O'A/O'D)} = \frac{(O'L)^2}{(O'A)^2}$$

If we neglect the no-load current, $O'L \simeq OL$ and $O'A \simeq OA$ [See Figs. 10.11 and 10.12].

$$\therefore \quad \frac{MJ}{AF} = \frac{(O'L)^2}{(O'A)^2} \simeq \frac{(OL)^2}{(OA)^2} = \frac{(OL)^2 \times R_{01}}{(OA)^2 \times R_{01}}$$

$$\therefore \quad \frac{MJ}{AF} \simeq \frac{\text{Copper losses on load}}{\text{Copper losses at standstill}}$$

Therefore, MJ is the copper loss (stator and rotor) for the point L on the circle diagram.

---

* Input power is proportional to the component of input current OL ($= I_1$) in phase with the applied voltage V_1 (*i.e.* $OL\cos\phi_1 = I_1\cos\phi_1$). Therefore, LK is the input power measured on power scale.

** The fixed losses in an induction motor are : (*i*) Iron losses (*ii*) Friction and windage losses. These are called fixed losses because they remain the same from no-load to full-load on the motor. Now, at no-load, the motor output is zero. Therefore, no-load input power is used to meet the fixed losses and very small Cu loss $I_0^2 R_1$ in the stator winding. The $I_0^2 R_1$ loss is negligible so that :

No-load input power = Fixed losses.

† In triangle $O'LJ$, $O'J/O'L = \cos\theta_1$. Therefore, $O'J = O'L\cos\theta_1$.

†† In triangle $O'AF$, $\cos\theta_2 = O'F/O'A$. Therefore, $O'F = O'A\cos\theta_2$.

10.7 COMPLETE CIRCLE DIAGRAM OF INDUCTION MOTOR

If we add torque line to the circle diagram shown in Fig. 10.11, we get the complete circle diagram shown in Fig. 10.13. *The line which separates the stator and rotor copper losses on the circle diagram is called the **torque line**.* Thus is Fig. 10.13, $O'E$ is the torque line. The vertical distance above this line up to the periphery of the circle is torque of the motor (on power scale). For the operating point L on the circle diagram, the vertical distance LN is the torque developed by the motor.

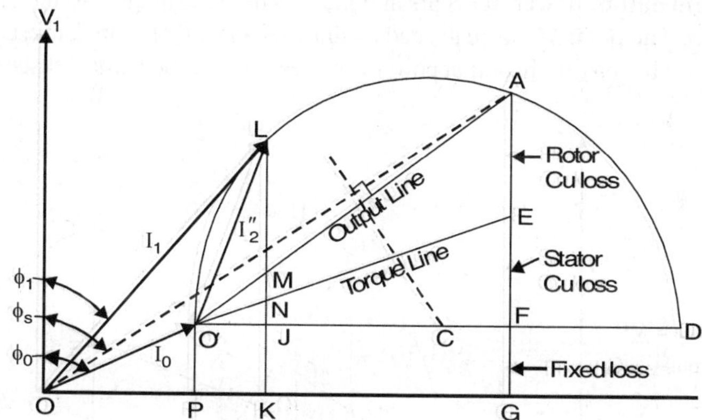

Fig. 10.13

How to draw torque line ? The torque developed by an induction motor is given by ;

$$T_g = \frac{\text{Rotor input}}{2\pi N_s}$$

For a given machine, N_s (synchronous speed) is constant.

$$\therefore \quad T_g \propto \text{Rotor input}$$

Therefore, to find torque from the circle diagram, we must find the rotor input. For this purpose, we should separate the stator Cu loss and the rotor Cu loss. This can be done for the standstill (blocked rotor) conditions. For blocked rotor conditions, *AF (See Fig. 10.13) is the sum of rotor Cu loss and stator Cu loss. The vertical distance AF is divided at E such that :

$$\frac{AE}{EF} = \frac{\text{Rotor Cu loss}}{\text{Stator Cu loss}}$$

The line joining points O' and E is the torque line. For the operating point L on the circle diagram, we have,

$$MN = \text{Rotor Cu loss}$$

$$NJ = \text{Stator Cu loss}$$

$$LN = \text{Motor torque}$$

Now $$LN = LM + MN$$

$$= \text{Output power} + \text{Rotor Cu loss} = \text{Rotor input}$$

$$\therefore \quad LN = \text{Rotor input}$$

* When the rotor is locked, the entire power supplied (= AG) to the motor is used up to meet the fixed losses (= $FG = O'P = JK$ = Core losses) and copper losses in stator and rotor. Note that when the rotor is locked, windage and friction losses are absent. However, under this condition, core loss in rotor is slightly increased. Therefore, $FG = O'P$.

$\therefore \quad AG - FG = AF$ = Copper losses in stator and rotor

Thus for point L on the circle diagram, the motor torque is LN *i.e.* vertical distance from point N upto the periphery of the circle. The line $O'E$ is the locus of such points as N and is called the *torque line*.

10.8 MAXIMUM QUANTITIES FROM CIRCLE DIAGRAM

Fig. 10.14 shows the circle diagram of an induction motor. We can find the maximum values of output power, torque and input power from this diagram.

(*i*) **Maximum output power.** It occurs at point M where the tangent to the circle is parallel to the output line $O'A$. The point M can be located by drawing a line CM from centre C perpendicular to the output line $O'A$. The maximum output power is represented by the vertical distance MP (on power scale).

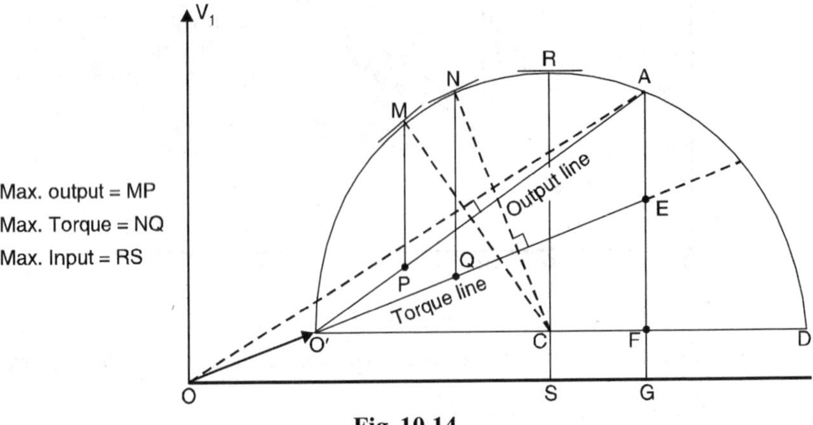

Fig. 10.14

(*ii*) **Maximum Torque (or Maximum Rotor Input).** It occurs at point N where the tangent to the circle is parallel to the torque line $O'E$. The point N can be located by drawing a line CN from centre C perpendicular to the torque line $O'E$. The maximum torque is represented by the vertical distance NQ (on power scale). Note that maximum torque is also called stalling torque or pull-out torque.

(*iii*) **Maximum Input Power.** It occurs at point R where the tangent to the circle is parallel to the horizontal. Obviously, R is the highest point of the circle. The maximum input power is represented by the vertical distance RS on power scale.

Note. It must be remembered that the method of drawing circle diagram of induction motor is based upon the approximations which have been used in developing the approximate equivalent circuit of the motor. Therefore, we should expect the approximate determination of the performance of the motor from the circle diagram. Nevertheless, the approximate information about the motor's performance is quite useful and handy in many practical situations.

Example 10.4. *Fig. 10.15 shows the circle diagram of an induction motor. The point P corresponds to the full-load conditions. What information is conveyed by this circle diagram?*

Solution. The vertical axis represents currents (on current scale) as well as powers (on power scale).

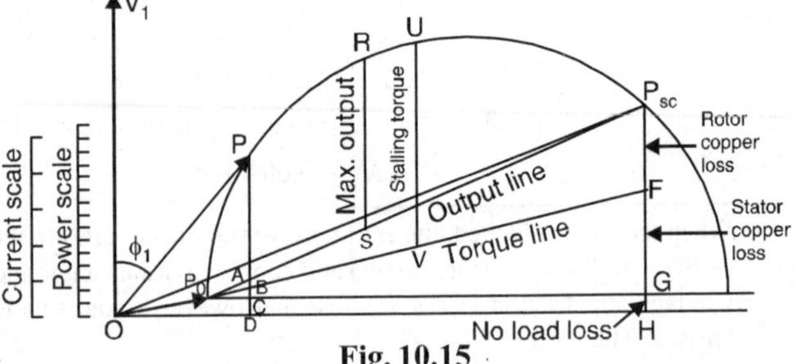

Fig. 10.15

This circle diagram conveys the following information :

(i) OP = Full-load stator input current

(ii) PD = Full-load input power to the motor

(iii) PA = Full-load output power of the motor

(iv) $\cos\phi_1 = PD/OP$ = Power factor of motor at full-load

(v) CD = Iron and Friction losses

(vi) AB = Rotor Cu loss

(vii) BC = Stator Cu loss

(viii) PB = Torque (Rotor input) developed by motor

(ix) PA/PD = Output/Input = Efficiency of the motor

(x) Motor slip, $s = \dfrac{\text{Rotor Cu loss}}{\text{Rotor input}} = \dfrac{AB}{PB}$

(xi) $\dfrac{RS}{PA} = \dfrac{\text{Max. output power}}{\text{Full-load output power}}$

(xii) $\dfrac{UV}{PB} = \dfrac{\text{Max. torque}}{\text{Full-load torque}}$

(xiii) $P_{sc}F$ = Starting torque

(xiv) AD = Total motor losses

(xv) $P_{sc}H$ = Input power under blocked rotor conditions

(xvi) $\dfrac{PA}{PB} = \dfrac{\text{Rotor output}}{\text{Rotor input}} = 1 - s = \dfrac{N}{N_s} = \dfrac{\text{Rotor speed}}{\text{Synchronous speed}}$

Note. In the following examples, due to reasons associated with the clarity of the figure, available size of the page of the book etc., the circle diagrams drawn may not be exactly according to mentioned scales.

Example 10.5. *A 400 V, 40 h.p., 50 Hz, 4-pole delta-connected induction motor gave the following test data :*

No-load test : 400 V, 20 A, 1200 W

Blocked rotor test : 100 V, 45 A, 2800 W

Draw the circle diagram and determine (i) full-load line current and power factor (ii) maximum output power (iii) maximum torque (iv) full-load efficiency and (v) full-load rotor speed. Assume stator and rotor Cu losses to be equal at standstill.

Solution. The data of the tests suggest that line values are given.

Construction of circle diagram. The circle diagram can be constructed from the data of no-load and blocked rotor tests.

$$\text{No-load p.f., } \cos\phi_0 = \frac{W_0}{\sqrt{3}V_1 I_0} = \frac{1200}{\sqrt{3}\times 400\times 20} = 0.0866$$

$$\therefore \quad \text{p.f. angle, } \phi_0 = \cos^{-1} 0.0866 = 85°$$

Power factor $\cos\phi_s$ under blocked rotor condition is

$$\cos\phi_s = \frac{W_s}{\sqrt{3}V_s I_s} = \frac{2800}{\sqrt{3}\times 100\times 45} = 0.359$$

$$\therefore \quad \phi_s = \cos^{-1} 0.359 = 69°$$

The input current and input power for blocked rotor conditions are at a reduced voltage of 100 V. These values are to be found at the normal or rated voltage (*i.e.* 400 V).

∴ Input current (at standstill) corresponding to normal voltage is

$$I_{SN} = 45\,(400/100) = 180\text{ A}$$

Input power (at standstill) corresponding to normal voltage is

$$W_{SN} = 2800\,(400/100)^2 = 44800\text{ W}$$

We now construct the circle diagram. Let us choose a convenient current scale *viz* 1 cm = 20 A. Represent voltage V_1 on the vertical axis. Draw horizontal line *OX* from *O* as shown in Fig. 10.16. Since current scale is 1 cm = 20 A, the no-load current of 20 A will have a length of 1 cm. Thus we can represent no-load current by a phasor *OO′* whose length is 1 cm and lagging 85° behind the vertical voltage axis V_1 as shown in Fig. 10.16. This locates the first point (*O′*) of the circle diagram. Similarly, I_{SN} (input current under blocked rotor condition corresponding to normal voltage) of 180 A is represented by a phasor *OA* of length 9 cm and lagging 69° behind the voltage axis V_1 as shown in Fig. 10.16. This locates the second point (*A*) of the circle diagram. Draw a horizontal line *O′D* from *O′* parallel to *OX*. Draw the perpendicular bisector of the line *O′A* (line joining points *O′* and *A*) to intersect the line *O′D* at point *C*. Then *C* is the centre of the circle. With *C* as centre and radius = *CO′*, draw the semicircle as shown in Fig. 10.16. Recall that *O′A* is the output line. Let us draw the torque line. From point *A* on the semicircle, draw a vertical line *AG* perpendicular to *OX*. Now *AG* cuts *O′D* at *F*. Therefore, for blocked rotor condition, *FG* = Fixed loss and *AF* = stator and rotor Cu loss. It is given that rotor and stator losses are equal at standstill. Therefore, divide the vertical line *AF* at *E* such that *AE* = *EF*. Then line *O′E* is the torque line. Thus we have the complete circle diagram. We now find the various quantities from this diagram.

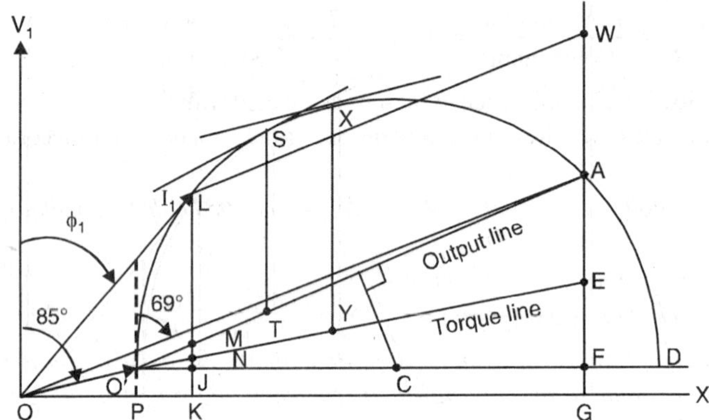

Fig. 10.16

Let us find power scale for this circle diagram. The vertical length *AG* represents input power under blocked rotor conditions corresponding to normal voltage (*i.e.* 44800 W in this case) and its length is 3.4 cm (by measurement).

∴ Power scale = 44800/3.4 = 13176 W/cm

or 1 cm = 13176 W

This is the power scale for the circle diagram.

(*i*) Full-load output of motor = 40 h.p. = 40 × 746 W = 29840 W. On power scale, this output power represents 29840/13176 = 2.26 cm. We are to locate the point *L* on the circle diagram such that vertical line *ML* above the output line is 2.26 cm. For this purpose, we raise the vertical line *FA* and *AW* is made equal to 2.26 cm. Now a line from *W* is drawn parallel to the output line *O′A* to meet the circle at point *L*. Thus *L* is the required load point. A vertical line *LM* is drawn on to the output line. Then *OL* represents the full-load input current (*OL* = 3.05 cm, by measurement).

∴ Full-load input current = 3.05 cm × 20 A = **61 A**

By measurement, the angle ϕ_1 which the full-load input current makes with vertical voltage axis V_1 is 34°.

$$\therefore \quad \text{Full-load p.f.} = \cos\phi_1 = \cos 34° = \mathbf{0.829}\ \textit{lagging.}$$

(*ii*) The maximum output power occurs at point *S* where tangent to the circle is parallel to the output line *O′A*. The vertical distance *ST* represents the maximum output power. Now *ST* measures 3.35 cm. Using power scale,

$$\text{Max. output power} = ST \text{ in cm} \times 13176 = 44140 \text{ W} = \mathbf{44.14\ kW}$$

(*iii*) The maximum torque occurs at point *X* where tangent to the circle is parallel to the torque line *O′E*. The vertical distance *XY* represents the maximum torque. Now *XY* measures 4.05 cm. Using power scale,

$$\text{Max. torque} = XY \text{ in cm} \times 13176 = \mathbf{53363\ Syn.\ W}$$

$$(iv)\ \text{Full-load efficiency} = \frac{\text{Output}}{\text{Input}} = \frac{LM}{LK} = \frac{2.23 \text{ cm}}{2.55 \text{ cm}} = \mathbf{87.4\%}$$

(*v*) To determine the full-load rotor speed (*N*), we first determine the full-load slip *s*.

$$\text{Full-load slip, } s = \frac{\text{Rotor Cu loss}}{\text{Rotor input}} = \frac{MN}{LN} = \frac{0.15 \text{ cm}}{2.32 \text{ cm}} = 0.0646$$

Now
$$N_s = \frac{120 f}{P} = \frac{120 \times 50}{4} = 1500 \text{ r.p.m.}$$

Also
$$s = \frac{N_s - N}{N_s}$$

or
$$0.0646 = \frac{1500 - N}{1500} \qquad \therefore\ N = \mathbf{1403\ r.p.m.}$$

Example 10.6. *Draw the circle diagram for a 5.6 kW, 400 V, 3-phase, 4-pole, 50 Hz, slip-ring induction motor from the following data :*

No-load test : 400 V, 6 A, $\cos\phi_0 = 0.087$

Blocked rotor test : 100 V, 12 A, 720 W

The ratio of primary to secondary turns = 2.62, stator resistance per phase is 0.67 Ω and of the rotor is 0.185 Ω. Calculate :

(i) full-load current (ii) full-load slip (iii) full-load power factor (iv) maximum power (v) max. torque/full-load torque.

Solution. The data of the tests suggest that line values are given.

Construction of circle diagram. The circle diagram can be constructed from the data of no-load and blocked rotor tests.

From no-load test : $I_0 = 6 \text{ A}\ ;\ \phi_0 = \cos^{-1}(0.087) = 85°$

From blocked rotor test : Power factor $\cos\phi_s$ under blocked rotor condition is

$$\cos\phi_s = \frac{W_s}{\sqrt{3}\, V_s I_s} = \frac{720}{\sqrt{3} \times 100 \times 12} = 0.347$$

$$\therefore \quad \phi_s = \cos^{-1}(0.347) = 69.7°$$

Input current (at standstill) corresponding to normal voltage is

$$I_{SN} = 12\,(400/100) = 48 \text{ A}$$

Input power (at standstill) corresponding to normal voltage is

$$W_{SN} = 720\,(400/100)^2 = 11.52 \times 10^3 \text{ W} = 11.52 \text{ kW}$$

We now construct the circle diagram. Let us choose a convenient current scale *viz.* 1 cm = 4 A. Represent voltage V_1 on the vertical axis. Draw horizontal line *OX* from *O* as shown in Fig. 10.17. Since the selected current scale is 1 cm = 4 A, the no-load current of 6A will have a length of 1.5 cm. Thus we can represent no-load current by a phasor *OA* whose length is 1.5 cm and lagging 85° behind the vertical voltage axis V_1 as shown in Fig. 10.17. This locates the first point (*A*) of the circle diagram. Similarly, I_{SN} (input current under blocked rotor condition corresponding to normal voltage) of 48 A is represented by a phasor *OB* of length 12 cm and lagging 69.7° behind the voltage axis V_1 as shown in Fig. 10.17. This locates the second point (*B*) of the circle diagram. Draw a horizontal line *AX′* from *A* parallel to *OX*. Draw the perpendicular bisector of the line *AB* to intersect the line *AX′* at *C*. Then *C* is the centre of the circle. With *C* as centre and radius = *CA*, draw the semicircle as shown in Fig. 10.17. Recall that *AB* is the output line.

How to draw torque line ? We now draw torque line on the circle diagram.

$$N_1/N_2 = 2.62 \; ; \; R_1 = 0.67\,\Omega \; ; R_2 = 0.185\,\Omega$$

$$\therefore \quad \frac{\text{Rotor Cu loss}}{\text{Stator Cu loss}} = \frac{I_2'^2 R_2}{I_1^2 R_1} = {}^*\left(\frac{N_1}{N_2}\right)^2 \times \frac{R_2}{R_1} = (2.62)^2 \times \frac{0.185}{0.67} = 1.9$$

$$\therefore \quad \frac{{}^{**}\text{Rotor Cu loss}}{\text{Total Cu loss}} = \frac{1.9}{2.9} = 0.655$$

Let us find power scale for this circle diagram. The vertical length *BD* represents input power under blocked rotor conditions corresponding to normal voltage (*i.e.* 11.52 kW in this case) and its length is 4.125 cm (by measurement).

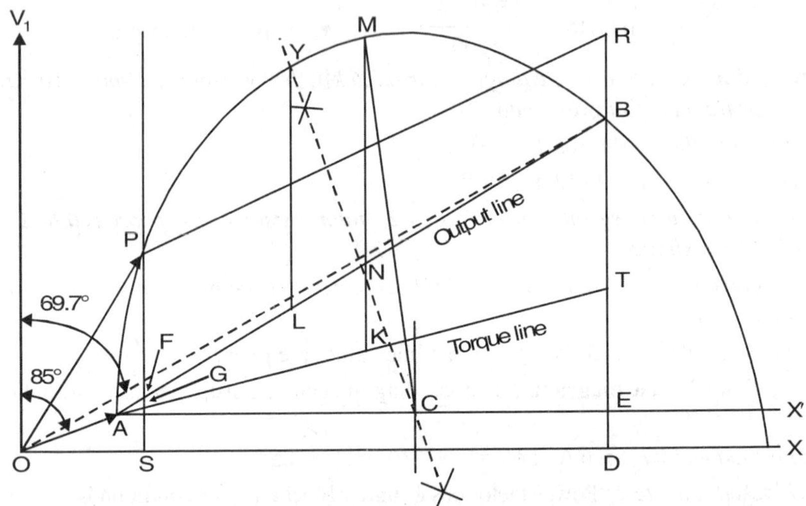

Fig. 10.17

$$\therefore \quad \text{Power scale} = 11.52/4.125 = 2.79 \text{ kW/cm}$$

or $\quad$ 1 cm = 2.79 kW

Now *BE* represents the total Cu loss (rotor and stator) and is divided at *T* in the ratio 1.9:1.

---

* $I_1 N_1 = I'_2 N_2 \quad \therefore \quad \dfrac{I'_2}{I_1} = \dfrac{N_1}{N_2}$

** Note this step. Add 1 to L.H.S. and R.H.S. of above equation and proceed.

$\therefore \quad BT = BE \times 1.9/2.9 = 4 \times 0.655 = 2.62$ cm

Thus point T on the vertical line BE is located. The line AT is the torque line.

(*i*) To find the full-load current, we proceed as follows. The full-load output = 5.6 kW. It is represented by a vertical line = 5.6/2.79 = 2 cm above the output line. Now vertical line DB is raised such that BR = 2 cm. Draw a line from R parallel to the output line AB to meet the circle at point P. Then OP represents the full-load line current. By measurement, OP = 2.87 cm.

$$\therefore \quad \text{F.L. current} = 2.87 \times 4 = \mathbf{11.48\ A}$$

(*ii*)
$$\text{F.L. slip} = \frac{FG}{PG} = \frac{0.1 \text{ cm}}{2.12 \text{ cm}} = \mathbf{0.047} \text{ or } \mathbf{4.7\%}$$

(*iii*)
$$\text{Power factor} = \frac{SP}{OP} = \frac{2.3 \text{ cm}}{2.87 \text{ cm}} = \mathbf{0.8} \textit{ lagging}$$

(*iv*) To find maximum output power, we proceed as follows. From centre C, draw a line perpendicular to the output line to meet the circle at point Y. Then vertical line YL represents the maximum output power.

$$\therefore \quad \text{Max. output power} = YL = 3.87 \text{ cm} = 3.87 \times 2.79 = \mathbf{10.8\ kW}$$

(*v*)
$$\frac{\text{Maximum torque}}{\text{Full-load torque}} = \frac{MK}{PG} = \frac{5.02 \text{ cm}}{2.12 \text{ cm}} = \mathbf{2.37}$$

Example 10.7. *The results of no-load and blocked rotor tests on a 400 V, 36.8 kW, 50 Hz, 3-phase, 4-pole induction motor are as under :*

No-load test : 400 V, 30 A, 1800 W

Blocked rotor test : 110 V, 80 A, 4000 W

The motor has a star-connected stator having a resistance of 0.1 Ω/phase. Draw the circle diagram and calculate :

(i) full-load stator current (ii) full-load power factor (iii) full-load slip (iv) full-load torque (v) full-load efficiency (vi) starting torque (vii) maximum torque (viii) maximum output power.

Solution. The data of the tests suggest that line values are given.

$$\cos\phi_0 = \frac{W_0}{\sqrt{3}\,V_1 I_0} = \frac{1800}{\sqrt{3} \times 400 \times 30} = 0.0866 \qquad \therefore \phi_0 = 85°$$

$$\cos\phi_s = \frac{W_s}{\sqrt{3}\,V_s I_s} = \frac{4000}{\sqrt{3} \times 110 \times 80} = 0.262 \qquad \therefore \phi_s = 74.8°$$

Input current (at standstill) corresponding to normal voltage is

$$I_{SN} = 80\,(400/110) = 290.9 \text{ A}$$

Input power (at standstill) corresponding to normal voltage is

$$W_{SN} = 4000\,(400/110)^2 = 52892 \text{ W}$$

Let us choose a convenient current scale *viz.* 1 cm = 25 A. The no-load current of 30 A will have a length of 1.2 cm and is represented by phasor OA lagging 85° behind the vertical voltage axis V_1 as shown in Fig. 10.18. Similarly, I_{SN} of 290.9 A is represented by the phasor OB of length 11.6 cm (290.9/25 = 11.6 cm) lagging 74.8° behind the voltage axis V_1 as shown in Fig. 10.18. The right bisector of the chord AB locates the centre P of the circle. With P as centre and radius = PA, the semicircle $AGSTBF$ can be drawn. Note that AB is the output line.

How to draw torque line ?

Stator Cu loss at rated voltage under blocked rotor condition

$$= {}^*3\, I_{SN}^2 R_1 = 3 \times (290.9)^2 \times 0.1 = 25386 \text{ W}$$

* Since the stator is star-connected, phase current = line current.

The power input at rated voltage under blocked rotor condition is calculated above as 52892 W and this goes to meet the Cu losses (rotor and stator).

∴ Rotor Cu loss at rated voltage under blocked rotor condition

$$= 52892 - 25386 = 27506 \text{ W}$$

$$\therefore \quad \frac{\text{Rotor Cu loss}}{\text{Stator Cu loss}} = \frac{27506}{25386} = 1.08$$

$$\therefore \quad \frac{\text{Rotor Cu loss}}{\text{Total Cu loss}} = \frac{1.08}{2.08} = 0.52$$

Let us find power scale for this circle diagram. The vertical length EB represents input power under blocked rotor conditions (*i.e.* 52892 W in this case) and its length is 2 cm (by measurement).

∴ Power scale = 52892/2 = 26446 W/cm

or 1 cm = 26446 W

Now BD represents the total Cu loss (rotor and stator) and is divided at C in the ratio 1.08:1.

$$\therefore \quad BC = BD \times 1.08/2.08 = 2 \times 1.08/2.08 = 1.04 \text{ cm}$$

Thus point C on the vertical line BE is located. The line AC is the torque line.

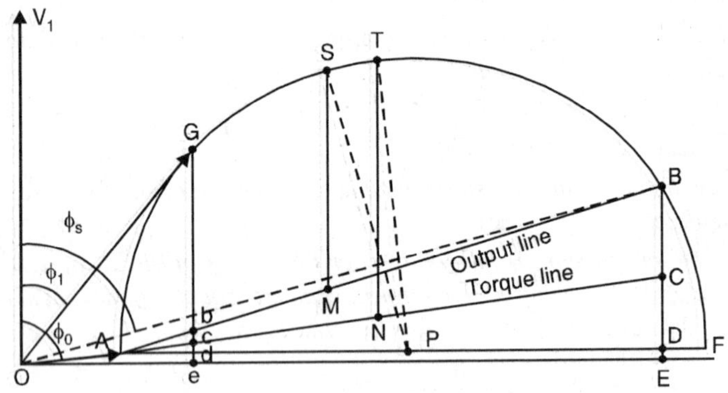

Fig. 10.18

(*i*) The full-load point G is located such that the vertical intercept Gb is equal to the full-load output power (= 36.8 kW) on power scale. Also Gb is approximately equal to the active component of full-load current on current scale. Its value is

$$= \frac{36.8 \times 10^3}{\sqrt{3} \times 400} = 53.1 \text{ A}$$

On current scale, the vertical length Gb should be = 53.1/25 = 2.12 cm. Therefore, vertical intercept Gb should be drawn to a length of 2.12 cm.

∴ Full-load stator current = OG = 3.4 cm = 3.4 × 25 = **85 A**

(*ii*) By measurement, full-load p.f. angle is ϕ_1 = 42°.

∴ Full-load p.f., $\cos \phi_1$ = cos 42° = **0.74** *lagging*

$$(iii) \quad \text{Full-load slip, } s = \frac{\text{Rotor Cu loss}}{\text{Rotor input}} = \frac{bc}{Gc} = \frac{0.11 \text{ cm}}{2.3 \text{ cm}} = \mathbf{0.05}$$

(*iv*) The active component of current corresponding to full-load torque is Gc and its length = 2.3 cm by measurement. Therefore, its value = 2.3 × 25 = 57.5 A.

$$\therefore \quad \text{Full-load torque} = \frac{\sqrt{3}\times 400\times 57.5}{2\pi\times 25^*} = \mathbf{253.7\ Nm}$$

(*v*) $$\text{Full-load efficiency} = \frac{\text{Output power}}{\text{Input power}} = \frac{Gb}{Ge}$$

$$= \frac{2.12\text{ cm}}{2.58\text{ cm}} = 0.82 = \mathbf{82\%}$$

(*vi*) The active component of current corresponding to starting torque is BC and its length is 1.1 cm by measurement. Therefore, its value = 1.1 × 25 = 27.5 A.

$$\therefore \quad \text{Starting torque} = \frac{\sqrt{3}\times 400\times 27.5}{2\pi\times 25} = \mathbf{121.3\ Nm}$$

(*vii*) From the centre P of the circle, draw perpendicular to the torque line. This perpendicular meets the circle at point T. The active component of current corresponding to maximum torque is TN and its length is 3.3 cm by measurement. Its value = 3.3 × 25 = 82.5 A.

$$\therefore \quad \text{Maximum torque} = \frac{\sqrt{3}\times 400\times 82.5}{2\pi\times 25} = \mathbf{364\ Nm}$$

(*viii*) From the cente P of the circle, draw perpendicular to the output line. This perpendicular meets the circle at point S. The active component of current corresponding to the maximum output power is SM and its length is 2.8 cm by measurement. Its value = 2.8 × 25 = 70 A.

$$\therefore \quad \text{Maximum output} = \sqrt{3}\times 400\times 70 = 48.5\times 10^3\text{ W} = \mathbf{48.5\ kW}$$

Example 10.8. *Draw the circle diagram of a 7.46 kW, 200 V, 50 Hz, 3-phase, slip-ring induction motor with a star connected stator and rotor, a winding ratio of unity, a stator resistance of 0.38 Ω/phase and a rotor resistance of 0.24 Ω/phase. The following are the test readings :*

No-load test : 200 V, 7.7 A ; cos $\phi_0 = 0.195$

Blocked rotor test : 100 V, 47.6 A ; cos $\phi_s = 0.454$

Find (i) starting torque (ii) maximum torque (iii) the maximum power factor (iv) the slip for maximum torque (v) the maximum output.

Solution. The data of the tests suggest that line values are given.

$$I_0 = 7.7\text{ A}; \cos\phi_0 = 0.195 \quad \therefore \phi_0 = 78.7°$$

Input current (at standstill) corresponding to normal voltage is

$$I_{SN} = 47.6\times(200/100) = 95.2\text{ A}$$

Now $$\cos\phi_s = 0.454 \quad \therefore \phi_s = 63°$$

Let us choose a convenient current scale *viz.* 1 cm = 10 A. The no-load current of 7.7 A will have a length = 7.7/10 = 0.77 cm and is represented by phasor OO' lagging 78.7° behind the vertical voltage axis V_1 as shown in Fig. 10.19. Similarly, I_{SN} of 95.2 A is represented by the phasor OA of length 9.52 cm (95.2/10 = 9.52 cm) lagging 63° behind the voltage axis V_1 as shown in Fig. 10.19. The right bisector of the chord $O'A$ locates the centre C of the circle. With C as centre, and radius = CO', the semicircle can be drawn. Note the $O'A$ is the output line.

---

* Frequency, $f = \dfrac{n_s P}{2}$ where n_s = synchronous speed in r.p.s.

$$\therefore n_s = \frac{2f}{P} = \frac{2\times 50}{4} = 25\text{ r.p.s.}$$

How to draw torque line ?

$$\frac{AH}{HB} = \frac{\text{Rotor Cu loss}}{\text{Stator Cu loss}} = \frac{0.24}{0.38}$$

$$\therefore \quad \frac{AH}{AB} = \frac{\text{Rotor Cu loss}}{\text{Total Cu loss}} = \frac{0.24}{0.62}$$

Let us find power scale for this circle diagram. The vertical length *AF* represents input power under blocked rotor conditions corresponding to normal voltage (*i.e.* $\sqrt{3} \times 200 \times 95.2 \times 0.454 =$ 14970 W in this case) and its length is 4.3 cm (by measurement).

$\therefore$ Power scale = 14970/4.3 = 3481 W/cm

or 1 cm = 3481 W

Now, $AB = 4.1$ cm by measurement.

$$\therefore \quad AH = AB \times \frac{0.24}{0.62} = 4.1 \times \frac{0.24}{0.62} = 1.6 \text{ cm}$$

(*i*) Starting torque = *AH* = 1.6 cm = 1.6 × 3481 = **5570 synch. watts**

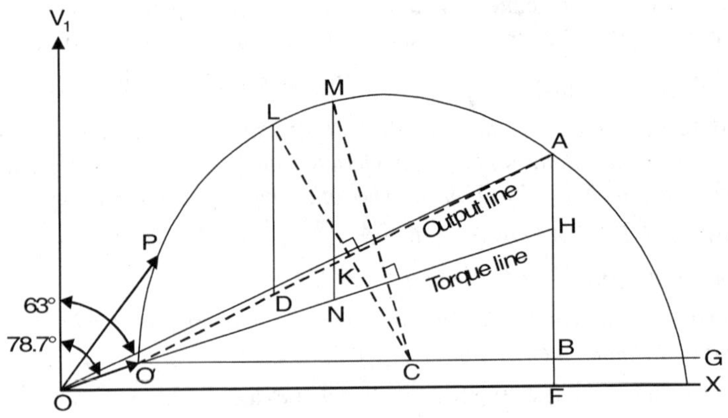

Fig. 10.19

(*ii*) From the centre *C* of the circle, draw perpendicular to the torque line *O'H*. This perpendicular meets the circle at point *M*. The intercept *MN* represents the maximum torque on power scale. Its length is 3.57 cm.

$\therefore$ Max. torque = *MN* = 3.57 cm = 3.57 × 3481 = **12427 synch. watts**

(*iii*) For finding maximum power factor, draw line *OP* tangent to the semicircle. By measurement, $\angle POV_1 = 28.5°$.

$\therefore$ Max. power factor = cos 28.5° = **0.879** *lagging*

(*iv*) Slip for maximum torque = *KN*/*MN* = 0.7 cm / 3.57 cm = **0.196**

(*v*) From the centre *C* of the circle, draw perpendicular to the output line *O'A*. This perpendicular meets the circle at point *L*. The intercept *LD* represents the maximum output power on power scale. Its length is 2.95 cm.

$\therefore$ Max. output power = *LD* = 2.95 cm = 2.95 × 3481 = **10270 W**

TUTORIAL PROBLEMS

1. Draw the circle diagram from no-load and short-circuit test on a 3-phase, 14.92 kW, 400 V, 6-pole induction motor from the following test results :

   No-load : 400 V, 11 A, p.f. = 0.2

   Short-circuit : 100 V, 25 A, p.f. = 0.4

   Rotor Cu loss at standstill is half the total Cu loss. From the circle diagram, find (*i*) line current, slip, efficiency and p.f. at full-load (*ii*) the maximum torque.

   [(*i*) 32.5 A ; 5.6% ; 80% ; 0.84 lag (*ii*) 27030 synch.watts]

2. Draw the circle diagram for a 5 h.p., 200 V, 50 Hz, 4-pole, 3-phase, star-connected induction motor from the following test data :

   No-load : 200 V, 5 A, 350 W

   Short-circuit : 100 V, 26 A, 1700 W

   From the circle diagram, find :

   (*i*) line current and power factor at full-load.

   (*ii*) maximum torque and starting torque in terms of full-load torque.

   The rotor Cu loss at standstill is half the total Cu loss. **[(*i*) 15 A ; 0.83 lag (*ii*) 180% ; 83%]**

3. A 29.84 kW, 440 V, 50 Hz, 3-phase induction motor gave the following test results :

   No-load test : 440 V, 16 A, p.f. = 0.15

   Short-circuit test : 100 V, 55 A, p.f. = 0.225

   Ratio of rotor to stator Cu losses on short-circuit = 0.9. Draw the circle diagram and find (*i*) full-load current and power factor (*ii*) pull-out torque (*iii*) maximum output power developed.

   [(*i*) 49 A ; 0.88 lag (*ii*) 78.5 synch. kW (*iii*) 701.2 kW]

4. A 415 V, 29.84 kW, 50 Hz, Δ-connected induction motor gave the following test data :

   No-load test : 415 V, 21 A, 1250 W

   Locked rotor test : 100 V, 45 A, 2730 W

   Construct the circle diagram and determine (*i*) line current and power factor for the rated output (*ii*) maximum torque.

   Assume stator and rotor Cu losses equal at standstill. **[(*i*) 104 A ; 0.819 lag (*ii*) 51980 synch. watts]**

5. Draw the circle diagram for a 3-phase, delta-connected, 30 h.p., 500 V, 4-pole, 50 Hz cage type induction motor from the following test results :

   No-load : 500 V, 8.3 A, + 2.85 kW and – 1.35 kW

   Short-circuit : 100 V, 32 A, – 0.75 kW and + 2.35 kW

   Find (*i*) full-load line current (*ii*) full-load p.f. (*iii*) full-load efficiency (*iv*) maximum output power.

   [(*i*) 33 A (*ii*) 0.9 lag (*iii*) 88% (*iv*) 68 h.p.]

10.9 SPEED CONTROL OF INDUCTION MOTORS

The slip of an induction motor is very small (< 3%) so that it is essentially a *constant-speed motor. Therefore, it is suitable for use in essentially constant-speed drive systems. However, many industrial applications require several speeds or a continuously adjustable range of speeds. Traditionally, d.c. motors have been used in such adjustable-speed applications. However, d.c. motors are expensive and require frequent maintenance of commutators and brushes. On the other hand, the induction motors (particularly the squirred cage motors) are cheap, rugged, have no commutators and are suitable for high-speed applications. The engineers have devised several methods to change the speed of induction motors. This has made it possible to use induction motors in variable-speed applications.

* When the load on an induction motor increases, its speed decreases by a very small amount. In fact, the change in slip from no-load to full-load is hardly 0.1% to 3% so that it is essentially a constant-speed motor.

The relation between motor speed (N), synchronous speed (N_s) and slip (s) is given by :

$$N = (1-s)\,N_s$$

or
$$N = (1-s)\,\frac{120\,f}{P} \qquad \ldots(i) \qquad \left[\because\ N_s = \frac{120\,f}{P}\right]$$

Eq. (i) reveals that the speed of an induction motor can be changed by the following methods :

1. By changing the number of stator poles (P).

2. By changing the line frequency (f).

3. By changing the slip (s) for a given load. The slip can be changed ;

(i) By changing the applied voltage.

(ii) By changing resistance in the rotor circuit.

(iii) By inserting foreign voltage of appropriate frequency in the rotor circuit.

The salient features of above *five* methods of speed control of induction motors are discussed in the following articles.

10.10 SPEED CONTROL BY CHANGING NUMBER OF STATOR POLES

We know that synchronous speed of induction motor $N_s = 120\,f/P$. Therefore, by changing the number of stator poles (P), the synchronous speed and hence the rotor speed (N) can be changed. This method is **easily applicable to squirrel cage motors but is not practicable for wound rotor motors.*

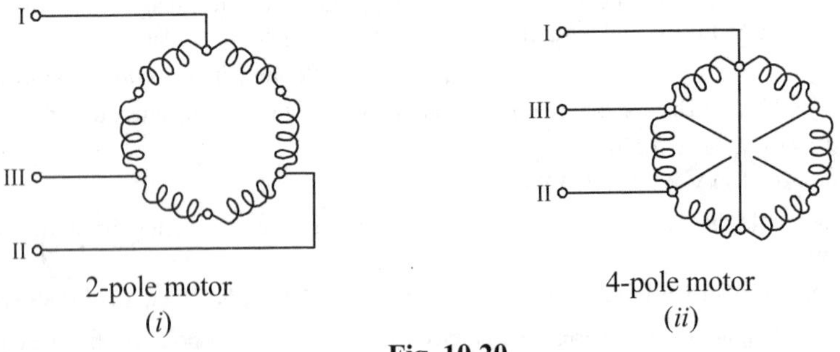

2-pole motor (i) 4-pole motor (ii)

Fig. 10.20

Fig. 10.20 shows the speed control of squirrel cage motor by pole changing method. The stator has six mesh-connected coils. By means of a change-over switch (not shown), the supply may be connected to alternate coil connections [See Fig. 10.20 (i)] to produce 2 poles. If the supply is connected to successive interconnections [See Fig. 10.20 (ii)], 4 poles are produced.

Applications. Squirrel cage motors designed for pole-changing control are commonly called multispeed motors. Because of the complications in the design and switching of the interconnections of the stator winding, it has been possible to obtain a maximum of only four synchronous speeds for any motor by this method. It is not possible to obtain gradual continuous speed control with this method. Nevertheless, it is very satisfactory for applications that require operation at only two to four approximately constant speeds. The pole-changing method has the advantages of simplicity, good speed regulation for each setting, high efficiency and moderate first cost and maintenance.

* For the satisfactory operation of the motor, the number of rotor poles must be equal to the number of stator poles. When the number of stator poles of squirrel cage motor is changed, the rotor acquires the same number of poles by induction. The number of rotor poles of wound rotor motor depends upon the design of the rotor circuit. It requires complex design of the rotor circuit to change the number of rotor poles to become equal to stator poles simultaneously.

10.11 SPEED CONTROL BY CHANGING LINE FREQUENCY

We know that synchronous speed of an induction motor is given by $N_s = 120\ f/P$. Therefore, by changing the line frequency f, the synchronous speed (N_s) of the motor and hence the running speed (N) can be changed. A major difficulty with this method is that it involves the use of 3-phase variable frequency power supply. The auxiliary equipment required for this purpose results in a high first cost, increased maintenance and lowering of the overall efficiency. However, the discovery of the thyristor (a controlled silicon rectifier), which is capable of providing large powers at variable frequency from a fixed frequency source, has solved this difficulty to a great extent. Although not suitable for general-purpose speed control, this method is useful for certain special applications due to many advantages including continuously variable speed range and good speed regulation.

When employing line-frequency control, the applied line voltage should be changed in direct proportion to the frequency *i.e.* if frequency is increased, the supply voltage must also be increased and if the frequency is decreased, the supply voltage must also be decreased proportionately. This is *necessary to maintain an approximately constant flux in the air-gap of the machine, otherwise the motor performance will not be satisfactory. Under these conditions, the maximum developed torque will remain approximately constant and the output power will vary approximately in direct proportion to the speed of the motor.

10.12 SPEED CONTROL BY CHANGING APPLIED VOLTAGE

We know that torque developed (T) by an induction motor is directly proportional to the square of applied voltage (V) *i.e.* $T \propto V^2$. Therefore, by changing the applied voltage, the torque and hence speed (or slip) of the motor can be changed. Fig. 10.21 shows the arrangement to control the speed of induction motor (squirrel cage or wound rotor motor) by changing the applied voltage.

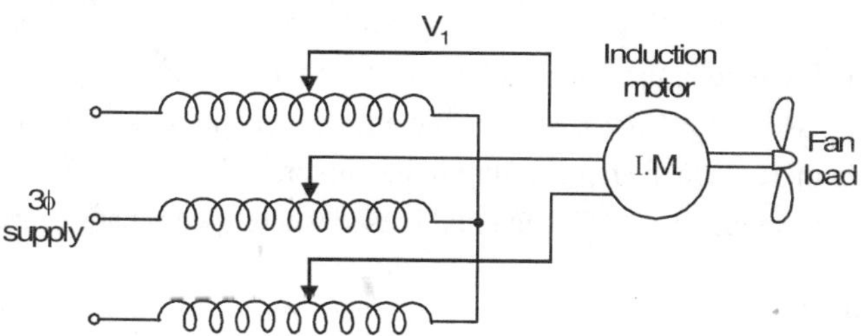

Fig. 10.21

The motor is supplied from 3-phase supply through a 3-phase auto-transformer. The motor drives fan load. The voltage applied to the motor can be changed by the autotransformer and hence desired motor speed can be obtained.

In fact, when we change the applied voltage, the slip (s) of the motor is changed *i.e.* motor speed changes. Let us illustrate this point. Suppose the applied voltage is V (say $V = 400$ volts)

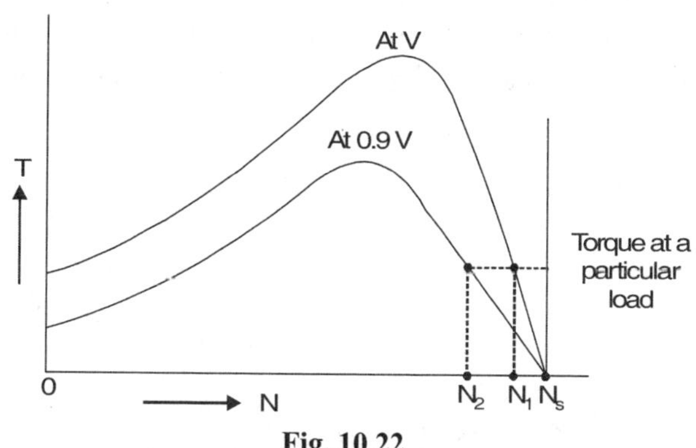

Fig. 10.22

* Applied voltage, $V \propto \phi f$. If V is fixed, $\phi \propto 1/f$. Therefore, for constant applied voltage V, if frequency (f) increases, the flux decreases and vice-versa. This change in flux will adversely affect the performance of the motor. For this reason, the frequency changing device should change freqeuncy and applied voltage simultaneously so that constant flux is obtained in the air-gap of the machine.

and the motor is running at a speed N_1. If the applied voltage is reduced to 0.9 V, the torque will be reduced from T to 0.81 T($\because$ Torque $\propto V^2$). Fig. 10.22 shows torque-speed curves for applied voltages V and 0.9 V. Since the torque is reduced, the rotor cannot continue to run at speed N_1. The rotor speed will decrease (*i.e.* slip will increase) until the increased rotor current produces the required load torque at a lower speed N_2.

Limitations. The stator voltage control method is the cheapest and the easiest method of speed control of induction motors. However, it is rarely used because of the following drawbacks :

(*i*) A large change in voltage is required for a relatively small change in speed.

(*ii*) The large change in voltage results in large change in the flux density. This affects the magnetic conditions and hence performance of the motor.

10.13 SPEED CONTROL BY CHANGING ROTOR CIRCUIT RESISTANCE

This method of speed control is suitable *only for *slip-ring motors*. The speed of the motor can be decreased by adding external resistance to the rotor as shown in Fig. 10.23. It was shown in Art. 9.18 that under normal running condition, the relation between torque (T) and slip (s) of an induction motor is given by :

$$T \propto s/R_2$$

Here R_2 is the rotor resistance/phase. It is clear from the above relation that for a given torque, $s \propto R_2$. Therefore slip can be increased (*i.e.* motor speed can be decreased) by increasing the rotor resistance.

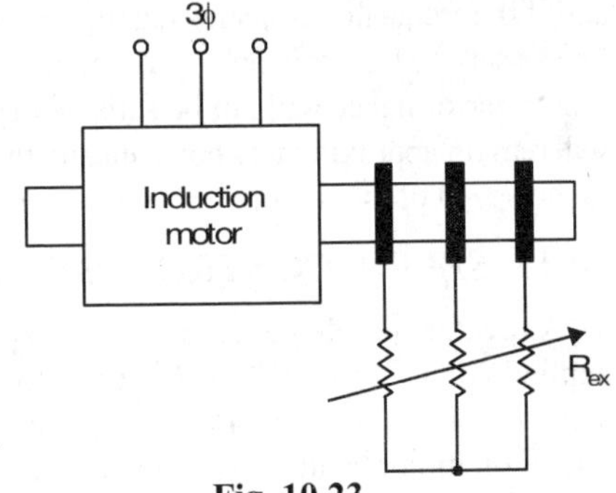

Fig. 10.23

Drawbacks.

(*i*) There is an increase in the rotor Cu losses due to the increased rotor circuit resistance.

(*ii*) Due to increased rotor Cu losses, the efficiency of the motor is decreased.

(*iii*) There is an increase in the temperature of the motor.

Because of the above drawbacks, this method is used where speed changes are required for short periods only.

Example 10.9. *The rotor of a 4-pole, 50 Hz slip-ring induction motor has a resistance of 0.30 Ω per phase and runs at 1440 r.p.m. at full-load. Calculate the external resistance per phase which must be added to lower the speed to 1320 r.p.m., the torque being the same as before.*

Solution. Suppose the required external resistance/phase to be added is R ohm.

$$\text{Synchronous speed, } N_s = 120\, f/P = (120 \times 50)/4 = 1500 \text{ r.p.m.}$$

For the first case : $N_s = 1500$ r.p.m. ; $N_1 = 1440$ r.p.m.

$$\therefore \quad \text{Slip, } s_1 = \frac{N_s - N_1}{N_s} = \frac{1500 - 1440}{1500} = 0.04$$

For the second case : $N_s = 1500$ r.p.m. ; $N_2 = 1320$ r.p.m.

$$\therefore \quad \text{Slip, } s_2 = \frac{N_s - N_2}{N_s} = \frac{1500 - 1320}{1500} = 0.12$$

Now, $$\text{Motor torque, } T = \frac{K\, s\, R_2}{R_2^2 + (sX_2)^2}$$ ....K is a constant

* Since the bars of a squirrel cage rotor are permanently short-circuited, the rotor circuit resistance cannot be changed.

Since X_2 is not given, it will be assumed to be zero.

$$\therefore \qquad T = \frac{K\,s\,R_2}{R_2^2} = \frac{Ks}{R_2}$$

Since torque remains the same in the two cases, $s \propto R_2$.

Now, $$s_1 \propto R_2 \text{ and } s_2 \propto R_2 + R$$

$$\therefore \qquad \frac{s_1}{s_2} = \frac{R_2}{R_2+R} \text{ or } \frac{0.04}{0.12} = \frac{0.3}{0.3+R} \qquad \therefore R = \mathbf{0.6\ \Omega/phase}$$

Example 10.10. *The rotor of a 6-pole, 50 Hz, 3-phase induction motor has a resistance of 0.2 Ω per phase and runs at 960 r.p.m. If the load torque remains unchanged, calculate the additional rotor resistance per phase which must be added to reduce the speed by 10%.*

Soluton. Suppose the required external resistance/phase to be added is R ohm.

Synchronous speed, $N_s = 120\,f/P = (120 \times 50)/6 = 1000$ r.p.m.

For the first case : $N_s = 1000$ r.p.m. ; $N_1 = 960$ r.p.m.

$$\therefore \qquad \text{Slip, } s_1 = \frac{N_s - N_1}{N_s} = \frac{1000-960}{1000} = 0.04$$

For the second case : $N_s = 1000$ r.p.m. ; $N_2 = 0.9\ N_1 = 0.9 \times 960 = 864$ r.p.m.

$$\therefore \qquad \text{Slip, } s_2 = \frac{N_s - N_2}{N_s} = \frac{1000-864}{1000} = 0.136$$

Since X_2 is not given, it will be assumed to be zero. Also torque in the two cases is the same. Therefore, as explained in Example 10.9, $s \propto R_2$.

Now, $$s_1 \propto R_2 \text{ and } s_2 \propto R_2 + R$$

$$\therefore \qquad \frac{s_1}{s_2} = \frac{R_2}{R_2+R} \text{ or } \frac{0.04}{0.136} = \frac{0.2}{0.2+R} \qquad \therefore R = \mathbf{0.48\ \Omega/phase}$$

Example 10.11. *A 4-pole, 3-phase, 50 Hz slip-ring induction motor rotates at 1440 r.p.m. with slip-ring terminals short-circuited. The per phase rotor resistance and standstill reactance are 0.1 Ω and 0.6 Ω respectively. If an extra resistance of 0.1 Ω/phase is added to the rotor circuit, what will be the new full-load speed ?*

Soluton.

Synchronous speed, $N_s = 120\,f/P = (120 \times 50)/4 = 1500$ r.p.m.

Full-load slip with rotor circuit resistance R_2 (= 0.1 Ω) is

$$s_1 = \frac{N_s - N_1}{N_s} = \frac{1500-1440}{1500} = 0.04$$

Suppose the full-load slip becomes s_2 when the rotor circuit resistance is R_2' (= R_2 + 0.1 = 0.1 + 0.1 = 0.2 Ω).

$$\text{For the first case, } T_1 = \frac{K\,s_1\,R_2}{R_2^2 + (s_1 X_2)^2} \text{ ; For the second case, } T_2 = \frac{K\,s_2\,R_2'}{R_2'^2 + (s_2 X_2)^2}$$

Since the load on the motor is unchanged, the full-load torque of the motor remains the same in the two cases *i.e.* $T_1 = T_2$.

$$\therefore \qquad \frac{K\,s_1\,R_2}{R_2^2 + (s_1 X_2)^2} = \frac{K\,s_2\,R_2'}{R_2'^2 + (s_2 X_2)^2}$$

$$\text{or} \qquad \frac{0.04 \times 0.1}{(0.1)^2 + (0.04 \times 0.6)^2} = \frac{s_2 \times 0.2}{(0.2)^2 + (s_2 \times 0.6)^2}$$

or $$0.378 = \frac{0.2\,s_2}{0.04 + 0.36\,s_2^2}$$

On solving, $s_2 = 0.08$

Now $$s_2 = \frac{N_s - N_2}{N_s} \quad \text{or} \quad 0.08 = \frac{1500 - N_2}{1500} \qquad \therefore N_2 = \mathbf{1380 \text{ r.p.m.}}$$

TUTORIAL PROBLEMS

1. The rotor of a 4-pole, 50 Hz, slip-ring induction motor has a resistance of 0.2 Ω per phase and runs at 1440 r.p.m. at full-load. Calculate the external resistance per phase which must be added to reduce the speed to 1200 r.p.m., the torque remaining the same in both the cases. **[0.8 Ω/phase]**
2. A 4-pole, 50 Hz, slip-ring induction motor fully loaded runs with a slip of 2.5%. Determine the value of series resistance to be connected in each phase of the rotor to reduce the speed by 12%. The resistance per phase of the rotor is 0.25 Ω. **[1.17 Ω/phase]**
3. The rotor of a 6-pole, 50 Hz, slip-ring induction motor has a resistance of 0.2 Ω per phase and runs at 960 r.p.m. on full load. Calculate the resistance per phase of a rotor rheostat such that speed is reduced to 800 r.p.m. for full-load torque. **[0.8 Ω/phase]**

10.14 SPEED CONTROL BY INJECTING VOLTAGE IN ROTOR CIRCUIT

The speed of a *silp-ring induction motor can also be changed by injecting an external or foreign voltage E_F in the rotor circuit. For proper operation of the motor, the frequency of E_F must be equal to the rotor frequency f' (= sf where f is the supply frequency and s is the motor slip). However, there is no restriction of the phase of the injected voltage. When external voltage E_F of rotor frequency is injected in the rotor circuit, the equivalent per phase circuit of the motor will be as shown in Fig. 10.24. Here R_F is the resistance of the external source (E_F) and X_F is the leakage reactance of source at stator frequency f.

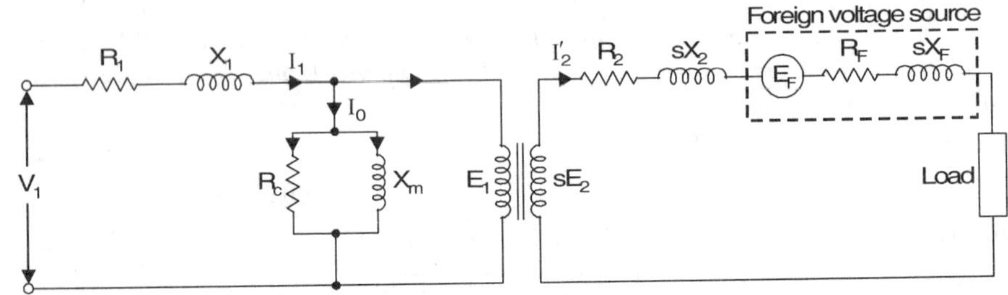

Fig. 10.24

(*i*) If the injected voltage E_F is in phase with the induced voltage in the rotor (sE_2), the rotor current increases. This is equivalent to decreasing the resistance of the rotor circuit. This decreases the slip and hence increases the motor speed.

(*ii*) If the injected voltage is in phase opposition to the induced voltage (sE_2) in the rotor, the rotor current decreases. This is equivalent to increasing the resistance of the rotor circuit. This increases the slip and hence decreases the motor speed.

Further control in speed can be achieved by varying the magnitude of the injected voltage E_F. Thus by injecting the correct slip frequency voltage in the rotor circuit of the slip-ring motor, the speed of the motor can be changed.

---

* This method cannot be used for squirred cage motor. It is because the bars of squirrel cage rotor are permanently short-circuited and external voltage cannot be injected into the rotor circuit.

10.15 SPEED CONTROL BY CASCADE CONNECTION

This method of speed control employs two induction motors coupled together to drive a common load. The rotor power of one motor (called *main motor* M_1) is fed to the stator of the other motor (called *auxiliary motor* M_2) as shown in Fig. 10.25. Therefore, the main motor M_1 must be of slip-ring type so that its rotor power can be fed to the stator of the auxiliary motor M_2. The auxiliary *motor M_2 may be a cage motor or wound rotor motor.

Operation. The stator winding of the main motor M_1 is connected to the 3-phase supply of frequency f_1 as shown in Fig. 10.25. The e.m.f. induced in the rotor of M_1 has a frequency of f_2 (= $s_1 f_1$ where s_1 is the slip of motor M_1). The rotor power of M_1 at frequency f_2 (= $s_1 f_1$) is fed to the stator of M_2. In practice, the combination is connected in **cumulative cascade** *i.e.* in such a way that the phase rotation of the stator fields of both the motors is in the same direction. Under this condition, it can be proved that the synchronous speed of the cascaded set is given by :

$$\text{Synchronous speed of the set, } N_{sc} = \frac{120\, f_1}{P_1 + P_2}$$

Here f_1 is the supply frequency and P_1 and P_2 are the number of poles of motor M_1 and motor M_2 respectively. Note that synchronous speed of the set or combination is equal to the synchronous speed of a single machine having ($P_1 + P_2$) poles.

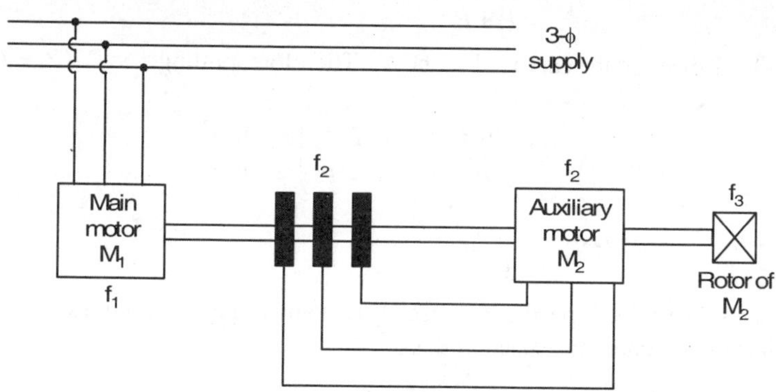

Fig. 10.25

Proof.

Let

f_1 = Supply frequency
P_1 = Number of poles of main motor M_1
P_2 = Number of poles of auxiliary motor M_2
s_1 = Slip frequency of motor M_1
s_2 = Slip frequency of motor M_2
f_2 = Rotor current frequency of M_1 = $s_1 f_1$
**f_3 = Rotor current frequency of M_2 = $s_2 f_2$

$$\therefore \text{ Synchronous speed of } M_1,\ N_{s1} = \frac{120\, f_1}{P_1}$$

* If the auxiliary motor M_2 is a wound rotor motor, its slip-rings must be short-circuited or connected to a rheostat. The resistance is used during starting conditions only. During operation, the slip rings of M_2 are short circuited.

** The stator current frequency of motor M_1 is f_1 (= supply frequency) and its rotor current frequency is f_2 (= $s_1 f_1$). The stator current frequency of motor M_2 is f_2 and its rotor current frequency is f_3 (=$s_2 f_2$).

$$\text{Synchronous speed of } M_2,\ N_{s2} = \frac{120\, f_2}{P_2}$$

$$\text{Actual speed of motor } M_1 = N_{s1}(1-s_1) = \frac{120\, f_1}{P_1}(1-s_1) \quad ...(i)$$

$$\text{Actual speed of motor } M_2 = N_{s2}(1-s_2) = \frac{120\, f_2}{P_2}(1-s_2) \quad ...(ii)$$

Since the two motors are mechanically coupled, their actual speeds are equal. Therefore, from eqs. (*i*) and (*ii*), we have,

$$\frac{120\, f_1}{P_1}(1-s_1) = \frac{120\, f_2}{P_2}(1-s_2)$$

or
$$\frac{120\, f_1}{P_1}(1-s_1) = \frac{120\, f_1 s_1}{P_2}(1-s_2) \quad ...(iii) \qquad (\because f_2 = s_1 f_1)$$

At no-load, the frequency f_3 of the rotor current of motor M_2 is nearly zero because $s_2 \simeq 0$. Putting $s_2 = 0$ in eq. (*iii*), we have,

$$s_1 = \frac{P_2}{P_1 + P_2}$$

Let the no-load speed of the cascaded set be N_0. Therefore, putting $s_1 = P_2/(P_1 + P_2)$ in exp. (*i*), we have,

$$N_0 = \frac{120\, f_1}{P_1}\left(1 - \frac{P_2}{P_1 + P_2}\right) = \frac{120\, f_1}{P_1 + P_2}$$

$$\therefore \quad N_0 = \frac{120\, f_1}{P_1 + P_2} \quad ...(iv)$$

Exp. (*iv*) gives the no-load speed of the cascaded set and is the *maximum speed that the set can attain. Hence it is the synchronous speed of the set.

$$\therefore \quad \text{Synchronous speed of set, } N_{sc} = \frac{120 f_1}{P_1 + P_2}$$

When the cascaded set is loaded, the actual speed (say N) of the set is slightly less than N_{sc}.

$$\therefore \quad \text{Slip of set, } s = \frac{N_{sc} - N}{N_{sc}}$$

It can be proved that slip s of the set is also given by the following relations :

(*i*) $$\text{Slip of set, } s = \frac{f_3}{f_1} = \frac{\text{Rotor current frequency of } M_2}{\text{Supply frequency}}$$

(*ii*) $$\text{Slip of set, } s = s_1 \times s_2$$

$$\text{where } s_1 = \frac{N_{s1} - N}{N_{s1}} \quad \text{and} \quad s_2 = \frac{N_{s2} - N}{N_{s2}}$$

---

* The reason is as follows. When the cascaded set is started, the supply voltage at frequency f_1 is applied to the stator of motor M_1. The same frequency (f_1) is induced in the rotor of M_1 ($\because$ the rotor is stationary to start with) and is supplied to the stator of M_2. Both motors then develop a forward torque. As the shaft speed rises, the rotor frequency of M_1 falls and so does the *synchronous speed* of motor M_2. The set comes to a stable speed when the shaft speed and the synchronous speed N_{s2} of M_2 are equal. For if the speed rose further, the rotor of M_2 would be running faster than its rotating field which is not possible.

Different shaft speeds available. The set can be operated in the following ways to get different synchronous speeds and hence the shaft speeds (N) :

(*i*) The main motor M_1 may be run separately from the supply to get a synchronous speed $N_{s1} = 120 f_1/P_1$ where f_1 is the supply frequency.

(*ii*) The auxiliary motor M_2 may be run separately (with motor M_1 being disconnected) to get a synchronous speed $N_{s2} = 120 f_1/P_2$.

(*iii*) The motors may be connected in cumulative cascade to get a synchronous speed given by ;

$$N_{sc} = \frac{120 f_1}{P_1 + P_2}$$

(*iv*) The two motors can also be connected in **differential cascade**. In this method, the phase rotation of stator field of the motor M_2 is opposite to that of the stator of motor M_1. This reversal of the phase of stator of motor M_2 is obtained by interchanging any of its two leads. By making analysis similar to that for cumulative cascade, it can be shown that synchronous speed of the set in differential cascade is given by ;

$$\text{Synchronous speed of set, } N_{sc} = \frac{120 f_1}{P_1 - P_2}$$

Here f_1 is the supply frequency.

Mechanical outputs of two machines. When the cascaded set is loaded, its speed falls from the synchronous speed by a small amount and normal load currents flow. The electrical power taken from the supply by the stator of main motor M_1 is partly used to meet stator Cu loss and core losses and the rest is given to its (M_2) rotor. The power given to the rotor is divided into two parts :

(*i*) One part, proportional to the speed of the set (N), is converted into mechanical power.

(*ii*) The other part, proportional to $(N_{s1} - N)$, is developed as electrical power at slip frequency and is passed on to auxiliary motor M_2 which uses it for producing mechanical power and losses.

Therefore, mechanical outputs of M_1 and M_2 are in the ratio $N : (N_{s1} - N)$ or $P_1 : P_2$ *i.e.*,

Output of M_1 : Output of M_2 $= P_1 : P_2$

Example 10.12. *Two motors of 12-poles (main motor M_1) and 8-poles (auxiliary motor M_2) are connected in cascade to 3ϕ, 50 Hz supply. How many synchronous speeds can be obtained ?*

Solution.

$$\text{For main motor alone, } N_{s1} = \frac{120 f_1}{P_1} = \frac{120 \times 50}{12} = \mathbf{500\ r.p.m.}$$

$$\text{For auxiliary motor alone, } N_{s2} = \frac{120 f_1}{P_2} = \frac{120 \times 50}{8} = \mathbf{750\ r.p.m.}$$

$$\text{For cumulative cascade, } N_{sc} = \frac{120 f_1}{P_1 + P_2} = \frac{120 \times 50}{12 + 8} = \mathbf{300\ r.p.m.}$$

$$\text{For differential cascade, } N_{sc} = \frac{120 f_1}{P_1 - P_2} = \frac{120 \times 50}{12 - 8} = \mathbf{1500\ r.p.m.}$$

Example 10.13. *A 4-pole induction motor and a 6-pole induction motor are connected in cumulative cascade. The frequency in the secondary circuit of the 6-pole motor is observed to be 1 Hz. Determine the combined speed of the set and the slip in each machine. Take supply frequency as 50 Hz.*

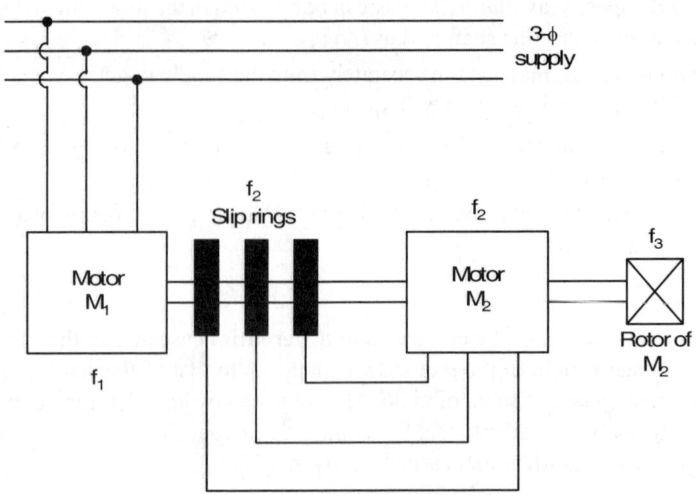

Fig. 10.26

Solution. Fig. 10.26 shows the conditions of the problem. The statement of the problem suggests that 4-pole motor is the main motor M_1 and 6-pole motor is the auxiliary motor M_2.

$$\text{Synchronous speed of set, } N_{sc} = \frac{120 f_1}{P_1 + P_2} = \frac{120 \times 50}{4+6} = 600 \text{ r.p.m.}$$

$$\text{Slip of the set, } s = \frac{\text{Rotor current frequency of } M_2}{\text{Supply frequency}} = \frac{1}{50} = 0.02$$

Now, $$s = \frac{N_{sc} - N}{N_{sc}} \quad \text{or} \quad 0.02 = \frac{600 - N}{600}$$

$\therefore$ $$\text{Speed of set, } N = 600 - 600 \times 0.02 = \mathbf{588 \text{ r.p.m.}}$$

$$\text{Synchronous speed of } M_1,\ N_{s1} = \frac{120 f_1}{P_1} = \frac{(120 \times 50)}{4} = 1500 \text{ r.p.m.}$$

$\therefore$ $$\text{Slip of } M_1,\ s_1 = \frac{N_{s1} - N}{N_{s1}} = \frac{1500 - 588}{1500} = \mathbf{0.608} \text{ or } \mathbf{60.8\%}$$

$$\text{Rotor current frequency of } M_1,\ f_2 = s_1 f_1 = 0.608 \times 50 = 30.4 \text{ Hz}$$

$$\text{Synchronous speed of } M_2,\ N_{s2} = \frac{120 f_2}{P_2} = \frac{120 \times 30.4}{6} = 608 \text{ r.p.m.}$$

$\therefore$ $$\text{Slip of } M_2,\ s_2 = \frac{N_{s2} - N}{N_{s2}} = \frac{608 - 588}{608} = \mathbf{0.033} \text{ or } \mathbf{3.3\%}$$

Example 10.14. *Two 50 Hz, 3-phase induction motors having six and four poles respectively are cumulatively connected, the 6-pole motor being connected to the main supply. Determine the frequency of the rotor currents and the slips referred to each stator field if the set has a slip of 2%.*

Solution. The statement of the problem suggests that 6-pole motor is the main motor M_1 and 4-pole motor is the auxiliary motor M_2.

$$\text{Synchronous speed of the set, } N_{sc} = \frac{120 f_1}{P_1 + P_2} = \frac{120 \times 50}{6+4} = 600 \text{ r.p.m.}$$

Actual speed of the set, $N = N_{sc}(1-s) = 600(1-0.02) = 588\text{ r.p.m.}$

$$\text{Synchronous speed of } M_1,\ N_{s1} = \frac{120 f_1}{P_1} = \frac{(120\times 50)}{6} = 1000\text{ r.p.m.}$$

$$\therefore\quad \text{Slip referred to this stator field, } s_1 = \frac{N_{s1}-N}{N_{s1}} = \frac{1000-558}{1000} = \mathbf{0.412} \text{ or } \mathbf{41.2\%}$$

Rotor current frequency of M_1, $f_2 = s_1 f_1 = 0.412 \times 50 = \mathbf{20.6\ Hz}$

The frequency f_2 (= 20.6 Hz) is the frequency of the stator currents of M_2.

$$\therefore\quad \text{Synchronous speed of } M_2,\ N_{s2} = \frac{120 f_2}{P_2} = \frac{(120\times 20.6)}{4} = 618\text{ r.p.m.}$$

$$\text{The slip as referred to } M_2,\ s_2 = \frac{N_{s2}-N}{N_{s2}} = \frac{618-588}{618} = \mathbf{0.0485} \text{ or } \mathbf{4.85\%}$$

Rotor current frequency of M_2, $f_3 = s_2 f_2 = 0.0485 \times 20.6 = \mathbf{1.0\ Hz}$

We can also find the frequency of rotor current of M_2 as under :

$$f_3 = \text{slip of set} \times \text{supply frequency}$$
$$= s \times f_1 = 0.02 \times 50 = \mathbf{1.0\ Hz}$$

Example 10.15. *A cascaded set consists of two motors A and B with 4-poles and 6-poles respectively. The motor A is connected to a 50 Hz supply. Find (i) synchronous speed of the set (ii) the electric power transferred to motor B when the input to motor A is 25 kW. Neglect losses.*

Solution. When not given, it is assumed that the motors are connected in cumulative cascade.

(*i*) Synchronous speed of set, $N_{sc} = \dfrac{120 f_1}{P_1+P_2} = \dfrac{120\times 50}{4+6} = \mathbf{600\ r.p.m.}$

(*ii*) We know that mechanical outputs of the motors are in the ratio of their number of poles *i.e.*,

Output of 4-pole motor : Output of 6-pole motor = 4 : 6

$\therefore$ Output of 4-pole motor $= 25\times\dfrac{4}{4+6} = \mathbf{10\ kW}$

Example 10.16. *Two 3-phase, 50 Hz, 200 V induction motors are mechanically coupled and connected in cumulative cascade. One machine is rated at 10 h.p. and has 4 poles. The other is rated at 15 h.p. and has 6-poles. Calculate (i) the synchronous speed of the set (ii) the greatest load that can be developed without either machine taking more than full-load current (iii) relative torques of machines at this load.*

Solution.

(*i*) Synchronous (N_{sc}) speed of the set is given by :

$$N_{sc} = \frac{120 f_1}{P_1+P_2} = \frac{120\times 50}{4+6} = \mathbf{600\ r.p.m.}$$

(*ii*) The maximum output power that can be developed without either machine taking more than full-load current is equal to the full-load power of the machine having lower power rating.

$\therefore$ Maximum power = **10 h.p.**

(*iii*) We have seen in Art. 10.15 that output mechanical power of the motors M_1 (main motor) and M_2 (auxiliary motor) are given as :

$$\text{Power of } M_1 : \text{Power of } M_2 = N : (N_{s1} - N) \quad \text{or} \quad P_1 : P_2$$

Suppose the torque of 4-pole motor is T_1 and that of 6-pole motor is T_2. Since torque is inversely proportional to speed,

$$T_1 : T_2 = 6:4 \text{ or } T_1 : T_2 = \mathbf{1.5:1}$$

10.16 ROTOR RHEOSTAT STARTER FOR SLIP-RING MOTORS

Slip-ring motors are invariably started by *rotor resistance starting.* In this method, a variable star-connected rheostat is connected in the rotor circuit through slip rings and full voltage is applied to stator winding. Fig. 10.27 shows one phase of the 3-phase rheostat starter. At starting, the entire starting resistance is in each phase of the rotor circuit. This reduces the starting current and at the same time starting torque is increased. As the motor picks up speed, the starting resistance in each phase is cut out gradually. When the motor attains normal speed, the whole starting resistance is cut out from the rotor circuit.

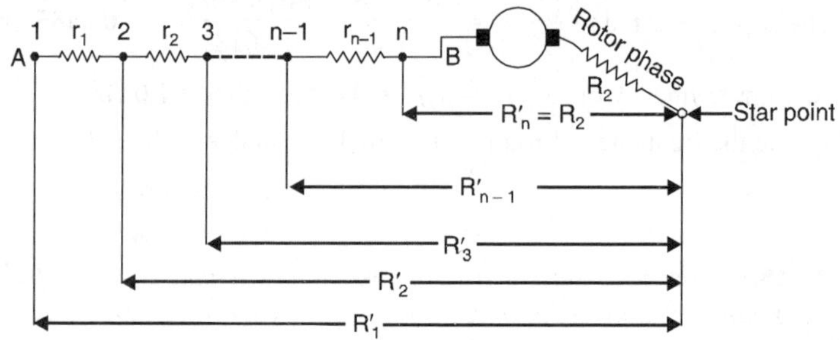

Fig. 10.27

Calculations of resistance steps. Fig. 10.27 shows one phase of the 3-phase rheostat having n steps [*i.e.,* there are $(n-1)$ resistance sections] and connected to one rotor phase. The rotor resistance per phase is R_2. Let $r_1, r_2 \ldots r_{n-1}$ be the resistances of the different $(n-1)$ resistance sections and $^*R_1'$, $R_2', R_3' \ldots R'_{n-1}$ and R_n' be the *total* resistance per phase in the rotor circuit on the 1st, 2nd, 3rd, ... $(n-1)$th and nth stud respectively as shown in Fig. 10.27. Let the values of slips be s_1 for R_1', s_2 for R_2', s_3 for R_3' and so on.

Let E_2 = Standstill rotor e.m.f. per phase

X_2 = Standstill rotor reactance per phase

As is usual in starter calculations, let us assume that the motor starts against a constant torque and the rotor current fluctuates between fixed maximum and minimum values of $^{**}I'_{2max}$ and I'_{2min} respectively. At the commencement of each step, the current is I'_{2max} and at the instant of leaving it, the current is I'_{2min}.

When the handle touches stud 1, the current rises to a maximum value I'_{2max} given by :

$$I'_{2max} = \frac{s_1 E_2}{\sqrt{\left(R_1'\right)^2 + (s_1 X_2)^2}} = \frac{E_2}{\sqrt{\left(R_1'/s_1\right)^2 + X_2^2}}$$

where s_1 = slip at starting (*i.e.,* $s_1 = 1$)

As the motor gains speed, the current decreases and so does the motor slip. At the instant of leaving stud No. 1, the current is reduced to I'_{2min} and slip falls from s_1 to s_2.

* Clearly, $r_1 = R_1' - R_2'$; $r_2 = R_2' - R_3'$ and so on.

** Note that in this book, we represent rotor current under running condition by I_2'.

$$\therefore \qquad I'_{2min} = \frac{E_2}{\sqrt{\left(R'_1/s_2\right)^2 + X_2^2}}$$

When the handle touches stud No. 2, the motor speed momentarily remains the same but current rises to I'_{2max} because some resistance in the rotor circuit is cut out.

$$\therefore \qquad I'_{2max} = \frac{E_2}{\sqrt{(R'_2/s_2)^2 + X_2^2}}$$

Now the motor speed increases further and the current is reduced to I'_{2min} and slip falls to s_3.

$$\therefore \qquad I'_{2min} = \frac{E_2}{\sqrt{(R'_2/s_3)^2 + X_2^2}}$$

Thus rotor current fluctuates between I'_{2max} and I'_{2min} as the handle moves over the studs. At the last stud (*i.e.*, nth stud), we have,

$$I'_{2max} = \frac{E_2}{\sqrt{(R_2/s_{max})^2 + X_2^2}}$$

Here s_{max} is the slip under normal running conditions (*i.e.*, rotor short-circuited) when the external resistance is completely cut out.

It is clear from the above mathematical treatment that :

$$I'_{2max} = \frac{E_2}{\sqrt{(R'_1/s_1)^2 + X_2^2}} = \frac{E_2}{\sqrt{(R'_2/s_2)^2 + X_2^2}} = \ldots = \frac{E_2}{\sqrt{(R_2^*/s_{max})^2 + X_2^2}}$$

$$\text{or} \qquad \frac{R'_1}{s_1} = \frac{R'_2}{s_2} = \frac{R'_3}{s_3} = \ldots = \frac{R'_{n-1}}{s_{n-1}} = \frac{R_2}{s_{max}} \qquad \ldots(i)$$

$$\text{Similarly,} \qquad I'_{2min} = \frac{E_2}{\sqrt{(R'_1/s_2)^2 + X_2^2}} = \frac{E_2}{\sqrt{(R'_2/s_3)^2 + X_2^2}} = \ldots = \frac{E_2}{\sqrt{(R'_{n-1}/s_{max})^2 + X_2^2}}$$

$$\text{or} \qquad \frac{R'_1}{s_2} = \frac{R'_2}{s_3} = \frac{R'_3}{s_4} = \ldots = \frac{R'_{n-1}}{s_{max}} \qquad \ldots(ii)$$

Combining eqs. (*i*) and (*ii*), we have,

$$\frac{s_2}{s_1} = \frac{s_3}{s_2} = \frac{s_4}{s_3} = \ldots = \frac{s_{max}}{s_{n-1}} = \frac{R'_2}{R'_1} = \frac{R'_3}{R'_2} = \frac{R'_4}{R'_3} = \ldots = \frac{R_2}{R'_{n-1}} = K\text{(say)} \qquad \ldots(iii)$$

From eq. (*i*), the total resistance R'_1 in the rotor circuit per phase at stud No. 1 is

$$R'_1 = R_2 \times \frac{s_1}{s_{max}} = \frac{R_2}{s_{max}} \qquad (\because s_1 = 1)$$

$$\therefore \qquad R'_1 = \frac{R_2}{s_{max}} \qquad \ldots(iv)$$

* Recall that R_2 is the rotor resistance per phase.

Thus R_1' is known in terms of R_2 (= rotor resistance/phase) and s_{max} (= motor slip under normal running conditions).

From eq. (*iii*), we have,

$$R_2' = K R_1' \;;\; R_3' = K R_2' = K^2 R_1' \text{ and so on.}$$

From eq. (*iii*), rotor resistance per phase (R_2) is given by :

$$R_2 = K R_{n-1}' = {}^*K^{n-1} R_1' = K^{n-1} \frac{R_2}{s_{max}} \qquad \left(\because R_1' = \frac{R_2}{s_{max}}\right)$$

$$\therefore \quad R_2 = K^{n-1} \frac{R_2}{s_{max}} \quad \text{or} \quad 1 = \frac{K^{n-1}}{s_{max}}$$

$$\therefore \quad K^{n-1} = s_{max} \quad \text{or} \quad K = (s_{max})^{\frac{1}{n-1}}$$

Thus we can find the value of K from the values of n and s_{max}.

The resistances of the various sections are found as under :

$$r_1 = R_1' - R_2' = R_1' - K R_1' = R_1' (1 - K)$$

$$r_2 = R_2' - R_3' = R_1' (K - K^2) = K r_1$$

$$r_3 = R_3' - R_4' = R_1' (K^2 - K^3) = K r_2 = K^2 r_1 \quad \text{etc.}$$

Example 10.17. *Calculate the steps in a 5-step rotor starter for a 3-phase induction motor. The slip at the maximum starting current is 2% with slip rings short-circuited and the resistance per rotor phase is 0.015 Ω.*

Solution. Here, $n - 1 = 5$; $s_{max} = 2\% = 0.02$; $R_1' = R_2/s_{max} = 0.015/0.02 = 0.75\ \Omega$

$$K = (s_{max})^{\frac{1}{n-1}} = (0.02)^{1/5} = 0.457$$

The resistances of the various sections are :

$$r_1 = (1 - K) R_1' = (1 - 0.457) \times 0.75 = \mathbf{0.40\ \Omega}$$
$$r_2 = K r_1 = 0.457 \times 0.40 = \mathbf{0.18\ \Omega}$$
$$r_3 = K r_2 = 0.457 \times 0.18 = \mathbf{0.08\ \Omega}$$
$$r_4 = K r_3 = 0.457 \times 0.08 = \mathbf{0.04\ \Omega}$$
$$r_5 = K r_4 = 0.457 \times 0.04 = \mathbf{0.02\ \Omega}$$

Example 10.18. *Calculate the steps in a 4-step starter for a 3-phase, slip-ring induction motor from the following data :*

Maximum starting current = Full-load current
Full-load slip = 2.5%
Rotor resistance/phase = 0.02 Ω

Solution. Here, $n - 1 = 4$; $s_{max} = 2.5\% = 0.025$; $R_1' = R_2/s_{max} = 0.02/0.025 = 0.8\ \Omega$

---

* ***How this step?*** At stud No. 3 ($n = 3$), the total rotor circuit resistance per phase is R_3' and is given by :

$$R_3' = K^2 R_1' \quad \text{or} \quad R_3' = K^{n-1} R_1' \qquad (\text{Here } n = 3)$$

At stud No. n (*i.e.*, $n = n$), the total rotor circuit resistance per phase is R_2 (= rotor resistance per phase).

$$\therefore \quad R_2 = K^{n-1} R_1'$$

$$K = (s_{max})^{\frac{1}{n-1}} = (0.025)^{1/4} = 0.398$$

The resistances of the various sections are :

$$r_1 = (1-K)R_1' = (1-0.398) \times 0.8 = \mathbf{0.48\,\Omega}$$
$$r_2 = K r_1 = 0.398 \times 0.48 = \mathbf{0.19\,\Omega}$$
$$r_3 = K r_2 = 0.398 \times 0.19 = \mathbf{0.08\,\Omega}$$
$$r_4 = K r_3 = 0.398 \times 0.08 = \mathbf{0.03\,\Omega}$$

10.17 CLASSES OF SQUIRREL-CAGE INDUCTION MOTORS

To meet the various starting and running requirements of a variety of industrial applications, 3-phase squirrel-cage induction motors have been divided into six different classes by assigning them the letters *A, B, C, D, E* and *F*. Each letter designation is intended to satisfy the requirements of a particular application.

1. Class *A* motors. A class *A* motor is considered a standard motor and is suitable for constant-speed applications. The rotor bars are placed close to the surface of the rotor laminations in order to reduce the leakage reactance. The important characteristics of class *A* motors are :

(*i*) The starting torque is 1.25 to 1.75 times the full-load torque.

(*ii*) The starting current at the rated voltage is 5 to 7 times the rated current.

(*iii*) The full-load slip is generally less than 5% because the rotor resistance is low.

(*iv*) The full-load speed regulation is 2% to 4%.

These motors drive low-inertia loads and possess high acceleration. They are used in such applications as fans, blowers, centrifugal type pumps and machine tools.

2. Class *B* motors. A class *B* motor is considered a general-purpose motor. The rotor resistance for a class *B* motor is somewhat higher than for a class *A* motor. The rotor conductors are placed deeper in the slots than for a class *A* motor. The increase in the rotor reactance reduces the starting torque whereas an increase in the rotor resistance increases the starting torque. Thus, the starting torque range for a class *B* motor is almost the same as that of a class *A* motor. The important characteristics of class *B* motors are :

(*i*) The starting torque is 1.4 to 1.6 times the full-load torque.

(*ii*) Owing to the increase in reactance, the starting current is about 4.5 to 5.5 times the full-load current.

(*iii*) The full-load slip is generally less than 5%.

(*iv*) The full-load speed regulation is 3% to 5%.

Most induction motors are class *B* motors and are employed to drive fans, centrifugal pumps, machine tools etc.

3. Class *C* motors. A class *C* motor usually has a double-cage rotor and is designed for full-voltage starting. Class *C* motors have high starting torque and low starting current. The important characteristics of class *C* motors are :

(*i*) The starting torque is 2 to 2.5 times the full-load torque.

(*ii*) Due to high-resistance rotor, the starting current is 3.5 to 5 times the full-load current.

(*iii*) The full-load slip is higher than that of class *A* and class *B* motors.

(*iv*) The full-load speed regulation is 4% to 5%.

Class *C* motors are used in applications that require high starting torque such as compression pumps, crushers, boring mills, conveyor equipment, textile machinery and wood-working equipment.

4. Class *D* motors. A class *D* motor is a high-resistance motor capable of developing a high starting torque. The high rotor resistance is created by using brass bars and by reducing the cross-sectional area of the bars. The following are the characteristics of class *D* motors :

(*i*) The starting torque is 2.5 to 3 times the full-load torque.

(*ii*) Depending upon the design, the starting current is 3 to 8 times the full-load current.

(*iii*) The full-load slip is high (8% to 15%) and, therefore, the motor efficiency is *low.

(*iv*) The full-load speed regulation is high (5% to 8%).

Class *D* motors are suitable for driving intermittent loads requiring rapid acceleration and high-inertia loads *e.g.*, punch presses, shearing machines, bulldozers, stamping machines, laundry equipment and hoists. In the case of high-inertia loads, a flywheel is connected to the system. The flywheel stores mechanical energy during idling period and releases it during the sharp impulse load.

5. Class *E* motors. A class *E* motor in general has low starting torque and operates at low slip at rated load. The starting current is relatively low for motors below 7.5 h.p. These motors may be started at rated voltage. However, for motors above 7.5 h.p., the starting current may be high enough to require a low-voltage starting circuit.

6. Class *F* motors. A class *F* motor is usually a double-cage motor and can be started by applying the rated voltage. It is a low-torque motor and requires the lowest amount of starting current of all motors. The important characteristics of class *F* motors are :

(*i*) The starting torque is about 1.25 times the full-load torque.

(*ii*) The starting current is 2 to 4 times the full-load current.

(*iii*) The full-load slip is quite small.

(*iv*) The full-load speed regulation is over 5%.

10.18 CRAWLING OF INDUCTION MOTORS

It has been found that induction motors, particularly the squirrel-cage type, sometimes show a tendency to run at speeds as low as one-seventh of their synchronous speed N_s. This peculiar behaviour of the cage motor at starting is known as *crawling* of an induction motor.

The above action is due to the fact that the a.c. winding of the stator produces a flux wave which is not a pure sine wave. It consists of a fundamental wave which revolves synchronously and odd harmonics like 3rd, 5th, 7th etc., which rotate either in the forward or backward direction at $N_s/3$, $N_s/5$ and $N_s/7$ respectively. These harmonic fields produce torques in the same way as the fundamental. The torque-speed curves of these harmonic torques have the same general shape as that due to the fundamental torque but with synchronous speed = N_s/n where n is the order of the harmonic. The magnitude of the harmonic torque is $1/n^2$ of the fundamental torque.

Since 3rd harmonic currents are absent in a balanced 3-phase system, the value of third harmonic torque is zero. Therefore, the total motor torque has the following three components :

(*i*) The fundamental torque T_f due to the fundamental flux rotating at a speed N_s.

(*ii*) The fifth harmonic torque T_5 due to the 5th harmonic flux rotating at a speed $N_s/5$.

(*iii*) The seventh harmonic torque T_7 due to the 7th harmonic flux rotating at a speed $N_s/7$.

Now the 5th harmonic currents have a phase difference of $-120°$ ($5 \times 120° = 600° = -120°$) in the three stator windings. Therefore, the 5th harmonic flux rotates at a speed $N_s/5$ in a direction opposite to that produced by the fundamental. The reverse torque produced by 5th harmonic is small and may be neglected.

* Recall that approximate efficiency of induction motor is $1 - s$.

Neglecting all higher harmonics, the resultant torque T_R is the sum of the fundamental torque (T_f) and the *7th harmonic torque (T_7). Fig. 10.28 shows the torque-speed characteristics due to the fundamental flux and the 7th harmonic flux as well as the resultant torque-speed characteristic. Note that the presence of the 7th harmonic flux has produced a dip in the normal torque-speed characteristic. The resultant curve and the load torque curve meet at point A which represents the stable operating point. Now point A represents a speed of $N/7$ where N is the normal speed of the motor. Therefore, the motor would not run up to full speed (N) but merely drive the load at a reduced speed of $N/7$. We say that the motor crawls at the speed $N/7$. The following points may be noted :

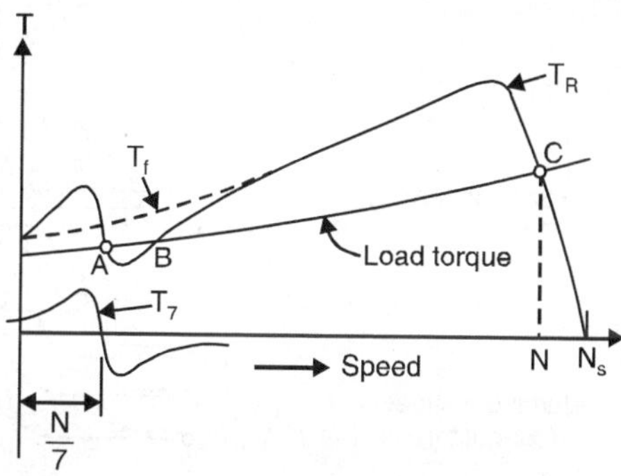

Fig. 10.28

(*a*) If the motor torque developed is due to the fundamental flux alone, the motor will accelerate to the point C [See Fig. 10.28] as the final operating speed (N) for the given load. However, the presence of the 7th harmonic flux has caused the motor to crawl at a speed corresponding to point A (*i.e.*, speed equal to $N/7$).

(*b*) To make the motor to run up to full speed, the load would have to be reduced to a value less than that of the minimum occurring between A and B [See Fig. 10.28].

(*c*) By proper choice of coil pitch and distribution of coils of stator winding, the harmonic fluxes in the air-gap of the motor can be reduced to a minimum. This eliminates the crawling effect.

10.19 COGGING

A squirrel-cage rotor may show a peculiar behaviour in starting for certain relationship between the number of stator slots (S_1) and rotor slots (S_2). If S_1 is equal to or an integral multiple of S_2, the motor may refuse to start. This phenomenon is known as *cogging* and is due to the magnetic locking between the stator and rotor teeth.

The reluctance of the magnetic path depends upon the positions of rotor teeth w.r.t. stator teeth. It is found that reluctance of the magnetic path is minimum when the rotor and stator teeth face each other. In such positions of minimum reluctance, there exists a strong alignment force between the stator and the rotor at standstill. The alignment force at the instant of start may become stronger than the starting torque. Consequently, the motor fails to start. To avoid cogging, the number of stator and rotor slots are never made to be equal or have an integral ratio.

10.20 LINEAR INDUCTION MOTOR (LIM)

So far we have seen that an induction motor (or any electric motor) produces rotation or circular motion. However, it is possible to design an induction motor in such a way that it produces linear motion. Such a motor is called linear induction motor (*LIM*). Fig. 10.29 shows the cross-sectional

* Note that 7th harmonic currents in the three stator windings have a phase difference of 120° (7 × 120° = 2 × 360° + 120° = 120°). Therefore, they set up a forward rotating field with a synchronous speed equal to 1/7th of the synchronous speed of the fundamental torque.

view of the rotary induction motor. Instead of a squirrel-cage rotor, a cylinder of conductor (usually made of metal) enclosing the rotor's ferromagnetic core is considered. If the rotary motor of Fig. 10.29 is cut along the line *xy* and laid flat, we get the linear induction motor as shown in Fig. 10.30. The stator is called primary and rotor the secondary of the linear induction motor. These terms are *appropriate because any member (primary or secondary) can be made movable. Note that the primary side consists of a magnetic core with a 3-phase winding and the secondary side is just a metal sheet. Thus a linear induction motor consists of a linear rotor and linear 3-phase stator with air-gap between them.

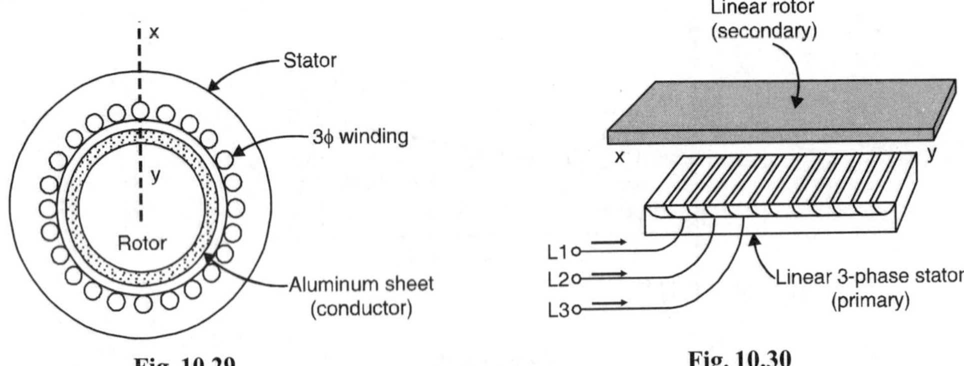

Fig. 10.29 **Fig. 10.30**

Operation. When a 3-phase supply is connected to the stator of a rotary induction motor, a revolving flux is produced in the air gap of the machine. Similarly, when a 3-phase supply is connected to the primary of a linear induction motor, the **flux produced moves at a constant speed in a straight line along the length of the primary. The speed of the travelling field is called the linear synchronous speed v_s. The travelling field will induce currents in the secondary. The induced currents will interact with the travelling field to produce a linear force F(or thrust). If one member is fixed and the other is free to move, the force will make the movable member to move. For example, suppose that the primary in Fig. 10.30 is fixed and the secondary is free to move. If the travelling flux moves from left to right, the secondary will also move to the right with linear velocity v, following the travelling field. Note that v (actual linear speed of rotor) will be less than v_s (linear synchronous speed of the travelling field).

The basic difference between a linear induction motor and a rotary induction motor is that the latter has ***endless air-gap while the former is open-ended due to the finite lengths of the primary and secondary sides. Also the angular velocity becomes linear velocity and the torque becomes the thrust or force in a linear induction motor. In order to maintain a constant thrust (force) over a considerable distance, one member is kept shorter than the other as shown in Fig. 10.30.

10.21 PROPERTIES OF LINEAR INDUCTION MOTOR

The properties of a linear induction motor are almost similar to those of a standard rotary induction motor. Consequently, the equations for slip, thrust, power etc. are also very similar.

(*i*) **Linear synchronous speed (v_s).** The linear speed of the travelling wave of flux is called linear synchronous speed. It is given by ;

---

* In a linear induction motor, the primary exciting winding may be static or moving part. This is not the case with rotary induction motor where the exciting winding is static. In other words, poles go on shifting their positions in a straight line.

** The travelling flux wave moves from one end of the linear stator to the other. Once the flux wave reaches the end of the linear stator, it restarts once more at the beginning.

*** The air-gap of rotary induction motor is circular.

$$v_s = 2wf$$

where v_s = Linear synchronous speed (m/s)

w = Width of pole-pitch (m)

f = Supply frequency (Hz)

Note that the synchronous speed (v_s) does not depend upon the number of poles but only on the width of pole-pitch. Moreover, the number of poles need not be an even number.

(*ii*) **Slip (*s*).** Similar to the rotary induction motor, the slip in a linear induction motor is defined as :

$$s = \frac{v_s - v}{v_s}$$

where v_s = Linear synchronous speed (m/s)

v = Linear speed of rotor (m/s)

(*iii*) **Thrust or Force (*F*).** The thrust or force developed by the induction motor is given by ;

$$F = \frac{P_r}{v_s}$$

where F = Thrust or force (N)

P_r = Power supplied to the rotor (W)

v_s = Linear synchronous speed (m/s)

(*iv*) **Active power flow.** The active power flows through a linear induction motor in the same way it does through a rotary induction motor. Thus, we have,

(*a*) Mechanical power P_m developed by motor is

$$P_m = P_r(1 - s)$$

(*b*) Rotor copper losses $= s\,P_r$

(*c*) $$\text{Efficiency, } \eta = \frac{\text{Output power}}{\text{Input power}}$$

The linear induction motor (*LIM*) requires a large air-gap (15-30 mm) whereas the air-gap for a rotary induction motor is small (1 – 1.5 mm). Therefore, the magnetising current is large and the power factor is low. For this reason, the efficiency of a linear induction motor is low.

(*v*) **Thrust-speed characteristic.** The thrust-speed characteristic of the linear induction motor has the same form as the torque-speed characteristic of a rotary induction motor as shown in Fig. 10.31. Note that velocity in a linear induction motor decreases rapidly with the increasing thrust.

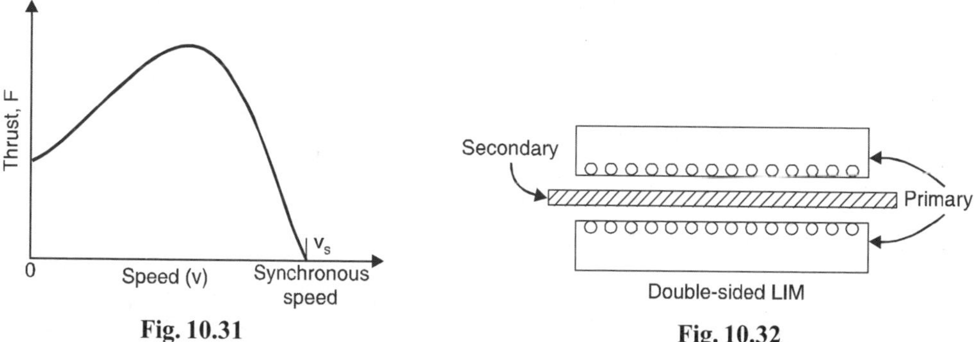

Fig. 10.31 **Fig. 10.32**

The *LIM* shown in Fig. 10.30 (Refer back) is called a *single-sided LIM* or *SLIM*. Another version is used in which primary is on both sides of the secondary as shown in Fig. 10.32. This is known as *double-sided LIM* or *DLIM*.

Applications. Theoretically, each type of rotating machine may find a linear counterpart. However, it is the linear induction motor that is being used in such industrial applications as high-speed ground transportation, sliding door systems, curtain pullers and conveyors.

Example 10.19. *A linear induction motor has 98-poles and a pole-pitch of 50 cm. Determine the synchronous speed and motor speed in km/h if the frequency is 50 Hz and slip is 0.25.*

Solution. Here, $w = 50 \text{ cm} = 0.5 \text{ m}$; $f = 50 \text{ Hz}$; $s = 0.25$

$$\text{Synchronous speed, } v_s = 2wf = 2 \times 0.5 \times 50$$

$$= 50 \text{ m/s} = \frac{50 \times 60 \times 60}{1000} \text{ km/h} = \mathbf{180 \text{ km/h}}$$

$$\text{Slip, } s = \frac{v_s - v}{v_s} \quad \text{or} \quad 0.25 = \frac{180 - v}{180}$$

$$\therefore \quad \text{Motor speed, } v = 180 - 180 \times 0.25 = \mathbf{135 \text{ km/h}}$$

Example 10.20. *The stator of a linear induction motor is excited from a 75 Hz electronic source. If the distance between consecutive phase groups of phase R is 300 mm, calculate the linear speed of the magnetic field.*

Solution. Pole-pitch, $w = 300 \text{ mm} = 0.3 \text{ m}$; Supply frequency, $f = 75 \text{ Hz}$

$$\text{Linear synchronous speed, } v_s = 2wf = 2 \times 0.3 \times 75 = 45 \text{ m/s}$$

$$= \frac{45 \times 60 \times 60}{1000} \text{ km/h} = \mathbf{162 \text{ km/h}}$$

Example 10.21. *An overhead crane in a factory is driven horizontally by means of two linear induction motors whose "rotors" are the two steel I-beams upon which the crane rolls. The 3-phase, 4-pole linear stators (mounted on opposite sides of the crane and facing the respective webs of the I-beams) have a pole pitch of 8 cm and are driven by a variable-frequency electronic source. During a test on one of the motors, the following results were obtained :*

Stator frequency : 15 Hz
Power to stator : 5 kW
Copper loss and iron loss in stator : 1 kW
Crane speed : 1.8 m/s

Calculate (i) synchronous speed and slip (ii) power to the rotor (iii) Cu loss in rotor (iv) mechanical power and thrust.

Solution. Here, Pole pitch, $w = 8 \text{ cm} = 0.08 \text{ m}$; Supply frequency, $f = 15 \text{ Hz}$

(*i*) $$\text{Linear synchronous speed, } v_s = 2wf = 2 \times 0.08 \times 15 = \mathbf{2.4 \text{ m/s}}$$

$$\text{Slip, } s = \frac{v_s - v}{v_s} = \frac{2.4 - 1.8}{2.4} = \mathbf{0.25}$$

(*ii*) $$\text{Input power to rotor, } P_r = \text{Power to stator} - \text{Cu loss and iron loss in stator}$$

$$= 5 - 1 = \mathbf{4 \text{ kW}}$$

(*iii*) $$\text{Rotor Cu loss} = s P_r = 0.25 \times 4 = \mathbf{1 \text{ kW}}$$

(*iv*) $$\text{Mechanical power developed, } P_m = P_r - \text{Rotor Cu loss} = 4 - 1 = \mathbf{3 \text{ kW}}$$

$$\text{Thrust, } F = \frac{P_r}{v_s} = \frac{(4 \times 10^3)}{2.4} = \mathbf{1.67 \times 10^3 \text{ N}}$$

Example 10.22. *The pole pitch of a linear induction motor is 0.5 m and the frequency of the applied 3-phase voltage is 60 Hz. The speed of the primary side of the motor is 200 km/h and the* *developed thrust is 100 kN. Calculate the power developed by the motor and the copper loss in the secondary side.*

Solution.

$$\text{Speed of the motor, } v = 200 \text{ km/h} = \frac{200 \times 10^3}{3600} = 55.55 \text{ m/s}$$

$$\text{Power developed by motor, } P_m = \text{Thrust} \times \text{Velocity} = F \times v$$

$$= (100 \times 10^3) \times 55.55 = 5555 \times 10^3 \text{ W} = \mathbf{5555\ kW}$$

$$\text{Synchronous speed of motor, } v_s = 2wf = 2 \times 0.5 \times 60 = 60 \text{ m/s}$$

$$\text{Slip for operating condition, } s = \frac{v_s - v}{v_s} = \frac{60 - 55.55}{60} = 0.074$$

$$\text{Input power to secondary, } P_r = \frac{P_m}{1-s}$$

$$\therefore \quad \text{Cu loss in the secondary} = s\,P_r = s \times \frac{P_m}{1-s} = 0.074 \times \frac{5555 \times 10^3}{1 - 0.074}$$

$$\simeq 444{,}000 \text{ W} = \mathbf{444\ kW}$$

SHORT ANSWER QUESTIONS

Q. 1. *What is a circle diagram ?*

Ans. Under appropriate circuit conditions, the locus of voltage, current, impedance and admittance of an a.c. circuit is circular. This is called a circle diagram. It is a convenient graphical method for circuit analysis.

Q. 2. *What is the importance of circle diagram ?*

Ans. The circle diagram of a circuit (*e.g.*, induction motor) shows us at a glance how current, power, power factor etc. change as load changes.

Q. 3. *What assumptions are made in drawing the circle diagram of an induction motor ?*

Ans. The following assumptions are made in drawing the circle diagram of an induction motor :

(*i*) The equivalent circuit of the motor is approximate equivalent circuit.

(*ii*) The leakage reactances are constant.

(*iii*) The saturation of magnetic circuit is neglected. This assumption affects the maximum output power and maximum torque.

Q. 4. *Why is circle diagram not an accurate method in determining the performance of an induction motor ?*

Ans. There are two principal reasons for it. First, it is a graphical method and, therefore, accuracy is limited. Secondly, many assumptions are made in drawing the circle diagram. Neverthe–less, it is a very simple and convenient method to study the performance of an induction motor under various load conditions.

Q. 5. *What data are required to draw the circle diagram of an induction motor ?*

Ans. In order to draw the circle diagram of an induction motor, we need two points of this circle. These two points are selected to represent the *no-load conditions* and *standstill conditions* respectively. The choice is appropriate because they represent the limiting conditions in the operation of the motor.

Q. 6. *Why is reduced stator voltage applied to perform the blocked-rotor test on a large induction motor ?*

Ans. In the case of small motors, full stator voltage may be applied for a brief period, sufficient to read current and power, the greatest care being taken not to overheat the winding. However, in case of large motors, it is generally not possible to provide test plant of sufficient capacity and the risk of damage to the motor is greater. Therefore, to perform blocked-rotor test on a large induction motor, the applied stator voltage is reduced to such a value that current drawn by the motor does not exceed 1.5 times the full-load current.

Q. 7. *What is the importance of slip in an induction motor ?*

Ans. The slip (s) of an induction motor is practically zero at no-load and its value is 1 when the rotor is locked. The slip plays an important role in the operation of the induction motor as is clear from the following facts :

(*i*) The approximate efficiency of an induction motor is $1 - s$.

(*ii*) Rotor Cu loss $= s\,P_r$

Here P_r is the rotor input power.

(*iii*) From no-load to full-load, the torque-slip characteristic is nearly a straight line so that $T \propto s$.

Q. 8. *Why is line-frequency control not used for general-purpose speed control of induction motors ?*

Ans. The major drawback of this method of speed control is that we have to provide the necessary variable-frequency supply. The auxiliary equipment required for this purpose results in a high first cost, increased maintenance and lowering of the overall efficiency.

Q. 9. *What are the disadvantages of speed control of induction motors by changing rotor circuit resistance ?*

Ans. This method of speed control is suitable *only for slip-ring motors.* The rotor-resistance control is simple, low in first cost and has low maintenance cost. However, this method has the disadvantages of low efficiency and poor speed regulation. We know that the approximate efficiency of induction motor is $(1 - s)$. Thus if the *speed of the motor developing rated torque were reduced to 50% of its normal value by rotor-resistance control, the efficiency would be less than 50%. Moreover, at low speeds, the speed regulation is poor.

Q. 10. *What is the effect of increasing the rotor resistance of the induction motor ?*

Ans. If we increase the rotor resistance, the torque-speed characteristic will change. The breakdown torque is not affected but it occurs at a lower speed. The other characteristics like starting torque, full-load torque etc. are also affected.

Q. 11. *What is the effect of high-inertia loads on induction motors ?*

Ans. High-inertia loads put a heavy strain on induction motors because they prolong the starting period. The starting current in both the stator and rotor is high during this interval so that overheating becomes a major problem. A prolonged starting may cause the line voltage to fall below for several seconds, thus affecting other connected loads. To relieve the problem, induction motors are often started on reduced voltage. This increases the starting time but it is generally not too important.

Q. 12. *What happens when an electric train driven by an induction motor begins to move downhill ?*

Ans. When the train begins to move downhill, the force of gravity causes the train to accelerate and its speed could easily exceed the synchronous speed. Under such conditions, the machine becomes induction generator. It converts mechanical energy it receives into electrical energy. This electrical energy is returned to the line.

---

* Note that in this method, applied voltage, supply frequency and synchronous speed are fixed.

Q. 13. *For the same output power and voltage rating, why are high-speed induction motors preferred to low-speed motors?*

Ans. This is due to the following advantages :

(*i*) The high-speed motors are smaller in size and consequently lighter in weight.

(*ii*) They have higher starting torque.

(*iii*) They have higher efficiencies.

(*iv*) At rated load, they have higher power factors.

(*v*) They are lower in cost.

Q. 14. *What is the fundamental difference between rotary induction motor and linear induction motor ?*

Ans. The fundamental difference between a conventional induction motor (producing rotary motion) and a linear induction motor (producing linear motion) is the *difference in their respective air gaps.* The rotating induction motor has a *closed* air gap. The linear induction motor has an *open* air gap with an *entry end* and an *exit end.*

Q. 15. *How does the weight of an induction motor vary as the output power increases?*

Ans. The dramatic increase in output power with increased weight is an inherent characteristic of all electrical machines. As a rule of thumb, we can say that for similar machines operating at the same speed, a doubling of the weight produces about 2.5 times as much power. Thus if a 10 h.p. motor weighs about 70 kg, a 25 h.p. motor of similar design and speed will weigh about 140 kg.

MULTIPLE-CHOICE QUESTIONS

1. Consider an *R-L* series a.c. circuit supplied from fixed voltage and fixed frequency. If inductive reactance is fixed and *R* is variable, the locus of current phasor will be a ....

(*i*) straight line (*ii*) circle

(*iii*) parabola (*iv*) none of above

2. In order to draw the circle diagram of an induction motor, we perform ....

(*i*) no-load test only

(*ii*) blocked rotor test only

(*iii*) both no-load and blocked rotor tests

(*iv*) none of above

3. The extreme operating conditions of an induction motor are ....

(*i*) no-load and standstill conditions

(*ii*) no-load and full-load conditions

(*iii*) full-load and standstill conditions

(*iv*) none of above

4. In a no-load test on a 3-phase induction motor, the input is measured by two-wattmeter method. The reading of the wattmeters are 5000 W and –3200 W. The no-load input power is ....

(*i*) 8200 W (*ii*) 5000 W

(*iii*) 4100 W (*iv*) 1800 W

5. The circle diagram of induction motor does not provide accurate results because it is ....

(*i*) a graphical method

(*ii*) uses approximate equivalent circuit

(*iii*) assumes leakage reactances to be constant

(*iv*) all of above

6. The pull-out torque is also called ....

(*i*) no-load torque

(*ii*) full-load torque

(*iii*) maximum torque

(*iv*) none of above

7. In a blocked rotor test on an induction motor, the mechanical output of the motor is ....

(*i*) zero (*ii*) maximum

(*iii*) quite large (*iv*) none of above

8. The power factor of an induction motor is 0.8. The machine is ....

(*i*) running at no-load
(*ii*) running at full-load
(*iii*) under blocked rotor condition
(*iv*) none of above

9. If the load torque in an induction motor exceeds the breakdown torque of the motor, the machine ....

(*i*) will suddenly stop
(*ii*) will start running at synchronous speed
(*iii*) will start running slowly
(*iv*) none of above

10. By changing the number of stator poles of a squirrel-cage motor, the number of synchronous speeds that can be obtained practically is ....

(*i*) 6 (*ii*) 4
(*iii*) 8 (*iv*) none of above

11. The speed of a slip-ring induction motor can be changed by ....

(*i*) changing the applied voltage
(*ii*) changing rotor circuit resistance
(*iii*) insertion and variation of a foreign voltage in the rotor circuit
(*iv*) all above

12. If the stator voltage of an induction motor drops by 10%, the torque of the motor will drop by ....

(*i*) 10% (*ii*) 40%
(*iii*) 20% (*iv*) 30%

13. The e.m.f. induced per phase in the rotor winding of a 3-phase induction motor is 100 V at standstill. Under full-load condition, this e.m.f. would be about ....

(*i*) 100 V (*ii*) 50 V
(*iii*) 4 V (*iv*) 0.2 V

14. Under blocked rotor condition, the frequency of rotor currents in a 50 Hz, 3-phase induction motor is ....

(*i*) 5 Hz (*ii*) 15 Hz
(*iii*) 30 Hz (*iv*) 50 Hz

15. At $s = 0$, the torque of an induction motor is ....

(*i*) zero
(*ii*) equal to full-load
(*iii*) very high
(*iv*) equal to maximum

16. The linear synchronous speed of a linear induction motor (*LIM*) does not depend upon ....

(*i*) supply frequency
(*ii*) pole pitch
(*iii*) number of poles
(*iv*) none of above

17. If line frequency drops by 20%, the speed of induction motor drops by ....

(*i*) 10% (*ii*) 20%
(*iii*) 5% (*iv*) 80%

18. A 50 Hz motor will operate well on 60 Hz line if the applied voltage is raised to .... of the nameplate rating.

(*i*) 20% (*ii*) 80%
(*iii*) 180% (*iv*) 120%

19. In the operating range of an induction motor, the relation between torque (T) and slip is (approximately) ....

(*i*) $T \propto s^2$ (*ii*) $T \propto \frac{1}{s}$
(*iii*) $T \propto s$ (*iv*) $T \propto s^{3/2}$

20. Between no-load and full-load conditions, the torque-speed characteristic of a 3-phase induction motor is essentially ....

(*i*) a straight line (*ii*) parabola
(*iii*) circle (*iv*) hyperbola

ANSWERS TO MULTIPLE-CHOICE QUESTIONS

1. (*ii*)	**2.** (*iii*)	**3.** (*i*)	**4.** (*iv*)	**5.** (*iv*)
6. (*iii*)	**7.** (*i*)	**8.** (*ii*)	**9.** (*i*)	**10.** (*ii*)
11. (*iv*)	**12.** (*iii*)	**13.** (*iii*)	**14.** (*iv*)	**15.** (*i*)
16. (*iii*)	**17.** (*ii*)	**18.** (*iv*)	**19.** (*iii*)	**20.** (*i*)

HINTS TO SELECTED MULTIPLE-CHOICE QUESTIONS

2. In order to draw the circle diagram, we need two points. For this reason, we perform two tests.

4. In no-load test on induction motor, the power factor is generally less than 0.5. Therefore, it becomes necessary to reverse the current coil or potential coil connections of that wattmeter which gives negative reading. The algebraic sum of wattmeter readings is the total input power.

7. Since the rotor is stationary, the mechanical output power of the motor is zero.

8. The power factor of an induction motor is low at no load as well as at blocked rotor condition.

10. Because of the complications in the design and switching of the interconnections of the primary, we can obtain a maximum of only four different synchronous speeds for any one motor.

12. The torque of an induction motor at any speed is directly proportional to the *square* of the applied voltage.

13. The full-load slip of an induction motor is 3% to 5%.

14. The supply frequency is $f = 50$ Hz. Under blocked rotor condition, the rotor is stationary so that slip, $s = 1$. Therefore, rotor current frequency is $f' = sf = 1 \times 50 = 50$ Hz.

11

Single-Phase Motors

INTRODUCTION

As the name suggests, these motors are used on single-phase supply. Single-phase motors are the most familiar of all electric motors because they are extensively used in home appliances, shops, offices etc. It is true that single-phase motors are less efficient substitute for 3-phase motors but 3-phase power is normally not available except in large commercial and industrial establishments. Since electric power was originally generated and distributed for lighting only, millions of homes were given single-phase supply. This led to the development of single-phase motors. Even where 3-phase mains are present, the single-phase supply may be obtained by using one of the three lines and the neutral. In this chapter, we shall focus our attention on the construction, working and characteristics of commonly used single-phase motors.

11.1 TYPES OF SINGLE-PHASE MOTORS

Single-phase motors are generally built in the fractional-horsepower range and may be classified into the following four basic types :

1. **Single-phase induction motors**
   (*i*) split-phase type (*ii*) capacitor type (*iii*) shaded-pole type
2. **A.C. series motor or universal motor**
3. **Repulsion motors**
   (*i*) Repulsion-start induction-run motor (*ii*) Repulsion-induction motor
4. **Synchronous motors**
   (*i*) Reluctance motor (*ii*) Hysteresis motor

11.2 SINGLE-PHASE INDUCTION MOTORS

A single phase induction motor is very similar to a 3-phase squirrel cage induction motor. It has (*i*) a squirrel-cage rotor identical to a 3-phase motor and (*ii*) a single-phase winding on the stator.

Unlike a 3-phase induction motor, a single-phase induction motor is not self-starting but requires some starting means. The single-phase stator winding produces a magnetic field that pulsates in strength in a sinusoidal manner. The field polarity reverses after each half cycle *but the field does not rotate.* Consequently, the alternating flux cannot produce rotation in a stationary squirrel-cage rotor. However, if the rotor of a single-phase motor is rotated in one direction by some mechanical means, it will continue to run in the direction of rotation. As a matter of fact, the rotor quickly accelerates until it reaches a speed slightly below the synchronous speed. Once the motor is running at this speed, it will continue to rotate even though single-phase current is flowing through the stator winding. This method of starting is generally not convenient for large motors. Nor can it be employed for a motor located at some inaccessible spot.

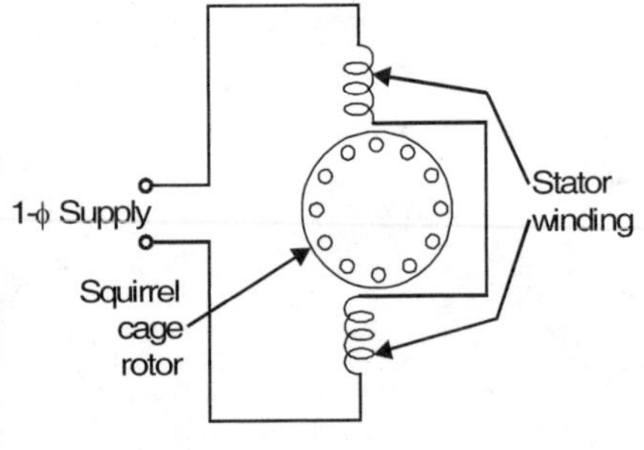

Fig. 11.1

Fig. 11.1 shows single-phase induction motor having a squirrel cage rotor and a single-phase *distributed stator winding. Such a motor inherently does not develop any starting torque and, therefore, will not start to rotate if the stator winding is connected to single-phase a.c. supply. However, if the rotor is started by auxiliary means, the motor will quickly attain the final speed. This strange behaviour of single-phase induction motor can be explained on the basis of double-field revolving theory.

11.3 DOUBLE-FIELD REVOLVING THEORY

The double-field revolving theory is proposed to explain this dilemma of no torque at start and yet torque once rotated. This theory is based on the fact that an alternating sinusoidal flux ($\phi = \phi_m \cos \omega t$) can be represented by two revolving fluxes, each equal to one-half of the maximum value of alternating flux (*i.e.*, $\phi_m/2$) and each rotating at synchronous speed ($N_s = 120 f/P$, $\omega = 2\pi f$) in opposite directions.

The above statement will now be proved. The instantaneous value of flux due to the stator current of a single-phase induction motor is given by ;

$$\phi = \phi_m \cos \omega t$$

Consider two rotating magnetic fluxes ϕ_1 and ϕ_2, each of magnitude $\phi_m/2$ and rotating in opposite directions with angular velocity ω [See Fig. 11.2]. Let the two fluxes start rotating from OX axis at $t = 0$. After time t seconds, the angle through which the flux vectors have rotated is ωt. Resolving the flux vectors along X-axis and Y-axis, we have,

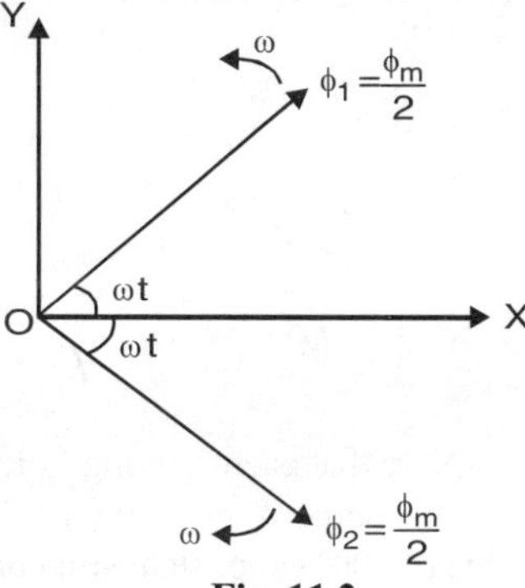

Fig. 11.2

$$\text{Total } X\text{-component} = \frac{\phi_m}{2} \cos \omega t + \frac{\phi_m}{2} \cos \omega t = \phi_m \cos \omega t$$

$$\text{Total } Y\text{-component} = \frac{\phi_m}{2} \sin \omega t - \frac{\phi_m}{2} \sin \omega t = 0$$

$$\therefore \quad \text{Resultant flux, } \phi = \sqrt{(\phi_m \cos \omega t)^2 + 0^2} = \phi_m \cos \omega t$$

Thus the resultant flux vector is $\phi = \phi_m \cos \omega t$ along X-axis. *Therefore, an alternating field can be replaced by two rotating fields of half its amplitude rotating in opposite directions at synchronous speed.* Note that the resultant vector of two revolving flux vectors is a stationary vector that oscillates in length with time along X-axis. When the rotating flux vectors are in phase [See Fig. 11.3 (*i*)], the resultant vector is $\phi = \phi_m$; when out of phase by 180° [See Fig. 11.3 (*ii*)], the resultant vector $\phi = 0$.

(*i*) (*ii*)

Fig. 11.3

Let us explain the operation of single-phase induction motor by double-field revolving theory.

(*i*) **Rotor at standstill.** Consider the case that the rotor is stationary and the stator winding is connected to a single-phase supply. The alternating flux produced by the stator winding can be represented as the sum of two rotating fluxes ϕ_1 and ϕ_2, each equal to one half of the maximum value of alternating flux and each rotating at synchronous speed ($N_s = 120 f/P$) in opposite directions as shown in Fig. 11.4 (*i*). Let the flux ϕ_1 rotate in anticlockwise direction and flux ϕ_2 in clockwise

* The stator winding is distributed in slots over the stator surface instead of being a concentrated coil.

direction. The flux ϕ_1 will result in the production of torque T_1 in the anticlockwise direction and flux ϕ_2 will result in the production of torque T_2 in the clockwise direction. At standstill, these two torques are equal and opposite and the net torque developed is zero. Therefore, single-phase induction motor is not self-starting. This fact is illustrated in Fig. 11.4 (*ii*).

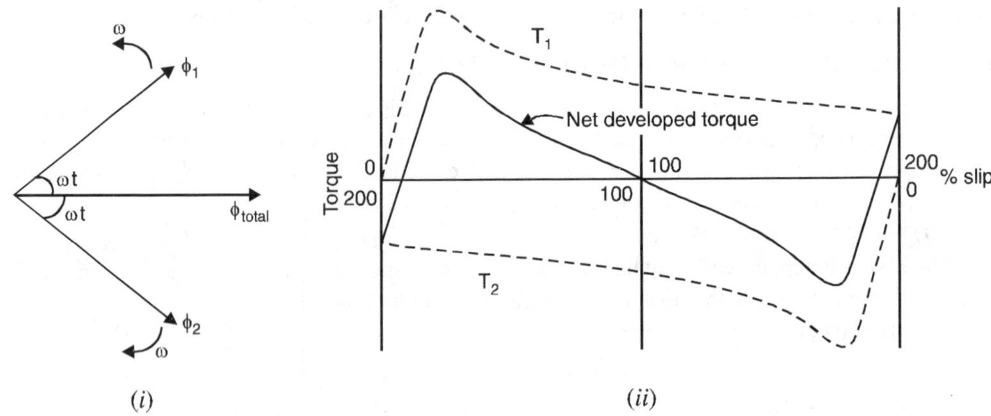

(*i*) (*ii*)

Fig. 11.4

Note that each rotating field tends to drive the rotor in the direction in which the field rotates. Thus the point of zero slip for one field corresponds to 200% slip for the other as explained later. The value of 100% slip (standstill condition) is the same for both the fields.

(*ii*) **Rotor running.** Now assume that the rotor is started by spinning the rotor or by using auxiliary circuit, in say clockwise direction. The flux rotating in the clockwise direction is the forward rotating flux (ϕ_f) and that in the other direction is the backward rotating flux (ϕ_b). The slip *w.r.t.* the forward flux will be

$$s_f = \frac{N_s - N}{N_s} = s$$

where N_s = synchronous speed

N = speed of rotor in the direction of forward flux

The rotor rotates opposite to the rotation of the backward flux. Therefore, the slip *w.r.t.* the backward flux will be

$$s_b = \frac{N_s - (-N)}{N_s} = \frac{N_s + N}{N_s} = \frac{2N_s - N_s + N}{N_s}$$

$$= \frac{2N_s}{N_s} - \frac{(N_s - N)}{N_s} = 2 - s$$

$\therefore$ $s_b = 2 - s$

Thus for forward rotating flux, slip is s (less than unity) and for backward rotating flux, the slip is $2 - s$ (greater than unity). Since for usual rotor resistance/reactance ratios, the torques at slips of less than unity are greater than those at slips of more than unity, the resultant torque will be in the direction of the rotation of the forward flux. Thus if the motor is once started, it will develop net torque in the direction in which it has been started and will function as a motor.

Fig. 11.5 shows the rotor circuits for the forward and backward rotating fluxes. Note that $r_2 = R_2/2$, where R_2 is the standstill rotor resistance *i.e.*, r_2 is equal to *half the standstill rotor resistance.

* A single-phase induction motor may be regarded as consisting of two motors having a common stator winding but with their respective rotors revolving in opposite directions. Hence resistance of each rotor is half the actual resistance of the original rotor.

Similarly, $x_2 = X_2/2$ where X_2 is the standstill rotor reactance. At standstill, $s = 1$ so that impedances of the two circuits are equal. Therefore, rotor currents are equal *i.e.*, $I_{2f} = I_{2b}$. However, when the rotor rotates, the impedances of the two rotor circuits are unequal and the rotor current I_{2b} is higher (and also at a lower power factor) than the rotor current I_{2f}. Their m.m.f.s, which oppose the stator m.m.f.s, will result in a *reduction of the backward rotating flux. Consequently, as speed increases, the forward flux increases, increasing the driving torque while the backward flux decreases, reducing the opposing torque. The motor quickly accelerates to the final speed.

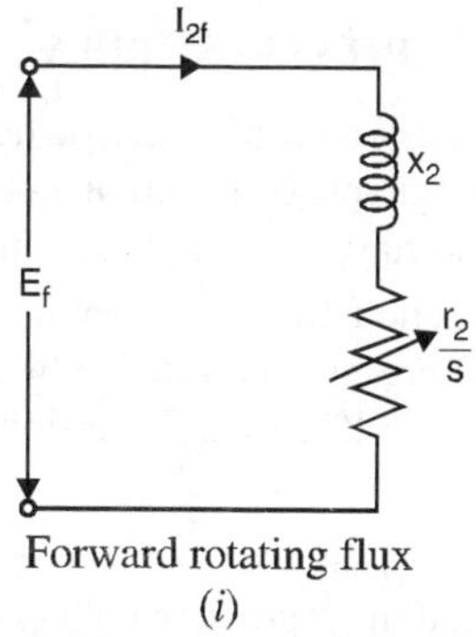

Forward rotating flux
(*i*)

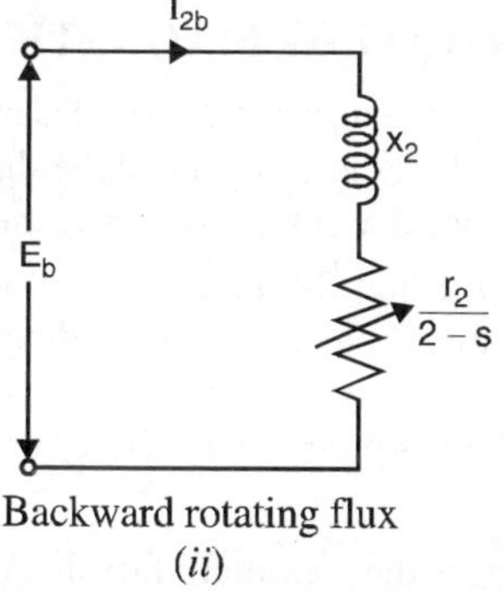

Backward rotating flux
(*ii*)

Fig. 11.5

11.4 MAKING SINGLE-PHASE INDUCTION MOTOR SELF-STARTING

The single-phase induction motor is not self-starting and it is undesirable to resort to mechanical spinning of the shaft or pulling a belt to start it. To make a single-phase induction motor self-starting, we should somehow produce a revolving stator magnetic field. This may be achieved by converting a single-phase supply into two-phase supply through the use of an additional winding. When the motor attains sufficient speed, the starting means (*i.e.*, additional winding) may be removed depending upon the type of the motor. As a matter of fact, single-phase induction motors are classified and named according to the method employed to make them self-starting.

(*i*) **Split-phase motors** – started by two phase motor action through the use of an auxiliary or starting winding.

(*ii*) **Capacitor motors** – started by two-phase motor action through the use of an auxiliary winding and a capacitor.

(*iii*) **Shaded-pole motors** – started by the motion of the magnetic field produced by means of a shading coil around a portion of the pole structure.

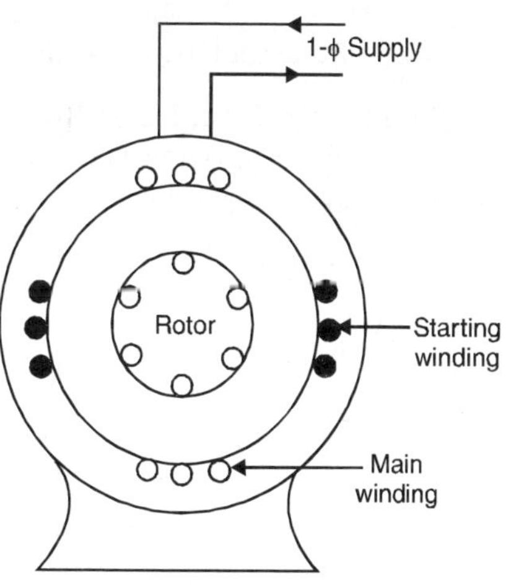

Fig. 11.6

Example 11.1. *A single-phase induction motor has an output torque of 0.25 Nm and a speed of 100 rad/s. If the input current is 0.6 A at 230 V and 0.6 lagging p.f., find (i) output power (ii) efficiency.*

---

* As the motor gathers speed, the rotor m.m.f. is composed of two parts : (*i*) an increasing backward m.m.f. since the current increases and (*ii*) a decreasing forward m.m.f. since the current decreases. It is the sum of these two rotor m.m.f.s that oppose the stator m.m.f. Just as in a transformer, the secondary m.m.f. opposes the primary m.m.f. leaving a net magnetising value, which produces the main flux. It follows that as the speed increases, the forward flux will increase since the rotor is providing a decreasing opposing m.m.f. On the other hand, the backward flux will decrease since the opposing m.m.f. is increasing.

Solution.

(*i*) Output power, $P_{out} = \dfrac{2\pi N\, T_{sh}}{60} = \omega T_{sh}$ $\qquad \left(\because \omega = \dfrac{2\pi N}{60}\right)$

$= 100 \times 0.25 = \mathbf{25\ W}$

(*ii*) Efficiency, $\eta = \dfrac{\text{Output power}}{\text{Input power}} \times 100 = \dfrac{25}{230 \times 0.6 \times 0.6} \times 100 = \mathbf{30.19\%}$

11.5 ROTATING MAGNETIC FIELD FROM 2-PHASE SUPPLY

As with a 3-phase supply, a 2-phase balanced supply also produces a rotating magnetic field of constant magnitude. With the exception of the shaded-pole motor, all single-phase induction motors are started as 2-phase machine. Once so started, the motor will continue to run on single-phase supply.

Let us see how 2-phase supply produces a rotating magnetic field of constant magnitude. Fig. 11.10 (*i*) shows 2-pole, 2-phase winding. The *phases *X* and *Y* are energised from a two-phase source and currents in these phases are indicated as I_X and I_Y [See Fig. 11.10 (*ii*)]. Referring to Fig. 11.10 (*ii*), the **fluxes produced by these currents are given by ;

$$\phi_Y = \phi_m \sin \omega t \quad \text{and} \quad \phi_X = \phi_m \sin(\omega t + 90^\circ) = \phi_m \cos \omega t$$

Here ϕ_m is the maximum flux due to either phase. We shall now prove that this 2-phase supply produces a rotating magnetic field of constant magnitude equal to ϕ_m.

(*i*) At instant 1 [See Fig. 11.10 (*ii*) and Fig. 11.10 (*iii*)], the current is zero in phase *Y* and maximum in phase *X*. With the current in the direction shown, a resultant flux is established toward the right. The magnitude of the resultant flux is constant and is equal to ϕ_m as proved under :

At instant 1, $\omega t = 0^\circ \quad \therefore \quad \phi_Y = 0$ and $\phi_X = \phi_m$

$\therefore$ Resultant flux, $\phi_r = \sqrt{\phi_X^2 + \phi_Y^2} = \sqrt{(\phi_m)^2 + (0)^2} = \phi_m$

ϕ_r

Fig. 11.7

(*ii*) At instant 2 [See Fig. 11.10 (*ii*) and Fig. 11.10 (*iii*)], the current is still in the same direction in phase *X* and an equal current flowing in phase *Y*. This establishes a resultant flux of the same value (*i.e.*, $\phi_r = \phi_m$) as proved under :

At instant 2, $\omega t = 45^\circ \quad \therefore \quad \phi_Y = \dfrac{\phi_m}{\sqrt{2}}$ and $\phi_X = \dfrac{\phi_m}{\sqrt{2}}$

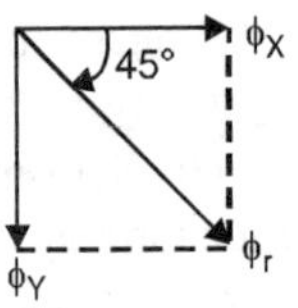

$\therefore$ Resultant flux, $\phi_r = \sqrt{(\phi_X)^2 + (\phi_Y)^2}$

$$= \sqrt{\left(\frac{\phi_m}{\sqrt{2}}\right)^2 + \left(\frac{\phi_m}{\sqrt{2}}\right)^2} = \phi_m$$

ϕ_X 45° ϕ_r ϕ_Y

Fig. 11.8

Note that resultant flux has the same value (*i.e.* ϕ_m) but turned 45° clockwise from position 1.

(*iii*) At instant 3 [See Fig. 11.10 (*ii*) and Fig. 11.10 (*iii*)], the current in phase *X* has decreased to zero and current in phase *Y* has increased to maximum. This establishes a resultant flux downward as proved under :

At instant 3, $\omega t = 90^\circ \quad \therefore \quad \phi_Y = \phi_m$ and $\phi_X = 0$

$$\therefore \quad \phi_r = \sqrt{\phi_X^2 + (\phi_Y)^2} = \sqrt{(0)^2 + (\phi_m)^2} = \phi_m$$

ϕ_r

Fig. 11.9

Note that resultant flux has now turned 90° clockwise from position 1.

The reader may note that in the three instants considered above, the resultant flux is constant and is equal to ϕ_m. However, this constant resultant flux is shifting its position (clockwise in this case). In other words, the rotating flux is produced. We shall continue to consider other instants to prove this fact.

* In two-phase system, the two windings (or phases) are 90° electrical apart.

** The flux produced by a current is in phase with the current that produces it.

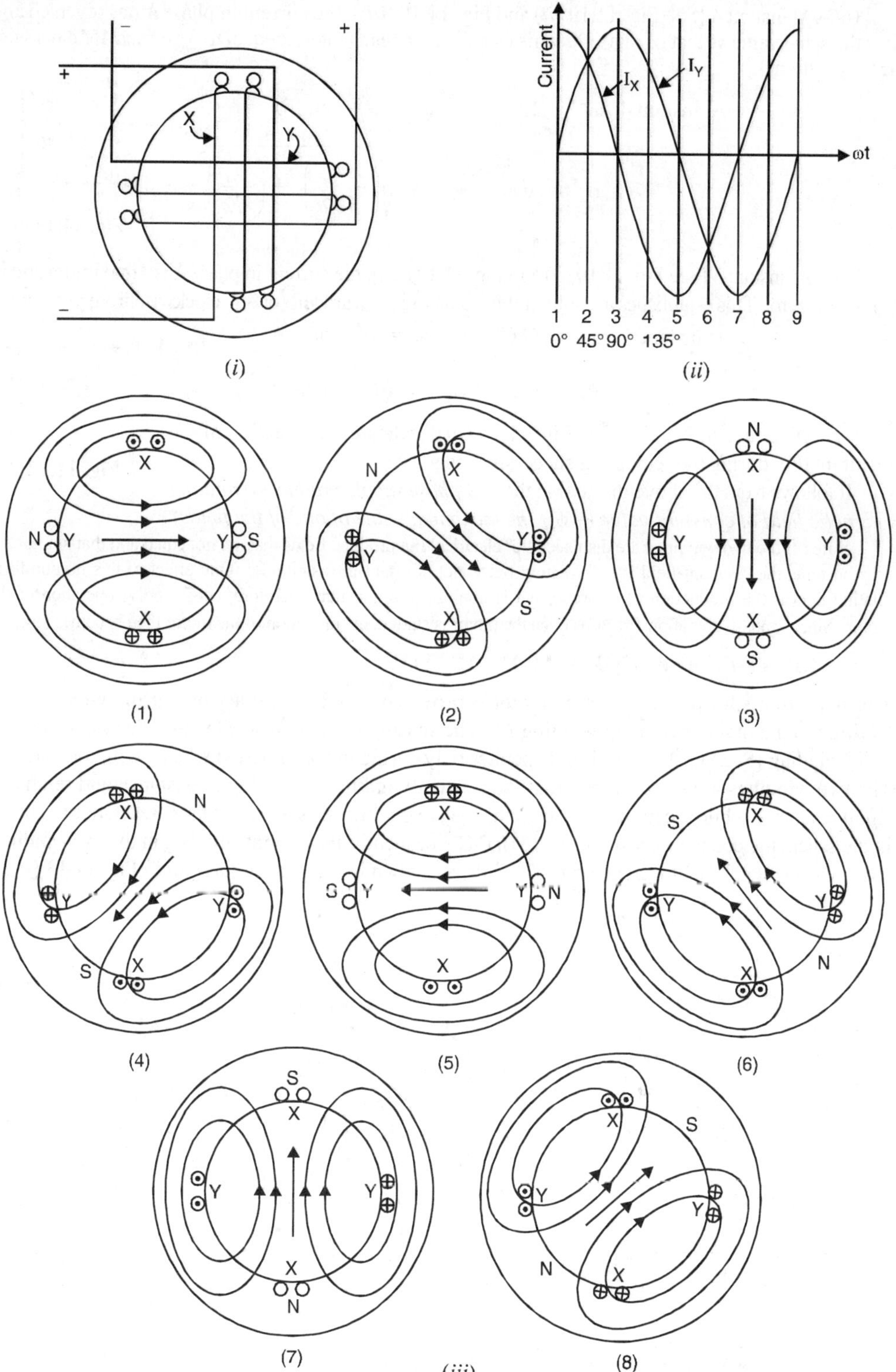

(*i*)

(*ii*)

(1) (2) (3) (4) (5) (6) (7) (8)

(*iii*)

Fig. 11.10

(*iv*) At instant 4 [See Fig. 11.10 (*ii*) and Fig. 11.10 (*iii*)], the current in phase *X* has reversed and has the same value as that of phase *Y*. This establishes a resultant flux equal to ϕ_m turned 45° clockwise from position 3.

$$\text{At instant 4, } \omega t = 135^\circ \quad \therefore \quad \phi_Y = \frac{\phi_m}{\sqrt{2}} \text{ and } \phi_X = -\frac{\phi_m}{\sqrt{2}}$$

$$\therefore \quad \phi_r = \sqrt{\phi_X^2 + \phi_Y^2} = \sqrt{\left(-\frac{\phi_m}{\sqrt{2}}\right)^2 + \left(\frac{\phi_m}{\sqrt{2}}\right)^2}$$

$$= \phi_m$$

Fig. 11.11

(*v*) At instant 5 [See Fig. 11.10 (*ii*) and Fig. 11.10 (*iii*)], the current in phase *X* is *maximum and in phase *Y* is zero. This establishes a resultant flux equal to ϕ_m toward left (or 90° clockwise from position 3).

$$\text{At instant 5, } \omega t = 180^\circ \quad \therefore \quad \phi_Y = 0 \text{ and } \phi_X = -\phi_m$$

$$\therefore \quad \phi_r = \sqrt{\phi_X^2 + \phi_Y^2} = \sqrt{(-\phi_m)^2 + (0)^2} = \phi_m$$

(*vi*) Diagrams 6, 7 and 8 [See Fig. 11.10 (*iii*)] indicate the direction of the resultant flux during the remaining successive instants.

Fig. 11.12

It follows from the above discussion that *a 2-phase supply produces a rotating magnetic field of constant value* (= ϕ_m, *the maximum value of one of the fields*).

Note : If the two windings are displaced 90° electrical but produce fields that are not equal and that are not 90° apart in time, the resultant field is still rotating but is not constant in magnitude. One effect of this non-uniform rotating field is the production of a torque that is non-uniform and that, therefore, causes noisy operation of the motor. Since 2-phase operation **ceases once the motor is started, the operation of the motor then becomes smooth.

11.6 SPLIT-PHASE INDUCTION MOTOR

The stator of a split-phase induction motor is provided with an auxiliary or starting winding *S* in addition to the main or running winding *M*. The starting winding is located 90° electrical from the main winding [See Fig. 11.13 (*i*)] and operates only during the brief period when the motor starts up. The two windings are so †designed that the starting winding *S* has a high resistance and relatively small reactance while the main winding *M* has relatively low resistance and large reactance as shown in the schematic connections in Fig. 11.13 (*ii*). Consequently, the currents flowing in the two windings have reasonable phase difference α (25° to 30°) as shown in the phasor diagram in Fig. 11.13 (*iii*).

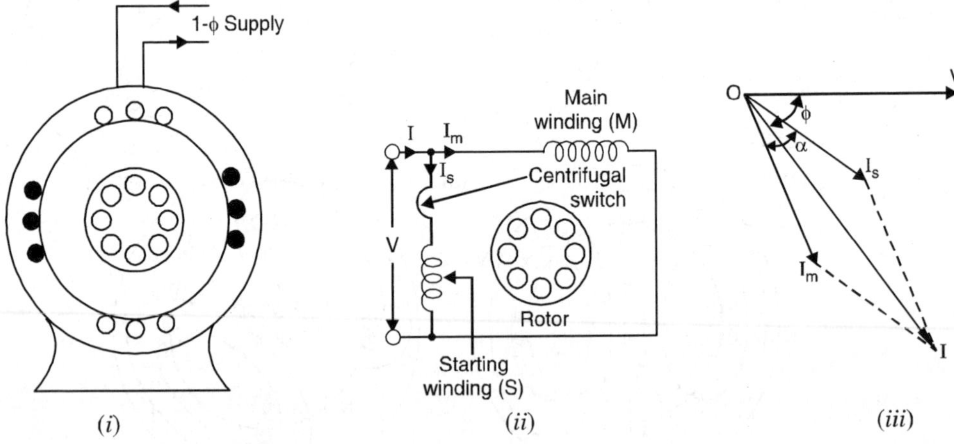

(*i*) (*ii*) (*iii*)

Fig. 11.13

* in the reverse direction.

** Once the motor is started, it operates on single-phase supply.

† This can be done by having main winding of large diameter and number of turns while starting winding of a fine wire of a small number of turns.

Operation.

(*i*) When the two stator windings are energised from a single-phase supply, the main winding carries current I_m while the starting winding carries current I_s.

(*ii*) Since main winding is made highly inductive while the starting winding highly resistive, the currents I_m and I_s have a reasonable phase angle α (25° to 30°) between them as shown in Fig. 11.13 (*iii*). Consequently, a weak revolving field approximating to that of a 2-phase machine is produced which starts the motor. The starting torque is given by ;

$$T_s = k\, I_m I_s \sin\alpha$$

where k is a constant whose magnitude depends upon the design of the motor.

(*iii*) When the motor reaches about 75% of synchronous speed, the centrifugal switch opens the circuit of the starting winding. The motor then operates as a single-phase induction motor and continues to accelerate till it reaches the normal speed. The normal speed of the motor is below the synchronous speed and depends upon the load on the motor.

Characteristics

(*i*) The starting torque is 1·5 to 2 times the full-load torque and the starting current is 6 to 8 times the full-load current.

(*ii*) Due to their low cost, split-phase induction motors are most popular single-phase motors in the market.

(*iii*) Since the starting winding is made of fine wire, the current density is high and the winding heats up quickly. If the starting period exceeds 5 seconds, the winding may burn out unless the motor is protected by built-in-thermal relay. This motor is, therefore, suitable where starting periods are not frequent.

(*iv*) An important characteristic of these motors is that they are essentially constant-speed motors. The speed variation is 2–5% from no-load to full-load.

(*v*) These motors are suitable where a moderate starting torque is required and where starting periods are infrequent *e.g.*, to drive :

(*a*) fans (*b*) washing machines (*c*) oil burners (*d*) small machine tools etc.

The power rating of such motors generally lies between 60 W and 250 W.

Example 11.2. *A resistance split-phase induction motor is rated at 187 W, 1725 r.p.m., 115V, 60 Hz. When the rotor is locked, a test at reduced voltage on the main and starting windings yields the following results :*

	Main winding	*Starting winding*
Applied voltage :	*23 volts*	*23 volts*
Current :	*4 A*	*1.5 A*
Active power (P) :	*60 W*	*30 W*

Calculate (i) the phase angle between I_m *and* I_s *(ii) the locked-rotor current drawn from the line at 115 V.*

Solution. (*i*) We shall first calculate the phase angle between I_m and the applied voltage V of the main winding.

$$\text{Apparent power, } S = V I_m = 23 \times 4 = 92 \text{ VA}$$

$$\therefore \quad \text{Power factor, } \cos\phi_m = P/S = 60/92 = 0.65 \qquad \therefore\ \phi_m = 49.2°$$

Therefore, I_m lags 49.2° behind V.

We now calculate the phase angle between I_s and V of the starting winding.

$$\text{Apparent power, } S = V I_s = 23 \times 1.5 = 34.5 \text{ VA}$$

$$\therefore \quad \text{Power factor, } \cos\phi_s = P/S = 30/34.5 = 0.87 \qquad \therefore\ \phi_s = 29.6°$$

Therefore, I_s lags V by 29.6°.

The phase angle between I_m and I_s, $\alpha = \phi_m - \phi_s = 49.2° - 29.6° = \mathbf{19.6°}$

(*ii*) Total active power, $P = P_m + P_s = 60 + 30 = 90$ W

$$\text{Total reactive power, } Q = Q_m + Q_s$$
$$= \sqrt{92^2 - 60^2} + \sqrt{34.5^2 - 30^2} = 86.8 \text{ VAR}$$
$$\text{Total apparent power, } S = \sqrt{P^2 + Q^2} = \sqrt{90^2 + 86.8^2} = 125 \text{ VA}$$

$\therefore$ Line current, $I = S/V = 125/23 = 5.44$ A (at 23 V)

$\therefore$ Current drawn at 115 V, $I_L = 5.44 \times (115/23) = \mathbf{27.2\ A}$

Example 11.3. *A 0.5 h.p., 230 V, single-phase induction motor (split-phase) takes a current of 4.2 A lagging the voltage by 10° for the starting winding and a current of 6.2 A lagging the voltage by 40° for its main winding. Find (i) total current and power factor at the time of starting (ii) total current and power factor during running (iii) phase angle between the main winding current and starting winding current (iv) power drawn by starting winding (v) power drawn by main winding (vi) total power drawn during starting (vii) total power drawn during running.*

Solution.

(*i*) Total current drawn by the motor is

$$I = I_m + I_s = 6.2 \angle -40° + 4.2 \angle -10°$$
$$= (4.75 - j\,3.986) + (4.136 - j\,0.73)$$
$$= (8.886 - j\,4.716) \text{ A} = 10.06 \angle -27.96° \text{ A}$$

$\therefore$ $I = \mathbf{10.06\ A}$; Power factor $= \cos(-27.96°) = \mathbf{0.883}$ *lag*

(*ii*) During running, the starting winding is not connected. Therefore, the total current drawn is equal to that drawn by the main winding.

$\therefore$ $I = I_m = \mathbf{6.2\ A}$; Power factor $= \cos(-40°) = \mathbf{0.766}$ *lag*

(*iii*) Phase angle between I_m and $I_s = 40° - 10° = \mathbf{30°}$

(*iv*) Power drawn by the starting winding is

$$P_s = V I_s \cos 10° = 230 \times 4.2 \times \cos 10° = \mathbf{951.32\ W}$$

(*v*) Power drawn by the main winding is

$$P_m = V I_m \cos 40° = 230 \times 6.2 \times \cos 40° = \mathbf{1092.38\ W}$$

(*vi*) Total power drawn during the starting of motor is

$$P = P_s + P_m = 951.32 + 1092.38 = \mathbf{2043.7\ W}$$

(*vii*) During running, only the main winding is in the motor circuit.

$\therefore$ Power drawn by the motor during running is

$$P = P_m = \mathbf{1092.38\ W}$$

Example 11.4. *A 50 Hz split-phase induction motor has a resistance of 5 Ω and an inductive reactance of 20 Ω in the main winding as well as starting winding. Determine the value of resistance and capacitance to be added in series with the starting winding to send the same current in each winding with a phase difference of 90°.*

Solution. Fig. 11.14 (*i*) shows the conditions of the problem.

$$Z_m = Z_s = (5 + j\,20)\ \Omega = 20.6 \angle 76°\ \Omega$$

$$\therefore \quad I_m = I_s = \frac{V \angle 0°}{20.6 \angle 76°} = \frac{V \angle -76°}{20.6} \text{ A}$$

Now $I_m = I_s$ and both lag behind the applied voltage V by 76° as shown in the phasor diagram in Fig. 11.14 (*ii*). It is desired that after the inclusion of *$R-C$ series circuit in the starting winding, the current in each winding remains the same but the phase difference between the two currents becomes 90°. This is possible only if the impedance of starting winding remains the same (*i.e.* $Z_s = 20.6\ \Omega$) and I_s leads V by 14° as shown in the phasor diagram in Fig. 11.14 (*ii*) *i.e.*

$$Z_s = 20.6\ \Omega \text{ and } \phi_s = 14°$$

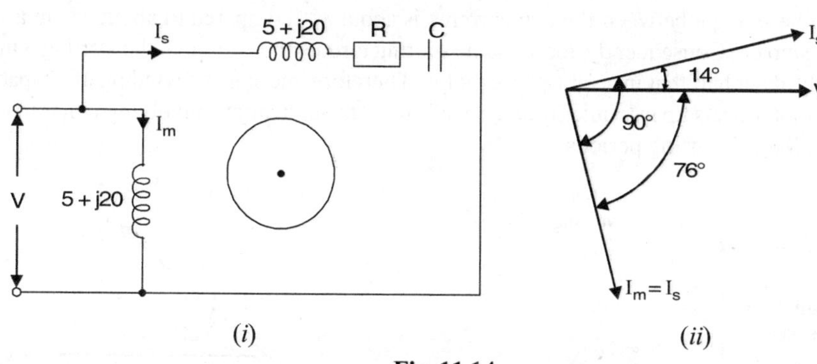

Fig. 11.14

$$\text{New resistance of starting winding, } R_s = Z_s \cos 14° = 20.6 \times \cos 14° = 19.99\ \Omega$$

$$\therefore \quad \text{Required resistance, } R = R_s - 5 = 19.99 - 5 = \mathbf{14.99\ \Omega}$$

$$\text{New reactance of starting winding, } X = Z_s \sin 14° = 20.6 \times \sin 14° = 4.98\ \Omega$$

Now $$^{**}X = X_C - X_L \text{ or } 4.98 = X_C - 20 \quad \therefore X_C = 24.98\ \Omega$$

$$\therefore \quad C = \frac{1}{2\pi f X_C} = \frac{1}{2\pi \times 50 \times 24.98} = 127.5 \times 10^{-6}\text{F} = \mathbf{127.5\ \mu F}$$

Example 11.5. *A single-phase split-phase motor is rated at 220 V, 175 W, 180 rad/s, 0.5 p.f. and 45% efficiency at rated load. Calculate its rated load (i) line current (ii) torque.*

Solution.

(*i*) $$\text{Input power, } P_i = \frac{\text{Output power}}{\eta} = \frac{175}{0.45} = 388.89 \text{ W}$$

$$\text{Line current, } I_L = \frac{P_i}{V \cos\phi} = \frac{388.89}{220 \times 0.5} = \mathbf{3.535\ A}$$

(*ii*) $$\text{Output power} = T\omega$$

or $$175 = T \times 180 \qquad \therefore T = \mathbf{0.972\ Nm}$$

11.7 CAPACITOR-START MOTOR

The capacitor-start motor is identical to a split-phase motor except that the starting winding has as many turns as the main winding. Moreover, a capacitor C is connected in series with the starting winding as shown in Fig. 11.15 (*i*). The value of capacitor is so chosen that I_s leads I_m by about 80° (*i.e.*, $\alpha \simeq 80°$) which is considerably greater than 25° found in split-phase motor [See Fig. 11.15 (*ii*)]. Consequently, starting torque ($T_s = k I_m I_s \sin\alpha$) is much more than that of a split-phase motor. Again, the starting winding is opened by the centrifugal switch when the motor attains about 75% of

* The purpose of C is to produce a phase difference between I_m and I_s. The purpose of R is to keep I_s equal to I_m even after the inclusion of C in the starting winding.

** Net reactance is capacitive.

synchronous speed. The motor then operates as a single-phase induction motor and continues to accelerate till it reaches the normal speed.

Characteristics

(*i*) Although starting characteristics of a capacitor-start motor are better than those of a split-phase motor, both machines possess the same running characteristics because the main windings are identical.

(*ii*) The phase angle between the two currents is about 80° compared to about 25° in a split-phase motor. Consequently, for the same starting torque, the current in the starting winding is only about half that in a split-phase motor. Therefore, the starting winding of a capacitor-start motor heats up less quickly and is well suited to applications involving either frequent or prolonged starting periods.

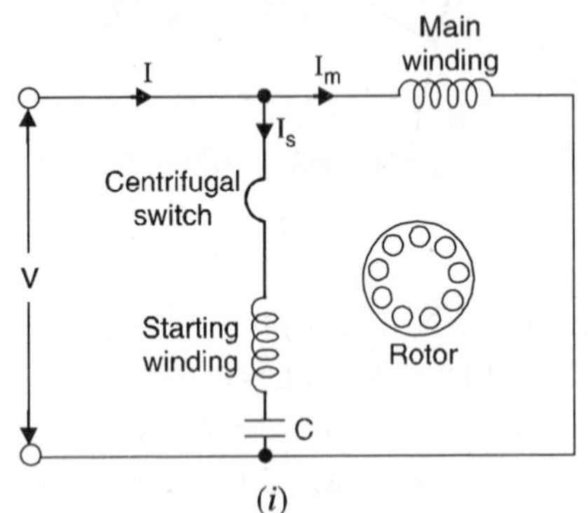

(*i*)

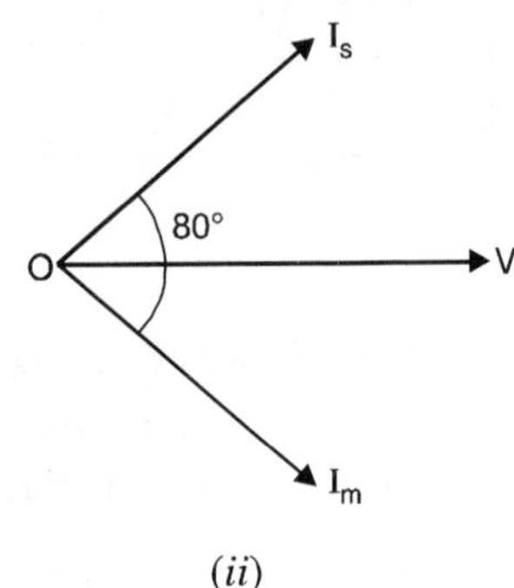

(*ii*)

Fig. 11.15

(*iii*) Capacitor-start motors are used where high starting torque is required and where the starting period may be long *e.g.*, to drive :

(*a*) compressors (*b*) large fans (*c*) pumps (*d*) high inertia loads

The power rating of such motors lies between 120 W and 7·5 kW.

Example 11.6. *A 4-pole, 50 Hz, single-phase induction motor is running with a slip of 3·4%. Calculate the speed of the motor.*

Solution.

$$\text{Synchronous speed, } N_s = 120f/P = 120 \times 50/4 = 1500 \text{ r.p.m.}$$

$$\text{Motor speed, } N = (1 - s)N_s = (1 - 0{\cdot}034)\,1500 = \mathbf{1449 \text{ r.p.m.}}$$

Example 11.7. *At starting, the windings of a 230 V, 50 Hz, split-phase induction motor have the following parameters :*

$$\textit{Main winding : } R = 4\,\Omega \; ; \; X_L = 7.5\,\Omega$$

$$\textit{Starting winding : } R = 7.5\,\Omega \; ; \; X_L = 4\,\Omega$$

Find (i) current I_m in the main winding (ii) current I_s in the starting winding (iii) phase angle between I_s and I_m and (iv) line current and (v) power factor of the motor.

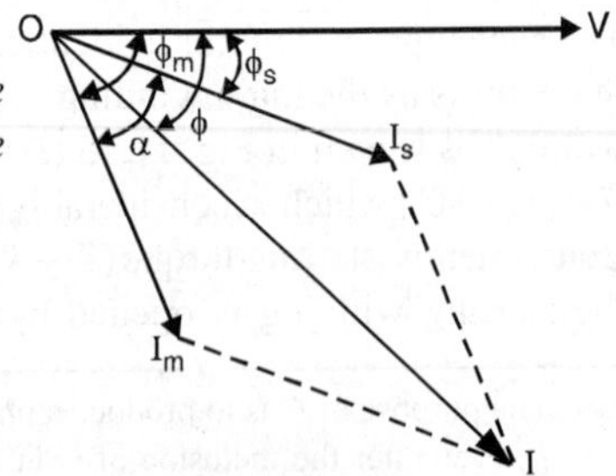

Fig. 11.16

Solution.

(*i*)
$$I_m = \frac{V}{Z_m} = \frac{230}{\sqrt{4^2 + (7 \cdot 5)^2}} = \mathbf{27\ A}$$

$$\phi_m = \tan^{-1} 7{\cdot}5/4 = 62°$$

(*ii*) $$I_s = \frac{V}{Z_s} = \frac{230}{\sqrt{(7 \cdot 5)^2 + 4^2}} = \mathbf{27\ A}$$

$$\phi_s = \tan^{-1} 4/7{\cdot}5 = 28°$$

(*iii*) Phase angle between I_s and I_m is

$$\alpha = \phi_m - \phi_s = 62° - 28° = \mathbf{34°}$$

(*iv*) Line current, $$I = \sqrt{(I_s)^2 + (I_m)^2 + 2\,I_s\,I_m \cos\alpha}$$

$$= \sqrt{(27)^2 + (27)^2 + 2 \times 27 \times 27 \times \cos 34°} = \mathbf{51{\cdot}6\ A}$$

(*v*) Resolving the currents along *OX*-axis, we get,

$$I \cos\phi = I_s \cos\phi_s + I_m \cos\phi_m$$

∴ Motor p.f., $$\cos\phi = \frac{27\cos 28° + 27\cos 62°}{I} = \frac{36 \cdot 51}{51 \cdot 6} = \mathbf{0{\cdot}7}\ lag$$

The resulting phasor diagram of split-phase induction motor is shown in Fig. 11.16.

Example 11.8. *A 200 W, 230 V, 50 Hz capacitor-start motor has the following winding constants*

Main winding : $R = 4{\cdot}5\,\Omega\ ;\ X_L = 3{\cdot}7\,\Omega$

Starting winding : $R = 9{\cdot}5\,\Omega\ ;\ X_L = 3{\cdot}5\,\Omega$

Find the value of starting capacitance that will result in the maximum starting torque.

Solution. Fig. 11.17 shows the phasor diagram of the motor. The current I_s in the starting winding leads the applied voltage V by ϕ_s while the current I_m in the main winding lags V by ϕ_m. The starting torque will be maximum when phase angle α between I_s and I_m is 90°.

$$\phi_m = \tan^{-1} 3{\cdot}7/4{\cdot}5 = 39.6°$$

$$\phi_s = 39{\cdot}6° - 90° = -50.4°$$

Let C farad be the capacitance of the starting capacitor to give maximum starting torque.

$$\tan(-50{\cdot}4°) = \frac{3 \cdot 5 - X_C}{9 \cdot 5}$$

or $$-1{\cdot}21 = \frac{3 \cdot 5 - X_C}{9 \cdot 5}$$

or $$X_C = 15\Omega$$

∴ $$C = \frac{1}{2\pi f X_C} = \frac{1}{2\pi \times 50 \times 15}$$

$$= 212 \times 10^{-6}\,\text{F} = \mathbf{212\ \mu F}$$

Fig. 11.17

Example 11.9. *A 4-pole, 250 W, 115 V, 60 Hz capacitor-start induction motor takes a full-load line current of 5·3 A while running at 1760 r.p.m. If the full-load efficiency of the motor is 64%, find (i) motor slip (ii) power factor and (iii) full-load torque.*

Solution.

(*i*) $$N_s = 120\,f/P - 120 \times 60/4 = 1800\ \text{r.p.m.}$$

$$s = \frac{N_s - N}{N_s} = \frac{1800 - 1760}{1800} = \mathbf{0{\cdot}022}$$

(*ii*) Input power $= 250/0{\cdot}64 = 390{\cdot}6$ W

Power factor $$= \frac{390 \cdot 6}{115 \times 5 \cdot 3} = \mathbf{0{\cdot}64}\ lag$$

(*iii*) Full load torque, $$T = 9{\cdot}55\,\frac{P_{out}}{N} = 9 \cdot 55\,\frac{250}{1760} = \mathbf{1{\cdot}35\ N\text{-}m}$$

Example 11.10. *A 0.25 h.p., 110 V, single-phase induction motor (split-phase) takes a current of 4 A lagging the supply voltage by 15° for the starting winding and a current of 6 A lagging the voltage by 40° for its main winding. Calculate (i) the total starting current and the power factor (ii) the phase angle between the main winding current and starting winding current.*

Solution.

(*i*) Starting winding current, $I_s = 4\angle -15°\text{ A} = (3.86 - j\,1.035)\text{ A}$

Main winding current, $I_m = 6\angle -40°\text{ A} = (4.60 - j\,3.86)\text{ A}$

Total starting current, $I = I_s + I_m = (3.86 - j\,1.035) + (4.60 - j\,3.86)$

$= (8.46 - j\,4.895)\text{ A} = 9.77\angle -30°\text{ A}$

∴ Total starting current = **9.77 A** ; Power factor = cos (– 30°) = **0.866** *lag*

(*ii*) Phase angle between I_m and I_s, $\alpha = 40° - 15° = \mathbf{25°}$

Example 11.11. *A capacitor is added to the starting winding of the above motor which causes the starting winding current to lead the supply voltage by 42°. The magnitude of currents in both the starting winding and the main winding remains the same, the latter lagging by the same amount as in problem 11.10. Calculate (i) the total starting current and the power factor (ii) the phase angle between the main winding current and starting winding current. Compare the results with those in problem 11.10.*

Solution.

(*i*) Starting winding current, $I_s = 4\angle 42°\text{ A} = (2.97 + j\,2.68)\text{ A}$

Main winding current, $I_m = 6\angle -40°\text{ A} = (4.60 - j\,3.86)\text{ A}$

Total starting current, $I = I_s + I_m = (2.97 + j\,2.68) + (4.60 - j\,3.86)$

$= (7.57 - j\,1.18)\text{ A} = 7.66\angle -8.9°\text{ A}$

∴ Total starting current = **7.66 A** ; Power factor = cos (– 8.9°) = **0.988** *lag*

(*ii*) Phase angle between I_m and I_s, $\alpha = 40° + 42° = \mathbf{82°}$

Comparison

(*a*) The starting current is reduced from 9.77 A to 7.66 A and power factor is raised from 0.866 to 0.988.

(*b*) The starting torque, $T_s \propto I_m I_s \sin\alpha$ where α is the angle between I_m and I_s. Therefore, the ratio of starting torques

$$= \frac{\text{Starting torque of capacitor – start motor}}{\text{Starting torque of resistance – start motor}}$$

$$= \frac{4\times 6\times \sin 82°}{4\times 6\times \sin 25°} = 2.35$$

Example 11.12. *A 600 W, 115 V, 60 Hz capacitor-start motor draws 13.8 A from the supply at rated load. If the efficiency is 65% and rated speed is 1750 r.p.m., calculate (i) input power at rated load (ii) power factor at rated load (iii) rated motor horsepower.*

Solution.

(*i*) Input power, $P_i = \dfrac{\text{Output power}}{\eta} = \dfrac{600}{0.65} = \mathbf{923.1\ W}$

(*ii*) Power factor, $\cos\phi = \dfrac{P_i}{VI} = \dfrac{923.1}{115\times 13.8} = \mathbf{0.582}$ *lag*

(*iii*) Motor horsepower $= P_{out}/746 = 600/746 = \mathbf{0.8\ h.p.}$

11.8 CAPACITOR-START CAPACITOR-RUN MOTOR

This motor is identical to a capacitor-start motor except that starting winding is not opened after starting so that both the windings remain connected to the supply when running as well as at starting. Two designs are generally used.

(*i*) In one design, a single capacitor *C* is used for both starting and running as shown in Fig. 11.18 (*i*). This design eliminates the need of a centrifugal switch and at the same time improves the power factor and efficiency of the motor.

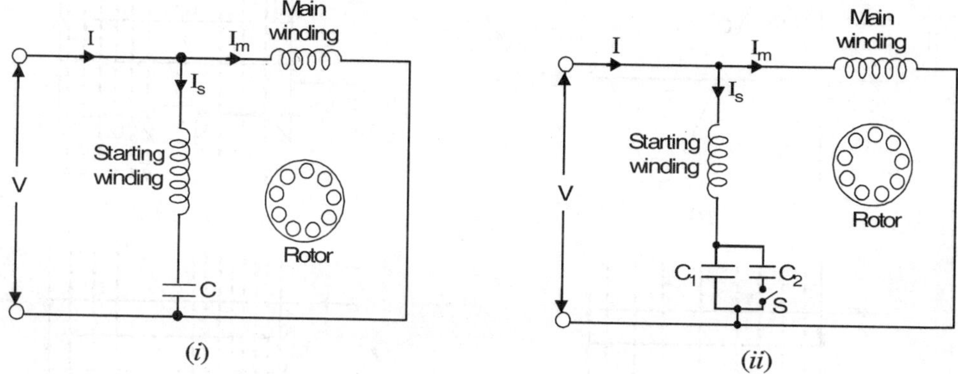

Fig. 11.18

(*ii*) In the other design, two capacitors C_1 and C_2 are used in the starting winding as shown in Fig. 11.18 (*ii*). The smaller capacitor C_1 required for optimum running conditions is permanently connected in series with the starting winding. The much larger capacitor C_2 is connected in parallel with C_1 for optimum starting and remains in the circuit during starting. The starting capacitor C_2 is disconnected when the motor approaches about 75% of synchronous speed. The motor then runs as a single-phase induction motor.

Characteristics

(*i*) The starting winding and the capacitor can be designed for perfect 2-phase operation at any load. The motor then produces a constant torque and not a pulsating torque as in other single-phase motors.

(*ii*) Because of constant torque, the motor is vibration free and can be used in :

(*a*) hospitals (*b*) studios and (*c*) other places where silence is important.

11.9 SHADED-POLE MOTOR

The shaded-pole motor is very popular for ratings below 0·05 h.p. ($\simeq$ 40 W) because of its extremely simple construction. It has salient poles on the stator excited by single-phase supply and a squirrel-cage rotor as shown in Fig. 11.19. A portion of each pole is surrounded by a short-circuited turn of copper strip called *shading coil*.

Operation. The operation of the motor can be understood by referring to Fig. 11.20 which shows one pole of the motor with a shading coil.

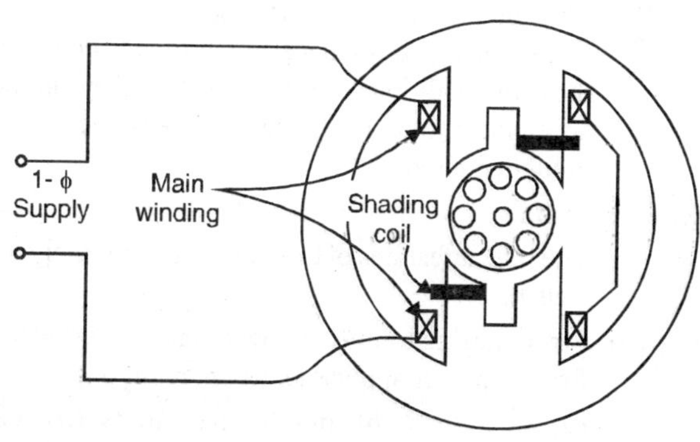

Fig. 11.19

(*i*) During the portion *OA* of the alternating-current cycle [See Fig. 11.20 (*i*)], the flux begins to increase and an e.m.f. is induced in the shading coil. The resulting current in the shading coil will be in such a direction (Lenz's law) so as to oppose the change in flux. Thus the flux in the shaded portion of the pole is weakened while that in the unshaded portion is strengthened as shown in Fig. 11.20 (*ii*).

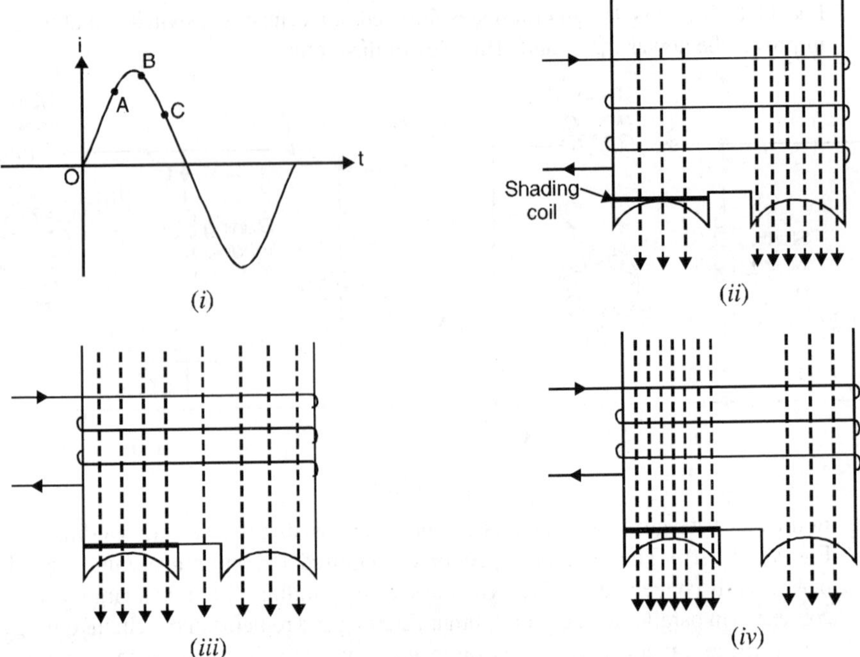

Fig. 11.20

(*ii*) During the portion *AB* of the alternating-current cycle, the flux has reached almost maximum value and is not changing. Consequently, the flux distribution across the pole is uniform [See Fig. 11.20 (*iii*)] since no current is flowing in the shading coil. As the flux decreases (portion *BC* of the alternating current cycle), current is induced in the shading coil so as to oppose the decrease in current. Thus the flux in the shaded portion of the pole is strengthened while that in the unshaded portion is weakened as shown in Fig. 11.20 (*iv*).

(*iii*) The effect of the shading coil is to cause the field flux to shift across the pole face from the unshaded to the shaded portion. This shifting flux is like a rotating weak field moving in the direction from unshaded portion to the shaded portion of the pole.

(*iv*) The rotor is of the squirrel-cage type and is under the influence of this moving field. Consequently, a small starting torque is developed. As soon as this torque starts to revolve the rotor, additional torque is produced by single-phase induction-motor action. The motor accelerates to a speed slightly below the synchronous speed and runs as a single-phase induction motor.

Characteristics

(*i*) The salient features of this motor are extremely simple construction and absence of centrifugal switch.

(*ii*) Since starting torque, efficiency and power factor are very low, these motors are only suitable for low power applications *e.g.*, to drive :

(*a*) small fans (*b*) toys (*c*) hair driers (*d*) desk fans etc.

The power rating of such motors is upto about 30 W.

11.10 EQUIVALENT CIRCUIT OF SINGLE-PHASE INDUCTION MOTOR

It was stated earlier that when the stator of a single-phase induction motor is connected to single-phase supply, the stator current produces a pulsating flux that is equivalent to two-constant-amplitude fluxes revolving in opposite directions at the synchronous speed (double-field revolving theory). Each of these fluxes induces currents in the rotor circuit and produces induction motor action similar to that in a 3-phase induction motor. Therefore, a single-phase induction motor can be imagined to be consisting of two motors, having a common stator winding but with their respective rotors revolving in opposite directions. Each rotor has resistance and reactance half the actual rotor values.

Let R_1 = resistance of stator winding

X_1 = leakage reactance of stator winding

X_m = total magnetising reactance

R'_2 = resistance of the rotor referred to the stator

X'_2 = leakage reactance of the rotor referred to the stator

We shall now develop the *equivalent circuit of single-phase induction motor using double-field revolving theory.

(*i*) **At standstill.** At standstill, the motor is simply a transformer with its secondary short-circuited. Therefore, the equivalent circuit of single-phase motor at standstill will be as shown in Fig. 11.21. The double-field revolving theory suggests that characteristics associated with each revolving field will be just one-half of the characteristics associated with the actual total flux. Therefore, each rotor has resistance and reactance equal to $R'_2/2$ and $X'_2/2$ respectively. Each rotor is associated with half the total magnetising reactance. Note that in the equivalent circuit, the core loss has been neglected. However, core loss can be represented by an equivalent resistance in parallel with the magnetising reactance.

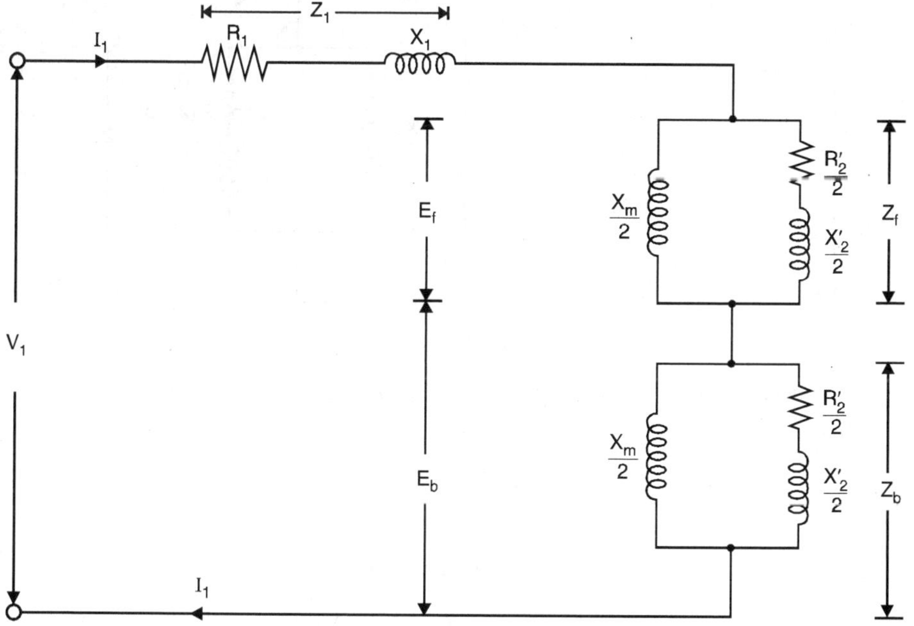

Fig. 11.21

* This equivalent circuit will be with motor operating with main winding only.

Now $E_f = 4{\cdot}44 f N \phi_f \;;\quad E_b = 4{\cdot}44 f N \phi_b$

At standstill, $\phi_f = \phi_b$. Therefore, $E_f = E_b$.

$$V_1 \simeq E_f + E_b = I_1 Z_f + I_1 Z_b$$

where Z_f = impedance of forward parallel branch

Z_b = impedance of backward parallel branch

(*ii*) **Rotor running.** Now consider that the motor is running at some speed in the direction of the forward revolving field, the slip being *s*. The rotor current produced by the forward field will have a frequency *sf* where *f* is the stator frequency. Also, the rotor current produced by the backward field will have a frequency of $(2 - s)f$. Fig. 11.22 shows the equivalent circuit of a single-phase induction motor when the rotor is rotating at slip *s*. It is clear from the equivalent circuit that under running conditions, E_f becomes much greater than E_b because the term $R'_2/2s$ increases very much as *s* tends toward zero. Conversely, E_b falls because the term $R'_2/2\,(2 - s)$ decreases since $(2 - s)$ tends toward 2. Consequently, the forward field increases, increasing the driving torque while the backward field decreases reducing the opposing torque.

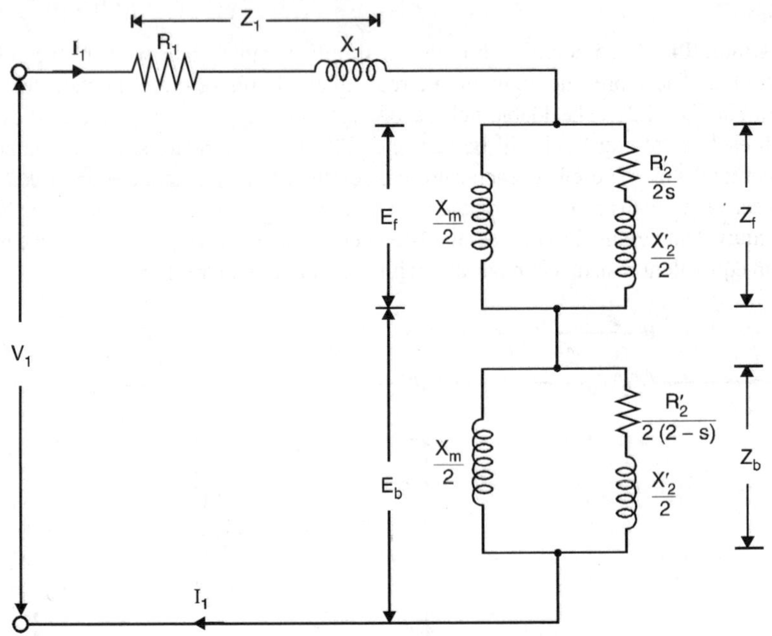

Fig. 11.22

Total impedance of the circuit is given by ; $\mathbf{Z_T} = \mathbf{Z_1} + \mathbf{Z_f} + \mathbf{Z_b}$

where $\mathbf{Z_1} = R_1 + jX_1$

$$\mathbf{Z_f} = \frac{j\dfrac{X_m}{2}\left(\dfrac{R'_2}{2s} + j\dfrac{X'_2}{2}\right)}{\dfrac{R'_2}{2s} + j\left(\dfrac{X_m}{2} + \dfrac{X'_2}{2}\right)}$$

$$\mathbf{Z_b} = \frac{j\dfrac{X_m}{2}\left(\dfrac{R'_2}{2(2-s)} + j\dfrac{X'_2}{2}\right)}{\dfrac{R'_2}{2(2-s)} + j\left(\dfrac{X_m}{2} + \dfrac{X'_2}{2}\right)}$$

$$\therefore \quad \mathbf{I_1} = \mathbf{V_1}/\mathbf{Z_T}$$

Example 11.13. *A 2-pole, 240 V, 50 Hz, single-phase induction motor has the following constants referred to the stator :*

$R_1 = 2{\cdot}2\,\Omega$; $X_1 = 3{\cdot}0\,\Omega$; $R'_2 = 3{\cdot}8\,\Omega$; $X'_2 = 2{\cdot}1\,\Omega$; $X_m = 86\,\Omega$

Find the stator current and input power when the motor is operating at a full-load speed of 2820 r.p.m.

Solution. Fig. 11.23 shows the equivalent circuit of single-phase induction motor with circuit values.

$$\text{Synchronous speed, } N_s = 120\,f/P = 120 \times 50/2 = 3000 \text{ r.p.m.}$$

$$\text{Slip, } s = \frac{N_s - N}{N_s} = \frac{3000 - 2820}{3000} = 0{\cdot}06$$

$$\therefore \quad \frac{3{\cdot}8}{2s} = \frac{3{\cdot}8}{2 \times 0{\cdot}06} = 31.67\,\Omega\,; \quad \frac{3{\cdot}8}{2(2-s)} = \frac{3{\cdot}8}{2\,(2-0{\cdot}06)} = 0{\cdot}979\,\Omega\,; \quad \frac{X_m}{2} = \frac{86}{2} = 43\,\Omega$$

$$\therefore \quad \mathbf{Z_f} = \frac{j\,43\,(31{\cdot}67 + j\,2{\cdot}1/2)}{31{\cdot}67 + j\,(43 + 2{\cdot}1/2)} = 25{\cdot}22\,\angle 37{\cdot}61^\circ\,\Omega = (19{\cdot}98 + j\,15{\cdot}39)\,\Omega$$

$$\mathbf{Z_b} = \frac{j\,43\,(0{\cdot}979 + j\,2{\cdot}1/2)}{0{\cdot}979 + j\,(43 + 2{\cdot}1/2)} = 1{\cdot}40\,\angle 48{\cdot}3^\circ\,\Omega = (0{\cdot}933 + j\,1{\cdot}05)\,\Omega$$

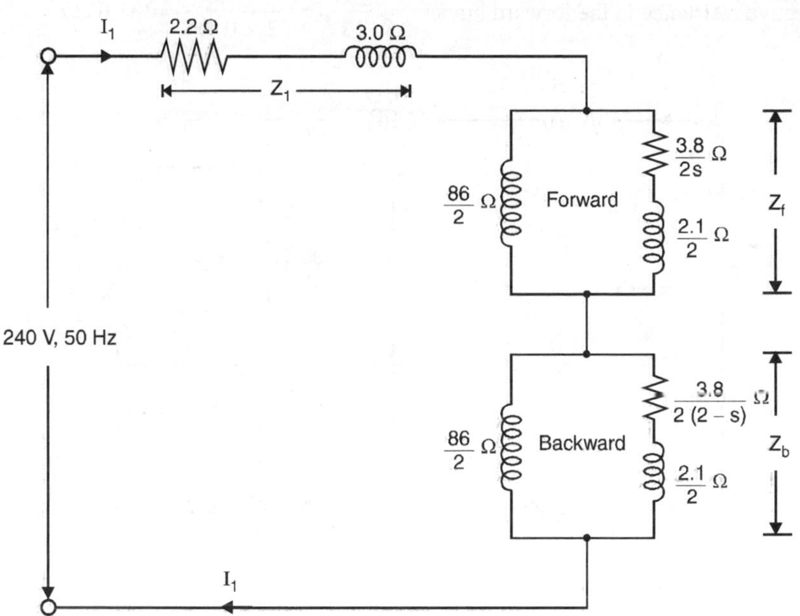

Fig. 11.23

$$\text{Total impedance, } \mathbf{Z_T} = \mathbf{Z_1} + \mathbf{Z_f} + \mathbf{Z_b}$$
$$= (2{\cdot}2 + j\,3{\cdot}0) + (19{\cdot}98 + j\,15{\cdot}39) + (0{\cdot}933 + j\,1{\cdot}05)$$
$$= (23{\cdot}11 + j\,19{\cdot}44)\,\Omega = 30{\cdot}2\,\angle 40.1^\circ\,\Omega$$

$$\therefore \quad \text{Stator current, } \mathbf{I_1} = \frac{\mathbf{V_1}}{\mathbf{Z_T}} = \frac{240\,\angle\,0^\circ}{30{\cdot}2\,\angle\,40{\cdot}1^\circ} = 7{\cdot}95\,\angle -40{\cdot}1^\circ\,\text{A}$$

$\therefore$ Magnitude of stator current, I_1 = **7·95 A** at 40·1° lagging p.f.

Input power = $V_1 I_1 \cos\phi = 240 \times 7{\cdot}95 \times \cos 40{\cdot}1^\circ$ = **1459·5 W**

Example 11.14. *In the above example, if the iron and friction losses amount to 50 W, find (i) total mechanical power developed (ii) useful shaft power (iii) shaft torque and (iv) efficiency.*

Solution. The equivalent circuit shown in Fig. 11.23 is reshown in Fig. 11.24 in the interest of clarity of the solution.

(*i*) The voltage across *AB* in Fig. 11.24 is

$$\mathbf{V_f} = \mathbf{I_1} \times \mathbf{Z_f} = 7{\cdot}95 \angle -40{\cdot}1^\circ \times 25{\cdot}22 \angle 37{\cdot}61^\circ = 200{\cdot}5 \angle -2{\cdot}5^\circ \text{ V}$$

The current in the resistive portion of the forward parallel branch is

$$\mathbf{I_f} = \frac{\mathbf{V_f}}{\dfrac{3\cdot 8}{2s} + j\dfrac{2\cdot 1}{2}} = \frac{200\cdot 5 \angle -2\cdot 5^\circ}{31\cdot 67 + j\, 1\cdot 05} = 6{\cdot}33 \angle -4{\cdot}4^\circ \text{ A}$$

The voltage across *BC* in Fig. 11.24 is

$$\mathbf{V_b} = \mathbf{I_1}\,\mathbf{Z_b} = 7{\cdot}95 \angle -40{\cdot}1^\circ \times 1{\cdot}40 \angle 48{\cdot}3^\circ = 11{\cdot}13 \angle 8.2^\circ \text{ V}$$

$$\mathbf{I_b} = \frac{\mathbf{V_b}}{\dfrac{3\cdot 8}{2\,(2-s)} + j\dfrac{2\cdot 1}{2}} = \frac{11\cdot 13 \angle 8\cdot 2^\circ}{(0\cdot 979 + j\, 1\cdot 05)}$$

$$= 7{\cdot}75 \angle -38{\cdot}8^\circ \text{ A}$$

Effective resistance in the forward branch $= \dfrac{3\cdot 8}{2\,s} = \dfrac{3\cdot 8}{2 \times 0\cdot 06} = 31.67\ \Omega$

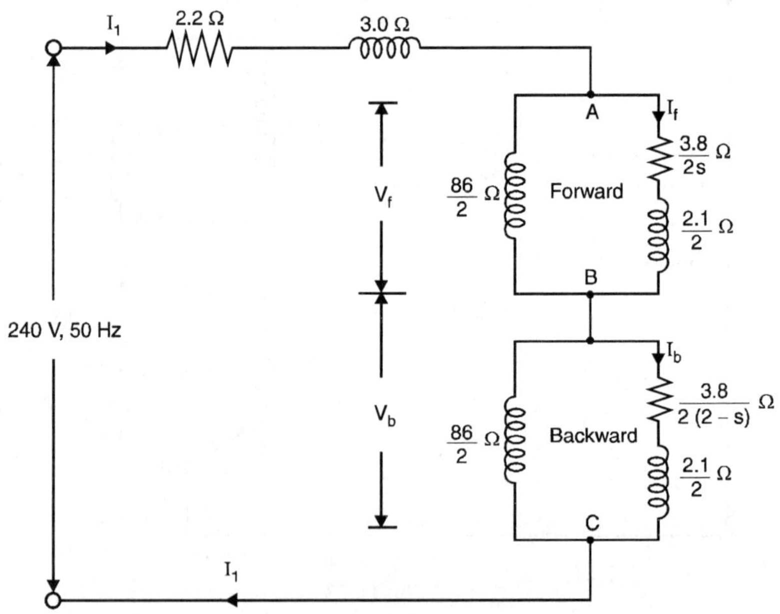

Fig. 11.24

The actual resistance existing in the branch is $= 3{\cdot}8/2 = 1{\cdot}9\ \Omega$. It follows that $(31{\cdot}67 - 1{\cdot}9)\ \Omega$ is the electrical equivalent of mechanical load.

$\therefore$ Forward branch power $= I_f^2 \times (31{\cdot}67 - 1{\cdot}9) = (6{\cdot}33)^2 \times (31{\cdot}67 - 1{\cdot}9) = 1193$ W

Effective resistance in backward branch $= \dfrac{3\cdot 8}{2\,(2-s)} = \dfrac{3\cdot 8}{2\,(2 - 0\cdot 06)} = 0{\cdot}979\ \Omega$

$\therefore$ Electrical equivalent of mechanical load $= (0{\cdot}979 - 1{\cdot}9)\ \Omega$

This is a negative value and indicates the production of a backward torque.

$\therefore$ Backward branch power $= I_b^2 \times (0{\cdot}979 - 1{\cdot}9) = (7{\cdot}75)^2 \times (0{\cdot}979 - 1{\cdot}9) = -55$ W

Total mechanical power developed is

$$P_m = (1193) + (-55) = \mathbf{1138\,W}$$

(*ii*) The total mechanical power developed includes iron and friction losses.

$\therefore$ Shaft power, $P_{out} = 1138 - 50 = \mathbf{1088\ W}$

(*iii*) Shaft torque, $T_{sh} = 9.55 \times \dfrac{P_{out}}{N} = 9{\cdot}55 \times \dfrac{1088}{2820} = \mathbf{3{\cdot}67\ N\text{-}m}$

(*iv*) Efficiency, $\eta = \dfrac{P_{out}}{P_{in}} \times 100 = \dfrac{1088}{1459{\cdot}5} \times 100 = \mathbf{74{\cdot}5\ \%}$

Example 11.15. *In the above example, find* (*i*) *two internal torques corresponding to the forward and backward rotating fields* (*ii*) *net torque* (*iii*) *lost torque* (*iv*) *output power.*

Solution.

Synchronous speed, $\omega_s = N_s/60 = 3000/60 = 50$ r.p.s. $= 100\,\pi$ rad/s

(*i*) Torque due to forward rotating field is

$$T_f = \frac{1}{\omega_s} \times I_f^2\, \frac{R_2'}{2s} = \frac{1}{100\,\pi} \times (6{\cdot}63)^2 \times \frac{3{\cdot}8}{2 \times 0{\cdot}06} = \mathbf{4{\cdot}038\ N\text{-}m}$$

Torque due to backward rotating field is

$$T_b = \frac{1}{\omega_s} \times I_b^2\, \frac{R_2'}{2(2-s)} = \frac{1}{100\,\pi} \times (7{\cdot}75)^2 \times \frac{3{\cdot}8}{2\,(2 - 0{\cdot}06)} = \mathbf{0{\cdot}187\ N\text{-}m}$$

(*ii*) Since the two torques are in opposition,

$\therefore$ Net or average torque, $T_g = T_f + T_b = 4{\cdot}038 + (-0{\cdot}187) = 3{\cdot}851$ N-m

(*iii*) Lost torque, $T_{loss} = 9{\cdot}55 \times \dfrac{\text{Iron and friction losses}}{N}$

$$= 9{\cdot}55 \times \frac{50}{2820} = \mathbf{0{\cdot}169\ N\text{-}m}$$

(*iv*) Shaft torque, $T_{sh} = T_g - T_{loss} = 3{\cdot}851 - 0{\cdot}169 = 3{\cdot}628$ N-m

Shaft power, $P_{out} = \dfrac{2\pi\, N\, T_{sh}}{60} = \dfrac{2\pi \times 2820 \times 3{\cdot}682}{60} = \mathbf{1087\ W}$

Note that we have used the method of internal and lost torques to solve the problem. The slight difference in results is due to approximations made in the course of calculations.

Example 11.16. *In a 6-pole, 50 Hz, single-phase induction motor, the gross power absorbed by the forward and backward fields are 160 W and 20 W respectively at a motor speed of 950 r.p.m. If the no-load frictional losses are 75 W, find the shaft torque.*

Solution.

Net input power, $P_i = 160 - 20 = 140$ W

Synchronous speed, $N_s = 120\,f/P = 120 \times 50/6 = 1000$ r.p.m.

$$\text{Slip, } s = \frac{N_s - N}{N_s} = \frac{1000 - 950}{1000} = 0.05$$

Mechanical power developed, $P_m = P_i\,(1 - s) = 140\,(1 - 0.05) = 133$ W

Mechanical power at shaft, $P_{out} = P_m - 75 = 133 - 75 = 58$ W

Now $P_{out} = \dfrac{2\pi N\, T_{sh}}{60}$ or $58 = \dfrac{2\pi \times 950 \times T_{sh}}{60}$ $\therefore T_{sh} = \mathbf{0.583\ Nm}$

TUTORIAL PROBLEMS

1. A 240 V, 50 Hz, single-phase capacitor start a.c. motor is loaded such that the slip is 6 %. The motor constants referred to the stator are :

Stator resistance = 2·2 Ω ; Rotor resistance = 3·8 Ω

Stator reactance = 3·0 Ω ; Rotor reactance = 2·1 Ω

Magnetising reactance = 86 Ω ; Iron and friction loss = 50 W

Draw the equivalent circuit and from it find the output power and efficiency at the given loading.

[1090 W ; 74·6 %]

2. Find the mechanical power output of the 185 W, 110 V, 60 Hz, single-phase induction motor whose constants referred to stator are :

Stator resistance = 1·86 Ω ; Stator reactance = 3·56 Ω

Rotor resistance = 3·56 Ω ; Standstill rotor reactance = 2·56 Ω

Magnetising reactance of stator winding = 53·5 Ω **[202 W]**

3. A 4-pole, single-phase, squirrel-cage induction motor operates from a 115 V, 50 Hz supply. The measured parameters are :

Stator impedance = $(1·9 + j\,2·7)\ \Omega$

Rotor impedance = $(4·4 + j\,2·6)\ \Omega$ referred to stator

Magnetising reactance = $j\,76\ \Omega$

Find the approximate output power and its corresponding power factor when the motor speed is 1400 r.p.m. Assume that iron and friction losses equal to copper losses at this speed.

[92 W ; 0·57 lagging]

11.11 A.C. SERIES MOTOR OR UNIVERSAL MOTOR

A d.c. series motor will rotate in the *same direction regardless of the polarity of the supply. One can expect that a d.c. series motor would also operate on a single-phase supply. It is then called an a.c. series motor. However, some changes must be made in a d.c. motor that is to operate satisfactorily on a.c. supply. The changes effected are :

(*i*) The entire magnetic circuit is laminated in order to reduce the eddy current loss. Hence an a.c. series motor requires a more expensive construction than a d.c. series motor.

(*ii*) The series field winding uses as few turns as possible to reduce the reactance of the field winding to a minimum. This reduces the voltage drop across the field winding.

(*iii*) A high field flux is obtained by using a low-reluctance magnetic circuit.

(*iv*) There is considerable sparking between the brushes and the commutator when the motor is used on a.c. supply. It is because the alternating flux establishes high currents in the coils short-circuited by the brushes. When the short-circuited coils break contact from the commutator, excessive sparking is produced. This can be eliminated by using high-resistance leads to connect the coils to the commutator segments.

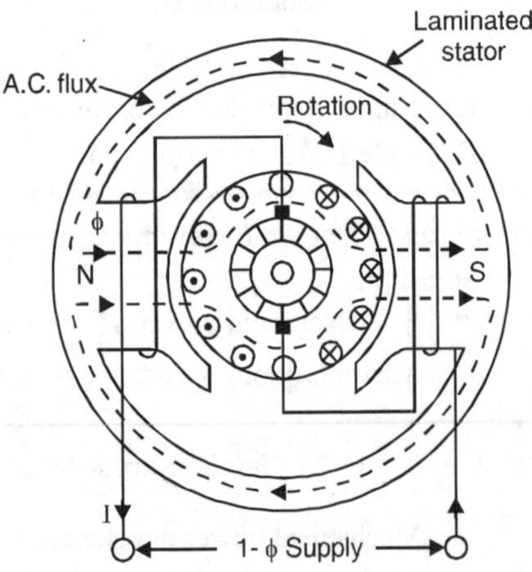

Fig. 11.25

* In a series motor, both armature and field windings are in series. Since the armature current and flux reverse simultaneously, the torque always acts in the same direction regardless of the polarity of the supply.

Construction. The construction of an a.c. series motor is very similar to a d.c. series motor except that above modifications are incorporated (See Fig. 11.25). Such a motor can be operated either on a.c. or d.c. supply and the resulting torque-speed curve is about the same in each case. For this reason, it is sometimes called a *universal motor*.

Operation. When the motor is connected to an a.c. supply, the same alternating current flows through the field and armature windings. The field winding produces an alternating flux ϕ that reacts with the current flowing in the armature to produce a torque. Since both armature current and flux reverse simultaneously, the torque always acts in the same direction. It may be noted that no rotating flux is produced in this type of machines; the principle of operation is the same as that of a d.c. series motor.

Characteristics. The operating characteristics of an a.c.series motor are similar to those of a d.c. series motor.

(*i*) The speed increases to a high value with a decrease in load. In very small series motors, the losses are usually large enough at no load that limit the speed to a *definite value (1500 – 15,000 r.p.m.).

(*ii*) The motor torque is high for large armature currents, thus giving a high starting torque.

(*iii*) At full-load, the power factor is about 90%. However, at starting or when carrying an overload, the power factor is lower.

Applications. The fractional horsepower a.c. series motors have high-speed (and corresponding small size) and large starting torque. They can, therefore, be used to drive :

(*a*) high-speed vacuum cleaners (*b*) sewing machines
(*c*) electric shavers (*d*) drills (*e*) machine tools *etc.*

Example 11.17. *A universal series motor has resistance of 30 Ω and an inductance of 0.5 H. When connected to a 250 V d.c. supply and loaded to take 0.8 A, it runs at 2000 r.p.m. Find the speed and power factor when connected to a 250 V, 50 Hz a.c. supply and loaded to take the same current.*

Solution.

A.C. Operation. Fig. 11.26 shows the circuit diagram as well as phasor diagram for a.c. operation.

$$X_a = 2\pi \times 50 \times 0.5 = 157\ \Omega$$
$$R_a = 30\ \Omega$$
$$I_a R_a = 0.8 \times 30 = 24\ \text{V}$$
$$I_a X_a = 0.8 \times 157 = 125.6\ \text{V}$$

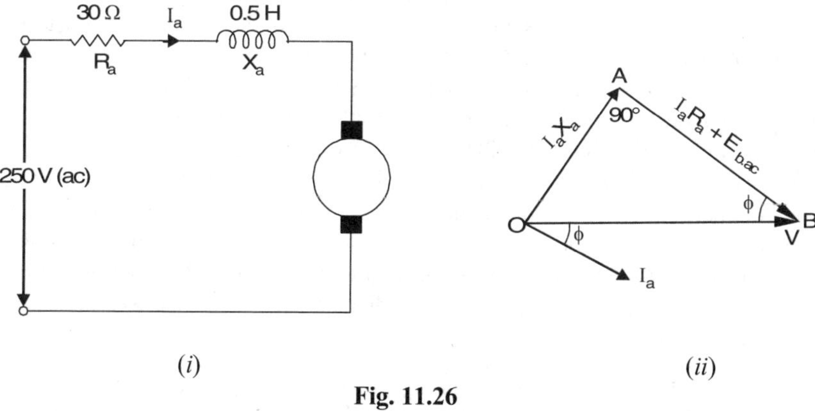

Fig. 11.26

* However, in a d.c. series motor, the speed at no load may rise to a dangerously high value.

Referring to the phasor diagram shown in Fig. 11.26 (*ii*), we have,

$$V^2 = (E_{b.ac} + I_a R_a)^2 + (I_a X_a)^2$$

or $$(250)^2 = (E_{b.ac} + 24)^2 + (125.6)^2$$

On solving, we have, $E_{b.ac} = 192.6$ volts

D.C. operation. For d.c. operation, $X_a = 0$.

$$\therefore \quad E_{b.dc} = 250 - 0.8 \times 30 = 226 \text{ volts}$$

Now $$\frac{E_{b.ac}}{E_{b.dc}} = \frac{N_{ac}}{N_{dc}} \text{ or } \frac{192.6}{226} = \frac{N_{ac}}{2000} \qquad \therefore N_{ac} = \mathbf{1700 \text{ r.p.m.}}$$

Referring to the phasor diagram in Fig. 11.26 (*ii*), we have,

$$\text{Power factor, } \cos\phi = \frac{E_{b.ac} + I_a R_a}{V} = \frac{216.6}{250} = \mathbf{0.866} \textit{ lag}$$

11.12 SINGLE-PHASE REPULSION MOTOR

A repulsion motor is similar to an a.c. series motor except that :

(*i*) brushes are not connected to supply but are short-circuited [See Fig. 11.27]. Consequently, currents are induced in the armature conductors by *transformer action.

(*ii*) the field structure has non-salient pole construction.

By adjusting the position of short-circuited brushes on the commutator, the starting torque can be developed in the motor.

Construction. The field of stator winding is wound like the main winding of a split-phase motor and is connected directly to a single-phase source. The armature or rotor is similar to a d.c. motor armature with drum type winding connected to a commutator (not shown in the figure). However, the brushes are not connected to supply but are connected to each other or short-circuited. Short-circuiting the brushes effectively makes the rotor into a type of squirrel cage. The major difficulty with an ordinary single-phase induction motor is the low starting torque. By using a commutator motor with brushes short-circuited, it is possible to vary the starting torque by changing the brush axis. It has also better power factor than the conventional single-phase motor.

Principle of operation. The principle of operation is illustrated in Fig. 11.27 which shows a two-pole repulsion motor with its two short-circuited brushes. The two drawings of Fig. 11.27 represent a time at which the field current is *increasing* in the direction shown so that the left-hand pole is *N*-pole and the right-hand pole is *S*-pole at the instant shown.

(*i*) In Fig. 11.27 (*i*), the brush axis is parallel to the stator field. When the stator winding is energised from single-phase supply, e.m.f. is induced in the armature conductors (rotor) by induction. By Lenz's law, the direction of the e.m.f. is such that the magnetic effect of the resulting armature currents will oppose the increase in flux. The direction of current in armature conductors will be as shown in Fig. 11.27 (*i*). With the brush axis in the position shown in Fig. 11.27 (*i*), current will flow from brush *B* to brush *A* where it enters the armature and flows back to brush *B* through the two paths *ACB* and *ADB*. With brushes set in this position, half of the armature conductors under the *N*-pole carry current inward and half carry current outward. The same is true under *S*-pole. Therefore, as much torque is developed in one direction as in the other and the armature remains stationary. The armature will also remain stationary if the brush axis is perpendicular to the stator field axis. It is because even then net torque is zero.

---

* In a.c. series motor, the field and armature windings are conductively coupled *i.e.*, the armature current is obtained by conduction from the supply. However, in a repulsion motor, the armature currents are obtained by induction *i.e.*, transformer action.

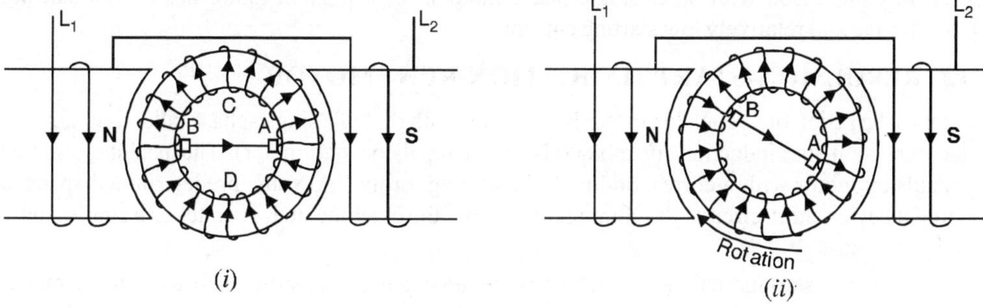

Fig. 11.27

(*ii*) If the brush axis is at some angle other than 0° or 90° to the axis of the stator field, a net torque is developed on the rotor and the rotor accelerates to its final speed. Fig. 11.27 (*ii*) represents the motor at the same instant as that in Fig. 11.27 (*i*) but the brushes have been shifted clockwise through some angle from the stator field axis. Now e.m.f. is still induced in the direction indicated in Fig. 11.27 (*i*) and current flows through the two paths of the armature winding from brush *A* to brush *B*. However, because of the new brush positions, the greater part of the conductors under the *N*-pole carry current in one direction while the greater part of conductors under *S*-pole carry current in the opposite direction. With brushes in the position shown in Fig. 11.27 (*ii*), torque is developed in the clockwise direction and the rotor quickly attains the final speed.

(*iii*) The direction of rotation of the rotor depends upon the direction in which the brushes are shifted. If the brushes are shifted in clockwise direction from the stator field axis, the net torque acts in the clockwise direction and the rotor accelerates in the clockwise direction. If the brushes are shifted in anti-clockwise direction as in Fig. 11.28, the armature current under the pole faces is reversed and the net torque is developed in the anticlockwise direction. Thus a repulsion motor may be made to rotate in either direction depending upon the direction in which the brushes are shifted.

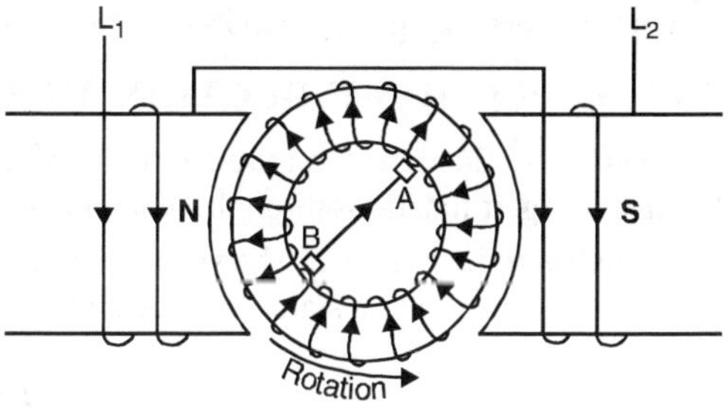

Fig. 11.28

(*iv*) The total armature torque in a repulsion motor can be shown to be

$$T_a \propto \sin 2\alpha$$

where α = angle between brush axis and stator field axis

For maximum torque, $2\alpha = 90°$ or $\alpha = 45°$

Thus adjusting α to 45° at starting, maximum torque can be obtained during the starting period. However, α has to be adjusted to give a suitable running speed.

Characteristics

(*i*) The repulsion motor has characteristics very similar to those of an a.c. series motor *i.e.*, it has a high starting torque and a high speed at no load.

(*ii*) The speed which the repulsion motor develops for any given load will depend upon the *position of the brushes.

* Shifting the brush position has the same effect as changing the number of turns of the series winding of an a.c. series motor *i.e.*, speed of the motor is affected.

(*iii*) In comparison with other single-phase motors, the repulsion motor has a high starting torque and relatively low starting current.

11.13 REPULSION-START INDUCTION-RUN MOTOR

Sometimes the action of a repulsion motor is combined with that of a single-phase induction motor to produce repulsion-start induction-run motor (also called repulsion-start motor). The machine is started as a repulsion motor with a corresponding high starting torque. At some predetermined speed, a centrifugal device short-circuits the commutator so that the machine then operates as a single-phase induction motor.

The repulsion-start induction-run motor has the same general construction of a repulsion motor. The only difference is that in addition to the basic repulsion-motor construction, it is equipped with a centrifugal device fitted on the armature shaft. When the motor reaches 75% of its full running speed, the centrifugal device forces a short-circuiting ring to come in contact with the inner surface of the commutator. This short-circuits all the commutator bars. The rotor then resembles squirrel-cage type and the motor runs as a single-phase induction motor. At the same time, the centrifugal device raises the brushes from the commutator which reduces the wear of the brushes and commutator as well as makes the operation quiet.

Characteristics

(*i*) The starting torque is 2·5 to 4·5 times the full-load torque and the starting current is 3·75 times the full-load value.

(*ii*) Due to their high starting torque, repulsion-motors were used to operate devices such as refrigerators, pumps, compressors etc.

However, they posed a serious problem of maintenance of brushes, commutator and the centrifugal device. Consequently, manufacturers have stopped making them in view of the development of capacitor motors which are small in size, reliable and low-priced.

11.14 REPULSION-INDUCTION MOTOR

The repulsion-induction motor produces a high starting torque entirely due to repulsion motor action. When running, it functions through a combination of induction-motor and repulsion motor action.

Construction. Fig. 11.29 shows the connections of a 4-pole repulsion-induction motor for 230 V operation. It consists of a stator and a rotor (or armature).

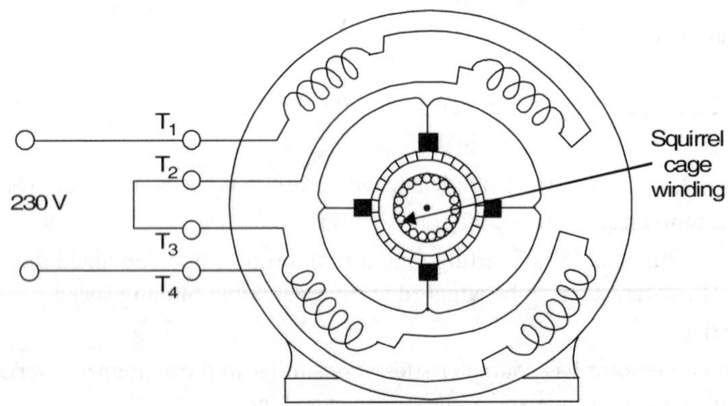

Fig. 11.29

(*i*) The **stator** carries a single distributed winding fed from single-phase supply.

(*ii*) The **rotor** is provided with two independent windings placed one inside the other. The inner winding is a squirrel-cage winding with rotor bars permanently short-circuited.

Placed over the squirrel cage winding is a repulsion commutator armature winding. The repulsion winding is connected to a commutator on which ride short-circuited brushes. There is no centrifugal device and the repulsion winding functions at all times.

Operation.

(*i*) When single-phase supply is given to the stator winding, the repulsion winding (*i.e.,* outer winding) is *active. Consequently, the motor starts as a repulsion motor with a corresponding high starting torque.

(*ii*) As the motor speed increases, the current shifts from the outer to inner winding due to the decreasing impedance of the inner winding with increasing speed. Consequently, at running speed, the squirrel cage winding carries the greater part of rotor current. This shifting of repulsion-motor action to induction-motor action is thus achieved without any switching arrangement.

(*iii*) It may be seen that the motor starts as a repulsion motor. When running, it functions through a combination of principle of induction and repulsion ; the former being predominant.

Characteristics

(*i*) The no-load speed of a repulsion-induction motor is somewhat above the synchronous speed because of the effect of repulsion winding. However, the speed at full-load is slightly less than the synchronous speed as in an induction motor.

(*ii*) The speed regulation of the motor is about 6 %.

(*iii*) The starting torque is 2·25 to 3 times the full-load torque ; the lower value being for large motors. The starting current is 3 to 4 times the full-load current.

This type of motor is used for applications requiring a high starting torque with essentially a constant running speed. The common sizes are 0·25 to 5 h.p.

11.15 SINGLE-PHASE SYNCHRONOUS MOTORS

Very small single-phase motors have been developed which run at true synchronous speed. They do not require d.c. excitation for the rotor. Because of these characteristics, they are called *unexcited single-phase synchronous motors*. The most commonly used types are :

(*i*) Reluctance motors (*ii*) Hysteresis motors

The efficiency and torque-developing ability of these motors is low. The output of most of the commercial motors is only a few watts.

11.16 RELUCTANCE MOTOR

It is a single-phase synchronous motor which does not require d.c. excitation to the rotor. Its operation is based upon the following principle :

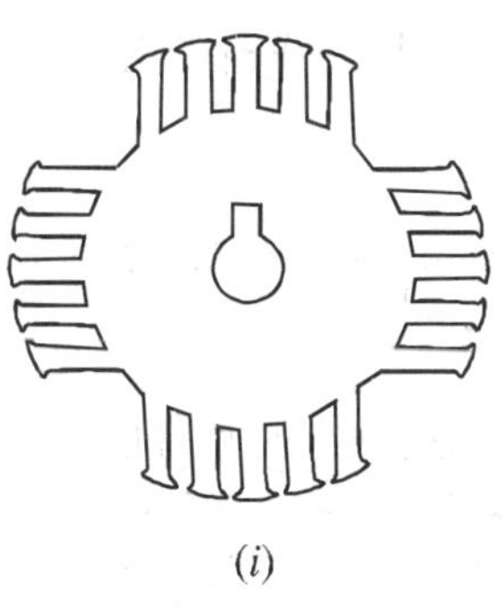

(*i*)

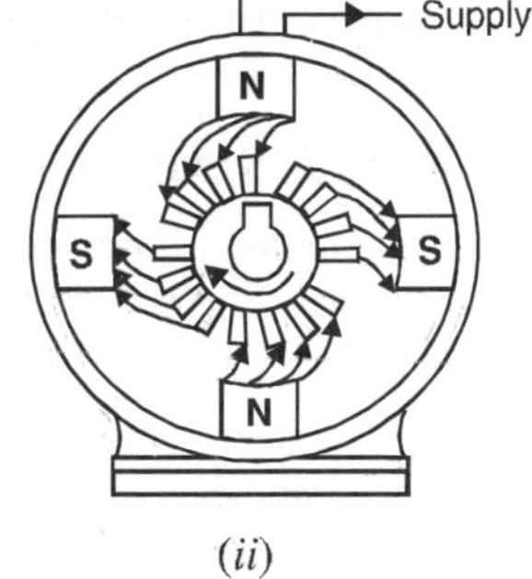

(*ii*)

Fig. 11.30

* The inner winding (*i.e.*, squirrel cage winding), being nearly surrounded by iron, has a high inductance. At starting, rotor frequency is equal to the supply frequency and the impedance of the outer winding is less than that of the inner winding. This results in a large proportion of rotor current flowing in the outer winding.

Whenever a piece of ferromagnetic material is located in a magnetic field, a force is exerted on the material, tending to align the material so that reluctance of the magnetic path that passes through the material is minimum.

Construction. A reluctance motor (also called synchronous reluctance motor) consists of :

(*i*) a **stator** carrying a single-phase winding along with an auxiliary winding to produce a synchronous-revolving magnetic field.

(*ii*) a **squirrel-cage rotor** having unsymmetrical magnetic construction. This is achieved by symmetrically removing some of the teeth from the squirrel-cage rotor to produce salient poles on the rotor. As shown in Fig. 11.30 (*i*), 4 salient poles have been produced on the rotor. The salient poles created on the rotor must be equal to the poles on the stator.

Note that rotor salient poles offer low reluctance to the stator flux and, therefore, become strongly magnetised.

Operation.

(*i*) When single-phase stator having an auxiliary winding is energised, a synchronously-revolving field is produced. The motor starts as a standard squirrel-cage induction motor and will accelerate to near its synchronous speed.

(*ii*) As the rotor approaches synchronous speed, the rotating stator flux will exert **reluctance torque* on the rotor poles tending to align the salient-pole axis with the axis of the rotating field. The rotor assumes a position where its salient poles lock with the poles of the revolving field [See Fig. 11.30 (*ii*)]. Consequently, the motor will continue to run at the speed of revolving flux *i.e.*, at the synchronous speed.

(*iii*) When we apply a mechanical load, the rotor poles fall slightly behind the stator poles, while continuing to turn at synchronous speed. As the load on the motor is increased, the mechanical angle between the poles increases progressively. Nevertheless, magnetic attraction keeps the rotor locked to the rotating flux. If the load is increased beyond the amount under which the reluctance torque can maintain synchronous speed, the rotor drops out of step with the revolving field. The speed, then, drops to some value at which the slip is sufficient to develop the necessary torque to drive the load by induction-motor action.

Characteristics

(*i*) These motors have poor torque, power factor and efficiency.

(*ii*) These motors cannot accelerate high-inertia loads to synchronous speed.

(*iii*) The pull-in and pull-out torques of such motors are weak.

Despite the above drawbacks, the reluctance motor is cheaper than any other type of synchronous motor. They are widely used for constant-speed applications such as timing devices, signalling devices etc.

11.17 HYSTERESIS MOTOR

It is a single-phase motor whose operation depends upon the hysteresis effect *i.e.*, magnetisation produced in a ferromagnetic material lags behind the magnetising force.

Construction. It consists of :

(*i*) a **stator** designed to produce a synchronously-revolving field from a single-phase supply. This is accomplished by using permanent-split capacitor type construction. Consequently, both the windings (*i.e.*, starting as well as main winding) remain connected in the circuit during running operation as well as at starting. The value of capacitance is so adjusted as to result in a flux revolving at synchronous speed.

---

* The reluctance torque arises from the tendency of the rotor to align itself in the minimum reluctance position *w.r.t.* the synchronously-revolving stator flux.

(*ii*) a **rotor** consisting of a smooth cylinder of magnetically hard steel, without winding or teeth.

Operation.

(*i*) When the stator is energised from a single-phase supply, a synchronously-revolving field (assumed in anticlockwise direction) is produced due to split-phase operation.

(*ii*) The revolving stator flux magnetises the rotor. Due to hysteresis effect, the axis of magnetisation of rotor will lag behind the axis of stator field by hysteresis lag angle α as shown in Fig. 11.31. Thus the rotor and stator poles are locked. If the rotor is stationary, the starting torque produced is given by :

$$T_s \propto \phi_s \, \phi_r \sin \alpha$$

where ϕ_s = stator flux

ϕ_r = rotor flux

From now onwards, the rotor accelerates to synchronous speed with a *uniform torque.

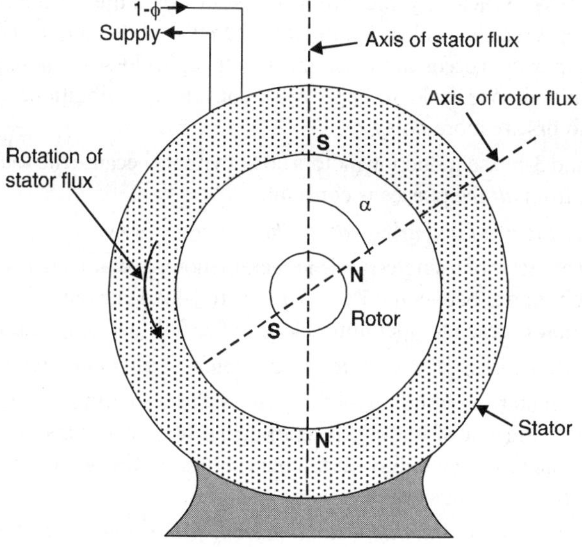

Fig. 11.31

(*iii*) After reaching synchronism, the motor continues to run at synchronous speed and adjusts its torque angle so as to develop the torque required by the load.

Characteristics

(*i*) A hysteresis motor can synchronise any load which it can accelerate, no matter how great the inertia. It is because the torque is uniform from standstill to synchronous speed.

(*ii*) Since the rotor has no teeth or salient poles or winding, a hysteresis motor is inherently quiet and produces smooth rotation of the load.

(*iii*) The rotor takes on the same number of poles as the stator field. Thus by changing the number of stator poles through pole-changing connections, wc can get a set of synchronous speeds for the motor.

Applications. Due to their quiet operation and ability to drive high-inertia loads, hysteresis motors are particularly well suited for driving (*i*) electric clocks (*ii*) timing devices (*iii*) tape-decks (*iv*) turn-tables and other precision audio-equipment.

---

* Now ϕ_s and ϕ_r are constant. When the rotor accelerates, the lag angle α remains constant since the angle α depends merely upon the hysteresis loop of the rotor and is independent of the rate at which the loop is traversed. The motor, therefore, develops constant torque from standstill to synchronous speed.

SHORT ANSWER QUESTIONS

Q.1. *Where do we require single-phase induction motors?*

Ans. Single-phase induction motors are required where

(*i*) 3-phase supply is not available (*ii*) efficiency is of lesser importance

(*iii*) equipment is portable (*iv*) rating is less than one h.p.

Q. 2. *Why are single-phase induction motors more noisy than 3-phase induction motors?*

Ans. It is due to the fact that a single-phase induction motor always receives a *pulsating* power whereas it delivers *constant* mechanical power. Consider a 250 W single-phase induction motor. The instantaneous input electric power is pulsating ; becoming positive, negative and zero depending upon resistance and inductive reactance of the motor circuit. However, whether the instantaneous electric power is positive, negative or zero, the mechanical power delivered is always 250 W. Obviously, the motor will slow down during the brief periods when the electric power it receives is less than 250 W. On the other hand, it will accelerate when the input electric power exceeds the mechanical output plus losses. These periods of acceleration and deceleration in one cycle of supply cause considerable vibrations and hence single-phase induction motors are more noisy.

Two-phase and 3-phase induction motors do not vibrate because the total instantaneous power they receive from *all* the phases is constant.

Q. 3. *Why is power factor of a single-phase induction motor low?*

Ans. The low power factor of a single-phase induction motor is mainly due to the large magnetising current which ranges between 60% to 70% of full-load current. Consequently, even at no-load, these motors reach temperatures close to the full-load temperature.

Q. 4. *How is starting winding disconnected when single-phase induction motor runs upto speed?*

Ans. The purpose of starting winding is that it provides a rotating flux in conjunction with the main winding at the time of starting. A centrifugal switch (which is speed-sensitive) is connected in series with the starting winding. It disconnects the starting winding when the motor reaches about 75% of synchronous speed.

Q. 5. *What is the function of centrifugal starting switch in a single-phase induction motor?*

Ans. The centrifugal switch is connected in series with the starting winding. The primary function of the starting winding is to produce rotating flux in conjunction with main winding at the start. When the motor has started and reaches nearly 75% of synchronous speed, it produces its own rotating field from the cross field effect. The starting winding now has no function to perform and is removed from the circuit by a centrifugally operated switch.

Q. 6. *What happens when the centrifugal starting switch fails to open?*

Ans. Most of the troubles in single-phase induction motors are traceable to the centrifugal switch connected in series with the starting winding. In the normal use, the contact gap of the switch changes due to arcing and pitting. If the starting switch fails to open when needed, the starting winding will almost always overheat and burn out. If *next* attempt is made to start the motor, it will result in a growling sound and no rotation. This can cause failure of main winding unless the fuse blows out.

Q. 7. *What happens when the centrifugal starting switch fails to close?*

Ans. If the centrifugal starting switch fails to close, the motor will overheat the main winding without any failure of the main winding.

Q. 8. *Why are resistance split-phase induction motors most popular?*

Ans. Due to their low cost, resistance split-phase induction motors are the most popular motors in the market. They are used where a moderate starting torque is required and where the starting periods are infrequent. They drive fans, pumps, washing machines, small machine tools etc. The power rating of such motors usually lies between 60 W and 250 W (1/12 h.p. to 1/3 h.p.).

Q. 9. *What is the drawback of resistance split-phase induction motor?*

Ans. In a resistance split-phase induction motor, the starting winding has a relatively small number of turns of fine wire. Consequently, its resistance is higher than that of the main winding. Owing to the fine wire used on the starting winding, the current density is high and the winding heats up quickly. If the starting period lasts for more than 5 seconds, the winding begins to smoke and may burn out unless the motor is protected by a built-in thermal relay.

Q. 10. *Why is the starting torque of a resistance split-phase motor not high?*

Ans. The starting torque T_s is given by ;

$T_s = k\, I_s\, I_m \sin \alpha$

(*i*) The angle α between I_s and I_m is small ($\simeq 25^{\circ}$) in a resistance split-phase motor. Due to small value of α, the starting torque is small.

(*ii*) Since currents I_m and I_s are not equal in magnitude, the rotating magnetic field is not uniform and the starting torque produced is small.

Q. 11. *Why is the starting torque of a capacitor start induction motor high?*

Ans. The capacitor C in the starting winding is so chosen that I_s leads I_m by about 75°. Since the starting torque is directly proportional to sin α, it is quite high in capacitor-start induction motors.

Q. 12. *The capacitor-start motor is also a split-phase motor. Why is the term split-phase not used?*

Ans. The capacitor-start motor is actually a split-phase motor but it is not designated as such. The reason for this is that resistance phase-splitting method was developed and put to use before the capacitor type. The use of the term split-phase, therefore, has become established for the resistance split-phase machine. To avoid confusion, the term split-phase is not used to designate the capacitor motor.

Q. 13. *Why do we use capacitor-start induction motors in applications requiring high starting torque in preference to repulsion induction motors?*

Ans. The wide use of capacitor-start motor in preference to repulsion-induction motor is a direct result of the availability of small, reliable and low cost electrolytic capacitors. Prior to the development of these capacitors, repulsion-induction motors had to be installed whenever a high starting torque was required. Repulsion-induction motors possess a special commutator and brushes that require considerable maintenance. Most manufacturers have stopped making them.

Q. 14. *Why is capacitor-start induction motor well suited for applications involving either frequent or prolonged starting periods?*

Ans. In a capacitor-start induction motor, the capacitor is so chosen that I_s (current in starting winding) leads I_m (current in main winding) by about 80° which is considerably more than the 25° found in resistance split-phase motor. Consequently, for equal starting torque, the current I_s in the starting winding is only about half that in a split-phase motor. Owing to the high starting torque and the relatively low value of I_s, the capacitor-start motor is well suited to applications involving either frequent or prolonged starting periods.

Q. 15. *What is the principle of operation of shaded-pole motor?*

Ans. A shaded-pole motor is basically a small single-phase squirrel cage motor in which the starting winding is composed of short-circuited copper ring (called shading coil) surrounding one-half of each pole. The effect of the shading coil is to cause a flux to sweep across the pole faces, from unshaded to shaded portion of the pole, producing a weak rotating magnetic field. Consequently, the rotor is set in motion due to induction principle.

MULTIPLE-CHOICE QUESTIONS

1. A single-phase induction motor employs .... rotor.
   (i) squirrel cage (ii) wound
   (iii) either squirrel cage or wound
   (iv) none of the above
2. For the same rating, the size of a single-phase induction motor is about .... that of a 3-phase induction motor.
   (i) 3 times (ii) less than
   (iii) 1·5 times (iv) 0·33 times
3. For the same rating, the p.f. of a single-phase induction motor is .... that of 3-phase induction motor.
   (i) the same as (ii) less than
   (iii) more than (iv) none of the above
4. For the same rating, the efficiency of a single-phase induction motor is .... that of 3-phase induction motor.
   (i) less than (ii) the same as
   (iii) more than (iv) none of the above
5. The starting winding of a single-phase induction motor has .... that of main winding.
   (i) more poles than (ii) less poles than
   (iii) same number of poles as
   (iv) none of the above
6. A 50 Hz, 4-pole, single-phase induction motor will have a synchronous speed of ....
   (i) 1500 r.p.m. (ii) 750 r.p.m.
   (iii) 1200 r.p.m. (iv) none of the above
7. A 4-pole, 50 Hz, single-phase induction motor has a slip of 5 %. The speed of the motor will be ....
   (i) 1500 r.p.m. (ii) 1425 r.p.m.
   (iii) 1200 r.p.m. (iv) none of the above
8. The purpose of starting winding in a single-phase induction motor is to ....
   (i) reduce losses
   (ii) limit temperature rise of the machine
   (iii) produce rotating flux in conjunction with main winding
   (iv) none of the above

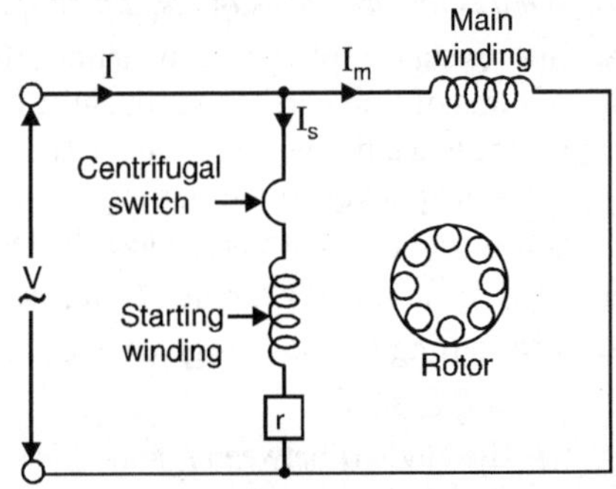

Fig. 11.32

9. In the resistance split-phase induction motor shown in Fig. 11.32, the main winding has ....
   (i) high resistance and high inductance
   (ii) high resistance and low inductance
   (iii) low resistance and high inductance
   (iv) low resistance and low inductance
10. At starting, the current I_s in the starting winding shown in Fig. 11.32 ....
    (i) lags V by 90° (ii) leads V by 90°
    (iii) is nearly in phase with V
    (iv) leads V by 75°
11. The resistance split-phase induction motor has .... starting torque.
    (i) high (ii) moderate
    (iii) very low (iv) none of the above
12. In the capacitor-start induction motor shown in Fig. 11.33, the angle α between I_m and I_s is ....
    (i) exactly 90° (ii) about 30°
    (iii) about 40° (iv) about 75°

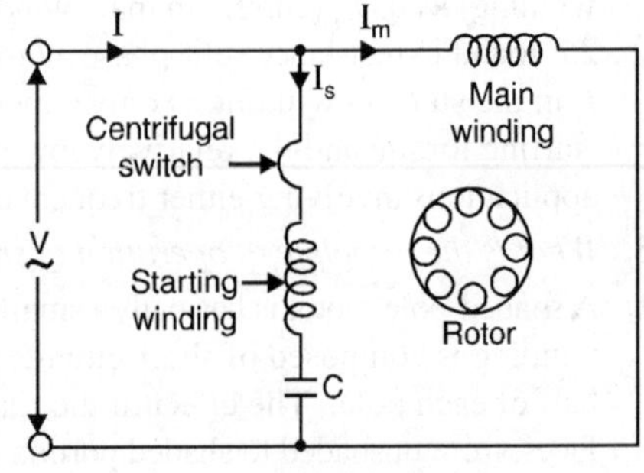

Fig. 11.33

13. The value of current I_s in capacitor-start motor shown in Fig. 11.33 is .... the current I_m.

(*i*) equal to (*ii*) less than
(*iii*) more than (*iv*) none of the above

14. The capacitor *C* used in the capacitor-start motor shown in Fig. 11.33 is always a .... capacitor.

(*i*) paper (*ii*) ceramic
(*iii*) mica (*iv*) electrolytic

15. A capacitor-start, capacitor-run motor has ....

(*i*) low power factor
(*ii*) high power factor
(*iii*) low efficiency
(*iv*) high starting torque

16. The capacitor-start, capacitor-run induction motor acts as a true 2-phase motor at ....

(*i*) starting (*ii*) no load
(*iii*) all loads (*iv*) full-load

17. The full-load efficiency of a shaded-pole motor is about ....

(*i*) 70 to 80 % (*ii*) 60 to 70 %
(*iii*) 30 to 35 % (*iv*) 5 to 10 %

18. The single-phase series motor can operate on ....

(*i*) a.c. only (*ii*) d.c. only
(*iii*) both a.c. and d.c.
(*iv*) none of the above

19. For the same rating, a d.c. series motor costs .... single-phase series motor.

(*i*) about the same as
(*ii*) more than (*iii*) less than
(*iv*) none of the above

20. The best suited motor to drive 1/4 h.p. fan in a hospital ward would be .... motor.

(*i*) shaded-pole
(*ii*) single-phase series
(*iii*) hysteresis
(*iv*) capacitor-run

21. The best suited motor to drive 3/4 h.p. air compressor would be .... motor.

(*i*) capacitor-start
(*ii*) single-phase series
(*iii*) shaded-pole
(*iv*) resistance split-phase

22. The least expensive fractional horse power motor is .... motor.

(*i*) shaded-pole (*ii*) capacitor-start
(*iii*) split-phase (*iv*) a.c. series

23. The washing machine generally employs .... motor.

(*i*) single-phase series
(*ii*) resistance split-phase
(*iii*) shaded-pole
(*iv*) hysteresis

24. The nameplate of a single-phase, 4-pole induction motor gives the following data :

Output 373 W ; 230 V ; 50 Hz
Input current 2·9 A ; Power factor 0·71
Speed 1410 r.p.m.

The full-load slip of the motor is

(*i*) 6 % (*ii*) 2 %
(*iii*) 4 % (*iv*) 3 %

25. In the above question, what is the efficiency of the motor ?

(*i*) 65·3% (*ii*) 78·7%
(*iii*) 89·2% (*iv*) 70·6%

ANSWERS TO MULTIPLE-CHOICE QUESTIONS

1. (*i*)	**2.** (*iii*)	**3.** (*ii*)	**4.** (*i*)	**5.** (*iii*)
6. (*i*)	**7.** (*ii*)	**8.** (*iii*)	**9.** (*iii*)	**10.** (*iii*)
11. (*ii*)	**12.** (*iv*)	**13.** (*ii*)	**14.** (*iv*)	**15.** (*ii*)
16. (*iv*)	**17.** (*iii*)	**18.** (*iii*)	**19.** (*iii*)	**20.** (*iv*)
21. (*i*)	**22.** (*i*)	**23.** (*ii*)	**24.** (*i*)	**25.** (*ii*)

HINTS TO SELECTED MULTIPLE-CHOICE QUESTIONS

1. Structurally, a single-phase induction motor resembles the 3-phase squirrel cage motor except that the stator winding is a single-phase winding.

5. The stator of a single-phase induction motor carries the *main winding* which creates a set of *N*-*S* poles. It also carries a smaller *starting winding* that only operates during the brief period when the motor starts up. The starting winding has the same number of poles as the main winding has.

6.
$$N_s = \frac{120\,f}{P} = \frac{120 \times 50}{4} = 1500 \text{ r.p.m.}$$

7.
$$N_s = \frac{120\,f}{P} = \frac{120 \times 50}{4} = 1500 \text{ r.p.m.}$$

$$s = \frac{N_s - N}{N_s} \times 100 \; or \; 5 = \frac{1500 - N}{1500} \times 100$$

$$\therefore \quad N = 1425 \text{ r.p.m.}$$

8. If a single-phase induction motor has main winding on the stator only, it will not be self-starting. To produce starting torque, we must somehow create a revolving field. This is done by placing an additional auxiliary winding on the stator. Both main and auxiliary windings (also called starting winding) are connected in parallel across the single-phase supply. The currents in the two windings have a phase difference and hence produce a weak revolving field.

9. The main winding of a resistance split-phase motor :

(*i*) is always made of relatively large wire to reduce I^2R losses

(*ii*) has a relatively large number of turns

Due to these reasons, the main winding has low resistance and high inductance.

10. The starting winding of a resistance split-phase induction motor has a relatively small number of turns of fine wire. Consequently, the starting winding has high resistance and low inductance. The result is that at starting, the current I_s in the starting winding is nearly in phase with the applied voltage *V*.

12. The capacitor *C* is so chosen that current I_s in the starting winding leads the current I_m in the main winding by about 75°.

13. In a capacitor-start motor, the starting winding has about as many turns as the main winding. Furthermore, a capacitor is connected in series with the starting winding. For this reason, the value of I_s is less than I_m at starting.

14. Because these capacitors are available in small size and are quite reliable. Moreover, they are also not expensive.

15. The capacitor-start, capacitor-run motor is similar to the capacitor-start type except that the starting winding remains in the circuit after the motor is started. Consequently, such a motor operates at a higher power factor.

17. The shading coils are not opened after the motor is running at normal speed. Therefore, considerable power loss takes place in the shading coils. For this reason, the efficiency of such a motor is very low.

19. The sparking at the brushes of an a.c. series motor is caused by the voltage induced by the alternating main field in the coils short-circuited by the brushes during commutation period. To reduce sparking in an a.c. series motor, the number of turns in each armature coil is kept small and high-resistance leads connect each coil to the commutator segments. As a result, the armature winding of an a.c. series motor has more coils and, therefore, more commutator segments than a comparable d.c. series motor. For this reason, a d.c. series motor costs less than the comparable a.c. series motor.

20. The capacitor-start, capacitor-run motor acts as a true 2-phase motor only when it operates on full load. The motor is then vibration-free. This particularly quiet motor is used to drive fixed loads in hospitals, studios and other places where silence is important.

21. The capacitor-start induction motor has a high starting torque and is, therefore, well suited to drive air compressors, large fans, pumps and high-inertia loads.

24. Synchronous speed, $N_s = 120f/P = 120 \times 50/4 = 1500$ r.p.m.

$$\text{Motor slip, } s = \frac{N_s - N}{N_s} = \frac{1500 - 1410}{1500} = 0{\cdot}06 \text{ or } 6\%$$

25.

$$\text{Input to motor} = VI\cos\phi = 230 \times 2{\cdot}9 \times 0{\cdot}71 = 474 \text{ W}$$

$$\text{Motor output} = 373 \text{ W}$$

$$\therefore \quad \text{Efficiency} = \frac{\text{Output}}{\text{Input}} = \frac{373}{474} = 0{\cdot}787 \text{ or } 78{\cdot}7\ \%$$

12

Alternators

INTRODUCTION

A.C. system has a number of advantages over d.c. system. These days 3-phase a.c. system is being exclusively used for generation, transmission and distribution of power. The machine which produces 3-phase power from mechanical power is called an alternator or synchronous generator. Alternators are the primary source of all the electrical energy we consume. These machines are the largest energy converters found in the world. They convert mechanical energy into a.c. energy. In this chapter, we shall discuss the construction and characteristics of alternators.

12.1 ALTERNATOR

An *alternator operates on the same fundamental principle of electromagnetic induction as a d.c. generator *i.e.*, when the flux linking a conductor changes, an e.m.f. is induced in the conductor. Like a d.c. generator, an alternator also has an armature winding and a field winding. But there is one important difference between the two. In a d.c. generator, the armature winding is placed on the rotor in order to provide a way of converting alternating voltage generated in the winding to a direct voltage at the terminals through the use of a rotating commutator. The field poles are placed on the stationary part of the machine. Since no commutator is required in an alternator, it is usually more convenient and advantageous to place the field winding on the rotating part (*i.e.*, rotor) and armature winding on the stationary part (*i.e.*, stator) as shown in Fig. 12.1.

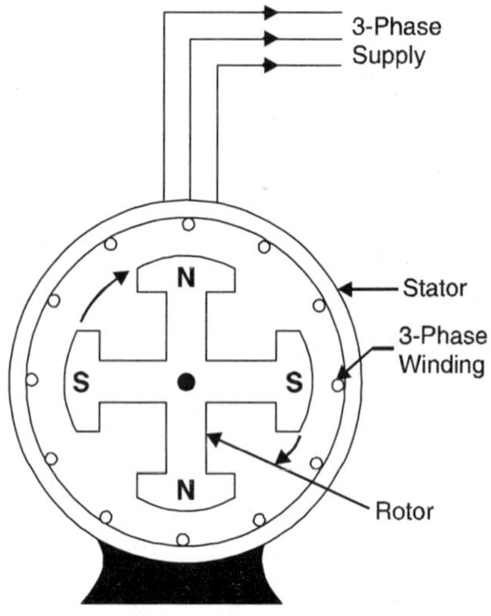

Fig. 12.1

Advantages of stationary armature. The field winding of an alternator is placed on the rotor and is connected to d.c. supply through two slip rings. The 3-phase armature winding is placed on the stator. This arrangement has the following advantages :

(*i*) It is easier to insulate stationary winding for high voltages for which the alternators are usually designed. It is because they are not subjected to centrifugal forces and also extra space is available due to the stationary arrangement of the armature.

(*ii*) The stationary 3-phase armature can be directly connected to load without going through large, unreliable slip rings and brushes.

(*iii*) Only two slip rings are required for d.c. supply to the field winding on the rotor. Since the exciting current is small, the slip rings and brush gear required are of light construction.

(*iv*) Due to simple and robust construction of the rotor, higher speed of rotating d.c. field is possible. This increases the output obtainable from a machine of given dimensions.

Note : All alternators above 5 kVA employ a stationary armature (or stator) and a revolving d.c. field.

---

* In this chapter, alternator means 3-phase alternator unless stated otherwise.

12.2 CONSTRUCTION OF ALTERNATOR

An alternator has 3-phase winding on the stator and a d.c. field winding on the rotor.

1. Stator. It is the stationary part of the machine and is built up of sheet-steel laminations having slots on its inner periphery. A 3-phase winding is placed in these slots and serves as the armature winding of the alternator. The armature winding is always connected in *star and the neutral is connected to **ground.

2. Rotor. The rotor carries a field winding which is supplied with direct current through two slip rings by a separate d.c. source. This d.c. source (called exciter) is generally a small d.c. shunt or compound generator mounted on the shaft of the alternator. Rotor construction is of two types, namely ;

(*i*) Salient (or projecting) pole type

(*ii*) Non-salient (or cylindrical) pole type

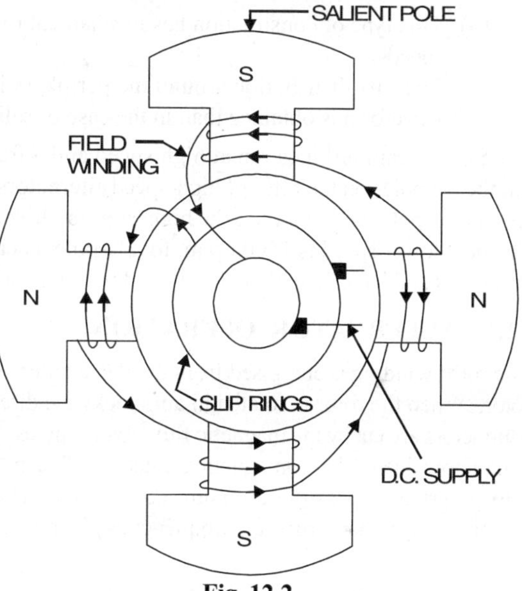

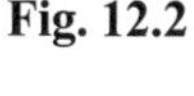

Fig. 12.2

(*i*) **Salient pole type.** In this type, salient or projecting poles are mounted on a large circular steel frame which is fixed to the shaft of the alternator as shown in Fig. 12.2. The individual field pole windings are connected in series in such a way that when the field winding is energised by the d.c. exciter, adjacent poles have opposite polarities.

Low and medium-speed alternators (120-400 r.p.m.) such as those driven by diesel engines or water turbines have salient pole type rotors due to the following reasons :

(*a*) The salient field poles would cause an excessive windage loss if driven at high speed and would tend to produce noise.

(*b*) Salient-pole construction cannot be made strong enough to withstand the mechanical stresses to which they may be subjected at higher speeds.

Since a frequency of 50 Hz is required, we must use a †large number of poles on the rotor of slow-speed alternators. Low-speed rotors always possess a large diameter to provide the necessary space for the poles. Consequently, salient-pole type rotors have large diameters and short axial lengths.

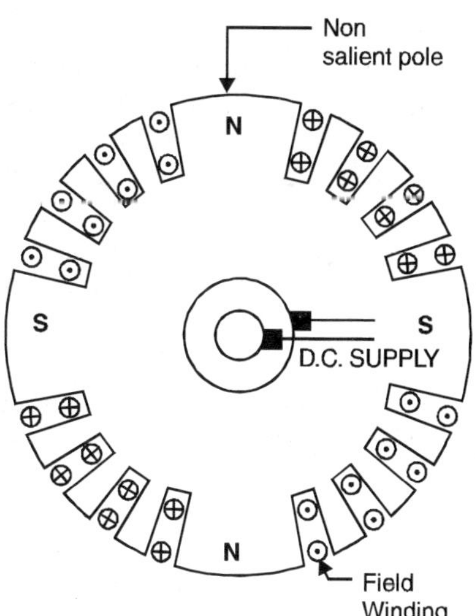

Fig. 12.3

(*ii*) **Non-salient pole type.** In this type, the rotor is made of smooth solid forged-steel radial cylinder having a number of slots along the outer periphery. The field windings are embedded in these slots and are connected in series to the slip rings through which they

* In star connection, phase voltage $V_{ph} = V_L/\sqrt{3}$. Since the induced e.m.f. in the phase winding of an alternator is directly proportional to the number of turns, a star-connected alternator will require less number of turns than a delta-connected alternator for the same line voltage.

** This permits to use protective devices (*e.g.*, relays) to protect the system in case of ground faults.

† $f = NP/120$

are energised by the d.c. exciter. The regions forming the poles are usually left unslotted as shown in Fig. 12.3. It is clear that the poles formed are non-salient *i.e.*, they do not project out from the rotor surface.

High-speed alternators (1500 or 3000 r.p.m.) are driven by steam turbines and use non-salient type rotors due to the following reasons :

(*a*) This type of construction has mechanical robustness and gives noiseless operation at high speeds.

(*b*) The flux distribution around the periphery is nearly a sine wave and hence a better e.m.f. waveform is obtained than in the case of salient-pole type.

Since steam turbines run at high speed and a frequency of 50 Hz is required, we need a small number of poles on the rotor of high-speed alternators (also called turboalternators). We can use not less than 2 poles and this fixes the highest possible speed. For a frequency of 50 Hz, it is 3000 r.p.m. The next lower speed is 1500 r.p.m. for a 4-pole machine. Consequently, turboalternators possess 2 or 4 poles and have *small diameters and very long axial lengths.

12.3 ALTERNATOR OPERATION

The rotor winding is energised from the d.c. exciter and alternate N and S poles are developed on the rotor. When the rotor is rotated in anticlockwise direction by a prime mover, the stator or armature conductors are cut by the magnetic flux of rotor poles. Consequently, e.m.f. is induced in the armature conductors due to electromagnetic induction. The induced e.m.f. is alternating since N and S poles of rotor alternately pass the armature conductors. The direction of induced e.m.f. can be found by Fleming's right hand rule and frequency is given by ;

$$f = \frac{NP}{120} \qquad \text{... See Art. 12.4}$$

where N = speed of rotor in r.p.m.

P = number of rotor poles

The magnitude of the voltage induced in each phase depends upon the rotor flux, the number and position of the conductors in the phase and the speed of the rotor.

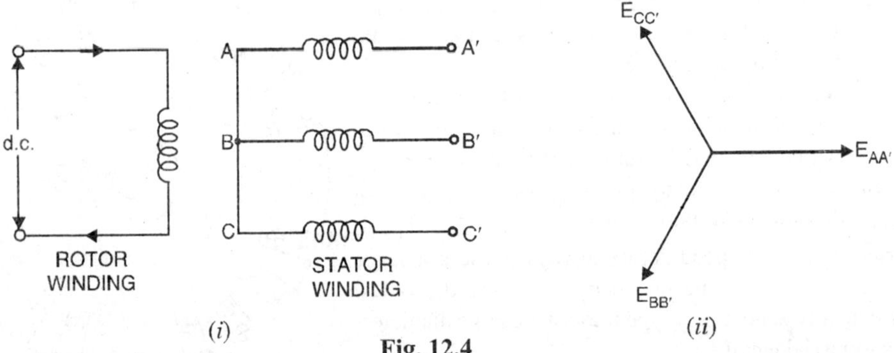

Fig. 12.4

Fig. 12.4 (*i*) shows star-connected armature winding and d.c. field winding. When the rotor is rotated, a 3-phase voltage is induced in the armature winding. The magnitude of induced e.m.f. depends upon the speed of rotation and the d.c. exciting current. The magnitude of e.m.f. in each phase of the armature winding is the same. However, they differ in phase by 120° electrical as shown in the phasor diagram in Fig. 12.4 (*ii*).

---

* The high speed of rotation produces strong centrifugal forces which impose an upper limit on the diameter. On the other hand, to build powerful alternators, we have to use massive rotors. Therefore, high-power and high speed rotors have to be very long.

12.4 FREQUENCY

The frequency of induced e.m.f. in the armature conductors depends upon speed and the number of poles.

Let N = rotor speed in r.p.m.

P = number of rotor poles

f = frequency of e.m.f. in Hz

Consider a stator conductor that is successively swept by the N and S poles of the rotor. If a *positive* voltage is induced when a N-pole sweeps across the conductor, a similar *negative* voltage is induced when a S-pole sweeps by. This means that *one complete cycle* of e.m.f. is generated in the conductor as a pair of poles passes it *i.e.*, one N-pole and the adjacent following S-pole. The same is true for every other armature conductor.

$\therefore$ No. of cycles/revolution = No. of pairs of poles = $P/2$

No. of revolutions/second = $N/60$

$\therefore$ No. of cycles/second = $^{*}(P/2)\,(N/60) = NP/120$

But number of cycles of e.m.f. per second is its frequency.

$$\therefore \quad f = \frac{NP}{120}$$

It may be noted that N is the synchronous speed and is generally represented by N_s. For a given alternator, the number of rotor poles is fixed and, therefore, the alternator must be run at synchronous speed to give an output of desired frequency. For this reason, an alternator is sometimes called *synchronous generator*.

12.5 A.C. ARMATURE WINDINGS

A.C. armature windings are always of the nonsalient-pole type and are usually symmetrically distributed in slots around the complete circumference of the armature. A.C. armature windings are generally **open-circuit type *i.e.*, both ends are brought out. An open-circuit winding is one that does not close on itself *i.e.*, a closed circuit will not be formed until some external connection is made to a source or load. The following are the general features of a.c. armature windings :

(*i*) A.C. armature windings are generally distributed windings *i.e.*, they are symmetrically distributed in slots around the complete circumference of the armature. A distributed winding has two principal advantages. First, a distributed winding generates a voltage wave that is nearly a sine curve. Secondly, copper is evenly distributed on the armature surface. Therefore, heating is more uniform and this type of winding is more easily cooled.

(*ii*) A.C. armature windings may use full-pitch coils or fractional-pitch coils. A coil with a ***span of 180° electrical is called a *full-pitch coil*. In this case, the two sides of the coil occupy identical positions under adjacent opposite poles and the e.m.f. generated in the coil is maximum. A coil with a span of less than 180° electrical is called a *fractional-pitch coil*. For example, a coil with a span of 150° electrical would be called a 5/6 pitch coil. Although e.m.f. induced in a fractional-pitch coil is less than that of a full-pitch coil, fractional-pitch coils are frequently used in a.c. machines for two main reasons. First, less copper is required per coil and secondly the waveform of the generated voltage is improved.

(*iii*) Most of a.c. machines use double layer armature windings. In a double layer winding, one coil side lies in the upper half of one slot while the other coil side lies in the lower half of another slot spaced about one-pole pitch from the first one. This arrangement permits simpler

---

* No. of cycles/second = No. of cycles/revolution × No. of revolutions/second

** D.C. armature windings must be of closed-circuit type in order to provide for the commutation of the coils.

*** The span of a coil is the electrical degrees between the two sides of the coils.

end connections and it is economical to manufacture.

(*iv*) Since most of a.c. machines are of 3-phase type, the three windings of the three phases are identical but spaced 120 electrical degrees apart.

(*v*) A group of adjacent slots belonging to one phase under one pole pair is known as *phase belt.* The angle subtended by a phase belt is known as *phase spread.* The 3-phase windings are always designed for 60° phase spread.

12.6 ARMATURE WINDING OF ALTERNATOR

With very few exceptions, alternators are 3-phase machines because of the advantages of 3-phase service for generation, transmission and distribution. The windings for an alternator are much simpler than that of a d.c. machine because no commutator is used. Fig. 12.5 shows a 2-pole, 3-phase, double-layer, full-pitch, distributed winding for the stator of an alternator. There are 12 slots and each slot contains two coil sides. The coil sides that are placed in adjacent slots belong to the same phase such as a_1, a_3 or a_2, a_4 constitute a *phase belt.* Note that in a 3-phase machine, phase belt is always 60° electrical. Since the winding has double-layer arrangement, one side of a coil, such as a_1 is placed at the bottom of a slot and the other side $- a_1$ is placed at the top of another slot spaced one pole pitch apart. Note that each coil has a span of a full pole pitch or 180 electrical degrees. Therefore, the winding is a full-pitch winding.

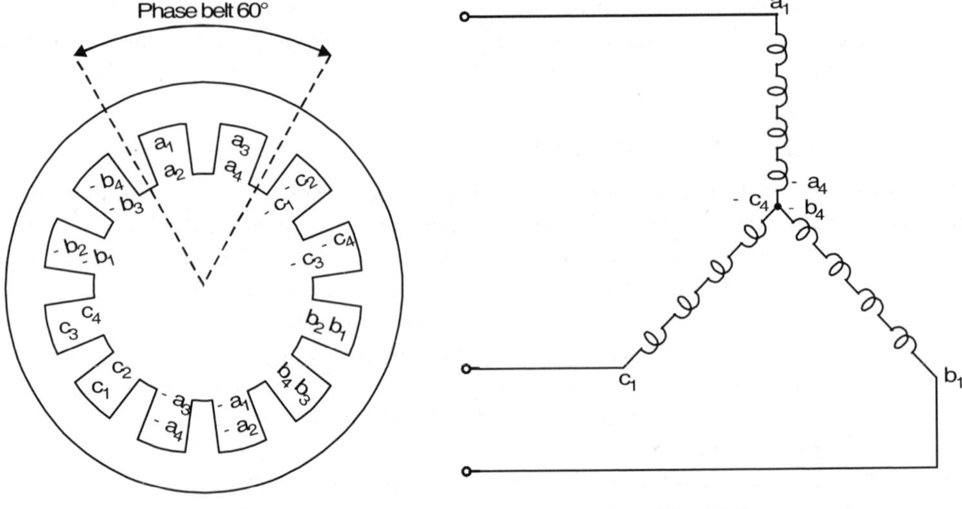

Fig. 12.5 **Fig. 12.6**

Note that there are 12 total coils and each phase has four coils. The four coils in each phase are connected in series so that their voltages aid. The three phases then may be connected to form *Y* or Δ-connection. Fig. 12.6 shows how the coils are connected to form a *Y*-connection.

12.7 WINDING FACTORS

The armature winding of an alternator is distributed over the entire armature. The distributed winding produces nearly a sine waveform and the heating is more uniform. Likewise, the coils of armature winding are not full-pitched *i.e.*, the two sides of a coil are not at corresponding points under adjacent poles. The fractional pitched armature winding requires less copper per coil and at the same time waveform of output voltage is improved. The distribution and pitching of the coils affect the voltages induced in the coils. We shall discuss two winding factors :

(*i*) Distribution factor (K_d), also called breadth factor

(*ii*) Pitch factor (K_p), also known as chord factor

(*i*) **Distribution factor (K_d).** A winding with only one slot per pole per phase is called a *concentrated winding. In this type of winding, the e.m.f. generated/phase is equal to the *arithmetic sum* of the individual coil e.m.f.s in that phase. However, if the coils/phase are distributed over several slots in space (distributed winding), the e.m.f.s in the coils are not in phase (*i.e.*, phase difference is not zero) but are displaced from each by the *slot angle* α (The angular displacement in electrical degrees between the adjacent slots is called slot angle). The e.m.f. /phase will be the phasor sum of coil e.m.f.s. The distribution factor K_d is defined as :

$$K_d = \frac{\text{e.m.f. with distributed winding}}{\text{e.m.f. with concentrated winding}}$$

$$= \frac{\text{phasor sum of coil e.m.f.s/phase}}{\text{arithmetic sum of coil e.m.f.s/phase}}$$

Note that numerator is less than denominator so that $K_d < 1$.

Expression for K_d

Let $\alpha = \text{slot angle} = \dfrac{180^\circ \text{ electrical}}{\text{No. of slots/pole}}$

n = slots per pole per phase

The distribution factor can be determined by constructing a phasor diagram for the coil e.m.f.s. Let $n = 3$. The three coil e.m.f.s are shown as phasors AB, BC and CD [See Fig. 12.7 (*i*)] each of which is a chord of **circle with centre at O and subtends an angle α at O. The phasor sum of the coil e.m.f.s subtends an angle $n\,\alpha$ (Here $n = 3$) at O. Draw perpendicular bisectors of each chord such as Ox, Oy etc [See Fig. 12.7 (*ii*)].

$$K_d = \frac{AD}{n \times AB} = \frac{2 \times Ax}{n \times (2\,Ay)} = \frac{Ax}{n \times Ay}$$

$$= \frac{OA \times \sin (n\,\alpha/2)}{n \times OA \times \sin (\alpha/2)}$$

$$\therefore \quad K_d = \frac{\sin (n\,\alpha/2)}{n \sin (\alpha/2)}$$

Note that $n\,\alpha$ is the phase spread.

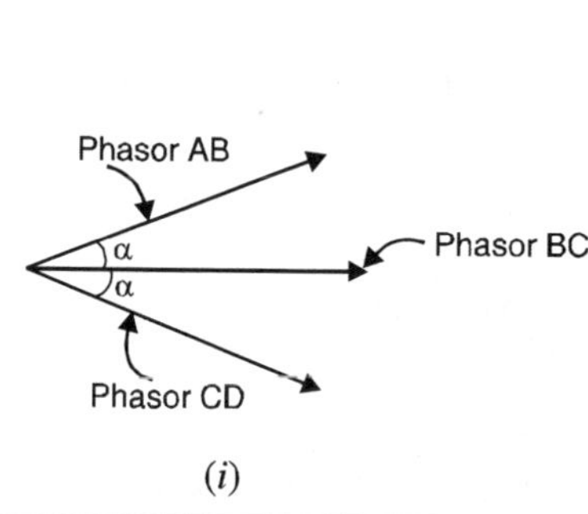

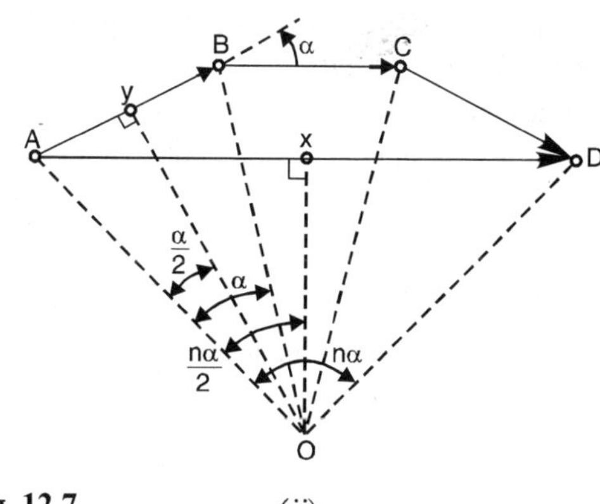

(*i*) **Fig. 12.7** (*ii*)

* A concentrated winding has the following drawbacks :
(*i*) In large machines, only a small portion of the stator surface is utilised.
(*ii*) There is a serious difficulty in producing a sine waveform of e.m.f. since in this case, waveform can be altered only by shaping the pole-faces.

** When n is large, which is usually so, then the curve $ABCD$ will become part of a circle of radius OA.

(*ii*) **Pitch factor (K_p).** A coil whose sides are separated by one pole pitch (*i.e.*, coil span is 180° electrical) is called a full-pitch coil. With a full-pitch coil, the e.m.f.s induced in the two coil sides are in phase with each other and the resultant e.m.f. is the arithmetic sum of individual e.m.f.s. However, the waveform of the resultant e.m.f. can be improved by making the coil pitch less than a pole pitch. Such a coil is called short-pitch coil. This practice is only possible with double-layer type of winding. The e.m.f. induced in a short-pitch coil is less than that of a full-pitch coil. The factor by which e.m.f. per coil is reduced is called pitch factor K_p. It is defined as :

$$K_p = \frac{\text{e.m.f. induced in short-pitch coil}}{\text{e.m.f. induced in full-pitch coil}}$$

Expression for K_p. Consider a coil *AB* which is short-pitch by an angle β electrical degrees as shown in Fig. 12.8. The e.m.f.s generated in the coil sides *A* and *B* differ in phase by an angle β and can be represented by phasors E_A and E_B respectively as shown in Fig. 12.9. The diagonal of the parallelogram represents the resultant e.m.f. E_R of the coil.

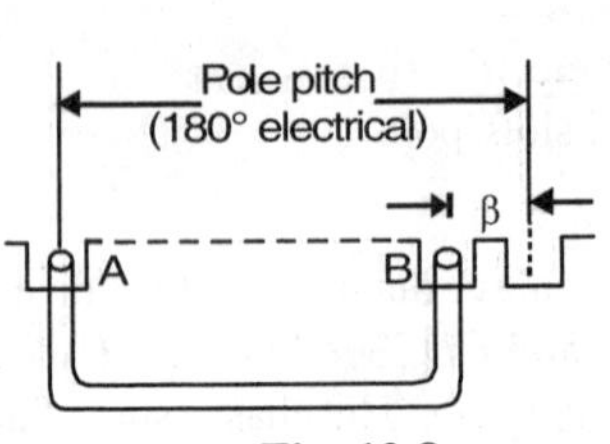

Fig. 12.8

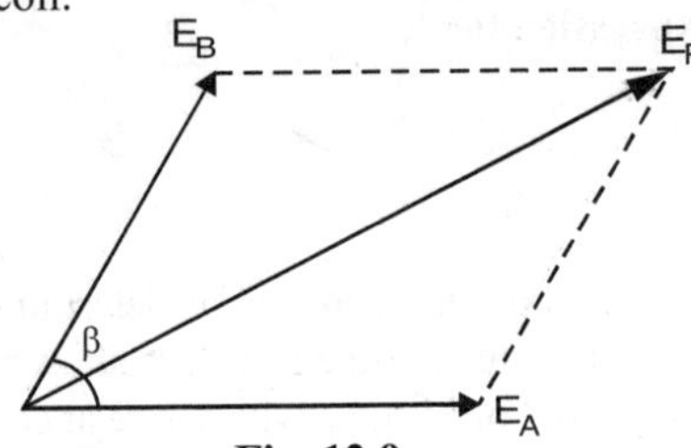

Fig. 12.9

$$\text{Since } E_A = E_B,\; E_R = 2\,E_A \cos \beta/2$$

$$\text{Pitch factor, } K_p = \frac{\text{e.m.f. in short-pitch coil}}{\text{e.m.f. in full-pitch coil}}$$

$$= \frac{2\,E_A \cos \beta/2}{2\,E_A} = \cos \beta/2$$

$$\therefore \quad K_p = \cos \beta/2$$

For a full-pitch winding, $K_p = 1$. However, for a short-pitch winding, $K_p < 1$. Note that β is always an integer multiple of the slot angle α.

Example 12.1. *Calculate the value of the distribution factor for a 3-phase winding of a 4-pole alternator having 36 slots.*

Solution.

$$\text{Number of slots/pole} = 36/4 = 9$$

$$\text{Slot angle, } \alpha = \frac{180^\circ}{\text{No. of slots/pole}} = \frac{180^\circ}{9} = 20^\circ \text{ (elect.)}$$

$$\text{Number of slots/pole/phase, } n = 9/3 = 3$$

$$\therefore \quad \text{Distribution factor, } K_d = \frac{\sin (n\,\alpha/2)}{n \sin (\alpha/2)} = \frac{\sin (3 \times 20^\circ/2)}{3 \sin (20^\circ/2)}$$

$$= \frac{\sin 30^\circ}{3 \sin 10^\circ} = \mathbf{0{\cdot}96}$$

Example 12.2. *The stator winding of an alternator has 48 slots. A 4-pole, 3-phase winding is made on the stator. Each coil spans 11 slots. Calculate the pitch-factor.*

Solution. Fig. 12.10 shows the conditions of the problem. The pole pitch is 48/4 = 12 slots. But the coil spans 11 slots.

$$\therefore \quad \text{Coil pitch} = \frac{11}{12} \times 180^\circ = 165^\circ \text{ (elect.)}$$

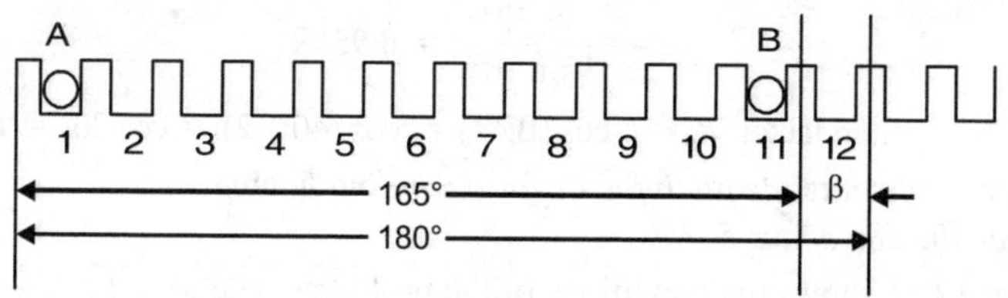

Fig. 12.10

$$\text{Coil is short-pitched by } \beta = 180° - 165° = 15°$$

$$\therefore \quad \text{Pitch factor, } K_p = \cos \beta/2 = \cos 15°/2 = \mathbf{0.99}$$

Example 12.3. *Calculate the distribution factor for a single-phase alternator having 6 slots/pole (i) when all the slots are wound and (ii) when only four adjacent slots per pole are wound, the remaining slots being unwound.*

Solution.

(*i*) *When all the slots are wound*

$n = 6$; Slot angle, $\alpha = 180°/6 = 30°$ (elect.)

$$\therefore \quad \text{Distribution factor, } K_d = \frac{\sin(n\,\alpha/2)}{n \sin(\alpha/2)} = \frac{\sin(6 \times 30°/2)}{6 \sin(30°/2)}$$

$$= \frac{\sin 90°}{6 \sin 15°} = \mathbf{0{\cdot}644}$$

(*ii*) *When only 4 adjacent slots per pole are wound*

$n = 4$; $\alpha = 30°$ (elect.)

$$\therefore \quad K_d = \frac{\sin(n\,\alpha/2)}{n \sin(\alpha/2)} = \frac{\sin(4 \times 30°/2)}{4 \sin(30°/2)}$$

$$= \frac{\sin 60°}{4 \sin 15°} = \mathbf{0{\cdot}837}$$

Example 12.4. *An alternator has 9 slots per pole. If each coil spans 8 slot pitches, what is the value of the pitch factor ?*

Solution.

$$\text{Coil pitch} = \frac{8}{9} \times 180° = 160° \text{ (elect.)}$$

$$\therefore \quad \text{Coil is short-pitched by } \beta = 180° - 160° = 20° \text{ (elect.)}$$

$$\therefore \quad \text{Pitch factor, } K_p = \cos \beta/2 = \cos 20°/2 = \cos 10° = \mathbf{0{\cdot}985}$$

Example 12.5. *The stator of a 3-phase alternator has 9 slots per pole and carries a balanced 3-phase, double-layer winding. The coils are short-pitched and the coil pitch is 7 slots. Find the distribution factor and pitch factor.*

Solution.

$$\text{Slot angle, } \alpha = \frac{180°}{\text{No. of slots/pole}} = \frac{180°}{9} = 20° \text{ (elect.)}$$

$$\text{No. of slots/pole/phase, } n = 9/3 = 3$$

$$\text{Coil pitch} = \frac{7}{9} \times 180° = 140° \text{ (elect.)}$$

$$\therefore \quad \text{Coil is short-pitched by } \beta = 180° - 140° = 40° \text{ (elect.)}$$

$$\text{Distribution factor, } K_d = \frac{\sin(n\,\alpha/2)}{n \sin(\alpha/2)} = \frac{\sin(3 \times 20°/2)}{3 \sin(20°/2)}$$

$$= \frac{\sin 30^\circ}{3 \sin 10^\circ} = \mathbf{0{\cdot}9598}$$

$$\text{Pitch factor, } K_p = \cos(\beta/2) = \cos(40^\circ/2) = \cos 20^\circ = \mathbf{0{\cdot}9397}$$

Example 12.6. *Calculate the pitch factor for a winding having 24 stator slots when the coil spans 5 slots.*

Solution. Fig. 12.11 shows the conditions of the problem. The pole pitch is 6 slots. But the coil spans 5 slots.

$$\therefore \quad \text{Coil pitch} = \frac{5}{6} \times 180^\circ = 150^\circ \text{ (elect.)}$$

Coil is short-pitched by ; $\beta = 180^\circ - 150^\circ = 30^\circ \text{ (elect.)}$

$$\therefore \quad \text{Pitch factor, } K_p = \cos \beta/2 = \cos 30^\circ/2$$

$$= \cos 15^\circ = \mathbf{0{\cdot}9659}$$

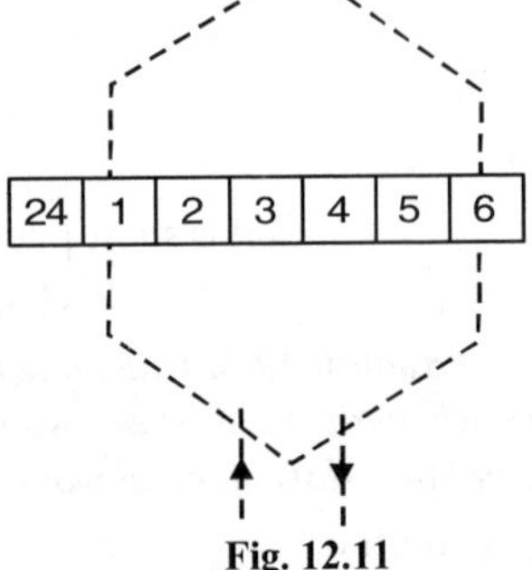

Fig. 12.11

TUTORIAL PROBLEMS

1. A certain alternator has 6 slots per pole and the coils are short-pitch by 1 slot (*i.e.*, coil pitch is 5 slot pitches). What is the value of pitch factor ? **[0·966]**
2. Calculate the value of distribution factor for a 3-phase alternator having 12 slots per pole. **[0·958]**
3. Calculate the distribution factor for a 36-slot, 4-pole, single-layer, 3-phase winding. **[0·96]**

12.8 E.M.F. EQUATION OF AN ALTERNATOR

Let Z = No. of conductors or coil sides in series *per phase*

ϕ = Flux per pole in webers

P = Number of rotor poles

N = Rotor speed in r.p.m.

In one revolution (*i.e.*, $60/N$ second), each stator conductor is cut by $P\phi$ webers *i.e.*,

$$d\phi = P\phi \; ; \; dt = 60/N$$

$\therefore$ Average e.m.f. induced in one stator conductor

$$= \frac{d\phi}{dt} = \frac{P\phi}{60/N} = \frac{P\phi N}{60} \text{ volts}$$

Since there are Z conductors in series per phase,

$$\therefore \quad \text{Average e.m.f./phase} = \frac{P\phi N}{60} \times Z$$

$$= \frac{P\phi Z}{60} \times \frac{120 f}{P} \qquad \left(\because N = \frac{120 f}{P}\right)$$

$$= 2 f \phi Z \text{ volts}$$

R.M.S. value of e.m.f./phase = Average value/phase × form factor

$$= 2 f\phi Z \times 1{\cdot}11 = 2{\cdot}22 f\phi Z \text{ volts}$$

$$\therefore \quad E_{r.m.s.}/\text{phase} = 2{\cdot}22 f\phi Z \text{ volts} \qquad ...(i)$$

If K_p and K_d are the pitch factor and distribution factor of the armature winding, then,

$$E_{r.m.s.}/\text{phase} = 2{\cdot}22\, K_p K_d f\phi Z \text{ volts} \qquad ...(ii)$$

Sometimes the turns (T) per phase rather than conductors per phase are specified. In that case, eq. (*ii*) becomes :

$$E_{r.m.s.}/\text{phase} = 4{\cdot}44\, K_p K_d f\phi T \text{ volts} \qquad ...(iii)$$

The line voltage will depend upon whether the winding is star or delta connected.

Example 12.7. *A 3-phase, 50 Hz, star-connected alternator has 180 conductors per phase and flux per pole is 0·0543 Wb. Find (i) e.m.f. generated per phase and (ii) e.m.f. between line terminals. Assume the winding to be full pitched and distribution factor to be 0·96.*

Solution.

(*i*) Generated e.m.f./phase, $E_{ph} = 2{\cdot}22\, K_p\, K_d\, Z f \phi$

$$= 2{\cdot}22 \times 1 \times 0{\cdot}96 \times 180 \times 50 \times 0{\cdot}0543 = \mathbf{1041{\cdot}5\ V}$$

(*ii*) Line voltage, $E_L = \sqrt{3}\, E_{ph} = \sqrt{3} \times 1041{\cdot}5 = \mathbf{1803{\cdot}93\ V}$

Example 12.8. *Find the number of armature conductors in series per phase required for the armature of a 3-phase, 50 Hz, 10-pole alternator. The winding is star-connected to give a line voltage of 11000 V. The flux per pole is 0·16 Wb. Assume $K_p = 1$ and $K_d = 0{\cdot}96$.*

Solution.

Generated e.m.f./phase, $E_{ph} = E_L/\sqrt{3} = 11000/\sqrt{3} = 6352$ V

Let Z be the number of conductors in series per phase.

$$E_{ph} = 2{\cdot}22\, K_p\, K_d\, Z f \phi$$

$$\therefore \quad Z = \frac{E_{ph}}{2{\cdot}22\, K_p\, K_d\, f\phi} = \frac{6352}{2{\cdot}22 \times 1 \times 0{\cdot}96 \times 50 \times 0{\cdot}16} = \mathbf{372{\cdot}5}$$

Example 12.9. *The armature of an 8-pole, 3-phase, 50 Hz alternator has 18 slots and 10 conductors/slot. A flux of 0·04 Wb is entering the armature from one pole. Calculate the induced e.m.f. per phase.*

Solution. Since the values of K_p and K_d are not given, they will be assumed 1.

Total no. of conductors $= 18 \times 10 = 180$

No. of conductors/phase, $Z = 180/3 = 60$

Induced e.m.f./phase, $E_{ph} = 2{\cdot}22\, K_p\, K_d f \phi$

$$= 2{\cdot}22 \times 1 \times 1 \times 60 \times 50 \times 0{\cdot}04 = \mathbf{266{\cdot}4\ V}$$

Example 12.10. *A 3-phase, 16-pole synchronous generator has a star-connected winding with 144 slots and 10 conductors per slot. The flux per pole is 0·03 Wb, sinusoidally distributed and the speed is 375 r.p.m. Calculate (i) the frequency and (ii) line induced e.m.f.*

Solution.

(*i*) Line frequency, $f = \dfrac{N_s P}{120} = \dfrac{375 \times 16}{120} = \mathbf{50\ Hz}$

(*ii*) Slot angle, $\alpha = \dfrac{180^\circ}{\text{No. of slots/pole}} = \dfrac{180^\circ}{144/16} = 20^\circ$ (elect.)

No. of slots/pole/phase, $n = \dfrac{144}{16 \times 3} = 3$

$\therefore$ Distribution factor, $K_d = \dfrac{\sin (n\,\alpha/2)}{n \sin (\alpha/2)} = \dfrac{\sin (3 \times 20^\circ/2)}{3 \sin (20^\circ/2)}$

$$= \frac{\sin 30^\circ}{3 \sin 10^\circ} = 0{\cdot}96$$

No. of turns/phase, $T = \dfrac{144 \times 10}{2 \times 3} = 240$

Assuming pitch factor $K_p = 1$ (since it is not given),

Induced e.m.f./phase, $E_{ph} = 4{\cdot}44\, K_p\, K_d\, T f \phi$

$$= 4{\cdot}44 \times 1 \times 0{\cdot}96 \times 240 \times 50 \times 0{\cdot}03 = 1534 \text{ V}$$

$\therefore$ Induced line voltage, $E_L = \sqrt{3}\, E_{ph} = \sqrt{3} \times 1534 = \mathbf{2657\ V}$

Example 12.11. *A 3-phase, star-connected alternator on open circuit is required to generate a line voltage of 3600 V at 50 Hz when driven at 500 r.p.m. The stator has 3 slots per pole per phase and 10 conductors per slot. Calculate (i) the number of poles and (ii) useful flux per pole. Assume all the conductors per phase to be connected in series and the coils to be full-pitch.*

Solution.

(*i*) $$f = \frac{N_s P}{120}$$

or $$50 = \frac{500 \times P}{120} \quad \therefore \; P = \mathbf{12}$$

(*ii*) $$\text{No. of slots/phase} = 3 \times 12 = 36$$

$$\text{No. of conductors/phase, } Z = 36 \times 10 = 360$$

$$\text{E.M.F./phase, } E_{ph} = \frac{3600}{\sqrt{3}} = 2080 \text{ V}$$

$$\text{No. of slots/pole/phase, } n = 3$$

$$\text{Slot angle, } \alpha = \frac{180^\circ}{\text{No. of slots/pole}} = \frac{180^\circ}{3 \times 3} = 20^\circ \text{ (elect.)}$$

$$\text{Distribution factor, } K_d = \frac{\sin(n\,\alpha/2)}{n \sin(\alpha/2)} = \frac{\sin(3 \times 20^\circ/2)}{3 \sin(20^\circ/2)}$$

$$= \frac{\sin 30^\circ}{3 \sin 10^\circ} = 0{\cdot}96$$

$$\text{Pitch factor, } K_p = 1 \qquad \text{...for full-pitch coils}$$

$$\text{Induced e.m.f./phase, } E_{ph} = 2{\cdot}22\, K_p K_d Z f \phi$$

or $$2080 = 2{\cdot}22 \times 1 \times 0{\cdot}96 \times 360 \times 50 \times \phi$$

$\therefore$ $$\phi = \mathbf{0{\cdot}0543\ Wb}$$

Example 12.12. *A 4-pole, 3-phase, 50-Hz, star-connected alternator has 60 slots with 4 conductors per slot. Coils are short-pitched by 3 slots. If the phase spread is 60°, find the line voltage induced for a flux per pole of 0·943 Wb, distributed sinusoidally. All the turns per phase are in series.*

Solution.

$$\text{No. of slots/pole/phase, } n = 60/4 \times 3 = 5$$

$$\text{Phase spread} = n\,\alpha$$

$\therefore$ $$\text{Slot angle, } \alpha = \frac{\text{Phase spread}}{n} = \frac{60^\circ}{5} = 12^\circ \text{ (elect.)}$$

$\therefore$ $$\text{Distribution factor, } K_d = \frac{\sin(n\,\alpha/2)}{n \sin(\alpha/2)} = \frac{\sin(5 \times 12^\circ/2)}{5 \sin(12^\circ/2)} = 0{\cdot}957$$

$$\text{Coil pitch} = \frac{12}{15} \times 180^\circ = 144^\circ \text{ (elect.)}$$

$$\text{Coil is short-pitched by } \beta = 180^\circ - 144^\circ = 36^\circ \text{ (elect.)}$$

$\therefore$ $$\text{Pitch factor, } K_p = \cos(\beta/2) = \cos(36^\circ/2) = \cos 18^\circ = 0{\cdot}95$$

$$\text{No. of conductors/phase, } Z = \frac{60 \times 4}{3} = 80$$

$$\text{No. of turns/phase, } T = Z/2 = 80/2 = 40$$

Now $$E_{ph} = 4{\cdot}44\, K_p K_d f T \phi$$

$$= 4{\cdot}44 \times 0{\cdot}95 \times 0{\cdot}957 \times 50 \times 40 \times 0{\cdot}943 = 7613 \text{ V}$$

$\therefore$ $$\text{Line voltage, } E_L = \sqrt{3}\, E_{ph} = \sqrt{3} \times 7613 = \mathbf{13185\ V}$$

Example 12.13. *A 12-pole, 3-phase, star-connected alternator has 72 slots. The flux per pole is 0·0988 Wb. Calculate :*

(*i*) *the speed of rotation if the frequency of the generated e.m.f. is 50 Hz*
(*ii*) *the terminal e.m.f. for full-pitch coils and 8 conductors per slot*
(*iii*) *the terminal e.m.f. if the coil span is reduced to 2/3 of the pole pitch*

Solution.

(*i*) $$\text{Synchronous speed, } N_s = \frac{120\, f}{P} = \frac{120 \times 50}{12} = \mathbf{500\ r.p.m.}$$

(*ii*) $$\text{Slot angle, } \alpha = \frac{180^\circ \text{ (elect.)}}{\text{No. of slots/pole}} = \frac{180^\circ}{72/12} = \frac{180^\circ}{6} = 30^\circ \text{ (elect.)}$$

$$\text{No. of slots/pole/phase, } n = \frac{72}{12 \times 3} = 2$$

$$\text{Distribution factor, } K_d = \frac{\sin (n\,\alpha/2)}{n \sin (\alpha/2)} = \frac{\sin (2 \times 30^\circ/2)}{2 \sin (30^\circ/2)}$$

$$= \frac{\sin 30^\circ}{2 \sin 15^\circ} = 0{\cdot}966$$

$$\text{Pitch factor, } K_p = 1 \qquad \text{...full-pitched coils}$$

$$\text{No. of slots/phase} = 72/3 = 24$$

$$\text{No. of conductors/phase, } Z = 24 \times 8 = 192$$

$$\therefore \quad E_{ph} = 2{\cdot}22\, K_p K_d Z f \phi$$

$$= 2{\cdot}22 \times 1 \times 0{\cdot}966 \times 192 \times 50 \times 0{\cdot}0988 = 2034 \text{ V}$$

$$\therefore \quad \text{Terminal e.m.f., } E_L = \sqrt{3}\, E_{ph} = \sqrt{3} \times 2034 = \mathbf{3523\ V}$$

(*iii*) $$\text{Coil pitch} = \frac{2}{3} \times 180^\circ = 120^\circ \text{ (elect.)}$$

$$\text{Coil is short-pitched by } \beta = 180^\circ - 120^\circ = 60^\circ \text{ (elect.)}$$

$$\therefore \quad \text{Pitch factor, } K_p = \cos (\beta/2) = \cos 30^\circ = 0{\cdot}866$$

$$\therefore \quad \text{Terminal e.m.f.} = 3523 \times 0{\cdot}866 = \mathbf{3051\ V}$$

Example 12.14. *Find the no-load phase and line voltage of a star-connected 3-phase, 6-pole alternator which runs at 1200 r.p.m., having flux per pole of 0.1 Wb sinusoidally distributed. Its stator has 54 slots having double layer winding. Each coil has 8 turns and the coil is chorded by 1 slot.*

Solution.

$$\text{Coil pitch} = (8/9) \times 180^\circ = 160^\circ$$

$$\text{Coil is short-pitched by } \beta = 180^\circ - 160^\circ = 20^\circ$$

$$\text{Pitch factor, } K_p = \cos \beta/2 = \cos 20^\circ/2 = 0.98$$

$$\text{Slot angle, } \alpha = \frac{180^\circ}{\text{No. of slots/pole}} = \frac{180^\circ}{54/6} = 20^\circ$$

$$\text{No. of slots/pole/phase, } n = 54/6 \times 3 = 3$$

$$\therefore \quad \text{Distribution factor, } K_d = \frac{\sin (n\alpha/2)}{n \sin(\alpha/2)} = \frac{\sin(3 \times 20^\circ/2)}{3 \sin(20^\circ/2)} = 0.96$$

$$\text{Line frequency, } f = NP/120 = 1200 \times 6/120 = 60 \text{ Hz}$$

$$\text{No of conductors/phase, } Z = {}^*54 \times 8/3 = 144$$

$$\text{No. of turns/phase, } T = 144/2 = 72$$

---

* Each coil occupies two slots so that the total number of conductors for three phases = 54 × 8.

$\therefore$ Induced e.m.f./phase, $E_{ph} = 4.44\, K_p K_d T f \phi$

$= 4.44 \times 0.98 \times 0.96 \times 72 \times 60 \times 0.1 =$ **1805 V**

Induced line voltage, $E_L = \sqrt{3}\, E_{ph} = \sqrt{3} \times 1805 =$ **3125 V**

Example 12.15. *A 50 Hz, 600 r.p.m., salient pole alternator has a sinusoidal flux density having a maximum value of 1 Wb/m². The alternator has 180 slots wound with 2-layer, 3-turn coils. The coil span is 15 slots and phase spread is 60°. The armature diameter is 1.25 m and core length 0.45 m. Find (i) flux/pole (ii) pitch factor (iii) distribution factor (iv) peak value of e.m.f. per conductor (v) peak value of e.m.f. per coil (vi) r.m.s. phase and line voltage if the machine is star-connected.*

Solution.

(*i*) No. of poles, $P = 120\, f/N_s = 120 \times 50/600 = 10$

Area under one pole pitch, $A = \dfrac{\pi D l}{P} = \dfrac{\pi \times 1.25 \times 0.45}{10} = 0.1767 \text{ m}^2$

Average flux density, $B_{av} = (2/\pi)\, B_m = (2/\pi) \times 1 = 2/\pi \text{ Wb/m}^2$

$\therefore$ Flux/pole, $\phi = B_{av} \times A = (2/\pi) \times 0.1767 =$ **0.1125 Wb**

(*ii*) Coil pitch $= (15/18) \times 180° = 150°$ (elect.)

Coil is short pitched by $\beta = 180° - 150° = 30°$

$\therefore$ Pitch factor, $K_p = \cos(\beta/2) = \cos(30°/2) =$ **0.966**

(*iii*) Slot angle, $\alpha = \dfrac{180°}{\text{No. of slots/pole}} = \dfrac{180°}{180/10} = 10°$ (elect.)

The phase spread is $n\,\alpha$ and its value is given to be 60°. Therefore, $n = 60°/10° = 6$. Note that n is the number of slots/pole/phase.

$\therefore$ Distribution factor, $K_d = \dfrac{\sin(n\alpha/2)}{n \sin(\alpha/2)} = \dfrac{\sin(60°/2)}{6 \sin(10°/2)} =$ **0.956**

(*iv*) R.M.S. e.m.f./conductor $= 2.22 f \phi = 2.22 \times 50 \times 0.1125 = 12.49$ V

Peak e.m.f./conductor $= \sqrt{2} \times 12.49 =$ **17.66 V**

(*v*) Each coil has 3 turns or 6 conductors.

$\therefore$ R.M.S. e.m.f./coil $= 12.49 \times 6 \times K_p = 12.49 \times 6 \times 0.966 =$ **72.39 V**

(*vi*) Number of coils/phase $= 180/3 = 60$

$\therefore$ R.M.S. value of phase voltage $= 72.39 \times 60 \times K_d$

$= 72.39 \times 60 \times 0.956 =$ **4152 V**

Line voltage $= \sqrt{3} \times 4152 =$ **7191.5 V**

Example 12.16. *An alternator on open circuit generates 360 V at 60 Hz when the field current is 3.6 A. Neglecting saturation, determine the open-circuit e.m.f. when the frequency is 40 Hz and field current is 2.4 A.*

Solution. From the e.m.f. equation of alternator, $E \propto \phi f$.

$$\therefore \quad \frac{E_1}{E_2} = \frac{\phi_1 f_1}{\phi_2 f_2}$$

Since saturation is neglected, $\phi \propto I_f$ where I_f is the field current.

$$\therefore \quad \frac{E_1}{E_2} = \frac{I_{f1} f_1}{I_{f2} f_2} \quad \text{or} \quad \frac{360}{E_2} = \frac{3.6 \times 60}{2.4 \times 40} \qquad \therefore\ E_2 = \mathbf{160\ V}$$

Example 12.17. *Find the no-load line voltage of a star-connected 4-pole, 50 Hz alternator from the following data :*

Flux per pole = 0.12 Wb ; No. of slots/pole/phase = 4

Conductors/slot = 4 ; Two layer winding with coil span 150°.

Solution.

$$\text{Total number of slots} = 4 \times 4 \times 3 = 48$$

$$\text{Total number of conductors} = 48 \times 4 = 192$$

$$\text{Total number of turns} = 192/2 = 96$$

$$\text{No. of turns in series, } T = 96/3 = 32$$

$$\text{No. of slots/pole/phase, } n = 4$$

$$\text{Slot angle, } \alpha = \frac{180°}{\text{No. of slots/pole}} = \frac{180°}{48/4} = 15° \text{ (elect.)}$$

$$\therefore \quad \text{Distribution factor, } K_d = \frac{\sin(n\alpha/2)}{n\sin(\alpha/2)} = \frac{\sin(60°/2)}{4\sin(15°/2)} = 0.958$$

$$\text{Coil is short pitched by } \beta = 180° - 150° = 30°$$

$$\therefore \quad \text{Pitch factor, } K_p = \cos(\beta/2) = \cos(30°/2) = 0.966$$

$$\therefore \quad \text{Induced e.m.f./phase, } E_{ph} = 4.44\, K_p\, K_d\, T f \phi$$

$$= 4.44 \times 0.966 \times 0.958 \times 32 \times 50 \times 0.12 = 789 \text{ V}$$

$$\therefore \quad \text{Induced line voltage, } E_L = \sqrt{3}\, E_{ph} = \sqrt{3} \times 789 = \mathbf{1366.6\ V}$$

TUTORIAL PROBLEMS

1. A 2-pole, 3-phase, 50 Hz, star-connected a.c. generator has 9 slots/pole and 6 conductors per slot. What is the flux/pole if the terminal voltage on open-circuit is 11000 V ? Take the coil span as unity. **[0·1656 Wb]**
2. Find the no-load phase and line-voltage of a star-connected, 3-phase, 6-pole alternator which runs at 1200 r.p.m. having flux per pole of 0·1 Wb sinusoidally distributed. Its stator has 54 slots having double layer winding. Each coil has 8 turns and the coil is chorded by one slot. **[1805 V ; 3125 V]**
3. A 3-phase synchronous generator has 4 poles and runs at 1500 r.p.m. The total number of armature conductors is 720 accommodated in 72 slots. If the flux per pole is 0·035 Wb, determine (*i*) the frequency of the e.m.f. (*ii*) the magnitude of e.m.f. for a star connection with coils short-chorded by two slot pitches. **[(*i*) 50 Hz (*ii*) 2650 V]**
4. Calculate the number of stator conductors per slot for a 3-phase, 50 Hz alternator if the winding is star-connected and has to give a line voltage of 13 kV when the machine is on open-circuit. The flux/pole is 0·15 Wb. Assume full-pitch coils and the stator to have 3 slots/pole/phase. The speed is 300 r.p.m. **[8]**
5. The resultant air-gap flux of a 16-pole, 3-phase alternator is 16×10^{-2} Wb/pole and is distributed sinusoidally over the pole. The stator has 2 slots/pole/phase and 4 conductors per slot are accommodated in two layers. The coil span is 150 electrical degrees. Calculate the e.m.f. generated in each phase when the machine runs at 375 r.p.m. **[795 V]**

12.9 ARMATURE REACTION IN ALTERNATOR

When an alternator is running at no-load, there will be no current flowing through the armature winding. The flux produced in the air-gap will be only due to the rotor ampere-turns. When the alternator is loaded, the three-phase currents will produce a rotating magnetic field in the air-gap. Consequently, the air-gap flux is changed from the no-load condition.

The effect of armature flux on the flux produced by field ampere-turns (i.e., rotor ampere-turns) is called **armature reaction.**

Two things are worth noting about the armature reaction in an alternator. First, the armature flux and the flux produced by rotor ampere-turns rotate at the same speed (synchronous speed) in the same direction and, therefore, the two fluxes are fixed in space relative to each other. Secondly, the modification of flux in the air-gap due to armature flux depends on the magnitude of stator current and *on the power factor of the load.* It is the load power factor which determines whether the armature flux distorts, opposes or helps the flux produced by rotor ampere-turns. To illustrate this important point, we shall consider the following three cases :

(*i*) When load p.f. is unity (*ii*) When load p.f. is zero lagging (*iii*) When load p.f. is zero leading

(*i*) **When load p.f. is unity.** Fig. 12.12 (*i*) shows an elementary alternator on no-load. Since the armature is on open-circuit, there is no stator current and the flux due to rotor current is distributed symmetrically in the air-gap as shown in Fig. 12.12 (*i*). Since the direction of the rotor is assumed *clockwise*, the generated e.m.f. in phase R_1R_2 is at its maximum and is towards the paper in the conductor R_1 and outwards in conductor R_2. No armature flux is produced since no current flows in the armature winding.

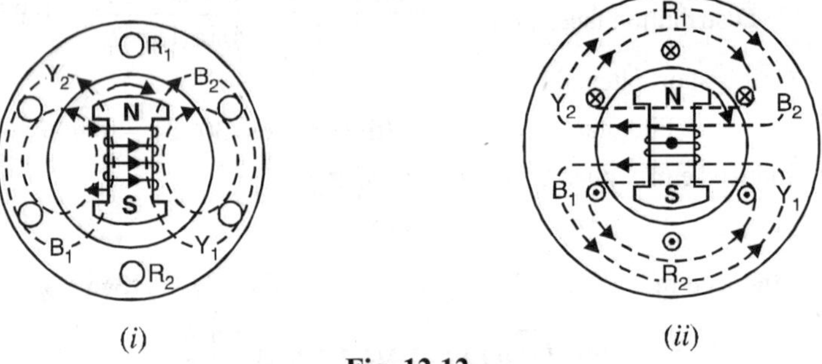

Fig. 12.12

Fig. 12.12 (*ii*) shows the effect when a resistive load (unity p.f.) is connected across the terminals of the alternator. According to right-hand rule, the current is "in" in the conductors under *N*-pole and "out" in the conductors under *S*-pole. Therefore, the armature flux is clockwise due to currents in the top conductors and anticlockwise due to currents in the bottom conductors. Note that armature flux is at 90° to the main flux (due to rotor current) and is behind the main flux. In this case, the flux in the air-gap is distorted but not weakened. *Therefore, at unity p.f., the effect of armature reaction is merely to distort the main field ; there is no weakening of the main field and the average flux practically remains the same.* Since the magnetic flux due to stator currents (*i.e.*, armature flux) rotates synchronously with the rotor, the flux distortion remains the same for all positions of the rotor.

(*ii*) **When load p.f. is zero lagging.** When a pure inductive load (zero p.f. lagging) is connected across the terminals of the alternator, current lags behind the voltage by 90°. This means that current will be maximum at zero e.m.f. and vice-versa.

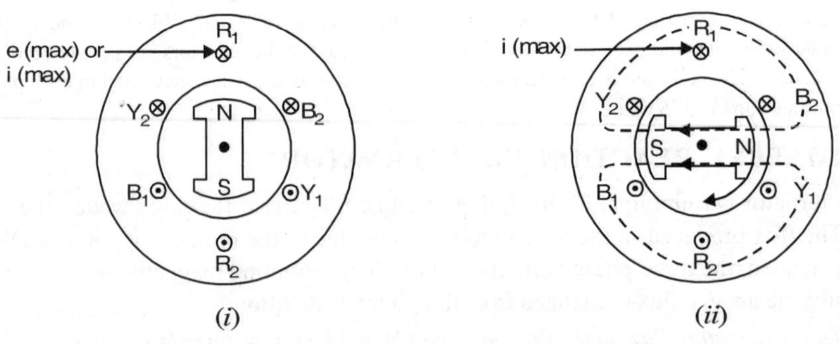

Fig. 12.13

Fig. 12.13 (*i*) shows the condition when the alternator is supplying resistive load. Note that e.m.f. as well as current in phase $R_1 R_2$ is maximum in the position shown. When the alternator is supplying a pure inductive load, the current in phase $R_1 R_2$ will not reach its maximum value until *N*-pole advanced 90° electrical as shown in Fig. 12.13 (*ii*). Now the armature flux is from right to left and field flux is from left to right. All the flux produced by armature current (*i.e.*, armature flux) opposes the field flux and, therefore, weakens it. In other words, armature reaction is directly demagnetising. *Hence at zero p.f. lagging, the armature reaction weakens the main flux*. This causes a reduction in the generated e.m.f.

(*iii*) **When load p.f. is zero leading.** When a pure capacitive load (zero p.f. leading) is connected across the terminals of the alternator, the current in armature windings will lead the induced e.m.f. by 90°. Obviously, the effect of armature reaction will be the reverse that for pure inductive load. Thus armature flux now aids the main flux and the generated e.m.f. is increased.

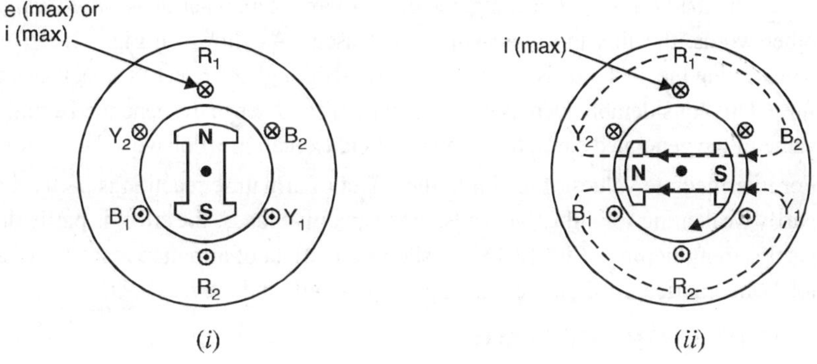

Fig. 12.14

Fig. 12.14 (*i*) shows the condition when alternator is supplying resistive load. Note that e.m.f. as well as current in phase $R_1 R_2$ is maximum in the position shown. When the alternator is supplying a pure capacitive load, the maximum current in $R_1 R_2$ will occur 90° electrical before the occurrence of maximum induced e.m.f. Therefore, maximum current in phase $R_1 R_2$ will occur if the position of the rotor remains 90° behind as compared to its position under resistive load. This is illustrated in Fig. 12.14 (*ii*). It is clear that armature flux is now in the same direction as the field flux and, therefore, strengthens it. This causes an increase in the generated voltage. *Hence at zero p.f. leading, the armature reaction strengthens the main flux.*

For intermediate values of p.f., the effect of armature reaction is partly distorting and partly weakening for inductive loads. For capacitive loads, the effect of armature reaction is partly distorting and partly strengthening. Note that in practice, loads are generally inductive.

Summary. When the alternator is loaded, the armature flux modifies the air-gap flux. Its angle (electrical) w.r.t. main flux depends on the load p.f. This is illustrated in Fig. 12.15.

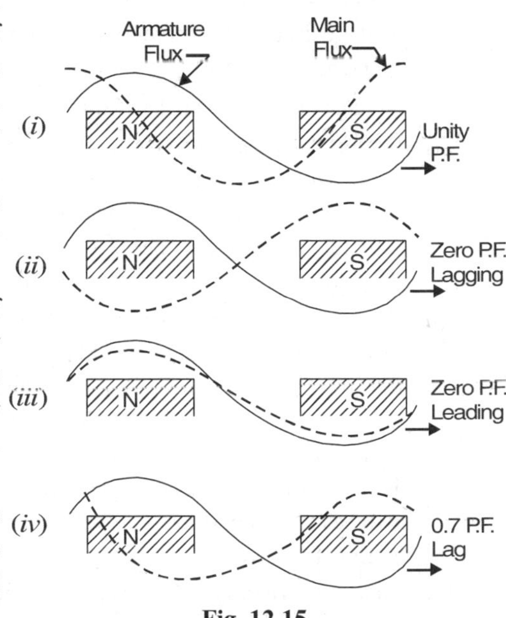

Fig. 12.15

(*a*) When the load p.f. is unity, the effect of armature reaction is *wholly distorting*. In other words, the flux in the air-gap is distorted but not weakened. As shown in Fig. 12.15 (*i*), the armature flux is 90° electrical behind the main flux. The result is that flux is strengthened at the trailing pole tips and weakened at the leading pole tips. However, the average flux in the air-gap practically remains unaltered.

(*b*) When the load p.f. is zero lagging, the effect of armature reaction is *wholly demagnetising*. In other words, the flux in the air-gap is weakened. As shown in Fig. 12.15 (*ii*), the wave representing the main flux is moved backwards through 90° (elect.) so that it is in direct opposition to the armature flux. This considerably reduces the air-gap flux and hence the generated e.m.f. To keep the value of the generated e.m.f. the same, the field excitation will have to be increased to compensate for the weakening of the air-gap flux.

(*c*) When the load p.f. is zero leading, the effect of armature reaction is *wholly magnetising*. In other words, the flux in the air-gap is increased. As shown in Fig. 12.15 (*iii*), the wave representing the main flux is now moved forward through 90° (elect.) so that it aids the armature flux. This considerably increases the air-gap flux and hence the generated e.m.f. To keep the value of the generated e.m.f. the same, the field excitation will have to be reduced.

(*d*) For intermediate values of load p.f., the effect of armature reaction is partly distorting and partly weakening for inductive loads. For capacitive loads, the effect is partly distorting and partly strengthening. Fig. 12.15 (*iv*) shows the effect of armature reaction for an inductive load. In practice, load on the alternator is generally inductive.

12.10 ALTERNATOR ON LOAD

Fig. 12.16 shows Y-connected alternator supplying inductive load (lagging p.f.). When the load on the alternator is increased (*i.e.*, armature current I_a is increased), the field excitation and speed being kept constant, the terminal voltage V (phase value) of the alternator decreases. This is due to

(*i*) Voltage drop $I_a R_a$ where R_a is the armature resistance per phase.

(*ii*) Voltage drop $I_a X_L$ where X_L is the armature leakage reactance per phase.

(*iii*) Voltage drop because of armature reaction.

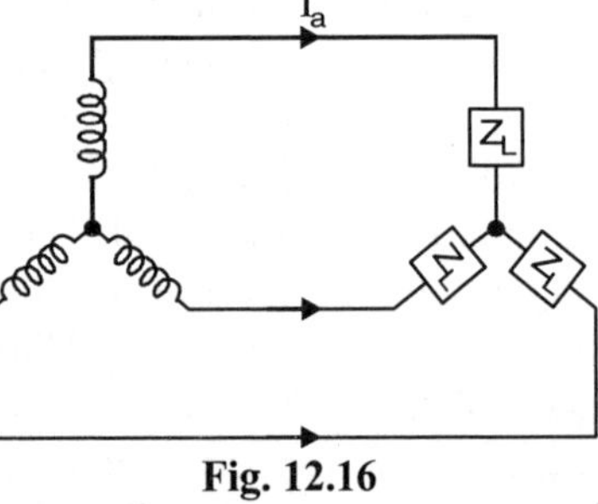

Fig. 12.16

(*i*) **Armature Resistance (R_a).** Since the armature or stator winding has some resistance, there will be an I_aR_a drop when current (I_a) flows through it. The armature resistance per phase is generally small so that I_aR_a drop is negligible for all practical purposes.

(*ii*) **Armature Leakage Reactance (X_L).** When current flows through the armature winding, flux is set up and a part of it does not cross the air-gap and links the coil sides as shown in Fig. 12.17. This leakage flux alternates with current and gives the winding self-inductance. This is called armature leakage reactance. Therefore, there will be I_aX_L drop which is also effective in reducing the terminal voltage.

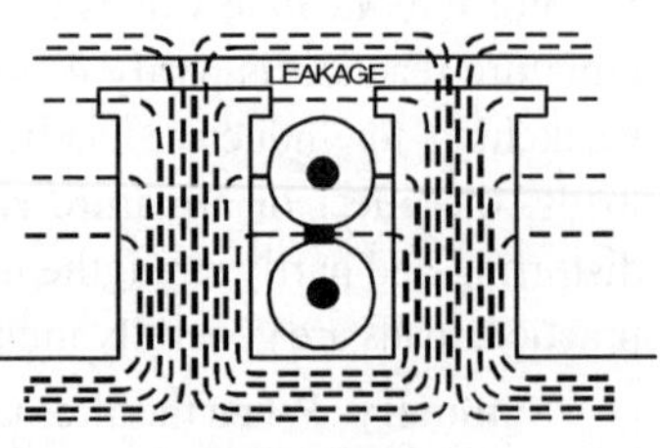

Fig. 12.17

(*iii*) **Armature reaction.** The load is generally inductive and the effect of armature reaction is to reduce the generated voltage. Since armature reaction results in a voltage effect in a circuit caused by the change in flux produced by current in the same circuit, its effect is of the nature of an inductive reactance. Therefore, armature reaction effect is accounted for by assuming the presence of a fictitious reactance X_{AR} in the armature winding. The quantity X_{AR} is called *reactance of armature reaction*. The value of X_{AR} is such that $I_a X_{AR}$ represents the voltage drop due to armature reaction.

Equivalent Circuit. Fig. 12.18 shows the equivalent circuit of the loaded alternator for one phase. All the quantities are per phase. Here

E_0 = No - load e.m.f.

E = Load induced e.m.f. It is the induced e.m.f. after allowing for armature reaction. It is equal to phasor difference of E_0 and $I_a X_{AR}$.

V = Terminal voltage. It is less than E by voltage drops in X_L and R_a.

$\therefore \quad \mathbf{E} = \mathbf{V} + \mathbf{I_a}\,(R_a + jX_L)$

and $\quad \mathbf{E_0} = \mathbf{E} + \mathbf{I_a}\,(jX_{AR})$

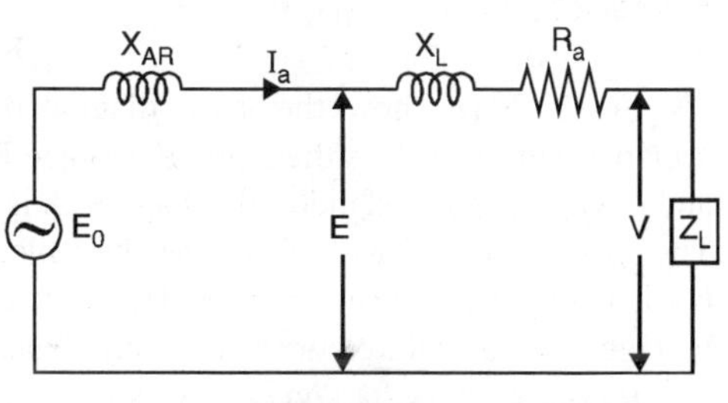

Fig. 12.18

12.11 SYNCHRONOUS REACTANCE (X_s)

The sum of armature leakage reactance (X_L) and reactance of armature reaction (X_{AR}) is called *synchronous reactance* X_s [See Fig. 12.19 (*i*)]. Note that all quantities are per phase.

$$X_s = X_L + X_{AR}$$

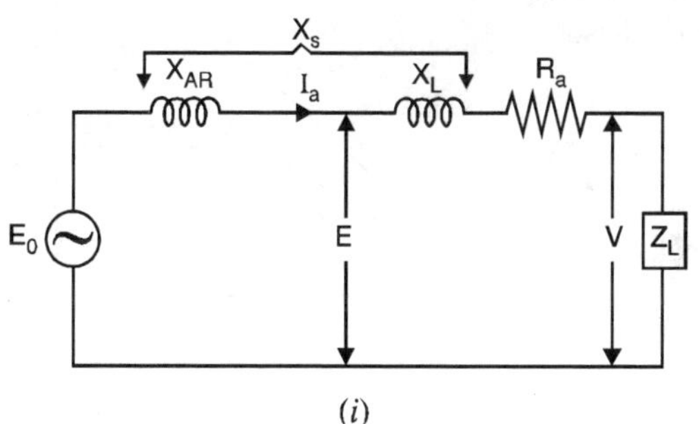

(*i*)

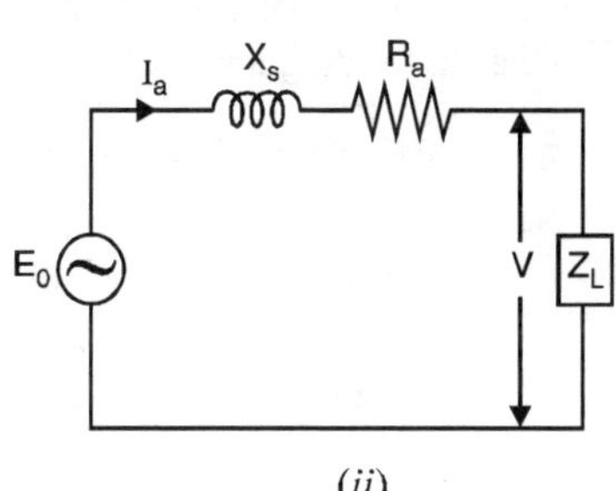

(*ii*)

Fig. 12.19

The synchronous reactance is a fictitious reactance employed to account for the voltage effects in the armature circuit produced by the actual armature leakage reactance and the change in the air-gap flux caused by armature reaction. The circuit then reduces to the one shown in Fig. 12.19 (*ii*).

$$\text{Synchronous impedance, } \mathbf{Z_s} = R_a + jX_s$$

The synchronous impedance is the fictitious impedance employed to account for the voltage effects in the armature circuit produced by the actual armature resistance, the actual armature leakage reactance and the change in the air-gap flux produced by armature reaction.

$$\mathbf{E_0} = \mathbf{V} + \mathbf{I_a}\,\mathbf{Z_s} = \mathbf{V} + \mathbf{I_a}\,(R_a + jX_s)$$

12.12 PHASOR DIAGRAM OF A LOADED ALTERNATOR

Consider a *Y*-connected alternator supplying inductive load, the load p.f. angle being ϕ. Fig. 12.20 (*i*) shows the equivalent circuit of the alternator per phase. All quantities are per phase.

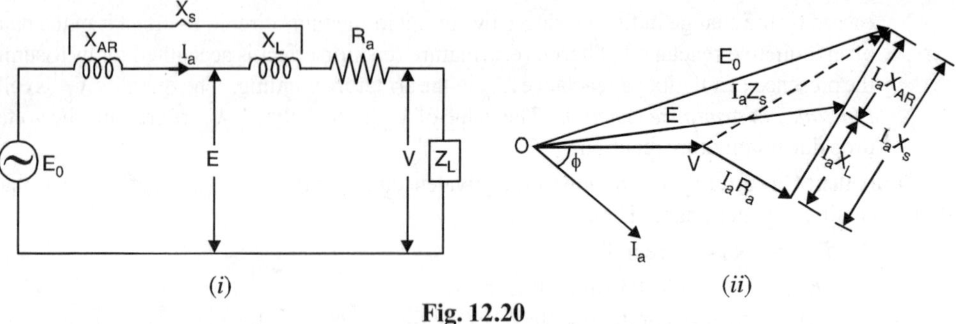

Fig. 12.20

Fig. 12.20 (*ii*) shows the phasor diagram of an alternator for the usual case of inductive load. The armature current I_a lags the terminal voltage V by p.f. angle ϕ. The phasor sum of V and drops $I_a R_a$ and $I_a X_L$ gives the load induced voltage E. It is the induced e.m.f. after allowing for armature reaction. The phasor sum of E and $I_a X_{AR}$ gives the no-load e.m.f. E_0. The phasor diagram for unity and leading p.f. is left as an exercise for the reader. Note that in drawing the phasor diagram either the terminal voltage (V) or armature current (I_a) may be taken as the reference phasor.

Example 12.18. *A 500 kVA, 3-phase, star-connected alternator has a rated line-to-line voltage of 3300 V. The resistance and synchronous reactance per phase are 0·3 Ω and 4 Ω respectively. Calculate the line value of the e.m.f. generated at full-load, 0·8 p.f. lagging.*

Solution.

$$\text{F.L. output current, } I_a = \frac{500 \times 10^3}{\sqrt{3} \times 3300} = 87{\cdot}5 \text{ A}$$

Fig. 12.21 shows the phasor diagram of the alternator. Note that current I_a has been taken as the reference phasor. From the phasor diagram, we have,

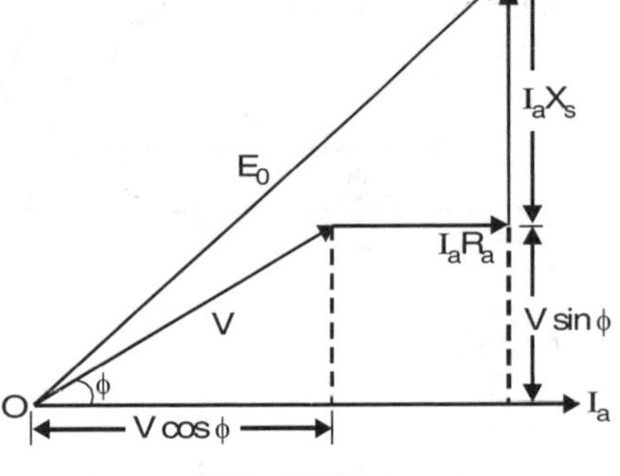

Fig. 12.21

$$E_0 = \sqrt{(V \cos\phi + I_a R_a)^2 + (V \sin\phi + I_a X_s)^2}$$

Now $\cos\phi = 0{\cdot}8 \quad \therefore \quad \sin\phi = 0{\cdot}6$

$$I_a R_a \text{ drop} = 87{\cdot}5 \times 0{\cdot}3 = 26{\cdot}25 \text{ volts}$$

$$I_a X_s = 87{\cdot}5 \times 4 = 350 \text{ volts}$$

Terminal voltage/phase is

$$V = 3300/\sqrt{3} = 1905 \text{ volts}$$

$$\therefore \quad E_0 = \sqrt{(1905 \times 0{\cdot}8 + 26{\cdot}25)^2 + (1905 \times 0{\cdot}6 + 350)^2} = 2152 \text{ volts}$$

$$\therefore \quad \text{Line e.m.f.} = \sqrt{3}\, E_0 = \sqrt{3} \times 2152 = \mathbf{3727 \text{ volts}}$$

Example 12.19. *A 1000 kVA, 2300 V, 3-phase, star-connected alternator has a resistance of 0·309 Ω/phase and a synchronous reactance of 3·31 Ω/phase. Calculate the change of line voltage when the rated output of 1000 kVA at power factor of 0·8 lagging is switched off. Assume the speed and the exciting current to remain unaltered.*

Solution.

$$\text{F.L. output current, } I_a = \frac{1000 \times 10^3}{\sqrt{3} \times 2300} = 251 \text{ A}$$

$$E_0 = \sqrt{(V \cos\phi + I_a R_a)^2 + (V \sin\phi + I_a X_s)^2}$$

Now $\cos\phi = 0{\cdot}8 \quad \therefore \quad \sin\phi = 0{\cdot}6$

$$I_a R_a \text{ drop} = 251 \times 0{\cdot}309 = 77{\cdot}6 \text{ volts}$$
$$I_a X_s \text{ drop} = 251 \times 3{\cdot}31 = 831 \text{ volts}$$
$$\text{Terminal voltage/phase, } V = 2300/\sqrt{3} = 1328 \text{ volts}$$
$$E_0 = \sqrt{(1328 \times 0{\cdot}8 + 77{\cdot}6)^2 + (1328 \times 0{\cdot}6 + 831)^2} = 1987 \text{ volts}$$
$$\text{Line value} = \sqrt{3}\, E_0 = \sqrt{3} \times 1987 = 3441 \text{ volts}$$
$$\text{Change in line voltage} = 3441 - 2300 = \mathbf{1141 \text{ volts}}$$

Example 12.20. *A 1500 kVA, 6.6 kV, 3-phase, star-connected alternator has effective armature resistance of 0·5 Ω/phase and a synchronous reactance of 5 Ω/phase. Find the percentage change in terminal voltage when the rated output of 1500 kVA at (i) unity p.f. (ii) 0·8 lagging p.f. (iii) 0·8 leading p.f. is switched off. The speed and excitation current remain unchanged.*

Solution.

$$\text{F.L. current, } I_a = \frac{1500 \times 10^3}{\sqrt{3} \times 6{\cdot}6 \times 10^3} = 131 \text{ A}$$
$$\text{Terminal voltage/phase, } V = 6{\cdot}6 \times 10^3/\sqrt{3} = 3810 \text{ volts}$$
$$I_a R_a \text{ drop} = 131 \times 0{\cdot}5 = 65{\cdot}5 \text{ volts}$$
$$I_a X_s \text{ drop} = 131 \times 5 = 655 \text{ volts}$$

(*i*) **Unity p.f.** In drawing the phasor diagram for unity p.f. [See Fig. 12.22], voltage has been taken as the reference phasor.

$$E_0 = \sqrt{(V + I_a R_a)^2 + (I_a X_s)^2}$$
$$= \sqrt{(3810 + 65{\cdot}5)^2 + (655)^2} = 3930 \text{ volts/phase}$$

∴ Percentage change in terminal voltage

$$= \frac{E_0 - V}{V} \times 100 = \frac{3930 - 3810}{3810} \times 100 = \mathbf{3{\cdot}15\ \%}$$

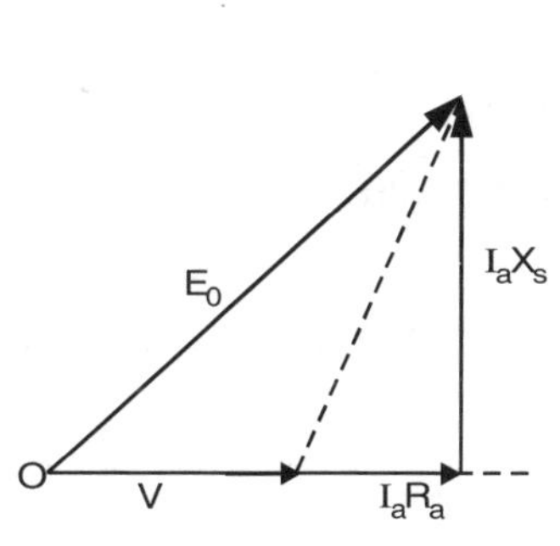

Fig. 12.22

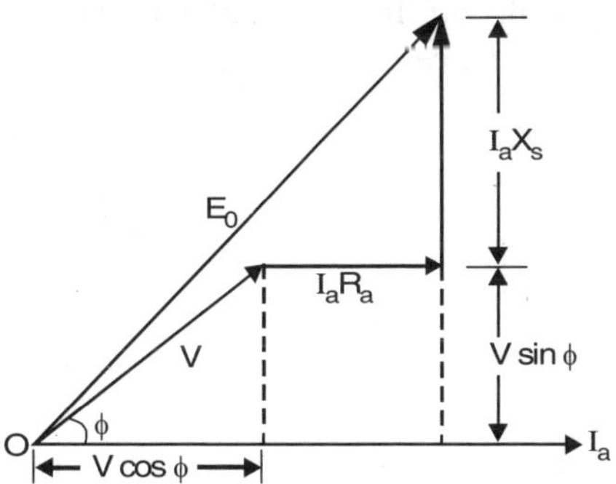

Fig. 12.23

(*ii*) **0·8 p.f. lagging.** In drawing the phasor diagram for 0·8 p.f. lagging [See Fig. 12.23], current has been taken as the reference phasor.

$$\cos\phi = 0{\cdot}8 \quad \therefore \quad \sin\phi = 0{\cdot}6$$
$$E_0 = \sqrt{(V\cos\phi + I_a R_a)^2 + (V\sin\phi + I_a X_s)^2}$$
$$= \sqrt{(3810 \times 0{\cdot}8 + 65{\cdot}5)^2 + (3810 \times 0{\cdot}6 + 655)^2} = 4283 \text{ volts/phase}$$

∴ Percentage change in terminal voltage

$$= \frac{E_0 - V}{V} \times 100 = \frac{4283 - 3810}{3810} \times 100 = \mathbf{12{\cdot}4\ \%}$$

(*iii*) **0·8 p.f. leading.** Fig. 12.24 shows the phasor diagram for 0·8 p.f. leading. Note that current has been taken as the reference phasor.

$$OB = OA + AB = V\cos\phi + I_a R_a$$

$$BC = BD - CD = V\sin\phi - I_a X_s$$

In right angled triangle OBC,

$$E_0 = \sqrt{(OB)^2 + (BC)^2}$$

$$= \sqrt{(V\cos\phi + I_a R_a)^2 + (V\sin\phi - I_a X_s)^2}$$

$$= \sqrt{(3810 \times 0{\cdot}8 + 65{\cdot}5)^2 + (3810 \times 0{\cdot}6 - 655)^2}$$

$$= 3515 \text{ volts/phase}$$

Fig. 12.24

∴ Percentage change in terminal voltage

$$= \frac{E_0 - V}{V} \times 100 = \frac{3515 - 3810}{3810} \times 100 = \mathbf{-7{\cdot}7\ \%}$$

Example 12.21. *A 60 kVA, 220 V, 50 Hz, single-phase alternator has effective resistance of 0·016 Ω and an armature leakage reactance of 0·07 Ω. Find the voltage induced in the armature when the alternator is delivering rated current at a load p.f. of* (*i*) *unity* (*ii*) *0·7 lagging and* (*iii*) *0·7 leading.*

Solution. Rated current, $I_a = \frac{60 \times 10^3}{220} = 272{\cdot}72$ A

Terminal voltage, $V = 220$ volts

$I_a R_a$ drop $= 272{\cdot}72 \times 0{\cdot}016 = 4{\cdot}3$ volts

$I_a X_L$ drop $= 272{\cdot}72 \times 0{\cdot}07 = 19$ volts

(*i*) **Unity p.f.** In drawing the phasor diagram for unity p.f. [See Fig. 12.25 (*i*)], current has been taken as the reference phasor.

$$E = \sqrt{(V + I_a R_a)^2 + (I_a X_L)^2}$$

$$= \sqrt{(220 + 4{\cdot}3)^2 + (19)^2} = \mathbf{225\ volts}$$

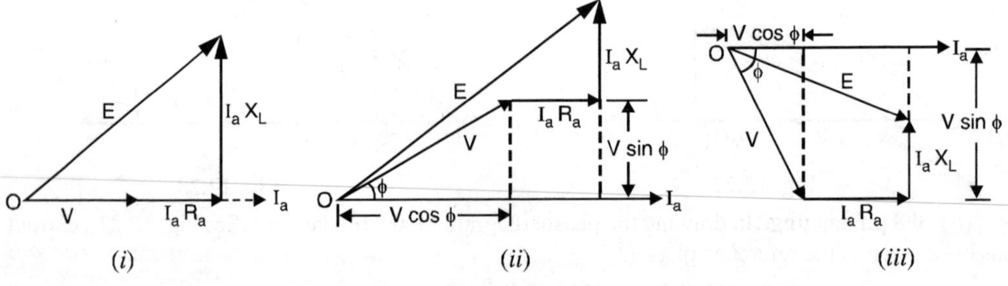

Fig. 12.25

(*ii*) **0·7 p.f. lagging** [See Fig. 12.25 (*ii*)].

$$\cos\phi = 0{\cdot}7 \quad \therefore \quad \sin\phi = 0{\cdot}7$$

$$E = \sqrt{(V\cos\phi + I_a R_a)^2 + (V\sin\phi + I_a X_L)^2}$$

$$= \sqrt{(220 \times 0{\cdot}7 + 4{\cdot}3)^2 + (220 \times 0{\cdot}7 + 19)^2} = \mathbf{234\ volts}$$

(*iii*) **0·7 p.f. leading** [See Fig. 12.25 (*iii*)].

$$E = \sqrt{(V\cos\phi + I_a R_a)^2 + (V\sin\phi - I_a X_L)^2}$$

$$= \sqrt{(220 \times 0{\cdot}7 + 4{\cdot}3)^2 + (220 \times 0{\cdot}7 - 19)^2} = \mathbf{208\ volts}$$

Example 12.22. *A 3-phase, 10 kVA, 400 V, 50 Hz, Y-connected alternator supplies the rated load at 0.8 p.f. lag. If armature resistance is 0.5 Ω and synchronous reactance is 10 Ω, find the voltage regulation.*

Solution. Fig. 12.26 shows the phasor diagram of the alternator at 0.8 p.f. lagging. Note that current has been taken as the reference phasor.

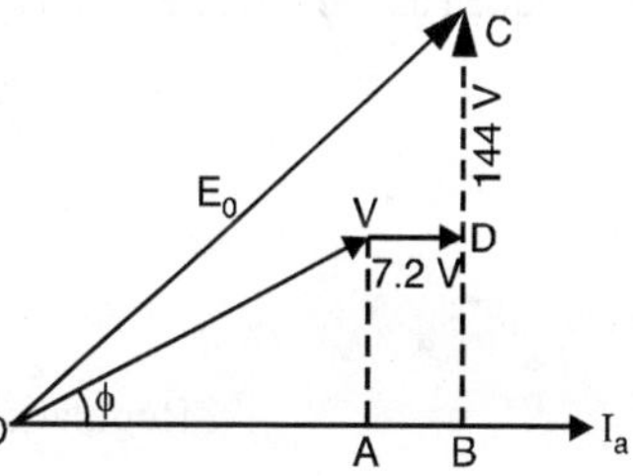

$$\text{Voltage/phase, } V = 400/\sqrt{3} = 231 \text{ volts}$$

$$\text{F.L. current, } I_L = \text{Armature current } I_a$$

$$= \frac{10 \times 10^3}{\sqrt{3} \times 400} = 14.4 \text{ A}$$

$$I_a R_a \text{ drop} = 14.4 \times 0.5 = 7.2 \text{ volts}$$

$$I_a X_s \text{ drop} = 14.4 \times 10 = 144 \text{ volts}$$

$$\cos\phi = 0.8 \text{ lag} \;;\; \sin\phi = 0.6$$

Fig. 12.26

$$\therefore \quad E_0 = \sqrt{(V\cos\phi + I_a R_a)^2 + (V\sin\phi + I_a X_s)^2}$$

$$= \sqrt{(231 \times 0.8 + 7.2)^2 + (231 \times 0.6 + 144)^2} = 342 \text{ volts}$$

$$\therefore \quad \%\text{ Regulation} = \frac{E_0 - V}{V} \times 100 = \frac{342 - 231}{231} \times 100 = \mathbf{48\%}$$

Example 12.23. *A 3-phase, 50 Hz, star-connected, 2000 kVA, 2300 V alternator gives a short-circuit current of 600 A for a certain field excitation. With the same excitation, the open circuit voltage was 900 V. The resistance between a pair of terminals was 0.12 Ω. Find the full-load regulation at 0.8 p.f. lagging.*

Solution.

$$\text{Synchronous impedance, } Z_s = \frac{\text{O.C. voltage/phase}}{\text{S.C. current/phase}} \quad \text{........for same excitation}$$

$$= \frac{900/\sqrt{3}}{600} = 0.866\ \Omega$$

The resistance between a pair of terminals is 0.12 Ω. It is the resistance of two phases connected in series. Therefore, d.c. armature resistance per phase = 0.12/2 = 0.06 Ω. The a.c. resistance is 1.5 times this value *i.e.*

$$\text{Armature resistance/phase, } R_a = 0.06 \times 1.5 = 0.09\ \Omega$$

$$\text{Synchronous reactance/phase, } X_s = \sqrt{Z_s^2 - R_a^2} = \sqrt{0.866^2 - 0.09^2} = 0.86\ \Omega$$

$$\text{F.L. Armature current, } I_a = \frac{2000 \times 10^3}{\sqrt{3} \times 2300} = 500 \text{ A}$$

$$I_a R_a = 500 \times 0.09 = 45 \text{ volts}$$

$$I_a X_s = 500 \times 0.86 = 430 \text{ volts}$$

Phase voltage, $V = 2300/\sqrt{3} = 1328$ volts

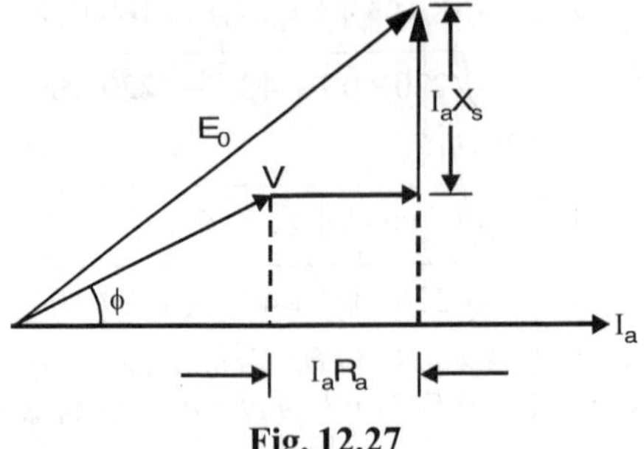

Fig. 12.27

Fig. 12.27 shows the phasor diagram of the alternator at 0.8 p.f. lagging. Note that current has been taken as the reference phasor.

$$E_0 = \sqrt{(V\cos\phi + I_aR_a)^2 + (V\sin\phi + I_aX_s)^2}$$

$$= \sqrt{(1328\times 0.8 + 45)^2 + (1328\times 0.6 + 430)^2} = 1653 \text{ V}$$

$$\therefore \quad \%\text{ Regulation} = \frac{E_0 - V}{V}\times 100 = \frac{1653 - 1328}{1328}\times 100 = \mathbf{24.5\%}$$

TUTORIAL PROBLEMS

1. A 3-phase, star-connected alternator is rated at 15000 kVA, 11 kV. The armature resistance is 1·5 Ω/phase and synchronous reactance is 30 Ω/phase. To what value will the terminal voltage rise when full-load at p.f. 0·8 lagging is switched off ? **[13970 V]**
2. A 3-phase, 11kV, star-connected alternator has armature resistance of 1 Ω/phase and a synchronous reactance of 20 Ω/phase. Find the percentage change in terminal voltage for a load of 1500 kW at a p.f. of (*i*) 0·8 lagging (*ii*) unity (*iii*) 0·8 leading. **[(*i*) 22 % (*ii*) 4·25 % (*iii*) –13·4 %]**
3. A single-phase alternator, having a synchronous reactance of 5·5 Ω and armature resistance of 0·6 Ω delivers a current of 100 A. Calculate the e.m.f. generated in the stator winding when the terminal voltage is 2000 V and the power factor of the load is 0·8 lagging. **[2412 V]**

12.13 VOLTAGE REGULATION

The voltage regulation of an alternator is defined as *the change in terminal voltage from no-load to full-load (the speed and field excitation being constant) divided by full-load voltage.*

$$\%\text{ Voltage regulation} = \frac{\text{No-load voltage} - \text{Full-load voltage}}{\text{Full-load voltage}}\times 100$$

$$= \frac{E_0 - V}{V}\times 100$$

Note that $E_0 - V$ is the arithmetic difference and not the phasor difference. The factors affecting the voltage regulation of an alternator are :

(*i*) I_aR_a drop in armature winding

(*ii*) I_aX_L drop in armature winding

(*iii*) Voltage change due to armature reaction

We have seen that change in terminal voltage due to armature reaction depends upon the armature current as well as power-factor of the load. For leading load p.f., the no-load

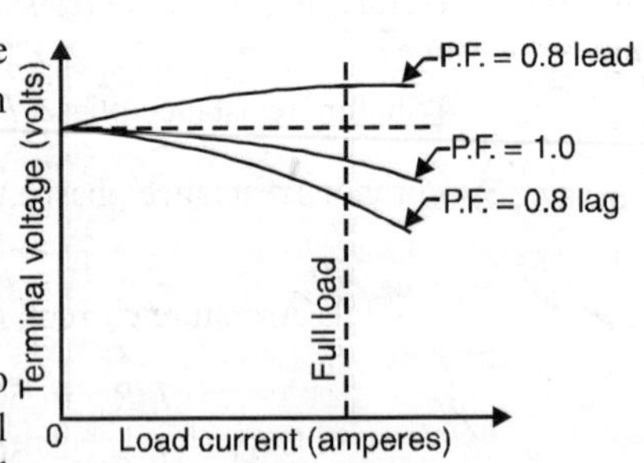

Fig. 12.28

voltage is less than the full-load voltage. Hence voltage regulation is negative in this case. The effects of different load power factors on the change in the terminal voltage with changes of load on the alternator are shown in Fig. 12.28. Since the regulation of an alternator depends on the load and the load power factor, it is necessary to mention power factor while expressing regulation.

12.14 DETERMINATION OF VOLTAGE REGULATION

The kVA ratings of commercial alternators are very high (*e.g.* 500 MVA). It is neither convenient nor practicable to determine the voltage regulation by direct loading. There are several indirect methods of determining the voltage regulation of an alternator. These methods require only a small amount of power as compared to the power required for direct loading method. Three such methods are :

1. *Synchronous Impedance or E.M.F. method*
2. *Ampere-turn or M.M.F. method*
3. *Zero power factor or Potier method*

For these methods, the following data are required :

(*i*) Armature resistance

(*ii*) Open-circuit characteristic (*O.C.C.*)

(*iii*) Short-circuit characteristic (*S.C.C.*)

[but with zero power factor lagging curve for Potier method]

(*i*) **Armature resistance.** The armature resistance R_a per phase is determined by using direct current and the voltmeter-ammeter method. This is the d.c. value. The effective armature resistance (a.c. resistance) is greater than this value due to skin effect. It is a usual practice to take the effective resistance 1·5 times the d.c. value ($R_a = 1{\cdot}5\, R_{dc}$).

(*ii*) **Open-circuit characteristic (O.C.C.).** Like the magnetisation curve for a d.c. machine, the open-circuit characteristic of an alternator is the curve between armature terminal voltage (phase value) on open circuit and the field current when the alternator is running at rated speed (synchronous speed).

Fig. 12.29 shows the circuit for determining the O.C.C. of an alternator. The alternator is run on no-load at the rated speed. The field current I_f is gradually increased from zero (by adjusting field rheostat) until open-circuit voltage E_0 (phase value) is about 50 % greater than the rated phase voltage. The graph is drawn between open-circuit voltage values and the corresponding values of I_f as shown in Fig. 12.30.

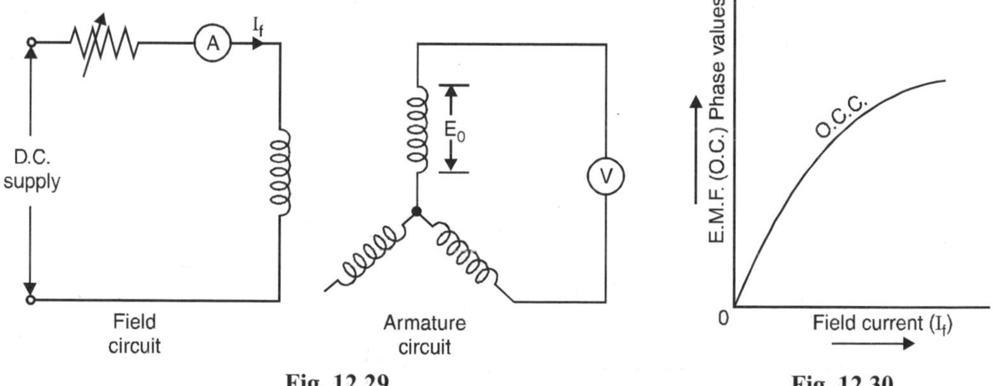

Fig. 12.29 Fig. 12.30

(*iii*) **Short-circuit characteristic (S.C.C.).** In a short-circuit test, the alternator is run at rated speed and the armature terminals are short-circuited through identical ammeters (See Fig. 12.31). Only one ammeter need be read; but three are used for balance. The field current I_f is gradually increased from zero until the short-circuit armature current I_{sc} is about twice the rated current. The graph between short-circuit armature current and field current gives the short-circuit characteristic (S.C.C.) as shown in Fig. 12.32.

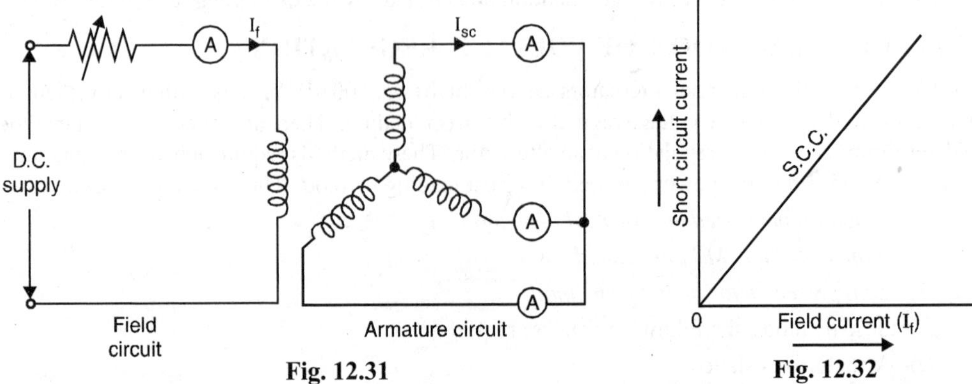

Fig. 12.31 Fig. 12.32

There is no need to take more than one reading because S.C.C. is a straight line passing through the origin. The reason is simple. Since armature resistance is much smaller than the synchronous reactance, the short-circuit armature current lags the induced voltage by very nearly 90º. Consequently, the armature flux and field flux are in direct opposition and the resultant flux is small. Since the resultant flux is small, the saturation effects will be negligible and the short-circuit armature current, therefore, is directly proportional to the field current over the range from zero to well above the rated armature current.

12.15 SYNCHRONOUS IMPEDANCE METHOD

In this method of finding the voltage regulation of an alternator, we find the synchronous impedance Z_s (and hence synchronous reactance X_s) of the alternator from the O.C.C. and S.C.C. For this reason, it is called synchronous impedance method. The method involves the following steps :

(*i*) Plot the O.C.C. and S.C.C. on the same field current base as shown in Fig. 12.33.

(*ii*) Consider a field current I_f. The open-circuit voltage corresponding to this field current is E_1. The short-circuit armature current corresponding to field current I_f is I_1. On short-circuit p.d. = 0 and voltage E_1 is being used to circulate the short-circuit armature current I_1 against the synchronous impedance Z_s. This is illustrated in Fig. 12.34.

$$\therefore \qquad E_1 = I_1 Z_s \quad \text{or} \quad Z_s = \frac{E_1 \text{ (Open-circuit)}}{I_1 \text{ (Short-circuit)}}$$

Note that E_1 is the phase value and so is I_1.

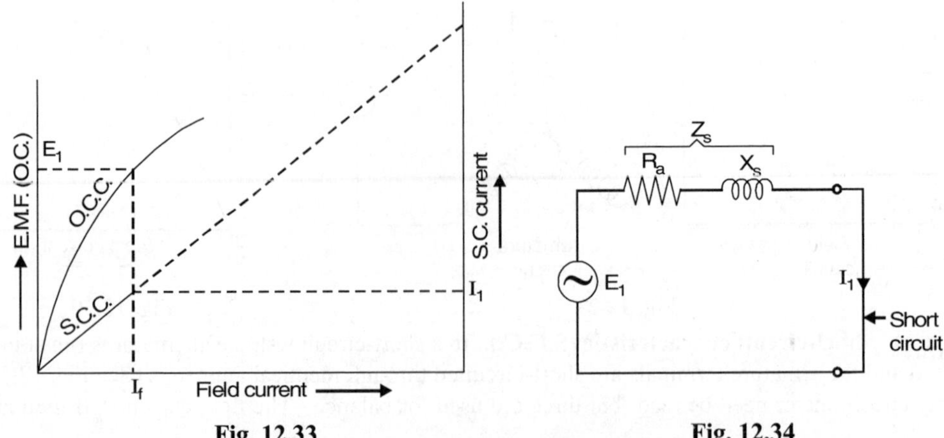

Fig. 12.33 Fig. 12.34

(*iii*) The armature resistance can be found as explained earlier.

$\therefore$ Synchronous reactance, $X_s = \sqrt{Z_s^2 - R_a^2}$

(*iv*) Once we know R_a and X_s, the phasor diagram can be drawn for any load and any p.f. Fig. 12.35 shows the phasor diagram for the usual case of inductive load; the load p.f. being cos ϕ lagging. Note that in drawing the phasor diagram, current I_a has been taken as the reference phasor. The $I_a R_a$ drop is in phase with I_a while $I_a X_s$ drop leads I_a by 90°. The phasor sum of V, $I_a R_a$ and $I_a X_s$ gives the no-load e.m.f. E_0.

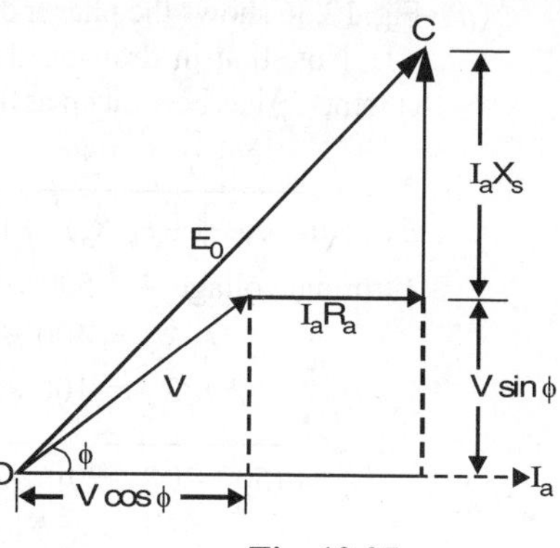

Fig. 12.35

$$E_0 = \sqrt{(OB)^2 + (BC)^2}$$

Now $\quad OB = V\cos\phi + I_a R_a \quad \text{and} \quad BC = V\sin\phi + I_a X_s$

$$\therefore \quad E_0 = \sqrt{(V\cos\phi + I_a R_a)^2 + (V\sin\phi + I_a X_s)^2}$$

$$\therefore \quad \%\ \text{voltage regulation} = \frac{E_0 - V}{V} \times 100$$

Drawback. This method is easy but it gives approximate results. The reason is simple. The combined effect of X_L (armature leakage reactance) and X_{AR} (reactance of armature reaction) is measured on short-circuit. Since the current in this condition is almost lagging *90°, the armature reaction will provide its worst demagnetising effect. It follows that under any normal operation at, say 0·8 or 0·9 lagging power factors will produce error in calculations. This method gives a value higher than the value obtained from an actual load test. For this reason, it is called *Pessimistic Method.*

Example 12.24. *A 50 kVA, 500 V, single-phase a.c. generator gave the following test results :*

Open circuit test : *A field current of 12 A produced an e.m.f. of 300 volts.*

Short-circuit test : *A field current of 12 A caused a current of 175 A to flow in the short-circuited armature.*

The effective armature resistance is 0·2 Ω.

(*i*) *Calculate the synchronous impedance and synchronous reactance.*

(*ii*) *If the a.c. generator is supplying full-load current of 100 A at 0·8 p.f. lagging, to what value would the terminal voltage rise if the load were removed ? Also find voltage regulation for this load and p.f.*

Solution.

(*i*) Synchronous impedance, $Z_s = \frac{\text{O. C volts}}{\text{S.C. current}}$ ...for same excitation

$$= \frac{300}{175} = \mathbf{1{\cdot}71\ \Omega}$$

Synchronous reactance, $X_s = \sqrt{Z_s^2 - R_a^2} = \sqrt{(1{\cdot}71)^2 - (0{\cdot}2)^2} = \mathbf{1{\cdot}69\ \Omega}$

---

* Because $X_s >> R_a$.

(*ii*) Fig. 12.36 shows the phasor diagram for lagging p.f. Note that in drawing the phasor diagram, current (I_a) has been taken as the reference phasor.

$$\cos\phi = 0{\cdot}8 \quad \therefore \quad \sin\phi = 0{\cdot}6$$

$$E_0 = \sqrt{(V\cos\phi + I_a R_a)^2 + (V\sin\phi + I_a X_s)^2}$$

Terminal voltage, $V = 500$ volts

$$I_a R_a = 100 \times 0{\cdot}2 = 20 \text{ volts}$$

$$I_a X_s = 100 \times 1{\cdot}69 = 169 \text{ volts}$$

$$\therefore E_0 = \sqrt{(500 \times 0{\cdot}8 + 20)^2 + (500 \times 0{\cdot}6 + 169)^2}$$

$$= \mathbf{630 \text{ volts}}$$

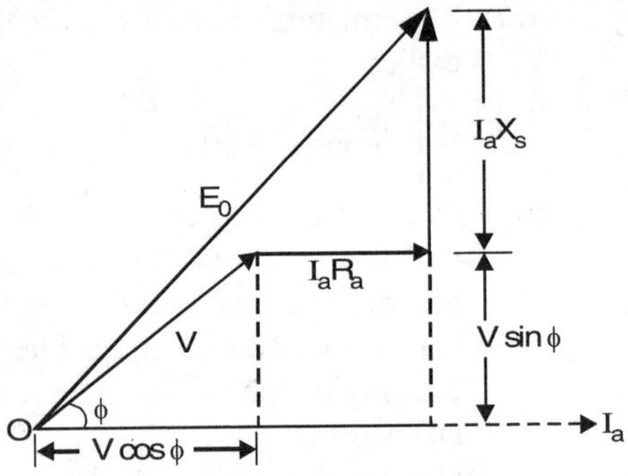

Fig. 12.36

Voltage regulation at p.f. 0·8 lagging

$$= \frac{E_0 - V}{V} \times 100 = \frac{630 - 500}{500} \times 100 = \mathbf{26\ \%}$$

Example 12.25. *The effective resistance of a 2200 V, 50 Hz, 440 kVA, 1-phase alternator is 0.5Ω. On short circuit, a field current of 40 A gives the full-load current of 200 A. The voltage on open-circuit with same field excitation is 1160 V. Calculate (i) synchronous impedance (ii) synchronous reactance.*

Solution.

(*i*) Synchronous impedance, $Z_s = \dfrac{\text{O.C. voltage}}{\text{S.C. current}}$ ...For same excitation

$$= \frac{1160}{200} = \mathbf{5.8\ \Omega}$$

(*ii*) Synchronous reactance, $X_s = \sqrt{Z_s^2 - R_a^2} = \sqrt{5.8^2 - 0.5^2} = \mathbf{5.78\ \Omega}$

Example 12.26. *A 1200 kVA, 3300 V, 50 Hz, three-phase, star-connected alternator has armature resistance of 0·25 Ω per phase. A field current of 40 A produces a short-circuit current of 200 A and an open circuit e.m.f. of 1100 V line-to-line. Find the voltage regulation on (i) full-load 0·8 p.f. lagging (ii) full-load 0·8 p.f. leading.*

Solution. F.L. current, $I_a = \dfrac{1200 \times 10^3}{\sqrt{3} \times 3300} = 210 \text{ A}$

Terminal voltage/phase, $V = 3300/\sqrt{3} = 1905$ volts

Synchronous impedance, $Z_s = \dfrac{\text{O.C. voltage/phase}}{\text{S.C. current/phase}}$ ...for same excitation

$$= \frac{1100/\sqrt{3}}{200} = 3{\cdot}17\ \Omega$$

Synchronous reactance, $X_s = \sqrt{Z_s^2 - R_a^2} = \sqrt{(3{\cdot}17)^2 - (0{\cdot}25)^2} = 3{\cdot}16\ \Omega$

(*i*) **p.f. 0·8 lagging.** Fig. 12.37 shows the phasor diagram.

$$E_0 = \sqrt{(V\cos\phi + I_a R_a)^2 + (V\sin\phi + I_a X_s)^2}$$

$I_a R_a$ drop $= 210 \times 0{\cdot}25 = 52{\cdot}5$ volts

$I_a X_s = 210 \times 3{\cdot}16 = 663{\cdot}6$ volts

$$\therefore \quad E_0 = \sqrt{(1905 \times 0{\cdot}8 + 52{\cdot}5)^2 + (1905 \times 0{\cdot}6 + 663{\cdot}6)^2} = 2398 \text{ volts/phase}$$

$$\therefore \quad \text{Voltage regulation} = \frac{E_0 - V}{V} \times 100 = \frac{2398 - 1905}{1905} \times 100 = \mathbf{25{\cdot}9\ \%}$$

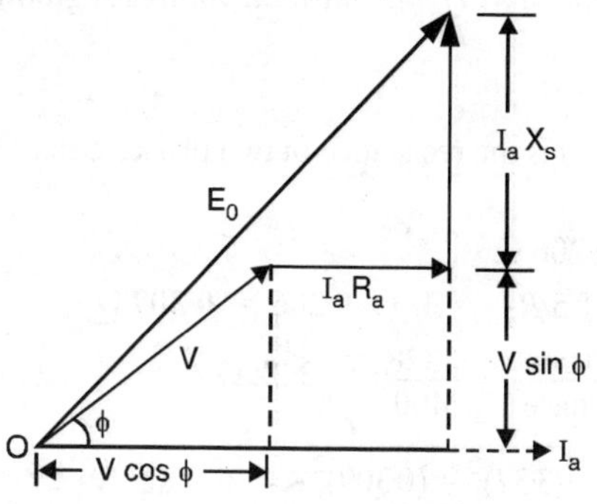

Fig. 12.37

Fig. 12.38

(*ii*) **p.f. 0·8 leading.** Fig. 12.38 shows the phasor diagram.

$$E_0 = \sqrt{(V\cos\phi + I_a R_a)^2 + (V\sin\phi - I_a X_s)^2}$$

$$= \sqrt{(1905 \times 0{\cdot}8 + 52{\cdot}5)^2 + (1905 \times 0{\cdot}6 - 663{\cdot}6)^2} = 1647 \text{ volts/phase}$$

$$\therefore \quad \text{Voltage regulation} = \frac{E_0 - V}{V} \times 100 = \frac{1647 - 1905}{1905} \times 100 = \mathbf{-13{\cdot}54\ \%}$$

Example 12.27. *A 3-phase, star-connected alternator is rated at 1600 kVA, 13500 V. The armature resistance and synchronous reactance are 1·5 Ω and 30 Ω respectively per phase. Calculate the percentage voltage regulation for a load of 1280 kW at 0·8 leading p.f.*

Solution. Fig. 12.39 shows the phasor diagram for leading p.f.

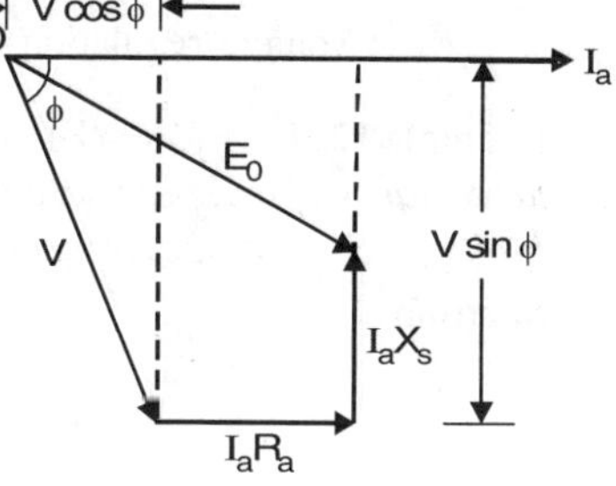

Fig. 12.39

$$\text{Load current, } I_a = \frac{1280 \times 10^3}{\sqrt{3} \times 13500 \times 0{\cdot}8} = 68{\cdot}4 \text{ A}$$

Terminal voltage/phase is

$$V = 13500/\sqrt{3} = 7795 \text{ volts}$$

$$E_0 = \sqrt{(V\cos\phi + I_a R_a)^2 + (V\sin\phi - I_a X_s)^2}$$

$$I_a R_a \text{ drop} = 68{\cdot}4 \times 1{\cdot}5 = 103 \text{ volts}$$

$$I_a X_s \text{ drop} = 68{\cdot}4 \times 30 = 2052 \text{ volts}$$

$$\therefore \quad E_0 = \sqrt{(7795 \times 0{\cdot}8 + 103)^2 + (7795 \times 0{\cdot}6 - 2052)^2} = 6861 \text{ volts/phase}$$

$$\therefore \quad \text{Voltage regulation} = \frac{E_0 - V}{V} \times 100 = \frac{6861 - 7795}{7795} \times 100 = \mathbf{-12\ \%}$$

Example 12.28. *If a field excitation of 10 A in a certain alternator gives a current of 150 A on short-circuit and a terminal voltage of 900 V (phase value) on open circuit, find the internal voltage drop with a load current of 60 A.*

Solution. Alternators are generally star-connected.

$$\text{Synchronous impedance, } Z_s = \frac{\text{O.C. volts/phase}}{\text{S.C. current/phase}} = \frac{900}{150} = 6\ \Omega$$

When the load current I_a = 60 A, then,

$$\text{Internal voltage drop} = I_a Z_s = 60 \times 6 = \mathbf{360\ V}$$

Example 12.29. *A 3-phase, 50 Hz, star-connected, 1000 kVA, 2300 V alternator gives a short-circuit current of 400 A for a certain field excitation. With the same excitation, the open-circuit voltage was 1328 V (phase value). The d.c. resistance between two lines is 0·412 Ω. Find (i)* *effective armature resistance (ii) synchronous reactance (iii) the full-load voltage regulation at 0·8 p.f. lagging.*

Solution.

(*i*) The resistance between two lines is 0·412 Ω. It is the resistance of two phases connected in series.

∴ D.C. resistance/phase, $R_{dc} = 0{\cdot}412/2 = 0{\cdot}206\ \Omega$

Effective armature resistance/phase, $R_a = 1{\cdot}5\ R_{dc} = 1{\cdot}5 \times 0{\cdot}206 =$ **0·309 Ω**

(*ii*)
$$Z_s = \frac{\text{O.C. volts/phase}}{\text{S.C. current/phase}} = \frac{1328}{400} = 3{\cdot}32\ \Omega$$

∴
$$X_s = \sqrt{Z_s^2 - R_a^2} = \sqrt{(3{\cdot}32)^2 - (0{\cdot}309)^2} = \mathbf{3{\cdot}31\ \Omega}$$

(*iii*)
$$\text{F.L. current, } I_a = \frac{1000 \times 10^3}{\sqrt{3} \times 2300} = 251\ \text{A}$$

$$\text{Terminal voltage/phase, } V = 2300/\sqrt{3} = 1328 \text{ volts}$$

$$E_0 = \sqrt{(V \cos\phi + I_a R_a)^2 + (V \sin\phi + I_a X_s)^2}$$

$$I_a R_a \text{ drop} = 251 \times 0{\cdot}309 = 77{\cdot}6 \text{ volts}$$

$$I_a X_s \text{ drop} = 251 \times 3{\cdot}31 = 831 \text{ volts}$$

∴
$$E_0 = \sqrt{(1328 \times 0{\cdot}8 + 77{\cdot}6)^2 + (1328 \times 0{\cdot}6 + 831)^2}$$

$$= \sqrt{(1140)^2 + (1628)^2} = 1987 \text{ volts/phase}$$

$$\text{Voltage regulation} = \frac{E_0 - V}{V} \times 100 = \frac{1987 - 1328}{1328} \times 100 = \mathbf{49{\cdot}6\ \%}$$

Example 12.30. *A 1500 kVA, 6600 V, 3-phase, star-connected alternator with a resistance of 0·4 Ω and reactance of 6 Ω per phase delivers full-load current at power factor 0·8 lagging and normal rated voltage. Estimate the terminal voltage for the same excitation and load current at 0·8 p.f. leading.*

Solution.

$$\text{F.L. current, } I_a = \frac{1500 \times 10^3}{\sqrt{3} \times 6600} = 131\ \text{A}$$

$$\text{Terminal voltage/phase, } V = 6600/\sqrt{3} = 3810 \text{ volts}$$

$$E_0 = \sqrt{(V \cos\phi + I_a R_a)^2 + (V \sin\phi + I_a X_s)^2}$$

$$I_a R_a \text{ drop} = 131 \times 0{\cdot}4 = 52{\cdot}4 \text{ volts}$$

$$I_a X_s \text{ drop} = 131 \times 6 = 786 \text{ volts}$$

∴
$$E_0 = \sqrt{(3810 \times 0{\cdot}8 + 52{\cdot}4)^2 + (3810 \times 0{\cdot}6 + 786)^2}$$

$$= 4364 \text{ volts/phase}$$

Since excitation remains constant, no-load e.m.f. in the second case will also be 4364 volts/phase. Let the new terminal voltage/phase be V'.

∴
$$E_0 = \sqrt{(V' \cos\phi + I_a R_a)^2 + (V' \sin\phi - I_a X_s)^2}$$

or
$$4364 = \sqrt{(V' \times 0{\cdot}8 + 52{\cdot}4)^2 + (V' \times 0{\cdot}6 - 786)^2}$$

$$\text{On solving, } V' = 4743 \text{ volts/phase}$$

$$\text{Line-to-line terminal voltage} = \sqrt{3} \times 4743 = \mathbf{8215\ volts}$$

Example 12.31. *A 3-phase, 10 kVA, 400 V, 50-Hz, Y-connected alternator supplies the rated load of 0·8 p.f. lagging. If the armature resistance is 0·5 Ω/phase and synchronous reactance is 10 Ω/phase, find (i) voltage regulation (ii) power angle.*

Solution. Fig. 12.40 shows the phasor diagram for 0·8 p.f. lagging. Note that in drawing the phasor diagram, current (I_a) has been taken as the reference phasor.

$$\text{F.L. current, } I_a = \frac{10 \times 10^3}{\sqrt{3} \times 400} = 14·4 \text{ A}$$

Terminal voltage/phase is

$$V = 400/\sqrt{3} = 231 \text{ volts}$$

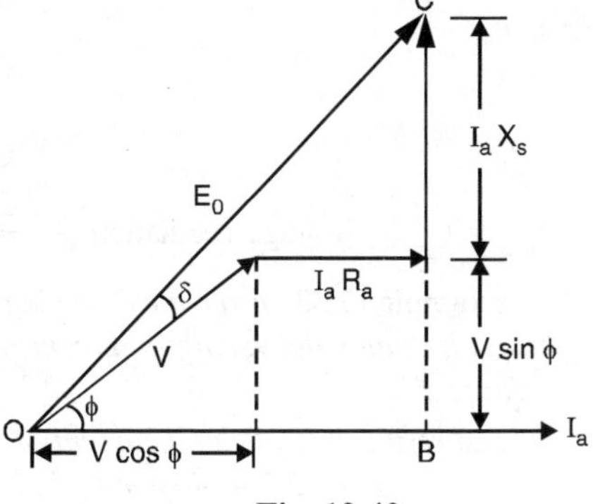

Fig. 12.40

(*i*) $E_0 = \sqrt{(V \cos\phi + I_a R_a)^2 + (V \sin\phi + I_a X_s)^2}$

$$I_a R_a \text{ drop} = 14·4 \times 0·5 = 7·2 \text{ volts}$$

$$I_a X_s \text{ drop} = 14·4 \times 10 = 144 \text{ volts}$$

$$\therefore \quad E_0 = \sqrt{(231 \times 0·8 + 7·2)^2 + (231 \times 0·6 + 144)^2} = 342 \text{ volts/phase}$$

$$\therefore \quad \text{Voltage regulation} = \frac{E_0 - V}{V} \times 100 = \frac{342 - 231}{231} \times 100 = \mathbf{48\ \%}$$

(*ii*) The power angle δ of the machine is the angle between E_0 and V. Referring to Fig. 12.40,

$$\tan(\phi + \delta) = \frac{BC}{OB} = \frac{V \sin\phi + I_a X_s}{V \cos\phi + I_a R_a} = \frac{231 \times 0·6 + 144}{231 \times 0·8 + 7·2}$$

$$= \frac{282·6}{192} = 1·472$$

$$\therefore \quad \phi + \delta = \tan^{-1} 1·472 = 55·8°$$

$$\therefore \quad \text{Power angle, } \delta = 55·8° - 36·9° = \mathbf{18·9°} \qquad (\because \phi = \cos^{-1} 0·8 = 36·9°)$$

Example 12.32. *A 10 MVA, 6·6 kV, 3-phase, star-connected alternator gave open-circuit and short-circuit test figures as under :*

Field current in amperes	:	*25*	*50*	*75*	*100*	*125*	*150*
Line e.m.f. in kV	:	*2·4*	*4·8*	*6·1*	*7·1*	*7·6*	*7·9*
Short-circuit current in A	:	*288*	*528*	*875*			

Find the voltage regulation on full-load at a p.f. of 0·8 lagging. Armature resistance/phase = 0·13 Ω.

Solution.

$$\text{F.L. current, } I_a = \frac{10 \times 10^6}{\sqrt{3} \times 6·6 \times 10^3} = 875 \text{ A}$$

$$\text{F.L. terminal voltage/phase, } V = \frac{6·6 \times 10^3}{\sqrt{3}} = 3810 \text{ volts}$$

The F.L. current of 875 A is produced by a field current of 75 A ; this latter current also produces an open-circuit line e.m.f. of 6·1 kV = 6100 volts.

$$\therefore \quad Z_s = \frac{\text{O.C. volt/phase}}{\text{S.C. current/phase}} = \frac{6100/\sqrt{3}}{875} = 4·02\ \Omega$$

Since R_a is small*, $X_s = Z_s = 4{\cdot}02\,\Omega$.

$$E_0 = \sqrt{(V\cos\phi + I_a R_a)^2 + (V\sin\phi + I_a X_s)^2}$$

$$= \sqrt{(3810 \times 0{\cdot}8 + 875 \times 0{\cdot}13)^2 + (3810 \times 0{\cdot}6 + 875 \times 4{\cdot}02)^2}$$

$$= \sqrt{(3162)^2 + (5803{\cdot}5)^2} = 6609 \text{ volts/phase}$$

$$\therefore \quad \text{Voltage regulation} = \frac{E_0 - V}{V} \times 100 = \frac{6609 - 3810}{3810} \times 100 = \mathbf{73{\cdot}5\ \%}$$

Example 12.33. *From the following test results, determine the voltage regulation of a 2000 V, 1-phase alternator delivering a current of 100 A at (i) unity p.f. (ii) 0.71 lagging p.f. and (iii) 0.8 leading p.f. :*

Test results : Full-load current of 100 A is produced on short-circuit by a field excitation of 2.5 A. An e.m.f. of 500 V is produced on open circuit by the same excitation.

The armature resistance is 0.8 Ω.

Solution.

$$\text{Synchronous impedance, } Z_s = \frac{\text{O.C. Voltage}}{\text{S.C. Current}} \quad \text{... for same excitation}$$

$$= 500/100 = 5\,\Omega$$

$$\text{Synchronous reactance, } X_s = \sqrt{Z_s^2 - R_a^2} = \sqrt{5^2 - 0.8^2} = 4.936\,\Omega$$

$I_a R_a$ drop = 100 × 0.8 = 80 volts, $I_a X_s$ drop = 100 × 4.936 = 494 volts.

(*i*) **At unity p.f.** (*i.e.* cos ϕ = 1). Fig. 12.41 (*i*) shows the phasor diagram of the alternator at unity p.f. Note that current has been taken as the reference phasor. Referring to phasor diagram in Fig. 12.41. (*i*), we have,

$$E_0 = \sqrt{(V + I_a R_a)^2 + (I_a X_s)^2}$$

$$= \sqrt{(2000 + 80)^2 + (494)^2} = 2140 \text{ volts}$$

$$\therefore \quad \% \text{ Regulation} = \frac{E_0 - V}{V} \times 100 = \frac{2140 - 2000}{2000} \times 100 = \mathbf{7\%}$$

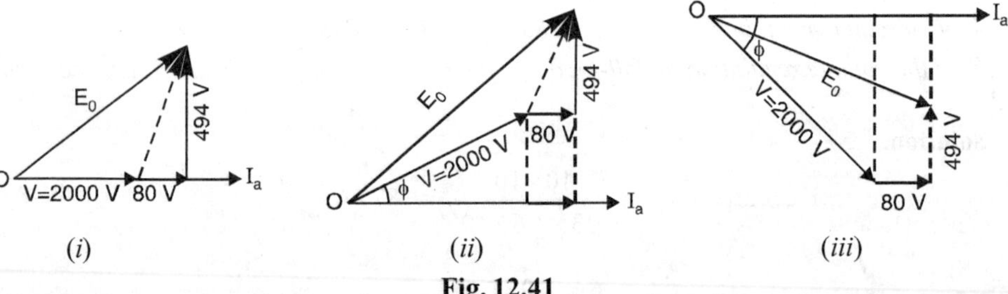

Fig. 12.41

(*ii*) **At 0.71 lagging p.f.** (*i.e.* cos ϕ = 0.71). Fig. 12.41 (*ii*) shows the phasor diagram of the alternator at 0.71 lagging p.f. Again, current is taken as the reference phasor. Referring to the phasor diagram in Fig. 12.41 (*ii*), we have,

---

* $X_s = \sqrt{Z_s^2 - R_a^2} = \sqrt{(4{\cdot}02)^2 - (0{\cdot}13)^2} = 4{\cdot}0179\,\Omega \simeq 4{\cdot}02\,\Omega$

In most large machines, $R_a \ll X_s$ so that $X_s \simeq Z_s$. For example, in the present case, the difference between X_s and Z_s is 0·0021 Ω or about 0·05 %. It is outside the accuracy limits of any likely measurement.

$$E_0 = \sqrt{(V\cos\phi + I_a R_a)^2 + (V\sin\phi + I_a X_s)^2}$$

$$= \sqrt{(2000 \times 0.71 + 80)^2 + (2000 \times 0.71 + 494)^2} = 2432 \text{ volts}$$

$$\therefore \quad \% \text{ Regulation} = \frac{E_0 - V}{V} \times 100 = \frac{2432 - 2000}{2000} \times 100 = \mathbf{21.6\%}$$

(*iii*) **At 0.8 leading p.f.** (*i.e.* $\cos\phi = 0.8$). Fig. 12.41 (*iii*) shows the phasor diagram of the alternator at 0.8 leading p.f. It is clear that :

$$E_0 = \sqrt{(V\cos\phi + I_a R_a)^2 + (V\sin\phi - I_a X_s)^2}$$

$$= \sqrt{(2000 \times 0.8 + 80)^2 + (2000 \times 0.6 - 494)^2} = 1820 \text{ volts}$$

$$\therefore \quad \% \text{ Regulation} = \frac{E_0 - V}{V} \times 100 = \frac{1820 - 2000}{2000} \times 100 = \mathbf{-9\%}$$

TUTORIAL PROBLEMS

1. A 3-phase, 600 kVA alternator has a rated terminal voltage of 3300 V (line value). The stator winding is star-connected and has a resistance of 0·37 Ω/phase and a synchronous reactance of 4·3 Ω/phase. Calculate the voltage regulation for full-load at a p.f. of (*i*) unity (*ii*) 0·8 lagging. **[(*i*) 4·7 % (*ii*) 17·3 %]**
2. A single-phase alternator has a rated output of 500 kVA at a terminal voltage of 3300 V. The stator winding has a resistance of 0·6 Ω and a synchronous reactance of 4 Ω. Calculate the voltage regulation for full-load at a p.f. of (*i*) unity (*ii*) 0·8 lagging (*iii*) 0·8 leading. **[(*i*) 4·4 % (*ii*) 14 % (*iii*) –7·3 %]**
3. A 3-phase, 50 Hz, star-connected, 2000 kVA, 2300 V alternator gives a short-circuit current of 600 A for a certain field excitation. With the same excitation, the open circuit voltage was 900 V (line value). The resistance between a pair of terminals was 0·12 Ω. Find full-load voltage regulation at (*i*) unity p.f. (*ii*) 0·8 p.f. lagging. **[(*i*) 7·3 % (*ii*) 23·7 %]**
4. A 3-phase, star-connected alternator has a resistance of 0·5 Ω/phase and a synchronous reactance of 5 Ω/phase. It is excited to give 6600 V (line) on open circuit. Determine the terminal voltage and voltage regulation on full-load current of 130 A when the load p.f. is (*i*) 0·8 lagging (*ii*) 0·6 leading. **[(*i*) 3340 V (*phase value*) ; 14·1 % (*ii*) 4260 V (*phase value*) ; 10·6 %]**
5. A 6600 V, 3-phase, star-connected alternator has a synchronous impedance of (0·4 + *j* 6) Ω per phase. Determine the voltage regulation of the machine when supplying a load of 1000 kW at normal voltage at p.f. (*i*) 0·866 lagging (*ii*) unity (*iii*) 0·866 leading. **[(*i*) 9·7 % (*ii*) 1·84 % (*iii*) – 6·03 %]**

12.16 AMPERE-TURN METHOD

This method of finding voltage regulation considers the opposite view to the synchronous impedance method. *It assumes the armature leakage reactance to be additional armature reaction.* Neglecting armature resistance (always small), this method assumes that change in terminal p.d. on load is due entirely to armature reaction. The same two tests (*viz* open-circuit and short-circuit test) are required as for synchronous reactance determination; the interpretation of the results only is different. Under short-circuit, the current lags by 90° (R_a considered zero) and the power factor is zero. Hence the armature reaction is entirely demagnetising. Since the terminal p.d. is zero, all the field *AT* (ampere-turns) are neutralised by armature *AT* produced by the short circuit armature current.

(*i*) Suppose the alternator is supplying full-load current at normal voltage V (*i.e.*, operating load voltage) and *zero p.f. lagging*. Then d.c. field *AT* required will be those needed to produce normal voltage V (or if R_a is to be taken into account, then $V + I_a R_a \cos\phi$) on *no-load* plus those to overcome the armature reaction.

Let AO = field *AT* required to produce the normal voltage V (or $V + I_a R_a \cos\phi$) at no-load

OB_1 = field *AT* required to neutralise the armature reaction

Then total field *AT* required are the phasor sum of *AO* and OB_1 [See Fig. 12.42 (*i*)] *i.e.*,

$$\text{Total field } AT,\ AB_1 = AO + OB_1 \quad \text{...phasor sum}$$

The *AO* can be found from O.C.C. and OB_1 can be determined from S.C.C. Note that the use of a d.c. quantity (field *AT*) as a phasor is perfectly valid in this case because the d.c. field is rotating at the same speed as the a.c. phasors *i.e.*, $\omega = 2\,\pi f$.

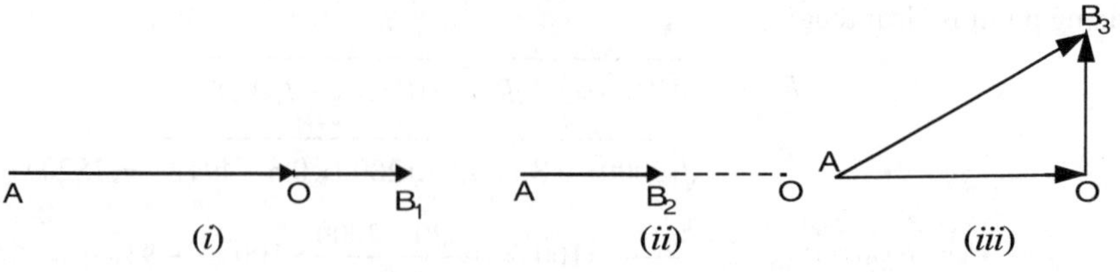

Fig. 12.42

(*ii*) For a full-load current of *zero p.f. leading*, the armature *AT* are unchanged. Since they aid the main field, less field *AT* are required to produce the given e.m.f.

$$\therefore \quad \text{Total field } AT,\ AB_2 = AO - B_2O \quad \text{...phasor difference}$$

where B_2O = field *AT* required to neutralise armature reaction

This is illustrated in Fig. 12.42 (*ii*). Note that again *AO* is determined from O.C.C. and B_2O from S.C.C.

(*iii*) Between zero lagging and zero leading power factors, the armature m.m.f. rotates through 180°. At unity p.f., armature reaction is cross-magnetising only. Therefore, OB_3 is drawn perpendicular to *AO* [See Fig. 12.42 (*iii*)]. Now AB_3 shows the required *AT* in magnitude and direction.

General case. We now discuss the case when the p.f. has any value between zero (lagging or leading) and unity. If the power-factor is cos ϕ lagging, then ϕ is laid off to the right of the vertical line OB_3 as shown in Fig. 12.43 (*i*). The total field *AT* required are AB_4 *i.e.*, phasor sum of *AO* and OB_4. If the power factor is cos ϕ leading, then ϕ is laid off to the left of the vertical line OB_3 as shown in Fig. 12.43 (*ii*). The total field *AT* required are AB_5 *i.e.*, phasor sum of *AO* and OB_5.

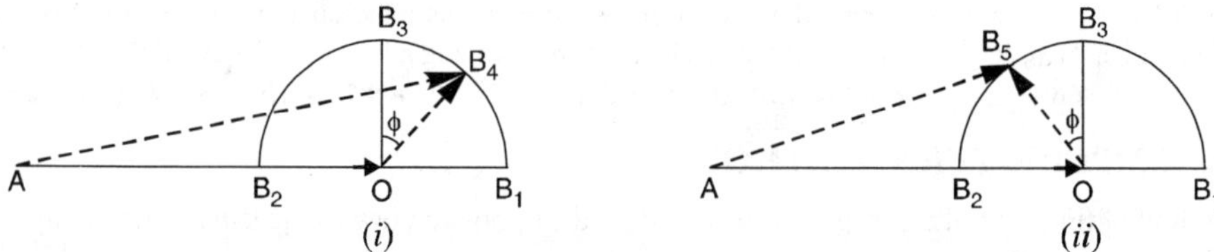

Fig. 12.43

Since current $\propto AT$, it is more convenient to work in terms of field current. Fig. 12.44 shows the current diagram for the usual case of lagging power factor. Here *AO* represents the field current required to produce normal voltage *V* (or $V + I_a R_a \cos\phi$) on no-load. The phasor *OB* represents the field current required for producing full-load current on short-circuit. The resultant field current is *AB* and is the phasor sum of *AO* and *OB*. Note that phasor *AB* represents the field current required for demagnetising and to produce voltage *V* and $I_a R_a \cos\phi$ drop (if R_a is taken into account).

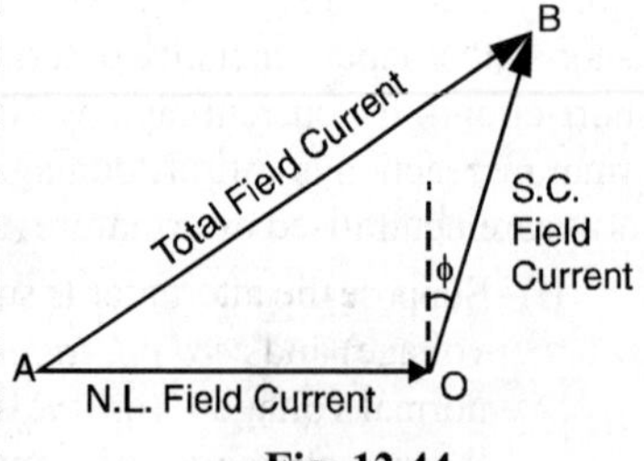

Fig. 12.44

12.17 PROCEDURE FOR AT METHOD

Suppose the alternator is supplying full-load current I_a at operating voltage V and p.f. cos ϕ lagging. The procedure for finding voltage regulation for AT method is as under :

(*i*) From the O.C.C., field current OA required to produce the operating load voltage V (or $V + I_a R_a \cos\phi$) is determined [See Fig. 12.45]. The field current OA is laid off horizontally as shown in Fig. 12.46.

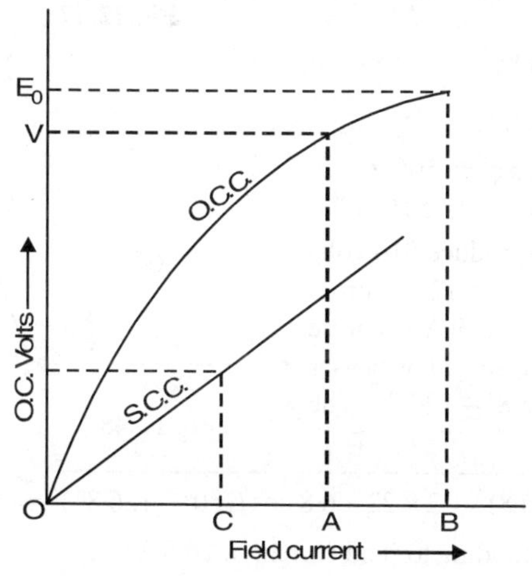

Fig. 12.45

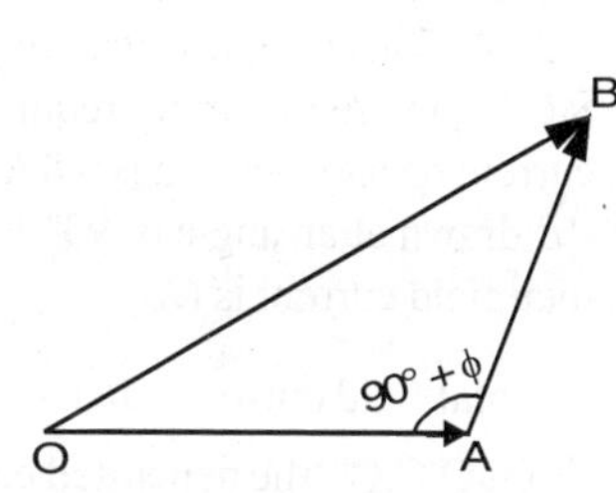

Fig. 12.46

(*ii*) From S.C.C., the field current OC required for producing full-load current I_a on short-circuit is determined. The phasor AB (= OC) is drawn at an angle of *(90° + ϕ) *i.e.*, $\angle OAB$ = (90° + ϕ) as shown in Fig. 12.46.

(*iii*) The phasor sum of OA and AB gives the total field current OB required. The O.C. voltage E_0 corresponding to field current OB on O.C.C. is the no-load e.m.f.

$$\therefore \quad \text{Voltage regulation} = \frac{E_0 - V}{V} \times 100 \qquad \text{...at }^{\dagger}\text{full-load}$$

This method gives a regulation lower than the actual performance of the machine. For this reason, it is known as *Optimistic Method*.

Example 12.34. *A 50 kVA, 500 V, single-phase a.c. generator gave the following results in the open and short-circuit tests :*

Field current (Amps) :	*5*	*10*	*15*	*20*	*25*	*30*
E.M.F. (volts) :	*125*	*250*	*370*	*480*	*566*	*640*
S.C. Armature current (Amps) :	*73*	*146*	*220*			

Using the ampere-turn method, find the full-load voltage regulation at (i) unity p.f. (ii) p.f. 0·8 lagging (iii) p.f. 0·8 leading. Effective armature resistance is 0·2 Ω.

Solution. From the given data, **plot O.C.C. and S.C.C. on the same field base.

$$\text{F.L. current, } I_a = \frac{50 \times 10^3}{500} = 100 \text{ A}$$

(*i*) **At Unity p.f.**

$$\text{E.M.F. at full-load current} = V + I_a R_a \cos\phi = 500 + 100 \times 0{\cdot}2 \times 1 = 520 \text{ volts}$$

* This is for lagging p.f. For leading p.f., the angle will be (90° – ϕ).

† The procedure is valid for any load current.

** The reader should plot these characteristics to verify the results.

From O.C.C., the field current required to produce 520 volts is 22·1 A. From S.C.C., the field current required to produce 100 A armature current on short-circuit is 6·8 A [See Fig. 12.47].

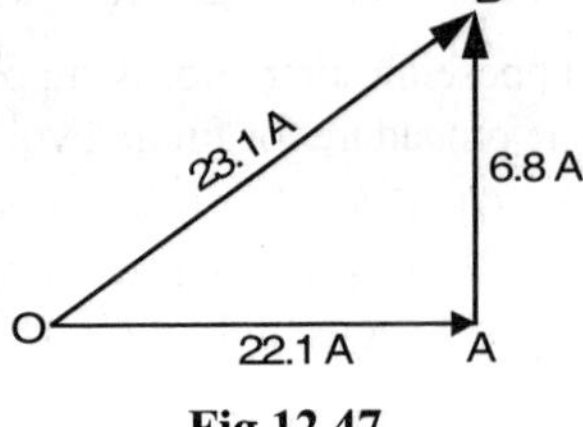

Fig.12.47

$$\therefore \quad \text{Total field current} = \sqrt{(22{\cdot}1)^2 + (6{\cdot}8)^2} = 23{\cdot}1 \text{ A}$$

From O.C.C., the generated e.m.f. corresponding to field current of 23·1 A is 538 volts *i.e.*, $E_0 = 538$ volts.

$$\text{Voltage regulation} = \frac{E_0 - V}{V} \times 100 = \frac{538 - 500}{500} \times 100 = \mathbf{7{\cdot}6\ \%}$$

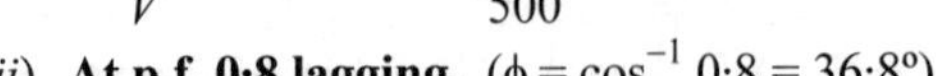

(*ii*) **At p.f. 0·8 lagging.** ($\phi = \cos^{-1} 0{\cdot}8 = 36{\cdot}8°$)

$$\text{E.M.F. at full-load current} = V + I_a R_a \cos\phi = 500 + 100 \times 0{\cdot}2 \times 0{\cdot}8 = 516 \text{ volts}$$

From O.C.C., the field current required to produce 516 volts is 22 A. This is represented by phasor *OA* in Fig. 12.48. From S.C.C., the field current required to produce 100 A armature current on short-circuit is 6·8 A. This is represented by phasor *AB* drawn at an angle of $90° + \phi = 90° + 36{\cdot}8° = 126{\cdot}8°$. The total field current is *OB*.

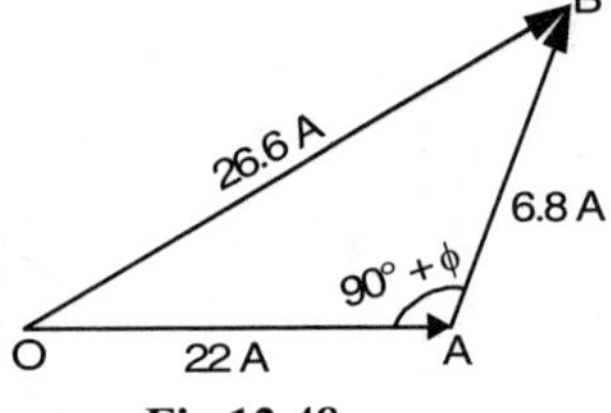

Fig.12.48

$$\text{Total field current, } OB = \sqrt{(22)^2 + (6{\cdot}8)^2 + 2 \times 22 \times 6{\cdot}8 \cos(180° - 126{\cdot}8°)} = 26{\cdot}6 \text{ A}$$

From O.C.C., the generated e.m.f. corresponding to field current of 26·6 A is $E_0 = 592$ volts.

$$\therefore \quad \text{Voltage regulation} = \frac{E_0 - V}{V} \times 100 = \frac{592 - 500}{500} \times 100 = \mathbf{18{\cdot}4\ \%}$$

(*iii*) **At p.f. 0·8 leading.**

$$\text{E.M.F. at full-load current} = V + I_a R_a \cos\phi = 500 + 100 \times 0{\cdot}2 \times 0{\cdot}8 = 516 \text{ volts}$$

From O.C.C., the field current required to produce 516 volts is 22 A. This is represented by phasor *OA* in Fig. 12.49. From S.C.C., the field current required to produce 100 A armature current on short-circuit is 6·8 A. This is represented by phasor *AB* drawn at an angle of $\phi = 36{\cdot}8°$ to the left of the vertical line *AC*. The total field current is *OB*.

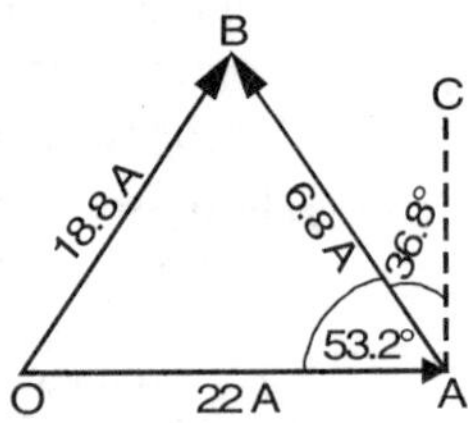

Fig.12.49

$$\text{Total field current, } OB = \sqrt{(22)^2 + (6{\cdot}8)^2 + 2 \times 22 \times 6{\cdot}8 \cos 126{\cdot}8°} = 18{\cdot}8 \text{ A}$$

From O.C.C., the generated e.m.f. corresponding to field current of 18·8 A is $E_0 = 455$ volts.

$$\therefore \quad \text{Voltage regulation} = \frac{E_0 - V}{V} \times 100 = \frac{455 - 500}{500} \times 100 = \mathbf{-\ 9\%}$$

Example 12.35. *A 3·5 MVA, Y-connected alternator rated at 4160 volts at 50 Hz has open-circuit characteristics given by the following data :*

Field current (amps) :	*50*	*100*	*150*	*200*	*250*	*300*	*350*	*400*
Line E.M.F. (volts) :	*1620*	*3150*	*4160*	*4750*	*5130*	*5370*	*5550*	*5650*

A field current of 200 A is found necessary to circulate full-load current on short-circuit of the alternator. Calculate by (i) synchronous impedance method and (ii) ampere-turn method the full-load voltage regulation at 0·8 p.f. lagging. Neglect armature resistance.

Solution. Plot the O.C.C. and S.C.C. on the same field base.

$$\text{F.L. current, } I_a = \frac{3{\cdot}5 \times 10^6}{\sqrt{3} \times 4160} = 486 \text{ A}$$

(*i*) **Synchronous impedance method.** From the O.C.C., the field current of 200 A produces O.C. voltage of 4750 volts (line value).

$$Z_s = \frac{\text{O.C. volts/phase}}{\text{S.C. current/phase}} = \frac{4750/\sqrt{3}}{486} = 5{\cdot}64\ \Omega/\text{phase}$$

$$\text{Since } R_a = 0 \ ; \ X_s = Z_s = 5{\cdot}64\ \Omega/\text{phase}$$

$$\therefore \quad I_a R_a = 0 \ ; \ I_a X_s = 486 \times 5{\cdot}64 = 2740 \text{ volts}$$

$$\text{F.L. voltage/phase, } V = 4160/\sqrt{3} = 2400 \text{ volts}$$

$$E_0 = \sqrt{(V\cos\phi + I_a R_a)^2 + (V\sin\phi + I_a X_s)^2}$$

$$= \sqrt{(2400 \times 0 \cdot 8 + 0)^2 + (2400 \times 0 \cdot 6 + 2740)^2} = 4600 \text{ volts}$$

$$\therefore \quad \text{Voltage regulation} = \frac{E_0 - V}{V} \times 100$$

$$= \frac{4600 - 2400}{2400} \times 100 = \mathbf{91{\cdot}7\ \%}$$

(*ii*) **Ampere-turn method.** From O.C.C., the field current required to produce normal voltage of 4160 volts (line value) is 150 A. This is represented by the phasor *OA* in Fig. 12.50. The field current required to produce F.L. armature current on short-circuit is 200 A. This is represented by the phasor *AB* drawn at an angle $\phi (= 36{\cdot}8°)$ to the right of the vertical line *AC*. The total field current is *OB*.

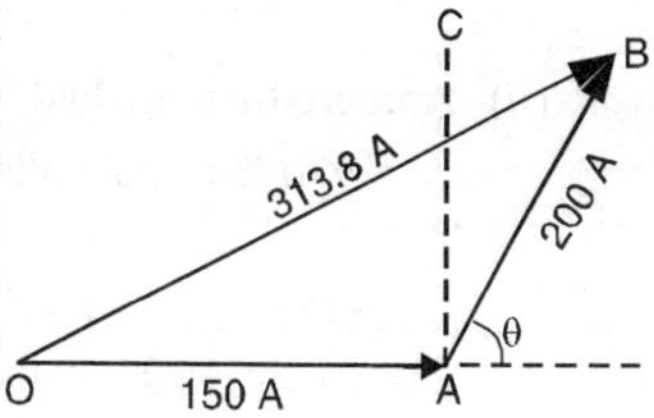

Fig. 12.50

Total field current is given by ;

$$OB = \sqrt{(150)^2 + (200)^2 + 2 \times 150 \times 200 \times \cos {}^*\theta}$$

$$= \sqrt{(150)^2 + (200)^2 + 2 \times 150 \times 200 \times \cos 53 \cdot 2°} = 313{\cdot}8 \text{ A}$$

From O.C.C., the generated e.m.f. corresponding to field current of 313·8 A is $E_0 = 5440$ volts (line value).

$$\therefore \quad \text{Voltage regulation} = \frac{5440 - 4160}{4160} \times 100 = \mathbf{30{\cdot}7\ \%}$$

There is wide difference between this and the value given by the synchronous impedance method. The true value lies somewhere between the two. Nevertheless, the synchronous impedance method is widely used for preliminary prediction calculations.

Example 12.36. *A 10 MVA, 6·6 kV, 3-phase, star-connected alternator has provided O.C.C. and S.C.C. test figures as under :*

Field current (Amps)	:	*25*	*50*	*75*	*100*	*125*	*150*	*175*	*200*	*225*
Line E.M.F. (kV)	:	*2·4*	*4·8*	*6·1*	*7·1*	*7·6*	*7·9*	*8·3*	*8·5*	*8·7*
S.C. current (Amps)	:	*288*	*582*	*875*						

Calculate by (*i*) *synchronous impedance method and* (*ii*) *ampere-turn method the full-load voltage regulation at 0·8 p.f. lagging. The armature resistance* $R_a = 0{\cdot}13\ \Omega/\text{phase}$.

Solution. Plot the O.C.C. and S.C.C. on the same field base.

$$\text{F.L. current, } I_a = \frac{10 \times 10^6}{\sqrt{3} \times 6 \cdot 6 \times 10^3} = 875 \text{ A}$$

$$\text{F.L. voltage/phase, } V = 6 \cdot 6/\sqrt{3} = 3{\cdot}81 \text{ kV} = 3810 \text{ volts}$$

---

* $\theta = 180° - (90° + 36{\cdot}8°) = 53{\cdot}2°$

(*i*) **Synchronous impedance method.** On short-circuit, the full-load armature current of 875 A is produced by field current of 75 A. Also field current of 75 A produces O.C. voltage of 6·1 kV (line value).

$$\therefore \quad Z_s = \frac{\text{O.C. volts/phase}}{\text{S.C. current/phase}} = \frac{6{\cdot}1 \times 10^3/\sqrt{3}}{875} = 4{\cdot}02\,\Omega$$

$$X_s = \sqrt{Z_s^2 - R_a^2} = \sqrt{(4\cdot 02)^2 - (0\cdot 13)^2} \simeq 4{\cdot}02\,\Omega$$

$$\cos\phi = 0{\cdot}8 \quad \therefore \quad \sin\phi = 0{\cdot}6$$

$$E_0 = \sqrt{(V\cos\phi + I_a R_a)^2 + (V\sin\phi + I_a X_s)^2}$$

$$= \sqrt{(3810 \times 0\cdot 8 + 875 \times 0\cdot 13)^2 + (3810 \times 0\cdot 6 + 875 \times 4\cdot 02)^2}$$

$$= \sqrt{(3162)^2 + (5803\cdot 5)^2} = 6609 \text{ volts/phase}$$

$$\therefore \quad \text{Voltage regulation} = \frac{E_0 - V}{V} \times 100 = \frac{6609 - 3810}{3810} \times 100 = \mathbf{73{\cdot}46\ \%}$$

(*ii*) **Ampere-turn method**

$$\text{Normal voltage} = V + I_a R_a \cos\phi = 3810 + 875 \times 0{\cdot}13 \times 0{\cdot}8$$
$$= 3901 \text{ volts/phase}$$

From O.C.C., the field current required to produce normal voltage of 3901 volts (phase value) is 98 A. This is represented by the phasor *OA* in Fig. 12.51. The field current required to produce F.L. armature current of 875 A on short-circuit is 75 A. This is represented by the phasor *AB* drawn at an angle ϕ (= 36·8°) to the right of the vertical line *AC*. The total field current is *OB*. Referring to Fig. 12.51, $\theta = 90° - 36{\cdot}8° = 53{\cdot}2°$.

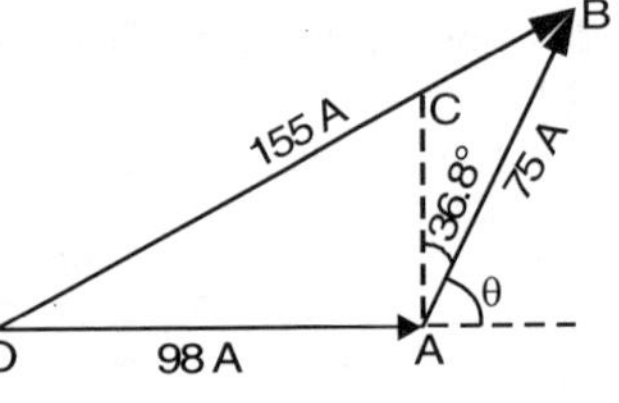

Fig. 12.51

$$\text{Total field current, } OB = \sqrt{(98)^2 + (75)^2 + 2 \times 98 \times 75 \times \cos 53\cdot 2°} = 155 \text{ A}$$

From O.C.C., the generated e.m.f. corresponding to field current of 155 A is E_0 = 4619 volts/phase.

$$\therefore \quad \text{Voltage regulation} = \frac{E_0 - V}{V} \times 100 = \frac{4619 - 3810}{3810} \times 100 = \mathbf{21{\cdot}2\ \%}$$

Example 12.37. *The no-load excitation of an alternator required to give rated voltage is 100 A. In a short-circuit test with full-load current flowing in the armature, the field excitation was 80 A. Determine the approximate excitation that will be required to give full-load current at 0.8 p.f. lag and at the rated terminal voltage.*

Solution. The field current required to give full-load rated voltage on open-circuit is 100 A. This is represented by the phasor *OA* in Fig. 12.52. The field current required to produce full-load current on short circuit in the armature is 80 A. This is represented by the phasor *AB* making an angle of (90° + ϕ) with *OA* where $\phi = \cos^{-1} 0.8 = 36.8°$. Now $90° + \phi = 90° + 36.8° = 126.8°$. The required field current is *OB* which is the phasor sum of *OA* and *AB*.

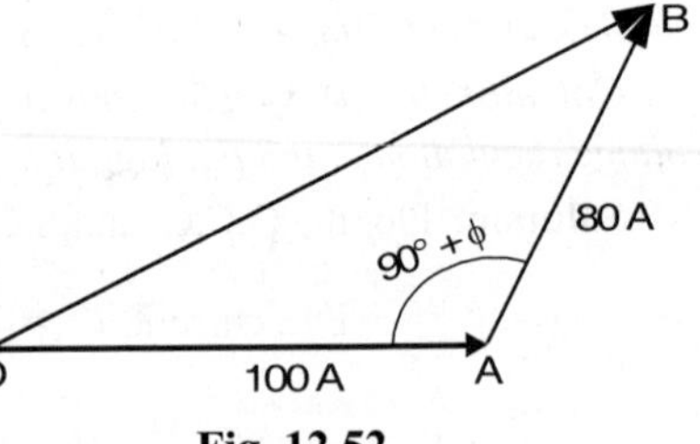

Fig. 12.52

$\therefore$ Required field current, $OB = \sqrt{(OA)^2 + (AB)^2 + 2 \times OA \times AB \cos(180° - 126.8°)}$

$$= \sqrt{(100)^2 + (80)^2 + 2 \times 100 \times 80 \times \cos(53.2°)} = \mathbf{161.2\ A}$$

Example 12.38. *The following test results are obtained on a 6600 V alternator :*

Open-circuit voltage :	*3100*	*5000*	*6600*	*7500*	*8300*
Field current (amp.) :	*16*	*25*	*37.5*	*50*	*70*

A field current of 20 A is found necessary to circulate full-load current on short-circuit of the armature. Using ampere-turn method, find the full-load regulation at 0.8 p.f. lagging.

Solution. Plot the O.C.C. from the given data as shown in Fig. 12.53. It is clear from the O.C.C. that field current required to produce the normal voltage of 6600 V on open circuit is 37.5 A. This is represented by the phasor OA in Fig. 12.54. The field current required to give full-load current on short circuit in armature is 20 A. This is represented by the phasor AB making an angle $(90° + \phi)$ with OA where $\phi = \cos^{-1} 0.8 = 36.8°$. Now $90° + \phi = 90° + 36.8° = 126.8°$.

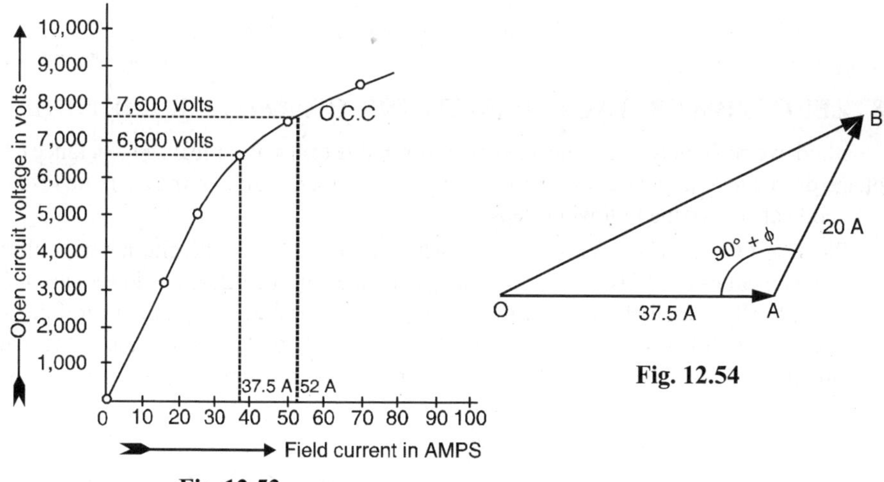

Fig. 12.53

Fig. 12.54

The phasor OB is the phasor sum of OA and AB and represents the field current required to produce a terminal p.d. of 6600 V at 0.8 p.f. lagging on *full-load.*

$$OB = \sqrt{(OA)^2 + (AB)^2 + 2 \times OA \times AB \times \cos(180° - 126.8°)}$$

$$= \sqrt{(37.5)^2 + (20)^2 + 2 \times 37.5 \times 20 \times \cos 53.2°} = 52\ \text{A}$$

From the O.C.C., the generated e.m.f. E_0 corresponding to this field excitation of 52 A is 7600 volts.

$$\therefore \quad \%\ \text{Regulation} = \frac{E_0 - V}{V} \times 100 = \frac{7600 - 6600}{6600} \times 100 = \mathbf{15.15\%}$$

TUTORIAL PROBLEMS

1. A 3-phase, 30 kVA, 50 Hz, 440 V alternator has a star-connected armature and on open-circuit and short-circuit test gave the following figures :

Field current (amps) :	1	3	5	9	12	15
Line E.M.F. (volts) :	80	222	335	500	565	600
S.C. current (amps) :	-	8	-	52	70	-

Calculate by (*i*) synchronous impedance method and (*ii*) ampere turn method the full-load voltage regulation at 0·8 p.f. lagging. The d.c. resistance measured between the line terminals was 0·27 Ω.

[(*i*) 53 % (*ii*) 33 %]

2. The open-circuit and short-circuit test readings for a 3-phase, star-connected, 1000 kVA, 2000 V, 50 Hz synchronous generator are :

Field amps	:	10	20	25	30	40	50
Line E.M.F. (volts)	:	800	1500	1760	2000	2350	2600
S.C. current (amps)	:	-	200	250	300	-	-

The effective armature resistance is 0·2 Ω/phase. Using ampere-turn method, find the full-load voltage regulation at (*i*) 0·8 p.f. lagging (*ii*) 0·8 p.f. leading. **[(*i*) 34·6 % (*ii*) – 6·4 %]**

3. A 25 kVA, single-phase, 500 V alternator gave the following open-circuit and short-circuit values :

Field current (amps)	:	2	4	6	8	10	12	14	16
E.M.F. (volts)	:	105	208	304	380	435	480	518	550
S.C. current (amps)	:	26	52·5	80	–	–	–	–	–

Effective armature resistance is 0·4 Ω.

Calculate by (*i*) synchronous impedance method (*ii*) ampere-turn method the full-load voltage regulation at a p.f. of 0·9 leading. **[(*i*) – 6 % (*ii*) – 0·8 %]**

12.18 ZERO POWER FACTOR METHOD OR *POTIER METHOD

In this method, we separately determine the voltage drop due to armature-leakage reactance ($= I_a X_L$) and voltage drop due to armature reaction ($= I_a X_{AR}$). Therefore, it gives more accurate results. The Potier method consists of the following steps :

(*i*) **Plotting O.C.C.** The open-circuit characteristic (O.C.C.) of the alternator is plotted by conducting no-load test on the alternator as explained in Art. 12.14. The lower part of O.C.C. is practically a straight line and when extended becomes the air-gap line. Therefore, air-gap line represents O.C.C. of the alternator if the reluctance of the iron portion of the magnetic circuit of the machine is neglected as compared to the reluctance of the air-gap.

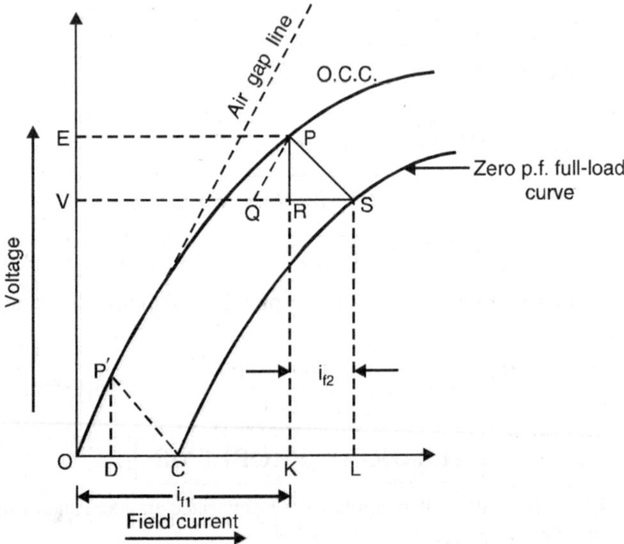

Fig. 12.55

* This method was suggested by Potier.

(*ii*) **Plotting zero p.f. (lagging) full-load curve.** This is the curve between terminal voltage and field current when the alternator is delivering its full rated current to a zero power factor (lagging) load. The test is carried out by running the alternator at synchronous speed and connecting a purely inductive 3-phase load to its terminals. The load is varied in steps and at each step, the field current is adjusted so that the armature current is equal to its rated value. There is no need to plot the full curve. Only two points *S* and *C* (See Fig. 12.55) are sufficient. The point *S* corresponds to a field current which gives the rated terminal voltage while the zero p.f. load is adjusted to draw rated armature current. The point **C* corresponds to the short-circuit conditions on the alternator (*i.e.* terminal voltage = 0) with the field current adjusted to give rated armature current. Since the armature resistance is negligible, the short circuit current lags behind the resultant induced e.m.f. by 90°. Therefore, point *C* constitutes a point on the zero p.f. curve.

(*iii*) **Constructing Potier Triangle.** Referring to Fig. 12.55, *OC* is the field current producing full-load armature current on short circuit (the current lagging by 90°). Therefore, field current *OC* must be sufficient to counter the demagnetising effect of armature reaction and armature leakage reactance drop at full-load. From *S*, draw *SQ* equal to and parallel to *OC*. From point *Q*, draw a line *QP* parallel to air-gap line. This line meets O.C.C. at point *P*. From point *P*, draw *PR* perpendicular to *QS* and meeting it at point *R*. The triangle *PRS* is known as **Potier triangle**. The following information is conveyed by the Potier triangle :

(*a*) The length *PR* represents the armature leakage reactance drop ($I_a X_L$).

(*b*) The length *RS* (= *KL* = i_{f2}) represents the field current to overcome the demagnetising effect of armature reaction at full-load. The length *RQ* represents the field current necessary to balance the armature leakage reactance drop ($I_a X_L$).

(*c*) The Potier triangle is the same for a given armature current and hence can be transferred anywhere to fit between the two characteristics. Thus the Potier triangle *PRS* is transferred to the base line as triangle *P′DC* and is identical to Potier triangle *PRS*.

Phasor diagram. Suppose the full-load terminal voltage of the alternator is *V*. Let the load p.f. be cos φ lagging. The phasor diagram for this condition is shown in Fig. 12.56. Here voltage *V* is taken as the reference phasor and current I_a lags behind *V* by φ. The $I_a R_a$ drop is drawn parallel to the current phasor and $I_a X_L$ drop is drawn perpendicular to it. The phasor *OM* represents the induced e.m.f. *E* (= phasor sum of *V*, $I_a R_a$ and $I_a X_L$). From O.C.C. in Fig. 12.55, the field current corresponding to induced e.m.f. *E* is i_{f1} (= *OK*). The field current i_{f1} is drawn **90° ahead of *E*.

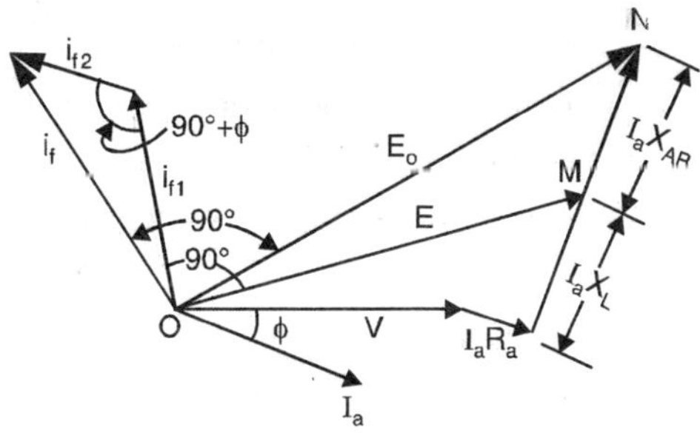

Fig. 12.56

The current phasor i_{f2} (= *RS* = *DC* in Fig. 12.55) represents the field current necessary to overcome the demagnetising effect of armature reaction at full load and is drawn parallel to the current phasor I_a. The phasor sum of i_{f1} and i_{f2} gives the total field current i_f required to produce E_0. The phasor E_0 (= *ON*) lags behind i_f by 90°. Note that phasor *MN* represents the voltage drop $I_a X_{AR}$ due to armature reaction.

$$\therefore \qquad \% \text{ Regulation} = \frac{E_0 - V}{V} \times 100$$

* This point is located from the short-circuit characteristic (S.C.C.).

** Recall that induced e.m.f. lags the flux producing it by 90°.

12.19 PROCEDURE FOR POTIER METHOD

For solving problems relating to Potier method, the following procedure is adopted :

(*i*) Suppose the terminal voltage per phase is V.

(*ii*) Find the armature leakage reactance drop ($= I_a X_L$) from the Potier triangle.

(*iii*) Add $I_a X_L$ (and $I_a R_a$ if given) vectorially to V to get E.

(*iv*) From the O.C.C., find the field current required to produce E. Let it be i_{f1}.

(*v*) From the Potier triangle, find the field current necessary for balancing armature reaction. Let it be i_{f2}.

(*vi*) Find the phasor sum of i_{f1} and i_{f2} to get the resultant field current i_f.

(*vii*) From the O.C.C., find the e.m.f. corresponding to field current i_f. This gives us E_0. Therefore, we can find the voltage regulation.

Limitations of Potier Method. The Potier method has the following drawbacks :

(*a*) The Potier triangle is based on the assumption that the armature leakage reactance is constant and that the O.C.C. of the alternator is the same under load as it is under open-circuit conditions. However, this is not correct. Therefore, the value of armature leakage reactance determined by the Potier triangle method is not absolutely accurate. In order to distinguish the value of leakage reactance thus determined from the absolutely correct value, the value of leakage reactance determined by the Potier triangle method is sometimes called *Potier leakage reactance.*

(*b*) In the Potier triangle method, some error is also introduced due to the fact that it is not possible to obtain exactly zero p.f. for the zero power factor load test.

(*c*) A greater amount of error is exhibited by machines of salient-pole construction than the nonsalient-pole construction.

Best results are obtained by determining the value of armature leakage reactance and the m.m.f. of armature reaction from a Potier triangle located as high as possible on the test curves.

Example 12.39. *A 3-phase, 6000V, star-connected alternator has the following O.C.C. at normal speed :*

Field amperes :	*14*	*18*	*23*	*30*	*43*
Terminal volts :	*4000*	*5000*	*6000*	*7000*	*8000*

With armature short circuited and full-load current flowing, the field current is 17 A and when the machine is supplying full-load of 20,000 kVA at zero p.f., the field current is 42.5 A and terminal voltage is 6000 V.

Determine the field current required when the machine is supplying the full load at 0.8 p.f. lagging.

Solution. From the given data, the O.C.C. is drawn in Fig. 12.57 with phase voltages (dividing line values by $\sqrt{3}$) which are :

2310, 2887, 3465, 4042, 4620

The full-load zero p.f. curve can be drawn because two points of this curve are given *viz* point C (17 A, 0V) and point S (42.5 A, 3465 V).

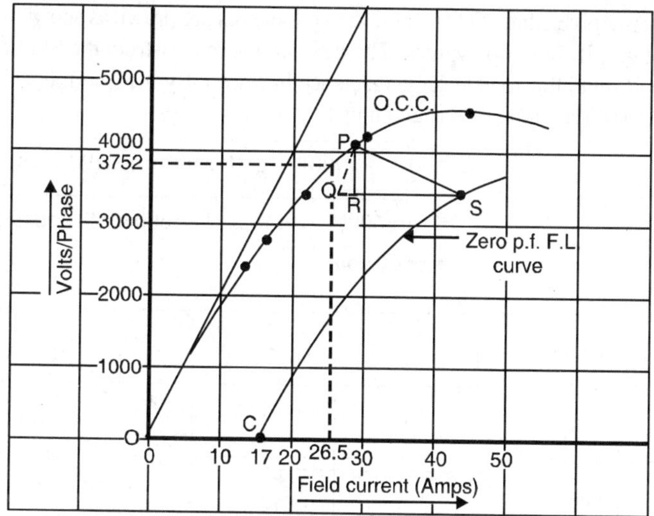

Fig. 12.57

In the Potier triangle *PRS*, length *PR* represents the leakage reactance drop ($I_a X_L$). By measurement, $PR = 450$ V. Since the armature resistance (R_a) is neglected, the value of *E* can be determined from Fig. 12.58.

$$E = \sqrt{(V\cos\phi)^2 + (V\sin\phi + I_a X_L)^2}$$

$$= \sqrt{(3465\times 0.8)^2 + (3465\times 0.6 + 450)^2} = 3752 \text{ volts}$$

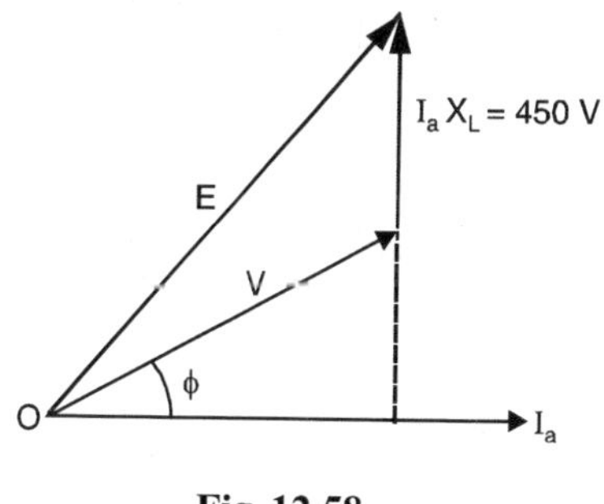

Fig. 12.58

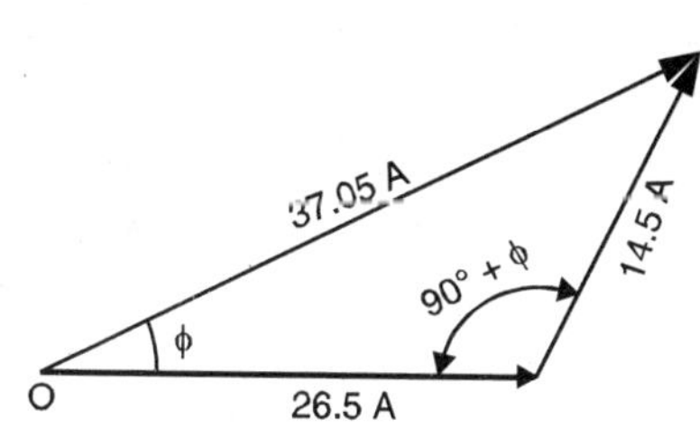

Fig. 12.59

From O.C.C. in Fig. 12.57, the field current corresponding to induced e.m.f. of 3752 V is 26.5 A. The length *RS* in the Potier triangle represents the field current required to overcome the armature reaction. By measurement, *RS* = 14.5 A. The total field current required to produce E_0 is the phasor sum of the two (See Fig. 12.59).

$\therefore$ Total field current $= \sqrt{(26.5)^2 + (14.5)^2 - 2\times 26.5\times 14.5\times\cos(90° + 36.8°)}$ = **37.05 A**

Example 12.40. *Find the voltage regulation by zero power factor method of 5000 kVA, 6600 V, 3-phase, 50 Hz, star-connected alternator at full load, unity p.f. having the following test data :*

Field current (A)	:	32	50	75	100	140
O.C. line voltage (V)	:	3100	4900	6600	7500	8300
Line volts, zero p.f. (V)	:	0	1850	4250	5800	7000

Neglect armature resistance.

Solution. From the given data, O.C.C. and zero p.f. curves are drawn as shown in Fig. 12.60. Note that voltages in the graph show line values. The next step is to construct the *Potier triangle *PRS*. In this triangle, *PR* represents full-load leakage reactance line drop. By measurement, *PR* = 900 volts. As shown in Fig. 12.61 (*i*), the value of *E* is given by ;

$$E = \sqrt{V^2 + (900)^2 - 2 \times V \times 900 \times \cos(90° + \phi)}$$

$$= \sqrt{(6600)^2 + (900)^2 - 2 \times 6600 \times 900 \times \cos 90°} \qquad (\because \phi = 0°)$$

$$= 6660 \text{ volts}$$

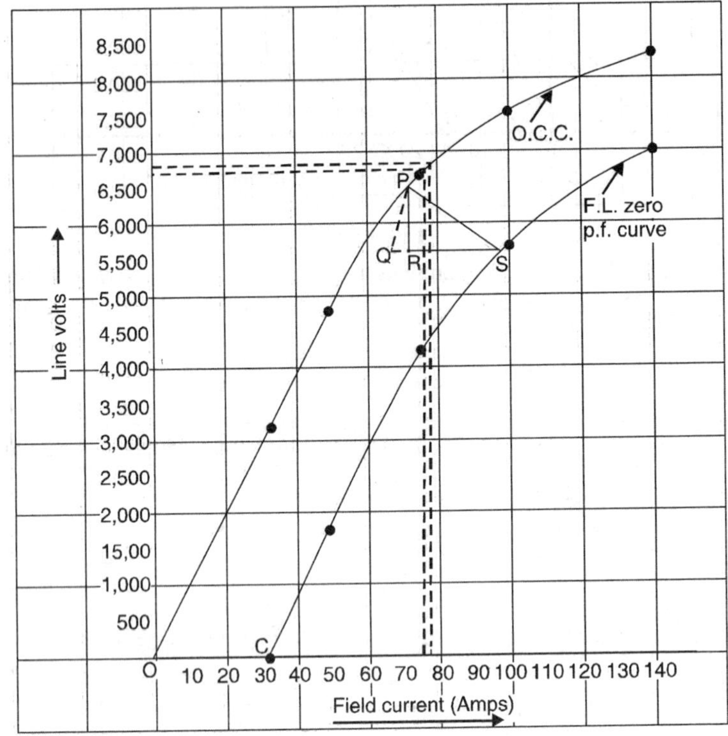

Fig. 12.60. Field current (Amps)

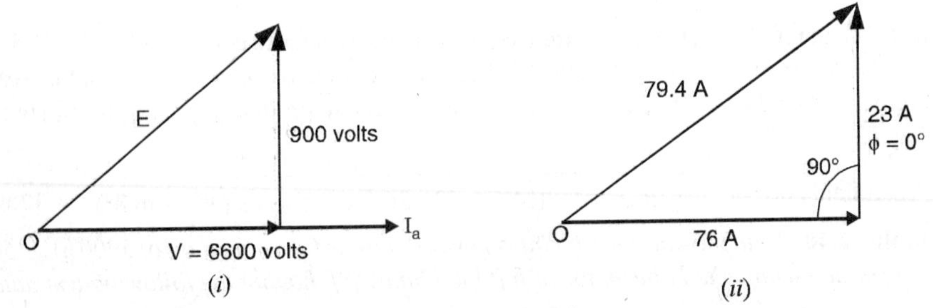

Fig. 12.61

* The steps of construction are as follows. From any point *S* on the zero p.f. curve, draw a line *SQ* parallel to and equal to *CO* (*i.e.* *SQ* = *CO*). From point *Q*, draw a line parallel to the initial slope of O.C.C. to meet the O.C.C. at point *P*. From point *P*, draw perpendicular to *QS*. Thus *PRS* is the Potier triangle.

From the O.C.C. in Fig. 12.60, the field current required to produce induced e.m.f. of 6660 volts is 76 A. The field current required to overcome the demagnetising effect of armature reaction on full load is represented by *RS* in the Potier triangle. By measurement, *RS* = 23 A. The total field current required to produce E_0 is the phasor sum of the two (See Fig. 12.61 (*ii*)).

$$\therefore \qquad \text{Total field current} = \sqrt{(76)^2 + (23)^2} = 79.4\text{A}$$

From O.C.C. in Fig. 12.60, the induced e.m.f. corresponding to field current of 79.4 A is $E_0 = 6820$ volts.

$$\therefore \qquad \%\ \text{Voltage regulation} = \frac{E_0 - V}{V} \times 100 = \frac{6820 - 6600}{6600} \times 100 = \mathbf{3.33\%}$$

Example 12.41. *Tests on a 15,000 kVA, 11,000 volts, 3-phase, 50 Hz, star-connected alternator gave the following results :*

Field AT in 10^3	:	*10*	*18*	*24*	*30*	*40*	*45*	*50*
O.C. line kV	:	*4.9*	*8.4*	*10.1*	*11.5*	*12.8*	*13.3*	*13.65*
Zero p.f., Full-load line kV	:	—	*0*	—	—	—	*10.2*	—

Find (i) the armature reaction ampere turns (ii) the armature leakage reactance per phase and (iii) voltage regulation for full load at 0.8 p.f. lagging. Neglect armature resistance.

Solution. From the given data, O.C.C. is drawn between line voltage (kV) and field *AT* as shown in Fig. 12.62. The full load zero p.f. curve can also be drawn because we are given its two points *viz* point *S* (45 × 10^3 *AT*, 10.2 kV) and point *C* (18 × 10^3 *AT*, *OV*). Therefore, we can construct the Potier triangle *PRS* [See Fig. 12.62].

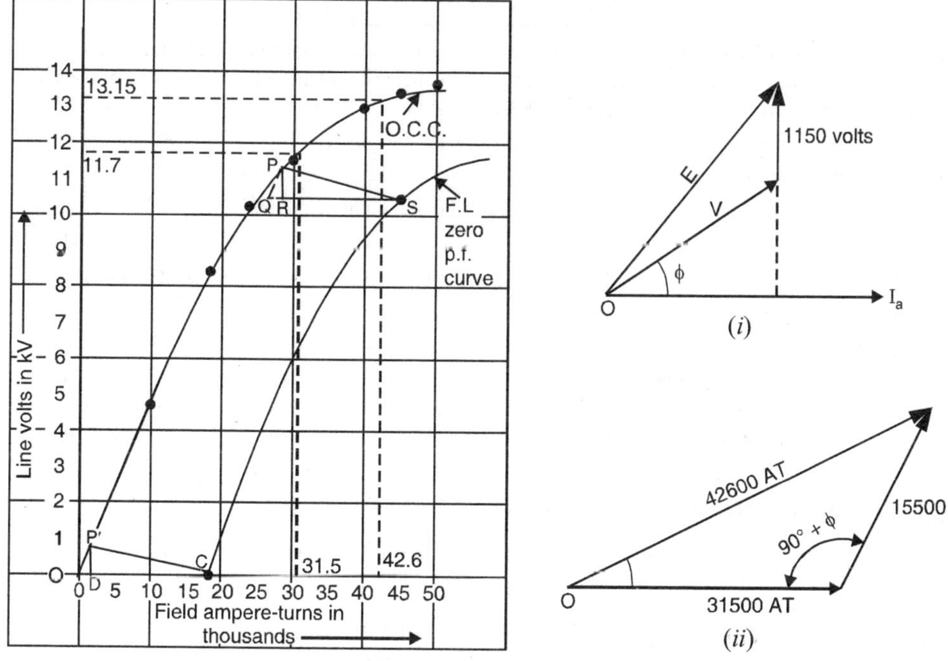

Fig. 12.62

Fig. 12.63

(*i*) In the Potier triangle *PRS*, the length *RS* represents the ampere-turns required to balance the effect of armature reaction. By measurement, *RS* = **15,500 *AT*.**

(*ii*) In the Potier triangle *PRS*, the length *PR* represents the leakage reactance line drop. By measurement, *PR* = 1.15 kV = 1150 volts.

$\therefore$ Leakage reactance drop/phase $= 1150/\sqrt{3}$ volts

$$\text{Full-load current, } I_a = \frac{15{,}000 \times 10^3}{\sqrt{3} \times 11{,}000} = \frac{5000\sqrt{3}}{11} \text{ A}$$

$$\therefore \quad \text{Leakage reactance/phase, } X_L = \frac{1150/\sqrt{3}}{I_a} = \frac{1150/\sqrt{3}}{5000\sqrt{3}/11} = \mathbf{0.843\ \Omega}$$

(*iii*) As seen in Fig. 12.63 (*i*), the value of E is given by ;

$$E = \sqrt{(V\cos\phi)^2 + (V\sin\phi + 1150)^2}$$

$$= \sqrt{(11000 \times 0.8)^2 + (11000 \times 0.6 + 1150)^2} = 11726 \text{ volts}$$

From the O.C.C. in Fig. 12.62, the field *AT* required to produce induced e.m.f. of 11726 are equal to 31500. The field *AT* required to balance armature reaction are equal to 15500 (calculated above). Therefore, total *AT* required are equal to the phasor sum of the two. As seen in Fig. 12.63 (*ii*),

$$\text{Total } AT \text{ required} = \sqrt{(31500)^2 + (15500)^2 - 2 \times 31500 \times 15500 \cos(90° + 36.8°)}$$

$$= 42600\ AT$$

From the O.C.C., the induced e.m.f. corresponding to 42600 *AT* is $E_0 = 13150$ volts.

$$\therefore \quad \%\text{ Voltage regulation} = \frac{E_0 - V}{V} \times 100 = \frac{13150 - 11000}{11000} \times 100 = \mathbf{19.55\%}$$

TUTORIAL PROBLEMS

1. An 11 kV, 1000 kVA, 3-phase, star-connected alternator has a resistance of 2Ω/phase. The open circuit and full-load zero power factor characterisitcs are given below. Find the voltage regulation of the alternator for full-load current at 0.8 p.f. lagging by Potier method.

Field current (A)	:	40	50	110	140	180
O.C.C. line voltage	:	5800	7000	12500	13750	15000
Line volts, zero p.f.	:	0	1500	8500	10500	12,200

[21.25%]

2. The open-circuit characteristic of a 6-pole, 440 V, 50 Hz, 3-phase, star-connected alternator is as under :

Field current (A)	:	2	4	6	7	8	10	12	14
O.C. Line volts	:	156	288	396	440	474	530	568	592

A field current of 7 A is needed to circulate the full-load armature current of 40 A under short circuit conditions. The field current for rated terminal voltage under full-load zero power factor conditions is 15 A. The armature resistance is 0.2 Ω per phase. Find voltage regulation at full-load current of 40 A at 0.8 p.f. lagging by Potier method. **[32.26%]**

12.20 POWER OUTPUT IN CYLINDRICAL ROTOR

Consider a star-connected cylindrical rotor alternator operating on infinite busbars.

Let V = busbars voltage/phase

E_0 = generated e.m.f./phase

I_a = armature current/phase delivered by the alternator

Z_s = synchronous impedance/phase $= R_a + jX_s$

$\cos\phi$ = lagging p.f. of the alternator

θ = internal angle $= \tan^{-1} X_s/R_a$

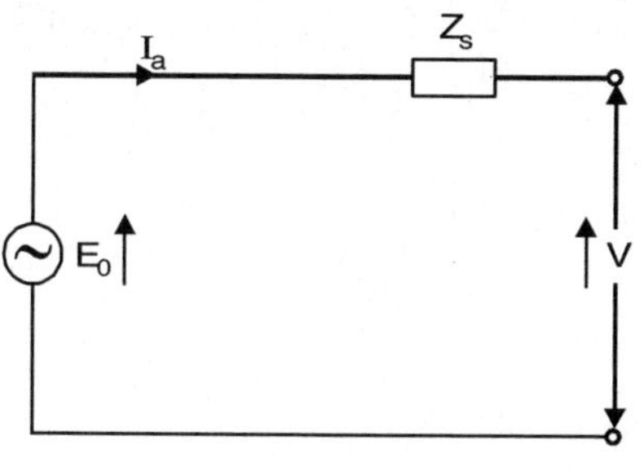

Fig. 12.64

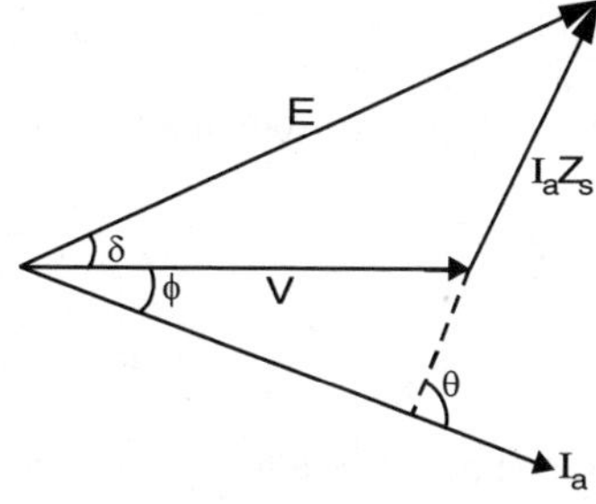

Fig. 12.65

Fig. 12.64 shows the equivalent circuit for one phase of the alternator whereas Fig. 12.65 shows the phasor diagram. Note that in drawing the phasor diagram, V has been taken as the reference phasor. We shall now derive an expression for the power delivered by the alternator. Referring to Fig. 12.65,

$$\boldsymbol{I_a} = \frac{E_0 \angle\delta - V \angle 0^\circ}{Z_s \angle\theta} = \frac{E_0}{Z_s} \angle(\delta - \theta) - \frac{V}{Z_s} \angle -\theta$$

$$\text{Power output/phase, } P = V \times \text{Real part of } \boldsymbol{I_a}$$

$$= V \times \left[\frac{E_0}{Z_s} \cos(\delta - \theta) - \frac{V}{Z_s} \cos(-\theta)\right]$$

$$= \frac{V}{Z_s} [E_0 \cos(\delta - \theta) - V \cos\theta]$$

$$= \frac{V^2}{Z_s} \left[\frac{E_0}{V} \cos(\delta - \theta) - \cos\theta\right]$$

$$\therefore \quad P = \frac{E_0 V}{Z_s} \cos(\theta - \delta) - \frac{V^2}{Z_s} \cos\theta \quad ...(i)$$

Eq. (*i*) gives the electrical output of the alternator in terms of E_0, V, Z_s, θ and load angle δ.

Maximum power output. For given E_0, V and frequency, the conditions for maximum power output can be obtained by differentiating eq. (*i*) *w.r.t.* δ and equating the result to zero *i.e.*,

$$\frac{dP}{d\delta} = 0$$

$$\text{or} \quad \frac{E_0 V}{Z_s} \sin(\theta - \delta) = 0$$

$$\text{or} \quad \sin(\theta - \delta) = 0$$

$$\text{or} \quad \theta = \delta$$

For constant busbars voltage V and fixed excitation (*i.e.*, E_0), the power output of the alternator will be maximum when $\theta = \delta$.

$$\therefore \quad P_{max}/\text{phase} = \frac{E_0 V}{Z_s} \cos 0^\circ - \frac{V^2}{Z_s} \cos\theta$$

$$= \frac{E_0 V}{Z_s} - \frac{V^2}{Z_s} \cos\theta \quad ...(ii)$$

Approximate expression. We have seen above that

$$\text{Power output/phase, } P = \frac{V^2}{Z_s} \left[\frac{E_0}{V} \cos(\theta - \delta) - \cos\theta\right]$$

$$= \frac{V^2}{Z_s}\left[\frac{E_0}{V}\cos(\theta - \delta) - \frac{R_a}{Z_s}\right] \qquad \left(\because \cos\theta = \frac{R_a}{Z_s}\right)$$

If $R_a << Z_s$, then $\theta = 90°$ and $Z_s = X_s$.

$$\therefore \qquad P = \frac{V^2}{X_s}\left[\frac{E_0}{V}\cos(90° - \delta) - 0\right]$$

or

$$P = \frac{E_0 V}{X_s}\sin\delta \qquad ...(iii)$$

Power output of the alternator will be maximum when $\delta = 90°$.

$$P_{max}/\text{phase} = \frac{E_0 V}{X_s} \qquad ...(iv)$$

Comments. We have seen above that if armature resistance R_a is very small (as is usually the case), then total power for three phases is given by ;

$$\text{Total power} = \frac{3 E_0 V}{X_s}\sin\delta \qquad ...(v)$$

When the alternator operates at a constant speed with constant field current, X_s and E_0 are both constant. Also the applied voltage/phase (*i.e.* V) is usually held constant. Therefore, power output of the alternator is directly proportional to sin δ where δ is the angle between V and E_0 and is called **power angle.** Eq. (v) is also known as the power-angle relation. This relation is shown in the power angle curve of Fig. 12.66. Note that in deriving eq. (v), we have assumed cylindrical rotor. For a salient-pole rotor, the curve is somewhat modified. In Art. 12.24, we shall develop relation between the power angle and the power output for salient-pole rotors.

The torque developed by a cylindrical rotor machine is given by ;

$$T_d = \frac{\text{Output power}}{\omega_s} = \frac{3 E_0 V}{\omega_s X_s}\sin\delta$$

where ω_s = angular velocity of the rotor

Since torque developed is also directly proportional to sin δ, the angle δ is also referred to as the **torque angle.** The torque developed by an alternator opposes the torque applied by the prime mover.

12.21 POWER/POWER ANGLE CHARACTERISTIC

The power output of an alternator is given by :

$$\text{Power output/phase, } P = \frac{E_0 V}{X_s}\sin\delta$$

$$\text{Total power output} = \frac{3 E_0 V}{X_s}\sin\delta$$

Note that power output varies sinusoidally with power angle δ. Fig. 12.66 shows the power angle characteristic of the alternator. The alternator delivers maximum power when $\delta = 90°$. If δ becomes greater than 90°, the machine will lose synchronism. The dotted portion of the curve refers to unstable operation *i.e.*, machine loses synchronism.

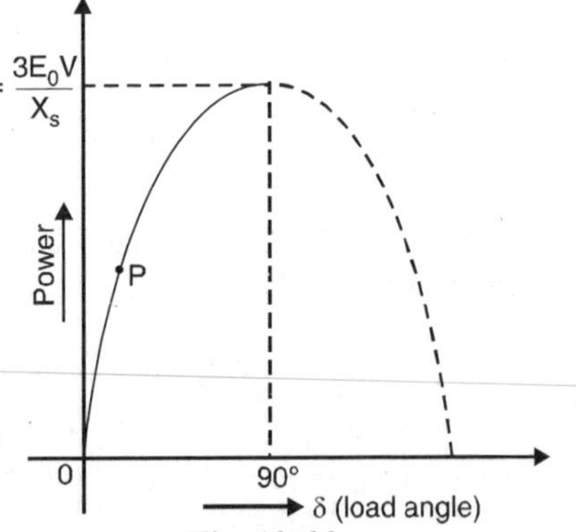

Fig. 12.66

Note that stability of the alternator is determined by the power/power angle characteristic. Suppose the operating position of the alternator is represented by point P on the curve. If unsteadiness occurs due to a transient spike of mechanical input, then load angle δ increases by a small amount. The additional electrical output caused by an increase in δ

produces a torque which is not balanced by the driving torque once the spike has passed. This torque causes retardation of the rotor and the alternator returns to the operating point *P*. The torque causing the return of the alternator to the steady-state position is called the synchronising torque and the power associated with it is known as synchronising power.

Example 12.42. *The rotor poles of an 8-pole alternator shift by 10 mechanical degrees between no-load and full-load.*

(i) Calculate the torque angle between E_0 and V at full-load.

(ii) Which voltage is leading ?

Solution.

(*i*) There is a direct relation between the mechanical angle α and torque angle δ (or power angle), given by ;

$$\delta = P\alpha/2$$

where δ = Torque angle between E_0 and V (electrical degrees)

P = Number of poles on the alternator

α = Mechanical angle between the centres of stator and rotor poles (mechanical degrees)

$\therefore \quad \delta = 8 \times 10/2 = \mathbf{40° \ (elect.)}$

(*ii*) When alternator delivers active power, **E_0 always leads V.**

Example 12.43. *A 3-phase, 11 kV, 5 MVA star-connected alternator has a synchronous impedance of (1 + j 10) Ω per phase. Its excitation is such that the generated line e.m.f. is 14 kV. The alternator is connected to infinite busbars. Determine the maximum output at the given excitation when (i) armature resistance is neglected (ii) armature resistance is considered.*

Solution.

$$\text{Busbars voltage/phase, } V = 11000/\sqrt{3} = 6352 \text{ volts}$$

$$\text{Generated e.m.f./phase, } E_0 = 14000/\sqrt{3} = 8083 \text{ volts}$$

$$\text{Synchronous impedance/phase, } Z_s = \sqrt{(1)^2 + (10)^2} = 10{\cdot}05\ \Omega$$

$$\text{Synchronous reactance/phase, } X_s = 10\ \Omega \ ; \ \cos\theta = \frac{R_a}{Z_s} = \frac{1}{10{\cdot}05}$$

(*i*) ***When R_a is neglected***

$$P_{max}/\text{phase} = \frac{E_0 V}{X_s} = \frac{8083 \times 6352}{10} = 5135 \times 10^3 \text{ W} = 5135 \text{ kW}$$

$$\text{Total } P_{max} = 3 \times 5135 = \mathbf{15405 \ kW}$$

(*ii*) ***When R_a is considered***

$$P_{max}/\text{phase} = \frac{V}{Z_s}(E_0 - V\cos\theta)$$

$$= \frac{6352}{10{\cdot}05}(8083 - 6352 \times 1/10{\cdot}05)$$

$$= 4711 \times 10^3 \text{ W} = 4711 \text{ kW}$$

$$\text{Total } P_{max} = 3 \times 4711 = \mathbf{14133 \ kW}$$

Example 12.44. *A 10,000 kVA, single phase alternator has an impedance of 10 Ω and a resistance of 1 Ω and operates in parallel with constant voltage 10,000 V busbars. Calculate the maximum external load that the machine can supply before dropping out of step when the machine is excited to give an e.m.f. of 11000 V. Also find the armature current and p.f. corresponding to this maximum load.*

Solution.

E_0 = 11000 volts ; V = 10,000 volts ; Z_s = 10 Ω ; cos θ = 1/10 = 0·1 ; sin θ = 0·995

Taking V as the reference phasor,

$$I_a = \frac{E_0 \angle\delta - V\angle 0^\circ}{Z_s \angle\theta}$$

When power output is maximum, δ = θ.

$$\therefore \quad (I_a)_{max} = \frac{E_0 \angle\theta - V\angle 0^\circ}{Z_s \angle\theta} = \frac{E_0}{Z_s} - \frac{V\angle -\theta}{Z_s}$$

$$= \frac{11000}{10} - \frac{10000}{10}\angle -\theta = 1100 - 1000\angle -\theta$$

$$= 1100 - 1000\,(\cos\theta - j\sin\theta) = 1100 - 1000\,(0{\cdot}1 - j\,0{\cdot}995)$$

$$= (1000 + j\,995)\text{A} = 1410\angle 44{\cdot}9^\circ\ \text{A}$$

∴ Current corresponding to P_{max} is **1410 A** and p.f. = cos 44·9° = **0·71** *lead*

$$P_{max} = V \times (I_a)_{max} \times \cos\phi = 10{,}000 \times 1410 \times 0{\cdot}71$$

$$\simeq 10{,}000 \times 10^3\ \text{W} = \mathbf{10{,}000\ kW}$$

Example 12.45. *A 3-phase, 11000 V, star-connected turbo-alternator delivers 200 A at unity p.f. when running on constant voltage and constant frequency bus-bars. If the induced e.m.f. be raised by 25 %, find the new current and p.f. at which the machine now works. Assume a constant steam supply and unchanged efficiency. Armature resistance is 0·5 Ω/phase and synchronous reactance is 8 Ω/phase.*

Solution.

$$\mathbf{Z_s} = (0{\cdot}5 + j\,8)\,\Omega\ ;\ \mathbf{I_a} = (200 + j\,0)\,\text{A}\ ;\ \mathbf{V} = \frac{11000}{\sqrt{3}}\angle 0^\circ\ \text{V}$$

$$\mathbf{E_0} = \mathbf{V} + \mathbf{I_a Z_s} = \frac{11000}{\sqrt{3}} + (200 + j\,0)\,(0{\cdot}5 + j\,8)$$

$$= 6351 + 100 + j\,1600 = (6451 + j\,1600)\ \text{volts}$$

$$\therefore \text{ Magnitude of induced e.m.f.} = \sqrt{(6451)^2 + (1600)^2} = 6646\ \text{volts}$$

When increased by 25 %, new e.m.f. E_1 = 1·25 × 6646 = 8308 volts

Now the steam supply is the same. Therefore, the wattful component (*i.e.*, 200 A) remains unchanged. The new current will be in the form

$$\mathbf{I_{a1}} = 200 + j\,x$$

$$\mathbf{E_1} = \mathbf{V} + \mathbf{I_{a1} Z_s} = 6351 + (200 + j\,x)\,(0{\cdot}5 + j\,8)$$

$$8308\angle\delta_1 = (6351 + 100 - 8x) + j\,(0{\cdot}5\,x + 1600)$$

$$\therefore \quad (8308)^2 = (6451 - 8x)^2 + (1600 + 0{\cdot}5\,x)^2$$

$$\text{or} \quad x^2 - 1581\,x - 387000 = 0$$

On solving, we have, x = – 215* A

∴ New current, $\mathbf{I_{a1}}$ = (200 – j 215) A = 294 ∠– 47° A

Magnitude of new current = **294 A**

New p.f. of the machine = cos (– 47°) = **0·68** *lag*

Example 12.46. *A 20,000 kVA, 11000 V, 3-phase, star-connected alternator running on constant voltage, constant frequency busbars has a resistance of 0·06 Ω/phase and a synchronous reactance of 1·8 Ω/phase. If excited to give a terminal voltage of 13000 V on open-circuit, find its maximum possible output power.*

* The positive value is ignored because it is not possible.

Solution.

$$\text{Open-circuit voltage/phase, } E_0 = 13/\sqrt{3}\ \text{kV}$$

$$\text{Bus-bars voltage/phase, } V = 11/\sqrt{3}\ \text{kV}$$

$$\text{Synchronous impedance/phase, } \mathbf{Z_s} = (0{\cdot}06 + j\,1{\cdot}8)\Omega = 1{\cdot}8\angle 88^\circ\,\Omega$$

$$\mathbf{I_a} = \frac{E_0\angle\delta - V\angle 0^\circ}{Z_s\angle\theta}$$

$$\text{At maximum output, } \delta = \theta.$$

$$\therefore \quad (\mathbf{I_a})_{\mathbf{max}} = \frac{E_0\angle\theta - V\angle 0^\circ}{Z_s\angle\theta} = \frac{E_0}{Z_s} - \frac{V}{Z_s}\angle -\theta$$

$$= \left(\frac{13}{\sqrt{3}\times 1{\cdot}8} - \frac{11}{\sqrt{3}\times 1{\cdot}8}\angle -88^\circ\right)\text{kA}$$

$$= (4{\cdot}17 - 0{\cdot}123 + j\,3{\cdot}526)\ \text{kA}$$

$$= (4{\cdot}047 + j\,3{\cdot}526)\ \text{kA} = (4047 + j\,3526)\ \text{A}$$

$$\therefore \quad P_{max} = \sqrt{3}\times V\times \text{Real part of } (\mathbf{I_a})_{\mathbf{max}}$$

$$= \sqrt{3}\times(11\times 10^3)\times 4047 = 77103\times 10^3\ \text{W} = \mathbf{77103\ kW}$$

Example 12.47. *A 36 MVA, 21 kV, 1800 r.p.m. alternator has a synchronous reactance of 5Ω per phase. If the exciting voltage is 12 kV (line to neutral) and the system voltage is 17.3 kV (line-to-line), calculate :*

(i) the power which the machine delivers when the torque angle δ is 30°.

(ii) the peak power which the machine can deliver before it falls out of step.

Solution.

Induced voltage/phase, $E_0 = 12\text{kV}$

Terminal voltage/phase, $V = 17.3/\sqrt{3} = 10\ \text{kV}$

(*i*) Power delivered by the alternator/phase is given by ;

$$P = \frac{E_0 V}{X_s}\sin\delta = \frac{12\times 10}{5}\sin 30^\circ = 12\ \text{MW}$$

The total power delivered by the three phases = 12 × 3 = **36 MW**.

(*ii*) The power delivered will be maximum when δ = 90°.

$$\therefore \quad P_{max}/\text{phase} = \frac{E_0 V}{X_s}\sin 90^\circ = \frac{12\times 10}{5}\times 1 = 24\ \text{MW}$$

The maximum power delivered by the alternator = 24 × 3 = **72 MW**.

Example 12.48. *A 9 kVA, 208 V, 1200 r.p.m., 3-phase, 60 Hz, star-connected generator has a field winding resistance of 4.5 Ω. The armature winding impedance is (0.3 + j 5) Ω per phase. When the generator operates at its full-load and 0.8 p.f. lagging, the field winding current is 5 A. The rotational loss is 500 W. Determine (i) voltage regulation (ii) efficiency of alternator (iii) torque applied by the prime mover.*

Solution.

(*i*) Terminal voltage/phase, $V = 208/\sqrt{3} \simeq 120$ volts

$$\text{Full-load current, } I_a = \frac{9\times 10^3}{3\times 120} = 25\ \text{A}$$

$$\therefore \text{ Generated voltage/phase, } \boldsymbol{E_0} = \mathbf{V} + \mathbf{I_a Z_s} = 120\angle 0° + 25\angle -36.8° \times (0.3 + j5)$$

$$= (120 + j0) + 25\angle -36.8° \times (0.3 + j5)$$

$$= 222.5\angle 25.4° \text{ V}$$

$$\therefore \quad \%\text{ Voltage regulation} = \frac{E_0 - V}{V} \times 100 = \frac{222.5 - 120}{120} \times 100 = \mathbf{85.45\%}$$

(*ii*)

$$\text{Power output, } P_{out} = 3\,VI_a \cos\phi = 3 \times 120 \times 25 \times 0.8 = 7200 \text{ W}$$

$$\text{Copper losses} = 3\,I_a^2 R_a = 3 \times (25)^2 \times 0.3 = 562.5 \text{ W}$$

$$\text{Power developed, } P_d = 7200 + 562.5 = 7762.5 \text{ W}$$

$$\text{Constant losses} = \text{Rotational loss} + \text{Excitation loss}$$

$$= 500 + (5)^2 \times 4.5 = 612.5 \text{ W}$$

$$\text{Total input power, } P_{in} = 7762.5 + 612.5 = 8375 \text{ W}$$

$$\therefore \quad \text{Alternator efficiency, } \eta = \frac{P_{out}}{P_{in}} \times 100 = \frac{7200}{8375} \times 100 = \mathbf{86\%}$$

(*iii*) Mechanical power input to prime mover is

$$P_m = P_d + \text{Rotational loss} = 7762.5 + 500 = 8262.5 \text{ W}$$

Angular velocity of prime mover is

$$\omega_s = \frac{2\pi N_s}{60} = \frac{2\pi \times 1200}{60} = 40\pi \text{ rad/s}$$

∴ Torque applied by prime mover is

$$T = \frac{P_m}{\omega_s} = \frac{8262.5}{40\,\pi} = \mathbf{65.75 \ Nm}$$

TUTORIAL PROBLEMS

1. Calculate the maximum load of a 5000 kVA, single phase alternator having an equivalent reactance of 5 Ω when connected to 6600 V busbars if its excitation is such that e.m.f. on open-circuit would be 6000 V. Also find the armature current and power factor at this load. **[7920 kW ; 1784 A ; 0·673 lead]**
2. An 11 kV, three-phase, star-connected turbo alternator has an armature synchronous impedance of (0 + *j* 10) Ω per-phase. It is connected to constant voltage and frequency busbars to which it delivers 6 MW at 0·8 p.f. lagging. If the current is increased 20 % by increasing the steam supply, with no change in excitation, what is the new output power and power factor ? **[6·56 MW ; 0·795 lag]**
3. A 3-phase, 11 kV, 10 MW, star-connected synchronous generator has synchronous impedance of (0·8 + *j* 8·0) ohms per phase. If the excitation is such that the open-circuit voltage is 14 kV, determine (*i*) the maximum output of the generator (*ii*) the current and p.f. at the maximum output. **[(*i*) 19·25 MW (*ii*) 1287 A ; 0·786 lag]**

12.22 EFFECT OF SALIENT POLES

The treatment developed so far is applicable only to cylindrical rotor machines. In these machines, the air-gap is uniform so that the reluctance of the magnetic path is the same in all directions. Therefore, the effect of armature reaction can be accounted for by one reactance - the synchronous reactance X_s. It is because the value of X_s is constant for all directions of armature flux relative to the rotor. However, in a salient-pole machine, the radial length of the air-gap varies [See Fig. 12.67]so that reluctance of the magnetic circuit along the polar axis (called *direct axis or d-axis*) is much less than the reluctance along the interpolar axis (called *quadrature axis or q-axis*). This is illustrated in Fig. 12.68.

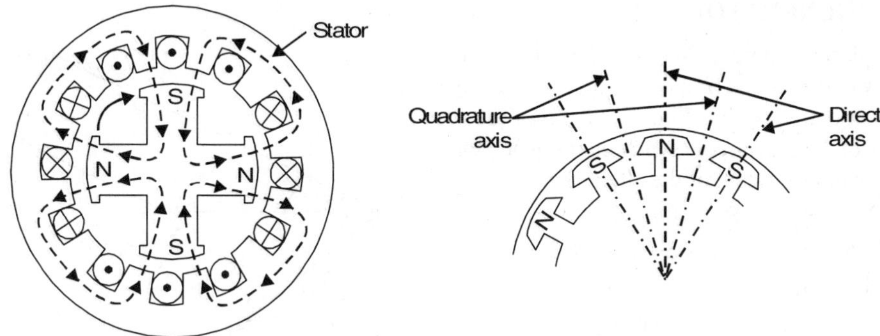

Fig. 12.67 **Fig.12.68**

Because of the lower reluctance along the polar axis (*i.e.*, *d*-axis), more flux is produced along *d*-axis than along the *q*-axis. Therefore, reactance due to armature reaction will be different along *d*-axis and *q*-axis. These are :

X_{ad} = direct axis reactance due to armature reaction

X_{aq} = quadrature axis reactance due to armature reaction

12.23 TWO-REACTANCE CONCEPT FOR SALIENT–POLE MACHINES

The effects of salient poles can be taken into account by resolving the armature current into two components ; I_d perpendicular to excitation voltage E_0 and I_q along E_0 as shown in phasor diagram in Fig. 12.69. Note that this diagram is drawn for an unsaturated salient-pole generator operating at a lagging power factor cos ϕ. With each of the component currents I_d and I_q, there is associated a component synchronous reactance X_d and X_q respectively.

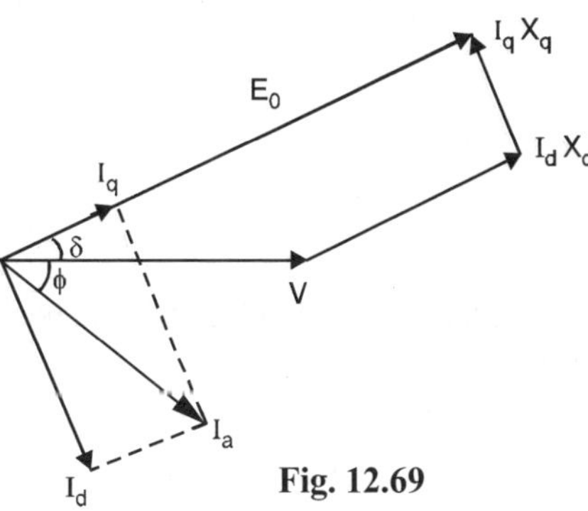

Fig. 12.69

X_d = direct axis synchronous reactance

X_q = quadrature axis synchronous reactance

If X_l is the armature leakage reactance and is assumed to be constant for direct and quadrature-axis currents, then,

$$X_d = X_{ad} + X_l \; ; \quad X_q = X_{aq} + X_l$$

Note that in drawing the phasor diagram, the armature resistance R_a is neglected since it is quite small. Further, all values are phase values. Here V is the terminal voltage/phase and E_0 is the e.m.f. per phase to which the generator is excited. Referring to Fig. 12.69,

$$I_q = I_a \cos(\delta + \phi) \quad \text{and} \quad I_d = I_a \sin(\delta + \phi)$$

The angle δ between E_0 and V is called the power angle or torque angle.

$$E_0 = V \cos\delta + I_d X_d$$

$$= V \cos\delta + I_a X_d \sin(\delta + \phi) \qquad [\because I_d = I_a \sin(\delta + \phi)]$$

$$\therefore \quad E_0 = V \cos\delta + I_a X_d (\sin\delta \cos\phi + \cos\delta \sin\phi) \qquad ...(i)$$

Also

$$V \sin\delta = I_q X_q$$

$$= I_a X_q \cos(\delta + \phi) \qquad [\because I_q = I_a \cos(\delta + \phi)]$$

$$\therefore \quad V \sin\delta = I_a X_q (\cos\delta \cos\phi - \sin\delta \sin\phi) \qquad ...(ii)$$

12.24 POWER DEVELOPED IN SALIENT-POLE SYNCHRONOUS GENERATOR

If we neglect armature resistance R_a (and hence Cu loss), then power developed (P_d) by an alternator is equal to the power output (P_{out}). Fig. 12.70 shows the phasor diagram of the salient-pole synchronous generator. The per phase power output of the alternator is

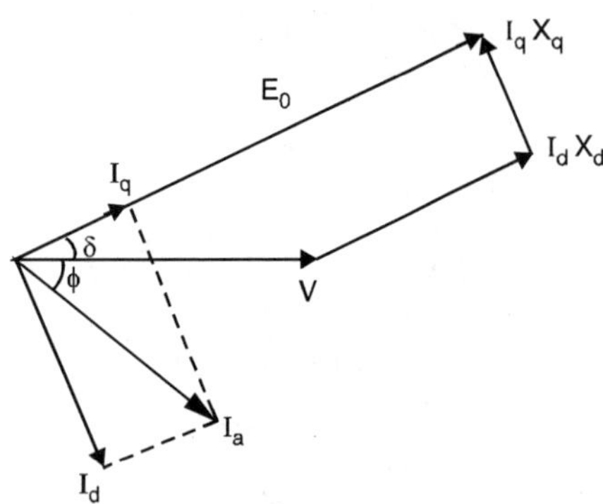

Fig. 12.70

$$P_{out} = P_d = V I_a \cos\phi \qquad ...(i)$$

Now $\quad I_a \cos\phi = I_q \cos\delta + I_d \sin\delta$

Also $\quad E_0 = V\cos\delta + I_d X_d$

$$\therefore \quad I_d = \frac{E_0 - V\cos\delta}{X_d}$$

And $\quad V\sin\delta = I_q X_q \qquad \therefore I_q = \frac{V\sin\delta}{X_q}$

$$\therefore \quad P_d = V[I_q\cos\delta + I_d\sin\delta] \qquad (\because I_a\cos\phi = I_q\cos\delta + I_d\sin\delta)$$

Putting the values of I_q and I_d, we have,

$$P_d = V\left[\frac{V\sin\delta}{X_q}\times\cos\delta + \frac{E_0 - V\cos\delta}{X_d}\times\sin\delta\right]$$

$$= \frac{E_0 V}{X_d}\sin\delta + V^2\left[\frac{\sin\delta\cos\delta}{X_q} - \frac{\sin\delta\cos\delta}{X_d}\right]$$

$$= \frac{E_0 V}{X_d}\sin\delta + \frac{V^2(X_d - X_q)}{X_d X_q}\times\frac{\sin 2\delta}{2}$$

$$\therefore \quad P_d = \frac{E_0 V}{X_d}\sin\delta + \frac{V^2(X_d - X_q)}{2X_d X_q}\sin 2\delta \text{ per phase} \qquad ...(ii)$$

The total power developed would be three times the above power. The following points may be noted :

(*i*) If there is no saliency, $X_d = X_q$.

$$\therefore \quad P_d = \frac{E_0 V}{X_d}\sin\delta \text{ per phase}$$

This is the power developed by cylindrical rotor machine (See Art. 12.20).

(*ii*) The second term in eq. (*ii*) above introduces the effect of salient poles. It represents the reluctance power *i.e.*, power due to saliency.

(*iii*) If $E_0 = 0$, then,

$$P_d = \frac{V^2(X_d - X_q)}{2X_d X_q}\sin 2\delta \text{ per phase}$$

The power obtained with zero excitation is called *reluctance power*. It should be noted that reluctance power is independent of the excitation.

Power angle characteristic. Fig. 12.71 shows the power-angle characteristic of a salient-pole machine. It is clear that reluctance power varies with δ at twice the rate of the excitation power. The peak

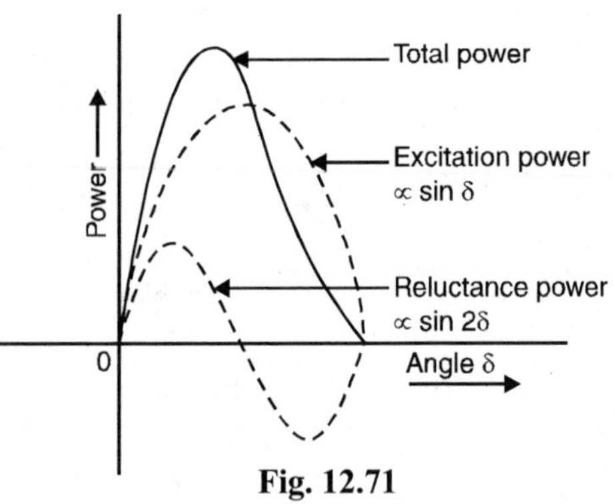

Fig. 12.71

power is seen to be displaced towards $\delta = 0$, the amount of displacement depends upon the excitation. In Fig. 12.71, the excitation is such that the excitation term has a peak value about 2·5 times that of the reluctance term. Under steady-state conditions, the reluctance term is positive because $X_d > X_q$.

Example 12.49. *A 3-phase alternator has a direct-axis synchronous reactance of 0·85 p.u. and quadrature-axis synchronous reactance of 0·55 p.u. Draw the phasor diagram for the alternator when operating on full-load at 0·8 p.f. lagging and determine therefrom (i) the load angle (ii) the no-load p.u. voltage (iii) per unit voltage regulation. Neglect armature resistance.*

Solution. Fig. 12.72 shows the phasor diagram.

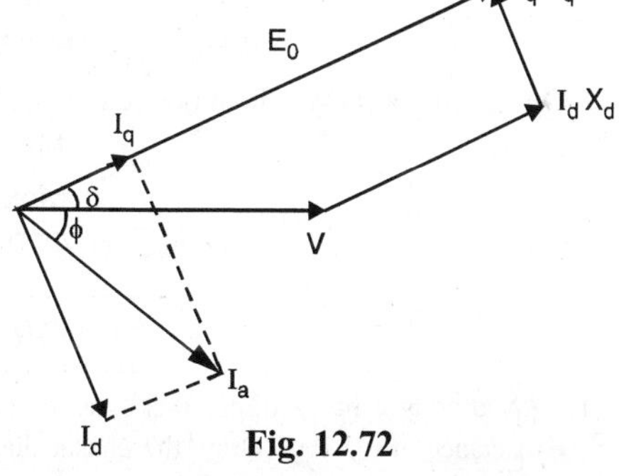

Fig. 12.72

$$V = 1 \text{ p.u.} \quad ; \quad I_a = 1 \text{ p.u.}$$

$$X_d = 0{\cdot}85 \text{ p.u.} \; ; X_q = 0{\cdot}55 \text{ p.u.}$$

(*i*) $\cos\phi = 0{\cdot}8 \quad ; \quad \sin\phi = 0{\cdot}6$

$$V \sin\delta = I_a X_q (\cos\delta\cos\phi - \sin\delta\sin\phi)$$

or $1 \times \sin\delta = 1 \times 0{\cdot}55 (\cos\delta \times 0{\cdot}8 - \sin\delta \times 0{\cdot}6)$

or $0{\cdot}44 \cos\delta = 1{\cdot}33 \sin\delta$

$$\therefore \quad \tan\delta = \frac{0{\cdot}44}{1{\cdot}33} = 0{\cdot}331$$

$\therefore$ Power angle, $\delta = \tan^{-1} 0{\cdot}331 = \mathbf{18{\cdot}3^\circ}$

(*ii*) $E_0 = V\cos\delta + I_a X_d (\sin\delta\cos\phi + \cos\delta\sin\phi)$

$$E_0 = 1 \times \cos\delta + 1 \times 0{\cdot}85 (\sin\delta \times 0{\cdot}8 + \cos\delta \times 0{\cdot}6)$$

or $E_0 = \cos\delta + 0{\cdot}85 (0{\cdot}8 \sin\delta + 0{\cdot}6 \cos\delta)$

Now $\cos\delta = \cos 18{\cdot}3^\circ = 0{\cdot}95 \quad ; \quad \sin\delta = \sin 18{\cdot}3^\circ = 0{\cdot}314$

$\therefore$ $E_0 = 0{\cdot}95 + 0{\cdot}85 \; (0{\cdot}8 \times 0{\cdot}314 + 0{\cdot}6 \times 0{\cdot}95) = \mathbf{1{\cdot}65 \text{ p.u.}}$

(*iii*) Voltage regulation $= \dfrac{E_0 - V}{V} = \dfrac{1{\cdot}65 - 1}{1} = \mathbf{0{\cdot}65 \text{ p.u.}}$

Example 12.50. *A 3-phase, Y-connected synchronous generator supplies a current of 10 A having phase angle of 20° lagging at 400 V (phase voltage). Find (i) the load angle (ii) components I_d and I_q of armature current (iii) voltage regulation. Given $X_d = 10\ \Omega$ and $X_q = 6{\cdot}5\ \Omega$. Neglect armature resistance.*

Solution. $V = 400$ volts ; $X_d = 10\ \Omega$; $X_q = 6{\cdot}5\ \Omega$; $I_a = 10$ A

$$\cos\phi = \cos 20^\circ = 0{\cdot}94 \quad ; \quad \sin\phi = 0{\cdot}342$$

(*i*) $V\sin\delta = I_a X_q (\cos\delta\cos\phi - \sin\delta\sin\phi)$

or $400 \sin\delta = 10 \times 6{\cdot}5 (\cos\delta \times 0{\cdot}94 - \sin\delta \times 0{\cdot}342)$

or $400 \sin\delta = 61{\cdot}1 \cos\delta - 22{\cdot}23 \sin\delta$

or $422{\cdot}23 \sin\delta = 61{\cdot}1 \cos\delta$

or $\tan\delta = \dfrac{61{\cdot}1}{422{\cdot}23} = 0{\cdot}1447$

$\therefore$ Load angle, $\delta = \tan^{-1} 0{\cdot}1447 = \mathbf{8{\cdot}23^\circ}$

(*ii*) $I_d = I_a \sin(\delta + \phi) = 10 \sin(8{\cdot}23^\circ + 20^\circ) = \mathbf{4{\cdot}73 \text{ A}}$

$$I_q = I_a \cos(\delta + \phi) = 10 \cos(8{\cdot}23^\circ + 20^\circ) = \mathbf{8{\cdot}81 \text{ A}}$$

(*iii*) $E_0 = V\cos\delta + I_a \sin(\delta + \phi) \times X_d = 400 \cos 8{\cdot}23^\circ + 4{\cdot}73 \times 10 = 443$ volts

$\therefore$ Voltage regulation $= \dfrac{E_0 - V}{V} \times 100 = \dfrac{443 - 400}{400} \times 100 = \mathbf{10{\cdot}75\ \%}$

Example 12.51. *A 3-phase star-connected salient pole synchronous generator is driven at a speed near synchronism with field circuit open and the stator is supplied from a balanced 3-phase supply. Voltmeter connected across the lines gave minimum and maximum readings of 2800 V and 2820 V. The line current fluctuates between 360 A and 275 A. Find the direct and quadrature axis synchronous reactances per phase. Neglect armature resistance.*

Solution. Direct axis synchronous reactance between lines

$$= \frac{\text{Maximum voltage}}{\text{Minimum current}} = \frac{2820}{275} = 10.25\ \Omega$$

$$\therefore \quad X_d/\text{phase} = 10.25/\sqrt{3} = \mathbf{5.92\ \Omega}$$

Quadrature axis synchronous reactance between lines

$$= \frac{\text{Minimum voltage}}{\text{Maximum current}} = \frac{2800}{360} = 7.778\ \Omega$$

$$\therefore \quad X_q/\text{phase} = 7.778/\sqrt{3} = \mathbf{4.5\ \Omega}$$

TUTORIAL PROBLEMS

1. An alternator has a direct-axis synchronous reactance of 0·8 p.u. and a quadrature-axis synchronous reactance of 0·5 p.u. Draw the phasor diagram for full load at a lagging p.f. of 0·8 and find (*i*) per unit open-circuit voltage (*ii*) per unit voltage regulation. **[(*i*) 1·6 p.u. (*ii*) 0·6 p.u.]**
2. The direct-axis and quadrature-axis synchronous reactance of a salient-pole synchronous generator are 1·0 p.u. and 0·6 p.u. respectively. Draw the phasor diagram for full-load 0·8 p.f. lagging and calculate the no-load voltage and load angle. Neglect the effect of saturation and armature resistance. **[1·773 p.u. ; 19·4°]**
3. An alternator has a direct-axis synchronous reactance of 0·8 p.u. and quadrature-axis synchronous reactance of 0·5 p.u. Draw the phasor diagram for full-load lagging power factor of 0·8 and find the p.u. value of the open circuit voltage with full-load excitation. Neglect armature resistance and saturation. **[1·0627 p.u.]**

12.25 V CURVES FOR ALTERNATOR

The graph between armature current (I_a) and field current (I_f) of an alternator for a constant output power is called its **V curve**. The curve is so called because it is *V*-shaped. Fig. 12.74 shows *V* curves for three different output powers P_1, P_2 and P_3 of the alternator.

Suppose a cylindrical rotor alternator is connected to an infinite bus bars and is delivering power at a lagging power factor (cos φ). If we neglect armature winding resistance, the power output of the alternator is given by ;

$$P = \frac{3\,E_0\,V}{X_s}\sin\delta = 3\,V\,I_a\cos\phi$$

Here V = Terminal voltage/phase ; E_0 = Generated e.m.f./phase

X_s = Synchronous reactance/phase ; δ = Power angle

(*i*) First, consider the output power equation :

$$P = \frac{3\,E_0\,V}{X_s}\sin\delta$$

Suppose that I_f is the field current necessary to generate voltage E_0. Let us now reduce the field current but keep the power output of the alternator the same as before. Since V is held constant by the infinite bus and X_s is constant, the same power output means that $E_0 \sin\delta$ must remain the same. However, reduction in field current must decrease generated e.m.f. E_0 so that power angle δ must increase (for the same output power) to maintain $E_0 \sin\delta$ constant. This means that tip of the generated e.m.f. phasor $\mathbf{E_0}$ must move along the horizontal line as shown in Fig. 12.73. As E_0 decreases, power angle δ increases so that $E_0 \sin\delta$ is constant.

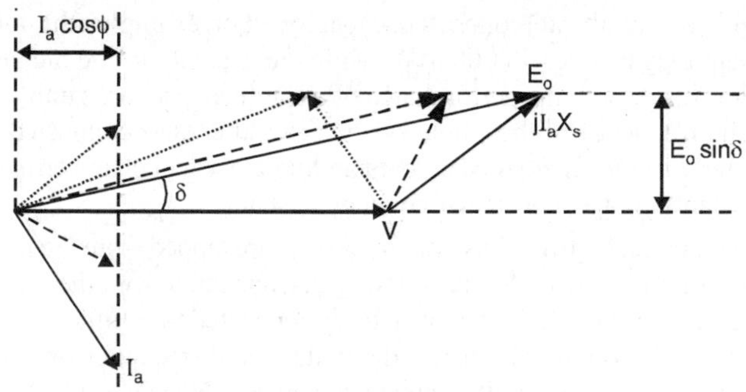

Fig. 12.73 Loci of E_0 and I_a for constant output power

(*ii*) Next, consider the output power equation :

$$P = 3\, V I_a \cos \phi$$

For the same power output (V is constant), $I_a \cos \phi$ must remain constant. This means that the projection of the current phasor $\boldsymbol{I_a}$ onto **V** must be the same even when I_f decreases. Therefore, the tip of the current phasor $\boldsymbol{I_a}$ must move along the vertical line in Fig. 12.73. The phasor diagram for the reduced field current is shown by the dashed lines.

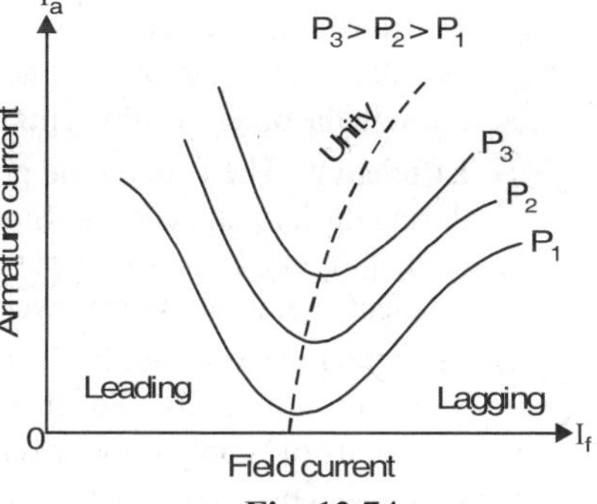

Fig. 12.74

From the above discussion, we see that for the constant power output of the alternator, if we go on reducing the field current (and hence E_0), the following sequence of events occur :

(*a*) The power angle δ goes on increasing until it becomes 90°.

(*b*) The p.f. angle (ϕ) continues changing from lagging to unity and finally to leading.

It is clear from Fig. 12.74 that as long as the power factor is changing from lagging to unity, the armature current is decreasing. The armature current is minimum when the power factor is unity. As the power factor changes from unity to leading, the armature current starts increasing again. Because of the shape of I_a/I_f graph, it is called *V curve.*

12.26 PARALLEL OPERATION OF ALTERNATORS

It is rare to find a 3-phase alternator supplying its own load independently except under test conditions. In practice, a very large number of 3-phase alternators operate in parallel because the various power stations are interconnected through the national grid. Therefore, the output of any single alternator is

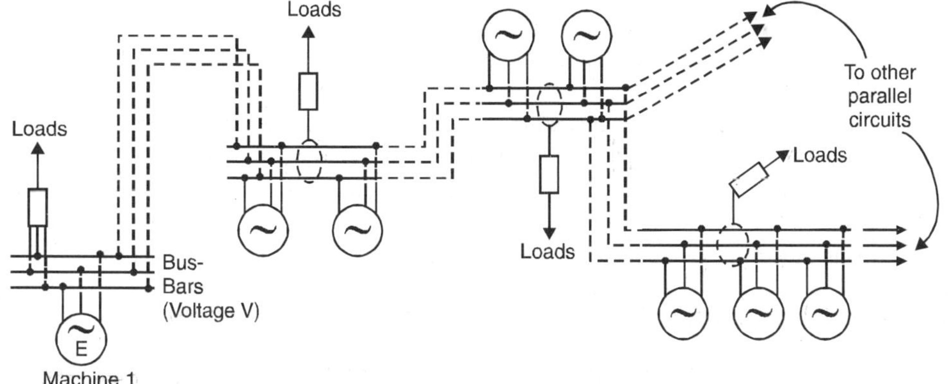

Fig. 12.75

small compared with the total interconnected capacity. For example, the total capacity of the interconnected system may be over 40,000 MW while the capacity of the biggest single alternator may be 500 MW. For this reason, the performance of a single alternator is unlikely to *affect appreciably the voltage and frequency of the whole system. An alternator connected to such a system is said to be connected to *infinite busbars*. The outstanding electrical characteristics of such busbars are that they are constant-voltage, constant-frequency busbars.

Fig. 12.75 shows a typical infinite bus system. Loads are tapped from the infinite bus at various load centres. The alternators may be connected to or disconnected from the infinite bus, depending on the power demand on the system. If an alternator is connected to infinite busbars, no matter what power is delivered by the incoming alternator, the voltage and frequency of the system remain the same. The operation of connecting an alternator to the infinite busbars is known as *paralleling with the infinite busbars*. It may be noted that before an alternator is connected to an infinite busbars, certain conditions must be satisfied.

12.27 ADVANTAGES OF PARALLEL OPERATION OF ALTERNATORS

The following are the advantages of operating alternators in parallel :

(*i*) **Continuity of service.** The continuity of service is one of the important requirements of any electrical apparatus. If one alternator fails, the continuity of supply can be maintained through the other healthy units. This will ensure uninterrupted supply to the consumers.

(*ii*) **Efficiency.** The load on the power system varies during the whole day; being minimum during the late night hours. Since alternators operate most efficiently when delivering full-load, units can be added or put off depending upon the load requirement. This permits the efficient operation of the power system.

(*iii*) **Maintenance and repair.** It is often desirable to carry out routine maintenance and repair of one or more units. For this purpose, the desired unit/units can be *shut down* and the continuity of supply is maintained through the other units.

(*iv*) **Load growth.** The load demand is increasing due to the increasing use of electrical energy. The load growth can be met by adding more units without disturbing the original installation.

12.28 CONDITIONS FOR PARALLELING ALTERNATOR WITH INFINITE BUSBARS

The proper method of connecting an alternator to the infinite busbars is called *synchronising*. A stationary alternator must not be connected to live busbars. It is because the induced e.m.f. is zero at standstill and a short-circuit will result. In order to connect an alternator safely to the infinite busbars, the following conditions are met :

(*i*) The terminal voltage (r.m.s. value) of the incoming alternator must be the same as busbars voltage.

(*ii*) The frequency of the generated voltage of the incoming alternator must be equal to the busbars frequency.

(*iii*) The phase of the incoming alternator voltage must be identical with the phase of the busbars voltage. In other words, the two voltages must be in phase with each other.

(*iv*) The phase sequence of the voltage of the incoming alternator should be the same as that of the busbars.

The magnitude of the voltage of the incoming alternator can be adjusted by changing its field excitation. The frequency of the incoming alternator can be changed by adjusting the speed of the prime mover driving the alternator.

---

* Since a large number of alternators operate in parallel, the system becomes a constant voltage source of very large inertia. Any change made in the operating conditions of one alternator will make insignificant change in voltage or frequency of the system because of its large size and inertia.

Condition (*i*) is indicated by a voltmeter, conditions (*ii*) and (*iii*) are indicated by synchronising lamps or a synchroscope. The condition (*iv*) is indicated by a phase sequence indicator.

12.29 METHODS OF SYNCHRONISATION

The method of connecting an incoming alternator safely to the live busbars is called *synchronising*. The equality of voltage between the incoming alternator and the busbars can be easily checked by a voltmeter. The phase sequence of the alternator and the busbars can be checked by a phase sequence indicator. Differences in frequency and phase of the voltages of the incoming alternator and busbars can be checked by one of the following two methods :

(*i*) By Three Lamp (one dark, two bright) method

(*ii*) By synchroscope

(*i*) **Three lamp method.** In this method of synchronising, three lamps L_1, L_2 and L_3 are connected as shown in Fig. 12.76. The lamp L_1 is straight connected between the corresponding phases (R_1 and R_2) and the other two are cross-connected between the other two phases. Thus lamp L_2 is connected between Y_1 and B_2 and lamp L_3 between B_1 and Y_2. When the frequency and phase of the voltage of the incoming alternator is the same as that of the busbars, the straight connected lamp L_1 will be dark while cross-connected lamps L_2 and L_3 will be equally bright. At this instant, the synchronisation is perfect and the switch of the incoming alternator can be closed to connect it to the busbars.

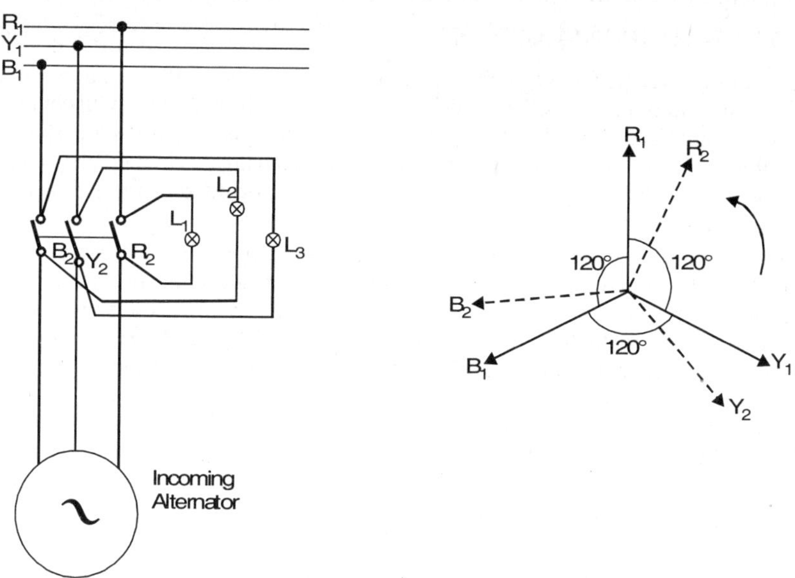

Fig. 12.76 **Fig.12.77**

In Fig. 12.77, phasors R_1, Y_1 and B_1 represent the busbars voltages and phasors R_2, Y_2 and B_2 represent the voltages of the incoming alternator. At the instant when *R_1 is in phase with R_2, voltage across lamp L_1 is zero and voltages across lamps L_2 and L_3 are equal. Therefore, the lamp L_1 is dark while lamps L_2 and L_3 will be equally bright. At this instant, the switch of the incoming alternator can be closed. Thus incoming alternator gets connected in parallel with the busbars.

(*ii*) **Synchroscope.** A synchroscope is an instrument that indicates by means of a revolving pointer the phase difference and frequency difference between the voltages of the incoming alternator

* Note that R_1 is in phase with R_2 with respect to external load. However, they are in phase opposition w.r.t. each other.

and the busbars (See Fig. 12.78). It is essentially a small motor, the field being supplied from the busbars through a potential transformer and the rotor from the incoming alternator. A pointer is attached to the rotor. When the incoming alternator is running fast (*i.e.*, frequency of the incoming alternator is higher than that of the busbars), the rotor and hence the pointer moves in the clockwise direction. When the incoming alternator is running slow (*i.e.*, frequency of the incoming alternator is lower than that of the busbars), the pointer moves in anticlockwise direction. When the frequency of the incoming alternator is equal to that of the busbars, no torque acts on the rotor and the pointer points vertically *upwards ("12 O' clock"). It indicates the correct instant for connecting the incoming alternator to the busbars. The synchroscope method is superior to the lamp method because it not only gives a positive indication of the time to close the switch but also indicates the adjustment to be made should there be a difference between the frequencies of the incoming alternator and the busbars.

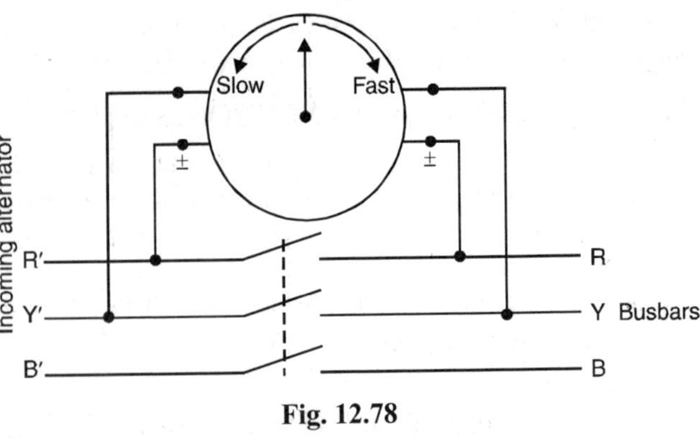

Fig. 12.78

12.30 SYNCHRONISING ACTION

When two or more alternators have been connected in parallel, they will remain in stable operation under all normal conditions *i.e.*, voltage, frequency, speed and phase equality will continue. In other words, once synchronised properly, the alternators will continue to run in synchronism under all normal conditions. If one alternator tries to fall out of synchronism, it is immediately counteracted by the production of a *synchronising torque* which brings it back to synchronism. This automatic action is called the *synchronising action* of the alternators.

Consider two similar single-phase alternators 1 and 2 operating in parallel as shown in Fig. 12.79 (*i*). For simplicity, let us assume that the alternators are at no-load. When in exact synchronism, the magnitudes of the e.m.f.s E_1 (machine 1) and E_2 (machine 2) are equal. These e.m.f.s are acting in the same direction with respect to the external circuit [See Fig. 12.79 (*ii*)]. But *in relation to each other*, these e.m.f.s are in phase opposition *i.e.*, if we trace the closed circuit formed by the two alternators we find that the e.m.f.s oppose each other [See Fig. 12.79 (*iii*)]. When the alternators are in exact synchronism, E_1 and E_2 are in exact phase opposition. Since $E_1 = E_2$ in magnitude, no current flows in the closed circuit formed by the two alternators.

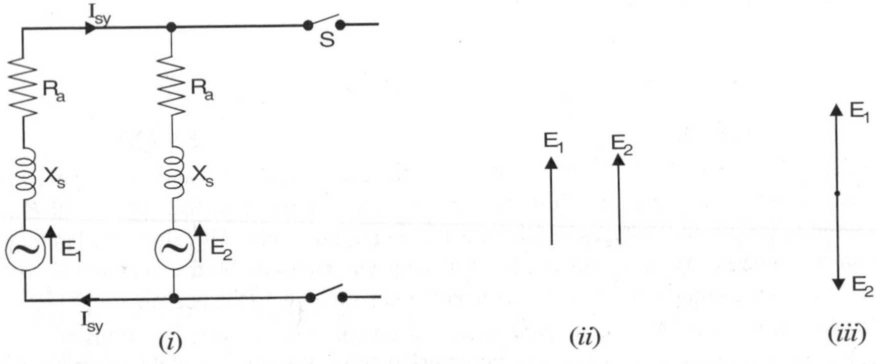

Fig. 12.79

* In this position, the voltages of the incoming alternator and that of busbars are in phase.

If one alternator drops out of synchronism, there is an automatic action to re-establish synchronism. Let us discuss this point.

(*i*) **Effect of speed change.** Suppose, due to any reason, the speed of machine 2 falls. Then e.m.f. E_2 will fall *back by a phase angle of α electrical degrees as shown in Fig. 12.80 (though still $E_1 = E_2$). There will be resultant e.m.f. E_r in the closed circuit formed by the two alternators. This e.m.f. E_r will circulate current (known as *synchronising current* I_{sy}) in this closed circuit.

$$\text{Synchronising current, } I_{sy} = \frac{E_r}{2\,Z_s} \qquad \text{... See Fig. 12.79 } (i)$$

The current I_{sy} lags behind E_r by an angle θ given by ;

$$\tan\theta = \frac{2\,X_s}{2\,R_a} = \frac{X_s}{R_a}$$

where R_a = armature resistance of each alternator

X_s = synchronous reactance of each alternator

Z_s = synchronous impedance of each alternator

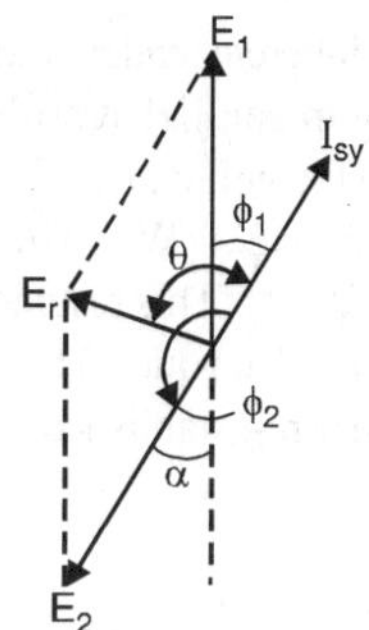

Fig. 12.80

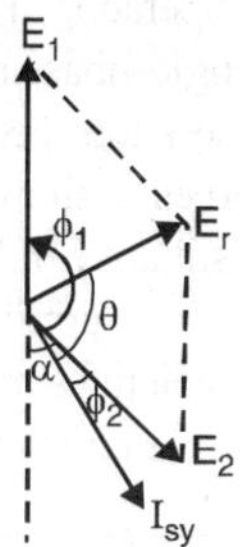

Fig. 12.81

Since R_a is very small as compared to X_s, θ is nearly 90° so that the current I_{sy} is almost in phase with E_1 and in phase opposition to E_2. This means that machine 1 is **generating and machine 2 is motoring. Consequently, machine 1 tends to slow down and machine 2, by accepting power, tends to accelerate. This restores the status quo *i.e.*, synchronism is re-established.

Conversely, if E_2 tends to advance in phase (See Fig. 12.81), the directions of E_r and I_{sy} are changed such that now machine 2 is generating and machine 1 is motoring. Once again the synchronism is restored.

(*ii*) **Effect of inequality of e.m.f.s.** The automatic re-establishment of synchronism of two alternators operating in parallel also extends to any changes tending to alter the individual e.m.f.s. When in exact synchronism, then $E_1 = E_2$ (magnitude) and they are in exact phase opposition as shown in Fig. 12.82 (*i*). Suppose due to any reason, e.m.f. E_1 increases. Then resultant e.m.f. E_r exists in the closed circuit formed by the two alternators. Then $E_r = E_1 - E_2$ and is in phase with E_1. The resultant e.m.f. E_r sends synchronising current I_{sy} in the closed circuit. Here again the current I_{sy} almost lags behind E_r by 90° ($\because Z_s \simeq X_s$) as shown in Fig. 12.82 (*ii*). Also I_{sy} lags almost 90° behind E_1 and leads E_2 almost by 90°. The power produced is practically zero ; just enough to overcome copper losses. The current I_{sy} lags behind E_1 and produces a demagnetising armature reaction effect on machine 1. At the same time, I_{sy} leads E_2 and produces magnetising armature reaction effect on

* The reader may recall that phasors rotate in anticlockwise direction.

** It is clear that I_{sy} is acting in the direction of E_1 but is in opposite direction to the direction of E_2. Therefore, machine 1 acts as a generator and machine 2 as a motor.

machine 2. Thus E_1 tends to fall and E_2 tends to rise. The result is that synchronism is re-established. The converse is true for $E_2 > E_1$ as shown in Fig. 12.82 (*iii*).

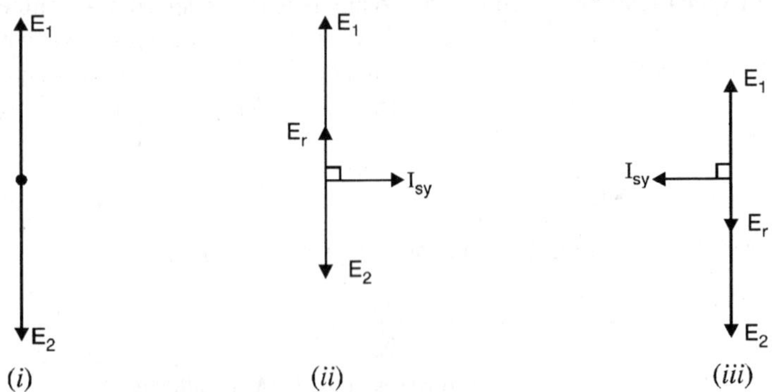

Fig. 12.82

12.31 SYNCHRONISING POWER

When two alternators are operating in parallel, each machine has an inherent tendency to remain synchronised. Consider two similar single-phase alternators 1 and 2 operating in parallel at no-load [See Fig. 12.83]. Suppose, due to any reason, the speed of machine 2 decreases. This will cause E_2 to fall back by a phase angle of α electrical degrees as shown in Fig. 12.84 (though still $E_1 = E_2$). Within the local circuit formed by two alternators, the resultant e.m.f. E_r is the phasor difference $E_1 - E_2$. This resultant e.m.f. results in the production of synchronising current I_{sy} which sets up synchronising torque. The synchronising torque retards machine 1 and accelerates machine 2 so that synchronism is re-established. *The power associated with synchronising torque is called* **synchronising power.**

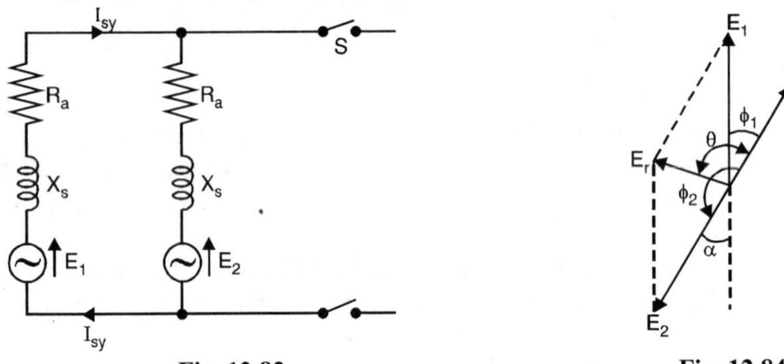

Fig. 12.83 **Fig. 12.84**

In Fig. 12.83, machine 1 is generating and machine 2 is motoring. The power supplied by machine 1 is called synchronising power. Referring to Fig. 12.84, we have,

$$\text{Synchronising power, } P_{sy} = E_1 I_{sy} \cos \phi_1 = E_1 I_{sy} \cos (90^\circ - \theta) = E_1 I_{sy} \sin \theta$$

$$= E_1 I_{sy} \qquad (\because \theta \simeq 90^\circ \text{ so that } \sin \theta \simeq 1)$$

The synchronising power goes to supply power input to machine 2 and the Cu losses in the local circuit of the two machines.

$$\therefore \quad E_1 I_{sy} = E_2 I_{sy} + \text{Cu losses}$$

$$\text{Resultant e.m.f., } E_r = 2\,E \cos \frac{(180^\circ - \alpha)}{2} \qquad [\because E_1 = E_2 = E \text{ (say)}]$$

$$= 2\,E \cos \left(90^\circ - \frac{\alpha}{2}\right) = 2\,E \sin \frac{\alpha}{2} = 2\,E \times \frac{\alpha}{2} \qquad [\because \alpha \text{ is small}]$$

$$\therefore \quad E_r = \alpha E \qquad ...(i)$$

Note that in this expression, α is in electrical radians.

$$\text{Synchronising current, } I_{sy} = \frac{E_r}{2\,X_s} = \frac{\alpha\,E}{2\,X_s} \quad \text{...}R_a \text{ of both machines is negligible}$$

Here X_s = synchronous reactance of each machine

$\therefore$ Synchronising power supplied by machine 1 is

$$P_{sy} = E_1\,I_{sy} \quad \text{...per phase}$$

$$= E \times \frac{\alpha\,E}{2\,X_s} = \frac{\alpha\,E^2}{2\,X_s} \quad \left(\because E_1 = E_2 \text{ and } I_{sy} = \frac{\alpha\,E}{2\,X_s}\right)$$

$$\therefore \quad P_{sy} = \frac{\alpha\,E^2}{2\,X_s} \quad \text{...per phase}$$

Total synchronising power for 3 phases

$$= 3\,P_{sy} = \frac{3\,\alpha\,E^2}{2\,X_s}$$

Note that this is the value of synchronising power when two alternators operate in parallel at no-load.

12.32 ALTERNATORS CONNECTED TO INFINITE BUSBARS

When an alternator is connected to an infinite busbars, the impedance (or synchronous reactance) of *only that alternator is considered.

$$\therefore \quad I_{sy} = \frac{E_r}{Z_s}$$

$$= \frac{\alpha\,E}{X_s} \quad \text{...if } R_a \text{ is negligible}$$

$$\therefore \quad P_{sy} = \frac{\alpha\,E^2}{X_s} \quad \text{...per phase}$$

Total synchronising power for three phases

$$= 3\,P_{sy} = \frac{3\,\alpha\,E^2}{X_s}$$

Synchronising Torque T_{sy}

Let T_{sy} be the synchronising torque in newton-metre (N-m).

(*i*) **When there are two alternators is parallel**

$$3\,P_{sy} = \frac{2\,\pi\,N_s\,T_{sy}}{60}$$

$$\therefore \quad T_{sy} = \frac{3\,P_{sy} \times 60}{2\,\pi\,N_s} \quad \left(\text{Here } P_{sy} = \frac{\alpha\,E^2}{2\,X_s}\right)$$

(*ii*) **When alternator is connected to infinite Busbars**

$$3\,P_{sy} = \frac{2\,\pi\,N_s\,T_{sy}}{60}$$

$$\therefore \quad T_{sy} = \frac{3\,P_{sy} \times 60}{2\,\pi\,N_s} \quad \left(\text{Here } P_{sy} = \frac{\alpha\,E^2}{X_s}\right)$$

---

* It is because the impedance of a very large number of machines in parallel tends to zero.

12.33 EFFECT OF LOAD ON SYNCHRONISING POWER

In this case, the approximate value of synchronising power is given by ;

$$P_{sy} = \frac{\alpha E V}{X_s} \quad \text{per phase}$$

where E = e.m.f. of alternator/phase

V = busbars voltage/phase

As already discussed, for a lagging p.f.,

$$E = \sqrt{(V \cos\phi + I_a R_a)^2 + (V \sin\phi + I_a X_s)^2}$$

Example 12.52. *A 2 MVA, 3-phase, 8-pole alternator is connected to 6000V, 50Hz busbars and has a synchronous reactance of 4 Ω per phase. Calculate the synchronising power and synchronising torque per mechanical degree of rotor displacement at no-load. Assume normal excitation.*

Solution. Assume the alternator to be star-connected.

E.M.F. per phase at no-load is given by ;

$$E = 6000/\sqrt{3} = 3465 \text{ volts}$$

Synchronous reactance per phase is given by ;

$$X_s = 4\,\Omega$$

Rotor displacement in electrical degrees

$$= \text{Rotor displacement in mechanical degrees} \times \frac{P}{2}$$

$$= 1 \times 8/2 = 4° \text{ (elect.)}$$

Rotor displacement in electrical radians is

$$\alpha = 4 \times \frac{\pi}{180} = \frac{\pi}{45} \text{ elect. radians}$$

$$\text{Synchronous speed, } N_s = \frac{120 f}{P} = \frac{120 \times 50}{8} = 750 \text{ r.p.m.}$$

$$\text{Synchronising power for 3-phases} = 3 P_{sy} = \frac{3 E^2 \alpha}{X_s} = \frac{3 \times (3465)^2 \times \pi}{4 \times 45}$$

$$= 628644 \text{ W} = \mathbf{628{\cdot}644 \text{ kW}}$$

$$\text{Synchronising torque, } T_{sy} = \frac{3 P_{sy} \times 60}{2\pi N_s} = \frac{628644 \times 60}{2\pi \times 750} = \mathbf{8000 \text{ N-m/mech. degree}}$$

Example 12.53. *A 3000 kVA, 6-pole alternator runs at 1000 r.p.m. in parallel with other machines on 3300 V busbars. The synchronous reactance is 25 %. Calculate the synchronising power for one mechanical degree of displacement and the corresponding synchronising torque.*

Solution. Since the alternator is operating in parallel with many machines, it may be considered to be connected to infinite busbars. Assuming the alternator to be star-connected,

$$\text{Voltage/phase, } E = 3300/\sqrt{3} = 1905 \text{ volts}$$

$$\text{F.L. current, } I_a = \frac{3000 \times 10^3}{\sqrt{3} \times 3300} = 525 \text{ A}$$

Now $I_a X_s$ = 25 % of 1905 $\therefore X_s = \dfrac{0{\cdot}25 \times 1905}{525} = 0{\cdot}9071\,\Omega$

Rotor displacement in electrical degrees

$$= \text{Rotor displacement in mech. degrees} \times \frac{P}{2}$$

$$= 1 \times 6/2 = 3^\circ \text{ (elect.)}$$

Rotor displacement in electrical radians is

$$\alpha = 3 \times \frac{\pi}{180} = \frac{\pi}{60} \text{ elect. radians}$$

$$\text{Synchronous speed, } N_s = 1000 \text{ r.p.m.}$$

Synchronising power for 3-phases

$$= 3\,P_{sy} = \frac{3\,\alpha\,E^2}{X_s} = \frac{3 \times \pi \times (1905)^2}{60 \times 0{\cdot}9071} = 628{\cdot}4 \times 10^3 \text{ W}$$

$$= \mathbf{628{\cdot}4\ kW}$$

$$\text{Synchronising torque, } T_{sy} = \frac{3\,P_{sy} \times 60}{2\pi\,N_s} = \frac{628{\cdot}4 \times 10^3 \times 60}{2\pi \times 1000} = \mathbf{6000\ N\text{-}m}$$

Example 12.54. *A 5·5 MVA, 50 Hz, 3-phase synchronous generator having a synchronous reactance of 0·30 p.u. is running at 1500 r.p.m. and is excited to give 11 kV. If the rotor deviates slightly from its equilibrium position, what is the synchronising torque in N-m per degree mechanical displacement ?*

Solution. Assuming the machine to be star-connected,

$$\text{E.M.F./phase, } E = \frac{11 \times 10^3}{\sqrt{3}} = 6350 \text{ volts}$$

$$\text{F.L. current, } I_a = \frac{5{\cdot}5 \times 10^6}{\sqrt{3} \times 11 \times 10^3} = 288{\cdot}7 \text{ A}$$

Now $I_a X_s = 0{\cdot}3 \text{ of } 6350 \quad \therefore \quad X_s = \dfrac{0{\cdot}3 \times 6350}{288{\cdot}7} = 6{\cdot}6\ \Omega$

$$\text{Synchronous speed, } N_s = 1500 \text{ r.p.m.}$$

$$\text{Number of poles, } P = \frac{120\,f}{N_s} = \frac{120 \times 50}{1500} = 4$$

Rotor displacement in electrical degrees

$$= \text{Rotor displacement in mech. degrees} \times P/2$$

$$= 1 \times 4/2 = 2^\circ \text{ (elect.)}$$

Rotor displacement in electrical radians

$$= 2 \times \pi/180 = \frac{\pi}{90} \text{ elect. radians}$$

Synchronising power for 3 phases

$$= 3\,P_{sy} = \frac{3\,\alpha\,E^2}{X_s} = \frac{3 \times \pi \times (6350)^2}{90 \times 6{\cdot}6} = 6{\cdot}4 \times 10^5 \text{ W}$$

$$\therefore \quad \text{Synchronising torque, } T_{sy} = \frac{3\,P_{sy} \times 60}{2\,\pi\,N_s} = \frac{6{\cdot}4 \times 10^5 \times 60}{2\pi \times 1500} = \mathbf{4074\ N\text{-}m}$$

Example 12.55. *A 5000 kVA, 10,000 V, 1500 r.p.m., 50 Hz alternator runs in parallel with other machines. Its synchronous reactance is 20 %. Find for (i) no-load (ii) full-load at power factor 0·8 lagging, synchronising power per unit mechanical angle of phase displacement and calculate the synchronising torque if the mechanical displacement is 0·5°.*

Solution.

$$\text{Voltage/phase, } V = 10{,}000/\sqrt{3} = 5775 \text{ volts}$$

$$\text{F.L. current, } I_a = \frac{5000 \times 10^3}{\sqrt{3} \times 10000} = 288{\cdot}7\,\text{A}$$

Synchronous reactance per phase is

$$X_s = \frac{20}{100} \times \frac{5775}{288{\cdot}7} = 4\,\Omega$$

$$\text{Number of poles, } P = \frac{120\,f}{N_s} = \frac{120 \times 50}{1500} = 4$$

$$\alpha = 1^\circ\,(\text{mech.}) = 1 \times P/2\,(\text{elect.}) = 1 \times 4/2 = 2^\circ\,(\text{elect.})$$

$$= 2 \times \pi/180 = \pi/90 \text{ elect. radian}$$

(*i*) **At no-load**

At no-load, V can be taken equal to E.

$\therefore$ Synchronising power for 3-phases

$$= 3\,P_{sy} = \frac{3\,\alpha\,E^2}{X_s} = 3 \times \frac{\pi}{90} \times \frac{(5775)^2}{4} = 873{\cdot}4 \times 10^3\,\text{W}$$

$$= \mathbf{873{\cdot}4\ kW}$$

$$\text{Synchronising torque, } T_{sy} = \frac{3\,P_{sy} \times 60}{2\,\pi\,N_s} = \frac{873{\cdot}4 \times 10^3 \times 60}{2\,\pi \times 1500} = 5564\,\text{N-m}$$

T_{sy} for 0·5° mechanical displacement

$$= 5564/2 = \mathbf{2782\ N\text{-}m}$$

(*ii*) **At Full-load 0·8 p.f. lagging.** Fig. 12.85 shows the phasor diagram.

Let $\mathbf{V} = 5775\,\angle 0^\circ \text{ volts} = (5775 + j\,0)$ volts

Then, $\mathbf{I_a} = 288{\cdot}7\,\angle{-36{\cdot}9^\circ}\,\text{A}$

$$\mathbf{E} = \mathbf{V} + \mathbf{I_a X_s}$$

$$\mathbf{I_a X_s} = 288{\cdot}7\,\angle{-36{\cdot}9^\circ} \times 4\,\angle 90^\circ$$

$$= 1154{\cdot}8\,\angle 53{\cdot}1^\circ\,\text{V} = (693{\cdot}4 + j\,923{\cdot}5)\,\text{V}$$

$$\therefore \quad \mathbf{E} = (5775 + j\,0) + (693{\cdot}4 + j\,923{\cdot}5)$$

$$= (6468{\cdot}4 + j\,923{\cdot}5)\,\text{V}$$

$$= 6533\,\angle 8{\cdot}1^\circ \text{ volts}$$

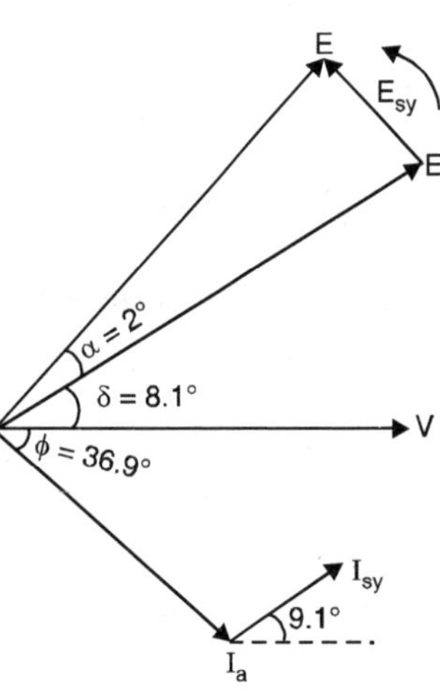

Fig. 12.85

As a result of mechanical displacement, an additional e.m.f. ($E_{sy} = 2\,E \sin \alpha/2$) will be produced. This will set up an additional current $I_{sy} = E_{sy}/X_s$.

$$E_{sy} = 2\,E \sin \alpha/2 = 2 \times 6533 \times \sin 2^\circ/2 = 228 \text{ volts}$$

$$\therefore \quad I_{sy} = E_{sy}/X_s = 228/4 = 57\,\text{A}$$

Now I_{sy} leads V by $(\delta + \alpha/2) = (8{\cdot}1^\circ + 2^\circ/2) = 9{\cdot}1^\circ$

$\therefore$ Synchronising power for 3-phases

$$= 3\,V\,I_{sy} \cos(\delta + \alpha/2) = 3 \times 5775 \times 57 \times \cos 9{\cdot}1^\circ$$

$$= 975 \times 10^3\,\text{W} = \mathbf{975\ kW}$$

Synchronising torque per degree mechanical displacement

$$= \frac{975 \times 10^3 \times 60}{2\pi \times 1500} = 6210\,\text{N-m}$$

Synchronising torque for 0·5 degree mechanical displacement

$$= 6210/2 = \mathbf{3105\ N\text{-}m}$$

Note. Approximate value of synchronising power/phase

$$= \frac{\alpha E V}{X_s} = \frac{\pi}{90} \times \frac{6533 \times 5775}{4} = 329{\cdot}3 \times 10^3 \text{ W/phase}$$

$$= 329{\cdot}3 \text{ kW/phase}$$

Example 12.56. *A 2-pole, 50 Hz, 3-ϕ turbo-alternator is excited to generate the busbars voltage of 11 kV at no-load. Calculate the synchronising power per degree of mechanical displacement of the rotor and the corresponding synchronising torque. The machine is star-connected and the short-circuit current for this excitation is 1200 A.*

Solution. E.M.F./phase, $E = \dfrac{11 \times 10^3}{\sqrt{3}} = 6350$ volts

Rotor displacement, $\alpha = 1°\text{(mech.)} = 1 \times P/2 \text{ (elect.)} = 1 \times 2/2 = 1°$ elect.

$= \pi/180$ elect. radian

Synchronous speed, $N_s = \dfrac{120\, f}{P} = \dfrac{120 \times 50}{2} = 3000$ r.p.m.

Synchronous reactance/phase, $*X_s = 6350/1200 = 5{\cdot}29\ \Omega$

Synchronising power for 3-phases $= \dfrac{3\,\alpha\, E^2}{X_s} = 3 \times \dfrac{\pi}{180} \times \dfrac{(6350)^2}{5{\cdot}29}$

$= 399 \times 10^3$ W $=$ **399 kW**

Synchronising torque, $T_{sy} = \dfrac{399 \times 10^3 \times 60}{2\pi \times 3000} =$ **1270 N-m**

TUTORIAL PROBLEMS

1. A 10 MVA, 3-phase alternator has an equivalent short-circuit reactance of 20 %. Calculate the synchronising power per mechanical degree of phase displacement when running in parallel on 10,000 V, 50 Hz busbars at 1500 r.p.m. **[1745 kW]**
2. A 2000 kVA, 3-phase, 8-pole alternator runs at 750 r.p.m. in parallel with other machines on 6000 V bus-bars. Find the synchronising power on full-load, 0·8 p.f. lagging per mechanical degree of displacement and the corresponding synchronising torque. The synchronous reactance is 6 Ω/phase. **[498·6 kW ; 6348 N-m]**
3. A 2 MVA, 3-phase, 8-pole alternator runs at 750 r.p.m. and operates in parallel with other machines on 6000 V bus-bars. Find the synchronising power on full-load at p.f. 0·8 lagging per mechanical degree of displacement and the corresponding synchronising torque. The synchronous reactance of the machine is 6 Ω/phase. **[505 kW ; 6400 N-m]**

12.34 SHARING OF LOAD CURRENTS BY TWO ALTERNATORS IN PARALLEL

Consider two alternators with identical speed/load characteristics connected in parallel as shown in Fig. 12.86.

Let $\mathbf{E_1}, \mathbf{E_2}$ = induced e.m.f.s per phase

$\mathbf{Z_1}, \mathbf{Z_2}$ = synchronous impedances per phase

$\mathbf{Z}$ = load impedance per phase

$\mathbf{I_1}, \mathbf{I_2}$ = currents supplied by two machines

$\mathbf{V}$ = common terminal voltage per phase

$$\mathbf{V} = \mathbf{E_1} - \mathbf{I_1}\mathbf{Z_1} = \mathbf{E_2} - \mathbf{I_2}\mathbf{Z_2}$$

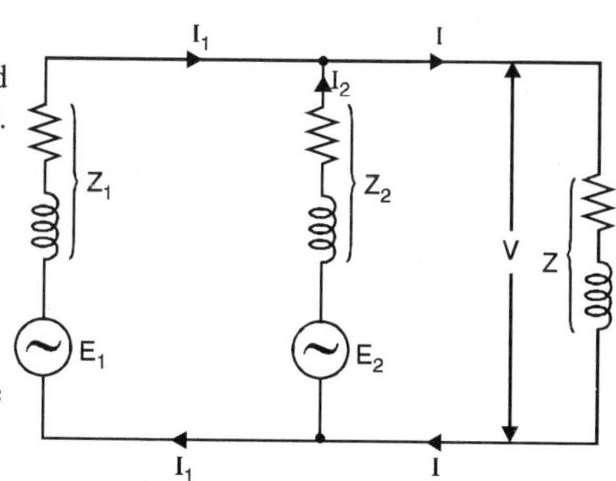

Fig. 12.86

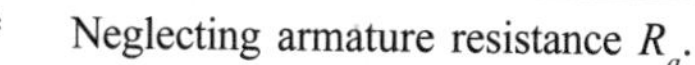
* Neglecting armature resistance R_a.

$$\therefore \quad \mathbf{I}_1 = \frac{\mathbf{E}_1 - \mathbf{V}}{\mathbf{Z}_1} \quad ; \quad \mathbf{I}_2 = \frac{\mathbf{E}_2 - \mathbf{V}}{\mathbf{Z}_2}$$

$$\mathbf{I} = \mathbf{I}_1 + \mathbf{I}_2 = \frac{\mathbf{E}_1 - \mathbf{V}}{\mathbf{Z}_1} + \frac{\mathbf{E}_2 - \mathbf{V}}{\mathbf{Z}_2}$$

$$\mathbf{V} = (\mathbf{I}_1 + \mathbf{I}_2)\,\mathbf{Z} = \mathbf{I}\,\mathbf{Z}$$

Circulating current on no-load is given by ;

$$\mathbf{I}_C = \frac{\mathbf{E}_1 - \mathbf{E}_2}{\mathbf{Z}_1 + \mathbf{Z}_2}$$

Example 12.57. *Two single-phase alternators are connected in parallel and supply current to a load at a terminal voltage of 10000 $\angle 0°$ volts. Alternator A has an induced e.m.f. of 13000$\angle 22{\cdot}6°$ volts and a reactance of 2 Ω ; alternator B has an e.m.f. of 12500 $\angle 36{\cdot}9°$ volts and a reactance of 3 Ω. Find the current supplied by each alternator.*

Solution. Current supplied by alternator A is given by ;

$$\mathbf{I}_A = \frac{\mathbf{E}_A - \mathbf{V}}{\mathbf{Z}_A} = \frac{13000\angle 22{\cdot}6° - 10000\angle 0°}{2\angle 90°}$$

$$= 6500\angle -67{\cdot}4° + j\,5000 = (2500 - j\,1000)\text{ A}$$

$$= \mathbf{2700\angle -21{\cdot}8°\ A}$$

Current supplied by alternator B is given by ;

$$\mathbf{I}_B = \frac{\mathbf{E}_B - \mathbf{V}}{\mathbf{Z}_B} = \frac{12500\angle 36{\cdot}9° - 10000\angle 0°}{3\angle 90°} = \frac{j\,7500}{j\,3}$$

$$= (2500 + j\,0)\text{A} = \mathbf{2500\angle 0°\ A}$$

Example 12.58. *Two 3-phase alternators A and B supply a sub-station through lines 1 and 2. The sub-station load is R = 50 Ω ; X = 40 Ω to neutral. If the generators e.m.f. are A, 10000 V and B, 12000 V and A leads B by 10° elect; find currents delivered by A and B.*

Generator A rating : 10,000 kVA, 11000 V, 10 % reactance

Generator B rating : 20,000 kVA, 11000 V, 15 % reactance

Reactance of line 1 = 3·8 Ω : Reactance of line 2 = 4·1 Ω

Solution. Fig. 12.87 shows the conditions of the problem.

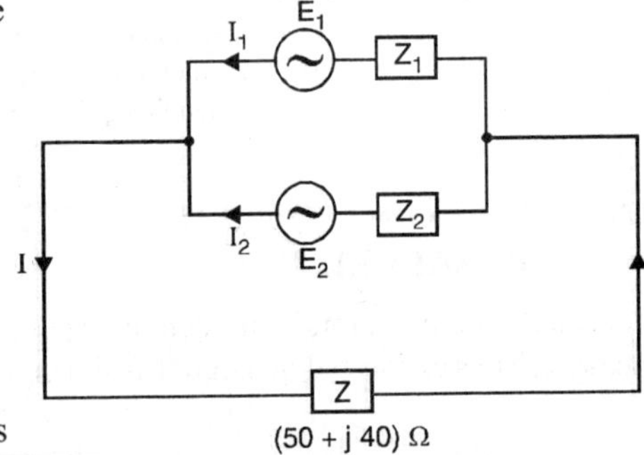

Fig. 12.87

F.L. current of generator A is

$$I_A = I_1 = \frac{10{,}000 \times 10^3}{\sqrt{3} \times 11000} = 525\text{ A}$$

F.L. current of generator B is

$$I_B = I_2 = \frac{20{,}000 \times 10^3}{\sqrt{3} \times 11000} = 1050\text{ A}$$

Let x_1 and x_2 be the respective per phase reactances of the generators in ohms.

$$\therefore \quad 525\,x_1 = 10\,\%\text{ of } 11000/\sqrt{3}$$

$$\therefore \quad x_1 = \frac{10}{100} \times \frac{11000}{\sqrt{3}} \times \frac{1}{525} = 1{\cdot}2\ \Omega$$

Also

$$1050\,x_2 = 15\,\%\text{ of } 11000/\sqrt{3}$$

$$\therefore \quad x_2 = \frac{15}{100} \times \frac{11000}{\sqrt{3}} \times \frac{1}{1050} = 0{\cdot}9\ \Omega$$

$$\therefore \qquad Z_1 = x_1 + 3{\cdot}8 = 1{\cdot}2 + 3{\cdot}8 = 5\,\Omega$$

$$\text{and} \qquad Z_2 = x_2 + 4{\cdot}1 = 0{\cdot}9 + 4{\cdot}1 = 5\,\Omega$$

$$\text{Now} \qquad \mathbf{Z_1} = (j\,5)\,\Omega\ ;\ \mathbf{Z_2} = (j\,5)\,\Omega\ ;\ \mathbf{Z} = (50 + j\,40)\,\Omega$$

$$\mathbf{E_1} = \frac{10{,}000}{\sqrt{3}}\angle 10^\circ = 5774\angle 10^\circ \text{ volts} = (5686 + j\,1003) \text{ volts}$$

$$\mathbf{E_2} = \frac{12000}{\sqrt{3}}\angle 0^\circ = (6928 + j\,0) \text{ volts}$$

Referring to Fig. 12.87, we have,

$$\mathbf{I_1 Z_1} + (\mathbf{I_1} + \mathbf{I_2})\mathbf{Z} = \mathbf{E_1} \qquad ...(i)$$

and

$$\mathbf{I_2}{}^*\mathbf{Z_1} + (\mathbf{I_1} + \mathbf{I_2})\mathbf{Z} = \mathbf{E_2} \qquad ...(ii)$$

Subtracting eq. (*ii*) from eq. (*i*), we have,

$$\mathbf{I_1} - \mathbf{I_2} = \frac{\mathbf{E_1} - \mathbf{E_2}}{\mathbf{Z_1}} = \frac{(5686 + j\,1003) - (6928 + j\,0)}{j\,5}$$

$$= \frac{-1242 + j\,1003}{j\,5} = (200 + j\,250)\,\text{A}$$

$$\therefore \qquad \mathbf{I_1} - \mathbf{I_2} = (200 + j\,250)\,\text{A} \qquad ...(iii)$$

Adding eqs. (*i*) and (*ii*), we have,

$$\mathbf{I_1} + \mathbf{I_2} = \frac{\mathbf{E_1} + \mathbf{E_2}}{2\,\mathbf{Z} + \mathbf{Z_1}} = \frac{(5686 + j\,1003) + (6928 + j\,0)}{2\,(50 + j\,40) + j\,5}$$

$$= \frac{12614 + j\,1003}{100 + j\,85} = \frac{12654\angle 4{\cdot}5^\circ}{131{\cdot}2\angle 40{\cdot}4^\circ}$$

$$= 96{\cdot}4\angle -35{\cdot}9^\circ\,\text{A} = (78{\cdot}1 - j\,56{\cdot}6)\,\text{A}$$

$$\therefore \qquad \mathbf{I_1} + \mathbf{I_2} = (78{\cdot}1 - j\,56{\cdot}6)\,\text{A} \qquad ...(iv)$$

On solving eqs. (*iii*) and (*iv*), we have,

$$\mathbf{I_1 = 169\,A} \quad ; \quad \mathbf{I_2 = 164\,A}$$

Example 12.59. *Two 3-phase, star-connected alternators 1 and 2 supply a total load of 18 MVA at 0·7 lagging p.f. at a line voltage of 6·6 kV. The two alternators are rated at 10 MVA, 6·6 kV. The machine 1 is operating on full-load at 0·8 lagging power factor. Find (i) the current supplied by machine 2 (ii) operating p.f. of machine 2 (iii) power delivered by each machine.*

Solution. Fig. 12.88 shows the conditions of the problem.

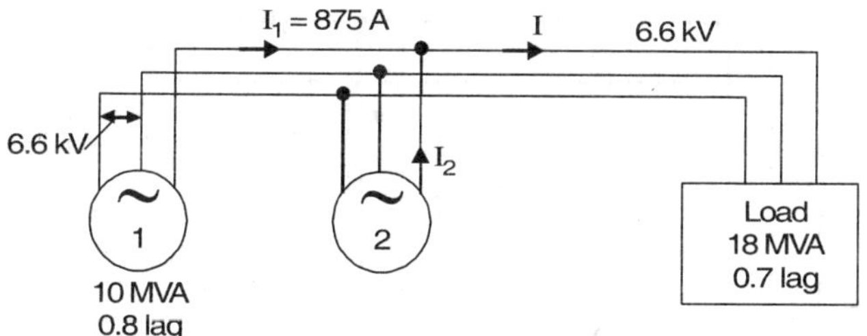

Fig. 12.88

F.L. current of machine 1 is $I_1 = \dfrac{10 \times 10^6}{\sqrt{3} \times 6{\cdot}6 \times 10^3} = 875\,\text{A}$

---

* $\mathbf{Z_1} = \mathbf{Z_2} = j\,5$

$$\text{Total load current is } I = \frac{18 \times 10^6}{\sqrt{3} \times 6 \cdot 6 \times 10^3} = 1575 \text{ A at } 0{\cdot}7 \text{ lag } (= 45{\cdot}6° \text{ lag})$$

Now $\mathbf{I_1} = 875 \angle{-36{\cdot}9°} \text{ A}$; $\mathbf{I} = 1575 \angle{-45{\cdot}6°} \text{ A}$

(*i*) $\mathbf{I} = \mathbf{I_1} + \mathbf{I_2}$

$\therefore$ $\mathbf{I_2} = \mathbf{I} - \mathbf{I_1} = 1575 \angle{-45{\cdot}6°} - 875 \angle{-36{\cdot}9°}$

$= 722{\cdot}3 \angle{-56{\cdot}2°} \text{ A}$

$\therefore$ Current supplied by machine 2 is I_2 = **722·3 A**

(*ii*) P.F. of machine 2 = $\cos(-56{\cdot}2°)$ = **0·556** *lag*

(*iii*) Power delivered by machine 1 is

$$P_1 = \sqrt{3}\, V I_1 \cos\phi_1 = \sqrt{3} \times 6{\cdot}6 \times 10^3 \times 875 \times 0{\cdot}8$$
$$= 8 \times 10^6 \text{ W} = \mathbf{8\ MW}$$

Power delivered by machine 2 is

$$P_2 = \sqrt{3}\, V I_2 \cos\phi_2 = \sqrt{3} \times 6{\cdot}6 \times 10^3 \times 722{\cdot}3 \times 0{\cdot}566$$
$$= 4{\cdot}6 \times 10^6 \text{ W} = \mathbf{4{\cdot}6\ MW}$$

The total power supplied = 8 + 4·6 = 12·6 MW. In fact the load was specified as 18 MVA at 0·7 lagging p.f. or = 18 × 0·7 = 12·6 MW.

Example 12.60. *Two 3-phase, star-connected alternators 1 and 2 supply a load of 3000 kW at 0·8 p.f. lagging and share the load equally. The excitation of machine 2 is so adjusted that it supplies 150 A at a lagging power factor. The per phase synchronous impedances of machines 1 and 2 are (0·4 + j 12) Ω and (0·5 + j 10) Ω respectively. Find current, power factor, induced e.m.f. and load angle of each machine. The terminal voltage is 6·6 kV.*

Solution. Fig. 12.89 shows the conditions of the problem.

$$\text{Terminal phase voltage, } V_{ph} = 6 \cdot 6 \times 10^3 / \sqrt{3} = 3810 \text{ volts}$$

Power supplied by each machine is

$$P_1 = P_2 = 3000/2 = 1500 \text{ kW}$$
$$\mathbf{Z_1} = (0{\cdot}4 + j\,12)\,\Omega \quad ; \quad \mathbf{Z_2} = (0{\cdot}5 + j\,10)\,\Omega$$

$$\text{Total load current, } I = \frac{3000 \times 10^3}{\sqrt{3} \times 6 \cdot 6 \times 10^3 \times 0 \cdot 8} = 328 \text{ A}$$

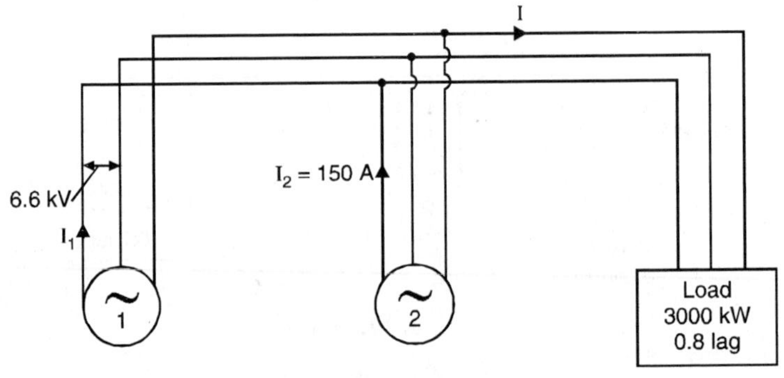

Fig. 12.89

Now $P_2 = \sqrt{3}\, V_L I_2 \cos\phi_2$

$$\therefore \quad \text{P.F. of machine 2, } \cos\phi_2 = \frac{P_2}{\sqrt{3}\, V_L\, I_2} = \frac{1500 \times 10^3}{\sqrt{3} \times 6 \cdot 6 \times 10^3 \times 150} = \mathbf{0{\cdot}875}\ \textit{lag}$$

Current supplied by machine 2 is

$$I_2 = \mathbf{150\ A}$$

$$\mathbf{I_2} = 150(\cos\phi_2 - j\sin\phi_2) = 150(0{\cdot}875 - j\,0{\cdot}485)$$

$$= (131{\cdot}2 - j\,72{\cdot}75)\,\text{A}$$

$$\mathbf{I} = 328\angle{-36{\cdot}9^\circ}\,\text{A} = (262{\cdot}4 - j\,196{\cdot}8)\,\text{A}$$

Now $\mathbf{I} = \mathbf{I_1} + \mathbf{I_2}$

$\therefore$ $\mathbf{I_1} = \mathbf{I} - \mathbf{I_2} = (262{\cdot}4 - j\,196{\cdot}8) - (131{\cdot}2 - j\,72{\cdot}75)$

$$= (131{\cdot}2 - j\,124{\cdot}05)\,\text{A} = 180{\cdot}6\angle{-43{\cdot}4^\circ}\,\text{A}$$

$\therefore$ Current supplied by machine 1 is $I_1 = \mathbf{180{\cdot}6\ A}$

Power factor of machine 1 is $\cos\phi_1 = \cos(-43{\cdot}4^\circ) = \mathbf{0{\cdot}726}$ *lag*

Induced e.m.f. of machine 1 per phase is

$$\mathbf{E_1} = \mathbf{V} + \mathbf{I_1 Z_1} = 3810\angle 0^\circ + (131{\cdot}2 - j\,124{\cdot}05)(0{\cdot}4 + j\,12)$$

$$= (5350 + j\,1520)\,\text{V} = 5560\angle 15{\cdot}9^\circ\,\text{V}$$

$\therefore$ Induced e.m.f. of machine 1 per phase is

$$E_1 = \mathbf{5560\ volts}$$

$$\text{Load angle} = \mathbf{15{\cdot}9^\circ}$$

Induced e.m.f. of machine 2 per phase is

$$\mathbf{E_2} = \mathbf{V} + \mathbf{I_2 Z_2} = 3810\angle 0^\circ + (131{\cdot}2 - j\,72{\cdot}25)(0{\cdot}5 + j\,10)$$

$$= (4600 + j\,1270)\,\text{V} = 4770\angle 15{\cdot}4^\circ\,\text{V}$$

$\therefore$ Induced e.m.f. of machine 2 per phase is

$$E_2 = \mathbf{4770\ volts}$$

$$\text{Load angle} = \mathbf{15{\cdot}4^\circ}$$

Example 12.61. *Two parallel-running alternators have e.m.f.s of 1000 V per phase. The synchronous impedances per phase are $Z_1 = (0{\cdot}1 + j\,2)\ \Omega$ and $Z_2 = (0{\cdot}2 + j\,3{\cdot}2)\ \Omega$. they supply a load of impedance $(2 + j\,1)\ \Omega$ per phase. Find their terminal voltage, load currents, power outputs and no-load circulating current for a phase divergence of 10° (elect.).*

Solution. Fig. 12.90 shows the conditions of the problem.

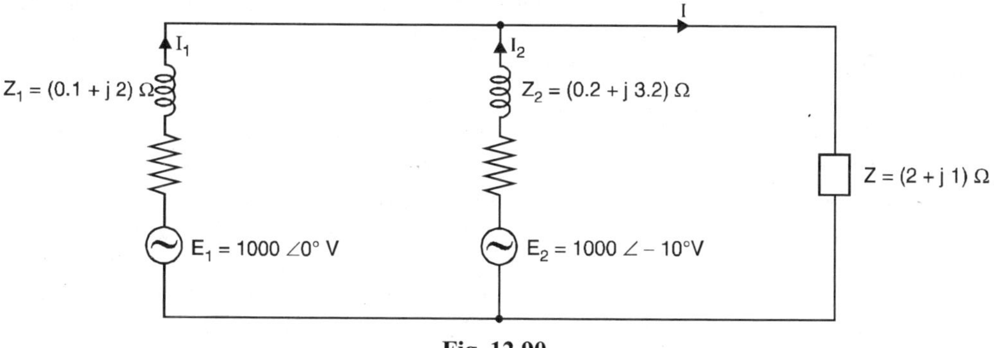

Fig. 12.90

$$\mathbf{E_1} = 1000\angle 0^\circ\,\text{V} = (1000 + j\,0)\,\text{V}$$

$$\mathbf{E_2} = 1000\angle{-10^\circ}\,\text{V} = 1000(\cos 10^\circ - j\sin 10^\circ) = (985 - j\,174)\,\text{V}$$

$$\mathbf{I_1} = \frac{(\mathbf{E_1} - \mathbf{E_2})\mathbf{Z} + \mathbf{E_1 Z_2}}{\mathbf{Z}(\mathbf{Z_1} + \mathbf{Z_2}) + \mathbf{Z_1 Z_2}}$$

$$= \frac{[(1000 + j\,0) - (985 - j\,174)]\,(2 + j\,1) + 1000\,(0\cdot2 + j\,3\cdot2)}{(2 + j\,1)\,[(0\cdot1 + j\,2) + (0\cdot2 + j\,3\cdot2)] + (0\cdot1 + j\,2)\,(0\cdot2 + j\,3\cdot2)}$$

$$= \frac{(15 + j\,174)\,(2 + j\,1) + 200 + j\,3200}{(2 + j\,1)\,[0\cdot3 + j\,5\cdot2] + (0\cdot1 + j\,2)\,(0\cdot2 + j\,3\cdot2)}$$

$$= \mathbf{224\,\angle{-44{\cdot}9^\circ}\,A} = (160 - j\,159)\,\text{A}$$

$$\mathbf{I}_2 = \frac{(\mathbf{E}_2 - \mathbf{E}_1)\,\mathbf{Z} + \mathbf{E}_2\,\mathbf{Z}_1}{\mathbf{Z}\,(\mathbf{Z}_1 + \mathbf{Z}_2) + \mathbf{Z}_1\,\mathbf{Z}_2}$$

$$= \frac{[(985 - j\,174) - (1000 + j\,0)]\,(2 + j\,1) + (985 - j\,174)\,(0\cdot1 + j\,2)}{(2 + j\,1)\,[(0\cdot1 + j\,2) + (0\cdot2 + j\,3\cdot2)] + (0\cdot1 + j\,2)\,(0\cdot2 + j\,3\cdot2)}$$

$$= \mathbf{106\,\angle{-64{\cdot}3^\circ}\,A} = (46 - j\,96)\,\text{A}$$

$$\text{Total current, } \mathbf{I} = \mathbf{I}_1 + \mathbf{I}_2 = (160 - j\,159) + (46 - j\,96)$$

$$= (206 - j\,255)\,\text{A} = \mathbf{327\,\angle{-51{\cdot}1^\circ}\,A}$$

$$\text{Terminal voltage, } \mathbf{V} = \mathbf{I\,Z} = (206 - j\,255)\,(2 + j\,1) = (667 - j\,304)\,\text{V} = \mathbf{730\,\angle{-24{\cdot}5^\circ}\,V}$$

The output of first machine is

$$P_1 = 730 \times 224 \times \cos(44{\cdot}9^\circ - 24{\cdot}5^\circ) = 153 \times 10^3\ \text{W/phase}$$

$$= \mathbf{153\ kW/phase}\ \textit{at p.f. 0·937 lagging}$$

The output of second machine is

$$P_2 = 730 \times 106 \times \cos(64{\cdot}3^\circ - 24{\cdot}5^\circ) = 60 \times 10^3\ \text{W/phase}$$

$$= \mathbf{60\ kW/phase}\ \textit{at p.f. 0·768 lagging}$$

No-load circulating current is

$$\mathbf{I}_C = \frac{\mathbf{E}_1 - \mathbf{E}_2}{\mathbf{Z}_1 + \mathbf{Z}_2} = \frac{(1000 + j\,0) - (985 - j\,174)}{(0\cdot1 + j\,2) + (0\cdot2 + j\,3\cdot2)}$$

$$= \frac{15 + j\,174}{0\cdot3 + j\,5\cdot2} = \mathbf{34\,\angle{-1{\cdot}68^\circ}\,A} = (34 - j\,1)\,\text{A}$$

12.35 ALTERNATOR ON INFINITE BUSBARS

Upto this point, we have considered only a single alternator working on an isolated load or two such machines operating in parallel. In practice, generating stations do not operate as isolated units but are interconnected by the national grid. The result is that a very large number of alternators operate in parallel. An alternator connected to such a network is said to be operating on *infinite busbars*. The behaviour of alternators connected to an infinite busbars is as under :

(*i*) Any change made in the operating conditions of one alternator will not change the *terminal voltage* or *frequency* of the system. In other words, terminal voltage (busbars voltage) and frequency are not affected by changing the operating conditions of one alternator. It is because of large size and inertia of the system.

(*ii*) The *kW output* supplied by an alternator depends *solely* on the mechanical power supplied to the prime mover of the alternator. An increase in mechanical power to the prime mover increases the kW output of the alternator and not the kVAR. A decrease in the mechanical power to the prime mover decreases the kW output of the alternator and not the kVAR.

(*iii*) If the mechanical power to the prime mover of an alternator is kept constant, then change in excitation will change the power factor at which the machine supplies changed current. In other words, change of excitation controls the kVAR and not kW.

The change of driving torque controls the kW output and not kVAR of an alternator. The change of excitation controls the kVAR and not the kW output of an alternator.

Note. An infinite busbars system has constant terminal voltage and constant busbars frequency because of its large size and inertia. However, the busbars voltage can be raised or lowered by increasing or decreasing simultaneously the field excitation of a large number of alternators. Likewise, system frequency can be raised or lowered by increasing or decreasing the speed of prime movers of a large number of alternators.

12.36 EFFECT OF CHANGE OF EXCITATION AND MECHANICAL INPUT

Consider a star-connected alternator connected to an infinite busbars as shown in Fig. 12.91. Note that infinite busbars means that busbars voltage will remain constant and no frequency change will occur regardless of changes made in power input or field excitation of the alternator connected to it.

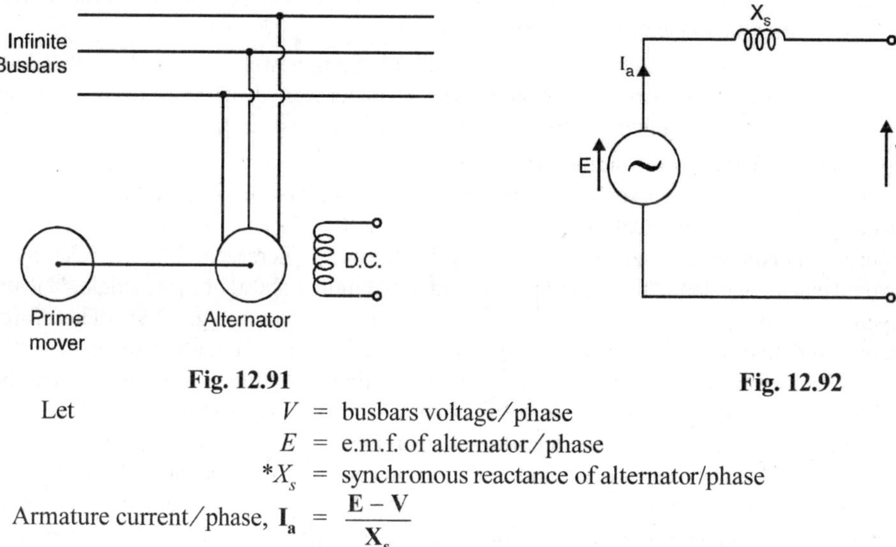

Fig. 12.91 Fig. 12.92

Let V = busbars voltage/phase

E = e.m.f. of alternator/phase

$^*X_s$ = synchronous reactance of alternator/phase

$$\text{Armature current/phase, } \mathbf{I_a} = \frac{\mathbf{E} - \mathbf{V}}{\mathbf{X_s}}$$

Fig. 12.92 shows the equivalent circuit of alternator for one phase.

(*i*) **Effect of change of field excitation.** Suppose the alternator connected to infinite busbars is operating at unity p.f. It is then said to be normally excited. Suppose that excitation of the alternator is increased (overexcited) while the power input to the prime mover is unchanged. The active power output (W or kW) of the alternator will thus remain unchanged *i.e.*, active component of current is unaltered. The overexcited alternator will supply lagging current (and hence lagging reactive power) to the infinite busbars. This action can be explained by the m.m.f. of armature reaction. When the alternator is overexcited, it must deliver lagging current since lagging current produces an opposing m.m.f. to reduce the overexcitation. Thus an overexcited alternator supplies lagging current in addition to the constant active component of current. *Therefore, an overexcited alternator will operate at lagging power factor.* Note that excitation does not control the active power but it controls *power factor* of the current supplied by the alternator to the infinite busbars. Fig. 12.93 shows the phasor

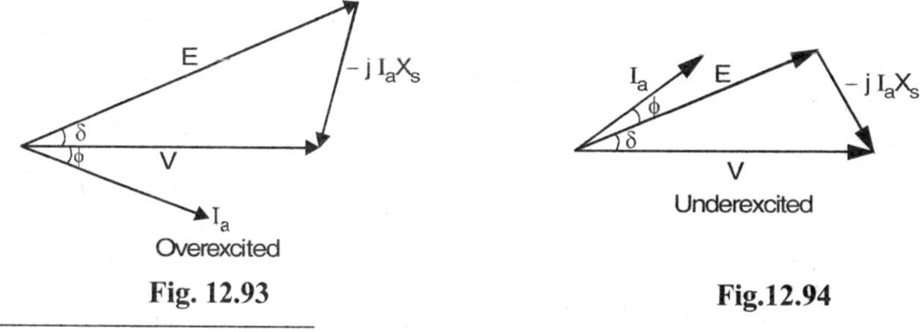

Fig. 12.93 Fig.12.94

* Since $R_a << X_s$, $Z_s = X_s$.

diagram of an overexcited alternator connected to infinite busbars. The angle δ between E and V is called *power angle*.

Now suppose that excitation of the alternator is decreased below normal excitation (under-excitation) while the power input to the prime mover is unchanged. Therefore, the active power output (W or kW) of the alternator will remain unchanged *i.e.*, active component of current is unaltered. The underexcited alternator supplies leading current (and hence leading reactive power) to the infinite busbars. It is because when an alternator is underexcited, it must deliver leading current since leading current produces an aiding m.m.f. to increase the underexcitation. Thus an underexcited alternator supplies leading current in addition to the constant active component of current. *Therefore, an under-excited alternator will operate at leading power factor.* Fig. 12.94 shows the phasor diagram of an underexcited alternator connected to infinite busbars.

Conclusion. An overexcited alternator operates at lagging power factor and supplies lagging reactive power to infinite busbars. On the other hand, an underexcited alternator operates at leading power factor and supplies leading reactive power to the infinite busbars.

(*ii*) **Effect of change in mechanical input.** Suppose the alternator is delivering power to infinite busbars under stable conditions so that a certain power angle δ exists between V and E and E leads V. The phasor diagram for this situation is depicted in Fig. 12.95. Now, suppose that excitation of the alternator is kept constant and power input to its prime mover is increased. The increase in power input would tend to accelerate the rotor and E would move further ahead of V *i.e.*, angle δ increases. Increasing δ results in larger I_a $(= E - V/X_s)$ and lower ϕ as shown in Fig. 12.96. Therefore, the alternator will deliver more active power to the infinite busbars. The angle δ assumes such a value that current I_a has an active power component corresponding to the input. Equilibrium will be re-established at the speed corresponding to the frequency of the infinite busbars with a larger δ. Fig. 12.96 is drawn for the same d.c. field excitation and, therefore, the same E as Fig. 12.95 but the active power output $(= VI_a \cos\phi)$ is greater than for the condition of Fig. 12.95 and increase in δ has caused the alternator to deliver additional active power to the busbars. Note that mechanical input to the prime mover cannot change the speed of the alternator because it is fixed by system frequency. Increasing mechanical input increases the speed of the alternator temporarily till such time the power angle δ increases to a value required for stable operation. Once this condition is reached, the alternator continues to run at synchronous speed.

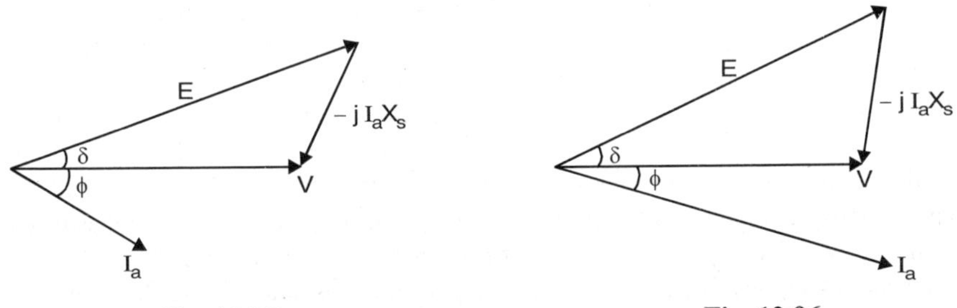

Fig. 12.95 Fig. 12.96

Conclusion. Increasing the mechanical input power to the prime mover will not change the speed ultimately but will increase the power angle δ. As a result, the change of driving torque controls the output kW and not the kVAR. When this change takes place, the power factor of the machine is practically not affected.

Example 12.62. *Two 20 MVA, 3-ϕ alternators operate in parallel to supply a load of 35 MVA at 0·8 p.f. lagging. If output of one machine is 25 MVA at p.f. 0·9 lagging, what is the output and p.f. of the other machine ?*

Solution.

$$\text{Load MW} = 35 \times \cos\phi_L = 35 \times 0{\cdot}8 = 28$$

$$\text{Load MVAR} = 35 \times \sin\phi_L = 35 \times 0{\cdot}6 = 21$$

$$\text{p.f. of first m/c, } \cos\phi_1 = 0{\cdot}9 \text{ lag} \quad ; \quad \sin\phi_1 = 0{\cdot}436$$

$$\text{MW of first m/c} = 25 \times \cos\phi_1 = 25 \times 0{\cdot}9 = 22{\cdot}5$$

$$\text{MVAR of first m/c} = 25 \times \sin\phi_1 = 25 \times 0{\cdot}436 = 10{\cdot}9$$

$$\therefore \quad \text{MW of second m/c} = 28 - 22{\cdot}5 = 5{\cdot}5$$

$$\text{MVAR of second m/c} = 21 - 10{\cdot}9 = 10{\cdot}1$$

$$\text{MVA of second m/c} = \sqrt{(5{\cdot}5)^2 + (10{\cdot}1)^2} = \mathbf{11{\cdot}5}$$

$$\text{p.f. of second m/c} = \frac{\text{MW of second m/c}}{\text{MVA of second m/c}} = \frac{5{\cdot}5}{11{\cdot}5} = \mathbf{0{\cdot}478}\ \textit{lag}$$

Example 12.63. *Two alternators A and B operate in parallel and supply a load of 8 MW at 0·8 p.f. lagging. The power output of A is adjusted to 5000 kW by changing its steam supply and its p.f. is adjusted to 0·9 lagging by changing its excitation. Find the p.f. of alternator B.*

Solution.

$$\text{*kVAR} = \text{kW} \times \tan\phi$$

$$\text{Load p.f., } \cos\phi_L = 0{\cdot}8 \text{ lag} \; ; \; \phi_L = 36{\cdot}9^\circ \text{ and } \tan\phi_L = \tan 36{\cdot}9^\circ = 0{\cdot}75$$

$$\text{p.f. of } A, \cos\phi_A = 0{\cdot}9 \text{ lag} \; ; \; \phi_A = 25{\cdot}84^\circ \text{ and } \tan\phi_A = \tan 25{\cdot}84^\circ = 0{\cdot}484$$

$$\text{Load kW} = 8 \times 10^3 = 8000$$

$$\text{Load kVAR} = 8000 \times \tan\phi_L = 8000 \times 0{\cdot}75 = 6000$$

$$\text{kW of } A = 5000 \; ; \; \text{kVAR of } A = 5000 \times \tan\phi_A = 5000 \times 0{\cdot}484 = 2420$$

$$\text{kW of } B = 8000 - 5000 = 3000 \; ; \; \text{kVAR of } B = 6000 - 2420 = 3580$$

$$\therefore \quad \tan\phi_B = \frac{\text{kVAR of } B}{\text{kW of } B} = \frac{3580}{3000} = 1{\cdot}19$$

$$\text{or} \quad \phi_B = \tan^{-1} 1{\cdot}19 = 50{\cdot}03^\circ$$

$$\therefore \quad \text{p.f. of } B = \cos\phi_B = \cos 50{\cdot}03^\circ = \mathbf{0{\cdot}64}\ \textit{lag}$$

Example 12.64. *Two alternators working in parallel supply a lighting load of 3000 kW and motor load amounting to 5000 kW at a p.f. of 0·71 lagging. One machine is loaded upto 5000 kW at 0·8 p.f. lagging. What is the load and p.f. of the other machine ?*

Solution.

$$\text{Lighting load p.f., } \cos\phi_l = 1 \quad \therefore \quad \tan\phi_l = 0$$

$$\text{Motor load p.f., } \cos\phi_M = 0{\cdot}71 \text{ lag} \quad \therefore \quad \tan\phi_M = \tan 44{\cdot}75^\circ = 0{\cdot}99$$

$$\text{Total load kW} = 3000 + 5000 = 8000 \text{ kW}$$

$$\text{Total load kVAR} = 3000 \times \tan\phi_l + 5000 \times \tan\phi_M$$

$$= 3000 \times 0 + 5000 \times 0{\cdot}99 = 4950$$

$$\text{p.f. of machine 1, } \cos\phi_1 = 0{\cdot}8 \text{ lag} \quad \therefore \quad \tan\phi_1 = \tan 36{\cdot}9^\circ = 0{\cdot}75$$

$$\text{kW of machine 1} = 5000$$

$$\text{kVAR of machine 1} = 5000 \times \tan\phi_1 = 5000 \times 0{\cdot}75 = 3750$$

$$\text{kW of machine 2} = 8000 - 5000 = \mathbf{3000}$$

$$\text{kVAR of machine 2} = 4950 - 3750 = 1200$$

$$* \quad \frac{\text{kVAR}}{\text{kW}} = \frac{VI \sin\phi}{VI \cos\phi} = \tan\phi \quad \therefore \quad \text{kVAR} = \text{kW} \times \tan\phi$$

$$\tan \phi_2 = \frac{\text{kVAR of machine 2}}{\text{kW of machine 2}} = \frac{1200}{3000} = 0{\cdot}4$$

$$\therefore \quad \phi_2 = \tan^{-1} 0{\cdot}4 = 21{\cdot}8^\circ$$

$$\therefore \quad \text{p.f. of machine 2} = \cos \phi_2 = \cos 21{\cdot}8^\circ = \mathbf{0{\cdot}928}\ \textit{lag}$$

Example 12.65. *Two identical 3-phase alternators work in parallel and supply a total load of 1500 kW at 11 kV at a power factor of 0·8 lagging. Each machine supplies half the total power. The synchronous reactance of each is 60 Ω per phase and the resistance is 2·6 Ω per phase. The field excitation of the first machine is so adjusted that its armature current is 45 A lagging. Find (i) the armature current of the second alternator (ii) the p.f. at which each alternator operates.*

Solution. Load current at 0·8 p.f. lagging is

$$I_L = \frac{1500 \times 10^3}{\sqrt{3} \times 11000 \times 0 \cdot 8} = 98{\cdot}4 \text{ A}$$

Wattful component of load current $= I_L \cos \phi_L = 98{\cdot}4 \times 0{\cdot}8 = 78{\cdot}72$ A

Wattless component of load current $= I_L \sin \phi_L = 98{\cdot}4 \times 0{\cdot}6 = 59$ A

When the conditions are identical, each machine supplies armature current = 98·4/2 = 49·2 A ; wattful current = 78·72/2 = 39·36 A and wattless current = 59/2 = 29·5 A.

Since the input power is not changed, the wattful current of each machine remains the same *i.e.*, 39·36 A. When excitation of first machine is changed, its armature current becomes 45 A.

∴ Wattless component of current of first machine after changing excitation

$$= \sqrt{(45)^2 - (39 \cdot 36)^2} = 21{\cdot}8 \text{ A}$$

∴ Wattless component of current of the second machine

$$= 59 - 21{\cdot}8 = 37{\cdot}2 \text{ A}$$

(*i*) Armature current of second machine

$$= \sqrt{(39 \cdot 36)^2 + (37 \cdot 2)^2} = \mathbf{54{\cdot}2\ A}$$

(*ii*) Power factor at which first machine operates

$$= \cos \phi_1 = 39{\cdot}36/45 = \mathbf{0{\cdot}8747}\ \textit{lag}$$

Power factor at which second machine operates

$$= \cos \phi_2 = 39{\cdot}36/54{\cdot}2 = \mathbf{0{\cdot}726}\ \textit{lag}$$

Example 12.66. *Two 50 MVA, 3-phase alternators operate in parallel. The settings of the governors are such that the rise in speed from full-load to no-load is 2 % in one machine and 3 % in the other, the characteristics being straight line in both cases. If each machine is fully loaded when the total load is 100 MW, what would be the load on each machine when the total load is 60 MW ?*

Solution.

Fig. 12.97 shows the parallel operation of the two alternators *A* and *B*. As per the conditions of the problem, the speed-load characteristics of alternators *A* and *B* are represented by straight lines *CA* and *DB* respectively as shown in Fig. 12.98. Since the machines are running in parallel, their frequencies must be the same. Let the total *load of 60 MW be represented by the horizontal line *EF* drawn through *G*. Then load supplied by machine *A* is *EG* and the load supplied by machine *B* is *GF* (= 60 – *EG*).

---

* In contrast with d.c. generators, parallel alternators must run at exactly the same steady-state speed. Consequently, the way in which active power (MW or kW) divides between them depends almost wholly on the load-speed characteristics of their prime movers (PM). All practical prime movers have drooping load-speed characteristics.

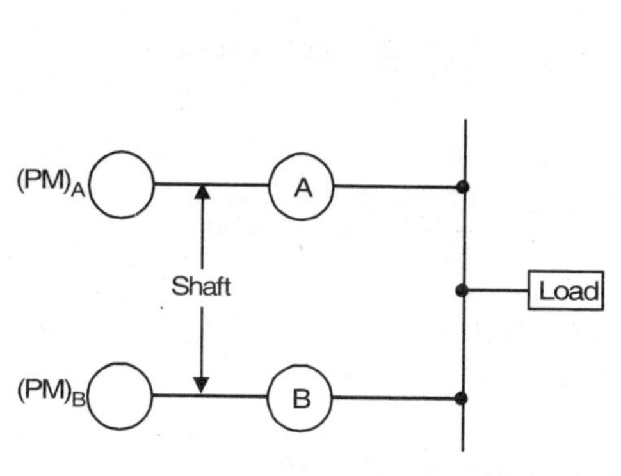

Fig. 12.97

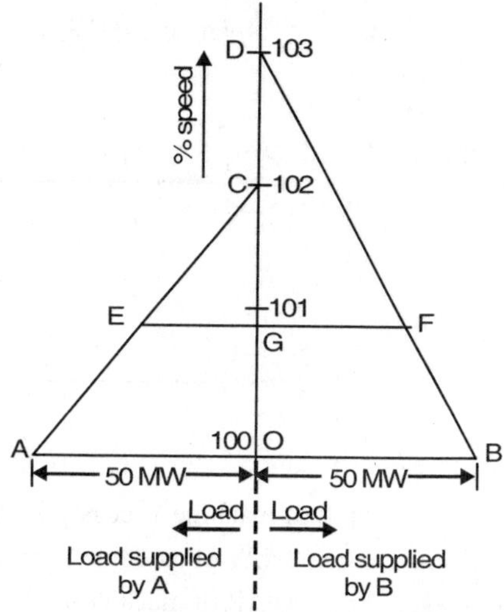

Fig. 12.98

Referring to Fig. 12.98, we have,

$$\frac{AO}{EG} = \frac{OC}{CG}$$

or $$EG = AO \times \frac{CG}{OC} = 50 \times \frac{OC - OG}{OC} = \frac{50}{2}\,(2 - OG)$$

∴ $$EG = 25\,(2 - OG)$$

Again referring to Fig. 12.98, we have,

$$\frac{BO}{GF} = \frac{OD}{DG}$$

or $$GF = BO \times \frac{DG}{OD} = 50 \times \frac{OD - OG}{OD} = 50 \times \frac{(3 - OG)}{3}$$

∴ $$GF = \frac{50}{3}\,(3 - OG)$$

But $$EG + GF = 60$$

or $$25\,(2 - OG) + \frac{50}{3}\,(3 - OG) = 60$$

∴ $$OG = 0{\cdot}96$$

∴ Load supplied by machine A

$$= EG = 25\,(2 - OG) = 25\,(2 - 0{\cdot}96) = \mathbf{26\ MW}$$

Load supplied by machine B

$$= GF = \frac{50}{3}(3 - OG) = \frac{50}{3}(3 - 0{\cdot}96) = \mathbf{34\ MW}$$

Example 12.67. *A lighting load of 2000 kW and a motor load of 4000 kW at a p.f. 0·8 lagging are supplied by two alternators running in parallel. One machine is loaded to 2400 kW at 0·95 p.f. lagging. What is the kW output and p.f. of the second machine?*

Solution. Fig. 12.99 shows the conditions of the problem.

Lighting load p.f., $\cos\phi_l = 1 \quad \therefore \quad \tan\phi_l = 0$

Motor load p.f., $\cos\phi_M = 0{\cdot}8$ lag $\quad \therefore \quad \tan\phi_M = 0{\cdot}75$

$$\text{Total load kW} = 2000 + 4000 = 6000$$

$$\text{Total load kVAR} = 2000 \times \tan\phi_l + 4000 \times \tan\phi_M$$

$$= 2000 \times 0 + 4000 \times 0{\cdot}75 = 3000$$

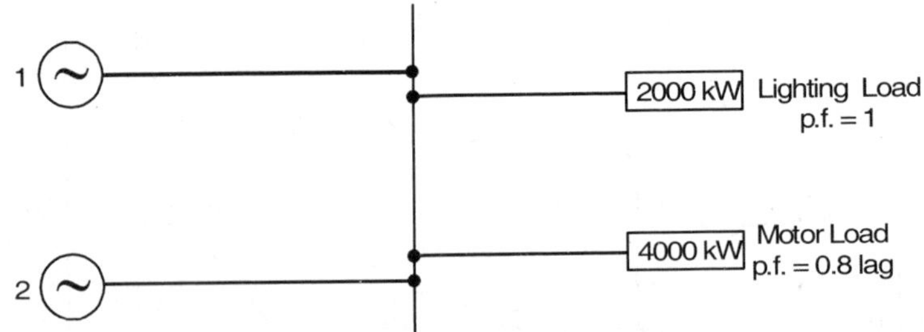

Fig. 12.99

$$\text{p.f. of machine 1, } \cos\phi_1 = 0{\cdot}95 \quad \therefore \quad \tan\phi_1 = 0{\cdot}329$$

$$\text{kW of machine 1} = 2400$$

$$\text{kVAR of machine 1} = 2400 \times \tan\phi_1 = 2400 \times 0{\cdot}329 = 789$$

$$\text{kW of machine 2} = 6000 - 2400 = \mathbf{3600}$$

$$\text{kVAR of machine 2} = 3000 - 789 = 2211$$

$$\tan\phi_2 = \frac{\text{kVAR of machine 2}}{\text{kWof machine 2}} = \frac{2211}{3600} = 0{\cdot}61$$

$$\therefore \quad \phi_2 = \tan^{-1} 0{\cdot}61 = 31{\cdot}4^\circ$$

$$\text{p.f. of machine 2} = \cos\phi_2 = \cos 31{\cdot}4^\circ = \mathbf{0{\cdot}85}\ \textit{lag}$$

Example 12.68. *Two identical 20 MW alternators operate in parallel. The governor of first machine is such that the frequency drops uniformly from 50 Hz on no-load to 48 Hz on full-load. The corresponding uniform speed drop of the second machine is 50 Hz to 47·5 Hz. How will the two machines share a load of 30 MW ?*

Solution. Fig. 12.100 shows the frequency/load characteristics of the two machines. As per the conditions of the problem, the frequency/load characteristics of machines 1 and 2 are represented by straight lines *AD* and *AB* respectively. Since the machines are operating in parallel, the frequency of the two machines must be the same at any load. Let the two machines share the load of 30 MW at frequency *x* measured from *A*. Then the horizontal line *GH* represents 30 MW load. The load supplied by machine 2 is *GI* and the load supplied by machine 1 is *HI*.

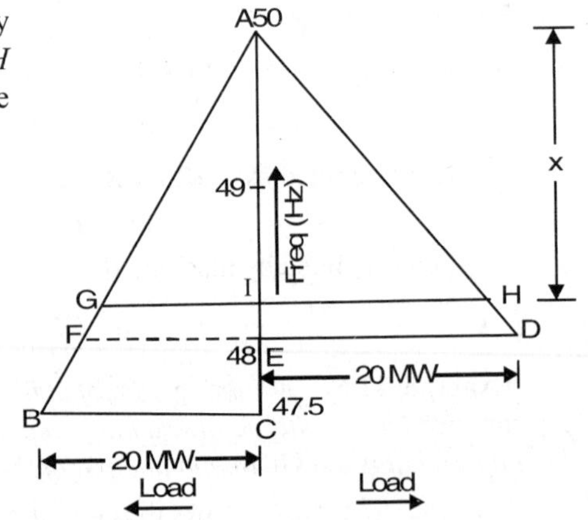

Fig. 12.100

From Δs *ABC* and *AGI*, we have,

$$\frac{GI}{BC} = \frac{x}{AC}$$

or
$$\frac{GI}{20} = \frac{x}{50 - 47{\cdot}5}$$

or
$$\frac{GI}{20} = \frac{x}{2{\cdot}5}$$

$$\therefore \quad GI = 8x$$

From Δs *AIH* and *AED*, we have,

$$\frac{HI}{ED} = \frac{AI}{AE}$$

$$\text{or} \quad \frac{HI}{20} = \frac{x}{50-48} \quad \text{or} \quad \frac{HI}{20} = \frac{x}{2}$$

$$\therefore \quad HI = 10x$$

$$\text{But} \quad GI + HI = 30$$

$$\text{or} \quad 8x + 10x = 30 \quad \therefore x = 5/3$$

$\therefore$ Load supplied by machine 2

$$= GI = 8x = 8 \times 5/3 = \mathbf{13{\cdot}33\ MW}$$

Load supplied by machine 1 $= HI = 10x = 10 \times 5/3 =$ **16·67 MW**

TUTORIAL PROBLEMS

1. A lighting load of 600 kW and a motor load of 707 kW at 0·707 lag are supplied by two alternators running in parallel. One of the machines supplies 900 kW at 0·9 p.f. lagging. Find the load and p.f. of the second machine. **[407 kW ; 0·7916 lag]**
2. A lighting load of 1500 kW and a motor load of 2200 kW at a p.f. of 0·707 lagging is supplied by two alternators running in parallel. One machine supplies 2000 kW at a p.f. of 0·8 lagging. Find the load and p.f. of the second machine. **[1700 kW ; 0·924 lagging]**
3. Two 750 kW alternators operate in parallel. The speed regulation of one machine is 100 % to 103 % from full-load to no-load and that of the other is 100 % to 104 %. How will the two alternators share a load of 1000 kW ? **[464·3 kW ; 535·7 kW]**
4. Two 3300 V star-connected alternators running in parallel supply the following loads :
   800 kW at unity p.f. ; 600 kW at p.f. 0·8 lagging ; 400 kW at p.f. 0·707 lagging
   The armature current of one machine is 150 A at p.f. 0·85 lagging. What is the armature current, output and p.f. of the other machine ? **[200 A ; 1071 kW ; 0·937 lagging]**
5. Two 50 MVA, 3-phase alternators operate in parallel. The settings of governors are such that the rise in speed from full-load to no-load is 2 % in one machine and 3 % in the other ; the characteristics being straight lines in both cases. If each machine is fully loaded when the total load is 100 MW, what would be the load on each machine when the total load is 60 MW ? **[26 MW ; 34 MW]**

12.37 HUNTING

Sometimes an alternator will not operate satisfactorily with others due to hunting. If the driving torque applied to an alternator is pulsating such as that produced by a diesel engine, the alternator rotor may be pulled periodically ahead of or behind its normal position as it rotates. This oscillating action is called *hunting*. Hunting causes the alternators to shift load from one to another. In some cases, this oscillation of power becomes cumulative and violent enough to cause the alternator to pull out of synchronism.

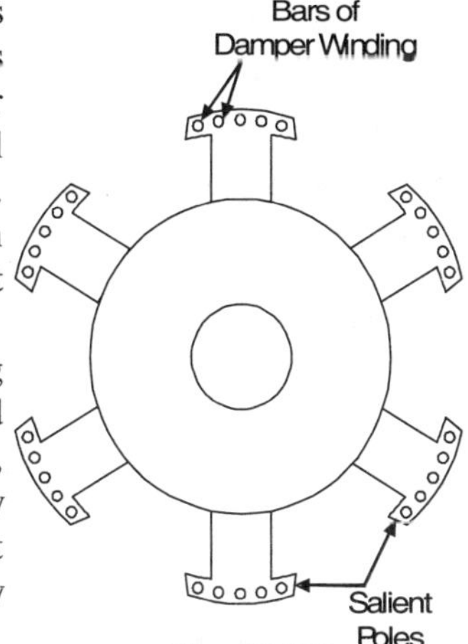

Fig. 12.101

In salient-pole machines, hunting is reduced by providing *damper winding*. It consists of short-circuited copper bars embedded in the pole faces as shown in Fig. 12.101. When hunting occurs, there is shifting of armature flux across the pole faces, thereby inducing currents in the damper winding. Since any induced current opposes the action that produces it, the hunting action is opposed by the flow of induced currents. The following points may be noted :

(*i*) Hunting generally occurs in alternators driven by engines because the driving torque of engines is not uniform.

(*ii*) Alternators driven by steam turbines generally do not have a tendency to hunt since the torque applied does not pulsate.

(*iii*) In cylindrical rotor machines, the damper windings are generally not used. It is because the solid rotor provides considerable damping.

Note. Under normal running conditions, damper winding does not carry any current because rotor runs at synchronous speed.

SHORT ANSWER QUESTIONS

Q. 1. *How does a 3-phase alternator differ from a d.c. generator?*

Ans. Like a d.c. generator, an alternator has also an armature and a field winding. However, there are following two important differences between the two :

(*i*) The armature of an alternator is an a.c. winding *i.e.*, it is *connected to produce 3-phase a.c. supply while the field winding is connected to d.c. supply. Therefore, no commutator is required in an alternator which makes its construction quite simpler than a d.c. generator.

(*ii*) In a d.c. generator, the armature winding rotates and the field system is stationary. But practically an alternator has armature winding on the stationary part called *stator* and field winding on the rotating part called *rotor*.

Q. 2. *Why are 3-phase alternators generally star-connected?*

Ans. Alternators are generally star-connected due to the following reasons :

(*i*) In a star-connection, phase voltage $V_{ph} = V_L/\sqrt{3}$ where V_L is the line voltage. Since the induced e.m.f. in the phase winding of an alternator is directly proportional to the number of turns, a star-connected alternator will require less number of turns than a delta-connected alternator for the same line voltage.

(*ii*) For the same line voltage, a star-connected alternator requires less insulation than a delta-connected alternator.

(*iii*) When an alternator is on load, the voltage per phase becomes distorted and the waveform is no longer sinusoidal. The distortion is mainly due to an undesired *third harmonic* voltage whose frequency is three times that of the fundamental frequency. With a Y-connection, the distorting line-to-neutral harmonics do not appear between the lines because they effectively cancel out each other. Consequently, the line voltages remain sinusoidal at all load conditions. However, in a delta-connection, third harmonic voltages do not cancel out but add up.

Q. 3. *Why do turboalternators use non-salient poles?*

Ans. Turboalternators or high-speed alternators have non-salient pole rotors because of the following reasons :

(*i*) For high speeds, it is difficult to build salient-pole construction strong enough to withstand the centrifugal forces.

(*ii*) Salient-pole rotors have large windage losses at high speeds.

Q. 4. *Why do we use hydrogen for cooling large turboalternators instead of air?*

Ans. For cooling large turboalternators, we use hydrogen instead of air due to the following reasons :

(*i*) Hydrogen being lighter than air causes less windage loss.

(*ii*) Hydrogen is a better heat conductor than air and thus provides better cooling.

(*iii*) It produces less noise and oxidation in the windings.

(*iv*) It reduces fire hazards.

Q. 5. *Why is the rotor of a turboalternator made cylindrical?*

Ans. The advantages of a cylindrical rotor for a turboalternator are :

* Although e.m.f. generated in the armature winding of a d.c. generator is also alternating but it is made unidirectional with the help of a commutator.

(*i*) It gives better balance.

(*ii*) The operation becomes quieter.

(*iii*) The windage losses are reduced.

Q. 6. *What are the advantages of short-pitched windings in an alternator?*

Ans. The advantages of short-pitched armature winding in an alternator are :

(*i*) The overhang, in which no e.m.f. is generated, is reduced in length so that less copper is required.

(*ii*) The waveform of the e.m.f. is made nearly sinusoidal. It is because the use of short-pitch coils eliminates harmonics from the e.m.f. of each coil.

Q. 7. *What do you mean by synchronous reactance of an alternator?*

Ans. Since the e.m.f. induced by the rotating flux in any one phase leads the current in that phase by 90°, the effect is similar to that of inductive reactance *i.e.,* the rotating magnetic flux produced by the stator currents may be regarded as being responsible for the reactance of the stator winding. Furthermore, since the rotating stator flux revolves synchronously with the poles, this reactance is referred to as synchronous reactance of the alternator.

Q. 8. *What is the effect of load power factor on armature reaction in an alternator?*

Ans. The load p.f. affects the armature reaction in the following manner :

(*i*) At unity load p.f., the effect of armature reaction is minimum ; its effect being merely to distort the main field flux.

(*ii*) At lagging power factors, the armature m.m.f. opposes the main field, resulting in a weakened field and a low generated voltage.

(*iii*) At leading power factors, the armature reaction aids or strengthens the main field, resulting in higher generated voltage.

Q. 9. *What factors govern the terminal voltage of an alternator when the load changes?*

Ans. As the load on an alternator changes, the change in its terminal voltage depends upon :

(*i*) design of the machine which determines the parameters of the armature circuit

(*ii*) amount of excitation (*iii*) magnitude of load current

(*iv*) power factor of the load

Q. 10. *Why is the knowledge of voltage regulation of an alternator important?*

Ans. The knowledge of voltage regulation of an alternator is important because of the following reasons :

(*i*) It indicates the range of field excitation required.

(*ii*) It indicates the magnitude of torque angle.

(*iii*) It indicates the performance of the machine when operating in parallel with others.

Q. 11. *What are the advantages of alternators having large ratings?*

Ans. The two main advantages of big alternators are :

(*i*) The efficiency of an alternator automatically increases when its power rating increases. For example, if an alternator of 1 kVA has an efficiency of 50%, a larger but similar model having a rating of 10 MVA may have an efficiency of about 90%.

(*ii*) The power output per kilogram of an alternator increases as its power rating increases. For example, if an alternator of 1 kW weighs 20 kg (*i.e.* 50 W/kg), a 10 MW machine of similar construction will weigh about 20,000 kg, yielding 500 W/kg. In other words, large alternators are cheaper.

Q. 12. *What difficulty is encountered when an alternator grows in size?*

Ans. Large alternators are very efficient and cheaper. However, as they grow in size, we run into serious cooling problems. Large machines produce high power loss per unit surface area. To prevent an unacceptable temperature rise, we must design efficient cooling system. The evolution of big alternators is dictated by the evolution of sophisticated cooling system.

Q. 13. *What is synchronising?*

Ans. When two or more alternators operating in parallel fulfil the desired conditions, they are said to be *in synchronism*. The operation of getting alternators in synchronism is called *synchronising*.

Q. 14. *What is the function of damper winding?*

Ans. The damper winding reduces the hunting in alternators operating in parallel. The damper winding consists of short-circuited copper bars embedded in the pole-faces of the salient-pole rotor. When hunting occurs, there is a shifting of the armature flux across the pole faces, thereby inducing currents in the damper winding. Since any induced current opposes the action that produces it (Lenz's law), the hunting action is opposed by the flow of induced currents.

Q. 15. *Why are alternators rated in kVA?*

Ans. The power $\sqrt{3}\, V_L I_L \cos\phi$ delivered by the alternator for the same value of current depends upon p.f. ($\cos\phi$) of the load. But the alternator conductors are calculated for a definite current and the insulation and magnetic system are designed for a definite voltage independent of p.f. of the load. For this reason, apparent power measured in kVA is regarded as the rated power of the alternator.

Q. 16. *Discuss the power flow in an alternator?*

Ans. For alternator operation, the rotor of the machine is supplied with mechanical energy from some prime mover. Some of the mechanical input energy is converted directly into heat energy through friction, windage and hysteresis and eddy current loss. The remainder of the mechanical input energy is converted into electrical energy in the armature circuit. Some of the electrical energy of the armature circuit is converted into heat energy through the I^2R loss in the armature circuit. The remainder becomes the useful electric output of the alternator.

Q. 17. *How is sine wave e.m.f. obtained in an alternator?*

Ans. The e.m.f. generated in a conductor at any instant is given by ;

$$e = B\, l\, v \text{ volts}$$

where B = instantaneous value of flux density in Wb/m^2

l = length of conductor in metres

v = speed of conductor in m/s

Since l and v are constant, $e \propto B$. Therefore, the e.m.f. wave for each conductor is of the same shape as the flux density wave.

(*i*) In the salient pole type of machine, the sine wave of flux density is closely obtained by suitably shaping the pole shoe so that the air gap gradually increases from the centre to the pole tips.

(*ii*) In the cylindrical type of rotor, the air gap is of constant length and the sine wave is approximated to by distributing the field winding in several slots.

A further improvement of waveform is obtained by making the coil span (or coil pitch) on the stator less than the pole pitch.

Q. 18. *How is the efficiency of an alternator effected by load power factor?*

Ans. The efficiency of an alternator depends not only on kVA output but also on p.f. of the load. For a given load, efficiency is maximum at unity p.f. and decreases as the p.f. falls.

Q. 19. *Is the efficiency of an alternator determined by direct loading?*

Ans. As with d.c. machines, the efficiency of an alternator is not determined by direct loading owing to the difficulty in finding a suitable load. The efficiency is generally determined from losses.

Q. 20. *How are iron and friction losses of an alternator measured?*

Ans. Iron and friction losses can be measured by coupling the alternator to a suitable calibrated d.c. motor (*i.e.,* one whose efficiency curve is known) and driving it at synchronous speed with normal excitation. Then,

Iron and friction losses = Output of motor in watts

MULTIPLE-CHOICE QUESTIONS

1. Majority of the alternators in use have ....

(*i*) revolving a.c. armature winding
(*ii*) stationary field type construction
(*iii*) revolving field type construction
(*iv*) none of the above

2. The stator of an alternator rarely uses .... slots.

(*i*) wide open type
(*ii*) semi-closed type
(*iii*) closed type
(*iv*) none of the above

3. The field winding of an alternator is .... excited.

(*i*) d.c. (*ii*) a.c.
(*iii*) both d.c. and a.c. (*iv*) none of the above

4. The a.c. armature winding of an alternator operates at .... the field winding.

(*i*) the same voltage as
(*ii*) much lesser voltage than
(*iii*) much higher voltage than
(*iv*) none of the above

5. An alternator is sometimes called .... generator.

(*i*) synchronous (*ii*) asynchronous
(*iii*) Rosenberg (*iv*) none of the above

6. A turboalternator uses ....

(*i*) salient-pole field structure
(*ii*) nonsalient-pole field structure
(*iii*) rotating a.c. armature winding
(*iv*) none of the above

7. High-speed alternators are driven by ....

(*i*) diesel engines
(*ii*) hydraulic turbines
(*iii*) steam turbines
(*iv*) none of the above

8. Turbo-alternators have rotors of ....

(*i*) small diameter and long axial length
(*ii*) large diameter and long axial length
(*iii*) large diameter and small axial length
(*iv*) same diameter and axial length

9. The speed at which a 6-pole alternator should be driven to generate 50 cycles per second is ....

(*i*) 1500 r.p.m. (*ii*) 1000 r.p.m.
(*iii*) 500 r.p.m. (*iv*) none of the above

10. The frequency of e.m.f. generated in an 8-pole alternator running at 900 r.p.m. is ....

(*i*) 50 Hz (*ii*) 60 Hz
(*iii*) 120 Hz (*iv*) none of the above

11. The synchronous reactance of an alternator is generally .... armature resistance.

(*i*) 5 times smaller than
(*ii*) 5 times greater than
(*iii*) 10 to 100 times greater than
(*iv*) 10 times smaller than

12. A 3-phase alternator generates an open-circuit voltage of 4000 V per phase when exciting current is 50 A ; the short-circuit current for the same excitation being 800 A. The synchronous reactance per phase is ....

(*i*) 80 Ω (*ii*) 5 Ω
(*iii*) 15 Ω (*iv*) none of the above

13. A 30 MVA, 15 kV alternator will have a per

phase nominal impedance of ....

(*i*) 9 Ω (*ii*) 15 Ω
(*iii*) 7·5 Ω (*iv*) none of the above

14. The full-load efficiency of an alternator .... with the size of the machine.

(*i*) increases (*ii*) decreases
(*iii*) remains unchanged
(*iv*) none of the above

15. In the armature winding of an alternator, the coil span falls short of full-pitch by 60° (electrical). The pitch factor is ....

(*i*) 0·866 (*ii*) 0·5
(*iii*) 0·25 (*iv*) none of the above

16. A 3-phase alternator has 3-slots per pole. The distribution factor of the winding is ....

(*i*) 0·866 (*ii*) 0·5
(*iii*) 1 (*iv*) none of the above

17. The disadvantage of a short-pitched coil is that ....

(*i*) harmonics are introduced
(*ii*) waveform becomes non-sinusoidal
(*iii*) voltage around the coil is reduced
(*iv*) none of the above

18. The per phase d.c. armature resistance of an alternator is 0·5 Ω. The effective a.c. armature resistance would be about ....

(*i*) 0·5 Ω (*ii*) 0·75 Ω
(*iii*) 4 Ω (*iv*) 0·25 Ω

19. The d.c. armature resistance of a Y-connected alternator measured across its two terminals is 0.5 Ω. The per phase resistance is ....

(*i*) 0·5 Ω (*ii*) 1 Ω
(*iii*) 0·25 Ω (*iv*) 1·5 Ω

20. The d.c. armature resistance of a delta-connected alternator measured across its two terminals is 1 Ω. The per phase d.c. resistance is ....

(*i*) 3 Ω (*ii*) 1·5 Ω
(*iii*) 0·33 Ω (*iv*) none of the above

21. When a number of alternators are operating in parallel, the power factor at which each operates is determined by ....

(*i*) power factor of the load
(*ii*) driving torque of the prime mover
(*iii*) its field excitation
(*iv*) none of the above

22. Alternators driven by .... do not have a tendency to hunt.

(*i*) diesel engines (*ii*) water turbines
(*iii*) steam turbines (*iv*) none of the above

23. A 3-phase, 16-pole alternator has a star-connected winding with 144 slots and 10 conductors per slot. The flux per pole is 30 mWb sinusoidally distributed and the speed is 375 r.p.m. The frequency of generated e.m.f. is

(*i*) 25 Hz (*ii*) 50 Hz
(*iii*) 100 Hz (*iv*) 200 Hz

24. In the above question, the line e.m.f. is

(*i*) 1120 V (*ii*) 3750 V
(*iii*) 1875 V (*iv*) 2770 V

25. For producing an e.m.f. of given frequency, an alternator will run at greatest speed for .... poles.

(*i*) 2 (*ii*) 4
(*iii*) 8 (*iv*) 12

26. A two-pole alternator is running at 1500 r.p.m. Its angular velocity is ....

(*i*) 118 rad/s (*ii*) 192 rad/s
(*iii*) 157 rad/s (*iv*) 212 rad/s

27. A 500 kVA, 1100 V, 50 Hz, Y-connected, 3-ϕ alternator has armature resistance of 0·1 Ω/phase and synchronous reactance/phase of 1·5 Ω. Find the generated e.m.f. per phase

(*i*) 769·2 V (*ii*) 832·6 V
(*iii*) 692·4 V (*iv*) 935·5 V

28. In the above question, what is the voltage regulation ?

(*i*) 15·6 % (*ii*) 29 %
(*iii*) 9·8 % (*iv*) 21·1 %

29. In Q. 27, if the load p.f. is 0·9 lagging, what is the voltage regulation ?

(*i*) 28·2 % (*ii*) 41·4 %
(*iii*) 32·6 % (*iv*) 18·7 %

30. In Q. 27, if the load p.f. is 0·8 leading, what is the voltage regulation ?

(*i*) – 5·8 % (*ii*) 12·7 %
(*iii*) – 15·9 % (*iv*) 21·5 %

ANSWERS TO MULTIPLE-CHOICE QUESTIONS

1. (*iii*)	**2.** (*iii*)	**3.** (*i*)	**4.** (*iii*)	**5.** (*i*)
6. (*ii*)	**7.** (*iii*)	**8.** (*i*)	**9.** (*ii*)	**10.** (*ii*)
11. (*iii*)	**12.** (*ii*)	**13.** (*iii*)	**14.** (*i*)	**15.** (*i*)
16. (*iii*)	**17.** (*iii*)	**18.** (*ii*)	**19.** (*iii*)	**20.** (*ii*)
21. (*iii*)	**22.** (*iii*)	**23.** (*ii*)	**24.** (*iv*)	**25.** (*i*)
26. (*iii*)	**27.** (*i*)	**28.** (*iv*)	**29.** (*ii*)	**30.** (*iii*)

HINTS TO SELECTED MULTIPLE-CHOICE QUESTIONS

2. The closed-type slots have the following disadvantages :

(*i*) They increase the inductance of the windings.

(*ii*) They are expensive in the initial cost.

(*iii*) They present a problem of end connections.

Due to these disadvantages, wholly closed slots are not used.

4. The field winding is supplied with direct current, usually at 125 V or 250 V. However, voltages as high as 33 kV are generated in the armature winding.

6. High-speed or turbo-alternators have nonsalient-pole (*i.e.*, cylindrical rotor) rotor to withstand the stresses at high speeds. Because the rotor is driven at a high speed, we require a small number of poles.

7. Steam turbines are smaller and more efficient when they turn at high speed. The same is true of alternators.

8. The turboalternators run at high speeds. The high speed of rotation produces strong centrifugal forces which impose an upper limit on the diameter of the rotor. In case of a rotor turning at 3600 r.p.m., the elastic limit of the steel requires the manufacturer to limit the diameter to a maximum of 1·2 m. We can double the diameter when the speed is 1800 r.p.m. but because of the transportation problems, we seldom go beyond 1·8 m.

9. $f = \dfrac{NP}{120}$ $\qquad \therefore N = \dfrac{120f}{P} = \dfrac{120 \times 50}{6} = 1000$ r.p.m.

10. $f = \dfrac{NP}{120} = \dfrac{900 \times 8}{120} = 60$ Hz

11. The value of synchronous reactance is 10 to 100 times greater than armature resistance. Consequently, we can neglect armature resistance unless we are interested in efficiency or heating effects.

12. Synchronous reactance, $X_s = \dfrac{E_{ph}}{I_{sc}} = \dfrac{4000}{800} = 5\ \Omega$

where E_{ph} = rated open-circuit phase voltage

I_{sc} = short-circuit current per phase using the same exciting current that was required to produce E_{ph}

13. Nominal impedance, $Z_n = \dfrac{E_L^2}{S} = \dfrac{(15000)^2}{30 \times 10^6} = 7{\cdot}5\ \Omega$

where E_L = rated line voltage in volts

S = rated power of alternator in VA

14. The efficiency of an alternator automatically improves as the power rating increases. For example, if an alternator of 1 kVA has an efficiency of 50%, a larger but similar model having a capacity of 10 MVA will have an efficiency of about 90%.

15. Pitch factor, $K_p = \cos(\beta/2) = \cos(60°/2) = 0{\cdot}866$
where β = electrical angle by which coil span falls short of full-pitch

16. Distribution factor, $K_d = \dfrac{\sin(n\,\alpha/2)}{n\sin(\alpha/2)}$

where n = slots/pole/phase

α = angular displacement between slots = 180°/slots per pole

In the present case, $n = 3/3 = 1$, $\alpha = 180°/3 = 60°$

$$\therefore \quad K_d = \frac{\sin\left(\dfrac{1 \times 60°}{2}\right)}{1 \times \sin\left(\dfrac{60°}{2}\right)} = 1$$

17. In a fractional pitch winding, the two sides of a given coil are not at the corresponding points under the adjacent poles. Hence the e.m.f. induced in the two sides is slightly out of phase and the coil e.m.f. is less than if full pitch were used with the same number of turns.

18. Since there are harmonics generated in individual coils of the alternator windings, some high frequencies are present. Due to skin effect, the effective a.c. resistance of armature winding is greater than its d.c. resistance. The factor of difference varies according to the base frequency and winding arrangement and is from 1·2 to 1·8. It is a usual practice to take a.c. armature resistance as 1·5 times the d.c. resistance for 50 Hz machine.

$$R_{ac} = 1{\cdot}5\,R_{dc} = 1{\cdot}5 \times 0{\cdot}5 = 0{\cdot}75\ \Omega$$

19. The d.c. resistance measured across any two terminals of Y-connected alternator is that of two phases. Therefore, per phase d.c. armature resistance is half of the measured value.

20. The d.c. resistance measured across any two terminals of a delta-connected alternator is equal to resistance of one phase *in parallel* with resistance of other two phases in series. Therefore, per phase d.c. armature resistance is 1·5 times the measured value.

21. The power factor of any a.c. distribution system depends upon the load. An alternator operating singly, then, must operate at the power factor of the load it supplies. However, when two or more alternators operate in parallel, the power factor at which each operates is determined by its field excitation.

22. Alternators driven by steam turbines do not have a tendency to hunt since the torque applied does not pulsate.

23. Frequency, $f = \dfrac{N\,P}{120} = \dfrac{375 \times 16}{120} = 50$ Hz

24. No. of slots/phase = 144/3 = 48, No. of conductors/phase, $Z = 48 \times 10 = 480$

Generated e.m.f./phase $= 2{\cdot}22\,f Z\,\phi = 2{\cdot}22 \times 50 \times 480 \times 30 \times 10^{-3} = 1600$ V

Line e.m.f. $= \sqrt{3} \times 1600 = 2770$ V

25. $f = \dfrac{N\,P}{120}$ or $N = \dfrac{120\,f}{P}$. Minimum number of poles = 2.

26. Angular velocity, $\omega = \dfrac{2\,\pi\,N}{60} = \dfrac{2\pi \times 1500}{60} = 157$ rad/s

27. $\sqrt{3}\,V_L I_L = 500 \times 10^3$ or $I_L = \dfrac{500 \times 10^3}{\sqrt{3} \times 1100} = 262$ A

$\therefore$ Armature current/phase, $I_a = 262$ A

$I_a R_a = 262 \times 0{\cdot}1 = 26{\cdot}2$ V ; $I_a X_s = 262 \times 1{\cdot}5 = 393$ V

Terminal voltage/phase, $V = V_L/\sqrt{3} = 1100/\sqrt{3} = 635$ V

Generated e.m.f./phase, $E_0 = \sqrt{(V + I_a R_a)^2 + (I_a X_s)^2} = \sqrt{(635 + 26\cdot2)^2 + (393)^2} = 769.2\text{ V}$

28. Voltage regulation $= \dfrac{E_0 - V}{V} \times 100 = \dfrac{769\cdot2 - 635}{635} \times 100 = 21\cdot1\,\%$

29. $\cos\phi = 0.9\,;\ \phi = 25.83^\circ\ \therefore\ \sin\phi = \sin 25.83^\circ = 0.436$

Generated e.m.f./phase, $E_0 = \sqrt{(V\cos\phi + I_a R_a)^2 + (V\sin\phi + I_a X_s)^2}$

$$= \sqrt{(635 \times 0\cdot9 + 26\cdot2)^2 + (635 \times 0\cdot436 + 393)^2} = 898\text{ V}$$

$\therefore$ Voltage regulation $= \dfrac{E_0 - V}{V} \times 100 = \dfrac{898 - 635}{635} \times 100 = 41.4\,\%$

30. $\cos\phi = 0\cdot8\,;\ \sin\phi = \sqrt{1 - (0\cdot8)^2} = 0\cdot6$

Generated e.m.f./phase, $E_0 = \sqrt{(V\cos\phi + I_a R_a)^2 + (V\sin\phi - I_a X_s)^2}$

$$= \sqrt{(635 \times 0\cdot8 + 26\cdot2)^2 + (635 \times 0\cdot6 - 393)^2} = 534\text{ V}$$

$\therefore$ Voltage regulation $= \dfrac{E_0 - V}{V} \times 100 = \dfrac{534 - 635}{635} \times 100 = -15\cdot9\,\%$

13

Synchronous Motors

INTRODUCTION

It may be recalled that a d.c. generator can be run as a d.c. motor. In like manner, an alternator may operate as a motor by connecting its armature winding to a 3-phase supply. It is then called a synchronous motor. As the name implies, a synchronous motor runs at synchronous speed ($N_s = 120f/P$) *i.e.*, in synchronism with the revolving field produced by the 3-phase supply. The speed of rotation is, therefore, tied to the frequency of the source. Since the frequency is fixed, the motor speed stays constant irrespective of the load or voltage of 3-phase supply. However, synchronous motors are not used so much because they run at constant speed (*i.e.*, synchronous speed) but because they possess other unique electrical properties. In this chapter, we shall discuss the working and characteristics of synchronous motors.

13.1 CONSTRUCTION

A synchronous motor is a machine that operates at synchronous speed and converts electrical energy into mechanical energy. It is fundamentally an alternator operated as a motor. Like an alternator, a synchronous motor has the following two parts :

(*i*) a **stator** which houses 3-phase armature winding in the slots of the stator core and receives power from a 3-phase supply (See Fig. 13.1).

(*ii*) a **rotor** that has a set of salient poles excited by direct current to form alternate N and S poles. The exciting coils are connected in series to two slip rings and direct current is fed into the winding from an external exciter mounted on the rotor shaft.

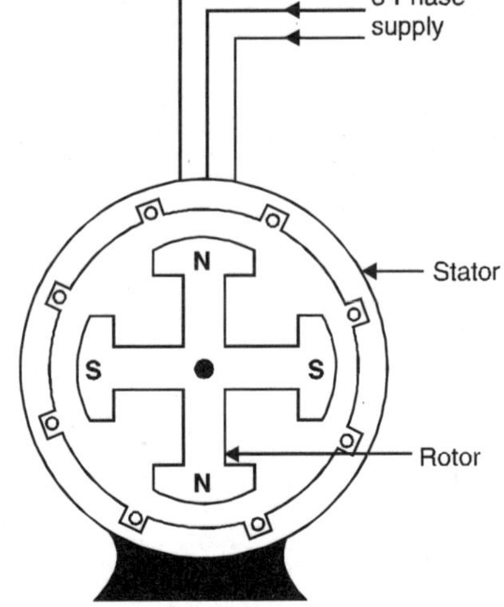

Fig. 13.1

The stator is wound for the same number of poles as the rotor poles. As in the case of an induction motor, the number of poles determines the synchronous speed of the motor :

$$\text{Synchronous speed, } N_s = \frac{120f}{P}$$

where f = frequency of supply in Hz

P = number of poles

An important drawback of a synchronous motor is that it is not self-starting and auxiliary means have to be used for starting it.

13.2 SOME FACTS ABOUT SYNCHRONOUS MOTOR

Some salient features of a synchronous motor are :

(*i*) A synchronous motor runs at synchronous speed or *not at all. Its speed is constant (synchronous speed) at all loads. The only way to change its speed is to alter the supply frequency ($N_s = 120f/P$).

* A synchronous motor develops torque only when running at synchronous speed.

(*ii*) The outstanding characteristic of a synchronous motor is that it can be made to operate over a wide range of power factors (lagging, unity or leading) by adjustment of its field excitation. Therefore, a synchronous motor can be made to carry the mechanical load at constant speed and at the same time improve the power factor of the system.

(*iii*) Synchronous motors are generally of the salient pole type.

(*iv*) A synchronous motor is not self-starting and an auxiliary means has to be used for starting it. We use either induction motor principle or a separate starting motor for this purpose. If the latter method is used, the machine must be run upto synchronous speed and synchronised as an alternator.

13.3 OPERATING PRINCIPLE

The fact that a synchronous motor has no starting torque can be easily explained.

(*i*) Consider a 3-phase synchronous motor having two rotor poles N_R and S_R. Then the stator will also be wound for two poles N_S and S_S. The motor has direct voltage applied to the rotor winding and a 3-phase voltage applied to the stator winding. The stator winding produces a rotating field which revolves around the stator at synchronous speed N_s (= $120\,f/P$). The direct (or zero frequency) current sets up a two-pole field which is stationary so long as the rotor is not turning. Thus, we have a situation in which there exists a pair of revolving armature poles (*i.e.*, $N_S - S_S$) and a pair of stationary rotor poles (*i.e.*, $N_R - S_R$).

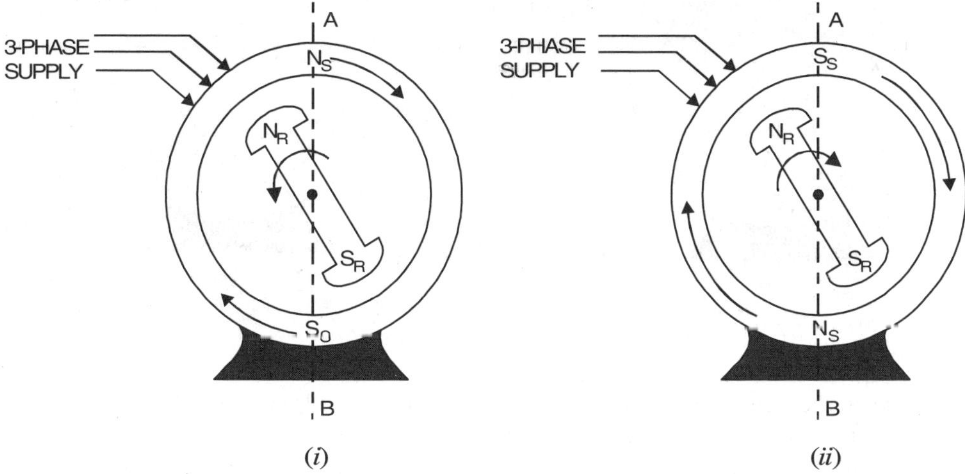

Fig. 13.2

(*ii*) Suppose at any instant, the stator poles are at positions *A* and *B* as shown in Fig. 13.2 (*i*). It is clear that poles N_S and N_R repel each other and so do the poles S_S and S_R. Therefore, the rotor tends to move in the anticlockwise direction. After a period of half-cycle (or $1/2f$ = 1/100 second), the polarities of the stator poles are reversed but the polarities of the rotor poles remain the same as shown in Fig. 13.2 (*ii*). Now S_S and N_R attract each other and so do N_S and S_R. Therefore, the rotor tends to move in the clockwise direction. Since the stator poles change their polarities rapidly, they tend to pull the rotor first in one direction and then after a period of half-cycle in the other. Due to high inertia of the rotor, the motor fails to start.

Hence, a synchronous motor has no self-starting torque i.e., a synchronous motor cannot start by itself.

How to get continuous unidirectional torque ? If the rotor poles are rotated by some external means at *such* a speed that they interchange their positions along with the stator poles, then the rotor will experience a continuous unidirectional torque. This can be understood from the following discussion :

(*i*) Suppose the stator field is rotating in the clockwise direction and the rotor is also rotated clockwise by some external means at such a speed that the rotor poles interchange their positions along with the stator poles.

(*ii*) Suppose at any instant the stator and rotor poles are in the position shown in Fig. 13.3 (*i*). It is clear that torque on the rotor will be clockwise. After a period of half-cycle, the stator poles reverse their polarities and at the same time rotor poles also *interchange their positions as shown in Fig. 13.3 (*ii*). The result is that again the torque on the rotor is clockwise. Hence a continuous unidirectional torque acts on the rotor and moves it in the clockwise direction. Under this condition, *poles on the rotor always face poles of opposite polarity on the stator* and a strong magnetic attraction is set up between them. This mutual attraction locks the rotor and stator together and the rotor is virtually pulled into step with the speed of revolving flux (*i.e.*, synchronous speed).

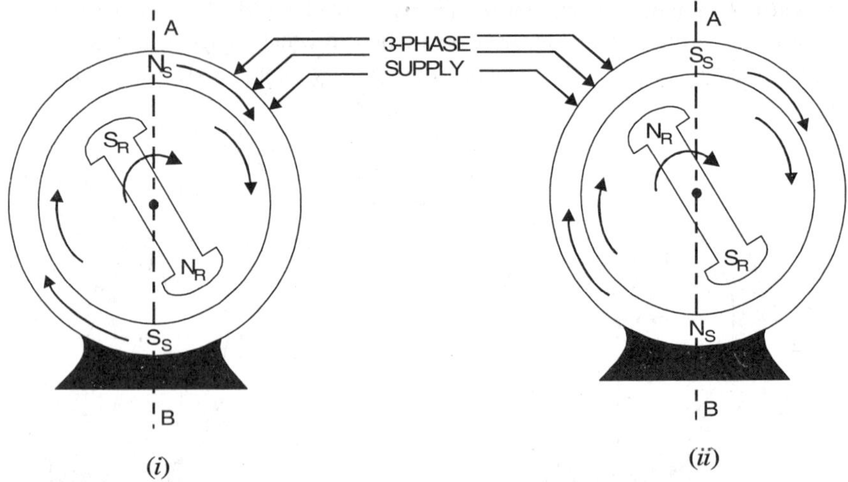

Fig. 13.3

(*iii*) If now the external prime mover driving the rotor is removed, the rotor will continue to rotate at synchronous speed in the clockwise direction because the rotor poles are magnetically locked up with the stator poles. It is due to this magnetic interlocking between stator and rotor poles that a synchronous motor runs at the speed of revolving flux *i.e.*, synchronous speed.

13.4 MAKING SYNCHRONOUS MOTOR SELF-STARTING

A synchronous motor cannot start by itself. In order to make the motor self-starting, a **squirrel cage winding (also called damper winding) is provided on the rotor. The damper winding consists of copper bars embedded in the pole faces of the salient poles of the rotor as shown in Fig. 13.4. The bars are short-circuited at the ends to form in effect a partial squirrel cage winding. The damper winding serves to start the motor.

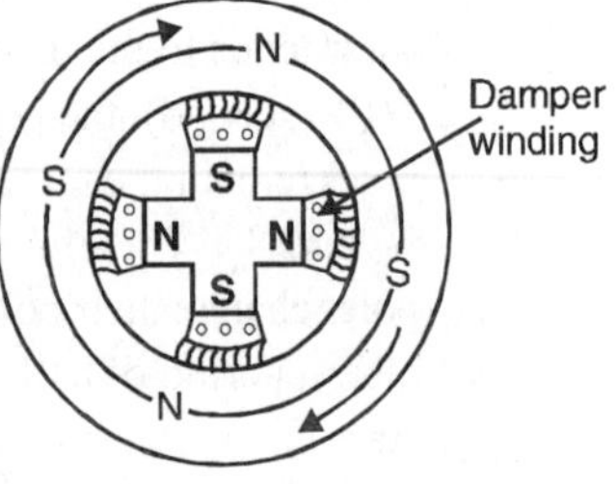

Fig. 13.4

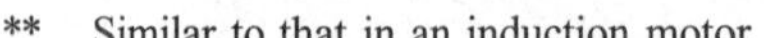

* Since the rotor poles are being rotated by an external means at such a speed that they interchange their positions along with the stator poles.

** Similar to that in an induction motor.

(*i*) To start with, 3-phase supply is given to the stator winding while the rotor field winding is left unenergised. The rotating stator field induces currents in the damper or squirrel cage winding and the motor starts as an induction motor.

(*ii*) As the motor approaches the synchronous speed, the rotor is excited with direct current. Now the resulting poles on the rotor face poles of opposite polarity on the stator and a strong magnetic attraction is set up between them. The rotor poles lock in with the poles of rotating flux. Consequently, the rotor revolves at the same speed as the stator field, *i.e.*, at synchronous speed.

(*iii*) Because the bars of squirrel cage portion of the rotor now rotate at the same speed as the rotating stator field, these bars do not cut any flux and, therefore, have no induced currents in them. Hence squirrel cage portion of the rotor is, in effect, removed from the operation of the motor.

It may be emphasised here that due to magnetic interlocking between the stator and rotor poles, a synchronous motor can only run at synchronous speed. At any other speed, this magnetic interlocking (*i.e.*, rotor poles facing opposite polarity stator poles) ceases and the average torque becomes zero. Consequently, the motor comes to a halt with a severe disturbance on the line.

Note : It is important to excite the rotor with direct current at the right moment. For example, if the d.c. excitation is applied when *N*-pole of the stator faces *N*-pole of the rotor, the resulting magnetic repulsion will produce a violent mechanical shock. The motor will immediately slow down and the circuit breakers will trip. In practice, starters for synchronous motors are designed to detect the precise moment when excitation should be applied.

13.5 EQUIVALENT CIRCUIT

Unlike the induction motor, the synchronous motor is connected to two electrical systems; a d.c. source at the rotor terminals and an a.c. system at the stator terminals.

1. Under normal conditions of synchronous motor operation, no voltage is induced in the rotor by the stator field because the rotor winding is rotating at the same speed as the stator field. Only the impressed direct current is present in the rotor winding and ohmic resistance of this winding is the only opposition to it as shown in Fig. 13.5 (*i*).

2. In the stator winding, two effects are to be considered ; the effect of stator field on the stator winding and the effect of the rotor field cutting the stator conductors at synchronous speed.

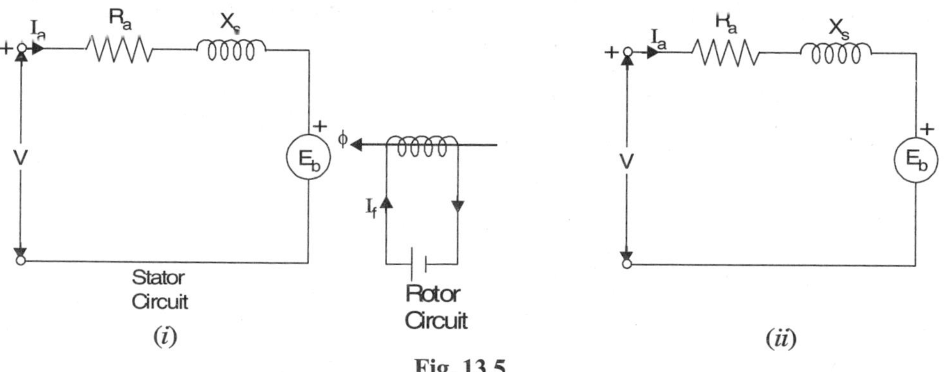

Fig. 13.5

(*i*) The effect of stator field on the stator (or armature) conductors is accounted for by including an inductive reactance in the armature winding. This is called *synchronous reactance* X_s. A resistance R_a must be considered to be in series with this reactance to account for the copper losses in the stator or armature winding as shown in Fig. 13.5 (*i*). This resistance combines with synchronous reactance and gives the synchronous impedance of the machine.

(*ii*) The second effect is that a voltage is generated in the stator winding by the synchronously-revolving field of the rotor as shown in Fig. 13.5 (*i*). This generated e.m.f. E_b is known as *back e.m.f.* and opposes the stator voltage V. The magnitude of E_b depends upon rotor speed and rotor flux ϕ per pole. Since rotor speed is constant, the value of E_b depends upon the rotor flux per pole *i.e.* exciting rotor current I_f.

Fig. 13.5 (*i*) shows the schematic diagram for one phase of a star-connected synchronous motor while Fig. 13.5 (*ii*) shows its equivalent circuit. Referring to the equivalent circuit in Fig. 13.5 (*ii*),

Net voltage/phase in stator winding is

$$E_r = V - E_b \qquad \text{... phasor difference}$$

Armature current/phase, $I_a = \dfrac{E_r}{Z_s}$

where $Z_s = \sqrt{R_a^2 + X_s^2}$

This equivalent circuit helps considerably in understanding the operation of a synchronous motor.

A synchronous motor is said to be *normally excited* if the field excitation is such that $E_b = V$. If the field excitation is such that $E_b < V$, the motor is said to be *under-excited*. The motor is said to be *over-excited* if the field excitation is such that $E_b > V$. As we shall see, for both normal and under-excitation, the motor has lagging power factor. However, for over-excitation, the motor has leading power factor.

Note : In a synchronous motor, the value of X_s is 10 to 100 times greater than R_a. Consequently, we can neglect R_a unless we are interested in efficiency or heating effects.

13.6 MOTOR ON LOAD

In d.c. motors and induction motors, an addition of load causes the motor speed to decrease. The decrease in speed reduces the counter e.m.f. enough so that additional current is drawn from the source to carry the increased load at a reduced speed. This action cannot take place in a synchronous motor because it runs at a constant speed (*i.e.*, synchronous speed) at all loads.

What happens when we apply mechanical load to a synchronous motor ? The rotor poles fall slightly behind the stator poles while continuing to run at synchronous speed. The angular displacement between stator and rotor poles (called torque angle α) causes the phase of back e.m.f. E_b to change *w.r.t.* supply voltage V. This increases the net e.m.f. E_r in the stator winding. Consequently, stator current I_a $(= E_r / Z_s)$ increases to carry the load.

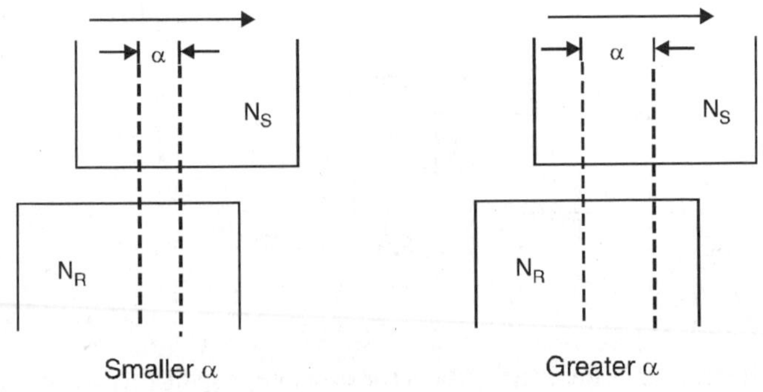

Fig. 13.6

The following points may be noted in synchronous motor operation:

(*i*) A synchronous motor runs at synchronous speed at all loads. It meets the increased load not by a decrease in speed but by the relative shift between stator and rotor poles, *i.e.*, by the adjustment of torque angle α.

(*ii*) If the load on the motor increases, the torque angle α also increases (*i.e.*, rotor poles lag behind the stator poles by a greater angle) but the motor continues to run at synchronous speed. The increase in torque angle α causes a greater phase shift of back e.m.f. E_b *w.r.t.* supply voltage V. This increases the net voltage E_r in the stator winding. Consequently, armature current I_a $(= E_r / Z_s)$ increases to meet the load demand.

(*iii*) If the load on the motor decreases, the torque angle α also decreases. This causes a smaller phase shift of E_b *w.r.t.* V. Consequently, the net voltage E_r in the stator winding decreases and so does the armature current I_a $(= E_r / Z_s)$.

13.7 PULL-OUT TORQUE

There is a limit to the mechanical load that can be applied to a synchronous motor. As the load increases, the torque angle α also increases so that a stage is reached when the rotor is pulled out of synchronism and the motor comes to a standstill. This load torque at which the motor pulls out of synchronism is called *pull-out or breakdown torque*. Its value varies from 1·5 to 3·5 times the full–load torque.

When a synchronous motor pulls out of synchronism, there is a major disturbance on the line and the circuit breakers immediately trip. This protects the motor because both squirrel cage and stator winding heat up rapidly when the machine ceases to run at synchronous speed.

13.8 MOTOR PHASOR DIAGRAM

Consider an under-excited star-connected synchronous motor ($E_b < V$) supplied with fixed excitation *i.e.*, back e.m.f. E_b is constant.

Let V = supply voltage/phase

E_b = back e.m.f./phase

Z_s = synchronous impedance/phase

(*i*) **Motor on no load.** When the motor is on no load, the torque angle α is small as shown in Fig. 13.7 (*i*). Consequently, back e.m.f. E_b lags behind the supply voltage V by a small angle *δ as shown in the phasor diagram in Fig. 13.7 (*iii*). The net voltage/phase in the stator winding is E_r.

Armature current/phase, $I_a = E_r / Z_s$

The armature current I_a lags behind E_r by $\theta = \tan^{-1} X_s / R_a$. Since $X_s >> R_a$, I_a lags E_r by nearly 90°. The phase angle between V and I_a is φ so that motor power factor is cos φ.

Input power/phase = $V I_a \cos \phi$

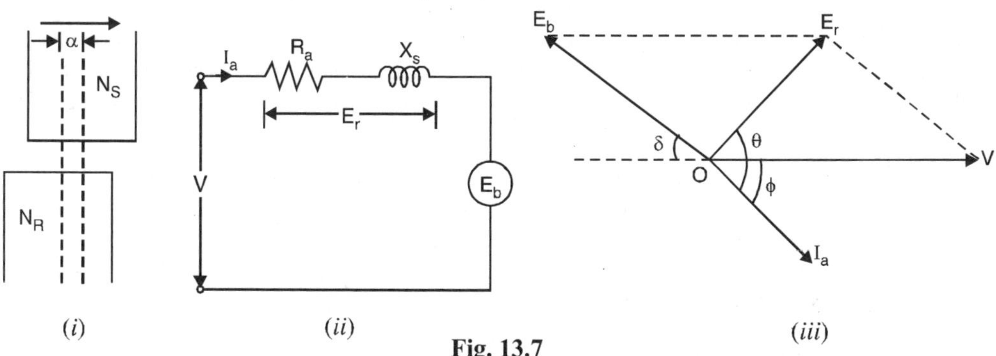

Fig. 13.7

Thus at no load, the motor takes a small power VI_a cos φ/phase from the supply to meet the no-load losses while it continues to run at synchronous speed.

* There is a definite relation between α and δ. Note that α is in mechanical degrees while δ is in electrical degrees.

(*ii*) **Motor on load.** When load is applied to the motor, the torque angle α increases as shown in Fig. 13.8 (*i*). This causes E_b (its magnitude is constant as excitation is fixed) to lag behind V by a greater angle as shown in the phasor diagram in Fig. 13.8 (*ii*). The net voltage/phase E_r in the stator winding increases. Consequently, the motor draws more armature current I_a (= E_r / Z_s) to meet the applied load.

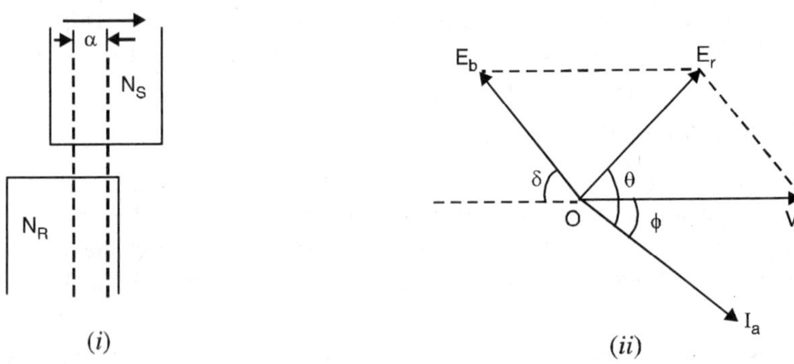

Fig. 13.8

Again I_a lags E_r by about 90° since $X_s >> R_a$. The power factor of the motor is cos φ.

$$\text{Input power/phase, } P_i = V I_a \cos \phi$$

Mechanical power developed by motor/phase is

$$P_m = E_b \times I_a \times \text{cosine of angle between } E_b \text{ and } I_a$$

$$= E_b I_a \cos (\delta - \phi)$$

13.9 EFFECT OF CHANGING FIELD EXCITATION AT CONSTANT LOAD

In a d.c. motor, the armature current I_a is determined by dividing the difference between V and E_b by the armature resistance R_a. Similarly, in a synchronous motor, the stator current (I_a) is determined by dividing voltage-phasor resultant (E_r) between V and E_b by the synchronous impedance Z_s.

One of the most important features of a synchronous motor is that by changing the field excitation, it can be made to operate from lagging to leading power factor. Consider a synchronous motor having a fixed supply voltage and driving a constant mechanical load. Since the mechanical load as well as the speed is constant, the power input to the motor (= 3 VI_a cos φ) is also constant. This means that the in-phase component I_a cos φ drawn from the supply will remain *constant. If the field excitation is changed, back e.m.f. E_b also changes. This results in the change of phase position of I_a *w.r.t.* V and hence the power factor cos φ of the motor changes. Fig. 13.9 shows the phasor diagram of the synchronous motor for different values of field excitation. Note that extremities of current phasor I_a lie on the straight line ***AB*.

(*i*) **Under-excitation.** The motor is said to be under-excited if the field excitation is such that $E_b < V$. Under such conditions, the current I_a lags behind V so that motor power factor is lagging as shown in Fig. 13.9 (*i*). This can be easily explained. Since $E_b < V$, the net voltage E_r is decreased and turns clockwise. As angle θ (= 90°) between E_r and I_a is †constant, therefore, phasor I_a also turns clockwise, *i.e.*, current I_a lags behind the supply voltage. Consequently, the motor has a lagging power factor.

---

* Input power = 3 $V I_a$ cos φ

Since V and input power are constant, I_a cos φ must be constant.

** Since I_a cos φ is constant, component of I_a along V is fixed.

† $\theta = \tan^{-1} X_s/R_a$. Since $X_s >> R_a$ and both are fixed, θ = 90° and is fixed.

(*ii*) **Normal excitation.** The motor is said to be normally excited if the field excitation is such that $E_b = V$. This is shown in Fig. 13.9 (*ii*). Note that the effect of increasing excitation (*i.e.*, increasing E_b) is to turn the phasor E_r and hence I_a in the anticlockwise direction, *i.e.*, I_a phasor has come closer to phasor V. Therefore, p.f. increases though still lagging. Since input power (= 3 $V I_a \cos \phi$) is unchanged, the stator current I_a must *decrease with increase in p.f.

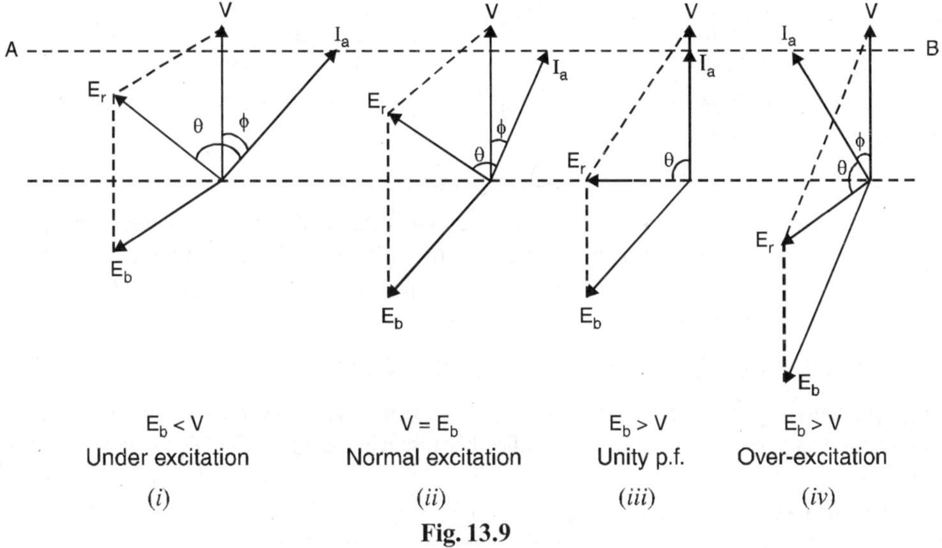

Fig. 13.9

Suppose the field excitation is increased until the current I_a is in phase with the applied voltage V, making the p.f. of the synchronous motor unity [See Fig. 13.9 (*iii*)]. For a given load, at **unity p.f. the resultant E_r and, therefore, I_a are miniumum.

(*iii*) **Over-excitation.** The motor is said to be over-excited if the field excitation is such that $E_b > V$. Under such conditions, current I_a leads V and the motor power factor is leading as shown in Fig. 13.9 (*iv*). Note that E_r and hence I_a further turn anticlockwise from the normal excitation position. Consequently, I_a leads V.

From the above discussion, it is concluded that if the synchronous motor is under-excited, it has a lagging power factor. As the excitation is increased, the power factor improves till it becomes unity at normal excitation. Under such conditions, the current drawn from the supply is minimum. If the excitation is further increased (*i.e.*, over-excitation), the motor power factor becomes leading.

Note. The armature current (I_a) is minimum at unity p.f. and increases as the power factor becomes poor, either leading or lagging.

13.10 PHASOR DIAGRAMS WITH DIFFERENT EXCITATIONS

Fig. 13.10 shows the phasor diagrams for different field excitations at constant load. Fig. 13.10 (*i*) shows the phasor diagram for normal excitation ($E_b = V$), whereas Fig. 13.10 (*ii*) shows the phasor diagram for under-excitation. In both cases, the motor has lagging power factor.

Fig. 13.10 (*iii*) shows the phasor diagram when field excitation is adjusted for unity p.f. operation. Under this condition, the resultant voltage E_r and, therefore, the stator current I_a are minimum. When the motor is over-excited, it has leading power factor as shown in Fig. 13.10 (*iv*). The following points may be remembered :

---

* $I_a = \sqrt{(I_a \cos \phi)^2 + (I_a \sin \phi)^2}$. Since $I_a \cos \phi$ is constant, the lagging reactive component $I_a \sin \phi$ will decrease.

** At unity p.f., $E_b > V$ but the difference is very small so that $E_b \simeq V$. Therefore, unity p.f. also corresponds to normal excitation.

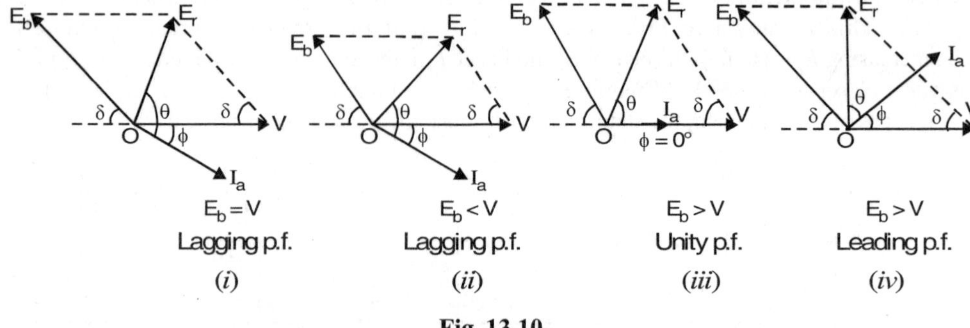

Fig. 13.10

(*i*) For a given load, the power factor is governed by the field excitation ; a weak field produces the lagging armature current and a strong field produces a leading armature current.

(*ii*) The armature current (I_a) is minimum at unity p.f. and increases as the p.f. becomes less, either leading or lagging.

13.11 V CURVES FOR SYNCHRONOUS MOTOR

The graph between armature current (I_a) and field current (I_f) of a synchronous motor for a constant load is called its **V curve.** The curve is so called because it is *V*-shaped. Fig. 13.11 shows *V* curves for three different loads.

When the level of excitation of a synchronous motor is changed gradually from under-excitation to over-excitation for a constant load, the following effects are observed :

(*i*) When the motor is under-excited (*i.e.* $E_b < V$), the power factor is lagging. In this case, the motor behaves like an inductive load.

(*ii*) When the motor is normally excited (*i.e.* $E_b = V$), the power factor is unity. In this case, the armature current is minimum and is in phase with the terminal voltage.

(*iii*) When the motor is over-excited (*i.e.* $E_b > V$), the power factor is leading. In this case, the motor behaves like a capacitive load. Therefore, an over-excited synchronous motor not only delivers load torque but also improves the power factor of the 3-phase power supply.

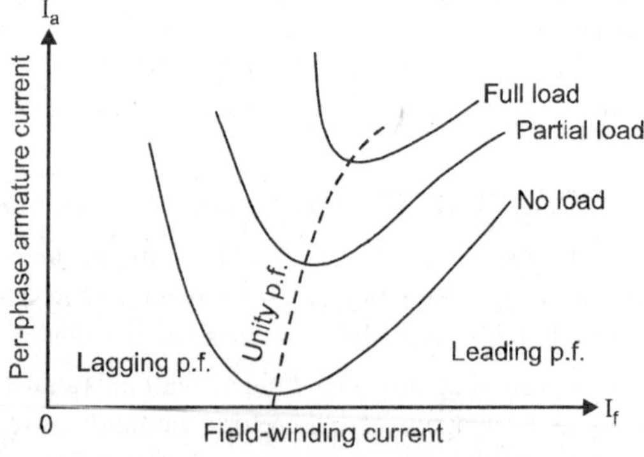

Fig. 13.11 V Curves For Synchronous Motor

When the magnitude of armature current (I_a) is plotted as a function of field current (I_f) for a constant load, we get the *V* curve as shown in Fig. 13.11. Note that these curves are similar to those obtained for an alternator in the previous chapter. The only difference is that the power factor is lagging when the motor is under-excited and leading when it is over-excited.

Example 13.1. *A 208 V, star-connected, 3-phase synchronous motor has a synchronous reactance of 4 Ω/phase and negligible armature winding resistance. At a certain load, the motor takes 7.2 kW at 0.8 p.f. lagging. If the power developed by the motor remains the same while the excitation voltage is increased by 50% by raising the field excitation, determine (i) the new armature current and (ii) the power factor.*

Solution.

$$\text{Supply voltage/phase, } V = 208/\sqrt{3} = 120 \text{ volts}$$

$$\text{P.F. angle, } \phi = \cos^{-1} 0.8 = 36.8°$$

$$\therefore \quad \text{Armature current, } I_a = \frac{7.2\times10^3}{3\,V\cos\phi} = \frac{7.2\times10^3}{3\times120\times0.8} = 25\,\text{A}$$

Taking supply voltage per phase as the reference phasor, we have,

$$\text{Generated voltage, } \mathbf{E_b} = \mathbf{V} - \mathbf{I_a}\mathbf{Z_s}$$

Here $\mathbf{V} = 120\angle 0°$ volts ; $\mathbf{I_a} = 25\angle{-36.8°}$ A ; $\mathbf{Z_s} = jX_s = 4\angle 90°\ \Omega$

$$\therefore \quad \mathbf{E_b} = 120\angle 0° - 25\angle{-36.8°} \times 4\angle 90°$$

$$= 120\angle 0° - 100\angle 53.2° = 100\angle{-53.13°} \text{ volts}$$

(*i*) $\quad$ New generated voltage, $E'_b = 1.5 \times E_b = 1.5 \times 100 = 150$ volts

The new torque angle δ′ can be found as under (See Art. 13.14).

$$P_m = \frac{3V E'_b}{X_s}\sin\delta' \quad \text{or} \quad 7.2\times10^3 = \frac{3\times120\times150}{4}\sin\delta'$$

$$\therefore \quad \sin\delta' = \frac{7.2\times10^3\times4}{3\times120\times150} = 0.533 \quad \text{or} \quad \delta' = 32.23°$$

$$\text{New armature current, } \mathbf{I'_a} = \frac{\mathbf{V} - \mathbf{E'_b}}{\mathbf{X_s}}$$

$$= \frac{120\angle 0° - 150 * \angle{-32.23°}}{j\,4} = \mathbf{20\angle 4.92°\ A}$$

(*ii*) $\therefore$ $\quad$ New power factor $= \cos 4.92° = \mathbf{0.996}$ *leading*

13.12 POWER RELATIONS

Consider an under-excited star-connected synchronous motor driving a mechanical load. Fig. 13.12 (*i*) shows the equivalent circuit for one phase, while Fig. 13.12 (*ii*) shows the phasor diagarm.

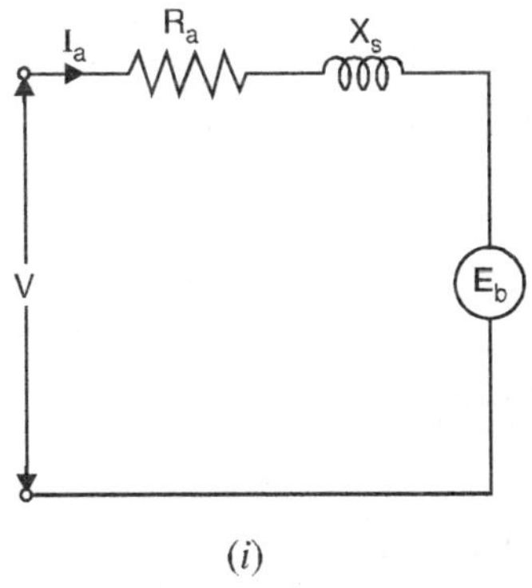

(*i*) $\quad$ **Fig. 13.12** $\quad$ (*ii*)

---

* Note E'_b lags behind V by δ′ (electrical).

(*i*) Input power/phase, $P_i = V I_a \cos \phi$

(*ii*) Mechanical power developed by the motor/phase is

$$P_m = E_b \times I_a \times \text{cosine of angle between } E_b \text{ and } I_a$$
$$= E_b I_a \cos (\delta - \phi)$$

(*iii*) Armature Cu loss/phase $= I_a^2 R_a = P_i - P_m$

(*iv*) Output power/phasor, $P_{out} = P_m$ – Iron, friction and excitation loss

Fig. 13.13 shows the power flow diagram of the synchronous motor.

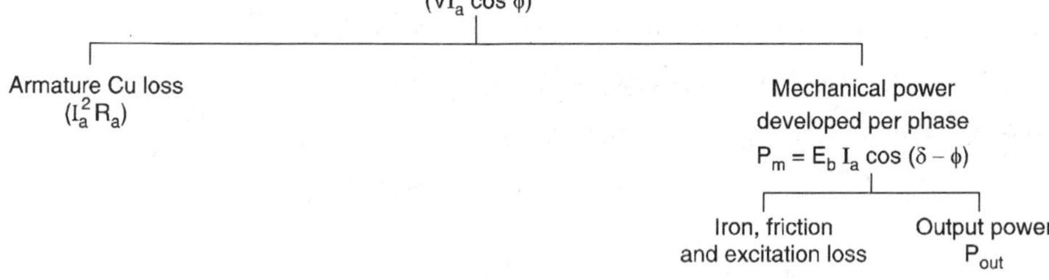

Fig. 13.13

13.13 MOTOR TORQUE

$$\text{Gross torque, } T_g = 9 \cdot 55 \frac{P_m}{N_s} \text{ N-m}$$

where P_m = Gross motor output in watts $= E_b I_a \cos (\delta - \phi)$

N_s = Synchronous speed in r.p.m.

$$\text{Shaft torque, } T_{sh} = 9 \cdot 55 \frac{P_{out}}{N_s} \text{ N-m}$$

It may be seen that torque is directly proportional to the mechanical power because rotor speed (*i.e.*, N_s) is fixed.

13.14 MECHANICAL POWER DEVELOPED BY MOTOR

(*Armature resistance neglected*)

Fig. 13.14 shows the phasor diagram of an under-excited synchronous motor driving a mechanical load. Since armature resistance R_a is assumed *zero, $\tan \theta = X_s / R_a = \infty$ and hence $\theta = 90°$.

$$\text{Input power/phase} = V I_a \cos \phi$$

Since R_a is assumed zero, stator Cu loss ($I_a^2 R_a$) will be zero. Hence input power is equal to the mechanical power P_m developed by the motor.

$$\text{Mech. power developed/ phase, } P_m = V I_a \cos \phi \quad ...(i)$$

Fig. 13.14

Referring to the phasor diagram in Fig. 13.14,

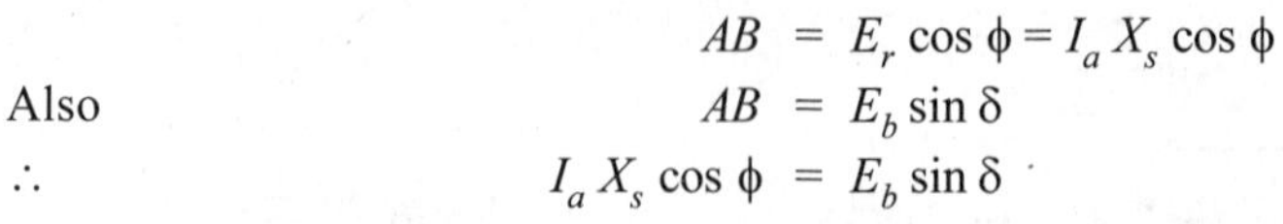

$$AB = E_r \cos \phi = I_a X_s \cos \phi$$

Also $AB = E_b \sin \delta$

$\therefore \quad I_a X_s \cos \phi = E_b \sin \delta$

* In a practical motor, the value of X_s is 10 to 100 times greater than R_a. Consequently, we can neglect R_a without significant error.

or $$I_a \cos\phi = \frac{E_b \sin\delta}{X_s}$$

Substituting the value of $I_a \cos\phi$ in exp. (*i*) above,

$$P_m = \frac{V E_b}{X_s}\sin\delta \quad \text{...per phase}$$

$$= \frac{3 V E_b}{X_s}\sin\delta \quad \text{...for 3-phase}$$

It is clear from the above relation that mechanical power increases with torque angle (in electrical degrees) and its maximum value is reached when $\delta = 90°$ (electrical).

$$\therefore \quad P_{max} = \frac{V E_b}{X_s} \quad \text{...per phase}$$

Under this condition, the poles of the rotor will be mid-way between *N* and *S* poles of the stator.

Example 13.2. *A 2000 V, 3-phase, star-connected motor has resistance and synchronous reactance per phase of 0·2 Ω and 1·9 Ω respectively. Calculate the generated (back) e.m.f. per phase with an input of 800 kW at p.f. 0·8 lagging.*

Solution.

Supply voltage/phase, $V = 2000/\sqrt{3} = 1155$ volts

Line current, $I_L = \dfrac{800\times 10^3}{\sqrt{3}\times 2000\times 0{\cdot}8} = 288\text{ A}$

Armature current/phase, $I_a = I_L = 288\text{ A}$

$$\theta = \tan^{-1} 1{\cdot}9/0{\cdot}2 = 84°$$

$$\phi = \cos^{-1} 0{\cdot}8 = 36{\cdot}8°$$

$$Z_s = \sqrt{(0\cdot 2)^2 + (1\cdot 9)^2} = 1{\cdot}91\ \Omega$$

$$E_r = I_a Z_s = 288\times 1{\cdot}91 = 550\text{ V}$$

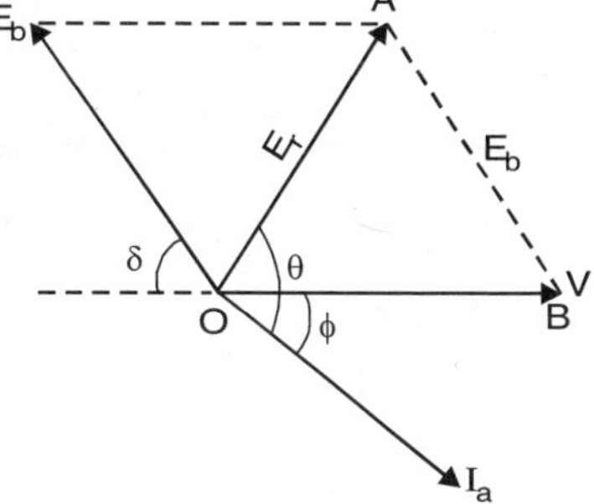

Fig. 13.15

Refer to the phasor diagram of the motor shown in Fig. 13.15.

Applying rule of trigonometry to Δ *AOB*,

$$AB^2 = OA^2 + OB^2 - 2\times OA\times OB\times\cos(\theta-\phi)$$

or $$E_b^2 = (550)^2 + (1155)^2 - 2\times 550\times 1155\times\cos(84° - 36{\cdot}8°)$$

$$= 773296$$

$$\therefore \quad E_b = \sqrt{773296} = \mathbf{879\ V/phase}$$

Alternatively. Taking applied voltage/phase (*V*) as the reference phasor, we have,

Generated (back) e.m.f., $\mathbf{E_b} = \mathbf{V} - \mathbf{I_a Z_s}$

Here $\mathbf{V} = 1155\angle 0°\text{ V}$; $\mathbf{I_a} = 288\angle -36.8°\text{ A}$

$$\mathbf{Z_s} = (0.2 + j\,1.9)\,\Omega = 1.91\angle 84°\,\Omega$$

$$\therefore \quad \mathbf{E_b} = (1155 + j\,0) - 288\angle -36.8° \times 1.91\angle 84°$$

$$= 1155 - 551\angle 47.2° = 1155 - (374.4 + j\,404.3)$$

$$= (780.6 - j\,404.3)\text{ V} = 879\angle -27.4°\text{ V}$$

Therefore, magnitude of E_b is **879 V/phase.**

Example 13.3. *A 3980 V, 50 Hz, 4-pole, star-connected synchronous motor generates back e.m.f. of 1790 V per phase. The resistance and synchronous reactance per phase are 2·2 Ω and 22 Ω respectively. The torque angle is 30° electrical. Calculate (i) the resultant armature voltage/phase (ii) armature current/phase (iii) power factor of the motor (iv) gross torque developed by the motor.*

Solution. Fig. 13.16 shows the phasor diagram of the motor. Since $V > E_b$, the motor power factor will be lagging.

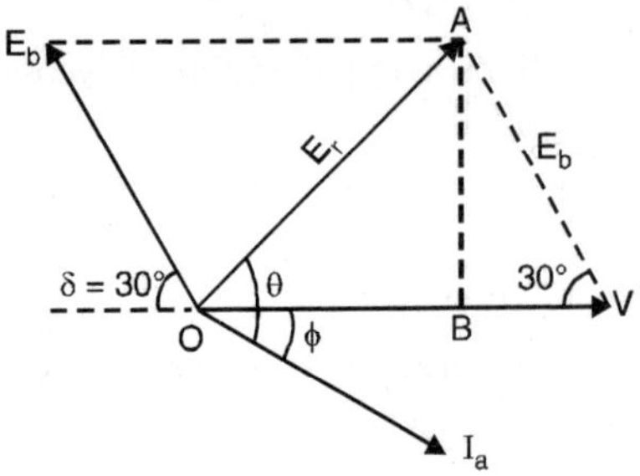

Fig. 13.16

Armature voltage/phase, $V = 3980/\sqrt{3} = 2300$ V

Back e.m.f./phase, $E_b = 1790$ V

(*i*) $E_r^2 = V^2 + E_b^2 - 2\,VE_b \cos \delta$

$= (2300)^2 + (1790)^2 - 2 \times 2300 \times 1790 \times \cos 30°$

$= 1363000$

$\therefore \quad E_r = \sqrt{1363000} = \mathbf{1167\ V}$

(*ii*) $Z_s = \sqrt{R_a^2 + X_s^2} = \sqrt{(2 \cdot 2)^2 + (22)^2} = 22 \cdot 1\ \Omega$

$I_a = E_r/Z_s = 1167/22 \cdot 1 = \mathbf{52 \cdot 8\ A}$

(*iii*) $\theta = \tan^{-1} X_s/R_a = \tan^{-1} 22/2 \cdot 2 = 84 \cdot 3°$

In right angled $\Delta\, ABO$, $\angle AOB = \theta - \phi$.

$$\therefore \quad \tan(\theta - \phi) = \frac{AB}{OB} = \frac{E_b \sin 30°}{V - E_b \cos 30°}$$

$$= \frac{1790 \sin 30°}{2300 - 1790 \cos 30°} = 1 \cdot 193$$

or $\theta - \phi = \tan^{-1} 1 \cdot 193 = 50°$

or $\phi = \theta - 50° = 84 \cdot 3° - 50° = 34 \cdot 3°$

$\therefore$ Motor p.f. $= \cos \phi = \cos 34 \cdot 3° =$ **0·826** *lag*

(*iv*) Motor input power, $P_i = \sqrt{3}\ V_L I_L \cos \phi$

$= \sqrt{3} \times 3980 \times 52 \cdot 8 \times 0 \cdot 826 = 300647$ W

Armature Cu loss $= 3\, I_a^2 R_a = 3 \times (52 \cdot 8)^2 \times 2 \cdot 2 = 18400$ W

Mechanical power developed, $P_m = 300647 - 18400 = 282247$ W

Synchronous speed, $N_s = 120 f/P = 120 \times 50/4 = 1500$ r.p.m.

Gross torque, $T_g = 9 \cdot 55 \dfrac{P_m}{N_s} = 9 \cdot 55 \dfrac{282247}{1500} =$ **1797 N-m**

Example 13.4. *A 3-phase synchronous motor has 12 poles and operates from 440 V, 50 Hz supply. Calculate its speed. If it takes a line current of 100 A at 0·8 power factor lead, what torque the motor will be developing ? Neglect losses.*

Solution.

Motor speed, $N_s = \dfrac{120\, f}{P} = \dfrac{120 \times 50}{12} = 500$ r.p.m.

Input power, $P_i = \sqrt{3}\ V_L I_L \cos \phi$

$= \sqrt{3} \times 440 \times 100 \times 0 \cdot 8 = 60966$ W

As losses are neglected, we have,

Output power, $P_{out} = P_i = 60966$ W

$\therefore$ Motor torque, $T = 9 \cdot 55 \dfrac{P_{out}}{N_s} = 9 \cdot 55 \dfrac{60966}{500} =$ **1165 N-m**

Example 13.5. *A 3-phase, 6000 kW, 4 kV, 180 r.p.m., 50 Hz motor has per phase synchronous reactance of 1·2 Ω. At full load, the torque angle is 20° electrical. If the generated back e.m.f./phase is 2·4 kV, calculate the mechanical power developed. What will be the maximum mechanical power developed ?*

Solution.

$$V = 4/\sqrt{3} = 2{\cdot}3 \text{ kV}$$

$$P_m = \frac{3 V E_b}{X_s} \sin \delta$$

$$= \frac{3 \times (2 \cdot 3 \times 10^3) \times (2 \cdot 4 \times 10^3)}{1 \cdot 2} \times \sin 20^\circ$$

$$= 4{\cdot}71 \times 10^6 \text{ W} = \mathbf{4{\cdot}71 \text{ MW}}$$

The mechanical power developed will be maximum when $\delta = 90°$.

$$\therefore \qquad P_{max} = \frac{3 V E_b}{X_s} = \frac{3 \times (2 \cdot 3 \times 10^3) \times (2 \cdot 4 \times 10^3)}{1 \cdot 2}$$

$$= 13{\cdot}8 \times 10^6 \text{ W} = \mathbf{13{\cdot}8 \text{ MW}}$$

Example 13.6. *A 400 V, 10 h.p. (7·46 kW), 3-phase synchronous motor has negligible armature resistance and a synchronous reactance of 10 Ω/phase. Determine the minimum armature current and the corresponding e.m.f. for full-load conditions. Assume an efficiency of 85 %.*

Solution. The armature current will be minimum when the power factor is unity *i.e.*, when cos $\phi = 1$. Under this condition, the phasor diagram will be as shown in Fig. 13.17.

Fig.13.17

$$\text{Voltage/phase, } V = 400/\sqrt{3} = 231 \text{ volts}$$

$$\text{Motor input, } P_i = 7460/0{\cdot}85 = 8775 \text{ W}$$

$$\text{Motor line current, } I_a = \frac{8775}{\sqrt{3} \times 400 \times 1} = \mathbf{12{\cdot}67 \text{ A}}$$

$$\text{Impedance drop} = I_a X_s = 12{\cdot}67 \times 10 = 126{\cdot}7 \text{ volts}$$

$$\text{e.m.f./phase, } E_b = \sqrt{V^2 + (I_a X_s)^2} = \sqrt{(231)^2 + (126 \cdot 7)^2}$$

$$= \mathbf{263{\cdot}4 \text{ volts}}$$

Example 13.7. *The input to an 11000 V, 3-phase, star-connected synchronous motor is 60 A. The effective resistance and synchronous reactance per phase are respectively 1 Ω and 30 Ω. Find (i) the power supplied to the motor at 0·8 p.f. lagging (ii) induced e.m.f. for a power factor of 0·8 leading (iii) the induced e.m.f. for a power factor of 0·8 lagging.*

Solution.

(*i*)

$$\text{Input current, } I_a = 60 \text{ A}$$

$$\text{Power supplied to motor} = \sqrt{3}\, V_L I_a \cos \phi = \sqrt{3} \times 11000 \times 60 \times 0 \cdot 8$$

$$= 915 \times 10^3 \text{ W} = \mathbf{915 \text{ kW}}$$

(*ii*) *Induced e.m.f. for 0·8 p.f. leading.*

$$\text{Supply voltage/phase, } V = 11000/\sqrt{3} = 6351 \text{ volts}$$

$$\text{Synchronous impedance/phase, } Z_s = \sqrt{R_a^2 + (X_s)^2} = \sqrt{(1)^2 + (30)^2} \simeq 30\ \Omega$$

$$\text{Internal angle, } \theta = \tan^{-1} \frac{X_s}{R_a} = \tan^{-1} \frac{30}{1} = 88{\cdot}1^\circ$$

$$\text{p.f. angle, } \phi = \cos^{-1} 0{\cdot}8 = 36{\cdot}9^\circ$$

$$\text{Impedance drop, } E_r = I_a Z_s = 60 \times 30 = 1800 \text{ volts}$$

Fig. 13.18 (*i*) shows the phasor diagram of the motor for a power factor of 0·8 leading.

$$E_b = \sqrt{V^2 + E_r^2 - 2\,V\,E_r\,\cos(\theta + \phi)}$$

$$= \sqrt{(6351)^2 + (1800)^2 - 2 \times 6351 \times 1800 \cos(88{\cdot}1^\circ + 36{\cdot}9^\circ)}$$

$$= 7529 \text{ volts}$$

$\therefore$ Line value of induced e.m.f. $= \sqrt{3} \times 7529$ = **13040 volts**

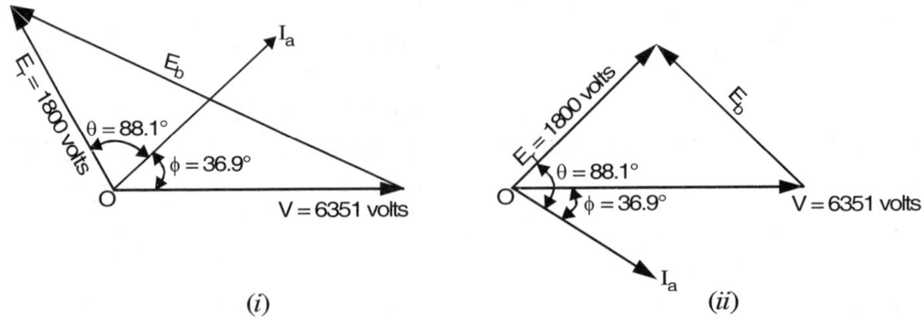

Fig. 13.18

(*iii*) *Induced e.m.f. for 0·8 p.f. lagging.* Fig. 13.18 (*ii*) shows the phasor diagram of the motor for a p.f. of 0·8 lagging.

$$E_b = \sqrt{V^2 + E_r^2 - 2\,V\,E_r\,\cos(\theta - \phi)}$$

$$= \sqrt{(6351)^2 + (1800)^2 - 2 \times 6351 \times 1800 \cos(88{\cdot}1^\circ - 36{\cdot}9^\circ)}$$

$$= 5408 \text{ volts}$$

$\therefore$ Line value of induced e.m.f. $= \sqrt{3} \times 5408$ = **9367 volts**

Example 13.8. *The synchronous reactance per phase of a 3-phase, star-connected, 6600 V synchronous motor is 10 Ω. For a certain load, the input is 900 kW and the induced e.m.f. is 8900 V (line value). Find the line current and power factor. Neglect resistance.*

Solution.

Supply voltage/phase, $V = 6600/\sqrt{3} = 3810$ volts

Back e.m.f./phase, $E_b = 8900/\sqrt{3} = 5140$ volts

Since the induced e.m.f. is greater than the supply voltage, the motor is operating with a leading power factor. Fig. 13.19 shows the phasor diagram of the motor.

Input power $= \sqrt{3} \times 6600 \times I_a \times \cos\phi$

or $900 \times 10^3 = \sqrt{3} \times 6600 \times I_a \cos\phi$

$$\therefore \quad I_a \cos\phi = \frac{900 \times 10^3}{\sqrt{3} \times 6600} = 78{\cdot}74 \text{ A}$$

Fig. 13.19

Referring to the phasor diagram in Fig. 13.19, we have,

$$OB = I_a X_s = I_a \times 10 = 10\,I_a$$

Also $BC = OB\cos\phi = 10\,I_a\cos\phi = 10 \times 78{\cdot}74 = 787{\cdot}4$ volts

Now $(AB)^2 = (AC)^2 + (BC)^2$

or $(5140)^2 = (AC)^2 + (787{\cdot}4)^2$ $\quad\therefore AC = 5079$ volts

$$\therefore \quad OC = AC - AO = 5079 - 3810 = 1269 \text{ volts}$$

In right angled Δ *OCB*, we have,

$$\tan\phi = \frac{OC}{BC} = \frac{1269}{787 \cdot 4} = 1{\cdot}612$$

$$\therefore \quad \phi = \tan^{-1} 1{\cdot}612 = 58{\cdot}2^{\circ}$$

$$\therefore \quad \text{p.f. of motor} = \cos\phi = \cos 58{\cdot}2^{\circ} = \mathbf{0{\cdot}527} \textit{ lead}$$

$$\text{Line current, } I_a = \frac{I_a \cos\phi}{\cos\phi} = \frac{78 \cdot 74}{0 \cdot 527} = \mathbf{149{\cdot}4\ A}$$

Example 13.9. *An 8-pole, 3300 V, 50 Hz, 3-phase, star-connected synchronous motor has armature resistance of 0·5 Ω per phase and synchronous reactance of 5 Ω/phase. While running on no-load, the excitation has been adjusted so as to make the e.m.f. numerically equal to and antiphase with the terminal voltage. With certain load torque applied, the rotor gets retarded by 3 mechanical degrees. Calculate the armature current and power factor of the motor.*

Solution.

Applied voltage/phase is

$$V = 3300/\sqrt{3} = 1905 \text{ volts}$$

$$\text{Back e.m.f./phase, } E_b = 1905 \text{ volts}$$

$$\text{Load angle, } \delta = \frac{P}{2} \times \text{angle of retardation in mech. degrees}$$

$$= (8/2) \times 3^{\circ} = 12^{\circ} \text{ (elect.)}$$

Fig. 13.20

Fig. 13.20 shows the phasor diagram of the motor.

$$E_r = \sqrt{V^2 + E_b^2 - 2\,V\,E_b \cos\delta}$$

$$= \sqrt{(1905)^2 + (1905)^2 - 2 \times 1905 \times 1905 \times \cos 12^{\circ}} = 400 \text{ volts}$$

Synchronous impedance/phase is given by;

$$Z_s = \sqrt{R_a^2 + X_s^2} = \sqrt{(0 \cdot 5)^2 + (5)^2} = 5{\cdot}025\,\Omega$$

$$\therefore \quad \text{Armature current is } I_a = E_r/Z_s = 400/5{\cdot}025 = \mathbf{79{\cdot}6\ A}$$

Referring to Fig. 13.20, from Δ *OAB*, we have,

$$\frac{AB}{\sin \angle AOB} = \frac{OB}{\sin \angle OAB}$$

or
$$\frac{E_b}{\sin(\theta - \phi)} = \frac{E_r}{\sin\delta}$$

or
$$\sin(\theta - \phi) = \frac{E_b}{E_r} \times \sin\delta = \frac{1905}{400} \times \sin 12^{\circ} = 0{\cdot}99$$

$$\therefore \quad \theta - \phi = \sin^{-1} 0{\cdot}99 = 82^{\circ}$$

Now
$$\theta = \tan^{-1}\frac{X_s}{R_a} = \tan^{-1}\frac{5}{0 \cdot 5} - 84 \cdot 3^{\circ}$$

$$\therefore \quad \phi = \theta - 82^{\circ} = 84{\cdot}3^{\circ} - 82^{\circ} = 2{\cdot}3^{\circ} \text{ lag}$$

$$\text{Motor p.f.} = \cos\phi = \cos 2{\cdot}3^{\circ} = \mathbf{0{\cdot}9992} \textit{ lag}$$

Example 13.10. *A 500 V, 1-phase synchronous motor is developing 10 h.p. and operates at 0·9 p.f. lagging. The effective resistance of armature is 0·8 Ω. Iron and frictional losses amount to 500 W and excitation losses are 800 W. Find (i) armature current (ii) total power supplied (iii) commercial efficiency.*

Solution.

$$\text{Motor input} = V I_a \cos\phi \ ; \ \text{Armature Cu loss} = I_a^2 R_a$$

If P_m is the total mechanical power developed (includes iron and friction losses), then,

$$P_m = V I_a \cos\phi - I_a^2 R_a$$

or $$I_a^2 R_a - V I_a \cos\phi + P_m = 0$$

(*i*) $\therefore$ $$I_a = \frac{V\cos\phi \pm \sqrt{V^2\cos^2\phi - 4 R_a P_m}}{2 R_a}$$

Here $P_m = 10 \times 746 + 500 = 7960 \text{ W}$; $R_a = 0{\cdot}8\,\Omega$; $\cos\phi = 0{\cdot}9$; $V = 500$ volts

$$\therefore \quad I_a = \frac{500 \times 0{\cdot}9 \pm \sqrt{(500 \times 0{\cdot}9)^2 - 4 \times 0{\cdot}8 \times 7960}}{2 \times 0{\cdot}8}$$

$$= \frac{450 \pm \sqrt{(450)^2 - 25472}}{1{\cdot}6} = \frac{450 \pm 420{\cdot}7}{1{\cdot}6}$$

$$= \frac{29{\cdot}3}{1{\cdot}6} = \mathbf{18{\cdot}31\ A}$$

(*ii*) Power supplied to stator $= V I_a \cos\phi = 500 \times 18{\cdot}31 \times 0{\cdot}9 = 8240$ W

Total power supplied to stator and rotor

$$= 8240 + 800 = \mathbf{9040\ W}$$

(*iii*) Commercial efficiency, $\eta_c = \dfrac{7460}{9040} \times 100 = \mathbf{82{\cdot}5\ \%}$

Example 13.11. *A 6-pole, 400 V, 3-phase, 50 Hz, star-connected synchronous motor has a resistance and synchronous impedance of 0·5 Ω and 4 Ω per phase respectively. It takes a current of 15 A at unity p.f. when operating with a certain field current. If the load torque is increased until the line current is increased to 60 A, the field current remaining unchanged, calculate the gross torque developed and the new power factor.*

Solution. First case. Fig. 13.21 shows the phasor diagram of the motor for the first case.

$$\text{Supply voltage/phase, } V = 400/\sqrt{3} = 231 \text{ volts}$$

$$\text{Synchronous reactance/phase, } X_s = \sqrt{Z_s^2 - R_a^2} = \sqrt{(4)^2 - (0{\cdot}5)^2} = 3{\cdot}968\,\Omega$$

$$\text{Internal angle, } \theta = \tan^{-1}\frac{X_s}{R_a} = \tan^{-1}\frac{3{\cdot}968}{0{\cdot}5} = 82{\cdot}8^\circ$$

$$\text{Impedance drop, } E_r = I_a Z_s = 15 \times 4 = 60 \text{ volts}$$

$$\text{Back e.m.f./phase, } E_b = \sqrt{V^2 + E_r^2 - 2 V E_r \cos\theta}$$

$$= \sqrt{(231)^2 + (60)^2 - 2 \times 231 \times 60 \times \cos 82{\cdot}8^\circ} = 231 \text{ volts}$$

Since the p.f. is unity, the armature current is minimum.

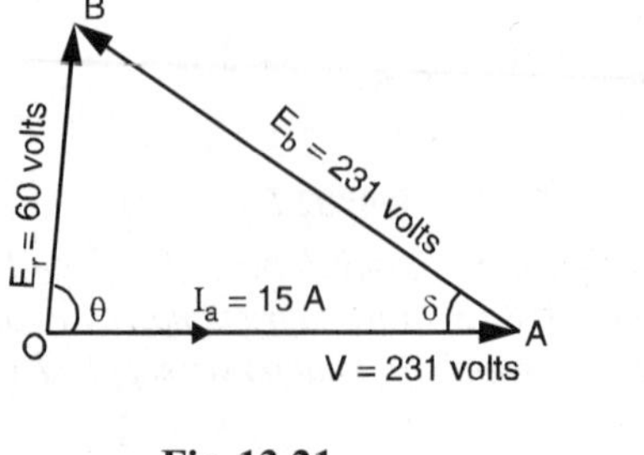

Fig. 13.21

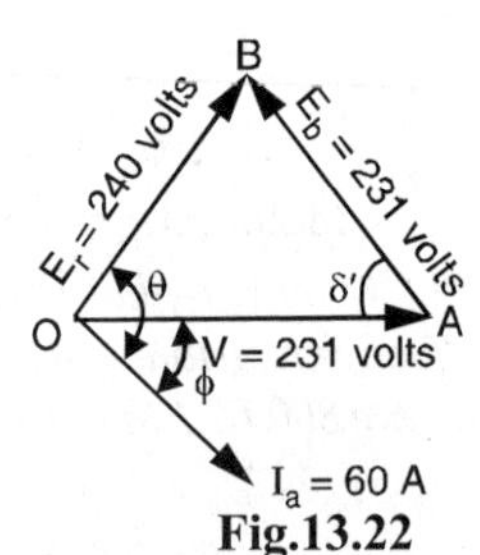

Fig.13.22

Second case. When load on the motor is increased, the load angle δ is increased to δ'. The phasor diagram for this case is shown in Fig. 13.22.

Input current, I_a = 60 A

Supply voltage/phase, V = 231 volts ... same as before

Back e.m.f./phase, E_b = 231 volts ... excitation is same

Impedance drop = $I_a Z_s = 60 \times 4 = 240$ volts

Referring to the phasor diagram in Fig. 13.22, we have,

$$E_b^2 = V^2 + E_r^2 - 2\,V E_r \cos \angle AOB$$

or
$$(231)^2 = (231)^2 + (240)^2 - 2 \times 231 \times 240 \times \cos \angle AOB$$

$$\therefore \quad \cos \angle AOB = \frac{(231)^2 - (231)^2 - (240)^2}{-2 \times 231 \times 240} = 0{\cdot}5195$$

$$\therefore \quad \angle AOB = \cos^{-1} 0{\cdot}5195 = 58{\cdot}7^\circ$$

$$\text{Internal angle, } \theta = \tan^{-1} \frac{X_s}{R_a} = 82{\cdot}8^\circ \qquad \text{...same as before}$$

$$\text{p.f. angle, } \phi = \theta - \angle AOB = 82{\cdot}8^\circ - 58{\cdot}7^\circ = 24{\cdot}1^\circ$$

New p.f. = $\cos\phi = \cos 24{\cdot}1^\circ =$ **0·913** *lag*

$$\text{New motor input} = \sqrt{3} \times 400 \times 60 \times 0{\cdot}913 = 38000 \text{ W} = 38 \text{ kW}$$

$$\text{Total armature Cu loss} = 3\, I_a^2 R_a = 3 \times (60)^2 \times 0{\cdot}5 = 5400 \text{ W} = 5{\cdot}4 \text{ kW}$$

Total mechanical power developed is

$$P_m = 38 - 5{\cdot}4 = 32{\cdot}6 \text{ kW}$$

$$\text{Synchronous speed, } N_s = 120 f/P = 120 \times 50/6 = 1000 \text{ r.p.m.}$$

$$\text{Gross torque developed, } T_g = 9{\cdot}55 \times \frac{P_m}{N_s} = 9{\cdot}55 \times \frac{32{\cdot}6 \times 10^3}{1000} = \mathbf{311 \text{ N-m}}$$

Example 13.12. *Consider a 3300 V delta connected synchronous motor having a synchronous reactance per phase of 18 Ω. It operates at a leading p.f. of 0·707 when drawing 800 kW from mains. Calculate its excitation e.m.f. and load angle.*

Solution. Fig. 13.23 shows the phasor diagram of the motor. Since R_a is negligible internal angle $\theta = 90^\circ$. The motor p.f. is 0·707 leading so that $\phi = 45^\circ$. The line current I_L is given by :

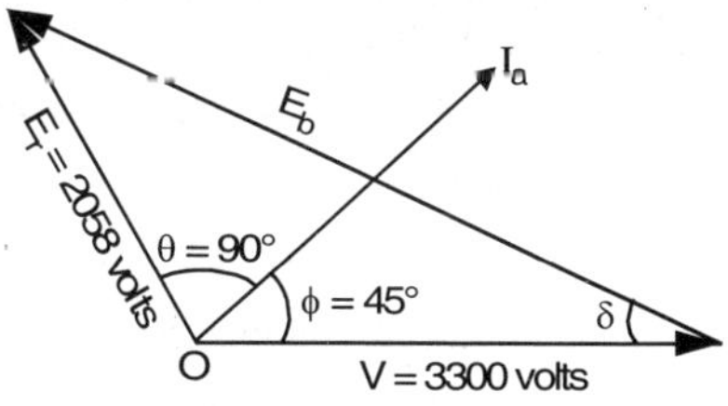

Fig. 13.23

$$I_L = \frac{800 \times 10^3}{\sqrt{3} \times 3300 \times 0{\cdot}707} = 198 \text{ A}$$

$$\text{Phase current, } I_a = 198/\sqrt{3} = 114{\cdot}3 \text{ A}$$

$$\text{Impedance drop, } E_r = I_a X_s = 114{\cdot}3 \times 18 = 2058 \text{ volts}$$

Supply voltage/phase, V = 3300 volts

Referring to the phasor diagram in Fig. 13.23, we have,

$$E_b = \sqrt{V^2 + E_r^2 - 2\,V\,E_r \cos(\theta + \phi)}$$

$$= \sqrt{(3300)^2 + (2058)^2 - 2 \times 3300 \times 2058 \cos(90^\circ + 45^\circ)}$$

$$= \mathbf{4973 \text{ volts}}$$

Also
$$\frac{E_r}{\sin\delta} = \frac{E_b}{\sin 135^\circ} \qquad \text{... sine rule}$$

$$\therefore \quad \sin\delta = E_r \times \frac{\sin 135^\circ}{E_b} = 2058 \times \frac{\sin 135^\circ}{4973} = 0{\cdot}293$$

$$\therefore \quad \text{Load angle, } \delta = \sin^{-1} 0{\cdot}293 = \mathbf{17^\circ}$$

Example 13.13. *A 10 h.p., 230 V, 60 Hz, 3-phase, star-connected synchronous motor delivers full load at a p.f. of 0.707 leading. The synchronous reactance of the motor is j5Ω/phase. The rotational loss is 230 W and the field winding loss is 70 W. Calculate (i) efficiency and (ii) generated voltage of the motor. Neglect armature winding resistance.*

Solution.

(*i*) Power output, $P_{out} = 10 \times 746 = 7460$ W ; Rotational loss = 230 W

$\therefore$ Power developed, $P_d = P_{out}$ + Rotational loss = 7460 + 230 = 7690 W

Since there is no copper loss in the armature winding, power supplied by the a.c. source is 7690 W.

$$\therefore \quad \text{Armature current, } I_a = \frac{7690}{\sqrt{3} \times 230 \times 0.707} = 27.3 \text{ A}$$

$$\text{Total input power, } P_{in} = 7690 + 70 = 7760 \text{ W}$$

$$\therefore \quad \text{Motor efficiency, } \eta = \frac{P_{out}}{P_{in}} \times 100 = \frac{7460}{7760} \times 100 = \mathbf{96.1\%}$$

(*ii*) Taking supply voltage/phase ($V = 230/\sqrt{3} = 132.8$ volts) as the reference phasor, we have,

$$\text{Generated voltage, } \mathbf{E_b} = \mathbf{V} - \mathbf{I_a X_s} \qquad (\because R_a = 0)$$

Here $\mathbf{V} = 132.8 \angle 0^\circ$ volts ; $\mathbf{I_a} = 27.3 \, * \angle 45^\circ$ A ; $\mathbf{X_s} = j\,5\,\Omega$

$$\therefore \quad \mathbf{E_b} = 132.8 \angle 0^\circ - 27.3 \angle 45^\circ \times 5 \angle 90^\circ$$

$$= 132.8 - 136.5 \angle 135^\circ$$

$$= 132.8 - (-96.5 + j\,96.5)$$

$$= (229.3 - j\,96.5) \text{ V} = \mathbf{248.8 \angle -22.8^\circ \text{ V}}$$

Example 13.14. *A 150 kW, 1200 r.p.m., 460 V, 3-phase synchronous motor has a synchronous reactance of 0.8 Ω per phase. If the excitation voltage is fixed at 300 V per phase, determine (i) the power versus torque angle δ curve (ii) the torque versus δ curve (iii) the pull-out torque.*

Solution.

Terminal voltage/phase, $V = 460/\sqrt{3} = 266$ volts

Induced voltage/phase, $E_b = 300$ volts

(*i*) Mechanical power/phase is given by;

$$P_{out} = \frac{E_b V}{X_s} \sin\delta = \frac{300 \times 266}{0.8} \times \sin\delta$$

$$\therefore \quad P_{out} = 99750 \sin\delta \text{ W}$$

$$\text{or} \quad P_{out} \simeq 100 \sin\delta \text{ kW}$$

For different values of δ, we can find the corresponding values of P_{out}. For example, if δ = 0°, then $P_{out} = 100 \sin 0^\circ = 0$. Similarly, if δ = 30°, then, $P_{out} = 100 \sin 30^\circ = 50$ kW. Likewise, if δ = 60°, then $P_{out} = 100 \sin 60^\circ = 86.6$ kW. The solid curve in Fig. 13.24 shows the curve between power/phase and torque angle δ. Note that maximum output power/phase of the motor is 100 kW.

---

* p.f. is 0.707 leading. Therefore, $\phi = \cos^{-1} 0.707 = 45^\circ$. The current I_a leads the voltage V by 45°.

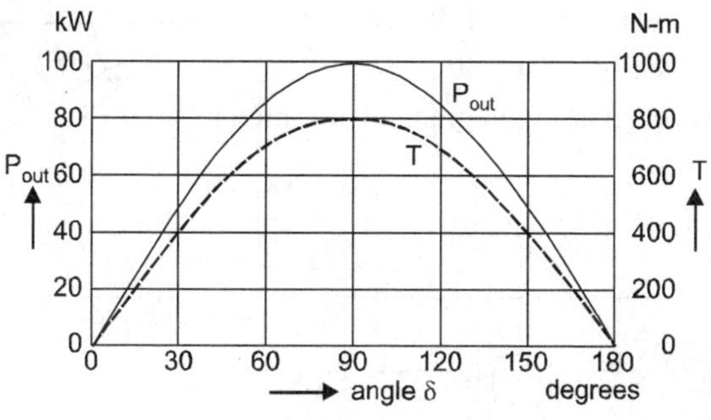

Fig. 13.24

(*ii*) $$\text{Torque/phase, } T = 9.55 \times \frac{P_{out}}{N_s} = 9.55 \times \frac{P_{out}}{1200}$$

$$\therefore \quad T = P_{out}/125 \text{ Nm}$$

Corresponding to different values of δ, P_{out} and hence T can be calculated. The dotted curve in Fig. 13.24 shows the graph between T and δ.

(*iii*) The pull-out torque (or maximum torque) coincides with the maximum power output.

$$\therefore \quad T_{max} = \frac{P_{max}}{125} = \frac{100 \times 10^3}{125} = 800 \text{ Nm}$$

Note that 800 Nm is the per phase torque. Since it is a 3-phase motor, the maximum torque = 3 × 800 = **2400 Nm.** Also maximum output power/phase = 100 kW so that the motor can develop maximum output power = 3 × 100 = 300 kW.

Example 13.15. *A 3-phase, 6000 kW, 4 kV, 180 r.p.m., 60 Hz synchronous motor has a synchronous reactance of 1.2 Ω per phase. At full-load, the rotor poles are displaced by a mechanical angle of 1° from their no-load position. If the line-to-neutral excitation voltage E_b = 2.4 kV, calculate the mechanical power developed.*

Solution.

$$\text{Induced voltage/phase, } E_b = 2.4 \text{kV}$$

$$\text{Terminal voltage/phase, } V = 4/\sqrt{3} = 2.3 \text{ kV}$$

$$\text{Number of poles, } P = \frac{120 f}{N_s} = \frac{120 \times 60}{180} = 40$$

$$\text{Torque angle, } \delta = \frac{P\alpha}{2} = \frac{40 \times 1}{2} = 20° \text{ (elect.)}$$

$$\text{Power developed/phase, } P_d = \frac{E_b V}{X_s} \sin \delta = \frac{2.4 \times 10^3 \times 2.3 \times 10^3}{1.2} \sin 20°$$

$$= 1.57 \times 10^6 \text{ W} = 1.57 \text{ MW}$$

$$\text{Total power} = 3 \times 1.57 = \mathbf{4.71 \text{ MW}}$$

Example 13.16. *A synchronous motor connected to a 3980 V, 3-phase line generates an excitation voltage E_b of 1790 V (per phase) when the d.c. exciting current is 25 A. The synchronous reactance is 22Ω / phase and the torque angle between V and E_b is 30°. Calculate (i) E_r (voltage across X_s) (ii) a.c. line current (iii) power factor of the motor.*

Solution.

(*i*) $$\text{Supply voltage/phase, } V = 3980/\sqrt{3} = 2300 \text{ volts}$$

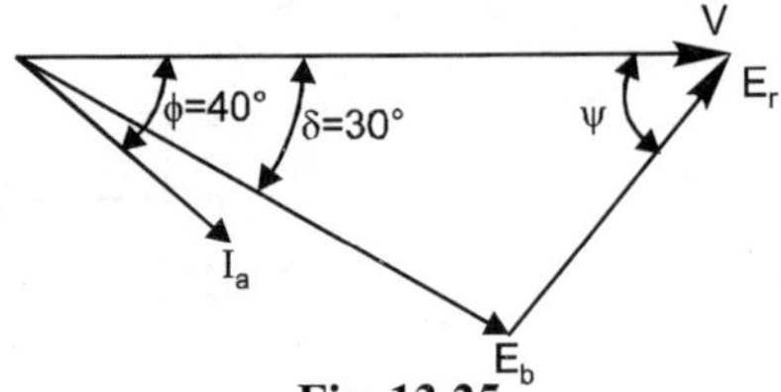

Fig. 13.25

Induced e.m.f./phase, E_b = 1790 volts

Torque angle, δ = 30°

Fig. 13.25 shows the phasor diagram of the motor. The value of E_r is given by :

$$E_r = \sqrt{V^2 + E_b^2 - 2V E_b \cos\delta}$$

$$= \sqrt{(2300)^2 + (1790)^2 - 2\times 2300\times 1790\times \cos 30°} = \textbf{1167 volts}$$

(*ii*) Line current, $I_a = E_r/X_s$ = 1167/22 = **53 A**

(*iii*) The angle Ψ between V and E_r can be found from the sine law as under :

$$\frac{E_b}{\sin\Psi} = \frac{E_r}{\sin\delta} \quad \text{or} \quad \frac{1790}{\sin\Psi} = \frac{1167}{\sin 30°}$$

$$\therefore \quad \sin\Psi = 0.7669 \quad \text{or} \quad \Psi = 50°$$

Since I_a lags *90° behind E_r, it follows that :

$$\phi = 90° - \Psi = 90° - 50° = 40°$$

$$\therefore \quad \text{Motor p.f.} = \cos\phi = \cos 40° = \textbf{0.766} \textit{ lag}$$

Example 13.17. *A 440 V, 3-phase, Δ-connected synchronous motor has a synchronous reactance of 36 Ω/phase. Its armature-winding resistance is negligible. When the motor runs at a speed of 188.5 rad/s, it consumes 9 kW and the excitation voltage is 560 V. Determine (i) power factor (ii) power angle and (iii) torque developed by the motor.*

Solution. Using Δ-Y transformation, the equivalent per phase reactance of Y-connected synchronous motor is X_s = 36/3 = 12 Ω.

Supply voltage/phase, $V = 440/\sqrt{3}$ = 254 volts

Generated voltage/phase, $E_b = 560/\sqrt{3}$ = 323.32 volts

Now
$$P_m = \frac{3V E_b}{X_s}\sin\delta \quad \text{or} \quad 9\times 10^3 = \frac{3\times 254\times 323.32}{12}\sin\delta$$

$$\therefore \quad \sin\delta = \frac{9\times 10^3\times 12}{3\times 254\times 323.32} = 0.438 \quad \text{or} \quad \delta = 26°$$

Taking supply voltage/phase as the reference phasor, we have,

$$\mathbf{V} = 254\angle 0° \text{ volts} \; ; \; \boldsymbol{E_b} = 323.32\angle{-26°} \text{ volts}$$

The phase current (or line current) in the equivalent star-connected synchronous motor is

$$\boldsymbol{I_a} = \frac{\mathbf{V} - \mathbf{E_b}}{jX_s} = \frac{254\angle 0° - 323.32\angle -26°}{12\angle 90°}$$

$$= \frac{254\angle 0°}{12\angle 90°} - \frac{323.32\angle -26°}{12\angle 90°} = 21.2\angle -90° - 26.9\angle -116°$$

$$= (11.8 + j3)\,\text{A} = 12.2\angle 14.3°\,\text{A}$$

(*i*) Motor power factor = cos ϕ = cos 14.3° = **0.97** *lead*

(*ii*) Power angle, δ = **– 26°** *electrical*

(*iii*) Torque developed, $T_d = \dfrac{P_m}{\omega_s} = \dfrac{9\times 10^3}{188.5}$ = **47.75 Nm**

* $I_a = E_r/X_s$. Therefore, I_a lags E_r by 90°.

TUTORIAL PROBLEMS

1. A 2200 V, 3-phase, 50 Hz, star-connected synchronous motor has a resistance of 0·6 Ω/phase and a synchronous reactance of 6 Ω per phase. Find the generated e.m.f. when the input is 200 kW at (*i*) a power factor of unity (*ii*) a power factor of 0·8 leading. **[(*i*) 2200 V (*ii*) 2640 V]**
2. A 3-phase, star-connected synchronous motor takes 20 kW at 400 V from 50 Hz supply. The synchronous reactance/phase is 4 Ω/phase and the effective resistance is negligible. If the exciting current is so adjusted that back e.m.f. is 550 V, find the power factor of the motor. **[0·872 *lead*]**
3. In a 3-phase, star-connected, 660 V synchronous motor, the synchronous reactance is 2 Ω/phase. When a certain load is applied, the input is 90 kW and the back e.m.f. is 890 V. Neglecting resistance, calculate the power factor. **[0·824 *lead*]**
4. A 6600 V, 3-ϕ, star-connected synchronous motor draws a full-load current of 70 A at 0·8 p.f. leading. The armature resistance/phase is 2 Ω and synchronous reactance/phase is 20 Ω. The rotational losses amount to 30 kW. Find (*i*) the induced e.m.f. (*ii*) output power of the machine and (*iii*) efficiency. **[(*i*) 8134 V (*ii*) 580·6 kW (*iii*) 90·72 %]**
5. A synchronous motor having 40 % reactance and negligible resistance is to be operated at rated load at (*i*) unity p.f. (*ii*) 0·8 p.f. lag (*iii*) 0·8 p.f.lead. What are the values of induced e.m.f.? **[(*i*) 108 V (*ii*) 82·5 V (*iii*) 128 V]**
6. A 220 V, single-phase synchronous motor has a synchronous impedance of 5 Ω and an effective armature resistance of 0·5 Ω. Calculate (*i*) the minimum armature current (*ii*) the angle of retard for a total load (including iron and friction losses) of 5 kW. **[(*i*) 24·04 A (*ii*) 29·9°]**
7. A 2300 V, 3-phase, star-connected synchronous motor has a resistance of 0·2 Ω per phase and a synchronous reactance of 2·2 Ω per phase. The motor is operating at 0·5 power factor leading with a line current of 200 A. Determine the value of generated e.m.f. per phase. **[1708 V]**
8. A 2000 V, 3-phase, star-connected synchronous motor has an effective per phase armature resistance and synchronous reactance of 0·2 Ω and 2·2 Ω respectively. The input is 800 kW at normal voltage and the induced line e.m.f. is 2500 V. Calculate the line current and power factor. **[254·15 A ; 0·909 *lead*]**

13.15 POWER FACTOR OF SYNCHRONOUS MOTORS

In an induction motor, only one winding (*i.e.*, stator winding) produces the necessary flux in the machine. The stator winding must draw reactive power from the supply to set up the flux. Consequently, induction motor must operate at lagging power factor.

But in a synchronous motor, there are two possible sources of excitation ; alternating current in the stator or *direct current in the rotor. The required flux may be produced either by stator or rotor or both.

(*i*) If the rotor exciting current is of such magnitude that it produces all the required flux, then no magnetising current or reactive power is needed in the stator. As a result, the motor will operate at unity power factor.

(*ii*) If the rotor exciting current is less (*i.e.*, motor is under-excited), the deficit in flux is made up by the stator. Consequently, the motor draws reactive power to provide for the remaining flux. Hence motor will operate at a lagging power factor.

(*iii*) If the rotor exciting current is greater (*i.e.*, motor is over-excited), the excess flux must be counterbalanced in the stator. Now the stator, instead of absorbing reactive power, actually *delivers* reactive power to the 3-phase line. The motor then behaves like a source of reactive power, as if it were a capacitor. In other words, the motor operates at a leading power factor.

To sum up, *a synchronous motor absorbs reactive power when it is under-excited and delivers reactive power to source when it is over-excited.*

* The rotor produces a d.c. magnetomotive force (m.m.f.). However, as far as the stator is concerned, d.c. m.m.f. appears as an a.c. m.m.f. because the rotor is constantly turning. Moreover, rotor m.m.f. has the same frequency as the stator m.m.f. because the rotor turns at synchronous speed. The total flux in the machine is, therefore, due to the combined action of the two m.m.f.s.

13.16 SYNCHRONOUS CONDENSER

A synchronous motor takes a leading current when over-excited and, therefore, behaves as a capacitor.

An over-excited synchronous motor running on no-load is known as **synchronous condenser**.

When such a machine is connected in parallel with induction motors or other devices that operate at low lagging power factor, the leading kVAR supplied by the synchronous condenser partly neutralises the lagging reactive kVAR of the loads. Consequently, the power factor of the system is improved.

Fig. 13.26 shows the power factor improvement by synchronous condenser method. The $3-\phi$ load takes current I_L at low lagging power factor $\cos\phi_L$. The synchronous condenser takes a current I_m which leads the voltage by an angle ϕ_m*. The resultant current I is the vector sum of I_m and I_L and lags behind the voltage by an angle ϕ. It is clear that ϕ is less than ϕ_L so that $\cos\phi$ is greater than $\cos\phi_L$. Thus the power factor is increased from $\cos\phi_L$ to $\cos\phi$. Synchronous condensers are generally used at major bulk supply substations for power factor improvement.

Advantages

(*i*) By varying the field excitation, the magnitude of current drawn by the motor can be changed by any amount. This helps in achieving stepless† control of power factor.

(*ii*) The motor windings have high thermal stability to short circuit currents.

(*iii*) The faults can be removed easily.

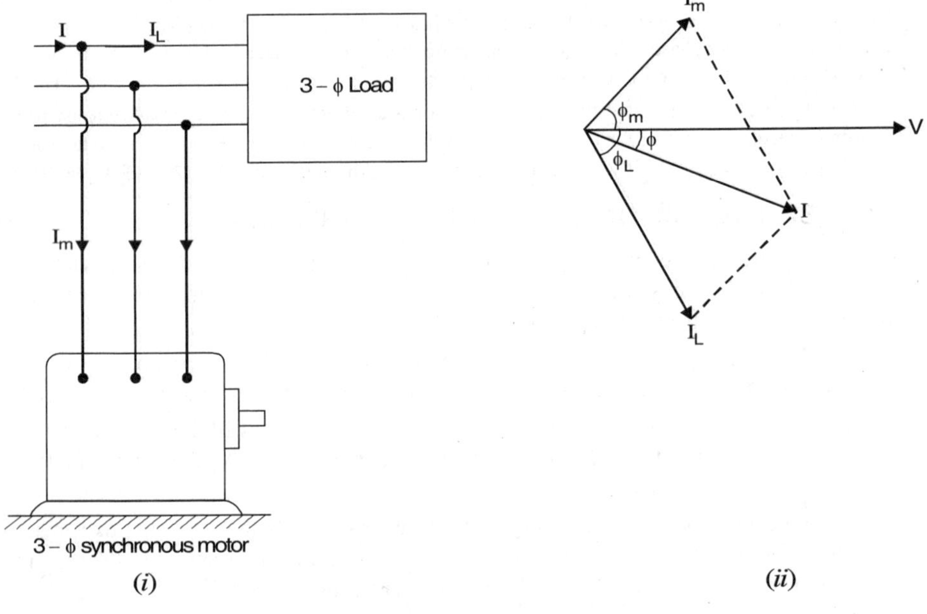

Fig. 13.26

Disadvantages

(*i*) There are considerable losses in the motor.

---

* If the motor is ideal *i.e.*, there are no losses, then $\phi_m = 90°$. However, in actual practice, losses do occur in the motor even at no load. Therefore, the current I_m leads the voltage by an angle less than 90°.

† The p.f. improvement with capacitors can only be done in steps by switching on the capacitors in various groupings. However, with synchronous motor, any amount of capacitive reactance can be provided by changing the field excitation.

(*ii*) The maintenance cost is high.

(*iii*) It produces noise.

(*iv*) Except in sizes above 500 kVA, the cost is greater than that of static capacitors of the same rating.

(*v*) As a synchronous motor has no self-starting torque, therefore, an auxiliary equipment has to be provided for this purpose.

Example 13.18. *A synchronous motor improves the power factor of a load of 200 kW from 0·8 lagging to 0·9 lagging. Simultaneously the motor carries a load of 80 kW. Find (i) the leading kVAR supplied by the motor (ii) kVA rating of the motor and (iii) the power factor at which the motor operates.*

Solution.

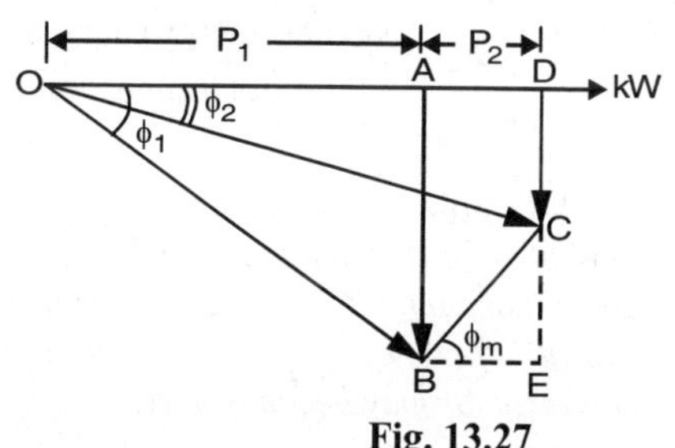

Fig. 13.27

$$\text{Load, } P_1 = 200 \text{ kW}$$

$$\text{Motor load, } P_2 = 80 \text{ kW}$$

$$\text{p.f. of load, } \cos\phi_1 = 0{\cdot}8 \textit{ lag}$$

$$\text{p.f. of combined load, } \cos\phi_2 = 0{\cdot}9 \textit{ lag}$$

$$\text{Combined load, } P = P_1 + P_2 = 200 + 80 = 280 \text{ kW}$$

In Fig. 13.27, Δ *OAB* is the power triangle for load, Δ *ODC* for combined load and Δ *BEC* for the motor.

(*i*) Leading kVAR supplied by the motor

$$= CE = DE - DC = AB - DC \qquad (\because AB = DE)$$

$$= P_1 \tan\phi_1 - P^* \tan\phi_2$$

$$= 200 \tan(\cos^{-1} 0{\cdot}8) - 280 \tan(\cos^{-1} 0{\cdot}9)$$

$$= 200 \times 0{\cdot}75 - 280 \times 0{\cdot}4843 = \mathbf{14{\cdot}4\ kVAR}$$

(*ii*) kVA rating of the motor $= BC = \sqrt{(BE)^2 + (EC)^2} = \sqrt{(80)^2 + (14\cdot4)^2} = \mathbf{81{\cdot}28\ kVA}$

(*iii*) p.f. of motor, $\cos\phi_m = \dfrac{\text{Motor kW}}{\text{Motor kVA}} = \dfrac{80}{81\cdot28} = \mathbf{0{\cdot}984}\ \textit{lead}$

Example 13.19. *A factory load consists of the following :*

(*i*) *an induction motor of 50 h.p. (37·3 kW) with 0·8 p.f. and efficiency 0·85.*

(*ii*) *a synchronous motor of 25 h.p. (18·65 kW) with 0·9 p.f. leading and efficiency 0·9.*

(*iii*) *lighting load of 10 kW at unity p.f.*

Find the annual electrical charges if the tariff is Rs 60 per kVA of maximum demand per annum plus 5 paise per kWh ; assuming the load to be steady for 2000 hours in a year.

Solution.

(*i*) Input power to induction motor = 37·3/0·85 = 43·88 kW

Lagging kVAR supplied by induction motor $= 43{\cdot}88 \tan(\cos^{-1} 0{\cdot}8) = 32{\cdot}91$

(*ii*) Input power to synchronous motor

$$= 18{\cdot}65/0{\cdot}9 = 20{\cdot}72 \text{ kW}$$

---

* In right angled Δ *OAB*, $AB = P_1 \tan\phi_1$

In right angled Δ *ODC*, $DC = OD \tan\phi_2 = (P_1 + P_2)\tan\phi_2 = P\tan\phi_2$

Leading kVAR supplied by synchronous motor

$$= 20{\cdot}72 \tan(\cos^{-1} 0{\cdot}9) = 10$$

(*iii*) Since lighting load works at unity p.f., its lagging kVAR = 0.

$$\text{Net lagging kVAR} = 32{\cdot}91 - 10 = 22{\cdot}91$$

$$\text{Total active power} = 43{\cdot}88 + 20{\cdot}72 + 10 = 74{\cdot}6 \text{ kW}$$

$$\text{Total kVA} = \sqrt{(74{\cdot}6)^2 + (22{\cdot}91)^2} = 78$$

$$\text{Annual kVA demand charges} = \text{Rs. } 60 \times 78 = \text{Rs. } 4{,}680$$

$$\text{Energy consumed/year} = 74{\cdot}6 \times 2000 = 1{,}49{,}200 \text{ kWh}$$

$$\text{Annual Energy charges} = \text{Rs. } 0{\cdot}05 \times 1{,}49{,}200 = \text{Rs. } 7{,}460$$

$$\text{Total annual bill} = \text{kVA demand charges} + \text{Energy charges}$$

$$= \text{Rs. } (4680 + 7460) = \textbf{Rs. 12,140}$$

Example 13.20. *A supply system feeds the following loads: (i) a lighting load of 500 kW (ii) a load of 400 kW at a p.f. of 0·707 lagging (iii) a load of 800 kW at a p.f. of 0·8 leading (iv) a load of 500 kW at a p.f. 0·6 lagging (v) a synchronous motor driving a 540 kW d.c. generator and having an overall efficiency of 90 %. Calculate the power factor of synchronous motor so that the station power factor may become unity.*

Solution. The lighting load works at unity p.f. and, therefore, its lagging kVAR is zero. The lagging kVAR are supplied by the loads (*ii*) and (*iv*), whereas loads (*iii*) and (*v*) supply the leading kVAR. For station power factor to be unity, the total lagging kVAR must be neutralised by the total leading kVAR. We know that kVAR = kW tan ϕ.

∴ Total lagging kVAR supplied by loads (*ii*) and (*iv*)

$$= 400 \tan(\cos^{-1} 0{\cdot}707) + 500 \tan(\cos^{-1} 0{\cdot}6)$$

$$= 400 \times 1 + 500 \times 1{\cdot}33 = 1066{\cdot}6$$

Leading kVAR supplied by the load (*iii*)

$$= 800 \tan(\cos^{-1} 0{\cdot}8) = 800 \times 0{\cdot}75 = 600$$

∴ Leading kVAR to be supplied by synchronous motor

$$= 1066{\cdot}6 - 600 = 466{\cdot}6 \text{ kVAR}$$

$$\text{Motor input} = \text{output/efficiency} = 540/0{\cdot}9 = 600 \text{ kW}$$

If ϕ is the phase angle of synchronous motor, then,

$$\tan\phi = \text{kVAR/kW} = 466{\cdot}6/600 = 0{\cdot}7776$$

$$\therefore \quad \phi = \tan^{-1} 0{\cdot}7776 = 37{\cdot}87^\circ$$

∴ p.f. of synchronous motor = cos ϕ = cos 37·87° = **0·789** *leading*

Therefore, in order that the station power factor may become unity, the synchronous motor should be operated at a p.f. of 0·789 leading.

Example 13.21. *An industrial load consists of (i) a synchronous motor of 100 metric h.p. (ii) induction motors aggregating 200 metric h.p., 0·707 power factor lagging and 82 % efficiency and (iii) lighting load aggregating 30 kW. The tariff is Rs 100 per annum per kVA maximum demand plus 6 paise per kWh. Find the annual saving in cost if the synchronous motor operates at 0·8 p.f. leading, 93 % efficiency instead of 0·8 p.f. lagging at 93 % efficiency.*

Solution. The annual power bill will be calculated under two conditions *viz.*, (*a*) when synchronous motor runs with lagging p.f. and (*b*) when synchronous motor runs with a leading p.f.

(*a*) **When synchronous motor runs at p.f. 0·8 lagging.** We shall find the combined kW and then calculate total kVA maximum demand using the relation :

$$\text{kVA} = \sqrt{(\text{kW})^2 + (\text{kVAR})^2}$$

$$\text{Input to synchronous motor} = \frac{100 \times 746}{0 \cdot 96 \times 1000} \simeq 79 \text{ kW}$$

*Lagging kVAR supplied by the synchronous motor

$$= 79 \tan(\cos^{-1} 0 \cdot 8) = 79 \times 0 \cdot 75 = 59 \cdot 25 \text{ kVAR}$$

$$\text{Input to induction motors} = \frac{200 \times 746}{0 \cdot 82 \times 1000} \simeq 179 \cdot 4 \text{ kW}$$

Lagging kVAR supplied by induction motors

$$= 179 \cdot 4 \tan(\cos^{-1} 0 \cdot 707) = 179 \cdot 4 \times 1 = 179 \cdot 4 \text{ kVAR}$$

Since lighting load works at unity p.f., its lagging kVAR is zero.

$$\therefore \quad \text{Total lagging kVAR} = 59 \cdot 25 + 179 \cdot 4 = 238 \cdot 65$$

$$\text{Total active power} = 79 + 179 \cdot 4 + 30 = 288 \cdot 4 \text{ kW}$$

$$\text{Total kVA} = \sqrt{(238 \cdot 65)^2 + (288 \cdot 4)^2} = 374 \cdot 4$$

$$\text{Annual kVA demand charges} = \text{Rs. } 100 \times 374.4 = \text{Rs. } 37{,}440$$

$$\text{Energy consumed/year} = 288 \cdot 4 \times 8760 = 25{,}26{,}384 \text{ kWh}$$

$$\text{Annual energy charges} = \text{Rs. } 0 \cdot 06 \times 25{,}26{,}384 = \text{Rs. } 1{,}51{,}583$$

$$\text{Total annual bill} = \text{Rs. } (37{,}440 + 1{,}51{,}583) = \text{Rs. } 1{,}89{,}023$$

(*b*) **When synchronous motor runs at p.f. 0·8 leading.** As the synchronous motor runs at leading p.f. of 0·8 (instead of 0·8 p.f. lagging), therefore, it supplies now 59·25 leading kVAR. The lagging kVAR supplied by induction motors are the same as before *i.e.*, 179·4.

$$\therefore \quad \text{Net lagging kVAR} = 179 \cdot 4 - 59 \cdot 25 = 120 \cdot 15$$

$$\text{Total active power} = \text{Same as before } i.e., 288 \cdot 4 \text{ kW}$$

$$\therefore \quad \text{Total kVA} = \sqrt{(120 \cdot 15)^2 + (288 \cdot 4)^2} = 312 \cdot 4$$

$$\text{Annual kVA demand charges} = \text{Rs } 100 \times 312 \cdot 4 = \text{Rs } 31{,}240$$

$$\text{Annual energy charges} = \text{Same as before } i.e., \text{ Rs } 1{,}51{,}583$$

$$\text{Total annual bill} = \text{Rs } (31{,}240 + 1{,}51{,}583) = \text{Rs. } 1{,}82{,}823$$

$$\therefore \quad \text{Annual saving} = \text{Rs. } (1{,}89{,}023 - 1{,}82{,}823) = \textbf{Rs. 6,200}$$

TUTORIAL PROBLEMS

1. A station supplies 250 kVA at a lagging power factor of 0·8. A synchronous motor is connected in parallel with the load. If the combined load is 250 kW with a lagging p.f. of 0·9, determine :
   (*i*) the leading kVAR supplied by the motor.
   (*ii*) kVA rating of the motor.
   (*iii*) p.f. at which the motor operates. **[(*i*) 28·9 kVAR (*ii*) 57·75 kVA (*iii*) 0·866 lead.]**
2. A generating station supplies power to the following :
   (*i*) a lighting load of 100 kW.
   (*ii*) an induction motor 800 h.p. (596·8 kW), p.f. 0·8 lagging, efficiency 92 %.
   (*iii*) a rotary convertor giving 150 A at 400 V at an efficiency of 0·95.
   What must be the power factor of the rotary convertor in order that power factor of the supply station may become unity ? **[0·128 leading]**
3. A 3-phase, 400 V synchronous motor having a power consumption of 50 kW is connected in parallel with an induction motor which takes 200 kW at a power factor of 0·8 lagging.

* Since the synchronous motor in this case runs at lagging p.f., it supplies lagging kVAR.

(*i*) Calculate the current drawn from the mains when the power factor of the synchronous motor is unity.

(*ii*) At what power factor should the synchronous motor operate so that the current drawn from the mains is minimum ? **[(*i*) 421 A (*ii*) 0·316 leading]**

13.17 STOPPING SYNCHRONOUS MOTORS

Owing to the large inertia of the rotor and load, large synchronous motors may take several minutes to stop after they are disconnected from the line. This long stopping time may be unacceptable in many practical situations. In order to reduce stopping time, we use the following braking methods :

(*i*) The full d.c. excitation is maintained and the armature is short-circuited.

(*ii*) The full d.c. excitation is maintained with armature connected to the external resistors.

(*iii*) Mechanical braking may be applied.

In methods (*i*) and (*ii*), the motor slows down because it functions as an alternator, dissipating its energy in the resistance of the circuit. Mechanical braking is usually applied after the motor has reached half speed or less in order to prevent undue wear of the brake shoes.

Example 13.22. *A 1500 kW, 4600 V, 600 r.p.m., 60 Hz synchronous motor possesses a synchronous reactance of 16 Ω per phase and a stator resistance of 0.4 Ω per phase. The excitation voltage per phase is 2400 V. We wish to stop the motor by short-circuiting the armature while keeping the d.c. rotor current fixed. Calculate :*

(i) the power dissipated in the armature at 600 r.p.m.

(ii) the power dissipated in the armature at 150 r.p.m.

Solution.

(*i*) **Motor running at 600 r.p.m.** Fig. 13.28 shows the per phase equivalent circuit of the motor when running at 600 r.p.m.

$$\text{Synchronous impedance/phase, } Z_s = \sqrt{R_a^2 + X_s^2} = \sqrt{(0.4)^2 + (16)^2} = 16\ \Omega$$

$$\therefore \quad \text{Current/phase, } I_a = \frac{E_b}{Z_s} = \frac{2400}{16} = 150\text{ A}$$

∴ Power dissipated in the 3 phases at 600 r.p.m. is given by;

$$P = 3\,I_a^2 R_a = 3 \times (150)^2 \times 0.4 = \mathbf{27000\ W}$$

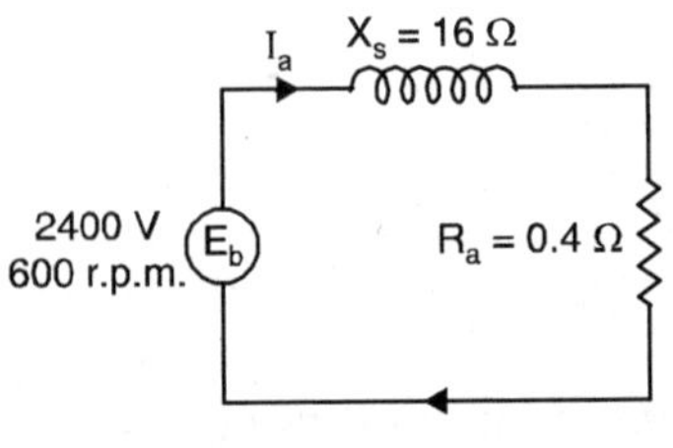

Fig. 13.28

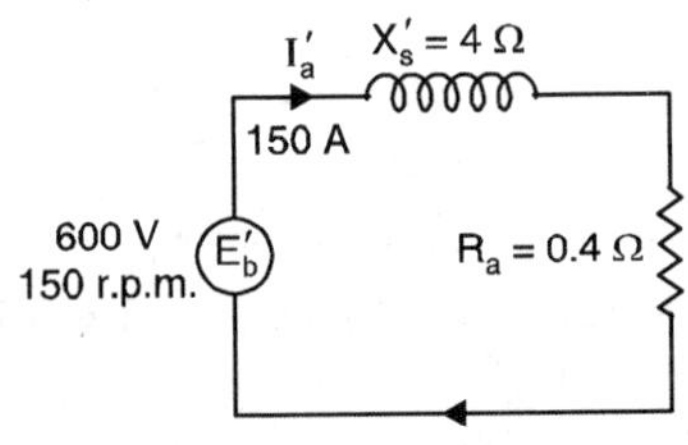

Fig. 13.29

(*ii*) **Motor running at 150 r.p.m.** Since the d.c. exciting current is fixed, the induced e.m.f. is proportional to the speed.

∴ Induced e.m.f. at 150 r.p.m. is given by;

$$E'_b = 2400 \times 150/600 = 600\text{ V}$$

The frequency of induced e.m.f. is proportional to speed.

∴ Frequency of induced e.m.f. at 150 r.p.m. is given by;

$$f' = f \times 150/600 = 60 \times 150/600 = 15 \text{ Hz}$$

The synchronous reactance at frequency f' (= 15 Hz) is

$$X_s' = X_s \times (15/60) = 16 \times (15/60) = 4\Omega$$

Therefore, at 150 r.p.m., the per phase equivalent circuit of the motor becomes as shown in Fig. 13.29.

$$\text{Synchronous impedance/phase, } Z'_s = \sqrt{R_a^2 + X_s'^2} = \sqrt{(0.4)^2 + (4)^2} = 4\Omega$$

$$\text{Current/phase, } I'_a = \frac{E_b'}{Z_s'} = \frac{600}{4} = 150 \text{ A}$$

∴ Power dissipated in the 3 phases at 150 r.p.m. is given by;

$$P' = 3 \; I_a'^2 R_a = 3 \times (150)^2 \times 0.4 = \mathbf{27000 \; W}$$

13.18 APPLICATIONS OF SYNCHRONOUS MOTORS

(*i*) Synchronous motors are particularly attractive for *low speeds (< 300 r.p.m.) because the power factor can always be adjusted to unity and efficiency is high.

(*ii*) Overexcited synchronous motors can be used to improve the power factor of a plant while carrying their rated loads.

(*iii*) They are used to improve the voltage regulation of transmission lines.

(*iv*) High-power electronic converters generating very low frequencies enable us to run synchronous motors at ultra-low speeds. Thus huge motors in the 10 MW range drive crushers, rotary kilns and variable-speed ball mills.

13.19 COMPARISON OF SYNCHRONOUS AND INDUCTION MOTORS

S.No.	*Particular*	*Synchronous Motor*	*3-phase Induction Motor*
1.	*Speed*	Remains constant (*i.e.*, N_s) from no-load to full-load.	Decreases with load.
2.	*Power factor*	Can be made to operate from lagging to leading power factor.	Operates at lagging power factor.
3.	*Excitation*	Requires d.c. excitation at the rotor.	No excitation for the rotor.
4.	*Economy*	Economical for speeds below 300 r.p.m.	Economical for speeds above 600 r.p.m.
5.	*Self-starting*	No self-starting torque. Auxiliary means have to be provided for starting.	Self-starting
6.	*Construction*	Complicated	Simple
7.	*Starting torque*	More	Less

SHORT ANSWER QUESTIONS

Q.1. *How does a synchronous motor start up ?*

Ans. A synchronous motor cannot start by itself; the squirrel cage winding (*i.e.* damper winding) on the rotor makes it self-starting.

(*i*) To start the synchronous motor, the rotor is de-energised and 3-phase supply is given to the stator. Due to the presence of damper winding, the motor starts up as a 3-phase induction motor.

* At low speeds, induction motors become heavy, costly and have relatively low power factor and low efficiency.

(*ii*) As soon as the motor approaches synchronous speed, the rotor is energised with direct current. Now the opposite polarity poles on the stator and rotor will attract. The mutual attraction locks the rotor and stator together and the rotor is literally pulled round at synchronous speed. The motor now runs at synchronous speed.

Q. 2. *During starting of a synchronous motor, when should the rotor excitation be applied?*

Ans. When the synchronous motor starts up as an induction motor upto nearly synchronous speed, its pull-in torque is quite powerful but d.c. excitation must produce the best effects. For example, if excitation is applied when the *N-S* poles of the rotor are opposite to *N-S* poles of the stator, the resulting magnetic repulsion produces a violent mechanical shock. The motor will immediately slow down and circuit breakers will trip, causing the motor to come to a halt.

In practice, starters for synchronous motors are designed to detect the precise moment when excitation should be applied. The rotor then pulls automatically and smoothly into step with the rotating field.

Q. 3. *How does a synchronous motor behave while in operation?*

Ans. A synchronous motor operates as under :

(*i*) When the synchronous motor is on no load, the rotor poles are directly opposite to the opposite-polarity poles of the stator and their axes almost coincide.

(*ii*) When we apply a mechanical load, the rotor poles fall slightly behind the stator poles, while continuing to run at synchronous speed. The mechanical angle α between the poles increases as we increase the load. Nevertheless, the magnetic attraction keeps the rotor locked to the revolving field (*i.e.*, motor continues to turn at synchronous speed) and motor develops more power torque. This is expected because a large angle corresponds to a bigger mechanical load and the increased power can only come from the a.c. source (*i.e.* stator winding).

Q. 4. *How does the behaviour of a synchronous motor differ from that of a 3-phase induction motor ?*

Ans. A synchronous motor cannot start by itself; consequently the rotor is equipped with a squirrel-cage winding (*i.e.*, damper winding) so that it can start up as a 3-phase induction motor. Once the synchronous motor runs at synchronous speed, no voltage is induced in the squirrel-cage winding. Consequently, the behaviour of a synchronous motor is entirely different from that of a 3-phase induction motor. Basically, a synchronous motor rotates because of the magnetic attraction between the poles of the rotor and the opposite poles of the stator.

Q. 5. *What is the origin of the name synchronous motor ?*

Ans. A synchronous motor moves in synchronism (*i.e.*, in time) with the rotating magnetic field. In other words, the motor runs at the speed of the rotating flux *i.e.*, synchronous speed. Hence the name synchronous motor.

Q. 6. *Why is the field winding of a synchronous motor shorted through a resistor at starting ?*

Ans. If voltage is impressed on the armature winding when the motor is at standstill with field winding unexcited, the revolving field produced by stator currents will cut across the field winding, thereby inducing a high voltage in the field winding. This induced voltage would be dangerous and would often result in the breakdown of insulation of field winding. Therefore, in starting a synchronous motor, the field winding is shorted through a suitable resistance. Consequently, the induced voltage is distributed throughout the whole winding and no part is subjected to high voltage. The resistance is removed and excitation applied to rotor when the motor attains 90-95% of synchronous speed.

Note : Under normal operating conditions, there is no relative motion between the stator flux and rotor flux. Consequently, no e.m.f. is induced in the field winding.

Q. 7. *What is the procedure of starting a synchronous motor ?*

Ans. The starting procedure of a synchronous motor is as under :

(*i*) Short the field winding through a suitable resistance.

(*ii*) Apply 3-phase supply to the stator winding. Since the rotor is equipped with squirrel-cage winding, the motor will start as a 3-phase induction motor.

(*iii*) When the motor approaches synchronous speed, remove the resistance and excite the field winding from d.c. supply. The rotor and stator poles will be locked, forcing the motor to run at synchronous speed.

(*iv*) Adjust the field current to give the desired power factor of operation of the motor.

Q. 8. *What happens when the mechanical load on a synchronous motor exceeds its pull-out torque?*

Ans. If the mechanical load exceeds the pull-out torque of a synchronous motor, the rotor is pulled out of synchronism *i.e.*, magnetic locking between rotor and stator poles ceases. Consequently, the motor comes to a halt. A synchronous motor that pulls out of synchronism creates a major disturbance on the line and circuit breakers immediately trip. This protects the motor because both damper and stator windings overheat rapidly when the machine ceases to run at synchronous speed.

Q. 9. *Show that synchronous motor is a variable power factor motor.*

Ans. The resultant flux ϕ in a synchronous motor is fixed in value because the stator voltage is fixed. The m.m.f. needed to create constant flux ϕ may be produced either by the stator or the rotor or by both.

(*i*) If the rotor excitation is zero, all the flux has to be produced by the stator. The stator must then absorb considerable reactive power from the 3-phase line. Consequently, the motor will have a low lagging power factor. If we excite the rotor with direct current, the rotor m.m.f. helps to produce part of the flux ϕ. Now the motor will draw less reactive power from the line and the motor will operate at improved lagging power factor. This condition is called *under-excitation.*

(*ii*) If we gradually increase the excitation, the rotor will eventually produce all the flux ϕ required by itself. The stator then draws no reactive power and the motor power factor becomes unity. This value of field excitation is called *normal excitation.*

(*iii*) If we excite the synchronous motor above normal excitation, the resultant flux is more than the required value ϕ. Consequently, the induced e.m.f. in the armature winding exceeds the applied stator voltage. The stator must now deliver reactive power to the line to restore the resultant flux to the original value ϕ. In other words, the motor now operates at leading power factor. This condition is called *over-excitation.*

Thus with the adjustment of field excitation, it is possible to make the synchronous motor to (*a*) draw reactive power from the line *i.e.*, operate at lagging power factor (*b*) draw no reactive power *i.e.* operate at unity p.f. and (*c*) deliver reactive power to the line *i.e.*, operate at leading p.f. Thus synchronous motor is a variable p.f. motor.

Q. 10. *What is a synchronous capacitor ?*

Ans. An over-excited synchronous motor running on no load is known as a synchronous capacitor. Under such conditions, the motor will take a leading current, though it is performing no shaft work. Most of the loads on the electric system are inductive in nature. This results in an overall lagging power factor which may often be of quite low value. Low lagging p.f. results in uneconomical use of copper of the system and poor voltage regulation. To improve the power factor, over-excited unloaded synchronous motors are connected to the system. The leading kVAR of the synchronous motor partly cancels the large lagging kVAR of the system, thus improving the p.f.

Q. 11. *How are large synchronous motors stopped?*

Ans. Due to inertia of the rotor and load, large synchronous motors may take several minutes to stop after they are disconnected from the line. In order to reduce the stopping time, we use the following braking methods :

(*i*) Maintain full d.c. excitation with armature in short circuit.

(*ii*) Maintain full d.c. excitation with armature connected to three external resistors.

(*iii*) Apply mechanical braking.

In methods (*i*) and (*ii*), the motor slows down because it functions as an alternator, dissipating energy in the resistive elements of the armature circuit. Mechanical braking is generally applied after the motor has reached half speed or less to prevent undue wear and tear of shoes.

Q. 12. *How does a synchronous motor differ from a d.c. motor ?*

Ans. A synchronous motor differs from a d.c. motor in the following respects :

(*i*) A synchronous motor requires two sources of supply *viz.,* a 3- phase source to supply power to armature and a d.c. source to excite the field. However, a d.c. motor requires a single d.c. source to supply the field exciting current as well as armature current.

(*ii*) In a synchronous motor, the stator current is directly proportional to the *phase difference* between the applied voltage and counter e.m.f. produced in the armature winding. However, the counter e.m.f. in a d.c. motor is approximately equal to the applied voltage; the small difference being necessary to drive current through the resistance of the armature winding.

(*iii*) A synchronous motor rotates because of the magnetic attraction between the poles of the rotor and the opposite poles of the stator. However, a d.c. motor operates on the principle that a mechanical force acts on a current carrying conductor placed in a magnetic field.

(*iv*) In a synchronous motor, the speed is constant (*i.e.*, synchronous speed) and a change in d.c. field excitation produces a change in counter e.m.f. in the armature winding. However, in a d.c. motor, the counter e.m.f. is practically constant and a change in field excitation produces a change in speed.

Q. 13. *What are the causes of faulty starting of a synchronous motor ?*

Ans. It can be due to the following causes :

(*i*) Voltage may be too low.

(*ii*) There may be open-circuit in one phase due to which motor may heat up.

(*iii*) Static friction may be large. This may be either due to high belt tension or too tight bearings.

(*iv*) Stator windings may be incorrectly connected.

(*v*) Field excitation may be too strong.

Q. 14. *What could be the reasons if a synchronous motor fails to start ?*

Ans. It may be due to the following reasons :

(*i*) Voltage may be too low.

(*ii*) Some faulty connection in auxiliary apparatus.

(*iii*) Too much starting load.

(*iv*) Open-circuit in one phase or short-circuit.

(*v*) Field excitation may be excessive.

Q. 15. *Under which conditions a synchronous motor will fail to pull into step ?*

Ans. This may be due to the following reasons :

(*i*) No field excitation.

(*ii*) Excessive load.

(*iii*) Excessive load inertia.

MULTIPLE-CHOICE QUESTIONS

1. The rotor of a synchronous motor is ....
   (*i*) salient-pole type
   (*ii*) nonsalient-pole type
   (*iii*) identical to that of a d.c. motor
   (*iv*) none of the above
2. The full-load slip of a synchronous motor is ....
   (*i*) 5% (*ii*) 1 %
   (*iii*) 2% (*iv*) zero
3. If the supply frequency of synchronous motor is 60 cycles/second, then rotor must revolve at ....
   (*i*) 25 cycles/second
   (*ii*) 60 cycles/second
   (*iii*) 100 cycles/second
   (*iv*) none of the above
4. When the synchronous motor runs at synchronous speed, the voltage induced in the damper winding is ....
   (*i*) maximum (*ii*) minimum
   (*iii*) zero
   (*iv*) none of the above
5. A synchronous motor runs at only one speed (*i.e.*, synchronous speed) because it ....
   (*i*) has no losses
   (*ii*) is a doubly fed machine
   (*iii*) has a damper winding
   (*iv*) none of the above
6. The speed of a synchronous motor can be changed by varying...............
   (*i*) mechanical load (*ii*) field excitation
   (*iii*) supply frequency
   (*iv*) none of the above
7. If the mechanical angle α between the stator and rotor poles is increased, then, stator current...............
   (*i*) is increased (*ii*) is decreased
   (*iii*) remains unchanged
   (*iv*) none of the above
8. The increased load demand in a synchronous motor is met by ....
   (*i*) reduction in speed
   (*ii*) increase in speed
   (*iii*) decrease in stator current
   (*iv*) relative shift between stator and rotor poles
9. The pull-out torque of a practical synchronous motor will occur when the torque angle is about...............
   (*i*) 0° (*ii*) 30°
   (*iii*) 45° (*iv*) 75°
10. For a given load, the normal field excitation of synchronous motor is that which gives...............power factor.
    (*i*) unity (*ii*) 0.8 lagging
    (*iii*) 0.8 leading.
    (*iv*) none of the above
11. For a given load, the armature current of a synchronous motor will be minimum for...............power factor.
    (*i*) 0.8 lagging (*ii*) unity
    (*iii*) 0.8 leading
    (*iv*) none of the above
12. An over-excited synchronous motor behaves as...............
    (*i*) a resistor (*ii*) an inductor
    (*iii*) a capacitor
    (*iv*) none of the above
13. An under-excited synchronous motor behaves as...............
    (*i*) an inductor (*ii*) a capacitor
    (*iii*) a resistor
    (*iv*) none of the above
14. At full-load, the rotor poles of a synchronous motor are displaced by a mechanical angle of 1° from their no-load position. If the machine has 40 poles, then torque angle is...............
    (*i*) 40° electrical (*ii*) 20° electrical
    (*iii*) 10° electrical
    (*iv*) none of the above
15. When the pull-out torque occurs in a synchronous motor, the poles of the rotor are...............
    (*i*) mid-way between *N* and *S* poles of

stator

(*ii*) coincident with stator poles

(*iii*) 45º (electrical)

(*iv*) none of the above

16. A synchronous capacitor is an over-excited motor running at...............

(*i*) full-load (*ii*) half full-load

(*iii*) no-load

(*iv*) none of the above

17. Leading power factor synchronous motors cost...............the unity p.f. motors.

(*i*) less than (*ii*) more than

(*iii*) about the same as

(*iv*) none of the above

18. When a unity p.f. synchronous motor is operated at leading p.f., it results in...........

(*i*) more than rated horsepower output

(*ii*) more than rated stator current

(*iii*) less than rated horsepower output

(*iv*) none of the above

19. The power factor correction capability of a loaded synchronous motor............... with the increase in mechanical load.

(*i*) decreases (*ii*) increases

(*iii*) remains unchanged

(*iv*) none of the above

20. A synchronous motor is cheaper than 3-phase induction motor for...............

(*i*) high speed, high output

(*ii*) high speed, low output

(*iii*) low speed, high output

(*iv*) none of the above

ANSWERS TO MULTIPLE-CHOICE QUESTIONS

1. (*i*)	**2.** (*iv*)	**3.** (*ii*)	**4.** (*iii*)	**5.** (*ii*)
6. (*iii*)	**7.** (*i*)	**8.** (*iv*)	**9.** (*iv*)	**10.** (*i*)
11. (*ii*)	**12.** (*iii*)	**13.** (*i*)	**14.** (*ii*)	**15.** (*i*)
16. (*iii*)	**17.** (*ii*)	**18.** (*ii*)	**19.** (*i*)	**20.** (*iii*)

HINTS TO SELECTED MULTIPLE-CHOICE QUESTIONS

2. A synchronous motor runs at one speed only *i.e.*, synchronous speed. Consequently , slip is zero.

3. The primary requirement is that frequency of rotation of the rotor should be such *that stator and rotor fields are stationary with respect to each other.* For this to happen, obviously, the rotor must revolve at 60 cycles/second.

4. When the motor runs at synchronous speed, there is no relative motion between rotor and revolving flux produced by stator currents. Hence e.m.f. in the damper winding placed on the rotor is zero.

5. The significant and distinguishing feature of synchronous motors in contrast to induction motors is that they are doubly fed. Electrical energy is supplied both to the field and the armature windings. When this is done, torque can only be developed at one speed, the *synchronous speed.* At any other speed, the torque is zero. In fact, it is this characteristic (*i.e.*, doubly fed) which enables the synchronous motor to develop non-zero torque at only one speed (*i.e.*, synchronous speed) and hence the name synchronous motor.

6. $f = \frac{N_s P}{120} \qquad \therefore \; N_s = \frac{120 f}{P}$

For a given machine, the value of P is fixed. Hence, the speed of a synchronous motor can only be changed by varying the supply frequency.

7. When synchronous motor is on no-load, the axes of the rotor and stator poles nearly coincide. As the motor is loaded, the rotor poles fall slightly behind the stator poles, while continuing to turn at synchronous speed. The increased power must come from the a.c. source (*i.e.,* stator winding). Hence, stator current increases.

9. The mechanical power developed by a synchronous motor per phase is given by :

$$P_m = \frac{V E_b}{X_s} \sin \delta$$

where P_m = mechanical power of motor per phase
E_b = phase voltage induced by excitation
V = phase voltage of a.c. source
X_s = synchronous reactance per phase
δ = torque angle (in electrical degrees between E_b and V)

The maximum torque that motor can develop is called the pull-out torque and will occur at $\delta = 90^o$. This is true for motors having smooth rotors. Most synchronous motors have salient poles; in this case the pull-out torque occurs at a torque angle of about 75° (electrical).

10. The resultant flux ϕ in a synchronous motor is fixed in value as the applied stator voltage is constant. Since synchronous motor is a doubly excited machine, both d.c. and a.c. sources may co-operate to produce this fixed resultant flux ϕ. For a given mechanical load, if the entire resultant flux is provided by d.c. excitation alone, the stator will absorb no reactive power (as it is not contributing to the resultant flux). Hence the p.f. of the motor will be unity.

11. Since at unity power factor the motor draws no reactive power, the stator current will be minimum.

12. An over-excited synchronous motor delivers reactive power to the 3-phase line. The motor then behaves like a source of power, as if it were a capacitor.

13. In an under-excited synchronous motor, the resultant fixed flux is (*i*) partly supplied by stator and (*ii*) partly by the rotor. As a result, motor absorbs reactive power from the 3-phase line. Consequently, it behaves as an inductor.

14. There is a precise relationship among the mechanical angle (α), torque angle (δ) and the number of poles (P).

$$\delta = \frac{P\alpha}{2} = \frac{40 \times 1^o}{2} = 20^o$$

15. Theoretically, pull-out torque occurs at $\delta = 90^o$. Clearly, the poles of the rotor should be midway between N and S poles of the stator.

16. An over-excited loaded synchronous motor supplies reactive power to the 3-phase line as well as absorbs active power (kW). But an unloaded over-excited synchronous motor performs only one function *i.e.,* delivers reactive power to the 3-phase line, like a capacitor. Hence, the name synchronous capacitor.

17. Since they must have large current carrying capacities.

18. This is because more than rated stator current must flow to carry the rated horsepower load at a reduced power factor.

19. If a synchronous motor is used to drive a load, over-exciting the motor causes it to operate with a leading power factor. The amount of power factor correction is limited by the rated current of the stator winding. Therefore, the more mechanical load a synchronous motor drives, the more its power factor correction ability is reduced.

20. Under such conditions, induction motors must use large amounts of iron in order that air-gap flux density does not exceed 0·7 Wb/m^2. In the synchronous machine, on the other hand, a value twice of this figure is permissible because of the separate excitation.

14

Special-Purpose Electric Machines

INTRODUCTION

In this chapter, we shall study some special-purpose electric machines. Although all electric machines have the same basic principle of operation, special-purpose machines have some features that distinguish them from conventional machines. For example, a stepper motor rotates by a specific number of degrees (*e.g.* 2°, 2.5°, 5° or 7.5°) in response to an input electrical signal and is widely used in digital control systems. Similarly, brushless d.c. motors eliminate mechanical commutator and brushes. Instead, these motors use electronic commutators. For applications that require constant-speed drives, brushless d.c. motors offer excellent characteristics. It is not possible to deal with all types of special-purpose machines in one chapter. However, we shall introduce the basic operating principles of some special-purpose machines that are being used extensively in homes, recreational and industrial applications.

14.1 STEPPER MOTOR

Stepper motors are also known as *stepping motors* or *step motors.* A stepper motor is an electromagnetic motor (*i.e.* it converts electrical energy into mechanical energy) that rotates by a specific number of degrees in response to an input electrical signal. Typical step sizes are 2°, 2.5°, 7.5° and 15° for each electrical pulse. Note that there is no continuous energy conversion (electrical to mechanical) so that the rotor does not rotate continuously as in a conventional electric motor. The stepper motor converts electrical pulses into proportionate mechanical movement. Each revolution of the stepper motor is made up of a series of definite individual steps. A step is defined as the angular rotation (in degrees) of the motor each time it receives the electrical pulse. Such a step control is required in many applicatons. Fig. 14.1 illustrates a simple application for a stepper motor. Each time the controller receives an input electrical signal, the paper is driven to a certain incremental distance. Stepper motors are relatively cheap and simple in construction and can be made to rotate in steps in either direction. These motors are excellent candidates for such applications as electric typewriters, control of floppy disc drives, numerical control of machine tools etc. The two most popular types of stepper motors are :

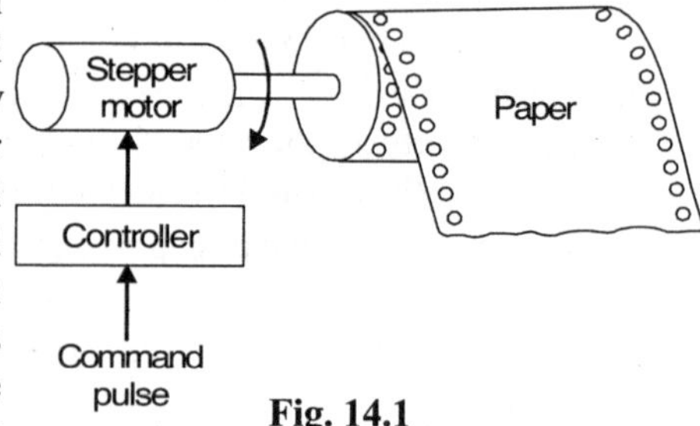

Fig. 14.1

(*i*) Permanent-magnet (PM) stepper motor

(*ii*) Variable-reluctance (VR) stepper motor

The stator of a stepper motor (PM or VR motor) carries stator windings which are energised from a d.c. source to create two or more stator poles. The stator poles are also called stator teeth. The rotor of a stepper motor may be a permanent magnet (in case of PM motor) or a soft-iron material (in case of VR motor). The rotor may have two or more than two poles. The rotor poles are also called rotor teeth.

The stator coils are energised in groups referred to as *phases*. The stator windings may be 2-phase, 3-phase or 4-phase windings. The phase windings are brought out to terminals for d.c. excitation. We shall now give introductory remarks about these motors.

(*i*) **PM Stepper Motor.** Fig. 14.2 shows a 2-pole, single phase permanent magnet (PM) stepper motor. When the stator is energised, the *excitation torque* acts on the rotor (permanent magnet). The rotor will move to a position where the excitation torque is zero *i.e.* the rotor will be aligned (parallel) with the stator field.

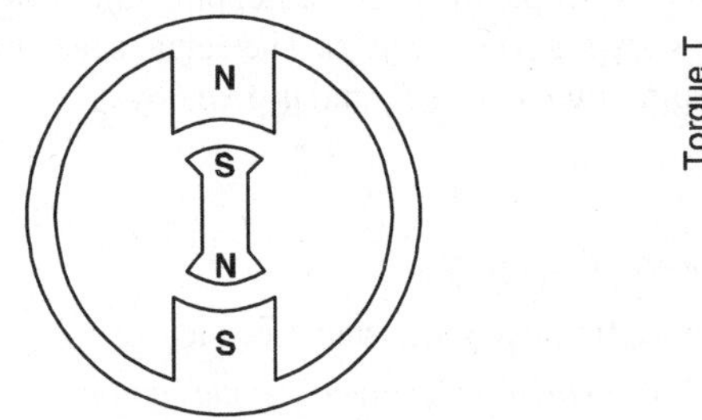

Fig. 14.2 **Fig. 14.3**

Fig. 14.3 shows how excitation torque varies with the rotor position for a PM rotor. Note that maximum torque is developed when the rotor is displaced from the stator field by either 90° or 270°. However, the torque is zero when the rotor is aligned (parallel) with the stator field.

(*iii*) **VR Stepper Motor.** Fig. 14.4 shows a 2-pole, single phase variable-reluctance (VR) stepper motor. When the stator is energised, *reluctance torque* acts on the rotor (soft-iron material). The rotor will move to a position where reluctance is *minimum* and air-gap flux is *maximum. This means that rotor teeth will align with the energised stator poles.*

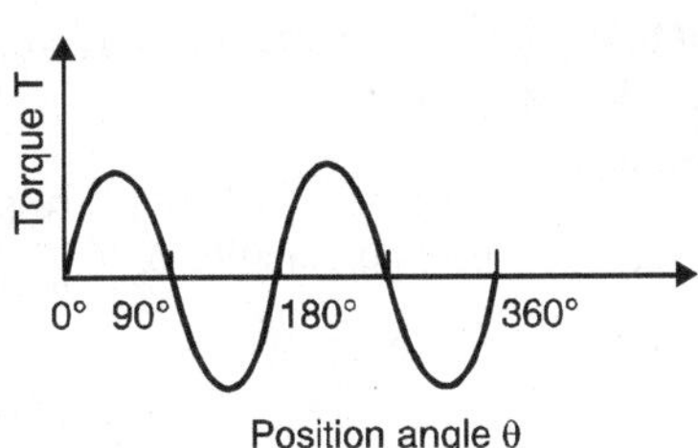

Fig. 14.4 **Fig. 14.5**

Fig. 14.5 shows how reluctance torque varies with the rotor position for a VR soft-iron rotor. With the rotor at 0° or 90°, no torque is developed. Maximum torque is developed at 45° and 135° which is the position where reluctance torque forces the rotor to move to position of minimum reluctance.

Step angle. The angle through which the motor shaft rotates for each command pulse is called *step angle.* It can be shown that for any PM or VR stepper motor, the step angle can be found from the following two relations :

(*i*) In terms of stator poles (N_s) and rotor poles (N_r), the step angle (α) is given by :

$$\text{Step angle, } \alpha = \frac{N_s - N_r}{N_s \times N_r} \times 360°$$

where α = Step angle in degrees

N_s = Number of stator poles (or teeth)

N_r = Number of rotor poles (or teeth)

(*ii*) In terms of stator phases (m) and rotor poles (N_r), the step angle is given by :

$$\text{Step angle, } \alpha = \frac{360°}{m N_r}$$

where α = Step angle in degrees

m = Number of stator phases

N_r = Number of rotor poles (or teeth)

Stepping rate. An important specification of a stepper motor is the stepping rate. The number of steps per second is known as *stepping rate* or *stepping frequency* (f). The actual speed of a stepper motor depends on the step angle (α) and stepping frequency (f) and is given by :

$$\text{Speed of stepper motor, } N = \frac{\alpha f}{6}$$

where N = Motor speed in r.p.m.

f = Stepping frequency *i.e.* steps/second

Example 14.1. *Determine the step angle of a variable-reluctance stepper motor with 12 teeth in the stator and 8 rotor teeth.*

Solution.

Number of stator teeth, $N_s = 12$

Number of rotor teeth, $N_r = 8$

$$\therefore \quad \text{Step angle, } \alpha = \frac{N_s - N_r}{N_s N_r} \times 360°$$

$$= \frac{(12-8)}{12 \times 8} \times 360° = \mathbf{15°/step}$$

Example 14.2. *Calculate the stepping angle for (i) a 3-phase, 16-tooth rotor VR motor (ii) a 3-phase, 24-pole PM motor.*

Solution.

$$\text{Stepping angle, } \alpha = \frac{360°}{m N_r}$$

(*i*) Here $m = 3$; $N_r = 16$

$$\therefore \quad \alpha = \frac{360°}{3 \times 16} = \mathbf{7.5°/step}$$

(*ii*) Here $m = 3$; $N_r = 24$

$$\therefore \quad \alpha = \frac{360°}{3 \times 24} = \mathbf{5°/step}$$

Example 14.3. *A stepper motor with a step angle of 15° has a stepping frequency of 300 steps/second. What is the motor speed ?*

Solution.

$$\text{Motor speed, } N = \frac{\alpha f}{6}$$

Here $\alpha = 15°$; $f = 300$ steps/second

$$\therefore \quad N = \frac{15° \times 300}{6} = \mathbf{750 \text{ r.p.m.}}$$

Example 14.4. *A stepper motor has a step angle of 10° and is required to rotate at 200 r.p.m. Determine the pulse rate (steps per second) for this motor.*

Solution.

$$\text{Motor speed, } N = \frac{\alpha f}{6}$$

$$\therefore \quad \text{Pulse rate, } f = \frac{6 \times N}{\alpha} = \frac{6 \times 200}{10} = \mathbf{120 \text{ steps/second}}$$

TUTORIAL PROBLEMS

1. Calculate the step angle of a VR stepper motor with 8 teeth in the stator and 6 rotor teeth. **[15°/step]**
2. A stepper motor has a step angle of 15° and is required to rotate at 400 r.p.m. Determine the pulse rate for this motor. **[160 steps/second]**
3. A stepper motor with a step angle of 10° has a pulse rate of 350 steps/second. What is the motor speed ? **[583.3 r.p.m.]**

14.2 PERMANENT-MAGNET (PM) STEPPER MOTOR

A permanent-magnet (PM) stepper motor is a popular type of stepper motor. It operates on the principle of interaction between a permanent-magnet rotor and an electromagnetic field.

Construction : The stator of a PM stepper motor is composed of steel laminations and carries stator windings. The stator windings are energised from a d.c. source to create two or more stator poles. The rotor of the motor is a permanent magnet made of high-retentivity steel alloy. The rotor has even number of poles. Fig. 14.6 shows a two-phase, 2-pole PM stepper motor. The motor has two rotor poles. The stator coils are grouped to form 2-phase winding *i.e.* phase-*A* winding and phase-*B* winding. The phase winding terminals are brought out for d.c. excitation.

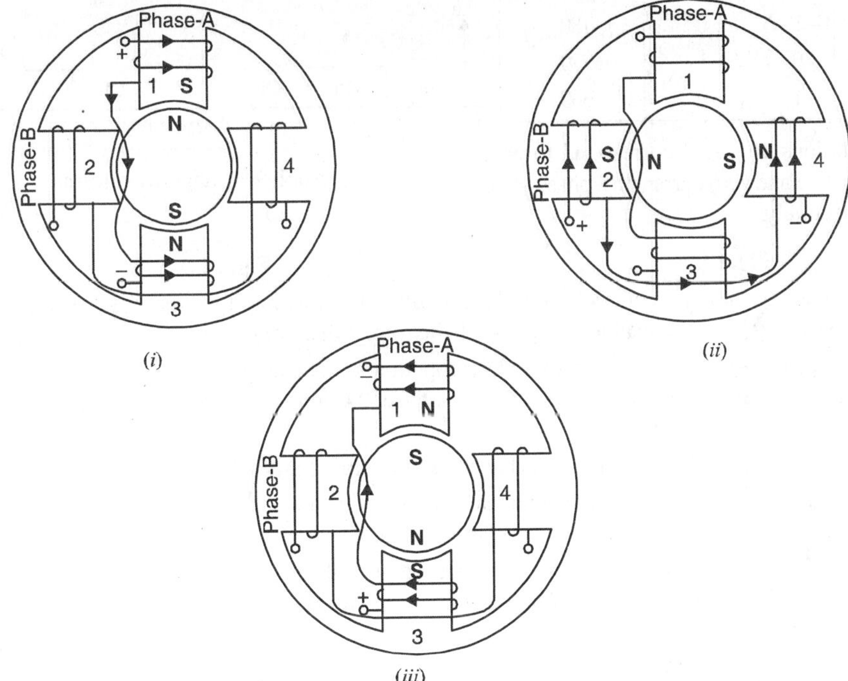

Fig. 14.6

Operation. For this PM stepper motor, the number of rotor poles, $N_r = 2$ and number of phases, $m = 2$.

$$\therefore \qquad \text{Step angle, } \alpha = \frac{360°}{m N_r} = \frac{360°}{2 \times 2} = 90°/\text{step}$$

(*i*) When *only* phase-*A* winding is excited by a constant current as shown in Fig. 14.6 (*i*), stator tooth 1 becomes south pole. This makes the north pole of the PM rotor to align (parallel) with the south pole (stator tooth 1) of the stator. The rotor will remain locked in this position as long as phrase-*A* winding remains energised. The first row of truth table in Fig. 14.7 shows that only phase-A winding is excited while phase-B winding is unexcited. Under this condition, step angle $\alpha = 0°$. The applied voltage waveforms in Fig. 14.8 also tally with facts shown in the truth table.

(*ii*) If phase-A winding is de-energised and phase-B winding is energised as shown in Fig. 14.6 (*ii*), stator tooth 2 becomes south pole. As a result, the north pole of the PM rotor aligns (parallel) with the south pole (stator tooth 2) of the stator. Thus the rotor has displaced 90° in the anticlockwise direction. The reader may see the truth table and applied voltage waveform for this switching sequence.

(*iii*) If phase-B winding is de-energised and phase-A is excited with reverse current [*i.e.* current in it is opposite to the case in (*i*) above], the rotor will further rotate 90° in anticlockwise direction as shown in Fig. 14.6 (*iii*). Now the north pole of PM motor aligns with the stator tooth 3.

Truth Table

Cycle	Phase		Position δ°
	A	B	
+	1	0	0
	0	1	90
–	– 1	0	180
	0	– 1	270
+	1	0	360

The notations 1, – 1, 0 correspond to positive, negative and zero current in a phase winding respectively.

Fig. 14.7

V_A: On, Off, On, Off, On (t)

V_B: Off, On, Off, On, Off, On (t)

Applied voltage waveforms

Fig. 14.8

(*iv*) So far the rotor has completed one-half revolution. However, if we continue the appropriate switching [See Truth table in Fig. 14.7], the rotor will complete one revolution in 90° steps.

We can change the step angle (α) of a PM stepper motor by changing the number of *rotor poles (N_r) and the number of phases (m). Thus for a 3-phase, 24-pole PM stepper motor, the step angle $\alpha = 360°/mN_r = 360°/3 \times 24 = 5°$/step.

Limitations : The PM stepper motors have the following drawbacks :

(*i*) It is difficult to make a small permanent magnet rotor with a large number of poles. Therefore, PM stepper motors are restricted to large step angles in the range of 30° to 90°.

(*ii*) The PM stepper motors have high inertia because of the permanent-magnet rotor. Therefore, these motors have slow acceleration. The maximum step rate (stepping frequency) is 300 steps/second.

(*iii*) The PM stepper motors have high rotational speed because of large stepping angle. Therefore, motor torque for a given output power is low.

14.3 VARIABLE-RELUCTANCE (VR) STEPPER MOTOR

The Variable-Reluctance (VR) stepper motor operates on the same principle as the reluctance motor. That is, when a piece of ferromagnetic material, free to rotate, is placed in a magnetic field, torque acts on the material to bring it to the position of minimum reluctance to the path of magnetic flux.

Construction : The stator construction of a VR stepper motor is the same as that of a PM stepper motor. The stator phase windings are wound on each stator tooth. The rotor is made of soft steel with

* We can make the permanent-magnet rotor multipolar.

teeth and slots. Fig. 14.9 shows the basic Variable-Reluctance stepper motor. In this circuit, the rotor is shown with fewer teeth than the stator. This ensures that only one set of stator and rotor teeth will align at any given instant. In Fig. 14.9, the stator has six teeth and the rotor has four teeth. The stator has three phases — *A*, *B* and *C* with teeth 1 and 4, 3 and 6 and 2 and 5 respectively. For this VR stepper motor,

$$\text{Step angle, } \alpha = \frac{N_s - N_r}{N_s N_r} \times 360° = \frac{6-4}{6 \times 4} \times 360° = 30°/\text{step}$$

Therefore, the rotor will turn 30° each time a pulse is applied.

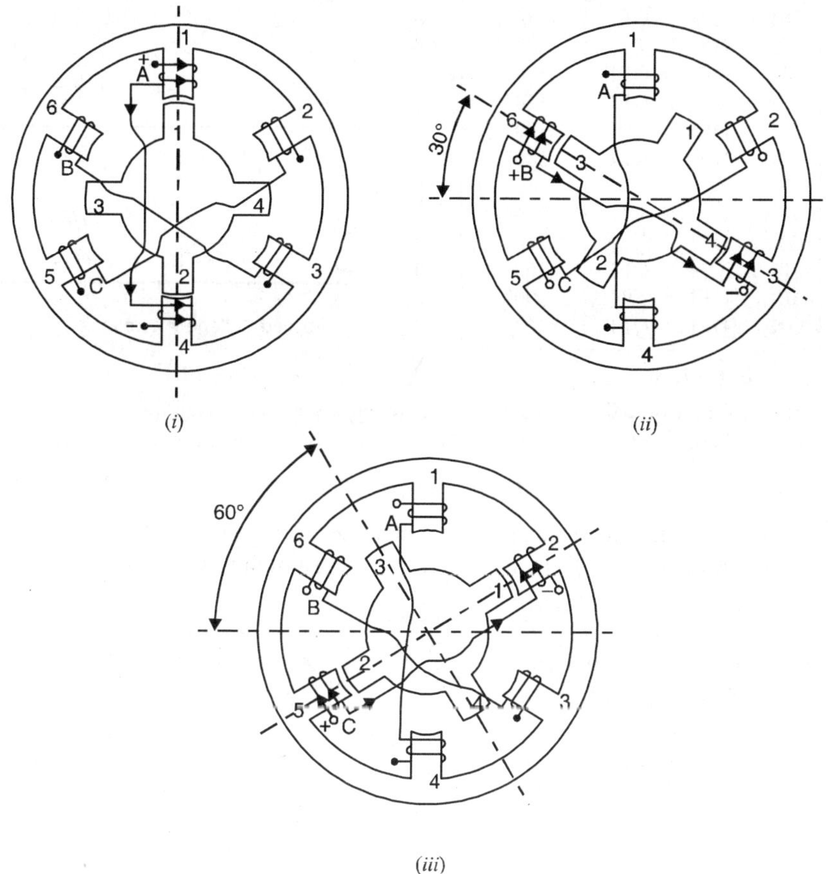

Fig. 14.9

Operation : When the phase winding is energised, the rotor teeth will align with the energised stator poles.

(*i*) Fig. 14.9 (*i*) shows the position of the rotor when phase *A* is energised with a constant current. *A*s long as phase *A* is energised, the rotor will be held stationary. Note that in this condition, the rotor teeth 1 and 2 are aligned with the *energised* stator teeth 1 and 4. The step angle $\alpha = 0°$. Also refer to truth table (Fig. 14.10) and applied voltage waveform (Fig. 14.11).

(*ii*) When phase *A* is switched off and phase *B* is energised, the rotor will turn 30° *clockwise* so that the rotor teeth 3 and 4 align with the energised stator teeth 6 and 3.

(*iii*) The effect of de-energising phase *B* and energising phase *C* is shown in Fig. 14.9 (*iii*). In this circuit, the rotor has further moved 30° clockwise so that rotor teeth 1 and 2 align with energised stator teeth 2 and 5.

(*iv*) After the rotor has displaced 60° (clockwise) from its starting point, the step sequence has completed one cycle. The truth table in Fig. 14.10 shows the switching sequence to complete a full 360° rotation for the motor with six stator poles and four rotor poles.

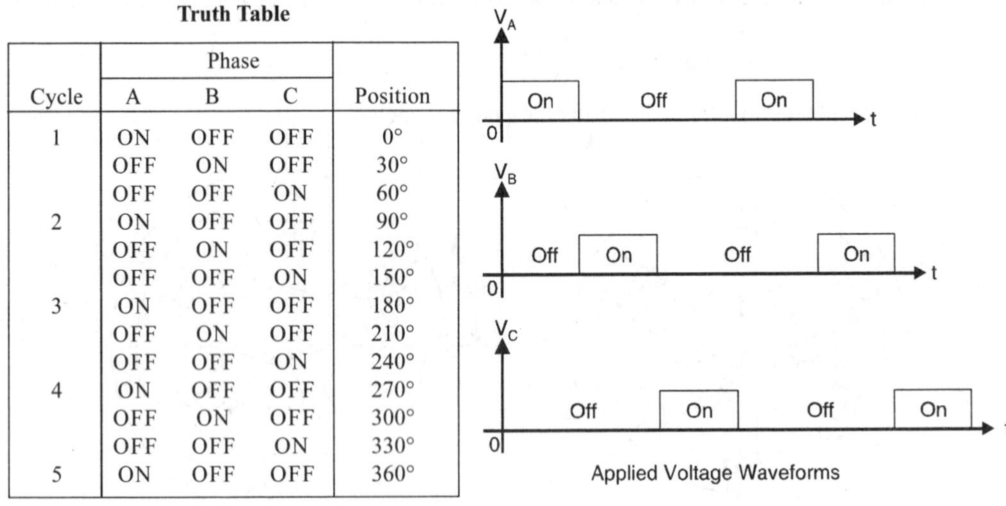

Truth Table

Cycle	Phase A	Phase B	Phase C	Position
1	ON	OFF	OFF	0°
	OFF	ON	OFF	30°
	OFF	OFF	ON	60°
2	ON	OFF	OFF	90°
	OFF	ON	OFF	120°
	OFF	OFF	ON	150°
3	ON	OFF	OFF	180°
	OFF	ON	OFF	210°
	OFF	OFF	ON	240°
4	ON	OFF	OFF	270°
	OFF	ON	OFF	300°
	OFF	OFF	ON	330°
5	ON	OFF	OFF	360°

Fig. 14.10

Fig. 14.11

The direction of rotation will be reversed if the switching sequence is in the order of *A*, *C* and *B*. For this particular motor, applied voltage must have at least five cycles for one revolution.

14.4 HYBRID STEPPER MOTOR

As the name implies, the *hybrid stepper motor combines the features of the PM and the VR stepper motors. The torque developed by this motor is greater than that of the PM or VR types.

Construction. Fig. 14.12 shows the basic construction of a hybrid stepper motor. The stator construction is similar to that of a VR or a PM motor. However, the rotor construction combines the design of the rotors of a VR and a PM stepper motor. The rotor of a hybrid stepper motor consists of two identical stacks of soft iron as well as an axially magnetised round permanent magnet. Soft iron stacks are attached to the north and south poles of the permanent magnet as shown in Fig. 14.12

The rotor teeth are machined on the soft iron stacks. Thus the rotor teeth on one end become the north pole and those at the other end become the south pole.

This rotor teeth of both north and south poles are displaced in angle for the proper alignment of the rotor pole with that of the stator as shown in Fig. 14.12.

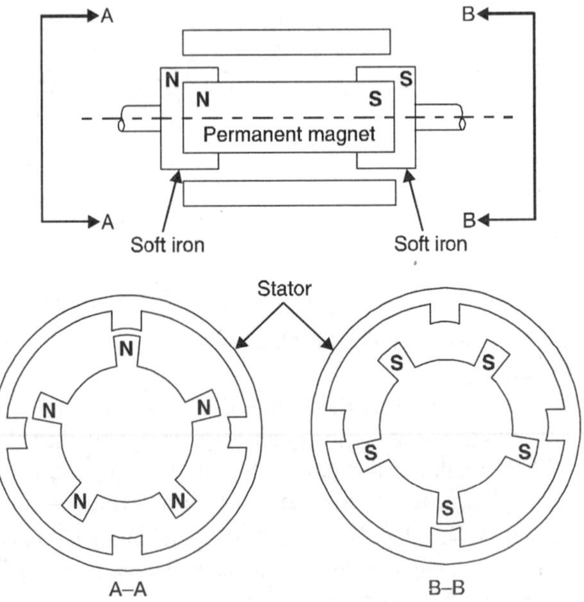

Fig. 14.12. Various views of Hybrid Stepper Motor.

* hybrid means mixed.

Operation. The operating mode of the hybrid stepper motor is very similar to that of a PM or VR stepper motor. The phase windings are energised in proper sequence and the rotor rotates in steps. Unlike the PM or VR stepper motors, the step angle of a hybrid stepper motor is independent of the number of stator phases and depends only on the number of rotor teeth (N_r). It is given by :

$$\text{Step angle, } \alpha = \frac{90°}{N_r} \text{ in degrees}$$

For a hybrid stepper motor having 5 rotor teeth, the step angle $\alpha = 90°/N_r = 90°/5 = 18°$/step. It means that for each change of stator excitation, the rotor will turn by a step of 18°.

It may be noted that a hybrid stepper motor operates under the combined principles of the permanent magnet and variable reluctance stepper motors. Therefore, the hybrid motor develops both excitation torque and reluctance torque. Consequently, the resultant torque developed by the hybrid stepper motor is greater than that of the PM or VR stepper motor.

Example 14.5. *A hybrid stepper motor has 50 variable-reluctance rotor teeth. Calculate the step angle in degrees.*

Solution.

$$\text{Step angle, } \alpha = \frac{90°}{N_r} = \frac{90°}{50} = \mathbf{1.8°/step}$$

Example 14.6. *A hybrid stepper motor has 36 variable reluctance rotor teeth and 4 stator phases. Calculate the step angle.*

Solution.

$$\text{Step angle, } \alpha = \frac{90°}{N_r} = \frac{90°}{36} = \mathbf{2.5°/step}$$

14.5 SERVOMECHANISM

The automatic control of a physical quantity (*e.g.* angular displacement or speed of a motor) is called *servomechanism.* The *automatic control system* or *servo system* has become necessary because human operator suffers from many disadvantages including fatigue, slow reaction time (about 0.3s) and limited power. The automatic control system (or servo system) involves the feedback of information about the physical quantity to be controlled in such a way that if the controlled quantity differs from the desired value, an error is observed. This error is transmitted to the control system which operates to reduce the error to zero. This type of control is called *closed-loop control.*

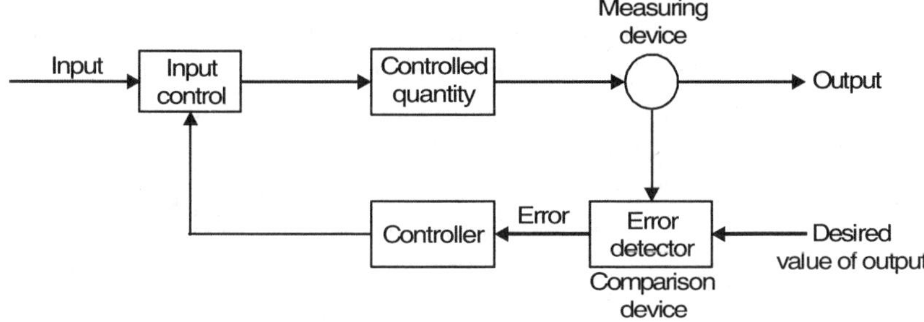

Fig. 14.13

The automatic control system (or servo system) can be represented by a block diagram shown in Fig. 14.13. The quantity to be controlled may be angular position or speed of a shaft or any other physical quantity. The actual output of the physical quantity is measured *electrically and compared

* By using a transducer.

with a desired value in the comparison device called *error detector*. The magnitude and sense of any difference between the desired and actual values of the output is an error signal. The error signal is fed to the controller (which normally incorporates a power amplifer). The controller actuates the correcting devices to reduce the error to zero.

Thus the essential elements of the automatic control system (or servo system) are :

(*i*) a measuring device.

(*ii*) a comparison device or error detector.

(*iii*) controller which normally incorporates power amplification.

(*iv*) a correcting device.

14.6 D.C. SERVOMOTORS

A d.c. motor that is used in servo mechanism is called a d.c. servomotor. The three main types of d.c. servomotors are :

1. Field-controlled d.c. servomotor
2. Armature-controlled d.c. servomotor
3. Permanent-magnet armature-controlled d.c. servomotor

1. Field-Controlled D.C. Servomotor. Fig. 14.14 shows the field-controlled d.c. servomotor. Here the field winding is controlled by an electronic amplifier. The armature is supplied by a constant current source. The error voltage represents the difference between the measured signal and the desired signal. The torque produced by this motor is zero when no field excitation is supplied by the d.c. error amplifier. Since armature current is *always constant,* the torque is directly proportional to the field flux and also it is directly proportional to field current upto saturation ($T \propto \phi I_a$). If the polarity of the field is reversed, the motor direction reverses. The control of field current by this method is used only in small servomotors because of the following reasons :

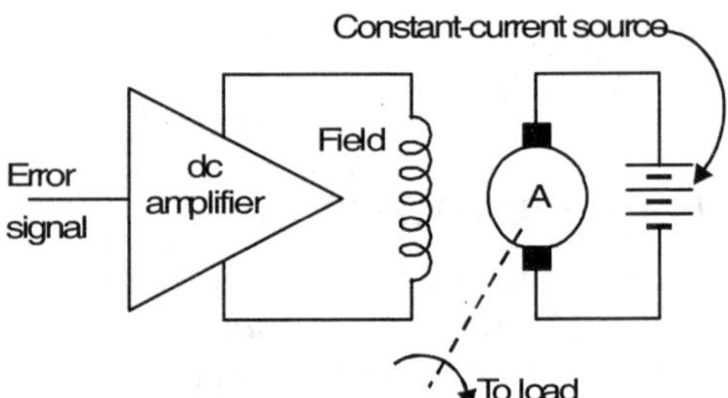

Fig. 14.14

(*i*) It is undesirable to supply a large and fixed armature current as would be required for large d.c. servomotors.

(*ii*) The time constant of the highly inductive field circuit is very long. Therefore, the dynamic response of the motor will be slow.

2. Armature-Controlled D.C. Servomotor. Fig. 14.15 shows the armature-controlled d.c. servomotor. Here the armature circuit is controlled by an electronic amplifier. The field winding

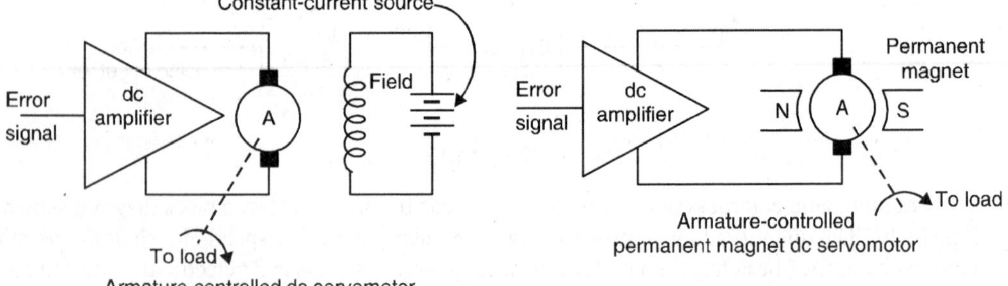

Fig. 14.15 **Fig. 14.16**

is supplied by a constant current source. The error voltage represents the difference between the measured signal and the desired signal. A sudden large or small change in armature voltage produced by an error signal will cause an almost immediate response in torque. It is because the armature circuit is essentially resistive compared to the highly inductive field circuit of field-controlled d.c. servomotor. If the error signal and the polarity of the armature voltage are reversed, the motor reverses its direction.

3. **Permanent-Magnet Armature-Controlled D.C. Servomotor.** Fig. 14.16 shows the permanent-magnet (PM) armature-controlled d.c. servomotor. It uses permanent magnets for constant field excitation instead of a constant current source. These types of motors use either alnico or ceramic magnets to produce the magnetic field. Permanent magnets have several advantages over wound-field servomotors including increased efficiency, reduced frame size and high accelerating torque. The speed of a PM d.c. servomotor is varied by changing the voltage applied to the armature. Conventional wound-field motors are often equipped with interpoles in order to improve commutation and minimise armature reaction. The high coercive force of permanent-magnet materials used in PM motors eliminates the need for interpoles due to their excellent commutation characteristics. PM servomotors are used in a wide variety of applications including electric vehicles and cordless power tools. The popularity of these motors is due to their high efficiency, compact design and good commutation.

14.7 A.C. SERVOMOTOR

An a.c. motor that is used in servomechanism is called an a.c. servomotor. Most a.c. servomotors are of two-phase, squirrel-cage rotor type induction motors.

Construction. Fig. 14.17 shows the schematic diagram of 2-phase a.c. servomotor. The squirrel cage rotor has high resistance and low inertia. The stator has two distributed windings which are displaced from each other by 90° (electrical). One stator winding called *main winding* is excited by a fixed a.c. voltage V_m. The other stator winding called *control winding* is fed by the ouput voltage V_a of an amplifier. The voltages V_m and V_a must be in synchronism which means that they must be derived from the same a.c. source. They must also be made to have a phase difference of 90° by connecting a suitable capacitor in series with the main winding as shown in Fig. 14.17.

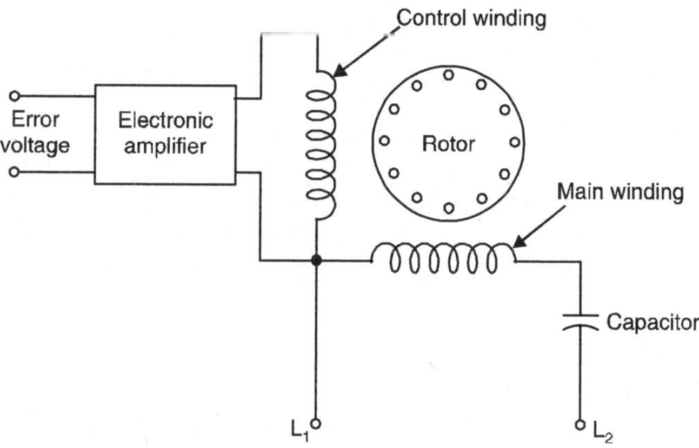

Fig. 14.17

Operation. With zero error voltage, the squirrel-cage rotor is at standstill. A small error signal is amplified by the a.c. amplifier and fed to the control winding. The servomotor rotates with a speed proportional to this voltage and in a direction so as to reduce the error signal. The motor ceases to rotate when zero error signal is produced at the control winding. Servomotors are designed to produce large values of torque at low speeds. Since these motors are widely used in positional control devices, it is important that the torque be reduced at high speeds to prevent the motor from overshooting its desired position. To reduce overshoot, the squirrel-cage rotor is made longer and thinner than in the ordinary induction motor. The rotor bars have a high value of resistance that help to prevent the inductive reactance of the motor affecting the torque. As a result, the torque of the motor decreases as the speed increases.

14.8 SWITCHED RELUCTANCE MOTOR (SRM)

A switched reluctance motor (SRM) operates on the same basic principle as a variable reluctance motor. Fig. 14.18 shows the schematic diagram of a switched reluctance motor. Unlike the conventional synchronous motor, both the stator and rotor of SRM have salient poles. The stator carries coils on each pole; the diametrically opposite coils are connected in series to form a phase. The four stator coils in Fig. 14.18 are grouped to form two phases *viz* phase *A* and phase *B*. The rotor is made of soft magnetic material and carries no windings. Both stator and rotor are laminated.

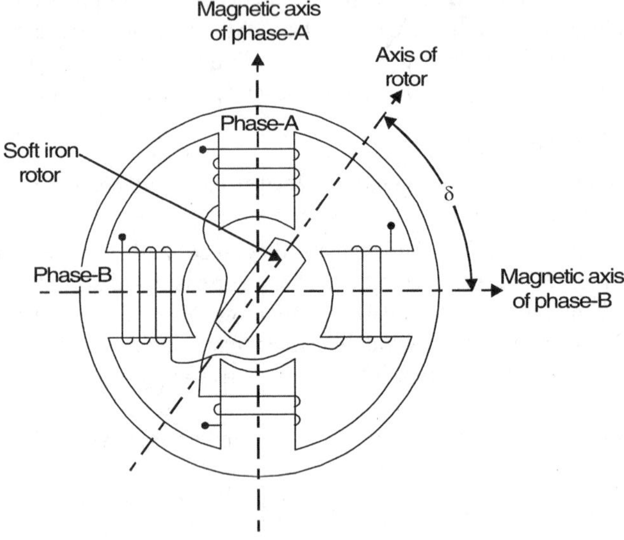

Fig. 14.18

Working. The stator coils are energised sequentially with a single pulse of current at high speed. The reluctance torque is developed due to the attraction between the rotor and stator poles. Consequently, the rotation of the rotor takes place. In order to develop torque in the motor, the rotor position should be determined by sensors so that excitation timing of the phase windings is precise. Very high speeds can be obtained by using high-frequency pulses to energise the poles sequentially.

Advantages

(*i*) The rotor has no windings or slip rings. This makes the motor very robust.

(*ii*) It has high efficiency (70% to 80%).

(*iii*) The motor works well in harsh environment where it may be exposed to high temperatures and severe vibrations.

(*iv*) It can be run at high speeds (10,000 to 30,000 r.p.m.).

14.9 PERMANENT-MAGNET DC MOTOR

A permanent-magnet (PM) d.c. motor is similar to an ordinary d.c. shunt motor except that its field is provided by permanent magnets instead of wound-field circuit. Although d.c. motors upto 75 h.p.

have been designed with permanent magnets, the major application of PM d.c. motors is confined to fractional-horsepower motors for economic reasons.

Construction. Fig. 14.19 shows the cross-sectional view of a 2-pole PM d.c. motor. The rotor (*i.e.* armature) has conductors in slots, commutator segments and brushes as in a conventional d.c. machine. The permanent magnets are supported by a cylindrical steel stator which also provides return path for the magnetic flux. The armature of a PM d.c. motor is larger than the wound-field machine because permanent magnets usually operate at a lower gap flux density. The overall size is not much changed because permanent magnets occupy less space.

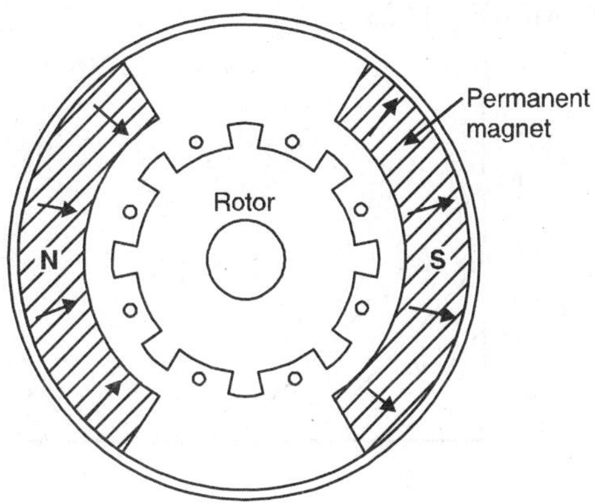

Fig. 14.19

Working. Most of PM d.c. motors are run on 12 V (or less) d.c. supply obtained from batteries or rectifiers. The interaction between the flux of permanent magnets and current-carrying conductors produces the driving torque. Consequently, the rotation of the rotor (*i.e.* armature) takes place. The stalling torque is normally about 6 times the full-load torque. This means that the motor cannot be run continuously anywhere near its maximum torque level. However, it is possible to start the motor directly from the supply without a conventional motor starter.

Motor analysis. The analysis of operation of PM d.c. motor is always comparable with ordinary d.c. shunt machine. Fig. 14.20 shows the schematic diagram of a PM d.c. motor. Note that there are no field connections. The flux is provided by the permanent magnets.

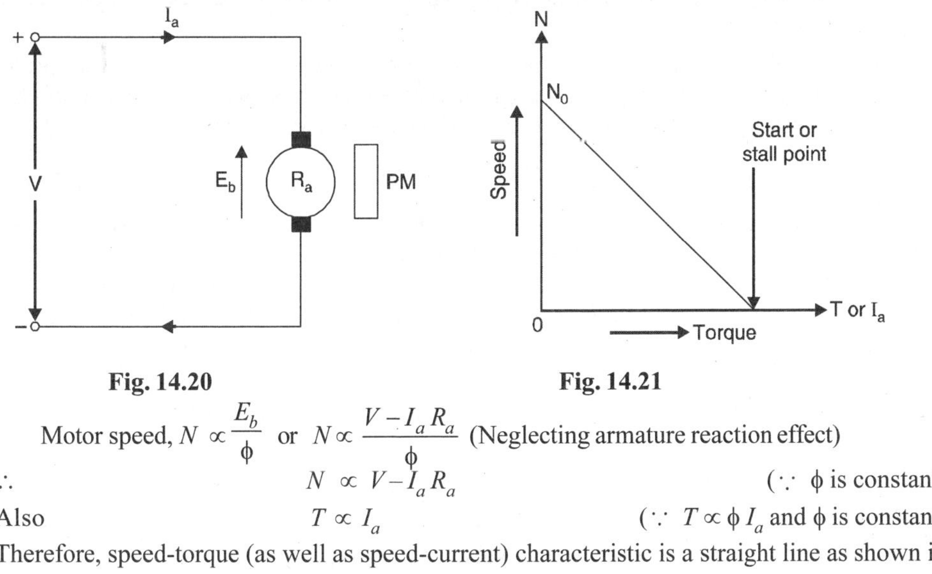

Fig. 14.20 **Fig. 14.21**

$$\text{Motor speed, } N \propto \frac{E_b}{\phi} \quad \text{or} \quad N \propto \frac{V - I_a R_a}{\phi} \quad \text{(Neglecting armature reaction effect)}$$

$$\therefore \qquad N \propto V - I_a R_a \qquad (\because \ \phi \text{ is constant})$$

$$\text{Also} \qquad T \propto I_a \qquad (\because \ T \propto \phi I_a \text{ and } \phi \text{ is constant})$$

Therefore, speed-torque (as well as speed-current) characteristic is a straight line as shown in Fig. 14.21.

The speed-torque characteristic of a PM d.c. motor can be controlled by changing either (*i*) the supply voltage V or (*ii*) the effective resistance of the armature circuit. The change in the supply voltage varies the no-load speed (N_0) of the motor without affecting the slope of the characteristic. Consequently, for different set of supply voltages (V_1, V_2, V_3 etc.), a set of parallel set of speed-torque characteristics are obtained as shown in Fig. 14.22. The change in the effective armature resistance changes the slope of the speed-torque characteristics without affecting the no-load speed (N_0) as shown in Fig. 14.23.

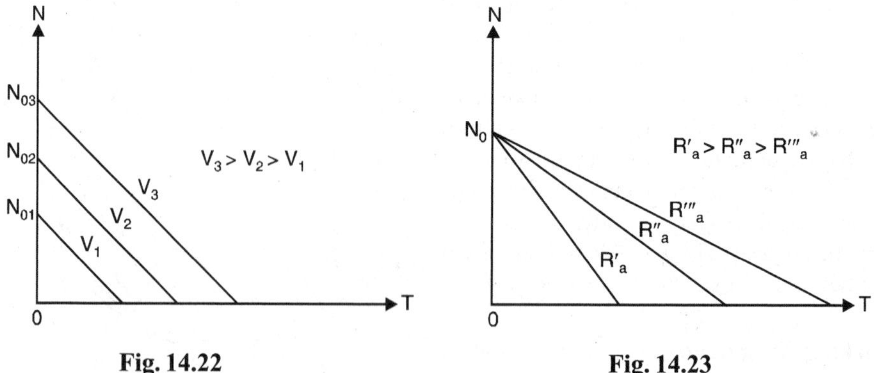

Fig. 14.22 **Fig. 14.23**

Salient Features

(*i*) For the same output and voltage, a PM d.c. motor is generally smaller in size than the wound-field d.c. motor.

(*ii*) Since field excitation is not required, the efficiency of such motors is generally higher than that of the wound-field motors.

(*iii*) The life of very small motors, determined by commutators and brushes, is limited to 1000–2000 hours.

(*iv*) These motors are mainly used for low voltages (< 12 V) and small power ratings (0.1 to 4 W). While designing these small motors, we make a compromise between initial cost, life span and working efficiencies. For battery-driven motors upto 4 W, the efficiency can be as high as 80%.

Applications. Millions of small PM d.c. motors operating on 12 V (or less), 0.1 – 4 W are produced every year. These motors are used for toys, shavers, cameras, portable vacuum cleaners, windshield wipers, automobile heaters, fans, radio antennas, wheel chairs, cordless power tools etc.

14.10 BRUSHLESS D.C. MOTORS

The conventional d.c. motors rely on mechanical commutation. These motors have two principal problems. First, they require periodic maintenance to repair and replace brushes. Secondly, electric arcs produced by the mechanical commutator brush arrangement limit the operatig speed and voltage of the motor. A motor that retains the characteristics of a d.c. motor but eliminates the commutator and the brushes is called a *brushless d.c. motor*. These motors use electronic commutators. An electronic commutator consists of transistors or thyristors and performs the function of its mechanical counterpart.

Advantages : The brushless d.c. motors have the following advantages over the mechanical commutator-brush type d.c. motors :

(*i*) They require little or no maintenance.

(*ii*) They have a longer operating life.

(*iii*) There is no arcing in these motors so that possibility of explosion is eliminated.

(*iv*) These are generally more efficient than the conventional d.c.motors.

(*v*) They can be used in situations where combustible fliuids and gases are present.

(*vi*) They are very reliable and their efficiencies may exceed 75%.

(*vii*) They are capable of rotating at speeds greater than 40,000 r.p.m.

Disadvantages : The disadvantages of brushless d.c. motors are :

(*i*) For the same output power, they are more expensive than the conventional d.c. motors.

(*ii*) They have greater overall size because additional equipment is required for electronic commutation.

Operating Principle. A brushless d.c. motor consists of a multiphase winding wound on a nonsalient stator and a permanent magnet (PM) rotor as shown in Fig. 14.24. Direct voltage or alternating voltage is applied to the individual phase windings through a sequential switching operation to achieve the necessary commutation to impart rotation of the motor. The switching is done electronically using power transistors or thyristors.

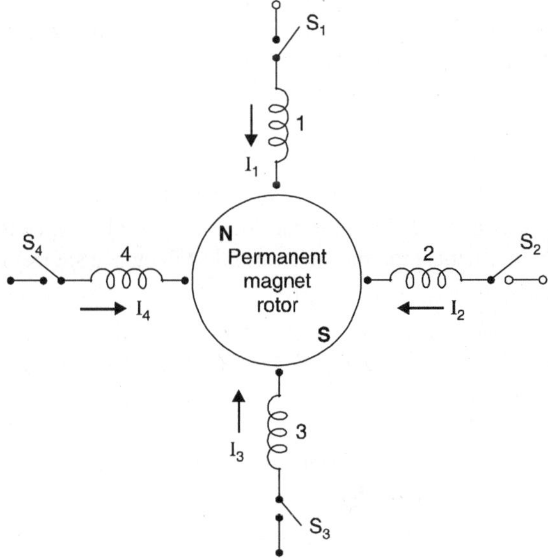

Fig. 14.24

Let us discuss the operation of brushless d.c. motor. If winding 1 is energised, the PM rotor aligns with the magnetic field produced by winding 1. When winding 1 is switched off and winding 2 is turned on, the rotor is made to rotate to line up with the magnetic field of winding 2. Thus by applying voltage to individual phase windings through a sequential switching circuitry, the motor rotation can be achieved. The control of the magnitude and the rate of switching of the phase currents essentially determine the speed-torque characteristic of a brushless d.c. motor.

Applications. The brushless d.c. motors are very popular in the biomedical and aerospace fields. Due to the high reliability and low maintenance requirements, these motors are used as artificial heart pump motors. In the aerospace industry, d.c. brushless motors are used in satellites, gyroscopes and high efficiency robotic systems. Other applications include disc drive motors, video recorders and tape transport systems.

SHORT ANSWER QUESTIONS

Q.1. *What is the origin of the name stepper motor ?*

Ans. It is called stepper motor because the motor rotates through a fixed angular step in response to each input current pulse.

Q.2. *Why is stepper motor very robust and reliable ?*

Ans. The only moving part in a stepper motor is its rotor which has no windings, commutator or brushes. This feature makes the motor quite robust and reliable.

Q.3. *What are the most common step sizes of stepper motors?*

Ans. Step motors are generally used in a range from 1 W to about 3 h.p. and their step sizes vary from 0.72° to 90°. However, the most common step sizes are 1.8°, 7.5° and 15°.

Q.4. *What happens to a stepper motor if pulse duration is very short?*

Ans. Since a stepper motor rotates when a series of pulses is applied to its phase windings, the duration of each pulse should be sufficiently long to accurately rotate the motor at the desired speed. If the pulse duration is too short, the rotor will miss the steps and be unable to follow the applied pulses accurately. Consequently, either the motor will not rotate or the required speed will not be achieved.

Q.5. *What is the main advantage of using a d.c. servomotor?*

Ans. The main advantage of using a d.c. servomotor is that its torque-producing ability is greater at low speeds than at high speeds. This is a tremendous advantage in precision control, since the servomotor is capable of accelerating the load rapidly from a standstill. Also, the d.c. servomotor is able to deliver a reverse torque to slow down the load as the load approaches its correct position.

Q.6. *What is the origin of the name hybrid stepper motor?*

Ans. The term *hybrid* is derived from the fact that the motor is operated under the combined principles of the permanent magnet and variable-reluctance stepper motors.

Q.7. *What is a brushless d.c. motor?*

Ans. A d.c. motor that uses electronic commutator is classed as a brushless d.c. motor. Such a motor has no mechanical commutator and brushes. Instead, electronic commutator is employed. An electronic commutator uses a few transistors (or thyristors) to provide commutation control.

Q.8. *What are the disadvantages of permanent magnet (PM) d.c. motors?*

Ans. A PM d.c. motor has the following disadvantages :

(*i*) The magnetic field of the motor remains active even when it is not in use. Therefore, these motors are made totally enclosed to prevent their magnets from collecting magnetic junk from neighbourhood. As a result, they heat up if used continuously.

(*ii*) The permanent magnets of the motor may be demagnetised by the armature reaction m.m.f. This may cause the motor to be inoperative.

Q.9. *What is the basic operating principle of d.c. stepping motors?*

Ans. In all types of stepping motors, rotation is produced by *switching* suitably connected *windings* in some *predetermined sequence* to produce *angular* discrete rotation *steps* that are essentially *uniform in magnitude.*

Q.10. *Why is the torque developed in hybrid stepping motor greater than that of PM or VR type stepping motors ?*

Ans. Since hybrid stepping motors develop both excitation torque (varying as the sine of displacement angle) and reluctance torque (varying as twice the sine of the displacement angle), the resultant torque developed is greater than that of either PM or VR types.

MULTIPLE-CHOICE QUESTIONS

1. A stepper motor is a ....

(*i*) d.c. motor

(*ii*) single-phase a.c. motor

(*iii*) two-phase motor

(*iv*) multiphase motor

2. If a hybrid stepper motor has a rotor pitch of 36° and a step angle of 9°, the number of its phases must be ....

(*i*) 4 (*ii*) 2

(*iii*) 3 (*iv*) 6

3. A two-phase a.c. servomotor has ....
   (*i*) wound rotor (*ii*) cage rotor
   (*iii*) wound or cage rotor
   (*iv*) a rotor similar to that in d.c. motors
4. A variable-reluctance stepper motor is constructed of .... material with salient poles.
   (*i*) paramagnetic (*ii*) diamagnetic
   (*iii*) ferromagnetic (*iv*) non-magnetic
5. In a brushless d.c. motor, we have, ...
   (*i*) no mechanical commutator
   (*ii*) no brushes
   (*iii*) no arcing (*iv*) all of above
6. The stepping angle for a 3-phase, 24-pole PM stepper motor is ....
   (*i*) 5°/step (*ii*) 10°/step
   (*iii*) 15°/step (*iv*) 2.5°/step
7. A hybrid stepper motor has 50 variable-reluctance rotor teeth. The stepping angle is ....
   (*i*) 2.16° (*ii*) 1.8°
   (*iii*) 3° (*iv*) 3.2°
8. Motors that use electronic commutators are classified as : ....
   (*i*) thyristor controlled
   (*ii*) servomotors
   (*iii*) brushless motors
   (*iv*) electronic motors
9. The most popular types of stepper motors are : ....
   (*i*) shunt and series
   (*ii*) shunt and compound
   (*iii*) PM and thin disc rotor
   (*iv*) PM and VR
10. The amount of torque required to move a stepper motor one full step is called ....
   (*i*) holding torque (*ii*) residual torque
   (*iii*) detent torque
   (*iv*) developed torque

ANSWERS TO MULTIPLE-CHOICE QUESTIONS

1. (*iv*)	**2.** (*i*)	**3.** (*ii*)	**4.** (*iii*)	**5.** (*iv*)
6. (*i*)	**7.** (*ii*)	**8.** (*iii*)	**9.** (*iv*)	**10.** (*i*)

INDEX